Reproduced here are three pages from the Codex Dresdensis, dated ca. 1200–1250 A.D. On the left is page 24 of Chapter 4, which consists of a table of the planetary motions (584-day synodical revolutions) of Venus. The Maya Venus tables are so accurate that only one day of error will be accumulated in 6000 years. The symbol for Venus as the morning star appears several times on this page. It has the appearance of a four-pointed star with a dot-centered circle in each lobe (see third column from left, about one-fourth distance from top of page).

In the center is page 53 from Chapter 5, which gives tables for moon phases and the dates on which solar and lunar eclipses might be visible. This page contains some dire predictions based on solar and lunar occultations. Thus, the upper part of page 53 shows the death god Cizin seated on his throne of bones. The moon hovers between light and shadow in the third glyph above his head. The picture below is of the moon goddess, hanging from the celestial band by a cord, her eye closed in death, her hands limp. The oval shell symbols (two appear on this page) represented the number zero in Maya calculations.

The right-hand frame, page 45, is probably the initial page of a missing almanac. The two columns in the upper left specify a date in time based on an accepted starting point in Maya history. The left column states that starting point. The second column tells the reader how many days must be added. A multiplication table (multiples of 364) occupies the remaining four columns. The year length was 364 days in the Maya calendar of the 1200s.

The middle and lower parts refer to another almanac or almanacs, probably related to weather and crops, and probably divinatory. The center shows a monster with upturned snout, cloven hoofs, and spots falling from the sky. The presence of axe glyphs above signifies impact from the sky, i.e., thunderbolts. Thus, the monster is identified as a "lightning beast." In the lower figure a god with torches in each hand sits on a deer dying of thirst, signifying drought. The presence of numbers refers to the time cycles of wet and dry weather.

From J.E.S. Thompson, A Commentary on the *Dresden Codex* (Philadelphia: American Philosophical Society, 1972), pp. 24, 45, and 53. (Courtesy of the New York Public Library.)

Handbook of Physical Testing of Paper

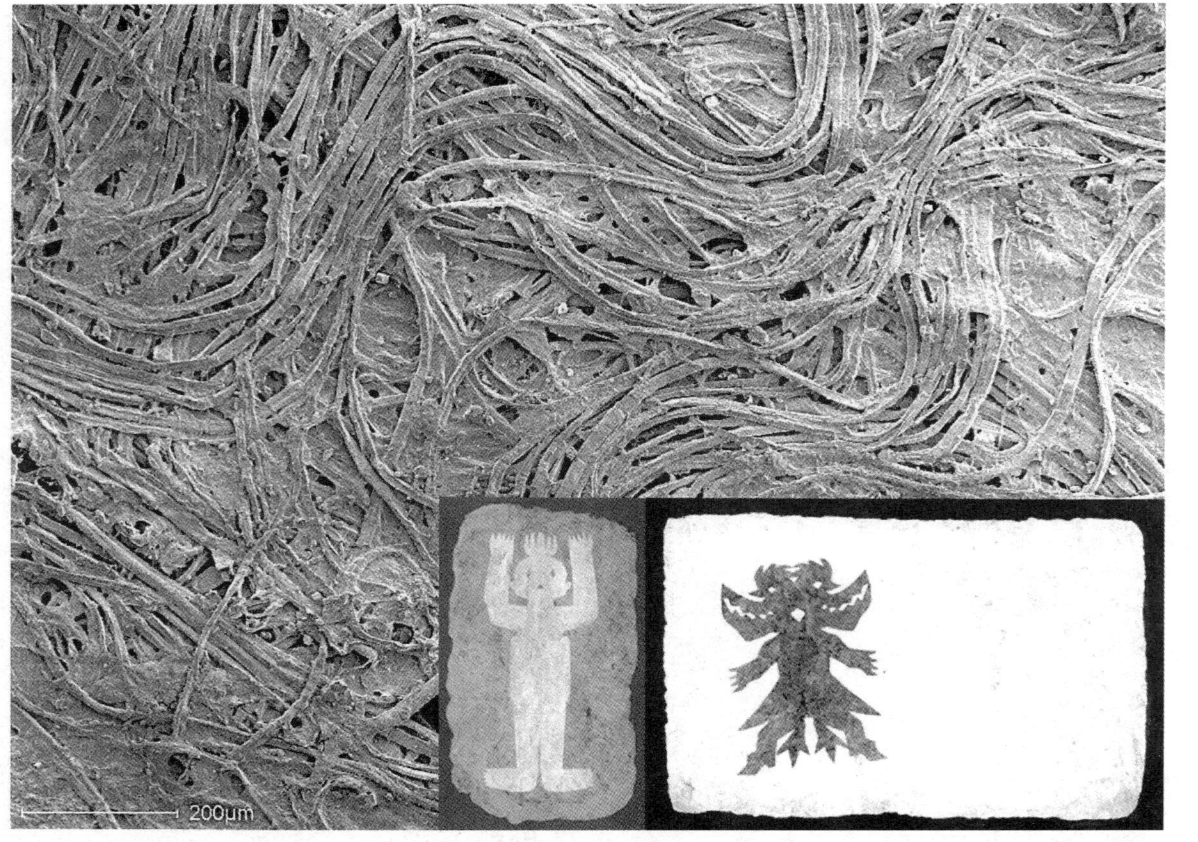

The main picture above (which also appears on the front cover of this volume) is an environmental scanning electron micrograph of uncoated ceremonial paper made by Otomi Indians in about 1940. Bar scale is 200 μm. Bark fibers of 4 Moraceae species are used by the Otomi: two *Ficus*, one *Urera*, and one *Morus*. Fibers are separated in boiling limewater, then washed, beaten, felted, couched, and sun-dried on boards much as in the manner of pre-Columbian Aztec and Maya papermakers. The two inserts are ceremonial paper figures pressed onto separate paper backing sheets. The left insert is approximately 1/5th of the original size; the right insert is approximately 1/4th of original size. Papers were provided courtesy of Dr. George E. Stuart, Center for Maya Research, Barnardsville, North Carolina. Micrographs courtesy of Dr. Susan Anagnost, N. C. Brown Center for Ultrastructure Studies, State University of New York College of Environmental Science and Forestry, Syracuse, New York.

Handbook of Physical Testing of Paper

Volume 1
Second Edition, Revised and Expanded

edited by
Richard E. Mark
Empire State Paper Research Institute
State University of New York
College of Environmental Science and Forestry
Syracuse, New York

Charles C. Habeger, Jr.
Institute of Paper Science and Technology
Atlanta, Georgia

Jens Borch
IBM Corporation
Boulder, Colorado

M. Bruce Lyne
International Paper
Tuxedo, New York

Associate Editor
Koji Murakami
Faculty of Agriculture
Kyoto University
Kyoto, Japan

MARCEL DEKKER, INC. NEW YORK • BASEL

The first edition of this book was published as *Handbook of Physical and Mechanical Testing of Paper and Paperboard*, R. E. Mark, ed., Marcel Dekker, Inc., 1983.

ISBN: 0-8247-0498-3

This book is printed on acid-free paper.

Headquarters
Marcel Dekker, Inc.
270 Madison Avenue, New York, NY 10016
tel: 212-696-9000; fax: 212-685-4540

Eastern Hemisphere Distribution
Marcel Dekker AG
Hutgasse 4, Postfach 812, CH-4001 Basel, Switzerland
tel: 41-61-261-8482; fax: 41-61-261-8896

World Wide Web
http://www.dekker.com

The publisher offers discounts on this book when ordered in bulk quantities. For more information, write to Special Sales/Professional Marketing at the headquarters address above.

Current printing (last digit):
10 9 8 7 6 5 4 3 2 1

PRINTED IN THE UNITED STATES OF AMERICA

PREFACE TO THE SECOND EDITION

In the mid-1990s, an agreement was concluded between the four editors and the publisher to produce a second edition of the *Handbook of Physical and Mechanical Testing of Paper and Paperboard.* The new version has a shorter title, *Handbook of Physical Testing of Paper*; but it is a substantial revision, modernization, and expansion of the previous work.

This edition contains 30 chapters (16 in Vol. 1, 14 in Vol. 2), one-third of which are entirely new. The other 20 chapters are more specifically "revised and expanded" from the previous edition. However, some of them have been totally rewritten to reflect the perspectives of the author(s) as well as to modernize the technical content. We hope that this edition will be at least as useful in today's world as the first edition has been over the past years.

We have learned that the creation of a technical book today involves vastly different problems from those encountered while preparing the first edition. But the positive experiences–with our authors, with the publisher's excellent professional staff, with Associate Editor Koji Murakami, with readers and critics, and with supportive employers—far outweigh the negatives. For these factors, we are most grateful.

Richard E. Mark
Charles C. Habeger, Jr.
Jens Borch
M. Bruce Lyne

PREFACE TO THE FIRST EDITION

We have tried to assemble here as many of the modern aspects of properties testing in the paper and paperboard field as possible. One of our objectives is to enable those concerned with planning, specifying, and evaluating the physical and mechanical testing of these materials to take advantage of the many advances and improvements that have taken place in recent years. We also feel that it is important and useful to codify the excellent pioneering work that has provided the essential base for these advances, which has been included to the maximum extent possible.

As with their associates in other aspects of the broad and dynamic paper industry, the pioneers in paper testing have come from many diverse areas of the world; the authors who have contributed to these volumes live on four continents and represent a spectrum of industrial, government, and educational institutions, bringing their own special insights to the areas covered in their respective chapters. It goes without saying that the very nature of the subject matter covered here has wide-ranging applicability; our intention has been to treat the topics in our various chapters in a holistic manner.

We wish to give some justly deserved praise to the contributors to this handbook and their affiliated organizations—academic, industrial, and governmental. Creation of a multiauthored book of this type requires more than professional expertise. It also requires a great deal of diligence, perseverance, and willingness to accept criticism, as well as laborious revision and adjustment to the coverage of related topics in other chapters. In short, it is hard work that calls for dedication to purpose. An editor fortunate enough to have become associated with this group of contributors is lucky indeed, for they have been exceptional in all these aspects, as has Associate Editor Koji Murakami. The organizations of which each contributor is a part must also be praised, for without their cooperation and support, even the most dedicated contributor would find it difficult if not impossible to do all that is required. Finally, we are all deeply indebted to the management and staff of Marcel Dekker, Inc., who have combined rigorous publishing professionalism with genuinely compassionate accommodation to the severe personal traumas that accompanied the preparation of these volumes. Thank you, all.

ACKNOWLEDGMENTS

The history of pre-Columbian paper and books in the New World is little known. Through the generous contributions of several leading companies in the paper industry, it has been possible to include color illustrations of several aspects of paper and papermaking in early Mesoamerica. We are indebted to these donors: Abitibi-Price, Inc., Crown Zellerbach Corporation, International Paper Company, Mead Corporation Foundation, St. Regis Paper Company, Scott Paper Company, Westvaco Corporation, and Weyerhaeuser Company.

Richard E. Mark

CONTRIBUTORS

Keith A. Bennett Corporate Research—Fiber Sciences, Weyerhaeuser Company, Federal Way, Washington

Curt A. Bronkhorst Corporate Research—Fiber Sciences, Weyerhaeuser Company, Federal Way, Washington

Leif A. Carlsson Department of Mechanical Engineering, Florida Atlantic University, Boca Raton, Florida

Benjamin C. Donner Corporate Research—Fiber Sciences, Weyerhaeuser Pulp, Paper, Packaging R&D, Tacoma, Washington

Christer Fellers STFI, Swedish Pulp and Paper Research Institute, Stockholm, Sweden

Charles C. Habeger, Jr. Department of Paper Physics, Institute of Paper Science and Technology, Atlanta, Georgia

Roger Hagen STFI, Swedish Pulp and Paper Research Institute, Stockholm, Sweden

Henry W. Haslach, Jr. Department of Mechanical Engineering, University of Maryland College Park, College Park, Maryland

Makio Hasuike Hiroshima Research and Development Center, Mitsubishi Heavy Industries Ltd., Hiroshima, Japan

José Iribarne Engineering and Technical Services, Solvay Paperboard, Syracuse, New York

D. Steven Keller Faculty of Paper Science and Engineering, Empire State Paper Research Institute, State University of New York College of Environmental Science and Forestry, Syracuse, New York

M. T. Kortschot Department of Chemical Engineering and Applied Chemistry and The Pulp & Paper Center, University of Toronto, Toronto, Ontario, Canada

Richard E. Mark* Empire State Paper Research Institute, State University of New York College of Environmental Science and Forestry, Syracuse, New York

Tsutomu Naito Yatsushiro Mill, Nippon Paper Industries, Kumamoto, Japan

Leena Paavilainen Finnish Forest Cluster Research Programme—Wood Wisdom, Helsinki, Finland

Richard W. Perkins, Jr. Department of Mechanical, Aerospace and Manufacturing Engineering, Syracuse University, Syracuse, New York

M. K. Ramasubramanian Mechanical and Aerospace Engineering, North Carolina State University, Raleigh, North Carolina

Elias Retulainen Pulp and Paper Technology, Asian Institute of Technology, Pathumthani, Thailand

Lennart Salmén STFI, Swedish Pulp and Paper Research Institute, Stockholm, Sweden

Rob Steadman Janus Research, Hamilton, New Zealand

Tetsu Uesaka Product Performance, Pulp and Paper Research Institute of Canada, Pointe-Claire, Quebec, Canada

John Frederick Waterhouse* Paper Physics–End Use and Converting, Institute of Paper Science and Technology, Atlanta, Georgia

*Retired.

CONTENTS

CONTENTS OF VOLUME 2

PAPER, BOOKS, AND PAPER TESTING: THE ORIGINS OF PAPER (AND BOOKS) IN THE NEW WORLD

Much has been written about the "humanity" of paper: its influence on civilization as its geographical and cultural distribution has widened; the increasing diversity in its use; and its central role in providing a convenient vehicle for the acquisition, storage, and dissemination of both tangible goods (as in packaging), and these same functions applied to intangible areas, such as communication and the infinitely diverse applications of human knowledge.

It is altogether in keeping with this humanistic, universal view of paper and civilization that we have acknowledged in our jacket design and elsewhere in these volumes, the role that paper played in the waxing and waning of human culture in several ancient societies not often associated with paper or papermaking.

The origin of paper as invented by T'sai Lun in China is a well-known story. The art of making paper also arose independently in Mesoamerica sometime before A.D. 660, and perhaps many centuries before that. As with their counterparts in the orient, Maya, Toltec, Aztec, and Zapotec papermakers utilized the bark fibers of trees of the Moraceae family. They devised their own techniques for fiber separation, washing, beating, felting, couching, sizing, drying, hot pressing, coating, and converting. They developed pigments, dyes, inks, and glues. Some of their writing was done in codices—handwritten books—fitted with covers of leather, jaguar skin, or wood, the latter often studded with decorative stones. From the few codices that survived the conquest, one senses that Mesoamerica may have been on the verge of printing when the conquistadores arrived, for the Mayas and Aztecs were already using wood, clay, and metal stamps for decorative stampings of ceramics and weavings.

Paper was of tremendous importance in pre-Columbian Mesoamerica. It was used in many rituals. It was a substantial article of commerce and served as a vehicle to sustain and enlarge commerce generally. The Maya originators of paper built libraries—guarded stone buildings—to house their records, documents, and sacred books, which were consulted before decisions were made in matters ranging from crop planting to war. The Aztecs, who improved on the techniques of papermaking that the Maya had started, employed paper for a much broader range of records keeping, including land surveys, engineering plans, tax rolls, and tribute lists, and they had even begun to use the medium for communication when the ships of Cortés arrived at their shores.

How important a role was assigned to paper in the governance of the Aztec domain can be deduced from the tribute lists that were housed in the libraries at Tenochtitlan. Areas of Mesoamerica that were under Aztec control were required to contribute substantial quantities of foodstuffs, spices, fabrics, blankets, skins, garments, shields, incense, jade, metals, firewood, hewn timbers, and other products of value, both natural and manufactured. The tribute lists were, of course, on paper. A handful of these tribute lists, such as those in the Codex Mendoza, have been preserved. From the lists in that codex we learn that two areas of present-day southern Mexico were required to contribute an amount of paper that seems impressive when we recall that each step of the operation, from the cutting of the trees to the carrying of the bundles of paper to the Aztec capital, was done by manual labor. The tribute towns had to deliver 16,000 *resmas* twice yearly—an annual contribution amounting to 480,000 sheets.

The Aztecs did not allow any paper or other tributes to be toted to their storehouses without prior approval of the ruler's inspectors. In order to carry out their responsibilities, the inspectors had standards of quality to verify; evidently some form of quality control testing was conducted in the papermaking towns, probably by both fabricators and inspectors.

Paper also played a role in communication in the empire that Cortés encountered in 1519—a situation he used as an aid, after minor skirmishing, to bring about the empire's downfall. The shrewd Spaniard noted that Indian artists were sketching scenes of the ships, the military deployment of the soldiers and horses, and the equipment that he had landed. What was being painted on the pads, he reasoned, would be the messages delivered to the capital. He then arranged for a mounted drill and demonstration firing of his cannon, which was duly recorded by the artists. As Cortés expected, the renditions of the scene on the beach caused great consternation in Tenochtitlan, for the Aztecs had no knowledge of horsemanship, gunpowder, or firearms, and could not be sure if they were dealing with men or gods. To act on either assumption was fraught with uncertainty and danger; while Moctezuma III temporized and, with lavish gifts, tried to induce the Europeans to leave, Cortés learned where all the disaffections and weaknesses, civil and military, lay in the restive, tribute-burdened satrapies.

The rest is history. The Aztec and all the other Indian nations of Mesoamerica were conquered militarily. Following that conquest came a spiritual assault on their customs and traditions, and a progressive disintegration of these once-proud societies ensued. It does not seem to be an overstatement to say that the loss of their paper records played a significant role in that disintegration. The chroniclers of the time duly recorded the extreme anguish of the Indians as the contents of their libraries were burned.

So complete was the destruction of the paper documents of the Maya ordered by Bishop Diego de Landa in 1561, that only four of the Maya codices have survived, and each of these relics has damaged and/or missing pages. Destruction of the Aztec libraries was not as total, and about 500 Aztec books were collected by a handful of individuals who recognized their value, monetary or otherwise. However, the man responsible for collecting most of them, the Chevalier Lorenzo Boturini, was shortly imprisoned on religious grounds. Partly "torn, pillaged, and dispersed" and partly stored with little care in a damp location for many years thereafter, the remains of Boturini's collection were finally auctioned off in 1804. At that time they were examined by the eminent scientist Alexander von Humboldt, who declared them to be so deteriorated that "there exists at present only an eighth part of the hieroglyphic manuscripts taken from the Italian traveler."

After succeeding vicissitudes and misadventures, there remain today less than 20 of the Aztec codices. Not a single Toltec or Zapotec book is known to have survived.

And so the accumulated centuries of writings by the Mesoamerican peoples—the paper record—has largely been lost and forgotten. Gone are their maps and hieroglyphic charts, their tribute lists and tax rolls, their herbals, their land surveys, the plans for their canals, roads, buildings, and other engineering and architectural works, the records of their wars and migrations, their histories, genealogies, charts of the constellations, calendars and almanacs, and other known writings on agriculture and crops, fishing, soils, astronomy and astrology, mythology, disease and medicine, mathematics, commerce, songs and chants, cosmogony (including the legendary *History of Heaven and Earth* of the Toltecs), and especially prophecy, religion, and sacred themes. On the front inside cover, we have reproduced a few pages of the most famous relict Maya book, the Dresden Codex.

Although giving recognition to the Amerindian paper-making societies constitutes only a minor (but we hope interesting and thought-provoking) part of these volumes, we have made one use of the number system of the Maya in the chapter texts. Their vigesimal system (see rear inside cover design) lends itself rather ideally to the problem of setting off itemizations (for example, lists of principles to follow in test design) from the rest of the textual material in a given section or subsection. We think it fulfills a real need in the publishing field, especially as regards technical books. Any comments on this usage will be welcomed by the editors.

Richard E. Mark

PART 1. THEORY AND TEST FOR MECHANICAL PARAMETERS

Emblem of Amatitlan

Amatitlan, the "Paper Place," was a town subjected to conquest under Moctezuma II in 1495. The conquest was represented by a burning, toppling temple in its symbol. Subsequently, Amatitlan was pressed to deliver paper rolls as tribute to the Aztecs, whence the roll of paper tied to the temple. The name is derived from *amatl* (paper) and *tlan* (a place where something is found in abundance). The town contributed paper to Tenochtitlan until the coming of Cortés.

1

MODELS FOR DESCRIBING THE ELASTIC, VISCOELASTIC, AND INELASTIC MECHANICAL BEHAVIOR OF PAPER AND BOARD

RICHARD W. PERKINS, JR.
Syracuse University
Syracuse, New York

I. INTRODUCTION

The objective of this chapter is to present a general treatment of material related to the problem of describing the stress–strain behavior of paper. Sections II and III are devoted to a general treatment of the problem of describing the deformation and stressing of materials. Certain of the concepts developed in these sections are necessary to enable a proper description of the constitutive relations that are presented in subsequent sections. Section IV treats the classical elastic and inelastic constitutive equations and their application to paper material. Section V provides a discussion of the prediction of both the elastic and inelastic reponse of paper in terms of the mechanical behavior of the fibers and the fiber network structure of paper.

II. MEASURES OF STRAIN AND DEFORMATION

The behavior of a solid material is often described in terms of a functional relationship between the stress applied to the material and the strain that takes place as a result of the applied stress. Consider an idealized experiment in which a strip of material is subjected to a tensile loading P as shown in Fig. 1. Assume that we focus attention at a central region in the strip located at some distance from the points of load application. Assume further that in this central region the strip deformation is homogeneous, that is, that the deformation of every point is essentially the same as that of any other point. The original dimensions L, W, and T change to new dimensions l, w, and t.

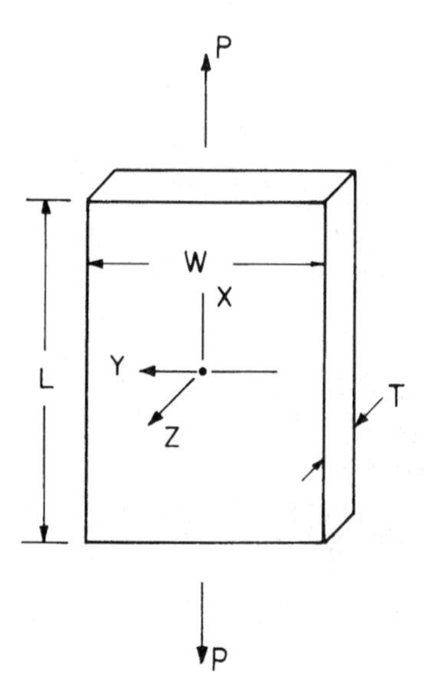

Fig. 1 An idealized strip of material of original length L, width W, and thickness T. The strip is subjected to tensile load P.

In order to quantatively describe the deformation of the strip, it is common practice to identify certain points, such as points O, A, B, C, and D shown in Fig. 2. The points are assumed to lie in the midplane of the strip, that is, at one-half the thickness. These points are moved to new locations O′, A′, B′, C′, and D′ by the deformation of the strip. Before deformation, the location of a point such as A is given by specifying its X, Y, Z coordinates relative to the origin O. After deformation, the location is given by specifying its coordinates x, y, z relative to the origin O′. Here, it is assumed that the point O was located at a position that does not move so that O′ is situated at the same point as O. For example, point O might be located at the center of strip while both grips of a testing machine are moved apart.

It is commonly assumed that the deformation of the material can be described by giving the functional relationship between the undeformed X, Y, and Z and the deformed x, y, z coordinates. This relationship can be considered to be a continuous function that changes from point to point within the material. For experimental purposes of determining the behavior of the material in question, it is desirable to have a test that is characterized by a uniform deformation, that is, a deformation that is the same at all points within the body. A deformation with this characteristic is called an affine deformation, and it can be represented by a matrix transformation of the form

$$\begin{pmatrix} x \\ y \\ z \end{pmatrix} = \begin{pmatrix} D_{XX} & D_{XY} & D_{XZ} \\ D_{YX} & D_{YY} & D_{YZ} \\ D_{ZX} & D_{ZY} & D_{ZZ} \end{pmatrix} \begin{pmatrix} X \\ Y \\ Z \end{pmatrix} \tag{1}$$

where the quantities in the matrix are constants. In the more general case, one can focus attention on a particular point X, Y, Z and a neighboring point $X + dX$,

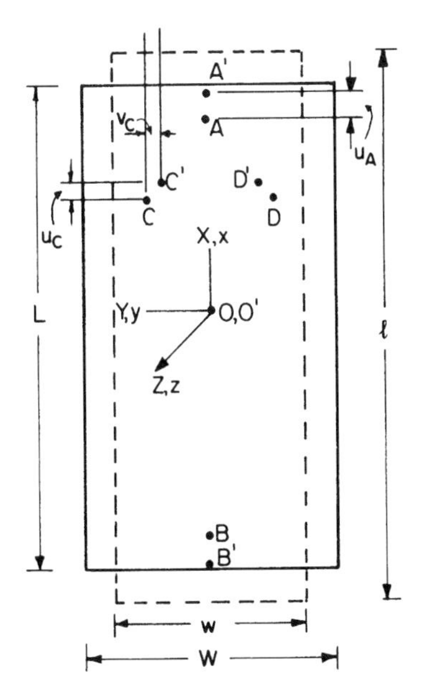

Fig. 2 Undeformed and deformed strip subjected to tensile loading in X direction.

$Y + dY$, $Z + dZ$. Then the distances dX, dY, dZ are changed as a result of the deformation into new elements dx, dy, dz. These quantities can be related to each other in the form

$$\begin{pmatrix} dx \\ dy \\ dz \end{pmatrix} = \begin{pmatrix} \frac{\partial x}{\partial X} & \frac{\partial x}{\partial Y} & \frac{\partial x}{\partial Z} \\ \frac{\partial y}{\partial X} & \frac{\partial y}{\partial Y} & \frac{\partial y}{\partial Z} \\ \frac{\partial z}{\partial X} & \frac{\partial z}{\partial Y} & \frac{\partial z}{\partial Z} \end{pmatrix} \begin{pmatrix} dX \\ dY \\ dZ \end{pmatrix} \tag{2}$$

and the elements of this matrix are termed the deformation gradients. Of course, in the case of the affine deformation, the deformation gradients would be the constants of Eq. (1).

As an example, consider the case of the ideal uniaxial tensile test. In this case, we have the situation that is depicted by Fig. 1, and the deformation is assumed to be a homogeneous affine deformation. The matrix $\{D\}$ that relates (x, y, z) and (X, Y, Z) or (dx, dy, dz) and (dX, dY, dZ) could be written as

$$\{D\} = \begin{pmatrix} \lambda & 0 & 0 \\ 0 & K_Y\lambda & 0 \\ 0 & 0 & K_Z\lambda \end{pmatrix} \tag{3}$$

The quantity λ, called the *stretch*, is the ratio of the element lengths dx and dX that are oriented in the direction of the load application in Fig. 1, that is,

$$dx = \lambda\, dX \tag{4}$$

Similarly,

$$dy = K_Y\lambda\, dY \tag{5}$$

$$dz = K_Z\lambda\, dZ \tag{6}$$

In order to determine the strain in the material it is important to be able to calculate the length of a line segment. In fact, the strain is defined by comparing the square of the length of a line segment before and after deformation. The square of the length of a line segment can be written

$$dS^2 = dX^2 + dY^2 + dZ^2 \tag{7}$$

After deformation, the square of the line segment is

$$ds^2 = dx^2 + dy^2 + dz^2 \tag{8}$$

By using relation (2), which describes the deformation of the material, both quantities dS^2 and ds^2 can be expressed in terms of the original coordinates x, y, z. For example, from relations (2),

$$\begin{aligned} dx^2 = {} & \left(\frac{\partial x}{\partial X}\right)^2 dX\, dX + \left(\frac{\partial x}{\partial Y}\right)^2 dY\, dY + \left(\frac{\partial x}{\partial Z}\right)^2 dZ\, dZ \\ & + 2\frac{\partial x}{\partial X}\frac{\partial x}{\partial T}\, dX\, dY + 2\frac{\partial x}{\partial X}\frac{\partial x}{\partial Z}\, dX\, dZ + 2\frac{\partial x}{\partial Y}\frac{\partial x}{\partial Z}\, dY\, dZ \end{aligned} \tag{9}$$

Specifically, suppose $dS^2 = dX^2$. Then,

$$
\begin{aligned}
dx^2 &= \left(\frac{\partial x}{\partial X}\right)^2 dX^2 \\
dy^2 &= \left(\frac{\partial y}{\partial X}\right)^2 dX^2 \\
dz^2 &= \left(\frac{\partial z}{\partial X}\right)^2 dX^2
\end{aligned}
\tag{10}
$$

The strain is defined by comparing ds^2 and dS^2, for if $ds^2 = dS^2$ then there will be no deformation of the body and the only motion possible is that associated with a rigid-body motion. The strains are defined by the relations

$$
\begin{aligned}
ds^2 - dS^2 &= \sum_{K=1}^{3}\sum_{L=1}^{3} 2E_{KL}\, dX_K\, dX_L \\
&= \sum_{K=1}^{3}\sum_{l=1}^{3} 2e_{kl}\, dx_{kl}\, dx_l
\end{aligned}
\tag{11}
$$

where a number subscript notation is used. Here, $dX = dX_1$, $dY = dX_2$, $dZ = dX_3$, $dx = dx_1$, $dy = dx_2$, $dz = dx_3$. It is common practice not to write the summation signs; summation is implied when the subscripts are repeated as, for example, in relation (12). Thus, it is equivalent to write

$$ds^2 - dS^2 = 2E_{KL}\, dX_K\, dX_L = 2e_{kl}\, dx_k\, dx_l \tag{12}$$

In the case of E_{KL}, called the components of the *Lagrangian strain tensor*, the originally undeformed coordinates are used. In the case of e_{kl}, the *Eulerian strain tensor*, the deformed coordinates are used.

As an example, consider the case where $dS^2 = dX^2 = dX_1^2$. Then

$$ds^2 - dS^2 = 2E_{11}\, dX_1\, dX_1 \tag{13}$$

or

$$\frac{1}{2}\left(\frac{ds^2 - dS^2}{dS^2}\right) = E_{11} \tag{14}$$

Thus, E_{11}, is the component of strain that represents one-half the ratio of the difference of the squares of the deformed and undeformed elements and the square of the undeformed element.

It is often convenient to use a representation of the deformation of the body that is closely related to the strain. The quantity defined by

$$C_{KL} = 2E_{KL} + \delta_{KL} \tag{15}$$

where δ_{KL} is the quantity that equals 1 if $K = L$ and 0 if $K \neq L$, is called the *Green deformtion tensor*. The quantity

$$c_{kl} = 2e_{kl} + \delta_{kl} \tag{16}$$

is called the *Cauchy deformation tensor*. These quantities can be used to express the undeformed length dS in terms of the deformed coordinates or the deformed length ds in terms of the undeformed coordinates. The expressions are

$$dS^2 = c_{kl}\, dx_k\, dx_l, \qquad ds^2 = C_{KL}\, dX_K\, dx_L \tag{17}$$

For theoretical reasons, it is important to define the strain in accordance with Eq. (12). This relation is not convenient, however, for experimental purposes because it is not possible to directly measure the square of a length such as dS^2 or ds^2. It is much more convenient to work directly with the lengths ds and dS rather than with the squares of these quantities. The ratio ds/dS of the lengths of two length elements is termed the *stretch*, $\Lambda_{(N)}$ or $\lambda_{(n)}$. The subscript N or n refers to the unit vector N or n that describes the direction of dS or ds, respectively. The equations for the calculation of the stretch are

$$\Lambda_{(N)} = \frac{ds}{dS} = \sqrt{C_{KL}N_K N_L} \tag{18a}$$

and

$$\lambda_{(n)} = \frac{ds}{dS}\frac{1}{\sqrt{c_{kl}n_k n_l}} \tag{18b}$$

Another measure of deformation based on ds and dS is the *extension*, $E_{(N)}$ or $e_{(n)}$, which can be calculated from the stretch and is defined by

$$E_{(N)} = e_{(n)} = \frac{ds - dS}{dS} = \Lambda_{(N)} - 1 \tag{19}$$

It is the extension, the change in length per unit length, that the experimentalist frequently measures. For example, a linear variable differential transformer (LVDT) or capacitive transducer is calibrated to directly sense $ds - dS$, and by knowing the gauge length dS one can calculate the extension. This experimental information can then be used to determine the strain components.

As an example, suppose dS is directed along the $X_1 = X$ axis (cf. Fig. 1). Then

$$\Lambda_{(1)} = \sqrt{C_{11}} = \sqrt{1 + 2E_{11}} \tag{20}$$

and

$$E_{(1)} = \sqrt{1 + 2E_{11}} - 1 \tag{21}$$

From these relations one can see that if the strain E_{11} is quite small, then a series approximation of the square root can be used, that is,

$$\sqrt{1 + 2E_{11}} \approx 1 + E_{11} - \tfrac{1}{2}E_{11}^2 + \cdots \tag{22}$$

When the strains are infinitesimal and approximations of the type in Eq. (22) are valid, then

$$E_{(1)} \cong E_{11} \tag{23}$$

In this case, the experimental measurement of extension is also a measurement of the strain component. On the other hand, when the material can be greatly deformed before failure, it would be necessary to use Eq. (3) to obtain the strain. Comparison of Eqs. (21) and (23), or other similar relations, can be used to determine whether the deformation is large enough to require the use of the finite strains or whether the approximate infinitesimal strains can be used.

The difference between the finite and infinitesimal strains can be easily appreciated by writing out the components of strain in terms of the displacement components, U and u, which are defined by the relations

$$x_1 = x_1 + U_1 = X_1 + u_1 \tag{24a}$$
$$x_2 = x_2 + U_2 = X_2 + u_2 \tag{24b}$$
$$x_3 = x_3 + U_3 = X_3 + u_3 \tag{24c}$$

The displacement components U_K are used when the displacement is described in terms of the undeformed coordinates, whereas the components u_k are used with the deformed coordinates. In terms of the displacements, the strains and the deformation tensors can be expressed as

$$2E_{KL} = C_{KL} - \delta_{KL} = \frac{\partial U_K}{\partial X_L} + \frac{\partial U_L}{\partial X_K} + \sum_{M=1}^{M=3} \frac{\partial U_M}{\partial X_K}\frac{\partial U_M}{\partial X_L} \tag{25a}$$

$$2e_{kl} = c_{kl} - \delta_{kl} = \frac{\partial u_k}{\partial x_1} + \frac{\partial u_1}{\partial x_k} + \sum_{m=1}^{m=3} \frac{\partial u_m}{\partial x_k}\frac{\partial u_m}{\partial x_1} \tag{25b}$$

For example,

$$\begin{aligned} 2E_{xx} &= C_{xx} - 1 = 2\frac{\partial U}{\partial X} + \left(\frac{\partial U}{\partial X}\right)^2 + \left(\frac{\partial V}{\partial X}\right)^2 + \left(\frac{\partial W}{\partial X}\right)^2 \\ 2E_{XY} &= C_{XY} = \frac{\partial U}{\partial Y} + \frac{\partial V}{\partial X} + \frac{\partial U}{\partial X}\frac{\partial U}{\partial Y} + \frac{\partial V}{\partial X}\frac{\partial V}{\partial Y} + \frac{\partial W}{\partial X}\frac{\partial W}{\partial Y} \end{aligned} \tag{25c}$$

As shown by Eqs. (25), the strain components can be broken down into first-degree terms and second-degree or product terms. When the derivatives of the displacements arer small the product terms can be neglected and

$$E_{xx} \approx \frac{\partial U}{\partial X}, \qquad E_{xy} \approx \frac{1}{2}\left(\frac{\partial U}{\partial Y} + \frac{\partial v}{\partial x}\right) \tag{26}$$

The strain components that are represented by the first-degree terms as in Eq. (26) are termed the *infinitesimal strain components*.

Ideal Uniaxial Tensile Test The deformation is uniform throughout the test material and can be described by Eq. (3) in terms of the stretch in the direction of uniaxial loading λ and the quantities K_Y, K_Z that correspond to the stretches in the Y and Z directions.

$$[C_{KL}] = \begin{bmatrix} \lambda^2 & 0 & 0 \\ 0 & K_Y^2\lambda^2 & 0 \\ 0 & 0 & K_Z^2\lambda^2 \end{bmatrix} = [c_{kl}^{-1}] \tag{27}$$

$$[c_{kl}] = \begin{bmatrix} \lambda^{-2} & 0 & 0 \\ 0 & (K_Y\lambda)^{-2} & 0 \\ 0 & 0 & (K_Z\lambda)^{-2} \end{bmatrix} = [C_{KL}^{-1}] \tag{28}$$

$$E_{11} = E_{(1)} + \tfrac{1}{2}E_{(1)}^2, \qquad E_{(1)} = \lambda - 1 \tag{29a}$$

$$E_{22} = E_{(2)} + \tfrac{1}{2}E_{(2)}^2, \qquad E_{(2)} = K_Y\lambda - 1 \tag{29b}$$

$$E_{33} = E_{(3)} + \tfrac{1}{2}E_{(3)}^2, \qquad E_{(3)} = K_Z\lambda - 1 \tag{29c}$$

Then the ratio of lateral strain to axially applied strain can be written as

$$\nu_{12} = -\frac{E_{(2)}}{E_{(1)}} = -\frac{K_Y\lambda - 1}{\lambda - 1} \tag{30a}$$

$$\nu_{13} = -\frac{E_{(3)}}{E_{(1)}} = -\frac{K_Z\lambda - 1}{\lambda - 1} \tag{30b}$$

These ratios are the *Poisson ratios* (the first subscript refers to the direction of applied strain; the second subscript refers to the direction of lateral strain).

The extension for a direction in the XY plane inclined at an angle θ from the $X = X_1$ direction can be calculated from

$$E_{(\theta)} = (C_{11}N_1N_1 + 2C_{12}N_1N_2 + C_{22}N_2N_2)^{1/2} - 1 \tag{31a}$$

$$E_{(\theta)} = \lambda\left(\cos^2\theta + K_y^2\sin^2\theta\right)^{1/2} - 1 \tag{31b}$$

By comparison, if $E_{(1)}$ is small,

$$E_{(\theta)} = E_{(1)}(\cos^2\theta - \nu_{12}\sin^2\theta) \tag{32}$$

Relations (31) and (32) can be used in any experimental situation to determine whether the deformation is large enough to require the use of finite strain measures or whether the use of infinitesimal strain components would suffice. For given experimental measurements of the extensions $E_{(1)}$ and $E_{(2)}$, for example, Eq. (30) can be used to determine λ, K_Y, and K_Z. Equations (31) and (32) can be compared as a function of θ. If the difference between the two quantities is small enough to be considered acceptable, then the use of infinitesimal strain components could be justified. For a thorough discussion of the deformation of a continuum, refer to Ref. 19 or 22.

In paper testing the strains are generally small enough that the strain components can be assumed to be infinitesimal components. In certain cases, however, it may be necessary to consider the nonlinear effects of finite strain. Situations for which the use of finite strains may be reqired are

- In-plane straining of highly extensible paper or nonwovens when the strains might be in excess of several percent.
- Strains involving the Z direction in paper where large strains may be observed due to the inherent flexibility of paper in this direction.

●●● Long-time loading involving creep deformation.
●●●● Bending where large deflections are involved; however, the existence of large deflections does not always result in large strains.

III. MEASURES OF STRESS

The stress at a point within a material can be defined by isolating a cubic element and considering the force that is transmitted through each surface. Consider the element shown in Fig. 3, where each side is assumed to have a unit area. The element shown is furthermore assumed to have been removed from the main body while in its loaded state. The force vector per unit area that acts on the side that has its normal directed in the x_1 direction is denoted T_1 and is called the *traction*. The components of the traction along the three coordinate directions, denoted σ_{11}, σ_{12}, σ_{13}, represent the force per unit area that is transmitted across the side. The quantities $\sigma_{kl}(k, l = 1, 2, 3)$ are called the *Eulerian* stress components.

For an elastic material, the components of stress and strain are related. In many cases the stress-free state is the natural state. The strains are most conveniently measured with respect to the undeformed configuration of the material. Then it would be desirable to relate the stresses in the undeformed configuration to the strains measured relative to the undeformed state. The Eulerian stresses that are defined relative to the deformed configuration may not always be appropriate

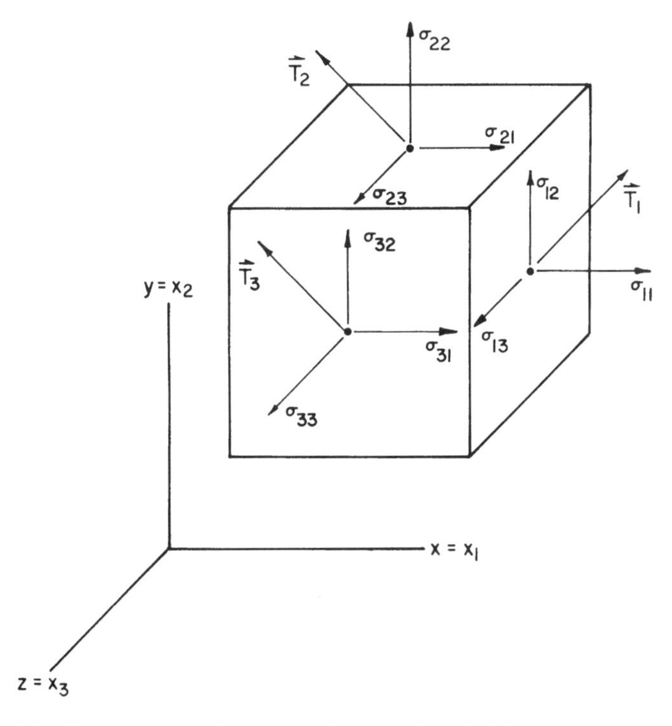

Fig. 3 Stresses acting on an element of material.

for purposes of expressing the material property relationship between stress and strain.

Referring to Figs. 1 and 2, the Eulerian stress component σ_{11} would be calculated from the relation

$$\sigma_{11} = \frac{P}{tw} \tag{33}$$

where P is the applied uniaxial load, and t and w are the deformed thickness and width dimensions. As an alternative stress measure, one could use the undeformed dimensions T and W and write the stress as

$$T_{11} = \frac{P}{TW} \tag{34}$$

Here the stress is the force per unit undeformed area.

There are basically two methods that can be used to relate the Eulerian stresses related to the deformed body to stresses in the undeformed body. Considering Fig. 3, a cubic element in the deformed body can be visualized as a deformed cube in the undeformed body. Likewise, the traction vectors $\vec{T}_1$, $\vec{T}_2$, and $\vec{T}_3$ will be transformed when they are visualized in the undeformed body. The methods for calculating the stresses in the undeformed body are related to assumptions regarding how the traction changes in passing from the deformed to the undeformed configurations.

Figure 4 illustrates a typical two-dimensional view of an element of material with sides dx_1, dx_2 in the loaded and deformed state. On one of the surfaces is shown the force $d\vec{T}$ that is transmitted across this surface. In the undeformed state this element would appear to have a different shape as shown in Fig. 4. The force vector $d\vec{T}$ is labeled $d\vec{T}_0$ in the undeformed state. The Lagrange hypothesis is that $d\vec{T}_0$ is equal to dT (Fig. 4A). The superscript L denotes Lagrange hypothesis. The Kirchhoff hypotheses state that force $d\vec{T}_0^K$ is related to the force $d\vec{T}$ by the same rule that governs the deformation of the body. Thus, if the deformed element length dx_i is given in terms of the undeformed coordinates dX_i,

$$dX_i = \frac{\partial X_i}{\partial x_j}\, dx_j \tag{35}$$

then

$$(dT_0^K)_i = \frac{\partial X_i}{\partial x_j} dT_i \tag{36}$$

In the deformed state, the force per unit surface area is described by the Cauchy stresses σ_{ij}. In the undeformed state, the force per unit area is given by the Lagrangian stresses T_{ij} or the Kirchhoff stresses S_{ij}. It can be shown that the three stress representations are related by the expressions [22]

$$T_{ji} = \frac{\rho_0}{\rho} \frac{\partial X_j}{\partial x_m} \sigma_{mi} \tag{37}$$

$$S_{ji} = \frac{\rho_0}{\rho} \frac{\partial X_i}{\partial x_m} \frac{\partial X_j}{\partial x_n} \sigma_{mn} \tag{38}$$

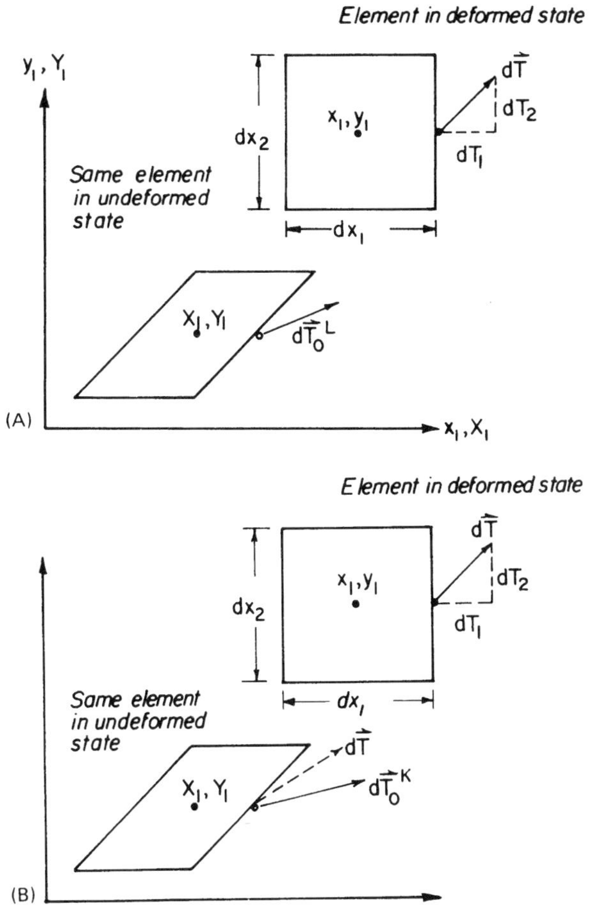

Fig. 4 Illustration of forces transmitted across the surface of an element of material in the deformed and undeformed states. (A) Lagrange case: $d\vec{T}_0^L = d\vec{T}$. Force acting on element surface is the same for deformed and undeformed element. (B) Kirchhoff case: $(dT_0^K)_i = \partial X_i/\partial x_j \, dT_j$. Force acting on element surface differs in the undeformed element from that of the deformed element according to the rule above. (From Ref. 22.)

and

$$\frac{\rho}{\rho_0} = \det\left|\frac{\partial X_i}{\partial x_j}\right| \tag{39}$$

As an example, consider the uniaxial loading case as described by Eq. (3), Thus,

$$\begin{bmatrix} dX \\ dY \\ dZ \end{bmatrix} = \begin{bmatrix} \frac{1}{\lambda} & 0 & 0 \\ 0 & \frac{1}{K_y\lambda} & 0 \\ 0 & 0 & \frac{1}{K_z\lambda} \end{bmatrix} \begin{bmatrix} dx \\ dy \\ dz \end{bmatrix}$$

and

$$\begin{bmatrix} L \\ W \\ T \end{bmatrix} = \begin{bmatrix} \frac{1}{\lambda} & 0 & 0 \\ 0 & \frac{1}{K_y\lambda} & 0 \\ 0 & 0 & \frac{1}{K_z\lambda} \end{bmatrix} \begin{bmatrix} l \\ w \\ t \end{bmatrix}$$

$$\frac{\rho}{\rho_0} = \frac{1}{\lambda^3 K_y K_z}$$

$$T_{11} = (\lambda^3 K_y K_z)\left(\frac{1}{\lambda}\right)\left(\frac{1}{wt}\right) = \left[\frac{K_y\lambda}{w}\left(\frac{K_z\lambda}{t}\right)\right]P = \frac{P}{WT}$$

$$S_{11} = (\lambda^3 K_y K_z)\left(\frac{1}{\lambda^2}\right)\left(\frac{P}{wt}\right) = \frac{P}{WT}\left(\frac{L}{l}\right)$$

Thus, if σ_{11} is the force P per unit deformed area wt, then T_{11} is the force P per unit undeformed area WT. S_{11} differs from T_{11} by the ratio L/l.

IV. RELATIONS BETWEEN LOADING AND DEFORMATION: CONSTITUTIVE RELATIONS

The relationship between loading and deformation that describes the characteristic behavior of a material is referred to as a constitutive relation. In fact, the load–deformation relation depends upon the magnitude of stress and strain, the rate of application of stress and strain, and such environmental factors as temperature and moisture content. From a practical point of view, it is not feasible to use one constitutive relation to describe all possible behavioral characteristics of a material. Rather, it is generally necessary to relate certain variables such as load and deformation with carefully stipulated restrictions on the magnitude and rate of application of load and deformation and the environmental conditions.

It is most common to attempt to describe behavior according to the following categories: (1) constant or time-varying environmental conditions of temperature and moisture content, (2) infinitesimal strain, (3) finite strain, (4) short time, (5) low rate of loading or deformation, (6) long time, (7) low stress, (8) moderate to high stress, and (9) strength.

A. Infinitesimal Linear Elasticity

When the duration of loading is short and the strain is small enough that the deformation can be described by the infinitesimal strain components and when the environmental conditions such as temperature and humidity are constant, the stress and strain can be assumed to be linearly related. Since the strains are infinitesimal, there is no distinction to be made between the Cauchy, Lagrangian, or Kirchhoff stresses or the deformed and undeformed coordinates. In the case of paper, the material can be considered to exhibit orthotropic symmetry (cf. Jones [2 .e stress–strain relation can be expressed as

$$\begin{bmatrix} e_{11} \\ e_{22} \\ e_{33} \\ e_{23} \\ e_{13} \\ e_{12} \end{bmatrix} = \begin{bmatrix} \frac{1}{E_1} & -\frac{\nu_{21}}{E_2} & -\frac{\nu_{31}}{E_3} & 0 & 0 & 0 \\ -\frac{\nu_{12}}{E_1} & \frac{1}{E_2} & -\frac{\nu_{32}}{E_3} & 0 & 0 & 0 \\ -\frac{\nu_{13}}{E_1} & -\frac{\nu_{23}}{E_2} & \frac{1}{E_3} & 0 & 0 & 0 \\ 0 & 0 & 0 & \frac{1}{2G_{23}} & 0 & 0 \\ 0 & 0 & 0 & & \frac{1}{2G_{13}} & 0 \\ 0 & 0 & 0 & & & \frac{1}{2G_{12}} \end{bmatrix} \begin{bmatrix} \sigma_{11} \\ \sigma_{22} \\ \sigma_{33} \\ \sigma_{23} \\ \sigma_{13} \\ \sigma_{12} \end{bmatrix} \tag{40}$$

The quantities E_1, E_2, E_3 are called the Young moduli in the 1, 2, and 3 directions. The ν_{ij} are the Poisson ratios for transverse strain in the j direction when a load is applied in the i direction. A condition of orthotropic symmetry is that

$$\frac{\nu_{12}}{E_1} = \frac{\nu_{21}}{E_2}, \qquad \frac{\nu_{13}}{E_1} = \frac{\nu_{31}}{E_3}, \qquad \frac{\nu_{23}}{E_2} = \frac{\nu_{32}}{E_3} \tag{41}$$

The G_{23}, G_{31}, and G_{12} are the shear moduli in the 23, 31, and 12 planes. The releations provided by (40) involve nine elastic constants.

The significance of the Young moduli and Poisson ratios can be made clear by considering the uniaxial tension or compression test. Suppose the load P is applied as in Figs. 1 and 2. Let $X = x_1$, $Y = x_2$, $Z = x_3$. Then $\sigma_{11} = P/wt$. $E_{11} = (l - L)/L$, $e_{22} = (w - W)/W$, $e_{33} = (t - T)/T$.

In accordance with the usual assumptions of small strain elasticity, the strains are assumed to be infinitesimal. The other stress components, $\sigma_{22}, \sigma_{33}, \sigma_{12}, \sigma_{23}$, are all assumed to be zero. As a consequence of Eq. (40), it will also be true that $e_{12} = e_{13} = e_{23} = 0$. These assumptions can be expected to be very realistic as long as the region shown in Figs. 1 and 2 pertains to the central portion of the uniaxial specimen that is a substantial distance away from the ends where the load is applied through the testing machine grips. From a practical viewpoint, the assumptions are nearly true when the distance from the grips is approximately equal to the largest dimension of the cross section of the specimen, for example, W.

For the stated assumptions, the relations (40) are

$$e_{11} = \frac{1}{E_1}\sigma_{11}, \qquad e_{22} = -\frac{\nu_{12}}{E_1}\sigma_{11}, \qquad e_{33} = -\frac{\nu_{13}}{E_1}\sigma_{11} \tag{42}$$

Thus, the Young modulus E_1 is the ratio of stress to strain. The Poisson ratios ν_{12} and ν_{13} can be expressed as

$$\nu_{12} = -\frac{e_{22}}{e_{11}}, \qquad \nu_{13} = -\frac{e_{33}}{e_{11}} \tag{43}$$

The relations (40) can be inverted to give the stresses in terms of the strains:

$$\begin{bmatrix} \sigma_{11} \\ \sigma_{22} \\ \sigma_{33} \\ \sigma_{23} \\ \sigma_{13} \\ \sigma_{12} \end{bmatrix} = \begin{bmatrix} C_{11} & C_{12} & C_{12} & 0 & 0 & 0 \\ C_{21} & C_{22} & C_{23} & 0 & 0 & 0 \\ C_{31} & C_{32} & C_{33} & 0 & 0 & 0 \\ 0 & 0 & 0 & 2C_{44} & 0 & 0 \\ 0 & 0 & 0 & 0 & 2C_{55} & 0 \\ 0 & 0 & 0 & 0 & 0 & 2C_{66} \end{bmatrix} \begin{bmatrix} e_{11} \\ e_{22} \\ e_{33} \\ e_{23} \\ e_{13} \\ e_{12} \end{bmatrix} \tag{44}$$

where

$$C_{11} = \frac{1 - \nu_{23}\nu_{32}}{E_2 E_3 \Delta}, \qquad C_{22} = \frac{1 - \nu_{13}\nu_{31}}{E_1 E_3 \Delta}, \qquad C_{33} = \frac{1 - \nu_{12}\nu_{21}}{E_1 E_2 \Delta}$$

$$C_{12} = C_{21} = \frac{\nu_{21} + \nu_{31}\nu_{23}}{E_2 E_3 \Delta} = \frac{\nu_{12} + \nu_{32}\nu_{13}}{E_1 E_3 \Delta}$$

$$C_{13} = C_{31} = \frac{\nu_{31} + \nu_{21}\nu_{32}}{E_2 E_3 \Delta} = \frac{\nu_{13} + \nu_{12}\nu_{23}}{E_1 E_2 \Delta}$$

$$C_{23} = C_{32} = \frac{\nu_{32} + \nu_{12}\nu_{31}}{E_1 E_3 \Delta} = \frac{\nu_{23} + \nu_{21}\nu_{13}}{E_1 E_2 \Delta}$$

$$C_{44} = G_{23'} \qquad C_{55} = G_{13'} \qquad C_{66} = G_{12}$$

$$\Delta = \frac{1 - \nu_{12}\nu_{21} - \nu_{23}\nu_{32} - \nu_{31}\nu_{13} - 2\nu_{21}\nu_{32}\nu_{13}}{E_1 E_2 E_3}$$

A laboratory-made paper is usually characterized by transversely isotropic symmetry. If it is assumed that the 1, 2 directions lie in the plane of the sheet and the 3 direction corresponds to the z direction of the sheet, the constitutive relations have the form of Eq. (40) but with $E_2 = E_1$, $\nu_{12} = \nu_{21} = \nu$. The transversely isotropic material is characterized by five independent constants.

To summarize, in the general case paper is assumed to have orthotropic symmetry. The complete description of the elastic properties of paper requires the determination of nine independent constants. In the case of handsheets made in a laboratory, there is no machine direction or cross-machine direction. Thus, the elastic properties in the plane of the paper do not depend on direction. For in-plane loading, the handsheet material exhibits isotropic symmetry. Clearly, this refers only to the in-plane loading. Paper is basically a layered structure, because the fibers usually lie essentially in the plane of the sheet. Because of this layered structure, the handsheet material is transversely isotropic, that is, it exhibits the same properties for all directions that lie in the plane of the sheet. On the other hand, the z-direction Young modulus and the interlaminar shear modulus are different than those associated with the plane of the sheet. In fact, as a consequence of the layered structure of paper, the z-direction Young modulus and the interlaminar shear modulus are approximately one to two orders of magnitude smaller than the corresponding in-plane moduli.

The elastic constants of (40) are independent. However, there are certain restrictions on their values that derive essentially from the concept that the strain energy stored in a loaded material should have the property that it is always greater than zero. It can be zero only if all the strains (or stresses) are zero. The strain energy function, therefore, is called positive definite. In the case of a truly isotropic material, the restriction imposed by positive definiteness of the strain energy is that the Poisson ratio lie between -1 and $+\frac{1}{2}$. One may refer to Fung [22] for a thorough discussion of these limits. Although it is theoretically possible to have a negative Poisson ratio, experimental measurements normally result in positive Poisson ratios when a material such as paper is loaded in the infinitesimal strain elastic region. The upper limit of $\frac{1}{2}$ can be understood as a consequence of having a positive bulk modulus associated with positive work being done when an isotropic material is subjected to pressure loading. In the more general case of an elastic material having orthotropic symmetry, Lempriere [37] provided restrictions on the Poisson ratios as

$$\nu_{12} < \left(\frac{E_1}{E_2}\right)^{1/2}, \qquad \nu_{21} < \left(\frac{E_2}{E_1}\right)^{1/2} \tag{45a}$$

$$\nu_{23} < \left(\frac{E_2}{E_3}\right)^{1/2}, \qquad \nu_{32} < \left(\frac{E_3}{E_2}\right)^{1/2} \tag{45b}$$

$$\nu_{13} < \left(\frac{E_1}{E_3}\right)^{1/2}, \qquad \nu_{31} < \left(\frac{E_3}{E_1}\right)^{1/2} \tag{45c}$$

As an example, for the case of the transversely isotropic handsheet, relations (45) give the restriction on the in-plane Poisson ratio $\nu < 1.0$, whereas for the z-direction Poisson ratios, $\nu_{13} < (E_1/E_3)^{1/2}$, $\nu_{31} < (E_3/E_1)^{1/2}$. For example, if $E_1/E_3 = 100$, then $\nu_{13} < 10$ and $\nu_{31} < 0.1$.

It should be noted that there is no universally used notation convention for the definition of Poisson ratios. See, for example, Whitney [96] or the ANSYS manual [35].

Stress–Strain Relations for In-Plane Loading In paper testing, the loading is often confined to the directions of the plane of the sheet, the z-direction normal and shear stresses being zero. This corresponds to a special case of loading commonly referred to as plane stress.

If $\sigma_{33} = \sigma_{13} = \sigma_{23} = 0$, then the stress–strain relation for σ_{11}, σ_{22}, σ_{23} and e_{11}, e_{22}, e_{12} can be given as

$$\begin{bmatrix} e_{11} \\ e_{22} \\ e_{12} \end{bmatrix} = \begin{bmatrix} \frac{1}{E_1} & -\frac{\nu_{12}}{E_1} & 0 \\ -\frac{\nu_{21}}{E_2} & \frac{1}{E_2} & 0 \\ 0 & 0 & \frac{1}{2G_{12}} \end{bmatrix} \begin{bmatrix} \sigma_{11} \\ \sigma_{22} \\ \sigma_{12} \end{bmatrix} \tag{46}$$

Rather than using the relations (44) for the inverse relation to (46), one can invert (46) directly. Then the relations for the stresses in terms of the strains can be written

$$\begin{bmatrix} \sigma_{11} \\ \sigma_{22} \\ \sigma_{12} \end{bmatrix} = \begin{bmatrix} \dfrac{E_1}{1-\nu_{12}\nu_{21}} & \dfrac{\nu_{12}E_2}{1-\nu_{12}\nu_{21}} & 0 \\ \dfrac{\nu_{21}E_1}{1-\nu_{12}\nu_{21}} & \dfrac{E_2}{1-\nu_{12}\nu_{21}} & 0 \\ 0 & 0 & 2G_{12} \end{bmatrix} \begin{bmatrix} e_{11} \\ e_{22} \\ e_{12} \end{bmatrix} \tag{47}$$

It is important to realize that it is assumed that $\sigma_{33} = \sigma_{23} = 0$ when relations (47) are to be employed.

In some cases it is convenient to express the stress–strain relations for coordinates that are not aligned with the directions of elastic symmetry. In the foregoing, it is assumed that the coordinate axes are directed along the axes of symmetry. These symmetry directions are generally identified as the machine direction, the cross-machine direction, and the z direction for paper. However, it is possible in some special cases that the elastic symmetry directions may not be aligned with the machine, cross-machine, and z directions. When the coordinate directions x, y for a plane stress loading are inclined at an angle θ to the 1, 2 directions of elastic symmetry, the stress–strain relations become

$$\begin{bmatrix} e_{xx} \\ e_{yy} \\ \gamma_{xy} \end{bmatrix} = \begin{bmatrix} \bar{S}_{11} & \bar{S}_{12} & \bar{S}_{16} \\ \bar{S}_{12} & \bar{S}_{22} & \bar{S}_{26} \\ \bar{S}_{16} & \bar{S}_{26} & \bar{S}_{66} \end{bmatrix} \begin{bmatrix} \sigma_{xx} \\ \sigma_{yy} \\ \sigma_{xy} \end{bmatrix} \tag{48}$$

where $\gamma_{xy} = 2e_{xy}$ and

$$\bar{S}_{11} = S_{11}\cos^4\theta + (2S_{12} + S_{66})\sin^2\theta\cos^2\theta + S_{22}\sin^4\theta \tag{49a}$$

$$\bar{S}_{12} = S_{12}\,(\sin^4\theta + \cos^4\theta) + (S_{11} + S_{22} - S_{66})\sin^2\theta\cos^2\theta \tag{49b}$$

$$\bar{S}_{22} = S_{11}\sin^4\theta + (2S_{12} + S_{66})\sin^2\theta\cos^2\theta + S_{22}\cos^4\theta \tag{49c}$$

$$\bar{S}_{16} = (2S_{11} - 2S_{12} - S_{66})\sin\theta\cos^3\theta - (2S_{22} - 2S_{12} - S_{66})\sin^3\theta\cos\theta \tag{49d}$$

$$\bar{S}_{26} = (2S_{11} - 2S_{12} - S_{66})\sin^3\theta\cos\theta - (2S_{22} - 2S_{12} - S_{66})\sin\theta\cos^3\theta \tag{49e}$$

$$\bar{S}_{66} = 2(2S_{11} + 2S_{22} - 4S_{12} - S_{66})\sin^2\theta\cos^2\theta + S_{66}(\sin^4\theta + \cos^4\theta) \tag{49f}$$

and

$$S_{11} = \frac{1}{E_1}, \qquad S_{22} = \frac{1}{E_2}, \qquad S_{12} = -\frac{\nu_{12}}{E_1} = -\frac{\nu_{21}}{E_2}, \qquad S_{66} = \frac{1}{G_{12}} \tag{50}$$

For further information concerning the stress–strain relations for coordinate directions that are not aligned with the elastic symmetry directions, refer to Jones [29].

The form of relations (48) has an important implication for the experimentalist. The existence of the terms with indices 16 and 26 implies a coupling between the extension and shearing deformations. If, for example, a tensile test were performed with loading in the x direction, then the tensile stress σ_{xx} could be expected to cause a shear deformation in addition to the normal strains e_{xx} and e_{yy}. Similarly, if a shear stress σ_{xy} were applied to a sample without normal stresses σ_{xx} and σ_{yy}, an extensional strain in the x and y directions would be observed in addition to shearing strain. Because of these complicated interactions, it is difficult to perform a stress–strain test on paper samples when the loading directions are not aligned with the directions of elastic symmetry.

Suppose uniaxial tensile or edgewise compression tests are carried out on paper samples for various orientations θ. This could be done by cutting rectangular samples from a sheet with the long axis of the sample inclined at the angle θ from the machine direction. In accordance with Eqs. (49), a test at $\theta = 0°$ would yield an experimental value for E_1. A test at $\theta = 90°$ would provide E_2. It would seem that if two other orientations were used, experimental values of G_{12} and ν_{12} or ν_{21} could be obtained. This is not possible, however, owing to the coupling of terms in the expressions. However, if one were to make experimental measurements for the extensional modulus along the machine and cross-machine directions along with a measurement of one of the two Poisson ratios, then Eq. (49) could be used to calculate the in-plane shear modulus. The use of this approach has been reported by Castagnede et al. [10,11]. This procedure must be used with caution because of the presence of induced shearing deformation under tensile loading. Special precautions must be taken to permit the shearing deformation to take place. In general, it is not too difficult to conduct tests of this type when the loading is tensile and the specimen length-to-width ratio is large. However, it is virtually impossible to ensure the proper conditions for edgewise compressive loading. Seo et al. [80] studied the errors involved in the determination of the in-plane shear modulus and proposed an optimization approach for minimizing the errors involved. For further information regarding the angular transformation relations, refer to Jones [29].

B. Other Kinds of Elasticity

The most common elasticity model is the one described in the preceding section, that is, linear elasticity. The linear elastic model of the behavior of paper is very useful in applications and can be considered to be a reasonable model for predicting the relations between stress and strain when the environmental conditions are constant, when the duration of loading or straining is relatively short, and when the stresses and strains are maintained at levels below the elastic limit. Unfortunately, the elastic limit is not easy to determine. The generally accepted requirements for linear elastic behavior are that the relation between stress and strain be linear and that the strain be zero (i.e., no residual deformation) when the stress is removed. There are three forms of elasticity that provide useful models for describing the behavior of paper. These are hypoelasticity, hyperelasticity, and viscoelasticity. The essential features of these models are outlined below.

Hypoelasticity A material that is hypoelastic is one that is characterized by a linear relationship between a time rate of change of stress and the deformation rate. Although a material may behave inelastically and therefore could not be considered to be linearly elastic, it may be reasonable to assume that in the vicinity of any deformed state increments in stress are linearly related to increments in deformation. This concept is most easily formulated by making the stress rate proportional to the deformation rate. In essence, one can say that a hypoelastic material is one that behaves elastically for incremental changes in stress and strain from a given state that may have been achieved through an inherently inelastic process. The constitutive relation for a hypoelastic material that exhibits orthotropic symmetry is

$$\begin{bmatrix} \overset{\nabla}{\sigma}_{11} \\ \overset{\nabla}{\sigma}_{22} \\ \overset{\nabla}{\sigma}_{33} \\ \overset{\nabla}{\sigma}_{23} \\ \overset{\nabla}{\sigma}_{13} \\ \overset{\nabla}{\sigma}_{12} \end{bmatrix} = \begin{bmatrix} C_{11} & C_{12} & C_{13} & 0 & 0 & 0 \\ C_{21} & C_{22} & C_{23} & 0 & 0 & 0 \\ C_{31} & C_{32} & C_{33} & 0 & 0 & 0 \\ 0 & 0 & 0 & C_{44} & 0 & 0 \\ 0 & 0 & 0 & 0 & C_{55} & 0 \\ 0 & 0 & 0 & 0 & 0 & C_{66} \end{bmatrix} \begin{bmatrix} V_{11} \\ V_{22} \\ V_{33} \\ 2V_{23} \\ 2V_{13} \\ 2V_{12} \end{bmatrix} \tag{51}$$

where $\overset{\nabla}{\sigma}_{11}$, etc. are stress rates that are defined below; V_{11}, etc. are deformation rates; and C_{11}, C_{22}, etc. represent incremental moduli. The incremental moduli may depend on the entire history of loading and deformation (i.e., the state and history of stress and strain), but they are independent of the stress and deformation rates. The deformation rate V_{ij} is defined as

$$V_{ij} = \frac{1}{2}\left(\frac{\partial v_i}{\partial x_j} + \frac{\partial v_j}{\partial x_i}\right) \tag{52}$$

where the terms v_i are rates of displacement with respect to time, i.e., the velocities. The rate of deformation times a small time increment dt is approximately equal to the infinitesimal strain tensor, i.e., $V_{ij}\,dt = \tilde{e}_{ij}$.

The stress rate $\overset{\nabla}{\sigma}_{ij}$ employed in relation (51) is linearly related to the rate of deformation. The rate of deformation has the property that it will always be zero whenever the body performs a rigid-body motion. In order for relation (51) to be a consistent and meaningful relationship, the stress rate $\overset{\nabla}{\sigma}_{ij}$ must also have the property that it will be zero for the case of a rotating body subjected to a constant state of stress (e.g., a disk rotating at constant angular velocity). The stress rate $\overset{\nabla}{\sigma}_{ij}$, called the Jaumann stress rate, properly takes the rotation into account and is defined as

$$\overset{\nabla}{\sigma}_{ij} = \frac{D}{DT}\sigma_{ij} - \sigma_{ip}\tilde{r}_{pj} - \sigma_{jp}\tilde{r}_{pi} \tag{53}$$

Here, $\tilde{r}_{ij}$ are the components of the spin given by

$$\tilde{r}_{ij} = \frac{1}{2}\left(\frac{\partial v_i}{\partial x_j} - \frac{\partial v_j}{\partial x_i}\right) \tag{54}$$

and D/DT is the material derivative that includes local time change and change due to convective effects. For a thorough explanation, refer to Fung [22].

The incremental moduli C_{11}, C_{22}, etc. depend on the details of the possibly inelastic process that leads up to the current state corresponding to relations (51). For example, if one could perform a uniaxial test with a prescribed V_{11} while other components were zero, then C_{11} would be the incremental tangent modulus. The tangent modulus E_T (refer to Fig. 5) depends on the value of the strain. At a particular point P, with stress σ_P and strain ε_P, the tangent modulus is the slope of the line AB drawn tangent to the stress–strain curve at P. The property of hypoelasticity requires that incremental increases or decreases in strain be accompanied by incremental increases or decreases in stress in accordance with

$$\Delta\sigma = E_T \Delta\varepsilon \tag{55}$$

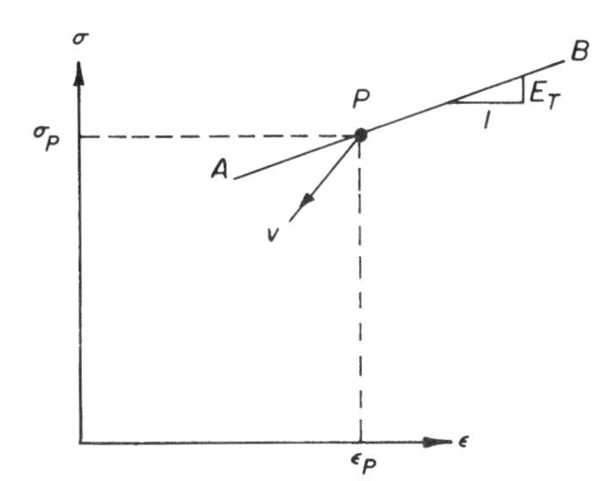

Fig. 5 Uniaxial stress–strain curve. Tangent modulus E_T.

In the uniaxial testing of a material that exhibits some inelastic effects, increases in stress and strain would be related by (51); however, for unloading, the curve might follow a different branch such as *PV* in Fig. 5. This type of material behavior could not be characterized as hypoelastic.

The idealization of hypoelasticity may be quite useful in the study of certain problems. For example, during the edgewise compression loading of paper, the material behaves elastically at first, then exhibits an inelastic reponse, and finally fails through the development of a localized material instability. The material response of the localized compressive instability can be analyzed with the assumption that the paper is incrementally elastic even though the time total response of loading to the instability point is inelastic. This approach to instability problems was extensively developed by Biot [8] for a wide variety of materials. For further information concerning the application to paper, refer to Perkins and McEvoy [62].

Hyperelasticity A hyperelastic material, as discussed by Fung [22], exhibits the special property of possessing a strain energy function per unit mass, W, that is a function of the strain components. Furthermore, this class of materials has the property that the rate of change of strain energy equals the rate of work done by the stresses,

$$\frac{D(W)}{Dt} = \frac{1}{\rho}\sigma_{ij}V_{ij} \tag{56}$$

and

$$W = W(e_{11}, e_{22}, \ldots, e_{12}) \tag{57}$$

The right-hand side of (56) represents the rate of work done by the stresses. If the functional form (57) that relates the strain components to the strain energy per unit mass is known, then by forming the material derivative of W in Eq. (56), the stress–strain law for the material can be obtained. This definition of elasticity includes the usual definition of linear elasticity. On the other hand, it is more general than linear elasticity, because hyperelastic behavior can be nonlinear. The behavior is linear only when the strain energy function is a second degree polynomial in the strain components.

The definition of hyperelasticity just given uses the Eulerian variables of stress and strain. The definition is also given by Fung [22] in terms of the Lagrangian variables, as

$$\frac{D(W)}{Dt} = \frac{1}{\rho_0} T_{ij} \frac{DE_{ij}}{Dt} \tag{58}$$

When the deformation is infinitesimal, the distinction between Eulerian and Lagrangian variables vanishes; then

$$V_{ij} = \frac{\partial e_{ij}}{\partial t} \tag{59}$$

Also, the material derivative can be replaced by the ordinary time derivative. Thus,

$$\frac{D(W)}{Dt} = \frac{dW}{dt} = \frac{\partial W}{\partial e_{ij}} \frac{\partial e_{ij}}{\partial t} \tag{60}$$

Then

$$\frac{\partial W}{\partial e_{ij}} \frac{\partial e_{ij}}{\partial t} = \frac{1}{\rho} \sigma_{ij} \frac{\partial e_{ij}}{\partial t} \tag{61}$$

Therefore,

$$\sigma_{ij} = \frac{\partial(\rho W)}{\partial e_{ij}} \tag{62}$$

When ρW is a known function of the strain components e_{ij}, the derivatives with respect to strain can be calculated, and Eq. (62) can be taken as the stress–strain law.

Relation (62) is of particular importance in the development of a theory of mechanical behavior of paper based on the network structure of the fibers. In this case, the strain energy stored in a typical fiber is calculated for arbitrary sheet strains. The total network strain energy is found by adding together the strain energies of the individual fibers in a unit mass. Finally, Eq. (62) is used to obtain the in-plane stress–strain behavior of paper. This approach is discussed more thoroughly later.

Johnson–Urbanik Hyperelasticity Model for Paper The use of a hyperelasticity model for describing the nonlinear behavior of paper was reported by Johnson and Urbanik [28]. In their work the complementary strain energy density is defined by the Legendre transformation as

$$H_C(T) = T_{ij} E_{ij} - H(E) \tag{63}$$

where $H(E) = \rho_0 W(E)$. In this definition the Lagrangian strains can be computed from the relation

$$E_{ij} = \frac{\partial H_C(T)}{\partial T_{ij}} \tag{64}$$

Johnson and Urbanik make the assumption that H_C depends only on the in-plane Lagrangian stress components T_{11}, T_{22} and T_{12}. Limiting attention to infinitesimal deformation and considering only the in-plane components of stress and strain, Johnson and Urbanik express the strain energy density in the form

$$H = \frac{\nu_2 E_1}{2(1 - \nu_1 \nu_2)} e \tag{65}$$

where

$$e = \frac{\varepsilon_1^2}{\nu_2} + \frac{\varepsilon_2^2}{\nu_1} + 2\varepsilon_1\varepsilon_2 + c\varepsilon_{12}^2 \tag{66}$$

and

$$c = \frac{4(1 - \nu_1 \nu_2)}{\nu_1 E_2} G \tag{67}$$

In the above, E_1, E_2, and G represent the two in-plane elastic moduli and the shear modulus. The Poisson ratios ν_1, ν_2 and the three elastic moduli are constants determined experimentally in the context of linear elasticity. The variable

$$e_g = \sqrt{e} \tag{68}$$

is a measure of the magnitude of the strain at a point in the plane of the material.

The in-plane infinitesimal stress–strain relations are given by the expressions

$$\sigma_{11} = H'(e)2\left(\frac{\varepsilon_1}{\nu_2} + \varepsilon_2\right) \tag{69a}$$

$$\sigma_{22} = H'(e)2\left(\frac{\varepsilon_2}{\nu_1} + \varepsilon_1\right) \tag{69b}$$

$$\sigma_{12} = H'(e)c\varepsilon_{12} \tag{69c}$$

where the prime notation $H'(e)$ represents the derivative of $H(e)$ with respect to the variable e. The elastic contstants and the functional form of $H(e)$ are determined by comparison with the experimentally observed stress–strain curve. The resulting model can be used to predict the nonlinear stress–strain behavior of paper for arbitrary in-plane deformation. One of the consequences of the model is that for uniaxial loading, e.g., $\sigma_{11} = 0$, the lateral strain $\varepsilon_1 = -\nu_2\varepsilon_2$, which is the same as that associated with linear elasticity. A similar result holds for the perpendicular material direction. Therefore, this model predicts that the Poisson ratios are constant even when the strains are sufficiently large as to exhibit nonlinear behavior.

The function $H(e)$ is determined by matching the uniaxial stress–strain curve corresponding to either the paper machine or cross-machine directions. For example, Johnson and Urbanik suggest using the two-parameter function

$$\sigma(\varepsilon) = c_1 \tanh\left(\frac{c_2 \varepsilon}{c_1}\right) \tag{70}$$

Using this function, it is found that

$$H(e) = \frac{c_1^2}{c_2} \log\cosh\left(\frac{c_2}{c_1}\sqrt{\frac{\nu_1 e}{1 - \nu_1 \nu_2}}\right) \tag{71}$$

This approach was used successfully to model the compressive behavior of linerboard where the strain to failure is less than 1%. Application to modeling behavior for more general loading conditions where failure strain is approximately 2% has not been reported. The hyperelasticity model can be expected to provide a convenient approach for modeling the nonlinear in-plane response of paper for proportional loading problems; however, the uniaxial stress–strain relation (70) may be replaced by alternative empirical functions when the loading includes tensile stress and strain that may lead to larger values of strain prior to failure. The hyperelasticity model is relatively simple to use in numerical work including finite element software.

Viscoelasticity When a material exhibits elastic behavior, the constitutive relation between stress and strain is not time-dependent. The sudden application of a constant value of stress is accompanied by an immediately observed constant level of strain. The assumption of elastic behavior is a realistic one when the strains are small, the duration of loading is relatively short, and the temperature and moisture content of the material are low. When these restrictions are not met, the relationship between stress and strain may be time-dependent.

For example, when a constant level of stress is suddenly applied to a material, the strain may immediately reach a certain level, but thereafter the strain may increase with time. This type of loading, in which the stresses are constant but the strains increase with duration of load application, is referred to as *creep loading*. On the other hand, when a constant strain is suddenly applied to the material, some stress is immediately observed; however, the level of stress decreases with the duration of time following the application of strain. This phenomenon is referred to as *stress relaxation*.

If the environmental conditions of temperature and relative humidity are constant with time and their values are moderate, and if the levels of applied stress or imposed strain are small (e.g., if the strains are restricted to be infinitesimal), then the material behavior can often be described by the linear viscoelasticity constitutive law. The assumption of linear viscoelasticity essentially requires that the Boltzmann superposition principle be valid. This is most easily understood in the context of a simple uniaxial strain test.

Suppose that a stress σ_0 is applied at a time $t_0 = 0$. The creep response is time-dependent as shown in Fig. 6A. This creep strain could be described by a relation of the form

$$e(t) = \sigma_0 S(t) \tag{72}$$

The entity $S(t)$, called the creep compliance, is a monotonically increasing function of time that provides the shape of the strain curve in Fig. 6A. Note that the strain in Eq. (72) is proportional to the level of applied stress. Referring now to Fig. 6B, Eq. (72) should hold until time $t = t_1$, at which point the stress is further increased. For times greater than $t = t_1$, the strain may be described by the relation

$$e(t) = \sigma_0 S(t) + (\sigma_1 - \sigma_0) S(t - t_1), \qquad t > t_1 \tag{73}$$

Similarly, for the case of Fig. 6C, the strain should be given by relation (72) for $t < t_1$. For $t > t_1$, the strain may be described by

$$e(t) = \sigma_0 S(t) - \sigma_0 S(t - t_1), \qquad t > t_1 \tag{74}$$

As shown by the form of Eqs. (73) and (74), it is assumed that the effects of stress change can be accounted for by superimposing the effects of load change at different times in the manner shown. It must be demonstrated by experiment that relations of the form (73) and (74) are valid. If this can be shown, then the

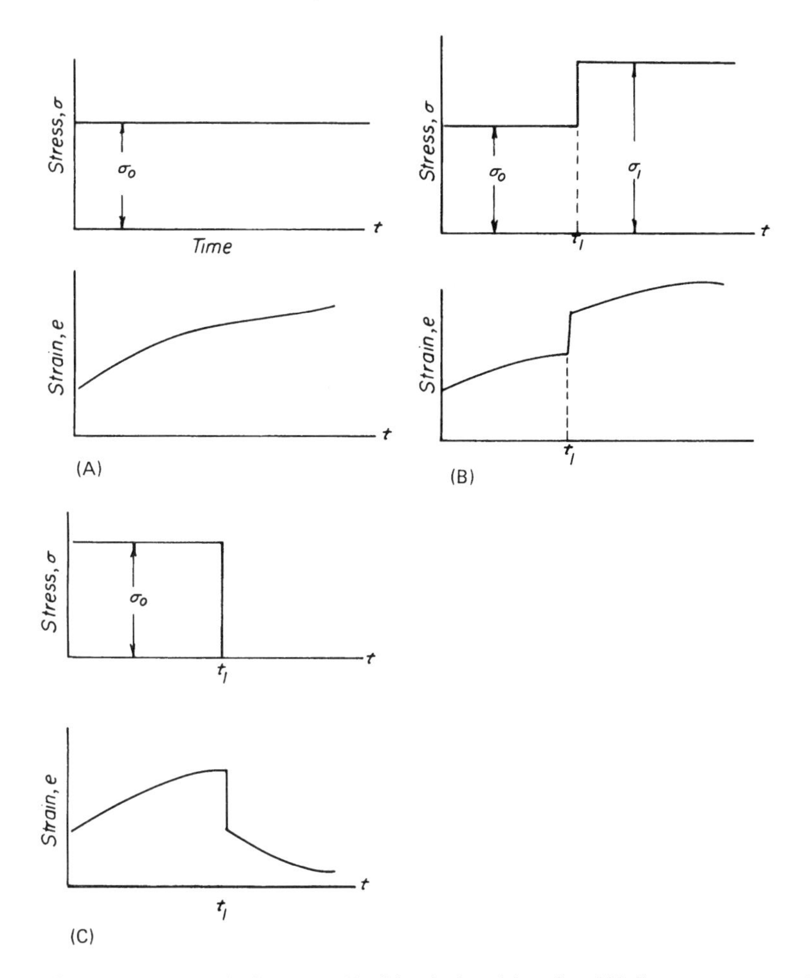

Fig. 6 Creep strain for prescribed load–time histories. (A) Step stress σ_0 applied at $t = 0$. (B) Step stress σ_0 applied at $t = 0$ followed by step stress σ_1 applied at $t = \tau_1$. (C) Step stress σ_0 applied at $\tau = 0$ and removed at $\tau = \tau_1$.

Boltzmann superposition principle is valid and the material behavior can be described by linear viscoelasticity. In the context of the foregoing discussion, it is assumed that environmental conditions remain constant. Furthermore, although the example described made the assumption of a uniaxial test, the principle must be demonstrated for general conditions of loading involving other components of stress and strain.

For a general history of stress application, the strain response in the uniaxial case can be given in the integral form

$$e(t) = \int_0^t S(t - \theta) \frac{d\sigma(\theta)}{d\theta} \, d\theta \tag{75}$$

Because of the special form of Eq. (75), which is a consequence of the assumed linearity of the system (Boltzmann superposition principle), the relationship between the Laplace transform of the stress and strain is the same in form as that for classical linear elasticity. Denote by a bar over the symbol its Laplace transform, for example,

$$\bar{f}(p) = \int_0^\infty e^{-pt} f(t)\, dt \tag{76}$$

Then the Laplace transform of (75) is

$$\bar{e}(p) = p\bar{S}(p)\bar{\sigma}(p) \tag{77}$$

The quantity

$$p\bar{S}(p) = \tilde{S} \tag{78}$$

(by definition) is called the *operational modulus*. Therefore,

$$\bar{e} = \tilde{S}\bar{\sigma} \tag{79}$$

The same principle applies to the general case of loading. Because paper exhibits orthotropic symmetry in the classic linear elasticity realm, the operational moduli exhibit the same symmetry for conditions consistent with the assumption of linear viscoelasticity. Therefore, referring to relation (44) for linear elasticity,

$$\begin{bmatrix} \bar{e}_{11} \\ \bar{e}_{22} \\ \bar{e}_{33} \\ \bar{e}_{23} \\ \bar{e}_{13} \\ \bar{e}_{12} \end{bmatrix} = \begin{bmatrix} \tilde{S}_{11} & \tilde{S}_{12} & \tilde{S}_{13} & 0 & 0 & 0 \\ \tilde{S}_{21} & \tilde{S}_{22} & \tilde{S}_{23} & 0 & 0 & 0 \\ \tilde{S}_{31} & \tilde{S}_{32} & \tilde{S}_{33} & 0 & 0 & 0 \\ 0 & 0 & 0 & \tilde{S}_{23} & 0 & 0 \\ 0 & 0 & 0 & 0 & \tilde{S}_{13} & 0 \\ 0 & 0 & 0 & 0 & 0 & \tilde{S}_{12} \end{bmatrix} \begin{bmatrix} \bar{\sigma}_{11} \\ \bar{\sigma}_{22} \\ \bar{\sigma}_{33} \\ \bar{\sigma}_{23} \\ \bar{\sigma}_{13} \\ \bar{\sigma}_{12} \end{bmatrix} \tag{80}$$

The inverse relation providing the stresses in terms of the strains can be obtained by the inversion of (80). Thus, if $[\tilde{C}_{ij}]$ represents the inverse of the $[\tilde{S}_{ij}]$ matrix in (80), then

$$\begin{bmatrix} \bar{\sigma}_{11} \\ \bar{\sigma}_{22} \\ \bar{\sigma}_{33} \\ \bar{\sigma}_{23} \\ \bar{\sigma}_{13} \\ \bar{\sigma}_{12} \end{bmatrix} = \begin{bmatrix} \tilde{C}_{11} & \tilde{C}_{12} & \tilde{C}_{13} & 0 & 0 & 0 \\ \tilde{C}_{21} & \tilde{C}_{22} & \tilde{C}_{23} & 0 & 0 & 0 \\ \tilde{C}_{31} & \tilde{C}_{32} & \tilde{C}_{33} & 0 & 0 & 0 \\ 0 & 0 & 0 & \tilde{C}_{23} & 0 & 0 \\ 0 & 0 & 0 & 0 & \tilde{C}_{13} & 0 \\ 0 & 0 & 0 & 0 & 0 & \tilde{C}_{12} \end{bmatrix} \begin{bmatrix} \bar{e}_{11} \\ \bar{e}_{22} \\ \bar{e}_{33} \\ \bar{e}_{23} \\ \bar{e}_{13} \\ \bar{e}_{12} \end{bmatrix} \tag{81}$$

To obtain the stress–strain relations in the time domain, the inverse Laplace transform of Eqs. (80) and (81) must be obtained. this can be achieved conveniently in the form of the convolution integral. For example,

$$e_{11}(t) = \int_0^t S_{11}(t-\theta)\frac{\partial \sigma_{11}}{\partial \theta}\,d\theta + \int_0^t S_{12}(t-\theta)\frac{\partial \sigma_{22}}{\partial \theta}\,d\theta$$
$$+ \int_0^t S_{13}(t-\theta)\frac{\partial \sigma_{33}}{\partial \theta}\,d\theta, \text{ etc.} \tag{82}$$

or,

$$\sigma_{11}(t) = \int_0^t C_{11}(t-\theta)\frac{\partial e_{11}}{\partial \theta}\,d\theta + \int_0^t C_{12}(t-\theta)\frac{\partial e_{22}}{\partial \theta}\,d\theta$$
$$+ \int_0^t C_{13}(t-\theta)\frac{\partial e_{33}}{\partial \theta}\,d\theta, \text{ etc.} \tag{83}$$

If the coordinate axes are not aligned with the axes of material symmetry, then constitutive equations can be obtained from Eqs. (81)–(83) by a transformation of exactly the same form as that for the linear elastic case. Thus, the creep functions S_{ij} can be represented in the form

$$C_{ij}(t) = C_{ij}^0 + \sum_k C_{ij}^k e^{-t/\rho_k} \tag{84}$$

$$S_{ij}(t) = S_{ij}^0 + \sum_{k=2} S_{ij}^k(1 - e^{-t/\rho_k}) \tag{85}$$

The various quatntities C_{ij}^k, S_{ij}^k in Eqs. (84) and (85) can be selected along with the relaxation times ρ_k and the retardation times τ_k to fit the experimental data. By selecting a sufficient number of parameters in (84) and (85), the desired degree of accuracy can be obtained between the representation (84) or (85) and the experimental data. Alternatively, when a large number of terms are needed in (84) or (85), the relaxation moduli and creep compliance may be represented by a continuous relaxation or retardation spectrum (cf. Ferry [21], Christensen [15], and Chapter 2 of this volume).

The parameters C_{ij}^k, ρ_k (or S_{ij}^k, τ_k) are dependent on temperature and the concentration of swelling agents such as water. Increases in temperature and moisture content can be expected to decrease the relaxation modulus or to increase the creep compliance. In certain materials, called thermorheologically simple materials, a special equivalence is observed between time and temperature. For example, consider the relation (84) for a material that is tested at two different constant temperatures T_0 and T. If the effect of the temperature change from T_0 or T is the same as the shift in time scale, then (84) can be given in the form

$$C_{ij}(\xi) = C_{ij}^0 + \sum_k C_{ij}^k e^{-\xi/\rho_k} \tag{86}$$

with

$$\xi = t/a_T \tag{87}$$

referred to as the reduced time. The quantity a_T, the temperature shift factor, is a decreasing function of temperature. A similar kind of simple relationship can be postulated to hold with regard to the concentration of a swelling agent such as water, in which case the material would also be moisture-rheologically simple. If the material exhibits this simple behavior with respect to temperature and moisture, the constitutive relations can be expressed in the form

$$\begin{aligned} e_i(\xi) = & \int_0^{\xi} S_{ij}(\xi - \xi') \frac{d\sigma_j}{\delta \xi'}\, d\xi' + \alpha_i^T + \alpha_i^c (C - C_0) \\ & + \int_0^{\xi} \phi_i^T(\xi - \xi', T - T_0) \frac{dT(\xi')}{d\xi'}\, d\xi' \\ & + \int_0^{\xi} \phi_i^c(\xi - \xi', C - C_0) \frac{dC(\xi')}{d\xi'}\, d\xi' \end{aligned} \tag{88}$$

and

$$\xi(t, T, C) = \int_0^{t} \frac{dt}{a_T(T, t) a_C(C, t)} \tag{89}$$

Here, T_0 and C_0 represent reference temperature and concentration and a_T and a_C represent the temperature and concentration shift factors. For further information refer to Halpin [23], Shapery [82], Christensen [15], and Ferry [21].

The assumption of rheologically simple behavior is not likely to be strictly valid for most materials. Surely it is not generally valid for paper testing, particularly when large changes in moisture content occur. Nonetheless, it may be an acceptable engineering approximation to use for certain purposes, and the validity of the approximation can be evaluated by laboratory test. Further discussion of the time, temperature, and moisture equivalence can be found in Chapter 2 of this volume and in the work of Bach and Pentoney [6]. A more general means for incorporating the effects of moisture and temperature can be achieved by the development of a series expansion of convolution integral terms involving temperature, moisture concentration, and components of stress and strain. For a discussion of such procedures, refer to Ranta-Maunus [71]. In general, as temperature and moisture content increase, the applicability of the assumption of system linearity diminishes. For further discussion of nonlinear effects, refer to Bach and Pentoney [6], Christensen [15], Halpin [23], Haslach [24], Pecht and Haslach [57], and Pecht and Johnson [58].

C. Plasticity

When the loading of a paper material is high relative to the loads that cause failure, the mechanical response becomes very nonlinear and is characterized by the occurrence of residual strains when the loads are removed. Nonlinear elastic models such as the hypoelasticity and hyperelasticity models discussed earlier can be used to account for the nonlinear stress–strain behavior; however, they cannot account for the nonrecoverable strains that remain after loading. The objective of a plasticity model is to predict the nonlinear stress–strain behavior on both loading and unloading of the material and to predict the complicated behavior of the material with loading and unloading cycles.

There does not appear to be much information in the literature regarding the application of plasticity approaches to paper materials. Therefore, it is of interest to review the methods that have been used for other material systems such as metals, soils, and cellular foams provided by Chen and Han [14], Chen and Baladi [13], and the *ABAQUS Theory Manual* [2].

The plasticity constitutive models are based on the concept of a yield surface for the onset of plastic behavior and on the existence of a flow rule for predicting the incremental strains. Plasticity constitutive models are time-independent. Plasticity approaches have been used extensively for metals, and Liang and Suhling [38] have shown that plasticity models developed for metals that exhibit anisotropic characteristics can also be used to model some aspects of the plastic response of paper materials.

In paper mechanics it is well recognized that the in-plane stress–strain response for uniaxial tensile loading is markedly different from the uniaxial response in compression. Paper exhibits plastic behavior in the three principal directions: MD, CD, and Z. The model proposed by Shih and Lee [83], although developed for the modeling of metals, has certain attributes that make it appropriate for paper modeling. This model acknowledges the difference between tensile and compressive loading and takes into account the orthotropic symmetry of the yielding and plastic response. Although this model does not appear to have been used for paper, it may be expected to provide a satisfactory approach for modeling the in-plane plastic behavior of paper. In the case of three-dimensional deformation, however, the approach may need to be modified for use in paper mechanics to remove the assumption of incompressible volumetric response.

V. RELATIONSHIPS BETWEEN FIBER PROPERTIES, SHEET STRUCTURE, AND PAPER MECHANICAL PROPERTIES

A. Methods for Predicting Paper Elastic Behavior in Terms of Network Structure

The mechanical behavior of a fiber network such as a paper sheet composed of pulped wood fibers depends on the fiber, fiber–fiber bond properties, and the geometrical structure of the bonded fibrous network. The principal objective of a mathematical model is to predict the mechanical properties of the fiber network in terms of the properties of the fibers, bonds, and the structure of the fiber network. In paper, the fibers are often collapsed, ribbonlike structures that are bonded together, primarily by hydrogen bonds formed when the sheet is pressed and subsequently dried.

A number of theories have been developed with the objective of predicting the elastic or inelastic response of paper based on the fibers and/or bonds. The approaches that have been used fall into the following three categories.

- *Network models* based on the analysis of a typical fiber coupled to the deformation of the macroscopic sheet. The first model of this type is attributed to

Cox [16], who assumed that the fibers are straight and that they are oriented in the sheet in accordance with an orientation distribution function expressed as a Fourier series. Cox was able to show how the elastic moduli of the sheet are dependent on the orientation distribution. Cox assumed that the transverse stress and strain components in the fiber can be neglected, thereby making the analysis of fiber stress and strain one-dimensional. The Cox model was modified by Qi [66] to account for transverse normal and shearing components of strain in the fiber by assuming that the fiber takes on the same strains as those of the sheet. Cox's original theory has been subjected to numerous refinements and extensions that take into account the structure of the fiber network, the modeling of the fibers, the effects of drying-induced strains, the modeling of bond deformation, the nonlinear constitutive effects of the deformation response of the fibers and bonds, and the effect of fiber buckling for fibers that experience compressive strains (cf. Refs. 53–55, 59, 60, 63, and 69).

●● *Continuum models* based on the treatment of the fiber and fiber–fiber bonded material as planar structures. Whereas the network models attempt to bring the geometrical aspects of the fiber network into account, they are limited by the assumption that the fibers are straight and that they carry only a uniaxial load. The network approach neglects the transverse loading of the fiber; therefore, it is also incapable of modeling the hygroexpansion effect. The application of the composite laminate approach to the prediction of the in-plane elastic behavior of paper was reported by Schulgasser and Page [79]. The laminate approach was further developed by Subramanian and Carlsson [87,88]. Subsequently, this approach was extended to incorporate a more general approach to the modeling of fiber segments, fiber–fiber bonds, and voids using a mosaic concept [39–41]. Uesaka [90] and Uesaka and Qi [92] applied a continuum approach to the problem of prediction of the hygroexpansion effect in paper.

●●● *Molecular model.* The concept that the mechanical properties of paper can be described in terms of the physics of the hydrogen bond was reported by Nissan and Batten [51]. Nissan and Batten [50] discussed the relationship of the hydrogen bond approach to the network and continuum approaches. The principal feature of the hydrogen bond approach lies in the ability to study the effects of moisture at a fundamental physics level. Apparently, however, the effects of the structure of the fiber and the physical geometry of the fiber–fiber bond as well as the network geometry of the paper system cannot be easily addressed with this approach.

Early work regarding the development of mathematical models can be found in Algar [1], Dodson [18], Kallmes [30,31], and Van den Akker [93,94]. General aspects of paper physics can be found in books by Niskanen [47] and Deng and Dodson [17]. A review of research involving the modeling of the mechanical behavior of paper is given by Ramasubramanian and Wang [70].

B. One-Dimensional Mesomechanics Models for Low Density Paper and Network Structure

The choice of the mesoelement that is used in the construction of a mathematical model depends on the density of the fiber sheet. In the case of low to medium density paper (to be defined subsequently), a typical mesoelement can be taken as an individual fiber along with the portions of crossing fibers that cross it and bond it to the fiber network. In order to develop a set of simple, closed-form expressions for the elastic moduli of the paper sheet, it is necessary to assume that the fibers are straight and that they lie in the plane of the paper. As will be discussed, these assumptions can be relaxed; however, it is then necessary to predict sheet properties by employing numerical techniques. Restricting attention to the low density system made up of straight fibers, a typical mesoelement of length λ and orientation θ is shown in Fig. 7. The mesoelements are coupled to the network by means of the crossing fibers. The strains in the sheet are presumed to be transmitted to the mesoelements by bending and shearing deformation of the crossing fibers and by shearing deformation of the fiber–fiber bonds. Thus the axial strain in the mesoelement fiber is not uniform but varies from the ends, where it is zero, to the middle, where it finds its maximum value. If the mesoelement is short enough and if the coupling is weak, the mesoelement strain is less than that associated with the sheet.

The model is further illustrated by Fig. 8, which shows a portion of a fiber that is coupled by two crossing fibers to the remainder of the network. The boundary between the element and the network is depicted by a dashed line. This boundary is assumed to be located a distance $l/2$ from the centerline of the primary fiber, where l represents the center-to-center distance between bonds, l_b is the bond length along the fiber, and w_f and t_f are the width and thickness of the fiber, respectively. If l is small in comparison with w_f, as would be expected in moderately dense paper, the coupling is primarily attributable to the shearing deformations of the fiber–fiber bonds. In a very low density system such as tissue paper, the bending and shearing deformation of the crossing fibers may be substantial.

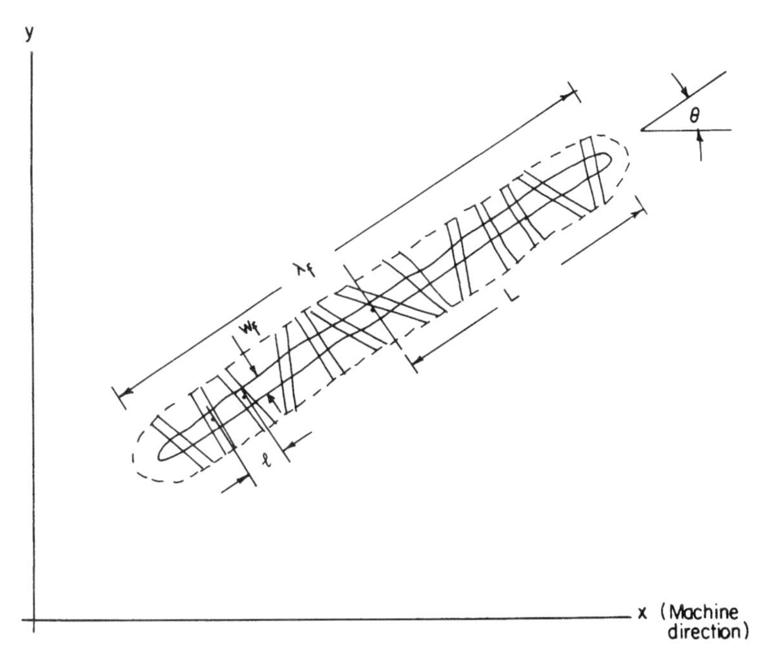

Fig. 7 Typical fiber of length λ_f and orientation θ with respect to the machine direction. (From Ref. 25.)

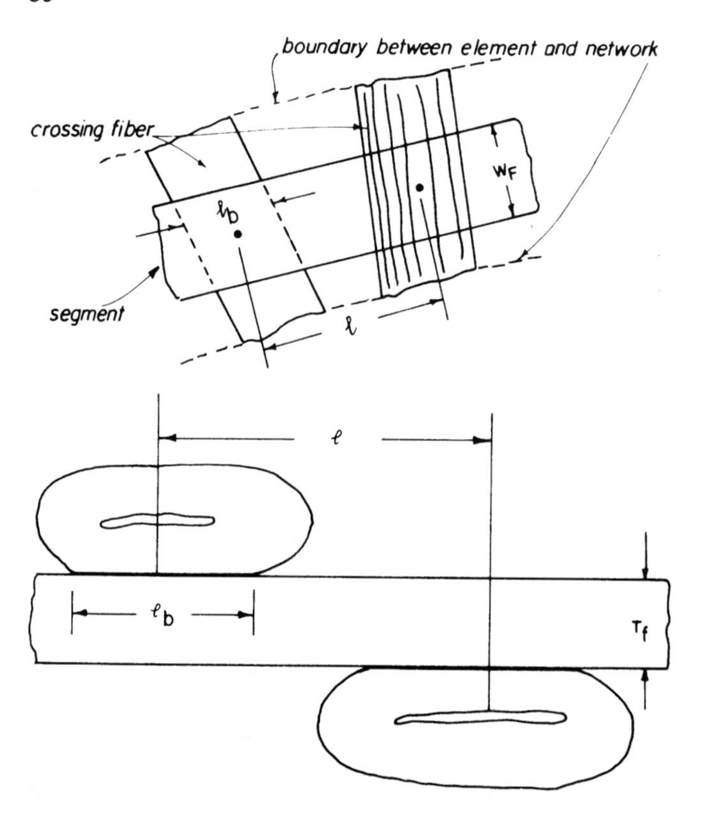

Fig. 8 Portion of an element of a fiber segment illustrating the bond length along a fiber, l_b, the center-to-center distance between bonds along a fiber, l, the (dashed line) boundary between the segment element and the remainder of the network, the fiber width w_f, and fiber thickness t_f.

On the assumption that the fiber is linearly elastic, the stress σ_f at any point along the fiber is related to the fiber strain ε_f at that point by

$$\sigma_f = E_{af}\varepsilon_f \tag{90}$$

where E_{af} is the effective axial modulus of the fiber. The effective axial modulus depends not only on the inherent properties of the fiber but also on how the fiber is bonded to other fibers in the network. Thus, there is a stiffening effect due to the reinforcement of other bonded fibers. E_{af} can be considered an inherent property of the paper system, or its value can be estimated in terms of an idealized paper network. For example, if the fibers are idealized (Figs. 8–10) as thin rectangular strips (e.g., perfectly collapsed thin-walled springwood fibers), and if the density of the network is low enough that on average only half (or less) of the fiber surface is in contact with other fibers in the network, then Perkins [60] has shown that

$$E_{af} = \frac{(l/l_b)(E_{fL} + E_{fT})}{C_B + \left(\dfrac{E_{fL} + E_{fT}}{E_{fL}}\right)\left(\dfrac{l - l_b}{l_b}\right)} \tag{91}$$

where

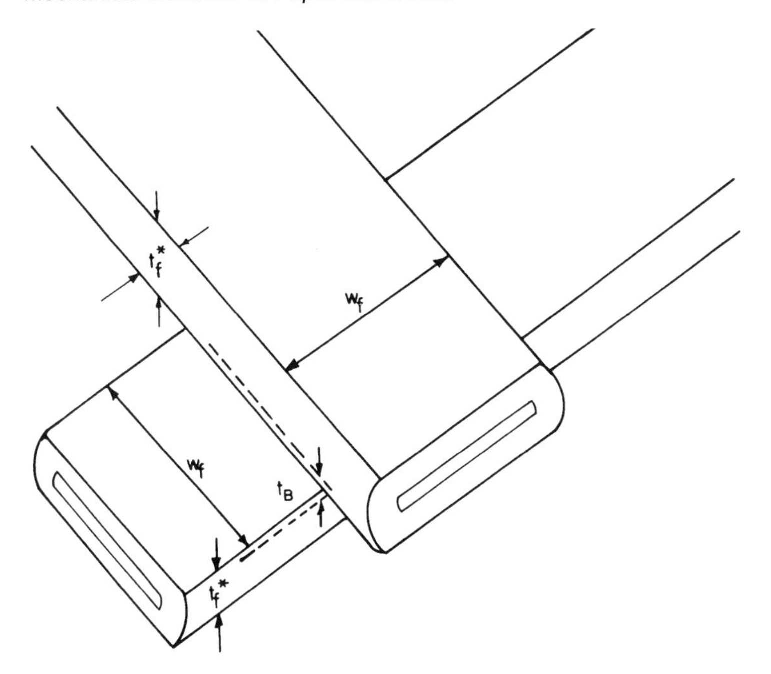

Fig. 9 Fiber–fiber bond. (From Ref. 25.)

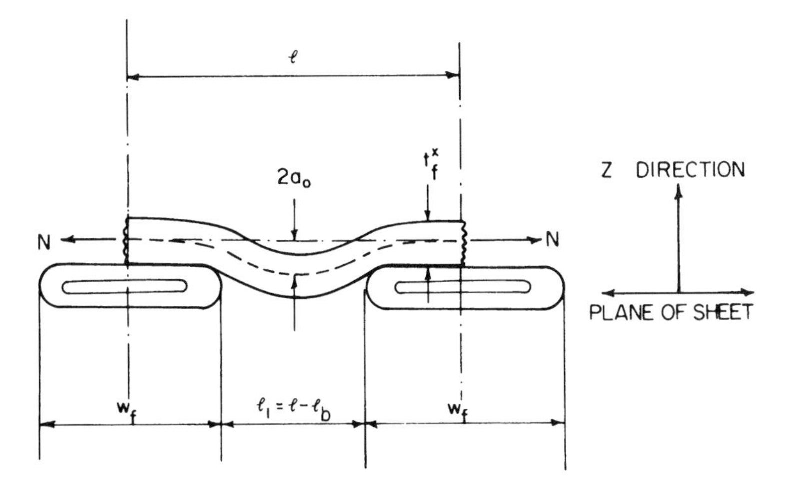

Fig. 10 Fiber segment situated between bond centroids. (From Ref. 25.)

$$C_B = 1 + \frac{E_{fT}}{E_{fL}}\left(\frac{\tanh B}{B}\right) \tag{92}$$

and

$$B = \frac{l_b}{2}\left[\frac{k_b}{t_f}\left(\frac{E_{fT} + E_{fL}}{E_{fT}E_{fL}}\right)\right]^{1/2} \tag{93}$$

Here E_{fL} and E_{fT} represent the elastic moduli in the axial and transverse directions of the fiber, respectively; t_f is the thickness of the collapsed fiber; and l and l_b are as previously defined.

The factor C_B determines the amount of reinforcement associated with fiber bonding. The constant k_b in the factor B [Eq. (93)] determines the stiffness of the fiber–fiber bond. It is expected that these factors are related to the degree of fiber refining, which roughens the fiber surface and causes damage or separation of the primary and S1 layers of the cell wall. As the bond stiffness becomes very small, C_B tends to a value of $1 + E_{fT}/E_{fL}$ and the reinforcing effect vanishes. At the other extreme, the bond becomes infinitely stiff and the factor C_B tends to unity.

The bond stiffness can be measured directly as shown by Thorpe et al. [89] where individual fibers are bonded to a fiber shive that is supported and the fiber is loaded by an external applied axial load. The value of k_b measured in this way is on the order of $1 \times 10^{13}\ \text{N/m}^3$. A further description of the fiber–fiber bond is found in Section V.D.

The modeling of the elastic properties of the fiber depends on a number of factors related to whether or not the fibers are thin-walled and collapsed in the paper sheet and whether or not it is considered necessary to include the effects of the S1 and S3 layers of the cell wall. Modeling of the fiber is discussed further in Section V.C. Note that the transverse fiber modulus E_{fT} should be computed based on the crossing angle that the crossing fibers make relative to the primary fiber as discussed in the following.

A fiber network representation of paper may also include "kinked" fibers. If it is assumed that axial load in a fiber cannot be transmitted through a kink, the kinked fiber can be modeled as a set of straight, independent segments as shown in Fig. 11.

The strain in the fiber ε_f is determined by the sheet strain and the degree of coupling between each fiber and the other fibers in the sheet. As shown in Fig. 7, a

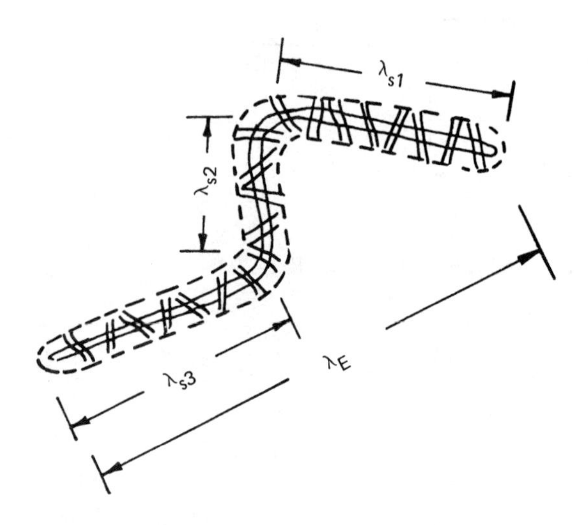

Fig. 11 Schematic of fiber having three distinct segments. Numerous crossing fibers are in contact with it.

fiber of length λ_f is coupled to the rest of the network through the crossing fibers. The equilibrium equation for a fiber mesoelement based on the equilibrium of a typical element along the fiber length (see Fig. 12) is

$$\frac{d}{d\xi}\left(E_{af}\frac{d\varepsilon_f}{d\xi}\right) - k_f\varepsilon_f = -k_f\varepsilon_s \tag{94}$$

where

$$\varepsilon_s = \varepsilon_x \cos^2\theta + \varepsilon_y \sin^2\theta + 2\varepsilon_{xy}\sin\theta\cos\theta \tag{95}$$

Equation (95) represents the normal component of strain in the sheet in the direction θ corresponding to the orientation of the fiber. The factor k_f is the effective stiffness of the coupling between the fiber of orientation θ and the other crossing fibers that couple the fiber to the network. The stiffness k_f includes the flexibility of the fibers in bending (which may be discernible for low density sheets) and the flexibility of the fiber–fiber bond. If the idealized system of rectangular cross section straight fibers is considered,

$$k_f = \frac{1}{l\left[\dfrac{(l-l_b)^3}{384E_{fL}I_f}\left(1+\dfrac{12fE_{fL}I_f}{G_fA_f(l-l_b)^2}\right)+\dfrac{1}{A_bk_b}\right]} \tag{96}$$

Here I_f represents the moment of inertia associated with the fiber cross section, G_f represents the cell wall shear modulus, f is a factor that depends on the fiber cross-sectional shape (it is 6/5 if the fiber is rectangular), A_b is the area of the fiber–fiber bond, and k_b is the bond stiffness.

Equations (91)-(93) and (96) can be used to estimate values of E_{af} and k_f in an idealized case. These quantities can also be considered fundamental parameters of

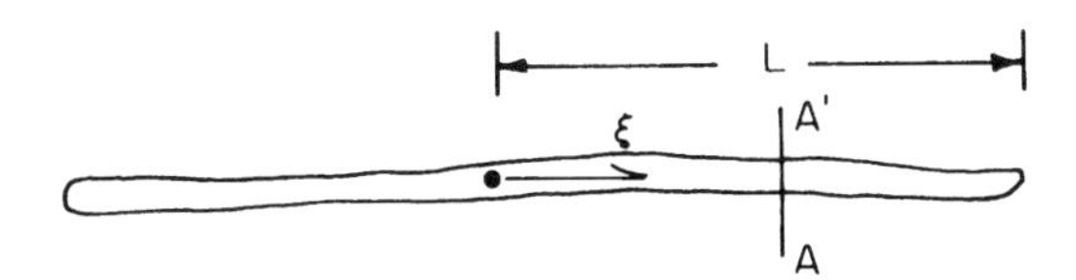

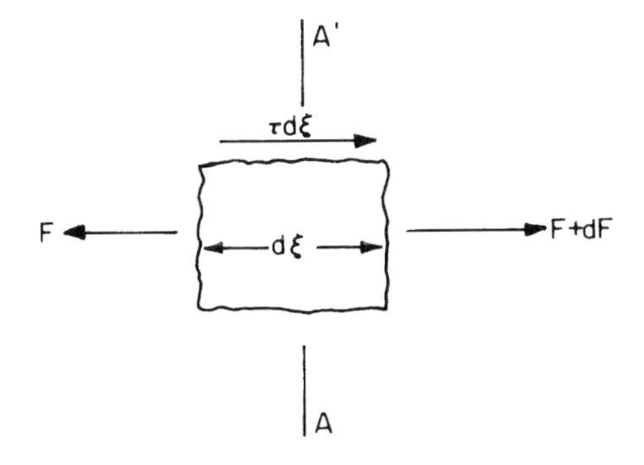

Fig. 12 Equilibrium of a typical element of the fiber. (From Ref. 25.)

the system that are to be determined by mechanical property testing. In the latter case, direct measurement of the paper's mechanical response is used with the theoretical property relations to calculate "experimental" values of E_{af} and k_f.

It should be noted that the values of the effective axial modulus E_{af} and the coupling stiffness k_f are dependent on the orientation angle θ inasmuch as the network lengths l, l_b and the bond area A_b are dependent on orientation angle. To permit the subsequent closed-form expressions for the moduli of the sheet it is necessary to take the network lengths as their average values for the sheet. These values are dependent on the orientation distribution of the fibers as shown below.

For a prescribed sheet strain given by ε_x, ε_y, ε_{xy}, the normal component of sheet strain in the direction of the fiber, ε_s, is known. Equation (94) can be solved by assuming an arbitrary value of ε_s. The factors A_f, E_{af}, and k_f should properly be considered as varying in some random fashion. In this case, the solution to Eq. (94) is very difficult. An alternative procedure is to select average values for A_f, E_{af}, and k_f and determine an average fiber response. When this procedure is followed, the solution of Eq. (94) is easily obtained as

$$\varepsilon_f = \varepsilon_s\left(1 - \frac{\cosh a\xi}{\cosh aL}\right) \tag{97}$$

where

$$a = \left(\frac{k_f}{A_f E_{af}}\right)^{1/2} \tag{98}$$

L is half the fiber length, and A_f is the cross-sectional area of the fiber.

The strain energy of a typical segment element is calculated as

$$W_e = \tfrac{1}{2} E_{af} A_f \lambda\, \varepsilon_s^2 \eta_L \tag{99}$$

where

$$\eta_L = 1 - \frac{\tanh aL}{aL} \tag{100}$$

and can be identified as a coupling efficiency. For strong coupling and long length, η_L approaches unity. At the other extreme of very short length or very weak coupling, η_L approaches the value of zero and the sheet elastic modulus would also approach the value of zero.

The energy of the system can be written

$$W = \int_0^{\lambda}\int_0^{\pi} D_0 f^*_{\theta\lambda} W_e \, d\theta \, d\lambda \tag{101}$$

where D_0 represents the number of segment elements per unit area and $f^*_{\theta\lambda}\, d\theta\, d\lambda$ represents the probability of finding a segment whose length is in the interval λ, $\lambda + d\lambda$ and whose orientation is in the interval θ, $\theta + d\theta$.

Suppose that $f^*_{\theta\lambda}$ can be expressed in the form

$$f^*_{\theta} = [f_\theta(\theta, a_1, a_2, a_3, \ldots, a_m)][f_\lambda(\lambda_s, b_1, b_2, b_3, \ldots, b_n)] \tag{102}$$

Here, f_θ describes the fiber orientation distribution as a function of θ depending on the m parameters, $a_1, a_2, \ldots, a_m$. Likewise, the distribution of fiber lengths λ depends on n parameters $b_1, b_2, \ldots, b_n$. In general, the parameters $a_1, a_2, \ldots, a_m$ are assumed to be functions of λ, whereas the parameters $b_1, b_2, \ldots, b_n$ are functions of θ.

The evidence available suggests that coupling between the two distributions is weak, and therefore as an approximation that is valid for at least certain papers it can be assumed that the length and orientation distributions are independent of each other. For this assumption, the parameters $a_1, a_2, \ldots, a_m$ and $b_1, b_2, \ldots, b_n$ are constants, and the strain energy per unit sheet area can be written in the form

$$W = \frac{1}{2}\left(\frac{\omega_s}{\rho_{af}}\right)\int_0^{\pi} \phi_\lambda E_{af}\varepsilon_s^2 f_\theta \, d\theta \tag{103}$$

where ω_s represents the basis weight, E_{af} represents the apparent fiber modulus, and

$$\phi_\lambda = \int_0^{\infty} \eta_L f_\lambda \, d\lambda \tag{104}$$

The length parameter ϕ_λ and the apparent fiber modulus E_{af} may depend upon the orientation direction θ as a result of the stresses imposed on the fibers during drying of the sheet. Analytically, it can be assumed that the apparent fiber modulus E_{af} depends upon the shrinkage and restraint conditions present during drying in the following way:

$$E_{af} = E_{af_0}(1 + H\varepsilon_{ND}) \tag{105}$$

where E_{af_0}, a constant, represents the apparent fiber modulus in the absence of drying restraint, H is a constant that predicts the magnitude of stiffening, and

$$\varepsilon_{ND} = (\alpha_x M + \varepsilon_{xD})\cos^2\theta + (\alpha_y M + \varepsilon_{yD})\sin^2\theta \tag{106}$$

Here $\alpha_x M$ and $\alpha_y M$ represent the sheet shrinkage strains during unrestrained drying in the x and y directions, and ε_{xD} and ε_{yD} represent the sheet strains applied or allowed during drying. Thus, for unrestrained shrinkage conditions, $\varepsilon_{xD} = -\alpha_x M$ and $\varepsilon_{yD} = -\alpha_y M$. It is evident that ε_{ND} is related to the magnitude of restraint that a fiber of orietation θ would be subjected to during the drying of the sheet. In the following analysis, it is further assumed that ε_{ND} is positive, that is, the fibers are loaded in tension as a result of any drying restraint.

Experimental results of the effects of drying restraint are given by Setterholm and Chilson [81]. Data taken from their work (cf. Perkins and Mark [63]) indicate that the parameter H has values that lie in the range of 10–30. Experimental evidence of the effect of drying restraint on the properties of the fibers in the network is presented by Wuu et al. [97].

With the help of Eqs. (104)–(106), integration of Eq. (103) can be easily carried out if the length factor ϕ_λ is independent of θ. In fact, ϕ_λ does depend on θ, because the coupling parameter a that appears in η_L depends on E_{af}. If one assumes that the influence of variation in E_{af} through θ in ϕ_λ has an insignificant effect on the prediction of the elastic moduli of the sheet, there is no serious analytical error commited in assuming that ϕ_λ is independent of θ.

The stress–strain relations for the sheet can be obtained from the strain energy function W after carrying out the integration through the relationships

$$\tau_x = \frac{\partial W}{\partial \varepsilon_x}, \qquad \tau_y = \frac{\partial W}{\partial \varepsilon_y}, \qquad \tau_{xy} = \frac{1}{2}\frac{\partial W}{\partial \varepsilon_{xy}} \tag{107}$$

where τ_x, τ_y, τ_{xy} represent the force per unit edge length of the paper sheet.

To conveniently express the resulting elastic moduli it is desirable to express the orientation distribution $f_\theta(\theta)$ in terms of its Fourier series expansion [16]. When the x direction is one of the axes of elastic symmetry, for example the machine direction, and $\theta = 0°$ corresponds to this direction, then

$$f_\theta(\theta) = \frac{1}{\pi}[1 + a_1 \cos 2\theta + a_2 \cos 4\theta + a_3 \cos 6\theta + \cdots + a_n \cos 2n\theta + \cdots] \tag{108}$$

After substitution and execution of the indicated steps, the Young moduli E_x^*, E_y^*, the shear modulus G_{xy}^*, and the Poisson ratios ν_{xy}, ν_{yx} corresponding to plain stress loading of the sheet are found to be

$$E_x^* = \frac{1}{16}\left(\frac{\omega_s}{\rho_{af}}\right)\phi_\lambda E_{af_0}\left[(1 + \langle e\rangle)(6 + 4a_1 + a_2) + \frac{e_\Delta}{2}(8 + 7a_1 + 4a_2 + a_3)\right] [1 - \nu_{xy}\nu_{xy}] \tag{109a}$$

$$E_y^* = \frac{1}{16}\left(\frac{\omega_s}{\rho_{af}}\right)\phi_\lambda E_{af_0}\left[(1 + \langle e\rangle)(6 - 4a_1 + a_2) + \frac{e_\Delta}{2}(8 - 7a_1 + 4a_2 - a_3)\right] [1 - \nu_{xy}\nu_{xy}] \tag{109b}$$

$$\nu_{xy} = \frac{(1 + \langle e\rangle)(2 - a_2) + (e_\Delta/4)(a_1 - a_3)}{(1 + \langle e\rangle)(6 - 4a_1 + a_2) - (e_\Delta/2)(8 - 7a_1 + 4a_2 - a_3)} \tag{109c}$$

$$\nu_{yx} = \frac{(1 + \langle e\rangle)(2 - a_2) + (e_\Delta/4)(a_1 - a_3)}{(1 + \langle e\rangle)(6 + 4a_1 + a_2) + (e_\Delta/2)(8 + 7a_1 + 4a_2 + a_3)} \tag{109d}$$

$$G_{xy}^* = \frac{1}{16}\left(\frac{\omega_s}{\rho_{af}}\right)\phi_\lambda E_{af_0}\left[(1 + \langle e\rangle)(2 - a_2) + \frac{e_\Delta}{2}(a_1 - a_3)\right] \tag{109e}$$

where

$$\langle e\rangle = \tfrac{1}{2}H[(\alpha_x M + \varepsilon_{xD}) + (\alpha_y M + \varepsilon_{yD})] \tag{110a}$$

$$e_\Delta = \tfrac{1}{2}H[(\alpha_x M + \varepsilon_{xD}) - (\alpha_y M + \varepsilon_{yD})] \tag{110b}$$

An inspection of Eqs. (109) indicates that the elastic constants depend on only the first three parameters, a_1, a_2, a_3, that occur in the Fourier series expansion for f_θ. Furthermore, if the influence of drying restraint is not incorporated or if the sheet is dried under conditions of no restraint, the elastic moduli depend only on the parameters a_1 and a_2. For this condition, Eqs. (109) are essentially the same as those given by Cox [16] for when $\theta = 0°$ corresponds to a direction of elastic symmetry in the sheet. [Note, however, that the effects of fiber–fiber bonding and the effect of finite fiber length are taken into account in Eqs. (109).]

The number of coefficients of the Fourier series expansion for f_θ that are necessary for predicting the elastic properties, therefore, depends on the functional relationship between the apparent fiber modulus E_{af} and orientation, which in turn depends on the procedures and conditions of drying. Incorporation of any other phenomenon affecting the angular dependence of the strain energy stored in the sheet may change the number of Fourier coefficients that appear in the elastic constants.

It is theoretically possible to describe any fiber orientation distribution in terms of the parameters $a_1, a_2, a_3, \ldots$ of the Fourier series expansion. However, a single-parameter function would simplify the expressions for the elastic moduli. The reduction to one parameter can be accomplished very simply by truncating the Fourier series and retaining the single parameter a_1 (Perkins [60]). However, reduction to a coefficient in the Fourier series expansion can have some significant effects. For example, for freely dried paper, the shear modulus should decrease with increased anisotropy, but this reduction effect will be predicted only if the coefficient a_2 is retained in the expansion of Eq. (108). A single-parameter function would predict the shear modulus to be independent of the degree of anisotropy of the paper.

An alternative approach is to employ a distribution function that has only one shape parameter. For a general discussion of fiber orientation distribution, refer to Perkins and Mark [63]. The most frequently used single-parameter distributions are the elliptical, the von Mises, and the normal distributions. The distributions and their corresponding values of a_1, a_2, and a_3 are given in Table 1.

The elliptical, von Mises, and normal distributions are illustrated in Fig. 13.

Application of Theory to Prediction of Elastic Moduli of a Paper Sheet Equations (109) can be used to estimate the effects of changes in structural parameters on the elastic properties of paper or to estimate from experimental measurement the parameters of the system. As an example, the values of E_x and E_y can be obtained from measurements of the Young moduli in the machine and cross-machine directions. Subsequently, the other in-plane elastic moduli for the sheet can be estimated.

To use Eqs. (109) in this example, it is necessary to investigate the influence of the drying restraint. The stiffening parameter H of Eq. (21) has been estimated by

Table 1 Parameters for the Elliptical and von Mises Fiber Orientation Distributions

	Parameter		
	a_1	a_2	a_3
Elliptical $f_\theta = \frac{\zeta}{\pi}\left(\frac{1}{\cos^2\theta + \zeta\sin^2\theta}\right)$	$2\left(\frac{\zeta-1}{\zeta+1}\right)$	$2\left(\frac{\zeta-1}{\zeta+1}\right)^2$	$2\left(\frac{\zeta-1}{\zeta+1}\right)^3$
Wrapped normal[a] $f_\theta = \frac{1}{\sigma\sqrt{2\pi}}e^{-\theta^2/2\sigma^2}$	2ρ	$2\rho^4$	$2\rho^9$
von Mises[b] $f_\theta = \frac{1}{\pi I_0(\kappa)}e^{\kappa\cos 2\theta}$	$2\frac{I_1(\kappa)}{I_0(\kappa)}$	$2\frac{I_2(\kappa)}{I_0(\kappa)}$	$2\frac{I_3(\kappa)}{I_0(\kappa)}$

[a] $\rho = \exp[-1/2\sigma^2]$.
[b] $I_n(\kappa)$ is a modified Bessel function of the first kind and order m.
Source: Refs. 42 and 43.

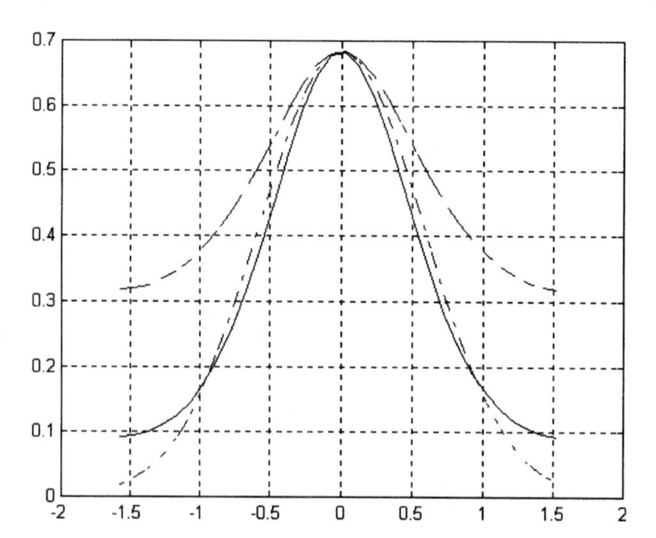

Fig. 13 Plot of von Mises distribution ($\kappa = 1$, solid line), elliptical distribution ($\zeta = 2.15$, dashed line), and normal distribution ($\sigma^2 = 0.34$, dash-dot line).

experiment [81] to lie in the range of 10–30. From a knowledge of the drying restraint conditions and sheet shrinkage behavior, the values of $\langle e \rangle$ and E_Δ can be calculated. The ratio of the experimental Young moduli for the machine and cross-machine directions can be equated to the ratio E_x^*/E_y^* from Eqs. (109a) and (109b). The elliptical or von Mises distribution can be written in a Fourier series form (Table 1). The first three coefficients of the expansion can then be taken as a_1, a_2, a_3, respectively. The inversion of this relation would yield a value of the concentration parameters for orientation ζ, or κ can then be used to obtain a_1, a_2, a_3, and subsequently the values of ν_{xy}, ν_{yx}, G_{xy}. The measurement of two Young moduli then permits estimation of all the in-plane elastic moduli for the sheet. The calculation is best performed for several different assumptions regarding the variable in Eq. (110) to ascertain the reliability of the assumptions.

The choice of fiber orientation distribution function can have significant effects on predictions obtained from Eqs. (109). For a given fiber orientation, predictions of ν_{xy}, ν_{yx}, G_{xy} from experimental measurements of E_x and E_y for the elliptical distribution may be different from those for the von Mises distribution. Although the elliptical distribution or the single-parameter cosine distribution is the easiest to apply, the von Mises distribution is considered to provide more accurate estimates [63].

In the case of fiber segment length distributions, the Erlang distribution has been found to be quite satisfactory. Thus, the function f_{λ_s} can be taken as

$$f_{\lambda_s} = \frac{(\lambda_s/b)^{c-1} \exp(-\lambda_s/b)}{b[(c-1)!]} \tag{111}$$

Here b and c are determined by fitting the distribution to the experimental data. The product bc is equal to the mean value of λ_s. The standard deviation for the distribution is $b\sqrt{c}$. For further information, refer to Perkins and Mark [63]. The Erlang distribution is illustrated in Fig. 14.

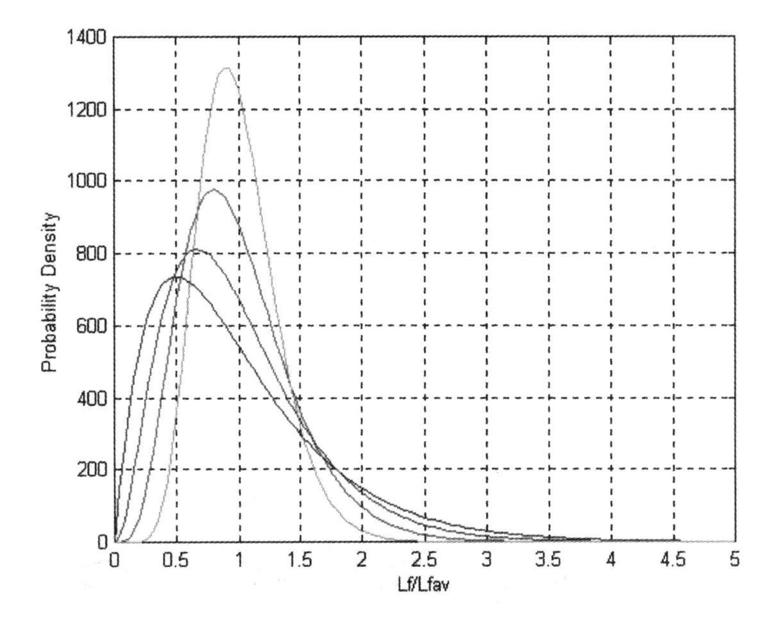

Fig. 14 The Erlang distribution function versus fiber length divided by average fiber length for values of $c = 2$, 3, 5, and 10.

Experimental investigations [53,73] have shown that the assumptions of the foregoing theory are reasonable for papers of sufficiently low density. Nonetheless, it is apparent that this theory is limited by many simplifying assumptions. As a result, theories based on the above approach have limited accuracy although they may be useful for assessing the approximate effect of the various parameters that influence the solution.

Network Geometry The approach outlined above depends on some knowledge of the structure of the network as related to fiber segment lengths between bonds. This information is difficult to obtain experimentally, so it is useful to have some approximate relations available. The distance between fiber crossings, fiber–fiber bonds, along a given fiber, $l(\theta)$, is an important property of the fiber network. Assuming that the fibers are all straight, the bond centroid distance depends on the direction of the fiber and the overall orientation distribution of the fibers in the network. The bond centroid distance is determined by the number of fibers that cross a given fiber. This problem was studied by Kallmes and Corte [32] and by Komori and Makashima [36], who showed that

$$n(\theta) = \frac{2D_f N \lambda^2}{V} J(\theta) \tag{112}$$

where

$$J(\theta) = \int_0^{\pi} f(\theta) \mid \sin(\theta - \alpha) \mid d\alpha \tag{113}$$

Here, D_f denotes the fiber diameter or, in the case of a flattened fiber, the Z-direction thickness of the fiber. N represents the number of fibers in a volume V. The function $J(\theta)$ represents the average value of the sine of the crossing angle, $\theta - \alpha$ for a fiber of orientation θ. The function $J(\theta)$ can be expressed in series form as shown by [12]

$$J(\theta) = \frac{2}{\pi}\left[1 - \sum_{n=1}^{\infty} \frac{a_n}{(2n-1)(2n+1)} \cos(2n\theta)\right] \tag{114}$$

The rationale for Eq. (112) follows from the observation that $2D_f\lambda(\lambda \sin\beta)$ represents the volume occupied by the two crossing fibers times the projected length of the crossing fiber as shown in Fig. 15. Thus, the average number of contacts is the number of fibers N in a volume V times the ratio of the volume associated with one crossing to the total volume of fiber material.

The average distance between bond centroids, $l(\theta)$, along a fiber of orientation θ is taken as the reciprocal of the average number of crossings per unit length. Therefore,

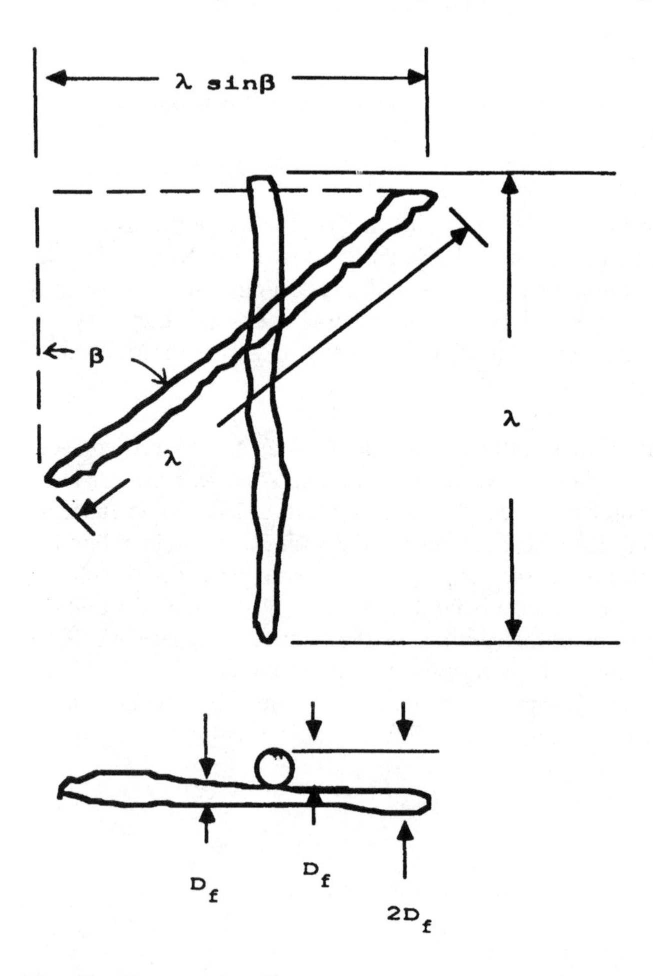

Fig. 15 Two crossing fibers.

$$l(\theta) = \frac{V}{2D_f N\lambda J(\theta)} \tag{115}$$

The average number of crossings considered over the set of all fibers is

$$\bar{n} = \frac{2D_f N\lambda^2}{V} I \tag{116}$$

where

$$I = \int_0^{\pi} f(\alpha) J(\alpha)\, d\alpha \tag{117}$$

or in series form as shown by Chang [12],

$$I = \frac{2}{\pi}\left[1 - \sum_{n=1}^{\infty} \frac{a_n}{(2n-1)(2n+1)}\right] \tag{118}$$

The overall distance between bond centroids for the network is

$$l = \frac{V}{2D_f N\lambda I} \tag{119}$$

The quantity $N\lambda$ represents the total length of fiber in a volume V. If $M = \rho_f A_f N\lambda$ represents the mass in volume V, then the expressions for bond centroid distance can be written

$$l(\theta) = \frac{\rho_f}{2\rho_s}\left(\frac{A_f}{D_f}\right)\left(\frac{1}{J(\theta)}\right) \tag{120}$$

$$l = \frac{\rho_f}{2\rho_s}\left(\frac{A_f}{D_f}\right)\left(\frac{1}{I}\right) \tag{121}$$

When the fibers are assumed to be completely flattened, then D_f is the thickness t_f and $A_f = w_f t_f$, and the bond centroid distances can be written

$$l(\theta) = w_f \frac{\rho_f}{2\rho_s}\left(\frac{1}{J(\theta)}\right) \tag{122}$$

and

$$l = w_f \frac{\rho_f}{2\rho_s}\left(\frac{1}{I}\right) \tag{123}$$

The fiber crossing angle can be obtained from the definition of $J(\theta)$ because this quantity represents the average value of the sine of the angle between two crossing fibers for a fiber of given orientation angle θ relative to the reference direction. Therefore, the average crossing angle β is given by

$$\beta = \arcsin(J(\theta)) \tag{124}$$

The bond area is calculated from the crossing angle. Therefore,

$$A_b(\theta) = \frac{w_f^2}{J(\theta)} \tag{125}$$

and the average bond area for fibers of all orientations is

$$A_b = \frac{w_f^2}{I} \tag{126}$$

These network properties are to be used in the calculations for the fiber coupling parameter k_f and the effective fiber modulus E_{af}. When the relations provided in Eq. (109) are to be used to predict the sheet elastic moduli, the values of l and A_b are to be taken from Eqs. (123) and (126) because these relations relate the average properties of the sheet over all directions. The portion of the fiber that is bonded, l_b, is defined using Eq. (126) as the apparent length necessary for the product $l_b w_f$ to be equal to the bond area. On the other hand, the methods leading to Eqs. (109) can be performed numerically with the angle dependence of bond centroid distance, bond length, and bond area taken into account by using Eqs. (122) and (125) in the expression for E_{af} and k_f. The numerical approach makes it possible to avoid the simplifying assumption of independence of angle-dependent and fiber length effects discussed above.

Using the foregoing relations it can be shown that when the sheet density, ρ_s, is equal to one-half the fiber density, ρ_f, the fibers will be in contact with adjacent fibers on one side of the fiber or the other and the unbonded length goes to zero. Given that the fiber density is 1500 kg/m^3, these network relations predict that fibers will have no unbonded or free length when the sheet density exceeds 750 kg/m^3. For paper sheets characterized by a density in excess of this figure, the mesoelement model depicted in Fig. 7 should be modified so that there is no free fiber length. This would suggest that the mesoelement should be taken as the fiber itself. Therefore, for sheet density in excess of 750 kg/m^3 the relations for the elastic moduli [Eqs. (109)] can still be used as long as the effective axial fiber modulus from Eq. (91) is taken as the axial fiber modulus, E_{fL}, and the network coupling stiffness, k_f, is taken as

$$k_f = \frac{A_b k_b}{l} \tag{127}$$

When the procedure described above is used for predicting the elastic moduli of sheets with sheet density in excess of 750 kg/m^3, it is clear that the prediction neglects the transverse and shearing strains in the fiber.

It is evident that the theoretical expressions that are discussed above regarding the network structure are very approximate and cannot reflect the actual structure of a paper network. Furthermore, they do not provide any information regarding the higher statistical moments of the distributions of fiber segment, bond area, etc. A more accurate analysis of the network structure can be obtained by computer simulation. This approach was taken by Subramanian and Carlsson [88]. A simulation of the three-dimensional structure was reported by Wang and Shaler [95].

Nonlinear Geometry and Constitutive Effects and Strength Models The assumption of linear elastic behavior of the fiber and bond materials can be relaxed while using essentially the same approach as that outlined above. The basic approach just outlined can be modified to take into account a number of nonlinear effects if one does not require that closed-form expressions be obtained for the prediction of mechanical response. In other words, the mesoelement approach can be extended by performing the calculations numerically rather than in algebraic form. The effects of fiber curl, fiber buckling, nonelastic constitutive models for the bond and fiber components, and certain aspects of failure at the mesoelement level can be studied when the mesoelement model is analyzed by the finite element method. The basic approach was presented by Perkins [61]. The advantage of this approach is that most of the simplifying assumptions regarding uniaxial loading of the fiber, fiber straightness, and linear elastic behavior of the fiber and bond can be relaxed.

The linear elasticity assumptions of the one-dimensional straight fiber network approach was extended by Ramasubramanian and Perkins [69], who assumed that the stress–strain curves for the fiber material and the bond materials could be approximated by two-slope relations. The initial slope represents the elastic modulus, whereas the second slope represents the material behavior after plastic response at the fiber cell wall or bond regions takes over. Ramasubramanian and Perkins derived closed-form expressions for the fiber and bond stress and strains as functions of applied load. This work also permitted the assumption that the fiber behavior in compressive loading is different from that for tensile loading. This behavior is expected owing to the presence of void regions in papers of moderate to low density. The results of this analysis showed how to predict the behavior of paper for both tensile and compressive uniaxial loading cases. It was also demonstrated that the apparent Poisson ratio for paper is not a constant as predicted by the elastic theory but rather depends on the plastic behavior of the fiber and bond materials. Fiber and bond plasticity lead to different expectations regarding the apparent Poisson ratios under uniaxial loading.

The one-dimensional fiber network analysis can also be used to predict some aspects of paper strength. McLaughlin and Batterman [46] developed a plastic limit analysis approach for fibrous materials, and Perkins [59] discussed the application of the McLaughlin and Batterman theory to paper materials. Feldman et al. [20] applied the limit analysis method to the prediction of the tensile strength of paper. Kärenlampi [33,34] developed closed-form expressions for the prediction of paper strength and applied this theory to the prediction of the tensile strength of paper. Jayaraman and Kortschot [26] provided a critical discussion of the use of closed-form network models for predicting the strength of paper materials.

C. Fiber Structure and Elastic Behavior

The elastic behavior of the fiber depends on a large number of factors related to the constituent properties of the cellulose, hemicellulose, and lignin components of the cell wall; the structure of the cell wall; and the existence of damaged zones in the cell wall. The model adopted in this chapter is based on the assumption that the fiber is a collapsed thin-walled fiber. The collapsed fiber walls as assumed to be bonded to one another in the lumen, giving rise to an antisymmetrical laminated structure. For further information, see Refs. 7, 9, 44, 47, 76, and Chap. 14 of this volume.

A typical layer of the cell wall, i.e., S1, S2, or S3, is composed of fibrils that are oriented at an angle to the axis of the fiber. The fibrils are in turn composed of cellulose chains in microfibrils embedded in a matrix of hemicellulose. The cellulose–hemicellulose composite is embedded in lignin. The cellulose and hemicellulose are generally assumed to be orthotropic, whereas the lignin is assumed to be isotropic. The cellulose elastic properties are assumed to be independent of moisture content. However, the hemicellulose and lignin elastic properties are dependent on the moisture content of the materials. Assumptions on percentages of cellulose, hemicellulose, and lignin within each of the cell wall layers and the elastic properties of each component are often made, for example in Ref. 7, but real data is available (Chap. 14).

The properties of the layers of cellulose embedded in hemicellulose can be modeled by using the approach often used to model the micromechanics of a composite material, general procedures for which can be found in Refs. 29 and 45. The cellulose–hemicellulose layers are treated as transversely isotropic, and the elastic properties are obtained with the help of the Halpin–Tsai relationship (Ref. 29, p.151),

$$\frac{M}{M_m} = \frac{1 + \zeta \eta V_f}{1 - \eta V_f} \tag{128}$$

where

$$\eta = \frac{M_f/M_m - 1}{M_f/M_m + \zeta} \tag{129}$$

M is the modulus of the composite, M_f is that of the reinforcement, M_m is that of the matrix, V_f is the volume fraction of the reinforcement, and ζ is a shape factor for the reinforcing elements. The property M can be taken as the extensional modulus parallel or perpendicular to the fibrils, the shear modulus, or the Poisson ratio.

When Eq. (128) is used to model the property along the reinforcement direction, the shape factor ζ is taken to have the value $2l/d$, where l/d is the length-to-diameter ratio of the reinforcing elements. As described in Refs. 7 and 75, the l/d

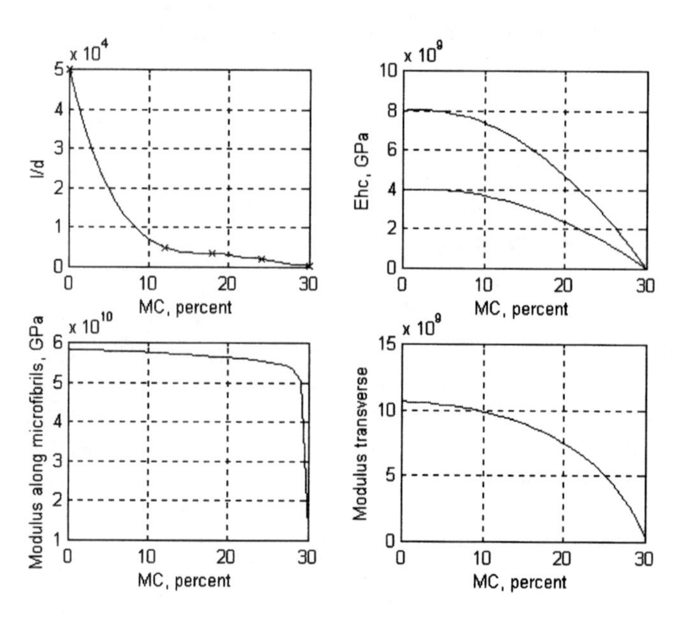

Fig. 16 Plots of l/d ratio for axial calculation, hemicellulose elastic moduli along and transverse to fibril direction (GPa), and S2 layer modulus (GPa) along and transverse to the microfibrils versus moisture content.

ratio can be modeled as a function of moisture content. Figure 16 illustrates the variation of properties with moisture content as provided by Berg and Gradin [7].

Following the work of Salmén [75], when Eq. (128) is used to model the transverse property, the l/d ratio is taken as unity based on the assumption that the cellulosic reinforcement is circular. The properties of the layer of the cell wall material are obtained by assuming a laminate structure consisting of a layer of isotropic lignin sandwiched between two special orthotropic layers of cellulose–hemicellulose composite.

The various parameters necessary for performing the calculations described are provided in Tables 2 and 3.

If one takes the 1,2 axes as the axes along the fibril direction and perpendicular to the fibril direction, respectively, the stress–strain relationship of the typical layer can be expressed as

Table 2 Fibril angle, Thickness, and Percent Relative Composition by Mass and Density of Cellulose, Hemiceullose, and Lignin in the Layers of the Cell Wall in Wood

Layer	Fibril angle (deg)	Thickness (μm)	Cellulose (%)	Hemicellulose (%)	Lignin (%)
ML	None	0.50	8	36	56
P	None	0.10	15	32	53
S1	70	0.30	28	31	41
S2	15	4.00	50	31	19
S3	70	0.04	48	36	16
ρ (kg/m^3)			1550	1500	1300

Source: Ref. 7.

Table 3 Elastic Properties of Cell Wall Components with Respect to Axes of Elastic Symmetry at 20°C at Various Moisture Contents

	Moisture content		
Component	0%	12%	30%
Lignin			
E (GPa)	4	2	2
ν	1/3	1/3	1/3
Hemicellulose			
E_1 (GPa)	8	7	0.02
E_2 (GPa)	4	3.5	0.01
G_{12} (GPa)	2	1.75	
0.005	ν_{12}	0.2	0.2
0.2	Cellulose		
E_1 (GPa)	134	134	134
E_2 (GPa)	27.2	27.2	27.2
G_{12} (GPa)	4.4	4.4	4.4
ν_{12}	0.1	0.1	0.1

Source: Ref. 7.

$$\begin{Bmatrix} \sigma_1 \\ \sigma_2 \\ \sigma_{12} \end{Bmatrix} = \begin{bmatrix} Q_{11} & Q_{12} & 0 \\ Q_{12} & Q_{22} & 0 \\ 0 & 0 & Q_{66} \end{bmatrix} \begin{Bmatrix} \varepsilon_1 \\ \varepsilon_2 \\ 2\varepsilon_{12} \end{Bmatrix} \tag{130}$$

where

$$Q_{11} = \frac{E_1}{1 - \nu_{12}\nu_{21}} \tag{131a}$$

$$Q_{22} = \frac{E_2}{1 - \nu_{12}\nu_{21}} \tag{131b}$$

$$Q_{12} = \frac{\nu_{21} E_1}{1 - \nu_{12}\nu_{21}} \tag{131c}$$

$$Q_{66} = G_{12} \tag{131d}$$

and

$$\frac{\nu_{12}}{E_1} = \frac{\nu_{21}}{E_2} \tag{131e}$$

The layers of the cell wall have different orientations relative to the axis of the fiber. By denoting the axes along and transverse to the fiber, respectively, as the x, y axes and the angle between the fiber axis and the fibril direction within the cell wall layers as θ, one can obtain the properties of the cell wall layers in the x, y system according to

$$\begin{Bmatrix} \sigma_x \\ \sigma_y \\ \sigma_{xy} \end{Bmatrix} = \begin{bmatrix} \bar{Q}_{11} & \bar{Q}_{12} & \bar{Q}_{16} \\ \bar{Q}_{12} & \bar{Q}_{22} & \bar{Q}_{26} \\ \bar{Q}_{16} & \bar{Q}_{26} & \bar{Q}_{66} \end{bmatrix} \begin{Bmatrix} \varepsilon_x \\ \varepsilon_y \\ 2\varepsilon_{xy} \end{Bmatrix} \tag{132}$$

where, with $m = \cos\theta$, $n = \sin\theta$,

$$\bar{Q}_{11} = Q_{11}m^4 + 2(Q_{12} + 2Q_{66})n^2m^2 + Q_{22}n^4 \tag{133a}$$

$$\bar{Q}_{12} = (Q_{11} + Q_{22} - 4Q_{66})m^2n^2 + Q_{12}(n^4 + m^4) \tag{133b}$$

$$\bar{Q}_{22} = Q_{11}n^4 + 2(Q_{12} + 2Q_{66})n^2m^2 + Q_{11}m^4 \tag{133c}$$

$$\bar{Q}_{66} = (Q_{11} + Q_{22} - 2Q_{12} - 2Q_{66})m^2n^2 + Q_{66}(n^4 + m^4) \tag{133d}$$

Expressions for the other coefficients are not provided because they are not used in the subsequent calculations. For their expressions, refer to Jones [29].

The properties of the fiber are obtained from laminate theory. Following the recommendation of Salmén and de Ruvo [76], the fiber is assumed to be a collapsed structure with the internal surfaces bonded to one another. The implication of this assumption is that the fiber is equivalent to an angle ply laminate, with the twisting deformation taken to be zero because the fibers are bonded together in the paper sheet. The effect of this is that the Q_{16} and Q_{26} terms in the expression for the composite cancel each other out. The final stress–strain relations for the fiber in the $-xy$ system is therefore specially orthotropic:

$$\begin{Bmatrix} \sigma_x \\ \sigma_y \\ \sigma_{xy} \end{Bmatrix} = \begin{bmatrix} Q^f_{11} & Q^f_{12} & 0 \\ Q^f_{12} & Q^f_{22} & 0 \\ 0 & 0 & Q^f_{66} \end{bmatrix} \begin{Bmatrix} \varepsilon_x \\ \varepsilon_y \\ 2\varepsilon_{xy} \end{Bmatrix} \tag{134}$$

where

$$Q^f_{11} = p_1 Q^{P+S1}_{11} + p_2 Q^{S2}_{11} + p_3 Q^{S3}_{11} \tag{135}$$

and similarly for the other coefficients and where p_1, p_2, and p_3 represent the proportion of P + S1, S2, and S3 layers in the cell wall.

The fiber elastic properties for the case of a fiber assumed to be thin-walled, perfectly collapsed, and bonded in the lumen, and with the S2 fibril angle assumed to be 15°, are shown in Fig. 17. As shown by this figure, the axial and transverse elastic moduli are found to be nearly equal. The fiber Poisson ratio (NUxyfib in Fig. 17) is found to be nearly constant until the moisture content approaches fiber saturation (assumed to be 30% MC), at which point the Poisson ratio is observed to decrease and become negative. This result is believed to be the result of using the values of the hemicellulose and lignin moduli provided in Ref. 7, which, as the authors state, are estimates and therefore may not be reasonable for the estimation of fiber elastic moduli.

D. Mechanical Properties of the Fiber–Fiber Bond

The physical structure of the fiber–fiber bond zone is not completely understood. However, a considerable amount of research has been devoted to this subject. Overviews of the structure and properties of the bond are given by Deng and

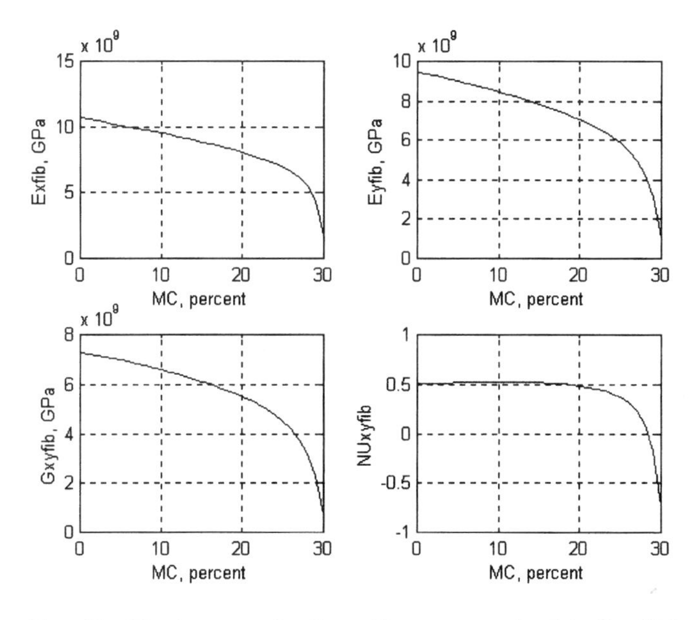

Fig. 17 Elastic properties for a fiber composed solely for S2 layer material with the fibril angle assumed to be 15°.

Dodson [17] and by Retulainen et al. [72]. See also Chapter 15 of this volume. Jayme and Hunger [27] and Page et al. [56] have provided considerable insight into the physical structure and aspects of the fiber–fiber bond that influence paper mechanical behavior.

It is generally believed that the fiber surfaces are bonded together by a network of hydrogen bonds that link the two mating surfaces. This view was used by Axelrad [5], who proposed modeling the interface zone by an array of hydrogen bonds that are characterized by a Morse potential function. It seems reasonable in terms of our limited knowledge of the structure of the bond zone, but it is difficult to use this approach in mechanics modeling owing to the fact that it is difficult to find experimental data for the number of hydrogen bonds per unit overlap area. Nissan and Batten [51] and Nissan et al. [52] provided a theoretical understanding of the properties of the hydrogen bond and its influence on the elastic behavior of paper, especially with regard to the effects of moisture. It has been understood that the stresses that are developed during the initial drying of the newly made paper sheet have an important effect on the mechanical behavior and strength of the paper [92]. (See also Chapter 10 of this book.) Some aspects of the modeling of the fiber–fiber bond in terms of hydrogen bonds and the effect of drying stresses are discussed by Salminen et al. [78].

The elastic properties of the fiber–fiber bond can be modeled by specifying the effective bond force spring constant per unit bond area. The approach is based on the experimental work reported by Thorpe et al. [89]. In this approach, a test pulp fiber is bonded to a shive taken from the same pulp system. The shive is supported while the end of the fiber is loaded along its axis. The displacements of end points of the bonded and unbonded zones are measured along with the applied load at failure. The width and length of the bonded and unbonded zones are recorded. If one assumes that the system behaves elastically and that the deformation of the shive is negligible, then a shear-lag analysis of the fiber–shive–bond system leads to the following expressions (see Fig. 18). The differential equation for all elements of the bonded fiber derives from the equilibrium of the typical element of the fiber–bond system (see Fig. 19).

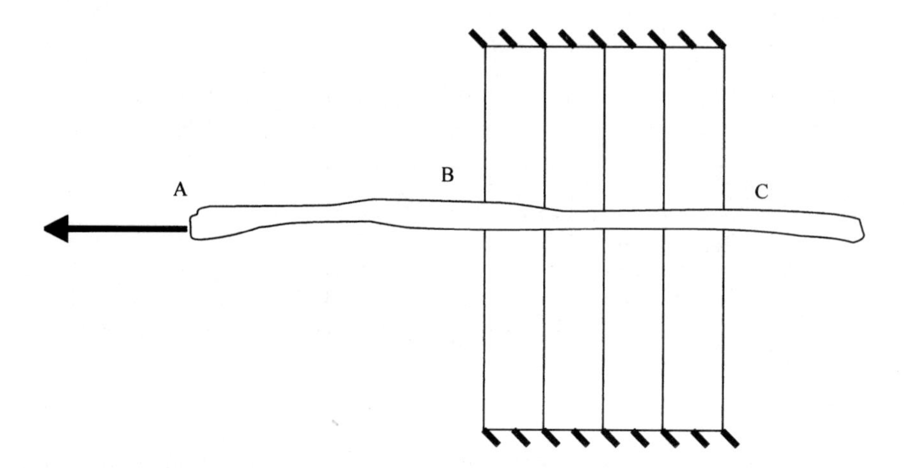

Fig. 18 Fiber–shive test setup. (From Ref. 89.)

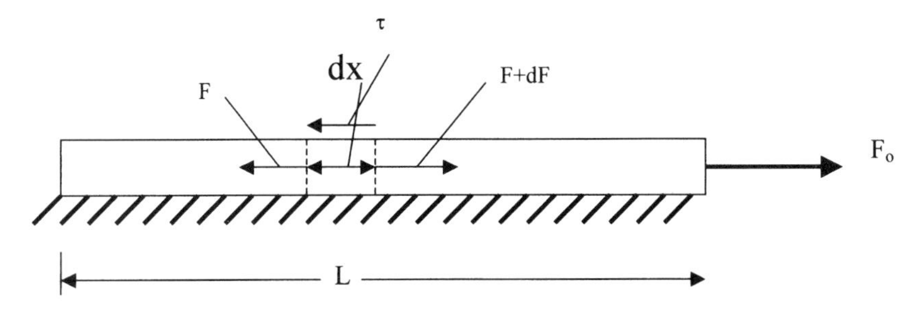

Fig. 19 Fiber–bond system.

$$dF + \tau w_f dx = 0 \tag{136}$$

where τ represents the shear stress exerted between the fiber and the shive through the bond. If we assume that the shear stress τ is proportional to the difference between the displacements of the fiber u_f, then we can write

$$\tau = k_b u_f \tag{137}$$

Furthermore, if the axial force transmitted by the fiber element, F, is expressed in the form

$$F = A_f E_f \frac{du_f}{dx} \tag{138}$$

where A_f is the fiber cross-sectional area and E_f is the fiber elastic modulus, then the equation of equilibrium can be written

$$\frac{d^2 u_f}{dx^2} - a^2 u_f = 0 \tag{139}$$

where

$$a = \left(\frac{k_b}{E_f t_f} \right)^{1/2} \tag{140}$$

and where it is assumed that the cross section is square with width w_f and thickness t_f.

The solution of Eq. (139) corresponding to the boundary conditions of a prescribed force F_0 applied at the right-hand end and zero stress at the left-hand end is

$$u_f = \frac{F_0 \cosh(ax)}{A_f E_f a \sinh(aL)} \tag{141}$$

In the case of the fiber–shive–bond test, the applied load and the displacement at $x = L$ were measured. Referring to the measured displacement as $u_f = \delta$, the following expression can be used to find parameter a, which in turn can be used to find k_b.

$$\tanh(aL) = \beta a L \tag{142}$$

where $\beta = \delta k_f / F_0$, where k_f is the axial stiffness of the fiber. The expression for k_b can be written in the form

$$k_b = \frac{k_f(aL)^2}{w_f L} \tag{143}$$

The maximum bond stress at maximum applied load is

$$\tau_{max} = \frac{k_f(aL)^2 \delta}{w_f L} \tag{144}$$

The strength of the bond can be identified as the maximum stress, τ_{max}, or as the maximum elastic strain energy, U_{max}, per unit bond area. In terms of the maximum bond stress and the bond stiffness, the bond energy can be given as

$$U_{max} = \frac{1}{2}\left(\frac{\tau_{max}^2}{k_b}\right) \tag{145}$$

Results from the paper by Thorpe et al. [89] were obtained with the foregoing expressions. These expressions are based on the assumptions that the fiber and bond material behaves elastically up to failure and that bond failure is catastrophic when the maximum stress at the end of the test specimen is reached. Furthermore, it is assumed that the deformation of the shive is negligible. The experimental materials used in the referenced study consisted of loblolly pine chlorite holocellulose fibers and Scotch pine thermomechanical pulp fibers. The results are shown in Table 4.

Stratton and Colson [86] studied the fiber–fiber bond using an experimental setup consisting of two isolated fibers that were bonded together. They measured the bond-breaking load and the bond area by using a polarized light scattering technique. Stratton and Colson investigated the bond strength of latewood and early wood fibers with varying levels of refining, and also the effect of polymeric strength additives. Their method does not permit the estimation of bond stiffness or bond energy at failure. However, the values of bond strength per unit area were found to be in the same range as those reported by Thorpe et al. [89] (see Table 4).

E. Linear Elastic Response of Medium-High Density Paper

In the work of Cox [16], the elasticity of the fibrous network is assumed to be attributed solely to the axial elastic properties of fibers that make up the network. Therefore, the transverse and shear deformations of the fibers are neglected. As discussed in Section V.A, a number of models have been developed that are based on a composite laminate approach with the intent of incorporating the transverse and shear deformations of the fibers [39–41,66,77,79,87,88,90]. The composite laminate approach, however, typically does not incorporate the effect of the fiber–fiber bond.

Table 4 Results for Bond Stiffness and Bond Strength

Material	k_b (N/m^3)	τ_{max} (N/m^2)	U_{max} (N/m)
Holocellulose chlorite bonded at room temperature	8.8×10^{12}	7.3×10^6	3.0
Thermomechanical pulp bonded at 110°C	5.7×10^{12}	6.4×10^6	3.6
Thermomechanical pulp bonded at 210°C	9.0×10^{12}	14.7×10^6	12.0

Source: Experimental data from Thorpe et al. [89].

The objective of the present section is to incorporate all of the in-plane normal and shear stress/strain components as well as the effect of the bond in transmitting load to fibers of arbitrary size. The fiber is assumed to be embedded in the paper network through a bond zone that covers the surface of the fiber and couples the fiber to the network. An approximate solution for the differential equation for the fiber, assuming elastic behavior, is solved by using the approach of classical elasticity.

The mesoelement for a high density system consists of a typical fiber of length $2L$, width $2W$, and thickness $2T$ that has an orientation θ with respect to the machine direction (MD) of the sheet as illustrated in Fig. 20. It is assumed that all fibers lie in the plane of the sheet and that they can be modeled as plane stress structures. It is assumed that the typical mesoelement fiber is bonded along its top and bottom surfaces by the other crossing fibers in the network. The proportion of the mesoelement fiber surface that is bonded to these crossing fibers is determined by the sheet density. The sheet density influences the distance between bonds as discussed in Section V.B. In addition, the bonded area depends on fiber orientation distribution because fiber orientation affects the fiber crossing angles, which in turn affects bond area.

The equilibrium equations are based on the equilibrium of a typical element of the fiber as shown in Fig. 21. The equilibrium equations are

$$\frac{\partial \sigma_x}{\partial x} + \frac{\partial \sigma_{xy}}{\partial y} + \frac{\sigma_{zx}}{2T} = 0 \tag{146}$$

$$\frac{\partial \sigma_{xy}}{\partial x} + \frac{\partial \sigma_y}{\partial y} + \frac{\sigma_{yx}}{2T} = 0 \tag{147}$$

The stress–strain relations for the fiber are taken to be in the form

$$\sigma_y = A_{12}\varepsilon_x + A_{22}\varepsilon_y - (A_{12}\alpha_x + A_{22}\alpha_y)M \tag{148a}$$

$$\sigma_x = A_{11}\varepsilon_x + A_{12}\varepsilon_y - (A_{11}\alpha_x + A_{12}\alpha_y)M \tag{148b}$$

$$\sigma_{xy} = A_{66}\varepsilon_{xy} \tag{148c}$$

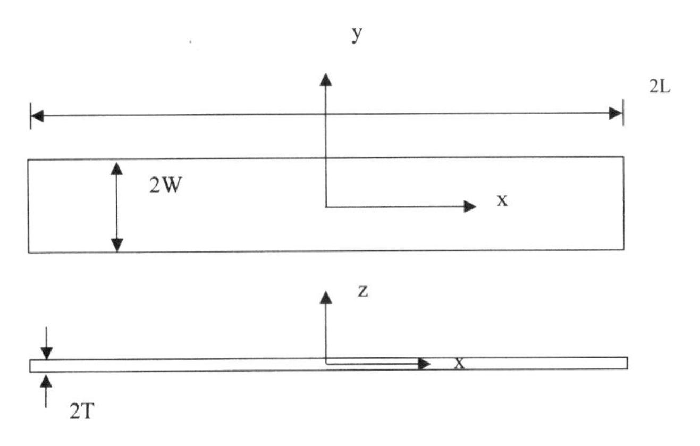

Fig. 20 Typical fiber of length $2L$, width $2W$, and thickness $2T$ oriented at angle θ relative to the machine direction.

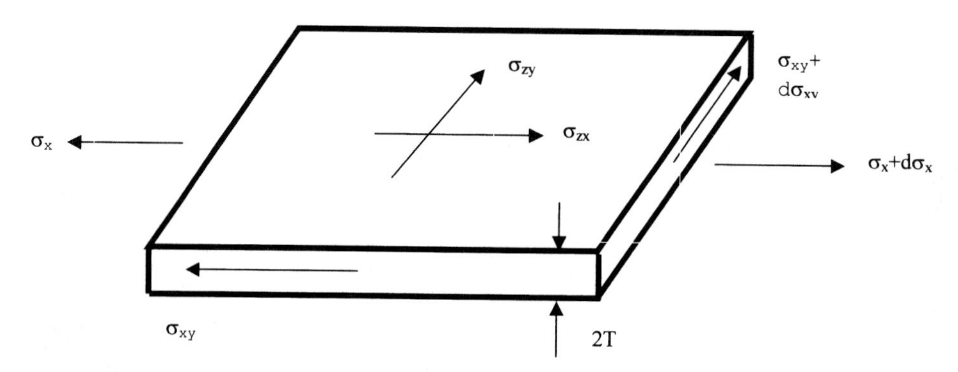

Fig. 21 Differential element of dimensions dx by dy and thickness $2T$ showing the in-plane stress components corresponding to the x direction along the transverse shear stress components acting on the top (bottom) surface that are associated with the fiber–fiber loading from the network.

where the A_{ij} are determined by the assumed layer structure of the collapsed fiber. Thus, the fiber is assumed to be a collapsed springwood fiber with the S3 layers bonded to one another to effectively make the fiber a ribbonlike laminate. The values of the A_{ij} depend on the elastic constants of the cell wall substance and on whether or not all three layers of the cell wall are incorporated in the laminate analysis. For further information, refer to Section V.D. It is convenient to write the equilibrium equations in terms of the assumed local displacement field on the assumption that the local strain field, ε_{ij}^s, is obtained by rotating the global strain field, ε_{ij}^g, along the axis of the fiber through the angle θ in accordance with

$$\varepsilon_x^s = \cos^2\theta\ \varepsilon_x^g + \sin^2\theta\ \varepsilon_y^g + \sin\theta\cos\theta\varepsilon_{xy}^g \tag{150a}$$

$$\varepsilon_y^s = \sin^2\theta\ \varepsilon_x^g + \cos^2\theta\ \varepsilon_y^g - \sin\theta\cos\varepsilon_{xy}^g \tag{150b}$$

$$\varepsilon_{xy}^s = 2\sin\theta\cos\varepsilon_x^g - 2\sin\theta\cos\theta\varepsilon_y^g + (\cos^2\theta - \sin^2\theta)\ \varepsilon_{xy}^g \tag{150c}$$

The shear components of strain are taken as the engineering shear strains. The deformation field surrounding the fiber is assumed to be an affine deformation field. Therefore, one can express the displacement components, u_s and v_s, in the form

$$v_s = \frac{1}{2}\varepsilon_{xy}^s x + \varepsilon_y^s y, \qquad u_s = \varepsilon_x^s x + \frac{1}{2}\varepsilon_{xy}^s y \tag{151}$$

One defines the fiber displacement components with the symbols u and v, and it is assumed that the shear stress components σ_{zx} and σ_{zy} are related to the relative displacement of the fiber and surrounding displacement field in the form

$$\sigma_{zx} = k_b(u_s - u), \qquad \sigma_{zy} = k_b(v_s - v) \tag{152}$$

Here, k_b is a shear constant that reflects the elastic coupling between the surrounding displacement field and the fiber displacement field through the fiber–fiber bond zone (cf. Section V.C). The equilibrium equations expressed in terms of the displacement components are

$$A_{11}\frac{\partial^2 u}{\partial x^2} + A_{66}\frac{\partial^2 u}{\partial y^2} + (A_{12} + A_{66})\frac{\partial^2 v}{\partial x\,\partial y} - Ku = -Ku_s \tag{153a}$$

$$A_{66}\frac{\partial^2 u}{\partial x^2} + A_{22}\frac{\partial^2 v}{\partial y^2} + (A_{12} + A_{66})\frac{\partial^2 u}{\partial x\,\partial y} - Kv = -Kv_s \tag{153b}$$

where the constant $K = k_b/2T$. At first it might appear that it would be a simple matter to solve these equations along with the boundary conditions of zero stress around the perimeter of the fiber. However, apparently no simple solution exists. A series solution approach can be used as proposed by Perkins [59]. The algebraic difficulties of using a solution of this type, however, obviate its use in a practical sense, because it would be difficult to achieve closed-form expressions for the elastic moduli. Further, as shown in Section V.F, an accurate numerical solution can be easily obtained using a finite element approach. Therefore, in order to derive closed-form expressions for the elastic moduli and hygroexpansion coefficients, an approximate solution is pursued.

Because fibers are generally much longer than their width dimension, it is not unreasonable to assume that $u_s = u$. Then

$$u = \varepsilon_x^s x + \varepsilon_{xy}^s y \tag{154}$$

$$\varepsilon_x = \frac{\partial u}{\partial x} = \varepsilon_x^s \tag{155}$$

$$\frac{\partial^2 u}{\partial x^2} = \frac{\partial^2 u}{\partial y^2} = \frac{\partial^2 u}{\partial x\,\partial y} = 0 \tag{156}$$

and the first of the displacement equilibrium equations reduces to

$$(A_{12} + A_{66})\frac{\partial^2 v}{\partial x\,\partial y} = 0 \tag{157}$$

which in turn leads to

$$(A_{12} + A_{66})v = \int f(x)\,dx + g(y) \tag{158}$$

We can therefore assume that

$$v = F(x) + G(y) \tag{159}$$

where $F(x)$ and $G(y)$ are functions that need to be determined. Substituting into the second equilibrium displacement equation leads to the form for $F(x)$ and $G(y)$ as

$$G(y) = B\sinh ay + \varepsilon_y^s y \tag{160}$$

$$F(x) = B_1 \sinh bx + \varepsilon_{xy}^s x \tag{161}$$

where

$$a = \sqrt{K/A_{22}} \tag{162}$$

$$b = \sqrt{K/A_{66}} \tag{163}$$

The constants B and B_1 are determined from the boundary conditions

$$\sigma_y(y = \pm L) = 0 \tag{164}$$

$$\sigma_{xy}(x = \pm L) = 0 \tag{165}$$

respectively, and their values are found to be

$$B = \frac{(A_{12}\alpha_x + A_{22}\alpha_y)M - A_{22}\varepsilon_y^2 - A_{12}\varepsilon_x^s}{A_{22}a\cosh aW} \tag{166}$$

$$B_1 = \frac{2\varepsilon_{xy}^s}{b\cosh bL} \tag{167}$$

In summary, the stress and strain components for the fiber are found to be

$$\varepsilon_x = \varepsilon_x^s x \tag{168}$$

$$\sigma_x = A_{11}\varepsilon_x^s + A_{12}\varepsilon_y - (A_{11}\alpha_x + A_{12}\alpha_y)M \tag{169}$$

$$\varepsilon_y = aB\cosh ay + \varepsilon_y^s \tag{170}$$

$$\sigma_y = A_{12}\varepsilon_x^s + A_{22}aB\cosh ay - (A_{12}\alpha_x + A_{22}\alpha_y)M \tag{171}$$

It is seen that the approximate solution provides all of the in-plane stress and strain components of the fiber, with the Poisson and shear effects all taken into account in accordance with the approximating assumption made regarding the displacement field for the fiber in the axial direction. Furthermore, the effects of the fiber–fiber bond and the in-plane dimensions of the fiber are also taken into account.

The elastic constant for the sheet can be calculated by using the solution provided above as follows. The strain energy density for the fiber is found by integrating over the domain of the fiber. Thus,

$$U_f = \frac{1}{4WL}\int_{-W}^{W}\int_{-L}^{L}\frac{1}{2}[\sigma_x\varepsilon_x + \sigma_y\varepsilon_y + \sigma_{xy}\varepsilon_{xy}]\,dW\,dL \tag{172}$$

Subsequently, the total strain energy for the paper sheet can be determined from

$$U = \frac{\rho}{\rho_f}\int_{-\pi/2}^{\pi/2}\int_{0}^{\infty}\int_{0}^{\infty} W_f f_\theta f_W f_L\,df_\theta\,df_W\,df_L \tag{173}$$

where f_θ, f_W, f_L are distribution functions for the fiber orientation, fiber width, and fiber length. The sheet stress–strain relations and the associated constants are determined from the relations

$$\sigma_x^{\text{sheet}} = C_{11}\varepsilon_x^g + C_{12}\varepsilon_y^g - \beta_1 M \tag{174}$$

where

$$\beta_1 = (C_{11}\alpha_x^s + C_{12}\alpha_y^s)M$$

$$\sigma_y^{\text{sheet}} = C_{12}\varepsilon_x^g + C_{22}\varepsilon_y^g - \beta_2 M \tag{175}$$

where

$$\beta_2 = (C_{12}\alpha_x^s + C_{22}\alpha_y^s)M$$

and

$$\sigma_x^{\text{sheet}} = C_{66}\varepsilon_{xy}^g \tag{176}$$

$$C_{11} = \frac{\partial^2 U}{\partial(\varepsilon_x^g)^2} \tag{177}$$

$$C_{12} = \frac{\partial^2 U}{\partial\varepsilon_x^g\,\partial\varepsilon_y^g} \tag{178}$$

$$C_{66} = \frac{\partial^2 U}{\partial(\varepsilon_{xy}^g)^2} \tag{179}$$

$$C_{22} = \frac{\partial^2 U}{\partial(\varepsilon_x^g)^2} \tag{180}$$

The von Mises and Erlang distributions are often used for the foregoing distribution functions. The general expressions for the elastic constants that result are cumbersome because of the large numbers of parameters. In the event that the fibers are assumed to be of constant width and length, and for the case that the orientation distribution is simply expressed in the series form

$$f_\theta = \frac{1}{\pi}[1 + a_1\cos 2\theta + a_2\cos 4\theta] \tag{181}$$

the elastic constants are easily found with the help of Maple or other symbolic software. The closed-form expressions for the elastic constants and hygroexpansion coefficients are provided in the Appendix. The analysis presented above neglects the strain energy in the bonds. An improved analysis can be developed with more general assumptions made regarding the form of the solution and with this inclusion of the bond strain energy.

F. Computer Simulation of Mesomechanics Models

There are a number of nonlinear effects that cannot be taken into account in any algebraic derivation analysis of the mesoelement such as the one given in Section V.E. For example, the definition of mesoelement geometry of the mesoelement for low density papers depends on the orientation angle, and geometric nonlinearity associated with large deformation exhibits an angle dependence. In addition, the algebraic analysis requires that one make very simplifying assumptions regarding the structure of the fiber such as uniformity along the fiber length and straightness of the fiber. Furthermore, the constitutive assumptions regarding the nature of the fiber and bond mechanical response are limited to linear elastic behavior in the algebraic analyses described above. To relax some of these simplifying assumptions, it is necessary to carry out the computations numerically rather than try to derive algebraic expressions for the mechanical response of low density paper.

A numerical approach that parallels the algebraic mesomechanics analysis provided in Section V.E can be carried out by making a model of the fiber and subjecting it to an affine displacement field. Because the analysis is completely numerical, it is necessary to devise a procedure that permits the equivalent of the closed-form analysis to be carried out without the use of calculus. The following procedure involving the use of the finite element method in the 2-D case has been used with success by Perkins [61]. In fact, the procedure was also used by Perkins and coworkers [64,65,69] to numerically simulate the uniaxial stress–strain behavior of paper sheets.

The first step is to construct a 2-D model of the fiber mesoelement, assuming plane stress. The mesoelement model depends on sheet density. For low density systems the models described in Section V.B would be used, whereas for medium to high density sheets, the model described in Section V.E would be used. Next, one must couple and load the mesoelement in accordance with the assumption of the affine strain field surrounding the mesoelement. The affine strain field depends on the fiber orientation and the applied strains for the sheet, so it is necessary to make a set of model analyses with different inputs. For example, assume that a finite element model is constructed of the mesoelement. One can simulate the uniaxial tensile test loading by prescribing the strain in the loading direction and assuming that the lateral strain is dependent on the assumed Poisson ratio. Thus, for an arbitrary prescribed uniaxial applied strain, the analysis is carried out for a set of assumed lateral strains corresponding to different Poisson ratios. These analyses are carried out for a set of different assumed orientations of the fibers relative to the fixed axes of the sheet. For example, one might use a set of six orientations of 18° increments from 0° to 90° relative to the machine direction in paper. In addition, one might employ six different assumed Poisson ratios 0, 0.1, 0.2, 0.3, 0.4, and 0.5 to control the lateral applied strains in the affine field surrounding the fiber. For each of this set of inputs, the total strain energy density or the work per unit volume of the mesoelement is calculated. The numerical procedures depend on whether the system is assumed to be linear elastic or whether nonlinear geometry, nonlinear constitutive models for the fiber and bond, and fiber or bond damage or failure are permitted to occur in the mesoelement.

Numerical Methods When the System is Assumed to be Linearly Elastic The strain energy density of the sheet in uniaxial tension, neglecting the swelling effect, can be expressed in the form

$$U = \tfrac{1}{2}\left[C_{11}(\varepsilon_x^g)^2 + 2C_{12}\varepsilon_x^g\varepsilon_y^g + C_{22}(\varepsilon_y^g)^2\right] \tag{182}$$

$$U = \tfrac{1}{2}\left[C_{11} - 2C_{12}\nu + C_{22}\nu^2\right](\varepsilon_x^g)^2 \tag{183}$$

where ν represents the assumed Poisson ratio. Here, the y component of strain is replaced by its equivalent in terms of the axial strain and the Poisson ratio. Of course, the true Poisson ratio is the one corresponding to the condition that the lateral stress be equal to zero. Therefore, the sheet stresses are given by

$$\sigma_x = \frac{\partial U}{\partial \varepsilon_x^g} = [C_{11} - 2C_{12}\nu + C_{22}\nu^2]\varepsilon_x^g \tag{184}$$

$$\sigma_y = \frac{\partial U}{\partial \varepsilon_y^g} = \frac{\partial U}{\partial \nu}\frac{\partial \nu}{\partial \varepsilon_y^g} = -\frac{\partial U}{\partial \nu}\frac{1}{\varepsilon_x^g} = -[-C_{12} + C_{22}\nu]\varepsilon_x^g = 0 \tag{185}$$

This implies that

$$\nu = C_{12}/C_{22} \tag{186}$$

which in turn implies that

$$\sigma_x = \left(C_{11} - \frac{C_{12}^2}{C_{22}^2} C_{22} \right) \varepsilon_x^g \tag{187}$$

The following numerical procedure is employed. We define a function

$$F_s(\nu) = \frac{2U}{(\varepsilon_x^g)^2} = C_{11} - 2C_{12}\nu + C_{22}\nu^2 \tag{188}$$

As an example, consider the simple case of uniaxially loaded fibers in a quasi-isotropic sheet. The macroscopic strain in the direction of the fiber axis is

$$\varepsilon_x^s = \varepsilon_x^g \big[(\cos\theta)^2 - \nu(\sin\theta)^2 \big] \tag{189}$$

Then the strain energy for the fiber is

$$U_f = \tfrac{1}{2} E_f (\varepsilon_x^s)^2 \tag{190}$$

Integrating the strain energy of the mesoelement multiplied by the orientation distribution function for an isotropic system with respect to orientation angle yields the strain energy of the sheet,

$$U_s = \tfrac{3}{16} E_f (\varepsilon_x^g)^2 (\nu^2 - 2\tfrac{1}{3}\nu + 1) \tag{191}$$

It is observed that the coefficient of the ν term is $-2C_{12}$. Therefore, the true value of ν is $1/3$. Furthermore, taking the second derivative of U_s with respect to the strain yields the value of the eleastic modulus, which is $E_f/3$, as expected. One can also observe that the minimum value of U_s corresponds to $\nu = 1/3$ and that the corresponding value of U_s divided by the applied strain squared is the elastic modulus, in this case $E_f/3$.

The strain energy information from the finite element analysis is computed by using appropriate distribution functions for orientation, fiber width, and fiber length. The strain energy for the set of orientation angles is curve-fit with a cubic spline. Subsequently, the curve-fit orientation information is integrated with respect to orientation angle in order to incorporate the actual fiber orientation distribution for the paper. This result is used to calculate the sheet strain energy U as a function of the computation variable ν. A second degree polynomial curve fit is used to fit the strain energy/strain squared data. Following the procedure outlined above, the sheet elastic modulus and Poisson ratio are taken as the ν and $2U(\varepsilon_x^g)^2$ values corresponding to the minimum values on the plot of these variables taken from the finite element analysis and curve-fitting processes. In accordance with the expression for $F_s(\nu)$ in Eq. (188); the curve fitting yields directly the elastic coefficients C_{11}, C_{12}, and C_{22}. All of the extensional in-plane elastic coefficients are simultaneously obtained from the numerical procedure.

The procedure just described is illustrated by carrying out the calculations for the strain energy of the mesoelement based on a finite element model. A plot of the strain energy for the mesoelement fiber versus orientation angle θ for the set of assumed values of ν is shown in Fig. 22. The plot of $2U/(\varepsilon_x^g)^2$ versus ν is shown in Fig. 23, where it is apparent that the Poisson ratio of the sheet is found to be 0.23. This method can be used in any approach that provides a numerical calculation of the strain energy of the mesoelement.

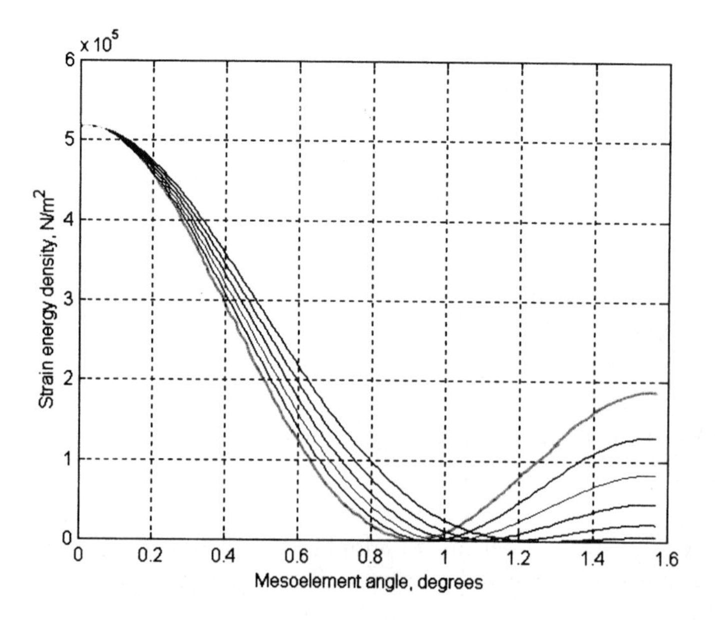

Fig. 22 Strain energy density versus mesoelement angle.

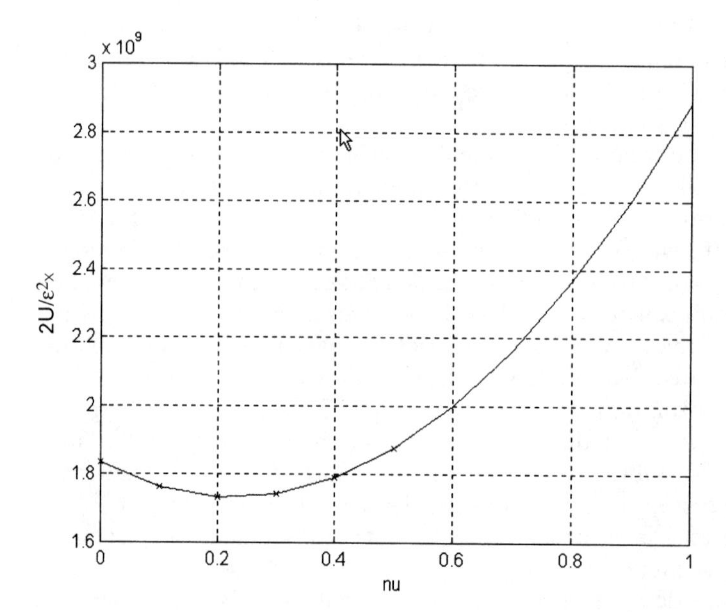

Fig. 23 Ratio of strain energy to strain squared versus calculated value of ν.

This method can be adapted for use in determining other elastic parameters as described next.

In-Plane Shear Modulus The in-plane shear modulus is easier to obtain. In this case,

$$\frac{2U}{(\varepsilon_{xy}^{g})^2} = C_{66} = G_{xy} \tag{192}$$

The procedure requires the determination of the fiber strain energy density for a set of fibers oriented at different directions to the machine direction, followed by an integration of the strain energy density multiplied by the fiber orientation distribution in order to determine the sheet strain energy density. Averaging with respect to other morphological variables is also performed when desired. Finally, Eq. (192) provides the shear modulus.

Moisture Swelling Coefficients The finite element numerical approach can also be used to obtain the moisture swelling coefficients. In this case, the sheet strain field is prescribed to be zero and the average values of the normal and transverse stress components are determined for the fiber. The sheet stress components are calculated by transforming the average fiber stresses to the global coordinate system and averaging with respect to orientation and size distribution functions. The global shear stress is always zero as a result of the averaging process for different orientations. Let σ_x^M and σ_y^M represent the calculated stresses corresponding to a particular value of the moisture content M. Then the swelling coefficients follow from

$$\begin{Bmatrix} \alpha_x \\ \alpha_y \end{Bmatrix} = \begin{bmatrix} C_{11} & C_{12} \\ C_{12} & C_{22} \end{bmatrix}^{-1} \begin{Bmatrix} \sigma_x^M/M \\ \sigma_y^M/M \end{Bmatrix} \tag{193}$$

Numerical Procedures When the System Exhibits Some Form of Nonlinearity Nonlinear effects are present when any of the following situations occur:

- Bond or fiber stress strain behavior is assumed to be nonlinear.
- Bond and/or fiber damage or failure is to be modeled.
- Finite strain effects are included, as in the case of very low density paper.

The general procedure that is described above can be used when nonlinear aspects are to be included. However, it is necessary to modify the assumption regarding the functional form of the strain energy for the system. Recall that in the case of the elastic procedure it can be assumed that the strain energy density is a parabolic function of the assumed lateral contraction ratio ν and that the strain energy density is proportional to the square of the axially applied strain ε_x. In the more general nonlinear case, we can only assume that the mesoelement strain energy density U_{meso} is a function of orientation direction, assumed lateral contraction ratio ν, and applied strain ε_x. Therefore, $U_{\text{meso}} = U_{\text{meso}}(\theta, \nu, \varepsilon_x)$. After a set of calculations for a prescribed value of ε_x have been performed numerically, one can define the apparent sheet strain energy density, $U_a(\nu, \varepsilon_x)$, as

$$U_a(\nu, \varepsilon_x) = \int_{-\pi/2}^{\pi/2} U_{\text{meso}}(\theta, \nu, \varepsilon_x) f_\theta \, d\theta \tag{194}$$

The minimum value of U_a is achieved for the value of the lateral contraction ratio ν^* that corresponds to the boundary condition of the uniaxial test that the transverse normal stress be zero. Thus,

$$U_a \mid_{\min} = U_a(\nu^*, \varepsilon_x) = U(\varepsilon_x) \tag{195}$$

Having determined $U(\varepsilon_x)$ for a set of values of ε_x, a curve fitting of this data is performed, which will then make it possible to determine the derivative of $U(\varepsilon_x)$ with respect to ε_x. Thus, the sheet stress corresponding to an arbitrary applied axial strain is obtained from

$$\sigma = \frac{dU(\varepsilon_x)}{d\varepsilon_x} \tag{196}$$

The resulting stress–strain curve will be expected to be nonlinear. Furthermore, the value of the solution lateral contraction ratio, ν^*, will be expected to vary with applied strain, ε_x.

Example of Numerical Calculations: Elastic Analysis of a Medium Density Sheet with Curled Fibers This example illustrates the effect of fiber curl for the case of a sheet of moderate density. The numerical results are based on an isotropic sheet having a density of $750\,\text{kg/m}^3$, comprising a set of uniform fibers having the parameters $\lambda = 1\,\text{mm}$, $w_f = 40\,\mu\text{m}$, $t_f = 4\,\mu\text{m}$, and for which the bond stiffness has the value $k_b = 1 \times 10^{13}\,\text{N/m}^3$. A typical fiber of 45° orientation is shown in Fig. 24. The figure shows the original and deformed shapes of the fiber corresponding to a uniaxial loading of 1% strain. Calculations for the sheet elastic modulus in the direction of uniaxial loading are shown in Fig. 25, where the dimensionless elastic modulus ratio is plotted as a function of fiber curl (defined as fiber radius of curvature divided by fiber length) and where

$$\rho_E = \frac{E_{\text{sheet}}}{E_f} \left(\frac{\rho_f}{\rho_s} \right) \tag{197}$$

The predicted Poisson ratio for the sheet is shown in Fig. 26. Observe that the sheet Poisson ratio does not approach the value of 1.3 as one would expect for a quasi-isotropic sheet as predicted by the closed-form expressions [e.g., Eq. (20), Section V.B]. This result is expected because the transverse and shear properties of the fibers are incorporated in the sheet elastic modulus calculation. The calculations shown were carried out for a flattened fiber consisting solely of S2 layer material. The fibril angle was selected to be 20°. The fiber moduli were calculated to be $E_{f,\text{axial}} = 3.3734 \times 10^{10}$, $E_{f,\text{transverse}} = 1.9197 \times 10^{10}$, $G_f = 6.0136 \times 10^9$, $\nu_{12} = 6.5780 \times 10^{-1}$, and $\nu_{21} = 3.7434 \times 10^{-1}$.

G. Computer Simulation of Fiber Networks Containing Many Fibers

A serious limitation of the mesomechanics models discussed above is the inherent presumed uniformity of the system. Paper, however, exhibits considerable nonuniformity at the network level. Fibers in a paper network are not straight, and the fibers that cross and become bonded do so in a distinctly nonuniform manner. The

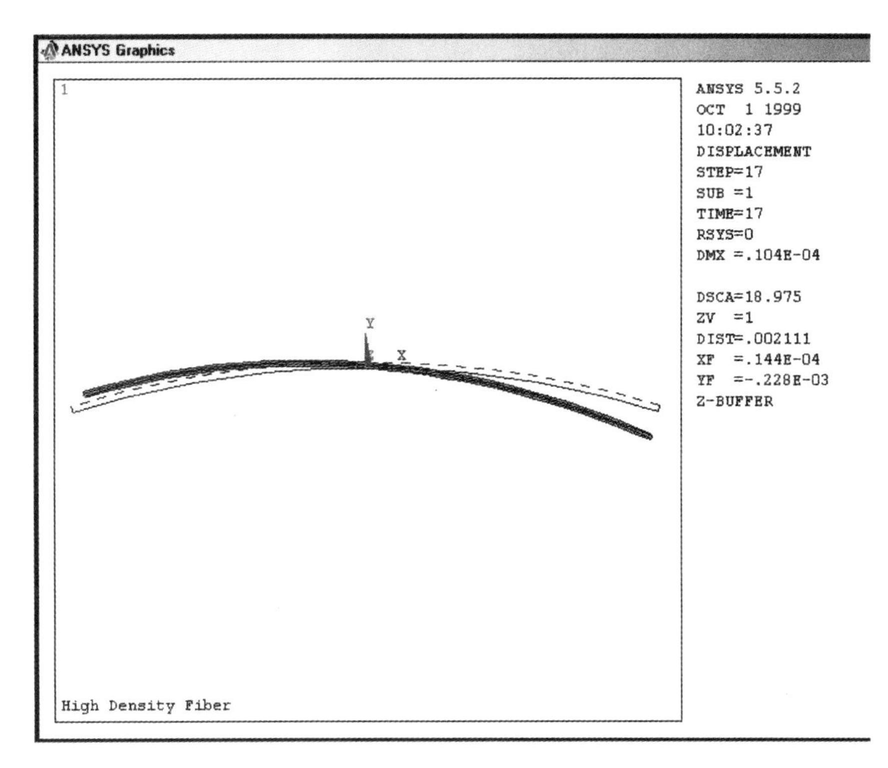

Fig. 24 Original and displaced curled fiber subjected to a 1% affine strain field. Fiber oriented between MD and CD in the sheet.

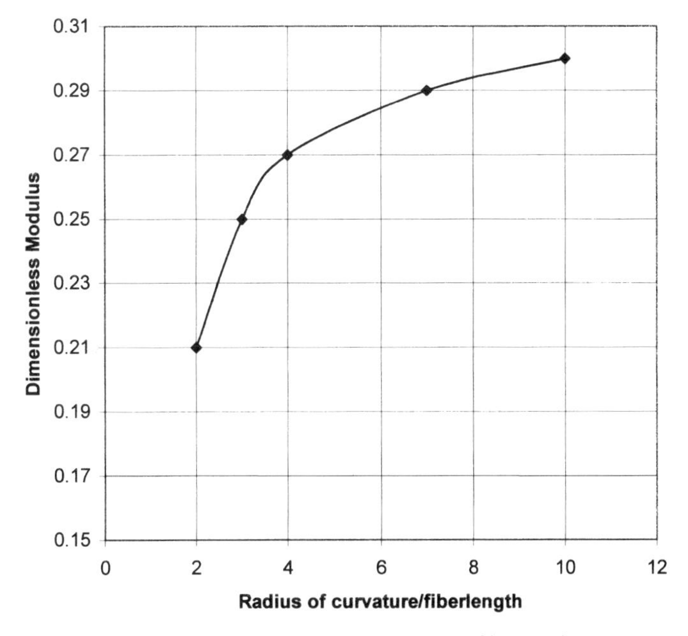

Fig. 25 Dimensionless modulus, ρ_E, versus fiber curl.

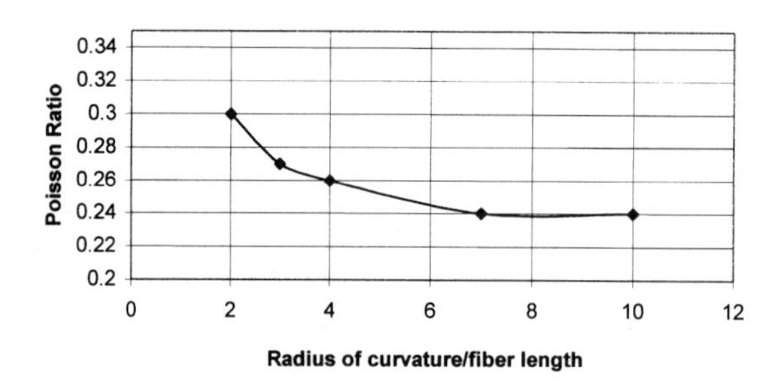

Fig. 26 Poisson ratio versus fiber curl.

existence of fiber ends that influence the local strain field in their vicinity and other factors give rise to a deformation field that is not uniform. The assumption of an affine deformation field is not valid, and this fact can be expected to be of considerable importance with regard to the stress and strain distributions within the fibers.

Rigdahl et al. [74] attempted to obtain some information regarding the interaction between fibers in a network by constructing a finite element model of a uniform rectangular array of fibers. Although this model is not at all like a paper network, those authors were able to demonstrate the importance of fiber interactions in a network.

It is not difficult to simulate a 2-D fiber network with a finite element model using beam elements to represent the fibers. At points where the fibers cross, a connection between fibers can be constructed to represent the effect of the fiber–fiber bond. Figure 27 illustrates a model of this type that was generated using the ANSYS finite element software. An inspection of Fig. 27 shows quite clearly the nonuniformity of the network structure, which appears to resemble a paper material rather well. The network in Fig. 27 was generated with distribution functions to control variation in fiber length, orientation, and curl.

One of the primary difficulties in the use of this type of 2-D network simulation lies in the uncertainty of the relationship between material density and the number and size of the fibers that constitute the model. One approach is to arbitrarily assign a thickness of the 2-D model to be twice the thickness of the fibers that compose the network. Based on this approach, the apparent density of the system in the simulation of Fig. 27 would be $750\,kg/m^3$. Thus, it is relatively easy to construct models with density levels that would be equivalent to a moderately dense paper sheet. On the other hand, the basis weight of a model of this type would be extremely low, e.g., on the order of $3\,g/m^2$. To be able to realistically model a paper sheet, it would be necessary to develop a 3-D network array having a much larger number of fibers than that illustrated in Fig. 27. Nonetheless, simulations with 2-D models can provide considerable insight into the mechanics of paper with the incorporation of nonlinear geometry effects in low density systems and nonlinear constitutive behavior for fiber and bond materials (including predictions of both fiber and bond failure).

Because the simulation model based on a 2-D network does not possess the attribute of a real thickness, it is not possible to specify the density of the resulting network. For this reason, Åström et al. [4] used the density of fibers in a unit area instead of the usual physical density. These authors used the variable q, defined as

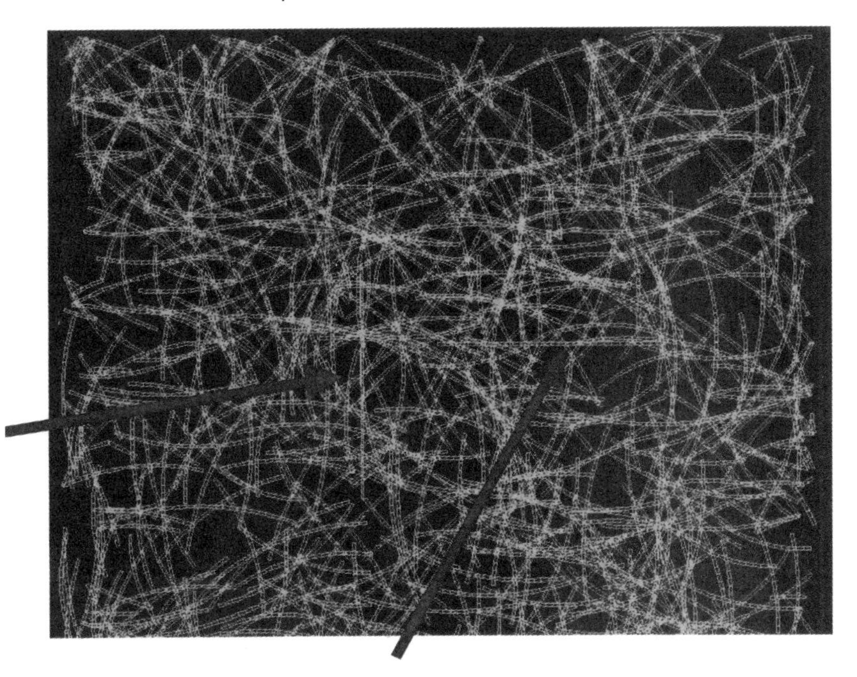

Fig. 27 A 2-D network of curved fibers. Approximate density equivalent is 750 kg/m^3. Arrows point to "test" fibers (one vertical and one horizontal) that were placed in the network for postprocessing purposes.

the total length of fiber in an area of λ_f^2, where λ_f is the fiber length. They also used the value of q_c, the critical length needed to develop a continuous network of interconnecting fibers. The critical q_c is also referred to as the "percolation" density. According to Åström et al., the value is

$$q_c = 5.71\lambda_f \tag{198}$$

Åström et al. further characterize the network in terms of the average segment length between bonds along a fiber, $\langle l \rangle$, by the relation

$$\frac{\langle l \rangle}{\lambda_f} = \frac{\pi}{11.42q/q_c} \approx \frac{q_c}{3.6q} \tag{199}$$

One of the important questions that arises in a study of network simulation concerns the extent to which the mesomechanics approach discussed in Section V.B can accurately predict the axial stress distribution along the fiber. Åström et al. [4] conclude that while the general shape of the distribution is similar to the hyperbolic cosine function as predicted by that theory, the magnitudes of the axial stresses are significantly different from those predicted by mesomechanics theory. They propose that the stress transfer is governed by network parameters that are related to fiber length more than to the attributes of fiber segment length as predicted by mesomechanics theory. The issue of the validity of the so-called shear-lag model was further studied by Räisänen et al. [68], who concluded that the shear-lag model does not apply to random fiber networks. A comparison of the mesomechanics model and the simulation model depicted in Fig. 27 serves to illustrate the nature of the difference between the two approaches. The arrows in Fig. 27 point to "test" fibers, one vertical straight fiber and one horizontal straight fiber. The vertical fiber is fairly uniformly bonded to other fibers in the network. Figure 28 illustrates a comparison of the axial strain along the vertical "test" fiber predicted by the mesomechanics model discussed in Section V.B [cf. Eq. (97)] and that of the simulation model shown in Fig. 27. It is of interest to note that the average segment length obtained from the simulation model by dividing the fiber length by the number of crossings was found to be 0.064 mm, whereas the average segment length calculated from the relations in Section V.B [Eq. (123)] was 0.060 mm. In the calculation for the "test" fiber, both the mesomechanics model and the simulation model used an assumed value of $1 \times 10^{13}\,\text{N/m}^3$ for the bond stiffness. The simulation for this example corresponds to the ratio $q/q_{\text{crit}} = 6.85$.

An extremely important aspect of the network simulation approach lies in the ability of this type of analysis to study the failure processes that take place as the applied loading is permitted to increase. Since the stress distribution predicted by the mesomechanics model is incapable of taking into account interactions between adjoining fibers as failure takes place, failure predictions based on mesomechanics are not expected to be very realistic. Some aspects of failure as predicted by network simulation have been published [3,4,49].

The network simulation approach is especially useful in the nonlinear analysis of fiber and bond constitutive behavior as well as the effects on mechanical behavior attributed to fiber curl. The effect of fiber curl was studied by Jangmalm and Östlund [25]. The effect of fiber nonlinear behavior was studied by Räisänen et al. [67], who concluded that the stress–strain behavior of a random network depends on the stress–strain behavior of the fibers in the network and does not depend on the

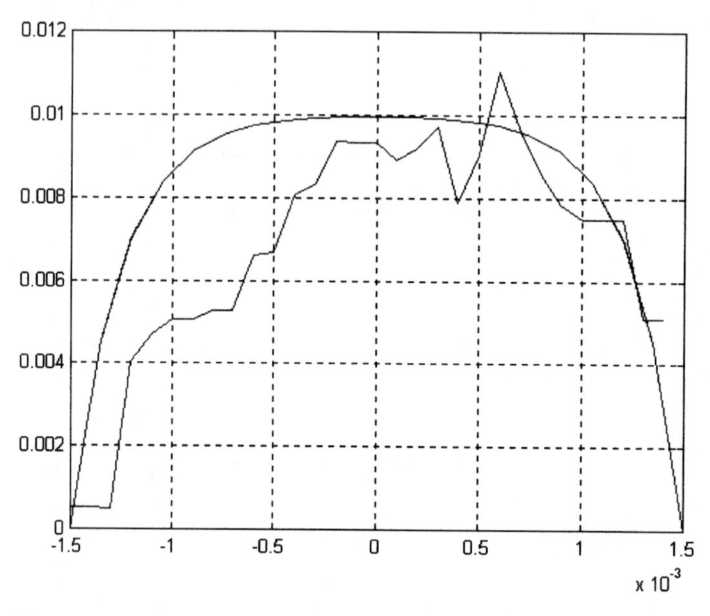

Fig. 28 Axial strain along the vertical "test" fiber shown in Fig. 27 and the axial strain along the fiber as predicted by the mesomechanics theory [cf. Eq. (97)].

density or bonding degree of the network. It should be noted, however, that the nonlinear behavior of the fiber–fiber bond does also have a significant effect on the network stress–strain behavior. This fact is illustrated by the results shown in Fig. 29, which were obtained from an analysis of the network depicted in Fig. 27. In this computer simulation it was assumed that the fibers were perfectly elastic and that therefore none of the network nonlinearity is attributable to nonlinear behavior of the fibers.

The extension of the 2-D simulation models to 3-D networks is necessary to obtain realistic predictions of paper mechanical behavior. Wang and Shaler [95] and Niskanen et al. [48] reported important progress in the development of 3-D network modeling. Neither of these works, however, endeavored to study the mechanical behavior aspects.

H. Plane Stress 2-D Nonlinear Constitutive Model for Paper

In problems involving the prediction of nonlinear inelastic response of paper structures characterized by a multiaxial state of stress and strain it is necessary to be able to carry out the analysis using commercial finite element software. The constitutive models used in commercial finite element software are typically designed for metals that exhibit similar response in tension and compression and incompressible plastic response and that are isotropic. These attributes are poor assumptions for paper inelastic behavior, where the stress–strain curve in compression is substantially dif-

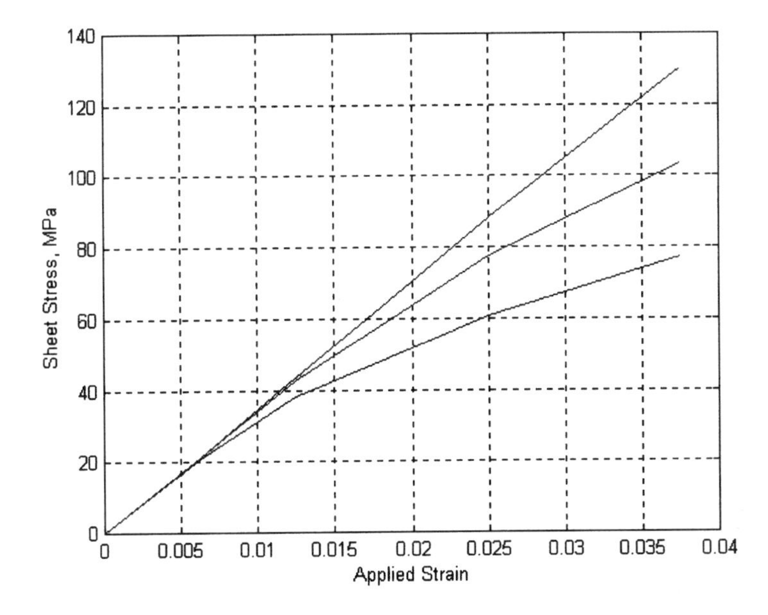

Fig. 29 Stress–strain curves from the uniaxial loading of a 1 cm by 1 cm network model approximating a density of 750 kg/m^3. The three curves correspond to different values of the critical displacement of the assumed two-slope spring constant that simulates the behavior of a nonlinear fiber–fiber bond (values are 1×10^{-6}, 2×10^{-6}, and infinity). The initial slope is equivalent to a bond modulus of 1×10^{13} N/m^3; the second slope is one-half the initial slope.

ferent from that in tension, where the material is porous and therefore is not incompressible, and where the response is strongly anisotropic. The hyperelasticity model used by Johnson and Urbanik [28] can be used in multiaxial stress problems to model the anisotropic behavior of paper. This model does not incorporate the aspect of different compression and tension response. The plasticity model presented by Shih and Lee [83], although designed specifically for the plastic behavior of metals, incorporates differences in compression and tension as well as anisotropic behavior. However, it does not incorporate compressibility. A 2-D approach based on mesomechanics that incorporates anisotropy and difference in tension and compression behavior was presented by Sinha and Perkins [85].

The mesoelement algebraic approach that is described in Section V.B can be extended to include certain nonlinear constitutive effects of the fiber and bond materials. Ramasubramanian and Perkins [69] derived expressions for predicting the nonlinear uniaxial stress–strain curve for paper for the case of straight fibers but allowing the fiber and bond to exhibit nonlinear behavior. This approach was also used by Perkins et al. [65] and by Perkins and Sinha [64]. The same approach can be used to derive a set of incremental expressions for predicting the 2-D response of the network for arbitrary in-plane applied strains. The following 2-D constitutive equations are based on an extension of the work of Sinha [84]. The method is based on the calculation of the average stress for a typical mesoelement. The average stress can be determined in closed form as was done by Sinha [84] or can be based on a finite element evaluation of a typical mesoelement. The approach is referred to as the method of mean effective mesoelements.

Method of Mean Effective Mesoelements In the following, the material is assumed to exhibit a rate-independent behavior. A typical mesoelement having an in-plane orientation angle θ relative to the machine direction in the sheet is characterized by its average stress $\langle\sigma(\varepsilon_s)\rangle$, a function of the sheet strain ε_s along its axis, where

$$\varepsilon_s = \varepsilon_x \cos^2\theta + \varepsilon_y \sin^2\theta + 2\varepsilon_{xy}\cos\theta\sin\theta \tag{200}$$

An incremental change in the local sheet strain is given by

$$\Delta\varepsilon_s = \Delta\varepsilon_x \cos^2\theta + \Delta\varepsilon_y \sin^2\theta + 2\Delta\varepsilon_{xy}\cos\theta\sin\theta \tag{201}$$

Dividing each term of Eq. (201) by $\cos^2\theta$, one obtains

$$\frac{\Delta\varepsilon_s}{\cos^2\theta} = \Delta\varepsilon_x + \Delta\varepsilon_y \tan^2\theta + 2\Delta\varepsilon_{xy}\tan\theta \tag{202}$$

The orientation angle that corresponds to having a mesoelement fiber strain of zero is given by

$$\tan\alpha = -\frac{\Delta\varepsilon_{xy}}{\Delta\varepsilon_y} \pm \left[\left(\frac{\Delta\varepsilon_{xy}}{\Delta\varepsilon_y}\right)^2 - \frac{\Delta\varepsilon_x}{\Delta\varepsilon_y}\right]^{1/2} \tag{203}$$

Thus, when θ is less than α, the incremental mesoelement strain $\Delta\varepsilon_s$ is greater than zero and the mesoelement loading is tensile, whereas for θ greater than α, the incremental mesoelement strain is less than zero and the mesoelement loading is compressive. It is assumed that in general mesoelements subjected to tensile loading will respond differently than those for which the loading is compressive.

The incremental change in the average stress in a mesoelement fiber due to the incremental change in local sheet strain depends on the previously accumulated local sheet strain, the incremental change in the local sheet strain, and the orientation angle of the mesoelement fibers. Let us write the incremental change in average mesoelement stress as

$$\Delta\langle\sigma\rangle = F(\varepsilon_s, \mathrm{sgn}(\Delta\varepsilon_s), \theta, \lambda_1, \lambda_2, \ldots)\, \Delta\varepsilon_s \tag{204}$$

where the function F represents the functional dependence of the average mesoelement stress on the accumulated mesoelement strain that results from the history of deformation prior to the increment under consideration. The function F can be assumed to be dependent on the orientation of the mesoelement and a set of shape and mesoelement constitutive parameters. However, if there is no directional dependence of these attributes with orientation, the function F can be taken as the average over all such parameters. The function F can either be found from the solution of the mechanics of the mesoelement or simply assumed to have a given functional form. Here the latter approach is used and F is assumed to be a bilinear function characterized by an initial slope, a critical strain corresponding to the change in slope, and a second slope. For unloading and reloading, changes are required to follow the first slope. The slope values and the critical strain are allowed to be different for tensile and compressive strains and to be dependent on moisture content. In this fashion, the behavior of the paper is determined by four moisture-dependent material parameters along with the mesoelement orientation distribution function. All of these parameters are obtained experimentally by comparison of standard tensile, compressive, and shear tests of the paper material with the model response. Assuming, therefore, that F depends only on the four material constants, orientation angle, and local sheet strain, the incremental stresses in the paper are given by

$$\begin{Bmatrix} \Delta\sigma_x \\ \Delta\sigma_y \\ \Delta\sigma_{xy} \end{Bmatrix} = \begin{bmatrix} k_{11} & k_{12} & 0 \\ k_{12} & k_{22} & 0 \\ 0 & 0 & k_{33} \end{bmatrix} \begin{Bmatrix} \Delta\varepsilon_x \\ \Delta\varepsilon_y \\ \Delta\varepsilon_{xy} \end{Bmatrix} \tag{205}$$

where

$$k_{ij} = \frac{\rho_s}{\rho_f} \int_{-\pi/2}^{\pi/2} G_{ij} f_\theta \, d\theta \tag{206}$$

and f_θ represents the orientation distribution,

$$G_{11} = Fm^4, \qquad G_{12} = Fn^2m^2, \qquad G_{22} = Fn^4, \qquad G_{33} = G_{12} \tag{207}$$

$n = \sin\theta$, $m = \cos\theta$.

Equation (205) forms the basis for a constitutive model that can be used in any finite element software that permits the user to incorporate a "user routine." The advantage of the approach outlined above is that it permits the development of a constitutive model that exhibits orthotropic plastic behavior having different responses for compressive loading and tensile loading. It is believed that these attributes are essential for modeling the behavior of paper materials.

APPENDIX. ELASTIC COEFFICIENTS FOR MODERATE TO HIGH DENSITY PAPER*

The elastic constants and the hygroexpansion coefficients from Eqs. (175) and (176) can be expressed in the form of a series involving the orientation coefficients $a_0 = 1$, a_1, and a_2 [cf. Eq. (181)] and a set of functions f_i, $i = 0, \ldots, 4$, defined as

$$C_{ij} = \sum_{m=0}^{m=2} d_{ij}^m a_m, \qquad i, j = 1, 2, 6 \tag{A.1}$$

where

$$d_{ij}^m = \sum_{n=0}^{n=4} b_{ijm}^n f_n \tag{A.2}$$

$$f_0 = 1 \tag{A.3a}$$

$$f_1 = \frac{\tanh(aW)}{aW} \tag{A.3b}$$

$$f_2 = \frac{\tanh(bL)}{bL} \tag{A.3c}$$

$$f_3 = \frac{1}{\cosh^2(aW)} \tag{A.3d}$$

$$f_4 = \frac{1}{\cosh^2(bL)} \tag{A.3e}$$

The hygroexpansion coefficients β_{I} can be expressed as the series

$$\beta_i = \sum_{m=0}^{m=2} e_i^m a_m \tag{A.4}$$

and

$$e_i^m = \sum_{n=0}^{n=4} g_{im}^n f_n \tag{A.5}$$

*Refer to Section V.E.

Table A.1 Coefficients b^n_{ijm} for the Calculation of C_{ij}, for $i, j = 1, 2, 6$, Using Eqs. (A.1), (A.2), and (A.3) and for the Coefficients g^n_{im} in Eqs. (A.4) and (A.5) Used in the Calculation of the β_i

		n				
	m	0	1	2	3	4
C_{11}	0	$3A_{11}/2 + A_{12} + 2A_{66}$	$A_{12} + 3A_{22}/4 + 9A_{12}^2/A_{22}$	$-3A_{66}$	$3A_{22}/4 + A_{12}/2 + 3A_{12}^2/(4A_{22})$	A_{66}
	1	A_{11}	$(A_{22} - A_{12}^2/A_{22})/2$	0	$-(A_{22} - A_{12}^2/A_{22})/2$	0
	2	$A_{11}/2 + 3A_{12} - 2A_{66}$	$-(3A_{12} - A_{22}/4 - 2A_{12}^2/3A_{22})$	$3A_{66}$	$3A_{12}/2 + A_{12}^2/4A_{22} + A_{22}/4)$	A_{66}
C_{12}	0	$3A_{11}/2 + A_{12} + 2A_{66}$	$-(A_{12} + 3A_{22}/4 + 9A_{12}^2/A_{22})$	$-3A_{66}$	$3A_{22}/4 + A_{12}/2 + 3A_{12}^2/4A_{22}$	A_{66}
	1	0	0	0	0	0
	2	$-A_{11}/4 + A_{22}/2 + A_{66}$	$-(A_{12}/2 - A_{22}/8 - 3A_{12}^2/8A_{22})$	$-3A_{66}$	$-(A_{22}/8 - A_{12}/4 + A_{12}^2/8A_{22})$	$A_{66/2}$
C_{22}	0	$3A_{11}/2 + A_{12} + 2A_{66}$	$-(A_{12} + 2A_{22}/3 + 9A_{12}^2/4A_{22})$	$-3A_{66}$	$2A_{22}/3 + A_{12}/2 + 2A_{12}^2/3A_{22}$	A_{66}
	1	$-A_{11}$	$-1/2(A_{22} - 3A_{12}^2/A_{22})$	0	$(A_{22} - A_{12}^2/A_{22})/2$	0
	2	$A_{11}/4 + A_{22} - A_{12}/2 - A_{66}$	$A_{12}/2 - A_{22}/8 - 3/8A_{12}^2/A_{22}$	$3A_{66}$	$-(A_{12}/4 + A_{22}/8 + A122/8A_{22})$	$2A_{66}$
C_{66}	0	$2A_{11} - 4A_{12} + 8A_{66}$	$4A_{12} - A_{22} - 3A_{12}^2/A_{22}$	$-12A_{66}$	$-2A_{12} + A_{22} - A_{12}^2/A_{22}$	$4A_{66}$
	1	0	0	0	0	0
	2	$4A_{66} + 2A_{12} - A_{11}$	$A_{22}/2 + 2A_{12}^2/2A_{22} - 2A_{12}$	$-6A_{66}$	$-A_{22}/2 - A_{12}^2/2A_{22} + A_{12}$	$2A_{66}$

Table A.1 Continued

	m	n = 0	1	2	3	4
β_1	0	$(A_{11}+A_{12})\alpha_x+(A_{22}+A_{12})\alpha_y$	$((A_{12}+2A_{12}^2/A_{22})\alpha_x+(A_{22}+2A_{12})\alpha_y)$	0	$(A_{12}+A_{12}^2/A_{22})\alpha_x+(A_{22}+A_{12})\alpha_y$	0
	1	$-((A_{11}-A_{12})\alpha_x-(A_{22}-A_{12}(\alpha_y)/2$	$-((A_{12}-2A_{12}^2/A_{22})\alpha_x+(A_{22}-2A_{12})\alpha_y)/2$	0	$((A_{12}-A_{12}^2/A_{22})\alpha_x+(A_{22}-A_{12})\alpha_y)/2$	0
	2	0	0	0	0	0
β_2	0	$(A_{11}+A_{12})\alpha_x+(A_{22}+A_{12})\alpha_y$	$-[(A_{12}+2A_{12}^2/A_{22})\alpha_x+(A_{22}+2A_{12}]\alpha_y)$	0	$(A_{12}+A_{12}^2/A_{22})\alpha_x+(A_{22}+A_{12})\alpha_y$	0
	1	$((A_{11}-A_{12})\alpha_x-(A_{22}-A_{12})\alpha_y/2$	$[(A_{12}-2A_{12}^2/A_{22})\alpha_x+(A_{22}-2A_{12})\alpha_y]/2$	0	$-((A_{12}-A_{12}^2/A_{22})\alpha_x+(A_{22}-A_{12})\alpha_y)/2$	
	2	0	0	0	0	0

REFERENCES

1. Algar, W. H. (1966). Effect of structure on the mechanical properties of paper. In: *Consolidation of the Paper Web*. Technical Section of the British Paper and Board Makers' Association, Cambridge, UK.
2. Anon. (1994). *ABAQUS Theory Manual*. Version 5.4. Hibbitt, Karlsson & Sorensen, Pawtucket, RI.
3. Åström, J., and Niskanen, K. (1991). Simulation of network fracture. 1991 International Paper Physics Conference, Kona, HI, TAPPI, Atlanta, GA.
4. Åström, J., Saarinen, S., Niskanen, K., and Kurkijarvi, J. (1994). Microscopic mechanics of fiber networks. *J. Appl. Phys.* *75*(5):2383–2392.
5. Axelrad, D. (1978). *Micromechanics of Solids*. Elsevier, New York.
6. Bach, L., and Pentoney, R. E. (1968). Nonlinear mechanical behavior of wood. *Forest Products J.* *18*(3):60–66.
7. Berg, J.-E., and Gradin, P. A. (1999). A micromechanical model of the deterioration of a wood fibre. *J. Pulp Paper Sci.* *25*(2):66–71.
8. Biot, M. A. (1965). *Mechanics of Incremental Deformation*. Wiley, New York.
9. Bristow, J. A., and Kolseth, P., eds. (1986). *Paper Structure and Properties*. (Int. Fiber Sci. Technol. Ser. Vol. 8, M. Lewin, ed.). Marcel Dekker, New York, p. 390.
10. Castagnede, B., Mark, R. E., and Seo, Y. B. (1989). New concepts and experimental implications in the description of the 3-D elasticity of paper. Part I: General considerations. *J. Pulp Paper Sci.* *15*(5):J178–J182.
11. Castagnede, B., Mark, R. E., and Seo, Y. B. (1989). New concepts and experimental implications in the description of the 3-D elasticity of paper. Part II: Experimental results. *J. Pulp Paper Sci.* *15*(6):J201–J205.
12. Chang, J. (1983). Modeling the behavior of paper with regard to fiber orientation and length distributions. Master's Thesis, Syracuse University, Syracuse, NY.
13. Chen, W. F., and Baladi, G. Y. (1985). *Soil Plasticity: Theory and Implementation*. Elsevier, New York.
14. Chen, W. F., and Han, D. J. (1988). *Plasticity for Structural Engineers*. Springer-Verlag, New York.
15. Christensen, R. M. (1971). *Theory of Viscoelasticity: An Introduction*. Academic Press, New York.
16. Cox, H. L. (1952). The elasticity and strength of paper and other fibrous materials. *J. Appl. Phys.* *3*:72–79.
17. Deng, M., and Dodson, C. T. J. (1994). *Paper: An Engineered Stochastic Structure*: TAPPI Press, Atlanta, MA.
18. Dodson, C. T. J. (1973). A survey of paper mechanics in fundamental terms. In: *The Fundamental Properties of Paper Related to Its Uses*. Technical Association of the British Paper and Board Makers' Association, Cambridge, UK.
19. Eringen, A. C. (1962). *Nonlinear Theory of Continuous Media*. McGraw-Hill, New York.
20. Feldman, H., Jayaraman, K., and Kortschot, M. T. (1996). A Monte Carlo simulation of paper deformation and failure. *J. Pulp Paper Sci.* *22*(10):J386–J391.
21. Ferry, J. D. (1980). *Viscoelastic Properties of Polymers*. 3rd ed. Wiley, New York.
22. Fung, Y. C. (1965). *Foundations of Solid Mechanics*. Prentice-Hall, Englewood Cliffs, NJ.
23. Halpin, J. C. (1968). Introduction to viscoelasticity. In: *Composite Materials Workshop*. S. W. Tsai, J. C. Halpin, and N. J. Pagano, eds. Technomic, Stamford, CT.
24. Haslach, H. W., Jr. (1996). A model for drying-induced microcompressions in paper: buckling in the interfiber bonds. *Composites* *27B*(1):25–33.
25. Jangmalm, A., and Östlund, S. (1995). Modelling of curled fibers in two-dimensional networks. *Nordic Pulp Paper Res. J.* *1995*(3):156–166.

26. Jayaraman, K., and Kortschot, M. T. (1998). Closed-form network models for the tensile strength of paper: A critical discussion. *Nordic Pulp Paper Res. J. 13*(3):233–242.
27. Jayme, G., and Hunger, G. (1962). Electron microscope 2- and 3-dimensional classification of fibre bonding. In: *Formation and Structure of Paper*. F. Bolam, ed. Tech. Sec, British Paper and Board Makers' Association, Inc., London, UK, p. 135.
28. Johnson, M. W., Jr., and Urbanik, T. J. (1984). A nonlinear theory for elastic plates with application to characterizing paper properties. *J. Appl. Mech. 51*(March):146–152.
29. Jones, R. M. (1999). *Mechanics of Composite Materials*. Taylor and Francis, Philadelphia, PA.
30. Kallmes, O. (1972). A comprehensive view of the structure of paper. In: *Theory of Design of Wood and Fiber Composite Materials*. B. A. Jayne, ed. Syracuse Univ. Press, Syracuse, NY, pp. 157–176.
31. Kallmes, O. (1972) The influence of nonrandom fiber orientation and other fiber web parameters on the tensile strength of nonwoven fibrous webs. In: *Theory of Design of Wood and Fiber Composite Materials*. B. A. Jayne, ed. Syracuse Univ. Press, Syracuse, NY, pp. 177–196.
32. Kallmes, O., and Corte, H. (1960). The structure of paper. I. The statistical geometry of an ideal two dimensional fiber network. *Tappi 43*(9):737–752.
33. Kärenlampi, P. (1995). Tensile strength of paper: A simulation study. *J. Pulp Paper Sci. 21*(6):J209.
34. Kärenlampi, P. (1995). Effect of distributions of fiber properties on tensile strength of paper: A closed-form theory. *J. Pulp Paper Sci. 21*(4):J138.
35. Kohnke, P., ed. *ANSYS User's Manual for Revision 5.0—Theory*, Vol. IV. Swanson Analysis Systems, Inc., 1992.
36. Komori, T., and Makishima, K. (1977). Number of fiber to fiber contacts in general fiber assemblies. *Textile Res. J. 47*(1):13–17.
37. Lempriere, B. M. (1968). Poisson's ratio in orthotropic materials. *AIAA J. 6*(11):2226–2227.
38. Liang, S., and Suhling, J. C. (1997). Finite element and experimental study of the embossing of low basis weight papers. R. W. Perkins, ed. ASME, Philadelphia, pp. 55–68.
39. Lu, W., and Carlsson, L. A. (1996). Micro-model of paper: Part 2: Statistical analysis of paper structure. *Tappi J. 79*(1):203–210.
40. Lu, W., Carlsson, L. A., and Andersson, Y. (1995). Micro-model of paper, Part 1. Bounds on elastic properties. *Tappi J. 78*(12):155–164.
41. Lu, W., Carlsson, L. A., and de Ruvo, A. (1996). Micro-model of paper, Part 3. Mosaic model. *Tappi J. 79*(2):197–205.
42. Mardia, K. V. (1972). *Statistics of Directional Data*. Academic Press, New York.
43. Mardia, K. V., and Jupp, P. E. (2000). *Directional Statistics*. Wiley, New York.
44. Mark, R. E., and Gillis, P. P. (1983). Mechanical properties of fibers. In: *Handbook of Physical and Mechanical Testing of Paper and Paperboard*. R. E. Mark, ed. Marcel Dekker, New York, pp. 409–495.
45. McCulloch, R. L. (1990). Micro-models for continuous materials—continuous fiber composites. In: *Micromechanical Materials Modeling*. L. Carlsson and J. Gillespie, eds. Technomic, Lancaster, PA, pp. 49–90.
46. McLaughlin, P. V., and Batterman, S. C. (1970). Limit behavior of fibrous materials. *Int. J. Solids Struct. 6*:1357–1376.
47. Niskanen, K. (1998). Paper physics. In: *Papermaking Science and Technology*, Vol. 16. J. Gullichsen and H. Paulapuro, eds. Fapet Oy, Helsinki, Finland, p. 324.
48. Niskanen, K., et al. (1997). KCL-PAKKA: Simulation of the 3-D structure of paper. In: *The Fundamentals of Papermaking Materials*. Transactions of the 11th Fundamental Research Symposium, Pira Int., Cambridge, UK.

49. Niskanen, K. J., Oleva, M. J., Seppala, E. T., and Astrom, J. (1999). Fracture energy in fibre and bond failure. *J. Pulp Paper Sci. 25*(5):167–169.
50. Nissan, A. H., and Batten, G. L. J. (1997). The link between the molecular and structural theories of paper elasticity. *Tappi 80*(4):153–158.
51. Nissan, A. H., and Batten, G. L. J. (1990). On the primacy of the hydrogen bond in paper mechanics. *Tappi* (February):159–164.
52. Nissan, A. H., et al. (1985). Paper as an H-bond dominated solid in the elastic and plastic regimes. *Tappi 68*(9):118–124.
53. Page, D. H., and Seth, R. S. (1980). The elastic modulus of paper. II. The importance of fiber modulus, bonding and fiber length. *Tappi 63*(6):113–116.
54. Page, D. H., and Seth, R. S. (1980). The elastic modulus of paper. III. The effects of dislocations, microcompressions, curl, crimps, and kinks. *Tappi 63*(10):99–102.
55. Page, D. H., Seth, R. S., and De Grâce, J. H. (1979). The elastic modulus of paper. I. The controlling mechanisms. *Tappi 62*(9):99.
56. Page, D. H., Tydeman, P. A., and Hunt, M. (1962). A study of fibre-to-fibre bonding by direct observation. In: *The Formation and Structure of Paper*. F. Bolam, ed. Technical Section of British Paper and Board Makers' Association, London, UK, p. 171.
57. Pecht, M., and Haslach, H. W. (1991). A viscoelastic constitutive model for constant rate loading at different relative humidities. *Mech. Mater. 11*:337–345.
58. Pecht, M. G., and Johnson, M. W. (1985). The strain response of paper under various constant regain states. *Tappi 68*(1):90–93.
59. Perkins, R. (1990). Micromechanics models for predicting the elastic and strength behavior of paper materials. In: *Materials Interactions Relevant to the Pulp, Paper and Wood Industries*. Materials Research Society, San Francisco, CA.
60. Perkins, R. W. (1979). Mechanical behavior of paper in relation to its structure. In: *Paper Science and Technology: The Cutting Edge*. Inst. Paper Chem., Appleton, WI.
61. Perkins, R. W. (1999). Mesomechanics analysis of medium to high density paper sheets. In: ASME Summer Mechanics Conference—Symposium on Mechanics of Cellulosic Materials. Am. Soc. Mech. Eng., Blacksburg, VA.
62. Perkins, R. W., and McEvoy, R. P. (1981). The mechanics of the edgewise compression strength of paper. *Tappi 64*(2):99–102.
63. Perkins, R. W., and Mark, R. E. (1981). Some new concepts of the relation between fibre orientation, fibre geometry and mechanical properties. In: *The Role of Fundamental Research in Papermaking*. Transactions of the Symposium Held at Cambridge, September 1981. Mech. Eng. Pub., Cambridge, UK.
64. Perkins, R. W., and Sinha, S. (1992). A micromechanics plasticity model for the uniaxial loading of paper materials. In: *Plastic Flow and Creep* (ASME Summer Mechanics Conference). H. M. Zbib, ed. ASME, Tempe, AZ, pp. 117–136.
65. Perkins, R. W., Sinha, S., and Mark, R. E. (1991). Micromechanics and continuum models for paper materials of medium to high density. In: 1991 International Paper Physics Conference. Kona, HI, TAPPI, Atlanta, GA.
66. Qi, D. (1997). Three-dimensional generalized anisotropic Cox-type structural model of a fiber network. *Tappi J. 80*(11):165–171.
67. Räisänen, V. I., Oleva, M. J., Nieminen, R. M., and Niskanen, K. J. (1996). Elastic-plastic behavior in fiber networks. *Nordic Pulp Paper Res. J. 1996*(4):243–248.
68. Räisänen, V. I., Oleva, M. J., Niskanen, K. J., and Nieminen, R. M. (1997). Does the shear-lag model apply to random fiber networks? *J. Mater. Res. 12*(10):2725–2732.
69. Ramasubramanian, M. K., and Perkins, R. W. (1988). Computer simulation of the uniaxial elastic-plastic behavior of paper. *J. Eng. Mater. Technol. 110*(April):117–123.
70. Ramasubramanian, M. K., and Wang, Y. Y. (1999). Constitutive models for paper and other ribbon-like nonwovens—A literature review. In: *Mechanics of Cellulosic Materials 1999*. R. W. Perkins, ed. ASME, New York, pp. 31–42.

71. Ranta-Maunus, A. (1975). The viscoelasticity of wood at varying moisture content. *Wood Sci. Technol. 9*:189–205.
72. Retulainen, E., Niskanen, K., and Nilsen, N. (1998). Fibers and bonds. In: *Paper Physics*. K. Niskanen, ed. Fapet Oy, Helsinki, Finland, pp. 55–87.
73. Rigdahl, M., Andersson, H. Westerlind, B., and Hollmark, H. (1983). Elastic behavior of low density paper described by network mechanics. *Fiber Sci. Technol. 19*:127–144.
74. Rigdahl, M., Westerlind, B., and Hollmark, H. (1984). Analysis of cellulose networks by the finite element method. *J. Mater. Sci. 19*(12):3945–3952.
75. Salmén, L. (1986). The cell wall as a composite structure. In: *Paper Structure and Properties*. J. A. Bristow and P. Kolseth, eds. Marcel Dekker, New York, pp. 51–73.
76. Salmén, L., and de Ruvo, A. (1985). A model for the prediction of fiber elasticity. *Wood Fiber Sci. 17*(3):336–350.
77. Salmén, L., et al. (1984). A treatise on the elastic and hygroexpansional properties of paper by a composite laminate approach. *Fibre Sci. Technol. 20*:283–296.
78. Salminen, L. I., Räisänen, V. T., Oleva, M. J., and Niskanen, K. J. (1996). Drying-induced stress state of inter-fiber bonds. *J. Pulp Paper Sci. 22*(10):J402–J407.
79. Schulgasser, K., and Page, D. H. (1988). The influence of transverse fibre properties on the in-plane elastic behavior of paper. *Compos. Sci. Technol. 32*(4):279.
80. Seo, Y. B., Castagnede, B., and Mark, R. E. (1992). An optimization approach for the determination of in-plane elastic constants for paper. *Tappi 75*(11):209–214.
81. Setterholm, V. C., and Chilson, W. A. (1965). Drying restraint: Its effect on the tensile properties of 15 different pulps. *Tappi 48*(11):634–640.
82. Shapery, R. A. (1968). Stress analysis in viscoelastic composite materials. In: *Composite Materials Workshop*. S. W. Tsai, J. C. Halpin, and N. J. Pagano, eds. Technomic, Stamford, CT.
83. Shih, C. F., and Lee, D. (1978). Further developments in anisotropic plasticity. *J. Eng. Mater. Technol. 100*(July):294–302.
84. Sinha, S. (1990). Finite element simulation of mechanical behavior of low basis weight crepe material. Ph.D. Thesis, Syracuse University, Syracuse, NY, p. 17.
85. Sinha, S. S., and Perkins, R. W. (1995). A micromechanics constitutive model for use in finite element analysis. In: *Mechanics of Cellulosic Materials*. R. W. Perkins, ed. ASME, New York.
86. Stratton, R. A., and Colson, N. L. (1990). Dependence of fiber/fiber bonding on some papermaking variables. In: *Materials Interactions Relevant to the Pulp, Paper, and Wood Industries*. D. F. Caulfield, J. D. Passaretti, and S. F. Sobczynski, eds. Mater Res Soc., Pittsburgh, PA, pp. 173–181.
87. Subramanian, L., and Carlsson, L. A. (1994). Influence of voids on the engineering constants of paper. Part 1: Continuum modeling. *Tappi 77*(11):209.
88. Subramanian, L., and Carlsson, L. A. (1994). Influence of voids on the engineering constants of paper. Part 2: Void modeling. *Tappi 77*(12):85.
89. Thorpe, J. L., Mark, R. E., Eusufzai, A. R. K., and Perkins, R. W. (1976). Mechanical proeprties of fiber bonds. *Tappi 59*(5):96–100.
90. Uesaka, T. (1990). Hygroexpansion coefficients of paper. In: *Mechanics of Wood and Paper Materials*. R. W. Perkins, ed. ASME, New York, pp. 29–35.
91. Uesaka, T., and Qi, D. (1994). Hygroexpansivity of paper—effects of fibre-to-fibre bonding. *J. Pulp Paper Sci. 20*(6):J175–J179.
92. Van den Akker, J. A. (1962). Some theoretical considerations on the mechancial properties of fibrous structures. In: *Formation and Structure of Paper*. F. Bolam, ed. Tech Section of the British Paper and Board Makers' Assoc, London, UK, p. 205.
93. Van den Akker, J. A. (1970). Structure and tensile characteristics of paper. *Tappi 53*(3):388–400.

94. Van den Akker, J. A. (1972). Large strains and inelastic behavior of paper. In: *Theory of Design of Wood and Fiber Composite Materials*. B. A. Jayne, ed. Syracuse Univ. Press, Syracuse, NY, pp. 197–218.
95. Wang, H., and Shaler, S. M. (1998). Computer-simulated three-dimensional microstructure and wood fiber composite materials. *J. Pulp Paper Sci.* *24*(10):314–319.
96. Whitney, J. M. (1987). *Structural Analysis of Laminated Anisotropic Plates*. Technomic, Lancaster, PA.
97. Wuu, F., Mark, R. E., and Perkins, R. W. (1991). Mechanical properties of "cut-out" fibers in recycling. In: 1991 International Paper Physics Conference, Kona, HI. TAPPi, Atlanta, GA.

2

VISCOELASTIC PROPERTIES

LENNART SALMÉN and ROGER HAGEN
STFI, Swedish Pulp and Paper Research Institute
Stockholm, Sweden

Partly based on the chapter "The Measurements of Viscoelastic Behavior for the Characterization of Time-, Temperature-, and Humidity-Dependent Properties," by P. Kolseth and A. de Ruvo, in the first edition of this book.

I. INTRODUCTION

The deformation of paper and paper products, like that of all polymeric materials, is dependent on the environmental conditions as well as on the time scale of the deformation and the loading pattern. Because paper material is highly hygroscopic, the relative humidity of the surroundings, and indeed moisture contact in general, affects the viscoelastic properties of paper to a great extent. Moisture uptake changes the properties of the paper so that although it is a relatively stiff and brittle material at low moisture contents it becomes a very ductile one at high moisture contents. The material properties of paper are strongly affected by process variables in the making of the paper in, for example, the pressing and drying operations; in converting operations such as folding, corrugating, and printing; and during its end use as a packaging or printed material. During all these processes, the paper is subjected to different and complicated loading patterns in terms of both stress levels and the time scale of the load. In order to model the performance of paper under all these conditions, it is first necessary to characterize its mechanical properties according to fundamental principles. Thus testing of paper with regard to its time-dependent behavior under a constant load (creep) or at a constant deformation (stress relaxation) or during cyclical deformation (dynamic behavior) at specific temperatures and relative humidities should provide a better understanding of the behavior of paper subjected to various loading situations.

All polymeric materials, including the wood polymers, exhibit to a greater or lesser extent a viscoelastic behavior [1,2]. This means that when the material is deformed it will not behave like an elastic spring but will partly show a viscous flow. For example, if the material is deformed at a slow rate, the material has more time to flow and will subsequently exhibit a lower stiffness. The material properties are dependent on the time scale of any loading type, such as the time under load (stress relaxation or creep) or the rate of deformation (frequency of the loading or rate of strain). The temperature also affects the degree of viscoelasticity, because the energy supplied in the form of an increase in temperature gives the polymers a higher energy level for movement. At certain temperatures characteristic of each polymer, this energy may exceed the energy required for large movements of the polymer chains, implying that the polymer shows a particularly large component of viscoelastic flow. The temperature at which this occurs is termed the glass transition temperature of the polymer (described in more detail later). At this temperature the polymer becomes particularly dependent on the time scale of mechanical testing.

II. CREEP AND LIFETIME

A. Creep Testing

Creep is defined as a time-dependent deformation under a constant load. A creep test is carried out by subjecting the test piece to a constant force or stress σ_0, then recording the strain $\varepsilon_c(t)$ as a function of time t (see Fig. 1). The loading may be followed by an unloading at time t_1, at which time the recovered strain $\varepsilon_r(t - t_1)$ is recorded as a function of time over a period of about 10 times the loading period. It is important that the recovered strain be measured as the difference between the projected creep strain at time t under the initial stress and the actual response. The residual strain $\varepsilon_p(t_1)$, is called the permanent set. Usually, either the strain ε_c is given versus $\log t$ or ε_r versus $\log(t - t_1)$. It is also possible to calculate $\varepsilon_c(t)/\sigma_0 = D(t)$, which is called the creep compliance. For a linear viscoelastic material, one in which the stress and strain are linearly proportional, the permanent set $\varepsilon_p(t_1)$ may be small, but for a nonlinear material there may be a considerable difference between the creep and the recovery. Often the initial rate of recovery from a given load is greater than the initial rate of creep under load.

The linearity in creep experiments can be expressed by the Boltzmann superposition principle. If a small stress is applied to the material at time equal to zero, a strain arises in response, $\varepsilon(t) = \sigma_0 D(t)$. If several stresses are applied in succession, the superposition principle states that the strain at the present time (t) is a linear superposition of the strains arising from the past stresses at times $t = u_1, u_2, \ldots, u_n$, each contributing independently of the others.

$$\varepsilon(t) = \sigma_0 D(t) + (\sigma_1 - \sigma_0)D(t - u_1) + \cdots + (\sigma_i - \sigma_{i-1})D(t - u_i) \tag{1}$$

or, more generally stated,

$$\varepsilon(t) = \sum_{i=1}^{n} \sigma_i D(t - u_i) \tag{2}$$

Equation (2) can be generalized in integral form as

$$\varepsilon(t) = \int_{-\infty}^{t} \frac{\partial \sigma(u)}{\partial u} D(t - u)\, du \tag{3}$$

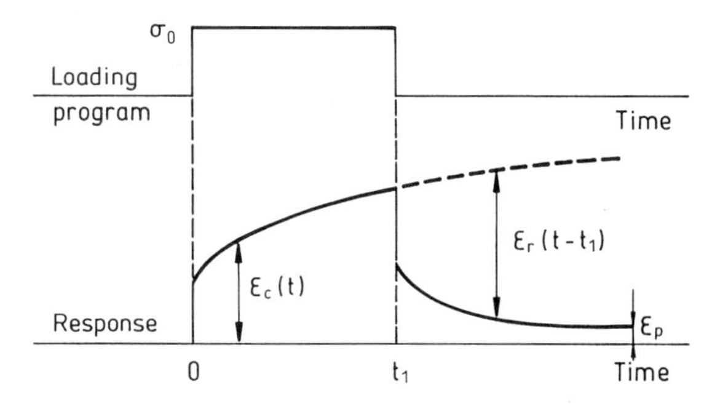

Fig. 1 Creep ε_c and recovery ε_r for a viscoelastic solid subjected to loading and unloading.

For a linear viscoelastic material, the creep compliance can be described in terms of a retardation function $D(\tau)$:

$$D(t) = D_i + \int_0^{\infty} D(\tau)(1 - e^{-t/\tau})\, d\tau \tag{4}$$

where D_i is the instantaneous elastic compliance and τ is the retardation time. Experience has shown that a logarithmic time scale is most convenient, and Eq. (4) then becomes

$$D(t) = D_i + \int_{-\infty}^{\infty} L(\tau)(1 - e^{-t/\tau})\, d \ln \tau \tag{5}$$

To a first approximation, the retardation spectrum can be found by differentiation of the creep curve:

$$L(\tau) \approx \left. \frac{dD(t)}{d \ln t} \right|_{t=\tau} \tag{6}$$

B. Evaluation of Creep Curves

A large number of descriptive equations have been used to model the creep of viscoelastic materials. Reviews of creep equations are given in several papers, e.g., Refs. 3–7.

One of the earlest creep laws is Andrade's equation [8],

$$\varepsilon_c(t) = \varepsilon_0 + ct^{1/3} \tag{7}$$

where ε_0 and c are material constants that depend on stress and temperature.

Another early creep law still in common use was formulated by Norton [9]:

$$\dot{\varepsilon} = c\left(\frac{\sigma}{\sigma_1}\right)^m \tag{8}$$

where $\dot{\varepsilon}$ is the creep rate $d\varepsilon/dt$, c and m are temperature-dependent material constants, and σ_1 is a constant used to adjust the scale of the graph. Plotting $\dot{\varepsilon}$ versus σ in a log-log diagram gives a straight line for materials that conform to Norton's law.

Other equations proposed to describe the creep behavior are [3]

$$\varepsilon(t) = \varepsilon_e \sinh\left(\frac{\sigma}{\sigma_0}\right) + \varepsilon_c \sinh\left(\frac{\sigma}{\sigma_c}\right) t^n \tag{9}$$

$$\varepsilon(t) = c\sigma^p t^n \tag{10}$$

which is the Nutting equation, mentioned in Findley et al. [3],

$$\varepsilon(t) = \varepsilon_0 \frac{\sigma}{\sigma_0} \exp\left(\frac{t}{t_0}\right)^n \tag{11}$$

given by Struik [10],

$$\frac{\varepsilon(t)}{\sigma} = b + ct^n \tag{12}$$

and

$$\frac{\varepsilon(t)}{\sigma} = b + c\log(t) \tag{13}$$

proposed by Brezinski [11] and mentioned by Pecht et al. [5], and

$$\varepsilon(t) = \frac{\sigma}{E} + a\sigma^m(1 - be^{-nt}) + c\sigma^p t \tag{14}$$

given by Marin-Pan (see Ref. 4). In Eqs. (9)–(14), a, b, c, m, n, and p are constants.

C. Isochronous and Isometric Curves

Isochronous stress–strain curves (stress versus strain at a given time) and isometric stress curves (stress versus time at a given strain) can be obtained from creep curves obtained at different stress levels. Isochronous curves are constructed from cross sections in the creep curves at different times (see Fig. 2), whereas isometric curves are constructed from cross sections at different strains, as shown in Fig. 3. Isochronous stress–strain curves are very useful for characterizing the viscoelastic properties of a material. For many polymeric materials, the stress is proportional to the strain at small strains characteristic of linear viscoelastic beahavior. Above the proportional limit, the behavior is nonlinear and viscoelastic and the Boltzmann superposition principle (see Section II.A) is not valid. The isometric curves are of interest in applications where design criteria are governed by a critical strain.

Isochronous stress–strain curves can also be constructed from a group of stress–strain curves obtained at considerably different rates of strain or stress loading. In the linear viscoelastic region, theory predicts identical results independently of the method used, but above the proportional limit significantly different isochrones are expected, because the nonlinear viscoelastic deformation processes are nonlinearly related to the level of stress.

D. Lifetime Prediction

Most creep testing is carried out under stresses that the test piece will sustain. However, if higher stresses are used, the creep test will come to an end after a while because the test piece breaks. Expressions to predict the lifetime from the loading conditions for viscoelastic materials can be found in the literature [12,13].

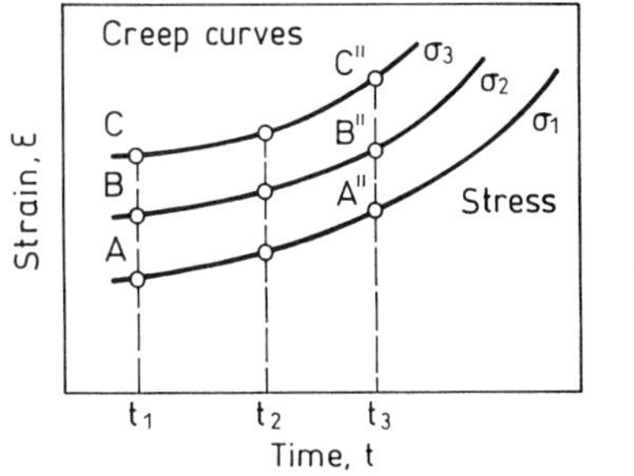

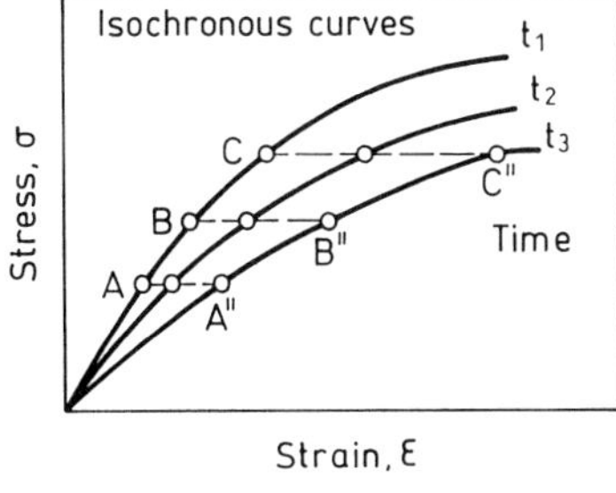

Fig. 2 Construction of isochronous curves from creep curves. (From Ref. 85.)

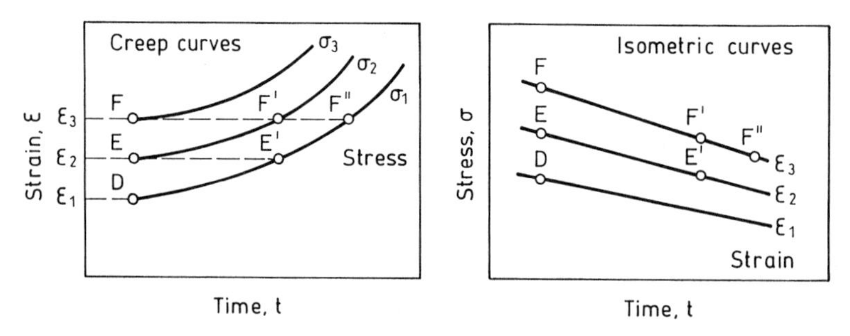

Fig. 3 Construction of isometric curves from creep curves. (From Ref. 85.)

The following semiempirical expression for time to fracture t_f was deduced by Zhurkov and Tomashevsky [13]:

$$t_f = t_0 e^{(U_0 - j\sigma)/RT} \tag{15}$$

or

$$\ln t_f = \ln t_0 + \frac{U_0 - j\sigma}{RT} \tag{16}$$

where U_0 is seen as the height of an energy barrier against fracture similar to an activation energy, σ is the applied stress, j is a stress intensity factor, t_0 is a constant that may be looked upon as the shortest possible time to fracture, and RT is the product of the gas constant [8.314 J/(mol K)] and the absolute temperature.

III. STRESS RELAXATION

A. Stress Relaxation

Stress relaxation is defined as the time-dependent dissipation of stress at a constant strain. In a stress relaxation experiment, the test piece is subjected to a strain ε_0, which is kept constant over time while the decay of stress $\sigma(t)$ is observed, as shown in Fig. 4. For linear viscoelastic behavior, the stress relaxation modulus may be defined as $E(t) = \sigma(t)/\varepsilon$. If the material is kept under strain for a sufficiently long time it will attain an equilibrium value, with σ_∞ defining E_∞. In the case of paper, this period may be several days or even weeks, but under certain conditions of low relative humidity it may not be possible to attain equilibrium during a practical period of testing.

The relaxation curve can be described by a distribution of relaxation times. For a linear viscoelastic material, the relaxation modulus can be given in terms of a relaxation function $E(\tau)$:

$$E(t) = E_\infty + \int_0^\infty E(\tau) e^{-t/\tau}\, d\tau \tag{17}$$

where E_∞ is the equilibrium modulus. On a logarithmic time scale, which is more convenient, Eq. (17) can be rewritten as

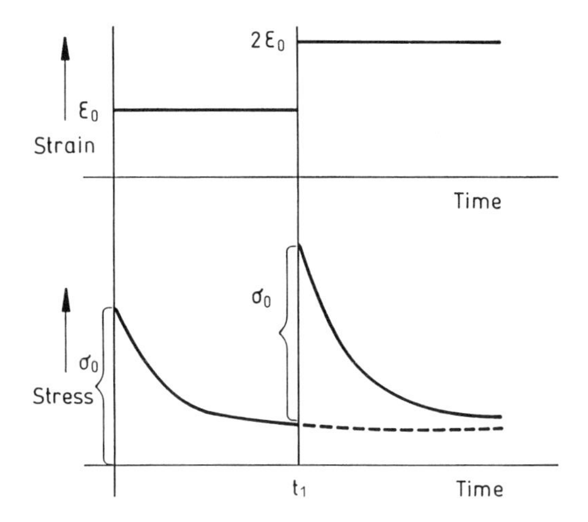

Fig. 4 Stress relaxation of a viscoelastic material.

$$E(t) = E_\infty + \int_{-\infty}^{\infty} H(\tau) e^{-t/\tau} d\ln\tau \tag{18}$$

To a first approximation, the relaxation spectrum $H(\tau)$ can be calculated from the expression

$$H(\tau) \approx -\left.\frac{dE(t)}{d\ln t}\right|_{t=\tau} \tag{19}$$

More accurate approximations can be found in Ferry's textbook [2].

Equations can be given for the relationship between $E(t)$ and $D(t)$ [2], but for simplicity it is sufficient to observe that a very good approximation is given by

$$E(t) = D(t)\frac{\sin m\pi}{m\pi} \tag{20}$$

where

$$m = \frac{d(\log D)}{d(\log t)} \tag{21}$$

B. Evaluation of Stress Relaxation Curves

An inherent experimental limitation of the relaxation experiment is that straining cannot be performed at infinite speed. It takes a certain time t_0 to reach the deformation. If the strain rate is constant, the relaxation modulus is expressed by the relation

$$E(t) = E_\infty + \int_0^\infty E(\tau) e^{-t/\tau} \frac{\tau}{t_0} \left(e^{t_0/\tau} - 1\right) d\tau \tag{22}$$

where t is the total elapsed time. If $e^{t_0/\tau}$ is expanded and all terms higher than the first power are neglected, Eq. (22) reduces to Eq. (17). This approximate form is applicable when the straining time t_0 is much smaller than the relaxation time τ. For 5% accuracy, this requires that t_0/τ be less than 0.1. To obtain reliable relaxation modulus data, the experimental time should thus be about 10 times that of the loading phase to the requried strain level.

Usually the stress relaxation test is considered to have started at the time t_0 at which the given constraint conditions were initially obtained. It has been argued, however, that, according to Boltzmann's superposition principle, the time should be measured from the beginning of extension. In practice, when a logarithmic time scale is used to compress the time scale, the relaxation curves obtained are not significantly different.

C. Evaluation of Internal Stress

During the processing and deformation of polymeric materials, internal stresses are introduced into the material. These internal stresss influence the mechanical properties of the material. The flow behavior of most solid materials is, for example, considered to be determined by the magnitude of the internal stress [14].

The internal stress in a material may be evaluated by stress relaxation measurements according to Kubát [14]:

$$-\left(\frac{d\sigma}{d\ln t}\right)_{\max} = c(\sigma_0 - \sigma_\infty) \tag{23}$$

where c is a constant. The internal stress is defined as being equal to the equilibrium stress σ_∞ reached after a sufficiently long relaxation time in a stress relaxation experiment. For practical reasons, it is not convenient to evaluate the final equilibrium stress, and instead indirect methods have been suggested.

If the main inflection slope $-(d\sigma/d\log t)_{\max}$ of the curve of σ versus $\log t$ (see Fig. 5 left) is plotted against the initial stress σ_0, the internal stress is obtained as the intercept on the σ_0 axis (see Fig. 5 right) [15,16].

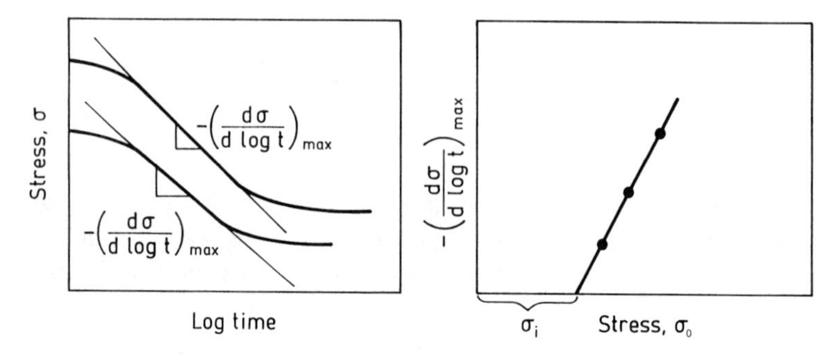

Fig. 5 The main inflection slope $-(d\sigma/d\log t)_{\max}$ of stress relaxation curves for σ vs. $\log t$ (left) is plotted against the initial stress, σ_0 (right). The intercept on the σ_0 axis gives the value of the internal stress, σ_i.

The internal stress can also be evaluated by using a method described by Li [17] based on a description of the flow rate according to the Norton-type power law, usually expressed as

$$\sigma - \sigma_i = c(t + a)^{-m} \tag{24}$$

where σ_i is the internal stress and c, a, and m are constants. According to this method, the relaxation rate $-d\sigma/d\log t$ is plotted against the corresponding stress σ, and the intercept on the σ axis is taken as a measure of the internal stress (see Fig. 6). If $-d\sigma/d\log t$ and σ values are taken from the same curve, a single experiment is sufficient. This method can, however, lead to difficulties if the curvature of the relaxation curve is not very pronounced.

At moderate initial stress levels, it has been shown using cold-drawn high density polyethylene that the two methods illustrated in Figs. 5 and 6 give the same internal stress value [18]. The internal stress evaluated by Li's method is, however, dependent on the initial stress [19] and is the sum of an induced and an irreversible internal stress [20]. The permanent or irreversible internal stress can be determined by using the maximum slope of the relaxation curve.

IV. CYCLICAL TESTING

A. Basic Concepts

Dynamic mechanical analysis or mechanical spectroscopy enables elastic, viscous, and viscoelastic properties to be characterized as a function of temperature, time, frequency, stress, and humidity. The mechanical properties of materials are determined in dynamic modes by applying an oscillating force or deformation and measuring the response of the materials. The term "dynamic mechanical property" generally applies to the linear viscoelastic response of a polymer to the externally applied force, which also means that the experimental techniques involve only very small strains.

Assuming that the applied stress is the force function according to

$$\sigma(t) = \sigma_0 \sin \omega t \tag{25}$$

where σ_0 is the maximum stress and ω is the angular frequency, the strain can be represented as

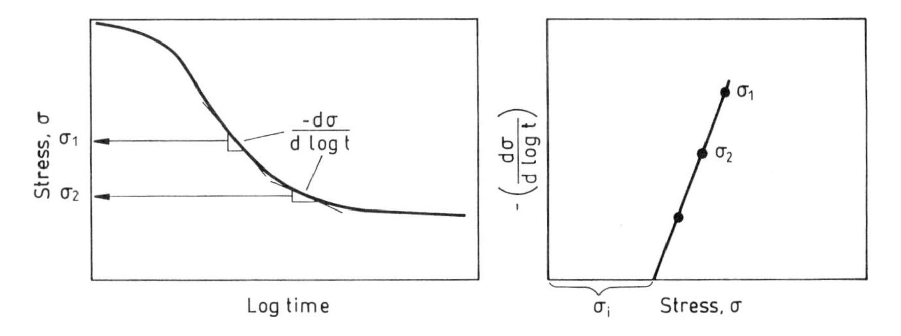

Fig. 6 Evaluation of the internal stress according to the method of Li. (From Ref. 17.)

$$\varepsilon(t) = \varepsilon_0 \sin(\omega t - \delta) \tag{26}$$

where ε_0 is the maximum strain and δ is the phase lag between stress and strain, being the angle between the in-phase and out-of-phase components in the cyclical motion.

The quantities most commonly determined by dynamic mechanical analysis are the storage modulus E', the loss modulus E'', and the loss factor $\tan\delta$ as functions of temperature or frequency. The storage and loss moduli are related to the complex modulus E^* by [2]

$$E^* = \frac{\sigma(t)}{\varepsilon(t)} = \frac{\sigma_0}{\varepsilon_0}\cos\delta + i\frac{\sigma_0}{\varepsilon_0}\sin\delta = E' + iE'' \tag{27}$$

where $i = \sqrt{-1}$. The real component of the complex modulus, the storage modulus E', is a measure of the ability of the material to store energy and is the in-phase component of the complex modulus. The imaginary part of the complex modulus, the loss modulus E'', is the ability of the material to dissipate mechanical energy by converting it to heat by molecular motion and is the 90° out-of-phase component of the complex modulus, i.e., the viscous behavior.

The loss modulus peak in a dynamic mechanical spectrum may be attributed to the natural frequency of chain motion that equals the external vibrational frequency [21]. The vectorial resolution of the components is shown in Fig. 7.

The mechanical loss factor or loss tangent, $\tan\delta$, can be defined as the tangent of the phase angle between the applied stress and strain. It is a dimensionless term and is also the ratio of the loss modulus to the storage modulus:

$$\tan\delta = \frac{E''}{E'} \tag{28}$$

The loss tangent is a measure of the relative dissipated energy, and its maximum is not directly associated with molecular motion related to molecular relaxation processes [22].

It is possible to obtain the components of the complex dynamic modulus from the relaxation modulus by Fourier transforms [2]:

$$E'(\omega) = E_e + \omega\int_0^\infty [E(t) - E_e]\sin\omega t\, dt \tag{29}$$

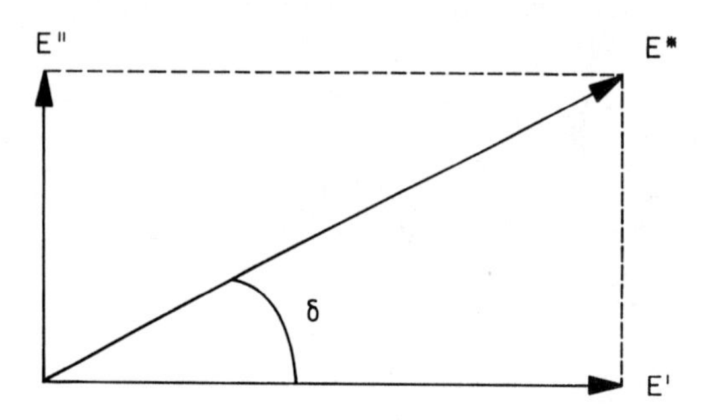

Fig. 7 Vectorial resolution of E^*, E', and E''.

$$E''(\omega) = \omega \int_0^\infty [E(t) - E_e] \cos \omega t \, dt \tag{30}$$

where E_e is equivalent to the long-term equilibrium relaxation modulus. The corresponding relations for the relaxation modulus $E(t)$ can be obtained in terms of E' and E'' as

$$E(t) = E_e + \frac{2}{\pi} \int_0^\infty \frac{E' - E_e}{\omega} \sin \omega t \, d\omega \tag{31}$$

and

$$E(t) = E_e + \frac{2}{\pi} \int_0^\infty \frac{E''}{\omega} \cos \omega t \, d\omega \tag{32}$$

B. Experimental Techniques

A number of commercial instruments for dynamic mechanical testing in different modes of operation are available on the market. These instruments can be divided into two groups according to whether they employ free vibration methods or forced vibration methods [23]. Free vibrations are always at a resonance frequency of the system, whereas forced vibrations permit either resonant or nonresonant conditions. The free torsion pendulum is an example of a resonant method that measures the natural frequency of vibration of a test piece [23]. Forced vibration techniques are more common, and a number of instruments are available that permit various modes of deformation and test geometries [23]. Forced vibration methods are to be preferred when the frequency and temperature dependence of the viscoelastic behavior are to be investigated. Common clamping geometries for testing in torsion, bending, and tension are illustrated in Fig. 8. Testing in torsion was the first technique developed for dynamic mechanical testing.

Depending on the stiffness, viscosity, and shape of the specimen, different modes of measurement may be appropriate. A paper, a polymer film, or a fiber may be more suitable for testing in tension, whereas a stiff polymer bar may be more suitable for testing in bending. Other dynamic mechanical testing modes are compression and shear (Fig. 8). Discrepancies have been reported among data on

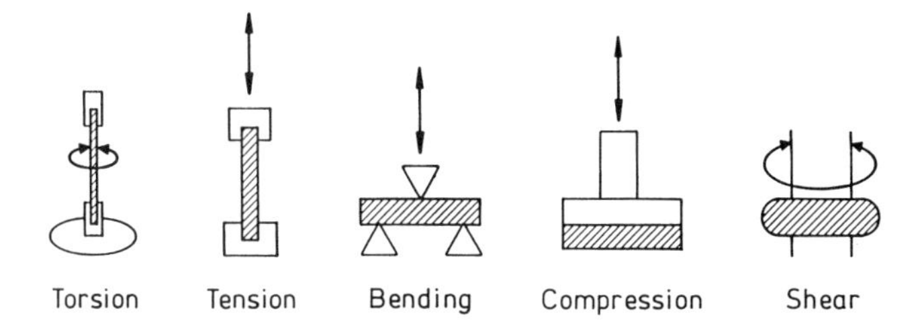

Fig. 8 Different deformation geometries for dynamic mechanical analysis.

polymers obtained by different dynamic mechanical instruments [24,25]. Such discrepancies may arise from limited frequency ranges, different types of deformation, limitations in data analysis, and overall test sensitivity to external influences. It is obvious that these factors must be considered when data obtained by different methods are compared.

Torsion Pendulum Dynamic mechanical studies in torsion exhibit a high response to the properties of the outermost part of the specimen [26], which means that experiments made on heterogeneous materials or layered composites may be more sensitive to the surface properties of the specimens.

The torsion pendulum based on free oscillation is a relatively simple instrument for dynamic mechanical testing. The experimental setup involves, in principle, only the recording of the free oscillation of the system. The technique has been described in more detail by, for example, McCrum et al. [27] and Kolseth [28]. Schematic diagrams of two different torsion pendulum instruments are shown in Fig. 9. The test piece—a bar, rectangular strip, or fiber—is clamped in its upper end, and an inertia disk is freely suspended in its lower end. The inertia disk and the specimen are set in oscillation by the influence of a magnetic field and are then permitted to oscillate freely at the resonance frequency of the system. The damped sinusoidal oscillation is then often measured optically. The period P required for one complete oscillation is related to the angular frequency ω by

$$\omega = \frac{2\pi}{P} \tag{33}$$

A variation of the torsion pendulum working with free oscillation is the inverted instrument, arrangement B in Fig. 9, where the inertia disk is supported

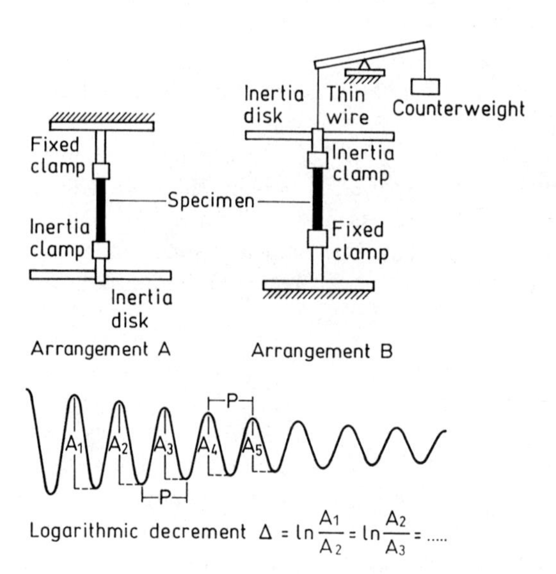

Fig. 9 Schematic diagram of the two torsion pendulum arrangements. A typical damping curve is illustrated at the bottom of the figure.

by a fine wire and the test piece is mounted beneath, clamped rigidly at its lower end [29].

The damping, i.e., the gradual decrease in amplitude, measured as the logarithmic decrement Δ, is used to calculate the dynamic mechanical loss factor, $\tan\delta$, of the material in free oscillation torsion using the expression

$$\Delta = \frac{1}{n}\ln\left(\frac{A_n}{A_{n+1}}\right) = \pi\tan\delta \tag{34}$$

where A_n and A_{n+1} are the amplitudes of successive swings of the pendulum [27]. The storage shear modulus in the plane normal to the axis for a specimen of circular cross section can be calculated from the period of oscillation, P; the shorter the period, the greater the modulus:

$$G' = \frac{8\pi LI}{R^4P^2} = \frac{\text{const.}}{P^2} \tag{35}$$

where L, R, and I are, respectively, the length, radius, and moment of inertia of the specimen [28]. This means that the frequency of vibration depends directly on the shear modulus of the specimen under test. The departure from a circular cross section, as for polymer films and paper strips, can be expressed using a shape factor [27]. In a freely suspended pendulum, the axial load on the specimen induced by the weight of the inertia disk may contribute to the torsional rigidity, because a tensile stress will oppose the warping of a noncircular cross section [30]. This means that it is difficult to measurer absolute values of the shear modulus of the specimen and the results are limited to relative stiffness values.

In a forced torsion pendulum, a force is applied at one end of the specimen, and the response to the imposed motion is measured by a transducer at the other end [28]. A nonresonance forced torsion pendulum allows measurements over a larger frequency range than is possible with the free oscillation torsion pendulum.

Torsional braid analysis (TBA) is a special kind of mechanical spectroscopy that uses torsion. An automatic TBA technique was developed by Gillham [31,32]. In the TBA method, a mechanically inert substrate is coated with a viscoelastic substance. The contribution of the substrate, usually a glass fiber braid, to the torsional properties of the viscoelastic substance is negligible, and it is therefore suitable as a support [32]. The TBA is suitable for test pieces in both the liquid and solid states and also for polymers at temperatures above the glass transition, T_g.

Forced Vibration Methods A number of dynamic mechanical instruments based on the forced vibration technique have been described in the literature [23,33–35]. The principle of operation is to measure the mechanical response of a material as it is deformed under a periodic stress. The sensitivity of dynamic mechanical measurements is high, which means that only very low stresses need be applied to a test piece and the response can be detected accurately without frictional errors. The high resolution of the signals from dynamic mechanical tests allows accurate data to be obtained even at strains where the dynamic amplitude is only 1–2 μm [33]. Typical frequency ranges for dynamic mechanical analysis (DMA) instruments are from 0.01 to 50–200 Hz. Tests performed at the higher frequencies are, however, often influenced by resonance effects. The temperature can be controlled in the range −150°C to +500°C in most instruments, which means that relaxation processes can be studied in most polymers. An instrument that has too low a rigidity itself, however, will give errors in modulus and stiffness values when stiff materials such as glassy polymers and paper materials are measured [25].

As already mentioned, the viscoelastic properties of polymer films and fibers are the most suitable for dynamic mechanical tests in tension. The high sensitivity of several DMA instruments and precision clamping possibilities make it possible to test single fibers and fragile films.

Dynamic mechanical testing in bending is, like the torsion method, particularly sensitive to the outermost parts of the specimen and is thus suitable for coated test pieces or sandwich structures [26]. Tests in bending can be performed in different ways. The simplest way is three-point bending, which does not require test piece clamping. It is typically used with high modulus materials such as thermoplastics and composites. Another common bending setup is the single or dual cantilever, where the specimen is clamped at one or both ends. The single cantilever is preferred to the dual cantilever for test pieces that exhibit considerable thermal expansion. Clamped fixtures are suitable for elastomers and polymers above their glass transition temperature.

Vibrating Reed The vibrating reed technique has been widely used for measuring the complex dynamic modulus and damping for paper and paperboard [36]. The apparatus is shown schematically in Fig. 10. A strip of paper (the reed) is clamped at one end and forced to vibrate transversely. This is achieved by attaching the clamp to an electromagnetic vibrator. The vibrator is driven by the amplified signal from a variable frequency generator. As the frequency of the vibrations is changed, the natural frequency of the reed will be reached, and the amplitude of the free end will pass through a maximum. The amplitude of vibration is measured optically. The frequency range is typically 200–1500 Hz [37]. The resonance frequency and the bandwidth are determined from an amplitude–frequency curve of the form shown in Fig. 11. The storage modulus E' and the mechanical loss factor $\tan\delta$ are calculated according to the equations

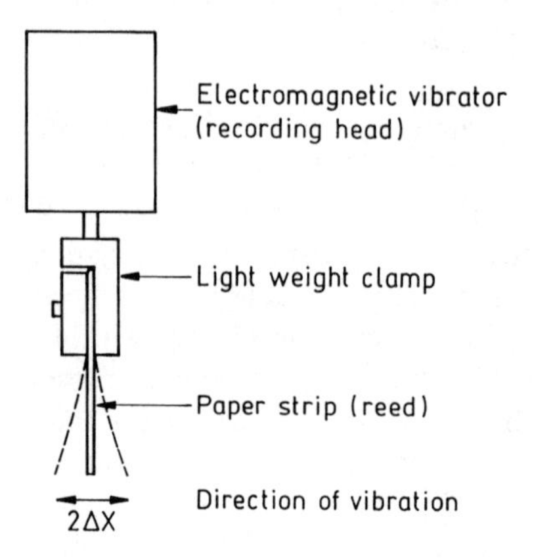

Fig. 10 A schematic diagram of a vibrating reed apparatus.

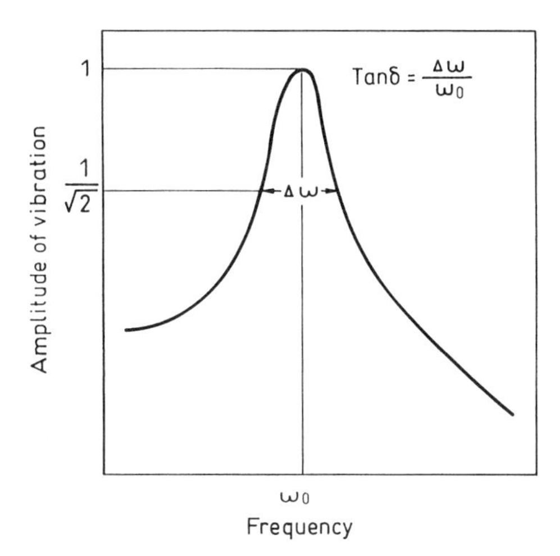

Fig. 11 The resonance curve as a function of frequency for forced vibrations of a viscoelastic material.

$$E' = \rho\left(\frac{A}{I}\right)\left(\frac{l^4}{a_0^4}\right)\omega_0^2 \tag{36}$$

and

$$\tan\delta = \frac{\Delta\omega}{\omega_0} \tag{37}$$

where l is the length of the reed, I is the moment of inertia of the cross-sectional area A, ρ is the density, ω_0 is the resonance frequency, and $\Delta\omega$ is the difference between the two frequencies above and below the resonance frequency where the displacement amplitude has dropped to $1/\sqrt{2}$ of its maximum (see Fig. 11).

Dynamic Rheological Methods Dynamic rheological experiments in shear are widely used for studying the viscoelastic properties of high viscosity fluids and polymers above T_g and in the melt state [38]. Examples of appropriate test geometries are the parallel-plate (see Fig. 8) and cone-and-plate methods. When being measured in shear, the test pieces are typically loaded between the plates, a dynamic shear force is applied from the upper plate, and the response is measured at the lower plate by a transducer. The dynamic shear modulus and the loss factor are often determined in order to characterize the rheological properties of the material. The relationship between Young's modulus E and the shear modulus G can be written

$$E = 2(1+\nu)G \tag{38}$$

where ν is the Poisson ratio. For an incompressible isotropic material, where $\nu = 0.5$, one obtrains $E \cong 3G$.

Wave-Propagating Methods Wave-propagating methods can be divided into two broad categories, sonic and ultrasonic methods. The sonic transducer is below 20 kHz, and the ultrasonic transducer is above 20 kHz. However, the ultrasonic method is the more commonly used. One transducer sends a sonic wave and another receives the sonic wave that propagates through the test material. The amplitude A of a longitudinal harmonic wave propagating in the x direction can be written

$$A = A_0 e^{-\alpha x} \cos \omega\left(t - \frac{x}{c}\right) \tag{39}$$

where A_0 is the initial amplitude, ω is the angular frequency of the wave, c is the velocity of propagation, and α is the attenuation factor. Substitution of Eq. (36) into the equation of motion gives the following equations relating wave propagation to mechanical properties of materials:

$$C' = \rho c^2 \frac{1 - r^2}{(1 + r^2)^2} \tag{40}$$

$$C'' = \rho c^2 \frac{2r}{(1 + r^2)^2} \tag{41}$$

where C' and C'' are the real and imaginary parts of the planar stiffness, r ($r = \alpha c/\omega$) is a dimensionless quantity, and ρ is the density of the material.

The acoustic propagation in a medium is studied by recording the logarithmic amplitude and the relative phase as a function of distance from the sound-transmitting device. The amplitude recording gives the attenuation in decibels per meter (dB/m), whereas the phase record gives the wavelength. In paper, however, the attenuation is usually so small that reflections from edges or from the receiver preclude measurements that use a continuous signal. Pankonin and Habeger [39] were, however, able to determine the loss tangent for paper by using a resonance technique. Usually, the propagation of sonic pulses has been studied, which permits only the determination of sonic velocity. For elastic materials, where the attenuation factor is negligible, the planar stiffness can be determined as

$$C = \rho c^2 \tag{42}$$

Equation (42) is a satisfactory approximation for paper in the absence of attenuation data [40]. For an isotropic planar material, e.g, handmade paper, the planar stiffness is given by

$$C = \rho c^2 (1 - \nu^2) \tag{43}$$

where ν is Poisson's ratio. For paper, the orientation and anisotropy due to drying stresses have been studied by determining the velocity of sound in various directions of the sheet [40,41].

V. VISCOELASTIC BEHAVIOR OF POLYMERS

A. Thermal Transitions

In polymeric materials and also in materials composed of the wood polymers (e.g., paper), the properties are highly dependent on the major thermal transitions, the primary transition being melting and secondary transitions being the glass transition and transitions in the glassy state below the glass transition temperature of the material. The main transition, the glass transition (T_g), is related to the onset of cooperative motion of the main chains of the polymer [23]. At the glass transition temperature, the free volume increases, and this facilitates molecular motion, which is an interpretation of the free volume theory of T_g [42]. As this temperature is passed, the properties of the polymer change drastically, resulting, for instance, in a 1000-fold reduction in the storage modulus for a completely amorphous material (see Fig. 12). With increasing crystallinity, the reduction decreases, because the transition occurs only in the amorphous part of the polymer. For crystalline polymers, the transition is also rather broad [43]. In general, polymers also have several minor transitions related to the movement of smaller groups of the polymer chain. The occurrence of such a transition below that of the main glass transition often signals the transition from a brittle to a more ductile material. Generally, the transitions are termed α, β, and γ in sequence, with the α transition most commonly the designation of the glass transition of the polymer [23].

The viscoelastic nature of polymeric materials also means that their mechanical properties depend on the time scale of the measurement, i.e., the time under load. Particularly in the transition region, when the material is changing from its glassy to its rubbery state, it exhibits a large viscous component and is thus more sensitive to the time scale of the measurement. In the case of measurements at different frequencies, the transitions thus appear at higher temperatures the higher the frequency, as exemplified in an Arrhenius diagram, where the logarithm of the frequency is plotted versus the reciprocal of the temperature (Fig. 13). For the β and γ transitions, the transitions related to the movement of different groups of the polymer, the frequency–temperature dependence follows an Arrhenius-type behavior [see Eq. (49)], where the activation energies of the processes, ΔH_β and ΔH_γ, are determined by the angle of inclination of the straight lines in the plots of log frequency versus $1/T$. The frequency dependence is greater for the γ transition, which means that its activation energy is lower than that of the β transition. The main transition, the α transition, i.e., the transition related to the glass transition, also termed the softening temperature, does not follow an Arrhenius type of dependence. Instead, it follows an empiri-

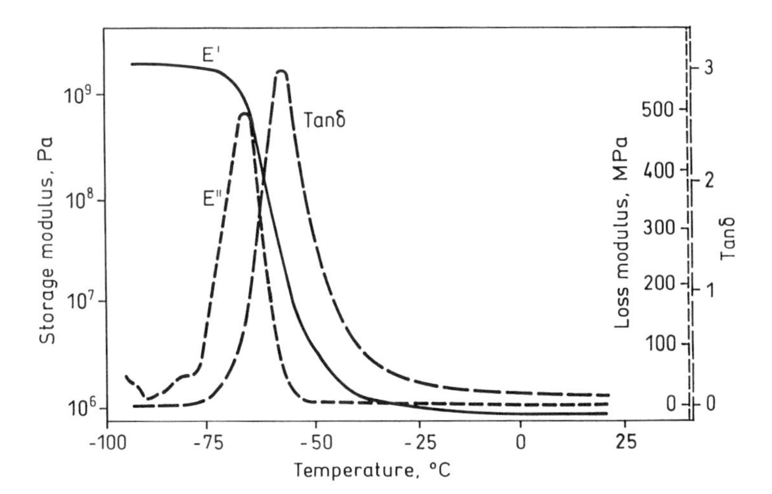

Fig. 12 The storage and loss modulus and $\tan\delta$ vs. temperature for an unfilled natural rubber at 1 Hz measured by dynamic mechanical analysis. (From Ref. 25.)

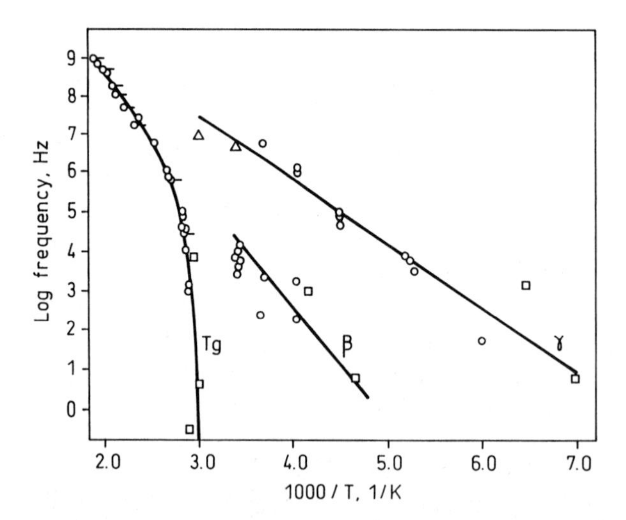

Fig. 13 Arrhenius diagram of the logarithm of frequency against the reciprocal of the temperature for α, β, and γ transitions of PA610. (Reprinted from Ref. 86.)

cal relation termed the WLF equation [see Section IV.D, Eq. (50)] [23]. The WLF-type behavior of the α transition in Fig. 13 is described by a curve, the tangent to which has a variable slope, that can be used to determine the apparent activation energy ΔH_α of the process.

B. Plasticizers

Plasticizers are low molecular weight components that, when mixed with a polymer, lower its softening temperature by interacting with the forces between the polymer chains and allowing more freedom of motion, thus increasing the free volume. In general, the effect of a plasticizer can be estimated from relations involving the free volume expansion coefficient [42] or molecular interaction parameters such as those given by an equation derived by Kaelble [44],

$$T_g = \frac{T_{g_p} X_p + (h_d/h_p) X_d T_{g_d}}{X_p + (h_d/h_p) X_d} \tag{44}$$

where X_p and X_d are molar fractions of monomer units of the polymer p and the diluent d, respectively, and h_p and h_d are corresponding parameters given by the relationship

$$h = \frac{2Z_g \, \Delta C_{Vg}}{R} \tag{45}$$

where Z_g is the lattice coordination number in the glassy state, ΔC_{Vg} is the change in specific heat capacity at T_g at a constant volume, and R is the gas constant. The parameter h can also be calculated from

$$h = \frac{2U}{RT_g} \tag{46}$$

where

$$U = \delta^2 v \tag{47}$$

U is the molar cohesive energy, δ is the solubility parameter, and v is the molar volume.

C. Mechanical Spectroscopy for Determining T_g

The softening of an amorphous polymer means that the storage modulus decreases by about three orders of magnitude, whereas the loss modulus and $\tan\delta$ go through maxima at T_g, as shown in Fig. 12. There are many different definitions of T_g from dynamic mechanical experiments, such as the temperature associated with the peak of $\tan\delta$, with the peak value of E'', with the first derivative, and with the onset of the modulus drop of E' [25,45]. The average of the peak temperatures of $\tan\delta$ and E'' has also been proposed [46]. The $\tan\delta$ peak is the most commonly used criterion in the literature for defining T_g [47]. When the T_g values derived according to different definitions are compared, large differences are often found due to the broad temperature range of the transition region and because they reflect different aspects of the same process. The $\tan\delta$ peak always occurs at a higher temperature than the E'' peak [48]. The frequency dependence of T_g for an amorphous polymer (unfilled natural rubber [25]) is shown in Fig. 14, where logarithm of frequency is plotted versus $1000/T_g$. The T_g values are here derived from the peak values of E'' and $\tan\delta$ and from the first derivative values of the onset of E'. There is clearly a large difference between the E'' data and the $\tan\delta$ data. A difference in T_g of approximately 8°C was obtained from dynamic mechanical measurements at 1 Hz. Peak values of E'' and derivative values of E' give similar values and almost the same frequency dependence.

Differential scanning calorimetry (DSC) is a commonly used technique for defining T_g of polymers [49]. A T_g value (at $-69.8°C = 4.92$, $1000/T_g$) obtained

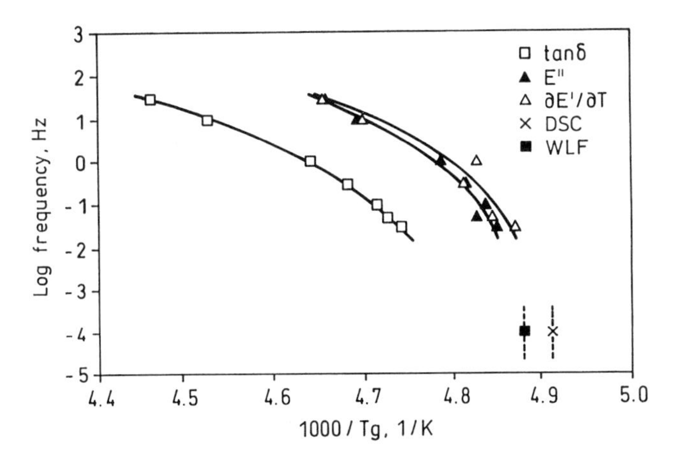

Fig. 14 Frequency–temperature map of the glass transition temperature of unfilled natural rubber. Different ways of determining T_g from the DMA data are compared. For comparison, DSC and WLF data are shown at 10^{-4} Hz. (From Ref. 25.)

from DSC data, taken from the heat flow midpoint at a heating rate extrapolated to 1°C/min, agrees well with the E'' value extrapolated to the same frequency. The frequency of 10^{-4} Hz for the DSC data is only an approximation [50], and dotted lines are therefore drawn in Fig. 14. A T_g value (at $-68.3°C = 4.89, 1000\,kg$), obtained from the dependence of the shift factor, $-\log a_T$, on the temperature at the intersection between Arrhenius and WLF behavior, is also reported in Fig. 14. The frequency for this definition of T_g is approximately 10^{-4} Hz [51]. The T_g values determined from $\tan\delta$ data, and not extrapolated to 10^{-4} Hz, have a poor correlation with the DSC and WLF data. The ratio between the loss modulus and the storage modulus, $\tan\delta$, is thus not comparable to the DSC and WLF data in the sense of a physical mechanism. However, $\tan\delta$ is very useful for determining T_g, because it is easy to determine and is independent of the dimensions of the specimen. It is thus important to specify both the measurement parameters and the method of definition when T_g data are reported.

D. Time–Temperature Superposition

The mechanical properties of a material may vary considerably, depending on the temperature and the climatic conditions. In addition, the time scale of the measurements has an influence on the properties of the material, as does also the loading pattern. To be able to predict how the material will behave in different situations, it is desirable for equations to be established to describe the property variations of the material. For some materials it is possible to establish such a relation where the response to a given load is independent of the response of the material to any load that is already applied to the test piece, the Boltzmann superposition principle [23]. A consequence of this principle is that the strain of a specimen is directly proportional to the applied stress when all strains are compared at equal times. The effects of different loads are additive, so that

$$\sigma_1(t) + \sigma_2(t) = c(\varepsilon_1(t) + \varepsilon_2(t)) \tag{48}$$

where c is a constant. Such materials are said to exhibit a *linear viscoelastic behavior*. For such materials it is possible to convert stress relaxation time spectra into creep retardation time spectra. Two conditions must be met:

1. The compliance must be independent of the applied stress.
2. The elongation due to a certain load must be independent of the elongation due to any previous load.

In practice, most polymers show an approximate linearity at strains of about 1%, whereas for paper or wood products strains below 0.1% have been shown to give only small nonlinear responses [52].

The influence of time and temperature can often be rationalized by using time–temperature superpositioning, which is based on the fact that an increase in temperature leads to an increase in molecular motion and thus brings the material more rapidly to equilibrium conditions by increasing the rate of the viscoelastic processes. Properties measured at a specific temperature and frequency, relaxation time, or retardation time can thus be shown to be the same at another combination of temperature and time. This effect of temperature is conveniently described in terms of a ratio a_T of the time at one temperature T to the time at a reference temperature T_0. In practice, data obtained as a function of time at one temperature can be shifted in relation to a selected reference curve so that a common relationship between the property and time is obtained. A vertical correction of the experimental data is also necessary to compensate for the change in density brought about by the change in temperature. Temperature correction is often made on the basis of the kinetic theory of rubber elasticity, but its applicability to systems other than those of pure amorphous polymers may be questioned. The complete correction factor can be written as $\rho_0 T_0/\rho T$. For amorphous polymers, the relationship between temperature and time represented by the shift factor a_T can often be twofold. Below the glass transition temperature, in the glassy region, an Arrhenius type of dependence is usually observed, given by

$$\log a_T = \frac{\Delta H}{2.303R}\left(\frac{1}{T} - \frac{1}{T_0}\right) \tag{49}$$

where ΔH is an activation energy and R is the gas constant [23].

At higher temperatures, i.e., above the glass transition temperature, the relaxation process is characterized by a cooperative action and can be represented by the equation

$$\log a_T = \frac{C_1(T - T_0)}{C_2 + (T - T_0)} \tag{50}$$

where the constants C_1 and C_2 are determined from the experiment. This equation is known as the Williams–Landel–Ferry (WLF) equation and functions in the temperature interval from T_g to $T_g + 100^\circ C$ [53]. The constants C_1 and C_2 are empirically determined, but they have been related to the fractional free volume f_g at T_0 and the volumetric expansion coefficient of the free volume α_f according to

$$C_1 = B/2.303 f_g \tag{51}$$

and

$$C_2 = f_g/\alpha_f \tag{52}$$

where B is a constant that is usually set equal to unity [54].

The time–temperature superposition is illustrated in Fig. 15 by the storage modulus measured on wet birch wood in the temperature range 23–130°C. The reference temperature was here set equal to $T_g = 62^\circ C$. The individual modulus curves measured at each temperature in the frequency range 0.6–20 Hz have been shifted along the frequency axis with respect to the reference curve at 62°C until they became superimposed. The resulting curve, called a *master curve*, covers several decades of frequency although the original experiments covered only two decades. The shift factor a_T as a function of temperature (Fig. 16) is described by the WLF equation, Eq. (50), in an interval from T_g and above. Usually the interval spans 100°C, but for wet wood test pieces this region is smaller [55]. Above and below this region, Arrhenius-type relations exist. From the shift factors of the master curve, a limiting apparent activation energy of the softening process, $\Delta H_{a(\mathrm{WLF})}$, at T_0 may be determined as

$$\Delta H_{a(\mathrm{WLF})} = 2.303R \left.\frac{d \log a_T}{d(1/T)}\right|_{T \to T_0} \tag{53}$$

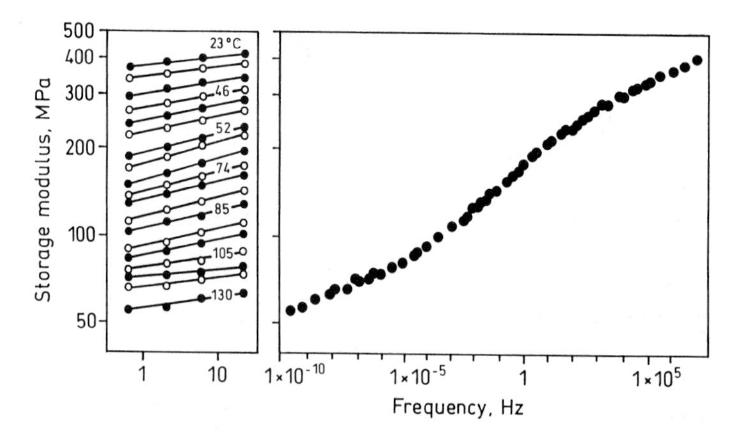

Fig. 15 Master curve for the storage modulus of birch wood tested under water-saturated conditions in the temperature range 23–130°C. (From Ref. 87.)

This apparent activation energy for the softening process of the glass transition of a polymer is found in the range of hundreds of kilojoules per mole, the activation energy increasing with T_g [56].

Master curves can also be constructed from creep or stress relaxation data. The angular frequency ω $(= 2\pi v)$ of a periodic experiment is roughly equivalent to the time $\tau = 1/\omega$ of a stress relaxation or creep experiment. Dynamic experiments may thus provide additional information corresponding to very short times when creep and stress relaxation are of interest. Master curves are of special importance for determining the distribution of relaxation and retardation times, because no single experiment can be carried out over a sufficiently long time span.

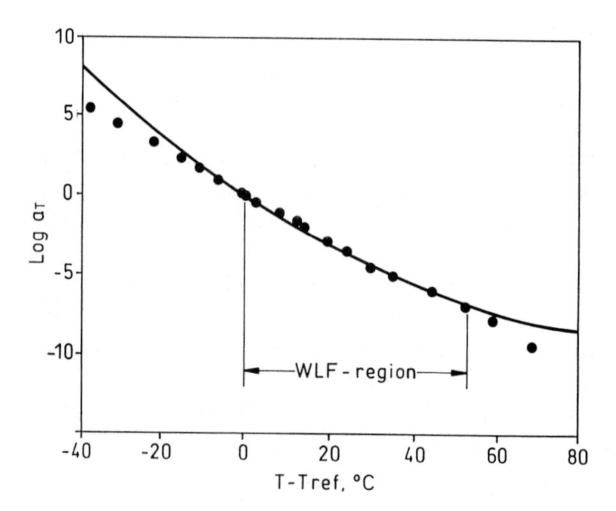

Fig. 16 The shift factor, $\log a_T$, versus the temperature difference, $T - T_0$, for the master curve of birch (Fig. 15). The solid line represents the calculated best fit of the WLF equation to the experiments. It is also evident that the WLF equation applies only in the $T - T_{ref}$ range from °C to 50°C. (From Ref. 87.)

Master curves can also be constructed on the basis of concepts other than the time–temperature equivalence. As mentioned earlier, the addition of a plasticizer, which lowers the glass transition temperature, is qualitatively equivalent to an increase in temperature, and consequently time–plasticizer concentration equivalence has also been suggested. This principle has been applied to cellulose–water [57] and nylon–water [58] systems. Water content–temperature superpositioning has also been applied to paper [59].

The construction of master curves usually requires that the Boltzmann superposition principle, which applies only to small stress and strain levels, be valid. Master creep curves have nevertheless been constructed on the basis of a principle of time–load equivalence (stress superposition for paper [11] and for pulp fibers [60]). In these cases, curves obtained at different stress levels of creep were shifted with respect to the logarithm of time. The result is reminiscent of a master curve from time–temperature superpositioning, but the empirical nature of the procedure must be recognized in interpreting data of this type.

E. Physical Aging Phenomena

Physical aging is a general phenomenon that occurs in all amorphous or partially amorphous polymers below the glass transition and is a fundamental characteristic of the glassy state [10]. When the polymer is cooled below its glass transition, the molecular configuration is in a nonequilibrium state, the degree of which depends on the rate of cooling. The aging process involves a slow movement of the polymer molecules toward an equilibrium configuration, with the aging rate dependent on the temperature relative to the glass transition temperature. Physical aging is a self-retarding process, so aging becomes less with time as the molecular configuration approaches equilibrium. At temperatures below the β transition, aging does not occur, because all significant moelcular motion has been frozen in. The aging process affects the viscoelastic properties of the material in that the elastic modulus increases and the creep and stress relaxation rates decrease, as illustrated in Fig. 17. The effect of physical aging is easily reversed by bringing the material above its glass transition temperature where large-scale molecular motion prevails (deaging of the material). This may be done either by increasing the temperature or by increasing the plasticizer concentration. For paper, for which the transitions depend on the relative humidity at room temperature, aging and deaging processes have a clear influence on properties [61].

VI. BEHAVIOR OF CELLULOSE AND PAPER

The basic constitutent in most papers is the pulp fiber, a fiber that is usually produced from wood. Wood pulp fibers are composed of the natural polymers cellulose, hemicelluloses, and lignin and are thus dependent on the characteristics relevant for all polymers. The hemicelluloses and the lignin are amorphous, whereas cellulose is semicrystalline, with a crystallinity in wood fibers of 60–70% as determined by X-ray diffraction [62].

The semicrystalline cellulose is arranged as extended chains in fibrils that are surrounded by a matrix of hemicelluloses and lignin. At least some of the hemi-

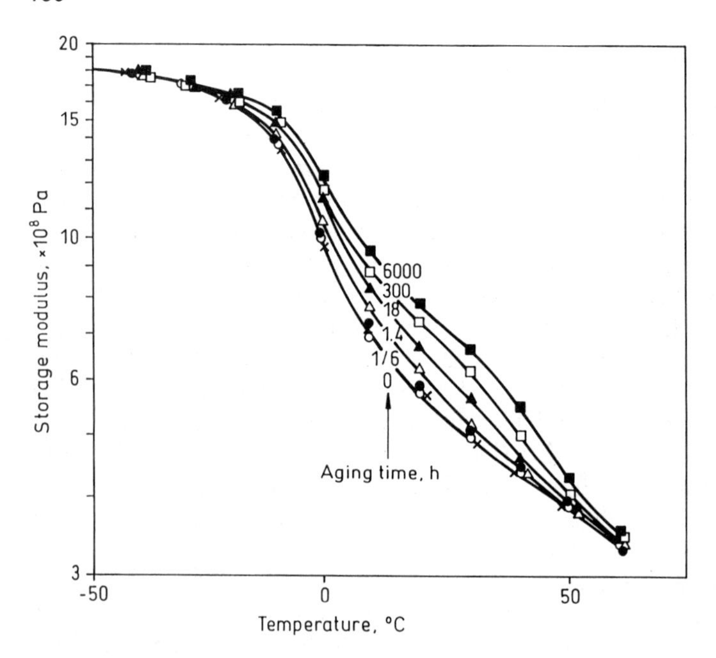

Fig. 17 The storage modulus in shear G at 1 Hz for polypropylene after different times of aging. (From Ref. 88.)

celluloses are arranged parallel to the cellulose fibrils [63], and there are also indications that the lignin may have such an orientation. Upon pulping, some of the dissolved hemicelluloses may reprecipitate in the fiber, and they may also crystallize [64].

A. Thermal Transitions of the Wood Polymers

In the dry wood polymers, the glass transitions occur at rather high temperatures, 180–220°C, mainly because of their high hydrogen-bonding capacity [65]. In dry cellulose, glass transition temperatures between 200°C and 250°C have been reported. In dry hemicelluloses, the glass transition is observed between 150°C and 220°C depending to a large extent on the technique of isolation. For native hemicelluloses, a transition in the region of 180°C is most probable. For dry native lignin it has been difficult to obtain a proper value of the glass transition due to the fact that degradation always occurs upon isolation, but a value of 205°C has been estimated as most probable [65].

All the wood polymers also exhibit secondary transitions. Of these, the most pronounced is a transition, generally termed the γ transition, at −90°C, 1 Hz, under dry conditions, associated with the rotation of the methylol group of the glucose units in cellulose [66,67] and glucomannan of the hemicellulose. At about −30°C, a transition assigned as the β transition appears. It has been suggested that this transition is due to methylol–water complexes [67]. In lignin, the rotation of the methoxyl group is associated with a transition in the −70 to −50°C interval [68].

B. Water as Plasticizer

In the wood polymers, plasticizing is a result of the interaction with water that lowers the softening temperature of all of the wood polymers, as indicated in Fig. 18, which is based on experimental data and on theoretical calculations based on Eqs. (44)–(47) [59]. The parameters used in these equations were estimated by Salmén and Back [69]. Water itself has been shown to have a glass transition temperature of −137°C [70]. For isolated hemicelluloses, the softening has been shown to be lowered to room temperature if sufficiently high relative humidities are reached [71]. This softening temperature thus lies in the interval of normal use of paper properties, as is evident in Fig. 19, and it is therefore of special interest when the viscoelastic properties of paper are evaluated. It has also been suggested that the amorphous part of the cellulose has its softening lowered to room temperature at very high water contents [72]. In the case of lignin, the softening is limited by the cross-linked structure of the lignin molecule. Thus, when saturated with water the lignin has a limiting softening temperature of 90°C at a frequency of 1 Hz [55]. The secondary transitions of the carbohydrates and the lignin are also affected by water acting as a softener, i.e., the transitions are shifted to lower temperatures [66]. Additional transitions may also occur as a result of the interaction between water and the side groups. A transition occurring at about −30°C, the temperature lowered with increasing moisture content, has been attributed to the rotation of methylol–water complexes [66]. The occurrence of this transition is sometimes interpreted as an antiplasticization, i.e., an increase of the transition temperatures at lower temperatures.

C. Linear Viscoelastic Behavior

In dynamic mechanical measurements, it is necessary to observe linear viscoelastic (LVE) behavior to obtain accurate and comparable data. Semicrystalline polymers

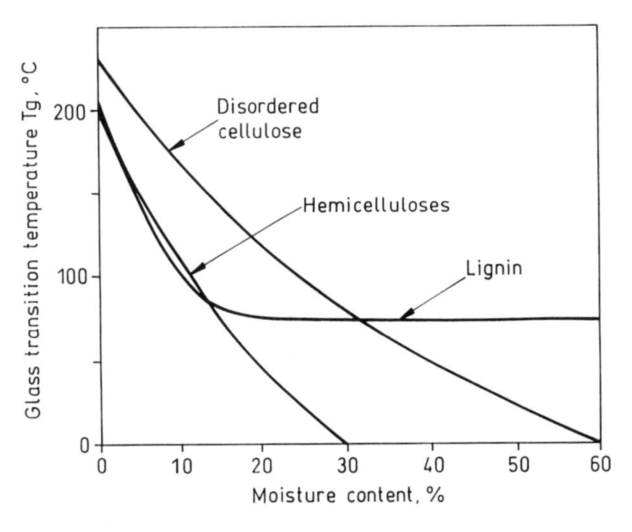

Fig. 18 Schematic picture of the glass transition temperatures of lignin, hemicelluloses, and amorphous cellulose as a function of moisture content at approximately 1 Hz, based on experimental data and theoretical calculations. (From Ref. 59.)

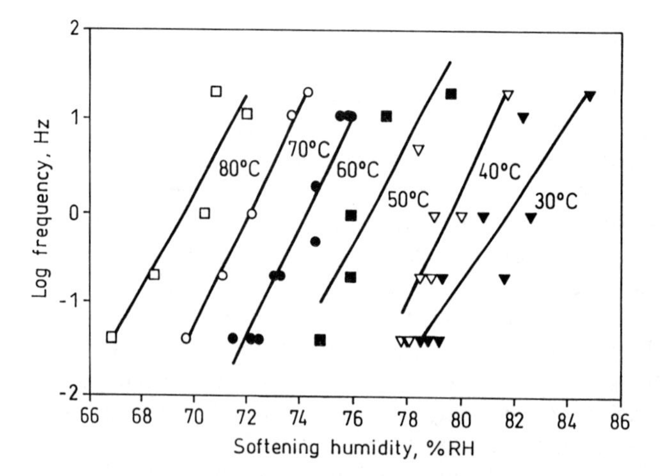

Fig. 19 Softening of birch xylan at different relative humidities as a function of frequency and temperature. The softening point or glass transition temperature was in these experiments determined from the onset of the decrease of the elastic modulus as a function of relative humidity from experiments at constant temperature and constant frequency. (From Ref. 89.)

are often nonlinear in their behavior even at very low levels of strain (below 0.1%), whereas amorphous polymers may often show linear viscoelastic behavior up to higher strains. It has been clearly shown that the effect of the strain amplitude must be taken into account when the dynamic mechanical properties of paper are measured [73]. Figure 20 shows two typical examples of the effect of strain amplitude on the complex modulus of a paper sheet. Increasing the strain amplitude evidently brings about a reduction in the absolute value of the complex modulus. The nonlinear viscoelastic behavior of paper may be an effect of the inherent polymeric properties of cellulose or of the network structure.

The influence of the strain amplitude on the mechanical loss factor, tan δ, for a paper sheet is shown in Fig. 21. The loss factor increases linearly with increasing

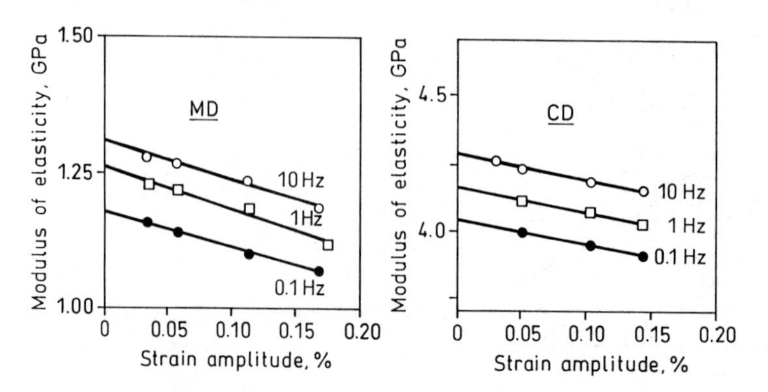

Fig. 20 The complex dynamic modulus $|E^*|$ plotted versus the strain amplitude at 0.1, 1, and 10 Hz. The left-hand figure refers to a freely dried sheet with a density of 326 kg/m^3 in the machine direction, and the right-hand figure to a restrained dried sheet in the cross-machine direction, with a density of 799 kg/m^3. (From Ref. 73.)

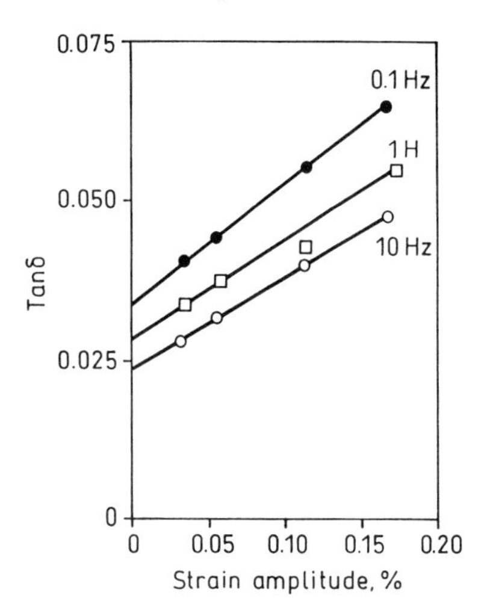

Fig. 21 The mechanical loss factor $\tan\delta$ plotted versus the strain amplitude at 0.1, 1, and 10 Hz for a freely dried sheet (MD) with a density of $326\,kg/m^3$. (From Ref. 73.)

applied strain amplitude. To compare data on different papers, values of $\tan\delta$ may be extrapolated to zero strain amplitude.

In the case of coated paper, the value of $(\tan\delta)_{max}$ associated with the glass transition of the coating polymer may be defined either as the absolute value including the damping from the paper, $(\tan\delta)^0_{max}$, or as a value calculated from a baseline taken as the damping level well outside the peak region to determine only the influence of the coating polymer, $(\tan\delta)^b_{max}$ [74], as shown in Fig. 22. The strain dependence of the two definitions of $(\tan\delta)_{max}$ is shown in Fig. 23, where $(\tan\delta)_{max}$ is plotted versus the dynamic strain for a coated paper with a coating containing 10 parts latex per 100 parts pigment (10 pph). It is evident that $(\tan\delta)^0_{max}$ is strain-dependent whereas

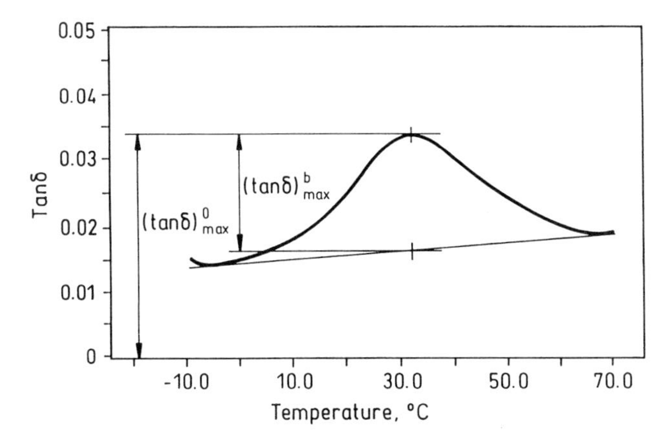

Fig. 22 Loss tangent ($\tan\delta$) for a coated paper with 10 pph latex polymer in the coating in the cross direction (CD) measured in tension. The peak maximum may be defined with respect to the zero level, $(\tan\delta)^0_{max}$, or with respect to a base line taken well outside the peak region, $(\tan\delta)^b_{max}$. (From Ref. 74.)

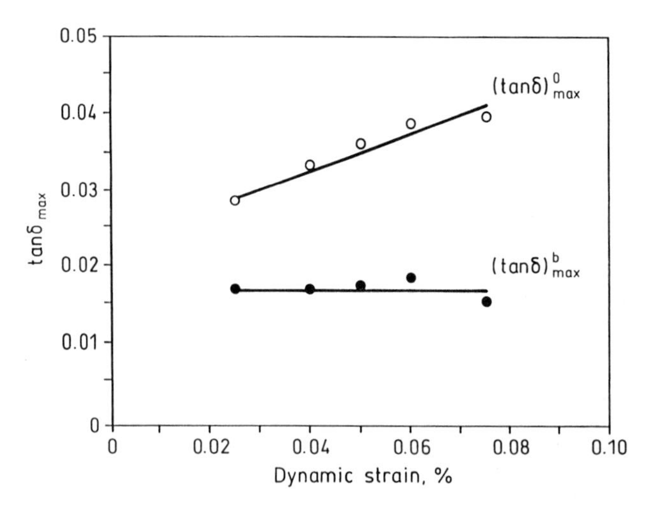

Fig. 23 The strain dependence of the peak maximum of $\tan\delta$ according to the two definitions in Fig. 22, $(\tan\delta)^0_{max}$ and $(\tan\delta)^b_{max}$. (From Ref. 74.)

$(\tan\delta)^b_{max}$ is strain-independent. The strain dependence appears to be associated with the nonlinear viscoelastic behavior of the paper. In torsion, the dynamic amplitude is so small that the strain dependence can be neglected.

D. Influence of Moisture and Temperature

As stated earlier, moisture and temperature influence the properties of paper to a great extent. For completely dry papers, the decrease in elastic modulus with increasing temperature is directly coupled to the thermal expansion of the paper and to the loss of strength of the hydrogen bonds with increasing temperature [75]. Softening of the wood polymers affects the properties only above 200°C [76] (see Fig. 24).

When moisture is present, the effect of temperature is somewhat more pronounced, increasingly so with increasing moisture content [77] (see Fig. 25). A softening region probably related to hemicelluloses and amorphous cellulose is shifted to lower temperatures the higher the moisture content. It is worth noting that although the moisture content at a given relative humidity varies depending on the way the paper has been conditioned, i.e., in absorption or desorption, the elastic properties are uniquely determined by its moisture content [77].

Drying stresses are often imposed on the paper during its manufacture, giving the sheet a residual internal stress. It has been shown [78] that the drying stress is directly related to an increase in the elastic modulus of the paper. Although the stress relaxation rate increases with increasing relative humidity, the internal stress level determined by the method described in Fig. 5 is not affected by moisture cycling, at least up to 80% RH [79]. However, at higher moisture contents, the internal stresses may be completely removed [80] as shown in Fig. 26. A decrease in bending stiffness caused by a release of dried-in strains has been found to be caused by the influence of water in the process of coating paper [81].

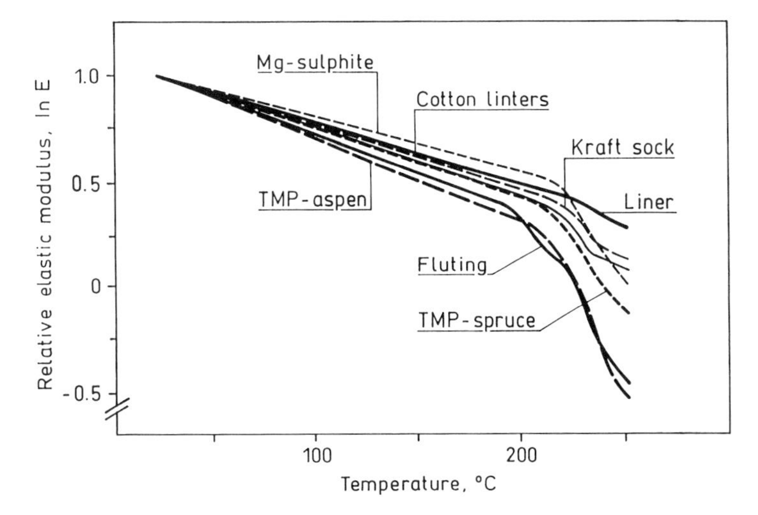

Fig. 24 The effect of temperature on dry papers illustrated by the natural logarithm of the modulus of elasticity divided by that at 20°C measured at a strain rate of 1.7 or 5.0 $10^{-3}\,s^{-1}$. (From Ref. 76.)

Because paper is normally handled under conditions close to the softening of the amorphous carbohydrates, the hemicelluloses, and the amorphous cellulose, the history of the paper also determines its viscoelastic properties. Thus, as is evident in Fig. 27, the relaxation modulus increases constantly with increasing storage time after having been exposed to a high relative humidity where it was deaged [61]. Thus, in order to measure the elastic properties of paper accurately, the paper must either

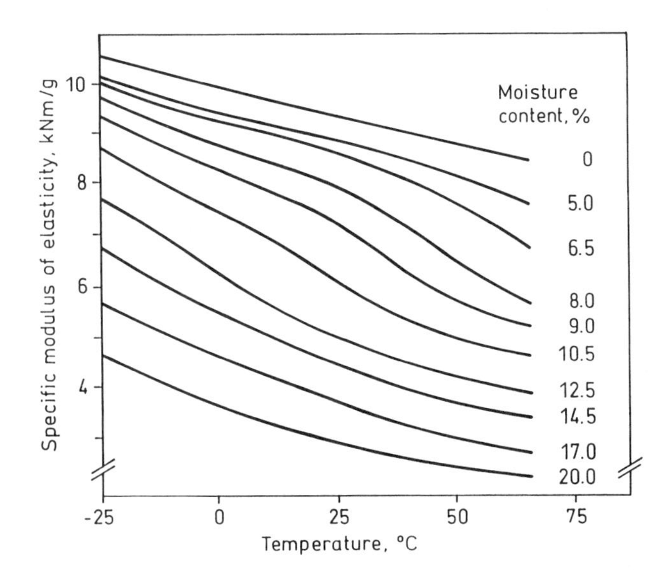

Fig. 25 The specific modulus of elasticity for a kraft sack paper versus temperature at different moisture contents (percent of moist paper). (From Ref. 77.)

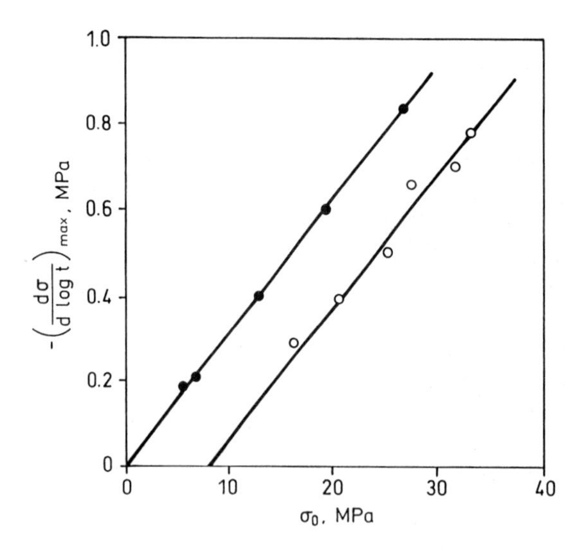

Fig. 26 The inflection slope $-(d\sigma/d\ln t)_{max}$ versus the initial stress σ_0 for untreated and water-treated handsheets of a pine kraft pulp. (From Ref. 80.)

be stored for a specified time in order to reach a quasi-equilibrium or be exposed to a high humidity environment immediate prior to the testing (a conditioning), in the same way as is often done when polymers are tested using temperature conditioning.

E. Influence of Time and Frequency

The properties of viscoelastic materials, e.g., cellulose and paper, are dependent on the time scale of testing. In dynamic mechanical testing, the properties vary with the

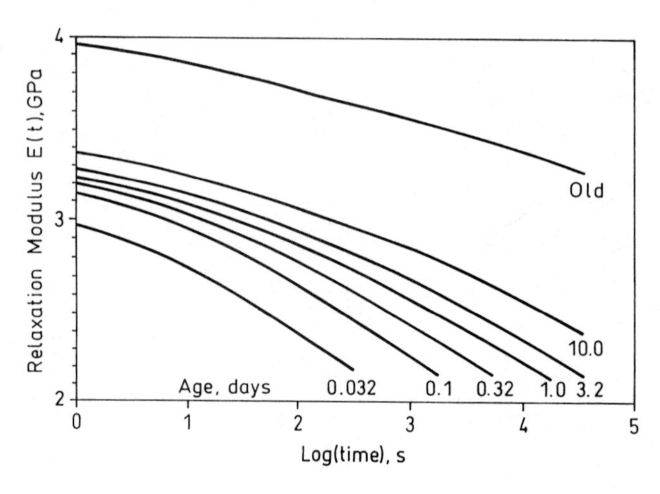

Fig. 27 Relaxation modulus at 23°C and 50% RH in the machine direction of a commercial linerboard, 210 g/m^2, as a function of storage time following a high moisture deaging (100% RH for 1 h). (From Ref. 61.)

oscillating frequency because at high frequencies there is insufficient time for the polymer chains to change their molecular configuration before the direction of deformation is reversed. A more rapid deformation or a high strain frequency results in a higher glass transition temperature (see, e.g., Fig. 14), a higher modulus (see, e.g., Fig. 15), and a more brittle polymer. The time scale of testing affects secondary transitions, the β and γ transitions, even more than the glass transition, owing to the lower activation energy (see Fig. 13).

The complex modulus and the mechanical loss factor tan δ are shown in Fig. 28 as a function of frequency in the 0.1–30 Hz range for a freely dried paper sheet with a density of 777 kg/m^3. With increasing frequency in this rather narrow frequency range in the absence of phase transitions, the modulus increases and the loss factor decreases somewhat. If a larger frequency range or a larger time scale are considered during the tests, phase transitions of polymers such as the glass transition can be located. To overcome the problem of the limited frequency range in a dynamic mechanical experiment, larger frequency ranges may be obtained by using ultrasonic and dielectric data or by constructing master curves as shown in Fig. 15.

Instead of changing the frequency of the experiment, the inverse of the frequency, i.e., the time, may be used to do measurements at constant temperature, i.e., stress relaxation or creep measurements. Because paper is viscoelastic, it displays both stress relaxation and creep behavior. Isochronous curves constructed from compression creep curves for a kraft liner in both the machine and cross-machine directions are shown in Fig. 29. The initial slope of the curves, the creep modulus, is higher for the short time test. Evidently, stress is proportional to strain in this region, and this means that the material exhibits a linear viscoelastic behavior. The higher creep modulus obtained at 1 s compared to that at 10^4 s is a consequence of the viscoelastic behavior of the material. It is also obvious that the modulus is higher in the machine direction than in the cross-machine direction, presumably because of the greater influence of the longitudinal fiber properties in the machine direction [82].

The time dependence of the properties of paper is also emphasized by the straining rate dependence of the stress–strain relationship. In normal stress–strain testing of paper, the tensile strength increases with increasing strain rate. The mod-

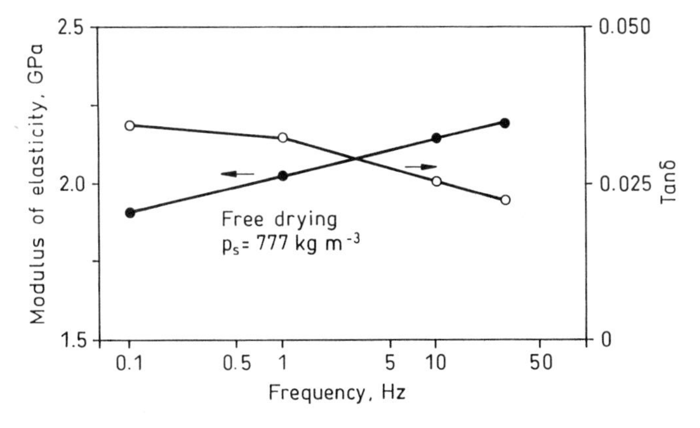

Fig. 28 The dynamic modulus and loss factor plotted versus the frequency for a freely dried sheet with a density of 777 kg/m^3. (From Ref. 73.)

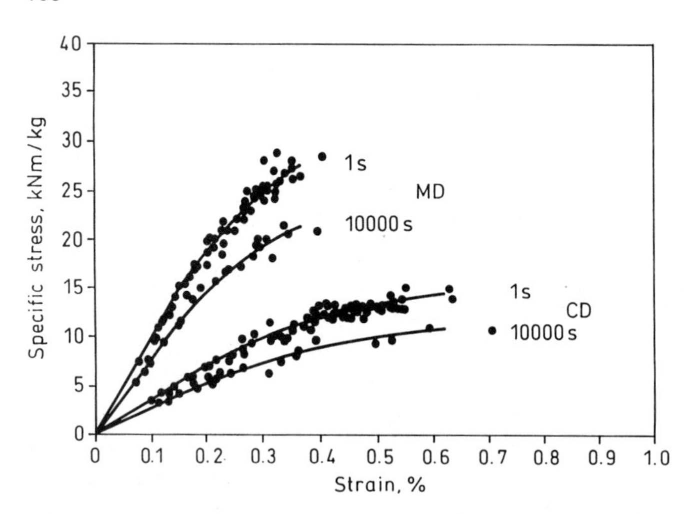

Fig. 29 Isochronous curves in compression for a kraft liner, $185\,g/m^2$, in machine and cross-machine directions, each at a short time (1 s) and a long time (10,000 s), at 50% RH and 23°C. (From Ref. 85.)

ulus of elasticity increases also, depending on the paper, as shown in Fig. 30, where strain rates covering two decades were used [83]. The stretch at rupture is, however, fairly unaffected by the strain rate [84].

The lifetime of paper products can be assessed by the construction of isometric curves, i.e., stress versus time at different strains, using the strain at break during creep. Predictions of paperboard lifetime ($185\,g/m^2$) in the machine and cross-machine directions are shown in Fig. 31, where it also is evident that a relatively small reduction in specific stress is sufficient to increase the lifetime. The strain at break of the paperboard was independent of this lifetime [85].

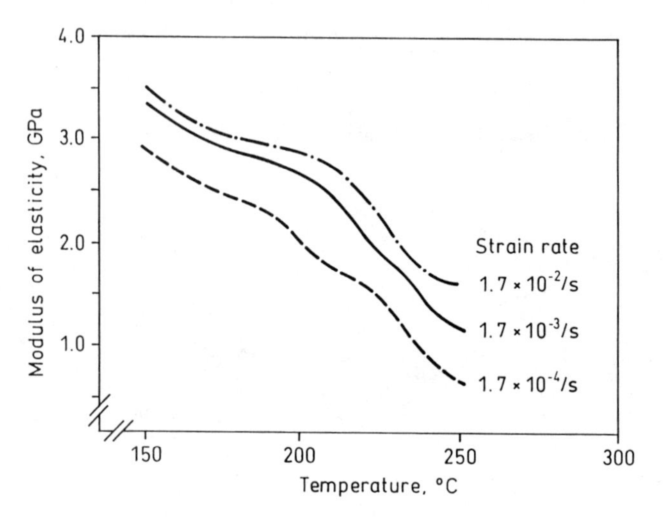

Fig. 30 Modulus of elasticity versus temperature at different strain rates for a dry fluting of $112\,g/m^2$ in the machine direction. (From Ref. 83.)

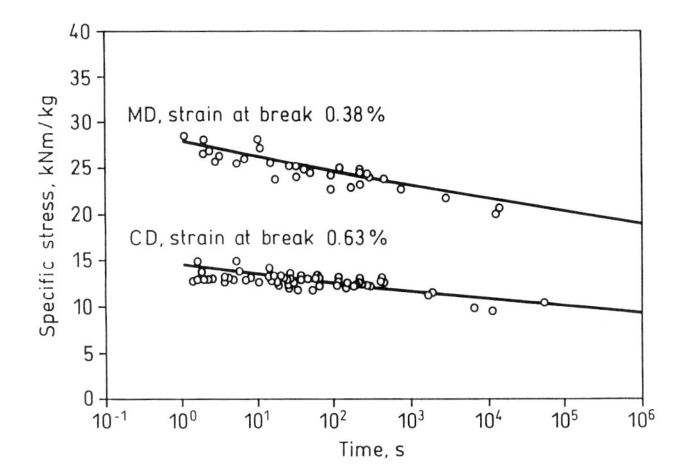

Fig. 31 Lifetime in compression in machine and cross-machine directions for a kraft liner, 185 g/m^2. The strains at break during creep are given in the figure. (From Ref. 85.)

REFERENCES

1. Aklonis, J. J., and MacKnight, W. J. (1983). *Introduction to Polymer Viscoelasticity*. 2nd ed. Wiley, New York.
2. Ferry, D. J. (1980). *Viscoelastic Properties of Polymers*. 3rd ed. Wiley, New York.
3. Findley, W. N., Lai, J. S., and Onaran, K. (1976). *Creep and Relaxation of Nonlinear Viscoelastic Materials*. North-Holland, Amsterdam.
4. Bodig, J., and Jayne, B. A. (1982). *Mechanics of Wood and Wood Composites*. Van Nostrand Reinhold, New York.
5. Pecht, M. G., Johnson, M. W., Jr., and Rowlands, R. E. (1984). Constitutive equations for creep of paper. *Tappi J. 67*(5):106–108.
6. Pecht, M. G., and Johnson, M. W., Jr. (1985). The strain response of paper under various constant regain states. *Tappi J 68*(1):90–93.
7. Tschoegl, N. W. (1989). *The Phenomenological Theory of Linear Viscoelastic Behavior: An Introduction*. Springer-Verlag, Berlin.
8. Andrade, E. N. (1910). On the viscous flow in metals, and allied phenomena. *Proc. Roy. Soc. (Lond.) A84*:1–12.
9. Norton, F. H. (1929). *Creep of Steel at High Temperatures*. McGraw-Hill, New York, p. 58.
10. Struik, L. C. E. (1977). Physical ageing in plastics and other glassy materials. *Polym. Eng. Sci. 17*(3):165–173.
11. Brezinski, J. P. (1956). The creep properties of paper. *Tappi J. 39*(2):116–128.
12. Caulfield, D. F. (1985). A chemical kinetics approach to the duration-of-load problem in wood. *Wood Fibre Sci. 17*(4):504–521.
13. Zhurkov, S. N., and Tomashevsky, E. E. (1968). An investigation of fracture processes of polymers by the electron spin resonance method. In: *Physical Basis of Yield and Fracture*. Institute of Physics, Conference Series No. 1, London, pp. 200–208.
14. Kubát, J. (1965). A similarity in the stress relaxation behaviour of high polymers and metals. Ph.D. Thesis. University of Stockholm, Stockholm.
15. Johansson, F., and Kubát, J. (1964). Measurements of stress relaxation in paper. *Svensk. Papperstidn. 67*(20):822–832.

16. Craven, B. D. (1962). Stress relaxation and work hardening in paper. *Appita 16*(2):57–70.
17. Li, J. C. M. (1967). Dislocation dynamics in deformation and recovery. *Can. J. Phys. 45*:493–509.
18. Kubát, J., Petermann, J., and Rigdahl, M. (1975). Internal stresses in polyethylene as related to its structure. *Mater. Sci. Eng. 19*:185–191.
19. Kubát, J., Seldén, R., and Rigdahl, M. (1978). Influence of strain rate on the stress relaxation behaviour of polyethylene and cadmium. *Mater. Sci. Eng. 34*(1):67–74.
20. Seldén, R. (1976). Stress relaxation and internal stresses in polyethylene. 7th International Congress on Rheology, Gothenburg, pp. 374–375.
21. Sperling, L. H., and Fay, J. J. (1990). Factors which affect the glass transition and damping capability of polymers. *Polym. Adv. Technol. 2*:49–56.
22. Gerard, J. F., Andrews, S. J., and Macosko, C. W. (1990). Dynamic mechanical measurements: Comparison between bending and torsion methods on a graphite-reinforced and rubber-modified epoxy. *Polym. Compos. 11*(2):90–97.
23. Nielsen, L. E., and Landel, R. F. (1994). *Mechanical Properties of Polymers and Composites*. 2nd ed. Marcel Dekker, New York.
24. Pournoor, K., and Seferis, J. C. (1991). Instrument-independent dynamic mechanical analysis of polymeric systems. *Polymer 32*(3):445–453.
25. Hagen, R., Salmén, L., Lavebratt, H., and Stenberg, B. (1994). Comparison of dynamic mechanical measurements and T_g determinations with two different instruments. *Polym. Testing 13*(2):113–128.
26. Lewis, T. B., and Nielsen, L. E. (1970). Dynamic mechanical properties of particulate-filled composites. *J. Appl. Polym. Sci. 14*:1449–1471.
27. McCrum, N. G., Read, B. E., and Williams, G. (1967). *Anelastic and Dielectric Effects in Polymeric Solids*. Wiley, New York.
28. Kolseth, P. (1983). Torsional Properties of Single Wood Pulp Fibers. Ph.D. Thesis Royal Inst. Technology, Stockholm.
29. Xiang, P. Z., Ansari, I., and Pritchard, G. (1985). Torsion pendulum automation by a single microcomputer. *Polym. Testing 5*:321–339.
30. Biot, M. A. (1939). Increase of torsional stiffness of a prismatical bar due to axial tension. *J. Appl. Phys. 10*:860–864.
31. Gillham, J. K. (1974). A semimicro thermomechanical technique for characterizing polymeric materials: Torsional braid analysis. *AIChE J. 20*(6):1066–1079.
32. Gillham, J. K. (1982). Torsional braid analysis (TBA) of polymers. In: *Developments of Polymer Characterisation-3*. J. V. Dawkins, ed. Applied Science, Barking, UK, pp. 159–227.
33. Cassel, B., and Twombly, B. (1991). Dynamic mechanical analysis of fibers and films. American Laboratory *23*(1):22–28.
34. Wetton, R. E. (1986). Dynamic mechanical thermal analysis of polymers and related systems. In: *Development in Polymer Characterization*. J. V. Fawkins, ed. Elsevier, London, pp. 179–221.
35. Wendlandt, W. W., and Gallagher, P. K. (1981). Instrumentation. In: *Thermal Characterization of Polymeric Materials*. E. A. Turi, ed. Academic Press, New York, pp. 1–90.
36. Riemen, W. P., and Kurath, S. F. (1964). The dynamic mechanical properties of paper. *Tappi J. 47*(10):629–633.
37. Ward, I. M., and Hadley, D. W. (1993). *An Introduction to the Mechanical Properties of Solid Polymers*. Wiley, Chichester, p. 334.
38. Barnes, H. A., Hutton, J. F., and Walters, K. (1989). *An Introduction to Rheology*. Elsevier, Amsterdam.

39. Pankonin, B., and Habeger, C. (1988). A strip resonance technique for measuring the ultrasonic viscoelastic parameters of polymeric sheets with application to cellulose. *J. Polym. Sci. B: Polym. Phys. 26*:339–352.
40. Craver, J. K., and Taylor, D. L. (1965). Nondestructive sonic measurement of paper elasticity. *Tappi J. 48*(3):142–147.
41. Berger, B. J., Habeger, C. C., and Pankonin, B. M. (1989). The influence of moisture and temperature on the ultrasonic viscoelastic properties of cellulose. *J. Pulp Paper Sci. 15*(5):170–177.
42. Sperling, L. H. (1986). *Introduction to Physical Polymer Science*. Wiley, New York, p. 439.
43. Boyer, R. F. (1975). Glassy transitions in semicrystalline polymers. *J. Polym. Sci. 50*:189–242.
44. Kaelble, D. H. (1971). *Physical Chemistry of Adhesion*. Wiley-Interscience, New York.
45. Cassel, B., and Twombly, B. (1991). Glass transition determination by thermomechanical analysis, a dynamic mechanical analyzer, and a differential scanning calorimeter. In: *Material Characterization by Thermomechanical Analysis*. A. T. Riga and C. M. Neag, eds. Am. Soc. Testing and Materials, Philadelphia, pp. 108–119.
46. Achorn, P. J., and Ferillo, R. G. (1994). Comparison of thermal techniques for glass transition measurements of polystyrene and cross-linked acrylic polyurethane films. *J. Appl. Polym. Sci. 54*:2033–2044.
47. Chartoff, R. P., Weissman, P. T., and Sircar, A. (1994). The application of dynamic mechanical methods to T_g determination in polymers: An overview. In: *Assignment of the Glass Transition*. R. J. Seyler, ed. ASTM, Baltimore, pp. 88–107.
48. Akay, M. (1993). Affects of dynamic mechanical analysis in polymeric composites. *Compos. Sci. Technol. 47*:419–423.
49. Turi, E. A., ed. (1981). *Thermal Characterization of Polymeric Materials*. Academic Press, New York.
50. Boyer, R. F. (1977). Transitions and relaxations. In: *Encyclopedia of Polymer Science and Technology*, Vol. 2. H. F. Mark, ed. Wiley, New York, pp. 745–839.
51. Leffingwell, J., and Bueche, F. (1968). Molecular motion in 2-chlorostyrene-styrene copolymers from dielectric measurements. *J. Appl. Phys. 39*(13):5910–5912.
52. Lif, J. O. (1997). Experimental methods and time dependent characterisation of paper for printing applications. Licentiate Thesis. Dept. of Solid Mechanics, Royal Inst. Technology, Stockholm.
53. Williams, M. L., Landel, R. L., and Ferry, J. D. (1955). The temperature dependence of relaxation mechanisms in amorphous polymers and other glass-forming liquids. *J. Am. Chem. Soc. 77*:3701–3707.
54. Smith, T. L. (1962). Stress–strain–time–temperature relationship for polymers. *ASTM Mater. Sci. Ser. 3. STP-325*:60–89.
55. Salmén, L. (1984). Viscoelastic properties of in situ lignin under water-saturated conditions. *J. Mater. Sci. 19*:3090–3096.
56. Lewis, A. F. (1963). The frequency dependence of the glass transition. *J. Polym. Sci. B-1*:649–654.
57. Kast, W., Meskat, W., Roseberg, O., and van der Vegt, A. K. (1956) Struhtur und mechanishe Eigenschaften von Faserstolten. In: *Physik der Hochpolymeren*. H. A. Stuart, ed. Springer Verlag, Berlin, Vol. 4 pp. 460–479.
58. Quistwater, J. M. R., and Dunell, B. A. (1959). Dynamic mechanical properties of nylon 66 and the plasticizing effect of water vapor on nylon. II. *J. Appl. Polym. Sci 1*(3):267–271.
59. Salmén, N. L. (1982). Temperature and water induced softening behaviour of wood fiber based materials. Ph.D. Thesis. Royal Inst. Technology, Stockholm.

60. Hill, R. L. (1967). The creep behaviour of individual pulp fibres under tensile stress. *Tappi J. 50*(8):432–440.
61. Padanyi, Z. V. (1993). Physical aging and glass transition: Effects on the mechanical properties of paper and board. In: *Products of Papermaking*. C. F. Baker, ed. Pira International, Leatherhead, pp. 521–545.
62. Nevell, T. P., and Zeronian, S. H., eds. (1985). *Cellulose Chemistry and Its Applications*. Ellis Horwood, Chichester, UK.
63. Liang, C. Y., Basset, K. H., McGinnes, E. A., and Marchessault, R.H. (1960). Infrared spectra of crystalline polysaccharides. VII. Thin wood sections. *Tappi 43*(12):1017–1029.
64. Marchessault, R., Settineri, W., and Winter, W. (1967). Crystallization of xylan in the presence of cellulose. *Tappi 50*(2):55–59.
65. Back, E. L., and Salmén, N. L. (1982). Glass transition of wood components hold implications for molding and pulping processes. *Tappi 65*(7):107–110.
66. Bradley, S. A., and Carr, S. H. (1976). Mechanical loss processes in polysaccharides. *J. Polym. Sci. Polym. Phys. 14*:111–124.
67. Klason, C., and Kubát, J. (1976). Thermal transitions in cellulose. *Svensk. Papperstidn. 15*:494–500.
68. Hatakeyama, H., Nakamura, K., and Hatakeyama, T. (1980). Studies on factors affecting the molecular motion of lignin and lignin-related polystyrene derivatives. *Can. Pulp Paper J. Trans. Techn. Sect. 6*(4):105–110.
69. Salmén, L., and Back, E. (1977). The influence of water on the glass transition temperature of cellulose. *Tappi 60*(12):137–140.
70. Rasmussen, D. H., and MacKenzie, P. (1971). The glass transition in amorphous water. Application of the measurement to problems arising in cryobiology. *J. Phys. Chem. 75*(7):967–973.
71. Cousins, W. J. (1978). Young's modulus of hemicellulose as related to moisture content. *Wood Sci. Technol. 12*:161–167.
72. Htun, M. (1984). Changes in in-plane mechanical properties during drying of handsheets. *Tappi 67*(9):124–127.
73. Rigdahl, M., and Salmén, L. (1984). Dynamic mechanical properties of paper: Effect of density and drying restraints. *J. Mater. Sci. 19*:2955–2961.
74. Hagen, R., and Salmén, L. (1995). Influence of the interaction zone between coating and paper on the dynamic mechanical properties of a coated paper. *J. Mater. Sci. 30*:2821–2828.
75. Nissan, A. H. (1957). The rheological behaviour of hydrogenbonded solids. *Trans. Faraday Soc. 53*:(5):700–721.
76. Salmén, N. L., and Back, E. L. (1978). Effects of temperature on stress–strain properties of dry papers. *Svensk. Paperstidn. 81*(10):341–346.
77. Salmén, N. L., and Back, E. L. (1980). Moisture-dependent thermal softening of paper evaluated by its elastic modulus. *Tappi J. 63*(6):117–120.
78. Htun, M., and de Ruvo, A. (1978). Correlation between the drying stress and the internal stress in paper. *Tappi J. 61*(6):75–77.
79. Salmén, L., Fellers, C., and Htun, M. (1987). The development and release of dried-in stresses in paper. *Nordic Pulp Paper Res. J. 2*(2):44–48.
80. Johansson, F., Kubát, J., and Pattyranie, C. (1967). Internal stresses, dimensional stability and deformation of paper. *Svensk. Papperstidn. 70*(10):333–338.
81. Joyce, M., Hagen, R., and de Ruvo, A. (1997). Mechanical consequences of coating penetration. *J. Coatings Technol. 69*(869):53–58.
82. Htun, M., and Fellers, C. (1986). The in-plane anisotropy of paper in relation to fiber orientation and drying restraints. In: *Paper Structure and Properties*. A. J. Bristow and P. Kolseth, eds. Marcel Dekker, New York, pp. 327–345.
83. Salmén, L., and Back, E. (1977). Simple stress–strain measurements on dry papers from −25°C to 250°C. *Svensk. Papperstidn. 80*(6):178–183.

84. Malmberg, B. (1964). Remslängd och töjningshastighet vid spänningstöjningsmätningar på papper. *Svensk. Papperstidn.* *67*(17):690–692.
85. Haraldsson, T., Fellers, C., and Kolseth, P. (1993). The edgewise compression creep of paperboard. In: *Products of Papermaking*. C.F. Baker, ed. Pira International, Leatherhead, pp. 601–637.
86. Törmälä, P., Weber, G., and Lindberg, J. J. (1980). Spin label and probe studies of relaxations and phase transitions in polymeric solids and melts. In: *Molecular Motion in Polymers by ESR*. R. F. Boyer and S. E. Keinath, eds. Midland Monographs Press Symp. Ser. 1. Harwood Academic Publ., Chur, pp. 81–114.
87. Olsson, A.-M., and Salmén, L. (1992). Viscoelasticity of in situ lignin as affected by structure: Softwood vs. hardwood. In: *Viscoelasticity of Biomaterals*. G. Glasser and H. Hatakeyama, eds. *ACS Symp. Ser. 489*, pp. 133–143.
88. Struik, L. C. E. (1982). The long-term physical aging of polypropylene at room temperature. *Plastics Rubber Process. Appl.* *2*(1):41–50.
89. Olsson, A.-M., and Salmén, L. (1997). Humidity and temperature affecting hemicellulose softening in wood. In: *Wood Water Relations*. P. Hoffmeyer, ed. COST E8. Tekst og Tryck, Copenhagen, pp. 269–280.

3

DIMENSIONAL STABILITY AND ENVIRONMENTAL EFFECTS ON PAPER PROPERTIES

TETSU UESAKA
Pulp and Paper Research Institute of Canada
Pointe-Claire, Quebec, Canada

I. INTRODUCTION

When relative humidity and temperature vary, paper dimensions and properties change, sometimes drastically. Effects of environmental conditions on paper properties and dimensions have been a major issue since the inception of papermaking.

Equipment used in converting and end use is becoming more sophisticated and complex and is operated at higher speeds with tighter tolerance than in years past. Accordingly, paper is subjected to tougher and more diverse environmental conditions. It is not unusual these days to see paper that is printed first in a multicolor offset press, subsequently in inkjet and laserjet printers, and then is subjected to perforation, folding, and cutting operations. The importance of properly taking environmental effects into account has been increasing ceaselessly in product design and process development. There is a strong need for paper engineers and scientists to have appropriate tools for characterizing the environmental effects on paper properties.

Testing in this area, however, is mostly nonroutine. As will be discussed later, the effects of moisture and temperature on paper properties and dimensions are generally nonlinear and history-dependent. In other words, these effects are not determined by the present values of relative humidity and temperature but are a function of how such conditions have been achieved through the entire time since the paper was manufactured. Because of this, readers are encouraged to read some background sections before designing any experiments. It is especially true in this specific area that careful planning saves months of time for experiments and mill trials later.

This chapter focuses on (1) general environmental effects on mechanical properties of paper and board and (2) dimensional stability, including linear dimensional changes in the machine, cross-machine, and thickness directions, curl, and cockle. For the environmental effects on optical, electrical, and other physical properties, readers can refer to the chapters of Baum, Bøhmer, and others in this handbook. Readers seeking information on commercial equipment and sensors may be able to find the latest information from Internet web sites by using appropriate key words.

II. BACKGROUND

A. Moisture Sorption and Desorption Characteristics

Because it is hygroscopic, paper sorbs or desorbs moisture according to the relative humidity of its surroundings. The moisture content of paper, however, is not determined by the present value of the relative humidity but rather depends on the history of the humidity changes in the surroundings. In testing the responses of paper to environmental changes and in relating such information to end-use behavior, it is extremely important to consider this history-dependent aspect of the moisture sorption/desorption characteristics of the material in question.

Figure 1 shows an example of history-dependent sorption/desorption characteristics [120]. The paper was subjected to a dynamic humidity change with a 4 h cycle (Fig. 1a). The moisture content was determined as a function of time. Figure 1b shows a plot of the moisture content against the corresponding relative humidity. The moisture content at a given relative humidity under nonequilibrium conditions generally depends on whether it is measured during the sorption stage or the desorption stage, the number of cycles that the paper underwent, the rate of humidity change, and the range of humidity change.

These charactersitics have certain implications for paper testing, some of which are as follows. First, paper can have significantly different moisture contents at the same relative humidity, and thus different dimensions and properties. The difference can be as much as 2%, depending on the previous moisture history. Second, paper can retain the same moisture content even when the relative humidity is changed. (This phenomenon often happens in paper mills and press rooms when paper is first dried to a low moisture content and then exposed to a higher relative humidity.)

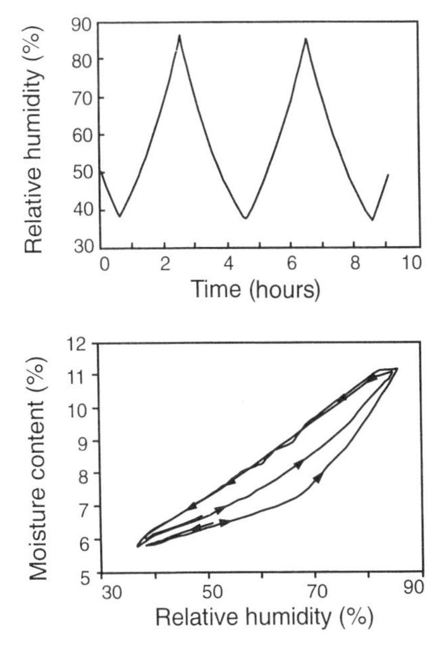

Fig. 1 Sorption of water vapor under nonequilibrium conditions. (From Ref. 120.)

Accordingly, it is a common practice to condition the paper first at a substantially lower humidity (e.g., 30% or lower) before it is tested in the standard condition in order to achieve an identical moisture content at a given relative humidity [131].

The understanding of interactions between paper and moisture has been the subject of intensive studies. The relationships (sorption isotherms) are obtained under equilibrium conditions where paper is exposed to a given relative humidity for at least 24 h. Figure 2 shows a typical sorption isotherm for paper [12,94]. The moisture sorption process is generally classified into three stages:

- • In the first stage where the relative humidity is low (less than about 20%), there is a steep change in moisture content with increasing relative humidity.
- •• In the second stage a steady moisture change takes place when the relative humidity lies between 20% and 65%.
- ••• In the third stage a very sharp change in moisture content occurs at a high relative humidity close to 100%.

Various sorption theories have been proposed to explain the sigmoidal shape of the sorption isotherm. For example, the surface sorption theory [14] considers monolayer and multilayer sorption of water molecules on internal surfaces with active hydroxyl or other polar groups. The solution theory [44], on the other hand, considers the water–polymer system a "solution" consisting of polymer (dry cellulose or fiber), hydrated polymer, and dissolved water [97]. Interested readers can refer to some excellent review papers on this subject [12,22,33,83].

Paper generally consists of several components, including cellulose, hemicellulose, lignin, fillers, and additives, each with different hygroscopicity. Therefore, depending on its constituents, paper exhibits different moisture sorption characteristics. Figure 3 shows sorption isotherms of papers made from mechanical

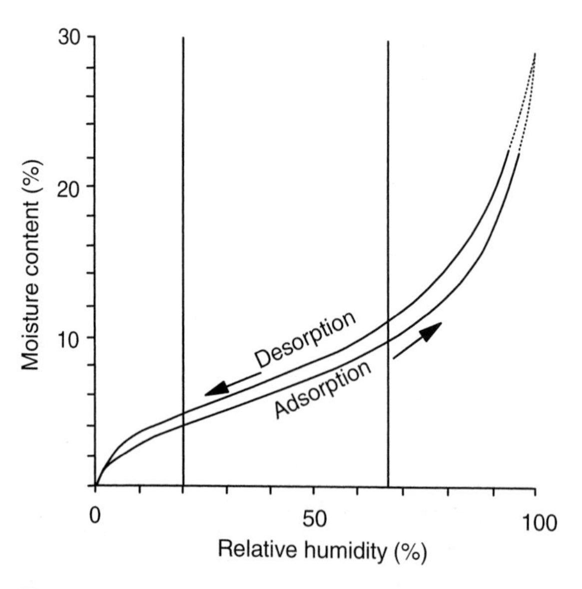

Fig. 2 Equilibrium sorption isotherm of water vapor by a sulfite pulp at 20°C. (From Ref. 12.)

pulps, chemical pulps, and cotton [83]. Because water is mainly adsorbed on the hydroxyl groups existing in the amorphous regions of hemicellulose, the equilibrium moisture content is expected to be higher for pulps with higher hemicellulose contents and lower for pulps with higher degrees of crystallinity. The results show that as hemicullose is removed (from mechanical pulp to chemical pulp) and as cellulose is purified (from chemical pulp to cotton), the equilibrium moisture content at a given relative humidity decreases. Bleaching also has some minor

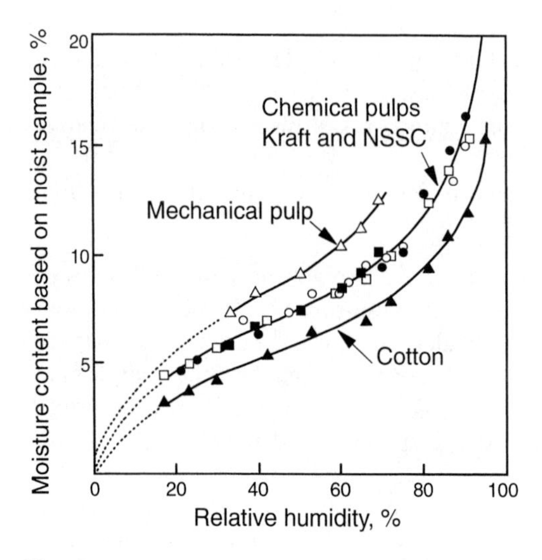

Fig. 3 Sorption isotherms at 23°C in absorption for papers made from mechanical pulp, chemical pulp, and cotton linters. NSSC: Neutral sulfite semichemical pulp. (From Ref. 83.)

effect—decreasing the equilibrium moisture content—because of the removal of hemicellulose [94].

In the wet state of a pulp, water uptake is known to be influenced by the ionic forms of the pulp [89]. The manifestation of this ionic effect depends on the amount of chemical groups such as carboyxlic acid groups. These groups are attached to the hemicellulose, become dissociated, and create counterions. Although this effect is most significant in the wet stage, it is also observable in the high relative humidity range of the moisture sorption curve of a dry sheet. Figure 4 shows sorption isotherms for a carboxylmethylated cellulose pulp with the ionic forms of H^+, Na^+, and Ca^{2+} [83]. The pulp in the sodium form sorbs significantly more water than pulp in the calcium form.

From a practical point of view, the most important effects related to the moisture sorption of paper are the irreversible loss of moisture absorbency and the hysteresis effect. When a wet sheet is first dried, the first desorption curve observed is always higher than the subsequent sorption and desorption curves, as illustrated in Fig. 5 [22]. This irreversible loss is most significant at higher humidities. On the assumption that the sorption behavior is strongly affected by the presence and size of micropores and capillaries, as presumed in the capillary condensation theory [97], one can speculate that the phenomenon is caused by the irreversible loss of some of these micropores and capillaries.

The hysteresis is common in many hygrosocopic materials of different structures [122]. We can expect such hysteresis phenomena by considering them as a thermodynamic process: (1) The moisture sorption/desorption process always involves diffusion; (2) the diffusion process, driven by the moisture gradient, creates the hygro stress gradient; (3) the stress gradient in turn influences the moisture gradient (thermodynamic coupling) and thus the local moisture content [31]; (4) the material is always viscoelastic and plastic to some extent, exhibiting time depen-

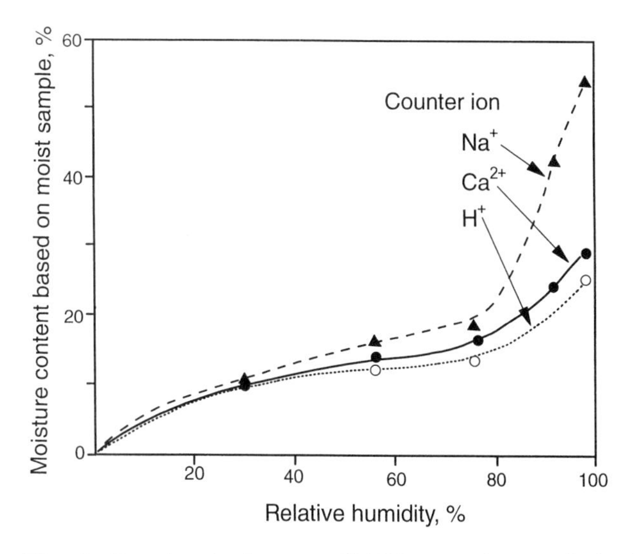

Fig. 4 Sorption, isotherms at 23°C for a carboxymethylated cellulose pulp (DS 0.29) in the ionic forms H^+, Na^+, and Ca^{2+}. (From Ref. 83.)

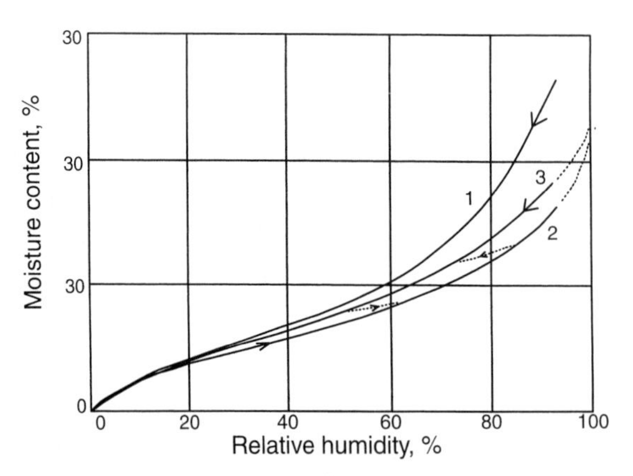

Fig. 5 Desorption and sorption isotherms of a bleached sulfite pulp. (From Ref. 22.)

dence and hysteresis in its stress–strain relationship; and (5) therefore, the sorption isotherm can also exhibit time dependence and hysteresis phenomena. Although the thermodynamic description of the sorption phenomena has not been fully developed, this thermodynamic coupling was already indicated in the early literature [5,6]. The dependence of "equilibrium" moisture content on the humidity history used to achieve "equilibrium" is demonstrated by Christensen and Kelsey [21]. It was shown that the equilibrium moisture content at 88% RH, achieved from 0% RH, is always higher when the relative humidity is changed in a single ramp step than when it is achieved in multiple ramp steps. Attempts to describe the hysteresis loop can be found in the recent literature [20,81].

Effects of papermaking variables are reported in the literature. Extensive beating of an unbleached sulfite pulp slightly increases moisture content (0.5% at 55% RH) [94]. Generally, the equilibrium moisture content increases with beating by at most 0.3–1.0% [11]. Drying restraint applied during the drying process was found to have no effect on the moisture isotherm [83]. Clay fillers have slight hygroscopicity, absorbing about 0.33% by weight when the relative humidity is raised from 33% to 66% [57]. Sizing chemicals have no effect on the moisture sorption characteristics (the vapor phase sorption) at a normal dosage level [11].

Adsorption of moisture generates heat [exothermic reaction). Therefore, the equilibrium moisture content at a constant relative humidity generally decreases with increasing temperature. Figure 6 shows an example of the temperature dependence of the equilibrium moisture content [8]. The decrease with temperature is much greater at 50% RH than at 65% RH. At a high humidity (95% RH), McKee and Shotwell [67], however, reported an *increase* in the equilibrium moisture content with temperatures above 30°C.

B. Moisture and Temperature Effects on Paper Properties

Moisture and temperature affect almost all paper properties and thus end-use performance. In this section the discussion is focused on effects on mechanical properties under equilibrium and nonequilibrium conditions of temperature and humidity.

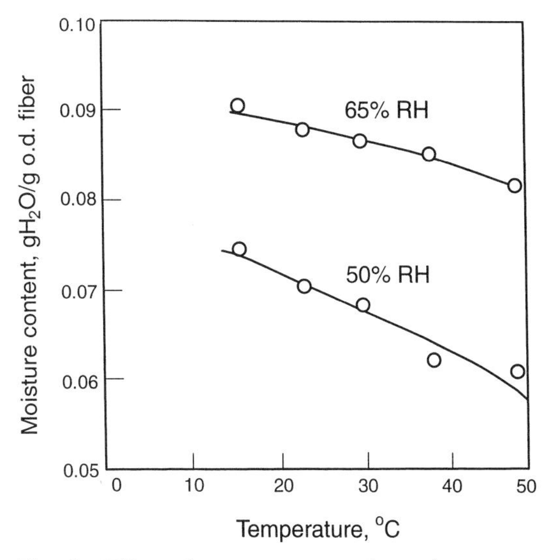

Fig. 6 Effect of temperature on the moisture content of kraft linerboards. o.d.: Oven dried. (From Ref. 8.)

Readers who are interested in effects on electrical, thermal, and other properties can refer to Volume 2 of this handbook.

Equilibrium Condition Relative humidity has a dramatic effect on tensile stress–strain behavior. Figure 7 shows a typical example of the effect for softwood unbleached kraft linerboards [8]. In both the machine direction (MD) and the cross-machine direction (CD), an increase in relative humidity decreases the initial slope of the curve (elastic modulus), yield stress (stress at the proportional limit), and ultimate tensile strength while increasing strain to failure (stretch). Among these properties, the yield stress showed the most sensitivity to changes in relative humidity [8]. The tensile energy absorption (TEA), on the other hand, showed a small peak at a moisture content of around 6–8%. In addition to the tensile properties, other mechanical properties are also affected by the moisture content; tear strength and folding strength increase with increasing moisture content. Folding strength generally shows the greatest change with moisture content change of all the paper mechanical properties routinely measured [131].

When paper is exposed to a higher humidity after being in the standard humidity condition, some paper properties exhibit significant irreversible changes. Table 1 shows the change in tensile strength, strain to failure, tensile energy absorption (TEA), and zero-span strength for various papers when they are exposed to the following cycle: 30% (preconditioning), 50% (control), 93%, 25%, and finally 50% RH [131]. Although tensile strength and zero-span strength showed minor changes, strain to failure and TEA increased significantly. Similar irreversible changes can be seen in dimensional stability, as discussed in subsection C of this section. These irreversible changes resulting from a high humidity excursion are closely associated with strains applied during drying processes and the time-dependent (viscoelastic) recovery of the strains. (See also Chapter 10 in this volume.)

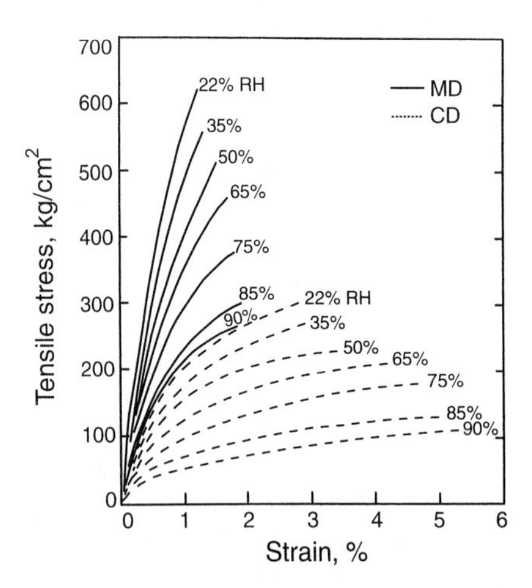

Fig. 7 Tensile stress–strain curves for various moisture levels. MD, machine direction; CD, cross-machine direction. (From Ref. 8.)

Table 1 Effect of Excursions to High and Low Relative Humidities on Tensile Strength, Strain to Failure, Tensile Energy Absorption, and Zero-Span Strength[a]

Sample	Tensile strength	Strain to failure	Tensile energy absorption	Zero-span strength
Sulfitel, MD	−3	7	3	2
Sulfite2, MD	−1	6	5	2
Sulfite3, MD	5	21	30	−1
Sulfite4, MD	−1	11	9	2
Sulfite5, MD	−2	14	15	
Sulfite5, CD	−3	6	2	
Linen writingl, MD	5	32	41	
Linen writingl, CD	−4	21	14	
Linen writing2, MD	−6	15	6	
Linen writing3, MD	−2	19	13	
Linen writing4, MD	−3	11	6	
Rag bond, MD	−3	4	0	
Rag bond, CD	1	14	12	
Newsprint, MD	−3	−1	−4	
Newsprint, CD	0	12	13	
Onionskin, MD	−2	31	31	
Onionskin, CD	2	33	35	
Parchment, MD	4	19	20	
Parchment, CD	−2	5	−1	
Kraft, MD	3	13	18	
Kraft, CD	1	9	12	

[a]The numbers represent percentage increases of the strength properties, based on values at 50% RH, after excursions to 93%, 25%, and 50% RH (23°C).
Source: Ref. 131.

Often in paper testing, moisture content and relative humidity are interchangeably used to indicate the environment effect. However, many paper properties display complex hysteresis phenomena when they are plotted against relative humidity. The phenomena are partly related to the irreversible change of the properties, as described above but are also caused by the hysteresis effect between moisture content and relative humidity, as discussed earlier. Therefore, in order to distinguish these two effects, it is preferable to monitor paper properties as a function of moisture content rather than relative humidity. Salmén and Back [84] showed that within the range of moisture content tested there is no significant difference in elastic modulus measured in the sorption and desorption stages. This one-to-one relationship between moisture content and static mechanical properties has also been observed by other workers for elastic modulus [10,46] and for edgewise compressive strength and tensile strength [1,30].

Figure 8 demonstrates relative effects of temperature on tensile properties of various papers at a "constant" moisture content ($6.1 \pm 0.1\%$). Increasing temperature decreased elastic modulus and tensile strength but increased strain to failure (stretch) dramatically [131]. Berger and coworkers [10] determined specific elastic modulus as a function of temperature and moisture content and found a linear dependence of elastic modulus on moisture and temperature within the range tested. The MD modulus was significantly less affected by temperature and moisture than the in-plane shear modulus and the CD modulus. The data also showed that an independent change in temperature by 10°C approximately corresponds to the same magnitude of change in specific elastic modulus when the moisture content is varied by 1%. (This estimate is considered a linear approximation of the relationship within the limited range. The elastic modulus–temperature–moisture content relationship is generally nonlinear [84], and the temperature and moisture effects may be coupled.)

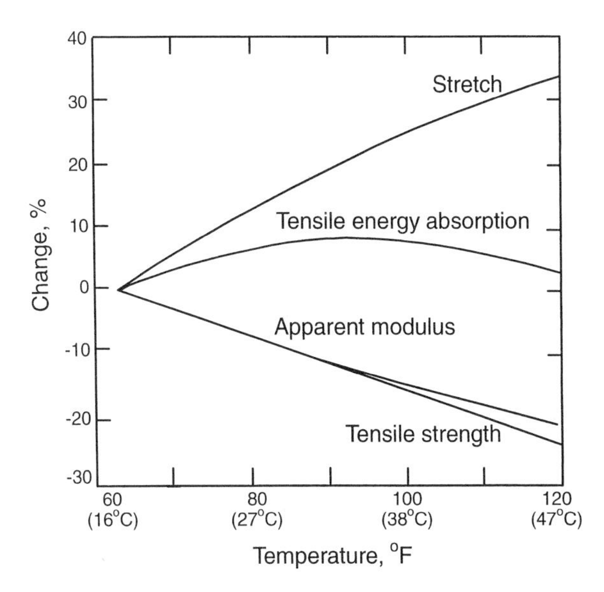

Fig. 8 Effect of temperature on physical properties of paper at a constant moisture content of 6.2%. (From Ref. 131.)

The basic mechanism of such temperature and moisture effects is explained in terms of the polymeric structures of the major pulp fiber components, i.e., cellulose, hemicellulose, and lignin (Fig. 9). As temperature increases (at the same moisture content), both hemicellulose and lignin are first "softened," because the softening temperature (T_g) of these polymers are lower than that of cellulose in a relatively low moisture range (< 20%) [83]. This softening reduces elastic modulus and tensile strength but increases strain to failure (or plasticity). Increasing moisture has a similar softening effect. As the moisture content increases, water molecules are predominantly sorbed in the amorphous or disordered regions linking crystalline regions and "soften" the whole structure (Fig. 9). Because of the orientation of the crystalline regions within the microfibrils and in the cell wall, the temperature and moisture effects on mechanical and dimensional properties of a fiber are highly directional.

In addition to this general softening effect, temperature and moisture affect the relaxation behavior of the molecular motion that results from stresses. As the struc-

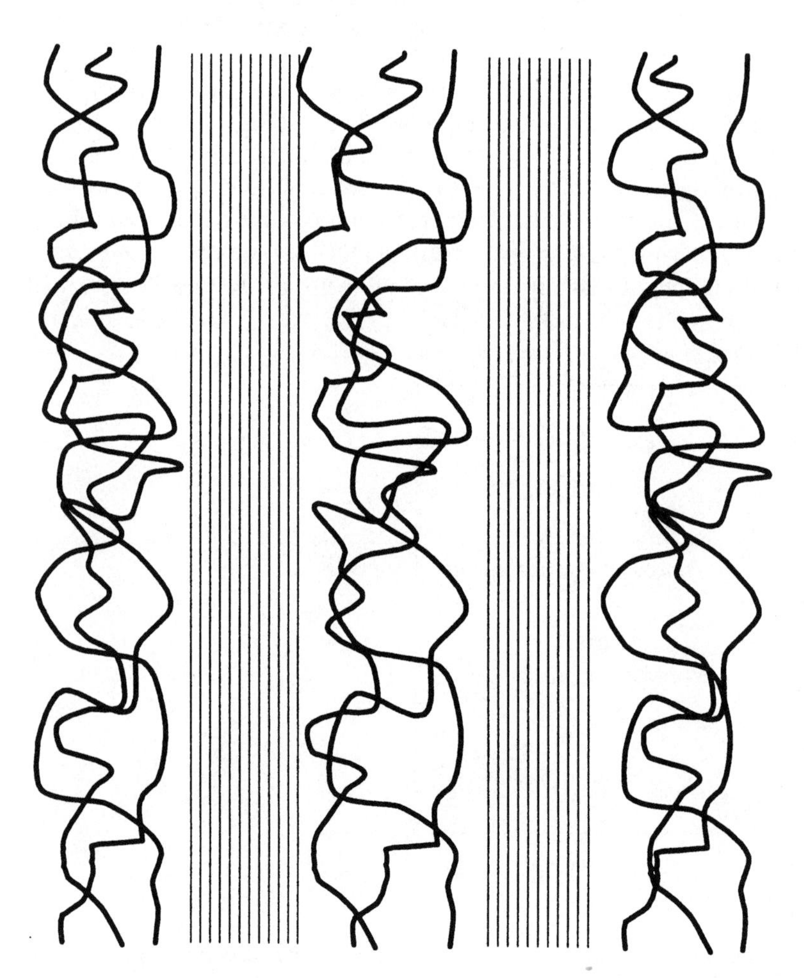

Fig. 9 A model polymeric structure.

ture is stressed, there are both immediate and delayed deformations, the latter being a result of the time-dependent reconfiguration of the structure (mainly in the amorphous or disordered regions). The time-dependent response may be illustrated by a series combination of an elastic spring and a dashpot (Maxwell model, Fig. 10), where the immediate deformation is represented by the spring and the delayed response is represented by the dashpot. The relaxation time τ ($= \eta/E$), defined as the time required for the stress to decay to $1/e$ of the initial value, is related to temperature by

$$\tau = \frac{h}{kT} \exp\left[\frac{G}{kT}\right] \tag{1}$$

where h is Planck's constant, k is the Boltzmann constant, T is the absolute temperature, and G is the free energy difference between the equilibrium and the activated states [10]. As this equation shows, increasing temperature T decreases the relaxation time. In other words, the temperature increase accelerates the process of realignment and reconfiguration of polymer molecules, causing faster stress relaxation or creep deformation. Moisture sorption has a similar effect through the interactions with hydrogen bonds in the noncrystalline region, creating free volume, lowering the energy barrier (free energy difference G), and thus reducing the relaxation time.

In real polymer systems, a number of relaxation mechanisms operate on different structural levels. The system is often represented (mathematically) by a parallel series of the Maxwell elements with different relaxation times. This distribution of relaxation times causes unique temperature and moisture dependencies of mechanical properties, dependent on the time scale for the viscoelastic measurements [10]. The theory and material characterization methods for linear viscoelastic materials have been fully developed. Interested readers can refer to Chapter 2 of this volume and to Ref. 10.

Nonequilibrium Condition When paper is exposed to varying humidity and temperature, it oftens shows unusual behavior that would not be predicted from a knowledge of its behavior under equilibrium humidity and temperature conditions. The phenomena are called "transient effects" or "mechanosorptive efffects." Precise descriptions of these phenomena, however, are still under active discussion.

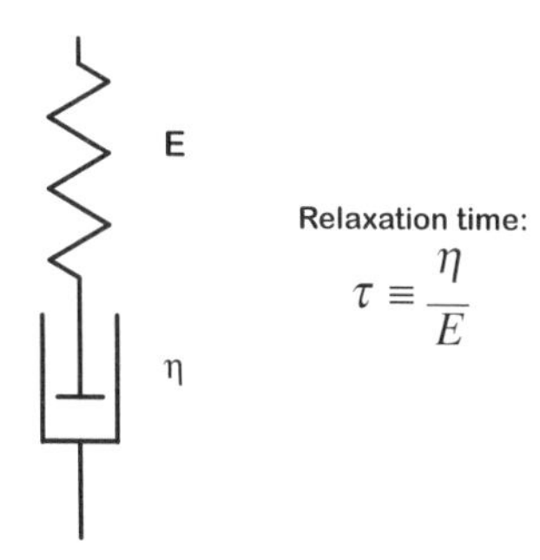

Relaxation time:

$$\tau \equiv \frac{\eta}{E}$$

Fig. 10 Maxwell model. Relaxation time $\tau = \eta/E$.

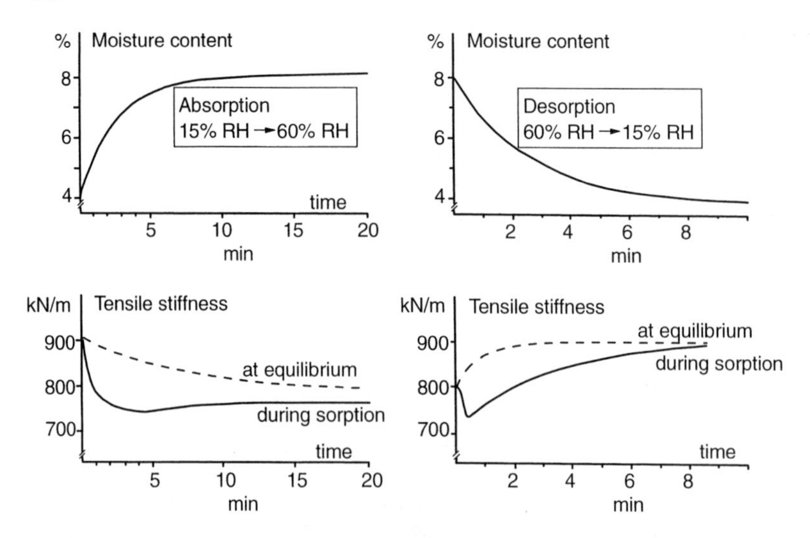

Fig. 11 Changes in tensile stiffness during (a) absorption (15% RH to 60% RH) and (b) desorption (60% RH to 15% RH) processes. (From Ref. 2.)

Figure 11 shows an example of the unusual softening behavior of kraft sack paper under nonequilibrium conditions [2]. The paper was exposed to both sorption (from 15% RH to 60% RH) and desorption (from 60% RH to 15% RH) processes, and tensile stiffness was measured as a function of time. During the sorption process the tensile stiffness measured (solid line) is much lower than the values predicted from the data obtained under the equilibrium conditions of the corresponding humidities (broken line). Similar behavior is seen in the desorption process. It should be noted that the difference between the equilibrium and nonequilibrium data points remained even after the moisture content reached almost a constant value.

A dramatic effect of the transient condition of humidity is seen in edgewise compressive creep behavior (Fig. 12) [15,42,99]. When linerboard is exposed to a cyclical humidity change between 35% and 90% RH under edgewise compression load, the board exhibits much greater creep deformation than the board exposed to a constant 90% RH, even taking into account the expected cyclical hygroexpansion and shrinkage resulting from the corresponding change in moisture content. This unusually large creep phenomenon has been called "accelerated creep" or "mechanosorptive creep." (A detailed discussion of accelerated creep is given in Chapter 4 of this volume.) A typical manifestation of accelerated creep in a practical environment is demonstrated by the behavior of a corrugated carton. When the box is exposed to varying humidities at a constant load, it collapses prematurely. Even if the box is safely designed to sustain the load at the highest possible humidity, it still fails under the varying humidity conditions after many cycles.

Basic observations in the literature of such transient phenomena are summarized as follows:

- The transient effect is more pronounced in bending and torsion modes than in tension and compression.
- The effect increases with increasing sorption rate.

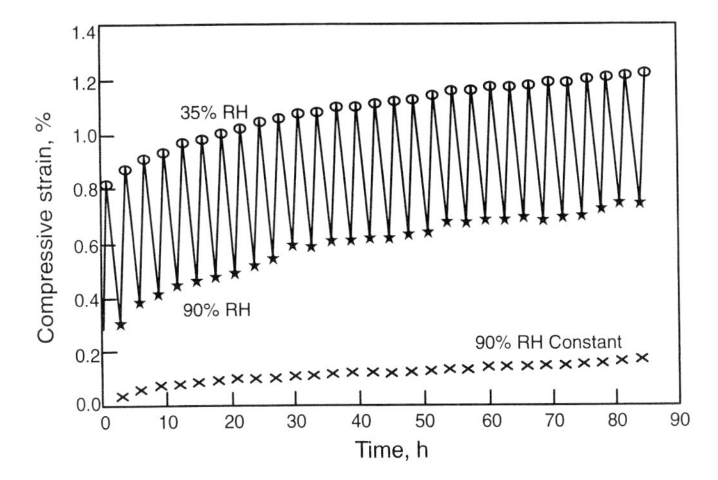

Fig. 12 Compressive creep curves for 205 g/m^2 linerboard. Cyclical and constant RH environment. (From Ref. 42.)

- ••• Ultrasonic measurements do not detect the transient effect [9].
- •••• The effect sometimes lasts for a very long time period.

The last effect has special significance in the measurement of viscoelastic properties under a (seemingly) equilibrium environmental condition. Figure 13 shows such an example [75]. A kraft linerboard was exposed to 100% RH for 2 h, after which it was reconditioned (aged) at 50% RH for different lengths of time. Then tensile creep behavior was determined at 30% of the failure load. An approximate moisture

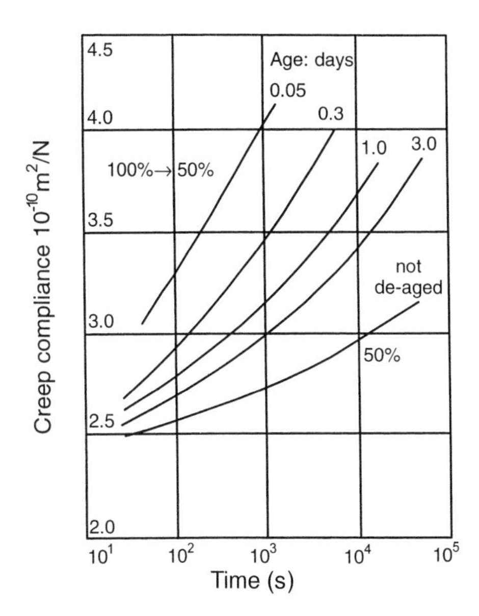

Fig. 13 Effect of "aging" time on tensile creep compliance for 210 g/m^2 linerboard that was exposed to 100% RH for 2 h. (From Ref. 75.)

equilibrium was achieved after 0.5 H (99% of the final moisture content). As a control, the creep behavior of the same sample was determined before the high humidity exposure. As seen in Fig. 13, the control sample shows significantly smaller creep deformation than samples that experienced the higher humidity excursion. Even after 3 days of reconditioning, the creep behavior of the exposed sample is still distinctively different from that of the control. Padanyi [75] showed that the same phenomenon occurs when the board is exposed to 0% RH. This results implies that in order to determine viscoelastic properties of paper and board under different equilibrium conditions of moisture and temperature, one needs a relatively lengthy conditioning time.

Explanations for these phenomena have been given mainly from two standpoints. One is a thermodynamic standpoint where stress–strain–temperature–moisture couplings are considered using basic thermodynamic principles. For example, Barkas [3,4] first considered the moisture change of a gel under tension or compression stresses, and his result was later generalized in the framework of equilibrium thermodynamics by Washburton [127] and Gurney [43]. In this approach, Gibbs' free energy for a material (gel) was assumed to be a function of the *present values* of moisture content, temperature, and stress (hygrothermal-*elastic* body). Further assuming that the change in the chemical potential of water in the material is equal to that in the surrounding air (i.e., equilibrium system) and that temperature is constant during the process (i.e., isothermal process), we can obtain a generalized form of Barkas's equation:

$$dm = \left(\frac{\partial P}{\partial m}\right)^{-1} \left(dP + \frac{1}{\rho v_v} \beta_{ij}\, d\sigma_{ij}\right) \tag{2}$$

where m is moisture content, P is the vapor pressure of water in the surrounding air, ρ is the density of the material, v_v is the specific volume of vapor, β_{ij} is the hygroexpansion coefficient, and σ_{ij} is stress. As can be seen in Eq. (2), increasing tension ($d\sigma_{ij} > 0$) increases the moisture content of the material, and increasing compression ($d\sigma_{ij} < 0$) decreases the moisture content. Gunderson [42] demonstrated for the first time that this phenomenon indeed occurs in paper, by carefully measuring moisture sorption and desorption under stresses. However, the magnitude of this coupling effect was found to be rather small and did not account for many of the observed transient phenomena reported in the literature.

Although the theory developed by Barkas, Washburton, and Gurney clearly demonstrated an important phenomenon of thermodynamic coupling, it was applied only to "equilibrium and isothermal" conditions, neither of which is the case for transient processes. In the general case where stress/strain gradients and temperature/moisture gradients exist, two types of thermodynamic coupling phenomena are expected. One is *constitutive coupling*, i.e., the material property itself becomes dependent on the gradient of field variables, such as the temperature gradient. The other is *field coupling*; in this case, the stress gradient induces heat and mass flow through plastic deformation, and in turn the heat/mass flow creates a stress gradient through thermal/hygro expansion. Uesaka [114] reviewed nonequilibrium thermodynamics approaches that could potentially explain the transient phenomena. By use of a simple example of a viscoelastic body having a single relaxation time, it was shown that under nonisothermal conditions the second law of thermodynamics requires the relaxation time to be dependent on the temperature gradient. This relation implies that viscoelastic behavior under nonisothermal conditions is inherently different from the viscoelastic behavior that occurs under equilibrium conditions. In addition, it was predicted that transient phenomena will not be observed when the elastic modulus is measured on a much shorter time scale than the relaxation time, such as in the case of ultrasonic measurements [9]. A similar dependence on moisture gradient (and its history) can be expected for a general viscoelastic body with distributed relaxation times.

The literature contains some qualitative explanations for transient phenomena from the molecular point of view [63,133]. In the fiber wall structure, such as that depicted conceptually in Fig. 9, the origin of the dissipation mechanism (or relaxation mechanism) is ascribed to the continuous breakage and creation of (secondary) bonds between the polymer chains in the amorphous region. Bond breakage and creation can induce small-scale conformational changes in the side chains of the polymer. These structural changes can occur on a relatively short time scale. However, large-scale segmental motions requiring long times can also be instigated by bond creation and breakage. In an equilibrium state where there is no heat and/or mass transfer, the probability of bond breakage is equal to the probability of bond creation. As heat and mass transfer occur, the local structure of the polymer network is distorted owing to the temperature and mass gradients, increasing the probability of bond breakage. This unbalance of bond breakage and creation temporarily makes the structure more mobile and thus results in temporal softening of the structure. This proposed mechanism is admissible to the nonequilibrium thermodynamic prediction described above.

The probability of bond breakage is generally a function of the end-to-end distance of the junction points [133]. Longer distances are less stable. Unstable structures are also generated by stresses [19], causing nonlinear viscoelastic behavior. Local distortions in the polymer network structure, such as might be created by temperature changes, moisture changes, and stresses, may require a long time period to readjust themselves toward a stable new structure. (These processes are often referred to as physical aging and deaging in the polymer chemistry area [75].) Therefore, in any characterization of paper involving temperature, humidity, and/or time dependence, one needs to pay special attention to these long-term transient phenomena. (See also Chapter 2 in this volume.)

C. Dimensional Stability

Hygroexpansion Behavior Paper expands or shrinks (hygroexpansion/shrinkage) as the relative humidity in the local environment changes. Figures 14 and 15 show an example of the hygroexpansion behavior of a copy paper in the machine (MD) and the cross-machine (CD) directions, respectively [120]. When hygroexpansion is plotted against relative humidity, paper generally shows complex hygroexpansion behavior, dependent on its humidity history before testing (conditioning RH, temperature, and time) and during testing (the speed of the change in relative humidity and the range of the humidity change). Such hygroexpansion behavior is partly attributable to the hysteresis effect of moisture absorption, as discussed in Section II.A. Therefore, in paper testing it is more sensible to plot hygroexpansion as a function of moisture content in the sheet, even though the

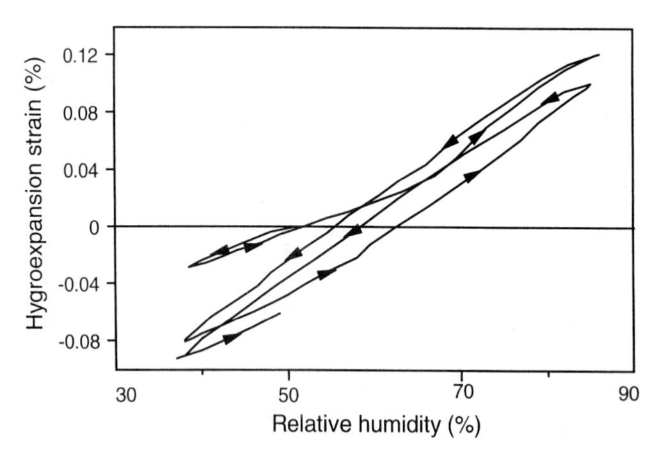

Fig. 14 Hygroexpansion as a function of ambient relative humidity of fine paper (MD). (From Ref. 120.)

determination of relative humidity is much easier than that of moisture content. Figures 16 and 17 are the corresponding plots for hygroexpansion versus moisture content. In the low moisture content range, the MD paper specimen shows approximately linear and reversible expansion/shrinkage behavior. As the paper experiences higher moisture for the first time after it was dried from the wet state, it starts showing nonlinear hygroexpansion behavior and irreversible shrinkage after the high humidity exposure, as shown in Fig. 16. After the same high humidity exposure has been repeated a few times, the expansion/shrinkage behavior finally becomes more reversible. These nonlinear and irreversible hygroexpansion characteristics of MD paper have been documented in the literature [12,118,120,131] and are generally attributed to the release of "dried-in strain" or the release of "internal stress" created during the drying process. (The term "internal stress" used in polymer science and paper science is, however, defined differently from the one used in applied mechanics. Interested readers can refer to Chapter 10 in this volume.) During the drying process, paper is under tension,

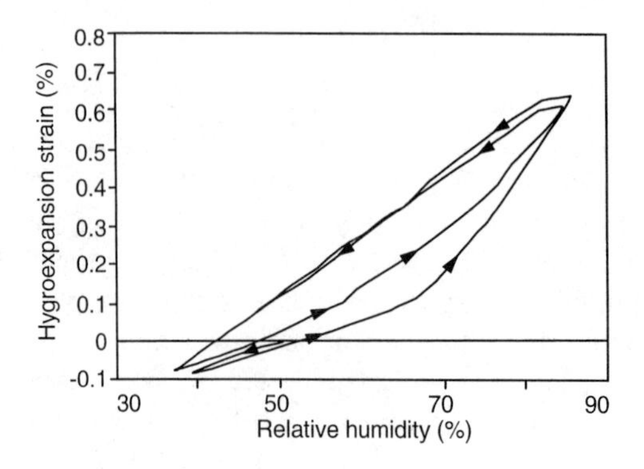

Fig. 15 Hygroexpansion as a function of ambient relative humidity of fine paper (CD). (From Ref. 120.)

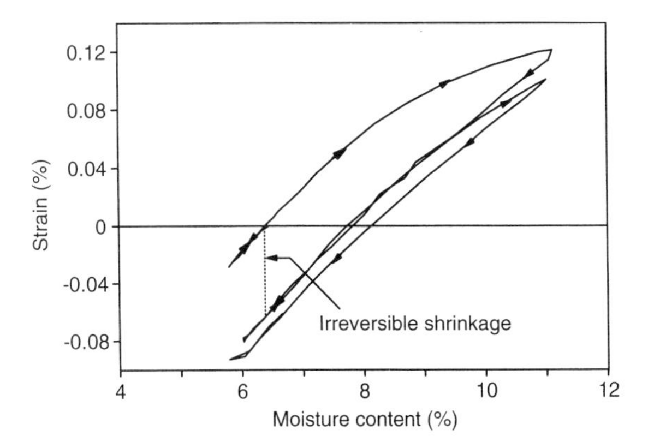

Fig. 16 Hygroexpansion as a function of moisture content of fine paper (MD). (From Ref. 120.)

partly due to the draw applied and partly due to the restraint of the drying shrinkage in the machine direction (Fig. 18). When the tension is released after the roll is finished, the paper shrinks back. This strain recovery continues to take place even months after release of the tension because of the viscoelastic nature of paper deformation. Increasing the temperature and the moisture content has the effect of accelerating this viscoelastic recovery process; the irreversible shrinkage is a manifestation of such strain recovery. (The transient conditions, as discussed before, have the same accelerating effect.) Therefore, if no draw (or very little draw) is applied in the papermaking process, irreversible shrinkage should not be observed and the hygroexpansion–moisture relationship should be linear. This is typically seen in the behavior of a CD specimen (Fig. 17). (It should be noted that, even for the CD specimen, if the moisture content is changed drastically as in the case of wetting, it develops an irreversible shrinkage.) If a very

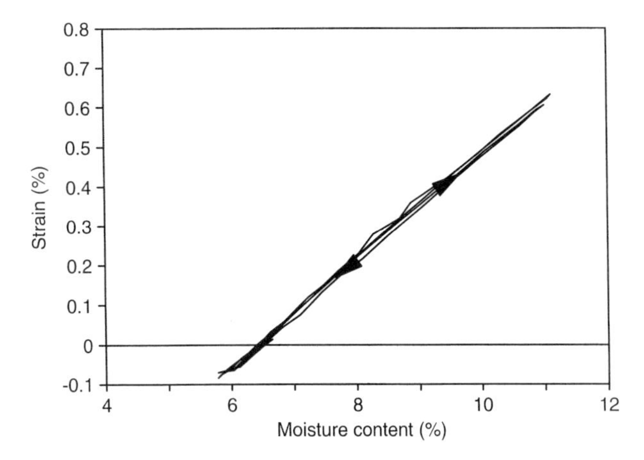

Fig. 17 Hygroexpansion as a function of moisture content of fine paper (CD). (From Ref. 120.)

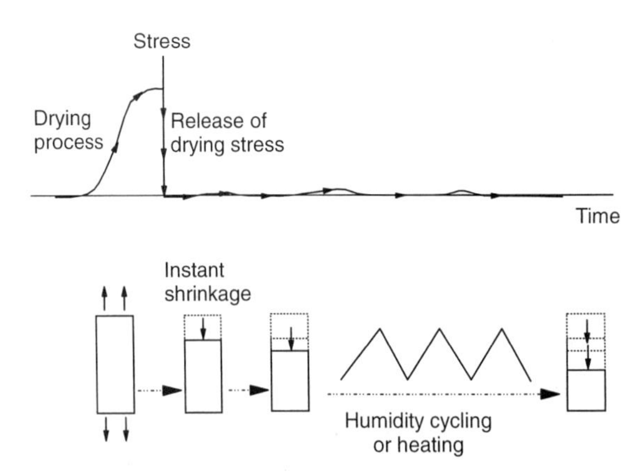

Fig. 18 Drying stress development and strain release. (a) Stress develops in paper during the drying process due to restraint drying or draw. After the paper is made, the stress is released and becomes nearly zero. (b) Associated strain change. After the release of the stress, the paper shrinks gradually. This shrinkage is accelerated by humidity/temperature cycling.

large draw is applied to paper, such as when newsprint is made on a high speed gap former, the paper sometimes shows shrinkage with increasing moisture content at the first exposure to a higher humidity [7].

To characterize the hygroexpansion of paper, we need to determine two aspects of the property: (1) irreversible shrinkage after the first high humidity exposure as a measure of the degree of drying restraint applied during the papermaking process and (2) reversible hygroexpansion in the low moisture content range as a measure of the inherent hygroexpansivity. Irreversible shrinkage is, however, a difficult property to determine because it depends on the time elapsed after the manufacturing of the paper, the whole humidity history between manufacturing and testing, and the highest moisture content achieved in the hygroexpansion test. Reversible hygroexpansivity can be characterized by determining the slope of the curve of hygroexpansion versus moisture content in the linear reversible range. The slope is called the *hygroexpansion coefficient*, a dimensionless parameter (percent strain/percent moisture content change) that is analogous to the thermal expansion coefficient. Relative humidity is sometimes used instead of moisture content to determine the hygroexpansion coefficient because of its convenience. In this case, the reversibility and the linearity should be carefully checked before testing, and the difference in the definition of the hygroexpansion coefficient should be noted.

Factors Affecting Hygroexpansivity In this section, effects of various papermaking factors on hygroexpansivity (hygroexpansion coefficient) are briefly summarized for readers who plan to conduct experiments of similar kinds. Detailed discussions can be found in the literature referenced below.

Increasing the wet pressing pressure, or increasing the degree of fiber–fiber bonding in general, is known to affect mechanical properties greatly, but hygroexpansivity responds differently [85,87,121]. Figure 19 shows such an example. The degree of bonding, as measured by specific elastic modulus, was changed either by

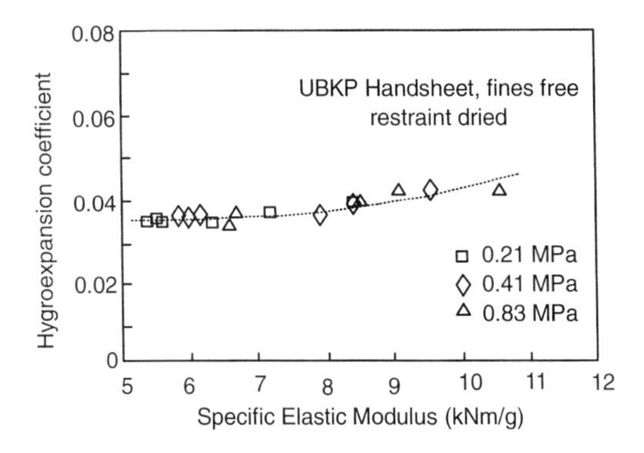

Fig. 19 Relationship between hygroexpansion coefficient and specific elastic modulus for fines-free restraint-dried handsheets of unbleached kraft pulp (UBKP) made from black spruce with various degrees of bonding (□) 0.21 MPa; (◇) 0.41 MPa; (△) 0.83 MPa. (From Ref. 121.)

wet pressing or by the use of a debonding agent. Hygroexpansivity increased only about 15% as the specific elastic modulus increased by almost 100%. This trend is most significant for sheets dried under restraint. When the measurement of hygroexpansion coefficient is extended to a higher humidity range, where irreversible shrinkage is involved, the apparent hygroexpansion coefficient sometimes shows a *decrease* with increasing density.

Among various papermaking variables, restraint applied during the drying process probably has the most significant effect on hygroexpansivity [11,38,73,87,88,121]. Figure 20 shows that hygroexpansivity for freely dried sheets is always higher than that for restraint-dried sheets. This difference increases with increasing sheet density; that is, higher density sheets exhibit more effect of drying restraint [121]. The effect of drying restraint on hygroexpansivity depends on the solids content at which the restraint is applied [88]. Nakayama et al. [69] extensively

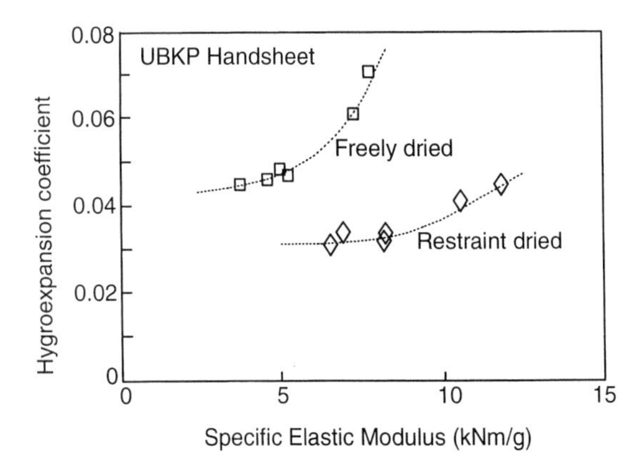

Fig. 20 Relationship between hygroexpansion coefficient and specific elastic modulus for (□) freely dried and (◇) restraint-dried handsheets of UBKP made from eastern Canada softwood species with varying degrees of fiber–fiber bonding. (From Ref. 121.)

studied the effects of drying restraint on hygroexpansivity, including both the hygroexpansion coefficient and the irreversible shrinkage. The hygroexpansion coefficient was found to be controlled by overall shrinkage between 40% solids content and the final dryness, whereas the irreversible shrinkage was mainly determined by the restraint (or draw) applied after a 70% solids content.

Fiber orientation in the sheet plane has another strong effect on hygroexpansivity [57,113,128]. With increasing degree of fiber orientation, as measured by the MD/CD ratio from a sonic modulus (extensional elastic stiffness) test, the CD hygroexpansivity increases dramatically, whereas the MD hygroexpansivity decreases slightly or is almost constant (Fig. 21). This fiber orientation effect depends on sheet density: Higher density sheets tend to be more affected by fiber orientation, particularly in the case of CD hygroexpansivity. In planning an experimental design for the effects of papermaking variables on hygroexpansivity, it is therefore crucial to always include sheet density as a variable.

The type of pulp does not have a great effect on the dimensional stability of paper in the dry state (hygroexpansivity), but it affects the wet-state dimensional stability (wet expansion and drying shrinkage). Table 2 shows that there is no consistent difference in hygroexpansivity between the mechanical pulps and chemical pulps, but drying shrinkage values for the mechanical pulps are much lower than those for chemical pulps. The difference results from the fact that mechanical pulps tend to have lower water retention values than chemical pulps. The wet dimensional stability properties, such as shrinkage and wet expansion, are determined mainly by the change in water content from the wet state to the dry state [71]. (It should be noted that much of the effect of pulp type is a combination of the effects of fiber shape, sheet density, fines content, and short fiber fractions, as described in this section.)

Web shrinkage in the cross-machine direction (as seen on the paper machine) is a function of shrinkage due to moisture loss, restraint applied by a dryer felt, and the draw applied. (With increasing draw in the machine direction, the web shrinks in the

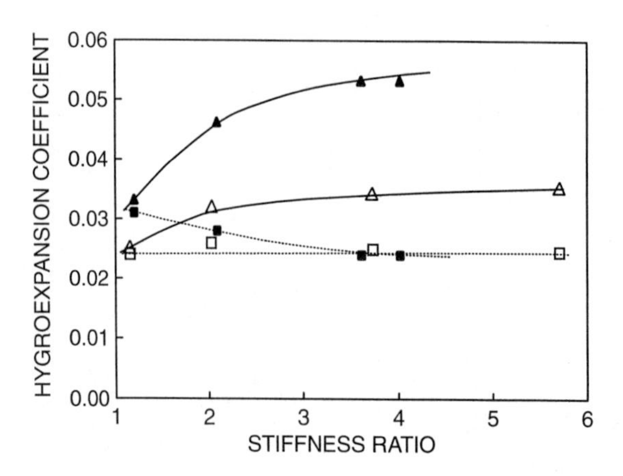

Fig. 21 Effect of fiber orientation on hygroexpansivity for restraint-dried sheets made from a fines-free, unbeaten, unbleached softwood kraft pulp. (- - -) Machine direction; (——) cross-machine direction. (▲, ■) High density: 419 kg/m^3. (△, □) Low density: 220 kg/m^3. (From Ref. 113.)

Table 2 Effects of Type of Pulp on Dimensional Stability[a]

	SGW	TMP	CTMP	LYS	Beaten LYS	Beaten BKP
Hygroexpansion coefficient (restraint-dried)	0.068	0.057	0.063	0.048	0.063	0.053
Hygroexpansion coefficient (freely dried)	0.086	0.074	0.087	0.075	0.123	0.092
Drying shrinkage (%)	1.50	1.40	1.99	2.35	4.60	3.23

[a]SGW; stone groundwood pulp; TMP, thermomechanical pulp; CTMP, chemi-thermo-mechanical pulp; LYS, low yield sulfite pulp; BKP, bleached kraft pulp.
Source: Ref. 71.

cross-machine direction by the tension applied in the machine direction.) A single felted dryer with vacuum rolls was shown to provide significant drying restraint, causing a reduction of drying shrinkage [132]. Web shrinkage generally has a strong correlation with wet expansion. However, the amount of web shrinkage that occurs in the wet part between the wire section and the second press is caused mainly by the wet draw and does not have a correlation with wet expansion. Todoroki et al. [111] demonstrated that the web shrinkage that occurs between the press and the reel controls the wet expansion.

Among furnish variables, the most important one affecting dimensional stability is fines content [71,85]. Figure 22 shows percentage increases in hygroexpansivity due to fines addition for three different pulps. A 10% fines addition increases hygroexpansivity about 10–20%, depending on the type of pulp. A well-known detrimental effect of beating or refining on hygroexpansivity [11] is caused mainly by this increase in fines content. Within the same furnish, shorter fiber fractions have significantly higher hygroexpansivity than longer fiber fractions [119]. Results of studies with cut fibers, however, show that hygroexpansivity increases very marginally with

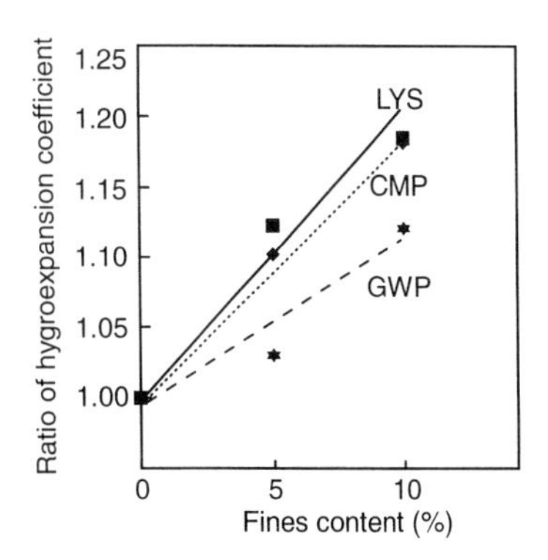

Fig. 22 Effect of fines on hygroexpansivity. LYS, low yield sulfite; CMP, chemimechanical pulp; GWP, groundwood pulp.

decreasing fiber length [119]. The fiber length effect seen for the fractionated pulps is attributed mainly to differences in the fiber wall structure rather than the geometric length [119].

Chemical modifications of cellulose, especially modification of hydroxyl groups, have been developed to improve dimensional stability [59,60,68,100,101]. Although these treatments have the capability to drastically change the wet dimensional stability, they often affect strength properties in negative ways. Wet-end additives and surface sizing agents have relatively minor effects on hygroexpansivity [57]. Fillers, including clay and calcium carbonates, tend to decrease hygroexpansion, but only at high addition levels, as shown in Fig. 23.

Hygroexpansion coefficients are often compared with thermal expansion coefficients. The ranges of values reported so far for these coefficients (for various papers) are

MD: 0.020–0.060; CD: 0.030–0.180

and for thermal expansion coefficient ($\times 10^{-6}\,K^{-1}$) [55],

MD: 2.0–7.5; CD: 7.9–16.2

That is, a 10% change in moisture content induces 0.2–1.8% hygroexpansion, whereas a 100°C temperature change causes 0.02–0.16% thermal expansion.

Structural Models to Explain Hygroexpansivity It is desirable to have an appropriate structural model for interpreting experimental results of hygroexpansivity and for designing new paper structures to optimize hygroexpansivity. A first attempt was made by Salmén et al. [86], who modeled paper as a laminated structure of the cell wall layers. The S2 wall layer was modeled as a unidirectional fiber-reinforced composite consisting of cellulosic microfibrils and a hemicellulose matrix. Its hygroexpansion coefficients were calculated by using Schapery's estimation of thermal expansion coefficients for a fiber-reinforced composite [91]. It was shown that the predicted anisotropy ratios (CD/MD) of hygroexpansivity are higher than the elastic anisotropy ratios (MD/CD), confirming some experimental observations. Later, Laurell Lyne [57] showed that the hygroexpansion coefficients calculated from this model are consistently higher than those experimentally determined. Although

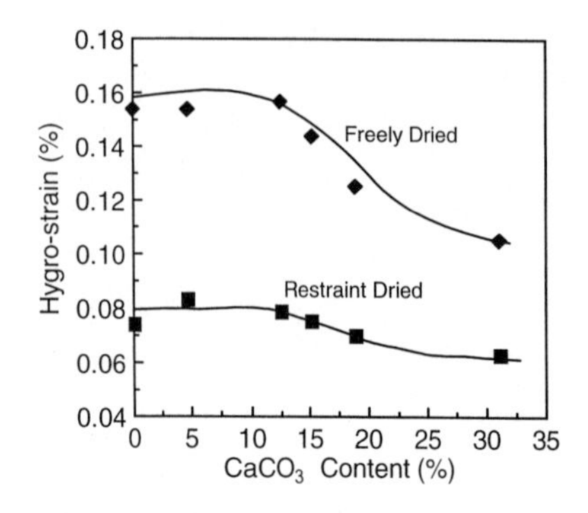

Fig. 23 Effect of filler content ($CaCO_3$) on hygroexpansion for freely dried and restraint-dried handsheets made from a bleached softwood kraft pulp.

the model still requires refinements, it is the first attempt to relate the properties of fiber constituents (cellulose microfibrils and hemicellulose) to paper hygroexpansivity.

Schulgasser [92,93] obtained an exact relationship between thermal (or hygro) expansion coefficients and elastic compliances of polycrystalline aggregates. The polycrystalline aggregates can be anisotropic, and therefore the model is considered applicable to high density papers with "no pore." He showed that under certain conditions the thermal expansion in a given direction can be inversely proportional to the elastic modulus in that direction. The prediction conforms to an earlier experimental finding of an inverse relationship between elastic modulus and the thermal expansion coefficient of paper [24]. However, a more recent study on the effects of anisotropy on hygroexpansivity did not show the inverse proportionality [113].

Perkins et al. [78] proposed an infinite fiber length model with a transverse fiber effect for moderate and high density sheets. The final expressions obtained for hygroexpansion can be further explored to depict effects of various fiber and network parameters on hygroexpansivity.

A general formula for hygroexpansion was derived by Uesaka [113], using an equation obtained by Dvorak and Benveniste [27]. Paper is modeled as a random three-dimensional fiber network. No special assumptions are made concerning fiber length and fiber shape (curliness). The physical parameters involved in the formula have clear physical significance, so one can interpret experimental results and predict effects of fiber properties and paper structure in a semiquantitative way. For example, in the case of transversely isotropic sheets, such as handsheets, the formula for in-plane hygroexpansion can be written as

$$\mu^*_{\text{in-plane}} = \mu^f_L + \gamma(\mu^f_T - \mu^f_L) \tag{3}$$

where μ^f_L and μ^f_T are hygroexpansions of a fiber (or a fiber segment when the fiber is curved) and γ is a parameter representing the relative efficiency of the stress transfer to the axial and transverse directions of the fiber, defined as

$$\gamma = \frac{\langle \sigma^f_{22} \rangle}{\langle \sigma^f_{11} \rangle} \tag{4}$$

where $\langle \sigma^f_{22} \rangle$ is an average stress transferred in the transverse direction of the fibers in the sheet when the sheet is subjected to uniaxial tension (Fig. 24), and $\langle \sigma^f_{11} \rangle$ is the corresponding stress in the fiber axis direction. The parameter γ is generally a function of fiber–fiber bonding, fiber (segment) orientation in the three dimensions, and other structural parameters. If the fibers become infinitely slender or highly anisotropic, the transverse stress component becomes negligibly small compared with the axial stress component ($\gamma \rightarrow 0$), and therefore

$$\mu^*_{\text{in-plane}} \cong \mu^f_L \tag{5}$$

Thus, for the $\gamma = 0$ condition hygroexpansion of paper is equal to the hygroexpansion of a fiber in the fiber axis direction. This condition creates a lower bound of the hygroexpansivity.

Because the transverse hygroexpansion of a fiber (μ^f_T) is expected to be much higher than the axial hygroexpansion (μ^f_L) [76,77], it is often speculated that the in-

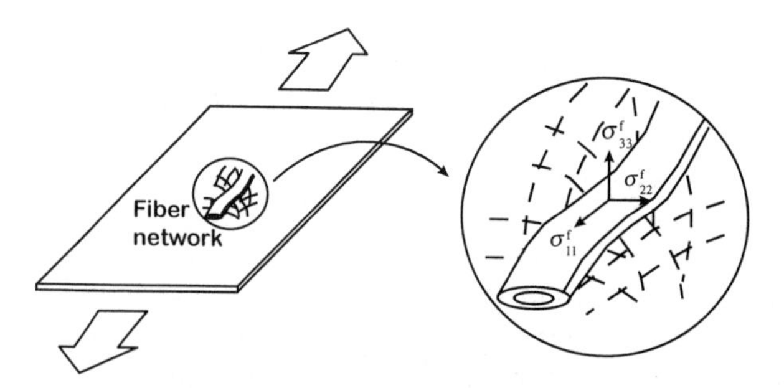

Fig. 24 Stresses transmitted to a typical fiber segment in the sheet under uniaxial tension. σ^f_{11} is the stress transmitted in the fiber axis direction; σ^f_{22}, the stress in the fiber width direction, and σ^f_{33}, the stress in the fiber thickness direction.

plane hygroexpansion of paper is dominated by the transverse hygroexpansion. However, some of the results seen in the previous section, such as the small effect of sheet density on hygroexpansivity for restraint-dried sheets, cannot be consistently explained by this speculation. According to Eq. (3), the effect of the transverse hygroexpansion of a fiber on paper hygroexpansion depends on the magnitude of the parameter γ. Uesaka and Qi estimated the magnitude of the parameter γ from both a shear-lag type of analysis [112] and finite element analysis [121]. When all fiber segments are aligned in the sheet plane direction, such as is shown in Fig. 25a, γ was found to be much smaller than 0.1. This result implies that even though the effect of fiber–fiber bonding on hygroexpansivity still exists, its contribution to paper hygroexpansivity [the second term of Eq. (3)] is very small in this case, and therefore the effect of wet pressing and fiber bonding on hygroexpansivity becomes minor. This also explains why the hygroexpansivity of restraint-dried handsheets is not sensitive to the change in the degree of fiber bonding (Fig. 19). However, when "out-of-plane orientation" of fiber segments is created in the bonded region, such as is shown in Fig. 25b, γ is increased drastically (by more than tenfold), and thus the hygroexpansivity of paper is increased [121]. The higher value of γ means that the effect of the degree of fiber bonding becomes more visible. This out-of-plane orientation effect explains why freely dried sheets show not only higher hygroexpansivity

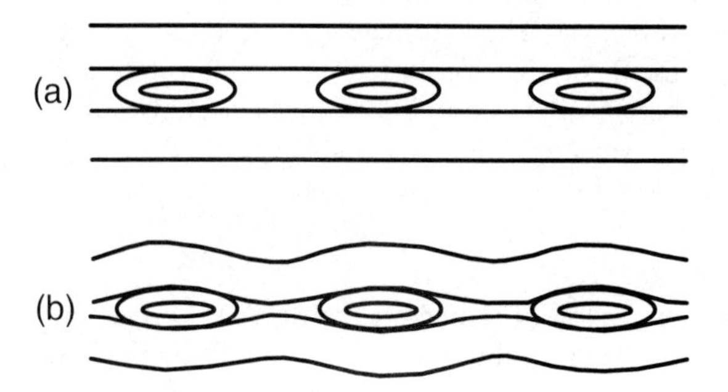

Fig. 25 Models for fiber segment orientations. (a) Fiber segments are oriented in the sheet plane direction; (b) a wavy structure is due to out-of-plane orientation of fiber segments.

but also more sensitivity in the hygroexpansivity to changes in fiber bonding than that exhibited by restraint-dried sheets (Fig. 20). One can also understand why the CD hygroexpansivity is more sensitive to the sheet density than the MD hygroexpansivity (Fig. 21).

Lu et al. [64–66] proposed a mosaic model for predicting hygroelastic properties of paper. A cell wall (S2 layer) model was first constructed from cellulose microfibrils embedded in a matrix of hemicellulose and lignin, and the hygroelastic properties of the cell wall were estimated using the Schapery–Salmén approach [86,91]. Paper is then modeled as a periodic array of mosaic elements consisting of a free segment material and a bond material. The overall response was calculated by a plane stress, finite element analysis. Substantial experimental scatter, however, prevented detailed comparisons with the prediction.

D. Curl

Curl is another type of dimensional instability phenomenon of paper. (In the paperboard and corrugated board field, the phenomenon is often called *warp*.) Although the basic cause of curl is simply a differential deformation of paper in its thickness direction, i.e., the in-plane deformation varies in the thickness direction, the processes causing the differential deformation are complex and variable, as will be described below.

Types of Curl Various types of curl typically seen in paper are shown in Fig. 26 [35,54]. The first two types are forms of cylindrical curl in which the axis of the cylinder, called the curl axis, is oriented either in the machine direction or in the cross-machine direction. Dependent on the side of the curvature (concave side), each form is further classified as top side (TS; felt side) or bottom side (BS; wire side) curl. For example, when the paper shows curl with the concave side on the bottom side and with the curl axis in the machine direction, the curl is designated as MDBS (or

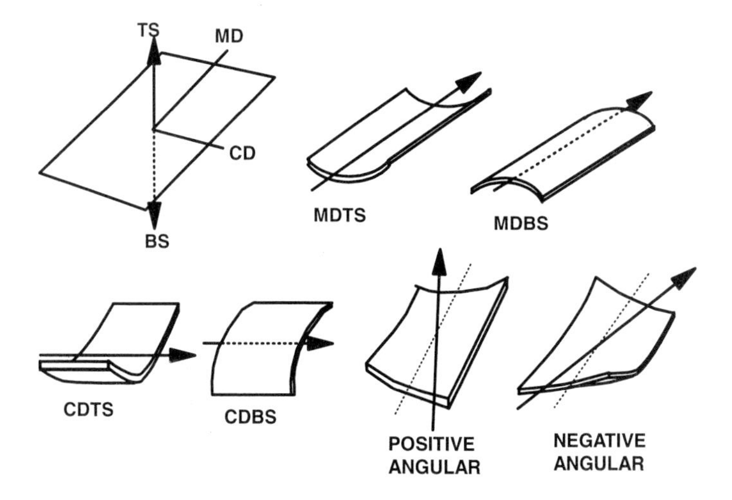

Fig. 26 Classification of different types of curl. (Based on Ref. 54.)

MDWS) curl. (In Europe, there is a different tradition: The same curl is designated as CDBS curl by focusing on the "direction" of the curvature rather than the curl axis.) The third type of curl is angular curl, according to the definition by Glynn et al. [35], where the curl axis is neither the machine nor the cross-machine direction. Dependent on the direction of the curl axis with respect to the machine direction, the angular curl is further classified as a positive angular or negative angular curl, as shown in Fig. 26. The types of curl shown in Fig. 26 do not necessarily represent accurate shapes of real curl but contain most of the important curl forms seen in practice. A more accurate description of curl shape is given in Section III. In the diagnosis of curl, it is extremely important to identify the curl axis and the concave side of the curl (top or bottom side), because each of the three types of curl has a different cause, as described later.

As with hygroexpansion behavior, curl depends on the moisture history to which the paper has been subjected. Figure 27 shows an example of curvature changes of fine paper subjected to cyclical humidity change [115]. As the sample is first brought to a higher moisture content, it develops top side curl (positive curvature). The subsequent moisture desorption process, however, further increases the top side curl. The repeated sorption and desorption processes gradually make the curvature change more reversible. On the other hand, Fig. 28 shows an example for the case in which the humidity cycling starts with desorption [115]. In both cases, the paper showed both a relatively reversible curvature change and an irreversible curvature change, the latter tending to be triggered by a higher humidity excursion similar to the hygroexpansion behavior. The transition toward reversible curl behavior depends on the type of paper; coated papers and film-laminated papers tend to require more humidity cycling to exhibit the reversible behavior.

Another important characteristic of curl is that the manifestation of a specific type of curl depends on the shape or size of the sample. For example (Fig. 29), paper may show simple cylindrical curl (MDTS) in the cut-size sheet form with the curl axis in the machine direction and the concave side on top. However, as the width of the sample becomes narrower, the paper can show a different curl shape (CDBS) with

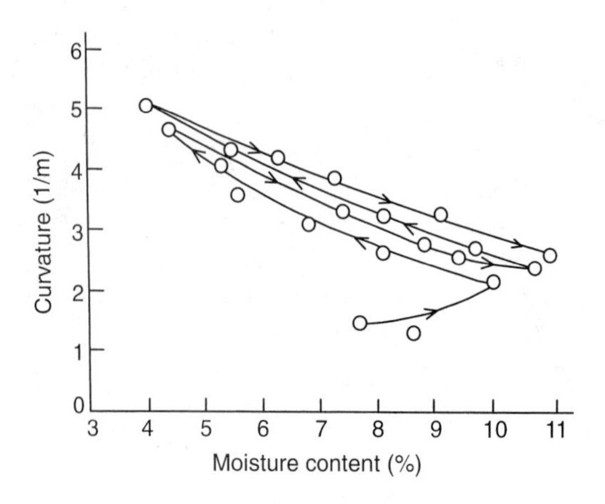

Fig. 27 Bending curvature response of fine paper to cyclical humidity changes when the process starts from the adsorption stage. (From Ref. 115.)

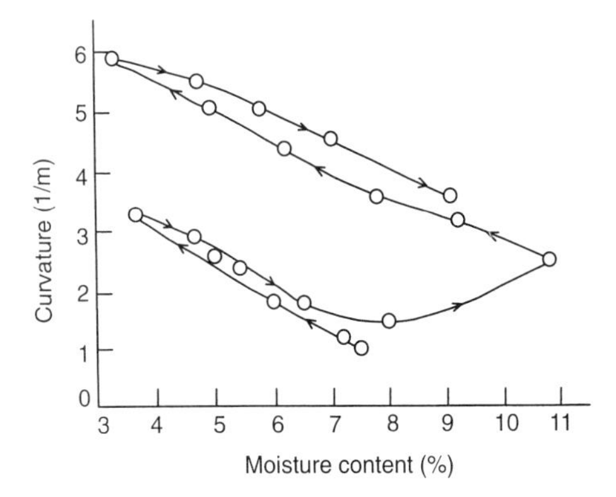

Fig. 28 Bending curvature response of fine paper to cyclical humidity changes when the process starts from the desorption stage. (From Ref. 115.)

the curl axis in the cross-machine direction and the concave side on the bottom. This change of curl shape with the specimen size or shape has the following reason. Paper usually has some hidden curvature components in the machine or cross-machine direction. Whether or not these curvature components become visible depends on the relative stiffness in the machine and cross-machine directions and on the size and shape of the sample. Paper tends to exhibit only the most stable curl at a given size and shape by minimizing the total strain energy required for curling. This stability of curl is well known in cross-ply laminates [48–50]. A cross-ply laminate normally exhibits a "saddle"-shaped curl when the specimen size is small. That is, the concave side is different in the two principal directions. As the size increases, reaching a critical value, the saddle-shaped curl suddenly becomes a cylindrical curl, which is energetically more stable than the saddle-shaped curl. Paper curl shows the same size and shape dependence [74].

Theories of Curl

Structural Curl (Reversible Curl) Curl in paper and paperbroad is generally dependent on the histories of moisture and temperature in the thickness direction of the sheet, as discussed earlier. Reversible curvature change is caused by structural nonuniformity of the sheet in the thickness direction and can be described by lamination theory. Carlsson [17] first formulated general out-of-plane dimensional instability phenomena, including cylindrical curl and twisting curl, using classical lamination theory. In his formulation, paper (or paperboard) is modeled as a multi-ply laminate.

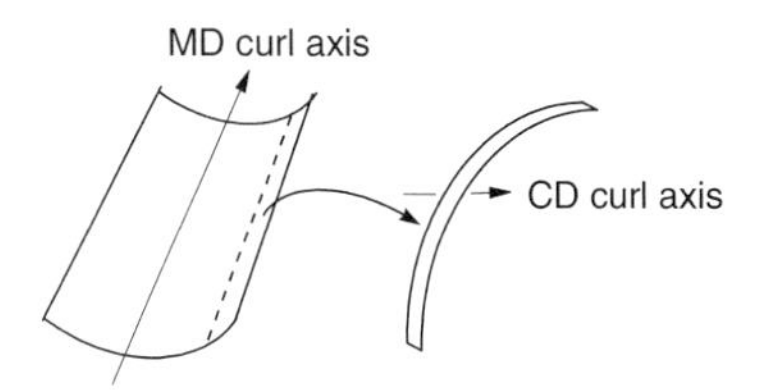

Fig. 29 Paper curl depends on the shape and size of a specimen.

Each ply has different hygroelastic properties. The model is essentially for a general multi-ply composite. Therefore the result can be equally applied to curl in coated papers, multi-ply boards, and laminated papers as long as the hygroelastic (or thermal-elastic) properties of each ply can be estimated.

Although the final expression for curl (three curvature components) is given in matrix form, it can easily be implemented in a computer program for conducting parametric studies. Attempts have been made to derive approximate formulas from the original equations [18,79,116]. Assuming a two-ply structure with plies of equal thickness (Fig. 30), Uesaka et al. [116] obtained approximate curl relations for two-sided papers. These formulas were obtained by taking only the first-order terms of the original equation:

$$k_1 = \frac{3}{2t}(H^B\beta_1^B - H^T\beta_1^T) \tag{6}$$

$$k_2 = \frac{3}{2t}(H^B\beta_2^B - H^T\beta_2^T) \tag{7}$$

$$k_3 = \frac{3}{t}\{\theta^B H^B(\beta_2^B - \beta_1^B) - \theta^T H^T(\beta_2^T - \beta_1^T)\} \tag{8}$$

where k_1 and k_2 are the curvature components in the x_1 (cross-machine) and x_2 (machine) directions, respectively, k_3 is the twisting curvature component, H is the change in moisture content of each layer, β is the hygroexpansion coefficient, θ is the misalignment angle of the principal direction of the property from the machine direction, and the superscripts T and B denote the top and bottom halves of the two-ply sheet with thickness t.

Expressions (6) and (7) show that the cylindrical curl is simply caused by the two-sidedness of the hygroexpansion of the sheet, as a first-order approximation. The elastic property effects exist only in the higher order terms [116]. The twisting curl (k_3), on the other hand, is a function of the two-sidedness of the orientation angle θ, the degree of anisotropy ($\beta_2 - \beta_1$), and the anisotropy two-sidedness. If anisotropy two-sidedness is insignificant, Eq. (8) for the twisting curl can be further simplified as

$$k_3 = \frac{3}{t}H(\beta_2 - \beta_1)(\theta^B - \theta^T) \tag{9}$$

That is, the twisting curl increases as the degree of anisotropy increases and the difference in the misalignment angle increases. (However, it is often the case in practical papermaking that the misalignment angle increases with decreasing anisotropy ratio if these parameters are controlled only by the jet–wire speed differential.)

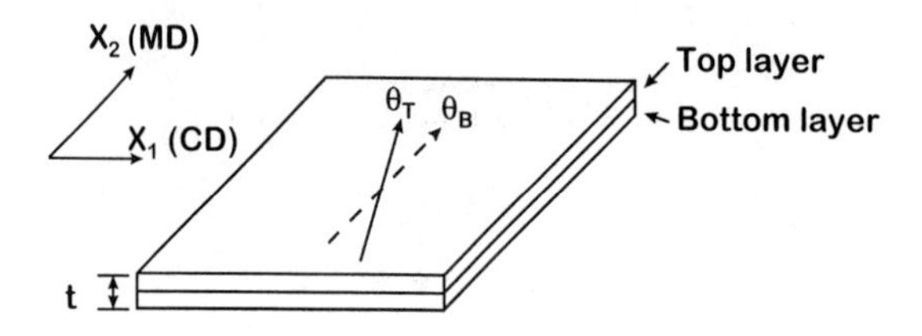

Fig. 30 A two-layered model of paper. Symbols are explained in text.

The classical lamination theory presented above is not accurate when the paper curl becomes large, because the theory assumes small out-of-plane deflection compared with the paper thickness. Using large deformation analysis, Nordström et al. [74] demonstrated that a normal size paper specimen does not show the saddle-shaped curl often predicted by the classical lamination theory, but a cylindrical curl. However, as was also found by Hyer [48–50], the curvature of the cylindrical curl (the curvature along the curl direction) was shown to be still very close to the value predicted from classical lamination theory.

Curl as a Result of Differential Dried-In Strain Release (Irreversible Curl) As shown in Figs. 27 and 28, when paper is exposed to high humidity (i.e., paper moisture content is raised), it shows a large irreversible change in its curvature. This type of curl results from the differential release of dried-in strain from the top and bottom sides of the sheet, similar to the irreversible shrinkage of paper seen in hygroexpansion behavior under a cyclical humidity condition (Fig. 16). During the papermaking process, paper is kept flat by the mechanical restraints exerted by drawing and by the dryer canvas. These restraints cause different (pseudo permanent) strains in the top and bottom sides of the sheet, which are later released (viscoelastic strain recovery) by exposure to high humidity or high temperature.

The origin of curl from differential dried-in strain is mainly the two-sided nature of the structure. This type of curl may be considered the latent component of the structural curl, originating from the wet end of the papermaking process. In addition, differential dried-in strain is created in the drying process, as will be discussed below.

Curl Caused by Different Moisture/Temperature Histories Between the Top and Bottom Side of the Sheet (Viscoelastic Curl) It is a general experience that paper often shows persistent curl, irrespective of its structural two-sidedness, when the top and bottom sides of the sheet are subjected to different moisture and/or temperature histories. Figures 31 and 32 illustrate this type of curl, where paper is assumed to be a viscoelastic body with no structural two-sidedness [118]. In this example, paper with a 5% moisture content is first exposed to a higher humidity, then gains moisture up to 15% on the bottom side of the sheet, is kept flat for 15 s, and is then exposed to the original environment without any restraint to regain the initial uniform moisture distribution through the thickness (Fig. 31). In this process, the paper first shows a large top-side curl because of the higher moisture content of the bottom side (Fig. 32). However, as the drying proceeds, it reduces the top-side curl and finally produces a large residual curl toward the bottom side, particularly in the cross-machine direction (x direction).

This type of curl is a manifestation of the viscoelastic relaxation of the hygro stress, which is caused by the elastic constraint within the sheet and/or by restraint of

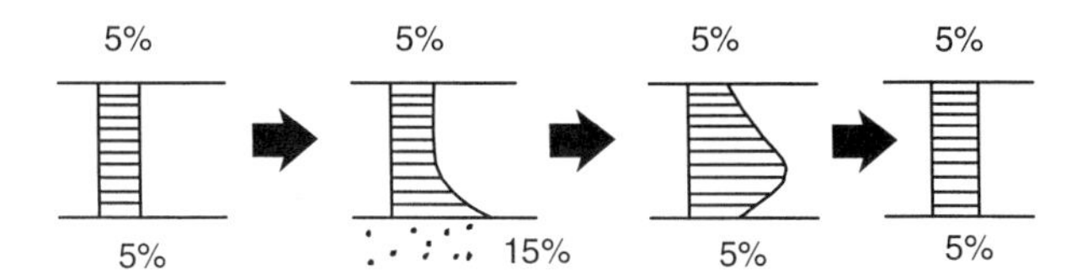

Fig. 31 Sequential moisture schedule for the simulation example of Fig. 32. (From Ref. 115.)

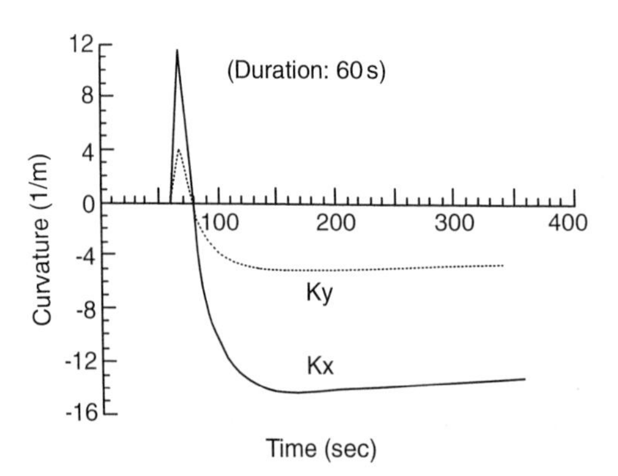

Fig. 32 Predicted curvature response to a nonuniform exposure of the top and bottom sides of the sheet to moisture. Ky: Machine direction curvature, Kx: Cross machine direction curvature. (From Ref. 115.)

bending deformation during the process. A parametric study showed that the major factors controlling this type of curl are the overall hygroexpansivity (not two-sidedness) and the amount of change in moisture content that takes place during the process [118].

Roll-Set Curl Roll-set curl is another type of curl related to viscoelastic bending stress relaxation. In this case, paper wound on a small diameter core or paper subjected to a decurler in the finishing operation exhibits curl. Because this type of curl is caused by viscoelastic relaxation of the bending stress or permanent deformation, higher temperature and higher moisture content both accelerate the roll-set process.

Curl in Practical Situations Curl observed in practical situations, such as drying, coating, printing, and electrophotographic processes, is generally a combination of the types of curl discussed above. For testing, diagnosing, and controlling curl, it is important to understand the types of curl involved in the practical process in question.

Drying Curl In the multicylinder dryer section, drying of the top and bottom sides of the sheet at different rates is known to cause curl. The papermakers' wisdom that paper curls toward the side that was dried last [34,36,126] reflects a typical manifestation of the curl caused by the viscoelastic relaxation of the hygro stresses. As shown in Fig. 33, the temperature gradient created by differential drying results in a moisture gradient (higher moisture on the lower temperature side), causing a process similar to the development of hygro stress seen in Figs. 31 and 32. However, if the drying temperature is low or the drying process is slow, so the moisture gradient is not large enough to create a hygro stress gradient, this drying curl may not be seen [39,40]. Papermakers often try to use this differential drying to counteract other curl tendencies created in the wet end of the paper machine. However, the structural curl tendency, which is important in electrophotographic printing processes, may not be easily changed by this action. Differential drying is most effective in influencing the irreversible curl tendency, as described earlier [126].

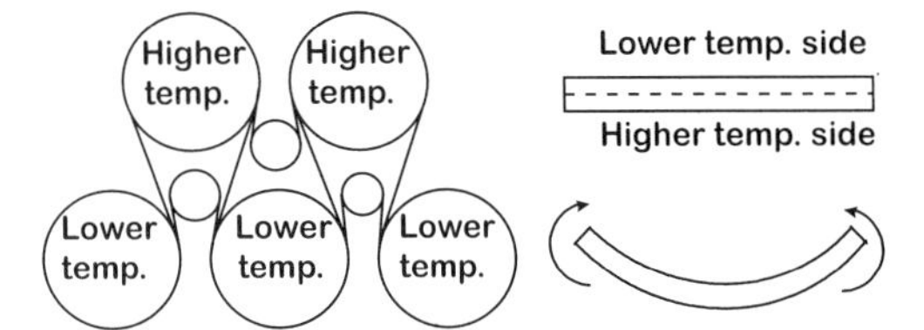

Fig. 33 Curl caused by differential drying in the multicylinder dryer.

Coating Curl In coating processes, the coating layer shrinks with increasing coating consolidation. This shrinkage creates strong curl tendencies toward the coating side (Fig. 34). The magnitude of the shrinkage depends on the type and amount of the binders used. To correct coating curl, a common practice is to remoisturize or coat on the back side of the sheet [61]. Such corrective actions are typical examples of applications of viscoelastic curl. Because the coating process involves wetting and drying, the structural curl and irreversible curl described in the previous section can be used to counterbalance the coating curl.

Offset Printing Curl In the offset printing process, paper picks up water from the blanket through the fountain solution and ink (water-emulsified ink). The paper is wetted on one side and then dried. The paper initially curls toward the drier side when the water is applied and later curls toward the other side (Fig. 35) [110,126]. This is another manifestation of the differential dried-in-strain release (irreversible curl) and the viscoelastic relaxation of the hygro stress (viscoelastic curl). The latter type of curl often becomes severe as the atmosphere in the press room becomes drier in winter. Control of this type of curl mainly depends on changing the structural curl in the wet end and the irreversible curl in the dryer. Waech [126] demonstrated that differential drying can counteract offset printing curl by the use of a pilot dryer.

Fusing Curl In an electrophotographic process, paper is heated on one side by a fuser roll to fix the toner image (Fig. 36). This heating results in moisture flow through the thickness of the sheet and an overall moisture loss. Therefore, fusing curl is a combination of viscoelastic curl, which tends to be toward the anti-image side, and structural curl that depends on the degree of sheet two-sidedness and moisture content. When the tendencies of viscoelastic curl and structural curl happen

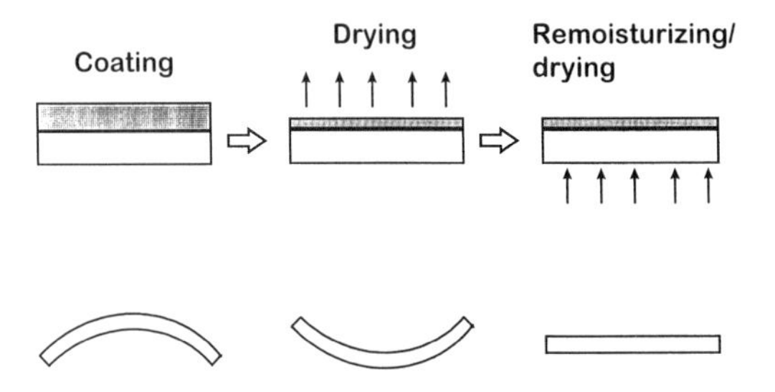

Fig. 34 Coating-induced curl. Left to right: Step 1, moisture in coating layer expands upper side; step 2, coating layer consolidates and shrinks; curvature reverses; step 3, curving is compensated for via remoisturizing.

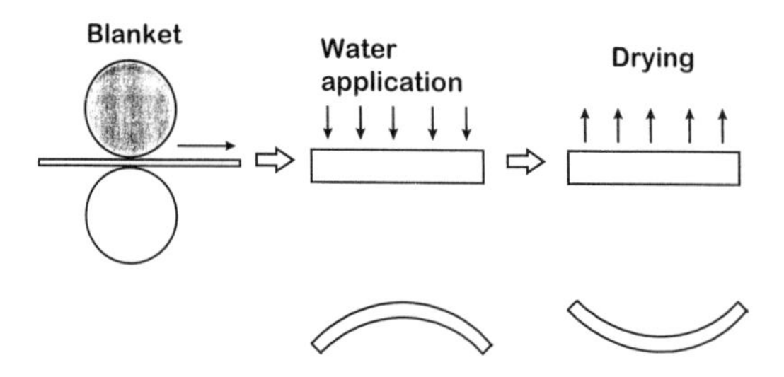

Fig. 35 Curl in offset printing.

to coincide with each other, a very large curl may be observed. Again, control of this type of curl depends mainly on control of the structural curl rather than a differential drying strategy. Basic factors of paper that affect fusing curl have been documented in the literature [37,39,40,47].

Papermaking Factors Affecting Curl The most important papermaking factor determining the extent of curl is two-sideness in the fiber orientation. Glynn et al. [35] demonstrated that there is indeed two-sideness of fiber orientation in paper made on a Fourdrinier machine (Fig. 37). (Determination of two-sidedness of fiber orientation has been done for many different types of former. Interested readers can refer to Chapter 16 in this volume.) In Fig. 37, the bottom side (wire side) shows a higher degree of fiber orientation than the top side. It has been shown by many researchers that two-sidedness in fiber orientation causes a corresponding two-sidedness in the hygroexpansion coefficient, resulting in structural curl [40,82,116]. An important characteristic of this fiber-orientation-induced curl is that strips cut from the machine and cross-machine directions exhibit curl toward opposite sides (see Fig. 38). This is a unique characteristic of the type of curl caused by two-sidedness in the fiber orientation and can be used for diagnosing the cause of structural curl.

When the principal directions of fiber orientation of the top and bottom sides are not the machine and cross-machine directions, paper tends to experience twisting curl as predicted by Eq. (9). This relation has been demonstrated experimentally by Carlsson [17] and by Whitsitt and Hoerschelman [130]. Twisting curl is most often seen in sheets taken from near the edge of the paper machine, because of the higher tendency toward off-axis orientation of the jet flow in that region.

Two-sidedness in fiber orientation (and thus curl) changes with the jet-to-wire speed ratio. Figure 39 shows an example of the curl response for a gap former [96]. In

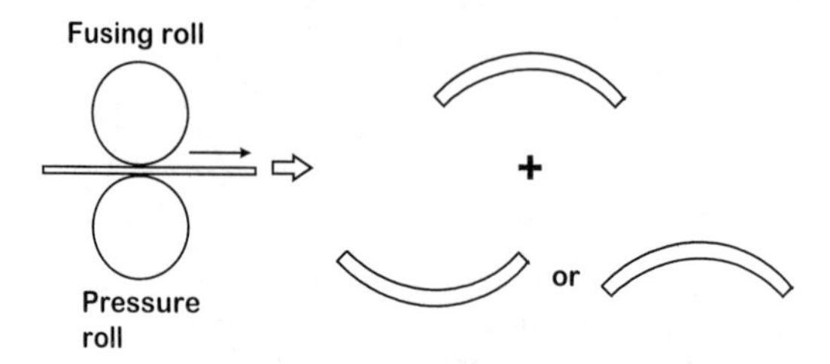

Fig. 36 Fusing curl in electrophotography.

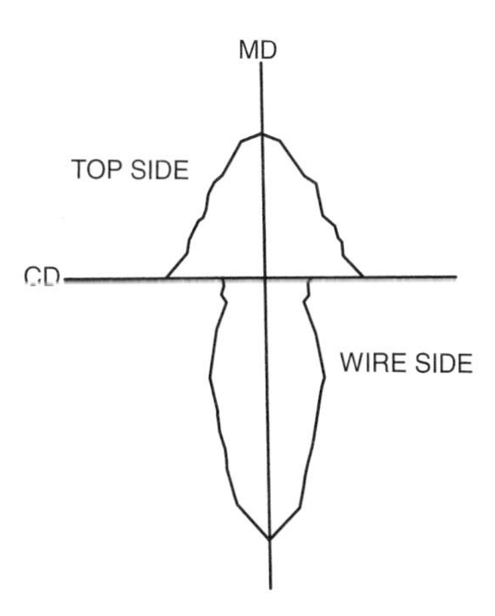

Fig. 37 Fiber orientation distribution on top and wire sides of paper in a Fourdrinier machine. (From Ref. 35.)

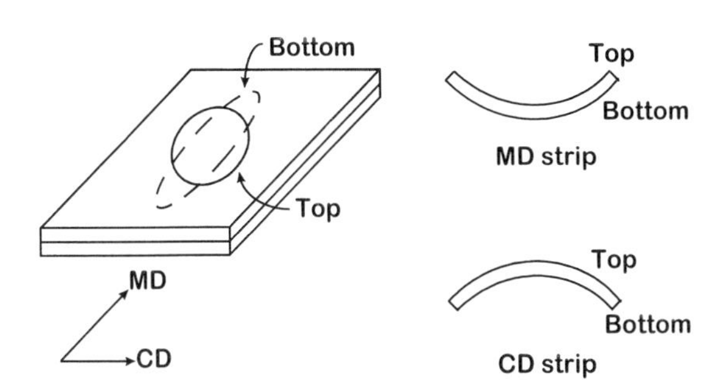

Fig. 38 Curl caused by two-sidedness of fiber orientation. In the example shown, the fibers on the bottom side are statistically more oriented in the machine direction (dashed ellipse) than the fibers on the top side (solid ellipse).

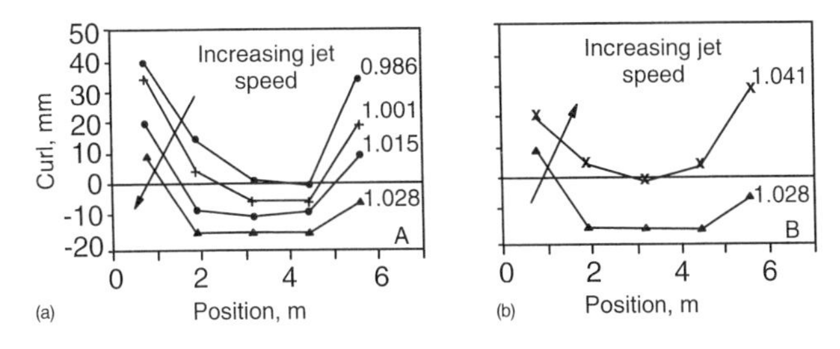

Fig. 39 Cross-machine curl profile for different jet/wire speed ratios. Wire speed 814 m/min. (a) Jet/wire speed ratio = 0.986, 1.001, 1.015, and 1.028. (b) Jet/wire speed radio = 1.028 and 1.041. Positive sign: curl toward side 1; negative sign; curl toward side 2. (From Ref. 96.)

this machine, the areas of the sheet near the edges have a strong tendency to curl toward the positive side (side 1), suggesting differential drying in the drying process. The overall curl level changes with the jet speed, as expressed by the jet/wire speed ratio; as the ratio increases from 0.986 to 1.028, the paper curl goes toward side 2. A further increase in the jet/wire speed ratio from 1.028 to 1.041 reverses the trend of the curl, and it heads toward side 1. A similar response has been observed for many gap formers and Fourdrinier-type machines with different turning points of the jet/wire speed ratio at which the curl trend is reversed.

Another factor affecting curl is two-sidedness in fines and filler distributions (see Fig. 40) [28]. The distributions of fines and fillers through the thickness are highly dependent on the type of former (drainage elements and their configuration) and the type of furnish. The fines-rich side tends to have higher hygroexpansivity than the other side, and the filler-rich side has slightly lower hygroexpansivity. Density two-sidedness is created in the press section. The side from which the water is removed (under dynamic conditions) tends to develop higher density than the other side (Fig. 41) [102]. This increasing density tends to increase hygroexpansivity on that side, but the effect is more evident in a CD strip than in an MD strip (Fig. 42). This tendency occurs because the CD hygroexpansivity is more sensitive to changes in the degree of fiber–fiber bonding (density) than the MD hygroexpansivity, as discussed earlier. The fact that MD and CD strips show different types of curl (Fig. 42), dependent on the causes, is being used effectively for curl diagnosis.

E. Cockle and Wrinkle

Cockle and wrinkle are other out-of-plane deformations of paper, but unlike curl, the deformations occur irregularly in the sheet plane. Cockle and wrinkle are sometimes seen well after drying has been completed, when the sheet is rewetted and dried without any restraint, or when the sheet is subjected to the fusing process of electrophotography. The shape and size of these deformations vary. The shape is often elongated toward the machine direction. A round deformation is often called cockle, and a highly elongated cockle is sometimes called wrinkle. The size varies from a few millimeters to ≈ 30 cm. Wrinkles discussed in this chapter differ in nature from other

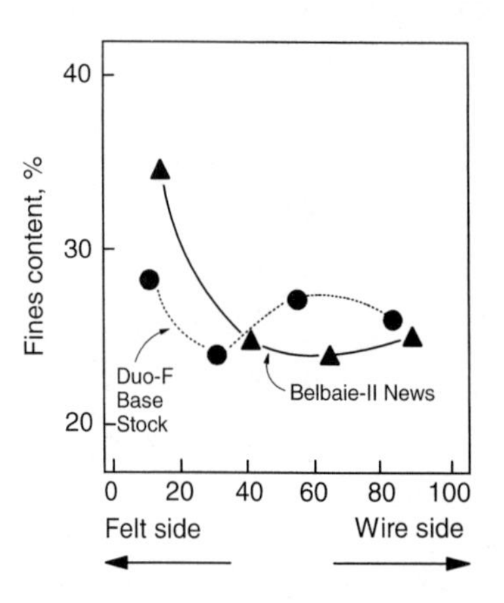

Fig. 40 Two-sidedness in fines content. (From Ref. 28.)

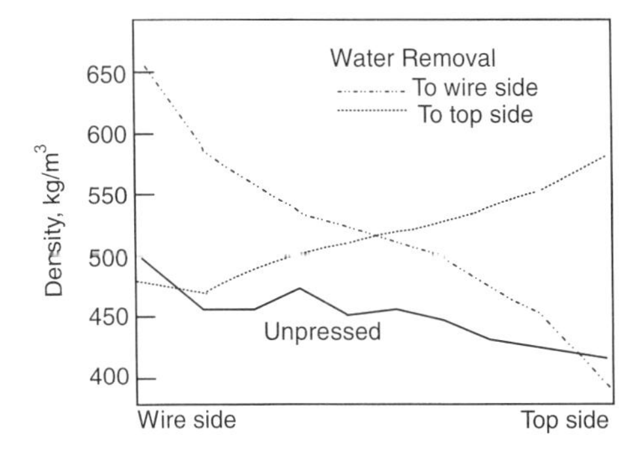

Fig. 41 Two-sidedness in density created by pressing. (From Szikla and Paulapuro, Ref. 102.)

wrinkles created by purely mechanical stresses, such as crepe wrinkle and tension wrinkle. In this chapter, we collectively call the out-of-plane deformations caused by dimensional instability "cockle."

The mechanism that creates drying cockle was first proposed by Smith [98] and demonstrated experimentally by Brecht et al. [13]. When there is moisture nonuniformity in the wet sheet (as shown in Fig. 43), the drier area starts shrinking earlier than the wetter area (shaded area) during the drying process. At this early stage of drying, because the rest of the area (wetter area) has not yet started shrinking, the drier area is under tension, which results in a small viscoelastic/plastic tensile deformation (dried-in strain). As drying proceeds and the moister area starts shrinking, the drier center part starts receiving compressive stresses from the surroundings. Because the paper is thin, this compressive stress causes local buckling, forming cockle.

Kajanto [52] found correlations between the height of cockle and formation, as changed by headbox consistency ($r = 0.85$), and between the cockle intensity and the local two-sidedness of fiber orientation angle, as affected by the jet/wire speed ratio ($r = 0.71$). He proposed two mechanisms for cockle: local buckling and local curl.

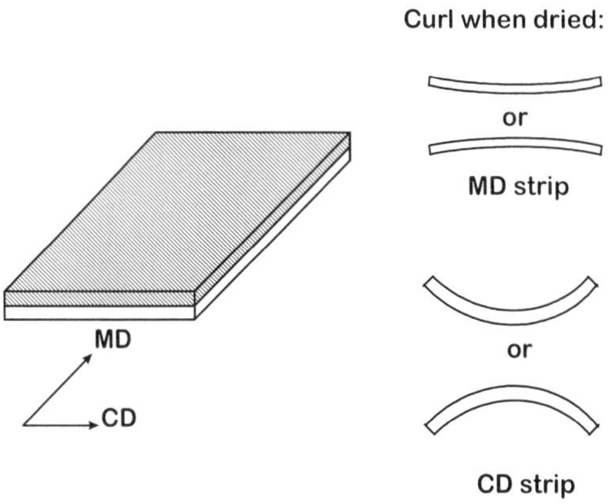

Curl when dried:
or
MD strip
MD
CD
or
CD strip

Fig. 42 Curl induced by two-sidedness in density and fines or filler contents. The MD strip tends to react less to this type of two-sideness than the CD strip.

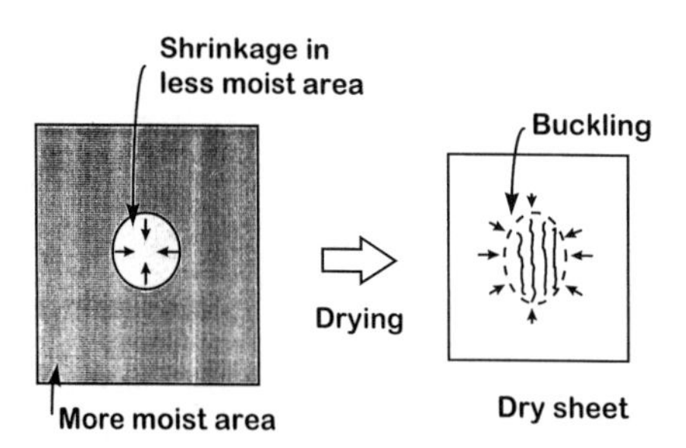

Fig. 43 A mechanism for creation of drying cockle.

An extensive mill investigation on cockle was carried out by Dittman and coworkers [26]. Using a frequency analysis of both cockle formation and hydraulic pulsation, the mill team identified several operational factors contributing to the cockles that were observed mainly on areas of the sheet formed near the edges of the paper machine. Those factors included the headbox fan pump and system resonance, headbox air content, primary centriscreen bearing damage, primary centriscreen foundation vibrations, and attenuation of hydraulic pulsations.

Because the basic cause of cockle is a nonuniform manifestation of hygroexpansion (reversible and irreversible) in the sheet plane, all factors that affect the magnitude of hygroexpansion also influence the intensity of cockle. For example, more shrinkage in the paper from areas near the edge of the machine and more extensive beating (producing fines) result in more cockling.

III. TESTING

A. Methods for Generating Controlled Temperature and Humidity

The determination of environmental effects on the physical and mechanical properties of paper requires the generation of various temperature–humidity conditions. For this purpose, various methods are used, using special solutions (saturated salt solutions, water–sulfuric acid solutions, and water–glycerin solutions) and special apparatuses (humidity generators) designed to control the temperature and humidity of the air. Special solution methods are simple and inexpensive but are not suitable for dynamic measurements. Humidity generators are more expensive but are more versatile for different kinds of environmental effects testing. In all cases, testing should be conducted within a standard constant temperature, constant humidity room, because the accuracy of temperature and humidity generation and control is strongly dependent on the surrounding atmosphere. (Readers who are interested in the design and performance of conditioned test rooms can refer to the literature [16,134] and to Chapter 3 of Volume 2 of this handbook.)

Solution Methods

Saturated Salt Solutions Data for the relative humidity values produced by a number of saturated solutions are extensively covered in the literature [124,129]. Table 3 lists the RH versus temperature value of some saturated salt solutions often used in testing laboratories [129]. An advantage of this method is that it is

Table 3 Saturated Salt Solution Values of Relative Humidity (%) Versus Temperature

Temperature (°C)	$LiCl{\cdot}H_2O$	$MgCl_2{\cdot}6H_2O$	$Na_2Cr_2O_7{\cdot}2H_2O$	$Mg(NO_3)_2{\cdot}6H_2O$	$NaCl$	$(NH_4)_2SO_4$	KNO_3	K_2SO_4
0	14.7	35.0	60.6	60.6	74.9	83.7	97.6	99.1
5	14.0	34.6	59.3	59.2	75.1	82.6	96.6	98.4
10	13.3	34.2	57.9	57.8	75.2	81.7	95.5	97.9
15	12.8	33.9	56.6	56.3	75.3	81.1	94.4	97.5
20	12.4	33.6	55.2	54.9	75.5	80.6	93.2	97.2
25	12.0	33.2	53.8	53.4	75.8	80.3	92.0	96.9
30	11.8	32.8	52.5	52.0	75.6	80.0	90.7	96.6
35	11.7	32.5	51.2	50.6	75.5	79.8	89.3	96.4
40	11.6	32.1	49.8	49.2	75.4	79.6	87.9	96.2
45	11.5	31.8	48.5	47.7	75.1	79.3	86.5	96.0
50	11.4	31.4	47.1	46.3	74.7	79.1	85.0	95.8

Source: Ref. 129.

relatively easy to prepare the solutions, because it only requires saturation of the solutions. Another important advantage, which can be applied to other solution methods, is that the humidity created is relatively insensitive to temperature changes, as shown in Table 3.

To achieve uniform relative humidity within a humidity chamber quickly, it is important to equip an appropriate fan for air circulation in the chamber and a stirring device for the tray containing the saturated solution. The motor driving the fan should be placed outside the chamber with appropriate thermal insulation, because the heat generated over the long testing period raises the temperatute and thus results in humidity drift in the chamber. These precautions apply equally to the other two solution methods.

Water–Sulfuric Acid Solutions This method requires different concentrations of sulfuric acid to control relative humidity. Table 4 lists the sulfuric acid solutions for generating different relative humidities at selected temperature [129]. One caution is that the concentration of the solution can change over a long time period and therefore a periodic check of the concentration is necessary, either by titration with a standard base or by measurement of specific gravity [129].

Water–Glycerin Solutions Table 5 shows relative humidity values obtained from a number of water–glycerin solutions [129]. An advantage of this method over the sulfuric acid method is that the solution is noncorrosive. The same precautions should be taken with the concentration change as in the previous method.

Humidity Generators The solution methods just discussed are best suited for creating a fixed humidity condition. However, for generating dynamic temperature–humidity conditions, specially designed humidity generators are useful. In particular, with the use of computer-based control systems, various data acquisition boards, and software, it is now possible to generate complex temperature and humidity conditions as a function of time in a very precise manner. Most of the generators

Table 4 Relative Humidity (%) Obtained from Water–Sulfuric Acid Solutions

% Sulfuric acid (w.w)	Specific gravity [a]	0°C	20°C	25°C	30°C	50°C	75°C
10	1.0661	95.9	95.6	95.6	95.6	95.6	95.6
20	1.1394	87.8	88.0	88.0	88.0	88.2	88.5
25	1.1783	81.8	82.4	82.5	82.6	83.1	83.6
30	1.2185	73.8	75.0	75.2	75.4	76.2	77.2
35	1.2599	64.6	66.0	66.3	66.6	67.9	69.5
40	1.3028	54.2	56.1	56.5	56.9	58.5	60.5
45	1.3476	44.0	45.6	46.1	46.6	48.5	50.8
50	1.3951	33.6	35.2	35.7	36.2	38.3	41.0
55	1.4453	23.5	25.3	25.8	26.3	28.5	31.1
60	1.4983	14.6	16.1	16.6	17.1	19.0	21.4
65	1.5533	7.8	9.2	9.7	10.1	11.8	14.0
70	1.6105	—	3.4	3.7	4.1	5.4	7.2

[a] Ratio of specific gravity at 20°C to that at 4°C.
Source: Ref. 129.

Table 5 Relative Humidity (%) Obtained from Water–Glycerin Mixtures

% Glycerin by weight	Specific gravity[a]	0°C	20°C	40°C	70°C
20	1.0470	90.0	94.2	93.9	93.1
40	1.0995	82.2	86.5	86.6	86.2
60	1.1533	70.0	73.8	73.8	74.4
80	1.2079	49.6	51.7	50.9	50.5
90	1.2347	32.4	32.0	30.9	30.3

[a]Ratio of specific gravity at 20°C to that at 4°C.
Source: Ref. 129.

used in the past were made in-house and designed for specific research and calibration purposes. A few commercial systems are now available attached to specific equipment (such as a hygroexpansimeter). The basic principles used for these generators include the two-pressure method, two-temperature method, divided-flow method, and combinations of these, each of which has some advantages and disadvantages.

Two-Pressure Method The two-pressure method uses high pressure air, which is first saturated with water vapor at pressure P_s in the saturator and then introduced into the testing chamber at a lower pressure P_t (normally atmospheric pressure) (Fig. 44). The relative humidity produced in the testing chamber is determined by the ratio of the two pressures [32,45,129]:

$$\mathrm{RH} = \frac{P_t}{P_s} \times 100 \tag{10}$$

Therefore, lower relative humidity in the testing chamber requires a higher pressure in the saturator.

The advantage of this method is the ease of obtaining a quick change in relative humidity in the testing chamber. On the other hand, in order to maintain a constant relative humidity over a long time period, it is imperative to achieve pressure stabilization at the pressure regulator position and a uniform and constant temperature

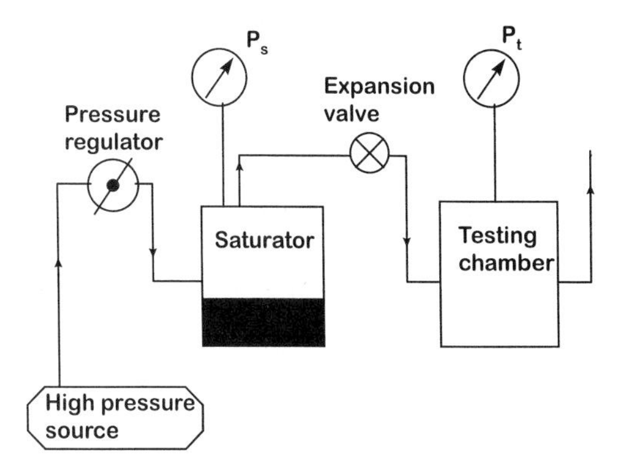

Fig. 44 Two-pressure method for generating controlled humidity air. (Based on Ref. 129.)

condition in both the saturator and the testing chamber (Fig. 44). The requirement of high pressure in the saturator for achieving a low humidity imposes a condition of the maximum flow rate, because the high pressure air expanding from the saturator sometimes carries mists that can damage samples in the test chamber. The range of the relative humidity obtainable by this method is normally 10–100%.

Two-Temperature Method In the two-temperature method, the air is saturated with water vapor at a given temperature T_s (lower than the test temperature) and then heated up to the test temperature T_t (Fig. 45). The relative humidity depends on the two temperatures through

$$\mathrm{RH} = \frac{p_s}{p_t} \times 100 \tag{11}$$

where p_s and p_t are the saturation pressures at temperature T_s and T_t, respectively. In the case for which the pressures in the saturator and the testing chamber are different, Eq. (11) can be modified to

$$\mathrm{RH} = \frac{p_s}{p_t}\left(\frac{P_t}{P_s}\right) \times 100 \tag{12}$$

This method is used in a closed system where the air is circulated through the saturator and the testing chamber. Because the temperature control in the saturator is accurate and the system is closed, the method allows very accurate and stable humidity control. It also allows the use of a high flow rate so that it is possible to supply controlled air to a chamber of relatively high volume. However, the temperature change in the saturator generally requires time, and therefore the two-temperature method is not suitable for changing the relative humidity very quickly (such as a change within a few seconds). Another drawback of this method is the difficulty in achieving a low relative humidity (< 30% RH) at room temperature, because in this case the water temperature in the saturator must be cooled down close to 0°C, at which temperature the efficiency of heat exchange in the cooling coil in the saturator decreases.

To compensate for the drawbacks of the two-pressure method and the two-temperature method, a combined method can be devised with the use of Eq. (12).

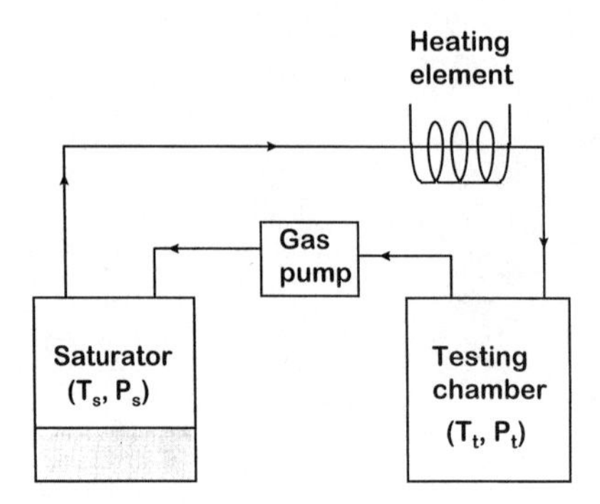

Fig. 45 Two-temperature method for generating controlled humidity air. (Based on Ref. 129.)

That is, by pressurizing the saturation chamber, as in the two-pressure method, a much lower relative humidity can be achieved.

Divided-Flow Method The divided-flow method is the most straightforward method for generating controlled air. The dry air is divided into two parts by a proportioning valve (Fig. 46); the fraction X of the dry air is saturated with water vapor in the saturation chamber and mixed with the other dry fraction $(1 - X)$ in the mixing chamber, and then the combined air is fed to the testing chamber. The relative humidity created depends on the fraction X and the pressures in the mixing chamber (P_c) and the saturator (P_s).

$$\mathrm{RH} = \frac{XP_c}{P_s - (1 - X)p_s} \times 100 \tag{13}$$

Because of this principle, the operation is very simple, and it is easy to change humidity very quickly (within a few seconds) if the testing chamber is small.

Readers interested in the system configurations of these humidity generators may refer to the comprehensive review paper by Wexler [129]. Automated humidity generators based on the foregoing principles are now commercially available. The following list is in no way complete but may help readers to find some examples of typical humidity generators available in the market.

General Eastern Instruments, 20 Commerce Way, Woburn, MA 01801
Parametric Generation & Control Inc., 1104 Old US 70 West, Black Mountain, NC 28711
Thunder Scientific Corp., 623 Wyoming S.E., Albuquerque, NM 87123-3198

B. Methods for Measuring Hygroexpansion, Shrinkage, and Wet Expansion

Hygroexpansion Hygroexpansion of paper is determined by changing its moisture content via a change in relative humidity. The range of relative humidity should be decided according to the end-use conditions (Fig. 47). If the end use is an electrophotographic process in which the paper is subjected to a high temperature and loses moisture in the process, the test should focus on the hygroexpansion behavior in the low relative humidity (low moisture content) range. If the end-use issue is the misregistration of paper in a multicolor lithographic process, one needs to cover a higher moisture content range.

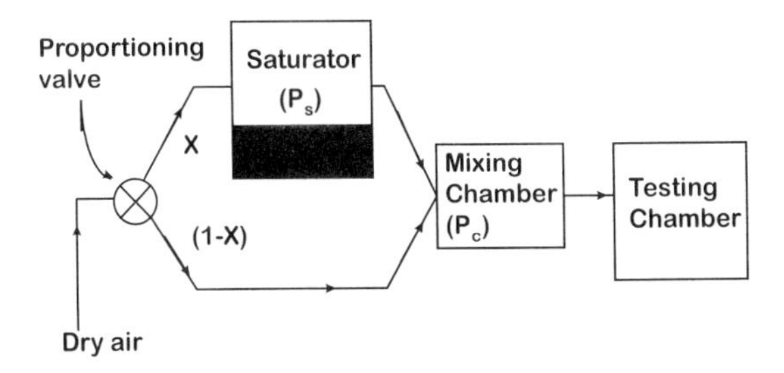

Fig. 46 Divided flow method for generating controlled humidity air. (Based on Ref. 129.)

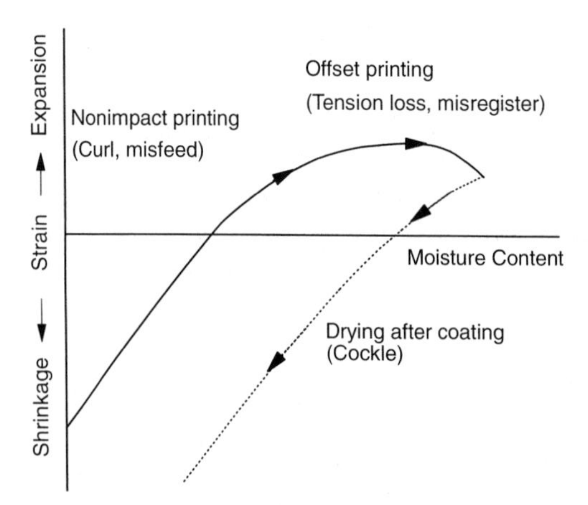

Fig. 47 Practical significance of hygroexpansion curve.

TAPPI Method for Measuring Hygroexpansion TAPPI Useful Method UM 549 (Hygroexpansivity of Paper) [104] recommends measurements in two RH ranges, i.e., from 25–50% and 50–86%. SCAN-P 28:88 [90] uses the RH ranges 33–66% and 33–84%. In both cases, the results are expressed as percent dimensional changes in the corresponding relative humidity ranges. Because the moisture content of the paper sample is not measured in either method, it is extremely important to precondition the sample at a relative humidity lower than the test humidity. For testing in the higher RH range, nonlinear hygroexpansion occurs, depending on the humidity history and the time elapsed since manufacture of the paper. Therefore, the determination of hygroexpansion behavior in a high RH range requires strict control of the whole humidity history of the sample after manufacturing.

The measurement of moisture content in addition to relative humidity is always recommended because of the moisture content–relative humidity hysteresis effect discussed earlier. This measurement can be achieved by placing a reference sample in the environmental chamber and simultaneously measuring the weight. Good design of the hygroexpansion chamber is crucial to maintaining uniform distribution of the humidity-controlled air. Literature listed in this section contains important information on chamber design.

Another important factor is the load applied to the paper sample during the test to prevent warping and cockling. SCAN-P 28:88 [90] recommends applying the load only at the time of the dimensional measurement and specifies the range of suitable load values for different basis weight samples. For continuous monitoring of hygroexpansion, a preliminary test is required in order to evaluate the effect of varying loads on the hygroexpansion measurement.

Other Methods for Measuring Hygroexpansion Neenah-type hygroexpansimeters have been used in the industry for years [80,123]. A paper strip is clamped at both ends; one is fixed, connected to a micrometer, and the other is movable. Expansion or contraction of the sample causes the displacement of the movable lower end, which is detected by the tilt of the vial attached to the lower clamp. Repositioning the vial by adjusting the micrometer attached to the upper clamp gives the displacement reading. Twenty specimens are accommodated in the chamber.

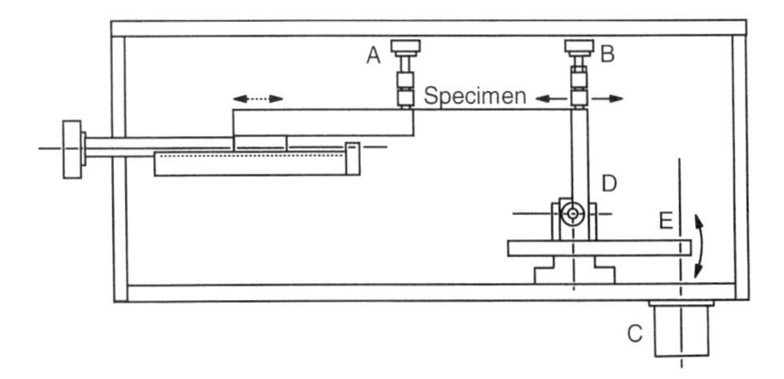

Fig. 48 Hygroexpansimeter (Paprican model). A, B, Specimen clamps; C, LVDT; D, balance beam; E, point of loading. (From Ref. 120.)

The basic design of the Neenah-type hygroexpansimeter is further developed in various research institutions by replacing the micrometer with a displacement or position sensor to detect hygroexpansion [7,23,51,95,120]. Figure 48 is an example of such a hygroexpansimeter [120]. The movable clamp (B) is attached to the top of a balance beam (D), and the balance beam displacement is detected by a linear variable differential transformer (C) (LVDT). Different loads can be placed at point E to remove slack or warp of the specimen. In the STFI model (Fig. 49) [58], the specimen is clamped with two razor blades, and the movable clamp, connected to an LVDT, is vertically supported by springs instead of a balance beam. The weight is applied to the paper surface before each measurement (not during the conditioning period). A hygroexpansimeter developed at the U.S. Forest Products Laboratory has a unique sample stage that forces the humidity-controlled air to go through the sample to accelerate the achievement of moisture equilibrium [23]. Ikuta and Hitosugi [51] designed a rotating disk on which 24 samples are continuously detected by a laser scanning position sensor. In most hygroexpansimeters, data for displacement, RH, and temperature are logged by computer. Some hygroexpansimeters are fully automated [51,95].

Digital correlation techniques have been applied to hygroexpansion measurements [53,62]. A speckle pattern is created on a paper surface, either by spraying colored particles or by using natural speckles (surface roughness or basis weight variation). Undeformed and deformed images are taken by a high resolution (charge-coupled device) (CCD) camera and digitized (Fig. 50). To measure a displacement at a specific point, two small subimages are taken, and a (two-dimensional)

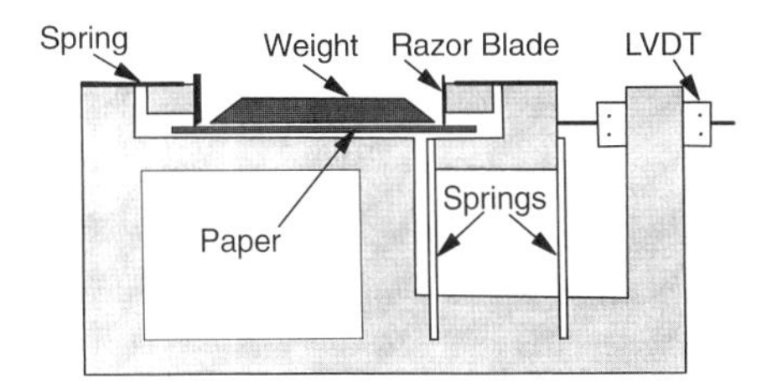

Fig. 49 Hygroexpanismeter (STFI model). The clamp with a razor blade is connected to an LVDT. (From Ref. 58.)

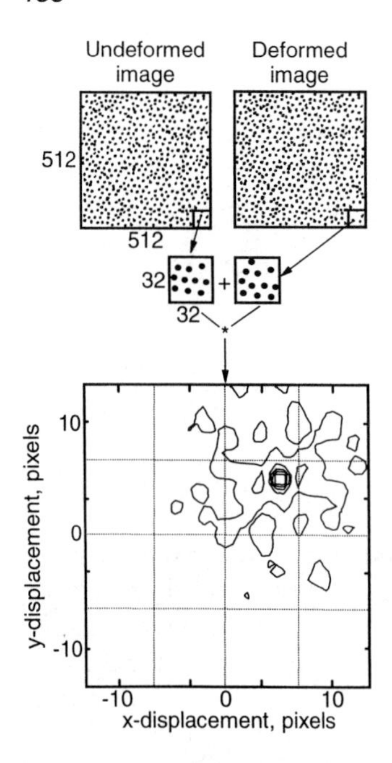

Fig. 50 Principle of electronic speckle photography (ESP). Cross-correlation technique is applied to the undeformed and deformed subimage areas. The total area (512 × 512 pixels) is divided into 16 × 16 subimage areas (each 32 × 32 pixels). (From Ref. 62.)

cross-correlation function is calculated. The position of the maximum of the cross-correlation function is the displacement of the point in the subimage. Figure 51 shows the calculated displacement field for a sample subjected to a climate change from 33% RH to 66% RH. From the displacement vectors, local strains, including normal and shear strain components, are calculated pointwise. To obtain a subpixel accuracy, various algorithms are employed (a bilinear interpolation of the gray level between pixels [53] and a Fourier series expansion of the discrete correlation surface [62]).

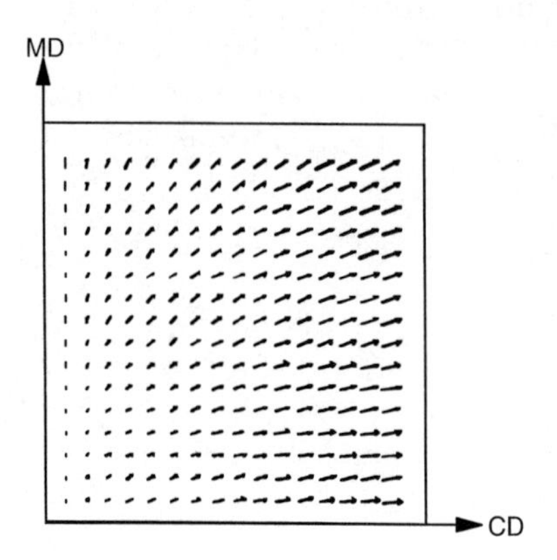

Fig. 51 A hygroexpansion displacement field after a climate change from 33% to 66% RH. (From Ref. 62.)

An advantage of this technique is that it is possible to obtain detailed information on two-dimensional hygrodeformation without physical contact. Therefore, there is no need to cut a number of samples in the machine and cross-machine directions. The calculation of the two-dimensional deformation field, however, still requires time, although the computational time can be shortened as more central processing unit (CPU) power is gained and the algorithm is improved. A disadvantage is that out-of-plane deformations such as cockle and warp that occur during the humidity change significantly deteriorate the accuracy of this technique.

Shrinkage and Wet Expansion The shrinkage of a pulp sheet during drying is often determined by measuring the dimensional change of the distance between marks on the sheet as it is dried without restraint. TAPPI Method T205 [103] describes the general sample preparation and measurement procedure. Drying without restraint (or with minimum restraint) may be achieved by placing the wet sheet between two Teflon screens with a small clearance corresponding to the thickness of the sheet. The dimensional change can be measured by a ruler or a traveling microscope. The shrinkage is affected not only by the swelling property of the furnish but also by some papermaking procedures. For example, as wet pressing is intensified, the shrinkage decreases because the sheet is expanded in the in-plane directions due to the Poisson effect (exclusion effect). At the same time, as seen in the case of hygroexpansivity, increased bonding by wet pressing increases shrinkage potential (percent shrinkage per 1% moisture content change) of the sheet. In addition, prolonged wet pressing or normal pressing of highly swollen pulps tends to decrease shrinkage because of the restraint of some shrinkage during wet pressing. Therefore, interpretation of the shrinkage results requires attention to the sample preparation procedure. Wet expansion is determined by using similar measuring devices.

On-line drying shrinkage has been recently measured on a commercial scale Fourdrinier machine [111]. Two non-contact laser scanning sensors mounted on a moving stage were placed in various positions on the paper machine to determine the change in width of the paper web. The shrinkage between the second press and the reel was found to have a strong correlation with wet expansion. Another on-line shrinkage measurement was attempted using CCD-captured images of the watermark [25].

C. Methods for Measuring Curl, Cockle, and Wrinkle

Curl Measurements In curl measurements, there are two important factors to be considered: (1) the size and shape of the specimen and (2) humidity history. As discussed earlier, the curl shape (curvature) is strongly dependent on the size and shape of the paper. In its end use, a normal size cut sheet tends to show only a cylindrical shape because of structural instability [48–50]: The curvature in one direction (often the machine direction) is suppressed; we see only the curvature in the other direction (the cross-machine direction). This shape does not necessarily mean that there is only one curvature component. A general curl shape, i.e., a deviation from a flat shape, is represented in terms of three curvature components [117] (Fig. 52):

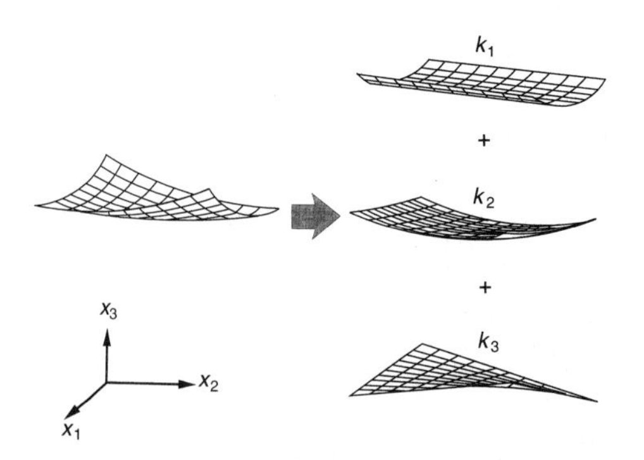

Fig. 52 Representation of a general curl shape by three curvature components. k_1, bending curvature in the x_1 direction; k_2, bending curvature in the x_2 direction; and k_3, twisting curvature. (From Ref. 117.)

$$w(x_1, x_2) = \frac{1}{2}(k_1 x_1^2 + k_2 x_2^2 + k_3 x_1 x_2) + a x_1 + b x_2 + c \tag{14}$$

where w is a deflection from a flat plane; k_1, k_2, and k_3 are the curvature components described earlier; and the last three terms simply describe a tilt of the reference plane. (In this equation, the deflection w is assumed to be small, so the curled surface is expressed in quadratic form.) The determinations of these three curvature components are crucial for curl diagnosis, because they can be used to evaluate the structural two-sidedness parameters through Eqs. (6), (7), and (9).

Humidity (or moisture) histories alter curl shape. In particular, for the simulation of end-use conditions such as converting and printing, we need to pay special attention to the entire humidity history to which the paper has been subjected. For the diagnosis of the causes of curl, a humidity schedule has to be chosen in such a way that the reversible curvature change can be determined. Measurement of the curvature change between the standard room humidity and a lower humidity is recommended for this purpose. (A better alternative is to first expose the paper to a higher humidity, then determine the relative curvature change from this high humidity to the lower humidity.) From the reversible curl test, "curl reactivity" is often determined. Curl reactivity is defined as the ratio of curvature change to relative humidity change [82]. To eliminate the effect of the hysteresis between moisture content and relative humidity, it is preferable to use the moisture change instead of the change in relative humidity.

Practical curl measurements normally determine only the dominant curl, because of the assumption of a cylindrical curl shape. Curl in cut-size sheets is measured either by laying the sheet on a flat plane or by suspending the sheet with the curl axis in the gravity direction. The degree of curl is evaluated by measuring the height of the curl with an engineer's vernier or by comparing the curl with a template consisting of a series of known curvature arcs (Fig. 53) [41]. TAPPI useful methods are practical methods for testing different grades of papers [105–109]. The ISO committee is currently in the process of approving two standards: "Paper and boards (cut-size office paper)—Measurement of curl in pack of sheets" and "Paper and board—Measurement of curl using a single vertically suspended test piece."

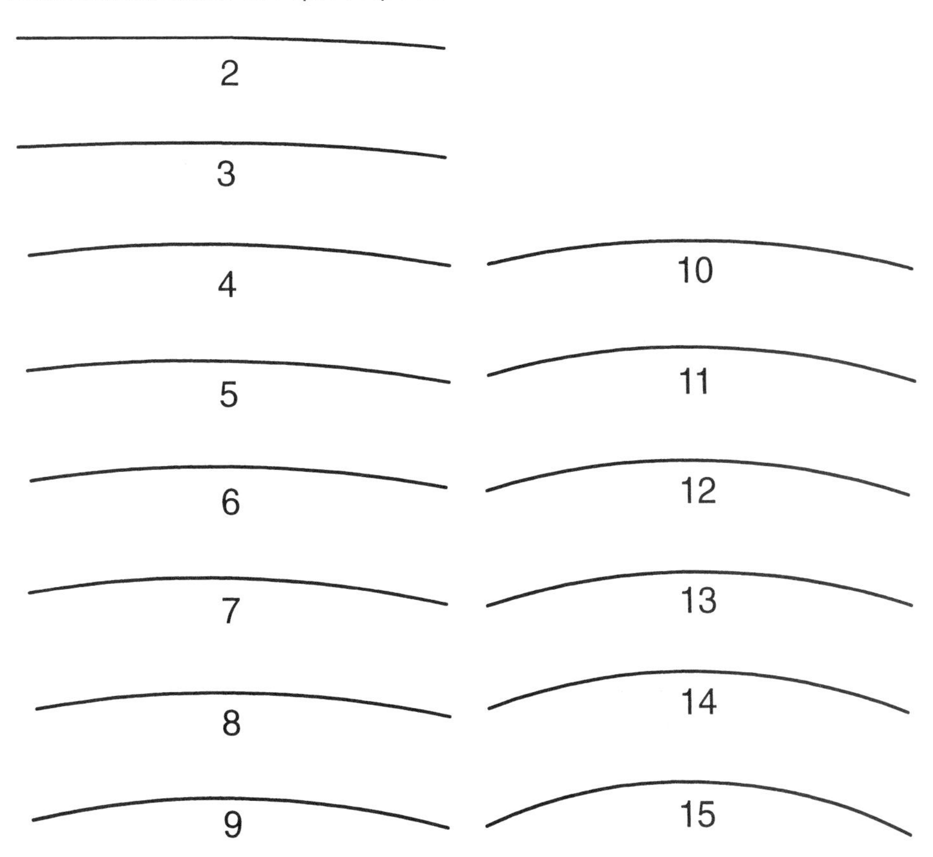

Fig. 53 Curvature template for curl measurements. (The numbers indicate the curvatures in m^{-1}.) (From Ref. 41.)

Another method that is often used in the mill is a cross-cut method, as shown in Fig. 54. The sheet center is cut in the diagonal directions, and the heights of the raised edges are measured. In this method, it is easy to identify a dominant side and the direction of the curl.

Methods that involve the measurement of the height of the edges of the sheet are always affected by gravity [29,125]. If the sheet lies in a horizontal plane, the deviation of the edges from the plane is given by [125]

$$w_{\text{edge}} = \frac{1}{2}\left(\frac{S_b}{gm}\right)K^2 \tag{15}$$

where K is the "dominant" curvature component, S_b is the bending stiffness, g is the gravitational constant ($9.8\ \text{m/s}^2$), and m is the basis weight. The basic assumptions in this equation are that the curvature and stiffness are moderate so that there is a flat area in the sheet and, more important, there is a dominant curl in either the machine or cross-machine direction. The edge height is sensitive to the curvature change (to the power of 2) but is also influenced by both the bending stiffness and the basis weight of the sheet.

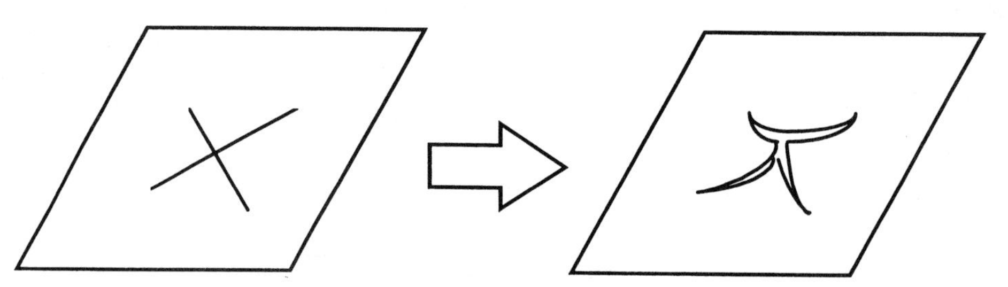

Fig. 54 Cross-cut method for curl measurements.

The idea of measuring the height of the edges has been applied for an on-line measurement of curl [56]. A combination of a laser beam and CCD camera detects a height deviation of the web edges, where the magnitude of curl is normally largest.

When paper shows mixed curl shapes or strong twisting curl, it is desirable to determine the three curvature components. The most straightforward method is to take a series of narrow strips (10–15 mm wide and 10–15 cm long) in the machine, cross-machine, and 45° directions (Fig. 55) [117] and then measure the curvature of each strip appropriately placed to avoid the effect of gravity, as shown in Fig. 56 [41]. The curvature component k_i in either the CD ($i = 1$) or MD ($i = 2$) can be determined as [117]

$$k_i = \frac{2\,\Delta y}{\Delta x^2 + \Delta y^2} \tag{16}$$

where Δx and Δy are defined in Fig. 55. Δx and Δy can be measured by using a template grid plate or an appropriate position sensor. The twisting curvature k_3 is determined from the curvatures measured in the machine, cross-machine, and 45° directions [$k(45°)$] by using the equation

$$k_3 = 2k(45°) - (k_1 + k_2) \tag{17}$$

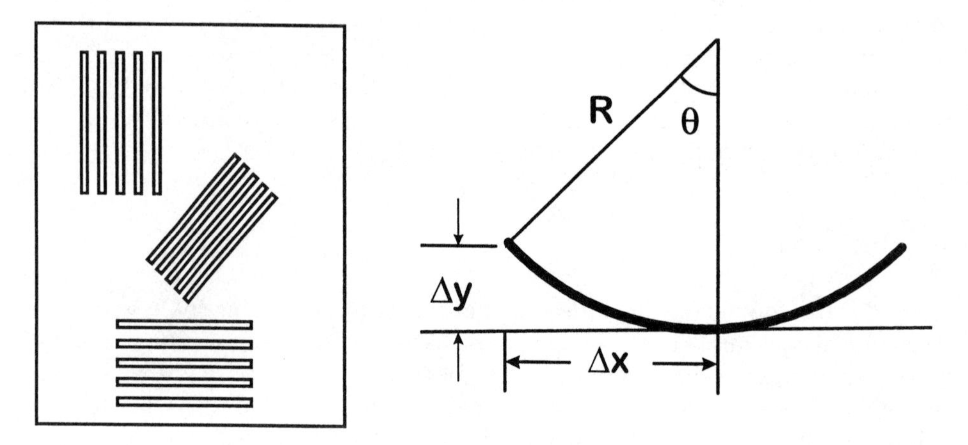

Fig. 55 Determination of three curvature components by paper strip method. (Based on Ref. 117.)

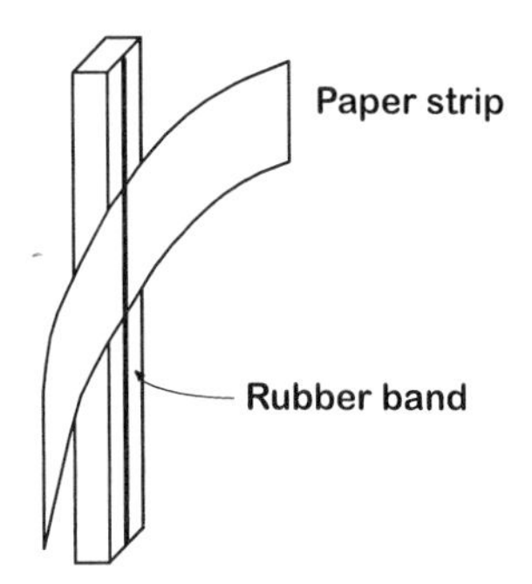

Fig. 56 Sample mounting to measure strip curvature. (Based on Ref. 41.)

Figure 57 shows an example of the CD variations of these curvatures for a 3750 mm wide Fourdrinier machine. The MD curvature (k_2) varied very little across the machine width, whereas the CD curvature (k_1) shows a bottom-side curl (wire-side curl, negative curvature) in several locations on the machine. The twisting curvature component (k_3) varied consistently from negative to positive, suggesting a systematic change in the two-sidedness of the fiber orientation direction across the machine width.

This method is suitable for measuring the "inherent" curl tendency of paper with a minimum influence of the shape-dependent curl, as discussed earlier. It is therefore suitable for curl diagnosis. However, a major drawback is that sample preparation is time-consuming.

An automatic curl measurement apparatus for paperboard was developed by the Swedish Packaging Research Institute [29]. Square samples (100 × 100 mm) are placed in an environmental chamber. The surface heights at selected points are determined by a non-contact distance sensor and are used for the determination of the three curvatures (Fig. 58). This apparatus has been used for routine quality control in board mills.

Another automated curl tester was developed at the Finnish Pulp and Paper Research Institute [72]. The device has horizontal shafts that hold 50 "round" speci-

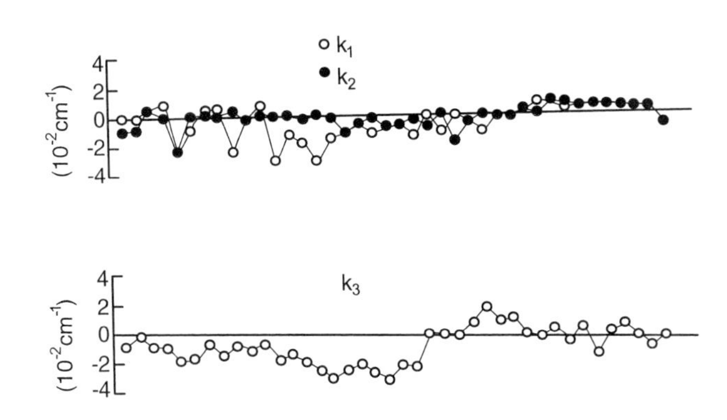

Fig. 57 Cross-machine direction curl profile. k_1, CD curvature; k_2, MD curvature; k_3, twisting curvature. (From Ref. 117.)

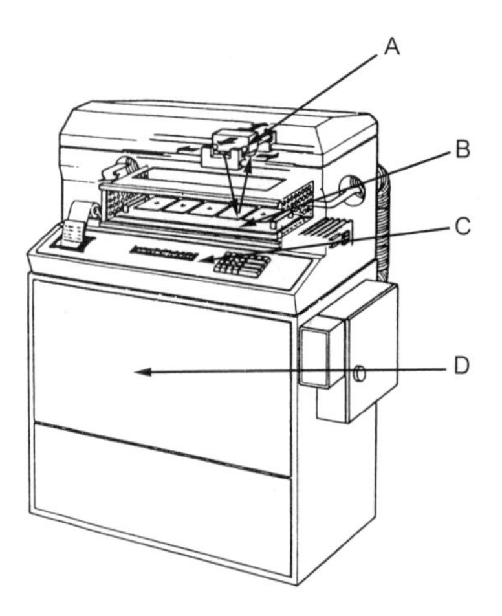

Fig. 58 The curl and twist tester. A, traversing, non-contact distance meter; B, test piece holder rack; C, microprocessor; D, conditioning unit. (From Ref. 29.)

mens (90 mm diameter) in a climate chamber. The specimen holder has three pin-holes in the center aligned vertically. As a shaft rotates, specimen surface displacement is measured by the position sensor at a constant radius (Fig. 59). The measurement is made twice at two different radii, and the curvature components are determined by applying the small surface displacement approximation. It was found that curl varies considerably in the machine and cross-machine directions, even within a short distance (25 cm), particularly for thin papers.

It should be noted that curvature components determined for different sizes and shapes can be different because of the structural stability or instability of the sheet. The sample size and shape, therefore, must be determined according to the purpose of the curl measurement. If the purpose is to find the shape of the curl in an end-use service condition, the shape and size of the specimen must be as close as possible to those of the actual paper sheet in that application. On the other hand, if

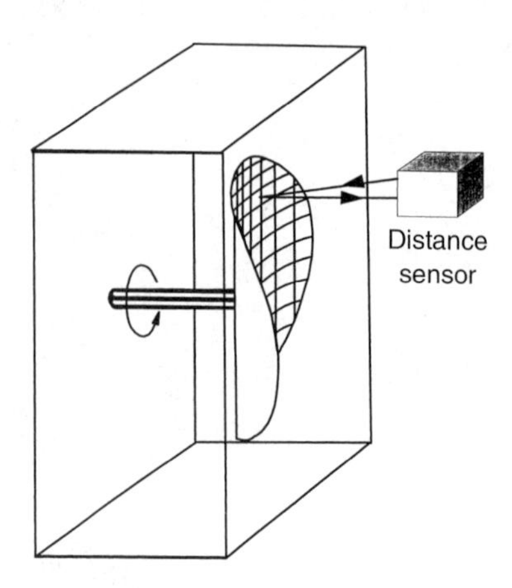

Fig. 59 Principle of curl tester (KCL model). (From Ref. 72.)

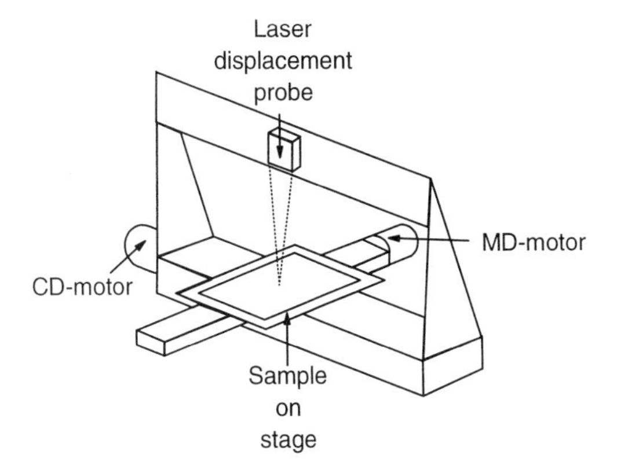

Fig. 60 Optical scanner for cockle measurements. (From Ref. 52.)

the purpose is to find the basic causes of curl (structural two-sidedness), then curvature measurements on narrow strip specimens provide adequate information on the inherent (potential) curl tendency of the paper.

Cockle and Wrinkle Measurements Cockle and wrinkle measurements are made by determining the surface deviation of a paper specimen from a flat surface. The surface deviation is detected either by an optical distance sensor or by employment of a digital image correlation technique in three dimensions.

An optical scanner has been developed at the Finnish Pulp and Paper Research Institute (Fig. 60). A laser displacement sensor measures the surface height variation of a paper sample placed on a stage moving in the machine and cross-machine directions [52]. A scanning time of 70 s is required with a size A4 paper sample. From the mapped image, the standard deviation of the surface height and characteristic length of the surface crossing the average surface height are determined. Figure 61 shows an example of the cockle measurements of samples taken from different locations across the paper machine in the cross-machine direction. The samples taken from the areas near the front and back edges of the machine tended to have higher cockle.

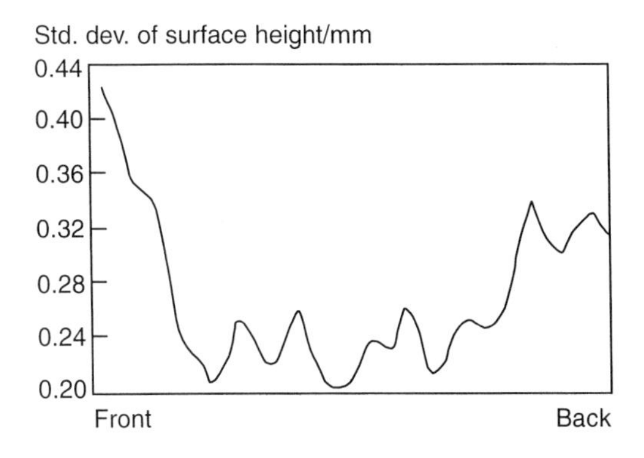

Fig. 61 Cross-machine direction profile of cockle. (From Ref. 52.)

Nam and Thorpe [70] determined cockles by employing a stereo viewing technique over the paper surface. The technique is an extension of the 2-D digital image correlation technique to determine the 3-D displacement components and thus strain fields. The advantage of this technique is that one can measure not only out-of-plane deformation (curl and cockle) but also in-plane deformation (hygroexpansion) simultaneously.

REFERENCES

1. Back, E. L. (1985). The relative moisture sensitivity of compression as compared to tensile strength. The Eighth Fundamental Research Symposium, Oxford, pp. 497–509.
2. Back, L., Salmén, L., and Richardson, G. (1983). Transient effects of moisture sorption on the strength properties of paper and wood-based materials. *Svensk Papperstidn.* *88*(6):R61–R71.
3. Barkas, W. W. (1942). Wood water relationships VII. Swelling pressure and sorption hysteresis in gels. *Trans. Faraday Soc.* *38*:194–209.
4. Barkas, W. W. (1950). The swelling of wood under stress. 3. *Svensk Papperstidn.* *53*(16):465–483.
5. Barkas, W. W. (1950). The swelling of wood under stress. 4. *Svensk Papperstidn.* *53*(17):509–517.
6. Barkas, W. W. (1950). The swelling of wood under stress. 3. *Svensk Papperstidn.* *53*(18):543–550.
7. Bennett, K. A. (1992). First cycle hygroexpansion behavior for commercial newsprint papers. 1993 Pan-Pacific Pulp and Paper Technology Conference, Tokyo, Japan, pp. 123–129.
8. Benson, R. E. (1971). Effects of relative humidity and temperature on tensile stress–strain properties of kraft linerboard. *Tappi* *54*(4):699–703.
9. Berger, B. J., and Habeger, C. C. (1989). Influences of non-equilibrium moisture conditions on the in-plane, ultrasonic stiffness of cellulose. *J. Pulp Paper Sci.* *15*(5):J160–J165.
10. Berger, B. J., Habeger, C. C., and Pankonin, B. M. (1989). The influence of moisture and temperature on the ultrasonic viscoelastic properties of cellulose. *J. Pulp Paper Sci.* *15*(5):J170–J177.
11. Brecht, V. W., Gerspach, A., and Hildenbrand, W. (1956). Drying tensions, their influence on some paper properties. *Das Papier* *10*:454–458 (in German).
12. Brecht, W. (1958). Beating and hygrostability of paper. In: *Fundamentals of Papermaking Fibers*, Vol. 1. F Bolam, ed. Tech. Sect. British Paper Board Makers' Assoc., London, pp. 241–262.
13. Brecht, W., Muller, F., and Weiss, H. (1955). About the blistering of papers. *Das Papier* *9*(4):133–142 (in German).
14. Brunauer, S., Emmett, P. H., and Teller, E. (1938). Adsorption of gases in multimolecular layers. *J. Am. Chem. Soci.* *60*:309–319.
15. Byrd, V. L. (1972). Effect of relative humidity changes during creep on handsheet paper properties. *Tappi* *55*(2):247–252.
16. Cairns, J. A. (1961). The design and control of test rooms and small ovens. *Tappi* *44*(6):191A–199A.
17. Carlsson, L. (1981). Out-of-plane hygroinstability of multi-ply paperboard. *Fiber Sci. Technol.* *14*:201–212.
18. Carlsson, L., Fellers, C., and Htun, M. (1980). Curl and two-sidedness of paper. *Svensk Papperstidn.* *83*(7):194–197.

19. Carreau, P. J. (1972). Rheological equations from molecular network theories. *Trans. Soc. Rheol. 16*(1):99–127.
20. Chatterjee, S. G., Ramarao, B. V., and Tien, C. (1997). Water-vapour sorption equilibria of a bleached kraft paperboard: A study of the hysteresis region. *J. Pulp Paper Sci. 23*(8):J366–J372.
21. Christensen, G. N., and Kelsey, K. E. (1959). The sorption of water vapor by the constituents of wood. *Holtz Roh Werkst. 17*(5):189–204 (in German).
22. Christensen, P. K., and Giertz, H. W. (1966). The cellulose–water relationship. In: *Consolidation of the Paper Web*, Vol. 1. F. Bolam, ed. British Paper and Board Makers' Assn., London, pp. 59–89.
23. Considine, J. M., and Bobalek, J. F. (1991). In-plane hygroexpansivity of postage stamp paper. TAGA (Technical Association of Graphic Arts) Conference, Rochester, NY, pp. 282–295.
24. de Ruvo, A., Lundberg, R., Martin-Löf, S., and Söremark, C. (1976). Influence of temperature and humidity on the elastic and expansional properties of paper and the constituent fiber. In: *The Fundamental Properties of Paper Related to Its Uses*, Vol. 2. F. Bolam, ed. British Paper and Board Makers' Assoc., London, pp. 785–806.
25. Di Mauro, E. C., Wadhams, K. R., and Wellstead, P. E. (1994). New on-line sensor for paper shrinkage measurement. Control Systems 94, Stockholm, Sweden, pp. 145–151.
26. Dittmann, R. L. (1991). The influence of hydraulic pulsation on cockles. 1991 Practical Aspects of Pressing and Drying Short Course, Atlanta, GA.
27. Dvorak, G. J., and Benveniste, Y. (1992). On transformation strains and uniform fields in multiphase elastic media. *Proc. Roy. Soc. Lond. A 437*:291–310.
28. Endo, Y. (1983). Operational experiences of Duoformer F. *J. Tappi 37*(1):56–61 (in Japanese).
29. Eriksson, L.-E., Cavilin, S., Fellers, C., and Carlsson, L. (1987). Curl and twist of paperboard: Theory and measurement. *Nordic Pulp Paper Res. 2*(2):66–70.
30. Fellers, C., and Bränge, A. (1985). The impact of water sorption on the compression strength of paper. The Eighth Fundamental Research Symposium, Oxford, pp. 529–539.
32. Fitts, D. D. (1962). *Non-Equilibrium Thermodynamics: A Phenomenological Theory of Irreversible Processes in Fluid Systems*. McGraw-Hill, New York.
32. Folland, C. K., and Sparks, W. R. (1975). A two-pressure humidity generator for calibrating electrical hygrometers used in meteorology. *J. Phys. E. Sci. Instrum. 9*:112–116.
33. Gallay, W. (1973). Stability and dimensions and form of paper. *Tappi 56*(11):54–63.
34. Glynn, P., and Gallay, W. (1962). Further studies of the mechanism of curl in paper. *Pulp Paper Mag. Can. 63*(8): T418–T424.
35. Glynn, P., Jones, W. H., and Gallay, W. (1959). The fundamentals of curl in paper. *Pulp Paper Mag. Can. 60*(10):T316–T323.
36. Glynn, P., Jones, W. H., and Gallay, W. (1961). Drying stresses and curl in paper. *Pulp Paper Mag. Can. 62*(1):T40–T48.
37. Green, C. (1981). Dimensional properties of paper structure. *Ind. Eng. Chem. Prod. Res. Dev. 20*:151–158.
38. Green, C. (1983). Relationships of dimensional stability to rheology of paper structure. TAPPI International Paper Physics Conf., Harwichport, MA.
39. Green, C. J. (1979). Characteristics of paper which contribute to curl. TAPPI International Paper Physics Conf., Harrison Hot Springs, BC, Canada, pp. 41–45.
40. Green, C. J. (1981). Curl properties of paper structure. *Ind. Eng. Chem. Prod. Res. Dev. 20*:147–150.
41. Green, C. J. (1984). Curl, expansivity, and dimensional stability. In: *Handbook of Physical and Mechanical Testing of Paper and Paperboard*, Vol. 2. R. E. Mark, ed. Marcel Dekker, New York, pp. 415–443.

42. Gunderson, D. E. (1981). A method for compressive creep testing of paperboard. *Tappi 64*(11):67–71.
43. Gurney, C. (1947). Thermodynamic relations for two phases containing two components in equilibrium under generalized stress. *Proc. Phys. Soc. A 59*:629–645.
44. Hailwood, A. J., and Horrobin, S. (1946). Absorption of water by polymers: Analysis in terms of a simple model. *Trans. Faraday Soc. 42B*:84–102.
45. Hasegawa, S., and Little, J. W. (1977). The NBS two-pressure humidity generator, Mark 2. *J. Res. Natl. Bur. Stand. 81A*(1):81–88.
46. Higgins, H. (1958). The structure and properties of paper. IX. Some critical problems. *Appita 12*(1):1–24.
47. Hosomura, H., and Makiyama, K. (1990). Evolution of non-impact printing with xerography and plain paper technology. *Jpn. Tappi 44*(4):417–432 (in Japanese).
48. Hyer, M. W. (1981). Calculations of the room-temperature shapes of unsymmetric laminates. *J. Compos. Mater. 15*:296–310.
49. Hyer, M. W. (1981). Some observations on the cured shape of thin unsymmetric laminates. *J. Compos. Mater. 15*:175–194.
50. Hyer, M. W. (1982). The room-temperature shapes of four-layer unsymmetric cross-ply laminates. *J. Compos. Mater. 16*:318–340.
51. Ikuta, S., and Hitosugi, F. (1996). New expansimeter for paper sample. 1996 Pulp and Paper Research Conference, Tokyo, Japan, pp. 70–73.
52. Kajanto, I. (1992). Analysis of paper cockling with an optical scanner. 1992 Paper Physics Seminar, Otaniemi, Finland, pp. 45–47.
53. Kajanto, I. M. (1996). Optical measurement of dimensional stability. Progress in Paper Physics—A Seminar, STFI, Stockholm, Sweden, 75–77.
54. Kijima, T. (1977). Dimensional stability of paper. In: *Paper Science*. I. Yoshino, T. Kadoya, and Y. Sumi, eds. Chugai-sangyo-chosakai, Tokyo, pp. 345–384 (in Japanese).
55. Kubát, J., Martin-Löf, S., and Söremark, C. (1969). Thermal expansivity and elasticity of paper and board. *Svensk Papperstidn. 23*:763–767.
56. Langevin, E. T., and Giguere, W. (1993). On-line curl measurement and control. 1993 Polymers, Laminations, and Coating Conference, Chicago, IL, pp. 125–130.
57. Laurell Lyne, Å. (1994). Hygroexpansivity of unfilled and filled laboratory-made fine papers. MS Thesis, Royal Institute of Technology, Stockholm, Sweden.
58. Laurell Lyne, Å., Fellers, C., and Kolseth, P. (1994). Three dimensional hygroexpansivity of unfilled and filled laboratory made fine papers. *Appita 47*(5):380–386.
59. LeBel, R., and Stradal, S. (1982). Control of fine paper curl in papermaking. *Pulp Paper Mag. Can. 83*(6):T183–T188.
60. LeBel, R. G., Schwartz, R. W., and Sepall, O. (1968). A novel approach to dimensional stabilization of paper. *Tappi 51*(9):79A–84A.
61. Leeman, D. A. (1993). A new approach to moisture and curl control on silicone release liners. 1993 Finishing and Converting Conference, New Orleans, LA, pp. 277–287.
62. Lif, J. O., Fellers, C., Söremark, C., and Sjodahl, M. (1995). Characterizing the in-plane hygroexpansivity of paper by electronic speckle photography. *J. Pulp Paper Sci. 21*(9): J302–J309.
63. Lodge, A. S. (1968). Constitutive equations from molecular network theories for polymer solutions. *Rheol. Acta 7*(4):379–392.
64. Lu, W., and Carlsson, L. A. (1996). Micro-model of paper. Part 2: Statistical analysis of the paper structure. *Tappi 79*(1):203–210.
65. Lu, W., Carlsson, L., and Anderson, Y. (1995). Micro-model of paper. Part 1: Bounds on elastic properties. *Tappi 78*(12):155–164.
66. Lu, W., Carlsson, L. A., and de Ruvo, A. (1996). Micro-model of paper. Part 3: Mosaic model. *Tappi 79*(2):197–205.

67. McKee, R. H., and Shotwell, S. G. (1932). Temperature effects of equilibrium between paper moisture content and humidity. *Paper Trade J. 94*:286–288.
68. Morton, J. L., and Bikales, N. M. (1959). Cyanoethylation as a means of improving the dimensional stability of paper. *Tappi 42*(10):855–858.
69. Nakayama, S., Toyama, T., and Kodaka, I. (1995). The influence of drying strategies on hygroexpansivity of paper—Effect of drying restraint conditions. *Jpn. Tappi 49*(8):1218–1226.
70. Nam, W.-S., and Thorpe, J. (1996). Deformation in copy paper with changing moisture conditions. Progress in Paper Physics—A Seminar, Stockholm, Sweden, pp. 65–68.
71. Nanri, Y., and Uesaka, T. (1993). Dimensional stability of mechanical pulps: Drying shrinkage and hygroexpansivity. *Tappi 76*(6):62–66.
72. Niskanen, K. (1996). Curl variations in paper and board. *Paperi Puu 78*(5):292–297.
73. Nordman, L. (1958). Laboratory investigations into the dimensional stability of paper. *Tappi 41*(1):23–30.
74. Nordström, A., Gudmundson, P., and Carlsson, L. A. (1998). Influence of sheet dimensions on curl of paper. *J. Pulp Paper Sci. 24*(1):18–25.
75. Padanyi, Z. V. (1991). Mechano-sorptive effects and accelerated creep in paper. TAPPI International Paper Physics Conference, Kona, HI, pp. 397–411.
76. Page, D. H., and Tydeman, P. A. (1962). A new theory of the shrinkage, structure and properties of Paper. In: *The Formation and Structure of Paper*, Vol. 2. F. Bolam, ed. Techn. Sect., British Paper and Boards Makers' Assoc., London, pp. 397–413.
77. Page, D. H., and Tydeman, P. A. (1966). Physical processes occurring during the drying phase. In: *Consolidation of the Paper Web*, Vol. 1. F. Bolam, ed. Tech. Sect., British paper and Board Makers' Assoc., London, pp. 371–396.
78. Perkins, R. W., Sinha, S., and Mark, R. E. (1991). Micromechanics and continuum models for paper materials of medium and high density. TAPPI International Paper Physics Conference, Kona, HI, pp. 413–435.
79. Pijselman, J., and Poustis, J. (1982). Curl of multi-ply papers: An analytical study. *Svensk Papperstidn. 85*(18):R177–R184.
80. Poustis, J., and Vidal, F. (1994). Paper hygroexpansion coefficient, an unknown parameter for corrugated box end users. Conference, "Moisture Induced Creep Behavior of Paper and Board," Stockholm, Sweden, pp. 181–190.
81. Ramarao, B. V., and Chatterjee, S. G. (1997). Moisture sorption by paper materials under varying humidity conditions. In: *The Fundamentals of Papermaking Materials*, Vol. 2. C. F. Baker, ed. Pira Int., Cambridge, UK, pp. 703–749.
82. Rutland, D. F. (1987). Dimensional stability and curl: A review of the causes. In: *Design Criteria for Paper Performance*. P. Kolseth, C. Fellers, L. Salmén, and M. Rigdahl, eds. STFI, Stockholm, pp. 29–50.
83. Salmén, L. (1993). Responses of paper properties to changes in moisture content and temperature. in: *Products of Papermaking*, Vol. 1. C. F. Baker, ed. Pira Int., Cambridge, UK, pp. 369–430.
84. Salmén, L., and Back, E. L. (1980). Moisture-dependent thermal softening of paper, evaluated by its elastic modulus. *Tappi 63*(6):117–120.
85. Salmén, L., Boman, R., Fellers, C., and Htun, M. (1987). The implication of fiber and sheet structure for the hygroexpansivity of paper. *Nordic Pulp Paper Res. J. 4*:127–131.
86. Salmén, L., Carlsson, L., de Ruvo, A., Fellers, C., and Htun, M. (1984). A treatise on the elastic and hygroexpansional properties of paper by a composite laminate approach. *Fiber Sci. Technol. 20*:283–296.
87. Salmén, L., Fellers, C., and Htun, M. (1985). The in-plane and out-of-plane hygroexpansional properties of paper. In: *Papermaking Raw Materials*, Vol. 2. V. Punton, ed. Mechanical Engineering Pub., London, pp. 511–527.

88. Salmén, L., Fellers, C., and Htun, M. (1987). The development and release of dried-in stress in paper. *Nordic Pulp Paper Res. J. 2*(2):44–48.
89. Scallan, A. M., and Tigerström, A. C. (1992). Swelling and elasticity of the cell walls of pulp fibers. *J. Pulp Paper Sci. 18*(5):J188–J192.
90. Scandanavian Pulp Paper and Board Testing Committe (1988). Standard SCAN-P 28:88. Hygroexpansivity.
91. Schapery, R. A. (1968). Thermal expansion coefficients for composite materials based on energy principles. *J. Compos. Mater. 2*(3):380–404.
92. Schulgasser, K. (1987). Moisture and thermal expansion of wood, particle board and paper. International Paper Physics Conf., Mont Gabriel, Quebec, CPPA Tech. Section, pp. 53–63.
93. Schulgasser, K. (1987). Thermal expansion of polycrystalline aggregates with texture. *J. Mech. Phys. Solids 35*(1):35–42.
94. Seborg, C. O., Simmonds, F. A., and Baird, P. K. (1938). Sorption of water vapor by papermaking materials: Irreversible loss of hygroscopicity due to drying. *Paper Trade J., Tappi Sect. 10*:223–228.
95. Serra-Tosio, J.-M., and Chave, Y. (1994). Automated measurement of dimensional variations and Young's moduli of paper and board under programmed variable conditions: The apparatus and some applications for process control and resarch. Int. Conf. "Moisture-Induced Creep Behavior of Paper and Board," STFI, Stockholm, Sweden, pp. 105–120.
96. Shands, J. A., and Genco, J. M. (1988). Cross-machine variation of paper curl on a twin-wire machine. *Tappi 71*(9):165–169.
97. Skaar, C. (1972). *Water in Wood.* Syracuse Univ. Press, Syracuse, NY.
98. Smith, S. F. (1950). Dried-in strains in paper sheets and their relation to curling, cockling and other phenomena. *Paper Maker 119*(3):185–192.
99. Söremark, C., and Fellers, C. (1993). Mechano-sorptive creep and hygroexpansion of corrugated board in bending. *J. Pulp Paper Sci. 19*(1):J19–J26.
100. Stamm, A. J. (1959). Dimensional stabilization of paper by catalyzed heat treatment and cross-linking with formaldehyde. *Tappi 42*(1):44–50.
101. Stamm, A. J., and Beasley, J. (1961). Dimensional stabilization of paper by acetylation. *Tappi 44*(4):271–275.
102. Szikla, Z., and Paulapuro, H. (1989). Changes in z-direction density distribution of paper in wet pressing. *J. Pulp Paper Sci. 15*(1):J11–J17.
103. TAPPI T205 om-88 (1988). Forming handsheets for physical tests of pulp.
104. TAPPI Useful Method UM549 (1991). Hygroexpansivity of paper.
105. TAPPI Useful Method UM425 (1991). Dry curl of coated paper (squares in changed humidity).
106. TAPPI Useful Method UM426 (1991). Dry curl of paper (normal and dry heat exposure).
107. TAPPI Useful Method UM427 (1991). Dry curl of paper (diagonal cuts).
108. TAPPI Useful Method UM428 (1991). Wet curl for printing paper (damp zinc plate).
109. TAPPI Useful Method UM434 (1991). Curl of cards.
110. Thalén, N. (1992). The influence of printing conditions on curl phenomena in newspapers. International Printing and Graphic Arts Conf., Pittsburgh, PA.
111. Todoroki, H., Tatara, T., Abe, Y., and Takeuchi, N. (1995). On-line monitoring system for web shrinkage of paper. 1995 Pulp and Paper Research Conf., Tokyo, Japan, pp. 110–113.
112. Uesaka, T. (1990). Hygroexpansion coefficients of paper. In: *Mechanics of Wood and Paper Materials.* ASME Vol. AMD-VOL. 112. R. W. Perkins, ed. American Society of Mechanical Engineers, New York, pp. 29–35.

113. Uesaka T. (1994). General formula for hygroexpansivity of paper. *J. Mater. Sci. 29*:2373–2377.
114. Uesaka, T. (1996). Transient effects on mechanical properties of paper: A review of non-equilibrium thermodynamic approaches. International Progress of Paper Physics Seminar, Stockholm, Sweden, pp. 61–63.
115. Uesaka, T., Ishizawa, T., Kodaka, I., and Okushima, S. (1989). Curl in paper. 3. Numerical simulation of history-dependent curl. *Jpn. Tappi 43*(7):689–696 (in Japanese).
116. Uesaka, T., Ishizawa, T., Kodaka, T., Okushima, S., and Fukuchi, R. (1987). Curl in paper. 2. Derivation of approximate curl formulas and their applicability. *Jpn. Tappi 41*(4):335–341 (in Japanese).
117. Uesaka, T., Kodaka, I., Okushima, S., and Fukuchi, R. (1985). Curl in paper. 1. A new approach to the evaluation of curl shape. *Jpn. Tappi 39*(10):953–958 (in Japanese).
118. Uesaka, T., Kodaka, I., Okushima, S., and Fukuchi, R. (1989). History-dependent dimensional stability of paper. *Rheol. Acta 28*:238–245.
119. Uesaka, T., and Moss, C. (1997). Effects of fiber morphology on hygroexpansivity of paper: A micromechanics approach. In: *The Fundamentals of Papermaking Materials*, Vol. 1. C. F. Baker, ed. Pira Int., Cambridge, UK, pp. 663–679.
120. Uesaka, T., Moss, C., and Nanri, Y. (1992). The characterization of hygroexpansivity of paper. *J. Pulp Paper Sci. 18*(1):J11–J16.
121. Uesaka, T., and Qi, D. (1994). Hygroexpansivity of paper: Effects of fiber-to-fiber bonding. *J. Pulp Paper Sci. 20*(6):J175–J179.
122. Urquhart, A. R., and Eckersall, N. (1930). The moisture relations of cotton. VII: A study of hysteresis. *J. Textile Inst. 21*(10):T499–T510.
123. Van den Akker, J. A., Root, C., and Wink, W. A. (1942). Multiple-specimen Neenah expansimeter. *Paper Trade J. 115*(24):33–36.
124. Van Eperen, R. H. (1986). Relative humidity and paper: The production and management of testing environments. *Tappi 69*(7):36–39.
125. Viitaharju, V., Kajanto, I., and Niskanen, K. (1997). "Heavy" papers and curl measurement. *Pap. Puu 79*(2):115–120.
126. Waech, T. G. (1992). Newsprint curl induced by offset printing. International Printing and Graphic Arts Conference, Pittsburgh, PA.
127. Washburton, F. L. (1946). Some thermodynamic relations of rigid hygroscopic gels. *Proc. Phys. Soc. A 58*:585–597.
128. Watanabe, M., and Abe, Y. (1990). Study on the heat contraction behavior of paper. 44th Appita Annual General Conference, Rotorua, New Zealand, pp. C1.1–C1.20.
129. Wexler, A. (1961). Humidity standards. *Tappi 44*(6):180A–191A.
130. Whitsitt, W. J., and Hoerschelman, E. (1990). Liner orientation effects of combined-board warp. *Tappi 73*(1):89–95.
131. Wink, W. A. (1961). The effect of relative humidity and temperature on paper properties. *Tappi 44*(6):171A–180A.
132. Yamaguchi, A. (1993). Production of fine paper by the BELRUN dryer technology and operating results. *Jpn. Tappi 47*(6):712–715 (in Japanese).
133. Yamamoto, Y. (1956). The viscoelastic properties of network structure. I. General formalism. *J. Phys. Soc. Jpn. 11*(4):413–421.
134. Zubryn, E. (1989). Conditioned test rooms. Design, layout and equipment. *Appita 42*(1):52–56.

4
MOISTURE-ACCELERATED CREEP

HENRY W. HASLACH, JR.
University of Maryland College Park
College Park, Maryland

I. INTRODUCTION

The mechanical response of paper in a variable relative humidity climate is quite different from that in constant relative humidity. Because water facilitates the bonding of the hydrophilic pulp fibers to form paper, moisture has a strong influence on the mechanical properties of paper. The mechanical response of paper is determined by the interaction of the external forces on the paper structure with the ambient relative humidity as well as temperature. A mechanosorptive effect is one arising from the interaction of loads with moisture sorption mechanisms. Moisture-accelerated creep occurs in the special case in which the load on the specimen is held constant while the ambient relative humidity is varied. Variations in the ambient conditions, such as relative humidity and temperature, magnify the mechanical creep strain response in paper compared to that measured at any constant ambient condition and can cause earlier failure. This behavior is observed under both tensile and compressive loads. Therefore, worst-case testing of paper products at high constant relative humidity is not a sufficient design strategy. This mechanosorptive problem has been studied by experimentalists, but there is as yet no universally accepted explanation of the physical mechanisms underlying such behavior.

Byrd [17] provided the first experiments on the acceleration of tensile creep due to variable ambient relative humidity in paper, although it had been studied earlier in wood by L. D. Armstrong and coworkers. One variable and two constant relative humidity (RH) tests are compared in Fig. 1 from Haslach [45]. The specimen was conditioned first at 23°C and 30% RH for 24 h to bring it to moisture equilibrium with the ambient air. The tensile creep tests of handsheets were conducted at a constant 30% RH, a constant 90% RH, and as the relative humidity linearly cycled from 30% to 90% to 30% at a rate of (1/3)% RH/min through four such cycles. One might have expected the constant 90% test to produce the largest creep strain at any given time. The variable relative humidity cycle, however, produces more strain at any time after the two curves cross. The varying relative humidity causes the tensile specimen to creep more than it would at a constant relative humidity equal to the highest relative humidity reached in the cycle. In this sense, the variation in the relative humidity accelerates the creep. A similar result was observed by Byrd [18] for the compressive loading of C-flute corrugated fiberboard (Fig. 2). The larger the creep load, the greater the difference between constant and variable relative humidity creep strains under identical creep stress. The creep strain seems to be magnified by variable relative humidity to a greater extent in compressive creep than in tensile creep.

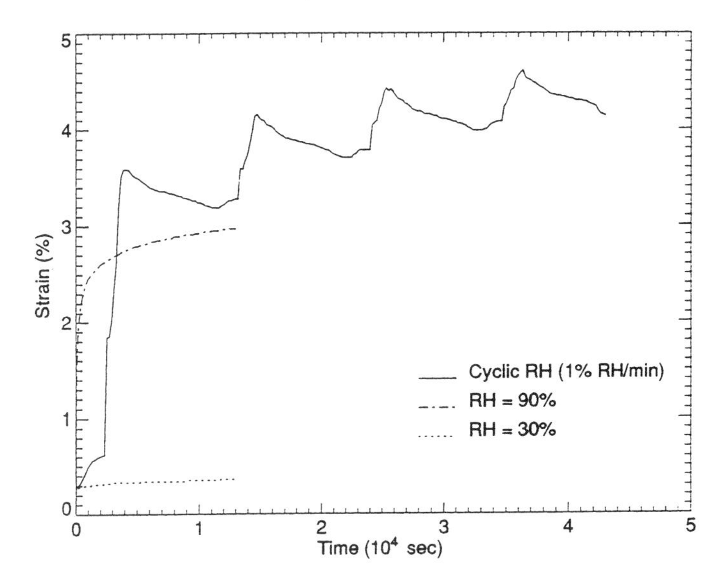

Fig. 1 Moisture-accelerated creep in a tensile specimen, with constant 30% RH, constant 90% RH, and relative humidity linearly cycled between 30% RH and 90% RH. (From Ref. 45.)

The problem is to understand and develop techniques to predict this variation in dimensional stability for the different types of paper products in order that controls can be designed. High levels or changes of humidity in the ambient air can cause sheet wrinkling or warping in printing and copying processes. Other undesirable results include the compressive buckling collapse of the walls of corrugated cartons stacked in warehouses lacking humidity control and the tensile creep of newly man-

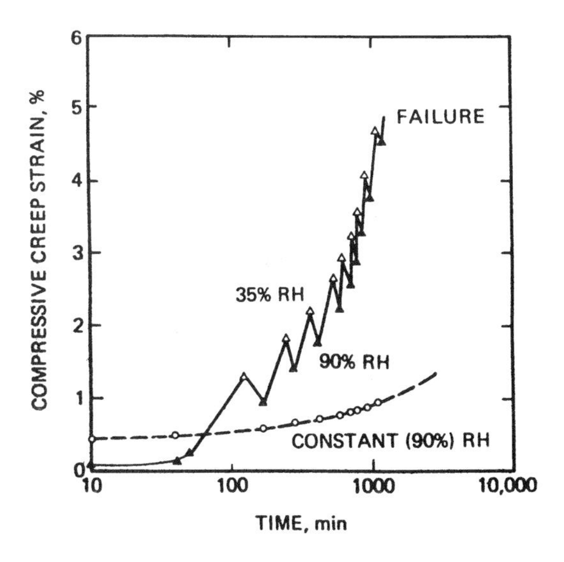

Fig. 2 Moisture-accelerated creep in a compressive specimen. (From Ref. 18.)

ufactured paper as the web is pulled through the press and dryer sections toward the reel. Misalignment in multirun offset printing processes can be induced by variations in sheet moisture content and the force on the sheet. Laminated paper products may warp due to the different responses of the paper components to changes in relative humidity. The edges of rolled sheets can wrinkle and cause unwinding problems. On the other hand, high levels of relative humidity can aid converting processes. Corrugating medium can be "softened" by variable humidity and temperature to prevent sheet tearing and cracking as the medium passes through the corrugating machine.

The mechanical response of paper to forces in a variable relative humidity climate is a nonequilibrium process, and therefore traditional equilibrium tests give only minimal information. In constant ambient relative humidity, the specimen is in equilibrium with its surroundings, but it is not in an equilibrium state if it is also loaded with an external force that induces creep. The key question is, What physical mechanisms come into play in variable relative humidity to drastically change the creep strain response from that observed in constant relative humidity under equal stresses?

The immediate goal of the various experiments on this issue is to identify the important parameters of the paper web or pulp that affect accelerated creep. The relationship between parameters should be represented in a mathematical model to aid design. Part of the response is hygroexpansion, the dimensional change of paper under zero load in response to variations in the ambient moisture conditions. Part is the reaction to the load. But the responses to simultaneous moisture variations and applied loads interact. A second goal is to discover design methods to control the loss of dimensional stability due to creep and any loss of strength due to fatigue induced by variable relative humidity. Better understanding of the underlying physics of this problem is required before useful design controls and descriptive mathematical models can be created.

The experimentals are described when possible with their motivating hypothesis about the physical mechanism underlying the phenomenon of moisture-accelerated creep. The few design techniques developed to guarantee that paper products such as corrugated containers survive their desired lifetime in a variable relative humidity environment are reviewed. This discussion concludes with some open questions, of which there are many, about this subject and proposed experiments that might help to resolve the questions.

II. MAJOR ISSUES

Paper responds mechanically to the interaction of external forces and ambient moisture. The contribution of each can be studied, but not isolated, by holding constant either the load or the relative humidity. A traditional stress-strain test results if the load is varied in a constant relative humidity. Such stress–strain tests [40] show that the response is more load rate dependent at higher constant humidities. The stress–strain curves can be fit by a variation on the classical hyperbolic tangent model or by a model based on the integral of a logarithmic function [88]. Such models have guided some proposed mathematical representations of moisture-accelerated creep phenomena. The complementary problem is the strain response of a specimen subjected to a constant load but variable relative humidity. Full knowledge of the interaction between load and ambient relative humidity will eventually require study of the consequences of simultaneous variation of both. The design of experiments on moisture-accelerated creep is difficult without a working hypothesis for the mechanism underlying the phenomenon. In formulating a hypothesis, one must balance the material, structural, and thermodynamic behavior involved in the phenomenon.

The experimental techniques for testing creep in paper have usually been devised in analogy to those used for metal creep and environmentally influenced polymer creep. The specimen is mounted on a load frame in a chamber in which the relative humidity can be controlled and varied. Before the test is conducted, the specimen is conditioned so that all tests begin with the specimen in the same state. Several methods have been used to condition the test specimen. Some researchers cycle the humidity for up to 10 cycles before applying the creep load; others only bring the specimen to temperature and moisture content equilibrium before beginning the test. Experimenters have cycled the relative humidity during the conditioning period for other types of tests in the belief that cycling removes residual stresses. Others also cycle the load on the specimen during conditioning, to remove some of the kinks and other distortions from the pulp fibers. A conditioning time depending on the specimen thickness must be chosen to ensure that moisture diffusion has time to bring the specimen to equilibrium with the ambient air. Some measurements have shown that 30 min is sufficient to attain moisture equilibrium in thin specimens. Other experimenters have conditioned for as long as 24 h. Clearly, the method of conditioning has a strong influence on the outcome of the test.

During a test, the sample adsorbs or desorbs moisture in response to changes in the ambient relative humidity. The paper specimen and the surrounding atmosphere form a two-phase thermodynamic system in which there is continual mass transfer of moisture in both directions between the two. At constant relative humidity, this transfer is in kinetic equilibrium, but it is not in equilibrium if the relative humidity varies. Creep behavior in variable relative humidity is a nonequilibrium process. To produce such a process, some experimenters have applied linear ramps, i.e., constant rates of change between the humidity extremes. Others have used step functions in which the humidity is as nearly as possible instantly varied between the extremes. Still others have used sinusoidal cycles so that the rate of change is also sinusoidal in time.

Both the load and variations in moisture content, or relative humidity variations, produce dimensional changes in a specimen. Because the hygroexpansive dimensional changes and the phenomenon of moisture-accelerated creep are so intertwined, a logical question is whether or not the same physical mechanisms underlie both hygroexpansion and moisture-accelerated creep. During variable relative humidity creep, can the hygroexpansive strain fluctuation due to variations in ambient relative humidity be distinguished from creep strain due to the load? The effect of the change of rate and of the period of cyclical relative humidity on the hygroexpansive strain and the time-dependent creep strain must also be investigated to determine if either is rate-dependent.

The response is at least partially determined by the material properties of the pulp. An important question is, How does the furnish affect the response of a fiber or a sheet to variable relative humidity? For example, as recycled fiber is used in more and more products, it is important to know if it is more or less susceptible to moisture-accelerated creep and consequent dimensional instability. A second important issue is the extent of influence of the fiber network structure of paper on its response to variable relative humidity. The behavior of single fibers loaded in variable relative humidity can be compared to that of a sheet. If moisture-accelerated creep appears only in a sheet, then the network structure of the sheet and therefore probably the interfiber bonds are an important deerminant. If the interfiber bonds are important, then the role and significance of hydrogen or other bonding must also be considered. A related idea is that moisture-accelerated creep depends on anisotropic fiber swelling and local stress gradients. An alternative to a structural explanation is that adsorbed moisture acts as a plasticizer to allow bond or intrafiber molecular slippage. The polymers that exhibit moisture-accelerated creep are semicrystalline and hydrophilic. A similar moisture-accelerated creep phenomenon has been observed in a few noncellulosic materials, such as concrete and various types of Kevlar.

If the magnified response of paper to variable relative humidity is a consequence of its being a nonequilibrium process, then other nonequilibrium processes may also accelerate creep. One such process is the response of paper to ambient thermal cycling. Paper is not a linear elastic material, but the measured apparent elastic modulus of paper decreases when either the moisture content or the temperature increases. The expansion due to temperature alone, as measured by the coefficient of thermal expansion under constant load, is analogous to the moisture-induced hygroexpansion in the sheet. This raises the question of whether or not a loaded specimen of paper responds to nonequilibrium temperature changes in the ambient environment in a manner similar to its response to relative humidity changes. The response of paper to a varying thermal environment has apparently not been closely studied, or at least not reported in the literature. This is surprising, because on the papermaking machine the web goes through temperature variations. The question of the influence of temperature on moisture-accelerated creep as well as on moisture sorption, and the analogy between thermally accelerated creep and moisture-accelerated creep, require that attention be paid to control of the temperature of the specimen during relative humidity cycling. Another issue is whether or not the fact that moisture adsorption is exothermic influences the response to a measurable extent.

Some of the more important features of moisture-accelerated creep have been identified by experimenters. It is not clear which of these are fundamental aspects of the phenomenon and which are merely symptoms and concomitant behavior. Agreement has yet to be reached on the proper manner in which to conduct experiments and on which simple parameters should be measured to predict behavior in a varying relative humidity ambient environment.

III. EXPERIMENTAL TECHNIQUES AND APPARATUS

A. Creep Testing

Most tests to date have been attempts to observe the interaction and evaluate the relative impotance of the various parameters affecting moisture-accelerated creep, under either a tensile or compressive load. Once this stage is completed, extensive tests will be required to build a constitutive model by quantifying the relationships between important parameters and their effect on paper response. The experimentalist's problem is to measure the strain while at the same time monitoring other important parameters in the paper specimen. For example, it may be desirable to measure the specimen moisture content and temperature during the test.

The test stand on which to mount the paper specimen must be constructed in such a way that it does not interfere with the measurements of the strain and other parameters or induce non-creep phenomena. The design of a test stand that permits isolation of the phenomena to be studied depends on the specimen type and its loading. Some tests focus on the behavior of corrugated containers, whose unexpected failure in variable relative humidity was the motivation for much of this work. Other tests employ paper sheets or even a single fiber as simpler specimens in an attempt to identify the underlying physical causes of moisture-accelerated creep. The type of sample to be tested affects affects the type of grips used to apply the load. For example, flat clamps satisfactorily hold the specimen in a tensile test of a paper sheet but not in one of a corrugated container board. The clamp would crush the corrugation. The clamps for a tensile test of a sheet might be rubber-coated to reduce the chances of specimen failure or slippage in the grip. Point or line contact grips can prevent slippage by applying greater pressures, but these grips risk generating high stress concentrations in the contact region of the specimen.

The type of specimen support required in a compression test depends on the specimen thickness. For example, corrugated containers have failed due to a combination of compressive creep and induced buckling, and one might wish to test the container components separately. To perform compression tests on thin sheets, a means must be found to prevent specimen buckling without affecting the strain observed under creep, either by the design of the supports or by specimen spatial orientation. A specimen under a compressive axial load might be best mounted in the vertical plane so that its own weight does not help initiate buckling. Gunderson [37] dealt with the difficulties of compressively loading a sheet by placing the specimen horizontally on an array of flexible vertical leaves transversely mounted so that only their edges contact the specimen. A vacuum system creates a pressure difference between the top and bottom faces of the specimen to hold it against the supporting leaves. This system prevents buckling in compression while allowing the humid ambient air to circulate around the specimen and reach all specimen surfaces. If a flat plate against one specimen surface were used to prevent buckling, humid air could not reach that surface, possibly setting up an undesirable moisture gradient within the specimen.

Most experimenters have performed axial tensile, axial compressive, or bending creep tests of paper. In each case, the state of stress is uniaxial. Most tensile or compressive axial creep tests are performed at a low percentage, say 25–50%, of the failure load in the stress–strain test conducted at a constant relative humidity of either 50% or 90%. The lower of such loads would be the one based on 90% relative humidity, because failure load decreases as relative humidity increases. This is done to ensure that the sample will not fail before passing through a sufficiently large number of humidity cycles to observe accelerated creep. The load is controlled in creep tests; for example, a constant axial load may be applied by a dead weight. An axial creep load was applied by a spring rather than a dead load by Gunderson and

Considine [39] so that the load could be easily changed during different periods of a creep test on a single specimen.

Some researchers have tried bending loads as a simpler means to create and measure compressive creep in paper or to allow testing of the same specimen in both tensile and compressive creep. One difficulty is that the creep stress in the specimen must be computed rther than directly measured. In order to use a linear elastic approximation for paper, a nonlinear material for which no known constitutive model exists, the strains were typically resricted to less than 0.1%, much smaller strains than those observed in axial creep tests. A four-point bending apparatus is often used to create a constant internal moment in the test section. The strain is compressive on one side of the neutral axis and tensile on the other. To overcome the problem of countervailing effects in tension and compression, Söremark and Fellers [98] placed a thin steel backing on the corrugated board so that the neutral axis would be essentially at the steel/paperboard interface. The bent corrugated paperboard was then completely in either tension or compression depending on which beam face the load was applied. Such difficulties do not arise with an axial member, assuming constant cross section, for which the stress is proportional to the dead load.

Strain measurement raises experimental difficulties. An electronic strain gauge glued to a paper sheet would certainly stiffen the specimen and change the specimen strain response. The gauge itself would also inhibit moisture transfer between the air and the specimen in exactly the region over which the strain is measured. An extensometer in contact with the specimen has been used to measure paper strain, but it can damage the specimen at the contact points because paper sheet specimens are thin. The strain would be most effectively measured without touching the specimen so that the gauge does not affect the strain response. The loads are quite small in most tests of paper so the grip deformation is negligible. The strain in the specimen can therefore be obtained from the change in grip separation. Frequently, an LVDT is used to measure the distance between grips. Linear encoder devices based on Moiré interference principles are also available for a similar application.

Experimentalists in several fields studying easily damaged materials, such as paper or biological tissues, have used optical non-contact strain measurement techniques. The inhomogeneity of a paper specimen may make it desirable to measure local strains in addition to an average strain over the specimen. The goal of these optical techniques is to calculate local strains in a small region of the specimen rather than a single average strain for the region between the grips. Video dimension analyzers (VDAs) measure the average normal strain in two perpendicular directions over a central region on the specimen. The VDA is a line-scanning device that produces a voltage proportional to the distance between lines in the video image. However, these instruments do not identify the local inhomogeneities in strain or determine the shear strain.

Alternatively, markers placed on the specimen are tracked by periodic camera images taken as the specimen deforms. These techniques depend on the ability to locate the new positions of the markers after deformation. The markers are lightweight to avoid influencing the strain response and are kept small for accuracy. In previous applications, the markers have been spaced 1–6 mm apart and within a test area 2.5–3.5 cm on a side. To measure the heterogeneous strain response of crepe paper, Chaves de Oliveira et al. sprayed a 7×5 dot matrix onto the specimen sheet with individual ink dots 4 mm apart [26]. Humphrey et al. [58] used four markers, each a vanilla bean speck wilth diameter of 250 μm, placed initially at the corners of a square to track the displacement. Their biaxial test stand loaded the specimen equally on both pairs of opposite edges so that the center of the square barely moved. This reduced the searching time for the image processing program devised to locate the marks.

The technique in Ref. 26 applied to paper required comparing and measuring, after the test is completed, photographs taken at discrete time intervals during the test. This method for measuring the displacement is slow and risks strain rate and relaxation effects coloring the results. A less labor-intensive technique automates the tracking of the markers by image processing software. If this can be done rapidly, feedback to obtain real-time experimental control, such as strain control over the loading, is possible.

An automated system using a video camera, which is slower than the VDA, was developed by Humphrey and coworkers [31,58] for membranes of soft biological tissue, but the technique would work equally well for paper. Software selects each video frame and locates the centroid of the markers by intensity measurements. The two-step search technique in Ref. 31 using a video camera, a charge-coupled device (CCD), follows the markers at a rate of 30 images/s. The first step searches a 35 × 35 pixel region centered on the last position of the marker. The region is divided into tiles. The tile with the optimal intensity averaged over its area is then searched for the new position of the marker. Once the new position is determined, the instantaneous strain is approximated from geometric computations based on the nonlinear strain–displacement relations,

$$\epsilon_x = \frac{\partial u}{\partial x} + \frac{1}{2}\left[\left(\frac{\partial u}{\partial x}\right)^2 + \left(\frac{\partial v}{\partial x}\right)^2\right] \tag{1a}$$

$$\epsilon_y = \frac{\partial v}{\partial y} + \frac{1}{2}\left[\left(\frac{\partial u}{\partial y}\right)^2 + \left(\frac{\partial v}{\partial y}\right)^2\right] \tag{1b}$$

$$\epsilon_{xy} = \frac{1}{2}\left[\frac{\partial u}{\partial y} + \frac{\partial v}{\partial x} + \frac{\partial u}{\partial x}\frac{\partial u}{\partial y} + \frac{\partial v}{\partial x}\frac{\partial v}{\partial y}\right] \tag{1c}$$

where u and v are the displacements in the x and y directions, respectively. The continuous displacement fields, u and v, are obtained by fitting interpolation functions to the measured discrete marker displacements. The accuracy of the strains obtained from the partial derivatives of these displacement fields depends on the order of the interpolation functions. This method was applied to biaxially loaded specimens and determines the shear as well as the normal strains.

The fundamental difficulty is that of tracking each marker. Two problems are that a false marker may be identified and that the specimen deformation may carry the markers out of the image processing search area. The latter is described by saying that the marker velocity is too high. Further, a good contrast between the markers and the specimen facilitates the image processing.

Electronic speckle photography (ESP) is a variation on tracking a few markers to determine displacements. Rather than following individual markers separately, the software is designed to identify changes in the random pattern of speckles on the specimen after deformation [15]. A fast Fourier transform analysis of the two-dimensional light intensity field reflected from the deformed speckled specimen is used to deduce the displacement field, from which the strains are calculated. If the field is resolved into 256 subimages, the analysis can take as long as an hour on a 166 MHz processor [15].

Moisture content has typically been determined by weighing a similar unloaded reference specimen subjected to the same environment as the specimen whose dimensional changes are measured in either hygroexpansion or creep tests [e.g., 17,103]. No one has reported a method to simultaneously load and weigh a single specimen. It is difficult to directly measure a change in moisture content by weighing the loading apparatus and specimen together because the apparatus weighs so much more than the specimen that sensitivity to small moisture variations is lost. Also, moisture may condense on the test stand during periods when the relative humidity is decreased.

A unique method of measuring the moisture content in a loaded specimen during a relative humidity cycle was developed by Gunderson and Tobey [41]. In their system, the moisture is added from wet sponges and removed with a desiccant contained in a removable capsule through which air is pumped to and from the test chamber, which is otherwise isolated from the atmosphere. This capsule is weighed to determine the moisture change in the system. Because the capsule shell weighed less than 40 g, the amount of moisture change could be computed within 0.01%. They report that it took at most 15 s to make the exchange of capsules. The chamber walls adsorb moisture, so this has to be accounted for when calculating the moisture content of the sample. About 20% of the specimen was clamped in the grips and assumed inaccessible to moisture in calculating the specimen moisture content.

Infrared techniques have been used to remotely measure the moisture content of the web on the papermaking machine. Infrared radiation at wavelengths between 1.92 and 1.94 μm is adsorbed by water in paper [14]. Detectors such as lead sulfide are used to measure the transmitted radiation, which varies with the amount of moisture in the paper. The resonant wavelength of 1.94 μm is easily transmitted by glass, so devices involving optical fibers can be used to direct the radiation to the specimen. Such radiation adsorption and reflection properties of water in paper might be applied to measure the moisture content of a loaded specimen, but the feasibility and accuracy of such a method have not been verified for moisture-accelerated creep or hygroexpansion. Infrared methods were not very sensitive in their early applications; the error was on the order of 0.1% moisture content. Furthermore, only the near surface region of the specimen affects the response to radiation, so only surface moisture content can be measured. A reflecting technique at a fixed wavelength is affected by surface roughness, fillers, and other surface features. The excitation of molecules by an imposed infrared radiation may not only raise the specimen temperature but may also directly affect the magnitude of the creep acceleration.

Nuclear magnetic resonance is a nondestructive and non-contact method to measure moisture content or the mobility of moisture in a solid. A magnetic field orients the hydrogen nuclei in a cellulose–water system [102]. The orientation is disturbed by radiant energy, and some of the enregy is adsorbed. This is a steady-state method because the nuclei are allowed to come to equilibrium with the magnetic field before being slightly disturbed. The change in adsorption is measured in one of two ways: Either the frequency is varied while the magnetic field is held constant or the magnetic field strength is varied under a constant-frequency energy.

Alternatively, the nuclei can be greatly disturbed, then monitored as they relax to equilibrium.

One problem with nuclear magnetic resonance is that the size of the sample is limited by the means of generating the magnetic field. In the preliminary study of Swanson et al. [102], the specimen size was restricted to a cross section of a few square inches. Also, difficulties arise if the method is to be applied to a variable relative humidity system. It takes time to scan the frequencies or amplitudes, so it is hard to get readings at a single point in time when the relative humidity varies. The adsorption of radiant energy may also excite the paper molecules and affect the creep response. An advantage is that this technique can detect very low moisture contents. In rayon, for example, Swanson et al. [102] were able to measure the steady-state moisture content corresponding to 0.03% relative humidity. They proposed using the method as a means to verify that a specimen is dry or to compare different drying procedures.

B. Humidity and Temperature Control

A system must be established to measure and control the relative humidity and temperature of the ambient environment in the test chamber. The relative humidity has been cycled with respect to time in step, ramp, and sinusoidal patterns. Any variation other than a step function requires equipment to regulate the relative humidity with time. In step function variations, the relative humidity is quickly changed from one extreme to the other. It is then held at the new position for a fixed time before it is quickly returned to the previous extreme. During the hold times at the extremes, the specimen begins to relax to an equilibrium state, and if the relative humidity is held constant long enough its moisture content eventually comes to equilibrium with the air. One justification given for this procedure is that the specimens of different paper can be compared only at equilibrium. However, such a method may lose some of the effect of exciting the specimen to a nonequilibrium state by the change in relative humidity. The ramp, or sawtooth, variation requires that the relative humidity be changed linearly with respect to time from one extreme to the other. Although in a strict sawtooth the relative humidity is not usually held at the extreme values for any period of time, some experimenters have done so. Sinusoidal variations might be the most useful, because a random variation can be resolved as a sum of sinusoidal functions. However, no one has examined whether or not the separate responses to two different humidity variation patterns can be superposed to obtain the same response as that due to the sum of the two humidity variations.

One technique to create air with a desired relative humidity is to mix streams of dry and saturated wet air. Sedlachek and Ellis [94,95] produced relative humidities ranging from 3% to 100% by combining dry air with wet air obtained by bubbling air through a 5 ft column of water. Padanyi [80] had a similar system in which the ratio of stream volumes could be manually selected to create changes in relative humidity in 5% increments.

The classical method used by chemists to maintain a desired relative humidity in a small test chamber is to use a combination of a desiccant, or drying element, to remove moisture and a moistening element such as a wet sponge to add moisture.

Such a system can produce stepwise changes in relative humidity. One problem with such a system is that the dessicant periodically becomes saturated.

Another technique to obtain rapid changes in relative humidity is to enclose the system so that a small volume of air has to be humidified or dehumidified. The air is circulated through a small glass tube containing calcium nitrate crystals to reduce the humidity from 100% to about 50% [10].

Commercially available temperature–humidity chambers use computer-controlled feedback systems to continuously vary the chamber temperature and relative humidity. A wet bulb–dry bulb sensor monitors the humidity, and a thermocouple monitors the chamber temperature. The measurements are compared by the computer to programmed setpoints of a temperature and humidity versus time regime. Any discrepancy between the actual and programmed humidity or temperature activates the controls to either raise or lower the temperature and humidity to bring them into agreement with the program. A steam system to add moisture and a refrigeration system to remove moisture by condensation are used to control the amount of moisture in the air at a given temperature and so to create the desired relative humidity. Excess moisture collects on the chamber floor and is removed through a drain. A fan mixes and circulates the air in the chamber. The humidity range actually attainable by these systems is not as wide as that possible with systems based on mixing moisture-saturated and dry air, often only extending from 30% to 90% maintainable over time periods of hours or days. These devices are only as good as their control systems.

IV. EXPERIMENTS ON STRUCTURAL RESPONSE

Experiments on moisture-accelerated creep fit into three broad categories. The first type investigates the response of a paper structural component including the pulp fiber internal substructure such as lamellae or the fibrils, a single pulp fiber, and the interfiber bonds. Another experimental strategy measures the response as reflected in thermodynamic variables such as the hygroexpansion, the strain, the strain rate, and the temperature. The third variety of tests emphasizes the chemistry involved, in particular the sorption reaction between moisture and cellulose, as measured by the changes in a mechanical modulus over time. Experimental results aimed at describing the structural response provide a foundation for analyses of moisture-accelerated creep.

A. Single-Fiber Creep

A single fiber is the fundamental structural component of paper. The creep behavior of the fiber in different types of ambient relative humidities should be an important determinant of the sheet behavior and should give clues to the physical mechanisms underlying moisture-accelerated creep of a sheet. A single fiber elongates under constant moisture content tensile creep. The graph of the tensile creep strain with respect to time for a single fiber has the form of a logarithmic curve, as does the graph of creep for the full paper sheet in constant relative humidity. The creep strain at a given time is greater in a constant 90% RH than in a constant 50% RH test. However, a single fiber shortens in tensile creep during periods of relative humidity increase, even though a paper sheet lengthens. This surprising result was originally observed by Byrd in 1972 [17]. The creep strain at the maximum relative humidity in the cycle is actually smaller after the first cycle in Byrd's single southern pine pulp fibers loaded to about 45% of the fracture stress and in a variable relative humidity atmosphere cycled between 35% and 90% RH. Byrd attributed this behavior to fibril angle changes. Some 20 years later, Sedlachek and Ellis [94,95] confirmed that after the first relative humidity cycle the axial length of a fiber under a creep load decreases as the humidity is rapidly increased from 50% to 90%. They suggested that this is due to anisotropic fiber swelling, which is much greater in the radial than in the longitudinal fiber direction. Anisotropic fiber swelling occurs with fibril angle increases.

The cross-sectional area of Byrd's fiber was 19,820 μm^2. The load of 3.76 g produced a stress of 1.861 MPa, which was about 45% of the failure fiber stress. Sedlachek's fiber had an area of 537 μm^2 for a stress of 368 MPa at a 20 g load, a stress almost 200 times that of Byrd's. Sedlachek and Ellis did not state what percentage of the failure load this was. This stress difference may at least partially explain the dissimilar creep behavior of their fibers. The stress in Byrd's fiber may not have been great enough to overcome the axial shrinkage in the increasing humidity periods. No one has tested one type of single fibers over a range of creep loads in variable relative humidity.

In either constant or variable relative humidity, the creep strain magnitudes observed by Byrd in a single fiber greatly differed from those in a sheet. Furthermore, Byrd's creep tests [17] showed that single fibers shorten more than fibers embedded in a sheet in a variable relative humidity atmosphere. The fiber network structure restricts the fiber strain. His press-dried handsheet specimens in constant relative humidity reached only a strain of nearly 1%, compared to a single fiber that strained over 4% after 1000 min and 13% after 10,000 min. In variable relative humidity, the sheet strained about 1.2%, whereas the single fiber contracted. The cycle took 20 min to change from 35% to 90% and 60 min to return. Byrd did, however, condition the handsheets by cycling the relative humidity several times to relieve drying stresses and then held the relative humidity at 50% before the load was applied. This may explain why it took more cycles for the moisture-accelerated creep strain in a sheet to exceed that in his constant 90% RH creep tests than in other experimenter's tests.

Sedlachek compared single-fiber creep of separate specimens conditioned at two different constant relative humidities. The fibers were dried under restraint and then conditioned at constant 50% or 90% RH with no load or humidity cycling in the conditioning stage. Sedlachek obtained tensile creep strain magnitudes in constant relative humidity of the same order as those obtained by Byrd. Specimens under a tensile creep load of 20 g were subjected to 20 min cycles between 50% and 90% RH. The relative humidity was changed from 50% to 90% and vice versa as rapidly as possible, within about 15 s, and held at each limit for 10 min, creating a step function relative humidity loading. If the specimen was conditioned at 50% RH, additional creep strain beyond that observed in constant relative humidity tests was observed, with most of the additional strain occurring in the first cycle. On the other hand, in those tested after a 90% RH conditioning period, no accelerated creep strain was observed. In both cases, the fiber under a creep load contracted as the humidity changed from 50% to 90%. Fiber shrinkage was observed during the cyclical creep tests, after conditioning at 50% RH, as the environment was transformed from 50% to 90% RH so that the strain at 90% was less than at 50% (Fig. 3). However, in contrast to Byrd's single-fiber results, the net strain in her variable humidity test was greater than that at 90% RH. Most of the creep in Sedlachek's tests occurred while the humidity was held constant, whereas in Byrd's tests the humidity was continuously varied.

The single-fiber creep strain rate is approximately the same at a given relative humidity whether the relative humidity is varying or constant in Sedlachek's tests [94]. She verified this by separately plotting the creep strains at 90% RH from three types of tests, a constant 90%, a cyclical 50% to 90%, and a cyclical 90% to 50% RH creep test. Time shifts of these three graphs produce a master curve. Her statistical analysis supports the proposition that the strain rate at the instant the relative humidity reaches 90% in a variable relative humidity creep test is no greater than that in a constant 90% RH creep test. It should be emphasized at this point that this result is for single fibers under fairly rapid humidity cycling. There is no guarantee that it also holds for creep tests of a sheet, a network of fibers. These results, as Sedlachek points out, do not include the behavior of the interfiber bonds. A sheet made of fibers of the same type as those used by Sedlachek did exhibit moisture-accelerated creep [28]. The handsheets, dried under restraint, had a creep strain rate that averaged only one-twentieth of that observed for the single-fiber creep.

Sedlachek [94] also verified that the strain magnitudes for creep tests of single fibers were the same for both constant and variable relative humidity creep tests. If a specimen is conditioned at 90% RH and the creep load is then applied while the relative humidity is cycled between 90% and 50% in steps, in those periods in which the humidity is 90% the strain coincides with that at constant 90% RH (Fig. 4). If the specimen is conditioned at 50% RH, then the cyclical creep strain curve, when cycling between 50% and 90% RH in steps, is greater than the constant 90% RH

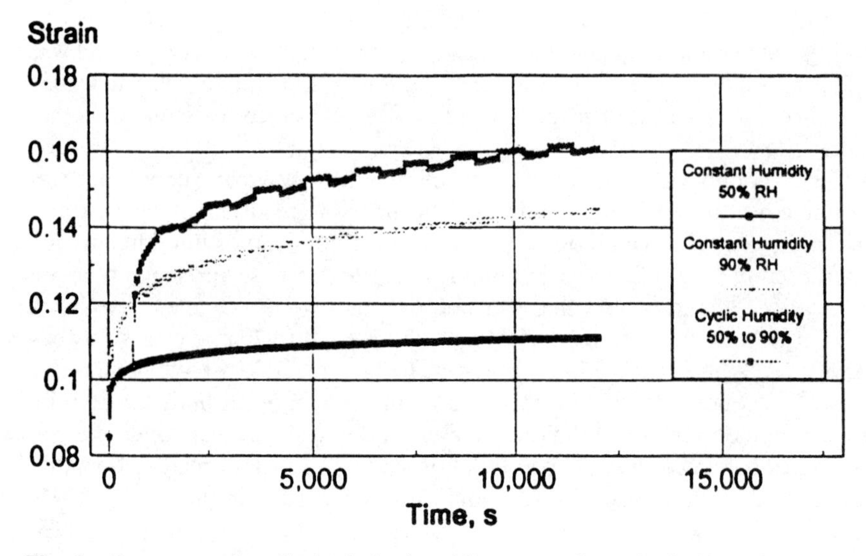

Fig. 3 Constant and cyclical relative humidity curves for a single fiber after conditioning at 50% RH. (From Ref. 95.)

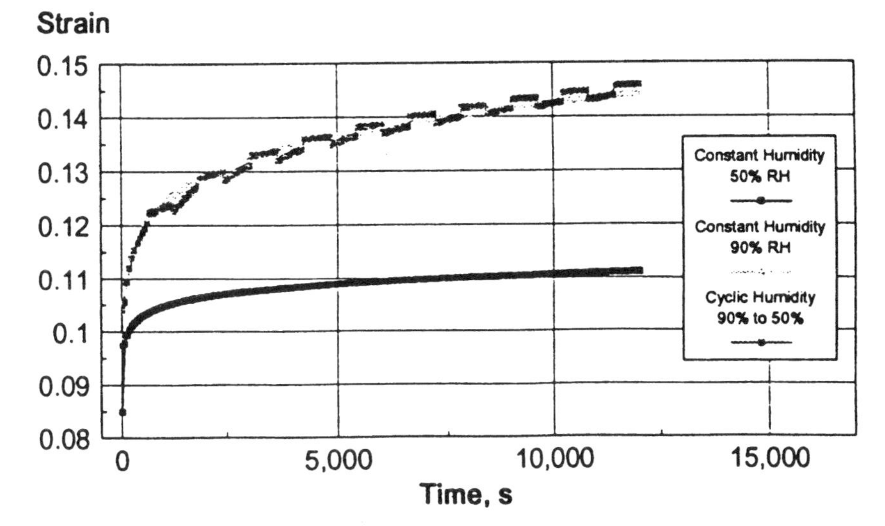

Fig. 4 Constant and cyclical relative humidity curves for a single fiber after conditioning at 90% RH. (From Ref. 95.)

creep curve. Most of the difference in strain occurs in the first step after the load is applied and the humidity is changed from 50% to 90% (Fig. 3). Sedlachek concluded that creep under a step change in relative humidity in a single fiber was not accelerated in either case because the strain rate was unaffected by moisture changes, in the sense that during the periods when the relative humidity was held constant the strain rate was the same in a constant relative humidity creep test. She suggested that the first cycle be examined more carefully to identify the cause of the large increment of strain during the first humidity change.

Because it was conjectured that hemicellulose exerts a large influence on tensile creep response, Sedlachek [94] raised the issue of whether the presence of hemicelluloses facilitates creep of the fiber. Her cyclical relative humidity tests of the creep behavior of southern pine (loblolly pine) single-fiber specimens with hemicellulose removed showed the same creep as the untreated fibers. The differences in comparison with those without the hemicellulose extracted were not as great as the scatter of response among different fibers of the same type. The presence of these structural polymers does not explain the additional observed creep in variable humidity.

If moisture-accelerated creep is not a single-fiber phenomenon, then its presence in a paper sheet must involve the interaction of fibers at their bond with other fibers.

B. Fiber Structure Behavior

The structure of the pulp fiber leads to the anisotropic swelling that experiments suggest is related to moisture-accelerated creep in materials made of networks of cellulosic fibers. The wood pulp fiber has primary and secondary structures. The secondary sublayers, denoted S1, S2, are made of helical fibrils. The S2 layer is by far the thickest. It contains about 80% of the total cellulose, but the total cellulose is no more than 50% of the dry weight of the fiber.

Byrd's test [17] shows that the fiber's structural behavior in variable moisture content creep is qualitatively different from that in constant moisture content creep. The fibril angle is the angle between the helically wound fibrils in the S2 layer and the longitudinal axis of the wood fiber. The contrasting creep strain behavior of a single fiber in constant versus variable humidity is consistent with the changes in fibril angle. To simplify, the fibril angle on a tube with a helix drawn on it increases if the tube shortens while its diameter increases and decreases if the tube lengthens as its diameter decreases. The fibril angle decreases in a constant moisture tensile creep test but increases in a cyclical relative humidity tensile creep test. The fibril angle typically ranges between 10° and 30°, although the fibril angle can be smaller. Byrd found that in constant moisture content creep of his southern pine bleached kraft handsheets at 90% RH, the fibril angle decreases from 19.2° to 16.8°. However, the fibril angle increased from 19.2° to 21.1° in variable moisture content creep of sheets in an environment in which the relative humidity was cycled between 35% and 90% RH. The magnitude of the change in both cases was around 10%. Prior to Byrd's tests, both Jentzen [63] and Hill [55] had concluded that single-fiber creep in constant relative humidity involves fibril motion. The radial dimension of the fiber decreases in constant relative humidity tensile creep but increases in variable relative humidity tensile creep. The process is more complex than the often proposed hypothesis that additional plasticizing layers of moisture molecules allow the polymer molecules to slip on one another more easily. The swelling of the fiber during variable relative humidity may be a consequence of free volume generated in the polymer fiber due to the excitation of molecules by the nonequilibrium ambient conditions. It is known that the free volume in polymers is minimized at equilibrium because the intermolecular distances are reduced in the equilibrium state.

Moisture-accelerated creep is found in materials made of fibers that exhibit anisotropic swelling. By anisotropic swelling is meant that the swelling of a fiber is greater in the fiber radial directions than in the fiber longitudinal direction. Page's observations [82,83] of the interfiber bonds show that as paper dries, each fiber is subject to up to 30% transverse shrinkage but only 1–2% longitudinal shrinkage. The secondary layers in wood pulp fibers, as well as other cellulosic fibers, are built up of lamellae, as evidenced by the observation of Page and De Grace [84] that beating causes the S2 layer to split into two to 10 layers. These concentric lamellae are themselves built up from fibrils. Lamellae separation may be a primary mechanism in fiber radial swelling and so may play a major role in moisture-accelerated creep by producing anisotropic swelling.

Some evidence for the role of lamellae lies in the observation of the behavior of binders in producing improved wet strength in paper. Formaldehyde is no longer in use as a binder owing to its health risks, but other small molecule binders may behave similarly. Covalent cross-links of formaldehyde in wood pulp paper restrict the swelling as moisture is adsorbed but do not seem to carry any of the load in the resulting higher wet strength. Caulfield proposed a model to reconcile these two effects. Caulfield [23] explained both the lack of swelling and the increased wet strength of formaldehyde-treated papers as being due to cross-linking between adjacent lamellae to prevent the entry of moisture and to hold existing hydrogen bonds close together. Caulfield [24] suggested that in the manufacture of the paper with binder, the moisture initially opens space between the fiber lamellae to allow formaldehyde to enter and link. The fiber is then dried, allowing cross-links to form and bringing hydroxyl sites on cellulose molecules close enough to hydrogen-bond. Subsequent increases in moisture will not again open spaces between the lamellae. These cross-links reduce the radial swelling of the fibers.

Other polymeric fibers exhibit a correlation between anisotropic swelling and accelerated creep. For example, Kevlar fibers swell anisotropically, and sheets made of Kevlar show moisture-accelerated creep [107]. However, drawn nylon fibers are hydrophilic but do not swell anisotropically, and variable ambient relative humidity does not accelerate the creep of nylon. For the cyclical relative humidity tests on nylon 6,6 fibers reported by Wang *et al.* [107, Fig. 8], the creep curve oscillated between and reached the creep curves obtained for the constant relative humidity at each extreme of the cycle. Not every hydrophilic material exhibits moisture-accelerated creep.

Kevlar is nearly 100% crystalline. According to Ericksen [32], it also has an axially oriented fibrillar structure in which the fibrils are tied together by other molecules. Kevlar 29 and 49 fibers seem to consist "of a system of sheets regularly pleated along their long axes and arranged radially" [32, p. 744]. These pleats lie in a plane transverse to the fiber axis. Ericksen could find no evidence that they play any role in Kevlar creep, although the pleats can be pulled out in a tensile test. The sheets in Kevlar lie in a radial plane containing the fiber axis, whereas the lamellae in wood pulp lie in circumferential sheets around the fiber axis.

Ericksen proposed that creep in Kevlar fibers is due to crystallite rotation with respect to the fiber axis by shear along the crystallite boundaries. The amount of shear depends on the motion of hydrogen and van der Waals bonds along the crystallite boundaries. The crystallites, which are 6–18 nm long, are bundled in a mutually parallel orientation into fibrils, much as in cellulosic fibers. The diameters of the crystallites are assumed to be a primary determinant of the amount of creep strain possible. There may be voids or amorphous material in the boundary regions. Ericksen computed that in a 12 μm diameter Kevlar fiber there may be as many as 1500 crystallites intersecting a cross-sectional plane, making a large amount of boundary surface available. A similar mechanism in metals, that of sliding along grain boundaries, also contributes to metal creep. Kevlar fibers increase in length during moisture-accelerated tensile creep, whereas paper fibers in tensile creep shorten during an increase in relative humidity. In tensile creep of a paper sheet, it is likely that the network structure of the pulp fibers prevents axial contraction of the fiber. Any hypothesis for the mechanism must account for these differences.

Experiments reported by Wang et al. [107] and by Ericksen [32] indicate that the potential for creep strain increases as the original crystallite angle with repect to the fiber axis increases. Their Kevlar 29, whose initial angle of 11° was the largest among the materials they tested, exhibited creep strains on the order of 2%. Kevlar 29 has a failure strain in tension of about 4%. The apparently analogous fibril angle in cellulose is around 20°, with strains ranging up to 3–4%. The stiffness of the Kevlar fiber also decreases with an increase in this orientation angle. In a varying relative humidity environment, the angle of inclination of the crystallites increases.

The creep strains in their nylon tests were greater than in Kevlar in the same time frame, about 6% strain, even though nylon does not undergo moisture-accelerated creep. Creep in drawn nylon fibers may be due to the breaking of hydrogen bonds in the amorphous region by moisture and stress and to the subsequent motion and re-forming of the bonds. However, nylon apparently does not have a fibril structure or lamellae that allow for anisotropic swelling. The component response in these different fibers suggests that creep behavior of a hydrophilic material in variable ambient relative humidity is probably governed by its supramolecular structure.

C. Formation of Microcompressions

The network structure of paper can influence moisture-accelerated creep through the behavior of the interfiber bonds. Microcompressions form in the interfiber bonds during the manufacture of paper from a pulp slurry because of the anisotropic contraction of the swollen wet fibers as they are dried. A weak interfiber bond is created when two fibers approach each other closely enough in the slurry that surface tensions can interact. As moisture is removed, the bond strength increases. The interfiber bonds initially form while the fibers are still immersed in water, well before the shrinking of the fiber in drying [86]. Because the bond is partially formed before shrinkage begins, the transverse contraction of a crossing fiber induces compressive stresses in the contact area between the two fibers and causes the fiber–fiber bond contact area to wrinkle, a local buckling. The wrinkles, called microcompressions, are approximately transverse to the longitudinal axis of the weaker fiber. The microcompressive wrinkles are typically a small portion of the bond area and do not always extend across the full fiber width. The wrinkling is initiated in the fiber with the least compression reistance at the fiber–fiber bond contact area. The S2 layer resists axial compression better than the S1 layer, so an S1–S2 bond can break when microcompression starts.

The microcompressions form unrelaxed sites in the fibers making up a dried sheet and induce delayed elastic behavior under loads. Usually a polymer relaxes to an equilibrium state when an applied load is removed. Delayed elastic strain occurs if the application of stress is required to produce relaxation from an unrelaxed state [2]. The polymer can be returned from the relaxed state to the original unrelaxed state by a "backstrain" from the surrounding material when the load is removed. If the microcompressions play a role in the physical mechanism of moisture-accelerated creep, then moisture-accelerated creep of paper involves delayed elastic behavior related to the release and re-formation of the microcompressions in the fiber–fiber bonds.

Microcompression formation can cause from 2% to 15% in-plane shrinkage of an unrestrained paper sheet. Drying handsheets under restraint reduces the number of microcompressions formed in the interfiber bonds and also affects the material properties. The apparent elastic modulus is increased, the tensile strength is increased, and the strain to failure is reduced [96]. One conjecture is that in paper made under drying restraint or wet straining, not only are fewer microcompresions formed, but also free fiber segments are straightened and help carry the load; stress is redistributed over more fibers. The extent of drying is measured by the percent solid content of the water–pulp slurry. The mechanical properties are most affected during the period of 30–60% solids content [61]. Combinations of free, restrained, and wet strain drying can be used to control the mechanical properties. Setterholm and Chilson [96] found that drying restraint affects the tensile strength of kraft pulp papers more than those of sulfite pulp paper. For a fixed density and under restraint, the tensile failure stretch was greater in unbleached kraft pulp paper than in bleached kraft pulp paper, which in turn was greater than in sulfite pulp paper. The reduction in failure strain was about 70% if the sheet was restrained in drying. Subsequent exposure to a constant 90% RH, environment did not significantly reduce the improvement in properties of the sheets dried under restraint. In general, kraft papers are stronger than sulfite paper. Up to 80% of the lignin and 50% of the hemicellulose is dissolved in a kraft cook. Cellulose can be more susceptible to hydrolysis after a sulfite pulping process. Bleaching may either remove the remaining lignin or brighten it. But the question is whether possible chemical reactions with moisture or mechanical properties like conformability and the ease of microcompression formation are more important for moisture-accelerated creep. Although the strength of the fibers and the sheet density also affect these material properties, the reduction in the number of fiber–fiber bond microcompressions may also contribute. Strain in a sheet may arise from release of microcompressions. Preventing the formation of microcompressions would then be expected to reduce the strain to failure. If so, microcompression behavior might also be expected to be part of the physical mechanism of the accelerated creep in a variable relative humidity environment. The same mechanism that allows a larger strain to failure in freely dried paper may also permit larger moisture-accelerated creep strains. In fact, the variable relative humidity creep rate is greater in sheets freely dried than in those dried under restraint [101]. An explanation may be that in a freely dried specimen, more microcompressions can form when the sheet is manufactured and thus more strain can be recovered from the release of microcompressions during variable relative humidity creep.

To see if the microcompressions could be recreated after creep, oven drying was used to remove all moisture remaining after a few cycles of variable relative humidity creep [51]. The idea was to mimic the drying of paper in manufacture. Oven drying does not return a paper specimen to its original condition after a variable relative humidity creep test. A handsheet of southern hardwood made of pulp beaten to 208 Canadian Standard Freeness was subjected to four linear 30% to 90% relative humidity cycles, oven dried for several hours at 105°C, and then subjected to four additional cycles. The 105°C temperature is well below the glass transition temperatures of hemicellulose and lignin, which are greater than 150°C. The magnitude of the strain on the second set of cycles was about 50% less than that on the first set. The initial increase in relative humidity on the second set again caused a large strain increase (Fig. 5). One explanation is that oven drying caused only a partial re-formation of the microcompressions present before the first set of relative humidity cycles. The drying may induce hornification, the process by which the formation of new bonds between microfibrils irreversibly reduces the ability of the fiber to swell [72, p. 104]. If so, these tests provide further evidence that fiber–fiber bond microcompressions resulting from fiber radial swelling behavior play a key role in moisture-accelerated creep.

D. Fiber–Fiber Bond Area

The total interfiber bonded area is the surface area of a sheet specimen less the free surface area of the fibers. The total interfiber bonded area is affected during moist-

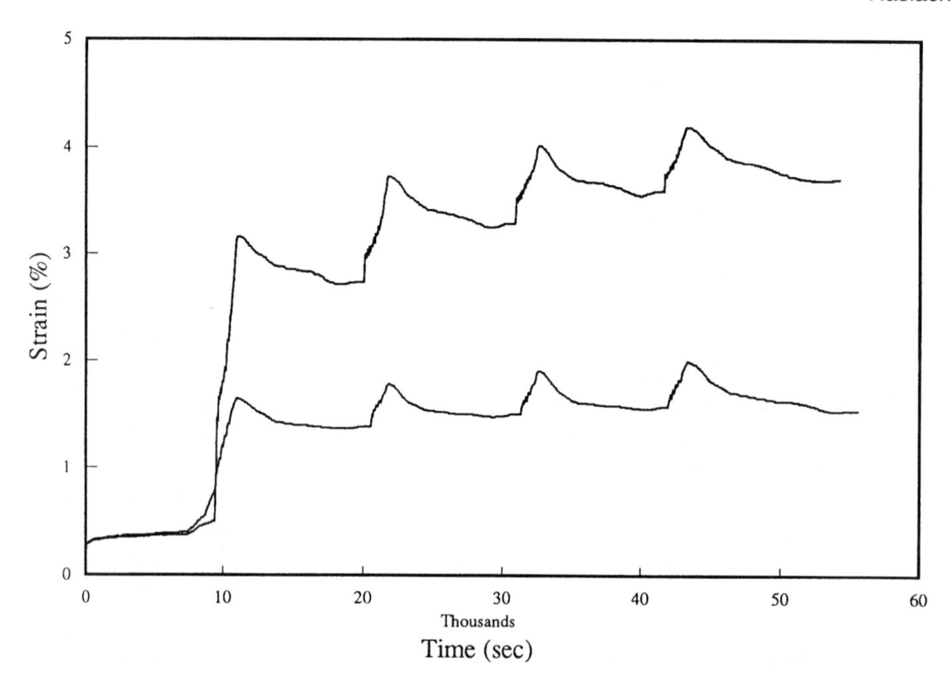

Fig. 5 Oven drying between two sets of relative humidity cycles. Lower curve after drying at 105°C. (From Ref. 51.)

ure-accelerated creep. The light scattered from the paper surface can be correlated to the creep strain under moisture-influenced creep [17]. The light scattering is also related to the behavior of the bonds using Nordmann's development of the Kubelka–Munk theory. The light scattered from a paper sheet is proportional to the free surface area in the sheet (see Nordman [77] and Nordman et al. [78] as well as Van den Akker's critique [106]). The light scattering coefficient is the slope of this linear relationship. The change in light scattering coefficient measures the amount of bond breaking in a loaded sheet. Constant relative humidity creep has a small, but detectable, effect on the light scattering coefficient in the sheet. In Byrd's constant 90% RH creep test, the scattering coefficient, when measured after 5000 min, decreased about 2% at 0.7% strain for a creep load equal to 40% of the 90% RH failure load, was slightly less than zero for a 55% load, and was slightly larger than zero for a 70% load. Viewing the scattering coefficient as a measure of unbonded area, this suggests that for low loads under constant relative humidity, the mechanism of creep may be the partial release of microcompressions with a resulting decrease in the proportion of unbonded area. The tensile creep load thins the fiber, as exhibited by the decrease in fibril angle, and may flatten some wrinkles in the fiber–fiber bonds. At higher loads, because the net change in unbonded area is nearly zero, such unwrinkling, which decreases the proportion of unbonded area, may be balanced with partial bond fracture, which increases it.

In variable relative humidity creep, the amount of light scattered depends on both the deformation and the creep load. The scattering coefficient increases as the deformation increases to 1.5%. Further increases in creep deformation produce a slight reduction in the scattering coefficient [17, p. 250]. Then cyclical relative humidity creep probably involves some partial bond failure, because the unbonded area increases. Byrd measured the scattering coefficient after 10 relative humidity cycles at various percentages of the failure load. For creep at 40% of the failure load, the scattering coefficient increased by about 7% at 1.1% strain. At the higher loads there is less increase in the scattering coefficient than at the lower loads for cyclical relative humidity. In fact, at the 55% load the change in scattering coefficient was 4% with a 1.5% creep strain, whereas at the 70% load the change was 2% but the final strain was 2.4%. The radial fiber swelling, as indicated by an increase in fibril angle, during variable relative humidity interacts with the tensile creep load. A possible mechanism to explain Byrd's results is that the radial fiber swelling increases the unbonded surface area. At higher loads, more of the wrinkles in the interfiber bonds are flattened, and the tension in the fiber resists radial swelling, thereby reducing the increase in the proportion of unbonded surface area. In variable relative humidity tensile creep, Byrd's single fiber first shortened, and then after several cycles its length remained constant. In a sheet, the interfiber bonds resist any individual fiber contraction. This resistance, in combination with the higher creep load, could induce greater tension on the bonded area. Byrd's variable relative humidity tensile creep experiments on handsheets produced a 10% thickening of the sheet [17], perhaps due to both fiber perimeter increases and some bond breaking. However, in constant relative humidity creep, the sheets thinned at larger deformations.

E. The Influence of the Pulp Furnish

The pulp finish affects the creep response of a sheet under variable humidity. Several experimenters have observed that recycled fiber paper exhibits more moisture-accelerated creep than similar virgin fibers but were unable to relate the difference to the amount or rate of moisture adsorbed. High yield pulp (63%) exhibits up to twice as much creep in a cyclical relative humidity environment as conventional yield pulp (50%) [21]. The higher lignin content leads to higher rates of moisture sorption in the high yield pulp. These results might suggest that a higher sorption rate induces more creep in a variable relative humidity environment. However, recycled fibers also exhibit about 25% more creep strain than virgin fibers, but the recycled fibers adsorb moisture at a reduced rate [21]. The compressive variable humidity creep strains in recycled fiber papers can be twice those in similar but virgin fiber papers. The difference between the strain response of a recycled and a virgin fiber handmade sheet under bending creep in 40% to 80% cyclical relative humidity is magnified at higher creep stresses [100]. Press-dried boards with high lignin content show less creep in variable relative humidity and about the same moisture sorption rate as virgin softwood low yield linerboard. Press drying would be expected to increase the number, and perhaps the strength, of interfiber bonds. Byrd conjectured that the excess creep in high yield pulp paper is related to the lower fiber–fiber bonding in the less conformable high yield fibers.

Either the drying during the recycling of fibers or press drying can induce hornification [72, p. 109]. Beating can also increase fiber conformability as well as sheet bond strength. However, tensile strength, which must depend on bond strength, is not an indicator of the magnitude of tensile creep response [41]. On the other hand, more comformable fibers might be more likely to form microcompressions in their fiber–fiber bonds. Beating might increase the radial lamellae separation in the fiber and so increase the anisotropic swelling that may be the basis of moisture-accelerated creep in paper. These results leave open the question of the relative importance of chemical effects of moisture sorption versus structural effects such as the conformability of the pulp fibers on moisture-accelerated creep.

F. Creep in Nonfibrous Cellulosic Materials

The relevance of the structure of paper as a network of fibers to its moisture-accelerated creep can be assessed by creep tests of cellophane, a nonfibrous cellulosic material. Cyclical relative humidity tests of cellophane do not exhibit significant accelerated creep, even though cellophane is made of cellulose [45]. The maximum creep strain does not exceed that at 90% constant relative humidity. In these tests, samples of 0.01 in. thick cellophane were subjected to a constant dead load at a constant 30% RH, a constant 90% RH, and four cycles linearly between 30% and 90% RH at a constant rate of change of 1% RH per minute. The cyclical 30% to 90% RH creep curve lies between the constant 30% and constant 90% RH curves (Fig. 6). During the cycles, the relative humidity was held constant for 60 min each time it reached 30% but was not held constant at 90%. The temperature was 23°C. The specimen was conditioned first at 30% RH for 24 h to allow it to come to moisture equilibrium. The dead creep load of 500 g used in these tests was one-sixth of the failure load in a uniaxial stress–strain test at 50% RH. These large strain creep tests [47] are in contrast to the small strain sonic tests performed by Berger and Habeger [12] on paper and cellophane.

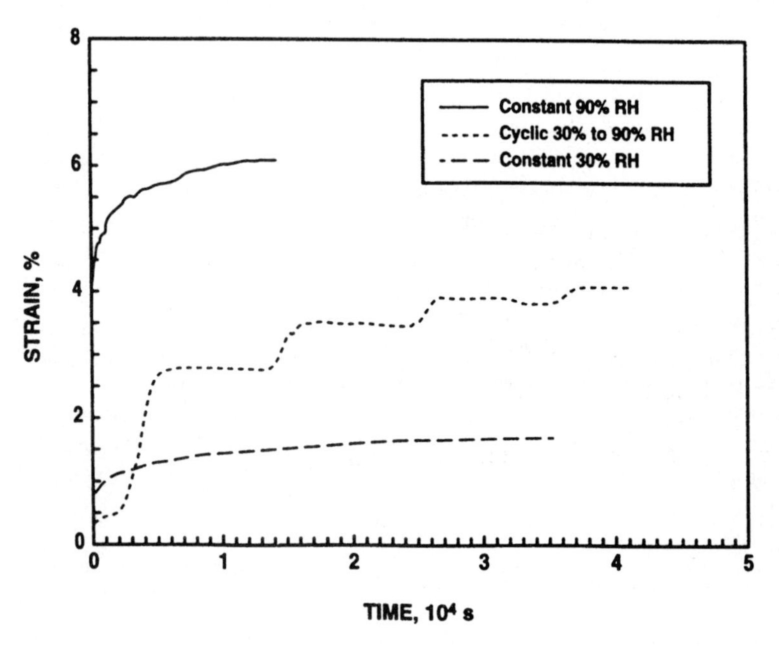

Fig. 6 Moisture-accelerated creep in a tensile cellophane specimen with constant 30% RH, constant 90% RH, and linear cycling between 30% RH and 90% RH. (From Ref. 46.)

Single rayon fibers also do not show moisture-accelerated creep [62]. Rayon is almost completely composed of reconstituted cellulose, although the crystalline structure of the cellulose is usually type II rather than the type I found in wood. Three types of rayon fibers were tested in a relative humidity changed stepwise between 10% and 80% under a load that was 40% of the tensile failure load to study whether moisture-accelerated creep depends only on the behavior of the cellulose polymer. In the tensile creep of all three fibers, the strains and rates during the 80% portion of the relative humidity cycle coincided with those obtained at a constant 80% relative humidity.

These results suggest that the network structure of paper is a significant factor in the phenomenon of variable moisture-accelerated creep. The adsorption of moisture in itself is not sufficient. Only with the network structure can anisotropic swelling play a role along with the microcompressions. Factors on a molecular scale probably have some influence because there is still a rapid increase in creep strain during the increasing portion of the initial 30% to 90% cycle on the cyclical creep curve for cellophane, as in the variable relative humidity creep of single fibers and paper sheets.

G. Other Hydrophilic Materials

A common experiment on α-keratin materials such as wool fibers or horsehair has been to calculate the behavior of a material modulus in the presence of a rapid change in moisture content. Dynamic moduli are computed under the assumption of linear viscoelasticity and a measurement of the loss tangent, tan δ (the lag between the strain response to a cyclical stress), during a rapid change in moisture content. Most of these materials show a maximum or minimum in moduli during transient moisture changes. This evidence that moisture changes affect the material properties for cyclical stress loading does not, however, prove that the material exhibits moisture-accelerated creep under constant load and variable relative humidity. This is so even in the limit obtained by taking longer periods even though low frequencies induce larger spikes.

A wool fiber hung vertically as a torsional pendulum with a small mass at its lower end and allowed to twist at very low frequencies in a stepwise changing relative humidity exhibits a minimum dynamic rigidity below that at equilibrium just after a humidity change from 0% to 61% [73]. A similar result was obtained with nuclear magnetic resonance techniques [97]. This response is reduced in smaller relative humidity changes [79]. The dip in modulus is reduced as the rate of relative humidity change is lowered [73]. This transient effect was attributed by Mackay and Downes [73] to the moisture initially adsorbed in the outer fiber layer. The differential swelling across the fiber radius, they proposed, sets up a stress that ruptures bonds in the fiber.

Dynamic uniaxial tests of a single fiber in a 4 μm extension at frequencies up to 200 Hz in wool and at 84 Hz for horsehair also showed a maximum in the loss tangent, tan δ, as the relative humidity was suddenly increased around the dry fiber [30]. The maximum in the phase angle δ is consistent with a minimum in the dynamic modulus, assuming that the material is actually linearly viscoelastic in this loading range.

Cotton poplin fabric has been shown to wrinkle more in an increasing relative humidity environment [71]. The specimens were conditioned at 30% RH, wrinkled during humidification to 65% RH, then allowed to come to equilibrium. The resulting wrinkle heights were greater than those obtained during wrinkling at a constant 65% RH. Cross-linking with binders to reduce the hygroscopicity of the material usually reduces the wrinkle height. Drying can cause recovery of the wrinkles.

Chain mobility in cotton and wool fibers as they were humidified from 0% to 50% RH was also measured by the nuclear magnetic resonance technique and shown to increase [97].

One explanation for the decrease in fiber rigidity offered in Ref. 71 is that diffusion builds up internal stresses that are relaxed in a second stage during which the material is very plastic and permits large permanent deformations. In the first stage, during which torsional rigidity decreases to a minimum below that of equilibrium, hydrogen bonds are broken or weakened. The bonds later re-form to create a gel state. Cross-linking helps reduce this drop in rigidity.

The loss tangent, tan δ, in paper and cellulose also attains a similar maximum as the relative humidity is abruptly changed. To make the measurements, the torsional pendulum test was performed at a frequency of 1 Hz in a relative humidity that began at 65%, was dropped to zero by creating a vacuum, and was then raised to 70% within a 10 min time interval [67].

That the existence of such transient behavior does not imply that the material exhibits moisture accelerated creep was demonstrated by Salmén and Fellers [93] in their axial extension tests of nylon 6,6 fibers and small paper samples. They calculated an instantaneous dynamic storage modulus and loss tangent, assuming linear viscoelasticity, by superposing a small cyclical stress on a constant stress. After conditioning by two stepwise 0% to 95% relative humidity cycles, the initial relative humidity of 95% was cyclically dropped to 0% around the loaded specimen. During these nearly instantaneous relative humidity changes, tan δ exhibited a sharp peak in both nylon and paper. The magnitude of the peak was lower at larger frequencies. However, nylon does not undergo moisture-accelerated creep (see also Ref. 107), whereas paper does. Jackson and Parker [62] suggest that the measured transients of tan δ are due to a rapid increase in the molecular motion in the amorphous region of the polymer due to moisture sorption. Further, this behavior is on a different structural level than that causing accelerated creep.

Both wood and concrete do exhibit moisture-accelerated creep. Armstrong and Kingston [6] observed that green wood specimens show accelerated bending creep when dried under a load. The compressive strains in bending were of larger magnitude than the tensile strains on the opposite face.

Dried wood specimens subjected to vapor at saturation pressure followed by drying in a stepwise cyclical pattern show large accelerated creep. If the load is removed in vapor at the saturation pressure, almost all of the strain is recovered [4]. Tests on specimens of different cross sections indicate that stresses due to non-uniform shrinkage are probably not the cause of the accelerated creep. Armstrong and Christensen [4] concluded that the deformation depends on the amount of moisture content change but not the rate at which it occurred. If the rate of imposition of saturated vapor to the specimen was reduced, the final deflection after saturation was not affected by the rate. Later bending tests of particle board showed that most of the accelerated creep occurred during the desorption portion of the humidity cycle after the first cycle [5]. The creep magnitude is larger if the specimen is initially dry.

Hearmon and Paton [54] found moisture-accelerated creep in bending tests of 2 × 2 mm cross section beech beams in which the relative humidity was varied between 30% and 90% in a stepwise fashion. The humidity was held constant at each stage for 1 day, so the specimen may have approached moisture equilibrium. The strains increased at 30% RH, a drying phase, and decreased during the wetting phase at 90% RH. The creep strain at failure after 14 complete cycles was about 25 times that of a specimen held in creep at a constant 93% RH. The load in both cases was three-eighths of the stress–strain test failure load at 93% RH.

To avoid the distribution of compressive to tensile stresses present in a bending specimen, Bažant and Meiri [10] performed compressive axial tests on hollow wood tubes made of spruce during changes between 50% and 100% relative humidity and also found that after the first moisture change the wood crept more during desorption than during adsorption. Again, these were step changes in relative humidity, so the rate of change of moisture content or even the rate of change of ambient relative humidity was not controlled. These tests of wood and particle board may involve a different mechanism of accelerated creep than a fiber network such as paper. The phenomenon of moisture-accelerated creep of wood has been captured in a mathematical model based on molecular deformation kinetics theory [43].

The first published observation of accelerated creep under variable moisture content in concrete preceded that of wood by 20 years. The bending tests of 2 in. × 2 in. cross section concrete beams 32 in. long with a transverse load at the midsection made by Pickett [90] showed that creep increased above that in constant conditions during alternate wetting in water and drying at 50% RH for periods of many days. This is called the *Pickett effect* by concrete specialists. The creep strain when either condition was held constant followed logarithmic shaped creep curves. The creep strain during constant 50% RH was greater at a given time than that in constant wetting in water [9, p. 315]. Paper behaves in the opposite manner, with its creep greater in higher relative humidities. Microcracking may be involved in the creep of dried concrete.

Viscoelastic constitutive models involving internal state variables have been developed for the combined shrinkage and creep of concrete including that accelerated during drying [9,11]. However, because concrete is an aging material, it is not clear how closely related this is to the similar phenomenon in wood or paper. The cement in concrete continuously combines with moisture in a chemical process called hydration, which is significantly reversible only under very high temperatures [9].

Bažant [8] suggests that the important characteristics of wood are that it is porous with a large range of pore sizes and that the material of the pore walls is hydrophilic. The magnitude of the moisture concentration gradient is not the cause of accelerated creep, because steady-state moisture diffusion of water does not accelerate creep in wood [3]. Moisture diffuses through the pores of least resistance. In steady state, the diffusion bypasses the micropores as it seeks the path of least resistance, because the flux varies with the cube of the pore diameter. Bažant proposes that a flux in micropores is induced by a non-steady diffusion, which then accelerates creep by locally breaking cross-linking bonds in the amorphous regions. Wood creep, in analogy with metal creep, results from the accumulation of dislocations due to the breakage of cross-links. There may be moisture flux in the micropores due to local diffusion gradients even when the global gradient is zero. However, this flux is not as significant as when the global gradient is nonzero. Bažant regards the inside of a wood fiber as a macropore; the micropores are in the fiber wall. Paper would not have such macropores if the fiber were collapsed to a nearly flat ribbon. Water in micropores can help support or transmit load, whereas that in macropores cannot. If the moisture in the micropores is disrupted, the deformation due to the applied load is affected.

A different mechanism underlying accelerated creep, due to either variable temperature or variable relative humidity, in concrete was suggested by Bažant [9, p. 316]. The accelerated creep during drying is a result of both stress-induced shrinkage or swelling and tensile cracking in combination with the aging of the concrete.

Whereas variable relative humidity accelerates the creep of wood, concrete, and paper, the mechanisms may not be the same in all of these materials.

V. EXPERIMENTS ON THERMODYNAMIC RESPONSE

Moisture-accelerated creep is a nonequilibrium process. One method to describe it is to monitor the time dependence of thermodynamic variables such as stress, strain rate, ambient relative humidity, and temperature in addition to the creep strain, which defines the process.

A. Separation of Creep and Hygroexpansive Strain

During a cyclical change in ambient relative humidity, a paper sheet undergoes in-plane swelling and deswelling, called hygroexpansion. The total axial strain under a creep load in variable relative humidity is a result of both the creep stress and in-plane swelling. To analyze variable moisture-accelerated creep, many researchers have tried to separate the experimental strain into creep and in-plane swelling components. The thickness dimensional changes that also occur are generally not measured during creep tests of thin paper specimens. Byrd [17] assumed that the creep and hygroexpansive strain superpose and "that the swelling deformation of a loaded specimen was approximately the same as for an unloaded specimen." A linear relationship between the equilibrium moisture content and the unloaded swelling strain was found for southern pine handsheets measured at five humidities between 35% and 90%. Byrd calculated the swelling strain from the companion specimen moisture content measured during a creep test. The creep component of strain was then derived by subtracting the computed swelling strain from the total strain when the relative humidity cycle reached 90%. Gunderson [38] later reported that the presence of a load on the paper specimen does not significantly affect the amount of moisture adsorbed. Other experiments have indicated that the presence of a load very slightly increases the moisture adsorbed at a given relative humidity [68]. However, the possibility that in the loaded case the swelling strain is not linearly related to moisture content cannot be rejected out of hand, especially if residual stresses and microcompressions in the fiber–fiber bonds are involved in sheet swelling. Byrd tried to remove such residual stresses by conditioning the paper with humidity cycles before performing tests. In fact, in cyclical humidity linearly varied between about 35% and 85% with a period of 4 h, Uesaka et al. [103] found linear relationships between moisture content and hygroexpansion only for handsheets dried under restraint and for fine papers in the cross-machine direction. Other papers in cyclical humidity conditions showed a nonlinear relationship dependent on the number of cycles and the humidity range, especially at higher humidities, at which irreversible shrinkage was observed. The hygroexpansion was measured on one sample, and the moisture content was determined by weighing an equivalent reference specimen.

If the swelling strain from an unloaded swelling test is subtracted from the variable relative humidity creep curves to obtain a candidate for the creep component, some implausible results can be obtained. The creep strain component curve, obtained by subtracting the unloaded swelling strain, can exhibit a temporary decrease in strain as the relative humidity increases from 30% to 90% in both a 30%–90%–30% and a 90%–30%–90% cycle in data reported in Ref. 52. Apparently, too much strain is removed if unloaded swelling is subtracted from the variable relative humidity creep curve. Some of the decrease could be accounted for by the lack of precise humidity control in the test chamber, but not all, because the decrease occurs over too long a time.

A different method of determining the creep portion of the total strain assumes that the conditioning relative humidity is the base on which the relative humidity cycles are superposed. In every cycle, when the relative humidity returned to the conditioning relative humidity of 30%, the strain at that time was assumed to be purely creep strain. In their tensile creep tests for humidity cycled between 30% and 90%, Haslach et al. [52] fit a logarithmic curve through the strain values at 30% RH. At any test time, this logarithmic curve was subtracted from the total strain; the remainder was assumed to be the in-plane swelling strain. Such a method does not require measurement of the current moisture content of the specimen. Even though this calculation is in effect the reverse of Byrd's, both methods produced a logarithmic mathematical model for the creep component of strain, $\epsilon(t) = a + b \ln(t)$, where t is time and a and b are material constants.

Another technique for computing the hygrosorptive strain during a variable relative humidity creep test was developed by Söremark and Fellers [98]. They write the total strain as the sum of an elastic component, a creep component, the hygroexpansive strain under zero load, and a stress-induced hygroexpansive strain term. The latter term, denoted ϵ_{SIH}, adjusts the hygroexpansion for the load. It is positive in the compressive case and negative in the tensile case. Its magnitude is on the order of 10–15% of the hygroexpansive strain measured under zero load [98]. Once the load is applied, this term is to remain constant. It can be estimated by comparing the amplitude of a hygroexpansive cycle under zero load to one under a creep load. These results indicate that the load affects hygroexpansion even though the influence of the load on moisture sorption is very small.

The success of a logarithmic model fit to the creep strain component is evidence that moisture-accelerated creep in paper involves delayed elastic behavior. Argon [2] summarizes the foundations of the laws of the time dependence of creep strain in materials with delayed elastic response. If the activation energies involved vary over a large range and the distribution of sites having the various activation energies is nearly constant, i.e., a nearly equal number of sites have each activation energy, then the strain is a logarithmic function of time. A candidate for the stress that is relaxed during delayed elastic creep of paper is the residual compressive stress in the fiber–fiber bond microcompressions induced by drying during manufacture of the paper sheet. The creep activation energy depends on the nature of the interfiber bonds in the paper sheet. The bonded area between each of the many pairs of bonded fibers varies over a large range. The network restraints on each fiber–fiber bond also vary greatly from fiber to fiber. Therefore there must be a large range of activation energies over the paper sheet. Paper fulfills the requirement for delayed elastic creep modeled by logarithmic functions.

B. The Creep Strain Relaxation Response

In classical uniaxial creep tests of polymers, after the material has been allowed to creep for some time under a constant load the load is removed so the relaxation behavior can be observed. The creep strain would be expected to relax to some equilibrium state by recovering strain. In this manner, one can determine whether or not the creep strain when relaxation is initiated had elastic and permanent strain components and can define concepts such as the relaxation time to characterize the material. A typical relaxation curve, under constant environmental conditions, is composed of an immediate drop in strain as the load is removed, followed by an exponential strain relaxation toward an equilibrium state. The immediate change in strain as the load is removed is called the *instantaneous elastic strain component*.

Such tests after the tensile creep of a single fiber in constant relative humidity have shown that only about 60% of the fiber strain is recovered after 12, 24, and 48 h creep loadings [95] when the load is fully removed and the humidity is held constant during relaxation. The amount of strain recovered after creep for these long times is not linear with respect to the time of the original creep test compared to that at very short creep times. This is another indication that either an important strain phenomenon occurs during the initial portion of a creep test or linear viscoelastic theories are not applicable.

Relaxation tests verify that some creep strain is recoverable during relaxation at constant ambient conditions and more of the strain is metastable in the sense that it can be recovered during variable ambient conditions. Sheet specimens of 26 lb virgin fiber corrugating medium were allowed to creep by Haslach [46] during four linear 30% to 90% to 30% relative humidity cycles. Not all of the creep strain is recovered when the load is removed and relaxation is allowed for a time equal to the creep time (Fig. 7). The recovered strain in constant relative humidity is on the order of the initial elastic strain when the load was applied. The relaxation behavior is strongly affected by the environmental conditions. At zero load, if the relative humidity continues to cycle during relaxation, more strain is recovered than if the relative humidity is held constant at 30% (Fig. 7), in the sense that more of the creep strain is recovered in a given time.

It might be parenthetically mentioned that relaxation in a variable relative humidity is reminiscent of the conditioning practice of cycling the relative humidity, which experimenters thought removed residual stresses. When one of the goals is to study the hypothesis that residual stresses influence the mechanical behavior of paper, the specimen should not be conditioned by cycling the ambient relative humidity.

A relaxation test also offers a means of estimating the average residual stress in a sheet. The standard creep relaxation test procedure is to completely remove the creep load and then measure the decrease in strain as the specimen relaxes. Because

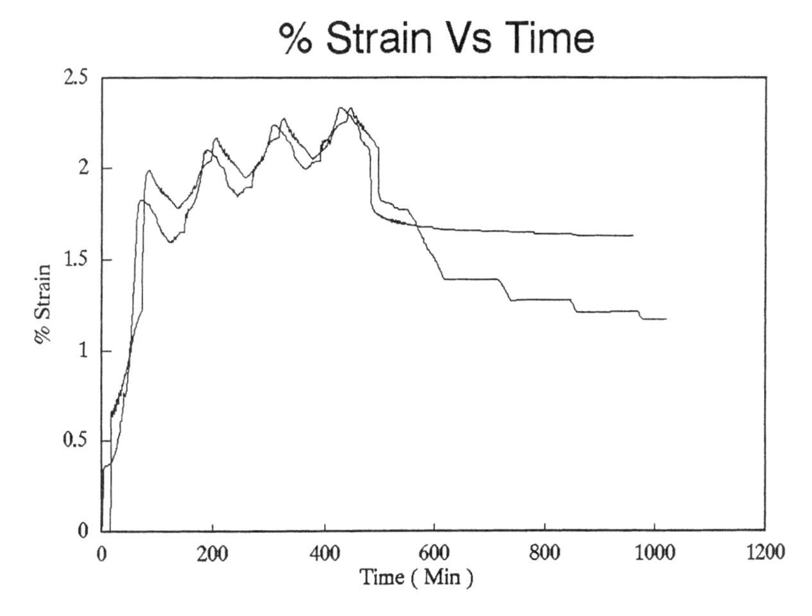

Fig. 7 Relaxation under zero load, with constant versus cyclical relative humidity during relaxation. (From Ref. 46.)

maintaining the creep load would cause the strain to increase and removing the load would cause it to decrease, in between the original creep load and zero load there may be a load reduction from the initial creep load at which the creep strain that has occurred then remains constant. This load will be called the *relaxation equilibrium load*. If the relaxation period of the test occurs in a variable relative humidity environment, the equilibrium relaxation load is that portion of the original creep load that maintains a constant mean strain over time during the relaxation period of the test.

An equilibrium load exists for paper and is the same whether the relative humidity is held constant or cycled after load has been reduced to the equilibrium load [46,48]. The existence of this load indicates that paper is not a linear viscoelastic material and cannot be represented by spring-and-dashpot models. The equilibrium relaxation load depends on only the initial creep load if the relative humidity was cycled during creep. In the tests of specimens of 26 lb virgin fiber corrugated medium, which were allowed to creep during four linear 30%–90%–30% RH cycles, the equilibrium load was about 55% of the initial creep load of either 5, 6, 7, or 9 lb (Fig. 8). This includes the case when the initial creep load was zero. When the relative humidity is held constant at 30% during relaxation at 55% of the initial creep load, the strain remains constant over time [48]. During relaxation at the equilibrium load under cyclical relative humidity, the strain cycles with the humidity in response to hygroexpansion and the amplitudes are not damped on subsequent cycles. The strain at the conditioning relative humidity remains constant in each cycle at the equilibrium relaxation load [48]. Smaller initial creep loads produce slightly larger hygroexpansive amplitudes during relaxation at the equilibrium load. Haslach did not investigate the dependence of such an equilibrium load on the creep time. If, on the other hand, the specimen is allowed to creep during several relative humidity

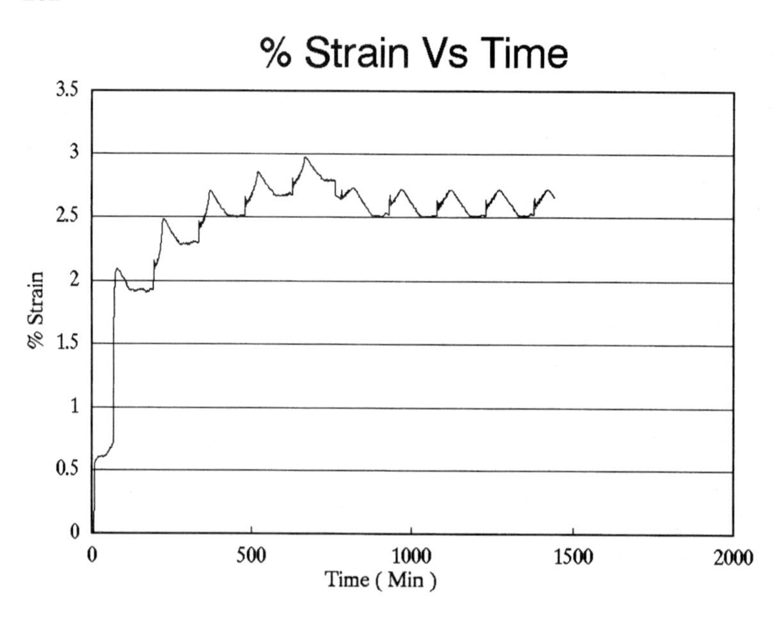

Fig. 8 The equilibrium relaxation load at 55% of the initial 9 lb creep load during cycling of relative humidity. (From Ref. 46.)

cycles and then the relative humidity is held constant for some time before initiating relaxation, the equilibrium load seems to increase slightly, perhaps because some fiber–fiber bond microcompressions re-form during the constant relative humidity interval.

Delayed elastic strain is due to stress-induced relaxation from an unrelaxed state. In the application here, the creep load relaxes the initial residual stresses in the paper. This is the opposite process from relaxation due to reducing the creep load. The relaxed stresses can be returned to their original unrelaxed state by a backstrain from the surrounding material when the load is removed. When the stress is removed, the material recovers some of the creep strain. This concept was originally used to describe the creep behavior of amorphous glasses [2]. The unrelaxed state might be represented by a relative minimum of an energy function from which the material can be removed by a stress that draws it over an energy barrier into a lower minimum. Therefore the behavior depends on the activation energy for the process under discussion. The logarithmic form of the creep strain component curve for paper is consistent with delayed elasticity, as previously mentioned. The physical manifestation of delayed elasticity when the creep load is reduced is an internal stress trying to force the material to recover its original metastable state, the microcompression stresses induced in the fiber–fiber bonds by drying the original paper specimen. Call this stress the "backstress" in analogy with viscoplasticity models of creep in metals [35]. The backstress is a residual stress that attempts to draw the specimen back into its original metastable state by re-forming microcompressions.

The amount of backstress is related to the microcompression release during the creep test. The backstress would therefore be expected to be an increasing function of the creep load and also to be affected by the radial fiber swelling induced by the rate of change of relative humidity. The backstress is zero in an unloaded specimen at constant relative humidity. If the load is reduced after creep, some recovery occurs and the backstress decreases.

The equilibrium load determined in a relaxation test is a measure of the backstress in the sheet at the time relaxation was initiated because the equilibrium load must balance the residual stress in the specimen to hold the strain constant. This type of test may offer a means to estimate the residual stress in the specimen after a period of creep.

C. Hygroexpansion and Moisture-Accelerated Creep

Hygroexpansion and moisture-accelerated creep certainly influence each other. The question is whether they arise from the same, or somehow complementary, mechanisms. The interaction of the load and moisture variation is not simple superposition. The total strain measured in a variable humidity creep test is not, as some have assumed, the sum of the creep strain and a hygroexpansive component equal to that measured on an equivalent but unloaded specimen. Insight into the relationship may be given by the difference between the hygroexpansive behavior during creep when a load is present and that during relaxation when either no load or the equilibrium load is present.

Determination of the hygroexpansive behavior during variable humidity creep requires the assumption of a method to calculate the creep component of the total creep strain. The hygroexpansive amplitude of each cycle will be taken as the difference between the maximum total strain and the strain at the conditioning relative humidity in that cycle. The latter strain will increase over time in creep and decrease in recovery. This definition is similar to the calculations for the creep strain component of strain by fitting a logarithmic model to the points in the relative humidity cycle corresponding to the conditioning relative humidity [52]. From this viewpoint, the hygroexpansive, or swelling, strain decreases in subsequent cycles in a tensile creep specimen as the relative humidity cycles are repeated. Apparently, the load "uses up" in creep some of the strain that would be available for swelling at any particular point in time. On subsequent cycles, swelling then appears damped by the load. Söremark and Fellers [98] found a similar variation of hygroexpansive strain amplitude in their tests of corrugated board with kraft liners in small strain bending in a 40% to 80% stepwise cyclical environment. In their tests, a tensile stress reduces the hygroexpansion and compressive stresses increase it. Amplitude variation is also visible in Urbanik and Lee's [105, Fig. 6] compressive creep data on corrugated containers in which the hygroexpansive strain amplitudes increase over many sinusoidal cycles (Fig. 9). This increase in amplitude occurs even though the relative humidity period is continuously varied and the container has a complex structure. Although a load may not affect the amount of moisture adsorbed, it can affect the swelling.

Furthermore, the amplitude of the hygroexpansive strain component on the same cycle of two different tests with the same linear relative humidity variation but different creep loads is smaller in the test with larger creep load. Hygroexpansive amplitude in variable relative humidity during the creep portion of the test is reduced as the creep load is increased [46,48].

The behavior of the hygroexpansion during creep relaxation under a reduced load is the opposite of that during creep. The load decrease reduces the strain

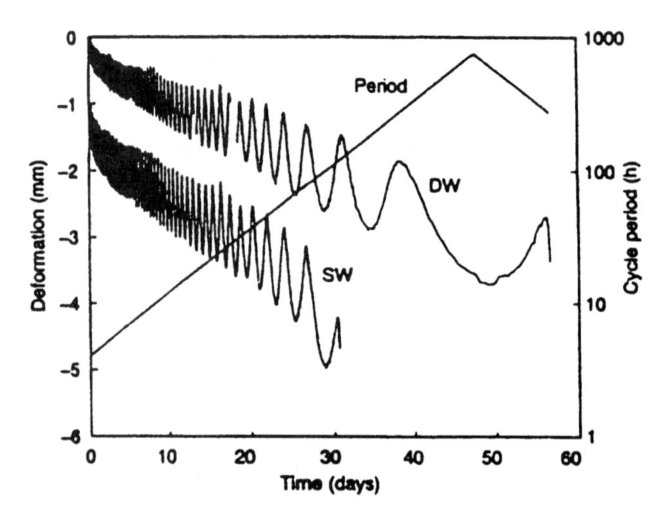

Fig. 9 Creep at variable sinusoidal relative humidity periods. (From Ref. 105.)

amplitude with respect to the conditioning humidity in comparison to the amplitudes during creep at the larger load. If the creep load is reduced to 85% of the original creep load and the relative humidity continues to cycle linearly, creep continues. More remarkably, the hygroexpansive strain cycles superposed on the creep seem to decrease in amplitude, say by 20%, after the load is reduced (Fig. 10). Alternatively, if the creep load is reduced to 25% of the initial creep load, the hygroexpansive amplitude is even smaller during relaxation and the reference creep strain at the conditioning relative humidity recovers [48]. One might conjecture that as the relaxation load is reduced to equilibrium, the amplitude will decrease until it matches the hygroexpansive strain amplitude in the unloaded specimen. The initial hygroexpansive strain amplitude must be much larger than the unloaded hygroexpansive strain amplitude and thus has room to be damped on subsequent cycles.

More strain is recovered in the relaxation portion of the test if the humidity continues to cycle rather than being held constant. The same nonequilibrium process that accelerates the creep in variable relative humidity seems to be at work. Furthermore, in cyclical relative humidity relaxation, reducing the initial creep load results in smaller amplitude hygroexpansive strain.

There is no hygroexpansive strain amplitude damping either when the specimen is unloaded or when it is under an equilibrium relaxation load. This could mean that damping of the hygroexpansive strain amplitude at the other loads occurs because there is an unbalanced backstress. During relaxation, the hygroexpansive strain amplitude increases with load because microcompressive residual stress has been released during creep. Hygroexpansive strain is resisted by the residual stress. If the initial creep load is reduced, the residual stress increases and lowers the hygroexpansive strain amplitude. On each subsequent cycle during the creep phase, there should be less residual stress.

The contrary effects of the creep load magnitude on the hygroexpansive strain amplitudes during creep and during relaxation could be explained by assuming that

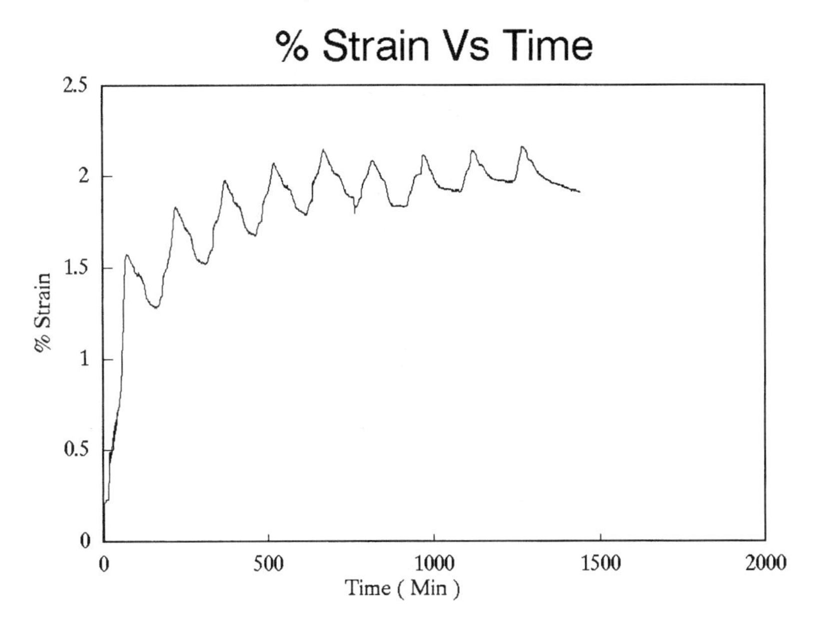

Fig. 10 Reduction of creep load by 15% causes reduction in hygroexpansion amplitude. (From Ref. 46.)

there is a maximum amount of strain that must be shared by the creep and hygroexpansive components. This strain limit might be determined by the maximum strain recoverable by the release of fiber–fiber bond microcompression. The limit could be estimated by comparing the shrinkage during the drying of unrestrained and restrained samples.

The mutual influence of creep and hygroexpansion supports the idea that in-plane sheet swelling and moisture-accelerated creep may have similar underlying physical mechanisms. The presence of a load on the paper specimen at most slightly affects the amount of moisture adsorbed. Therefore, variation in moisture content alone does not explain the relationship between hygroexpansion and accelerated creep. A structural phenomenon, such as fiber–fiber bond microcompresion release, is likely involved in both responses.

D. Rate Effects

The quantity of moisture adsorbed in wood is more important than the rate of change of moisture content for moisture-accelerated creep, according to the interpretation of Armstrong and coworkers of their results under rapid relative humidity changes [3,4,6]. A logical question is whether or not this finding carries over to paper creep in slower rate changes. For a series of tests in variable relative humidity of the same amplitude, the rate of change of the relative humidity depends on the period of the relative humidity cycle. Presumably the rate of change of moisture content is related in some way to the rate of change of relative humidity. As the period of the relative humidity cycle becomes longer and longer, the creep curve should approach that obtained at a constant relative humidity equal to that at the beginning of the chosen cycle. Therefore, the creep strain should decrease with the rate of change of the relative humidity. The amount of creep strain per cycle, the creep rate, the time to failure, and the amplitude of the hygroexpansive strain component might be affected by the period of the ambient relative humidity cycles.

One possibility is that the creep response to variable relative humidity depends on the magnitude of the relative humidity change but is independent of the rate of change. Gunderson and Tobey [41] raise the question of whether on a cycle-per-cycle basis the creep strain is independent of the rate of humidity change. The strain per cycle was nearly the same in the 70 min and 140 min ramps in tensile creep tests of virgin fiber 26 lb corrugating medium at different periods by Haslach et al. [50]. The different periods also imply different rates of change of relative humidity because the humidity was always linearly cycled between 30% and 90%. At a given time from the beginning of the test, the creep strain was smaller in the test with the longer period. Gunderson and Tobey [41] tested virgin Lake States softwood unbleached kraft paperboard in tensile creep. By rescaling the time coordinate of the creep strain versus time data so that each cycle takes the same interval on the time axis, they found that in cases in which the period was long, with ramp times of 12 and 84 h between 30% and 90% relative humidity, the strain per cycle was nearly the same. However, for the shorter ramp times of 10 min and 1 h, the creep strain per cycle was not independent. In fact, as the ramp time is reduced, the strain per cycle is also reduced [41, Fig. 7]. It has not been verified that the cycle-by-cycle strain is unaffected by the rate of relative humidity cycling.

The effect on creep strain of bringing the specimen to equilibrium by forcing a hold time at constant humidity at each of the extremes in the relative humidity cycle is not clear [41]. The small sample of experiments by Gunderson and Tobey indicate that for the same hold time, 2 h, the longer ramp time of 1 h produces more strain per cycle than the shorter ramp time of 2 min. A ramp time of 2 min with a hold time of 12 h produced nearly the same creep strain per cycle as the 1 h ramp with a hold time of 2 h. Much more work is needed to determine whether comparing the creep strain per cycle is a useful method of investigating moisture-accelerated creep. If it were, then the area under each cycle of the creep curve should have a physical interpretation.

One suggestion for the cause of moisture-accelerated creep is that a stress gradient is set up in either the web or the fiber by a gradient in moisture content resulting from a cyclical relative humidity environment. Gunderson and Tobey [41] interpreted their longer period experiments as showing that this did not appear to be the case because if such gradients were important a fast ramp would accelerate the creep more than a slow ramp.

Creep Stiffness Index Isochronous stress–strain curves can be obtained from a series of creep tests performed at several different creep loads but all in the same humidity environment. A time from the beginning of the creep test is fixed. The strains, at the chosen time in each test, are then plotted against their corresponding creep stresses to create a creep analog to a stress–strain curve. The *creep stiffness index* is the initial slope of such a curve extended to zero strain [101]. The index is well defined at all stresses if the isochronous curve is linear. This modulus is dependent on the choice of time. The smaller the index, the more rapidly the creep strain increases at the chosen time. There is no known relationship between the creep stiffness and the classical stiffness, the elastic modulus, obtained from a stress–strain curve. Isochronous curves were first applied to the creep of paper for the case of constant ambient relative humidity [64].

Isochronous stress–strain curves can also be obtained for a series of creep tests in variable relative humidity [100,101]. The isochronous curves are linear for small creep stresses in variable as well as constant relative humidity, in the bending tests of corrugated board by Söremark and Fellers. The creep stiffness index distinguishes between constant and variable relative humidity creep. As expected, the constant relative humidity test has a larger creep stiffness than the test in which the specimen was subjected to a stepwise cycle betwen 40% and 80% RH (Fig. 11a). The tensile and compressive indices in constant relative humidity are the same [101]. However, in variable relative humidity, the creep rate of the board is greater in compression than in tension, and both are greater than the strain rate at constant relative humidity. Recall that a single fiber has the same tensile creep strain rates in constant and variable relative humidity [94]. The indices in the machine direction under both constant and variable relative humidity are greater than those in the cross-machine direction [101, Fig. 9]. Boards made of freely dried sheets had higher creep strain rates in variable relative humidity than the boards dried under restraint in both the tension and compression tests [101] (Fig. 11b). Drying under restraint reduces the accelerated creep of paper. However, there is no guarantee that the isochronous curves obtained from higher creep loads and strains will be linear. The creep stiffness can be used as tool to compare the response of different types of paper to variable relative humidity environments.

Cycling Period and Rate Effects A slower rate or longer relative humidity period increases the failure time of a single sheet of corrugated medium in tensile creep at 40°C [50]. A specimen under a 140 min ramp failed in the first cycle, whereas one under a 280 min ramp survived for an additional cycle at the same creep load. The relative humidity period has the opposite effect on the failure time for the compressive creep of corrugated containers. With the humidity cycled between 50% and 90% and periods of 12, 24 and 48 h, the specimens of Leake and Wojcik [70] subjected to the longer period cycles failed first, but those subjected to a constant 90% relative humidity survived longer than any of those tested in cyclical humidity. The extra time needed for moisture diffusion in the corrugated board has been offered as an explanation of the difference between paper sheets and corrugated board [105].

The length of the relative humidity cycle period influences compressive moisture-accelerated creep of corrugated containers. The goal of the experiments of Urbanik and Lee [105] was to propose a relative humidity schedule that could serve as a standard test of the tendency toward moisture-accelerated creep of a particular type of paper. The specimens were loaded to either 17% or 25% of their maximum static compressive strengths. In other tests, a single-wall specimen was loaded to 13% and a double-wall specimen to 17% of its compressive strengths. Those at 25% did not show a primary region, so the lower load was used. One must raise the question of whether or not the ideas of primary and secondary creep, taken from metals, have meaning for a material that is a network of composite polymeric fibers. The idea of a secondary creep region is of

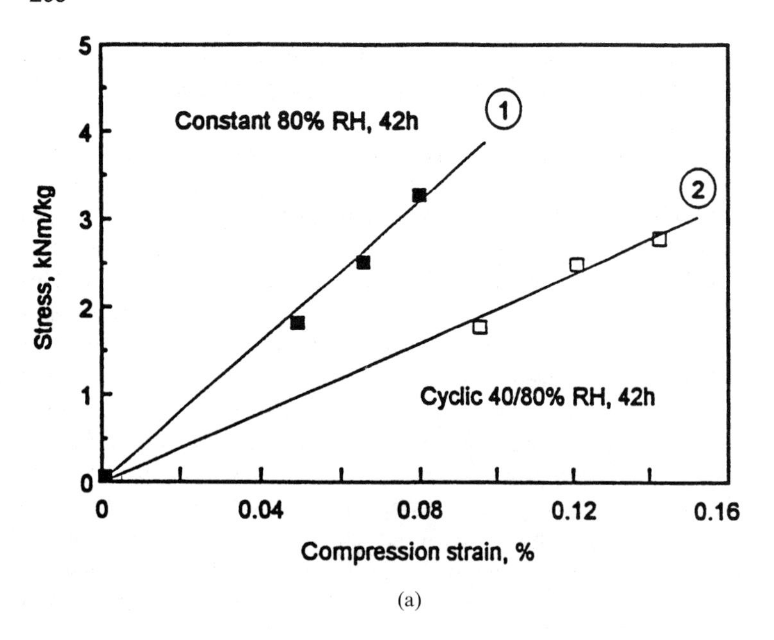

(a)

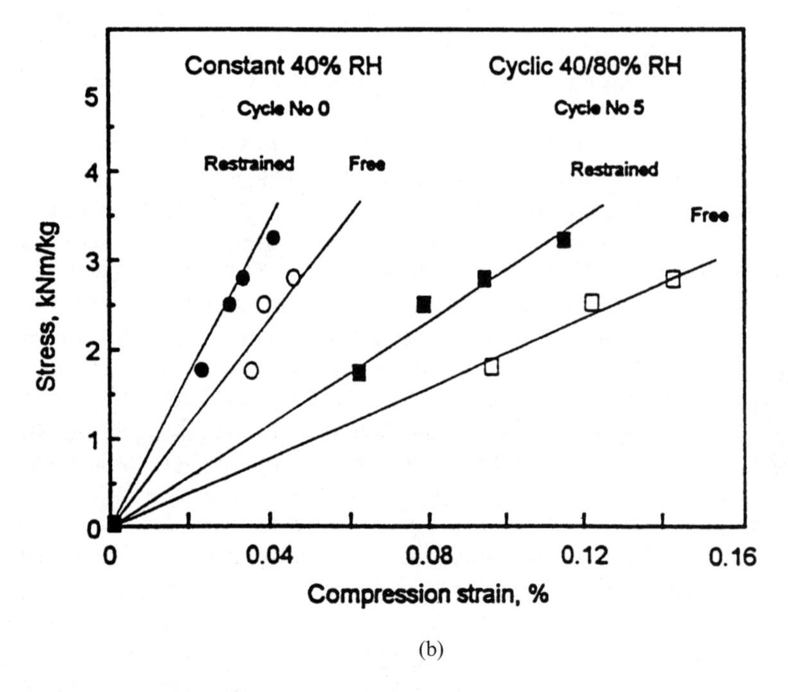

(b)

Fig. 11 Isochronous curves (a) at 42 h for freely dried linerboard and (b) for freely dried and restrained dried linerboard in compressive bending creep under constant and cycled 40%–80% RH. (From Ref. 101.)

use only when the creep strain rate is constant there; there is no evidence that paper has such a region. Dislocation climb, the primary mechanism of creep in metals, has no analog in paper.

To produce a single, efficient test of the consequences of changing the period length, Urbanik and Lee varied the ambient relative humidity according to what they call a swept sine schedule in which the frequency, in radians per unit time, is a logarithmic function of time. In such a case, as the frequency changes with time, the number of cycles in a fixed frequency difference interval is held constant. This control tends to keep fatigue effects independent of frequency over the various tests and presumes that the frequency change per cycle is a relevant parameter. The amplitude of the sinusoidal relative humidity variation was held constant by cycling beween 50% and 90% RH. The temperature was held constant at 23°C. Corrugated fiberboard tubes were placed under a constant longitudinal compressive load parallel to their sides. Two different flute constructions were tested. The specimen interiors were filled to about three-fourths of their volume with soybeans so that the sides would slightly bulge outward. In the language of buckling theory, the structure was given a small initial imperfection. The idea was to model the buckling of stacked corrugated containers. In fact, the postbuckling behavior induced by the structure of the corrugated specimen likely affects the test as much as the material properties of the paper.

The creep per cycle increased rapidly for periods longer than 1 day. In these tests, the deformation of the specimen depends on the cycle frequency in that the creep rate increases as the cycle frequency decreases or the period increases. Urbanik and Lee suggest that this increase would occur until a critical period was reached. At that point, they conjecture, the creep strain rate would decrease with period until it reached that of a constant relative humidity test. Gressel [36], for example, found that once the relative humidity period is greater than 96 h for the material tested, the 21 day creep is the same for all periods.

The corrugated board specimen deformation is out of phase with, and lags, the cycled ambient relative humidity [105]. Others have found that at a constant relative humidity frequency, hygroexpansive strain in an unloaded specimen closely tracks the variations in ambient relative humidity. However, over large frequency changes, Urbanik and Lee observed that the hygroexpansion and ambient relative humidity could be as much as 56° out of phase (Fig. 12). At high relative humidity frequencies, the hygroexpansion during desorption was further out of phase with the ambient relative humidity than in the adsorption phase; the relation reversed at low relative humidity frequencies. It is not clear what physical mechanism the phase lag measures. The adsorption and desorption phase lags are greater in the double-walled than in the single-wall specimen, espeically at smaller periods. They conjectured that this is due to differences in moisture diffusion, but it could also be due to structural differences.

E. Thermal Effects and Thermal Cycling

Almost all published tests on cyclical relative humidity creep have been conducted at room temperature. One exception that shows that raising the temperature increases the creep strain magnitude is the tensile creep tests of Haslach et al. [50] at two different constant temperatures. A virgin fiber 26 lb corrugated medium was sub-

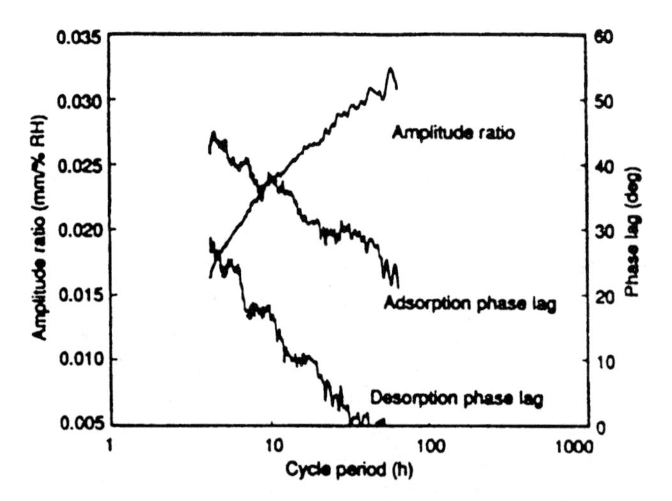

Fig. 12 Variation of amplitude and phase angle with period. (From Ref. 105.)

jected to linear humidity cycles between 30% and 90% RH with a ramp time of 70 min and a hold time of 2 h at each extreme. The load was 6.87 lb, which was 66% of the ultimate tensile strength at 90% RH and 40°C. At 40°C, cycling the relative humidity caused the specimens to fail in the first cycle; however, the same test run at 20°C found that the specimens had not failed after the third cycle. The strain in the first cycle at 40°C was greater than that at 20°C.

If creep is accelerated by the nonequilibrium cycling of the ambient relative humidity, other ways of creating nonequilibrium states such as thermal cycling should also produce analogous behavior. Tensile creep tests were performed by Haslach and Abdullahi [49] on a virgin fiber 26 lb corrugated medium around which the relative humidity is held constant at 50% while the temperature linearly cycles between 25°C and 55°C at constant rates of either 0.25°C/min, 0.5°C/min, or 1°C/min. The maximum strains in the 4 h test period were greater than the 0.5% maximum reached in the constant temperature test, and some reached nearly 1.0%. These tests indicate that the creep strain is, in fact, accelerated in a variable temperature environment (Fig. 13). The largest change in strain occurs in the rising portion of the first temperature cycle. This is analogous to the response under constant temperature and cycled relative humidity. The creep strain due to temperature changes is smaller than the creep response due to cycled relative humidity.

In each test, after the creep load was applied through four cycles, the specimens were allowed to relax under zero load either while allowing the temperature to cycle or while holding the temperature constant at the lowest value in the cycle. The strain relaxes exponentially to an equilibrium state under constant temperature. For cyclical temperature during relaxation, the relaxation curve appears to be a sinusoidal curve owing to thermal expansion and contraction superimposed on a typical exponential relaxation curve. The sample regains at least half its strain when the temperature is cycled during the creep relaxation phase.

The accelerated creep response due to variable temperature may have the same underlying mechanism as that for the response due to variable relative humidity. A

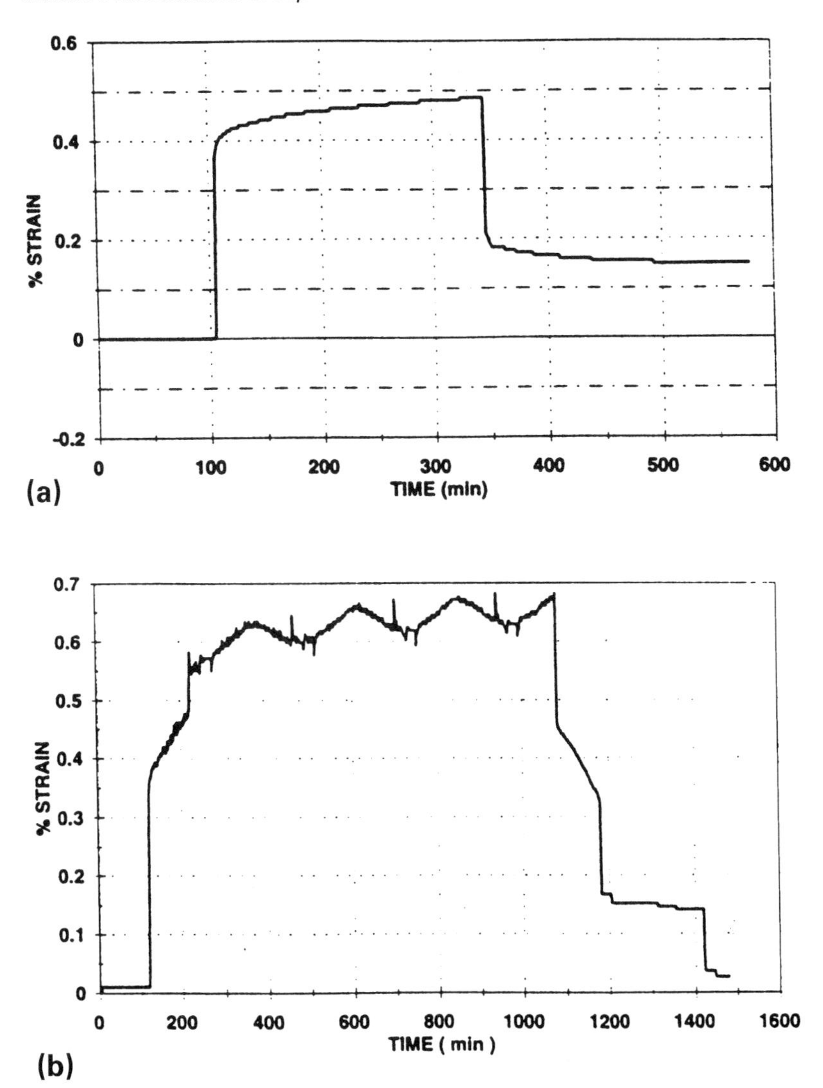

Fig. 13 Creep strain at 6.6 lb (a) under constant 55°C with relaxation and (b) under linear thermal cyclng between 25°C and 55°C with relaxation. (From Ref. 49.)

possible explanation is that the nonequilibrium state in the paper sheet caused by the variable temperature generates free volume, although apparently much less than that created by variable relative humidity. As the long-chain molecules are excited from equilibrium by the variations in temperature, the distance between them increases. This free volume induces a small radial fiber swelling that partially releases some microcompressions existing in the fiber–fiber bonds to permit accelerated creep. The fact that variation in temperature can do little else than generate free volume in a polymer by exciting the molecules supports the idea that the process is one of interfiber bond microcompresion-induced delayed elasticity, as Haslach [46] conjectured governs the behavior of moisture-induced creep.

VI. EXPERIMENTS IN MOISTURE CHEMISTRY

A paper specimen interacts with the ambient environment to adsorb or desorb moisture. The sorption chemistry can be measured through mechanical moduli such as the apparent stiffness, dynamic loss moduli, or relaxation times. Small-strain tests seem to give more information about how water bonds with cellulose under various relative humidity or temperature conditions than about the mechanical response of the structural components of paper or of the sheet itself. For example, mechanical behavior is not the only phenomenon influencing measured moisture-induced transients in dynamic loading tests of a specimen in a very small strain range. Moisture affects the activation energy and relaxation of the cellulose transitions described by Berger et al. [13]. These tests provide information on how the water molecule bonds to the amorphous region cellulose or other substructures but little on the behavior of the supramolecular structure. Moisture-accelerated creep is visible at small strains, perhaps because large amounts of free volume are generated immediately when the relative humidity begins to change.

A. Stiffness

Paper is not a linearly elastic material. The elastic modulus of a material with nonlinear elastic stress–strain response must be defined as the slope of the tangent line to the stress–strain curve at the origin and therefore depends only on the small strain behavior. Because of the long experience with linearly elastic metals, the idea of an elastic modulus was carried over to paper, but the physical meaning of such a modulus for paper is not clear. The measured small-strain apparent elastic modulus of paper decreases with increasing moisture content. Nissan [74], on the basis of his analysis of hydrogen bonding in an isotropic material continuum, proposed that the elastic modulus is related to moisture content m by $E(m) = E_0 \exp(a - bm)$, where E_0, a, and b are positive material constants. The number of hydrogen bonds in a cellulosic material can be estimated from Nissan's theory [75, p. 250] for the elastic modulus E of a hydrophilic material. Approximating the material as isotropic, Nissan derived the relation for the number N of effective hydrogen bonds as $N = (E/k)^3$ bonds/cm^3, where for cellulose $k = 8000$ [76]. On the other hand, the apparent elastic modulus is the same in both tension and compression at a fixed relative humidity up to about 50%, but at higher constant relative humidities the apparent elastic modulus of corrugated board is much greater in compression than in tension [25].

One of the earliest hypotheses to explain moisture-accelerated creep in paper was that a decrease, due to variable relative humidity, in the postulated elastic modulus allows more strain per unit time [7]. Back et al. [7] found that at the same moisture content the apparent elastic modulus at a variable moisture content is lower than that at a constant moisture content. Berger and Habeger [12] did not observe the same behavior in their small-strain sonic tests. Temperature changes of up to 5°C occurred in the first 2 min of their variable moisture content tests. They showed that there is no experimental apparent elastic modulus difference between constant and variable moisture content tests if the results are compared at a common moisture content and if the temperature changes in the paper induced by sorption are accounted for. Even so, no evidence was found that temperature change accounts for any measured decrease in apparent elastic modulus. Stress–strain measurements suggesting a decrease in elastic modulus may merely reflect other structural responses of the paper fiber. In wood pulp fibers, the apparent elastic stiffness decreases with an increase in fibril angle [85]. Recall that an increase in fibril angle is associated with a variable relative humidity ambient environment.

Another method to measure the current apparent stiffness during a cyclical humidity creep test of paper was devised by Considine et al. [29]. At any desired time during a creep test, the current apparent elastic modulus is measured by superposing a sawtooth load on the creep load; they chose one with amplitude 0.2 kN/m and duration 0.5 s. The slope of the strain response during the increasing portion of this small sawtooth load was assumed to be the current elastic modulus. In compressive creep tests of paperboard, this measured current elastic modulus decreased to as little as one-third of its initial value by the time of specimen failure.

B. Physical Aging

The variation of the dynamic moduli of paper in a variable relative humidity environment also depends only on small strains. The dissipation under dynamic loads in a solid, like paper, is measured by the loss tangent, the tangent of the phase angle lag, δ, between the applied sinusoidal strain (or stress) and the stress (or strain) response. In linear viscoelastic models, the loss modulus, E_l, is defined by requiring that the ratio of the loss modulus to the elastic modulus equal the loss tangent, $E_l/E = \tan\delta$. The loss tangent is zero when the dissipation is zero, so the material is in an elastic state. There is no reason to believe that paper is a linear viscoelastic material; however, its constitutive model might be quantitatively approximated as linear at small strains. An apparent elastic modulus and the loss tangent can be measured, independently of the choice of a constitutive model for paper, in an ambient environment in which the relative humidity is changed in a stepwise manner [80].

Padanyi [80] measured the dynamic moduli of 210 g/m^2 kraft linerboard specimens in dynamic bending under strain control with a small sinusoidal strain amplitude of 0.36%. The specimens were 2 mm long, 10 mm wide, and 0.3 mm thick. Moisture equilibrium was reached in less than 30 min, for these specimens, as verified by weighing. Bending tests were chosen over tensile tests in the belief that mechanosorptive behavior is more pronounced in bending than in axial tests. The relative humidity was controlled by using the technique of mixing dry and wet air streams into the test chamber. The relative humidity was rapidly changed in increments of 25% between 0% and 100% RH and held constant for 30 min after each change. As might be expected from other researchers' results, the apparent elastic modulus decreased with increasing increments of relative humidity and vice versa. The apparent elastic modulus also varied while the relative humidity was held constant. During desorption steps, the apparent elastic modulus increased at the lower constant relative humidities and increased at the higher constant relative humidities. During adsorption increments, the apparent elastic modulus increased slightly during the time intervals in which the relative humidity was held constant. The loss tangent, on the other hand, shows a very large jump at the time the relative humidity was changed and was generally higher at the higher relative humidities. During the periods of constant relative humidity the loss tangent decreased. This information could be used to predict creep behavior only if a constitutive relation between these moduli and the creep compliance were known, and verified, for paper.

Padanyi suggested that such loss tangent behavior results from the amorphous cellulose region relaxation lagging the generation of free volume due to the variation in relative humidity [62,80]. The long-chain molecules forming a polymer are more separated when the material is in a nonequilibrium state than when it is in an equilibrium state. The portion of the volume of a solid containing no polymer molecules is called the free volume. Padanyi [80] proposed that variations in the free volume of the polymeric components of paper cause moisture-accelerated creep. The excess volume is presumed to increase molecular mobility in the material and thereby allow additional creep.

Aging is the variation of mechanical properties over time. In a paper specimen brought to a nonequilibrium state, aging occurs as the polymer relaxes to an equilibrium state along with a concurrent increase in relaxation time and reduction in free volume. Increasing the free volume by some means would reverse the relaxation to equilibrium; Padanyi calls such a process deaging. Relaxation times increase continuously in constant moisture and constant temperature [80]. In this sense, the material ages. On the other hand, in variable moisture, the relaxation times decrease. The decrease is more pronounced with more rapid moisture changes.

Molecular mobility and therefore the rate of decrease in free volume are greatly reduced if the temperature of the solid is below the glass transition temperature. In many polymers, aging occurs if the material is at least partially amorphous and its temperature is below its glass transition temperature. Free volume is increased when the material is taken above its glass transition temperature, T_g, by raising its moisture content. Aging due to the loss of free volume over time as the polymer molecules relax to their equilibrium positions and the resulting reduction in molecular mobility is reversed by quenching the material from above its glass transition temperature to preserve the generated free volume. The glass transition temperature of paper decreases with increasing moisture content [92] and can be in the range of room temperature at high moisture contents. The glass transition temperature also depends on the degree of crystallinity of the cellulose. Increasing the relative humidity at constant temperature could cause the glass transition temperature to be exceeded and create additional free volume. It is therefore likely that the generation of significant free volume in variable relative humidity is a consequence of the material nonequilibrium state induced by the lack of kinetic moisture balance between the paper and the ambient atmosphere.

The response of paper often depends on how the sample is conditioned. Padanyi [80] refers to certain conditioning processes as deaging the material. The specimens were first conditioned, or deaged by increasing moisture content, by placing the specimens in a 100% RH environment for 2 h, then brought to the test conditions by placing them in a 50% RH environment for 30 min. Before the creep tests were begun, each specimen was allowed to age from the equilibrium state at 50% RH for a fixed time period. Tensile creep tests were performed on 160 mm × 15 mm specimens of kraft linerboard at 23°C and 50% RH under a load of 30% of the breaking strength at these ambient conditions. The creep compliance at a given time during the creep test decreased as a function of increasing aging. A similar pattern was found when the deaging was accomplished by placing the specimens in a 0% RH environment for 1 h to generate free volume by desorption. However, at a given aging period, the compliance curve was below that for those deaged at 100% RH. A larger creep compliance means accelerated creep. Padanyi therefore concluded that deaging accelerates creep.

The relaxation modulus $E(t)$ determined from a dynamic test and the relaxation time τ are often assumed to be related by the Kohlrausch–Williams–Watts (KWW) law for stress relaxation,

$$E(t) = E(0)\exp\left[-\left(\frac{t}{\tau}\right)^{\beta}\right] \tag{2}$$

where β is a material constant. The relaxation time predicted by this model is directly proportional to the age. To investigate the validity of this relation for paper under various relative humidities, Padanyi [81] placed a specimen in a 100% RH environment for 1 h to deage it. Then after various times to allow it to age, the relaxation modulus $E(t)$ was determined from a dynamic test. Humidity changes were made in 5% increments rather than continuously, then the humidity was held constant while the dynamic moduli were measured. This law was found by Padanyi to apply only for times less than the specimen age from the end of its conditioning at 100% RH. Unfortunately, the value of β had to be changed, to fit the data, after times of 31.4 days; it is not a material parameter.

C. Sheet Moisture Content Variations

One of the earliest hypotheses proposed for the mechanism of moisture-accelerated creep in a paper sheet is that the moisture acts as a plasticizer by breaking some load-bearing hydrogen bonds and thereby allowing slip. For example, Byrd [21] suggested that perhaps "humidity cycling reduces the number of load-bearing elements in the fiber network through accumulation of water molecules in the cellulose lamellae." It is not made clear whether this was thought to occur only within the fiber, or in the interfiber bond, or in both. Nor is it clear how this mechanism functions differently when the relative humidity is merely raised to a higher constant value compared to the case when the relative humidity is cycled. The role of moisture content has been investigated by measuring the specimen moisture content during various creep tests.

A corollary of this idea is that moisture-accelerated creep is due to an accumulation of sorbed moisture on each subsequent cycle. To examine this, Gunderson and Tobey [41] used their total-moisture-in-the-system technique described in Section III to monitor the moisture sorbed by a paperboard specimen during variable relative humidity tensile creep. A single material had the same amount of creep strain per cycle in two different cyclical relative humidity, schedules, but the one whose period was four times longer adsorbed 34% more moisture. On the other hand, a sample treated with a binder and one of the same material untreated both adsorbed the same amount of moisture, but the one with binder crept much less. This paper was a virgin liner made of Lake States softwood unbleached kraft furnish. Its density was 670 kg/m^3, and its tensile strength at 50% RH in the cross-machine direction (CD) was 8.07 kN/m. The specimens were loaded in the cross-machine direction by 1.98 kN/m, which was about 25% of the tensile strength. The specimens were tested in an environment in which the humidity cycled linearly between 30% and 90%. The specimens were conditioned by cycling the relative humidity before the creep load was applied. Gunderson and Tobey concluded that no clear relationship exists between the creep deformation and the amount of moisture sorbed. The results contradict the hypothesis that moisture acts as a plasticizer and that this serves as the complete explanation for moisture-accelerated creep.

The load does not significantly affect the gross moisture sorption in the specimen [38]. Thermodynamics suggests that it could. Therefore variation in total moisture content is not the explanation for moisture-accelerated creep. Gunderson and Tobey also found no statistically significant differences in moisture sorption between when the load is applied and when it is not, at least for a load of 25% of the tensile strength. They point out that these results do not eliminate the possibility that load may affect moisture sorption in specific local sites in the sheet. Sites of particular interest would, of course, be the fiber–fiber bonds. No experiments on this question are discussed in the literature.

VII. DESIGN FOR A VARIABLE RELATIVE HUMIDITY ENVIRONMENT

Design criteria for paper products intended to survive a variable relative humidity environment have concentrated on the magnitude of the creep strain, the creep strain rate, and lifetime estimates. Several proposals have been made to predict the lifetime, i.e., either the time or the number of cycles to failure, of a paper sheet or a corrugated container. One technique is to compare the tensile strengths of two containers to predict which would have the longer lifetime in a cyclical relative humidity environment, but the one with the greater tensile strength does not always have the longer life.

The lifetime can be related to a strain rate and a rate of change of stiffness in some papers. A stiffness was periodically measured during a creep test, as the slope of the strain response to a superposed small additional linear impulse load at 50% RH, by Considine et al. [29]. Values were plotted as a function of the time at which they were measured. The minimum slope of this curve is the minimum current stiffness rate loss, $\dot{E}$, as defined in Ref. 29. The minimum compressive creep rate, $\dot{\epsilon}$, is similarly defined from the creep strain versus time curve. Both are correlated to the time to failure, t_f, under varying relative humidity creep. Equations of the form

$$\log(t_f) = a - b\log(\dot{E}) \qquad (3a)$$

$$\log(t_f) = c - d\log(\dot{\epsilon}) \qquad (3b)$$

where a, b, c, d are material constants, were obtained in Ref. 29 for two types of paperboard in compresive creep at 29% of the 50% RH stress–strain failure load and in sinusoidal humidity cycles between 50% and 90% RH with a 10 min period. The specimen conditioning included 10 relative humidity cycles before the creep load was applied. It is not known how strongly the correlation depends on the application of this cyclical conditioning. The failure time can be eliminated from Eqs. (3) to give a relationship between the current stiffness rate loss and the minimum compressive creep rate.

A. Endurance Limit

It might be tempting or helpful to think of design problems for paper in variable relative humidity in terms analogous to those used to avoid fatigue in solids such as metals. In metal fatigue design, an *S–N* diagram is used to determine the maximum stress *S* allowed for a given lifetime as measured in the number of cycles to failure, *N*. The uniaxial *S–N* diagrams obtained for steel seem to show a stress below which the metal appears to have infinite fatigue life. This value is called the *endurance limit*. However, only body-centered cubic (bcc) metals have been shown to have such an apparent endurance limit under which they will survive for an infinite number of cycles. Designers today choose a working load to produce a specified lifetime, a desired minimum number of cycles to failure, rather than overdesigning for an infinite life. Also, it is unlikely that an endurance limit exists for paper. However, until plots of the number of relative humidity cycles to failure versus the creep stress are available for designers, estimates must be made.

An increase in creep load magnitude can drastically reduce the number of cycles to failure, the lifetime, of paper in a cyclical relative humidity environment. Varying the creep load as a parameter verifies that there is a load at which there is a large decrease in the number of cycles to failure. Gunderson and Considine [39] performed compressive creep tests on a softwood handsheet in humidity cycled between 50% and 90% with a time period of 10 min per cycle. They conditioned the specimen by cycling the relative humidity nine times before applying the creep load. Their test with a creep load of 22% of compressive strength at 50% RH failed in 27 cycles, but a test at 17% of the compressive load had not failed after 274 cycles. There is therefore a load between these two at which the cyclical creep becomes particularly destructive to the strength of the paper. This critical load, which they call the working strength, is to be taken as the design load for the paper, because below it the endurance of the specimen is extremely long. This idea is analogous to that of an endurance limit in fatigue studies of metals.

Gunderson and Considine [39] proposed a novel experimental method to quickly determine the working strength of a specimen in cyclical relative humidity by periodically increasing the creep load. To locate a working load, the relative humidity is cycled at each stepwise increment of load 10 times with a period of 10 min; this process is continued until the specimen fails. They increased the load from zero in steps of 0.0945 kN/m on a softwood handsheet of basis weight 129 g/m^2 and density of 566 kg/m^3. Failure occurred at 32% of the 50% RH compressive strength after a total of 103 cycles with a 10 min period and amplitude between 50% and 90% RH. The data presented are for the shortest period of those tested, ranging from 10 min to 4 h, apparently because the effects should be most visible in the more rapid rate of change of ambient relative humidity. A working strength is estimated by inspection of the test curve of strain versus time. Gunderson and Considine expect it to be the load near failure, where the strain rate increases exponentially. No one has fully examined whether increasing the load in such a stepwise manner causes failure at either a lower load or in a lower number of cycles than the application of a single large load.

B. Corrugated Containers

The cross-machine direction hygroexpansive strain is greater than that in the machine direction, especially at high rates of change of ambient relative humidity.

Corrugated containers carry the compressive load in their walls in the cross-machine direction. Component hygroexpansivity is a good indicator of container performance, as measured by creep rate and total creep strain, under cyclical creep [69]. The boards made of components with lower hygroexpansivity tested by Kuskowski et al. [69] had less total creep and lower creep rates than those with high hygroexpansivity. However, strength is not a good predictor of the response to variable relative humidity. Standard compression tests produced little difference in the 50% RH strength of the various boards. Kuskowski et al. suggest that corrugated containers could be ranked for their survivability in variable relative humidity by measuring the hygroexpansivity of their components.

The creep of a corrugated board can also be compared to the creep of its components. Byrd [19] found that a corrugated short column crept two to five times faster than predicted by the creep measured in the individual components. This result raises the question of whether the adhesive or the structure itself accounts for the increase in creep under variable relative humidity. To study the influence of the adhesive on creep, Byrd [20] joined paperboard strips into short columns with two types of adhesive. The columns were loaded at 25% of the 90% RH strength in a constant 23°C environment. The humidity was linearly cycled from 90% to 35% RH in 1 h and then from 35% to 90% in 2 h. The creep rates were determined at the 90% RH portion of the creep curve in the time range of 18–72 h. The board whose interface was joined by a water-sensitive adhesive crept at almost three times the rate of boards joined by water-insensitive adhesives. Moisture adsorption by the adhesive probably accounts for most of the increase in creep.

The lifetime of corrugated containers under linearly varying compressive loads in constant relative humidity, a complementary problem to moisture-accelerated creep, has been thought to depend on the load, the load rate, and the compression strength in a constant humidity. Forsberg [34] examined the assumption that the slope of the stacking load versus logarithm of lifetime curve and the slope of the compression strength versus the logarithm of the load rate curve have the same absolute value. Her experiments on boxes made of either virgin or recycled fibers showed this relationship to be valid within a 90% confidence interval for a constant relative humidity of 50% but not for a constant relative humidity of 90%.

Bronkhorst [16] developed an empirical relationship between container lifetime under a compressive dead load and the secondary creep rate in changing relative humidity. The relative humidity was changed from 38% to 93% and back to 38% relative humidity. The humidity was held constant at each extreme for 24 h. It took 20 min for the ambient air to change from 38% to 93% RH and 2 h to make the 93% to 38% RH transition. A 320 N compressive creep load, which was 30% of the 90% RH failure strength, was placed on an aluminum platform that sealed the container top. Simultaneously, the moisture content was measured in an unloaded matched container and the temperature and humidity were monitored inside an unloaded but similarly sealed matched container. The temperature increased about 2°C inside the container as the humidity outside was raised from 38% to 93% and eventually came to an equilibrium owing to heat transfer through the container wall. The temperature similarly decreased initially as the relative humidity was lowered from 93% to 38%. The relative humidity inside the container slowly approached and nearly reached the outside humidity over each 24 h cycle. However, the unloaded moisture content of the container varied in a repeatable manner over each stepwise cycle. Although not mentioned by Bronkhorst, the measurements of the internal humidity and temperature give a means of estimating the moisture diffusion through the container wall.

These container tests produced creep curves with both primary and secondary regions, as occurs for metals. The strain was due to structural changes such as out-of-plane deformation of the container walls as well as material behavior. Container lifetime, t_L, in days is defined as the time until the container fails to carry the load and is modeled by

$$t_L = A\frac{1}{\dot{\epsilon}_2} \tag{4}$$

where $\dot{\epsilon}_2$ is the strain rate in the secondary region. The value $A = -0.0292$ was determined from a set of 66 tests involving three types of containers. This empirical model is a variation of that used by Koning and Stern [66] to combine data from tests at various constant relative humidities and loads in a relationship of the form

$$t_L = A\frac{1}{|\dot{\epsilon}_2|^n} \tag{5}$$

Recasting the data of Ref. 66 in terms of Eq. (4) resulted in a positive value of A that was about one-sixth of the variable relative humidity value for tests conducted in a constant relative humidity and an intermediate value less than one-half their value for tests involving one relative humidity change. Varying the relative humidity accelerates the secondary creep and reduces container life.

VIII. PRELIMINARY CONSTITUTIVE MODELS

Mathematical models can both help delineate the relationship between various parameters and aid in the design of products. Only a few tentative constitutive models that include the effect of relative humidity on the mechanical response of paper have been proposed. Continuum mechanics models exist that attempt to represent the consequences of moisture transport into a solid from an ambient atmosphere. However, such models may not be applicable, because paper is a fiber network, not a continuum.

Several models for creep of paper have been devised in analogy with stress–strain models. The classical empirical model proposed by Andersson and Berkyto [1] for paper in tension was $p = C_1 \tanh(C_2\epsilon/C_1) + C_3\epsilon$. This was later modified for compressive stress–strain to $p = C_1 \tanh(C_2\epsilon/C_1)$, where p is the load and C_1, C_2 are material constants [40]. However, the hyperbolic tangent alone has an asymptote and therefore may not satisfactorily model a stress–strain curve. To avoid such an asymptote, the expression can be inverted to write the strain in terms of the stress and then substitute the hyperbolic sine for the hyperbolic tangent [87].

The hyperbolic sine uniaxial creep model was extended to explicitly include the stress rate by the Pecht and Haslach [88] integral model for the stress–strain response of paper at various constant temperatures.

$$\epsilon(t) = \frac{1}{K}\sinh\left(\frac{K\sigma(t)}{E}\right) + B\dot{\sigma}_0^4\int_0^t s^3 \ln\left[1 + \left(\frac{(t-s)}{\tau}\right)^n\right]ds \tag{6}$$

where $\dot{\sigma}_0$ is the constant stress rate, E and K are functions of moisture content, B is a function of moisture content and stress rate, the durability τ is a function of moisture content and stress, and n is a material constant. For small strains, the hyperbolic sine term is approximated as $\sinh[K\sigma(t)/E]/K = \sigma(t)/E$. This model generalizes rheologically simple creep models. At extremely slow rates, the hyperbolic term approaches zero and the creep behavior is represented by the time-dependent integral. It has a superficial resemblance to the Eyring rate of flow hyperbolic sine creep model [33,42,57] for textile fibers, which was used by Sedlachek [94] to discuss single-fiber creep.

The classical hyperbolic tangent constitutive stress–strain model proposed for paper has been modified [44] to relate the creep strain and the creep stress in isochronous curves in the case of both constant and variable humidity creep,

$$\sigma = \left(\frac{t}{t_o}\right)^n c_1 \tanh\left(\frac{c_2}{c_1}\epsilon\right) \tag{7}$$

where σ is the creep stress, ϵ is the creep strain, t is time, and t_o, n, c_1, c_2 are material constants.

A logarithmic curve represents the creep component of moisture-accelerated creep [17,52].

$$\epsilon(t) = a + b\ln(t) \tag{8}$$

But no physical meaning was assigned to the coefficients a and b. These coefficients must be functions of at least the creep stress, the relative humidity, and the material. Furthermore, such a model cannot account for the initial creep strain, because it is undefined at $t = 0$ unless a function of the form $\ln(1 + t)$ is substituted for $\ln(t)$. The creep curves of paper at constant relative humidity also have the qualitative form of logarithmic curves.

Urbanik's model [104] for the creep strain in a cyclical relative humidity environment relates hygroexpansion and mechanosorptive deformation. It assumes that the deformation is the sum of a hygroexpansion deformation, X_h, due to moisture content and a creep, X_c, resulting from the "cumulative ratcheting of swelling effect with each change in specimen moisture content." In this model, the hygroexpansion rate determines the creep rate in the sense that

$$\frac{dX_c}{dt} = \mu\left|\frac{dX_h}{dt}\right| \tag{9}$$

where μ is the experimentally determined creep coefficient, which may not be a constant. This relation indicates that the creep rate is zero at the instant that the ambient relative humidity cycle changes direction, i.e., has a zero velocity. The absolute value implies that the creep strain rate depends only on the fact that the moisture content is changing, not on whether it is increasing or decreasing, i.e., in sorption or desorption. The creep deformation is determined by integrating the creep rate with respect to time.

The model presented in Ref. 104 is developed in detail for a sinusoidal relative humidity cycle. It would not apply to continuous cycling by a sawtooth linear function between extremes for which the rate of relative humidity change is constant. The relative humidity changes at different rates during one sinusoidal cycle. According to the data of Ref. 105, the hygroexpansion in an unloaded specimen follows the sinusoidal form but with a phase lag behind the relative humidity,

$$X_h(t) = A_o + A \sin \alpha \tag{10}$$

where $\alpha = \omega t - \theta$ if ω is the angular frequency of the relative humidity change and θ is the measured phase lag. The model does not depend on the rate of change of the period, $d\omega/dt$.

In this model, the creep deformation in each quarter-cycle is the same. Therefore an average creep rate can be defined as $\overline{R} = 4A\mu/P$, where P is the cycle period. This implies that the accelerated creep rate is greater for shorter period cycles, as was experimentally shown for paper, the opposite of the results for corrugated containers.

A general relationship between hygroexpansion and the ambient relative humidity is not known. In applying this model to data obtained by varying the period of the sinusoidal relative humidity change [105], it was found that the coefficient μ is not a constant. Therefore the average rate had to be calculated over time intervals small enough to apply the relation assuming that μ is constant, because its functional form was not determined. In each interval, the average rate and μ were assumed to satisfy a decaying exponential function of time. The average rate is integrated over a given time interval to determine the predicted average creep, excluding the contribution from hygroexpansion. The resulting model deformation predictions seem to match experimental data.

A model for small strain tests in which the relative humidity is varied in a stepwise manner was proposed by Söremark and Fellers [98]. For a given change in relative humidity over a time interval, the total change in strain, $\Delta\epsilon_t$, is the sum of the elastic, $\Delta\epsilon_e$, and creep, $\Delta\epsilon_c$, strain changes with respect to constant relative humidity as well as the hygroexpansive strain at zero stress, ϵ_h, and the stress-induced strain due to humidity, ϵ_{SIH}.

$$\Delta\epsilon_t = \Delta\epsilon_e + \Delta\epsilon_c + \epsilon_h + \epsilon_{\text{SIH}} \tag{11}$$

The latter term, which measures the difference between constant and variable relative humidity behavior, is the change in hygroexpansive strain due to the presence of a stress. It is greater than the sum of the constant relative humidity elastic and creep strains from constant relative humidity creep tests. The change in the sum of the elastic and creep strain is obtained by subtracting the constant relative humidity creep strain at the original relative humidity from that at the final relative humidity in the chosen interval. The total strain is obtained from the variable relative humidity creep test and is the difference between the strains at the beginning and end of the humidity change. The hygroexpansive strain is the difference in strain over the change in relative humidities, measured from a strain–relative humidity curve obtained at zero stress. Small strain bending theory is used to relate the load and measured deflection and to write the elastic strain as proportional to the stress, accounting for the change in paper cross-sectional area due to the swelling and strain.

No one has yet reached the point of proposing a fully developed constitutive model. Such a model would suggest experiments to determine any empirical coefficients needed in the model.

IX. METHODS TO INHIBIT MOISTURE-ACCELERATED CREEP

An important reason to study moisture-accelerated creep is the need to develop design methods to control the dimensional instability that results from such creep. One strategy is to treat the paper in some manner to reduce moisture-accelerated creep.

It is known that binders, such as the now banned carcinogenic melamine formaldehyde, can inhibit hygroexpansion and increase wet strength. Formaldehyde compounds were formerly used for this purpose in U.S. paper currency. Given the apparently close relationship between hygroexpansion and moisture-accelerated creep, a logical question concerns whether or not binders can reduce moisture-accelerated creep. This idea was raised by Back et al. [7] when they reviewed experiments with low molecular weight urea resin forming cross-links in textile fibers that indicated that a binder may reduce the effects of moisture-accelerated creep.

The idea that moisture-accelerated creep is reduced by minimizing moisture sorption was tested by Gunderson and Tobey [41]. One set of specimens were wax-dipped to prevent moisture diffusion, and two sets were cross-linked, one with butane-tetracarboxylic acid (BTCA) at 7.7% by dry fiber weight and the other with formaldehyde. The loads were 25% of the tensile strength at 50% relative humidity. These specimens and similar but untreated specimens were subjected to humidity cycling from 30% to 90% RH with a 12 h ramp and no hold time at the extremes. The BTCA and wax specimens exhibited about half as much creep strain as the untreated. But all specimens showed the characteristic large initial strain increase associated with the beginning of the first humidity cycle. However, the response of the waxed specimen to the humidity cycling was damped. The formaldehyde specimen had about 1.5% bound formaldehyde. Its creep was less than one-third that of the untreated specimen. The binder does not change the variable relative humidity creep strain pattern. A large amount of creep occurred during the initial increasing component of the first humidity cycle followed by slower creep. The formaldehyde specimen had a tensile strength only about 80% of that of the untreated specimen. Clearly, binders inhibit moisture-accelerated creep. However, no one has definitively identified which sites with cross-links are the most important to control accelerated creep or the mechanism by which the reduction occurs. No correlation was obtained between moisture sorption and moisture-accelerated creep. A more likely possibility is that binders inhibit moisture-accelerated creep by reducing swelling.

Press-drying can reduce moisture-accelerated creep, as observed by Byrd [21] in compression tests of press-dried oak high yield (63%) linerboard under relative humidity cycles between 35% and 90%. He suggested that the improvement of interfiber bonding caused the reduction in creep. The creep rate was comparable to that of a virgin fiber normal yield (50%) linerboard. The moisture sorption of the high yield press-dried board was also reduced to a level near that of normal yield linerboard. Further support is given by Gunderson [37], whose compressive creep tests of press-dried, 205 g/m^2, oak linerboard showed only about 0.5% strain under relative humidity cycles between 35% and 90%. This reduction in strain accompanying press-drying supports the idea that the nature of the fiber–fiber bonds affects moisture-accelerated creep.

X. CONCLUSION

Experimental work has established that moisture-accelerated creep is a real phenomenon. The experiments, however, leave many open questions about the underlying physical mechanism of moisture-accelerated creep. Design criteria must be developed, along with the best experimental techniques to apply such criteria, to ensure that a paper application will avoid the detrimental effects of moisture-accelerated creep.

The experiments have uncovered some important characteristics of moisture-accelerated creep in paper. However, when describing and analyzing this phenomenon, one has to be careful about generalizing ideas such as the elastic modulus, endurance limit, and primary and secondary regions of creep from metals to the network structure of composite polymeric fibers that forms paper. In metals, creep behavior is believed to be a consequence of mechanisms involving lattice dislocation behavior, such as dislocation climb, but there is no a priori reason why this model should hold for paper. A large portion of the accelerated creep strain in paper occurs in the initial increasing portion of the first relative humidity cycle, when the specimen is conditioned at the low extremity of the cyclical humidity range. This behavior is seen in most cyclical humidity situations including single-fiber creep, virgin or recycled fiber sheet creep, creep of papers treated with binder, and after a specimen has been cycled under a load and then redried. This large initial strain jump is not primary creep. To pass through a putative primary creep region, a specimen can be allowed to creep for 2 h after the load is applied while the humidity is held constant at the conditioning relative humidity. If the humidity is then cycled to a higher relative humidity, the strain exhibits the same rapid increase in the initial increasing portion of the cycle (Fig. 14). This test casts doubt on the validity of a primary–secondary region creep model for paper.

Certain paper substructures behave differently during constant relative humidity creep than during variable relative humidity creep. The fibril angle gets smaller during constant relative humidity creep and larger during variable relative humidity tensile creep. The fiber therefore swells radially during variable, and contracts during constant, relative humidity tensile creep. A single fiber in variable relative humidity tensile creep also contracts longitudinally during the increasing humidity portion of relative humidity cycles. Moisture-accelerated creep in fibrous materials is correlated to anisotropic individual fiber swelling in the presence of moisture.

The experiments show, as expected, that load as well as the period and amplitude of a cyclical relative humidity environment affect creep behavior. The nature of the pulp furnish has an important influence on the strain magnitudes. Thermal cycling also accelerates creep. However, no mathematical model exists that successfully quantifies these relationships. It is not easy to compare the published tests, because they were performed at various creep loads. A testing standard should be developed to compare moisture-accelerated creep in different papers.

Many additional tests are needed, because neither the underlying physical mechanism nor methods to control moisture-accelerated creep are fully understood. The types of tests already reported need to be repeated at other parameter values. Tests investigating different aspects are needed to follow up on questions raised both by the tests already completed and by the hypotheses that have been proposed to explain existing experimental data.

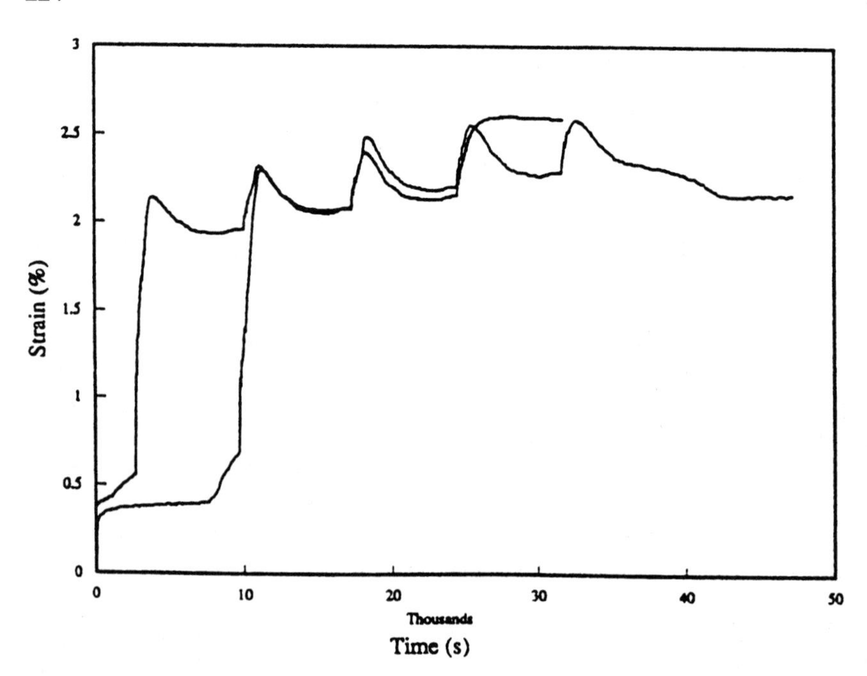

Fig. 14 Application of a cyclical relative humidity after 2 h at a constant relative humidity. (From Ref. 51.)

The several hypotheses presented for the physical mechanisms causing moisture-accelerated creep include the idea that the accumulation of moisture in the paper reduces the number of load-bearing elements [e.g., 21]. The paper is plasticized. High constant relative humidity does increase creep. This idea does not explain why variable relative humidity increases the strain so drastically beyond that observed in high constant relative humidities.

Another hypothesis is that the changing moisture content during cyclical relative humidity sets up a stress gradient within individual fibers that accelerates the strain. However, no one has shown how such stressed fibers would interact in the sheet. No experiment has been conceived of that would allow the direct measurement of such a stress gradient; there are no stress gauges. Measuring the strain distribution would be instructive but would not permit calculation of the stresses, because no one has developed a verified constitutive stress–strain model for either paper or a pulp fiber in the presence of varying relative humidity.

A third hypothesis is that aging and deaging of the paper through the creation and relaxation of free volume during the nonequilibrium process of loading in variable relative humidity causes the moisture-accelerated creep [80,81]. The resulting reduction of relaxation times as the paper passes above and below its moisture-dependent glass transition temperature increases the creep strain. These investigations of the variation of dynamic moduli need to be related to studies of the glass transition in cellulose such as Ref. 13. Whereas the nonequilibrium process of moisture-accelerated creep likely creates free volume as it would in any polymer, the direct connection to increased creep strain is not developed in this hypothesis. Other factors must also be involved, because not all hydrophilic polymeric materials exhibit moisture-accelerated creep. The measured mechanical transients in variable relative humidity most likely result from moisture phenomena that occur in the amorphous regions of the fibers.

The hypothesis I favor [45] says that moisture-accelerated creep is a consequence of the network structure of paper and the anisotropic swelling of the individual fibers. The nonequilibrium state during variable relative humidity generates free volume and so induces radial swelling of the pulp fiber. This nonequilibrium induced radial swelling is what distinguishes variable and constant relative humidity creep in which the tensile creep load reduces the fiber radius. The microcompressions formed during drying in the fiber–fiber bond are released more easily because of the excess radial swelling in the nonequilibrium environment of cyclical relative humidity. This process allows additional creep compared to the constant relative humidity sheet in kinetic equilibrium with its environment. The tendency of such fiber–fiber bond microcompressions to re-form causes the relaxation observed when the creep load is sufficiently reduced.

Additional creep tests under both constant and variable relative humidity on sheets dried under restraint would give information about the role of microcompressions in moisture-accelerated creep, presuming that restraint in drying reduces the formation of interfiber bond microcompressions. Drying under restraint, which reduces hygroexpansion, might be a possible technique in paper design to inhibit moisture-accelerated creep as well. A related matter would be to determine whether or not press-drying reduces the formation of fiber–fiber bond microcompression in addition to strengthening the fiber–fiber bonds. Processes such as beating that tend to make the fibers more comformable might be investigated from the point of view of asking if they encourage the formation of interfiber bond microcompressions and if this is reflected in the amount of moisture-accelerated creep observed. This would help decide whether bond strength or comformability is a better indicator of the magnitude of moisture-accelerated creep.

The relationship between hygroexpansion and moisture-accelerated creep should be studied further to decide whether or not they have a common underlying physical mechanism. More information about the possible role of delayed elasticity could be obtained from a set of tests of creep followed by relaxation that would attempt to quantify the dependence, on creep load and relative humidity period, of the separate creep and hygroexpansive strain components during creep and during relaxation. Variable and constant relative humidity creep relaxation tests at different loads should be compared while relaxing both under zero load and under the equilibrium relaxation load in variable and constant relative humidity. Such a variety of tests might allow quantification of the mutual influence of the hygroexpansive and creep strains. It seems doubtful that hygroexpansion and moisture-accelerated creep can be treated as independent phenomena whose contributions to the total strain can be simply superposed. In any case, better means are needed for distinguishing the hygroexpansive effect from the creep strain during variation of either relative humidity or temperature.

The roles of the amount or rate of change of moisture content in moisture-accelerated creep are ambiguous. Certainly, equilibrium moisture contents are not particularly relevant to the study of a nonequilibrium moisture transfer problem. More precision might be obtained if a technique could be devised to monitor the moisture content of the loaded specimen itself rather than using a reference specimen or other indirect method. If non-contact techniques such as infrared radiation are to be used, the creep in a specimen subjected to radiation must be compared to one that is not. The radiation itself could generate free volume and drive the specimen into a nonequilibrium state that would accelerate creep.

Experimental studies of the behavior of the fiber–fiber bond during creep in both constant and variable relative humidity are needed. A movie of the interfiber bond behavior in variable relative humidity might possibly be made with an environmental electron microscope. One goal would be to measure the interaction of fibril angle changes in two bonded fibers and the formation or release of microcompressions in the fiber–fiber bond during the cyclng of relative humidity. A second goal would be to measure and compare the bond strength to the swelling stresses in the bond induced by variable relative humidity to determine if the response is indeed delayed elasticity.

Additional measurements of the effect of binders, such as melamine formaldehyde or more modern, less carcinogenic binders, on hygroexpansion and creep during either relative humidity or thermal variation would be helpful in developing techniques to control the strain.

Sorption is exothermic, but it is not clear if the heat released has a measurable effect on moisture-accelerated creep. For that matter, very little study has been completed on the significant effect of the ambient temperature and its variation on the creep response of paper. A series of creep tests is easily visualized in which the ambient relative humidity and the temperature are varied both in phase and out of phase.

Experimenters have cycled the ambient relative humidity in a stepwise, a linear, and a sinusoidal fashion. To prepare the ground for an investigation of the behavior under random variations in relative humidity, an investigation is needed of whether or not the superposition of the responses to two different relative humidity variations gives the response due to their sum. In the event that this is the case, a sum of responses to sinusoidal relative humidity variations would give the general case, because an arbitrary function can be written as a Fourier series. The nonlinear behavior of the accelerated creep response makes it unlikely that such a superposition is valid.

Almost all measurements of strains associated with moisture-accelerated creep have been in uniaxial tests. Information about the interaction of the strains in the machine and cross-machine directions in a sheet might be uncovered by measurement of the biaxial strains under biaxial creep loads during changes in either relative humidity or temperature. Such a biaxial test, for example, could compare the effect of drying restraints in the machine and cross-machine directions.

Much work remains to be done on the scientific study of paper mechanics to control paper behavior in response to the external influences of loads and ambient conditions including variable humidity and temperature.

REFERENCES

1. Andersson, O. and Berkyto, E. (1951). Some factors affecting the stress–strain characteristics of paper. *Svensk Papperstidn.* *54*(13):437–444.
2. Argon, A. S. (1968). Delayed elasticity in inorganic glasses. *J. Appl. Phys.* *30*(9):4080–4086.

3. Armstrong, L. D. (1972). Deformation of wood in compression during moisture movement. *Wood Sci.* *5*(2):81–86.
4. Armstrong, L. D., and Christensen, G. N. (1961). Influence of moisture changes on deformation of wood under stress. *Nature 191*:869–870.
5. Armstrong, L. D., and Grossman, P. U. A. (1972). The behavior of particleboard and hardwood beams during moisture cycling. *Wood Sci. Technol. 6*:128–137.
6. Armstrong, L. D., and Kingston, R. S. T. (1960). Effect of moisture changes on creep in wood. *Nature 185*:862.
7. Back, E. L., Salmén, L., and Richardson, G. (1983). Transient effects of moisture sorption on the strength properties of paper and wood-based materials. *Svensk Papperstidn. 86*:R61–R71.
8. Bažant, Z. P. (1985). Constitutive equation of wood at variable humidity and temperature. *Wood Sci. Technol. 19*:159–177.
9. Bažant, Z. P., and Kaplan, M. F. (1996). *Concrete at High Temperatures*. Longman, Harlow, Essex, England.
10. Bažant, Z. P., and Meiri, S. (1985). Measurement of compression creep of wood at humidity changes. *Wood Sci. Technol. 19*:179–182.
11. Bažant, Z. P., and Wu, S. T. (1974). Creep and shrinkage law for concrete at variable relative humidity. *J. Eng. Mech. Div., Proc. ASCE 100*:1183–1209.
12. Berger, B. J., and Habeger, C. C. (1989). Influences of non-equilibrium moisture conditions on the in-plane ultrasonic stiffnesses of cellulose. *J. Pulp Paper Sci. 15*(5):J160–J165.
13. Berger, B. J., Habeger, C. C., and Pankonin, B. M. (1989). The influence of moisture and temperature on the ultrasonic viscoelastic properties of cellulose. *J. Pulp Paper Sci. 15*(5):J170–J177.
14. Beutler, A. J. (1965). An infrared backscatter moisture gage. *Tappi 48*(9):490–493.
15. Bostrom, B. (1997). New perspectives on test methods for dimensional stability. In: *Proceedings of the 3rd International Symposium: Moisture and Creep Effects on Paper, Board and Containers*. I. R. Chalmers, ed. PAPRO, Rotorua, New Zealand.
16. Bronkhorst, C. A. (1997). Towards a more mechanistic understanding of corrugated container creep deformation behavior. *J. Pulp Paper Sci. 23*(4):J174–J181.
17. Byrd, V. L. (1972). Effect of relative humidity changes during creep on handsheet paper properties. *Tappi 55*(2):247–252.
18. Byrd, V. L. (1972). Effect of relative humidity changes on compressive creep response of paper. *Tappi 55*(11):1612–1613.
19. Byrd, V. L. (1984). Edgewise compression creep of fiberboard components in a cyclic-relative-humidity environment. *Tappi J 67*(7):86–90.
20. Byrd, V. L. (1986). Adhesives influence on edgewise compression creep in a a cyclic relative humidity environment. *Tappi J. 69*(10):98–100.
21. Byrd, V. L. (1988). Effect of cyclic moisture changes on paperboard performance in a service environment. Proceedings of the 1988 Appita General Conference, Hobart, Tasmania, Australia.
22. Byrd, V. L., and Koning, J. W., Jr. (1978). Edgewise compression creep in cyclic relative humidity environments. *Tappi 61*(6):35–37.
23. Caulfield, D. F. (1980). Interactions at the cellulose-water interface. In: *Paper Science and Technology: The Cutting Edge*. Institute of Paper Chemistry, Appleton, WI, pp. 70–88.
24. Caulfield, D. F. (1987). Dimensional stability of paper: Papermaking methods and stabilization of cell walls. Wood Science Seminar 1, Dec. 15–16, East Lansing, MI, pp. 87–98.
25. Chalmers, I. R. (1997). The effect of humidity on packaging grade paper elastic modulus. In: *Proceedings of the 3rd International Symposium: Moisture and Creep Effects on*

Paper, Board and Containers. I. R. Chalmers, ed. PAPRO, Rotorua, New Zealand, pp. 201–213.

26. Chaves de Oliveira, R., Mark, R. E., and Perkins, R. W. (1990). Evaluation of the effects of heterogeneous structure on strain distribution in low density papers. In: *Mechanics of Wood and Paper Materials.* R. Perkins, ed. AMD-Vol. 112, MD Vol.. 23. ASME, New York, pp. 37–61.
27. Christensen, R. M. (1971). *Theory of Viscoelasticity.* Academic Press, New York.
28. Coffin, D. W., and Boese, S. B. (1997). Tensile creep behavior of single fibers and paper in a cyclic humidity environment. In: *Proceedings of the 3rd International Symposium: Moisture and Creep Effects on Paper, Board and Containers.* I. R. Chalmers, ed. PAPRO, Rotorua, New Zealand, pp. 39–52.
29. Considine, J. M., Thelin, P., Gunderson, D. E., and Fellers, C. (1989). Compressive creep behavior of paperboard in a cyclic humidity environment: Exploratory experment. *Mechanics of Cellulosic and Polymeric Materials*, The Third Joint ASCE/ASME Mechanics Conference, La Jolla, CA, July 9–12. ASME, AMD-Vol 99, ASME, New York, pp. 149–156.
30. Danilatos, G., and Feughelman, M. (1976). The internal dynamic mechanical loss in α-keratin fibers during moisture sorption. *Textile Res. J. 46*:845–846.
31. Downs, J., Halperin, H. R., Humphrey, J., and Yin, F. (1990). An improved video-based computer tracking system for soft biomaterials testing. *IEEE Trans. Biomed. Eng. 37*(9):903–907.
32. Ericksen, R. H. (1985). Creep of aromatic polyamide fibers. *Polymer 26*:733–746.
33. Eyring, H. (1936). Viscosity, plasticity, and diffusion as examples of absolute reaction rates. *J. Chem. Phys. 4*:283–291.
34. Forsberg, K. (1994). Predicting long term stacking strength of corrugated boxes. *Proceedings of the Moisture-Induced Creep Behavior of Paper and Board Conference*, Stockholm, Sweden, Dec. 5–7, 1994, pp. 191–203.
35. Freed, A. D., and Walker, K. P. (1993). Viscoplasticity with creep and plasticity bounds. *Int. J. Plasticity 9*:213–242.
36. Gressel, V. P. (1986). A proposal for consistent experimental principles for conducting and evaluating creep tests. *Holz. Roh-Werkst. 44*(4):133–138.
37. Gunderson, D. E. (1981). A method for compressive creep testing of paperboard. *Tappi 64*(11):67–71.
38. Gunderson, D. E. (1989). Method for measuring mechanosorptive properties. *Mechanics of Cellulosic and Polymeric Materials*, The Third Joint ASCE/ASME Mechanics Conference, La Jolla, CA, July 9–12. AMD-Vol 99, ASME, New York, pp. 157–166.
39. Gunderson, D. E., and Considine, J. M. (1986). Measuring the mechanical behavior of paperboard in a changing humidity environment. *1986 International Process and Materials Quality Evaluation Conference.* TAPPI Press, Augusta, GA, pp. 245–251.
40. Gunderson, D. E., Considine, J. M., and Scott, C. T. (1988). The compressive load-strain curve of paperboard: Rate of load and humidity effects. *J. Pulp Paper Sci. 14*(2):J37–J41.
41. Gunderson, D. E., and Tobey, W. E. (1990). Tensile creep of paperboard: Effect of humidity change rates. Materials Research Society, 1990 Spring Meeting, San Francisco, CA, Apr. 18–20.
42. Halsey, G., White, H. J., Jr., and Eyring, H. (1945). Mechanical properties of textiles, I. *Textile Res. J. 15*(9):295–311.
43. Hanhijärvi, A. (1995). Deformation kinetics based rheological model for time-dependent and moisture induced deformation of wood. *Wood Sci. Technol. 29*: 191–195.

44. Haraldsson, T., Fellers, C., and Kolseth, P. (1994). Modelling of creep behavior at constant relative humidity. *Proceedings of the Moisture-Induced Creep Behavior of Paper and Board Conference*, Stockholm, Sweden, Dec. 5–7, 1994, pp. 139–147.
45. Haslach, H. W., Jr. (1994). The mechanics of moisture accelerated creep in paper. *Tappi J. 77*(10):179–186.
46. Haslach, H. W., Jr. (1994). Relaxation of moisture accelerated creep and hygroexpansion. *Proceedings of the Moisture-Induced Creep Behavior of Paper and Board Conference*, Stockholm, Sweden, Dec. 5–7, 1994, pp. 121–138.
47. Haslach, H. W., Jr. (1996). A model for drying-induced mirocompressions in paper: Buckling in the interfiber bonds. *Composites 27B*(1):25–33.
48. Haslach, H. W., Jr. (1997). Relaxation of moisture accelerated creep, backstress and hygroexpansion in paper. In: *Mechanics of Cellulosic Materials—1997*. R. W. Perkins, ed. AMD-Vol. 221, MD-Vol. 77. Joint Applied Mechanics and Materials Summer Conference, ASME AMD-MD, Evanston, IL, June 30, 1997, ASME, New York, pp. 45–53.
49. Haslach, H. W., Jr. and Abdullahi, Z. (1995). Thermally cycled creep of paper. Joint Applied Mechanics and Materials Summer Conference, Los Angeles, CA, June 28–30, 1995. In: *Mechanics of Cellulosic Materials*—1995. R. Perkins, ed. AMD-Vol. 209, MD-Vol. 60, ASME, New York. pp. 13–22.
50. Haslach, H. W., Jr., Khan, S., Mohammad, S., and Pecht, M. G. (1989). Behavior of Virginia fiber paper as influenced by relative humidity. Symposium on the Mechanics of Cellulosic and Polymeric Materials, 3rd ASCE-ASME Mechanics Conference, LaJolla, CA, July 9–12, 1989. *Mechanics of Cellulosic and Polymeric Materials*. AMD-Vol 99, MD-Vol 13. ASME, New York, pp. 167–172.
51. Haslach, H. W., Jr., Pecht, M. G., and Wu, X. (1990). A viscoelastic model for variable humidity loading in creep. Symposium on Mechanics of Wood and Paper Materials, 1990 ASME Winter Annual Meeting, Dallas, TX, Nov. 25–30, 1990. *Mechanics of Wood and Paper Materials*. AMD-Vol. 112, MD-Vol 23. ASME, New York, pp. 1–7.
52. Haslach, H. W., Jr., Pecht, M. G., and Wu, X. (1991). Variable humidity and load interaction in tensile creep of paper. Proceedings of the 1991 International Paper Physics Conference, Kona, HI. TAPPI Press, Atlanta, GA, pp. 219–224.
53. Haslach, H. W., Jr. and Wu, X. (1992). Mechanisms of moisture accelerated tensile creep in paper. ASME Symposium on the Mechanics of Cellulosic Materials, Nov. 8–13, 1992, Anaheim, CA. *Mechanics of Cellulosic Materials*, R. W. Perkins, ed. AMD-Vol. 145, MD-Vol. 36, ASME, New York, pp. 39–47.
54. Hearmon, R. F. S., and Paton, J. M. (1964) Moisture content changes and creep of wood. *Forest Products J. 14*:357–359.
55. Hill, R. L. (1967). The creep behavior of individual pulp fibers under tensile stress, *Tappi 50*(8):432–440.
56. Hoffmeyer, P., and Davidson, R. W. (1989). Mechano-sorptive mechanism of wood in compression and bending. *Wood Sci. Technol. 23*:215.
57. Holland, H. D., Halsey, G., and Eyring, H. (1946). Mechanical properties of textiles. VI. A study of creep of fibers. *Textile Res. J. 16*(5):201–210.
58. Humphrey, J. D., Vawter, D. L., and Vito, R. P. (1987). Quantification of strains in biaxially tested soft tissues. *J. Biomech. 20*(1):59–65.
59. Hunt, D. G. (1980). Prediction of sorption and diffusion of water vapour by nylon-6,6. *Polymer 21*:495–501.
60. Hunt, D. G., and Darlington, M. W. (1980). Creep of nylon-6,6 during concurrent moisture changes. *Polymer 21*:502–508.
61. Htun, M. (1986). The control of mechanical properties by drying restraints. In: *Paper: Structure and Properties*. J. A. Bristow and P. Kolseth, eds. Marcel Dekker, New York, pp. 311–326.

62. Jackson, T., and Parker, I. (1997). Accelerated creep in rayon fibres? In: *Proceedings of the 3rd International Symposium: Moisture and Creep Effects on Paper, Board and Containers.* I. R. Chalmers, ed. PAPRO, Rotorua, New Zealand, pp. 53–67.
63. Jentzen, C. A. (1964). The effect of stress applied during drying on some of the properties of individual pulp fibers. *Tappi 47*(7):412–418.
64. Kolseth, P., and de Ruvo, A. (1983). The measurement of viscoelastic behavior for the characterization of time-, temperature-, and humidity-dependent properties. In: *Handbook of Physical and Mechanical Testing of Paper and Paperboard,* Vol. 1. R. E. Mark, ed. Marcel Dekker, New York, pp. 255–322.
65. Kolseth, P., and de Ruvo, A. (1986). The cell wall components of wood pulp fibers. In: *Paper: Structure and Properties.* J. A. Bristow and P. Kolseth, eds. Marcel Dekker, New York, pp. 1–25.
66. Koning, J. W., and Stern, R. K. (1977). Long-term creep in corrugated fiberboard containers. *Tappi 60*(12):128–131.
67. Kubat, J., and Lindbergson, B. (1965). Damping transients in polymers during sorption and desorption. *J. Appl. Polym. Sci. 9*:2651–2654.
68. Kubat, J., and Nyborg, L. (1962). Influence of mechanical stress on the sorption equilibrium of paper *Svensk Papperstidn. 65*(18):698–702.
69. Kuskowski, S., Considine, J., and Lee, S. (1994). Corrugating components and their relationship to combined board performance. Proceedings of the Moisture-Induced Creep Behavior of Paper and Board Conference, Stockholm, Sweden, Dec. 5–7, 1994, pp. 223–231.
70. Leake, C., and Wojcik, R. (1992). Humidity cycling rates: How they affect container life spans. Proceedings, Cyclic Humidity Effects on Paperboard Packaging. Forest Products Laboratory, Madison, WI, pp. 134–144.
71. Liljemark, N. T., Åsnes, H., and Kärrholm, M. (1971). The sensitivity of cotton fabrics to wrinkling during changing moisture regain and its dependence on setting and cross-linking parameters. *Textile Res. J. 41*:526–533.
72. Lindström, T. (1986). The porous lamellar structure of the cell wall. In: *Paper: Structure and Properties,* J. A. Bristow and P. Kolseth, eds. Marcel Dekker, New York, pp. 99–120.
73. Mackay, B. H., and Downes, J. G. (1959). The effect of the sorption process on the dynamic rigidity modulus of the wool fiber. *J. Appl. Polym. Sci. 11*(4):32–38.
74. Nissan, A. H. (1959). A molecular approach to the problem of viscoelasticity. *Nature 185*:1477.
75. Nissan, A. H. (1967). The significance of hydrogen bonding at the surfaces of cellulose network structures. In: *Surface and Coatings Related to Paper and Wood.* R. H. Marchessault and C. Skaar, eds. Syracuse Univ. Press, Syracuse, NY, pp. 221–268.
76. Nissan, A. H. (1976). H-bond dissociation in hydrogen bond dominated solids. *Macromolecules 9*(5):840–850.
77. Nordman, L. S. (1958). Bonding in paper sheets. In: *Fundamentals of Papermaking Fibers.* F. Bolam, ed. British Pulp and Paper Manuf. Assoc., Kenley, UK, pp. 333–347.
78. Nordman, L. S., Gustafsson, C., and Olofsson, G. (1955). Optical measurement of bond breaking during a tensile test. *Tappi 38*(12):724–727.
79. Nordon, P. (1962). Some torsional properties of wool fibers. *Textile Res. J 32*:560–568.
80. Padanyi, V. (1991). Mechano-sorptive effects and accelerated creep in paper. *1991 International Paper Physics Conference Proceedings.* TAPPI Press, Atlanta, GA, pp. 397–411.
81. Padanyi, V. (1994). Reversible age dependence of tensile creep and stress relaxation in paperboard. *Proceedings of the Moisture-Induced Creep Behavior of Paper and Board Conference*, Stockholm, Sweden, Dec. 5–7, 1994, pp. 67–87.
82. Page, D. H. (1969). The structure and properties of paper. Part 1. The structure of paper. *Trend 15*:7.

83. Page, D. H. (1971). The structure and properties of paper. Part 2. Shrinkage, dimensional stability and stretch. *Trend 18*:6–11.
84. Page, D. H., and De Grace, J. H. (1967). The delamination of fiber walls by beating and refining. *Tappi 50*(10):489–495.
85. Page, D. H., El-Hosseiny, E., Winkler, K., and Bain, R. (1972). The mechanical properties of single wood-pulp fibres. Part 1. A new approach. *Pulp Paper Mag. Can. 83*(8):T198–T203.
86. Page, D. H., and Tydeman, P. A. (1961). In: *Formation and Structure of Paper*. F. Bolam, ed. Transactions of the Fundamental Reserach Symposium, Oxford, p. 397.
87. Pecht, M. G. (1985). Creep of regain rheologically simple hydrophilic polymers. *J. Strain Anal. 20*(3):179–181.
88. Pecht, M. G., and Haslach, H. W., Jr. (1991). A viscoelastic model for constant rate loading at different relative humidities. *Mech. Mater. 11*:337–345.
89. Perkins, R. W., and Ramasubramanian, M. K. (1989). Concerning micromechanics models for the elastic behavior of paper. *Mechanics of Cellulosic and Polymeric Materals*, The Third Joint ASCE/ASME Mechanics Conference, La Jolla, CA, July 9–12. AMD-Vol 99. ASME, New York, pp. 23–33.
90. Pickett, G. (1942). The effect of change in moisture content on the creep of concrete under a sustained load. *J. Am. Concrete Inst. 29/54*(10):857–864.
91. Salmén, N. L. (1986). The cell wall as a composite structure. In: *Paper: Structure and Properties*. J. A. Bristow and P. Kolseth, eds. Marcel Dekker, New York, pp. 51–73.
92. Salmén, N. L., and Back, E. L. (1980). Moisture-dependent thermal softening of paper, evaluated by its elastic modulus. *Tappi 63*(6):117–120.
93. Salmén, L., and Fellers, C. (1996). Moisture induced transients and creep of paper and nylon 6,6: A comparison. *Nordic Pulp Paper Res. J. 11*(3):186–191.
94. Sedlachek, K. M. (1995). The effect of hemicelluloses and cyclic humidity on the creep of single fibers. Ph.D. Thesis, Institute of Paper Science and Technology, Atlanta, GA.
95. Sedlachek, K., and Ellis, R. (1994). The effect of cyclic humidity on the creep of single fibers of southern pine. *Proceedings of the Moisture-Induced Creep Behavior of Paper and Board Conference*, Stockholm, Sweden, Dec. 5–7, 1994, pp. 29–49.
96. Setterholm, V. C., and Chilson, W. A. (1965). Drying restraint. *Tappi 48*(11);634–640.
97. Shishoo, R., and Lundell, M. (1972). NMR studies of cotton and wool in sorption processes. *Textile Res. J. 42*:285–291.
98. Söremark, C., and Fellers, C. N. (1991). Mechano-sorptive creep and hygroexpansion of corrugated board in bending. *1991 International Paper Physics Conference Proceedings*. TAPPI Press, Atlanta, GA, pp. 549–559.
99. Söremark, C., and Fellers, C. N. (1993). Mechano-sorptive creep and hygroexpansion of corrugated board in bending. *J. Pulp Paper Sci. 19*(1):J19–J26.
100. Söremark, C., and Fellers, C. N. (1994). Assessing cyclic creep behavior of different paper grades. *Proceedings of the Moisture-Induced Creep Behavior of Paper and Board Conference*, Stockholm, Sweden, Dec. 5–7, 1994, pp. 275–285.
101. Söremark, C., Fellers, C. N., and Henriksson, L. (1993). Mechano-sorptive creep of paper: Influence of drying restraint and fibre orientation. In: *Products of Papermaking*. Transactions of the tenth fundamental research symposium held at Oxford, Sept. 1993 Vol.1, C. F. Baker (ed) PIRA International, Leatherhead, Surrey, United Kingdom, pp 547–574
102. Swanson, T., Stejskal, E. O., and Tarkow, H. (1962). Nuclear magnetic resonance studies on several cellulose-water systems. *Tappi 45(12):929–932*.
103. Uesaka, T., Moss, C., and Nanri, Y. (1991). The characterization of hygroexpansivity in paper. *1991 International Paper Physics Conference Proceedings*. TAPPI Press, Atlanta, GA, pp. 613–622.

104. Urbanik, T. J. (1995). Hygroexpansion-creep model for corrugated fiberboard. *Wood Fiber Sci. 27*:134–140.
105. Urbanik, T. J., and Lee, S. K. (1995). Swept sine humidity schedule for testing cycle period effects on creep. *Wood Fiber Sci 27*:68–78.
106. Van den Akker, J. A. (1969). An analysis of the Nordman "bonding strength." *Tappi 52*(12):2386–2389.
107. Wang, J. Z., Dillard, D. A., Wolcott, M. P., Kamke, F. A., and Wilkes, G. L. (1990). Transient moisture effects in fibers and composite materials. *J. Compos. Mater. 24*:994–1009.

5

BENDING STIFFNESS, WITH SPECIAL REFERENCE TO PAPERBOARD

CHRISTER FELLERS
STFI, Swedish Pulp and Paper Research Institute
Stockholm, Sweden

LEIF A. CARLSSON
Florida Atlantic University
Boca Raton, Florida

I. INTRODUCTION

Bending stiffness is an important property for both paper and paperboard in both converting and end-use situations. For thin papers, high bending stiffness is extremely important during converting operations such as feeding in fast presses, folders, copying machines, and sack-forming machines. On the other hand, low bending stiffness is a requirement in, for instance, label papers and papers in books and magazines in order to obtain good conformability and folding properties [11].

For heavier papers such as carton board for packaging, bending stiffness is the most important mechanical property in the forming of cartons, for the bulging of the panels, for protection of the contents, and for stacking strength in end-use situations [2]. Thus, it is essential for the manufacturer of carton board to have a good understanding of the engineering principles involved in the evaluation of bending stiffness of multi-ply board and a proper appreciation of the relative importance of the properties of the different plies. The focus is then on the producer of the material to meet the specified bending stiffness level in the most economical way.

Many instruments have been developed over the years for testing of bending stiffness. Review articles on the subject are presented by, e.g., Luey [19], Ranger [22,23], Hohmann [12–14], Carlsson and Fellers [3], Koran and Kamden [18], Franklin [11], and Fellers [7].

The following discussion and examples deal with the theoretical background of bending stiffness evaluation and the principles of pulp characterization with respect to bending stiffness in single-ply and multi-ply board.

II. THEORETICAL BACKGROUND: PURE BENDING OF BEAMS

Consider a homogeneous beam with a rectangular cross section subjected to a bending moment M^b per unit width (Fig. 1). Because no shear forces are acting on the beam, the bending moment is constant along the beam. Longitudinal elements on the convex side of the beam are subjected to tension, and those on the concave side to compression. The plane within the beam that remains undeformed during bending,

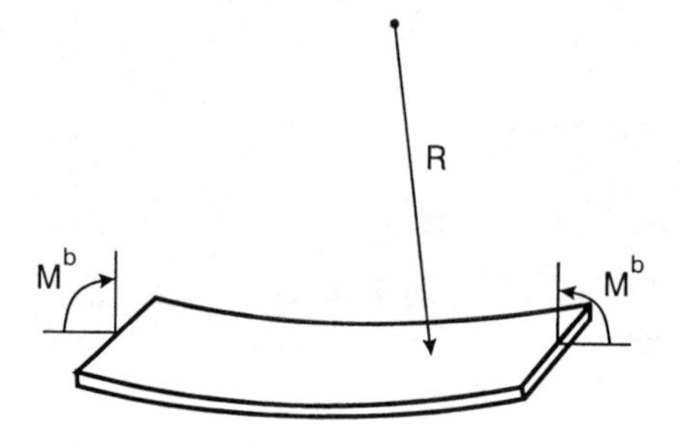

Fig. 1 Beam subjected to pure bending, where M^b is the bending moment per unit width and R is the radius of curvature of the neutral surface.

i.e., neither compressed nor extended, is called the neutral plane. The radius of curvature of the beam is denoted by R.

In this analysis, it is assumed that plane cross sections are initially straight and normal to the neutral plane and remain so after bending. Under these conditions the beam adopts a circular deflection shape with constant curvature. The strain ε of an element at a distance z from the neutral plane for a given beam curvature $1/R$ is given by

$$\varepsilon = \frac{z}{R} \tag{1}$$

Figure 2 shows the strain variation through the thickness of the beam in the bent state. The x axis is aligned with the neutral plane. Thus, the strains of the longitudinal fibers are proportional to the distance z from the neutral plane and inversely proportional to the radius of curvature. According to Hooke's law, the stress σ is proportional to the strain ε in the elastic region of the material, the constant of proportionality E being called the elastic modulus, the modulus of elasticity, or Young's modulus:

$$\sigma = E\varepsilon \tag{2}$$

Equations (1) and (2) yield

$$\sigma = \frac{Ez}{R} \tag{3}$$

This equation expresses the variation of stress through the thickness of the beam in the elastic region. By analogy with the strain variation, the stress variation illustrated in Fig. 2 is linear.

The bending moment per unit width of the beam, M^b, is obtained by integration of the stresses through the thickness of the beam:

$$M^b = \int \sigma z \, dz \tag{4}$$

Substitution of Eq. (3) into Eq. (4) yields

$$M^b R = \int E z^2 \, dz \tag{5}$$

The product $M^b R$ generally defines the bending stiffness S^b, i.e.,

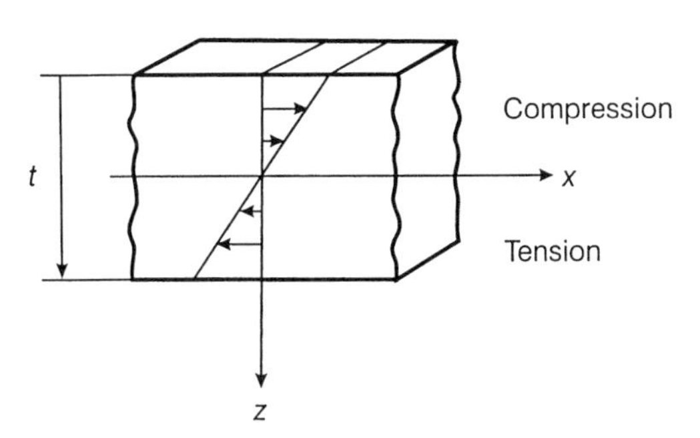

Fig. 2 Strain variation in the elastic region across the thickness of the beam.

$$S^b = M^b R \tag{6}$$

For a homogeneous beam of thickness t and rectangular cross section, we obtain from Eq. (5) the following well-known equation for the bending stiffness per unit width:

$$S^b = \frac{Et^3}{12} \tag{7}$$

The foregoing one-dimensional plane stress analysis ignores the two-dimensional behavior of paper during bending. A more detailed analysis that considers the theory of bending of an elastic plate is given in the Appendix. Depending on the constraints of the paper in the direction perpendicular to the bending direction, we may get different bending stiffness values. In reality one may, however, not expect bending stiffness values to exceed the simple plane stress analysis above by more than 5% (see Appendix).

III. EVALUATION OF BENDING STIFFNESS

The two-point, three-point, and four-point methods and the resonance method are discussed in this section. The theoretical background and some practical limitations are also mentioned. The starting point for bending stiffness evaluation is Eq. (6). The bending stiffness can be obtained by evaluating the product $M^b R$ for small deformations of the strip in a bending test. A method for measuring the pure bending properties of paper, where the strip forms a circular arc, has been reported [9]. Because of the difficulty of measuring the radius at small deformations, measurement of the deflection is commonly preferred.

A. The Four-Point Method

The principle of the four-point method is illustrated in Fig. 3. The strip is subjected to a pure bending moment between two symmetrically located inner loads and adopts a circular deflection shape. For small deflections, the relationship between

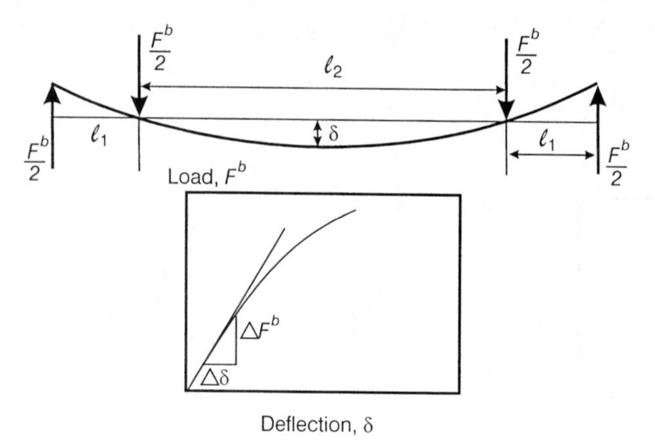

Fig. 3 The four-point method.

the radius of curvature R, the length l_2 between the inner loads, and the midpoint deflection δ is [12]

$$R = \frac{l_2^2}{8\delta} \tag{8}$$

The bending moment per unit width is

$$M^b = \frac{F^b l_1}{2} \tag{9}$$

where F^b is the force per unit width of the strip and l_1 is the distance between each inner load and the corresponding outer load. Combining Eqs. (6), (8), and (9) gives the bending stiffness,

$$S^b = \frac{F^b}{\delta}\left(\frac{l_1 l_2^2}{16}\right) \tag{10}$$

In practice, the steepest slope $\Delta F^b/\Delta\delta$ of the curve in Fig. 3 is evaluated:

$$S^b = \frac{\Delta F^b}{\Delta\delta}\left(\frac{l_1 l_2^2}{16}\right) \tag{11}$$

B. The Two-Point Method

Figure 4 shows the principle of the two-point method, where the strip is clamped at one end and subjected to a concentrated load at the other end. As distinct from the four-point method, the bending moment decreases linearly as the distance x from the clamped end increases:

$$M^b = F^b(l - x) \tag{12}$$

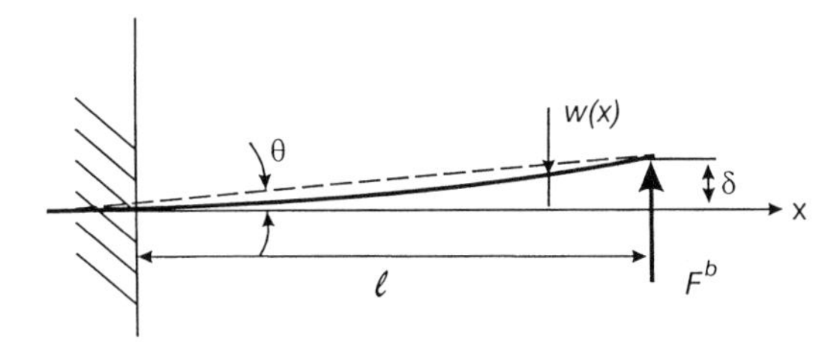

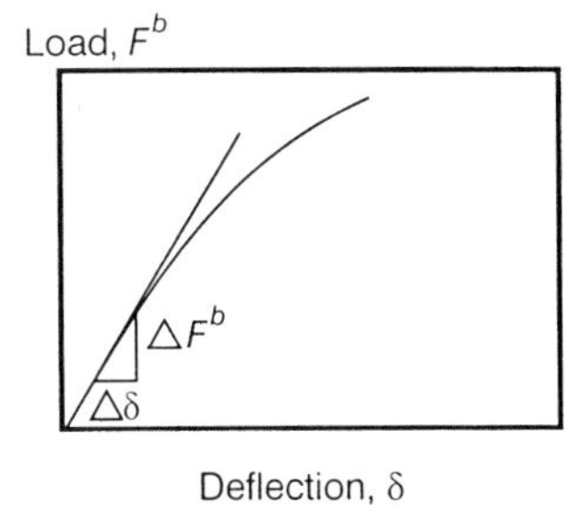

Fig. 4 The two-point method.

The radius of curvature consequently increases with the distance x from the clamped end. In order to obtain a relation between deflection δ and curvature $1/R$, the following differential equation [27] is used:

$$\frac{1}{R} = \frac{d^2w/dx^2}{[1 + (dw/dx)^2]^{3/2}} \tag{13}$$

where $w(x)$ is the deflection of the strip at an arbitrary distance x from the clamped end (Fig. 4).

For bending stiffness measurements, only small deflections and rotations of the strip are considered. In such cases, the term $(dw/dx)^2$ in Eq. (13) is small in comparison with unity and can therefore be neglected.

$$\frac{1}{R} = \frac{d^2w}{dx^2} \tag{14}$$

When Eqs. (6), (12), and (14) are combined for the case of the two-point test, the result is

$$\frac{d^2w}{dx^2} = \frac{F^b(l - x)}{S^b} \tag{15}$$

Integration of this equation and consideration of the boundary conditions yields

$$w = \frac{F^b}{2S^b}\left(\frac{(l - x)^3}{3} + l^2x - \frac{l^3}{3}\right) \tag{16}$$

The deflection δ at the point of application of the load is given by

$$\delta = \frac{F^b l^3}{3S^b} \tag{17}$$

The bending stiffness is thus

$$S^b = \frac{F^b}{\delta}\left(\frac{l^3}{3}\right) \tag{18}$$

As in the four-point method, the bending stiffness can be evaluated from the steepest slope of the curve in Fig. 4:

$$S^b = \frac{\Delta F^b}{\Delta\delta}\left(\frac{l^3}{3}\right) \tag{19}$$

The deflection δ can be measured directly or can be obtained from the angle θ of the clamp in Fig. 4.

$$\delta = l\theta \tag{20}$$

If θ is measured in degrees, the bending stiffness is given by

$$S^b = \frac{\Delta F^b}{\Delta\theta}\left(\frac{60l^2}{\pi}\right) \tag{21}$$

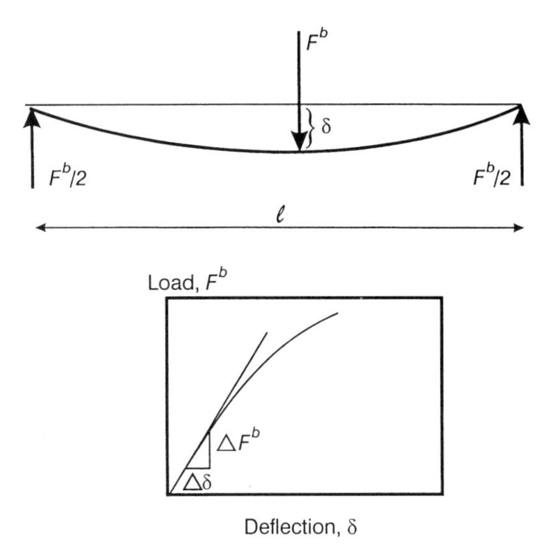

Fig. 5 The three-point method.

C. The Three-Point Method

Figure 5 shows the three-point method, which can be treated basically as two mirror-imaged two-point methods (the center of the strip corresponding to the clamp). The equations derived for bending stiffness for the two-point method can consequently be used for the three-point method by substituting loads and strip lengths in Eq. (19), yielding

$$S^b = \frac{\Delta F^b}{\Delta \delta}\left(\frac{l^3}{48}\right) \tag{22}$$

D. The Resonance Method

The principle of the resonance method is shown in Fig. 6. The strip is clamped at one end and oscillates perpendicular to the plane of the strip at a constant frequency f of 25 Hz. The span l is varied. At a given length, the strip becomes a vibrating system of standing waves with resonance, which can be detected by the maximum deflection δ.

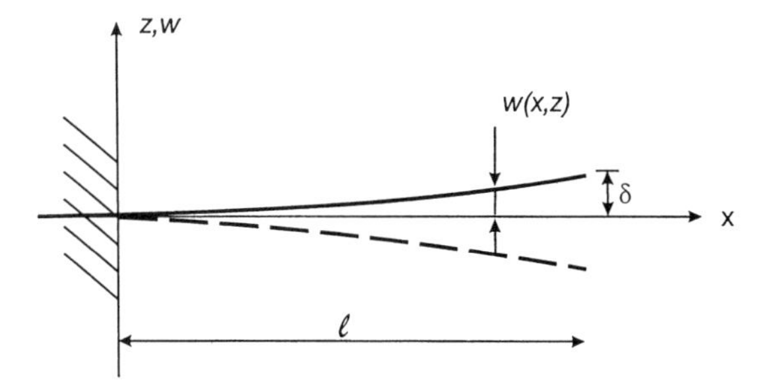

Fig. 6 The resonance method. Resonance corresponds to maximum deflection δ.

Consider the vibrating strip in Fig. 6. The equation of motion of an elastic strip of density ρ and thickness t is [28]

$$S^b \frac{d^4}{dx^4} w(x, t) + t\rho \frac{d^2}{dt^2} w(x, t) = 0 \tag{23}$$

The general solution of Eq. (23) for sinusoidal vibrations is

$$w(x, t) = (A \cosh \mu x + B \, sinh \mu x + C \cos \mu x + D \sin \mu x) \sin \omega t \tag{24}$$

where

$$\mu^2 = \left(\frac{t\rho}{S^b}\right)^{0.5}, \qquad \omega = 2\pi f \tag{25}$$

When the boundary conditions are considered, it can be shown that the solution for the first vibrational mode gives the resonant eigenfrequency or the fundamental frequency f_r [28],

$$f_r = \frac{1}{2\pi}\left(\frac{(1.875)^2}{l^2}\right)\left(\frac{S^b}{\rho t}\right)^{0.5} \tag{26}$$

which results in

$$S^b = \frac{4\pi^2 l^4 \rho t f_r^2}{(1.875)^4} \tag{27}$$

Applying the relation $w = \rho t$, where w is the basis weight, we obtain

$$S^b = 3.19 l^4 w f_r^2 \tag{28}$$

The ISO Standard 5628 specifies a frequency of 25 Hz.

IV. SOURCES OF ERROR IN THE EVALUATION OF THE BENDING STIFFNESS OF PAPER

In the preceding section, expressions for the bending stiffness according to four different testing principles were derived. Certain geometrical conditions must, however, be fulfilled to give correct values of the bending stiffness. The relative importance of these conditions is examined and specified in this section.

A. Effect of Deviations from Hooke's Law

Deformation during a bending test must be small in order to satisfy Hooke's law in Eq. (2), which means that the beam must deform elastically. The strain limit ε_{pl} for elastic deformation can be estimated in a stress–strain test by determining the point where the stress–strain curve starts to deviate from a straight line, that is, the proportional limit.

It should be pointed out that the nonlinear character of the stress–strain curve is more pronounced in compression than in tension [4]. This means that the proportional limit in a bending test depends primarily on the proportional limit in compression. The largest strain ε_0 in a bent strip will, according to Eq. (1), be at the outer surfaces of the strip:

$$\varepsilon_0 = \frac{t}{2R} \tag{29}$$

where t is the thickness and R is the radius of curvature. To satisfy Hooke's law, ε_0 must not exceed ε_{pl} in a bending test.

From the equations in the preceding sections, it is possible to obtain expressions for maximum allowable deflections δ_a to satisfy Hooke's law for the two-point, three-point, and four-point methods.

For the *four-point method*, the deflection δ obtained from Eq. (8) is

$$\delta = \frac{l_2^2}{8R} \tag{30}$$

where l_2 is defined in Fig. 3. Equations (29) and (30) give the connection between maximum allowable deflection δ_a and ε_{pl}.

$$\delta_a = \varepsilon_{\text{pl}} \frac{l_2^2}{4t} \tag{31}$$

For the *two-point method*, the expressions for maximum allowable deflection δ_a and angle θ_a are obtained from Eqs. (14), (15), (17), (20), and (29):

$$\delta_a = 2\varepsilon_{\text{pl}} \frac{l^2}{3t} \tag{32}$$

$$\theta_a = 120\varepsilon_{\text{pl}} \frac{l}{\pi t} \tag{33}$$

For the three-point method,

$$\delta_a = \varepsilon_{\text{pl}} \frac{l^2}{6t} \tag{34}$$

Figure 7 shows maximum allowable deflections and angles for the two-point method as a function of thickness for a span of 50 mm and for different values of proportional limit strains ε_{pl}. According to Hohmann [12,13], who investigated a number of carton board grades with a large variation in properties, the proportional limit strain in tension ranged from 0.01 to 0.17%. For a board of thickness 0.5 mm at a span of 50 mm tested in the two-point method, this range corresponds to maximum allowable bending angles between 0.5° and 7.5° or maximum allowable deflections between 0.5 and 6 mm.

Standard methods such as those of Taber and Lorentzen and Wettre work according to the two-point principle, where the load at a certain angle is registered. The correct value of the bending stiffness should, however, be evaluated from the steepest slope of the load versus bending angle curve. If, instead, the load at a certain angle is used for bending stiffness evaluation, a lower apparent value of the bending stiffness may be obtained. In some unfortunate cases, the ranking of bending stiffness between two boards will be reversed on account of bending beyond the elastic region of the material, as is shown schematically in Fig. 8 [13,14]. Paperboard A has a higher bending stiffness than B. As a result of strong nonlinear response, evalua-

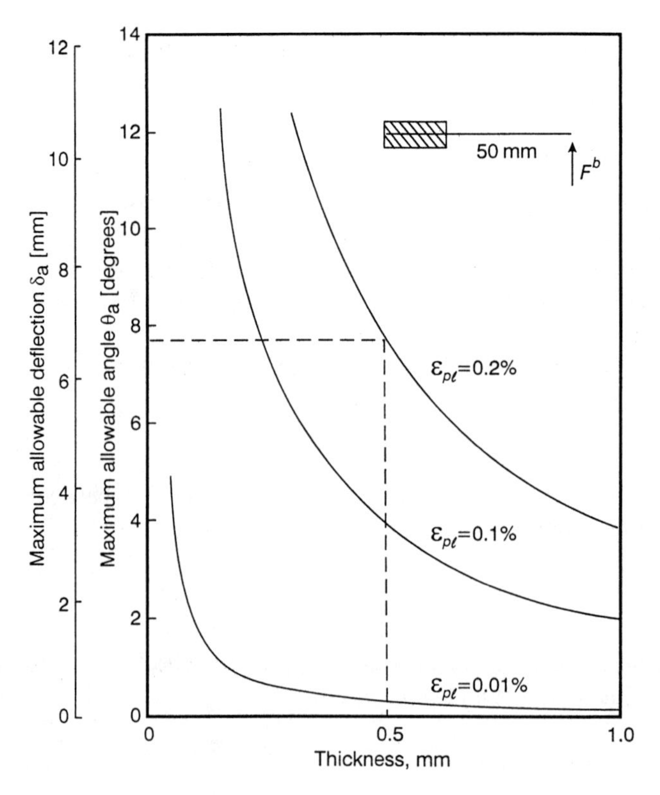

Fig. 7 The maximum allowable deflection δ_a and angle θ_a in the two-point test as a function of thickness for different values of proportional limit strain. Refer to Eqs. (32) and (33).

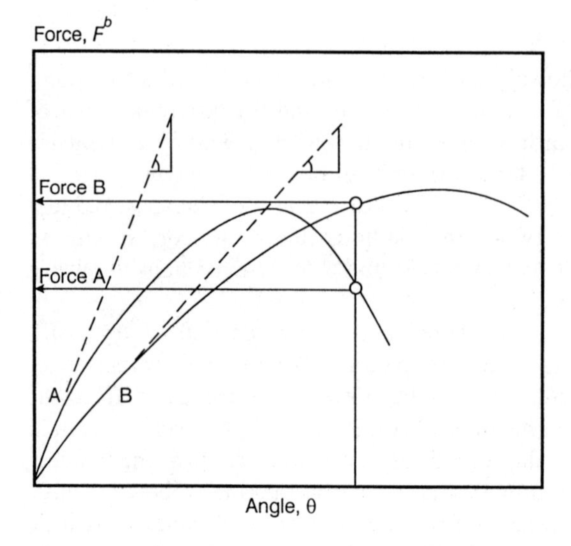

Fig. 8 Evaluation of the bending load at large angles results in underestimation of the correct bending stiffness in the two-point method.

tion of the bending load at a large angle will give a lower apparent bending stiffness for A than for B.

B. Effect of Slope of the Deflected Beam

The derivations of the bending stiffness formulas (21) and (22) for the two-point and three-point methods, respectively, are based on the assumption of small deflections and rotations in order to fulfill the approximation in Eq. (14). This means that $(dw/dx)^2$ must be small compared with unity, where dw/dx is the slope of the strip. The relative importance of the slope is illustrated next.

The deflection $w = w(x)$ of the strip in the two-point method was derived earlier [Eq. (16)]. Differentiation of this equation gives the slope as a function of x:

$$\frac{dw}{dx} = \frac{F^b[l^2 - (l - x)^2]}{2S^b} \tag{35}$$

The maximum slope, which occurs at the point $x = l$ is, for a given load F^b,

$$\left(\frac{dw}{dx}\right)_0 = \frac{F^b l^2}{2S^b} \tag{36}$$

Equations (17) and (36) are combined to obtain the relation between maximum allowable deflection δ_a and maximum allowable slope:

$$\delta_a = \frac{2}{3} l \left(\frac{dw}{dx}\right)_a \tag{37}$$

Equation (14) may be used as an approximation for Eq. (13) only if the slope dw/dx found from Eq. (37) is less than the permissible limit. For $(dw/dx)_a = 0.1$, the relative error in $1/R$ is 1.5%, which may be regarded as small.

If, as recommended in the DIN standards [5], the board is bent through 7.5° in the two-point method test, the corresponding value of $(dw/dx)_a$ is 0.2, which gives a relative error of approximately 6%. The allowable deflection then, for $(dw/dx)_a =$ 0.2 in the two-point method is, according to Eq. (37),

$$\delta_a = 0.132l \tag{38}$$

The corresponding deflection in the *three-point method* is

$$\delta_a = 0.066l \tag{39}$$

For the four-point method, there is no simple limit in maximum deflection that is related to the slope, although geometrical nonlinearities will influence the evaluation of bending stiffness for this method also.

C. Effect of Interlaminar Shear

In contrast to the four-point method, the applied load in the two-point and three-point methods causes shear stresses in the strip. These stresses cause initially plane cross sections of the strip to be warped, as illustrated in Fig. 9. The effect of shear stresses has been considered by Timoshenko [27]. The total deflection, including the contribution due to shear, is, for the two-point method,

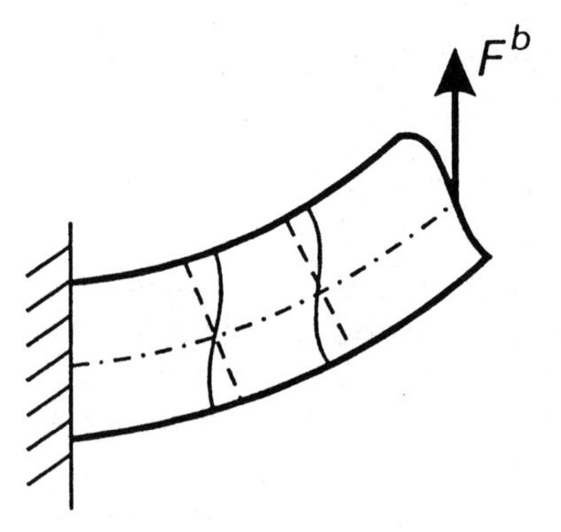

Fig. 9 Warping of initially plane cross section as a result of a shear load. (From Ref. 27.)

$$\delta_{\text{tot}} = \frac{F^b l^3}{3S^b}\left[1 + 0.3\frac{t^2}{l^3}\left(\frac{E}{G}\right)\right] \tag{40}$$

where t is the thickness of the strip, l is the span, and G is the interlaminar shear modulus.

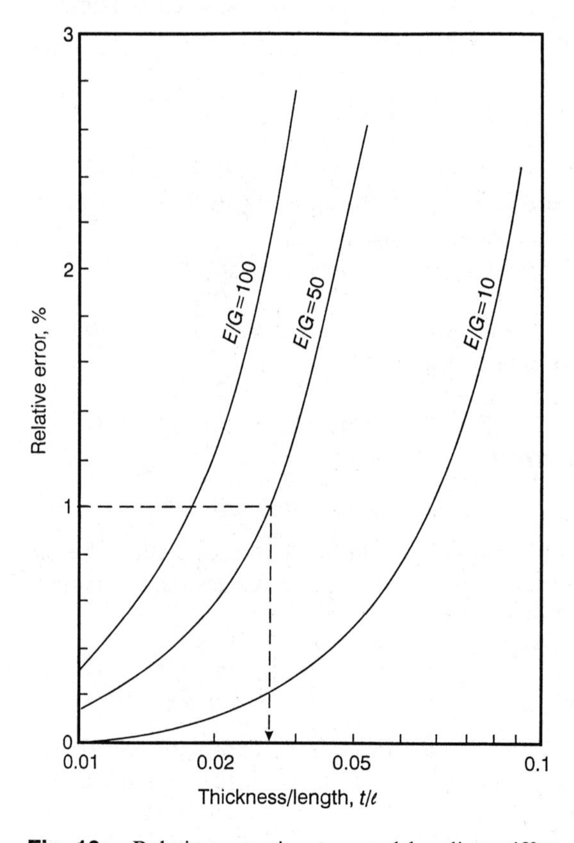

Fig. 10 Relative error in measured bending stiffness, according to the two-point method, due to shear deformation as a function of the thickness-to-length ratio of the strip for different E/G ratios.

Equation (17) was derived on the assumption that shear deformation could be neglected. A comparison of Eqs. (17) and (40) shows the magnitude of the effect of shear deformation on the deflection δ_{tot}. If shear is neglected, as in Eq. (17), there will be an error in the measured bending stiffness. This is illustrated in Fig. 10, where the influence of the ratios t/l and E/G on the relative error in S^b is shown. Measured values of E/G for paper are on the order of 50 [6]. Consequently, for a maximum relative error of 1%, the ratio of thickness to span must be less than 0.025; i.e., the span must be greater than 40 times the thickness. This condition is generally fulfilled in practice.

D. Effects of Deformation of the Paper in the Clamp in the Two-Point Method

In the preceding section it was shown that the influence of shear deformation on the evaluation of bending stiffness can be neglected at spans greater than approximately 40 times the thickness. Experiments show that the results are nevertheless dependent on the span [9]. As mentioned earlier, the three-point emthod can be regarded as two mirror-image two-point tests, which implies that the span dependence should theoretically be the same. The span l in both methods is defined here as the free span to simplify comparison. As shown in Fig. 11, measurements of bending stiffness on a solid bleached board of thickness 0.5 mm and with a basis weight of 380 g/m^2 by the

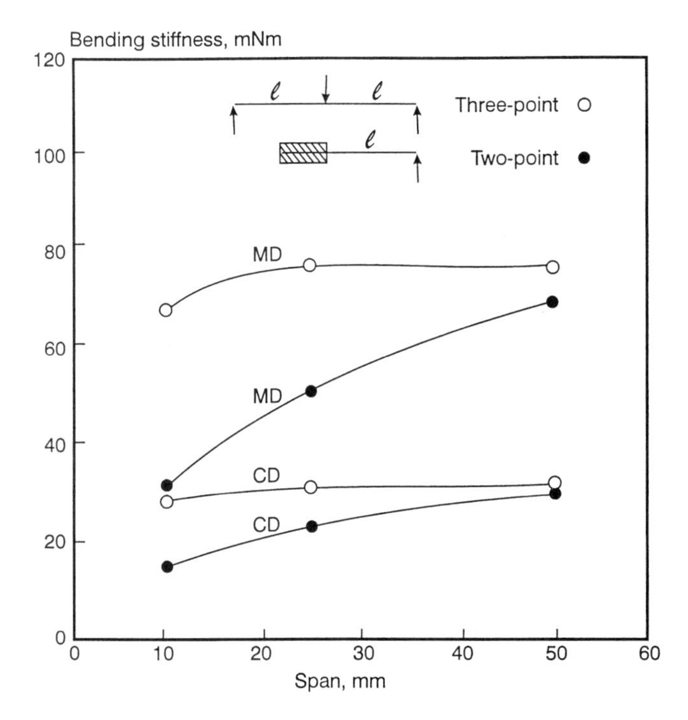

Fig. 11 The influence of span on measured bending stiffness evaluated from the (●) two-point and (○) three-point methods.

two-point method show a marked dependence on span, as distinct from the three-point method.

The greater dependence on span of the two-point method has been attributed to slippage between the clamp and paper allowing shear deformation of the paper strip inside the clamp [9]. Such slip deformations do not occur in the three-point method in the middle section; this can be concluded from symmetry considerations.

The deformation in the clamp in the two-point method is expected to depend on the clamping pressure. This is demonstrated for a 340 g/m^2 cardboard in the machine direction (MD) in Fig. 12. The bending stiffness increases up to approximately 100 kPa, above which it remains essentially constant. Measurements on several papers show that this pressure is sufficient and that the risk of reducing the bending stiffness by the use of excessively high pressure is minimal.

E. Discussion

The most comprehensive standards concerning testing conditions for bending stiffness evaluations by the strip methods of paper and board are ISO Standard ISO-5628 1990 based on the German DIN standards (DIN 1973), summarized in Table 1. Because paper is a viscoelastic material, the mechanical properties depend on the deformation speed. Table 1 gives recommended speed values. According to these standards, the bending stiffness should be evaluated by bending to small deformations with large spans to avoid effects of plastic deformation, shear, and too steep a slope of the tangent to the strip. The resonance method is also standardized by ISO-5629 [16] and many national standards.

Both theory and experiment show that is is appropriate to register the load–deflection curve on an XY recorder or digitally and calculate the bending stiffness from the steepest slope of the curve. With an apparatus that instead of registering the curve, records the load at a certain angle, errors are likely to occur as a result of curled strips or bending in the plastic region of the material. Even if the bending stiffness is calculated from the initial slope of the load–deformation curve, errors due to curled strips and clamping effects may occur, particularly in the two-point

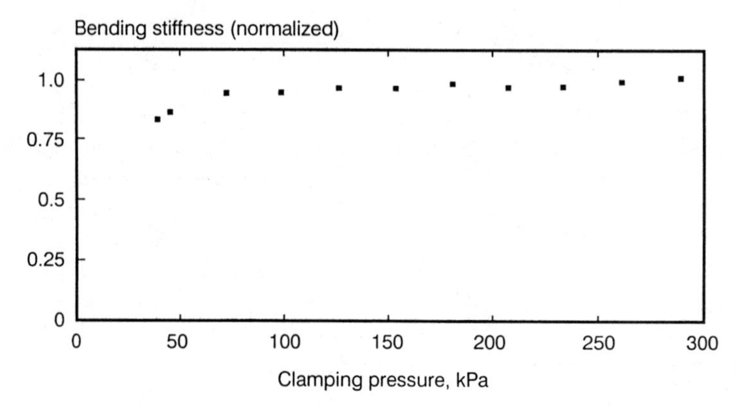

Fig. 12 Bending stiffness in the machine direction of a 340 g/m^2 board as a function of the clamping pressure. Bending angle is 7.5° and span length is 50 mm. The bending stiffness scale is normalized from 0 to 1. (From Ref. 21.)

Table 1 Recommended Testing Conditions (DIN 1973)

	Two-point, Fig. 4	Three-point, Fig. 5	Four-point, Fig. 3
Bending stiffness equations	$S^b = F^b l^3/3\ \delta$ $S^b = 60F^b l^2/\pi\theta$	$S^b = F^b l^3/48\ \delta$	$S^b = F^b l_1 l_2^2/16\ \delta$
Recommended spans	$l = 20, \quad t < 0.1$ $l = 50, \quad t > 0.1$	$l = 20, \quad t < 0.3$ $l = 50, \quad 0.3 < t < 0$ $l = 100, \quad t > 0.8$	$l_2 = 10, l_1 = 10, \quad t < 0.3$ $l_2 = 30, l_1 = 35, \quad 0.3 < t < 0.8$ $l_2 = 50, l_1 = 50, \quad t > 0.8$
Allowable deflections and angles to obey Hooke's law	$\delta_a = \varepsilon_{pl} l^2/150t$ $\theta_a = 1.2\varepsilon_{pl} l/\pi t$	$\delta_a = \varepsilon_{pl} l^2/600t$	$\delta_a = \varepsilon_{pl} l_2^2/400t$
Allowable deflections and angles that will keep the slope of the tangent to the strip sufficiently small	$\delta_a = 0.132l$ $\theta_a = 7.5°$	$\delta_a = 0.066l$	
Test rates	$\dot{\delta} = \dot{\varepsilon} l^2/150t$ $\dot{\Theta} = 1.2\dot{\varepsilon} l/\pi t$ $(0.2 < \dot{\varepsilon} < 1.0)$	$\dot{\delta} = \dot{\varepsilon} l^2/600t$ $(0.2 < \varepsilon < 1.0)$	$\dot{\delta} = \dot{\varepsilon} l_1(l_1 + 1.5l_2)/150t$ $(0.2 < \varepsilon < 1.0)$

S^b = bending stiffness, mN/m; F^b = force per unit width, N/mm; δ = deflection, mm; t = thickness, mm; θ = angle, deg; l, l_1, l_2 = span, mm; δ_a = allowable deflection, mm; θ_a = allowable angle, deg; ε_{pl} = strain at the proportional limit, percent; $\dot{\delta}$ = rate of linear deflection, mm/min; $\dot{\theta}$ = rate of angular deflection, deg/min; $\dot{\varepsilon}$ = strain rate, percent/min.

method, if the ratio of span to thickness of the strip is too small. The span should be at least 100 times the thickness of the strip and the clamping pressure on the strip at least 100 kPa to avoid or minimize clamping effects. The effect of clamping, however, is not considered by the DIN standards even though in many cases it is the dominant one. Special care should be taken to ensure that the measurements are not affected by deviations from flatness of the test pieces.

Commercial methods for bending stiffness evaluation such as those of Lorentzen and Wettre, Taber, Ohlsen, and Kodak Pathé, and also among the two-point, three-point, and four-point methods have been compared by Hohmann [14]. Good agreement among the methods was obtained when the tests were performed according to the DIN standards. If the testing conditions were changed, for example, bending to larger deflections, the agreement became significantly poorer.

V. CALCULATION OF BENDING STIFFNESS OF MULTI-PLY SHEETS

A. General Background

The preceding text focused mainly on test methods for the evaluation of bending stiffness of paper or paper board. However, it is often an advantage if the bending stiffness can be calculated. The ability to make such a calculation is beneficial for the optimization of the product with regard to choice of raw material, basis weight, and ply layup in the multi-ply sheet.

Few theoretical and experimental investigations of the bending stiffness of multi-ply sheets have been reported. Luey [19] discusses the bending stiffness contributions from different plies in multi-ply boxboard with numerical examples. The calculations are based on a formula for bending stiffness applicable when the elastic modulus and the thickness of each constituent ply are known. The Luey formula, however, has the drawback of being difficult to use due to the fact that the position of the neutral plane must first be calculated. An alternative formula has been derived from classical lamination theory by Carlsson and Fellers [3]. With this procedure, the neutral plane need not be considered. A presentation of the formula and a description of how to use it follow in the next section.

B. Procedure for the Calculation of Bending Stiffness

From a two-dimensional treatment of a multi-ply sheeet (Fig. 13) in bending, the following expression for the bending stiffness can be derived [3].

$$S^b = D - \frac{B^2}{A} \tag{41}$$

where

$$A = \sum_{k=1}^{N} (E_x)_k (z_k - z_{k-1}) \tag{42}$$

$$B = \frac{1}{2} \sum_{k=1}^{N} (E_x)_k (z_k^2 - z_{k-1}^2) \tag{43}$$

$$D = \frac{1}{3} \sum_{k=1}^{N} (E_x)_k (z_k^3 - z_{k-1}^3) \tag{44}$$

and $(E_x)_k$ is the elastic modulus of ply k in the x direction, which may be either MD or CD.

The ply coordinates z_k in Fig. 13 ($k = 0, 1, 2, \ldots, N$) are calculated from Eqs. (45) and (46):

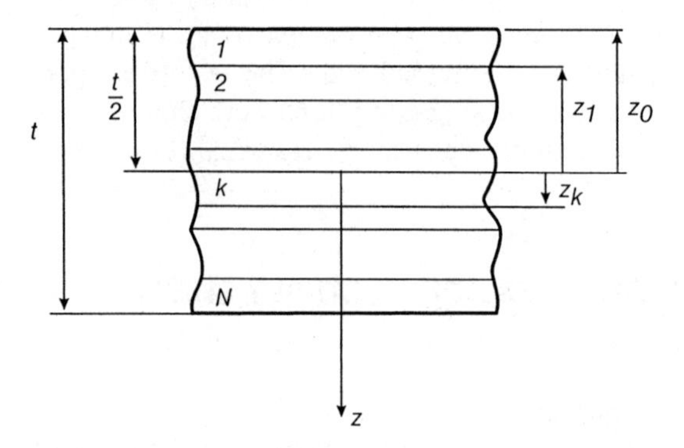

Fig. 13 The ply coordinates z_k of a multi-ply sheet.

Table 2 Properties of the Plies

Ply number k	Basis weight w (g/m^2)	Tensile stiffness index E^w (MN m/kg)	Density ρ (kg/m^3)	Elastic modulus $E = E^* \rho$ (N/mm^2)	Ply thickness $t = w/\rho$ (mm)
1	80	7.5	800	6000	0.1
2	80	2.5	400	1000	0.2
3	50	4.0	500	2000	0.1
4	70	7.14	700	5000	0.1

$$z_0 = \frac{-t}{2}, \qquad k = 0 \tag{45}$$

$$z_k = z_{k-1} + t_k, \qquad k = 1, 2, \ldots, N \tag{46}$$

where t is the total thickness and t_k is the thickness of ply k.

To illustrate the calculation of bending stiffness, a four-ply board is considered with properties as given in Table 2. As will be discussed later in this chapter, the calculations are based on basis weight, tensile stiffness index, and density.

The total thickness of the sheet is calculated as follows:

$$t = \sum t_k = 0.1 + 0.2 + 0.1 + 0.1 = 0.5 \qquad \text{(From Table 2)}$$

$$z_0 = -0.5/2 = -0.25 \qquad \text{(From Eq. (45))}$$

$$z_1 = z_0 + t_1 = -0.25 + 0.1 = -0.15 \qquad \text{(From Eq. (46))}$$

$$z_2 = 0.05, \qquad z_3 = 0.15, \qquad z_4 = 0.25$$

Table 3 gives the squares and cubes of the ply coordinates.

A simple way of performing the summations in Eqs. (42)–(44) is to insert in these equations the contributions from each ply according to Table 4. The values thus obtained for A, B, and D are substituted into Eq. (41) to obtain S^b:

$$S^b = 48.25 - \frac{(-10)^2}{1500} = 48.18 \text{ mN m}$$

The B term in Eq. (41) is zero for symmetrical sheets. It can be seen directly in Eq. (41) that the optimum bending stiffness is obtained by making a board with $B = 0$,

Table 3 Data Used for Eqs. (42)–(46)

k	z_k (mm)	z_k^2 (mm^2)	z_k^2 (mm^3)
0	−0.25	0.0625	-15.625×10^{-3}
1	−0.15	0.0225	-3.375×10^{-3}
2	0.05	0.0025	0.125×10^{-3}
3	0.15	0.0225	3.375×10^{-3}
4	0.25	0.0625	15.625×10^{-3}

Table 4 Calculation of A, B, and D

Ply k	$E_k(z_k - z_{k-1})$	$E_k(z_k^2 - z_{k-1}^2)$	$E_k(z_k^3 - z_{k-1}^3)$
1	$6000[(-0.15-(-0.25)] = 600$	$6000(0.0225 - 0.0625) = -240$	$600[-3.375-(-15.625)]\times 10^{-3} = 73.5$
2	$1000[0.05-(-0.15)] = 200$	$1000(0.0025 - 0.0225) = -20$	$1000[0.125 - (-3.375)] \times 10^{-3} = 3.5$
3	$2000(0.15–0.05) = 200$	$2000(0.0225-0.0025) = 40$	$2000(3.375 - 0.125) \times 10^{-3} = 6.5$
4	$5000(0.25-0.15) = 500$	$5000(0.0625-0.0225) = 200$	$5000(15.625 - 3.375) \times 10^{-3} = 61.25$
	$A = \sum = 1500$	$b = \frac{1}{2}\sum = -10$	$D = \frac{1}{3}\sum = 48.25$

that is, a symmetrical sheet, with the plies arranged to achieve the highest possible D value, i.e., high modulus plies on the surfaces.

Carlsson and Fellers [3] obtained excellent agreement between measured and calculated bending stiffness for sheets subjected to controlled drying conditions. In this investigation the tensile stiffness index of the constituent plies was measured in a tensile tester in which an extensometer gauge was devised to measure the strain. The thickness was measured by a mercury immersion technique [10].

It was pointed out by Carlsson and Fellers [3] and Markström [21] that the use of a standard thickness measuring method, such as the TAPPI Standard-T411 os-76 [26], results in an underestimation of the density and consequently gives poor agreement between theory and practice. Methods for thickness and density evaluation for paper have been summarized and discussed by, e.g., Fellers et al. [10].

VI. INFLUENCE OF PULP AND PAPERMAKING VARIABLES ON BENDING STIFFNESS

This section considers the factors of importance for the bending stiffness of different sheet constructions and how pulps may be chosen and papermaking variables adjusted to achieve optimum bending stiffness.

A. Bending Stiffness of a Single-Ply Sheet

The bending stiffness of a homogeneous single-ply sheet is given in Eq. (7) as a function of elastic modulus E and thickness t. Alternative expressions of Eq. (47) use either the tensile stiffness E^b and thickness t or the tensile stiffness index $E^w = E/\rho$ [25], basis weight w, and density $\rho = w/t\rho$.

$$S^b = \frac{Et^3}{12} = \frac{E^b t^2}{12} = \frac{E^w w^3}{12\rho^2} \tag{47}$$

where t is the thickness (m), E^b is the tensile stiffness (N/m), E^w is the tensile stiffness index (Nm/kg), w is the basis weight (kg/m^2), and ρ is the density (kg/m^3).

The tensile stiffness index is related to density according to the following empirical relationship [8,20].

$$E^w = k\rho^a \tag{48}$$

Luner et al. [20] found that $a = 1$ for handsheets when the density was increased by beating. Results of Fellers et al. [8] show that the exponent depends on fiber properties and process conditions during papermaking. From Eqs. (47) and (48) we obtain:

$$S^b = kw^3 \rho^{a-2} \tag{49}$$

Values of a greater than 2 indicate that the bending stiffness increases with increasing density. Under constant manufacturing conditions (beating, pressing, and drying) and constant tensile stiffness index, the bending stiffness increases in proportion to the third power of the basis weight. An increase in density resulting from beating at constant basis weight and under constant drying conditions may or may not increase bending stiffness depending on the value of the exponent $a - 2$. Increased density achieved by wet pressing seldom increases the bending stiffness. Restrained shrinkage results in a greater bending stiffness because it leads to an increase in tensile stiffness index, whereas the density may be considered to change marginally. Dry calendering always results in a diminished bending stiffness, because E^w is usually unaffected or decreased whereas ρ is increased.

To compare the bending stiffness of sheets of different basis weights, the following equation derived from Eq. (47) is used to define a property that is independent of basis weight, i.e., the bending stiffness index S^w [24]:

$$S^w = \frac{S^b}{w^3} = \frac{1}{12}\left(\frac{E^w}{\rho^2}\right) \tag{50}$$

Equation (50) may thus be used to normalize bending stiffness values measured on sheets with various basis weights.

B. Bending Stiffness of a Multi-Ply Sheet

If the elastic modulus and thickness of each separate ply are known, the bending stiffness of a multi-ply paper or board can be calculated according to Eqs. (41)–(46). To simplify the analysis, a specific multi-ply sheet will be treated—the symmetrical three-ply sheet in Fig. 14, consisting of a thick middle ply of thickness t_c and thin surface plies of thickness t_s. The total thickness is denoted by t.

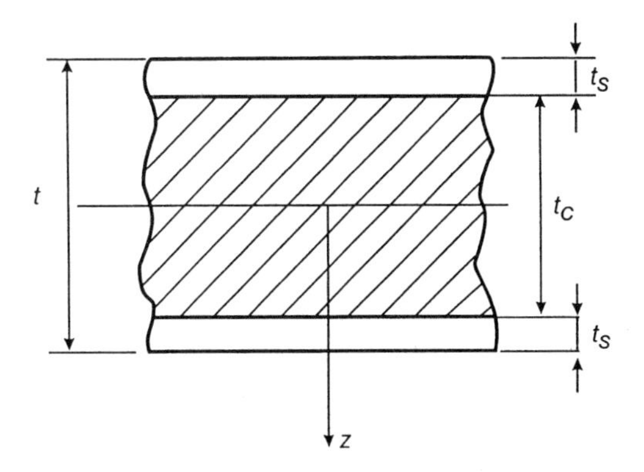

Fig. 14 Cross section of a three-ply sheet.

Surface Plies Each surface ply provides, according to Eqs. (41)–(44), a contributiotn S_s^b to the total bending stiffness $(B = 0)$:

$$S_s^b = \frac{1}{3}E_s\left[\left(t_s - \frac{t}{2}\right)^3 - \left(-\frac{t}{2}\right)^3\right] = E_s\left[\frac{t_s^3}{3} - t_s^2\left(\frac{t}{2}\right) + t_s\left(\frac{t}{2}\right)^2\right] \tag{51}$$

The thickness of the surface ply, t_s, is usually much less than t, which means that t_s^2 and t_s^3 in Eq. (51) can, to a first approximation, be neglected and S_s^b becomes

$$S_s^b = E_s^b t_s\left(\frac{t}{2}\right)^2 = E_s^w w_s\left(\frac{t}{2}\right)^2 \tag{52}$$

It is thus evident that for a surface ply, tensile stiffness $E_s^b = E_s t_s$ or the tensile stiffness index $E_s^2 = E_s^b/w_s$ and the square of the distance from the neutral plane, $\approx (t/2)^2$, are important. Using Eq. (48) in Eq. (52) yields

$$S_s^b = k\rho_s^a w_s\left(\frac{t}{2}\right)^2 \tag{53}$$

An increase in the density of the surface ply due, for example, to increased beating or pressing, is always advantageous because the exponent a is positive.

The bending stiffness contribution of a surface ply in a multi-ply sheet is thus proportional to the tensile stiffness $E_s^b = E_x t_w$ or tensile stiffness index $E_s^w = E_s^b/w_s$ of the surface ply.

The tensile stiffness index is consequently the critical property that characterizes pulp for use in a surface ply. The usefulness of the tensile stiffness index for pulp characterization has also been discussed by Ranger [22,23].

Middle Ply As mentioned in the preceding section, the distance from the neutral plane to the surface ply is an important factor for the bending stiffness contribution of the surface ply. The middle ply thus contributes to the total bending stiffness partly through its own bending stiffness and partly through its function of separating the surface plies. An analysis shows, however, that the contribution of the middle ply to the total bending stiffness of the multi-ply sheet through its own bending stiffness is negligible compared with the contribution resulting from its ability to separate the surface plies. The most important property of the middle ply is consequently a low density, i.e., a large thickness for a given basis weight [8]. For this reason, pulps for use in middle plies should have a high yield and a low degree of beating. The degree of beating of the pulp, however, is often determined by other demands such as delamination resistance and surface evenness.

Optimal Bending Stiffness The previous sections have shown that the requirements with regard to pulp properties vary, depending on the position of the ply within the sheet. The question then arises as to how the total bending stiffness of a sheet can be optimized for a given basis weight. The following example illustrates the importance of composition and construction for bending stiffness.

Consider a sheet made from chemical pulp and mechanical pulp, where the relative amounts of each pulp are varied while the total basis weight is kept constant. Figure 15 shows how the calculated bending stiffness at constant basis weight varies for three constructions: (1) a symmetrical three-ply sheet with surface plies made from chemical pulp; (2) a two-ply sheet with a chemical pulp on one side and a mechanical pulp on the other; and (3) a homogeneous sheet where the pulps are mixed. As expected, the bending stiffness is highly dependent on both board construction and composition. For the homogeneous sheet, the highest value is obtained with 100% mechanical pulp. A comparison of the two multi-ply sheets shows that the symmetrical three-ply sheet is superior to the two-ply sheet. Optimum bending stiffness is obtained for approximately 35% chemical pulp in the three-ply sheet. This type of analysis should be valid for newsprint, magazine paper, linerboard, and carton board.

In summary, this simple analysis shows that a combination of low density in the middle ply and high density in the surface plies optimizes or maximizes bending stiffness. In practice, these conditions are obtained by choosing high yield pulps with low beating level for the middle ply and low yield pulps with high beating level for the surface plies.

APPENDIX

Consider a thin element of paper subjected to bending moments M_x^b and M_y^b as in Fig. A1. When the element is bent by M_x^b only, it will bend also in the y direction due to the Poisson's ratio effect, Fig. A2. This is called the *unconstrained condition.*

If the paper is prevented from bending perpendicular to the bending direction, we call this situation the *constrained condition.* The following analysis is made to estimate the difference in bending stiffness between the two extreme cases.

Paper normally has different elastic properties in the machine and cross-machine directions. The governing equations for orthotropic bending are [17]

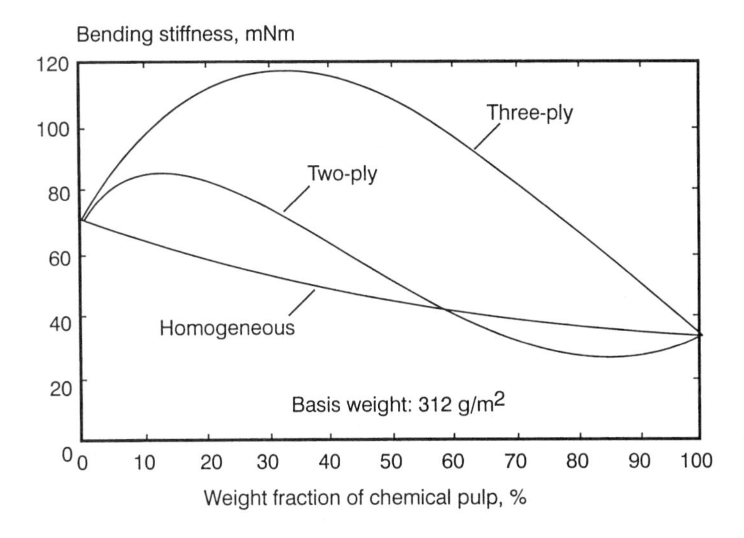

Fig. 15 Bending stiffness for homogeneous, two-ply, and three-ply sheets with different weight fractions of chemical pulp. Chemical pulp: $E^w = 7.69$, $\rho = 826$. Mechanical pulp: $E^w = 3.02$, $\rho = 400$. (Units as in Table 2.)

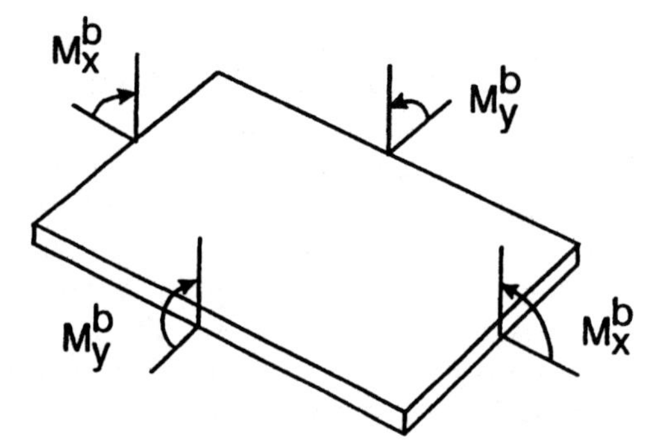

Fig. A1 An element of paper subjected to bending moments M_x^b and M_y^b.

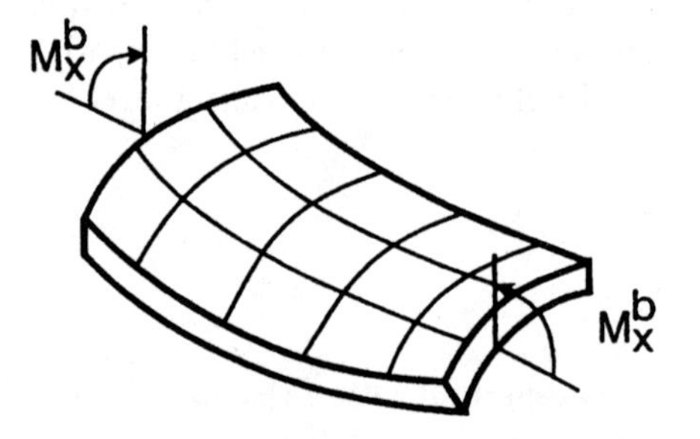

Fig. A2 An element of paper subjected to bending moments M_x^b.

$$\begin{bmatrix} M_x^b \\ M_y^b \end{bmatrix} = \begin{bmatrix} D_{11}^b & D_{12}^b \\ D_{21}^b & D_{22}^b \end{bmatrix} \begin{bmatrix} k_x \\ k_y \end{bmatrix} \tag{A1}$$

where D_{ij}^b are stiffness components per unit width b, and k_x and k_y are bending curvatures.

For *unconstrained bending* by M_x^b only (Fig. A2),

$$M_x^b = D_{11}^b k_x + D_{12}^b k_y \tag{A2a}$$

$$M_y^b = 0 = D_{21}^b k_x + D_{22}^b k_y \tag{A2b}$$

$$D_{12}^b = D_{21}^b \tag{A2c}$$

Equations (A2) combined yield

$$M_x^b = D_{11}^b k_x - \frac{(D_{12}^b)^2}{D_{22}^b} k_x = \left[D_{11}^b - \frac{(D_{12}^b)^2}{D_{22}^b} \right] k_x \tag{A3}$$

Hence, the *unconstrained bending* stiffness per unit width S_{UC}^b is

$$S_{UC}^b = D_{11}^b - \frac{(D_{12})^b}{D_{22}^b} \tag{A4}$$

Noting that

$$D_{11}^b = \frac{E_1 t^3}{12(1 - v_{12}v_{21})} \tag{A5a}$$

$$D_{12}^b = v_{21} D_{11}^b = v_{12} D_{22}^b \tag{A5b}$$

and

$$D_{22}^b = \frac{E_2 t^3}{12(1 - v_{12}v_{21})} \tag{A5c}$$

Eqs. (A4) and (A5) yield

$$S_{\mathrm{UC}}^b = \frac{E_1 t^3}{12} \tag{A6}$$

which is the same formula as for a rectangular beam under plane stress.

For *constrained* bending, $k_y = 0$, the paper attains a cylindrical shape. Equation (A1) yields

$$M_x^b = D_{11}^b k_x = \frac{E_1 t^3}{12(1 - v_{12}v_{21})} k_x \tag{A7}$$

The *constrained bending stiffness* per unit width S_C^b is

$$S_C^b = D_{11}^b \tag{A8}$$

and we get

$$S_C^b = \frac{E_1 t^3}{12(1 - v_{12}v_{21})} \tag{A9}$$

The product of Poisson's ratios [Eq. (A9)] is 0.293^2 for paper [1], which yields a 10% larger bending stiffness for the constrained bending than for the unconstrained bending. The actual bending stiffness of a paper panel is expected to lie between the two extremes in Eqs. (A6) and (A9). One would expect the difference to be on the order of 5%.

REFERENCES

1. Baum, G. A., Brennan, D. C., and Habeger, C. C. (1981).Orthotropic elastic constants of paper. *Tappi J. 64*(8):97–101.
2. Buchanan, J. S. (1968). Stiffness: Its importance and its attainment. In: 12th Eucepa Conference: Multiply Board, Berlin, Germany, pp. 8–16.
3. Carlsson, L., and Fellers, C. (1980). Flexural stiffness of multi-ply paperboard. *Fibre Sci. Technol. 13*:213–223.
4. Carlsson, L.,Fellers, C., and de Ruvo, A. (1980). The mechanism of failure in bending of paperboard. *J. Mater. Sci. 15*:2636–2642.
5. DIN. (1973). DIN 53121. Testing of paper and board; determination of the bending stiffness by the beam method (in German).
6. Fellers, C. (1977). Procedure for measuring the interlaminar shear properties of paper. *Svensk Papperstidn. 80*(3):89–93.
7. Fellers, C. (1997). Bending stiffness of paper and paperboard: A round robin study. *Nordic Pulp Paper Res. J. 1997*(1):42–45.

8. Fellers, C., de Ruvo, A., Htun, M., Carlsson, L., Engman, C., and Lundberg, R. (1983). *Carton Board. Profitable Use of Pulps and Processes.* STFI, Stockholm.
9. Fellers, C., and Carlsson, L. (1979). Measuring the pure bending properties of paper. *Tappi J. 62*(8):107–109.
10. Fellers, C., Andersson, H., and Hollmark, H. (1986). The definition and measurement of thickness and density. In: *The Definition and Measurement of Thickness and Density*. J. A. Bristow and P. Kolseth, eds. Marcel Dekker, New York, pp. 151–167.
11. Franklin, F. (1989). Paper properties and testing: Stiffness. *Paper Focus*, August, p. 20.
12. Hohmann, H. J. (1968). Bending stiffness of board. 1. Measurement bases for the determination of bending stiffness of board (in German). *Verpack.-Rundsch. 19*(4):25–32.
13. Hohmann, H. J. (1970). Bending stiffness of board. 2. Critical investigations of a few common bending test methods to determine their suitability for determination of the bending stiffness of boards (in German). *Verpack.-Rundsch. 21*(1):1–10.
14. Hohmann, H. J. (1977). Advantage of a unified method for the determination of bending stiffness of board (in German). *Das Papier 31*(8):338–345.
15. ISO. (1990). ISO-5628. Paper and board: Determination of bending stiffness by static methods—General principles.
16. ISO. (1983). ISO-5629. Paper and board: Determination of bending stiffness—Resonance method.
17. Jones, R. M. (1975). *Mechanics of Composite Materials.* McGraw-Hill, New York.
18. Koran, Z., and Kamden, D. P. (1989). The bending stiffness of paperboard. *Tappi J. 72*(6):175–179.
19. Luey, A. T. (1963). Stiffness of multiply box board. *Tappi J. 46*(11);159–162A.
20. Luner, P., Kärnä, A. E. U., and Donofrio, C. P. (1961). Studies in interfiber bonding of paper. The use of optically bonded areas with high yield pulps. *Tappi J. 44*(6):409–414.
21. Markström, H. (1994). Comparison Between Measured Bending Stiffness and Calculated Bending Stiffness. AB Lorentzen & Wettre Research Report, 6 June 1994.
22. Ranger, A. E. (1967). Evaluation of fibrous materials for boardmaking and converting. *Paper Technol. 8*(8):245–250.
23. Ranger, A. E. (1967). Flexural stiffness of multiply box board. *Paper Technol. 8*(1):51–56.
24. SCAN-P64:90 (1990). Paper and board: Bending stiffness—Resonance method.
25. SCAN-P67:93 (1993). Tensile strength, strain at break, tensile energy absorption and tensile stiffness.
26. TAPPI. (1976). Tappi-Standard-T411 os-76. Thickness (caliper) of paper and paperboard.
27. Timoshenko, S. (1970). *Theory of Elasticity*. 3rd ed. McGraw-Hill, New York.
28. Timoshenko, S., and Young, D. H. (1955). *Vibration Problems in Engineering*. 3rd ed. Van Nostrand, Princeton, NJ.

6

ULTRASONIC DETERMINATIONS OF PAPER STIFFNESS PARAMETERS

CHARLES C. HABEGER, JR.
Institute of Paper Science and Technology
Atlanta, Georgia

I. INTRODUCTION

I plan to examine ultrasound in paper within a limited scope. I will review in-plane and out-of-plane ultrasonic diagnostics of paper. I will outline the mechanics underlying these techniques. I will conjecture on physical meaning. I will address issues of experimental difficulty, and I will compare techniques. To my knowledge, there is no other unified exposition of these important topics as applied to paper. I believe this leaves me an opening for a contribution. On the other hand, I do not include on-machine measurements. I omit the influences of papermaking variables on ultrasonic velocity and the relationships between ultrasonic stiffnesses and other physical properties. I will not talk about ultrasonic processing in the pulp and paper industry. These subjects are adequately entertained elsewhere. I acknowledge that my narrowed perspective excludes important works that would be referenced in a more even account of ultrasound in paper. I rationalize my convenient lapses by appeal to the charge of the book. The objective of these volumes is to critique experimental methods. The intent is not to emphasize papermaking or paper physics.

In a semblance of fairness, I offer to any innocent reader the following warning. One purpose of this chapter is to advance my own idiosyncratic points of view. Not everyone versed in the subject will agree with everything I say. Nonetheless, I do not pay equal attention to all opinions. I will cite other works in those cases in which counterexplanations are published. When, from personal experience, I know that someone is clenching his fists over a particular passage, I will cue the reader. When you see something like "I believe" or "In my opinion", you should reflect on the author's fallibility. Extrapolating from previous misadventures, I am sure that I will regret my excesses. But, from my heart, I can only do it the way I feel it, now.

II. HISTORY

The first investigators of sound in paper relied on techniques developed by the textile research community. Therefore, a proper history begins with a rundown of the early fiber literature. In 1936, Meyer and Lotmar [64] stirred up enthusiasm for textile acoustic measurements with an odd musical demonstration. They mounted a single test fiber (or film strip) in a tensioning apparatus and rubbed it so as to induce a longitudinal disturbance. Listening carefully, they detected a half-wavelength resonance in the ringing fiber. A standard fiber of known modulus and density was excited likewise in another apparatus. They adjusted the length of the standard fiber until it sounded like the test fiber. The ratio of fiber lengths at the same resonance frequency was set equal to the ratio of velocities. In the 1940s, Ballou and Silverman [4] transformed the Meyer–Lotmar curiosity into an objective measurement. They fastened one end of a long fiber (or film strip) to a 10 kHz Rochelle salt piezoelectric crystal. This transducer was electrically driven at resonance. A receiving Rochelle salt crystal was mounted to a sliding stand and administered to the fiber at adjustable separations from the transmitter. The receiver partially reflected the sonic energy, producing a standing wave pattern between the transducers. They determined the wavelength of the longitudinal wave by measuring the distance between receiver positions at standing wave maxima. They reported modulus numbers for rayon fibers, regenerated cellulose films, and cellulose acetate films in the range of modern measurements. Film velocity was demonstrated to depend upon the humidity, the temperature, and the strip orientation relative to the machine direction. Latter, by recording the decay in the standing wave maxima, Ballou and Smith [5] determined an attenuation coefficient and calculated the imaginary part of the Young's modulus. By comparing sonic results to moduli determinations at lower frequencies, they confirmed that polymer stiffness increases with frequency.

Pulsed time-of-flight experiments on fibers and film strips were first conducted by Hamburger and coworkers [39] in 1948. A pair of Rochelle salt crystal transducers were applied, at separate locations, along a strip. One of the transducers was strobed with an electric pulse, and the time lapse between its stimulation and the response of the second transducer above a threshold was measured electronically. The velocity was taken as the transducer separation distance divided by the delay time. Chaikin and Chamberlain [23] pointed out inadequacies in the Hamburger technique, and in 1955 they offered an improved time-of-flight measurement. One end of a fiber was tied to a Rochelle salt transmitter, and the other end was attached to the front of a condenser microphone. The transmitter was pulsed, and the sweep of a homemade oscilloscope was triggered. The microphone output was displayed as a function of time on the oscilloscope cathode ray tube. The microphone trace along with timing markers was photographed. The string was shortened repeatedly, and new photographs were captured. By examining the photographs through a traveling microscope, the time of arrival of a common feature in each microphone response was measured. The velocity was the regressed slope of the plot of string length versus time delay. The Chaikin and Chamberlain transducers and diagnostic electronic equipment were, by modern standards, very crude. Nonetheless, with ingenuity, they were able to engineer an excellent (though tedious) experiment.

Also in 1955, the Morgan tester patent [77] was granted. The Morgan tester was a commercial sonic velocity measurement instrument. It was less accurate than a Chaikin–Chamberlain apparatus. However, it could sample sheets as well as strips, and measurements were much quicker and easier. Two transducers were slide-mounted on a rule. A sheet or strip sample was placed below the mount, and the transducers were gravity loaded to the sample. One transducer was intermittently pulsed, and special circuitry was employed to output a voltage proportional to the time lag between the excitation and the ascendance of the output of the second transducer above a threshold. The transducer spacing was incremented, and further outputs were recorded. The velocity was calculated as the regressed slope of the distance–time line. This is a big improvement over the Hamburger technique. Time measurements at multiple transducer separations allow for an automatic compensation of non-sample delay times and of variations in pulse shape resulting from sample-dependent and load-dependent phase shifts between sample and transducer. However, the signal amplitude falls off with separation, and a constant signal threshold level corresponds to different events in the received signal at different separations. The search for a common receiver feature, as was incorporated in the Chaikin–Chamberlain work, was omitted. The introduction of ceramic bimorph transducers was an important innovation of the Morgan tester. Ceramic transducers are more rugged than Rochelle salt crystals. Transducers of bimorph construction oscillate in a beam-bending mode. This generates a low impedance mechanical motion that is efficiently coupled to and from thin polymer strips and sheets. Moreover, the transducer strips can be rotated in their mounts. Thus, the polarization of the detected radiation can be oriented either parallel or perpendicular to the direction of propagation. The velocities of transverse, as well as longitudinal, motions are determined. To this day, bimorph transducers, slightly modified from the original Morgan tester design, are the mainstays of in-plane ultrasonic characterizations of paper.

Craver and Taylor [24,87] are the pioneers of ultrasonic velocity measurement in paper. During the mid-1960s, they published the first reports on submissions of paper samples to the Morgan tester. More important, they communicated a proper understanding of the subtle mechanical, acoustical, and viscoelastic issues that influence the interpretation of ultrasonic measurements on thin polymer sheets. This limited the unavoidable confusion that accompanied the introduction of a new diagnostic technique to the paper industry. Taking advantage of the versatility of the ceramic bimorph transducers, they made determinations of shear moduli as well as normal planar stiffnesses. They presented curves of quasi-transverse and quasi-longitudinal velocity versus angle to the machine direction. They headed off potential misunderstanding by emphasizing the difference between Young's modulus and planar stiffness. Appreciating Musgrave's [67,68] important distinction between wave surface velocity and plane wave velocity, they described a technique for calculating Poisson's ratios from plots of quasi-longitudinal velocity versus angle to the machine direction. They noted that sonic stiffness coefficients are greater than those calculated from normal load-elongation experiments, and they attributed this (in what I believe is the correct manner) to the reduced time for viscoelastic transitions in short-time sonic pulse experiments.

In 1971, Luukkala et al. [56] introduced a continuous wave technique that they applied to paperboard. An extended sheet sample was mounted in an open frame. An air-coupled transducer was placed on each side of the sheet. The transducers were fixed to a common arm, which rotated relative to the frame while keeping the transducer faces parallel to each other. In this way, the transducers were forced to always communicate through the sample, but the angle between the sheet normal and the line between transducers could be altered by rotating the arm. One transducer was driven sinusoidally, and the signal at the other transducer was monitored. They determined the phase velocity of plate waves in the sample by noting peaks in the received signal as the transducer arm was rotated or the driving frequency was swept.

In the late 1970s and throughout the 1980s a host of experimenters (Baum, Bornhoeft, Forbes, Habeger, Mann, Pankonin, Treleven, Van Zummeren, Wink, Young) at The Institute of Paper Chemistry continued the development and interpretation of techniques for analyzing paper with ultrasound. In-plane Poisson's ratios were calculated using an off-axis quasi-shear velocity measurement [6]. Out-of-plane longitudinal and transverse velocity measurements were introduced [31,59–61]. A mathematical model for plate waves in highly anisotropic materials was shown to be consistent with Luukkala method results on paperboard [31]. Out-of-plane Poisson's ratios were estimated from the velocity difference between longitudinal bulk waves and low frequency symmetrical plate waves [59,61]. Paper-specific transduction methods were devised for examination of out-of-plane longitudinal [32,33] and transverse [34] waves. The Luukkala plate technique was enhanced by transducer refinement and the addition of a lock-in amplifier to the detection circuitry [28,59]. Up-to-date electronic diagnostic and computing methods allowed the introduction of automated measuring equipment [36,90]. Ceramic bender transducers were modified for better sensitivity, broader bandwidth, and improved modal purity [36]. Morgan tester style time delays were calculated with cross-correlation rather than threshold detection analysis [90]. A strip resonance apparatus for finding real and imaginary Young's moduli was constructed [75]. A Musgrave interpretation of the in-plane, quasi-longitudinal velocity versus angle curve allowed an assessment of the orthotropic assumption and an estimation of the shear coupling coefficients [35].

Piezoelectric technologies continued to advance. Hall [38] and Brodeur et al. [20] describe a fluid-filled-wheel technique for measuring of out-of-plane longitudinal velocity in paperboard. A fluid-filled wheel consists of a broadband transducer mounted to a stationary axle, a thin rubber tire, and a coupling fluid occupying the space between transducer and tire. A paper sample is fed into the nip between two such wheels. One transducer is pulsed, and a series of pulses, corresponding to acoustic paths of different lengths, are detected at the other transducer. By examining the times of flight and the amplitudes of the pulses, the velocity and attenuation of the out-of-plane longitudinal wave in paper are calculated.

We have benefited from adaptations of laser instrumentation to the excitation and detection of ultrasonic waves in paper. In the late 1980s, patents were issued to Leugers, Pace, and Salama for in-plane [53] and out-of-plane [74] pulsed laser generations of paper ultrasonic waves. These laser pulses are of very short duration, thereby initiating a more broadband disturbance than piezoelectric transduction methods. They do not contact the sample, which is of particular advantage when analyzing a moving web non-contact. Broadband, laser-based reception techniques followed in the 1990s. These include a direct detection method from Luukkala [53]; interferometric techniques from Johnson et al. [21,44,45], Walker et al. [93], and Lafond et al. [49,50]; and full-image holographic approaches from Olofsson and Kyosti [73] and Hale et al. [37].

Three instruments for automated in-plane longitudinal measurements are available commercially. All adopt the Hamburger approach (single spacing time of flight taken from a threshold detection). The original commercial instrument [69] consists of two standard (not bimorph) ceramic transducers and a turntable sample holder. A measurement cycle consists of applying the transducers to the sample, taking a velocity measurement, raising the transducers, and rotating the transducers for the next cycle. Times of flight at different angles to the machine direction are calculated, and the final output is a curve of longitudinal velocity versus angle. Two later designs incorporate a circular array of transducers. One [88] uses bimorph transducers, whereas the other [70] does not. The array is lowered onto the sample once. The transmitters are sequentially fired. At each firing, a signal is detected by the receiver at the opposite side of the trasnducer array. Times of flight are calculated and displayed as a plot of velocity versus angle.

III. BASIC SOLID MECHANICS

An understanding of solid wave propagation is a prerequisite to a recognition of the opportunities and limitations of ultrasonic testing of paper. However, I will not make a rigorous presentation of all the necessary physics. Instead, I present a cursory review emphasizing insights and intuitions that help to rationalize the phenomenon of mechanical wave motion in paper. For those intent on a deeper study, I will identify my favorite texts.

To appreciate the principles of wave propagation in solids, you first need a background in continuum mechanics. The first chapter of this book gives a fine description of the subject, and I will refer to it as necessary. For the complete treatment, I recommend a textbook authored by Fung [29]. If you are only interested in ultrasonic testing, there can be some holes in your understanding of continuum mechanics. You can ignore the discourses on large strain behavior and on nonlinear responses. The strains encountered in diagnostic ultrasound are extremely small, and linearity is always an excellent assumption. Linear viscoelasticity, however, is an important topic. Ferry's book [27] on viscoelasticity in polymers is a well-known general textbook on the subject, whereas Chapters 1 and 6 in this volume provide more specific information on cellulosic materials. For an overall coverage of mechanical wave propagation, I recommend *Theoretical Acoustics* by Morse and Ingard [66]. For plate and bulk wave propagation in isotropic solids, you should look to *Wave Motion in Elastic Solids* by Graff [30]. Laser excitation and detection of ultrasound are discussed extensively by Scruby and Drain [79]. The propagation of plane waves in anisotropic media is discussed by Auld [3], Kolsky [48], Borngis [16], and Love [55]. Musgrave [67,68] contributed an important distinction between the off-axis phase velocity of a plane wave and the group velocity of a "wave surface" generated by a point source excitation. Auld [3] provides a clear, modern discussion of this topic.

A. Continuum Mechanics

In rigid-body mechanics, one assumes that an object moves as a whole. The separations between all particles that make up a rigid body are fixed. The positions of the center of mass and the attitude of the body axes are all that is necessary to locate a rigid body in space. Application of Newton's laws to any assemblage of particles yields two corollaries: The acceleration of the center of mass is equal to the net external force divided by the total mass, and the time rate of change of angular momentum relative to the center of mass is equal to the torque about the center of mass. For rigid bodies, these rules (along with initial boundary conditions) are sufficient to determine motion for all time. In actuality, no substance is rigid, and mechanics is not so easy. At first the complication seems unfortunate, but, by generalization to the mechanics of nonrigid bodies, physics uncovers new and interesting phenomena. My present subject, the propagation of solid mechanical waves, is a classic example.

Once rigidity is rejected, one needs to independently account for all of the individual parts of the body. The center-of-mass laws still apply; however, they are no longer capable of describing the motion in detail. The tracking process is simplified if continuity clauses restrict the composition and the movement. For a continuum, the mass of material in any infinitesimal volume is assumed to be proportional to the volume; i.e., the mass density (ρ) is a continuous function of the material coordinates. Motion is referenced from an "undeformed" or zero external force state. Continuous reference coordinates ($X_i, i = 1, 3$) are constructed on the undeformed body. A second set of coordinates (x_i) are assumed to be admissible on the deformed body. For now, Cartesian coordinates will be employed in both cases. The motion of the deformed body is defined when x_i's are given as continuous functions of X_i's and time (t). To properly apply straightforward continuum mechanics, the discrete features of the actual medium must not be of consequence. The continuum requirements are valid for wave motion analysis only if the wavelength of the elastic wave is much greater than the scale of the microscopic structure of the material. There will be situations in which application of the standard interpretation of ultrasound to paper is in jeopardy because the wavelength becomes of the order of the microstructure.

The displacements (U_i) are the differences between the deformed coordinates and the reference coordinates: $U_i \equiv x_i - X_i$. In the Euler convention [29] (see also Chapter 1, this volume), small strains are defined in terms of the partial derivatives of the displacement with respect to the coordinates in the deformed body, i.e.,

$$\varepsilon_{ij} \equiv \frac{1}{2}\left(\frac{\partial U_i}{\partial x_j} + \frac{\partial U_j}{\partial x_i}\right), \qquad i = 1, 2, 3 \tag{1}$$

The strains are local measures of the deformation of the body. Note that ε_{ij} is equal to ε_{ji} and that there are only six independent strain components at each location in the body. Consider ε_{11}, which equals $\partial U_1/\partial x_1$. If ε_{11} is positive, displacement in the x_1 direction is increasing with x_1. This means that a line segment in the 1 direction is locally elongated. To be exact, ε_{11} is the ratio of the increase in length to the length of an infinitesimal line segment in the x_1 direction. Likewise, ε_{22} and ε_{33} are the normal strains in the 2 and 3 directions. The off-axis terms are the shear strains. They are measures of distortion. For example, ε_{12} equals one-half of the change in angle between the 1 and 2 axes. So, on-diagonal (or normal) strains specify relative changes in length, whereas the off-diagonal (or shear) strains represent changes in angle.

In the absence of external forces, solid bodies maintain their undeformed size and shape. Deformation requires a load. The local loading intensity is expressed in terms of stresses (σ_{ij}). A particular σ_{ij} is defined as the force in the j direction per unit area perpendicular to the i axis. Imagine an infinitesimal volume ($dx_1\, dx_2\, dx_3$) in a deformed body. To maintain the distortion, a differential force equal to $\sigma_{in}\, dx_j\, dx_k$ ($i \neq j \neq k$) s applied by the remainder of the body in the n direction to the face of the infinitesimal volume on the positive i side. This is balanced by an opposite, but equal-magnitude, force on the negative i side. So, a positive on-diagonal (or normal) stress (σ_{ii}) represents a material locally in tension in the i direction. The off-diagonal (or shear) stresses pertain to forces in the plane of the surface. By balancing the moments on an infinitesimal element it can be shown that, like strain, the stress tensor is symmetrical [29], i.e., $\sigma_{ij} = \sigma_{ji}$.

B. Linear Elasticity

In elasticity theory, stresses and strains are related through constitutive equations. It is assumed that at small strain there is a one-to-one linear relationship between the stresses and the strains. That is, for a linear elastic material, each stress is proportional to the immediate strains (or vice versa), irrespective of loading history. There are six independent stresses and six independent strains; therefore, the linear relationships can be expressed through a 6×6 bulk "stiffness" matrix [29] (also Chapter 1, this volume). If the material possesses orthotropic symmetry, symmetry and physical arguments reduce the number of independent stiffness coefficients from 36 to 9 [29], and in a coordinate frame aligned with planes of orthotropic symmetry the constitutive relations are expressed in the following matrix notation:

$$\begin{vmatrix} \sigma_{11} \\ \sigma_{22} \\ \sigma_{33} \\ \sigma_{23} \\ \sigma_{13} \\ \sigma_{12} \end{vmatrix} = \begin{vmatrix} C_{11} & C_{12} & C_{13} & 0 & 0 & 0 \\ C_{12} & C_{22} & C_{23} & 0 & 0 & 0 \\ C_{13} & C_{23} & C_{33} & 0 & 0 & 0 \\ 0 & 0 & 0 & 2C_{44} & 0 & 0 \\ 0 & 0 & 0 & 0 & 2C_{55} & 0 \\ 0 & 0 & 0 & 0 & 0 & 2C_{66} \end{vmatrix} \begin{vmatrix} \varepsilon_{11} \\ \varepsilon_{22} \\ \varepsilon_{33} \\ \varepsilon_{23} \\ \varepsilon_{13} \\ \varepsilon_{12} \end{vmatrix} \qquad (2)$$

Orthotropic symmetry means that the material has three mutually perpendicular planes of reflectional symmetry. Paper is generally assumed to be orthotropic, and the symmetry planes are perpendicular to the thickness direction (ZD) and usually perpendicular to the machine direction (MD) and the cross-machine direction (CD).

The constitutive equations are often written equivalently in terms of a bulk "compliance" matrix:

$$\begin{vmatrix} \varepsilon_{11} \\ \varepsilon_{22} \\ \varepsilon_{33} \\ \varepsilon_{23} \\ \varepsilon_{13} \\ \varepsilon_{12} \end{vmatrix} = \begin{vmatrix} S_{11} & S_{12} & S_{13} & 0 & 0 & 0 \\ S_{12} & S_{22} & S_{23} & 0 & 0 & 0 \\ S_{13} & S_{23} & S_{33} & 0 & 0 & 0 \\ 0 & 0 & 0 & S_{44}/2 & 0 & 0 \\ 0 & 0 & 0 & 0 & S_{55}/2 & 0 \\ 0 & 0 & 0 & 0 & 0 & S_{66}/2 \end{vmatrix} \begin{vmatrix} \sigma_{11} \\ \sigma_{22} \\ \sigma_{33} \\ \sigma_{23} \\ \sigma_{13} \\ \sigma_{12} \end{vmatrix} \qquad (3)$$

You can derive the exact relationships between compliances and stiffnesses by inverting the matrix in Eq. (2). For on-diagonal compliances, the results are $S_{ii} = (C_{jj}C_{kk} - C_{jk}C_{jk})/\det(C_{jk})$ ($i \neq j \neq k$ and $i, j, k = 1, 2,$ or 3) and $S_{ii} = 1/C_{ii}$($i = 4, 5,$ or 6).

In engineering parlance, the compliances are communicated in the modulus convention (Perkins, this volume). The Young's modulus in the i direction (E_{ii}) is defined as $1/S_{ii}$. A Poisson's ratio (ν_{ij}), $i \neq j$, is minus the strain in the j direction divided by the strain in the i direction when normal stress is applied only in the i direction. From Eq. (3), ν_{ij} is seen to equal $-S_{ij}/S_{ii}$. The shear modulus in the 1,2 plane (G_{66}) is defined as $1/S_{66}$, G_{55} is $1/S_{55}$, and G_{44} is $1/S_{44}$.

For later application, I need to emphasize the differences between normal stiffnesses (C_{ii}) and Young's moduli ($1/S_{ii}$). A stiffness coefficient is the ratio of stress to strain with no lateral strains, whereas a modulus is the ratio of stress to strain with no lateral stresses. In a modulus experiment, the material is free to move laterally. In a stiffness determination lateral movement is prevented. The material is easier to deform without lateral constraint, and normal stiffnesses are greater than the corresponding Young's moduli. The exact relationship depends on the off-axis coefficients.

Bulk stiffnesses and bulk compliances, as defined above, are the common parameters for expression of three-dimensional linear elasticity. However, to study in-plane ultrasound propagation in paper, it is convenient to use a third, simplified convention. For in-plane ultrasonic testing, as for most other in-plane experiments, the surfaces of the sheet are unrestrained. This means that at the sheet surfaces σ_{33}, σ_{13}, and σ_{23} are zero. Two points must be separated by at least the order of a wavelength for there to be instantaneous differences in stress. Therefore, under the special condition that the sheet is thin compared to the wavelength of ultrasound in the out-of-plane direction, it is fair to assume that σ_{33}, σ_{13}, and σ_{23} are always zero throughout the sheet. This is called the "plane stress" condition. Usually, plane stress is a good assumption for in-plane wave propagation. However, as discussed later, it breaks down at lower frequencies than you might expect for thick paperboards. With out-of-plane stresses set to zero, Eq. (3) is considerably shortened. Ignoring the out-of-plane strains, in-plane strains in an orthotropic thin plate are related to in-plane stresses through bulk compliances as

$$\begin{vmatrix} \varepsilon_{11} \\ \varepsilon_{22} \\ \varepsilon_{12} \end{vmatrix} = \begin{vmatrix} S_{11} & S_{12} & 0 \\ S_{12} & S_{22} & 0 \\ 0 & 0 & S_{66}/2 \end{vmatrix} \begin{vmatrix} \sigma_{11} \\ \sigma_{22} \\ \sigma_{12} \end{vmatrix} \tag{4}$$

When the expressions in Eq. (4) are inverted, the in-plane stresses, under the plane stress condition, in terms of in-plane stresses are

$$\begin{vmatrix} \sigma_{11} \\ \sigma_{22} \\ \sigma_{12} \end{vmatrix} = \begin{vmatrix} Q_{11} & Q_{12} & 0 \\ Q_{12} & Q_{22} & 0 \\ 0 & 0 & 2Q_{66} \end{vmatrix} \begin{vmatrix} \varepsilon_{11} \\ \varepsilon_{22} \\ \varepsilon_{12} \end{vmatrix} \tag{5}$$

where $Q_{11} = S_{22}/(S_{11}S_{22} - S_{12}S_{12})$, $Q_{22} = S_{11}/(S_{11}S_{22} - S_{12}S_{12})$, $Q_{12} = -S_{12}/(S_{11} S_{22} - S_{12}S_{12})$, and $Q_{66} = 1/S_{66} = G_{66}$. Putting the compliances in terms of engineering coefficients, I have $Q_{11} = E_{11}/(1 - \nu_{12}\nu_{21})$ and $Q_{22} = E_{22}/(1 - \nu_{12}\nu_{21})$. The Q coefficients are called "planar stiffnesses" [89]. A normal planar stiffness (Q_{ii}) is the ratio of stress to strain with no out-of-plane stresses and no lateral in-plane strain. In magnitude, it lies between the corresponding Young's modulus and bulk stiffness. If the geometric mean of the Poisson's ratios is about 0.3, a planar stiffness is roughly 10% larger than the corresponding Young's modulus. In performing normal load-elongation experiments, the ratio of stress to strain equals a Young's modulus when (as normally is the case) the jaw span is much greater than the width of the sample. However, if instead the measurement were made with a wide sample over a short span, the ratio would equal a planar stiffness.

There is a troublesome issue indigenous to the measurement of the in-plane elastic parameters of paper. Standard modulus, stiffness, and compliance calculations require a stress input. For example, consider load-elongation experiments. Here, the stress equals the load (F) divided by the sample width (w) multiplied by its thickness (h). Unfortunately, the thickness of paper is poorly defined. Paper has a rough surface, and it is unusually compressible. Caliper measurements are therefore highly dependent on the test load and the surface properties of the contact platens. Absolute modulus determinations are no better than caliper measurements, and these are very ambiguous. The standard trick for circumventing this difficulty is to redefine the constitutive parameters so that the suspect thickness measurement is eliminated. This is achieved by normalizing the stresses and the stiffnesses to the mass density (ρ). In a load-elongation test example, the mass specific modulus (E/ρ) can be calculated as the ratio of $F/\rho wh$ to the strain. The density is the mass of the sample divided by the volume (M/whL), where L is the sample length. Canceling the thickness terms, F/pwh becomes $(F/w)/M/wL$), which is the ratio of force to width divided by basis weight (mass per unit area). So, by normalizing to density, a mass measurement replaces a caliper measurement in the determination of the constitutive parameters. The basis weight of paper is a better-defined property. Mass-specific elastic coefficients are divorced from the vagaries of paper caliper measurements. Therefore, I think they are, in almost all cases, the preferred convention of expression for paper in-plane elasticity. As you will see later, wave propagation velocities in solids are directly related to mass specific stiffnesses. They fall straight out of the analysis, no tricks.

C. Linear Viscoelasticity

Paper is partially composed of amorphous polymers. Amorphous polymers are not linearly elastic at any strain level. The polymer backbones and side groups are subject to conformational change. Driven by random thermal excitations, they are continually wriggling around to different positions and orientations. Distinct polymer conformations are separated by energy barriers. In order to experience a conformational change, a polymer subgroup must gain enough thermal energy to push it over an energy barrierr. The rate at which reconfigurations occur increases exponentially with the ratio of Boltzmann's constant times temperature to the height of the potential barrier [41]. At equilibrium, the relative populations of rival conformations depend on the ground-state energy differences. The low energy conformations are overpopulated, but the disparity falls off with temperature.

When such a polymeric substance is step-loaded, there is, as discussed above, an immediate elastic response. However, this is not the whole story. The stress alters the potential energy differences between conformations. A new equilibrium distribution ensues and, with it, a further compliance to the load. However, this does not happen at once. The new molecular distribution is established only after the random thermal motion is able to energize the repartition. Each "thermal relaxation" has a different activation energy and thereby a different time frame for compliance. Thus, strain trails stress. After the immediate elastic compliance, a series of viscoelastic relaxations follow with a potentially wide range of time lags. The thermal relaxations in cellulosic materials are classified as one of two secondary transitions [1,18,46,47,71,80,86] denoted as γ and β or as the glass (α) transition [18,47]. The rapid secondary relaxations are associated with the redistribution of side group conformations, whereas the longer time glass transition requires backbone movement. Moisture generally, but not always [7,18,43,75], "plasticizes" [78] the thermal relaxations in cellulose. This means that moisture reduces the potential barrier and/or adds to the number of sites capable of reconfiguration. In this way, moisture decreases the relaxation times and increases the relaxation intensities. In other words, hydrophilic amorphous polymers have significant time-delayed compliances that are highly sensitive to moisture and temperature.

As just pointed out, the ratio of stress to strain in a viscoelastic material depends on the history of the deformation. There is no longer a simple proportionality between instantaneous stress and strain. The elasticity assumption must be abandoned, but, if strains are small, linearity is still a valid approximation. This means that the total time-dependent response from a combination of stimuli equals the superposition of the time-dependent responses to the individual stimuli. The stress–strain constitutive relationships for a linear viscoelastic material are expressed as linear differential equations rather than as simple proportionalities. Linear viscoelasticity adds complexity. However, if only time sinusoidal (harmonic) forcing functions are considered, there is a reprieve: The constitutive relations are only slightly complicated. The steady-state solution to a linear differential equation with a sinusoidal forcing function is another sinusoid at the same frequency. Thus, if a small harmonic stress is applied to a linearly viscoelastic material, the resulting steady-state strain is also a sinusoid. In this case, the relationships between stresses and strains can be specified in terms of amplitude ratios and phase shifts. A stiffness can be expressed as a complex number with an amplitude representing the ratio of stress amplitude to strain amplitude and a loss angle (θ) denoting the phase difference between the stress and the strain, that is, $C_{ij} = |C_{ij}|e^{i\theta}$.

The complex stiffnesses are functions of frequency. As frequency increases, there is less time available in one cycle for viscoelastic relaxation. When frequency is well above a particular transition, there is no contribution from this transition to the compliance, whereas well below the transition there is full contribution. Therefore, the amplitudes of the complex stiffnesses increase monotonically with frequency. At frequencies near the inverse of the time required for thermal relaxation, the phases of the complex stiffnesses reach maxima, and the amplitudes change rapidly with frequency. The transition frequency increases with temperature and most often with moisture addition.

Using Fourier analysis, any arbitrary forcing function can be written as a sum of sinusoids. Once a forcing function is broken down into its frequency components, the response to each component can be determined from the complex stiffnesses at the proper frequency. Because the system is linear, the total response is the sum of the responses to each frequency component. Thus, a knowledge of the complex stiffnesses as a function of frequency is sufficient to determine the response of a linear material to any disturbance. So, to account for viscoelasticity, the constitutive coefficients must become frequency-dependent complex numbers that relate sinusoidal stresses to sinusoidal strains. All the foregoing discussions of stiffness and compliance still apply when the consstitutive concept is generalized to complex parameters and specialized to a single frequency.

In paper testing, the dominant frequency of the ultrasonic burst is about 50 kHz. The time frame of the modulus measurements on a tensile machine is about 0.1 s. Ultrasonically determined stiffnesses are made at around 50 kHz, whereas standard mechanical ones are at about 10 Hz. There is over three orders of magnitude more time available for viscoelastic relaxation in the mechanical measurements. Therefore, the ultrasonic stiffnesses of papers (and all other viscoelastic polymers) are greater than those taken from low frequency mechanical measurements. For paper, the frequency dependence of stiffness is greater in the cross-machine direction, because in this direction more load is carried by the amorphous regions of the polymers. Zauscher et al. [97] put forth a counterexplanation.

IV. WAVE PROPAGATION IN SOLIDS

I will present the principles of solid wave propagation through a series of examples. I will progress from the conceptually simple to the more difficult. All the topics will have relevance to paper science, and I will digress briefly as a preview of later paper-specific discussions.

A. Longitudinal Waves in Elastic Strips

In my first example (a minor modification of an acoustic argument from Chapter 6 of Morse and Ingard [66]), I will try to persuade you that the velocity of an elastic wave should equal the square root of a mass-specific stiffness. Consider a semi-infinite strip of elastic material as in Fig. 1. The strip is narrow and thin, and it has a small cross-sectional area (a). Its elastic modulus is E, and its mass density is ρ. At time $t = 0$, I pull with a finite force (F) at the end of the strip. The total strip has infinite mass, and it cannot accelerate as a whole under a finite force. Nonetheless, the force must influence the strip locally. An elastic body can deform; it does not have to move rigidly as a whole. Momentum transfer can be accommodated by a nonuniform, time-dependent deformation of the body. Motion would surely originate from the point of force application. As time progresses and additional momentum is transferred to the strip, more and more of the strip could be shocked into motion. Assume that the scenario proceeds as follows. The step function force suddenly jerks the material near the strip end to a constant particle velocity (v). A moving boundary develops. It divides the strip between a near-end portion moving at velocity v and a far-end portion that is undisturbed. This boundary progresses along the strip with a velocity c. To avoid nonlinear and supersonic complications, I apply a small F, and $|v|$ is guaranteed to be much less than c. In other words, assume that the force step causes a shock wave (a boundary between motion and no motion) to move down the strip at velocity c. What does the application of Newton's second law reveal about c, the velocity of elastic wave propagation?

As assumed, the strip, at a distance x from the end, will start moving in the x direction at velocity v at a time x/c. Thus, if t is greater than x/c, the deformation as a function of time and position $[U_1(x, t)]$ is $(t - x/c)v$. The strain (ε_{11}) is defined as $\partial U_1/\partial x$, which equals $-v/c$ on the back side of the shock wave. I assumed that the strip was thin and narrow and that no forces were applied at the lateral boundaries. Therefore, I take all lateral stresses to be zero throughout the strip, and from Eq. (3)

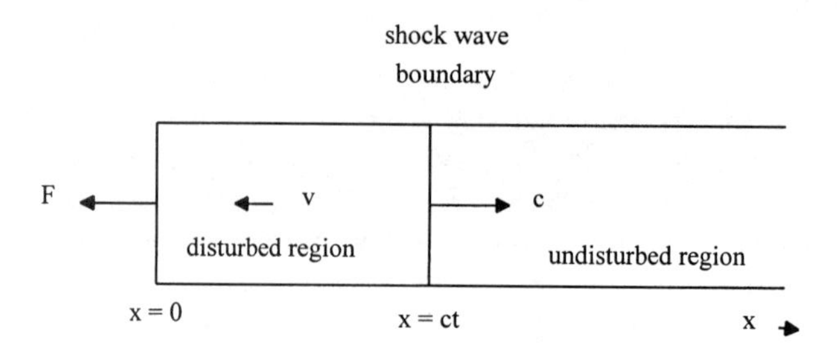

Fig. 1 Longitudinal shock wave propagation in a thin strip.

the elastic constitutive relationship becomes $\varepsilon_{11} = S_{11}\sigma_{11}$ or $\sigma_{11} = E\varepsilon_{11}$. In terms of v, the external force $(-\sigma_{11}a)$ can now be written as Eav/c. The strip momentum, which equals the mass of the moving part of the strip multiplied by v, is $\rho actv$. Setting the time rate of change of momentum equal to the force gives me $Eav/c = \rho acv$, or, solving for c,

$$c = (E/\rho)^{1/2} \tag{6}$$

The postulated shock wave will satisfy Newton's second law if its wave velocity is as given in Eq. (6).

Using Eq. (6) to put c in terms of E and ρ, the constitutive equation can now be rewritten as

$$\sigma_{11}/v = -(E\rho)^{1/2} \tag{7}$$

The quantity $I \equiv (E\rho)^{1/2}$ is called the intrinsic mechanical impedance of a longitudinal wave. If you would repeat this exercise with the strip extending in the negative x direction, you would find that, for a left-hand traveling wave, $c = -(E/\rho)^{1/2}$ and $\sigma/v = (E\rho)^{1/2}$.

Note that the wave velocity and impedance are material properties. Regardless of the strip cross section or how hard I pull on the strip, the shock wave velocity squared equals the mass specific modulus, and the absolute value of the ratio of stress to particle velocity is $(E\rho)^{1/2}$. Intuitively, this all makes sense. Mass resists acceleration, and mechanical disturbances should be expected to progress slowly in dense materials. According to Newton's second law, a wave will move quickly through a stiff material, because a large stress is necessary to achieve a given particle velocity. Materials of high modulus and density require a large ratio of stress to particle velocity because modulus and density both enter the analysis as multipliers in a force equality. In this exercise, I directed the force along the strip. The displacement was collinear with the direction of wave propagation; thus I produced a "longitudinal" wave. Actually, there are lateral Poisson strains; however, the strip is very thin and narrow and lateral displacements are small compared with displacements along the strip. Flexural and torsional waves can also be excited in rods and strips [30], but I do not discuss them.

Now, let me nudge this example one small step ahead. Cut the strip at some place ($x = d$) along its length, dispose of the strip to the right of the cut, and attach another infinite strip with a different density (ρ'), modulus (E'), and cross-sectional area (a'). What will happen when a longitudinal shock wave hits the boundary? Well, obviously, a step force function will be applied to the second strip, and the shock wave will continue, but at a rate of c', along the second strip. There may also be a reflection at the boundary. A single shock wave, neatly handed off across the boundary, can happen only if the two forward-traveling shock waves coincidentally meet the boundary conditions at $x = d$. The boundary conditions are that velocity and force must be continuous. So $a\sigma_{11}$ must equal $a'\sigma'_{11}$, and v must equal v'. However, remember that σ_{11}/v is a fixed material property. In general, $a\sigma_{11}/v = -a(E\rho)^{1/2}$ will not equal $a'\sigma'_{11}/v' = -a'(E'\rho')^{1/2}$. If $a(E\rho)^{1/2}$ just happens to be the same as $a'(E'\rho')^{1/2}$, the two strips are "impedance-matched." A single longitudinal shock will pass smoothly across the interface. In general, good acoustic coupling between systems occurs if radiation propagates in both with nearly the same ratio of force to particle velocity. Otherwise, the boundary conditions can be met only if a reflected shock wave, beginning at $t = d/c$, travels from the boundary back into the first strip. The impedance of the backward traveling wave is the inverse of the impedance of the forward traveling wave, and combinations of waves in two directions in the first strip can match the boundary conditions for a single forward wave in the second strip. For details, see the text by Graff [30]. If the values of the area times the impedance of the strips are nearly equal, most energy will be transmitted across the boundary. If the two are far apart, most energy will be reflected. Note that if $a = a'$, it is the intrinsic impedances of the two media that must be matched to avoid wave reflection at the boundary.

I included the above demonstration because it gets to the basics of solid wave propagation with a minimum of mathematics. However, to progress, I need a slightly more sophisticated outlook. Consider the same thin, narrow strip of deformable material. This time I do not specialize to an idealized shock wave; I look at arbitrary continuous disturbances along the strip. Any imaginable longitudinal displacement history can be represented by $U_1(x, t)$, the displacement as a function of position and time. However, not every continuous function of x and t is admissible as a dynamic state of the strip. Newton's second law restricts the displacement history. To see how, look to Fig. 2, which is a free-force diagram of a representative infinitesimal slice of the strip. The force exerted on the left-hand side of the infinitesimal section is $-a\sigma_{11}(x)$, whereas the right-hand-side force is $a\sigma_{11}(x + dx)$. Using the elastic constitutive equation for a thin narrow strip, the total force on the infinitesimal section becomes $dF = aE[\partial U_1(x + dx, t)/\partial x - \partial U_1(x, t)/\partial x]$. In the limit as dx approaches zero, this simplifies to $dF = aE[\partial^2 U_1(x, t)/\partial x^2]dx$. To first order in U, the differential time rate of change of momentum of the slice is $a\rho[\partial^2 U_1(x, t)/\partial t^2]dx$. Equating the two differential terms, I find that any acceptable longitudinal disturbance in an elastic, deformable strip must be a solution to the differential equation

$$\frac{\partial^2 U_1(x, t)}{\partial t^2} = \frac{E}{\rho}\frac{\partial^2 U_1(x, t)}{\partial x^2} \tag{8}$$

So the second derivative with respect to time equals E/ρ multiplied by the second derivative with respect to distance. This is a "wave equation." Using the chain rule, you can verify that functions of the form $F(x - ct)$ and $G(x + ct)$ are solutions to Eq. (8) if $c^2 = E/\rho$. Imagine any $F(x - ct)$ plotted out as a function of position at some time t_0. Note that $F(x - ct)$ is unchanged if you increment time by an amount Δt and position by an amount $c\Delta t$. There is enough information in $F(x - ct_0)$ to supply a similar plot at any time. To find $F(x - ct)$ at a new time

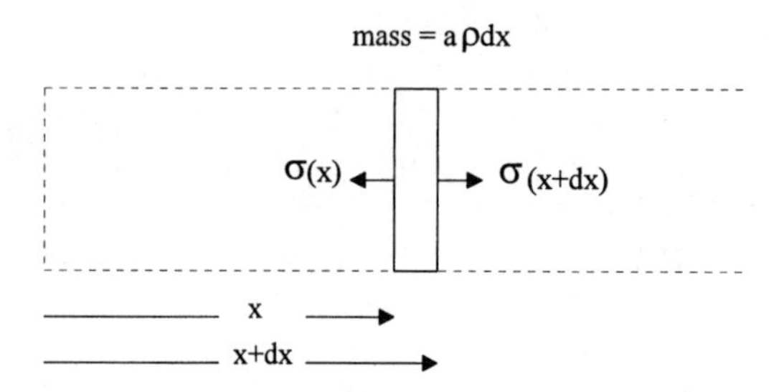

Fig. 2 Free force diagram of an infinitesimal section of a thin strip.

(t), you merely shift $F(x - ct_0)$ to the right by a distance equal to $c(t - t_0)$. Thus, $F(x - ct)$ is a forward traveling wave; it marches to the right undistorted with a velocity $c = (E/\rho)^{1/2}$. Likewise, $G(x + ct)$ is a backward traveling wave; it marches to the left undistorted with a velocity equal to $-(E/\rho)^{1/2}$. Once a pure traveling wave is generated in an elastic strip, one can measure its propagation velocity and determine the mass specific modulus of the strip. You might recall that this was the approach of Hamburger [39] and Chaikin and Chamberlain [23] for fibers and film strips. The wave equation is linear; therefore, linear combinations of solutions are also solutions. Standing waves are mixtures of forward and backward traveling waves.

B. Longitudinal Waves in Viscoelastic Strips

Longitudinal waves in linearly elastic strips travel along undistorted at a velocity squared of E/ρ. However, paper is partially composed of amorphous polymers, and they are linearly viscoelastic (not linearly elastic) at small strains. How does this change the discussion? Well, the wave equation is the consequence of a force balance and a constitutive equation. Newton's second law holds regardless of the subject material, but we need to carefully reconsider the constitutive equation. Remember that the simple linear constitutive equation is good for viscoelastic materials only for sinusoidal forcing functions. In this case, the viscoelastic parameters become complex, and they depend on frequency. So we can continue with Eq. (8) only if the time dependence is sinusoidal and the elastic modulus concept is generalized to complex notation ($E = E' + iE''$). For a function of time and position to be a proper solution of Eq. (8), the time portion must be harmonic ($e^{i\omega t}$), and the entire function must depend on x plus or minus a constant times t. This happens only if the position dependence is also sinusoidal. That is, constant frequency solutions of Eq. (8) are in the form $Ae^{i(\omega t - kx)}$ or $Be^{(i(\omega t + kx)}$. Here, ω is the angular frequency, and the wavenumber (k) is an as yet undetermined function of frequency and material properties. To get a fix on the wavenumber, I insert $U(x, t) = Ae^{(i(\omega t - kx)}$ into Eq. (8). This is valid (i.e., $Ae^{(i(\omega t - kx)}$ is a solution to the viscoelastic wave equation) if

$$\omega^2/k^2 = (E' + iE'')/\rho \tag{9}$$

This equation plays only if k is complex. Let me write k as $k = k' - ik''$. In terms of the real and imaginary parts of k, the traveling wave in the viscoelastic strip becomes $Ae^{(i(\omega t - k'x)}e^{-k''x}$. This is an attenuating sinusoidal traveling wave. The wave amplitude drops off exponentially in the direction of propagation. The phase velocity c is ω/k' and the attenuation coefficient is k''. Viscoelastic relaxation is an irreversible process; therefore, mechanical energy is degraded to heat as a mechanical wave propagrates through a viscoelestic strip. If c and k'' are measured, the real and imaginary parts of the mass specific Young's modulus can be calculated by equating the real and imaginary parts of Eq. (9). The results are

$$\frac{E'}{\rho} = \frac{c^2(1 - r^2)}{(1 + r^2)^2} \tag{10}$$

and

$$\frac{E''}{\rho} = \frac{2rc^2}{(1+r^2)^2} \tag{11}$$

where r is an abbreviation for $k''/k' = k''c/\omega$. In almost all cases, r is considerably less than 1, and we can safely ignore terms of second order and higher in r. The real and imaginary parts of Young's modulus then simplify to the following functions of propagation parameters:

$$E'/\rho \approx c^2 \tag{12}$$

and

$$E''/\rho \approx 2k''c^3/\omega \tag{13}$$

Of course, $Be^{i(\omega t+k'x)}e^{k''x}$ is also a solution to Eq. (9). It represents an attenuated traveling wave moving in the negative x direction. Again, standing waves can be constructed by combining traveling waves in opposite directions. In the case of Ballou and Smith [5], the receiving transducer generated reflections that set up a standing wave in a section of a viscoelastic strip. By measuring the distance between peaks in the standing wave pattern and the amplitude of the standing wave pattern as a function of position, they determined E' and E''. Pankonin and Habeger [75] calculated the complex modulus by observing the frequency response of a resonant strip. Strips are convenient structures for making in-plane loss measurements. The wave is confined to one direction, and viscoelastic loss, not wave spreading, is the dominant factor in amplitude change with position.

The stiffness of a viscoelastic polymer is monotonically increasing function of frequency; compliance increases as more time is allowed for thermal relaxations. This means that not only do waves in viscoelastic strips have a frequency-dependent attenuation, but also the phase velocity increases with frequency. Consider a finite pulse of mechanical energy propagating in one direction along a viscoelastic strip. In order to describe the progression of the pulse, we must break it down into Fourier components that can be handled individually as sinusoidal traveling waves. The Fourier components will have different phase velocities and attenuation coefficients. Thus, the pulse will spread out and die off as its propagates. Viscoelastic materials are "dispersive" in wave propagation. This is in contrast with elastic media, which are nondispersive. As already argued, thin strip elastic pulses in one dimension propagate without distortion.

C. Longitudinal Waves in Thin Plates and in Bulk Materials

Because the strip was thin and narrow and because I administered no lateral constraints, I was able to reasonably assume that strip waves propagated with no lateral stresses. My only nonzero stress was σ_{11}, and normal stress divided by normal strain equaled Young's modulus. Even though there were no lateral stresses, from Eq. (3) there were lateral strains ($S_{12}\sigma_{11}$ and $S_{13}\sigma_{11}$). By Poisson expansion and contraction, the strip pulsed laterally in and out as a traveling wave progressed. Now, imagine an infinite number of strips laid out side by side in the y direction. The strips are miraculously fastened together at their edges to form an infinite thin plate. I apply an identical normal stress regime to the end of each strip. A traveling wave propagates down the plate, but each strip can no longer freely Poisson contract and expand. Because they are now attached, attempted y-direction lateral expansion (or contraction) in a strip works against lateral expansion (or contraction) in an adjacent strip. In fact, if one strip did y-direction Poisson expansion, its neighbor would have to contract or shear. However, by symmetry, each strip must behave in the same manner. Therefore (unlike in a single strip), there can be no y-direction strain when a uniform normal stress regime is applied to the end of a thin plate. This changes my elastic constitutive equation. The plate faces are still free; therefore, I can use the plane stress constitutive relations of Eq. (5). However, this time e_{22} is zero, and the pertinent elastic constitutive equation is $s_{11} = Q_{11}e_{11}$. The wave propagation of longitudinal waves in thin plates becomes $(Q_{11}/\rho)^{1/2}$, and the impedance is $(Q_{11}\rho)^{1/2}$. Since free boundaries have been eliminated, it is harder to deform a plate than a strip. Thus, longitudinal waves have higher velocities and higher impedances in thin plates than in narrow thin strips.

As you probably guessed, I am going to push this construction up into the third direction. Let's say that I stack an infinite number of thin plates and rigidly bond the faces of next-door neighbors. Now, by the same argument, a longitudinal wave propagates without strain in either lateral direction. the stress–strain relation from Eq. (2) is $\sigma_{11} = C_{11}\varepsilon_{11}$. Longitudinal bulk waves are faster than longitudinal plate waves; they have a velocity of $(C_{11}/\rho)^{1/2}$ at an impedance of $(C_{11}\rho)^{1/2}$. So longitudinal elastic waves propagated in strips, plates, and bulk materials travel with a velocity equal to the square root of a mass specific stiffness. Differences in lateral boundary conditions cause waves in the different structures to move at different speeds and with different impedances. The viscoelastic concept can be incorporated into plate and bulk wave motion, just as it was with strips: the stiffnesses become complex and frequency-dependent.

D. Transverse Waves

So far, I have dealt only with longitudinal wave motion. Now I consider transverse waves. This time I want to propagate shear stresses. Therefore, instead of applying a normal force to induce wave motion, I aim the force in the cross-sectional plane. For the strip, this causes some technical problems, and I am going to sidestep the issue. Imagine reorienting the driving force on the strip to the y direction in hopes of generating shear stress (σ_{12}) waves. Unfortunately, I have lateral, stress-free boundary conditions on the strip. This means that σ_{21}, σ_{22}, and σ_{23} are zero at the y-direction limits of the strip. Remember that to balance the moments on infinitesimal elements, σ_{12} must always equal σ_{21}. Thus, σ_{12} must also be zero at the y-direction boundaries. I am interested in thin plates, and I am not prepared to address stress or strain gradients in the lateral directions. I cannot apply a uniform shear stress to the strip cross sections. Therefore, I give up and skip along to transverse waves in thin plates.

If I try to align my plate stress in the z direction, I bump into the same contradiction that I encountered with the strip. Nonetheless, I can get away with a uniform shear stress in the xy plane of a thin plate. There are no longer any stress-free, y-direction boundary conditions. I can envision a uniform σ_{12} on the entire plate cross section, and I can postulate a uniform y-direction deformation as a function of x and

t, $U_2(x, t)$. Now, the elastic constitutive equation becomes $\sigma_{12} = G\,\partial U_2(x, t)/\partial x$, and the new wave equation is

$$\frac{\partial^2 U_2(x,t)}{\partial t^2} = \frac{G}{\rho}\frac{\partial^2 U_2(x,t)}{\partial x^2} \tag{14}$$

The velocity of transverse wave propagation equals $(G/\rho)^{1/2}$, and the impedance is $(G\rho)^{1/2}$. So, in a thin isotropic plate, I have two kinds of simple, uniform stress wave motions: a longitudinal wave with particle motion in the direction of propagation, and a generally slower, lower impedance transverse wave with particle motion in the plane of the plate but normal to the direction of propagation.

In the bulk material, all of the stress-free boundary conditions are eliminated. Lateral stresses can be uniformly applied in any direction. Transverse waves in isotropic bulk materials propagate with velocity $(G/\rho)^{1/2}$ and impedance $(G\rho)^{1/2}$, and they can be "polarized" in any direction in the plane perpendicular to the wave propagation.

E. Plane Waves in Thin Orthotropic Plates

My next step is to extend the strip wave equation [Eq. (8)] to thin orthotropic plates [48,67]. This time, there are two spatial coordinates; an arbitrary displacement is specified by $U_i(x, y, t), i = 1, 2$. As shown in Fig. 3, the free-force diagram is drawn on a representative segment that is infinitesimal in the x and y directions. There are normal and shear stresses on all four out-of-plane faces. The small rectangular section is centered at coordinates (x, y), and it extends a distance dx in the x direction and dy in the y direction. The x and y axes are aligned in the principal directions of orthotropic symmetry. The plate thickness is h. The arrows corresponding to each stress point in the direction of the resulting force on the section assuming positive stress. As with the strip exercise, it is the differences between like stresses on opposite sides of the section that lead to net forces. The problem is just more complex here, because we have two displacements, four sides, and four stresses to deal with. Assuming small strains, force equals mass times acceleration in the x direction and in the y direction can be written as in Eqs. (15) and

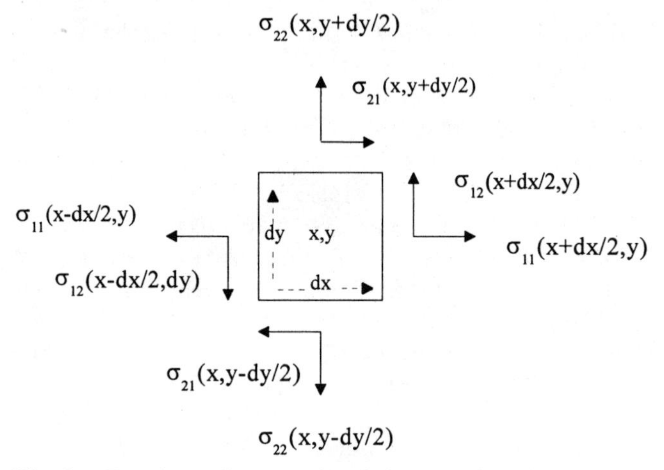

Fig. 3 Free force diagram of an infinitesimal section of a thin plate.

$$\left[\sigma_{11}\left(x+\frac{dx}{2}y,t\right)dy-\sigma_{11}\left(x-\frac{dx}{2},yt\right)dy+\sigma_{21}\left(x,y+\frac{dy}{2},t\right)dx-\sigma_{21}\left(x,y-\frac{dy}{2},t\right)dx\right]h=\frac{\partial^2 U_1(x,y,t)}{\partial t^2}\rho h\,dx\,dy \tag{15}$$

and

$$\left[\sigma_{22}\left(x,y+\frac{dy}{2},t\right)dy-\sigma_{22}\left(x,y-\frac{dy}{2},t\right)dy+\sigma_{12}\left(x+\frac{dx}{2},y,t\right)dx-\sigma_{12}\left(x-\frac{dx}{2},y,t\right)dx\right]h=\frac{\partial^2 U_2(x,y,t)}{\partial t^2}\rho h\,dx\,dy \tag{16}$$

In the limit as dx and dy go to zero, these become

$$\frac{\partial\sigma_{11}(x,y,t)}{\partial x}+\frac{\partial\sigma_{21}(x,y,t)}{\partial y}=\rho\frac{\partial^2 U_1(x,y,t)}{\partial t^2} \tag{17}$$

and

$$\frac{\partial\sigma_{22}(x,y,t)}{\partial y}+\frac{\partial\sigma_{12}(x,y,t)}{\partial x}=\rho\frac{\partial^2 U_2(x,y,t)}{\partial t^2} \tag{18}$$

Using the plane stress constitutive equations [Eq. (5)] to replace stresses with strains and expressing the strains in terms of the derivatives of deformations, I get the following coupled linear differential equations for U_1 and U_2:

$$Q_{11}\frac{\partial^2 U_1}{\partial x^2}+Q_{12}\frac{\partial^2 U_2}{\partial x\,\partial y}+Q_{66}\frac{\partial^2 U_1}{\partial y^2}+Q_{66}\frac{\partial^2 U_2}{\partial x\,\partial y}=\rho\frac{\partial^2 U_1}{\partial t^2} \tag{19}$$

$$Q_{12}\frac{\partial^2 U_1}{\partial x\,\partial y}+Q_{22}\frac{\partial^2 U_2}{\partial y^2}+Q_{66}\frac{\partial^2 U_1}{\partial x\,\partial y}+Q_{66}\frac{\partial^2 U_2}{\partial x^2}=\rho\frac{\partial^2 U_2}{\partial t^2} \tag{20}$$

Now, please look back to Eq. (8) and compare it with Eqs. (19) and (20). Equation (8) is a simple one-dimensional wave equation: The second spatial derivative equals a constant multiplied by the second time derivative. Traveling waves, which are functions of the form $F(x \pm ct)$, are solutions, providing c^2 equals the mass specific stiffness. Equations (19) and (20) are not so different. We now have two coupled deformations and four separate stiffness coefficients, but second spatial derivatives are still proportional to second time derivatives. Some sorts of traveling waves will surely make solutions. This time, however, we have to be prepared for traveling waves heading in any xy plane direction. Consider a function of the form $F(x\cos(\theta) + y\sin(\theta) - ct)$, where θ is any angle to the x axis. Note that this function is unchanged as you move in a plane perpendicular to a line at an angle θ to the x axis. At a given time, it has one value for each plane perpendicular to θ. It is also unchanged as you increment t by an amount Δt and move a distance $c\,\Delta t$ along θ. It is a "plane wave" that travels with a velocity c along θ. Let's see if there are traveling plane wave solutions to Eqs. (19) and (20).

Assume solutions of the form $U_1 = F_1(x\cos(\theta) + y\sin(\theta) \pm ct)$ and $U_2 = F_2(x\cos(\theta) + y\sin(\theta) \pm ct)$. Both functions represent a plane wave traveling at a speed equal to c in the θ direction. Letting F_i'' denote the second derivative of

F_i, substitution of the plane wave trial solutions into Eqs. (19) and (20) gives, using the chain rule, Eqs. (21) and (22):

$$\cos^2\theta Q_{11}F_1'' + \cos\theta\sin\theta\, Q_{12}F_2'' + \sin^2\theta\, Q_{66}F_1'' + \cos\theta\sin\theta\, Q_{66}F_2'' = \rho c^2 F_1'' \quad (21)$$

$$\cos\theta\sin\theta\, Q_{12}F_1'' + \sin^2\theta\, Q_{22}F_2'' + \cos\theta\sin\theta\, Q_{66}F_1'' + \cos^2\theta\, Q_{66}F_2'' = \rho c^2 F_2'' \quad (22)$$

The above two equations can be rewritten in matrix notation as

$$\begin{vmatrix} \cos^2\theta Q_{11} + \sin^2\theta Q_{66} - \rho c^2 & \cos\theta\sin\theta(Q_{12}+Q_{66}) \\ \cos\theta\sin\theta(Q_{12}+Q_{66}) & \sin^2\theta Q_{22} + \cos^2\theta Q_{66} - \rho c^2 \end{vmatrix} \begin{vmatrix} F_1'' \\ F_2'' \end{vmatrix} = \begin{vmatrix} 0 \\ 0 \end{vmatrix} \quad (23)$$

Nontrivial solutions are possible only if the determinant of the coefficient in the above matrix is zero. Note that setting the determinant to zero provides a quadratic equation in c^2. It is

$$c^4 - (A+B)c^2 + (AB - H^2) = 0 \quad (24)$$

where $A \equiv \cos^2\theta Q_{11}/\rho + \sin^2\theta Q_{66}/\rho$, $B \equiv \sin^2\theta Q_{22}/\rho + \cos^2\theta Q_{66}/\rho$, and $H \equiv \cos\theta\sin\theta\,(Q_{12}/\rho + Q_{66}/\rho)$. Writing out the solutions to the quadratic equation, I find that the two possible values of c^2 are

$$c^2 = 0.5\{A + B \pm [(A-B)^2 + 4H^2]^{1/2}\} \quad (25)$$

The coefficients A, B, and H depend on the angle of propagation of the traveling wave and on the mass specific planar stiffnesses. At every angle, Eq. (25) provides two values of c^2 for which there are plane wave solutions to the coupled wave equations [Eqs. (19) and (20)]. Once one of these values is chosen from Eq. (25), the relations in Eq. (23) give the ratio of F_1'' to F_2'' and the polarization of the deformation of the traveling wave. In the principal directions of orthotropic symmetry ($\theta = 0$ or $\pi/2$), H becomes zero, and Eq. (25) simplifies considerably to $c^2 = A$ and $c^2 = B$. For $\theta = 0$, the acceptable speeds are $c^2 = Q_{11}/\rho$ and $c^2 = Q_{66}/\rho$. At $\theta = \pi/2$, they are $c^2 = Q_{22}/\rho$ and $c^2 = Q_{66}/\rho$. From Eq. (23), the waves with velocity determined by a normal stiffness are found to be along the direction of propagation; they are longitudinal waves. The other two are transverse waves: Deformation is perpendicular to the direction of propagation. The longitudinal waves have different speeds in the two principal directions, whereas transverse waves are of equal speed in both directions. In the principal directions, this more elaborate exposition coincides with the earlier thin plate analyses: There exist purely longitudinal and purely transverse waves that propagate with different velocities. Beware that Eq. (25) is not equivalent to a stiffness rotation. The off-axis quasi-longitudinal velocity does not equal the mass specific normal stiffness at the angle of propagation; likewise for the quasi-transverse wave.

Excluding isotropic materials ($Q_{11} = Q_{22}$ and $Q_{12} = Q_{11} - 2Q_{66}$), the situation is more difficult when θ is not equal to 0 or $\pi/2$ [3,48,67]. At each angle, there are still two types of plane waves. Neither, however, is exactly parallel or perpendicular to the direction of propagation. If the medium is not overly anisotropic, the solution coming from the plus sign in Eq. (25) will be polarized close to the direction of propagation, and the minus sign solution will be nearly perpendicular to propagation. The polarizations of the two solutions are still perpendicular to each other. To emphasize this complexity, an off-angle plane wave is classified as quasi-longitudinal or quasi-transverse. A plot of quasi-longitudinal wave speed [the plus alternative in

Eq. (25)] as a function of angle will peak in the principal direction of greater stiffness and decrease monotonically to the lower stiffness principal direction. The quasi-transverse speed is the same in both principal directions, but off-axis it can deviate (up or down) from $(Q_{66}/\rho)^{1/2}$.

Harmonic plane traveling waves are plane waves of particular interest. Functions of the form $\exp\{\pm i[k\cos(\theta)x + k\sin(\theta)y \pm \omega t]\}$ represent plane waves that are sinusoidal in time and displacement and move at θ to the x axis. The wavelength (the distance along θ at which the wave repeats) is $\lambda = 2\pi/k$. The acceptable values of the phase velocity squared ($\omega^2/k^2 = \lambda^2 f^2 = c^2$) are restricted by Eq. (25). By Fourier superposition, any general plane wave at θ can be written as a summation of sinusoidal plane waves at θ. In general, a plane wave could have a wide range of frequency components. In application, however, narrowband transducers are employed, and the diagnostic pulse expression is bunched in the frequency domain. In this case, even though the wave is not sinusoidal, an approximate wavelength (that at the peak in Fourier superposition) can be used to characterize the pulse.

Now consider viscoelastic thin plates. Sinusoidal traveling waves are time harmonic, and one can use the complex constitutive equations to relate stress to strain when a single-frequency disturbance is present in a viscoelastic medium. Equation (25) (with $c^2 = \omega^2/k^2$) becomes complex, and k takes on an imaginary part. Attenuating plane sinusoidal waves now become solutions of Eqs. (19) and (20). Like the stiffnesses, the phase velocity and the attenuation are frequency-dependent. Consequently, finite pulses, which are combinations of plane waves at different frequencies, disperse.

F. Near Field Issues

The excitation of in-plane ultrasonic waves in paper is often accomplished with small, almost point-contact, transducers. These will not make plane waves. A plane wave function is a good model of a disturbance field far from a point source or when the lateral extent of the transducer is much greater than the transducer separation. In out-of-plane testing, the diameter of the transducer faces dwarfs the sheet thickness, and the plane wave picture is good. However, for in-plane work, the transducer-to-paper contact dimensions are often small compared to transducer separation. For the sake of interpreting in-plane measurements, I need to take a hard look at the disturbances near point sources.

To get an idea of the wave structure encountered in actual measurements, imagine an isotropic point-source longitudinal wave generator on an isotropic thin plate. Assume that it radiates uniformly so that the particle motion is in the r direction and is a function only of r, the radial distance from the source. Now, the wave is spreading away from the source. By conservation of energy, the amplitude of the wave must decrease as it moves out from the source. In fact, energy per unit volume goes as the square of the amplitude, and a progressing circular wave increases in volume in proportion to the radius. Therefore, in a conservative (elastic) medium, you might expect amplitude squared times radius to remain constant. In this circumstance, the amplitude would be proportional to $1/r^{1/2}$. For the moment, there is no reason to expect the phase velocity of the radial wave to be different from that of a plane wave. Thus (for a sinusoidal forcing function), an intuitive guess for an out-going radial longitudinal wave solution would be $U_r(r, t) = (A/r^{1/2})e^{\pm i(kr-\omega t)}$, with $\omega^2/k^2 = c^2 = Q_{11}/\rho$. This disturbance, as intended, drops off in intensity in proportion to the distance from the source and has the established traveling wave form at the right phase velocity.

To test this hypothetical solution, I need the governing equation for radial wave propagation. This time the representative infinitesimal section is a radial slice between $\theta - d\theta/2$ and $\theta + d\theta/2$ and between r and $r + dr$ (see Fig. 4). As before, I sum the forces on the four sides of the section and set that sum equal to mass times acceleration. To first order, the forces in the r direction from the two radial sides are $-\sigma_{rr}(r, t)hr\ d\theta$ and $\sigma_{rr}(r + dr, t_h)h(r + dr)\ d\theta$. Because the coordinates are curvilinear, the normal stresses on the θ faces have a small component in the r direction. To first order, each θ face contributes a force of $-\sigma_{\theta\theta}(r, t)\sin(d\theta/2)\ h\ dr$ in the r direction. The sum of these four forces must equal the time rate of change of momentum, which to first order is $\rho hr\ d\theta\ dr\ \partial^2 U_r(r, t)/\partial t^2$. In the limit as dr and $d\theta$ go to zero, this becomes

$$\frac{\partial(r\sigma_{rr})}{\partial r} - \sigma_{\theta\theta} = \rho r \frac{\partial^2 U_r(r, t)}{\partial t^2} \tag{26}$$

As before, I will use the constitutive relations to put the stresses in terms of the deformation: $U_r(r, t)$. Since I have a thin plate under plane stress, $\sigma_{rr} = Q_{11}\varepsilon_{rr} + Q_{12}\varepsilon_{\theta\theta}$ and $Q_{\theta\theta} = Q_{12}\varepsilon_{rr} + Q_{11}\varepsilon_{\theta\theta}$. Particle motion is in the r direction, and ε_{rr} is $\partial U_r(r, t)/\partial r$. At first it seems that $\varepsilon_{\theta\theta}$ would be zero. After all, there is no deformation in the θ direction. Nevertheless, consider a hypothetical ring of material at an initial, undisturbed radius r. As the ring moves toward and away from the source, its radius increases and decreases. This changes the circumference of the ring, resulting in normal strain in the θ direction. The circumference is proportional to the radius; thus $\varepsilon_{\theta\theta} = U_r(r, t)/r$. Putting the strain definitions into the constitutive equations, the constitutive equations into the force balance equation [Eq. (26)], and making some simplifications, I arrive at the following (not quite Bessel) differential equation for the deformation of a longitudinal radial wave in a thin isotropic plate:

$$\frac{Q_{11}\partial}{\partial r}\left(\frac{\partial U_r(r, t)}{\partial r} + \frac{U_r(r, t)}{r}\right) = \rho \frac{\partial^2 U_r(r, t)}{\partial t^2} \tag{27}$$

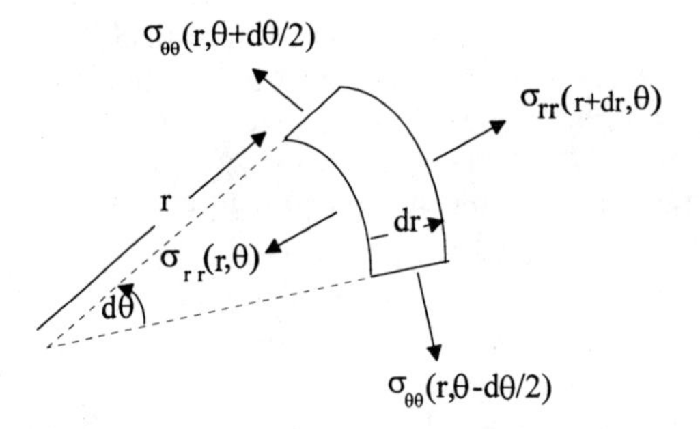

Fig. 4 Cylindrical coordinate free force diagram of an infinitesimal section of a thin plate.

The next step is to insert the test function and see if it is a solution. Putting $U_r(r, t) = (A/r^{1/2})e^{+i(kr-\omega t)}$ into Eq. (27), I discover a necessary relationship between k and ω: $k^2 = \omega^2/c^2 - 3/4r^2$, or if you like,

$$\frac{\omega^2}{k^2} = \frac{c^2}{1 - 3c^2/4r^2\omega^2} \tag{28}$$

where c^2 represents the mass specific planar stiffness Q_{11}/ρ. Remember, Q_{11}/ρ equals the speed squared of the old plane wave. Well, this almost worked. If it wasn't for that pesky $3/4r^2$ term, the phase velocity ω/k would equal c independent of r, and everything would be fine. All the same, note that as r^2 becomes large compared to c^2/ω^2, the problem term evaporates. Far from the source, the trial solution is approximately correct; the radial wave is almost like a plane wave whose amplitude decreases as the square root of r. Here, phase velocity can be measured and used to determine the specific stiffness in the normal way. However, near the source (in the near field) there are effectively radial dependent modifications to the phase velocity. Roughly and inexactly speaking, longitudinal phase velocity is greater near the source, because hoop stresses provide an extra restoring force. I also did this calculation for a radial transverse wave [$U_\theta(r, t)$], and I derived again Eq. (28) (except, of course, $c^2 = G/\rho$). Please note that $U_r(r, t) = (A/r^{1/2})e^{\pm i(kr-\omega t)}$ with $k^2 = \omega^2/c^2 - 3/4r^2$ is not an exact solution to Eq. (27). It gives the first-order correction to $(A/r^{1/2})e^{\pm ik(r-ct)}$ as the origin is approached. Further in, higher order terms arise to account for the derivative of k with respect to r. When r^2 becomes small compared to c^2/ω^2, Eq. (28) is no longer a good estimate of phase velocity.

Clearly, there will be serious consequences if one naively tries to make a velocity measurement close to a small source. In practice, with nonisotropic plates and nonsymmetrically pulsating disturbances, the near field may not have exactly this form. Near field headaches will still persist, and Eq. (28) will give the right order of magnitude correction. Obviously, I will have to return to this subject when I discuss specific measurement schemes.

G. Wave Surfaces

This is the most difficult argument that I have yet to make. Again, I am interested in velocity measurements on thin plates using small sources. This time, I stay many wavelengths away from the generator so that I can forget about the near field. However, I generalize to anisotropic thin plates. My new concern is the off-axis velocity of quasi-longitudinal and quasi-transverse waves measured by transducers that are small compared to a wavelength. A first rational guess would be that the velocity of a radially diverging wave equals that of a plane wave in the same direction. That is, in a thin plate, a locus of constant phase emanating from a point source would be a curve whose distance from the origin would be proportional to a velocity from Eq. (25). Along a principal axis of orthotropic symmetry, this very logical assumption holds, but off-axis, especially for longitudinal waves in very anisotropic plates, it does not. This point was originally made for general anisotropic media in the 1950s by Musgrave [67,68]. It is presented in modern terms by Auld [3]. Straudt and Cook [84,95] present an impressive experimental demonstration of the relationship between phase velocity of a plane wave and the group velocity of an energy pulse. Craver and Taylor [24] recognized the importance of this distinction early in the history of paper testing. The IPC group [35] demonstrated that it is important to account for the differences in the proper determination of general elastic constants.

I will try to explain this intuitively with reference to Fig. 5. For rigor, refer to Musgrave [67,68] or, better yet, Auld [3]. In Fig. 5, I depict the result of a hypothetical purely quasi-longitudinal generating point source in an orthotropic thin plate. The source oscillates sinusoidally. At a given time, as I move away from the source, the phase is retarded. As drawn, a line of equal phase (a wave surface) will have an oblong shape. The wavelength is greatest along the principal axis of high stiffness and least in the low stiffness direction. At an in-between angle, the phase progresses at an intermediate velocity. The velocity of the wave surface, $r(\alpha)$, is proportional to the distance from the source to a point on the wave surface. A particular r at an angle α to the minor principal axis is emphasized in the figure. As time advances, this wave surface will progress out to a larger, but similar, figure. This is not plane wave propagation; therefore, I cannot directly apply my known relations between the angle and the phase velocity of plane waves. However, if I appeal to Huygen's principle of wave propagation, all is not lost.

A Huygen's wave construction [15,30] uses an existing wave front as the source for future wave fronts. In a short time interval dt, the drawn wave surface will generate a new wave surface slightly and proportionally larger at all angles. I want to take a close look at this process at a particular angle α; therefore, I blow up the region at α and show the wave surfaces at t and $t + dt$. Now, in the limit as dt approaches zero, the local wave surfaces are flat, and I do have plane wave propagation. As argued above, plane waves propagate with a velocity $c(\theta)$ perpendicular to the plane. Note that since the plate is anisotropic, α and θ cannot always be equal. Along the principal axis, r and c are equal, but elsewhere the effective plane wave can move in a different direction from the wave surface. Nonetheless, the plane wave

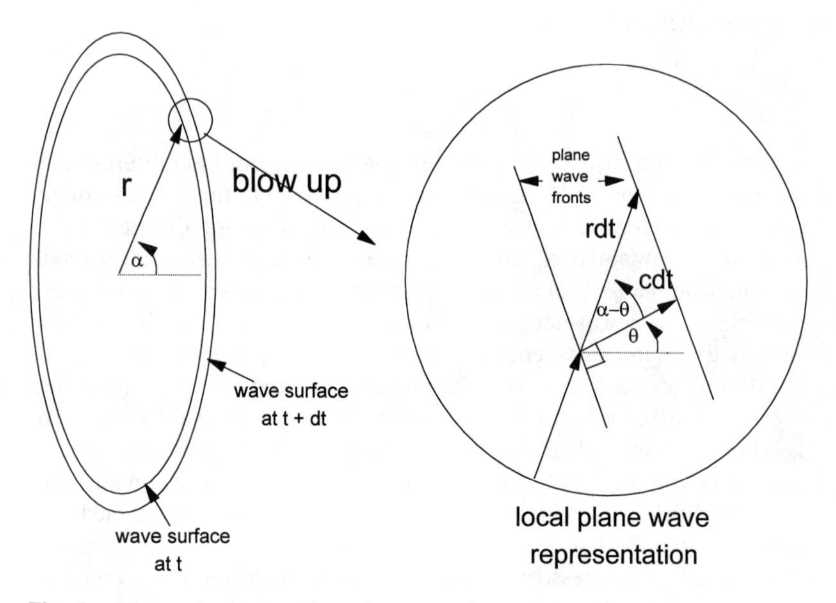

Fig. 5 Huygen's construction of wave surface propagation.

propagation at θ determines the wave surface propagation at α. In fact, from Fig. 5, you can verify that $c(\theta) = r(\alpha)\cos(\alpha - \theta)$.

Using point source transducers, I can measure the quasi-longitudinal group velocities as a function of angle and draw an $r(\alpha)$ curve. I have a direct relationship between elastic parameters and $c(\theta)$ but not between the elastic parameters and $r(\alpha)$. Therefore, in any effort to calculate elastic constants from off-axis velocity measurements, I need a way to convert $r(\alpha)$ into information concerning constitutive relations. I might measure $r(\alpha)$, determine the angle (θ) of the normal to the wave surface, and calculate a phase velocity along the normal by multiplying $r(\alpha)$ by $\cos(\alpha - \theta)$. In this way, I could reconstruct $c(\theta)$ from $r(\alpha)$. If the wave surface is well behaved (no lacunas [67] or no caustics [62], depending on your point of view), there is a shortcut that avoids the task of determining wave surface normals. It is easier to calculate $c(\theta)$ as the maximum projection of all $r(\alpha)$ to θ. To convince yourself that this is a valid course, consider a unit vector in the θ direction. Define $P(\alpha, \theta)$ as the projection of $r(\alpha)$ along θ for all values $\alpha : P(\alpha, \theta) = \mathbf{r}(\alpha) \cdot \mathbf{n}(\theta)$. The derivative of $P(\alpha, \theta)$ with respect to α is the dot product of $\mathbf{n}(\theta)$ with a vector tangential to $\mathbf{r}(\alpha)$. This will be zero, because $\mathbf{n}(\theta)$ is normal to the $r(\alpha)$ curve. Thus, the projection of $\mathbf{r}(\alpha)$ has a extremum when $\mathbf{n}(\theta)$ is normal to $\mathbf{r}(\alpha)$. For well-behaved wave surfaces (such as quasi-longitudinal waves in orthotropic plates) the maximum of $\mathbf{r}(\alpha) \cdot \mathbf{n}(\theta)$ will equal $c(\theta)$.

The distinction between these two velocities is not a trivial point for paper samples. For quasi-longitudinal waves in machine-made paper, there are large ($\approx 10\%$) off-axis differences between the $r(\alpha)$ and $c(\theta)$ curves. For quasi-transverse waves, the velocities in the principal directions are equal, and the paper wave surfaces are almost circular. The wave surface and plane wave propagation angles, within experimental error, are the same. We can forget about the Musgrave construction for quasi-transverse waves in paper but not for quasi-longitudinal waves.

H. Flexural Waves in Thin Plates and Tension Waves

The longitudinal and transverse in-plane waves that I have considered so far are "symmetric": The in-plane motion is symmetrical about the midplane of the plate. The longitudinal wave does include some minor out-of-plane Poisson contraction and expansion, but the low frequency symmetric waves are basically in-plane motions propagating in plane. There also are "antisymmetric" modes. Bending of a thin plate is an example of an antisymmetrical deformation: One face of the plate is in compression, whereas the other is in tension. Thin plates have flexural rigidities, and elastic forces arise in proportion to the deviation of the bent plate from its flat equilibrium state. So, to extend our understanding of plate modes, let us consider wave motion in which the flexural rigidity provides a restoring force. At low frequencies in thin plates, the Bernoulli–Euler theory of beams can be justly applied. Here it is assumed that the plane cross sections initially normal to the axis of the plate remain plane and normal to the neutral axis during bending. Under these conditions, the dynamic state of the plate is specified by the out-of-plane displacement $z(x, y, t)$. Just so things are not too simple, let's put the plate under a uniform, finite in-plane tension. To achieve out-of-plane deformation, work must be done against this added in-plane restraint. So the macroscopic tensile force will assist the first-order bending forces in their struggle to maintain a flat plate. As a simpli-

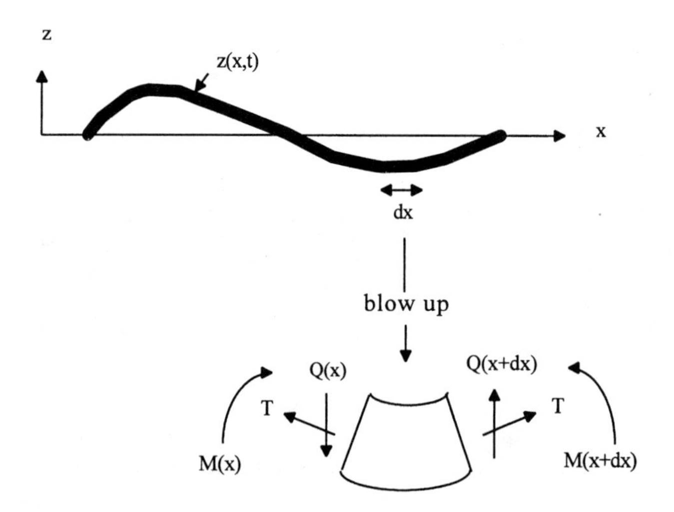

Fig. 6 Free force diagram of an infinitesimal section of a bent thin plate.

fication, consider out-of-plane deformations that are independent of the y direction. Now the dynamic state is represented by $z(x, t)$.

I am looking for out-of-plane waves that propagate in the x direction. Figure 6 is a force–moment diagram of an infinitesimal y-direction section of the plate. The vertical force on the infinitesimal section per unit dimension on the y direction is represented by $Q(x)$, T is the tensile force per y-direction extension, and $M(x)$ is the moment per y-direction extension. The tension is a finite quantity that is present regardless of the deformation, whereas Q and M are small quantities proportional to the posited infinitesimal deformations. A force summation in the x direction reveals that, to first order in $z(x, t)$, T is independent of x and t. However, due to plate deformation, there is a first-order out-of-plane component to the tension ($T\ \partial z/\partial x$). Equating the z-direction component of first-order forces on the small section to the first-order time rate of change of momentum gives

$$\frac{\partial Q}{\partial x} + T\frac{\partial^2 z}{\partial x^2} = \rho h \frac{\partial^2 z}{\partial t^2} \tag{29}$$

where h stands for the thickness of the plate and ρ is the mass density. Doing a moment balance on the section tells me that $Q(x) = -\partial M/\partial x$. The Bernoulli–Euler theory relates the bending moment to the deformation: $M = D\ \partial^2 z/x^2$. where $D = Q_{11}h^3/12$ is the plate's flexural rigidity per unit y-direction extension. Inserting these relations into Eq. (29) provides the equation of motion for $z(x, t)$:

$$-D\frac{\partial^4 z}{\partial x^4} + T\frac{\partial^2 z}{\partial x^2} = \rho h \frac{\partial^2 z}{\partial t^2} \tag{30}$$

Seeking harmonic plane wave solutions with phase velocity c equal to ω/k, I insert a trial function of the form $z(x, t) = e^{i(\omega t - kx)}$ into Eq. (30). This will work if the wavenumber k is related to ω as in Eq. (31).

$$\frac{Q_{11}h^3}{12}k^4 + Tk^2 = \rho h \omega^2 \tag{31}$$

Solving Eq. (31) for k^2 yields the following propagating solution:

$$k^2 = \frac{6T}{Q_{11}h^3}\left[\left(1 + \frac{\rho\omega^2 Q_{11}h^4}{3T^2}\right)^{1/2} - 1\right] \tag{32}$$

Equation (32) has two asymptotic solutions: one in the tension-dominant regime $(\rho\omega^2 Q_{11}h^4/3T^2 \ll 1)$ and one in which tension is insignificant $(\rho\omega^2 Q_{11}h^4/3T^2 \gg 1)$. When the tension is large, we find in the limit of Eq. (32) that nondispersive, drumlike waves propagate as in a membrane. The phase velocity $(c = \omega/k)$ is

$$c = (T/\rho h)^{1/2} \tag{33}$$

Here, phase velocity is independent of frequency. It depends only on the tension and the basis weight.

More often thin-plate antisymmetrical propagation is best described as a flexural wave. Here, tension is small and bending forces dominate the restorative process; the phase velocity approaches

$$c = (Q_{11}h^2\omega^2/12\rho)^{1/4} \tag{34}$$

Now, the phase velocity increases as the square root of the frequency. A finite pulse will disperse as it propagates because the phase velocity is different for different Fourier components of the pulse. Flexural rigidity and basis weight are the sheet properties that affect propagation.

Between the two extremes, both tension and flexural rigidity contribute to wave propagation, and the full complexity of Eq. (32) is necessary.

I. Plane Waves in Thick Orthotropic Plates

Up to now, the in-plane discussions hang on a thin strip or thin plate assumption. Let's take a closer look at the plate. Remember, I started by pointing out that zero stress boundary conditions apply on the faces. Then I slyly told you that, because the plate was thin, I could extrapolate the absence of out-of-plane shear and normal stresses at the faces to the entire plate. This allowed me to use plane stress constitutive equations, and I proceeded to derive velocities as functions of material plate properties. So, under what conditions does the thin-plate concept break down, and what are the consequences? On a fine scale, the plate is a three-dimensional material, and any disturbance is a combination of bulk waves that meet the stress-free boundary conditions on the faces. Consider harmonic motion in a three-dimensional orthotropic material. Along a principal axis, there are longitudinal and transverse waves whose velocity squared, c^2, equals a mass specific bulk stiffness (C_{ii}/ρ). The wavelength of a particular mode is its velocity divided by the frequency. In order for there to be a difference, at a given time, between stress and strain at two different points, they must be separated by a distance on the order of a wavelength in the direction of propagation. If the out-of-plane stress conditions are to vary across the plate, there must be a component of out-of-plane bulk wave motion and the plate must be at least the order of a wavelength thick for out-of-plane motion. If the plate thickness

(h) is much less than an out-of-plane wavelength, the plane stress assumption is acceptable. That is, the thin plate condition is

$$h \ll c/f \tag{35}$$

where c is the velocity of the slowest out-of-plane bulk wave and f is the frequency of the radiation. If a simple pulsed excitation is employed, a prominent Fourier component of the pulse has a wavelength equal to about the pulse width, and the thin-plate condition is that the transit time through the plate of the slowest out-of-plane mode must be much less than the pulse width. Looking at Eq. (35), it is apparent that the thin-plate assumption will be invalid for thick papers at high frequency or at short time pulses. Paper is a very anisotropic medium: Out-of-plane stiffnesses are hundreds of times less than in-plane ones. Therefore, the relevant c in Eq. (35) is extremely small, and the thin-plate criterion becomes invalid at smaller thickness and lower frequencies than one might expect. As you will see later, we must be extra careful.

In-plane transverse bulk waves meet the free-face boundary conditions naturally. Pure in-plane transverse waves can exist in a plate regardless of frequency or thickness; there is no fundamental concern as the paper becomes thicker and the frequency gets higher. In-plane thin-plate transverse waves and bulk traverse waves are the same motion with the same velocity $(C_{66}/\rho)^{1/2}$. On the other hand, in-plane and out-of-plane bulk longitudinal waves and out-of-plane transverse waves require out-of-plane stresses, and they will not propagate independently in a plate. They have to be combined to build harmonic plate waves. In a thin plate this results in the simple, nondispersive, symmetrical plate waves and dispersive antisymmetrical plate waves that I discussed above. However, at higher frequencies it is not so easy. There are modes of harmonic propagation, and as the frequency increases so does the number of modes. These modes are traveling waves in plane, but they have different standing wave patterns out of plane. They are dispersive; the phase velocity changes with frequency. This time, dispersion is not due to loss processes; it is a result of the frequency dependence of the mixture of the elastic bulk waves that meet the plate boundary conditions. The plate modes are neatly divided into symmetric and antisymmetric modes depending on the symmetry of in-plane motion about the plate center. At low frequency the zeroth-order symmetric mode (S_0) is nondispersive and identical to longitudinal thin-plate wave, and the zeroth-order antisymmetric wave (A_0) is identical to the flexual wave. High frequency plate waves play a role in paper diagnostics. I will briefly introduce measurement technique in Section VII; however, I do not undertake a detailed exposition. If you are interested, there are discussions available of general isotropic plate waves [30,91], general anisotropic plate waves [3,83], plate waves in isotropic paper [56], and plate waves in orthotropic paper [31]. Otherwise, for in-plane analyses, be resolved to play within the confines of Eq. (35).

V. IN-PLANE MEASUREMENT TECHNIQUES

For this discussion, the purpose of wave propagation examinations in paper is the calculation of linear mechanical constitutive parameters. Ultrasonic determinations complement low-strain stiffnesses from other techniques (load elongation, cyclical straining, vibrating reed, torsion pendulum). Like all the rest, they have advantages, shortcomings, and peculiarities. In selling ultrasonic characterization, one should first point out that (unlike the others) it is performed nondestructively on sheet-sized samples. Also, it is by far the most rapid technique. It is practical to scan the sample making repeat measuerments and establishing averages and deviations. These are important distinctions, but the shining point of ultrasonic testing is the completeness of the stiffness determinations. Other methods provide one or two stiffness coefficients. Ultrasound routinely gives all of the in-plane parameters (normal stiffnesses, shear stiffnesses, Poisson coefficients, and even shear-coupling coefficients) as well as the out-of-plane normal stiffness and the two out-of-plane shear coefficients. The big downside is that ultrasound in paper is valid over a high and narrow frequency range. In-plane work can be done only around 60 kHz, whereas out-of-plane numbers are possible only near 1 MHz. These are well above the effective frequencies of the other methods. Because paper is viscoelastic and there is less time for viscoelastic relaxations at high frequency, paper is stiffer under the ultrasonic straining. Paper scientists desire stiffness coefficients to assess compliance under commonplace, practical stress–strain applications. Almost always, conditions of interest have longer time frames than the ultrasonic tests; therefore, they elicit a more compliant material. A sufficient characterization of a viscoelastic response requires measurement over all operative frequencies; the ultrasonic stiffnesses are at the very high end of practical concern.

Now, in light of the fundamentals already presented, I would like to discuss the details of present ultrasonic techniques. I begin with the velocity measurements of nondispersive symmetric waves through thin plates. In later sections, I continue with thin-plate flexural waves, dispersive high frequency plate waves, and out-of-plane bulk waves.

A. Transduction

Perhaps it is not an exaggeration to claim that transducers are the critical elements of every mechanical measurement. Proper transduction surely is paramount in diagnostic ultrasonics that use piezoelectric transducers. Almost always, an ultrasonic experiment begins and ends with an electrical signal. For the electrical signals to be indicative of mechanical properties, there must be agents to transfer an electrical signal to mechanical motion and back again. These agents are the transducers. Often they rely on the piezoelectric properties of their active components. In piezoelectric materials, electric fields induce small molecular level rearrangements and small concomitant overall changes in shape. Likewise, when a piezoelectric material is strained, the molecular charge distribution is disturbed and small electric fields arise. Efficient, sensitive utlrasonic measurements are achieved by using materials with large piezoelectric coefficients. These couple mechanical and electric energy sufficiently well to permit electrical stimulation and detection of mechanical disturbances.

Quartz crystals and Rochelle salt crystals were the early choices for active elements in ultrasonic transducers. Quartz is very stable and durable, but its piezoelectric parameters are relatively small. Rochelle salt crystals have large coupling coefficients, but they are piezoelectric over only a narrow temperature range. The widespread introduction of ferroelectric ceramics in the 1950s expanded the selection of piezoelectric materials. A ferroelectric ceramic is poled and made piezoelectric by maintaining an electric field as it is cooled through its Curie temperature. The ceramic can be fired in an arbitrary shape and poled at any direction. Various forms of lead zirconate titanate (PZT) and barium titanate are common ferroelectric ceramics used in transducer construction. Today, ceramics dominate the market because they are rugged, sensitive, inexpensive, and versatile.

Frequency response is a major consideration in transducer construction. The preferred behavior depends on the type of experiment. Flat response versus frequency curves are best for pulsed time-of-flight measurements. Well-defined pulses are composed of a wide range of Fourier components. Thus, to faithfully reproduce a pulse, one wants sensitivity to be unaffected by frequency. Pulsed time-of-flight transducers should be "broadband." On the other hand, continuous wave (cw) experiments are driven at a single frequency. Anomalous responses at other frequencies are undesired. A sharp, resonant, behavior is ideal. Swept wave techniques present intermediate specifications. The response should be flat over the sweep band, but the high frequency response needed to reproduce sharp-edge pulses is unnecessary. Generally, there are trade-offs between sensitivity and bandwidth. Transducers often resonate. They are particularly responsive over narrow frequency bands. A transducer's sensitivity at its center resonance frequency is inversely related to resonance peak width. Bandwidth can be increased by mechanical or electrical damping; however, damping degrades the signal. Operation far from all resonance peaks improves bandwidth, but a transducer that doesn't ring could yield poor signal-to-noise ratio.

Transducer-to-sample mechanical coupling is another important issue. A transducer is more than a bridge between electrical and mechanical energy. Mechanical action in the transducer must be transferred into and out of the sample. When a traveling wave is incident on an interface between objects, some energy is transferred and some is reflected. One should design the transducers to minimize reflections at the transducer/sample boundary. In our case, transduction will be specialized to accommodate the distinctive nature of paper. This will require proper design of the transducers and of the coupling between the transducers and the paper.

We perform ultrasonic experiments on paper to determine linear constitutive coefficients. These are related to the parameters of wave propagation. However, there are different modes of propagation with different relations to constitutive coefficients. We are advised to generate a simple, pure disturbance if we expect to decode constitutive information from an ultrasonic signal. For example, if we want to determine the in-plane shear modulus of paper, we had best excite an in-plane transverse wave in the sample. Transverse wave propagation is directly related to the shear modulus. Any anomalous longitudinal component of the signal will taint the measurement. The transducer constructions and sample couplings must be selected to disturb and sense transverse motion. In general, the experimental design must accommodate the particular constitutive relation of interest.

So: Transducers should be sensitive; they should be rugged; their frequency response must be appropriate for the measurement; they have to be mechanically matched to the sample; and they need to couple predominantly a pure, well-defined motion into the sample. Let's see how this all shakes out in the case of time-of-flight measurements on thin-plate waves in paper. I begin by establishing the desired frequency characteristics. For the sake of improved time resolution, I would like to operate at high frequency. Sharp pulse fronts are needed to make accurate time interval measurements, and sharp fronts require high frequency Fourier components. Unfortunately, there are high frequency limits. First of all, I must keep the thin-plate condition in force. When the plate is no longer thin, there are extra modes of propagation and the modes are dispersive (phase velocity depends on frequency). Pulse shapes cannot be maintained, and flight times are not easily deduced. Rigorously, all samples should be very thin compared to the out-of-plane wavelength of sound. In practice, if the thickness is less than one-quarter wavelength, there is little dispersion, and no higher order modes are encountered. Wavelength is equal to velocity divided by frequency; therefore, the maximum allowed frequency component is $f_{max} = c_{min}/4t_{max}$, where c_{min} is the slowest out-of-plane velocity and t_{max} is the maximum sample thickness. Paper has an unusually high degree of in-plane to out-of-plane anisotropy. In comparison to other solid media, the out-of-plane stiffnesses of paper are very small. To one not initiated in the peculiarities of paper, c_{min} is unexpectedly small, and dispersive plate wave action initiates at a surprisingly low frequency. A typical out-of-plane longitudinal bulk wave in paper has a velocity of only about 300 m/s. The mechanical properties of paperboard grades are particularly important, and I would like to include paperboard in the testing range. Accordingly, I take 0.75 mm as the maximum sample caliper. Therefore, I should keep my frequency response below about 100 kHz. For thinner samples, I might move to higher frequencies. However, plate wave attenuation increases rapidly with frequency in paper. For narrow pulses the waveshape becomes dependent on transducer spacing, and time-of-flight determination becomes precarious. So, at high frequency, there are practical, as well as fundamental, limitations. On the low frequency end, I must stay away from near-field problems. (I will talk more about the near field later.) This means that the transducer spacing should be greater than about one-half the wavelength of the fastest in-plane wave. I would like to do my test on normal size sheets (around 20 cm square). So transducer separations will be about 10 cm. The fastest in-plane disturbance, the longitudinal wave in the machine direction, has a velocity of about 3000 m/s. Thus, the minimum acceptable frequency component is at 15 kHz. An in-plane instrument for time-of-flight measurement on nondispersive thin-plate waves over the full range of papers and paperboards is confined to a narrow frequency range. Giving myself a small safety factor, I restrict such an instrument to the 20–80 kHz frequency band. Undesirable out-of-band responses should be electronically filtered away. The transducers are expected to have only a strong, flat response over the band.

A straightforward transduction scheme is to apply, through a face plate, a piezoelectric ceramic directly to the surface of the paper sample. However, with one exception [69,70], laboratory experimenters nowadays eschew this approach because of its inherent shortcomings. Consider a ceramic-based transducer whose active face is in contact with the surface of a paper sample. The transducer is polarized so that the main motion is out of plane with respect to the paper. There will be coupling of mechanical energy between the transducer and thin-plate waves in the sample; however, the coupling is inefficient and impure. The dominant face-plate motion is not conducive to exciting transverse or longitudinal sample motion. There is no out-of-plane motion in a transverse plate wave and very little in a longitudinal one. Also, the transducer-to-sample impedance match is poor. Ferroelectric ceramics have high densities (about 7500 kg/m^3) and high stiffnesses (about 1.2×10^{11} kg/

ms^2). Thus, they are very high intrinsic impedance materials. The out-of-plane stiffness of paper is very small (about 9×10^7 kg/ms^2), and paper density is moderate (about 1000 kg/m^3). Out-of-plane motion in paper is of very low impedance. Transducer and sample are terrifically mismatched, and the interfacial reflection coefficient is large. Remember that force and velocity are continuous across a tight transducer–sample connection. To get good energy transfer, the ratio of force to velocity should be nearly the same in both elements. The ceramic has a large force/velocity ratio, whereas there is much motion for little force in the sample. Regardless of the inefficient matching design, the pounding of a ceramic transmitter will induce some plate wave motion in the sample away from the transducer interface and vice versa. In general, the sample disturbance will be a mixture of longitudinal and shear modes. Transducer geometry, sample characteristics, and coupling load can be adjusted empirically to force one motion to predominate in particular directions, but purity is poor and dependent on coupling and sample vagaries.

I believe that in-plane transduction is improved when the transducers have lower mechanical impedance and the transducer motion is in the plane of the sample. Ceramic bimorph (bender) transducers [36,77] meet these requirements. Benders are made by bonding two thin ceramic piezoelectric plates together. There are two varieties of benders: parallel and series. In the series configuration, the two layers are of opposite polarity and metal electrodes are attached only to the opposite faces of the stack. When a voltage drop is applied across the stack, one layer expands whereas the other contracts. Similar to the response of a thermal bimetal strip, the stack bends, and the motion at the bender edge is perpendicular to the piezoelectric plates. Likewise, when the tip of a clamped bender is displaced normal to the plane, one layer contracts, the other expands, and a voltage develops between the outer electrodes. The two layers in a parallel bender are aligned with the same polarity. A third electrode is sandwiched between the ceramic layers. The outer electrodes are generally grounded, whereas the middle electrode is live. As before, voltage application leads to bending, and bending generates a voltage. The parallel design is functionally superior because the active electrode is electrically shielded by the surrounding ground planes. However, the electrical connection to the center plate of a parallel bender is a manufacturing challenge.

In my preferred in-plane paper transduction embodiment, 0.5 mm thick bimorph plates are scribed into small rectangular plates about 3 mm long and 3 mm wide [36]. One end of a plate is clamped in the transducer housing, leaving a free span that is about 1.5 mm long and 3 mm wide. The free end is rounded for repeatable application to the sample. When administered in this fashion, the bender provides a low impedance coupling for matching to plate waves in paper. In a bending mode, the interfacial force/velocity ratio is reduced by the leveraging action of the cantilevered plate. Also, the bending motion is in the plane of the sample and conducive to symmetric plate wave generation. Longitudinal waves radiate predominantly along a line that is perpendicular to the excited bender face, whereas transverse waves propagate mainly parallel to the plane of the transducer. The same transducers can be used for longitudinal and transverse measurement. Transducers mounted on the sample surface and aligned face to face are suitable for longitudinal plate wave inspection. Rotating each transducer 90° about its center axis perpendicular to the sample surface orients the benders along the line of transducer separation and positions the benders for transverse wave detection.

We can classify the low frequency vibrational modes of a thin cantilevered beam as bends, twists, or wobbles. For broadband response the transducer plates are cut small, and the resonance frequencies of all vibrational modes are above 100 kHz. The transducers are particularly effective at generating, detecting, and coupling bending motions; however, unwanted secondary twists and wobbles can complicate the simplicity of the first-order bending motions and result in the mixing of transverse and longitudinal plate wave motion in the sample. The rounding off of the business end of the bender plate reduces the influence of the twisting motions. These are pronounced at the sides of the beam, and the beam sides are not in contact with the sample. The relatively large width/length ratio of the flat beam was chosen to push the resonance frequency of lowest order wobble far above the 20–80 kHz band and thereby minimize wobble contamination. Properly constructed transducers will give much stronger signals along a principal direction of an orthotropic plate when the receiver and transmitter are polarized in the same direction (both for longitudinal or both for transverse) than when they are cross-polarized.

Non-contact laser-based techniques for paper ultrasonic diagnosis are in their early stages of development. They have inherent advantages over the contact piezoelectric methods just discussed and will, in time, become prominent. It was established long ago that pulsed lasers generate mechanical shock waves in paper [53,74]. Then as now, the pulse energy is in the tens of millijoules range, the pulse width is less than a 1.0 μs, and the spot size is under 0.5 mm. When a paper surface is startled by a jolt of laser energy, there is extreme local trauma. Out-of-plane momentum transfer, heat generation, and molecular level disruptions roil the targeted speck of paper. As a side effect, mechanical shock waves, which can be detected by piezoelectric means, emanate from the point of encounter. Laser pulses generate particularly strong signals when operated in the "ablation" mode, and paper is vaporized locally, producing pits about 100 μm in diameter. The damage is superficial but there are cosmetic consequences, and the method is no longer, strictly speaking, nondestructive. A "thermoelastic" excitation regime exists at lower power and/or larger spot size. Here, there is no visible damage to the sample; however, signal strengths are much attenuated. Optimizing laser pulses in terms of energy, temporal pulse width, wavelength, and excitation geometry to generate maximum signal with minimum sample impact provides an ongoing challenge [44,45].

Laser detection schemes are also workable, allowing one to make fully non-contact measurements. In fact, there are many ways for laser interferometers to sense ultrasonic displacements on smooth, regular surfaces [65,79]. The rough, diffusely scattering characteristics of paper cause special problems, but some methods have been adapted for paper wave detection. Generally, they monitor the influence of sample motion on the interference between coherent laser beams whose phases are influenced differently by sample motion. In straightforward applications, laser interferometers are most sensitive to out-of-plane motion. As discussed later, they are best suited for detection of the flexural motion of the low frequency A_0 mode. However, they can also be used for in-plane sensing. Compared to piezoelectric methods, laser ultrasound excitation is very broadband. When both laser excitation and reception are employed, the pulse is highly localized in time and space. Full pulse width for the low frequency S_0 mode are on the order of a microsecond, whereas they can extend to over 100 μs in purely piezoelectric approaches. This is a huge advantage for laser ultrasonics in time-of-flight work where time resolution is inherently linked to pulse width.

The first laser-based detection of S_0 waves in paper samples was reported by Johnson et al. [44,45]. They used a heterodyne Mach–Zehnder interferometer. The output of a continuous wave argon laser was split into two paths. One, the reference beam, was diffracted through a Bragg cell, giving it a 40 MHz frequency shift. For out-of-plane work, the other, the signal beam, was focused normally to a spot on the sample displaced from the excitation point. The light scattered from the sample was gathered and mixed with the reference beam on the face of a photodetector. Without sample motion, an interference fringe pattern oscillated on the face of the detector at 40 MHz. After the sample was strobed by a separate, pulsed laser, a packet of motion passed by the continuously illuminated spot. While the spot is in motion, the scattered light is Doppler phase shifted by an amount proportional to the out-of-plane velocity of the spot, and the detector signal is frequency modulated. An FM discriminator examines the RF signal and produces a signal proportional to the velocity of the sample. For in-plane detection, both beams are focused obliquely, at opposite angles, to the face of the sample. This time, the radio frequency fringe pattern appears on the sample. The mixed radiation, scattered from the spot on the sample, is focused to the face of the photodetector. The Doppler effect from out-of-plane motion frequency shifts the scattered radiation from both illumination components equally, and there is no net shift in the frequency of oscillation of the fringe pattern. The in-plane velocity component, however, shifts the frequencies of the two components of scattered radiation in opposite directions and is manifested as a frequency modulation of the detector signal.

Confocal Fabry–Perot self-referencing interferometry has been demonstrated for ultrasound detection on paper by Walker et al. [93]. This is primarily suitable for detection of the out-of-plane motion, but some weak S_0 signals have been recorded. A Fabry–Perot interferometer is a pair of parallel glass plates with highly reflective coatings on their adjacent faces. The coated surfaces and the intervening space comprise a high Q optical cavity. Light, incident to one of the glass plates, will pass through the cavity only if it is very near a resonance frequency of the cavity. The interferometer functions as a narrowband spectrometer. For ultrasound detection, a coherent laser source is reflected off the surface of the sample and directed to the interferometer. The glass plate separation is adjusted slightly off the transmission peak for the laser source. This makes the output of the interferometer very sensitive to the laser frequency. When a mechanical pulse passes by the laser spot on the sample, the reflected light is Doppler frequency shifted, modulating the response of a photodetector at the exit of the interferometer.

Photorefractive methods [12,96] have also been applied to paper. Again, they work better for flexural waves, but longitudinal S_0 motion has been sensed by adjusting the orientation of the optics to the sample. The index of refraction of a photorefractive material depends on light intensity. When two coherent light beams pass through a photorefractive crystal at different angles, they establish an interference pattern. The index of refraction of the crystal responds, forming an optical grating corresponding to the interference pattern. Because the grating is established by the two incoming beams, it is efficient at transferring light from one beam to the other. The two beams are said to be mixed in the photorefractive crystal. For ultrasonic detection purposes, the signal beam is scattered from the reception point on the sample, whereas the reference beam is directed straight to the crystal. Oscillation of the sample causes phase shifts in the signal beam, changing diffraction coefficients. Longitudinal waves in paper have been detected by monitoring intensity changes in the transmitted signal beam [49] and by monitoring the voltage changes across the crystal resulting from movements in the interfernece pattern [50].

B. Time-of-Flight Techniques

For a time-of-flight measurement with piezoelectric transducers, the paper sample is placed on a substrate and transducers are administered to the paper surface. One or more transducers (transmitters) are excited with an electrical pulse. This generates a shock wave that is electrically detected by one or more other transducers (receivers). As discussed below, the time of flight is determined in various ways from the differences in signals at different transmitter–receiver separations. One is interested in characteristics of the sample, not the substrate. There must be minimal wave propagation induced in the substrate. Since the days of Morgan, foam rubber [77] and soft neoprene rubber [87] have been the substrates of choice. As a demonstration that the substrate is nonparticipatory, you can adjust a pair of benders to get a strong pulse through a paper sample. If you then remove the sample and place the transducer directly on the rubber substrate, the pulse will completely disappear. For time-of-flight measurements of longitudinal waves with the laser interferometer apparatus [44,45], the sample is suspended and there are no substrate issues. Signals are recorded with different transmitter placements and compared to calculate a time of flight.

I prefer to classify the different time-of-flight techniques according to their velocity calculation processes. In my lexicon, the basic approach is the "Hamburger" method. It requires the simplest, most economical means; it has by far the shortest measurement cycle; and it outputs the least reliable velocities. The only readily available commercial instruments [70,88] use Hamburger techniques because the market puts premiums on price and quick results. Hamburger [39] introduced the single-spacing, threshold-detection, time-of-flight technique for paper and polymer films. Two transducers contact the sample. A clock is triggered at the same time a pulse excites one of the transducers. The clock stops when the signal at the second transducer exceeds a voltage threshold. Ostensibly, extraneous time delays attributed to electronics, transducers, and time to reach threshold can be calibrated out with a standard. A pulse through a standard is timed; a true time of flight is calculated from the transducer separation and the known velocity of sound propagation in the standard; and the true time is subtracted from the measured time to establish a time correction. The velocity of sound in test samples is estimated by dividing the transducer separation by the clocked time less the time correction.

As a consequence of near field effects, the phase velocity does not equal the square root of the mass specific stiffness near a point source transmitter. If the receiver is well out of the near field range, the near-source propagation craziness is calibrated out. However, if $(c/r\omega)^2$ is not small, the near field perturbation will depend on the velocity, and it will be different for the sample and the standard. When the separation equals the wavelength of the radiation, $(c/r\omega)^2$ is about 0.01. Therefore, for paper velocity measurements, one wavelength is a fair estimate of the practical extent of the near field. Assuming a paper longitudinal velocity of 3000 m/s,

Hamburger transducer separations should be greater than 15 cm for 20 kHz signals and greater than 4 cm for 80 kHz signals. There is a real advantage in going to higher frequencies to reduce near field errors.

The ultrasonic modeling of a transducer loaded on a planar sample has reactive (imaginary) components as well as real parts. In this case, the reactive terms come from such things as elastic restorative forces between transducer, sample, and substrate and from the near field response of the sample. If all the impedance terms were real, the wave transmitted into the sample would be in phase with the incident wave. Reactive impedance terms cause a phase shift in the transmitted signal. The shift is frequency-dependent, and pulse shapes are altered in transfer from transducer to sample and from sample to transducer. The effective impedance match varies with the loading of the transducer to the sample, with transducer construction, with the basis weight of the sample, with the density of the sample, with all the elastic coefficients of the sample, and with orientation of the sample to the polarization of the transducer. The pulse shape can be different for every sample, at every transducer loading, in every orientation. This impedance uncertainty cannot be calibrated out with a single time correction from a standard sample. Hamburger errors on the order of 10% due to sample–standard impedance differences are often encountered.

The definition and subsequent determination of the arrival time of a pulse is always a ticklish issue. The Hamburger design makes it particularly troublesome. Here, one establishes an arbitrary voltage threshold that is just comfortably above the noise level of the receiver. The time-of-flight measurement starts when the transmitter is triggered and stops when the receiver signal exceeds this threshold. The time of flight, so defined, is measured for a standard of known propagation velocity. In order to reproduce the correct velocity for the standard, a "dead time" is subtracted from the time of flight measured on the standard. As a calibration procedure, this same dead time is subtracted from the times of flight of samples tested at the same transducer spacing. As discussed above, the pulse shape is inconstant and the threshold error cannot be entirely calibrated away in this manner. But even if the pulse shapes were identical for the standard and all samples, there would still be problems with the Hamburger approach. Differences in signal strength would cause different pulse features (at the same absolute threshold) to terminate the time-of-flight counter. At first it seems that one could minimize this problem by increasing the signal-to-noise ratio and catching the pulse as soon as possible. However, the front end of a propagated wave pulse does not resemble an abrupt transition. Surprisingly, the slope of a rising pulse becomes more gradual as you progress to lower thresholds at earlier times. You can actually compound the threshold error if you try to catch the pulse too soon. Brillouin [19] discusses this precursor phenomenon in the analogous case of electromagnetic pulse propagation.

I call threshold time determinations with two (or more) transducer separations Morgan measurements [24,77]. Calibration standards are not necessary. Threshold times of flight are recorded for one separation of the transmitter and receiver; the transducer separation is changed; and another time of flight is taken. The velocity is calculated as the difference in separation divided by the difference in time of flight. Extraneous time delays cancel in the time-of-flight subtraction. If the transducers are applied to the sample in exactly the same way at both separations, the sample–transducer couplings are constant. Mounting the probes vertically in linear bearings and dead weight loading them is a good way to maintain repeatable coupling [36].

Now the pulse distortion due to coupling is the same near and far, and one of the Hamburger shortcomings is overcome. The signal, however, is weaker at the far spacing, and a common threshold level appears farther into the far pulse. Errors due to the dependence of threshold times of flight on signal strength remain. There is also a three-transducer version of the Morgan method. Here, time-consuming probe manipulation is avoided by having either two transmitters or two receivers at different sample locations. This adaptation makes for easier and faster measurement. However, it is difficult to establish and maintain identical transducers and transducer-to-sample couplings, and the three-transducer Morgan methods are more error-prone.

Resorting to more sophisticated signal analysis procedures, Chaikin and Chamberlain (CC) methods alleviate the Morgan threshold identification headache. This time, the shapes of the received signals are recorded at different separations, and the time difference measurement is made relative to common, distinctive features in the recorded pulses [23,25,36,90]. With the CC method, one is looking at the entire front end of the pulses and lining them up by structure. Since reflections from the edges of the sample eventually mix with the straight-through signal, the analysis must stick to the beginning of the pulse. A CC practitioner must be careful not to place the transducers too close to the sample boundaries and not to analyze too deeply into the pulses. The pulse analysis domain should be considerably shorter than the time required for the quickest wave to go from a transducer to the nearest edge of the sheet. Once the need to align pulses is accepted, one must decide how to do the matching. With a dispersive medium, this is not a clear-cut decision. There are distinct approaches for extracting velocities from the shape of pulses propagating in dispersive media [2,13,14,17]. Bloch [11] discusses the various alternatives and makes a spirited defense of the cross-correlation technique. The cross-correlation method, with necessary modification, is commonly used in CC paper testers [90]. Here, to get a finite pulse at predominantly one frequency, the transmitter is excited with a single-cycle sine wave. The near and far receiver signals are digitally recorded. The near signal is terminated at the first zero crossing after pulse detection. The far signal is displaced in time relative to the near signal, and the two signals are multiplied together, point by point. This process is repeated for a series of near to far signal offsets. The time-of-flight difference between the two signals is defined as the offset time that generates a maximum in the point-by-point multiplication. In this way, only the first half-wavelength front ends of the signals are considered, and the time-of-flight determination without prejudice considers all of the signal in the front end. The interferometer also does a cross-correlation determination for thin-plate longitudinal waves [44,45]. This broadband technique has the advantage of producing a much shorter pulse. Since any edge reflections would arrive long after the straight-shot pulse is complete, the full bursts can be used for the cross-correlation calculation. This is effectively a higher frequency technique; the peak Fourier components of the pulses are in the megahertz range. Thus, the longitudinal mode becomes dispersive at a lower sample caliper threshold. The simple relationship between stiffness and velocity is lost on thick samples.

Near field worries persist in Morgan and CC determinations. With reasonable transducer separations, the very near field phenomena cancel in the time subtraction, and Eq. (28) can be manipulated to provide a first-order estimate of the near field generated error. If the two transducer separations are represented by r_1 and r_2 and k_0

signifies the far field wavenumber (ω/c), the relative near field error in velocity for a two-spacing measurement is

$$\mathrm{Err} = 1 - (r_2 - r_1)^{-1} \int_{r_1}^{r_2} \left(1 - \frac{3}{4k_0^2 r^2}\right)^{1/2} dr \tag{36}$$

For a longitudinal wave with a phase velocity of 3000 m/s at a frequency of 80 kHz and transducer separations of 3.0 and 6.5 cm, the relative error is about 0.7%. If the frequency drops to 20 kHz, the error balloons to 12% at the same separations.

C. Time-of-Flight Transverse Velocity Determinations

The transverse velocity is the same in both principal directions of a thin orthotropic plate. At first it seems that transverse measurements in the two directions would have equal validity. However, the measurement along the minor principal axis (the one with the smaller longitudinal velocity) is preferred. Let me explain. No matter how well one constructs transducers, there will be a mixture of longitudinal and transverse components in every received signal. For longitudinal determinations, this is less consternating. In time-of-flight measurements, only the front end of the pulse is of interest. The longitudinal wave is faster than the transverse one. For a longitudinal analysis, little of the transverse contamination has arrived by the time the front end of the pulse is complete. For the transverse signal, the reverse is true; the longitudinal portion is up and ringing before the transverse front end is done. The longitudinal wave is slower along the minor principal axis; thus it will perturb the transverse wave less in that direction. But that is not the only reason to do the transverse work along the minor axis. There is a better impedance match between bender transducers and paper for motions along the stiffest paper direction. That is, longitudinal energy is coupled best for motion along the major principal axis, and transverse energy is coupled best for motion along the minor principal axis. The major longitudinal signal is stronger than the minor longitudinal one, whereas the minor transverse signal is stronger than the major transverse one. Therefore, the ratio of the desired transverse signal to the unwanted longitudinal one is much better along the minor axis.

Comparing the transverse velocities in the two principal axes is a good way to assess the transducer quality in terms of modal purity. The test is particularly severe if highly anisotropic paper samples are examined. For paper sheets with principal axes stiffness ratios of around 4, the original Morgan transducers [77] and the new shorter, wider transducers [35] give nearly equal tranverse velocities along the minor axis. However, the difference in minor and major direction transverse velocities is twice as great for the Morgan transducers. I submit this, along with the argument in the above paragraph, as evidence for the superiority of the minor transverse measurement and the better modal purity of the short, wide benders.

D. Off-Axis Velocity Determinations

A thin orthotropic plate has four independent in-plane elastic constants (Q_{11}/ρ, Q_{22}/ρ, Q_{12}/ρ, and Q_{66}/ρ) [see Eq. (5)]. Thus, to complete the stiffness matrix, one is required to perform four independent velocity measurements. The two normal stiffnesses ($Q_{11/\rho}$ and Q_{22}/ρ) come from the longitudinal velocities in the principal directions. The shear coefficient (Q_{66}/ρ) equals the transverse velocity squared in either of the two principal directions. So the simple, on-axis measurements are exhausted, and we are shy one coefficient. Off-axis measurements are needed to calculate a Poisson coefficient and finish the elastic characterization of the plate. Taylor and Craver [87] very early on described a valid method for determining Q_{12}/ρ. They suggested making a polar plot of the quasi-longitudinal wave surface velocity. The quasi-longitudinal plane wave phase velocity at 45° to the principal axis is then found by the Musgrave construction. This is inserted into Eq. (25) with the plus sign, and Eq. (25) is solved for Q_{12}/ρ. The Poisson coefficient can be inferred from off-axis quasi-transverse as well as quasi-longitudinal measurements. The quasi-longitudinal approach has experimental advantage. Off-axis the quasi-transverse and quasi-longitudinal motions are neither exactly perpendicular with nor parallel to the direction of propagation. Therefore, it is difficult to line up bender transducers properly for best modal purity. Signals taken off-axis generally have more of the unwanted mode. This is a larger problem for the quasi-transverse work as the quasi-longitudinal wave is the faster moving disturbance. In ultrasonics, as with other methods, accurate Poisson determinations are difficult. The coefficient is calculated from differences in velocities, and uncertainties in velocities are magnified in the Poisson calculation [6,22,61].

The first actual reports of ultrasonic Poisson's coefficients came from the IPC group [6,40]. This time, the quasi-transverse velocity was measured off-axis at 45°. This quantity was inserted into Eq. (25) with the minus sign, and Q_{12}/ρ was calculated. Remember that since the quasi-transverse polar plot is nearly circular there is no significant difference between the wave surface velocity and the plane wave velocity at the same angle. Thus, the Musgrave construction is not necessary for transverse determination of the Poisson coefficient. On normal machine-made papers, the longitudinal and transverse techniques give the same result, within experimental error. However, if one naively asumes that the quasi-longitudinal wave surface velocity at 45° equals the 45° quasi-longitudinal plane wave velocity, very different Poisson ratios are found. For anisotropic samples, the Musgrave construction is mandatory for quasi-longitudinal Poisson measurements.

A later IPC methd [35] reverted to the quasi-longitudinal measurements. This time, the full quasi-longitudinal wave surface was measured. The Musgrave construction transformed the wave surface into a 360° plane wave velocity polar plot. The shear stiffness was determined by the transverse velocity along the minor principal axis. The rest of the stiffnesses were optimized to best reproduce the experimental plane wave polar plot with Eq. (25). If the sample is orthotropic, the velocity polar plots will have two perpendicular axes of reflectional symmetry. Otherwise, nonzero shear coupling coefficients (Q_{16}/ρ and Q_{26}/ρ) will be needed to characterize the elastic response of the plate. The above optimization was generalized to include shear coupling coefficients. For all paper samples studied (including some with fiber orientation and drying stress intentionally oblique), the shear coupling coefficients were not significant. That is, the paper samples were orthotropic within experimental error. On the other hand, some oriented polymer films were found to be clearly nonorthotropic.

E. Strip Waves

Ultrasound diagnostic techniques are neatly divided into three categories according to the form of the excitation. They are driven by either a continuous wave (cw), swept wave, or pulse wave generator. Time-of-flight measurements are the preferred methods for everyday paper characterization. These employ pulsed excitations, and pulsed methods therefore receive the brunt of my attention. However, cw and swept wave experiments have been practiced on paper, they have unique merit, and they also deserve attention. Their neglect arises, not from measurement deficiencies, but as a consequence of experimenter impatience. They are time-consuming and often tedious. A major appeal of ultrasound is its ease and rapididity of measurement. It is disconcerting, once drawn to ultrasound, to be asked to forgo the incentive for conversion.

For cw and swept wave work, the signal excitation and detection times are long compared to the time required for the transit of sound across the sample. There is a large impedance mismatch at the sample edges and a concomitant large reflection coefficient. The disturbance at the rceiver can be a complex interference of the signal directly from the transmitter and the signals reflected off the sample edges and between transducers. To rationalize the signal and to relate it to sample properties, all signal paths must be considered. To make this a tractable undertaking, the boundary conditions should be simple and well defined. If a two-dimensional sheet-like sample is examined, there will be many paths for transverse and longitudinal waves contributing to the received signal. The output will be a complicated function of the sample's mechanical properties and of sample size and shape. It will be nearly impossible to decode the constitutive parameters from the jumble. There are two ways to maneuver around this difficulty. One is to make the sample so large that reflections are greatly attenuated before they return to the area of investigation. The high frequency plate wave technique of Luukkala et al. [56], which I discuss in Section VII, takes advantage of this simplification.

The other approach is to analyze a strip of width much less than one wavelength. This gets rid of transverse waves and leads to great simplifications in the boundary conditions. One needs to account only for reflections at the strip ends and at the transducer contact points. Propagation is in one dimension, and the standing wave pattern in any free section is a simple combination of longitudinal traveling waves in opposite directions. Consider the old-time cw work of Ballou et al. [4,5]. Here the end of a long strip is attached to a cw-driven transducer. Another transducer contacts the strip at variable distances from the transmitter. The strip is cut long enough that reflections from the free end are strongly attenuated at the receiver. One monitors the receiver signal as a function of transducer separation and records the standing wave pattern. Because only reflections at the transducers need be considered, phase velocity and attenuation coefficient will follow from a straightforward analysis of the standing wave pattern.

The real and imaginary normal stiffness coefficients have also been determined in a swept wave experiment [75]. Here the ends of a short strip are bonded to separate broadband transducers. One transducer is driven with a frequency-swept, sinusoidal electrical signal, while the resulting signal amplitude of the other is monitored as a function of drive frequency. A resonance peak is observed in the output when the strip is an integral number of half-wavelengths long. Phase velocity is determined from the resonance frequency and the strip length, whereas measurement of the width-to-height ratio of the resonance peak provides the loss tangent.

The simplicity and purity of longitudinal wave propagation in strips allows the determination of the loss coefficients. The more convenient sheet time-of-flight methods are developed only for velocity measurement. The major attraction of strip work is the opportunity to make good loss determinations.

VI. FLEXURAL WAVE MEASUREMENT TECHNIQUES

Piezoelectric bender transducers are unsuited for flexural wave detection. Low frequency flexural waves are slow-traveling out-of-plane motions. The benders are insensitive to out-of-plane motion, and any flexural disturbance arrives long after reflected longitudinal and transverse signals begin to influence the received signal. Laser techniques with their bias for out-of-plane motion are more appropriate. Laser blasters naturally impart more out-of-plane disturbances, and laser interferometers are suited to detect out-of-plane motion. The laser techniques described in Section V all work well for flexural waves in paper. Mass specific bending stiffness is determined by recording signals at two different source-to-receiver displacements, doing a Fourier analysis of both signals, and finding phase velocity as a function of frequency from the phase differences of the Fourier components [44,45].

In addition to single-point detections, full-field imaging methods are applied to paper. In one embodiment, a series of double-exposure holograms [73] are transformed by computer manipulation into a plot of the overall deformation of a thin plate a fixed time delay after a pulsed laser excitation. The experiment begins with a pulsed laser excitation. A portion of the laser pulse is diverted from the excitation route and expanded into two broad flash beams. One beam illuminates the back side of the paper, whereas the other shines directly on a holographic film. The pulse is so short ($\approx$25 ns) that there is practically no paper movement over its duration. Thus, the interference of the reference beam and the light reflected from the sample surface produce a hologram of the undeformed sheet. A fixed time later (1–800 μs), a second pulse is triggered. A rotating mirror slightly shifts the angle of the reference beam to the holographic plate, and a second interference pattern is recorded on the film. Sufficient time has elapsed for the first laser blast to perturb the sample, and a double-exposure hologram of the paper, undeformed and deformed, develops. The next step is to decode the paper disturbance at the time of the second pulse from the hologram. Two reference beams, at angles equal to those in the exposure process, illuminate the hologram, producing an interferogram of the sample in the two states. The image is recorded by a CCD camera. This is repeated a series of times as the phase of one of the reference beams is shifted. From the recorded images, the computer calculates the interference phase of the two exposures at each location of the image. Finally, the wavelength of the laser pulse is inserted into an unwrapping algorithm to recreate the surface of the deflected plate. The deformation map is then compared to the solution of the flexural wave equation forced by a delta function excitation. Finally, the flexural rigidity is determined from the comparison without intermediary phase velocity calculations.

Full-field images [37] are also made of oscillating paper samples [26] in real time. A mechanical oscillator is applied at a point near the middle of an extended paper sample. It is excited to ring continuously at a steady frequency ω_1. The output of a continuous wave laser is slit into two paths. The signal beam is expanded and reflected off the area of the vibrating sample to be imaged. The reference beam is frequency shifted by ω_2 and expanded. The beams are directed to a photorefractive crystal, producing a hologram of the vibrating sheet and two-wave mixing the beams. A polarizer on the exit side of the signal beam selects the component of the reference beam diffracted along the signal beam path. This is imaged on the face of a CCD camera, producing a recording of the paper motion. The intensity of light imaged at a given spot on the camera depends on the phase difference between oscillation at the corresponding point on the sample and the phase of the reference beam. Therefore, the image displayed beats at a frequency equal to $\omega_2 - \omega_1$. If $\omega_2 - \omega_1$ is on the order of 1.0 Hz, the progression of traveling waves away from the source is conveniently viewed in slow motion.

Laser methods have also made possible the measurement of large-amplitude tension waves in paper [58]. Output from a low power continuous laser is shone obliquely on the surface of a paper web under tension. The reflected light is focused to the face of a position-dependent light detector. As tension waves push the web up and down, the focused spot translates across the detector, and an electrical signal representative of the out-of-plane motion is recorded.

VII. THICK PLATE WAVE MEASUREMENT TECHNIQUES

I have reserved this section for discussion of high frequency dispersive plate waves. All plate waves are harmonic disturbances that meet the free stres boundary conditions on the sheet surface. At low frequencies, symmetrical mode wave propagation is nondispersive and amenable to time-of-flight examination. At high frequencies or for thick plates [$\omega h \gg (C_{33}/\rho)^{1/2}$], all modes are dispersive and straightforward time-of-flight measurements are inappropriate. In this realm, we can use the resonance technique developed by Luukkala et al. [56]. Here, radiation is transferred into and out of the sample via a pair of air-coupled transducers positioned on opposite sides of the sample. The sample is suspended in an open frame. The transducers are fixed to a common bar with their faces parallel. The transducer bar can rotate with respect to the paper. A variable-frequency generator excites one transducer, and the amplitude of the resulting signal is monitored at the other. The response of the receiver is recorded either as the frequency is swept or as the frame is rotated. Peaks in the receiver output occur when the wavenumber of the airborne radiation as projected onto the paper surface is equal to the wavenumber of a paper plate mode. That is, a peak corresponds to the fulfillment of the following relationship between phase velocities in air and paper:

$$C_p = C_0/\sin\theta \qquad (37)$$

where C_p is the velocity of a plate wave mode at the frequency of operation, C_0 is the velocity of sound in air, and θ is the angle between the sample normal and the line between transducer faces. By noting the frequencies of the peaks at a series of transducer rotations, one can find the phase velocity as a function of frequency for a number of different plate wave modes.

The technique is very successful, but it has limitations. The Luukkala method is ill-equipped for low frequency, symmetrical, thin-plate waves because these produce little out-of-plane motion. Because sin θ is always less than 1, only plate wave modes with velocities greater than sound in air are excited. This precludes the detection of low frequency antisymmetrical plate waves. The Luukkala method is best suited for high frequency, dispersive plate wave modes. Sound in air becomes very lossy above a few hundred kilohertz. Plate waves in paperboard become dispersive above about 150 kHz. The thin-plate condition persists for thinner paper to much higher frequencies. Thus, the method gives good results only on thick papers and only over a narrow frequency range. In addition, the experiment is foiled by minor air currents and is best conducted in a glove box. One increases the signal-to-noise ratio when electret transducers replace the original condenser microphones (especially if the new transducers are scavenged from the range finders of Kodak cameras) and when a lock-in amplifier conditions the receiver signal [28].

Through thin plates, the antisymmetric mode is a flexural wave obeying a low frequency dispersion equation, $c = (Q_{11}h^2\omega^2/12\rho)^{1/4}$. As frequency increases, this mode approaches a Rayleigh wave [30,91]. That is, it exhibits predominantly out-of-plane motion, which is concentrated at the surface of the plate. This is a non-dispersive wave moving with velocity slightly less than that of the out-of-plane bulk shear wave. Using its high frequency capabilities, the laser interferometer [44,45] is successful in detecting antisymmetric waves above the frequency of applicability of the flexural dispersion equation. By extrapolating the measured dispersion curve to higher frequencies, the researchers obtain the velocity of the Rayleigh wave and an estimate of out-of-plane shear stiffness.

VIII. OUT-OF-PLANE MEASUREMENT TECHNIQUES

A. Introduction

In-plane and out-of-plane velocity measurements in paper have little in common. The transduction techniques are entirely different, the frequency ranges are limited by different concerns, and different physical phenomena plague the interpretations. For in-plane deformations there are solid definitions for the stiffnesses. The material is linear over a wide range, and the ambiguity of paper caliper can be finessed with the mass specific stiffness concept. Out of plane, there is no such luck. Consider a sheet of paper sandwiched between two platens. Apply a variable load to the platens and plot the platen separation as a function of pressure. At extremely small loads, the platens make contact at only the outermost fiber excursions. There is very little resistance to deformation, and the ratio of stress to strain is abnormally small. As the load increases, more of the fibrous structure bears the load, and additional paper deformation is more difficult. The slope of the stress–strain curve increases over many orders of magnitude of load. There is no linear region in the stress–strain curve over which tangent stiffness has a singular definition. The low pressure stiffness is surface-dependent and does not account for loading of the entire structure, whereas at high pressure the slope is indicative of cell wall properties independent of the fiber structure and bonding of the sheet. In between, the slope evokes structure, fiber, and bonding properties. However, the tangent stiffness is highly load-dependent, and any standard load definition of stiffness is arbitrary. There is no clear linear regime that characterizes the important features of the out-of-plane mechanical properties of paper.

Please reconsider the same experiment with a time-of-flight ultrasonic measurement added. This time the platens are ultrasonic transducers. At each pressure, one is pulsed, and a time of flight is determined from the signal received at the other transducer. An out-of-plane ultrasonic velocity is calculated as the platen separation divided by the time of flight, and the out-of-plane velocity squared provides a normal mass specific stiffness (C_{33}/ρ). Paper is viscoelastic, and the ultrasonic modulus will be greater than the slope of the stress–strain curve because the ultrasonic experiment has a much shorter time frame. Nonetheless, like the tangent stiffness, the ultrasonic stiffness always increases with pressure, and there is no convenient plateau region for stiffness definition. The mass specific stiffness as a function of pressure would be a proper ultrasonic accounting of out-of-plane sound propagation, but that would require a tedious, time-consuming experiment. Instead, out-of-plane velocity measurements are customarily made at the TAPPI standard load for caliper measurement (50 kPa). This is in the region where most of the sheet is engaged and sheet structure and bonding have influence. It is admissible and useful to make comparisons between sheets, but out-of-plane stiffness, so determined, is not an indicator of paper compliance in a linear regime.

Unlike the in-plane case, caliper measurement is an integral part of the out-of-plane velocity determination. Here, velocity is the caliper divided by the time of flight. All of the machinations of the caliper measurement corrupt out-of-plane velocity calculations. Caliper depends not only on platen pressure but also on platen surface characteristics and the time of load application [82,94]. Softer platens conform better to the surface of paper and give significantly smaller calipers on rough papers than do hard platens under the same load regime. At constant load, the paper yields and platen separation slowly decreases. Platen surface and load time regime must be fixed to get repeatable caliper readings. One can make a feeble attempt to divorce the out-of-plane ultrasonic determination from the caliper measurement. Instead of measuring velocity by dividing caliper by time, you could divide basis weight by time of flight. This would provide the density times a velocity that equals an intrinsic acoustic impedance, e.g., $(\rho C_{33})^{1/2}$. So if you adopted out-of-plane impedance as the prime ultrasonic parameter, you could avoid a direct caliper measurement in the ultrasonic characterizations. However, time of flight also depends on load and platen surface. The exchange of impedance for velocity marginally reduces caliper-generated ambiguity, but one is forced to deal with a more arcane parameter. As a consequence, acoustic impedance is a seldom-quoted quantity.

Regardless of the arbitrary definitions of the out-of-plane parameters, one thing is clear. Because of the preferred in-plane alignment of fibers, paper ultrasonic velocities are much smaller out of plane than in plane. Adhering to any reasonable standard condition, one finds that the out-of-plane longitudinal velocities are a full order of magnitude less than in-plane longitudinal velocities. This disparity translates into two orders of magnitude difference in stiffness. Paper is an unusually mechanically anisotropic medium. Compared with normal solid materials, paper, out of plane, transmits sound very slowly. In fact, the velocity of sound in lightly bonded papers is well below even the velocity of sound in air (330 m/s). This leads to a curious observation when out-of-plane longitudinal velocity is plotted versus loading pressure on very poorly bonded sheets.

Paper is a porous medium. It is a mixture of fiber and air. Wave propagation constitutes a disturbance in the gas phase as well as in the solid phase. The consequence of the gas-phase motion is appreciated by reference to Biot's model of wave propagation in porous media [8–10]. In a two-phase medium, there are two independent longitudinal plane wave modes that propagate at different velocities with different impedances. If, as in the case of paper, there is a large intrinsic impedance mismatch between phases, the phases are decoupled. That is, the energy of one mode is concentrated in one phase, whereas the other mode is concentrated in the other phase. Out of plane in paper, one mode is fiber-dominant and of relatively high impedance: It is basically the vacuum mode of the fiber matrix with some extra loss due to viscous drag of the air. The low impedance mode is envisioned as propagation in the combination of air passages between fibers. In dense sheets it is slow and highly attenuated by the viscous drag along the interfiber pathways. In very low density sheets it travels at near the velocity of sound and is less attenuated. In most cases, the Biot generalization is of no importance to our ultrasonic characterization of paper. Transducers are better matched to the higher impedance solid-phase disturbances, and the matrix is dense enough to rapidly attenuate any air phase motion. However, in very poorly bonded papers the airborne disturbance can be detected through a sheet. Consider a low density sheet suspended between two time-of-flight transducers. Without sample contact, a distinct ultrasonic pulse with nearly the velocity of sound in air is detected. The transducer spacing is slowly decreased. Just after the sample is engaged, the transducers are not mated well to the fiber matrix, and the air phase mode is not overly attenuated. The airborne mode remains dominant. However, as separation decreases further, the matrix becomes better coupled to the transducers and the air phase signal becomes slower and more attenuated. The time-of-flight velocity decreases. Eventually the air phase portion is inconsequential, and the fiber mode dominates the received signal. As pressure further increases, the structure becomes stiffer. Velocity increases. The plot of apparent time-of-flight velocity versus load begins at about 330 m/s, dips to a minimum, and then rises.

There is a common misconception concerning the propagation of ultrasound in paper. Some people imagine a maze of fibers individually channeling radiation through the sheet. In fact, the wavelengths used in diagnostic ultrasound are too large to resolve fiber structure. Wave propagation is insensitive to structural features much smaller than a wavelength. Parameters calculated from ultrasonic experiments represent the overall behavior of the material on wavelength and greater scales. If one does increase frequency to the point that the wavelength approaches the order of the inhomogeneities, the structure begins to scatter radiation. This increases rapidly as wavelength further decreases. Attenuation due to scattering limits plane wave penetration, and diagnostic ultrasonic evaluations become impossible. Scattering imposes a practical high frequency limit. For in-plane studies, this is unimportant. The scale of the microstructure in paper is about 30 μm, and the lowest in-plane velocities encountered are about 1500 m/s. Thus, the maximum available frequency ($f = c/\lambda$) is in the neighborhood of 50 MHz. Remember, time-of-flight work is already confined below 150 kHz by plate wave considerations, so the scattering restriction is purely academic.

For out-of-plane work, we are liberated from the plate wave limitations and near field worries. The lateral dimensions of the transducer faces are much greater than the paper thickness, and plane wave propagation is an excellent approximation. Attenuation due to scattering now imposes the high frequency boundary. Assuming that the lowest velocity encountered will be 250 m/s and allowing the 30 μm structure to be as much as a quarter-wavelength, I expect to experience excessive attenuation with signals above 2 MHz. To finely resolve time of flight, we always want to go to the highest possible frequency. This is especially true out of plane, because the transducer spacing is only one paper thickness and the time of flight is consequently very short. An average writing paper has a caliper of around 100 μm. With an out-of-plane velocity of 300 m/s, the pulse transit time through this sheet is just 0.33 μs. Because of fiber scattering losses, the pulse width cannot be much less than a microsecond. The transit time will be less than the pulse width, making good time-of-flight resolution difficult. To compound the problem, reflections from the transducer/sample interface can interfere with the straight-through signal. Consider the partial signal passing through the sample, reflected at the receiver, reflected back at the transmitter, and transmitted to the receiver on the second try. This arrives and contributes to the time-of-flight signal two sample transit times after the signal of interest. If the time allotted for time-of-flight signal analysis is greater than two transit times, reflected signals will participate. Consequently, out-of-plane velocity measurements on thin papers are inadvisable. This condition ameliorates when the samples are thick papers or paperboards. The nuisance signal, traveling through an extra two thicknesses of sample, occurs later and is weaker than the main pulse. The transit times have increased with respect to the pulse widths, and there is more confidence in the time-of-flight determinations. As will be established quantitatively later, out-of-plane measurements are proper only on paperboards and thick papers. After all, a sample must be homogeneous at some level if we are to assign overall properties to it. We are interested in the bulk stiffnesses, so we use a probe that does not resolve fiber structure. If the sample thickness is not much greater than the fiber width, sheet properties cannot be accurately determined with such a coarse probe.

B. Transduction

Figure 7 is a schematic of transducer application in an out-of-plane velocity measurement. It depicts all of the generic parts for transduction. Individual elements are omitted in some embodiments. A thin sample is sandwiched between transmitter and receiver transducers. A transducer consists of an active piezoelectric element, a ballast, a delay line, and a coupling layer. The piezoelectric elements accomplish the transfer of electrical to mechanical energy in the transmitter and mechanical to electrical energy in the receiver. The ballast is intended to eliminate back-side reflection and increase transducer bandwidth. It is a mechanically lossy material that is impedance matched to the piezoelectric element. Signals propagating into the ballast rapidly attenuate, and little energy returns to the signal path. When the ballast is properly selected and well coupled to the piezoelectric element, along-axis resonances are muted, and interferences from back-side reflections are inconsequential. The purpose of the delay line is to temporally isolate the diagnostic pulse. The transit time through the delay line is long compared to the pulse analysis time, and signals internally reflected in the delay line reach the receiver element long after the straight-through pulse analysis is complete. Also, transverse waves progress more slowly through the delay line than do longitudinal waves. Thus, the transmitter delay line

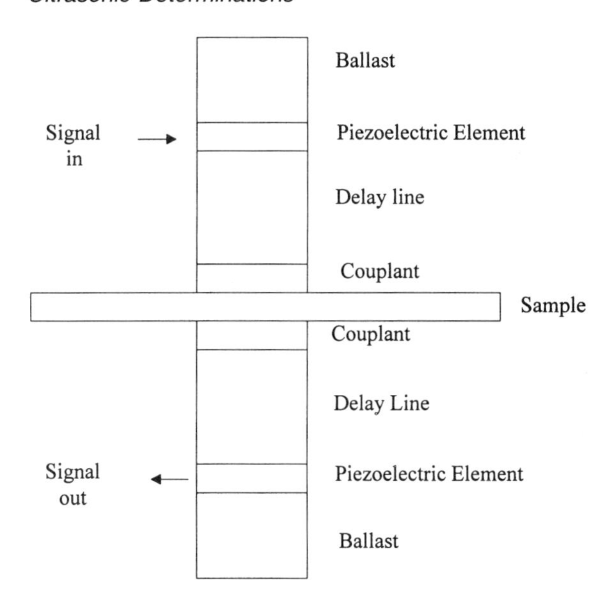

Fig. 7 General schematic of direct contact, out-of-plane transduction.

separates in time off-mode pulses unintentionally generated by the active element. The combination of ballast, active element, and delay line is intended to transmit a mechanical signal that accurately reproduces the electrical pulse stimulation. In the receiver, a mechanical pulse traveling down the delay line will be transferred into a similar electrical pulse with little reverberation or distortion. Another purpose of the delay line is to impedance match the active element to the coupling layer. The impedance of the delay line should lie between that of the active element and that of the coupling layer. The best case is when the impedance of the delay line is the root mean square of the active element impedanace and the couplant impedance. This provides optimum energy transfer between the active element and the coupling layer. Likewise, the couplant impedance should be positioned between the sample impedance and the delay line impedance. The coupling layer has the difficult task of efficiently coupling energy between the delay line and the paper sample.

There are two designs for solid-state longitudinal transducers. The original instrument uses a PZT ceramic disk as the active element [34]. The high impedance ceramic element needs a lossy, high impedance ballast. This is accomplished by mixing powdered tungsten in a silicone potting compound, curing the mixture under pressure to form a pellet, and bonding the pellet to the back-side of the ceramic element. The delay line is a fused silica rod. This is a low-loss medium with a somewhat lesser impedance than the active element. The impedance of the silica delay line [1.3×10^7 kg/(m^2 s)] is still almost two orders of magnitude greater than the impedance of an out-of-plane longitudinal wave in paper [$\sim 3.5 \times 10^5$ kg/(m^2 s)]. For best transfer, the couplant impedance should be around 2×10^6 kg/(m^2 s). The coupling layer is a soft (5–10 durometer) 0.8 mm thick neoprene disk. It has an impedance of around 1.6×10^6 kg/(m^2 s), thereby providing a good match from silica to paper. Rubber, with its very small shear stiffness, is a poor propagator of transverse radiation. Therefore, neoprene layers improve the modal purity of a longitudinal measurement. The main function of the neoprene layer is to mate the transducer to the sample. In general ultrasonic practice, one either bonds the transducer to the sample or intercedes with a fluid coupling layer. Both of these methods are unacceptable for paper. A fluid coupling or an adhesive application would penetrate the porous paper mat, perhaps seep into the fibers, and undoubtedly alter the mechanical properties of the sheet. Without a coupling layer, large reflection coefficients arise from rigid surface contact regardless of impedance matching precautions. The thin neoprene disk conforms to the rough paper surface and accomplishes good ultrasonic mating without contaminating the sample. In addition, the neoprene disk is identical to the soft platen of the IPC soft platen caliper gauge [94]. Thus, the thickness inserted into the velocity calculation complies with a standard caliper measurement.

The introduction of sensitive piezoelectric polymers [51,81,85] precipitated improvements in longitudinal out-of-plane transduction [32]. Polyvinylidene fluoride (PVDF) films, when electrically poled, develop large piezoelectric coefficients. They have much lower impedances than ferroelectric ceramics; their mechanical loss tangents are larger, and they are effective in thin layers. The reduced impedance of the PVDF films narrows the impedance gap between the active element and the sample, thereby opening the door to transducer designs with less overall loss from interfacial reflections. The relatively thick ceramic disks have natural resonances in the megahertz frequency range. The tungsten ballast tones down the ringing; however, interfacial reflections still build up to unwanted resonances in this low-loss medium. The PVDF films are much thinner, and the active element resonances are well above the excitation band. In addition, PVDF has a large mechanical loss tangent, and active element resonances are highly damped. The design of an out-of-plane transducer with PVDF active elements benefits from ballast and the delay line modifications. A machined disk of PVDF is a good ballast choice. This is a high-loss medium that closely impedance matches the PVDF film. The delay line is made of polystyrene because polystyrene is a low-loss plastic whose impedance lies between PVDF and neoprene. The coupling layer is made of the same neoprene rubber as that used with the ceramic-based transducer. However, to accommodate longer pulses without accounting for interference from neoprene interface reflections, it was made from a 3.2 mm thick sheet rather than a 0.8 mm sheet. Brass retaining rings are added to constrain lateral expansion of the thicker neoprene disks and bring caliper readings into agreement with those from the IPC soft platen caliper gauge. The PVDF transducers, so made, are more braodband than their PZT counterparts and exhibit better signal-to-noise ratios. Pulse-echo work is made easier with a double-element PVDF transducer [33].

Wheel transducers also perform out-of-plane longitudinal transduction [20,38]. To construct a fliud-filled wheel transducer, a commercial immersion transducer is mounted to the axle of a wheel with a rubber tire, and the tire is filled with water. The paper sample is fed into the nip between a transmitter and a receiver wheel. The transduction schematic of Fig. 7 still applies, but now there is a liquid delay line, and the rubber coupling layer can rotate with respect to the transducer. The immersion transducer is a broadband PZT transducer with a front-end impedance-matching layer optimized for water coupling. The rubber tire is not as soft as the platens of the IPC caliper gauge, but it still conforms to the paper and provides sufficient ultrasonic mating.

Out-of-plane transverse transduction requires different active elements. Excitation of a transverse element should predominantly create movement in the plane of the transducer front face. This is perpendicular to the out-of-plane motion of a longitudinal transducer. For a longitudinal transducer, the exciting electric field in the transmitter application and the sensed electric field in the receiver application are in the direction of the electric field used to polarize the piezoelectric material. To couple transverse motion, the transducing electric field must be perpendicular to the polarizing electric field. Commercial transverse transducers have been adapted to out-of-plane measurements in paper [34]. They are modified by the attachment of polystyrene delay lines to their front faces. Unfortunately, neoprene disks cannot be used for relaying transverse ultrasound to and from the sample. The shear modulus of rubber is almost zero, and it is a very poor carrier of transverse radiation. (This is the reason neoprene works well as the sample support for in-plane testers.) In the everyday testing of thick, smooth paperboards, one can live with the consequences of direct coupling between the sample and the delay line. However, this is unacceptable for very rough papers such as corrugated medium. Hard-earned mating improvement comes if the practitioner resorts to the "pillow method" [34]. A pillow is a viscous fluid encapsulated in a thin polymer bag. It transfers transverse radiation between a rough paper sample and a flat delay line without contaminating the sample.

C. Instrumentation

With the exception of the fluid-filled-wheel apparatus, out-of-plane velocity-measuring instruments are modified caliper gauges. The caliper gauge platens are fitted with ultrasonic transducers as shown in Fig. 7. As with caliper gauges, it is important to maintain a repeatable platen application and testing regime. The two flat transducers are accurately aligned, and one is deadweight-loaded atop the other. A motor-driven hoist lifts the upper transducer, making a gap for sample insertion. With sample in place, the motor is reversed, and the sample is squeezed between transducer jaws. After a fixed application time, displacement measurement and ultrasonic pulsing proceed simultaneously. The neoprene and the paper sample exhibit a time-dependent compliance to the load. With longitudinal neoprene coupling, 5 s is allowed for rubber and paper flow to level off and for coupling to stabilize [32]. One second is sufficient for hard-platen transverse measurements [34]. The fluid-filled wheels determine caliper without a direct displacement measurement [38]. Pulses in the received signal from different reflections are identified. The total water delay line transit time decreases when the sample is inserted by a time equal to the sample thickness divided by the velocity of sound in water. Thus, the caliper can be calculated from reflected path transit times with and without the sample inserted.

Time-of-flight techniques are commonly used for longitudinal and transverse out-of-plane velocity measurements in paperboard. For a clean, straightforward pulse analysis, interference from internal transducer reflections and from sample–transducer reflections must arrive after the analysis time window. With the just-described transduction schemes, reflections back and forth across the neoprene disks and reflections back and forth across the sample are the first signals to reach the receiver after the straight-through signal. Thus, to avoid complications from multiple reflections, the analysis window must be less than twice the transit time through the sample and less than twice the transit time through the neoprene. The PZT longitudinal transducer [32] maintains the 0.8 mm neoprene thickness adopted by the IPC soft platen caliper gauge. The neoprene transit time is only about 0.5 μs, and the analysis time is thereby limited to about a microsecond. The PZT longitudinal transducers are not broadband enough to generate a pulse of less than a few microseconds in duration. Therefore, time-of-flight analysis is restricted to the front portion of the pulse regardless of sample caliper. In practice, the PZT transducer-based instrument does a first half-wavelength cross-correlation determination identical to that of the in-plane time-of-flight measurement. The PVDF transducers [34] are more broadband and their neoprene disk is four times thicker. Pulse widths of under 2.0 μs are feasible [33], and the entire first pulse will elapse before neoprene cross-reflections can interfere. If the sample is thick enough, the full pulse can be used for cross-correlation time-of-flight determination. Reverberations across the sample must also cause minimal interference. Paper is highly attenuating to ultrasound in the megahertz frequency range; thus the concern over sample reflections is eased rapidly with increasing sample thickness. The reflections arrive later and are relatively weaker in thicker papers. As a rule of thumb, when using a 1.5 MHz single-cycle excitation, first half-wave times of flight should be a greater than 0.25 μs and full-pulse times of flight should be greater than 0.5 μs. This generally limits this style of out-of-plane work to thick papers and paperboards.

For in-plane velocity measurements, extraneous time delays were calibrated away by measuring times of flight at different transducer separations. Out of plane, there is only one possible sample path: across the entire sheet. Using the hard platen transverse technique, the non-sample delays are eliminated by defining time of flight as the time difference between the pulse arrival with and without the sample in place. However, it is not so simple in the soft platen longitudinal case. The soft rubber coupling disks deform differently when compressed in direct contact than when a sample is sandwiched between them. In direct contact, there is a single perimeter bulge curve resulting from lateral Poisson expansion. With the sample, the lateral expansion is restrained at the sample interface, and each disk has a lateral bulge maximum at its center. Because it is unconstrained by the sample, the rubber thicknesses and transit times are smaller when the sample is absent. To maintain a common rubber condition, the reference state is defined with a thin sheet of aluminum foil replacing the sample [32]. Now the rubber is laterally restrained in the reference and sample states. The time delay through the foil is much less than that through the paper, but there is still a necessary correction. Because the foil transit time is small compared to pulse width, one must account for foil-to-neoprene reflections in the correction. When the neoprene faces are properly cleaned, a valid correction can be made [32]. Sometimes, with very clean surfaces it is necessary to place thin threads across the foil interfaces to prevent air entrapment between the rubber and the aluminum.

As discussed in Section V, there are inevitable ambiguities in the alignment of the near and far pulses in simple time-of-flight determinations. McSkimmin [63] devised a delay line method that has been adapted to make clear-cut out-of-plane phase velocity measurements in paper [34]. Here, the electrical excitation is a train of around 20 sinusoidal pulses. The delay line must be long enough to contain the pulse and prevent transducer reflections from interfering with the first receiver train and the first sample-reflected transmitter train. The sample and coupling layer transit times, on the other hand, are small compared to the train duration. Toward the middle of the train, the sample and coupling layer portion of the response approaches that of an infinitely long sinusoidal pulse. Transmission line equations predict the fate of sinusoidal radiation incident on plane stacks of materials with different propagation parameters. Applying transmission line equations (delay line to neoprene to paper to neoprene to delay line) to the middle train phase and amplitude relations of the first transmitted and first reflected pulses allows one to infer sample phase velocity and attenuation coefficient at the carrier frequency. Transducer–paper coupling is too poor for the McSkimmin technique to give a valid loss measurement; however, mating worries do not derail the phase velocity calculation [34]. Generally, the large uncertainties in sample–transducer mating loss confound the adaptation of stnadard attenuation measurement techniques to paper. Fluid-filled-wheel enthusiasts get around this difficulty by invoking the Kramers–Kronig relationship [20,72]. This allows the calculation of loss coefficients from the change in phase velocity with frequency and precludes a mating-contaminated amplitude measurement.

IX. MISCELLANY

Linna, Moilanen, and Luukkala calculate tensions in paper webs from measurements of the time-of-flight velocities of low frequency antisymmetric waves [54,57]. Resonance techniques are also reported [92]. As I discussed earlier, there are two restoring forces for low frequency out-of-plane disturbances propagating in the plane of a thin plate under tension. If the plate is thin, the frequency is low, and the web is under high tension, then membrane waves propagate, and the velocity of out-of-plane waves is fixed by Eq. (33), $c = (T/\rho h)^{1/2}$. Otherwise, the low frequency antisymmetrical mode is a flexural wave. These authors intentionally operate in the tension-dominant regime. They mount a loudspeaker above a stretched web. They pulse the loudspeaker, which responds over a 200–400 Hz frequency range. Receiver transducers (microphones [54] or position-dependent light detectors focused on laser spots [57]) sense the resulting web motion. Time of flight comes from a cross-correlation of receiver signals at different separations from the transmitter. Velocity is the receiver separation divided by the time of flight, and web tension is velocity squared multiplied by basis weight.

Recently, an interesting technique for the ultrasonic determination of paper basis weight was published [42,52]. Low frequency ($\approx$ 40 kHz) air-coupled transducers are mounted on both sides of a suspended thin sheet. One transducer is excited with a continuous wave, and the resulting signal at the other is monitored. Assuming that the sample is much less than a wavelength in thickness and that is provides only an inertial loading, the basis weight of the sample is calculated [42,52] from the ratio of the received signals with and without the sample in place. The very limited data reported indicate good results for thin paper samples [42].

REFERENCES

1. Abdel Moteleb, M. M., Naoum, H. G., Shinouda, H. G., and Rizk, H. A. (1982). Some of the dielectric properties of cotton cellulose and viscose. *J. Polym. Sci. Polym. Chem. Ed. 20*:765–774.
2. Anderson, D. G., and Askne, J. I. H. (1974). Wave packets in strongly dispersive media, *Proc. IEEE 62*(11): 1518–1523.
3. Auld, B. A. (1990). *Acoustic Fields and Waves in Solids*, Vols. 1 and 2, 2nd ed. Krieger, Malabar, FL.
4. Ballou, J., and Silverman, S. (1944). Determination of Young's modulus of elasticity for fibers and films by sound velocity measurements. *Text. Res. J. 14*:282–292.
5. Ballou, J., and Smith, J. (1949). Dynamic measurements of polymer physical properties. *J. Appl. Phys. 20*:493–502.
6. Baum, G., and Bornhoeft, L. (1979). Estimating Poisson ratios in paper using ultrasonic techniques. *Tappi 62*:87–90.
7. Berger, B. J., Habeger, C. C., and Pankonin, B. M. (1989). The influence of moisture and temperature on the ultrasonic viscoelastic properties of cellulose. *J. Pulp Paper Sci. 15*(5): J170–J177.
8. Biot, M. A. (1956). Theory of propagation of elastic waves in a fluid-saturated porous solid. I. Low-frequency range. *J. Acoust. Soc. Am. 28*(2):168–178.
9. Biot, M. A. (1956). Theory of propagation of elastic waves in a fluid-saturated porous solid. II. Higher-frequency range. *J. Acoust. Soc. Am. 28*(2):179–191.
10. Biot, M. A. (1962). Generalized theory of acoustic propagation in porous dissipative media. *J. Acoust. Soc. Am. 34*(9):1254–1264.
11. Bloch, S. (1977). Eighth velocity of light. *Am. J. Phys. 45*:538–549.
12. Blouin, A., and Monchalin, J. (1994). Detection of ultrasonic motion of a scattering surface by two-wave mixing in a photorefractive GaAs crystal. *Appl. Phys. Lett 65*(8):932–934.
13. Bonnet, G. (1983). Beyond the group velocity: Signal and wave velocities, Part I. *Ann. Telecommun. 38*(9–10):345–366.
14. Bonnet, G. (1983). Beyond the group velocity: Signal and wave velocities, Part II. *Ann. Telecomm. 38*(11–12):471–487.
15. Born, M., and Wolf, E. (1980). *Principles of Optics*. 6th ed. Pergamon Press, Oxford.
16. Borngis, F. E. (1955). Specific directions of longitudinal wave propagation in anisotropic media. *Phys. Rev. 98*(4):1000–1005.
17. Bradford, H. M. (1976). Propagation and spreading of a pulse or wave packet. *Am. J. Phys. 44*(11):1058–1063.
18. Bradley, S. A., and Carr, S. H. (1976). Mechanical loss processes in polysaccharides. *J. Polym. Sci. Phys. Ed. 14*:111–124.
19. Brillouin, L. (1960). *Wave Propagation and Group Velocity*. Academic Press, New York.
20. Brodeur, P. H., Hall, M. S., and Esworthy, C. (1993). Sound dispersion and attenuation in the thickness direction of paper materials. *J. Acoust. Soc. Am. 94*(4):2215–2225.
21. Brodeur, P. H., Johnson, M. A., Berthelot, Y. H., and Gerhardstein, J. P. (1997). Noncontact laser generation and detection of Lamb waves in paper. *J. Pulp Paper Sci. 23*(5):J238–J243.
22. Castagnede, B., Mark, R. E., and Seo, Y. B. (1989). New concepts and experimental implications in the description of the 3-D elasticity of paper. Part I. *J. Pulp Paper Sci. 15*(5):J178–J182.
23. Chaikin, M., and Chamberlain, N. (1955). The propagation of longitudinal stress pulses in textile fibres, Part I. *J. Text. Inst. Trans. 46*: T25–T43.
24. Craver, J., and Taylor, D. (1965). Nondestructive sonic measurement of paper elasticity. *Tappi 48*(3):142–147.

25. Dimond, P. M., unpublished work, Weyerhaeuser Co.
26. Deason, V. A., Telschow, K. L., Schley, R. S., and Watson, S. M. (1999). Imaging the anisotropic elastic properties of paper with the INEEL laser ultrasonic camera. Proc. 26th Review of Progress in Quantitative NDE, Montreal, July 25–30.
27. Ferry, J. D. (1980). *Viscoelastic Properties of Polymers*. 3rd ed. Wiley, New York.
28. Forbes, M. (1986). Ultrasonic characterization of layered composite systems. Ph.D. Thesis. The Institute of Paper Chemistry, Appleton, WI.
29. Fung, Y. C. (1965). *Foundations of Solid Mechanics*. Prentice-Hall, Englewood Cliffs, NJ.
30. Graff, K. F. (1975). *Wave Motion in Elastic Solids*. Dover, New York.
31. Habeger, C., Mann, R., and Baum, G. (1979). Ultrasonic plate waves in paper. *Ultrasonics 17*:57–62.
32. Habeger, C., Wink, W., and Van Zummeren, M. (1988). Using neoprene-faced PVDF transducers to couple ultrasound into solids. *J. Acoust. Soc. Am. 84*(4):1388–1396.
33. Habeger, C., and Wink, W. (1990). Development of a double-element pulse echo, PVDF transducer. *Ultrasonics 28*:52–54.
34. Habeger, C., and Wink, W. (1986). Ultrasonic velocity measurements in the thickness direction of paper. *J. Appl. Polym. Sci. 32*:4503–4540.
35. Habeger, C. (1990). An ultrasonic technique for testing the orthotropic symmetry of polymeric sheets by measuring their elastic shear coupling coefficients. *J. Eng. Mater. Technol. Trans. ASME 112*:366–371.
36. Habeger, C., Van Zummeren, M., Wink, W., Pankonin, B., and Goodlin, R. (1989). Using a robot-based instrument to measure the in-plane ultrasonic velocities of paper. *Tappi 72*(7):171–175.
37. Hale, T. C., Telschow, K. L., and Deason, V. A. (1997). Photorefractive optical lock-in vibration spectral measurement. *Appl. Opt. 36*(31):8248–8258.
38. Hall, M. S. (1990). On-line ultrasonic measurement of paper strength. *Sensors 7*(7):13–20.
39. Hamburger, W. (1948). Mechanics of elastic performance of textile materials. II. *Text. Res. J. 18*:704–743.
40. Hardacker, K. W. (1981). Instrument and specimen shape for biaxial testing of paper. *J. Phys. E: Sci. Instrum. 14*:593–596.
41. Hill, T. L. (1960). *Introduction to Statistical Thermodynamics*. Addison-Wesley, Reading, MA.
42. Imano, K., Okuyama, D., and Chubachi, N. (1993). A noncontact thickness measurement of thin samples using 40 kHz ultrasonic waves. *IEICE Trans. Fundam. E76–A 10*:1861–1862.
43. Jackson, W. J., and Caldwell, J. R. (1965). Antiplasticization. II. Characteristics of antiplasticizers. *J. Appl. Polym. Sci. 11*:211–244.
44. Johnson, M. A. (1996). Investigation of the mechanical properties of copy paper using laser generated and detected Lamb waves. Ph.D. Thesis, Georgia Institute of Technology, Atlanta, GA.
45. Johnson, M. A., Berthelot, Y. H., Brodeur, P. H., and Jacobs, L. A. (1996). Investigation of laser generation of Lamb waves in copy paper, *Ultrasonics* (April).
46. Kimura, M., and Nakano, J. (1976). *Polym. Sci. Polym. Lett. 14*:741–745.
47. Klason, C., and Kubat, J. (1976). Thermal transitions in cellulose. *Svensk Papperstidn. 9*(15):494–500.
48. Kolsky, H. (1963). *Stress Waves in Solids*. Dover, New York
49. Lafond, E. F., Brodeur, P. H., Gerhardstein, J. P., Habeger, C. C., and Telschow, K. L. (1999). Photorefractive interferometer for ultrasonic measurements on paper. Proc. 26th Review of Progress in Quantitative NDE, Montreal, July 25–30.
50. Lafond, B. F., Gerhardstein, G. D., Klein, M. B., and Brodeur, P. H. (1999). Non-contact characterization of static paper materials using a photorefractive interferometer.

SPIE Conf. on Process Control and Sensors for Manufacturing, Newport Beach, March, pp. 30–41.

51. Lancee, C. T., Souquet, J., Ohigashi, H., and Bom, N. (May 1985). Ferro-electric ceramic versus polymer piezoelectric materials. *Ultrasonics 23*:138–142.
52. Lefebvre, J., Bruneel, C., Delebarre, C., Lutgen, P., and Ecker, T. (1988). Remote ultrasonic measurement of the thickness of thin films. *J. Acoust. Soc. Am. 84*(3):1094–1096.
53. Leugers, M. A. (1986). Laser induced acoustic generation for sonic modulus. U.S. Patent 4,622,853.
54. Linna, H., and Moilanen, P. (1988). Comparison of methods for measuring web tension, *Tappi 71*(10):134–137.
55. Love, A. E. H. (1944). *A Treatise on the Mathematical Theory of Elasticity*. Dover, New York.
56. Luukkala, M., Heikkila, P., and Surakka, J. (1971). Plate wave resonance: A continuous test method. *Ultrasonics 9*(3):201–208.
57. Luukkala, M. (1990). Tenscan, an acoustic NDE device to measure tension in a moving paper web. In: *Review of Progress in Quantitative Nondestructive Evaluation*, Vol 9. D. O. Thompson and D. E. Chimenti, eds. Plenum Press, New York, pp. 1987–1991.
58. Luukkala, M., and Marttinon, T. (1989). Method and apparatus for noncontacting tension measurement in a flat foil and especially in a paper web. U.S. Patent 4,833,928.
59. Mann, R. (1978). Elastic wave propagation in paper. Ph.D. Thesis. The Institute of Paper Chemistry, Appleton, WI.
60. Mann, R., Baum, G., and Habeger, C. (1979). Elastic wave propagation in paper. *Tappi 62*(8):115–118.
61. Mann, R., Baum, G., and Habeger, C. (1980). Determination of all nine orthotropic elastic constants for machine-made paper. *Tappi 63*(2):163–166.
62. Maznev, A. A., and Every, A. G. (1995). Focusing of acoustic modes in thin anisotropic plates. *Acta Acustica 3*:387–391.
63. McSkimmin, H. J. (1951). *J. Acoust. Soc. Am. 23*(4):429.
64. Mayer, K., and Lotmar, W. (1936). The elasticity of cellulose. *Helv. Chim. Acta 19*:68–86 (in French).
65. Monchalin, J. P. (1986). Optical detection of ultrasound. *IEEE Trans. UFFC 33*:485.
66. Morse, P. M., and Ingard, K. U. (1968). *Theoretical Acoustics*. McGraw-Hill, New York.
67. Musgrave, M. (1954). On the propagation of elastic waves in aeolotropic media. I. General principles. *Proc. Roy. Soc. Lond. A226*:339–355.
68. Musgrave, M. J. P. (1954). On the propagation of elastic waves in aeolotropic media. II. Media of hexagonal symmetry. *Proc. Roy. Soc. Lond. A226*:356–366.
69. Normuri Shoji Co. Sales brochure for SST-250 Sonic Tester, Japan.
70. Normuri Shoji Co. Sales brochure for SST-3000 Sonic Tester, Japan.
71. Nishinari, K., and Fukada, E. J. (1980). Viscoelastic, dielectric, and piezoelectric behavior of solid amylose. *J. Polym. Sci. Polym. Phys. Ed. 18*:1609–1619.
72. O'Donnel, M., Jaynes, E., and Miller, J. (1981). *J. Acoust. Soc. Am. 69*(3):696–701.
73. Olofsson, K., and Kyosti, A. (1994). Stiffness and stiffness variation in paper measured by laser-generated and laser-recorded bending waves. *J. Pulp Paper Sci. 20*(11):J328–J3332.
74. Pace, S. A., and Salama, S. S. (1987). Laser induced acoustic generation for sonic modulus. U.S. Patent 4,674,332.
75. Pankonin, B. M., and Habeger, C. C. (1988). A strip resonance technique for measuring the ultrasonic viscoelastic parameters of polymeric sheets with application to cellulose. *J. Polym. Sci. Polym. Phys. Ed. 26*:339–352.
76. Rayleigh, J. (1945). *The Theory of Sound*, Vol. 2. Dover, New York.
77. Rich, S. (1955). Modulus determining system. U.S. Patent 2,706,906.

78. Salmen L., and Back, E. (1977). The influence of water on the glass transition of cellulose. *Tappi 60*(12):137–140.
79. Scruby, C. B., and Drain, L. E. (1990) *Laser Ultrasonics: Techniques and Applications*. Adam Hilger, Bristol, UK.
80. Seidman, R., and Mason, S. G. (1954). Dielectric relaxation in cellulose containing sorbed vapors. *Can. J. Chem. 20*:744–762.
81. Sessler, G. M. (1981). Piezoelectricity in polyvinylidene fluoride. *J. Acoust. Soc. Am. 70*(6):1596–1608.
82. Setterhom, V. C. (1974). A new concept in paper thickness measurement. *Tappi 57*(3):164.
83. Solie, L. F., and Auld, B. A. (1973). Elastic waves in free anisotropic plates. *J. Acoust. Soc. Am. 54*:50–65.
84. Staudt, J. H., and Cook, B. D. (1965). Visualization of quasilongitudinal and quasitransverse elastic waves. *J. Appl. Phys. 36*:759–768.
85. Stiffler, R., and Henneke, E. G. (1983). The application of polyvinylidene fluoride as an acoustic emission transducer for fibrous composite materials. *Mater. Eval. 41*:956–960.
86. Stratton, R. A. (1973). Dependence of the viscoelastic properties of cellulose on water content. *J. Polym. Sci. Polym. Chem. Ed. 11*:535–544.
87. Taylor, D., and Craver, J. (1966). Anisotropic elasticity of paper from sonic velocity measurements. In: *Consolidation of the Paper Web*. The British Paper and Board Makers' Association, London, pp. 852–872.
88. Titus, M. (1993). Ultrasonic technology: Measurements of paper orientation and elastic properties. TAPPI 1993 Process and Product Quality Conference, pp. 117–122.
89. Tsai, S., and Hahn, H. (1980). *Introduction to Composite Materials*. Technomic, Lancaster, PA.
90. Van Zummeren, M., Young, D., Habeger, C., Baum, G., and Treleven, R. (1987). Automatic determination of ultrasound velocities in planar materials. *Ultrasonics 25*:288–294.
91. Viktorov, L. (1967). *Rayleigh and Lamb Waves*. Plenum Press, New York.
92. Wallbaum, H. H., and Lisnyansky, K. (1983). Review of process control instruments for measuring paper quality variables. *Paper Trade J. 167*(14):34–40.
93. Walker, J. B., Telschow, K. L., Pufahl, B. M., Gerhardstein, J. P., Habeger, C. C., Brodeur, P. H., and Lafond, E. F. (1999). Fabry-Perot laser ultrasonic elastic anisotropy measurements on a moving paper web. Proc. 26th Review of Progress in Quantitative NDE, Montreal, July 25–30.
94. Wink, W. A., and Baum, G. A. (1983). A rubber platen caliper gage. *Tappi 66*(9):131–133.
95. Wolfe, J. P., and Hauser, M. R. (1995). Acoustic wavefront imaging. *Ann. Phys. 4*:99–126.
96. Yeh, P. (1993). *Introduction to Photorefractive Nonlinear Optics*. Wiley, New York.
97. Zauscher, S., Caulfield, D. F., and Nissan, A. H. (19??), Influence of water on elastic modulus of paper; Extension of the H-bond theory. *Tappi 79*(12):178–182.

7
DEFORMATION AND FAILURE BEHAVIOR OF PAPER

CURT A. BRONKHORST and KEITH A. BENNETT
Weyerhaeuser Company
Federal Way, Washington

I. INTRODUCTION

The focus of this chapter is the deformation response of paper to imposed mechanical loading up to the point of ultimate failure. Ultimate failure is implicitly defined here as the point of loading at which the material can no longer offer significant resistance to further deformation or at which a peak functional applied load has been reached. As is the case for most materials, ultimate failure is a sequence of failure events beginning at a small size scale. These events then either escalate in magnitude or coalesce to the largest size scale of the material being tested. Failure events that occur before ultimate failure are sometimes termed damage. Generally, then, ultimate failure is not a singular event but rather the final result of an accumulation of smaller scale deformation and failure events occurring during the mechanical deformation of the material. This chapter acknowledges this fact and attempts to describe the material behavior leading up to ultimate failure of paper for some primary as well as practical modes of loading.

II. DEFINITIONS AND CONCEPTS

The term *stress* is defined as a force per unit area. A force applied across an area can be resolved into three components that are aligned with the chosen coordinate system. Division of the force magnitude of these components by the area over which the force acts defines a normal stress and two shear stresses, as illustrated in Fig. 1. Normal and shear stresses will be designated by the symbols σ and τ, respectively, followed by two subscripts that indicate the direction of the force component and the direction of the normal to the area. For any arbitrary small cube of material (six sides), there are 18 possible stress components (one normal and two shear stress components on each of the six sides of the cube). For bodies at rest, only nine stress components are independent, because equilibrium of forces demands that they occur in equal and opposite pairs. Moreover, three of these remaining nine stress components can also be shown to occur in pairs on account of moment equilibrium arguments. Therefore, for a body at rest, only six stress components are required to characterize the state of stress on any cubic volume surrounding a given point in the material in that body. The definition of stress is well established, and more detailed discussions are given in many texts, e.g., those of Sokolnikoff [165], McClintock and Argon [110], Crandall et al. [35], Gurtin [62], and Malvern [105]. The state of stress at a *point* is considered to be the *limit* of the components as the volume tends to zero. This infinitesimal point is generally termed a *material point*.

For a sheet material with no transversely applied forces the state of stress is two-dimensional, and only three components are needed to characterize such a state, as shown in Fig. 2. This state of material stress is generally called *plane stress*. In the paper literature, in-plane forces are sometimes divided by the length, rather than the area, of the edge. This, of course, is not a stress but a force per unit length even though this measure is generally referred to as *stress*. This practice is sometimes undertaken because of difficulties associated with reliably measuring the thickness of paper. In this chapter the term stress is used in its strict sense.

It should be noted that the definition of stress depends on the choice of orientation for the (Cartesian) coordinate system. In a state of plane stress, if the stress

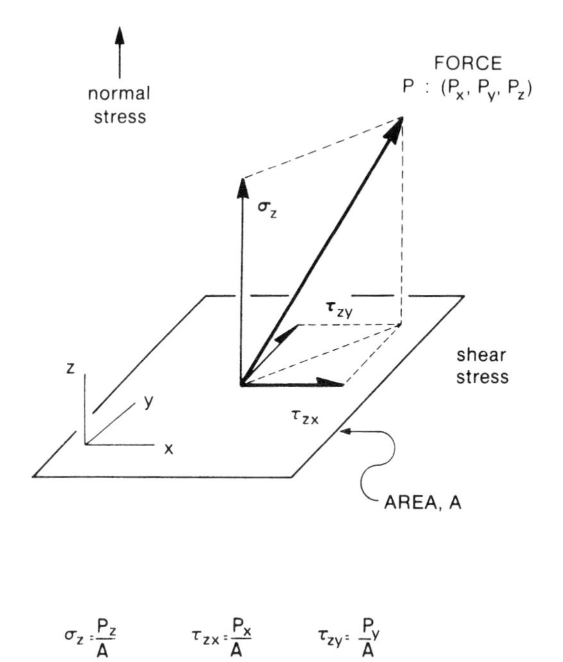

Fig. 1 A traction force resolved into its shear and normal stress components.

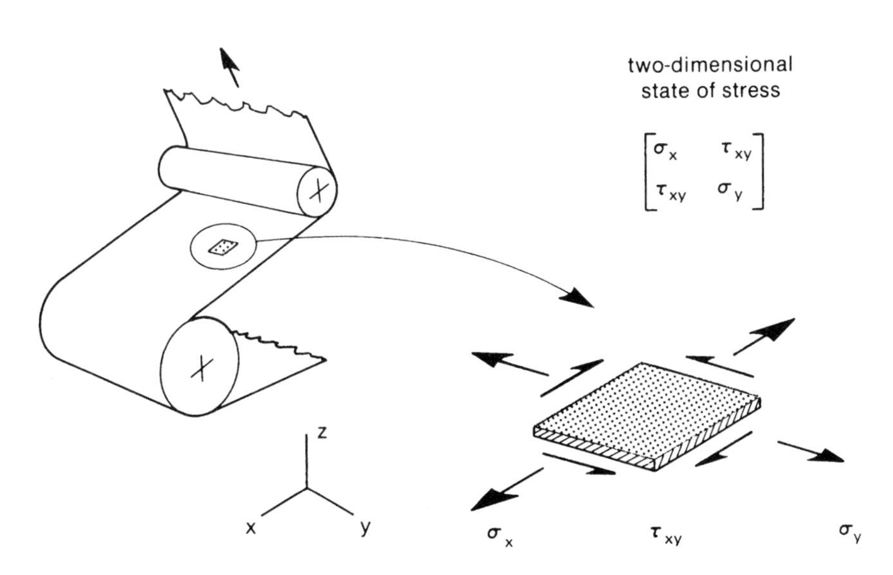

Fig. 2 The state of stress existing under plane stress conditions (applicable for many cases involving paper).

components in one Cartesian system are known, then the components for any other system can be calculated using the following equations:

$$\sigma_{x'} = \frac{\sigma_x + \sigma_y}{2} + \frac{\sigma_x - \sigma_y}{2}\cos(2\theta) + \tau_{xy}\sin(2\theta) \tag{1a}$$

$$\tau_{x'y'} = -\frac{\sigma_x - \sigma_y}{2}\sin(2\theta) + \tau_{xy}\cos(2\theta) \tag{1b}$$

$$\sigma_{y'} = \frac{\sigma_x + \sigma_y}{2} - \frac{\sigma_x - \sigma_y}{2}\cos(2\theta) - \tau_{xy}\sin(2\theta) \tag{1c}$$

where θ is the angle between the two coordinate systems in the plane of the sheet. These equations can be represented graphically by a diagram known as Mohr's circle (Fig. 3). The arrow indicator of this diagram is simply rotated twice the angular distance between the coordinate system in which the components are known and another coordinate system (primed). This diagram shows that it is always possible to select a particular coordinate system in which the state of stress has no shear component. The two normal stresses σ_1 and σ_2 in such a coordinate system are called the

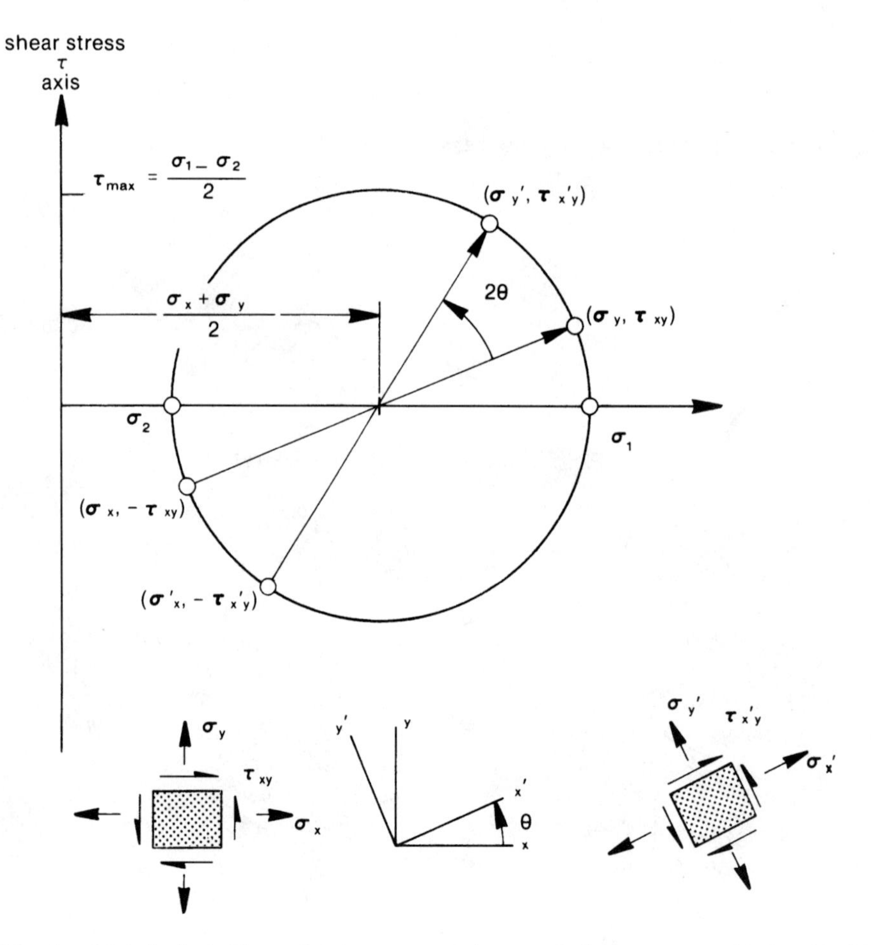

Fig. 3 Mohr's circle illustrating how a general state of plane stress can be described in any coordinate system as well as principal stresses σ_1 and σ_2.

principal stresses. In a similar manner, the state of stress in three dimensions can be characterized by three principal stresses: σ_1, σ_2, and σ_3.

A measure of deformation must be defined. First, consider the concept of displacement. The position of a material in an undeformed, unstressed body is specified by the coordinates X, Y, and Z relative to a fixed coordinate system. If the body is moved, the same material point will have different coordinates x, y, and z, and the difference between the two sets defines three displacements:

$$u = x - X, \qquad v = y - Y, \qquad w = z - Z \tag{2}$$

Measures of deformation, strains, are defined as rates of change of displacement relative to their positions. Normal strains are defined as

$$\varepsilon_x = \frac{\partial u}{\partial x}, \qquad \varepsilon_y = \frac{\partial v}{\partial y}, \qquad \varepsilon_z = \frac{\partial w}{\partial z} \tag{3}$$

These measures are called infinitesimal strains and serve well for small displacements. For large displacements, it has proven useful to use a length ratio of an arbitrary line segment $\lambda_x = \Delta x/\Delta X$ before and after deformation as a measure of deformation. These ratios, one for each direction, are called stretch ratios. Normal strains or stretch ratios are associated with volumetric changes as illustrated in Fig. 4. Shear strains are associated with angular distortions, also shown in Fig. 4. Two definitions are generally encountered:

Mathematical shear strains:

$$\varepsilon_{xy} = \frac{1}{2}\left\{\frac{\partial v}{\partial x} + \frac{\partial u}{\partial y}\right\}, \qquad \varepsilon_{xz} = \frac{1}{2}\left\{\frac{\partial w}{\partial x} + \frac{\partial u}{\partial z}\right\}, \qquad \varepsilon_{yz} = \frac{1}{2}\left\{\frac{\partial w}{\partial y} + \frac{\partial v}{\partial z}\right\} \tag{4}$$

Engineering shear strains:

$$\gamma_{xy} = 2\varepsilon_{xy}, \qquad \gamma_{xz} = 2\varepsilon_{xz}, \qquad \gamma_{yz} = 2\varepsilon_{yz} \tag{5}$$

Engineering shear strain, twice the mathematical shear strain, is the total angular deformation. The importance of the distinction is that mathematical strains can be transformed to a new coordinate system using the same equations as the stresses by replacing σ_x, τ_{xy}, and σ_y with ε_x, ε_{xy}, and ε_y, respectively, in Eqs. (1). In two dimensions, then, it is always possible to find two principal strains ε_1 and ε_2 that will characterize the state of strain at a point. The concept of strain, along with stress, is also well established in mechanics, and the reference texts mentioned for more detailed discussions of stress can also be consulted regarding strain.

Suppose a planar material is subjected to an in-plane homogeneous stress state. By homogeneous, we mean that the state of stress is the same at each material point. It has been demonstrated that the state of plane stress at a point can be monitored by the two principal stresses. If we were to plot the path of stress states or the trajectory of σ_1, σ_2 in a principal stress plane during the loading history (as illustrated in Fig. 5), we could arrive at a principal stress surface that identifies those combinations of principal stresses that cause ultimate failure of the material. If the material does not fail, all points on the trajectory are safe. If the material does fail, then the principal stress point on the loading trajectory marking the initiation of failure defines the boundary between the safe zone and the failure zone. Subjecting the material to all possible failure trajectories establishes the failure surface. In three dimensions, the failure surface would be evaluated in principal stress space defined by σ_1, σ_2, and σ_3.

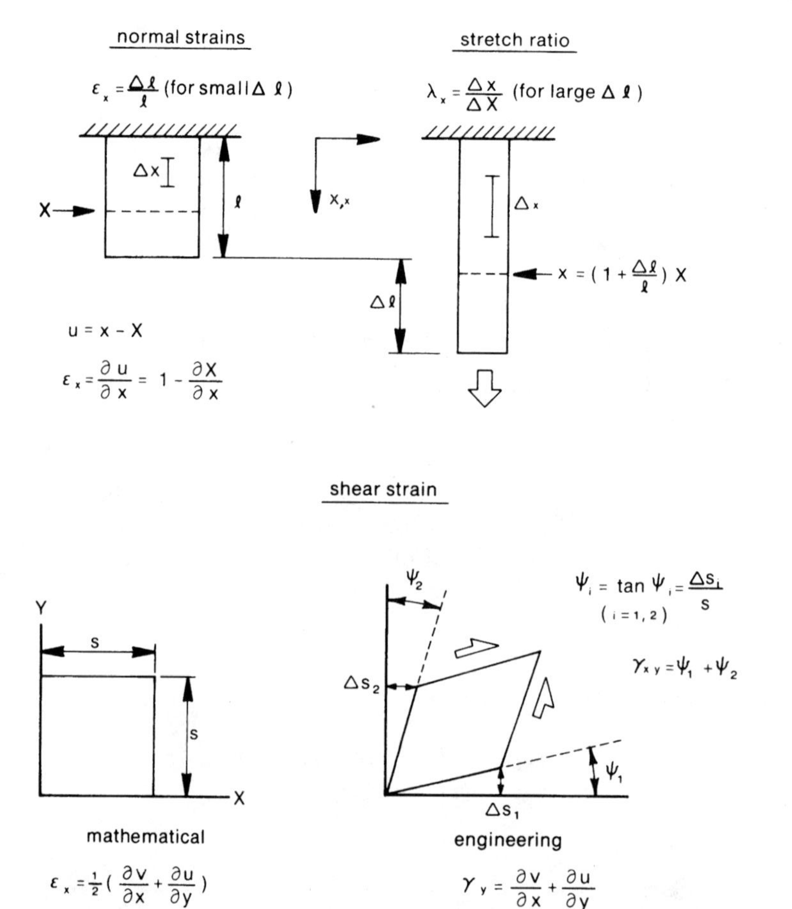

Fig. 4 A description of normal and shear strain for small displacements.

It is important to note that the events that indicate ultimate failure are not the same for each principal stress combination on the failure surface. In tension, for example, the failure event can be complete rupture, whereas in compression the event might be the development of a crease or the appearance of a kink.

As we discussed earlier, stress and strain define both the kinetics and kinematics, respectively, of deformation. Most properly instrumented mechanical test systems can accurately measure both ultimate stress and ultimate strain. As an example, consider a uniaxial tension test. An idealized plot of a stress–strain curve for a material is shown in Fig. 6. At some point during loading, the material fails catastrophically. The failure can be characterized by a maximum (or ultimate) stress and a maximum (or ultimate) strain at the breakpoint. This point is indicated by a star in Fig. 6.

It will be noticed that the stress–strain curve is not entirely a straight line but becomes nonlinear at some point called the proportional limit (located by a small circle on the curve in Fig. 6). This point on the stress–strain curve is also termed the *yield* point and is generally the stress level at which inelastic or irreversible deforma-

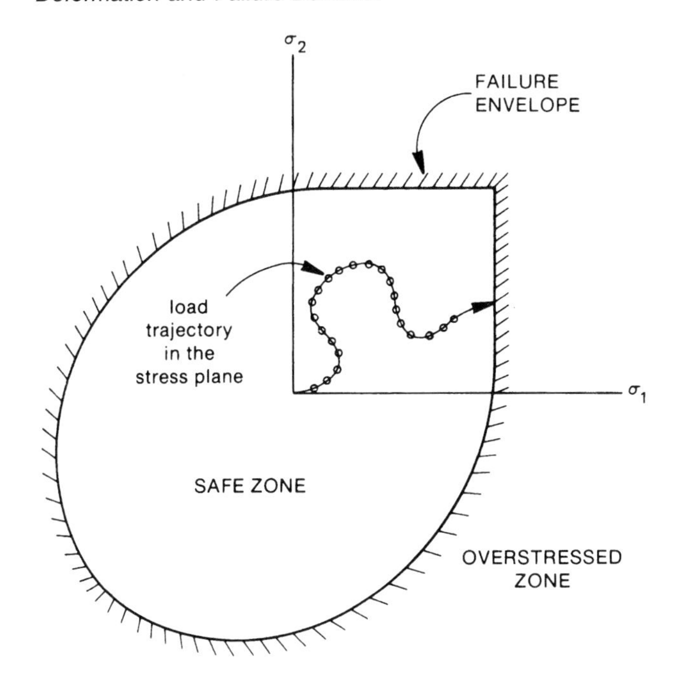

Fig. 5 A hypothetical failure envelope in the principal plane stress plane.

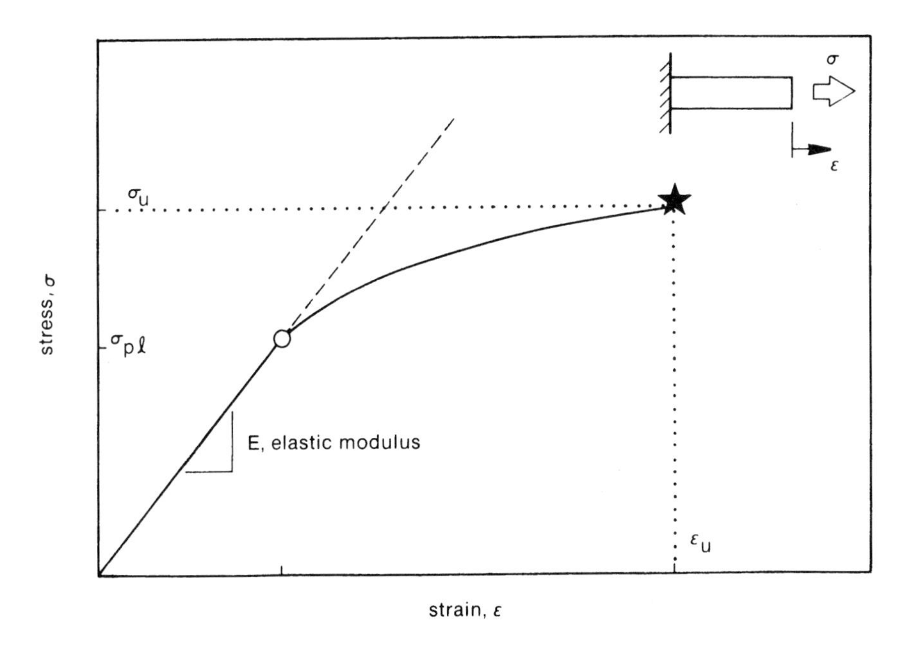

Fig. 6 An idealized uniaxial tensile stress–strain curve showing both the proportional limit (circle) and point of ultimate failure (star).

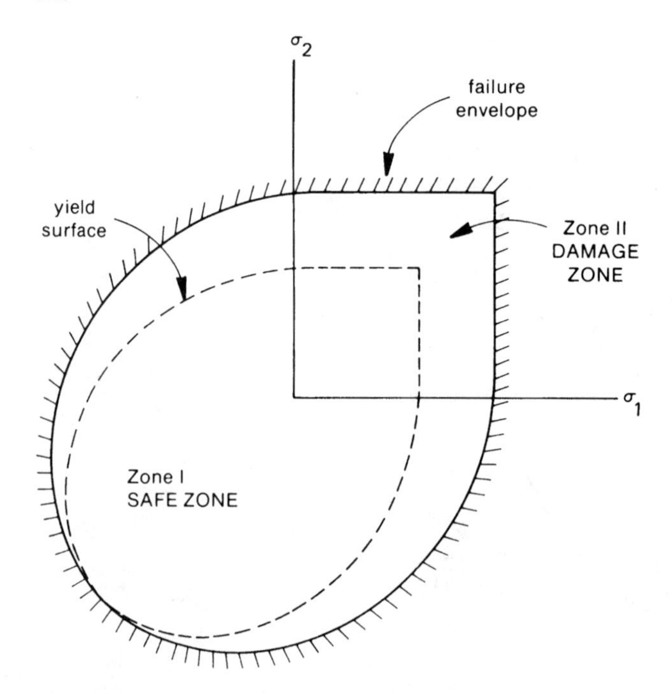

Fig. 7 The hypothetical failure envelope shown in Fig. 5 with a hypothetical yield surface added in the principal plane stress plane. The region between the two surfaces is suggested to be the region where localized failure or damage can occur in the material, leading to ultimate failure.

tion of the material begins. If we evaluate the yield point of a material for all principal stress combinations, we would arrive at a *yield surface*. This yield surface can be thought of as the point at which localized damage in the material is possible for any load trajectory. The region between the yield surface and the failure surface can then be thought of as the *damage zone* as illustrated hypothetically in Fig. 7. Within the damage zone, the rate at which damage occurs is highly nonlinear with deformation. The amount of localized damage accelerates rapidly as the failure envelope is approached. This is particularly true for paper, where damage is generally intrafiber and interfiber hydrogen bond failure.

III. PAPER AS A MATERIAL

Paper is a structural composite material. It is composed of ribbonlike load-bearing elements: collapsed (or partially collapsed) wood pulp fibers or other types of cellulosic fibers. The fibers form a network in which externally applied loads must be transmitted to the individual fiber segments through the bonded contact areas between the fibers. For the most part these fibers are laid down with their axis parallel to the plane of the sheet (or at least nearly so); the interfiber bond areas are also generally oriented with their normal directions perpendicular to the plane of the sheet. Moreover, commercial paper is manufactured in such a way that there are more fiber axes aligned parallel with than perpendicular to the flow of the paper through the paper machine. This organization of the microstructure, combined with machine direction web tension and cross-machine direction drying restraint, produces a material that is orthotropic in its response to mechanical deformation. The three, mutually perpendicular, principal directions are referred to as the machine direction (MD), cross-machine direction (CD), and transverse (through-the-thickness or *Z*) direction (ZD) as illustrated in Fig. 8. Paper deformation and strength behavior must be considered in terms of this anisotropy.

Deformation and failure in paper are highly dependent on the direction of the applied loads relative to the principal material directions. For example, it is obvious that the microscopic response to tensile forces applied in each of the principal directions will be different due to the orientation of the fiber segments and the bonded areas connecting them. Forces acting on a hypothetical diamond-shaped portion of an aligned fiber network are shown in Fig. 9. It can be seen that the resultant forces in the fiber elements are quite different if loaded in the machine direction than if loaded in the cross-machine direction. The bonded areas, consequently, will be subject to different shear stresses.

Paper is also a heterogeneous material. This heterogeneity is difficult to characterize, because it depends on the scale of observation. If a sheet of ordinary writing paper is held up to a light source, a variation is seen to exist in the sheet density with a characteristic size on the order of 10–20 times the sheet thickness. Closer observations of paper reveal the heterogeneous nature of the fiber segments and interfiber voids, and examination of the fibers reveals a multilayer cylindrical cellulose microfibril composite.

A simplified model can be used to illustrate how heterogeneity influences strength. A system of three parallel arrays of three springs in series is shown in Fig. 10. The stiffness of the system is a function of all the individual spring stiffnesses

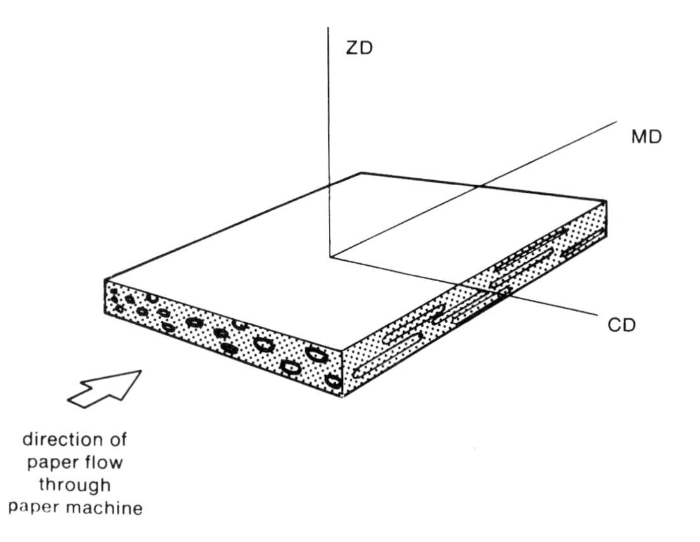

Fig. 8 The principal material directions for paper: machine direction (MD), cross machine direction (CD), and transverse (ZD) direction.

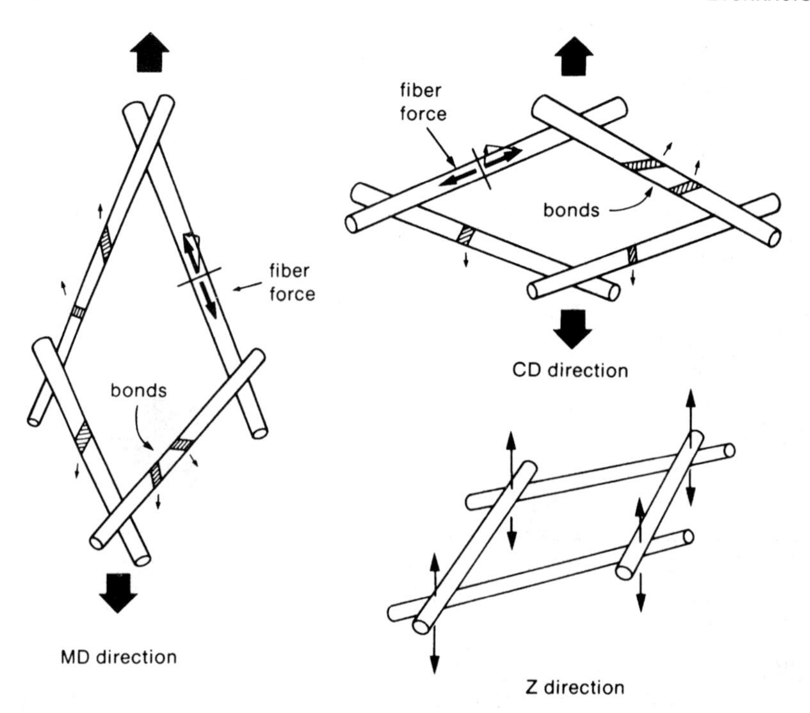

Fig. 9 Highly simplified force distributions in fiber networks for tensile loading along the principal material directions. The hatched areas are surfaces where other fibers are attached.

k_{ij}. The system can be thought of as an idealized replacement of a real system in which physical and mechanical properties change from point to point. Suppose all of the springs in Fig. 10 except the one in the center have the same stiffness k_{I}. Let the center spring have a stiffness of k_{II}, and let the ratio of the two be given by $\beta = k_{\mathrm{II}}/k_{\mathrm{I}}$. The overall stiffness of the system becomes

$$k = \frac{1}{3}\left\{\frac{7\beta + 2}{2\beta + 1}\right\}k_{\mathrm{I}} \tag{6}$$

For a homogeneous system ($k_{\mathrm{I}} = k_{\mathrm{II}}$, or $\beta = 1$), the system stiffness is identical to the individual spring stiffness $k = k_{\mathrm{I}}$. If a gap is left in the center ($\beta = 0$), the system becomes more flexible, $k = (2/3)k$; if a rigid rod is placed at the center ($\beta = \infty$), the system becomes slightly stiffer, $k = (7/6)k_{\mathrm{I}}$.

Changing the stiffness ratio, however, produces unequal load sharing among the three parallel arrays:

$$P_1 = \left(\frac{2\beta + 1}{7\beta + 2}\right)P = P_3, \qquad P_2 = \frac{3\beta}{7\beta + 2}P \tag{7}$$

If the strengths of the two spring types are S_{I} and S_{II}, then there are three possible ways to calculate the potential load-bearing force of the system. When the force in each of the springs of the first and third arrays is equal to its strength, the overall force acting on the system will be

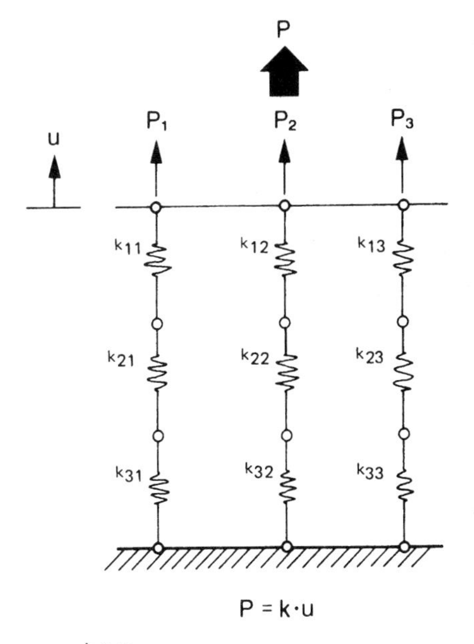

where

$$k = \left\{ \frac{1}{\sum_{1}^{3} \left(\frac{1}{k_{i1}}\right)} + \frac{1}{\sum_{1}^{3} \left(\frac{1}{k_{i2}}\right)} + \frac{1}{\sum_{1}^{3} \left(\frac{1}{k_{i3}}\right)} \right\}$$

Fig. 10 A simple spring model used to illustrate the effect of heterogeneity on the mechanical response of a network.

$$P = \left(\frac{7\beta + 2}{2\beta + 1}\right) S_{\text{I}} \tag{8}$$

When the force in the center spring of the central array is equal to its strength, the overall force will be

$$P = \left(\frac{7\beta + 2}{3\beta}\right) S_{\text{II}} \tag{9}$$

and when the forces in the other two springs of the central array are equal to their strengths, the overall force will be

$$P = \left(\frac{7\beta + 2}{3\beta}\right) S_{\text{I}} \tag{10}$$

Assume, for the sake of simplicity, that the ratio of the strengths of the two types of springs is equal to the stiffness ratio, that is

$$\frac{S_{\text{II}}}{S_{\text{I}}} = \frac{k_{\text{II}}}{k_{\text{I}}} = \beta \tag{11}$$

A plot of the three potential load-bearing forces of the system is shown as a function of the stiffness ratio in Fig. 11. For a given stiffness ratio, the ultimate strength of the system will be the minimum of the three strengths. It is clear that the homogeneous

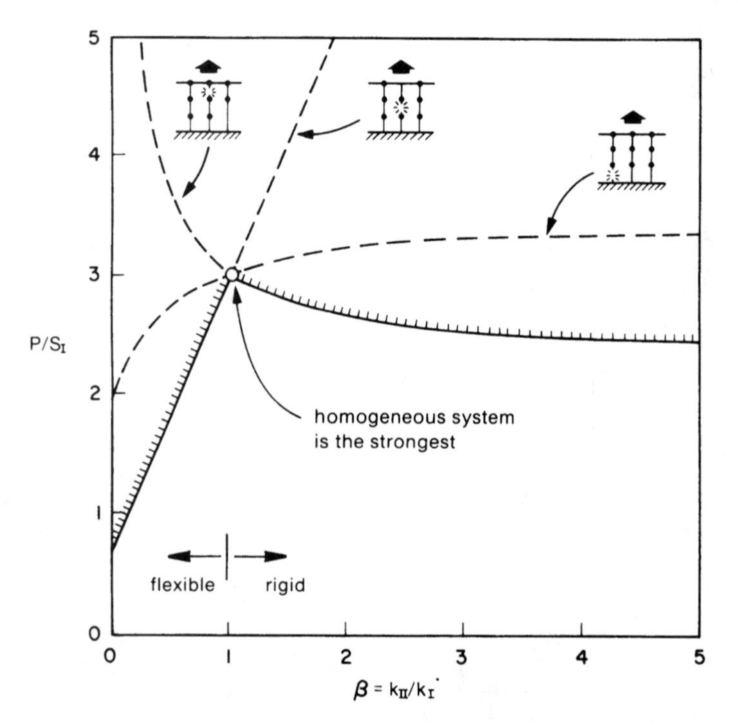

Fig. 11 The strength of the array of springs illustrated in Fig. 10 as a function of the ratio of spring constants.

system is the strongest. In the flexible case ($\beta < 1$), the weakest link is the center spring of the central array; in the stiffer system ($\beta > 1$), the outer springs of the central arrays are critical.

It is not hard to imagine a large network of springs representing a heterogeneous paper system in which the properties change from point to point, such as would occur in a highly flocculated system, for example. The analysis of strength would be conceptually similar to the simplified model just considered but would be much more complicated. A more comprehensive two-dimensional representation of paper is illustrated in Fig. 12, where each line segment is a free fiber segment. In such an analytical framework, the spatial distribution of the fibers as well as the mechanical behavior of the fibers themselves can be defined independently. If we deform such a network in uniaxial tension and examine the state of stress at the center of each fiber segment, we can see how inhomogeneously such a system is loaded. This is illustrated in Fig. 13, where each point is the axial stress in a single fiber segment. Note that the axial stress state in a fiber segment is a function of its orientation with respect to the loading direction. For a good fundamental overview of paper mechanics, the reader is urged to consult the book by Niskanen [121].

In the following sections, we will discuss a few of the basic stress states and some important test methods in use to characterize the deformation and strength behavior of paper. We shall begin with tests involving basic stress states for sheet material. Tests involving more complex stress distributions and tests for composite paper material and paper structures will follow. Conventional tests such as tensile,

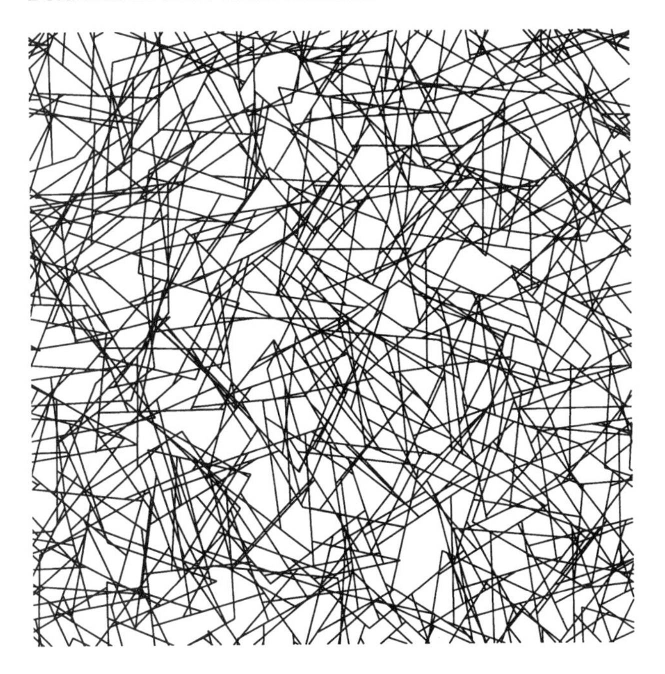

Fig. 12 An example of a numerical two-dimensional random fiber network. Each line segment represents a fiber segment.

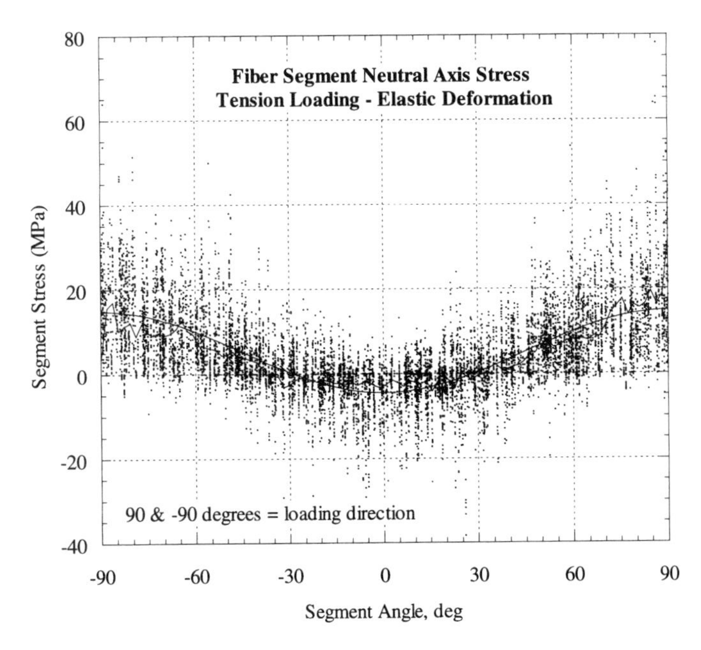

Fig. 13 The stress in each fiber segment presented in the network of Fig. 12 as a function of its orientation relative to the loading direction of the network for an imposed elastic uniaxial tension deformation. Each point represents a single fiber segment. The direction of applied tension is at $+90^\circ$ and -90°, whereas 0° is the direction perpendicular to the loading direction.

burst, and tear strength tests are intermixed with nonconventional tests such as those for transverse and in-plane shear and biaxial deformation. Background information, test procedures, and analysis of results will be discussed for each test or state of stress. The paper literature is filled with attempts to correlate paper strength measurements with fiber and sheet properties. A number of models have been proposed to explain the results. Although an important endeavor, this is not the objective of the present chapter.

A. Tensile Strength

Background Perhaps the most common strength test for paper is the tensile test. Two tension test specimen geometries are typically used: a necked-down "dog-bone" specimen or a constant width strip. These are both illustrated in Fig. 14. To properly perform a uniaxial tensile test, the sample should be loaded in a state of pure tension. The most common method is to grip the ends of a test specimen and pull,

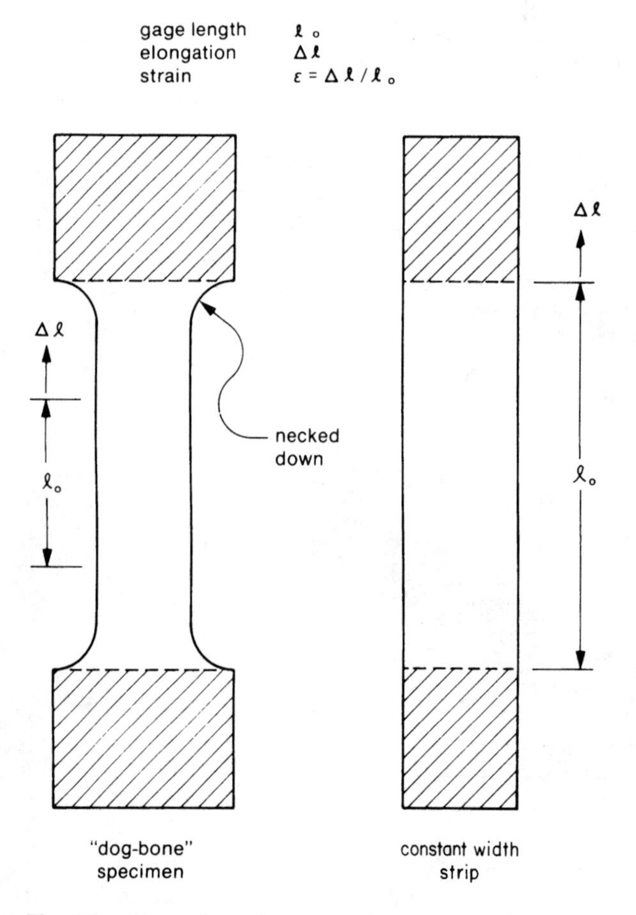

Fig. 14 General tensile test specimen geometries. Note that the radius of curvature of the necked-down region of the dog-bone geometry is variable. ℓ_0 = gauge length; $\Delta\ell$ = elongation; strain $\varepsilon = \Delta\ell/\ell_0$.

using either a mechanical or hydraulic testing machine, while constantly measuring the load applied and the imposed elongation of the gauge section. The grips are used to clamp onto the paper specimen by compressing the paper through the thickness, and a tensile load is applied to the material between the grips as the grips are driven apart (Fig. 15). Note that the contact between the grips and sample is frictional, which leads to a shear traction stress across the surfaces of the paper inside the grips (Fig. 15). Natural lateral contraction, which takes place for most materials during tensile loading, is prevented from occurring at the grips, and because of this restraint an in-plane lateral tensile stress is imposed near the grips. Therefore, in the vicinity of the grips the paper is in a rather complex state of stress—not uniaxial tension.

Speckle interferometry was used by Lyne and Bjelkhagen [103] to study the strain pattern of paper and polyester film in the vicinity of the grips. It can be seen that the strain pattern is not uniform near the grips for either the paper or the polyester film (Fig. 16). The influence of heterogeneity is clearly portrayed in the paper specimen in comparison with the more homogeneous polyester. Seo et al. [150] also examined grip area mechanics for paper and showed numerically that the stress state at the grip is not uniaxial tension. They also showed that the gauge section of a sample will not remain flat during loading unless a necked-down specimen geometry is used, particularly for low basis weight materials. Bodig and France [8] also suggested the use of a necked-down specimen to minimize problems associated with failures occurring at the grip.

The dominant resultant force acting on the sample, however, is still the tensile force pulling the two ends of the sample apart. An interesting principle of solid mechanics—St. Venant's principle—states that stress disturbances in proximity to a dominant stress state will decay over some characteristic distance measured from the source of the perturbation. This principle implies that a state of pure tensile stress

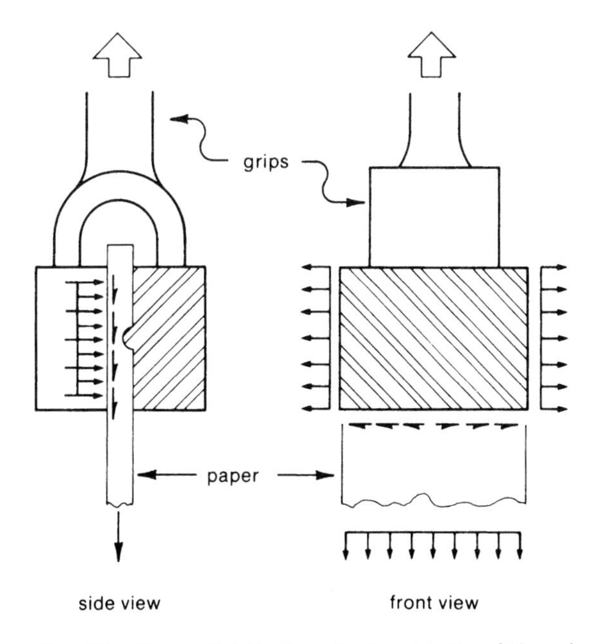

Fig. 15 Force distributions in the vicinity of the grips.

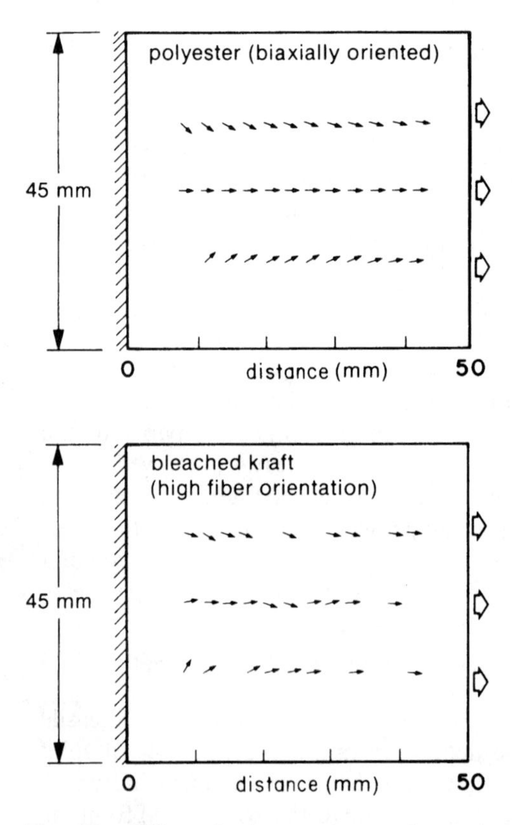

Fig. 16 Effect of grip restraint on the strain distribution of a tensile test sample. (From Ref. 103.)

exists in the middle of the sample if it is sufficiently long. Therefore, the material at equal distance from each grip experiences the closest thing to a uniaxial state of stress.

The most general result from a tension test is the force required to elongate the material of unit cross-sectional area (stress) versus the imposed fractional elongation (strain). Note that strain (Fig. 14) is simply the magnitude of elongation divided by the original length of the measured area. Strain is reported as fractional (as in Fig. 17) or as percentage elongation (fractional value multiplied by 100). Typical tensile material property measures are indicated in Fig. 17.

In addition to problems associated with the grips, the size of the tensile specimens themselves can influence tensile results. Larger samples, for example, will give lower measured strengths. The probability of finding a weak link in the heterogeneous structure is increased if larger samples are used. The statistical nature of failure in relation to microstructural voids and other microscopic parameters has been studied by Fong et al. [50].

The strain rate imposed on a material during mechanical testing is also important. Tensile strength generally increases with increasing strain rate. Therefore the rate of deformation should be kept constant for experiments in a study if comparisons are to be made. In addition, the rate of deformation should be reported so that mechanical measurements can be reproduced and compared. The effect of strain rate on the tensile behavior of newsprint in the machine and cross-machine directions can

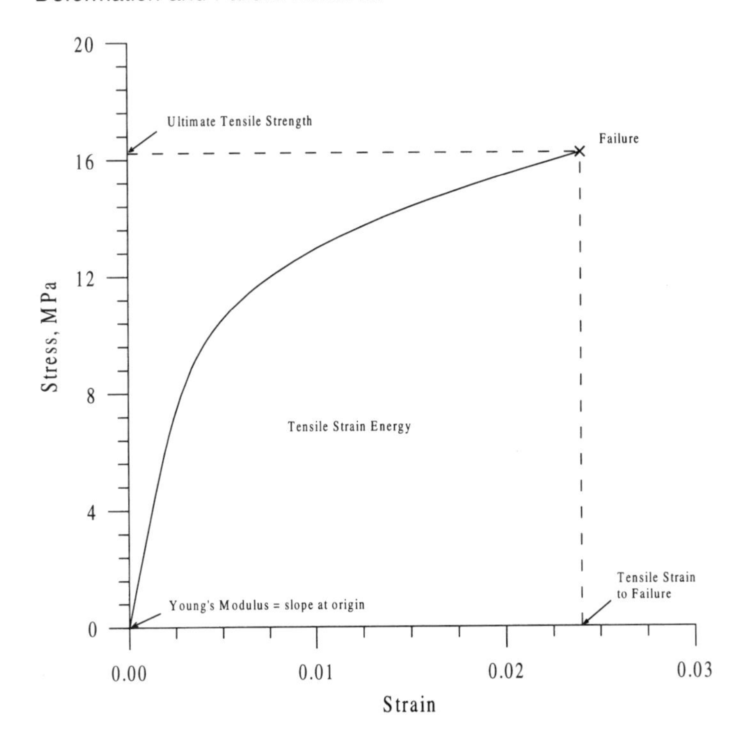

Fig. 17 A typical tensile stress–strain curve for paper showing the typical quantities of interest. Note that the tensile strain energy is the area under the curve to failure.

be seen in Fig. 18. There are three curves for both MD and CD. The upper and lower curves for each were obtained at a constant strain rate. The curve between is taken from a strain rate jump test where the strain rate was increased during the test. Strain rate jump tests are sometimes used to measure the strain rate sensitivity of a material. Note that the curves in Fig. 18 suggest that the tensile strength measured quasi-statically is not necessarily the tensile strength of the material being used in high speed operations, where the rates of deformation can be quite high.

In general, moisture content and temperature are important state variables of paper. The effects of moisture content [4] and temperature [147] on tensile properties of paper are shown in Figs. 19 and 20, respectively. Generally, tensile strength of paper will decrease with increasing moisture content and temperature. Although violated in Fig. 19, it is important to keep in mind that moisture content is the material state variable, not relative humidity (RH) of the surrounding air (RH defines the moisture state of the air, not that of the paper). When reporting results, moisture content of the paper and the relative humidity and temperature of the environment should be recorded.

Paper is a fibrous composite with cellulosic fibers that are hydrogen bonded to each other. Tensile failure of paper is therefore a process that involves breakage of the bonds between fibers and rupture of the fibers themselves. The issue of whether bond failure or fiber failure is most responsible for the ultimate tensile strength of paper is still a topic of debate [3,120]. Generally speaking, however, the tensile failure of paper is a combination of fiber and interfiber bond failure. The proportion of

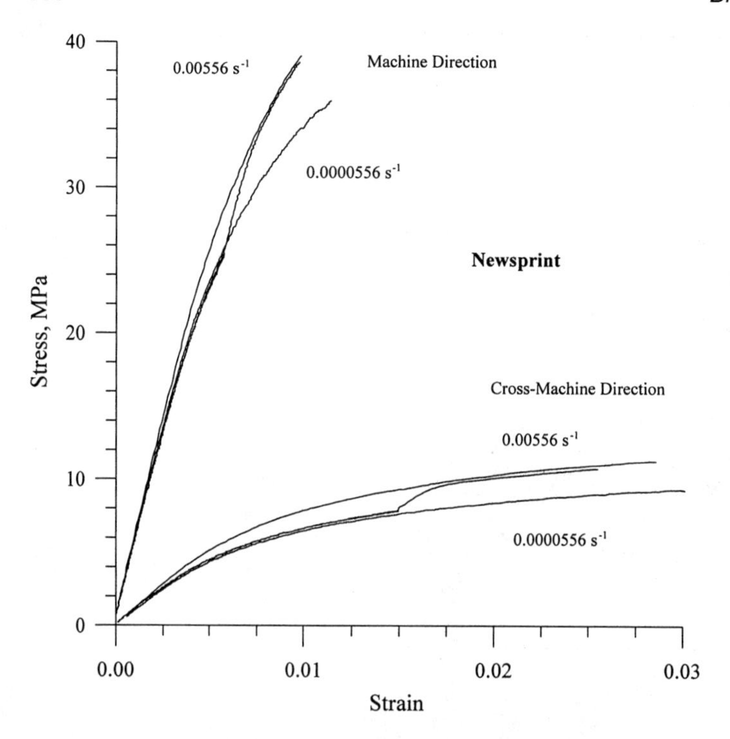

Fig. 18 Stress-strain curves for machine-made newsprint showing the effect of strain rate. The strain rates used are noted and differ by a factor of 100. A strain rate jump test was also performed in each direction, and the results are shown between each of the two constant deformation rate curves.

these two mechanisms involved in failure is likely determined by a combination of variables such as network geometry, fiber strength, degree of fiber latency, interfiber bond area, and interfiber bond specific strength. For example, a paper sample that has a low density and therefore a low interfiber bond area will likely fail in tension primarily by failure of interfiber bonds. On the other hand, a paper sample that has a high density (by wet press densification) and high specific bond strength made with fibers that are weak will likely fail primarily by fiber failure. The tensile failure of paper is still not well understood from a micromechanical standpoint [3,120].

The foregoing brief discussion also demonstrates the importance of density to the tensile strength of paper materials. Increasing the density of paper by wet pressing will increase the tensile strength on account of the increase of interfiber bond area [3]. A more comprehensive account of the tensile response of paper is given by Niskanen and Kärenlampi [122].

Procedure The TAPPI standard test method T494 om-88 [177] contains a procedure to determine the tensile strength of paper. This procedure is summarized as follows.

First, the samples must be prepared, including appropriate temperature and moisture conditioning of the sample material. From representative sources, 10 test strips are cut that are 25 ± 1 mm wide and long enough to be clamped in jaws when the span between the jaws is 180 ± 5 mm. The test specimens should be aligned and

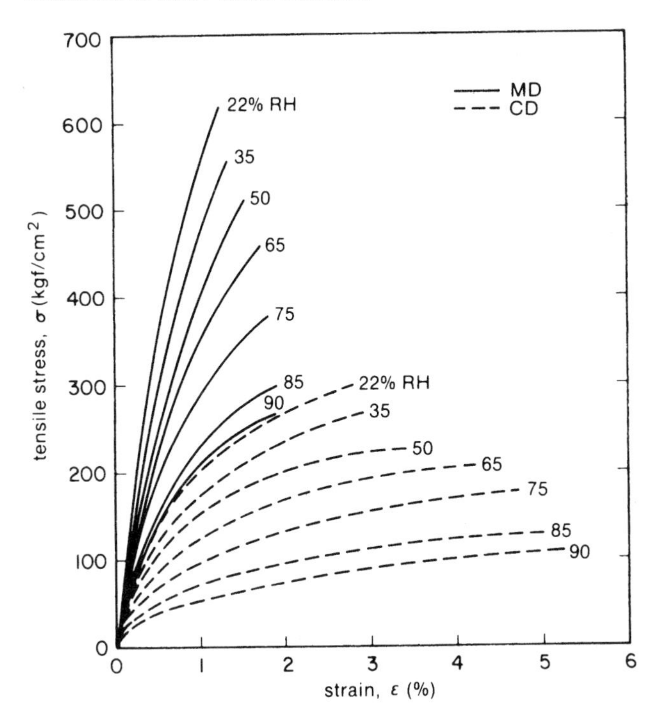

Fig. 19 The effect of moisture content on the tensile stress–strain curves of kraft linerboard. (From Ref. 4.)

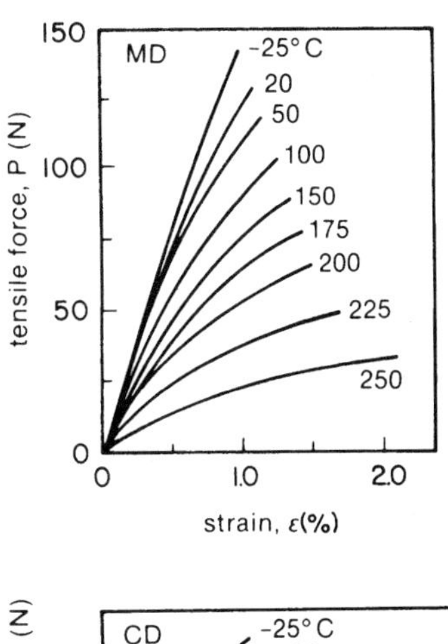

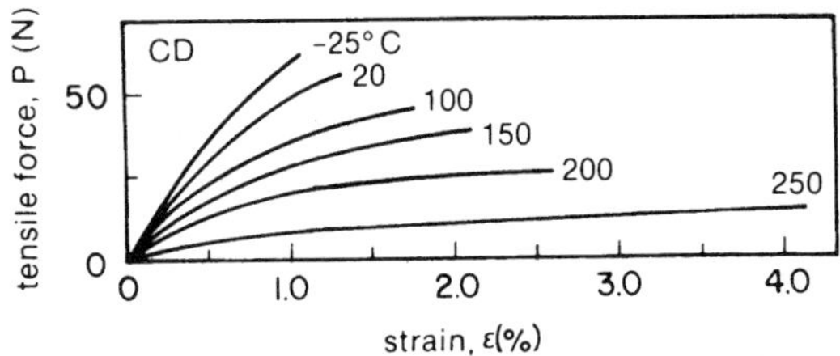

Fig. 20 The effect of temperature on tensile test results of NSSC fluting. (From Ref. 147.)

clamped in the jaws with a line contact grip (Fig. 15). Finger contact with the specimen test area should be prevented, because it would change the temperature and moisture state of the sample. The specimens are inserted into the jaws of a test machine whose speed of grip separation is set at a constant $25 \pm 5\,\text{mm/min}$. A test result is rejected if significant slippage in the jaws occurs or if the specimen breaks near or within the grips. The test procedure calls for three properties to be reported for each sample: the maximum stress (force per unit width), the maximum strain or stretch, and tensile energy absorption (TEA) (work done to rupture divided by the facial area of the sample).

Although the standard procedure prescribes the use of constant width specimens, the tensile specimens for most materials are dog-bone shaped (Fig. 14). In this way the size difference between the gripped and gauge sections can be chosen such that the stress state in the gripped portion remains smaller than in the gauge section so that failure occurs somewhere in the gauge section. If strain is of interest, the strain near the center of the gauge section is generally measured with a strain gauge extensometer (hence the term "gauge section"). This is a small device that clips onto the sample and measures the elongation of the center 1 in. or so of the gauge section. For low basis weight materials in particular, the use of a strain gauge extensometer can be difficult. On this account non-contact extensometers are also commercially available that use high resolution video cameras or a laser to measure the distance between two materials points on a line parallel to the loading direction. With the use of either type of extensometer, the strain is measured over the portion of the specimen that is experiencing the closest thing to uniaxial tension, i.e., the central part of the gauge section. A stress–strain curve, as shown in Fig. 17, is the most common way to represent tension test results. Ultimate tensile strength, strain to tensile failure, and tensile energy absorption can all be derived from the stress–strain curve.

Analysis of Results Once again, a typical stress–strain curve is shown in Fig. 17, in which four mechanical behavior measures are illustrated. The ultimate tensile strength is the stress at failure and has units of Newtons per square meter (Pascals, Pa). The tensile strain to failure is the magnitude of strain at failure and has units of meters per meter; or percent. The Young's modulus of elasticity is the slope of the stress–strain curve at the origin and has units of stress, Newtons per square meter (Pa). The tensile strain energy (or tensile energy absorption) is the area under the stress–strain curve and has unit of joules per cubic meter.

Because measurement of the thickness of paper has an uncertainty associated with it [143,161,175,209], a practice of expressing strength as a force per unit width rather than a force per unit area has evolved. Furthermore, it has become customary to report the maximum (or ultimate) tensile force to rupture of the test strip in terms of a fictitious length—the length of a strip of the same material whose weight is equivalent to the force that would break it. This is known as the breaking length. The ultimate tensile load is also reported as a tensile index that is equivalent to specific strength σ_u^*, defined as the actual ultimate tensile strength σ_u divided by the apparent density ρ of the material. it is easy to show that the specific strength σ_u^* is equal to the maximum load per unit width, $s_u = P_{\max}/b$, divided by the basis weight m_A (mass per unit area):

$$\sigma_u^* = \frac{\sigma_u}{\rho} = \frac{P_{\max}/hb}{m/hA} = \frac{S_u}{m_A} \tag{12}$$

where it is seen that the thickness h cancels. Consequently, if a tensile strip is made long enough, its weight could be made equivalent to the force required for failure, given as

$$L = \frac{P_{\max}}{bm_A g} \tag{13}$$

where g is the acceleration due to gravity. This measure of strength is termed the breaking length, L. Thus, specific strength is proportional to the breaking length:

$$\sigma_u^* = gL \tag{14}$$

Therefore, breaking length L can be determined from specific strength and the acceleration due to gravity, g. Conversion factors required to make these calculations depend on the values of g in the different systems as well as on the definition of the forces themselves. Such conversions are illustrated in Table 1, where unit specific strengths are converted to breaking lengths for some commonly used units.

Tensile strain energy to failure is represented by the area under the stress–strain curve. Since work done is force times the distance through which the force acts, the area under the curve represents total energy consumed per unit volume of material. Units of this quantity are J/m^3 (SI), dyn cm/cm^3 or kg_f cm/cm^3 (cgs), and lb_f in./in.^{3} (English).

B. Zero-Span Tensile Strength

Background If the bonds between fibers are sufficient to withstand any level of stress transfer from one fiber to the next, it is intuitively obvious that fiber tensile strength will be related to sheet tensile strength. In 1925, Hoffman-Jacobsen [68] introduced the idea for a zero-span tensile test to distinguish the difference between

Table 1 Conversion of Specific Strength to Breaking Length

System of units	Specific strength[a]		$(\sigma_u^* = S_u/m_A)$	÷	Acceleration due to gravity (g)	=	Breaking length L
SI	$1\,\frac{\text{Nm}}{\text{kg}}$	or	$\frac{(\text{kg m/s}^2)\,\text{m}}{\text{kg}}$	÷	$9.81\ \text{m/s}^2$	=	0.102 m
cgs	$1\,\frac{\text{dyn m}}{\text{g}}$	or	$1\,\frac{(\text{g cm/s}^2)\ \text{cm}}{\text{g}}$	÷	$981\ \text{cm/s}^2$	=	0.00102 cm
	$1\,\frac{\text{kg}_\text{f}\ \text{cm}}{\text{kg}}$	or	$981\,\frac{(\text{kg cm/s}^2)\ \text{cm}}{\text{kg}}$	÷	$981\ \text{cm/s}^2$	=	1 cm
English	$1\,\frac{\text{lb}_\text{f}\ \text{in.}}{\text{lb}}$	or	$32.2\,\frac{(\text{lb in./s}^2)\ \text{in.}}{\text{lb}}$	÷	$32.2\ \text{ft/s}^2$	=	1 in.

[a] 1N is equivalent to $1\ \text{kg} \times 1\ \text{m/s}^2$; 1 dyn is equivalent to $1\ \text{g} \times 1\ \text{cm/s}^2$; $1\ \text{kg}_\text{f}$ is equivalent to $1\ \text{kg} \times 981\ \text{cm/s}^2$; $1\ \text{lb}_\text{f}$ is equivalent to $1\ \text{lb}_\text{m} \times 32.2\ \text{ft/s}^2$.

the strength of the fiber network and the strength of the fibers themselves. In this test, the grips are brought together and clamped to the specimen in such a manner that the resultant tensile force is applied across a plane through the thickness (Fig. 21). By pulling the paper apart along a line (zero gauge length, hence the name of the test), it is assumed that the fibers are pulled apart and consequently fiber strength is determined. This idea of using a zero-span test method to measure the strength of fibers has a lengthy and somewhat controversial history. Clark [30] reviewed the work prior to the 1940s and concluded that the zero-span tension test results may not be a fundamental measure of ultimate fiber strength but that they are probably fairly good indicators of this property. Van den Akker et al. [193], on the other hand, redesigned the testing apparatus and concluded, on the basis of experimental data and theoretical reasoning, that the zero-span test was a valid measure of fiber strength.

Fiber pull-out is particularly important to zero-span tensile strength because fibers not clamped in both jaws cannot bear load and hence the tensile force required to cause sample failure will be reduced. Pullout is affected by both fiber length and interfiber bonding. Shorter fibers have a higher percentage of fiber ends that do not cross the clamping line compared with long-fiber pulps. Poorly bonded networks will exhibit more fiber pullout then well-bonded networks. A technique to correct for the effects of fiber pullout was demonstrated by Boucai [11]. The concept is to conduct a series of short-span tests with a zero-span apparatus that defines zero-span tensile strength as the limiting value of the sequence of strengths as the span length, suitably

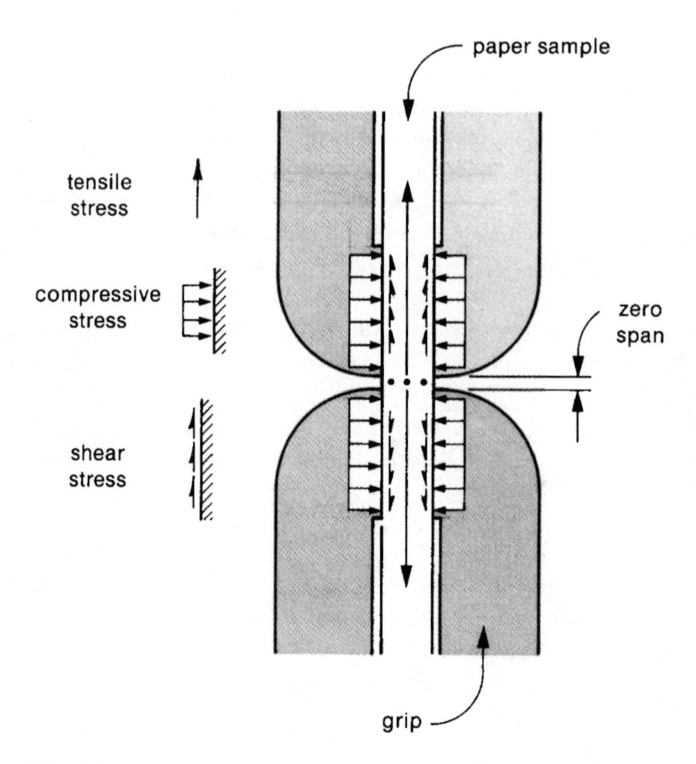

Fig. 21 Gripping arrangement for the zero-span tensile test.

adjusted, approaches zero. A residual span is added to the actual span, the rationale for this adjustment being that some straining must occur before failure. Cowan published a small monograph [32] that deals in detail with the empirical modifications that will lead to consistent results. Additional sources of information can be found in this useful publication.

In some studies it has been found that zero-span tensile strength changes with interfiber bonding and is thus affected by sheet structure. To account for this effect, Cowan suggests the use of wet or rewetted sheets based on the premise that wet sheets should effectively remove all bonds. Wet sheets, however, tend to give results somewhat lower than those of dry sheets. Gurnagul and Page [61] present data that shows lower zero-span tensile test results for wet sheets than for dry sheets. They argue that the drop in zero-span strength from dry to wet cannot be caused by loss of bonding, because the sheets with the highest bonding as indicated by finite-span tensile strength do not necessarily give the greatest loss in zero-span tensile strength.

The difference between dry and wet zero-span strength seems to be more closely related to pulp type. It is argued by Gurnagul and Page [61] that wetting of sheets to remove the effect of bonding reduces the strength of individual fibers that have been degraded by chemical or mechanical treatments. They hypothesize that the magnitude of the loss in strength is controlled by chemical differences in the hemicellulose–lignin matrix holding the cellulose fibrils together in the cell wall of the fiber. They found that, in general, standard unbleached kraft pulps of softwoods and hardwoods change very little in strength upon wetting. Unbleached sulfite pulps (spruce) were found to be weakened more upon wetting than kraft pulps. Upon bleaching, all kraft and sulfite pulps showed a lower wet zero-span strength than the corresponding unbleached pulps. And, in general, sheets made from beaten pulps showed a larger decrease in rewetted strength than the corresponding unbeaten pulps.

According to Mohlin [116] and Mohlin et al. [117], zero-span tensile test results are influenced by fiber structural deformities such as kink and curl (see Chapters 13 and 14). Fiber deformities are "industrial" fiber properties, which means that extra care must be taken when they are studied in the laboratory. Pulp handling in the laboratory does not introduce fiber deformities to the same extent as in the mill, and the commonly used laboratory beating procedures often remove those that are there. In industrial pulps, fiber deformities introduced during pulping and bleaching thus have an effect on paper quality that is hidden by laboratory beating. Experimental results suggest that an additional reason for the differences observed between wet and dry zero-span tensile strengths may be that fines could bridge a deformed or weak section of the fiber in the dry state and thus improve its load-transmitting ability.

In recent work, Seth [154] argues that fiber bonding and fiber length have little effect on zero-span tensile strength. He further argues that because the zero-span tensile strength of commercial papermaking fibers is found to increase with beating or refining, it has been erroneously concluded that fibers are better gripped and carry load more effectively because of better bonding. The increase in zero-span tensile strength with beating or refining comes rather from straightening of fibers and removal of kinks and crimps. In light of the strong evidence that bonding has little effect on zero-span tensile strength, it is argued that the notion that there is a need for rewetting dry sheets to eliminate the effects of bonding is unsupported.

Zero-span tensile test results are most often corrected for differences in basis weight, in much the same fashion as common tensile test results are converted to tensile index or breaking length. What is often overlooked in zero-span tensile testing is that the number of fibers participating in the failure zone must be taken into account before inferences about fiber strength can be made. For a given pulp, the zero-span tensile strength is directly proportional to the number of load-bearing fibers participating in the test. The number of fibers, N, clamped in a test strip of known width L (mm) is calculated from measurements for sheet basis weight BS (g/m^2) and fiber coarseness C (mg/100 m), according to the relationship $N = 100 \times L \times \mathrm{BW}/C$. It is important to mention that not all of these N fibers share the load uniformly or fail at the same time, owing to differences in fiber orientation as well as variation in fiber strength. It would therefore be necessary to correct the zero-span tensile strength results for these effects.

The fiber orientation distribution of the test sheet will also affect the zero-span tensile strength value. Statistically, the strength of a randomly oriented sheet should be only 37.5% of that of a completely oriented sheet. Only fibers loaded to ultimate stress will fail and subsequently shift their loss in load-bearing ability to neighboring fibers. This concept can be illustrated with a slight modification to the models already proposed. In Fig. 22, a random array of line segments is shown being intersected by two parallel lines representing the zero-span grips. Without too much difficulty, it can be shown that the stress-strain diagram for the ensemble of N fiber segments is given by

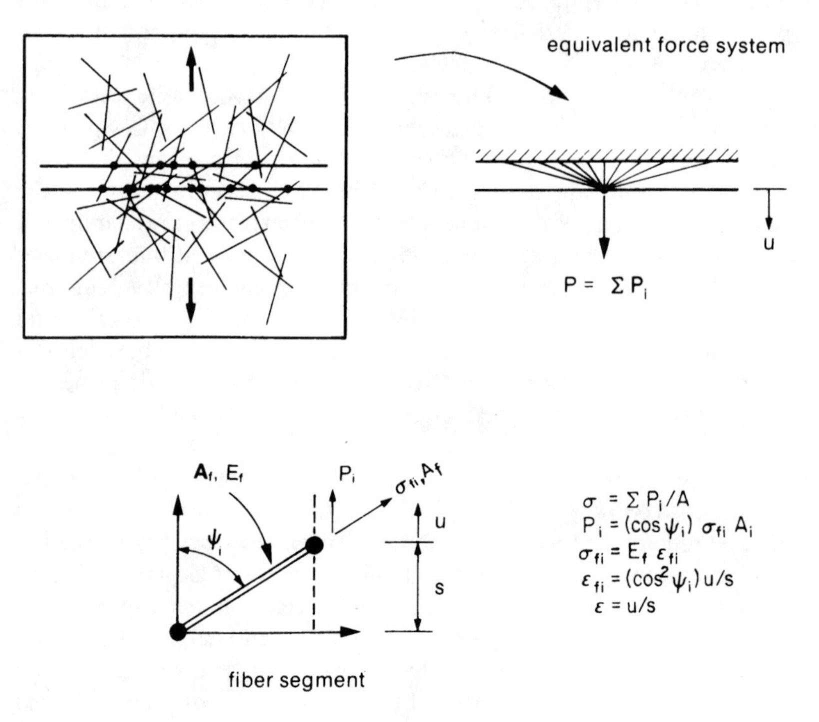

Fig. 22 Model for pulling apart a random array of elements along a line.

$$\sigma = \left(E_f \frac{A_f}{A} \sum_{i=1}^{N} \cos^3 \psi_i \right) \varepsilon \tag{15}$$

where σ and ε represent the effective stress and strain of the sheet, A the cross-sectional area of the sheet, and E_f, A_f the elastic modulus and cross-sectional area, respectively, of the fibers. The fibers are assumed to be identical and to fail at a particular ultimate strain ε_u. Because the largest strains are found in the fibers whose axes are perpendicular to the grip edges, these will fail first. Once failed, they will be removed from the ensemble; N is decreased. Consequently, the stiffness will decrease, slowly at first, then rapidly as the stress is shifted to fiber segments with larger orientation angles The orientation distribution of the fibers plays a major role in this model. El-Hosseiny and Bennett [44] also developed a more comprehensive model to account for the statistical properties of the fibers.

Whether or not the zero-span test measures what is purported to be measured, that is, individual fiber strength, is a question that has yet to be convincingly resolved. Theoretical arguments developed to date appear to be somewhat simplified and perhaps a bit weak, but experimental data obtained on individual fibers and zero-span results from handsheets made from these fibers suggest that there is a relationship. For example, Kellogg and Wangaard [86], Van den Akker et al. [193], and Cowan and Cowdrey [33] use two-dimensional models to derive expressions for expected behavior. However, the overall tensile load is transferred into the sheet by shear stresses acting on the grip face. A sufficient amount of pressure must be applied to the faces to ensure this transfer, but these stresses are not accounted for in the theories. Lateral restraint by the grips adds another stress component to the already complex stress state that exists in the sample.

Although the philosophical basis of the test may be in doubt, it has been found that the zero-span tensile strength of pulp can be related to end-use performance test results [32]. In fact, Page [128] developed an expression relating zero-span tensile strength to uniaxial tensile strength results. A relationship of this nature can be useful in quality control procedures.

From this brief discussion it can be concluded that although the idea of pulling a material apart along a line so as to measure the intrinsic strength of the microstructural elements has an intuitive appeal, it is, in fact, fraught with theoretical difficulties. From a practical point of view, however, the results of a properly conducted zero-span test can be used to characterize pulp quality, which in turn can serve as an index to monitor expected end-use performance once a relationship has been established.

Procedure The test method for the zero-span tensile test is given in the TAPPI standard Zero-Span Breaking Length of Pulp, T231 su-70 [177]. Boucai [11] discusses procedures for short-span testing. A sample size of 6–18 mm wide and any convenient length over 15 mm is used. It is very important that the jaws be perfectly aligned when they are clamped on the specimen. Clamping pressure should be adjusted to a level that ensures no slippage at the grips but does not affect the results. Wink and Van Eperen [205,206] found that microscopic burrs along the jaws of the grips caused stress concentrations and reduced the zero-span tensile strength substantially. The jaws should also be free of any traces of oil or other friction-reducing agents. Van den Akker et al. [193] maintain that a wide range of clamping pressures

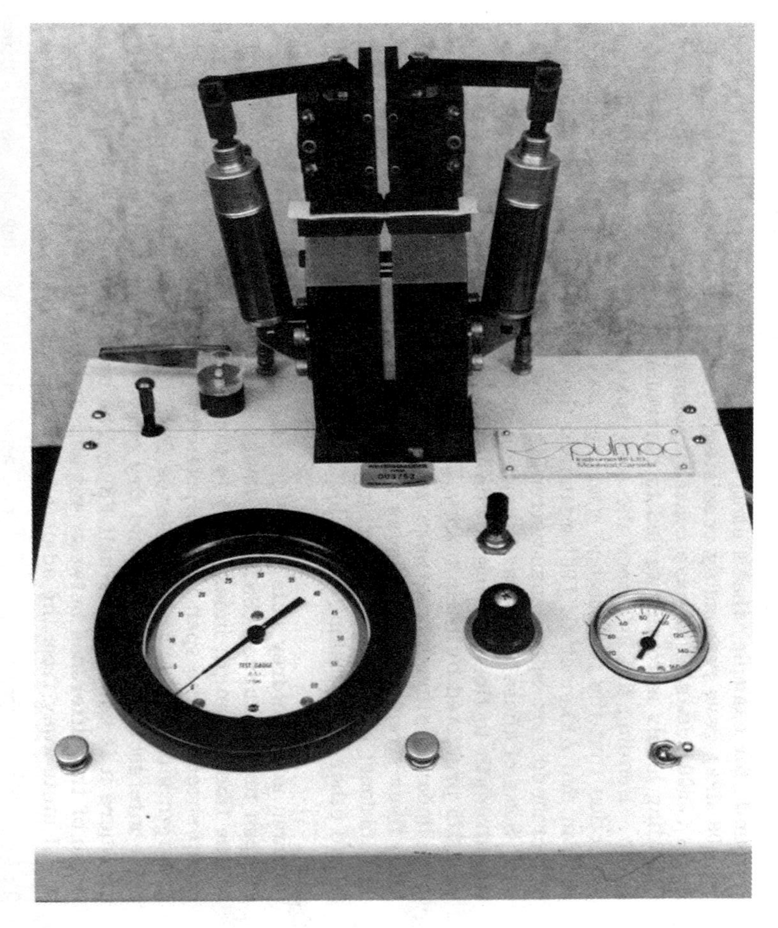

Fig. 23 Two types of zero-span test apparatuses.

are acceptable. An experiment can always be run to verify the consistency of breaking load with clamping pressure if in doubt.

The zero-span tensile test is performed rather rapidly (2.5 s), and the final breaking load is recorded. The breaking load divided by cross-sectional area will yield a nominal ultimate zero-span tensile strength or, if divided by sample width, a maximum force per unit width. The breaking load can be converted to breaking length in a manner similar to tensile testing discussed previously. The units are the same as those of tensile strength. Two types of zero-span apparatuses are shown in Fig. 23.

It is up to each investigator's discretion whether or not to use wet, rewet, or dry samples and whether or not to run tests at machine zero span or to use multiple spans and then supply Boucai's "extrapolation" method to estimate true zero span for which the fiber pullout effect is minimized. Some people prefer to run multiple samples with increasing fiber bonding as a way of minimizing fiber pullout. Other people prefer to pretreat pulps with moderate PFI refining (500 revolutions) to minimize fiber curl effects on zero-span tensile strength results. Finally, when pulps known to differ in fiber coarseness are tested, some attention should be given to the number of fibers, N, involved in the zero-span tensile test.

Analysis of Results Wink and Van Eperen [206] employed a number of procedural variables to establish operating characteristics of the zero-span test method. They found, for example, that a clamping pressure range of 6–11 kg/mm^2 (59–108 MPa) gave reasonably consistent results; loading rates over 4 kg/(cms) [3900 N/(m s)] also gave constant values. Basis weight, formation, and bonding levels were also varied to determine their effect on zero-span strength. Bonding did influence the zero-span results, which runs contrary to the notion that the test method measures only fiber strength. On the other hand, Van den Akker et al. [193] measured the strength of individual fibers and determined zero-span tensile strengths from handsheets made from these fibers. Some of these values are shown in Table 2. The average zero-span tensile strength/fiber strength ratio is 0.38, which approximates the value of 0.375 predicted by their theory. Kellogg and Wangaard [86] obtained a

Table 2 Comparison of Individual Fiber Strengths, Zero-Span Tensile Strengths, and Tensile Strength of Handsheets

Fiber type[a]	Average fiber length (mm)	Tensile strength of handsheets[b] (MPa)	Zero-span tensile strength[c] (MPa)	Fiber strength (MPa)	Ratio of zero-span to fiber strength
Douglas fir (U)	3.7	78	257	627	0.410
Southern pine (B)	3.3	83	207	590	0.351
Sweetgum (B)	1.7	46	227	526	0.432
Sweetgum (U)	1.7	8	217	665	0.326

[a] B = beaten; U = unbeaten.
[b] Approximately the same basis weight: 62–65 g/m^2.
[c] Computed on the basis of equivalent area of cellulose, using a density of 1.55 for the fiber.
Source: Data from Ref. 193.

linear relationship between zero-span breaking length and individual fiber strength with 5 km as the intercept and a slop of 0.00019 km/psi. Additional information concerning the relationship between zero-span tensile tests and other papermaking variables is given in the literature cited by Cowan [32].

Cowan and Cowdrey [33] developed four indices based on the short-span tensile curves (SSTCs) obtained from wet and dry measurements (Fig. 24). The fiber strength index (FSI) is the value of the intercept of wet and dry pulp sheets as the curves are extrapolated beyond the nominal zero span. The fiber length index (FLI) is the distance from the "true" span, associated with the intersection of the wet and dry curves, to the position on the wet curve where it is equal to one-half FSI. The bonding index (BI) is equal to the ratio of the difference between wet and dry strength evaluated at the span associated with the fiber length index to one-half the FSI expressed as a percentage. The orientation index (OI) is the ratio of FSI in the machine direction to the FSI in the cross-machine direction. Ionides and Mitchell [73] used these indices to predict the results of end-use performance tests and pointed out a few limitations of the procedures. Law et al. [98] further refined the use of these indices.

It appears that the zero-span test has evolved into a rather practical but somewhat arbitrary test method that is finding usefulness in quality control applications. It is an example of a test whose theoretical basis is much more complex than the simple idea from which it originated but whose usefulness may not depend on whether the theory is understood.

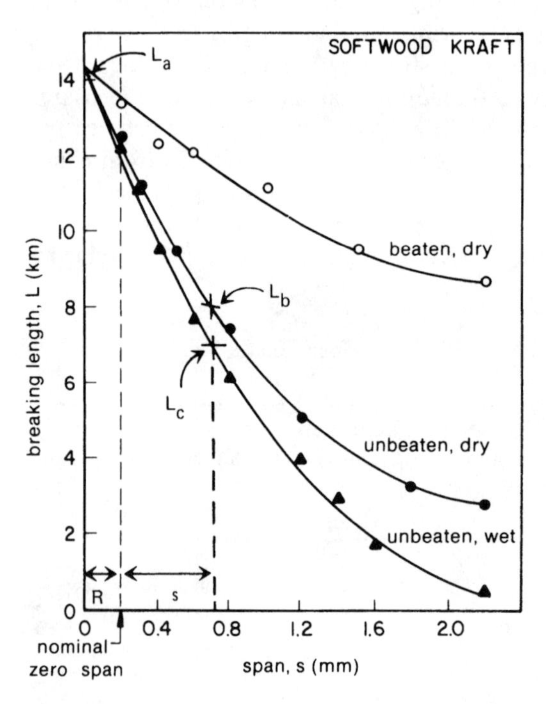

Fig. 24 Variation of zero-span strength with initial span separation. Fiber strength index (FSI) $=L_a$; fiber length index (FLI) $= R + S$; bonding index (BI) $= 100(L_b - L_c)/L_c$; orientation index (OI) $= L_a(\text{MD})/L_a(\text{CD})$. *Note*: $L_c = (1/2)L_a$. (From Ref. 32.)

C. Through-the-Thickness Tensile Strength (*Z*-Direction Tensile Strength)

Background Adequate fiber–fiber bonding is required for stress transfer between fibers in a sheet of paper. A perfect bond, not necessarily of infinite strength, would resist transfer stresses, allowing the fibers themselves to be weak links in the system. The surface area of a fiber is much greater than its cross-sectional area. For a hypothetical fiber, collapsed to a rectangular ribbonlike shape with dimensions 0.01 mm thick, 0.02 mm wide, and 1 mm long, the ratio of surface area to cross-sectional area is 300. Hence, the bond strength would need to be 0.3% of the fiber strength to be considered perfect.

This analysis implies, of course, that all of the surface area is involved in stress transfer and that there are no stress concentrations. In reality, only a portion of the surface area, the relative bonded area (RBA), is involved in this stress transfer; accordingly, a bond strength much greater than 0.3% of the fiber strength is required in practice. Unequal stressing at various bond sites while keeping the cross section as before would also increase the requirement. On the other hand, increasing the fiber length while keeping the cross section as before would decrease the required bond strength. From this cursory analysis it would appear that adequate bond strength does not have to be very large to ensure stress transfer from fiber to fiber.

Because most of the bond areas are roughly coplanar with the surface of the sheet, a tensile test that pulls paper apart through the thickness would be one possible way to characterize the extent and quality of the bonding. Tensile strength through the thickness is referred to as *Z*-direction tensile strength or, in certain cases, internal bond strength. Two types of test methods have been used to measure this strength (Fig. 25). In the first test method (*Z*-tensile), highly polished metal blocks are glued to the surfaces of the sheet of paper and subsequently the blocks are pulled apart. In this case, failure is normally reported in units of failure stress. In the second test method

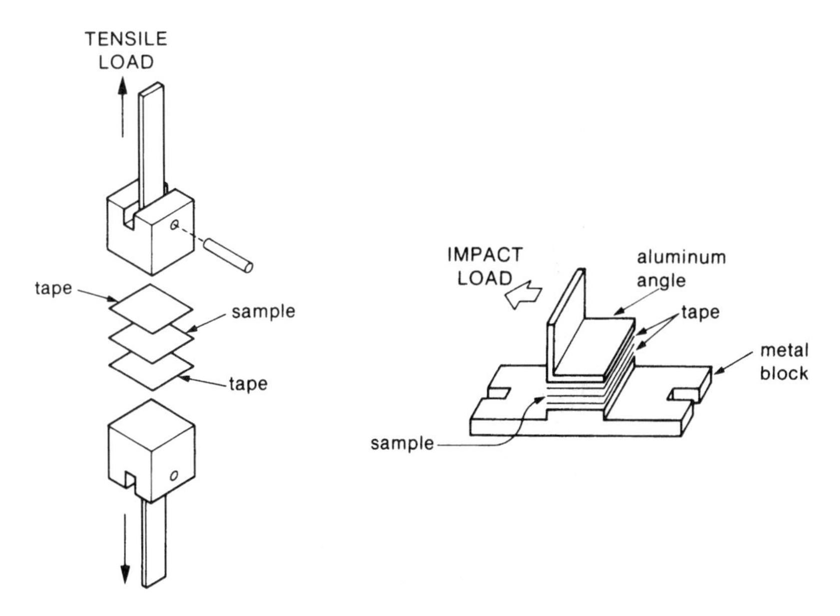

Fig. 25 Two types of fixtures for measuring through-the-thickness tensile strength.

Scott bond (Fig. 26), an impact load is applied to one arm of a right-angle metal bracket. the other arm, glued to the faces of the paper test specimen, pulls the sample apart while rotating about a pivot point. This method reports results in units of energy. Neither method measures the true intrinsic fiber–fiber bond strength, because the number and size of bonds involved in the fracturing process cannot be readily determined. The fiber walls themselves could also be torn apart, which negates the assumption that the fracture process involves only bond failure.

There are considerable experimental difficulties associated with through-the-thickness testing. The most difficult problem in the first method is that of ensuring the application of uniform load over the entire surface of the sample as it is pulled apart. The tensile forces must be perfectly aligned through the center of the planar

Fig. 26 Scott bond test apparatus.

area of the sample. Even if the alignment is satisfactory, nonparallel separation of the grip faces may occur because of natural heterogeneity in the sample. The method of attaching the samples to the grip faces represents another problem. The sample is attached to the test blocks with adhesive glue or double faced tape. It is important that the adhesive system not penetrate the paper sample and provide unwanted reinforcement.

As in the case of the zero-span tensile test, the gauge length of the sample used for measurement of Z-direction tensile strength is very small; it is just the thickness of the sheet. One would expect, therefore, considerable local perturbations of tensile stress through the sample. Lateral contraction is also restrained to various degrees in this test; thus, a transverse stress will exist. Both the basis weight of the paper and the mechanical properties of the adhesive system will therefore have an effect on the contribution from lateral contraction. These effects are discussed by Wink and Van Eperen [206].

In the second method (Scott bond), a very complex state of stress exists in the sample at the moment of impact. A nonuniform tensile stress distribution is induced along the surface of the sample as well as a nonuniform shear distribution. Because the maximum tensile force occurs at the end of the arm, the paper will begin to separate at this position. Fracture proceeds into the paper as the angular bracket rotates in response to the impact load. The energy absorbed during this process is used to characterize strength in this test, in contrast to the first method for which a maximum stress is used.

Procedure The Z-tensile strength test is the oldest test used to measure internal bond strength. This is a quasi-static test that can be carried out according to TAPPI Method T541 om-9 [177] or CPPA Standard Method D 37P. The Z-tensile strength is generally defined as the force required to produce a unit area fracture perpendicular to the plane of the paper sample and is expressed in units of kilopascals. In the case of Scott bond [7], a pendulum apparatus is used to apply an impact load to the test assembly (Fig. 26). A pointer attached to the pendulum indicates, on a scale, the energy absorbed during the rupture process. The absorbed energy divided by the cross-sectional area is used as the measure of internal bond. The units are the same as those of TEA.

Analysis of Results Typical Z-direction stress–strain curves are shown in Fig. 27 [195]. The total nonlinearity is evident. It does not appear that the concept of a well-identified safe zone is applicable to tensile behavior in the Z direction. Continuous damage takes place during the entire load history and might indicate a transfer from one type of failure to another. Part of the reason for this behavior might be the nonuniformity of stress, resulting in a wedgelike strain pattern across the face of the sample. The effect of specimen size on the tensile strength was reported by Wink and Van Eperen [206] (Fig. 28). It appears that a sample with an area greater than 400 mm^2 will yield a constant tensile strength.

Because the stress distribution is not known in the Scott bond impact test and the total amount of energy consumed is the only quantity measured, the test results are of limited value with respect to providing fundamental information about through-the-thickness failure. This is an example of a test that was ostensibly designed to obtain internal bond strength but that actually measures the summation

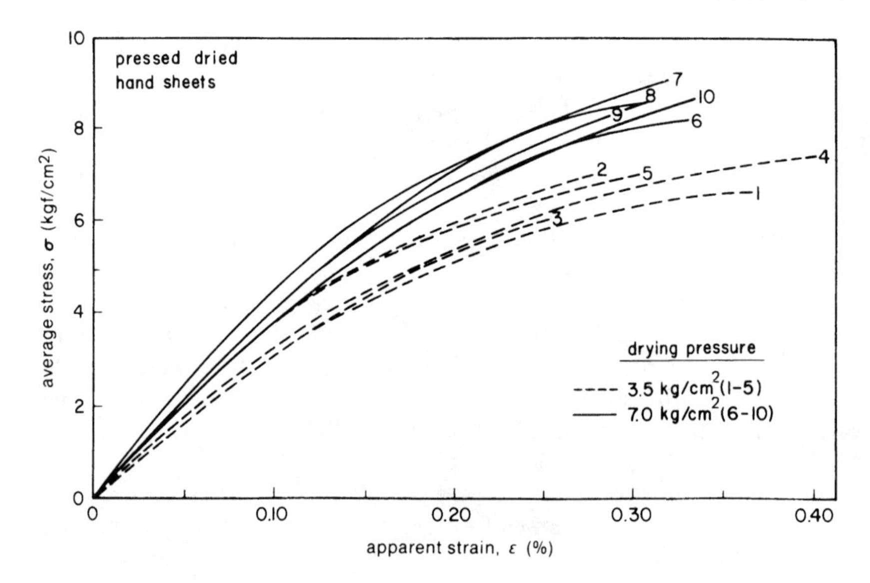

Fig. 27 Typical stress–strain behavior of Z-direction tensile tests. (From Ref. 195.)

of a complex sequence of failure events that occur during a very short period of time. The recorded values should be considered as indices of bond impact strength. Reynolds [141] presented results showing the influence of various papermaking variables on this bond impact index. For example, mixtures of pine and hardwood kraft pulps show an increase in bond strength as higher percentages of pine are used in the mixtures (Fig. 29).

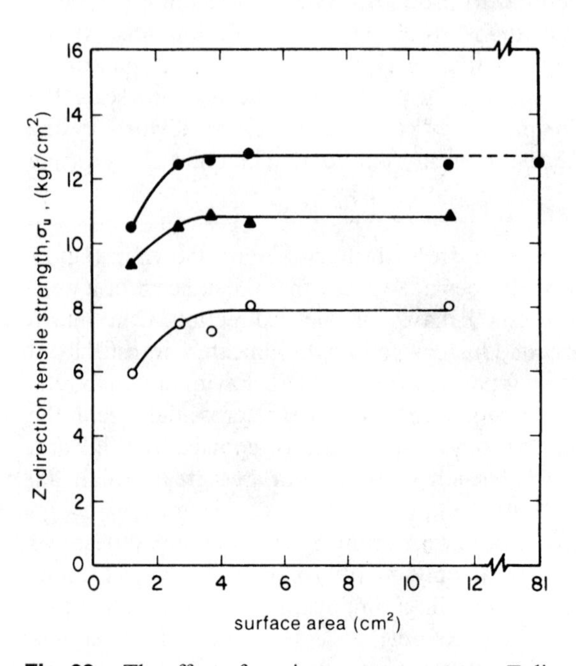

Fig. 28 The effect of specimen contact area on Z-direction tensile strength. (From Ref. 206.)

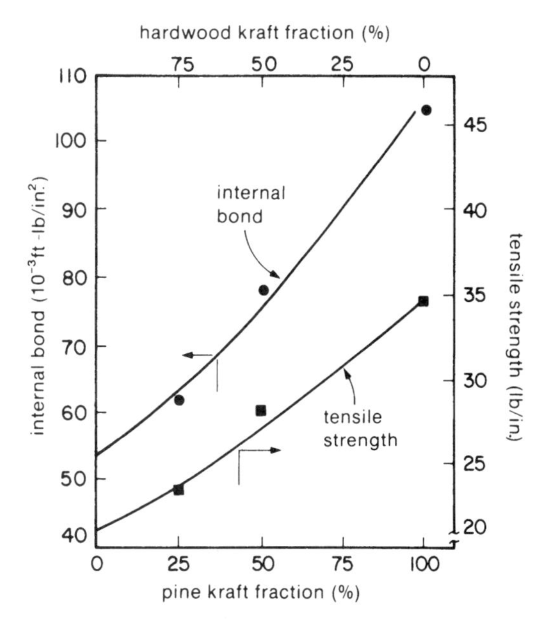

Fig. 29 Relationship between Scott internal bond strength and uniaxial tensile strength as a function of hardwood–softwood fiber mixtures. (From Ref. 141.)

A comparison of these two tests is given by Koubaa and Koran [96]. Comparisons of two methods, even using the same unit of measure, show that although results are highly correlated these tests do not measure the same thing. The authors' analysis shows that the *Z*-direction tensile strength test is better suited for measurement of internal bond strength.

D. Edgewise Compression Tests

Background It is recognized that for in-plane loading paper and paperboard are generally weaker in compression than in tension. Experimental data demonstrating this are given in Fig. 30 for commercially manufactured linerboard. It is important to consider this strength difference when designing structures from paper and paperboard so that states of tensile stress are designed into load-bearing elements of the system as much as possible. Often, though, it is not entirely possible to do that, in which case compressive deformation behavior and strength of paperboard may dominate a structure's performance. For example, corrugated shipping containers can be exposed to several modes of loading such as external compression on more than one face, internal pressure or force from contents, or both. Such forces can lead to biaxial states of stress in the side wall materials (linerboard and corrugating medium), where tensile and compressive stresses coexist. Even though tensile stresses may exist, compressive stresses usually dominate, because loads imposed onto corrugated containers are generally from the outside where only externally applied compression loads are usually possible. The fact that the compression strength of paperboard is lower than its tensile strength makes in-plane compression strength

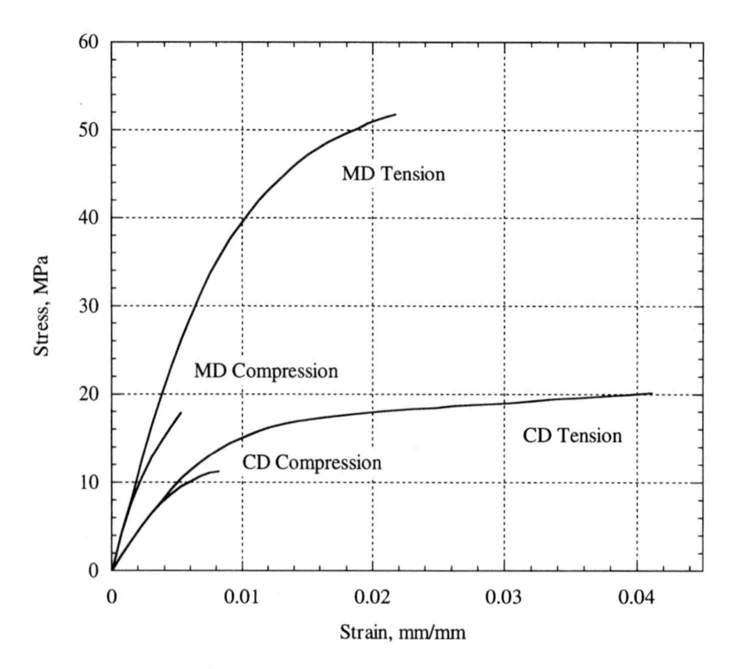

Fig. 30 An example of the difference in strength between compression and tension of paper. The initial slopes of the tension and compression stress–strain curves are the same. (From Ref. 40.)

one of the more important material properties for the performance of linerboard and corrugating medium.

Unlike tension, the mode of paper failure in compression generally depends upon the relative dimensions of the sample. A strip of paper or paperboard loaded in compression may be thought of as a column of length ℓ, thickness h, and width b. If $b > h$ (as it usually is), then a slenderness ratio λ can be defined as

$$\lambda = \frac{\ell}{\sqrt{I/A}} = 2\sqrt{3}\frac{\ell}{h} \tag{16}$$

The variables $I = bh^3/12$ and $A = bh$ are the area moment of inertia and cross-sectional area, respectively, of the strip of paper.

A typical relationship between maximum stress in edgewise (in-plane) compression and slenderness ratio is given in Fig. 31. The slenderness ratio range in Fig. 31 has been divided into three regions:

I.	Long-column height	$100 < \lambda$
II.	Medium-column height	$30 < \lambda < 100$
III.	Short-column height	$10 < \lambda < 30$

One might even consider the range $\lambda < 10$ as a fourth region of column height and behavior [63]. Experimental data in support of Fig. 31 can be found in Figs. 32 and 33 for linerboard and bleached paperboard, respectively. Note that the numerical divisions between regions is approximate because it depends upon the properties of

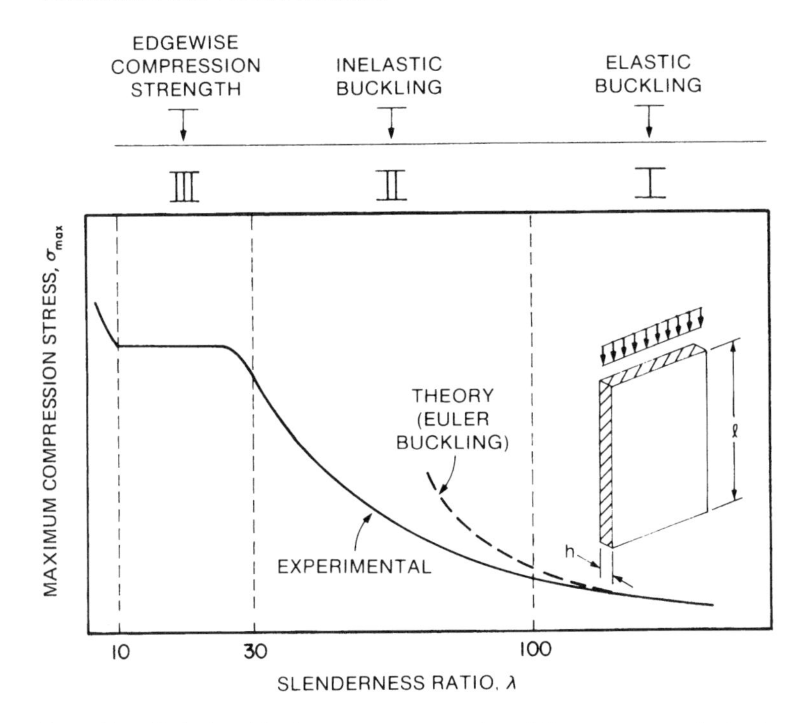

Fig. 31 Relationship between compression failure stress and slenderness ratio. (From Ref. 170.)

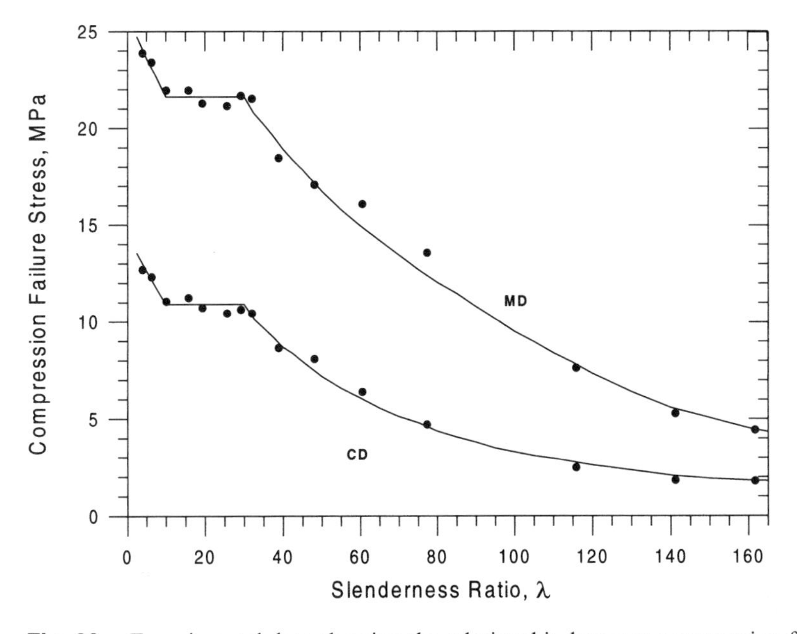

Fig. 32 Experimental data showing the relationship between compression failure stress and slenderness ratio. The measurements were made on machine-made linerboard with basis weight 205 g/m^2 and thickness 0.28 mm. The experiments were performed with a long-span method with lateral support. (From Ref. 170.)

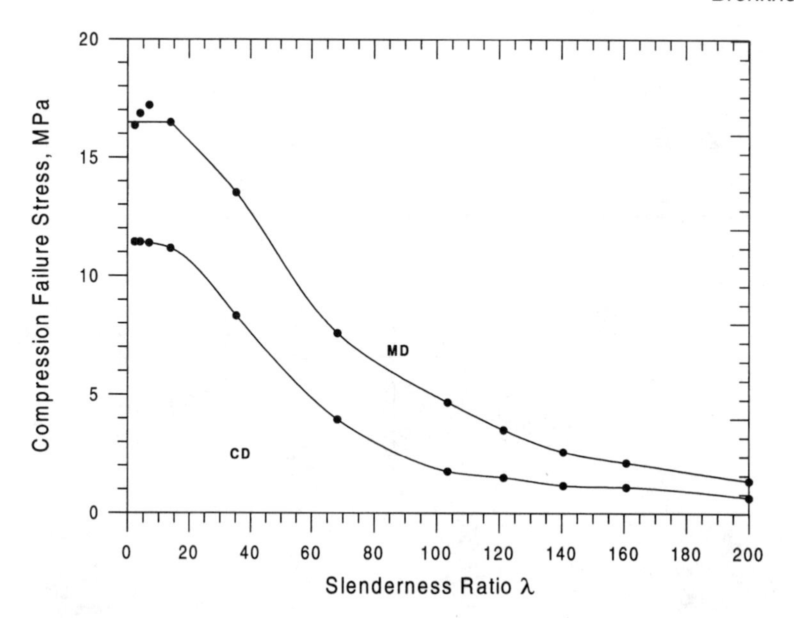

Fig. 33 Experimental data showing the relationship between compression failure stress and slenderness ratio. The measurements were made on machine-made bleached paperboard with basis weight 300 g/m^2 and thickness 0.49 mm. The experiments were performed with a long-span method with lateral support. (From Ref. 28.)

the material under examination. Division of slenderness ratio into regions is instructive, because the paper or paperboard will deform differently in reach region.

In region I, the sample becomes unstable and buckles in the direction out-of-plane to the sample well before the intrinsic compression strength is achieved. This is a mode of structural failure, not of material failure. The material at the level of the fibers has only deformed elastically. The failure mode is one of excessive out-of-plane displacement of the paper sample. The sheet simply finds an alternative equilibrium configuration that minimizes its potential energy. In region I, the displacement is similar to that experienced by a long handsaw if one places the end of the saw on the floor and pushes down on the handle—it bows outward elastically but returns to its original configuration once the pressure is released (assuming the handle is not pushed too far). The maximum attainable stress σ_1 in this region can be described by the classical Euler equation, given as

$$\sigma_1 = \frac{C_1 \pi^2 E}{\lambda^2} \tag{17}$$

where E is the Young's modulus of elasticity, C_1 is a constant depending on the end conditions, and λ is the slenderness ratio defined above. This maximum stress is achieved immediately before the sample buckles. Once the sample buckles, the force required to continue the deformation would be lower. Because the maximum load-bearing capacity is a function of the sample geometry and Young's modulus in this region, the term *edge crush resistance*, rather than strength, is sometimes given to this property.

Edge crush resistance is an appropriate term for the behavior found in region II also. Medium-height columns fail as a result of inelastic buckling and will not return to their original shape on removal of the applied load. This is to say that out-of-plane buckling is the mechanism of structural deformation that limits greater levels of stress from being reached in the material in both regions I and II. The slenderness ratio of the column in region II is smaller than in region I, so the critical level of stress required to initiate out-of-plane buckling in region II is greater than in region I. By the time a region II column has reached its critical stress level for out-of-plane buckling, it has already been deformed plastically, and once the column buckles the buckling deformation also requires that the paperboard deform plastically. If, after buckling, the load is released from a region II sample, it will not return to its original shape.

The phenomenon of inelastic or plastic buckling is more involved than elastic buckling. Different forms of analytical relationships have been derived for the inelastic buckling stress of columns [e.g., 34,63,80]. The simplest relationship for the maximum attainable stress σ_{II} in region II takes the same basic form as Eq. (17) [80]:

$$\sigma_{\mathrm{II}} = \frac{C_{\mathrm{II}}\pi^2 E_T}{\lambda^2} \tag{18}$$

where E_T is now the tangent modulus of the material at failure, C_{II} is once again an end conditions constant, and λ is the slenderness ratio [Eq. (16)].

The tangent modulus of a material is simply the slope of the stress–strain curve. The tangent modulus changes with strain because the general stress–strain response of paper for compression loading is nonlinear. If the stress–strain curve for edgewise compression can be characterized by the function Σ,

$$\sigma = \Sigma(\varepsilon) \tag{19}$$

then the compressive tangent modulus for the material is

$$E_T = \frac{d\Sigma(\varepsilon)}{d\varepsilon} \tag{20}$$

Note that $E = E_T|_{\varepsilon=0}$, so $E_T \leq E$ always.

Even though the concept of tangent modulus is simple, Eq. (18) is not as straightforward as it seems. The tangent modulus used in Eq. (18) is the slope of the material's stress–strain curve at compression failure. This information is not known a priori. Therefore, an actual region II compression test would have to be performed to evaluate E_T for each material.

As we have seen, failures of the column in regions I and II are structural failures. Only when the column height is relatively short (region III) does the material fail. The strength achieved in slenderness ratio region III is often called the intrinsic compression strength or edgewise compression strength of the paperboard. This strength criterion relies only on the ability of the fibers and interfiber bonds to resist externally applied compressive loads without failing. Failure in this region is typically characterized by a well-defined localized shear band produced through the thickness of the sheet at maximum load. This shear band is usually at or near 45° to the loading direction. An example of this shear band formed in a post-failure compression specimen is shown in a photomicrograph in Fig. 34. Unlike failure in tension, which is catastrophic (complete rupture), material failure in edgewise com-

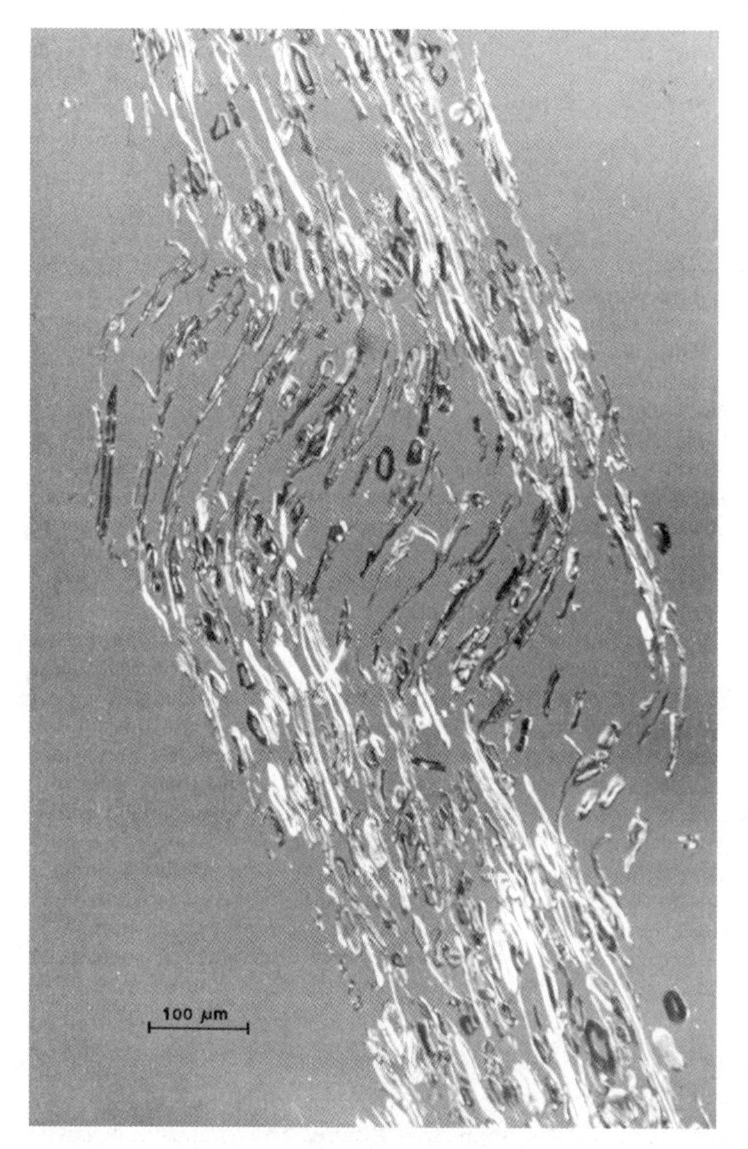

Fig. 34 Characteristic compression failure zone found in samples tested in slenderness ratio region III.

pression is not. Although a sample that has failed in compression is no longer able to support increased compression load, it can be loaded in tension to as much as 80% of its original tensile strength, which suggests that compression failure in region III is perhaps a structural failure of the material at the size scale of a single fiber [28,40,48,145,207].

For slenderness ratios less than 10, it has been reported that the measured compression strength increases beyond that measured in region III [169,170].

Much of the more recent literature on the topic of in-plane compression strength has dealt with behavior of paperboard in region III. This emphasis is consistent with the direction of industry as well. It is certainly well established that sheet density (achieved through wet pressing or refining) is an important variable that heavily influences region III compression strength [207] and implies the importance of interfiber bonding in compression. In addition, as sheet density is increased by means of wet pressing, not only does the level of interfiber bonding increase, but also the free-span fiber length is decreased. It follows that the slenderness ratio existing within the material at the size level of the single fiber is reduced, thereby increasing the potential structural strength of the fibers that are loaded under axial compression. The combined role of interfiber bonding and fiber strength has been discussed before [160,207] in the context of in-plane compression loading.

Some of the more interesting recent work has suggested that region III compression failure of paperboard is an interlaminar shear phenomenon directly related to out-of-plane shear properties [63,107]. Both of these papers put forth mathematical relationships between region III compression strength of paperboard and out-of-plane sheet properties. Although more work is needed, the implications are important for how we think about in-plane compression loading and failure and how paperboard is manufactured.

The topic of in-plane compression strength of paperboard is also covered in Chapter 9 in this volume.

Procedures The test configurations shown in Fig. 35 correspond to the following test methods:

FCT: Fluted edge crush test
RCT: Ring crush test
SSC: Short-span compression test

Details concerning each of these test methods can be found in the following TAPPI standards [177]: FCT, TAPPI Standard T824 om-93; RCT, TAPPI Standards T822 om-93 and T818 om-87; SSC, TAPPI Standard T826 pm-92.

For the edge crush tests FCT and RCT, the specimen is held on edge and compressed between two parallel plates in a standard benchtop compression device at a constant rate of loading or a constant rate of displacement of the top loading plate relative to the bottom loading plate. Ring crush method T822 specifies a constant platen displacement rate of 10 mm/min. Ring crush method T818 and fluted edge crush method T824 both specify a constant loading rate of 111 ± 22 N/s. The sample size for all three tests is 152.4 mm long and 12.7 mm wide. The sample is loaded edgewise in the direction of the 12.7 mm dimension. In all three cases, the maximum achieved load or the maximum achieved load per unit length (kN/m) is recorded as the edge crush resistance. Sometimes this value is converted to units of breaking length in a manner similar to the procedure for the tensile test. In all cases, it is the upper free edge of the test specimen that is loaded by the movement of the top loading plate. Because all three edge crush tests involve direct edge loading of the paperboard, it is exceedingly important that uniform contact between sample and plate be established. Nonuniform loading will significantly reduce the measured edge crush resistance. Uniform contact is more probable when die-cut samples with straight, parallel edges are used and if it is made certain that loading plates are flat and coplanar. If we neglect the fact that the paperboard is not tested in a flat configuration (i.e., ring or fluted), the slenderness ratio of samples tested with these

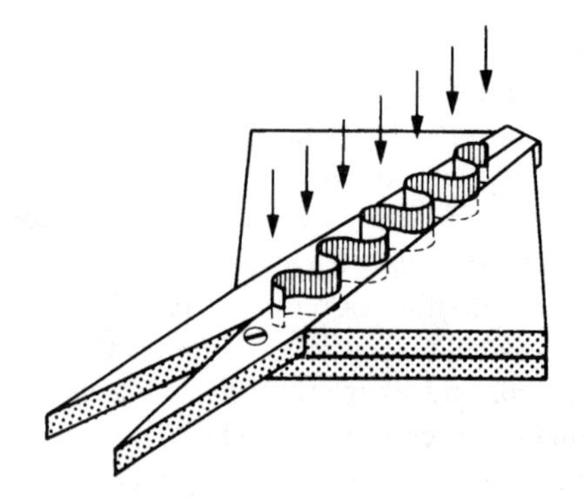

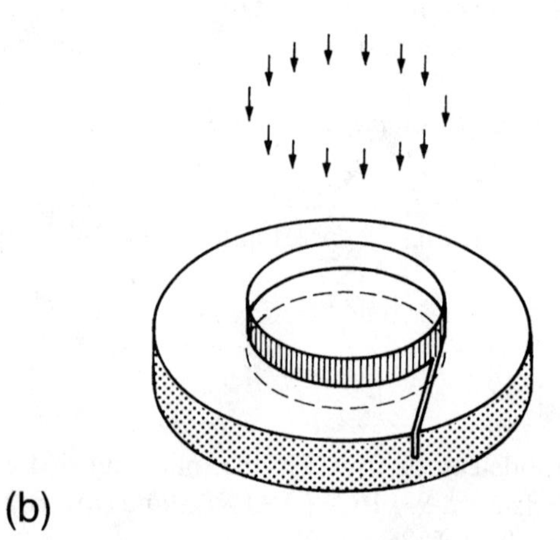

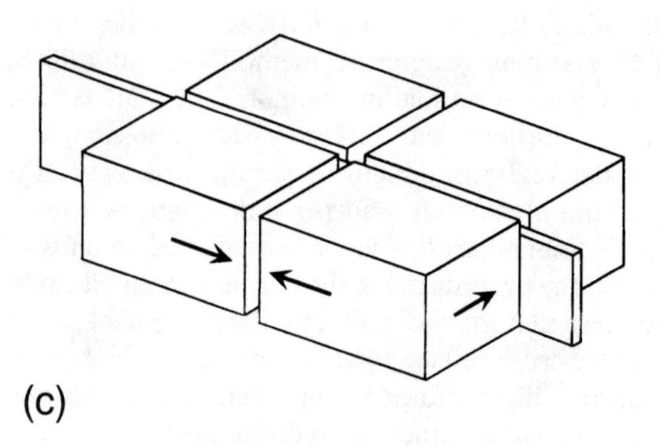

Fig. 35 Standard TAPPI tests: (a) Fluted edge crush test (FCT); (b) ring crush test (RCT); (c) short-span compression test (SSC).

two methods is in the neighborhood of 100. By fluting the paperboard or forming a ring, the sample is rendered structurally more stable so its effective slenderness ratio would be something less than 100. These test methods have found criticism in the literature for not operating in region III and therefore being unreliable [151].

The short-span compression test SSC [28,47] uses a sample of width 15 mm and length of at least 60 mm (Fig. 35). The sample is placed between two pairs of grip blocks, and the paperboard is held with a gripping force of 2300 ± 500 N between each pair of blocks. The grip blocks are 30 mm long and slightly greater than 15 mm tall. There is a gap of length 0.7 mm between the two pairs of grip blocks. Once the gripping force has been applied, the two pairs of grips are forced toward each other at a rate of 3 ± 1 mm/min. The maximum achieved load before specimen failure in the 0.7 mm zone between the grips is the recorded strength (kN/m). This is a very simple and straightforward test to perform. Because the gauge section of the test specimen is very short, proper alignment of the grips relative to each other and the direction of travel is very important. The slenderness ratio of samples tested with this method is typically in the neighborhood of 10.

The test methods discussed above are those designed to be used in a production environment where only the ultimate compression strength of a paperboard is of interest. Each test is very easy and quick to perform. If one is interested in the entire stress–strain response of paperboard, then other measurement techniques are required.

Over the years, several concepts have been introduced. The central theme of each experimental technique or specimen design is that of offering lateral support to the paperboard specimen while it is being deformed so that it does not deform or fail by transverse buckling (as in slenderness ratio regions I and II), but rather the intrinsic compression deformation response of the material is measured (as in slenderness ratio region III). The two test methods discussed briefly here (Fig. 36) are the FPL vacuum restraint test [56] and the cylinder test [37,108,162,185] The two techniques are quite different, but both offer lateral restraint to the paperboard and easy access to measuring strain with an external extensometer during testing.

The FPL vacuum restraint test device [56] consists of slender rods ($3 \times 3 \times 115$ mm) on their ends, spaced closely together (0.7 mm) and embedded in a relatively thin elastomeric base (Fig. 36b). This assembly is mounted inside an enclosure of square cross section that is open at the top. Paperboard samples are placed on top of the surface created by the free ends of the rods, and a vacuum is applied through the metal box (the open area not covered by paperboard is covered by sheet metal to develop a pressure differential across the sample). The pressure differential between the top and bottom sides of the paperboard forces the paperboard against the ends of the rods. The paperboard is then deformed between a pair of grips in either tension or compression. The length of the rods and the compliance of the elastomeric base allow the free ends of the rods to move easily with the paperboard as it is deformed while offering constant lateral support.

The cylinder test [37,108,162,185] is a type of specimen design rather than the design of a test system (Fig. 36c). Samples of paperboard are formed into cylinders of diameter and length that depend on the basis weight and thickness of the material of interest. The vertical seam made by wrapping the paperboard sample into a cylinder is usually glued by a lap or scarf joint. A specimen design has also been used in which the vertical seam is restrained from the outside by a simple restraint

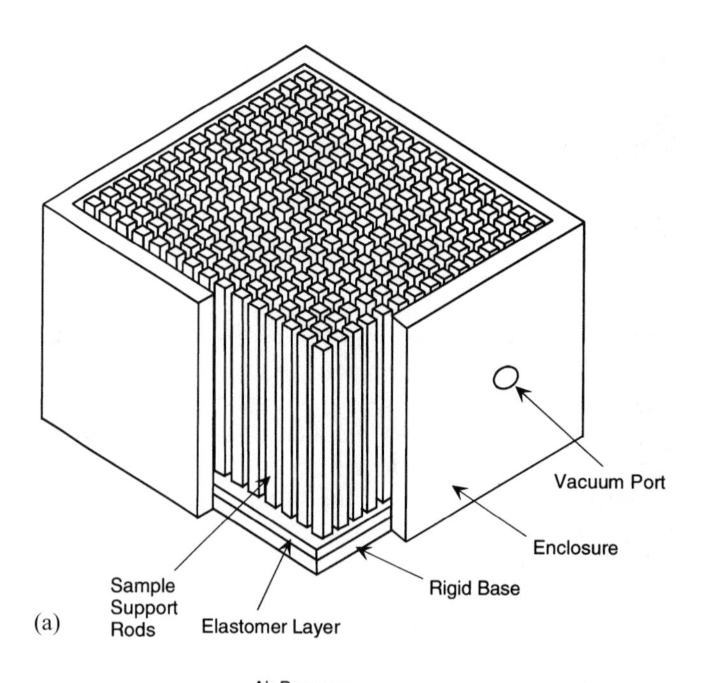

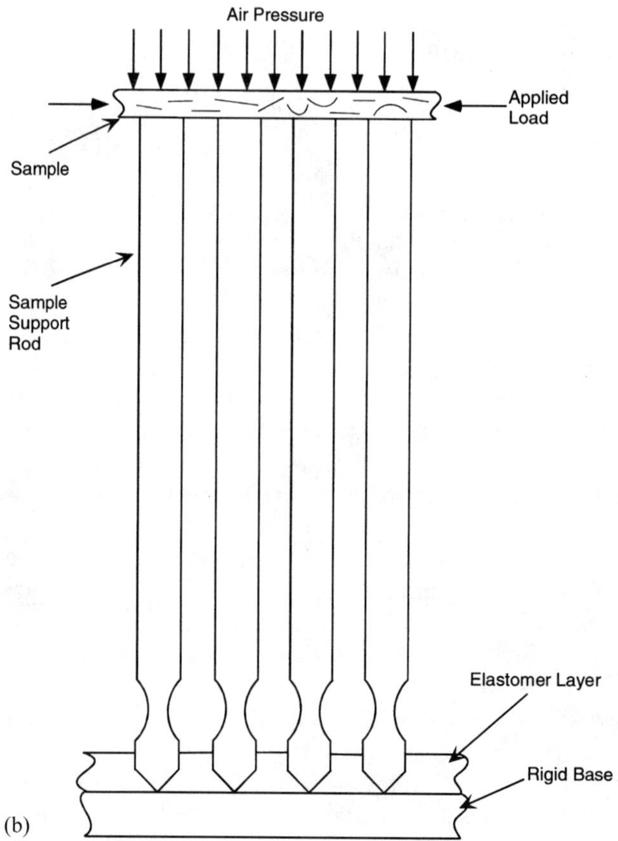

Fig. 36 (a) The lateral restraint system used for the Forest Products Lab vacuum restraint test; (b) a closer view of the sample support rods used in the lateral restraint system; (c) specimen geometry used for the cylinder test.

(c)

device [16]. The ends of the cylinders are always reinforced by gluing extra layers of paperboard or even stainless steel rings to the sample material. The cylinder is also supported from the inside, usually by a close-fitting cylinder or tube slid inside the test cylinder after it has been formed. Once again, the entire stress–strain response of this type of specimen can be easily measured using a strain gauge extensometer. In addition, the cylinder test can be performed by anyone with access to a standard mechanical test system and a few simple additional pieces of hardware.

Analysis of Results Among the TAPPI standard tests [177] for edgewise compression strength of paperboard, the ring crush and short-span methods are by far the most widely used. Compression strength as measured by the short-span method is significantly greater than that measured by the ring crush method (Fig. 37). This relationship has been examined in some detail [28,41,151,158,203]. Collectively there were five reasons stated for the discrepancy between the two test methods:

1. A short-span compression test specimen is much smaller in surface area [41].
2. The short-span compression test clamps exert a restraining influence [28,41] (recall that the slenderness ratio is around 10).
3. The ring crush sample buckles [151,203].
4. The top or bottom edges of the sample are crushed inside the ring crush clamps [151,203].
5. The short-span test strain rate is greater (nominally $-0.071\,s^{-1}$ for SSC and $-0.013\,s^{-1}$ for RCT).

Each of these five reasons would act to increase the measured short-span compression strength relative to the measured ring crush compression strength. Reasons 1, 2, and 5 increase measured short-span strength, and reasons 3 and 4 decrease measured ring crush strength.

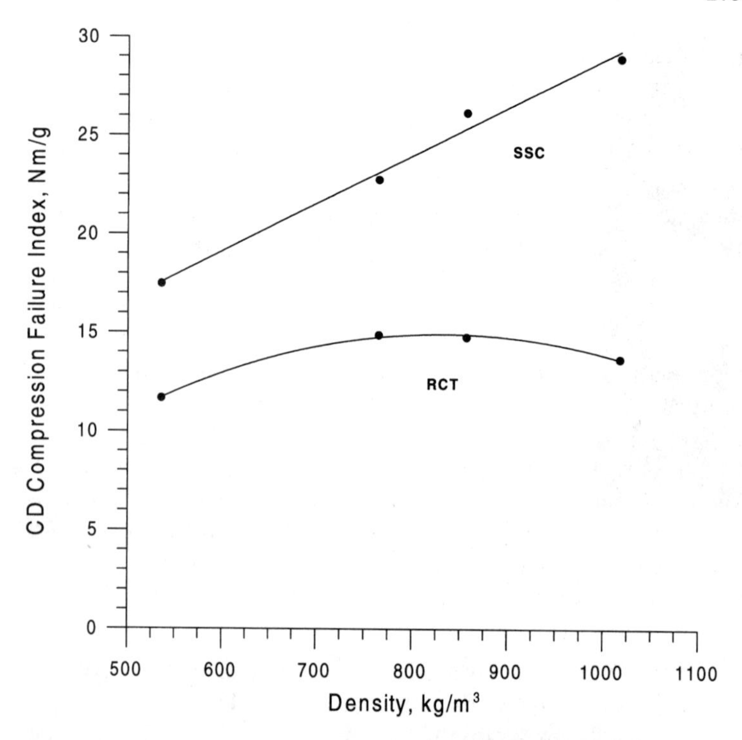

Fig. 37 Cross-machine direction compression failure index versus sheet density measured with both the short-span compression test and ring crush test using a 195 g/m^2 corrugated medium. (From Ref. 203.)

The only known comparison between the FPL vacuum restraint test and the cylinder test was presented by Mark et al. [108]. They reported that there was good qualitative and quantitative similarity between the two compression test methods. In their examination, however, the same material was not tested by each technique. A clear difference between the two test methods is that for the vacuum restrain test the paperboard sample is tested flat, whereas in the cylinder test the paperboard sample must be deformed into the shape of a cylinder before testing. Some researchers form the samples while the material is at an elevated moisture content to improve sample compliance and reduce the tendency for material damage during forming. The deformation induced during sample forming can be minimized by proper selection of cylinder diameter for the basis weight and density of the paperboard being tested. A practical difference between the two techniques is that the FPL vacuum restrain test requires an extensive amount of specialized equipment whereas the cylinder test requires far less special equipment, other than a standard mechanical test system.

E. Transverse Compression Tests (*Z*-Direction Compression)

Background The performance of numerous paper and paperboard products is, in some measure, dependent on the transverse compression behavior for paper and paperboard. Paper is delivered and stored in the form of wound rolls. During the twin-drum winding process the paper is compressed transversely. The transverse compressibility of the paper in part controls the structure quality of the roll. Contact printing operations such as offset and rotogravure also transversely compress the paper between rolls. Standard methods for measuring paper thickness apply some transverse compressive force to the sheet of paper; hence we typically measure apparent sheet thickness and work with apparent sheet density. Density is an important paper property, so accurate measurement of sheet thickness is affected by the transverse compressibility of paper. Hence a more compressible paper will render an apparent density that is more in error than one that would be less compressible. Although performed while the paper is still wet, wet pressing operations transversely deform the matted network a great deal, and the mechanical compressibility of that mat contributes to the densification of the paper.

Another important papermaking operation that transversely compresses the paper repeatedly at high pressure and temperature is calendering. In contrast with wet pressing, the sheet is calendered after drying when the moisture content is rather low. Because of the high level of transverse pressure applied to the web with both pressing and calendering, there is the potential to apply transverse pressure to the mat or sheet that might cause the fibers or interfiber bonds to be damaged.

Ting et al. [180] present transverse compression stress-strain responses for laboratory paper starting at a relative density of 0.24 (apparent density 360 kg/m^3). A selection of their data are plotted in Fig. 38, showing both loading and unloading response. These data are interesting because the paper used in the

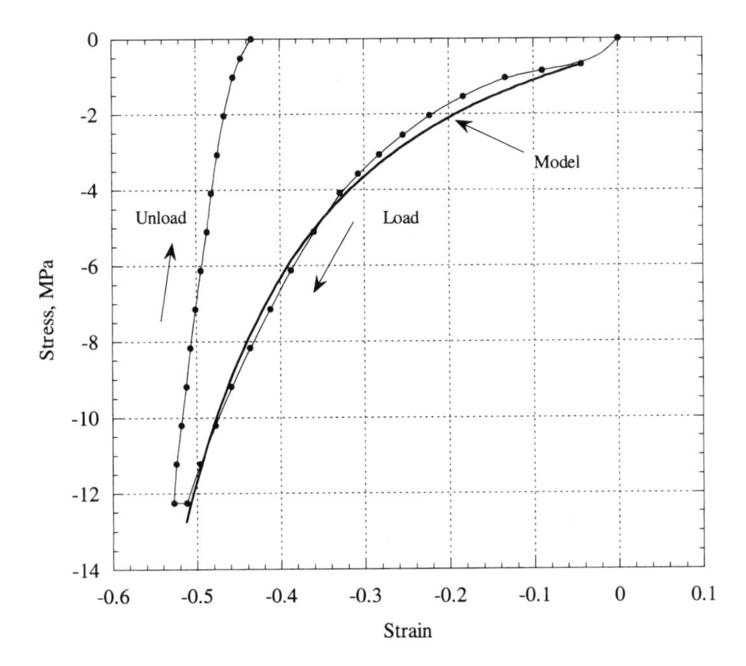

Fig. 38 Load and unload transverse compression data for low density bleached pulp handsheets. The bold curve is the representation of transverse compression by Eq. (21). (From Ref. 180.)

experiment begins at an apparent density that is low enough to classify it as a cellular solid. The loading curve in Fig. 38 is very similar to those for wood and polymeric foams [55]. Wood and foams display three distinct regions of behavior when compressed: (1) linear elastic; (2) plateau, and (3) densification. At the beginning of loading, the walls of the cells are not loaded enough to buckle the cell walls, so the material responds linear-elastically. As the load is increased, the cell walls are loaded to a point where they begin to buckle. As this occurs the material is deformed with little increase in load. This is the plateau region. As deformation proceeds, the cell walls begin to increase contact and load rapidly increases with deformation. This is the densification region. The data in Fig. 38 show an initial linear-elastic portion followed by plastic buckling of the fiber cell walls and then perhaps a very brief plateau region followed by densification as the slope of the loading curve increases rapidly. We know that the behavior of the fibers is plastic because the unload curve does not retrace the loading curve but is substantially linear. In general, machine-made papers or papers with an initial apparent density typical of machine-made papers would not show a linear-elastic portion of the loading curve, because by wet pressing or calendering to apparent density of 600–700 kg/m^3 the fibers will already be collapsed and will have been taken beyond the plastic buckling plateau stages, so only the densification region will be observed.

Gibson and Ashby [55] present a semiempirical description of the stress–strain response of polymeric foams in the densification region—the region in which paper will almost always exist. Using this relationship as motivation, we get the following relationship between stress and imposed transverse strain for loading:

$$\sigma = \sigma_0 \left[\left(\frac{\varepsilon_D}{\varepsilon_D - \varepsilon} \right)^m - 1 \right] \tag{21}$$

where

$$\varepsilon_D = \frac{\rho_0}{\rho_f} - 1 \tag{22}$$

The quantities σ_0 and m are constants. The parameters ρ_f and ρ_0 are the densities of the fiber cell wall ($\cong 1540\,kg/m^3$) and the undeformed density of the paper sample, respectively. The material parameter ε_D as defined by Eq. (22) (densification strain [55]) is simply the strain required to achieve a density in the sheet that is the same as the cell wall density of the fibers. This is an asymptotic quantity as demonstrated in Eq. (21). For the paper data plotted in Fig. 38, Eq. (22) gives $\varepsilon_D = -0.75$. Then curve fitting Eq. (21) to the loading curve presented in Fig. 38 (after subtracting out the elastic strains), we get $\sigma_0 = -2.9637$ MPa and $m = 1.6466$. The resulting description of the data presented by Ting et al. [180] is plotted in Fig. 38.

Note that the transverse strain ε in Eq. (21) is related to the sheet density (ρ) as follows:

$$\varepsilon = \frac{\rho_0}{\rho} - 1 \tag{23}$$

Therefore, as the sheet density increases, the slope of the stress–strain curve increases and the paper becomes more difficult to compress. This is relevant

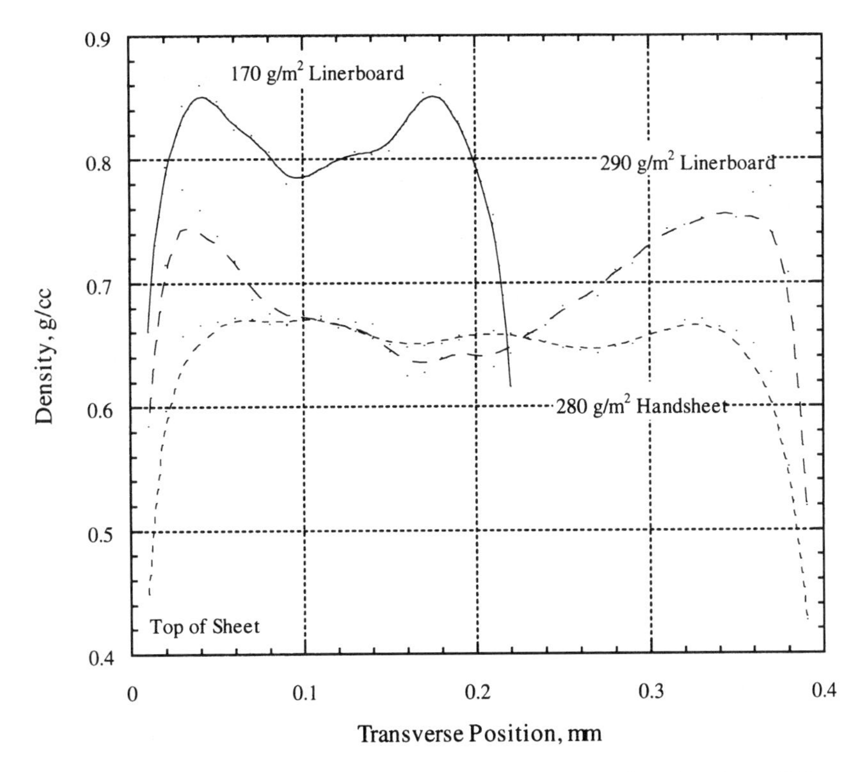

Fig. 39 Paper density versus distance from the top surface for three different materials. These measurements were made using an X-ray transmission method.

because the density of paper, in particular machine-made paper, is not uniform through the thickness. Density versus out-of-plane position data for three different paper samples are graphically presented in Fig. 39. Note that handsheet material is rather uniform near the center but the density drops rapidly near both surfaces. The surface density phenomenon has been known for some time [82,88,89,106,148,174,178,208]. This is generally attributed to surface roughness and interfiber bond disruption near the surface. The machine-made paperboards also show rapid reduction of density at both surfaces; however, the density of the center also changes with position. This has also been shown by Kimura and Mark [88,89,106]. Therefore if a paper sample does not begin with a uniform density through the thickness, its transverse compressibility will also not be uniform, so that upon compression the less dense regions will deform more than the higher density regions [148].

For some studies such as those concerned with the development of internal stresses in rolls of paper, it is more appropriate to measure the transverse compression behavior of stacks of paper rather than that of individual sheets [136]. It is very important to note that the transverse compression behavior of a single sheet is different than that of a stack of several sheets. In general, a single sheet is more transversely deformable than a stack of more than one sheet [119,148,178]. That is to say, at a given transverse stress level the corresponding strain will be greater for a single sheet of paper. This difference is attributed to the interaction of the paper surfaces with each other, thereby reducing the impact of the low density surfaces as shown in Fig. 39.

For the deformation discussed so far, little damage is thought to occur to the fibers and interfiber bonds. Much of the deformation is believed to be plastic deformation of the fibers, much like that demonstrated by Elias [45]. As the fibers are brought closer to one another, they are forced to conform to their nearest neighbors and are thus deformed enough to do so irreversibly. We know that the deformation is largely plastic, because the unloading curve plotted in Fig. 38 is quite linear and far from the loading curve. However, if we were to increase loading to the magnitude generally used during calendering operations, the deformation would be large enough to induce damage to the fibers and interfiber bonds. The transverse stresses found in calendering operations are generally several times greater than the maximum stress attained in Fig. 38 [142]. Although the deformation in paper imposed by passing it between two cylinders under high load is not simply transverse compression, it is still dominantly transverse compression. Although calendering does not necessarily cause large numbers of fiber cells and interfiber bonds to fail, the potential for damage is real. Berger [5] demonstrated that by densifying machine-made paperboard on a laboratory calender using different roll loadings and temperatures. Data of Berger are presented in Figs. 40 and 41 for short-span compression strength and tensile strength, respectively. In this case, exposing paperboard to high transverse compressive stresses probably caused failures to occur in the fibers and inter-

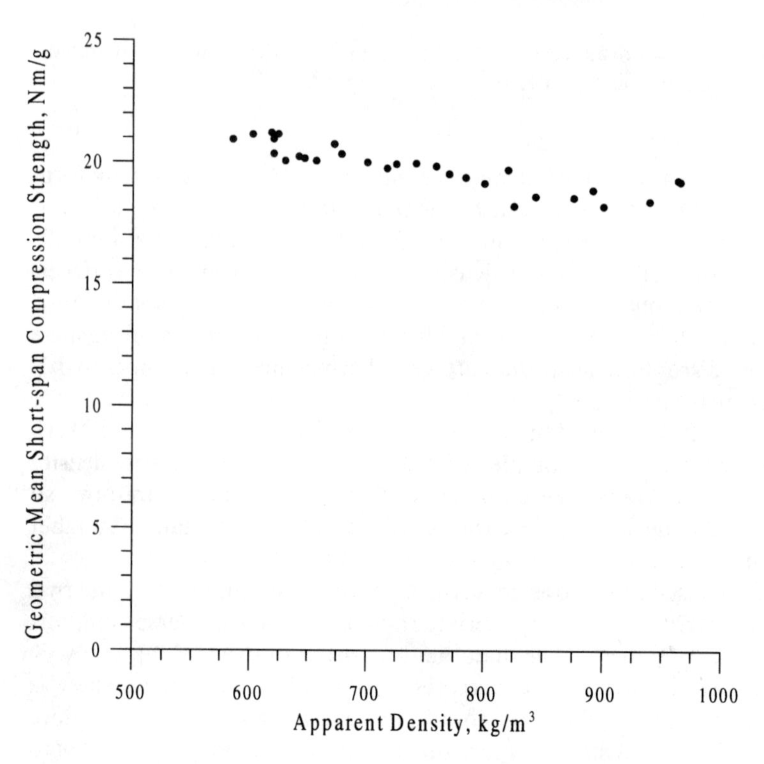

Fig. 40 Geometric mean short-span compression strength data for machine-made paperboard, densified by laboratory calendering. (From Ref. 5.)

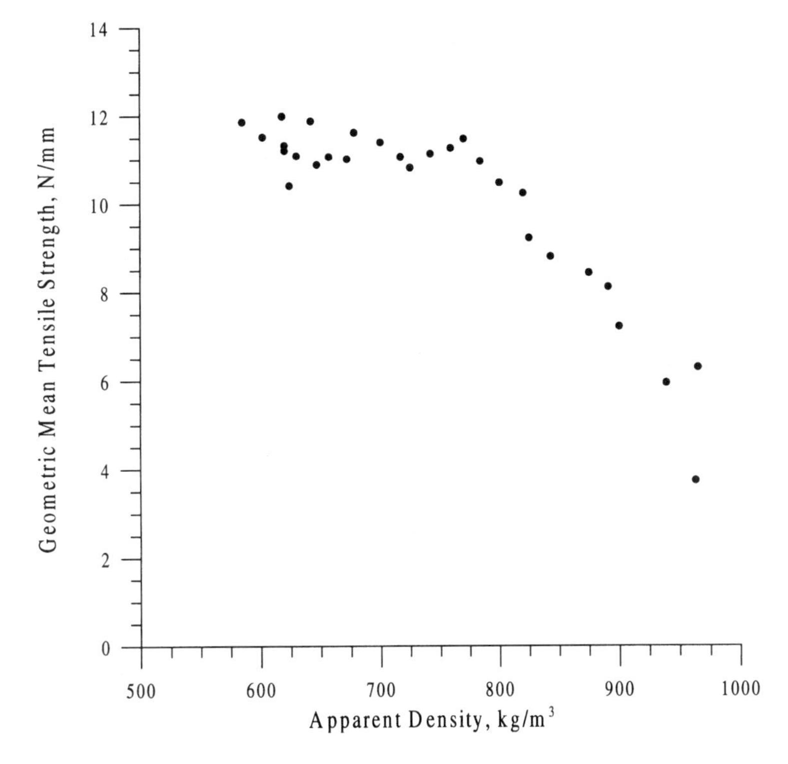

Fig. 41 Geometric mean tensile strength data for machine-made paperboard densified by laboratory calendering. (From Ref. 5.)

fiber bonds. Indeed, paper blackening during calendering has been attributed to local areas of substantial fiber failure, which changes the optical properties in those areas of damage [196].

Procedure There is no industry standard test method or test apparatus for evaluation of transverse compression deformation behavior of paper and board. However, uniaxial compression is an extremely common measurement made on many materials. To perform such a test, it is necessary to have a mechanical test system that has coplanar compression platens, one of which is connected to a load cell. The compression platens must be rigidly fixed except for allowance of imposed motion of the platens in the direction perpendicular to their loading surface. Most mechanical test systems have displacement measurement devices built into the machine (such as an LVDT connected to the crosshead or hydraulic piston), but the resident LVDT generally does not provide the required accuracy for measurement of such small displacements. For acceptable accuracy, one must measure displacements as close as possible to the sample. This can usually be done by measuring the relative displacement between the compression platens by attaching an extensometer between them (using appropriate fixtures).

For compression testing of solid materials, the interface between the material and the compression surfaces is lubricated, because there is considerable lateral expansion with uniaxial compression. This is generally not done with paper; because the material is not solid, much of the imposed transverse compression will be used for densification rather than volume-preserving lateral expansion.

F. In-Plane Shear

Background Just as tension and compression define elementary states of normal stress, shear is an elementary state of stress. Compared with tension and compression deformation testing, the in-plane shear properties of paper are more difficult to measure directly. Planar methods, such as those used for simple shear deformation of metals [17] (as illustrated in Fig. 42), develop undesirable edge effects that are manifested as in-plane tensile and compressive stresses. These edge effects are illustrated schematically in Fig. 43. The error in measured shear stress can be minimized if the size of the material gauge section is substantially greater in the direction of loading than it is perpendicular to loading ($L \gg w$). For this planar approach, the shear stress and shear strain are given by

$$\tau = \frac{F}{2Lh} \qquad \text{and} \qquad \gamma = \frac{\delta}{w} \tag{24}$$

where L, w, and h are the length, width, and thickness, respectively, of the gauge section, F is the applied force, and δ is the shear displacement. For many materials, these edge effects can be eliminated by using thin-walled cylindrical tubes and deforming them in torsion. Such a short gauge section specimen is illustrated in Fig. 44 for metals [17,202]. For paper materials, procedures based on this approach have been developed by Setterholm et al. [163], Suhling et al. [175], and Uesaka et al. [184]. For a thin-walled cylinder as illustrated in Fig. 45, the in-plane shear stress and strain are

$$\tau = \frac{T}{2\pi\bar{r}^2 h} \qquad \text{and} \qquad \gamma = \frac{\bar{r}\theta}{L} \tag{25}$$

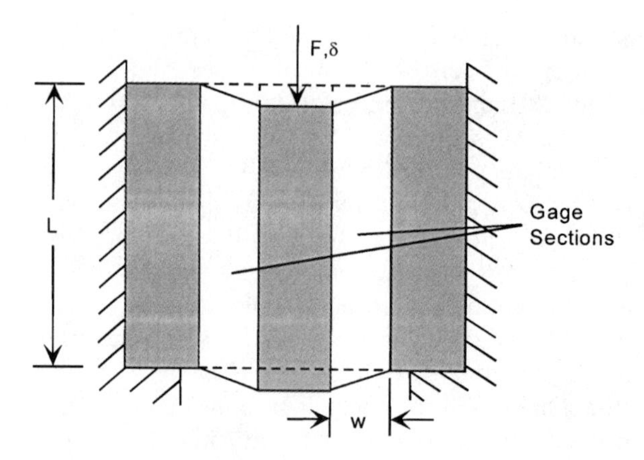

Fig. 42 Schematic representation of a planar specimen used to measure the inelastic simple shear deformation behavior of materials. Note that the specimen is symmetrical to balance forces. Generally $L \gg w$ and w is taken as approximately three times the thickness of the gauge section. (From Ref. 17.)

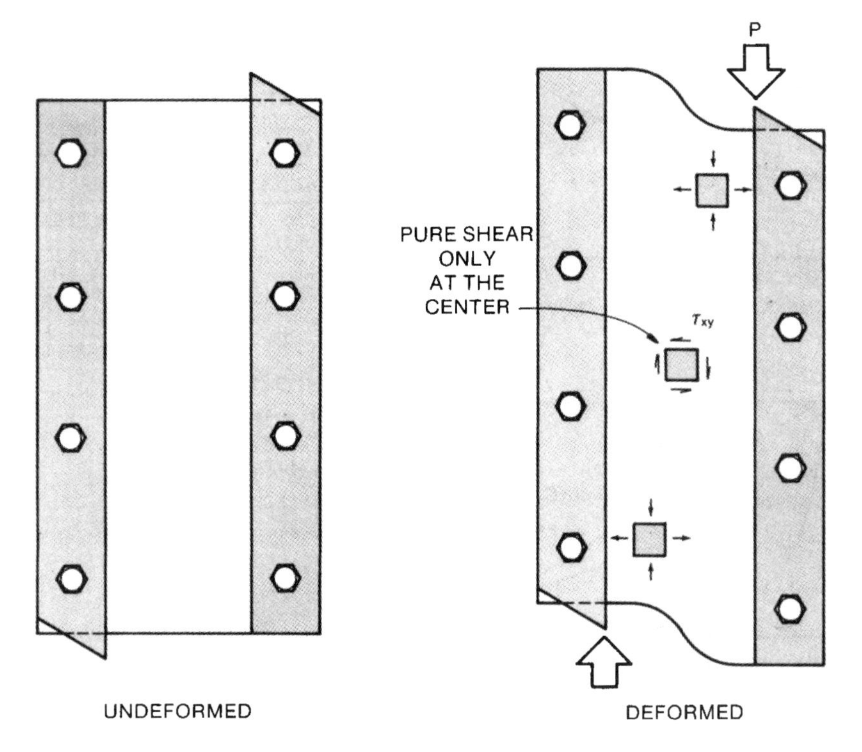

Fig. 43 Illustration of the edge effects active in the planar simple shear approach. The edge effects can be minimized by choosing $L \gg w$ as noted in Fig. 42.

where $\bar{r}$ is the mean gauge section cylinder radius, h is the wall thickness, L is the gauge section length, T is the applied torque at the end of the tube, and θ is the imposed angle of twist over the entire gauge section [149].

As Setterholm et al. [163] and Suhling et al. [175] discovered, it is not generally possible to achieve shear failure in thin-walled tubular paper specimens. The specimen will first fail by compression buckling in the direction of maximum compressive stress in the wall of the gauge section. This problem will exist with planar specimens also. Setterholm et al. suppressed this by applying axial tension to the tubes, although when this is done a state of pure shear stress no longer exists. In their work on the biaxial strength of paperboard, Gunderson and coworkers [57,59] also discovered that only tensile or compressive failures were possible and that the magnitude of in-plane shear stress was significant only to the extent that it changed the principal stress magnitudes.

More recently, Qiu et al. [138] used the Iosipescu test to measure the in-plane shear deformation response of paperboard laminate materials. This approach apparently allows deformation to failure in shear although it is not suitable for measurement on single sheets.

Procedure The thin-walled tube torsion approach to the in-plane shear deformation of paper has been used more than any other technique [163,175,184]. Measurements of this type are not easy to perform. In addition, attaining a

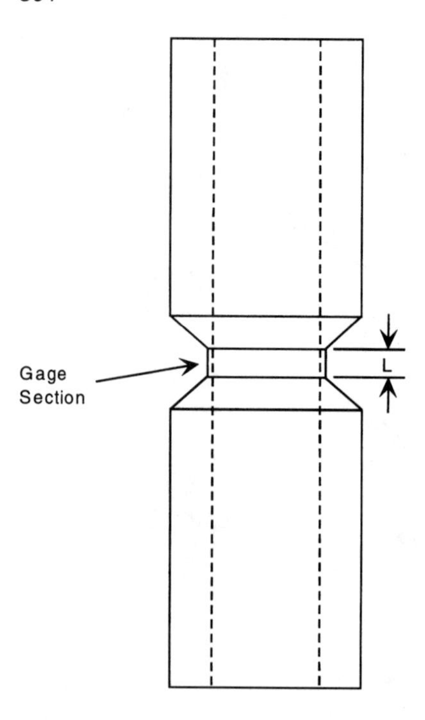

Fig. 44 Schematic representation of a thin-walled tubular specimen used to measure the inelastic simple shear deformation behavior of materials. The length of the gauge section, *L*, is taken as approximately three times the thickness of the wall of the gauge section. (After Refs. 17 and 202.)

reliable stress–strain curve to failure will not generally be possible. Therefore, this procedure should be used only for evaluation of elastic in-plane shear properties. In the procedure used by Setterholm et al. [163] and Suhling et al. [175], preparation for testing consists of five distinct steps. A schematic of the relationship between the metal cylinders and the paper sample is given in Fig. 46. The first step involves cleaning and drying the metal cylinders around which the paper sample is to be wrapped. Next the paper specimen is prepared by carefully cutting it to size, keeping track of the desired orientation of shear stress during testing. It is marked for adhesive application and measured for thickness. Next, adhesive is applied to the cylinder areas to which the paper is to be bonded, taking care to leave a tacky glue surface. The paper specimen is then wrapped around the cylinder, the vertical overlap is glued, and the entire assembly is set aside to dry. The final preparation step consists of conditioning the sample under standard TAPPI temperature and humidity conditions. Following preparation, the assembly is placed in a test system and loaded. Combined linear and torsion servohydraulic test systems, which are commercially available, are most suitable for these measurements. Rotational displacement should be measured on the material in the gauge section (see, e.g., Suhling et al. [175]). Using a rotational LVDT resident in the test system is generally not suitable because it will also measure the deformation of the test frame.

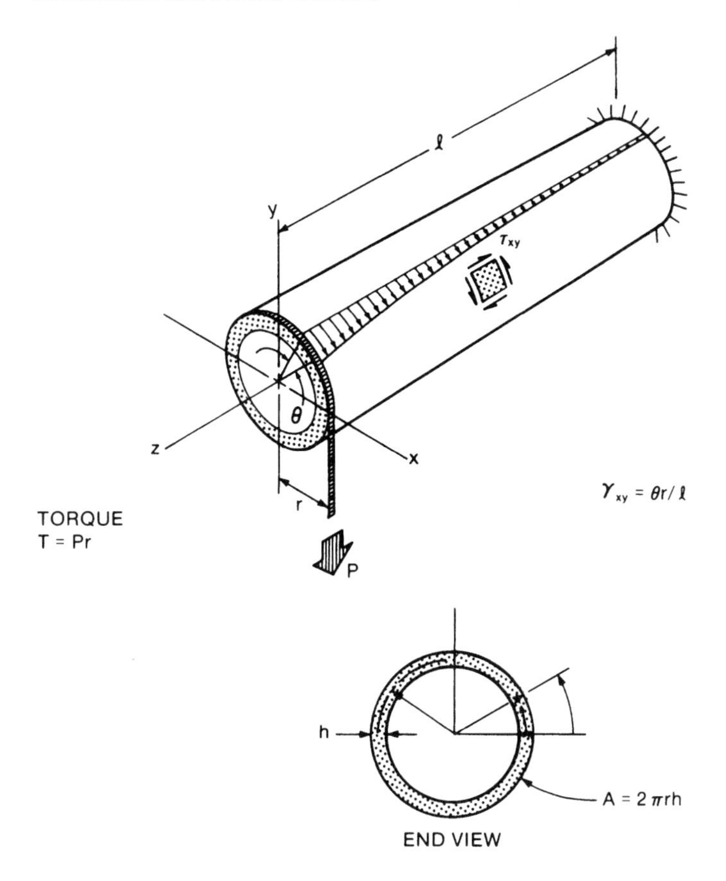

Fig. 45 Twisting of a thin-walled cylinder to produce simple shear deformation.

Currently, there is no experimental method for measurement of the general in-plane shear stress-strain response of paperboard that is fully satisfactory. Perhaps work should be done to assess the suitability of experimental approaches characterized by Figs. 42 and 44 with gauge section geometries chosen small enough that the principal compressive stress is not so large as to cause buckling failure. This approach to experimental design would be similar to that which has been done for the problem of determining the optimum gauge length for measurement of inherent compression strength of paperboard.

Analysis of Results A typical stress–strain curve from a torsion test of a thin-walled tube by Suhling et al. [175] is shown in Fig. 47. The initial linear portion can be used to calculate the in-plane shear modulus. Despite the elegance of the thin-walled tube methodology, a major limitation is that compressive buckling failure is still the dominant failure mode (as pointed out in Fig. 47 and earlier). This compressive buckling type of failure can be inhibited by adding a tensile axial load to the specimen while torque is being applied, although the torsional strains are then no longer simple shear deformations. Gunderson and coworkers [57,59] suggested that shear failure in paper is far subordinate to tensile and compressive failures.

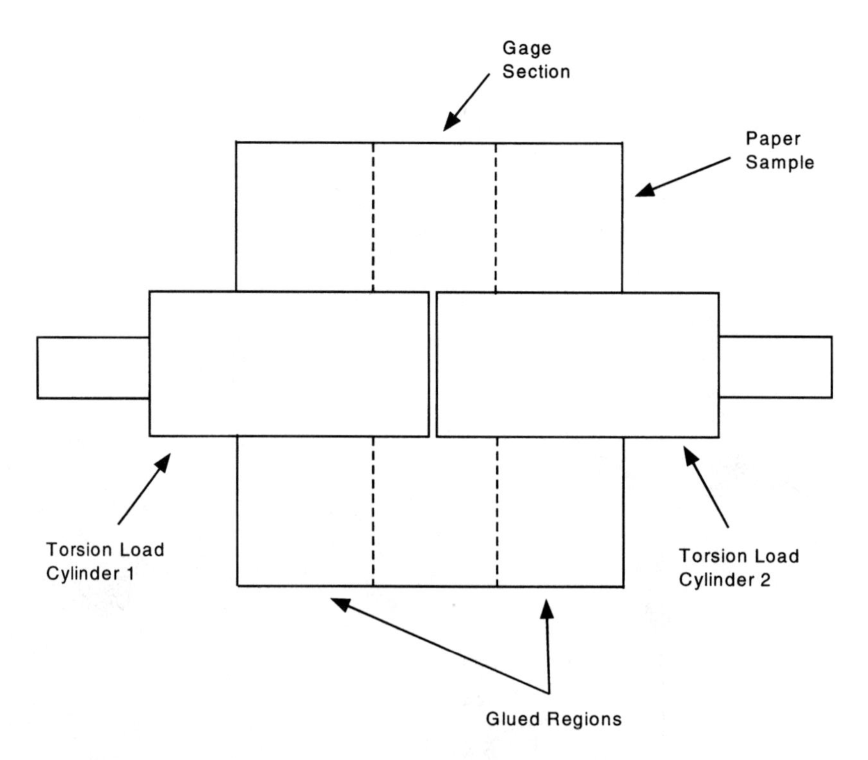

Fig. 46 Schematic representation of the sample preparation procedure used by Setterholm et al. [163].

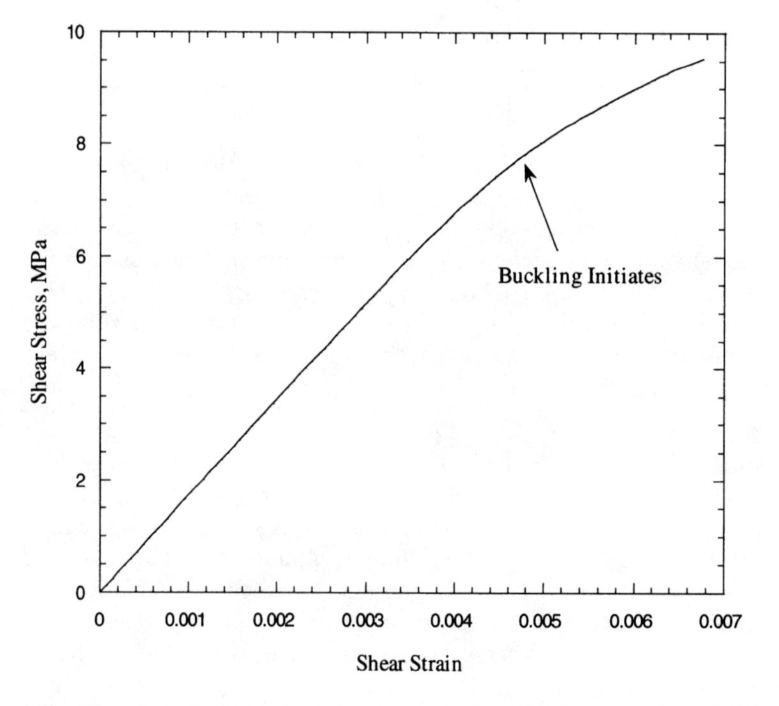

Fig. 47 A typical simple shear stress–strain curve for paperboard. Note that the sample buckles very near the proportional limit of the material. (From Ref. 175.)

G. Interlaminar Shear

Background Because paper is essentially a laminar fibrous composite material, it is especially susceptible to out-of-plane shear deformations. Thus, interlaminar (or through-the-thickness) shear strength is an important physical property. Low shear strength is a disadvantage in many end uses, such as during corrugation and printing, although it is occasionally an advantage. Some packaging materials need to be delaminated during scoring and bending without significant loss in tensile strength. Interlaminar shear stiffness and strength are required in end uses involving bending of paper or paperboard. Interlaminar shear properties of linerboard also contribute to its edgewise compression strength [63,107,131]. Knowledge of these properties allows for rational design of paper products for particular applications.

The goal of any technique employed is to deform the material in simple shear and measure the applied load and displacement so that the shear stress–strain response of the material can be determined (Fig. 48). Shear strain is determined as follows:

$$\gamma = \frac{\delta}{h} \tag{26}$$

where δ is the imposed shear displacement and h is the material thickness. Shear stress is determined simply as

$$\tau = \frac{F}{A} \tag{27}$$

where F is the applied load over the surface area of contact A.

The experimental determination of interlaminar shear strength properties of paper has been the subject of investigation by Byrd et al. [24], Fellers [46], Waterhouse [198,199], and Heckers and Göttsching [64]. Each study used an experimental procedure involving shear load applied through parallel loading plates bonded to the top and bottom surfaces of a paper specimen by adhesive. Both Byrd et al. and Fellers used techniques that employed rectangular samples and linear loading, whereas Waterhouse used circular samples and torsional loading. The geometry used in each of these studies is shown schematically in Figs. 49, 50, and 51, respectively.

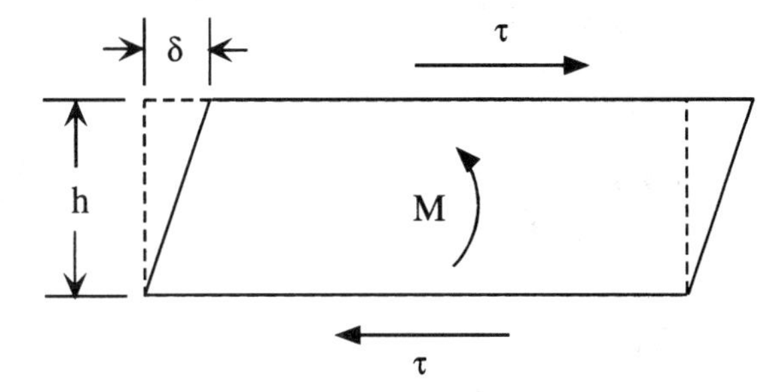

Fig. 48 Kinematics of simple shear deformation. The applied shear stress is τ, and M is the moment required to balance the imposed stress.

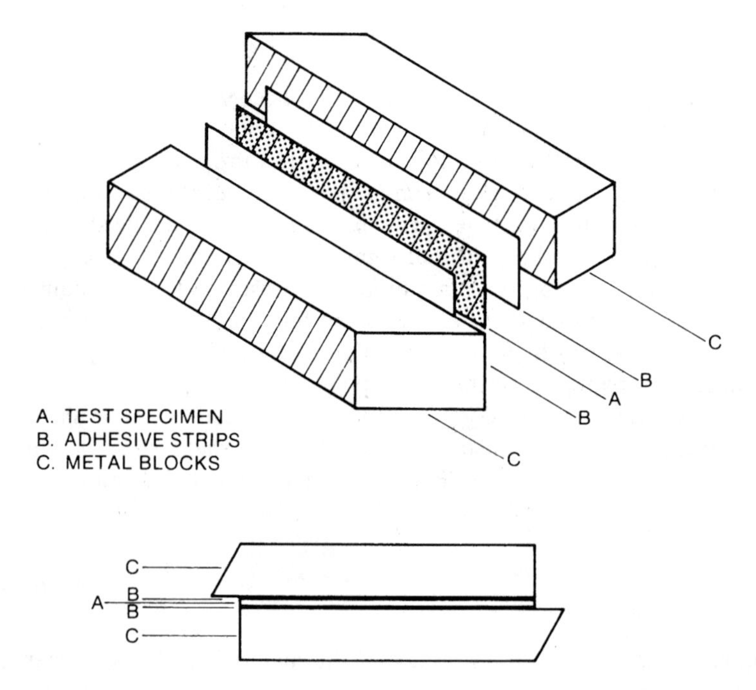

Fig. 49 The linear interlaminar shear experimental approach used by Byrd et al. [24].

As discussed in the previous section on in-plane shear, using a rectangular specimen, as by Byrd et al. [24], Fellers [46], and Heckers and Göttsching [64] did, does produce edge effects. However, edge effects should not be significant here, because the thickness of the sample material is generally much smaller than the dimension of the sample in the direction of loading. Of greater concern is that the loading surfaces must be kept exactly coplanar during testing. The free-body diagram in Fig. 48 shows a segment of material exposed to a simple shear deformation field (as is applied in all three techniques referenced above). Note that in order to prevent rotation of the sample, an external moment must be applied. Because of this required moment, it is more difficult for the surfaces of the loading blocks to remain coplanar and at a constant distance from each other. This is particularly true for the linear deformation approaches (Figs. 49 and 50). If the fixtures are sufficiently rigid, the balanced approach of Heckers and Göttsching [64] can minimize the effect of the unbalanced moment. Given that undesirable out-of-plane tensile and compressive stresses can develop during shear deformation, erroneous results can be obtained if the measurements are not taken with great care.

The torsion technique illustrated in Fig. 51 does not have a problem with edge effects, because the sample is continuous in the direction of deformation. In addition, the external moment is balanced automatically because of the circular geometry as long as the testing fixture is infinitely rigid compared with the sample material. A significant shortcoming of the torsional measurement technique is that machine and cross-machine direction shear stress–strain curves cannot be easily measured independently. This technique is best applied to random handsheets. In

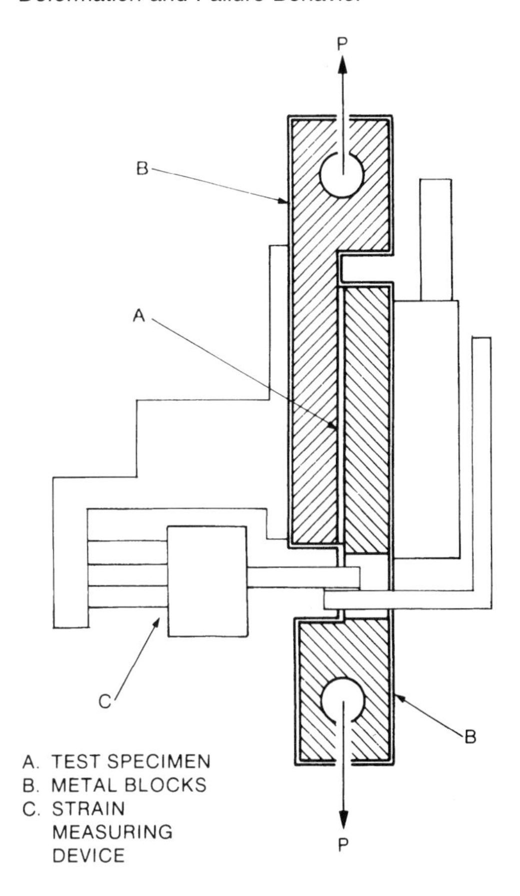

Fig. 50 The linear interlaminar shear experimental approach used by Fellers [46].

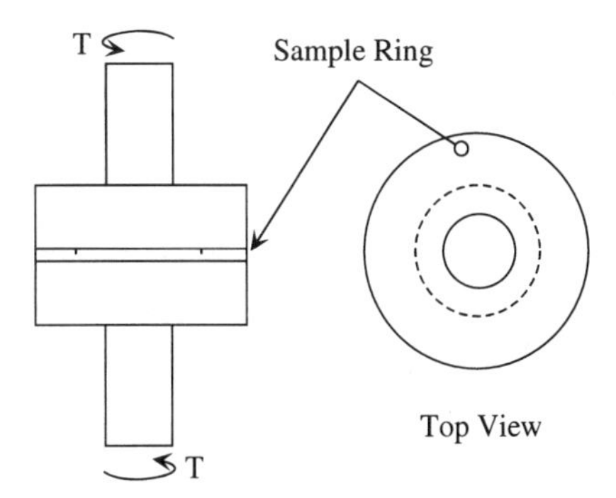

Fig. 51 The torsion interlaminar shear experimental approach used by Waterhouse [198,199]. T is the applied torque. A ring of the test material is mounted between the two torsion platens.

addition, the applied shear strain is a function of radial position in the deforming ring according to

$$\gamma = \frac{r\theta}{h} \tag{28}$$

where r is the radial distance from the center of rotation, θ is the rotation angle in radians, and h is the thickness of the material. Therefore, shear strain will be the greatest at the outer perimeter of the testing ring. The way to minimize this shortcoming is to use a sample ring with inner and outer radii that are both as large as possible.

Procedure The Byrd et al. [24] test configuration is illustrated in Fig. 49. Paper specimens of 19 × 51 mm are attached to metal blocks by photographic film adhesive strips. This adhesion procedure requires heating at 120°C for 45 min while transverse pressure is applied to secure the samples to the blocks, then cooling and equilibration to a standard test environment. Surface roughness of the paper has been found to give erratic results; consequently, the specimens must be surface-ground.

Regardless of which technique is used to measure the interlaminar shear response of paper, the thickness of paper is small and therefore accurate measurement of the very small shear displacements* is the most critical. These displacements, either linear or rotational, should be measured as close to the material as possible (as shown in Fig. 50) using a device designed for small displacement measurements (an extensometer, for example). Measurement of displacements through LVDTs resident in most standard test systems is generally unacceptable. In addition, because the test material is attached to the metal fixtures by using a film of adhesive, it is important to subtract the compliant displacement of the adhesive layers from the measured displacement. This correction is rather straightforward and is covered in sufficient detail in the literature [24,46,198,199]. Generally there is no correction for the measured load, which can be used directly to determine shear stress.

Byrd et al. [24] and Fellers [46] both found that it was not possible to determine the interlaminar shear response of materials with thickness of less than 0.1 mm. Results from Fellers are given in Fig. 52.

Analysis of Results A typical interlaminar shear stress–strain curve from the work of Fellers [46] is shown in Fig. 53. For the western Hemlock handsheet material used (apparent density of 600 kg/m^3), the ratio of tensile strength to interlaminar shear strength was 24 and the ratio of Young's modulus to interlaminar shear modulus was 35. Byrd [20] examined the effect of level of restraint during the drying of oriented handsheets for the range from freely dried to 2% wet stretch. He showed a reduction in interlaminar shear strength with increasing degree of restraint during drying. Waterhouse [198,199] examined the effect of interlaminar shear deformation on residual in-plane compression strength. He saw no clear effect on compression strength of initial shear deformation up to 5%. These results are shown in Fig. 54.

---

* For a shear strain of 5% and sample thickness of 0.2 mm, the resulting shear displacement is only 10 μm.

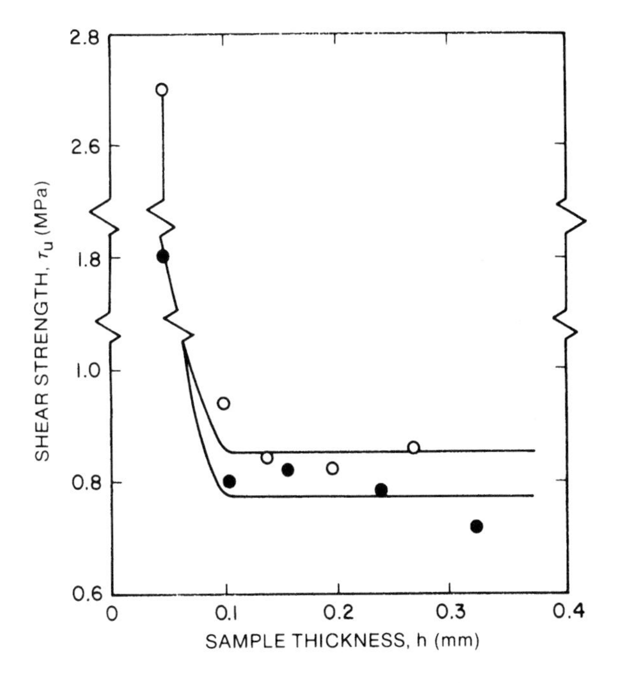

Fig. 52 The dependence of interlaminar shear strength on sample thickness. (From Ref. 46.)

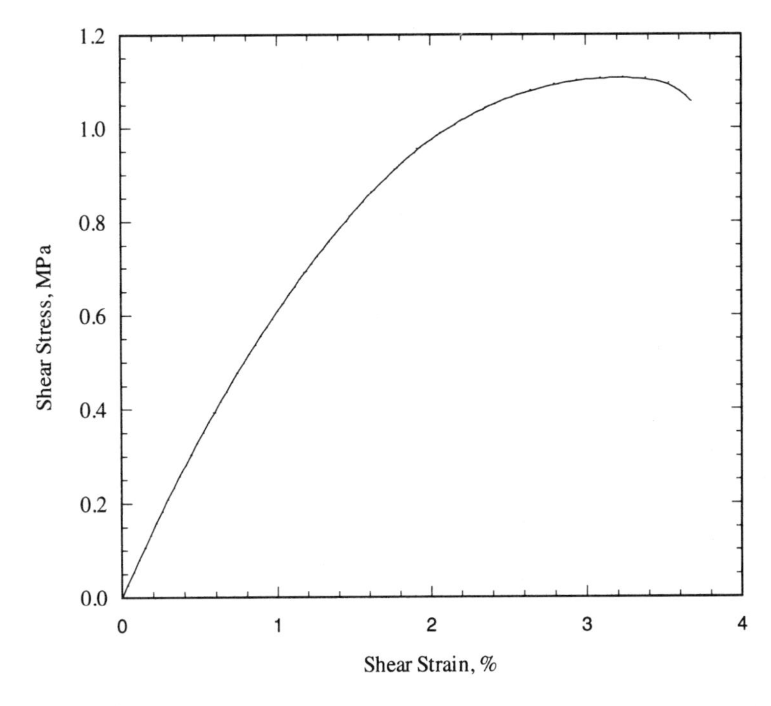

Fig. 53 A typical interlaminar shear stress–strain curve as presented by Fellers. (From Ref. 46.)

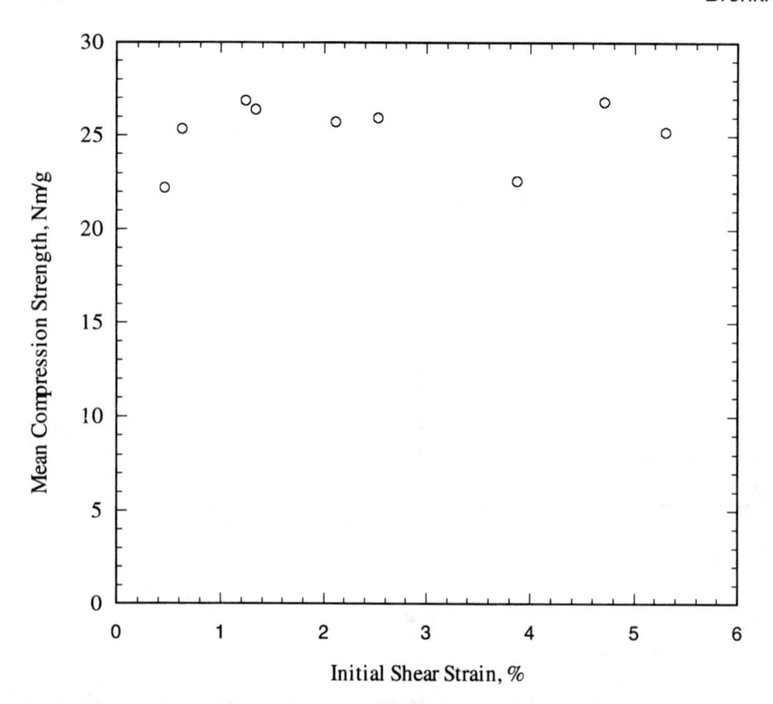

Fig. 54 Data showing the residual compression strength of paperboard after being deformed in interlaminar shear to different levels. (Data from Refs. 198 and 199.)

H. Biaxial Stress Test

Background During the manufacture of paper and in most end-use applications, paper is subjected to a complex state of stress involving superposition of the two elementary stress states: normal (tension, compression) and shear. Examples include deformation in an open draw or printing operation, the constituents in a corrugated container simultaneously subjected to internal pressure and external loading due to stacking, a filled grocery bag being carried, the burst test, and the stress state around a shive or hole in a web under tension. These are all examples of complex biaxial loadings. In the previous sections, test methods were discussed that attempt to induce in the test specimen a pure state of a particular stress component. In this section, test methods will be reviewed that induce a known, but uniform, arbitrary state of biaxial stress.

In its simplest form the biaxial plane stress state in sheet material can be thought of as arising from two perpendicular uniaxial loadings being performed simultaneously. A particular example has already been discussed in the section on edgewise shear. In fact, in all cases where cylindrical specimens are subjected to axial compression, axial tension, or combinations thereof, the state of plane stress can be resolved into two normal stresses (σ_1 and σ_2) and one shear stress (τ_{12}) at each material point along the machine and cross-machine directions. Failure in a biaxial state of stress is generally represented by a single point in two-dimensional principal stress space (constant shear stress). An example of such a collection of points for paper is given in Fig. 55 for zero shear stress [59,144]. Note that there

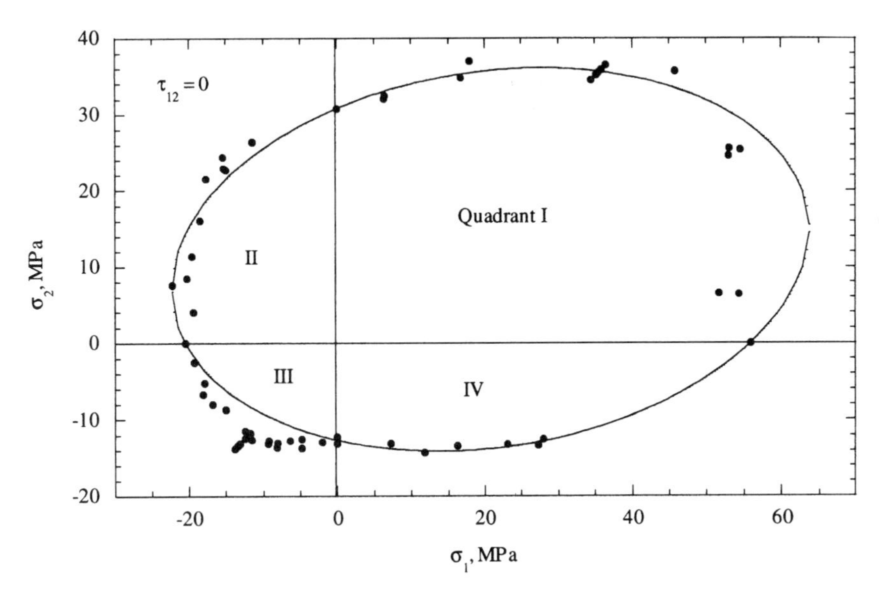

Fig. 55 Biaxial strength data of Gunderson and Rowland [57] and Gunderson et al. [59] at zero shear stress. The four quadrants are also indicated. The curve is for Eq. (31) with $F_{12} = -2.373 \times 10^{-4}\ \text{MPa}^{-2}$ as determined from the work of Suhling et al. [176].

are four normal stress quadrants and the failure behavior for paper is unique to each. That is,

Quadrant I	$\sigma_1 > 0;\ \sigma_2 > 0$ (both tension)
Quadrant II	$\sigma_1 < 0;\ \sigma_2 > 0$
Quadrant III	$\sigma_1 < 0;\ \sigma_2 < 0$ (both compression)
Quadrant IV	$\sigma_1 > 0;\ \sigma_2 < 0$

Numerous phenomenological failure theories have been developed in an attempt to represent the strength of materials under combined stress states from the results of fundamental strength tests involving pure and biaxial stress states. Particularly useful for metals have been yield criteria proposed by von Mises [197] and Hill [66]. The von Mises theory postulates a critical value for elastic distortional energy for isotropic material. The Hill theory is a variation of the von Mises theory for orthotropic material. Mathematically, the Hill [67] relationship takes the form

$$F(\sigma_2 - \sigma_3)^2 + G(\sigma_3 - \sigma_1)^2 + H(\sigma_1 - \sigma_2)^2 + 2L\tau_{23}^2 + 2M\tau_{31}^2 + 2N\tau_{12}^2 = 1 \qquad (29)$$

where F, G, H, L, M, and N are experimentally evaluated material constants: σ_1, σ_2, and σ_3 are normal stress components; and τ_{12}, τ_{23}, and τ_{31} are shear stress components. Note that for plane stress conditions (which is the case for paper in most situations), all terms with subscript 3 are taken as zero.

These two widely used theories illustrate the general nature of phenomenological failure envelope characterization. Although the Hill theory is designed for orthotropic material, without modification it cannot represent the difference in tensile and compressive behavior as is the case for paper and many other cellular solids [55]. Rowlands et al. [144] used the following form of the Hill equation:

$$F_{11}\sigma_1^2 + 2F_{12}\sigma_1\sigma_2 + F_{22}\sigma_2^2 + F_{66}\tau_{12}^2 = 1 \tag{30}$$

where the F_{ij} are material constants; σ_1, σ_2 are normal stress components in the machine and cross-machine directions, respectively; and τ_{12} is the in-plane shear stress component. They evaluated how this relationship, with different functional forms for F_{12}, represented biaxial experimental data of Gunderson et al. [57,59,60,144] for zero and nonzero values of shear stress. Note that the coefficient F_{12} attenuates the product $\sigma_1\sigma_2$, which is positive in quadrants I and III and negative in quadrants II and IV. Suhling et al. [176] were more successful in representing the data of Gunderson et al. by using the tensorial relationship

$$F_{11}\sigma_1^2 + 2F_{12}\sigma_1\sigma_2 + F_{22}\sigma_2^2 + F_1\sigma_1 + F_2\sigma_2 + F_{66}\tau_{12}^2 = 1 \tag{31}$$

Equation (31) reasonably represents all the experimental data when $F_{12} = 0$. The results of Suhling et al. [176] and the experimental data of Gunderson et al. [57,59,60,144] are presented in Fig. 56 at four levels of shear stress.

Most recently, Tryding [181] built upon earlier work [39,49] by proposing the relationship

$$\begin{aligned}\left(p + q - \frac{p\sigma_1 + q\sigma_2}{P}\right)\left(\frac{\sigma_1\sigma_2}{T_1T_2}\right) + F_{11}\sigma_1^2 + 2F_{12}\sigma_1\sigma_2 + F_{22}\sigma_2^2 + F_{66}\tau_{12}^2 \\ + F_1\sigma_1 + F_2\sigma_2 = 1\end{aligned} \tag{32}$$

where p, q, P, T_1 and T_2 are additional material constants. This relationship gave reasonable results compared to experimental data, although it was not attempted with biaxial strength data at nonzero values of shear stress. In general, one should expect improved accuracy of representation as the mathematical degrees of freedom become increasingly greater. Which phenomenological model is used depends upon the failures being represented and the degree of accuracy desired.

Procedure As one might imagine, evaluation of the biaxial failure surface for paper is not a straightforward matter. Several approaches have been taken over the years. De Ruvo et al. [38] used 220 mm long tubes, with 40 mm on each end being attached with epoxy to circular clamps. The diameter of the tube was 50 mm. A glued seam of 5 mm converted the planar sheet into a closed tube that was pressurized using a thin, flexible plastic bag as an interior lining. The measurements were performed with a conventional testing machine. As illustrated in Fig. 57, the circumferential stress σ_x (hoop stress) is related to internal pressure p by

$$\sigma_x = p\left(\frac{r}{h}\right) \tag{33}$$

Since internal pressure pushes the ends apart (in the way the experiments were performed), the longitudinal stress is a combination of stresses due to the internal pressure and the axial loading P:

$$\sigma_y = p\left(\frac{r}{2h}\right) + P\left(\frac{1}{2\pi rh}\right) \tag{34}$$

(a)

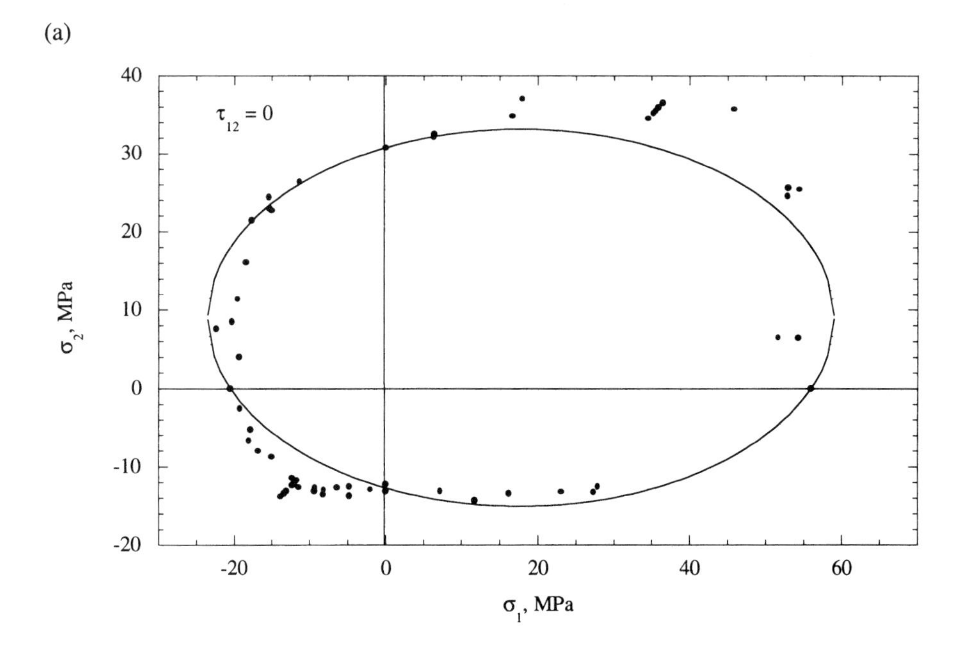

(b)

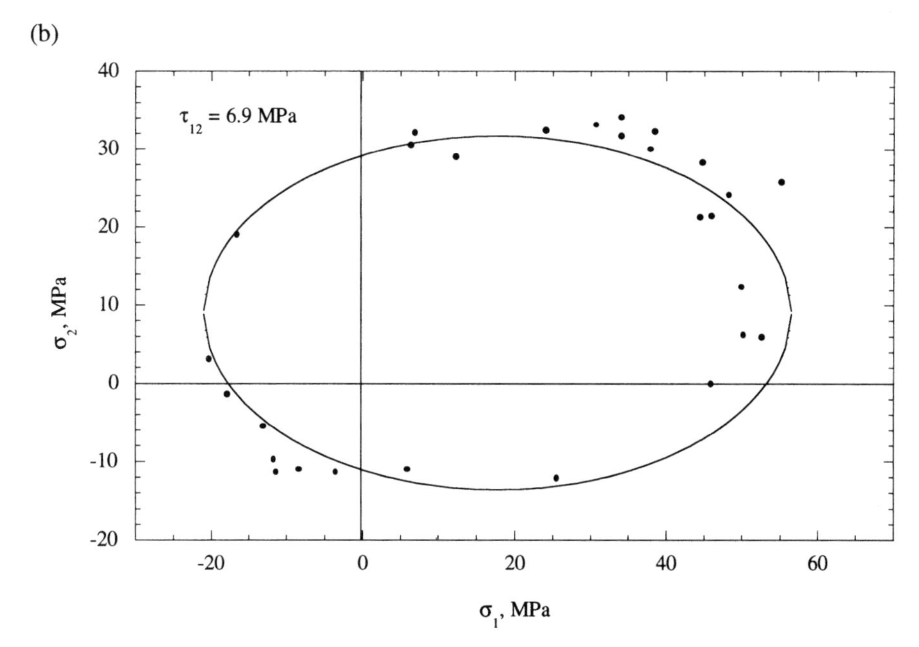

Fig. 56 Failure envelopes of paperboard at four different constant shear stress levels: (a) 0 MPa; (b) 6.9 MPa; (c) 10.3 MPa; (d) 15.9 MPa. The curves are for Eq. (31) with $F_{12} = 0$. (From Ref. 176.)

(c)

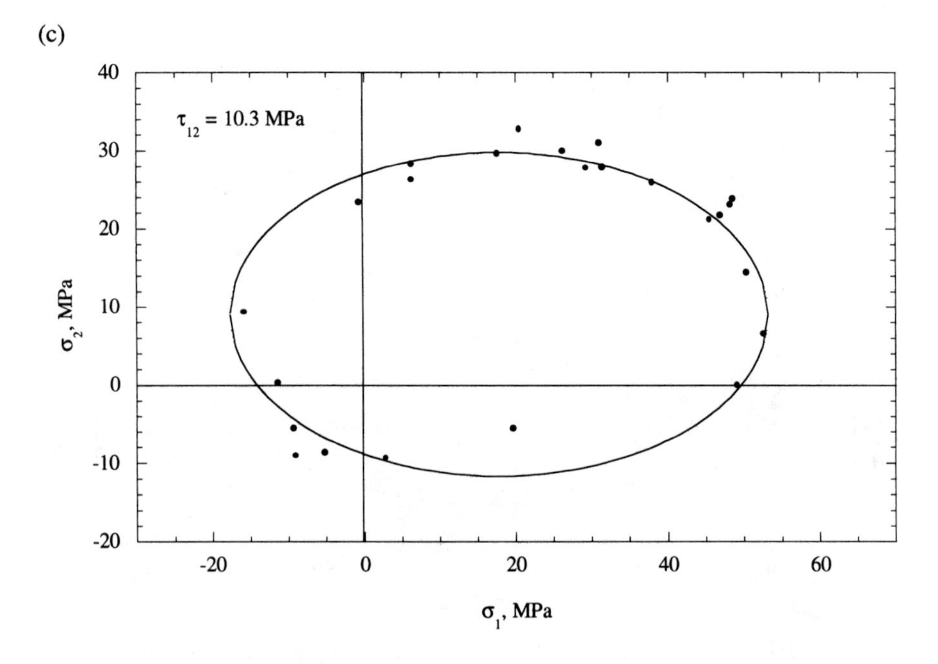

(d)

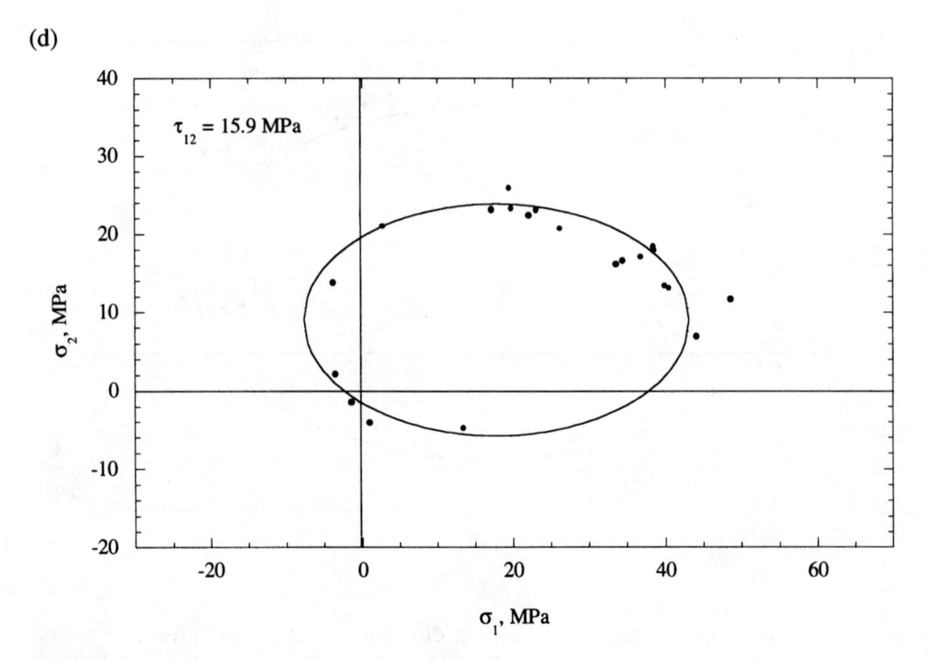

Fig. 56 *(continued)*

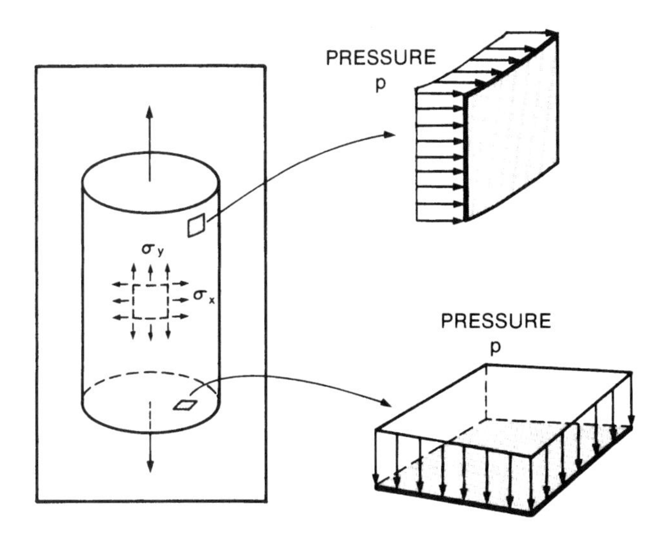

Fig. 57 A cylindrical sample internally pressurized and axially extended.

where r is the tube radius and h is paper thickness. Expressions (33) and (34) can be converted to force per unit width by multiplying both sides of the equation by thickness:

$$s_x = pr \tag{35}$$

and

$$s_y = \frac{1}{2}\left(pr + \frac{P}{\pi r}\right) \tag{36}$$

A schematic diagram of the planar biaxial tensile tester used by Uesaka et al. [183] is shown in Fig. 58. With this device, both strains ε_x and ε_y can be varied independently. Roller bars and fixed load detectors allow for measurement of forces without rigid-body restraints acting on the samples. Sample sizes on the order of 100 mm × 100 mm are used.

Brezinski and Hardacker [12] introduced a planar specimen design meant strictly for biaxial testing. The geometry is represented schematically in Fig. 59. Gunderson and Rowlands [57] used both cylindrical and planar (similar to the Brezinski and Hardacker design) specimens to evaluate the entire four-quadrant failure surface at four different levels of shear stress as presented in Fig. 56. These different levels of shear stress were developed by rotating the principal material directions relative to the loading directions. The shear stress and normal stress components were then resolved in the machine and cross-machine directions from the measured stress values using the Mohr's circle technique. For the planar measurements they used the method of lateral specimen support developed by Gunderson [56] (Fig. 36a,b). This work [57,59,60,144,176] represents the most comprehensive biaxial strength data set and analysis available in the literature.

More recently, the Brezinski–Hardacker type of specimen was used with new mechanical and optical equipment dedicated to biaxial testing by measurement of the planar displacement field using CCD video cameras [190]. Equipment was also

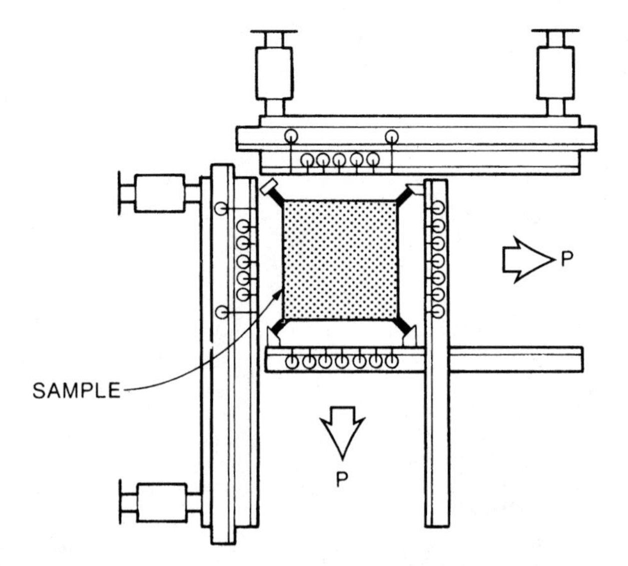

Fig. 58 Planar biaxial tension test system used by Uesaka et al. [183].

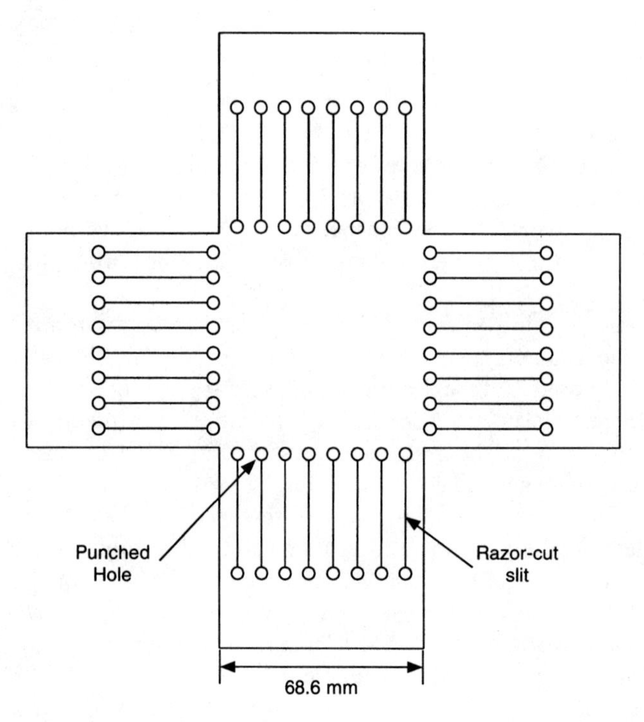

Fig. 59 Specimen configuration used for biaxial testing in the work of Brezinski and Hardacker [12].

recently developed (J. Tulonen, TJT-Technik AB, Järfälla, Sweden) to measure the biaxial stress development in paper during drying [102].

Analysis of Results The data of Gunderson et al. [57,59,60,144] and the analytical results of Suhling et al. [176] are given in Fig. 56. The curves in Fig. 56 represent the fit of Eq. (31) with $F_{12} = 0$. Although most of the work done in this area has been from a phenomenological perspective, some thought has been given to explaining the general shape of the failure surfaces [39,49,59] from a mechanistic perspective.

It is interesting to note the first-quadrant response of paper where biaxial tension is applied. Generally it has been found that in the first quadrant the biaxial strength is greater than both machine and cross-machine direction uniaxial tensile strengths. As loading transitions from uniaxial tension to biaxial tension, fewer fiber segments are loaded in compression [14,192]. Gunderson et al. [59] developed a mechanistic model for paper failure and compared the suggested results with those of their experiments. The model that best described the failure envelope for paper conformed to the following criteria.

Failure does *not* occur if

1. The normal strain in any direction is algebraically greater than the uniaxial compressive strain at failure in that direction (compressive strain is negative).
2. Tensile stress in any direction is not greater than the maximum tensile strength in that direction during biaxial testing.
3. Normal stress in any direction is algebraically greater than or equal to the uniaxial compressive strength in that direction (compressive stress is negative).
4. The combined probability of failure in any two perpendicular directions is not greater than 0.5.

The resulting failure surface for zero shear stress was then presented, showing the criterion that was violated (failure mode) at various stress space positions. This figure is reproduced in Fig. 60. It is interesting to note that the compressive strain limitation (criterion 1) is responsible for much of the transition between quadrants I and III. Gunderson and Rowlands [57] point out that the model discussed above is conceptually similar to the one proposed by Ranger and Hopkins [140] where the compressive failure of fibers plays a prominent role in the failure of paper. They also point out that in-plane shear stress is significant only in that it alters the magnitude of normal stresses in other than the principal stress directions. Although out of scope for this brief treatment, Patel [129] examined the biaxial stress deformation response of corrugated board.

I. Bursting Strength

Background One of the oldest most widely used of the strength tests for paper, paperboard, and corrugated fiberboard is the bursting test, usually called the Mullen test after one of the early instruments development around 1900. It was undoubtedly developed to simulate end-use conditions. The primary function of the test is to indicate the resistance of fiber products to rupture in use. The test essentially consists

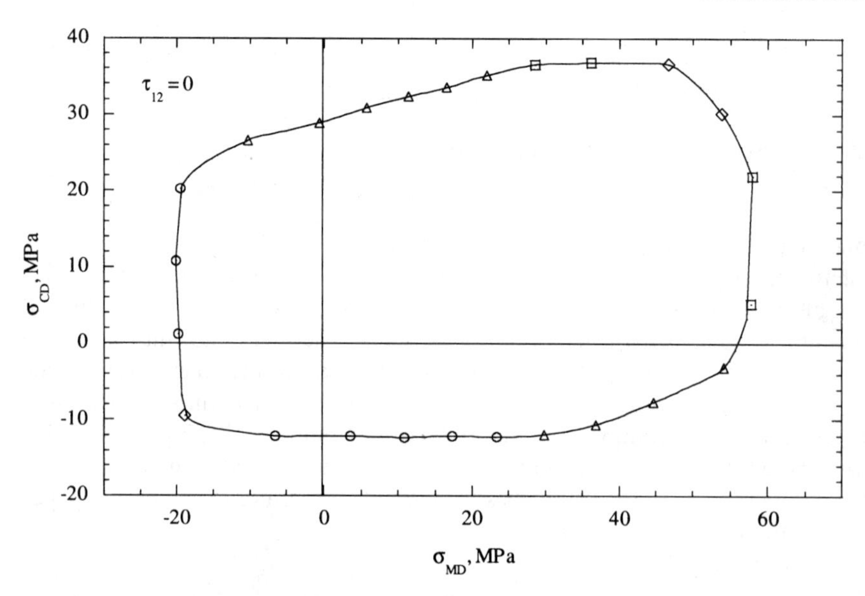

Fig. 60 The mechanistic failure envelope as proposed by Gunderson et al. [59]. There are four failure criteria: (△) compressive strain (criterion 1); (□) tensile strength (criterion 2); (○) compressive strength (criterion 3); (◇) combined probability (criterion 4).

of applying a pressure to one side of a clamped sample until it breaks. Bursting strength is defined as the hydrostatic pressure required to produce rupture.

Understanding the significance of burst test results is difficult because of the complex stress and strain fields present in the test specimen. It is likely that bursting strength is related to tensile strength, tensile strain, and shear characteristics of the material. Because the burst results depend on so many variables, the burst strength is not an intrinsic material property. It is commonly assumed that forces are applied biaxially during the test and that bursting strength is proportional to the square root of strain in the machine direction (MD) and the arithmetic mean of MD and cross-machine direction (CD) tensile strength.

To illustrate the complexity of the test, consider a circular flat sheet of paper that is deformed into a surface of revolution (Fig. 61) by a pressure acting on one surface. Assuming that membrane stresses are uniform through the thickness, the external pressure can be related to the in-plane principal stresses [179] by

$$p = \left(\frac{h}{r_1}\right)\sigma_1 + \left(\frac{h}{r_2}\right)\sigma_2 \tag{37}$$

where r_1 and r_2 are the principal radii of curvature and h is the thickness of the sheet. If the surface of revolution is assumed to be a spherical cap, then $r_1 = r_2 = r$. In this case, the strain components will be equal, $\varepsilon_1 = \varepsilon_2 = \varepsilon$, and related to the geometry by

$$\theta = \frac{a}{r}(\varepsilon + 1) \tag{38}$$

where a is the radius of the circular boundary of the clamp and θ is one-half the subtended angle shown in Fig. 61.

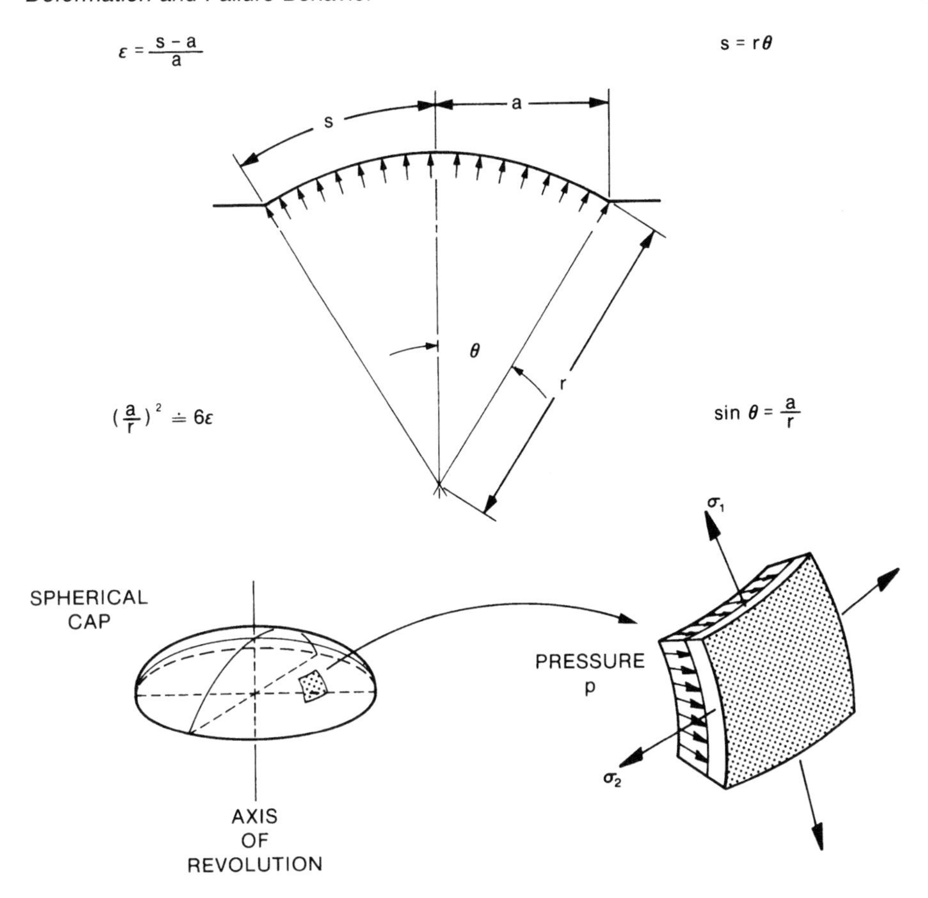

Fig. 61 Model for analysis of stress in a burst test.

$$\frac{a}{r} = \sin\theta = \theta - \frac{\theta^3}{3!} + \frac{\theta^5}{5!} - \cdots \tag{39}$$

Then, using the first two terms of the Taylor series for $\sin\theta$ in conjunction with Eq. (38),

$$\left(\frac{a}{r}\right)^2 = \frac{6\varepsilon}{(1+\varepsilon)^3} \tag{40}$$

which, for small strains, reduce to

$$\frac{a}{r} = \sqrt{6}\sqrt{\varepsilon} \tag{41}$$

Consequently, the pressure required to produce the bulge is related to the strain in the paper by

$$p = \sqrt{6}\frac{h}{a}(\sigma_1 + \sigma_2)\sqrt{\varepsilon} \tag{42}$$

This equation states that the results of a particular biaxial test—one in which strain is applied equally in both directions (regime I, Fig. 62)—should predict the burst strength.

The anisotropic nature of the results from burst tests can be illustrated by assuming that paper obeys a linear orthotropic stress-strain law:

$$\sigma_M = \frac{E_M}{1 - \nu_{CM}\nu_{MC}}\{\varepsilon_M + \nu_{CM}\varepsilon_C\} \tag{43}$$

$$\sigma_C = \frac{E_C}{1 - \nu_{CM}\nu_{MC}}\{\nu_{MC}\varepsilon_M + \varepsilon_C\} \tag{44}$$

where E_M and E_C are the Young's moduli in the machine and cross-machine directions, respectively, and ν_{MC} and ν_{CM}, are the in-plane Poisson ratios. Substituting Eqs. (43) and (44) ($\sigma_1 = \sigma_M$ and $\sigma_2 = \sigma_C$) into Eq. (42) and recalling that $\nu_{MC}E_C = \nu_{CM}E_M$ yields

$$p = \sqrt{6}\frac{h}{a}E_M\Psi_{MC}\sqrt{\varepsilon} \tag{45}$$

where

$$\Psi_{MC} = \frac{1 + \nu_{MC} + \nu_{CM} + \nu_{CM}/\nu_{MC}}{1 + \nu_{CM}\nu_{MC}} \tag{46}$$

Obviously, bursting pressure is related not only to the strain at rupture but also to the stiffness and lateral contraction behavior in both the machine and cross-machine directions.

Bohmer [9], IPC [72], and Stenitzer [168], who developed the ideas covered in the preceding discussion, took the analysis one step further. Although the machine direction strength from a uniaxial tensile test is greater than the corresponding cross-machine direction value, the strain at failure for the machine direction is smaller than the cross-machine counterpart. This can be readily seen in Fig. 18 from the tensile

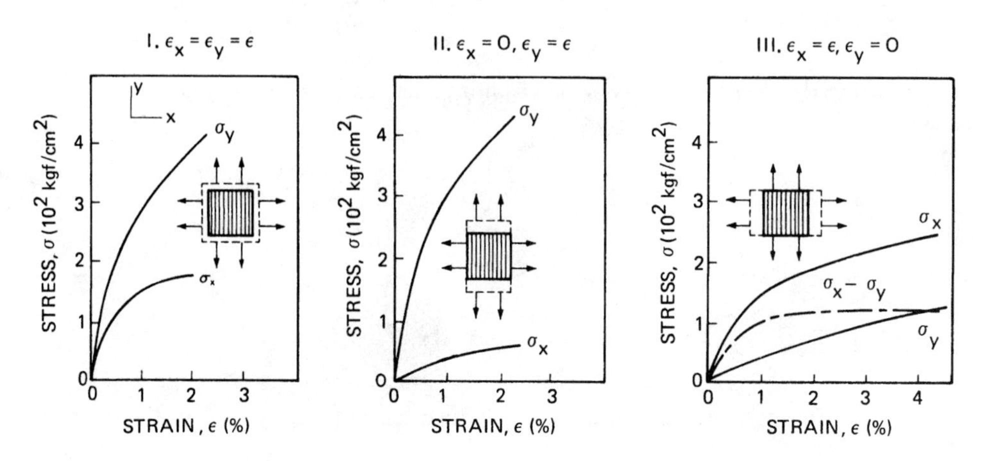

Fig. 62 Biaxial stress–strain diagrams from the planar biaxial tester for kraft paper corresponding to three loading regimes I, II, and III (machine direction parallel to y). (From Ref. 183.)

testing section. If the spherical cap assumption for the shape of the bulge is correct, which seems to be valid based on the work of Campbell [25], then failure in a burst test should initiate at machine direction strain at failure. Following Bohmer [9] and Stenitzer [168], the stress in both directions is assumed to be proportional to the square root of strain,

$$\sigma_M = k_M\sqrt{\varepsilon_M}, \qquad \sigma_C = k_C\sqrt{\varepsilon_C} \tag{47}$$

where k_M and k_C are empirical constants that can be expressed in terms of the stress and strain components at failure (Fig. 63):

$$k_M = \frac{\sigma_{M,u}}{\sqrt{\varepsilon_{M,u}}}, \qquad k_C = \frac{\sigma_{C,u}}{\sqrt{\varepsilon_{C,u}}} \tag{48}$$

Failure in a burst test will occur when the strain in the sample equals the ultimate strain in the machine direction. If lateral contraction effects are ignored, the stress in the cross-machine direction, σ_C, will be

$$\sigma_C = k_C\sqrt{\varepsilon_{M,u}} = \sqrt{\frac{\varepsilon_{M,u}}{\varepsilon_{C,u}}}\sigma_{C,u} \tag{49}$$

Consequently, burst can be related to the results from the uniaxial tension tests by

$$P_{\max} = \sqrt{6}\frac{h}{a}\left(\sigma_{M,u} + \sqrt{\frac{\varepsilon_{M,u}}{\varepsilon_{C,u}}}\sigma_{C,u}\right)\sqrt{\varepsilon_{M,u}} \tag{50}$$

In the investigation of burst behavior by Bohmer, Van den Akker, and Stenitzer, it is assumed that the shape of the deformed surface is a spherical cap. By assuming this

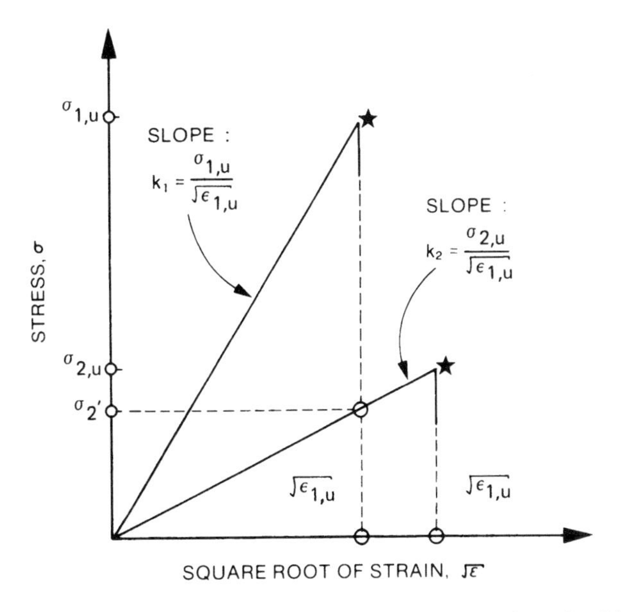

Fig. 63 Plot of stress versus square root of strain. [Assumption used to predict burst strength from uniaxial tensile tests: (1) machine direction; (2) cross-machine direction.]

shape and also assuming that paper is a linear elastic orthotropic material, these researchers related burst strength to the machine direction uniaxial tensile failure. However, Strikwerda and Considine [172] presented an analysis based on membrane theory that allows for unlimited deformation and no shape specification when the deformation of the specimen is greater than the specimen thickness. Further, Suhling [173,174] found that the shape of the burst deformation of a paperboard specimen was not spherical but more nearly parabolic.

Two other factors should be mentioned in conjunction with the burst test. One is the effect of sample thickness, which may increase the influence of bending as thickness increases. The other, rigidity of the clamps, will change the bending stress distribution if significant bending stresses are present. Thus, for thick specimens, a combined state of bending and membrane stress will exist through the sample thickness (Fig. 64). If significant bending is present, then initiation of failure would be expected to occur on one side of the sheet. Moreover, p_{max} will depend on whether the clamps are rigid or flexible, with rigid clamps yielding lower burst strength. It is important to remember, therefore, that when significant bending is present, as may be the case for thick paperboard, the combined stress will not remain constant through the thickness of the sample. This factor may explain why the wire and felt sides of paperboard sometimes show a two-sided effect.

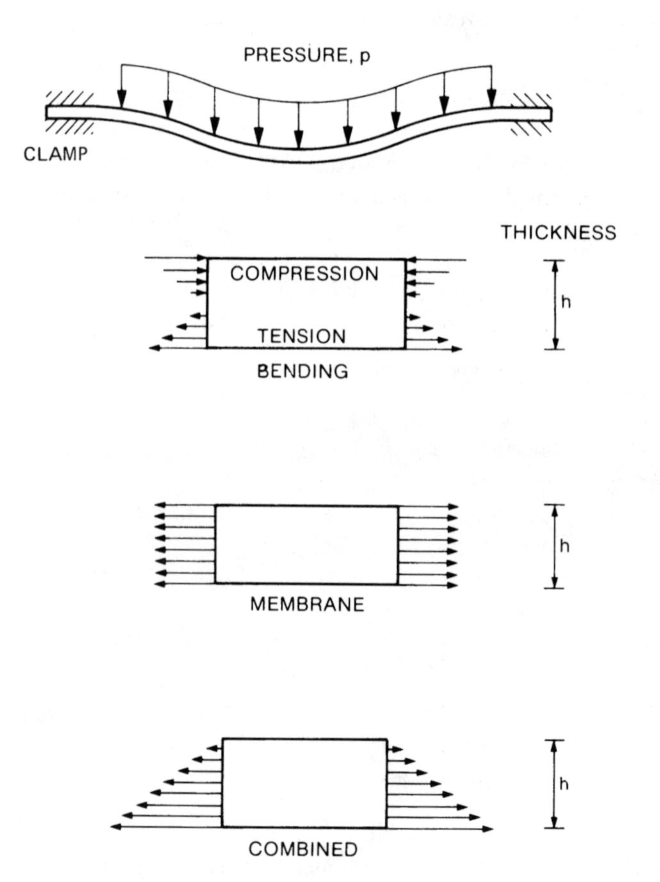

Fig. 64 Combined bending and membrane stresses in a thick burst sample.

Procedure The burst test is conducted by clamping a sample, 100 × 100 mm, between two annular plane surfaces with circular openings in their centers and fine concentric tool markings in the clamping region to minimize slippage of the test specimen during testing. The upper clamping ring is connected to the clamping mechanism through a swivel joint to facilitate an even clamping pressure. The specimen is initially flat and held rigidly at the circumference of the circular test area. During testing, the circular test area is subjected to an increasing hydrostatic pressure as an expandable gum rubber diaphragm is distended at a constant rate through the lower platen by means of hydraulic fluid.

The test specimen is thus deflected laterally and continues to bulge until rupture ensues. The maximum hydraulic pressure, recorded when the specimen ruptures, is defined as bursting strength. It is reported in units of kilopascals (kPa) or pounds per square inch (psi). Sometimes the burst index is reported. It is expressed as bursting strength divided by basis weight in units of kPa m^2/g or psi ft^2/lb. Average, maximum, and minimum values, the standard deviation, and the number of tests should always be reported.

A burst testing machine is shown in Fig. 65. The burst test is a common test used throughout the paper, paperboard, and corrugated fiberboard industries. Accordingly, TAPPI standards T403 os-76, T807 os-75, and T810 su-66 [177], respectively, are given for these categories. These standards specify calibration and maintenance of the test equipment as well as preparation and testing of burst samples.

A new tester quoted by Lorentzen and Wettre [109] was designed to measure both bursting strength and bursting energy absorption. Measurement of the amount of paper strain as well as stress during the course of bursting, in order to obtain the burst energy, is a completely new feature that has been integrated into Lorentzen and Wettre's classical burst strength tester. It is argued, for example, that the burst strength value by itself cannot differentiate between sack paper and liner. Sack paper is tough and stretches, whereas liner feels stiff. Despite the obvious differences between these two grades, they may have the same burst strength value. The burst energy absorption value for sack paper, on the other hand, will be substantially higher.

Analysis of Results A schematic illustration of a typical H-shaped crack pattern that develops at failure in a burst specimen is shown in Fig. 66. The midsection of the sample appears to fail perpendicular to the machine direction first; then the crack spreads to the sides, where it branches on both sides and pulls apart perpendicular to the cross-machine direction. This pattern confirms the fact that the MD stress has reached the level of stress corresponding to the maximum allowable strain in the cross-machine direction.

Perhaps the greatest cause of erroneous results in burst testing is improper clamping of the burst sample. Kill and Maltenfort [87], Aldrich [2], and Snyder [164] have shown that variations in clamping pressure can cause up to 15–20% error in results for bursting strength (Fig. 67). Too low a clamping pressure will cause slippage of the sample and excessive bulging during testing; this leads to inflated values for bursting strength. As with most testing procedures, the boundary or end conditions can significantly affect test results. It is also known that loading rates can affect burst; consequently, the rate of strain must be maintained effectively constant to obtain reproducible results.

Fig. 65 Burst tester.

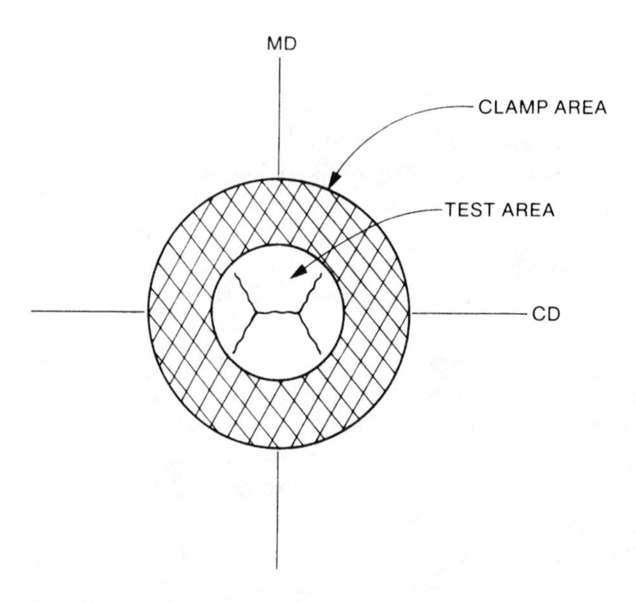

Fig. 66 Typical failure pattern from a burst test (From Ref. 1.)

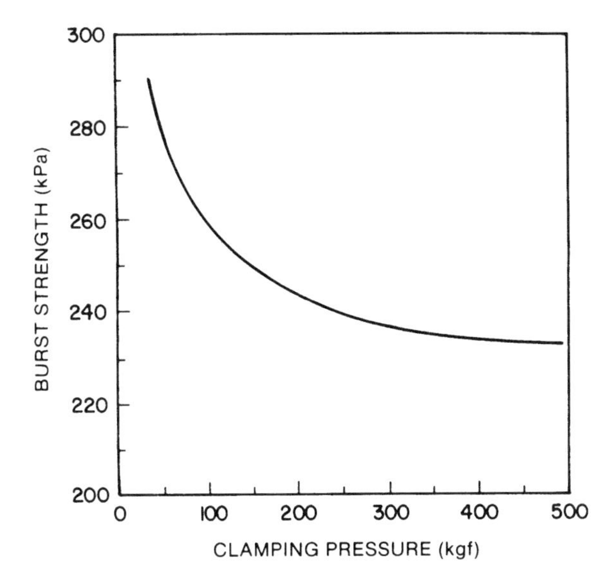

Fig. 67 Variation of burst strength with clamping pressure. (From Ref. 1.)

Because burst strength is the result of a combination of several factors, it will depend on the type, proportion, preparation, and amount of fiber present. For example, bursting strength is known to increase with increase in basis weight, sheet densification, and degree of pulp refining [6,69]. Obviously, the degree of bonding and individual fiber strength contribute to bursting strength. Furthermore, formation, distribution of flaws, internal sizing, surface treatments, distribution of fines, multiple plies, and wire-felt sides all have an impact on bursting strength.

In recent work by El-Hosseiny and Anderson [43], the authors discuss the effect of fiber length and fiber coarseness on bursting strength. For randomly oriented sheets, theoretically derived models provide an invariant relationship in which bursting strength is proportional to tensile strength and the square root of strain to failure. The constant of proportionality in this relationship reflects the particular geometry of the burst tester, which is customarily presumed to be independent of furnish variables. El-Hosseiny and Anderson argue that this constant is affected by fiber length and fiber coarseness. Increased fiber length increases the constant, and increased coarseness decreases it.

Any variation in atmospheric relative humidity will affect the moisture content of the sample. This moisture has a pronounced effect on the stress–strain response of the material and consequently affects bursting strength. A maximum burst strength has been obtained for samples conditioned at about 40% relative humidity by Abrams [1] and Douty [42] (see Fig. 68). This is an example of the effect of maximum strain to failure on burst strength. It is highly recommended, therefore, that TAPPI Standard T402 be used to condition each sample prior to testing. Although the burst test has found widespread use, the correlation of results among different laboratories and using different instruments leaves much to be desired [27,139]. These results emphasize the importance of standardization and vigilance in keeping the instrument in good condition and the test procedures carefully controlled.

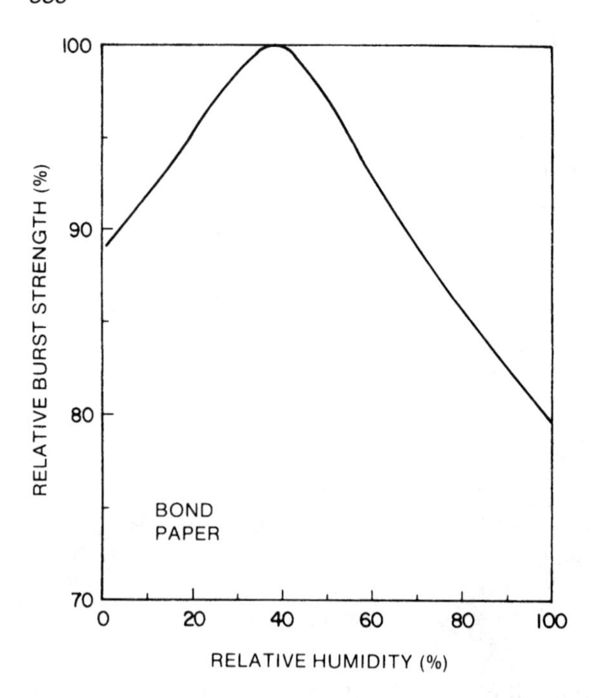

Fig. 68 Variation of burst strength with moisture content. (From Ref. 42.)

J. Tear Strength

Background Tear tests are classical examples of test methods developed pragmatically for evaluating a particular kind of paper strength. Tearing of paper consumes energy, and the force needed to continue the tearing is assumed to correlate with important end-use properties. With this in mind, the Elmendorf test, introduced in the 1940s, was developed and standardized. The test method involves applying an impact load to pull paper apart perpendicular to its faces. The energy absorbed by the sample from a swinging pendulum is used as a measure of tear strength. An in-plane tear test, similar to a fracture resistance test, was developed by Van den Akker et al. [194], but the principles of fracture mechanics were apparently not used as guidelines. The results from both tear tests have been used as indices of paper performance.

Although the Elmendorf tear strength tester is commonly used in North America and elsewhere for measuring the tearing resistance of paper, many European customers use the Brecht–Imset tear tester. The Brecht–Imset tear tester is also a pendulum-type instrument.

Attempts have been made to correlate the strength indices obtained with these instruments with various end-use situations, such as evaluating web runnability, controlling quality of newsprint, and characterizing the toughness of packaging papers where the ability to absorb shocks is essential.

It is instructive to compare the failure modes of the tear tests with the principal modes of failure distinguished in the study of fracture (Fig. 69): cleavage (mode I), transverse shear (mode II), and tear (mode III). These three modes are differentiated by the manner in which the breaking forces are applied relative to the crack area.

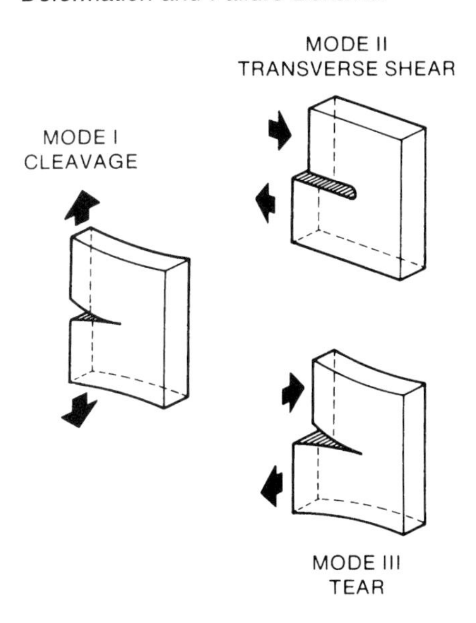

Fig. 69 The three principal modes of failure.

Because paper is an orthotropic material, a number of characteristic modes are possible as defined by the orientation of the failure surface, direction of crack propagation, and direction of the applied load. The failure mode of the in-plane tear test is essentially one of cleavage, or mode I (Fig. 70). (Note that it is not a mode I test from the point of view of fracture mechanics.) To confine the cleavage forces to the region in front of the crack tip, the specimen is purposely clamped in a misaligned configuration (α being the misalignment angle). Unfortunately, this causes wrinkling of the sample, and the test is conducted with the sample in a buckled state. For this reason, as well as others related to specimen width and speed of testing, the in-plane tear results cannot be directly related to the true fracture resistance.

The Elmendorf tear test involves out-of-plane loading (Fig. 71); however, the plane of fracture changes during the test. The difference between a true fracture-tearing test (mode III) and an out-of-plane paper tear test is shown schematically in Fig. 72. Scissors apply shear stresses through the thickness of the paper and rupture the fibers along a well-defined surface perpendicular to the paper faces. Pulling paper apart by out-of-plane forces, such as those encountered in the Elmendorf test, leads to a mixed mode of failure. It is apparent that the weak plane (parallel to the faces) dominates the behavior exhibited by this type of loading. It is evident that the fundamentals of both tear tests are complex, and test results must be interpreted with great care.

Procedure The Elmendorf tear test requires a sample 76 mm wide. It is clamped in split jaws in such a way that the unclamped length is 63 mm. A 20 mm slit is made in the sample with a razor blade, leaving a 43 mm segment through which the tear will propagate. Samples are tested by applying an impact load through a swinging pendulum (Fig. 73). The pendulum of a standard Elmendorf tester will deliver a maximum energy of 1.35 J. A scale on the apparatus is constructed in such a manner that

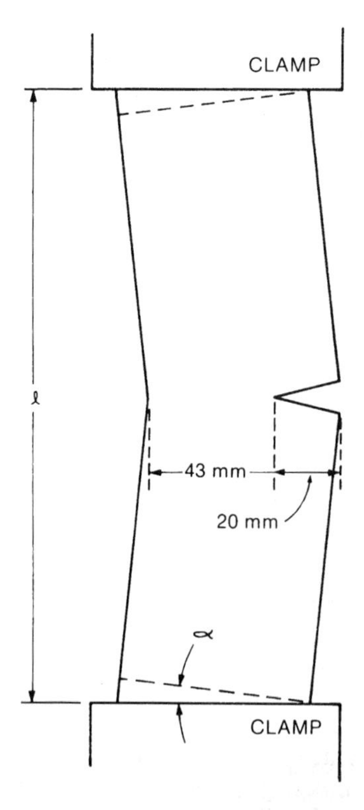

Fig. 70 In-plane tear testing of paper or paperboard.

if this maximum energy were required to completely tear 16 sheets (total tear length of 16 sheets × 4.3 cm/side × two sides = 137.6 cm) then the average amount of energy consumed to tear each sheet per unit crack length would be indicated as 0.98 J/m.

The scale on the apparatus indicates energy consumption per unit crack length assuming that 16 sheets are simultaneously tested. If fewer (or more) than 16 sheets are tested, the scaled reading is multiplied by the ratio of 16 to the actual number of sheets tested.

According to the Elmendorf test procedure recommended by the CPPA (Technical Section Standards D.9 and D.12), four plies of paper are clamped together and torn simultaneously in each test. Pendulums of different capacities depending on the sample strength, are used. The procedure recommended by TAPPI (test methods T 414 om-88 and T 220 om-88) is somewhat different. Although the capacity of the pendulum is fixed, the number of plies torn simultaneously is varied so as to obtain readings within a certain range on the tester scale. With both procedures, the test instrument provides the average amount of work consumed to tear each sheet per unit tear length (J/m). Tear strength is often normalized with respect to sheet basis weight (grammage); tear index is calculated by dividing the tear strength by sheet basis weight; e.g., joules per meter per kilogram per square meter [(J/m)/(kg/m^2)] equals joule-meters per kilogram (J m/kg).

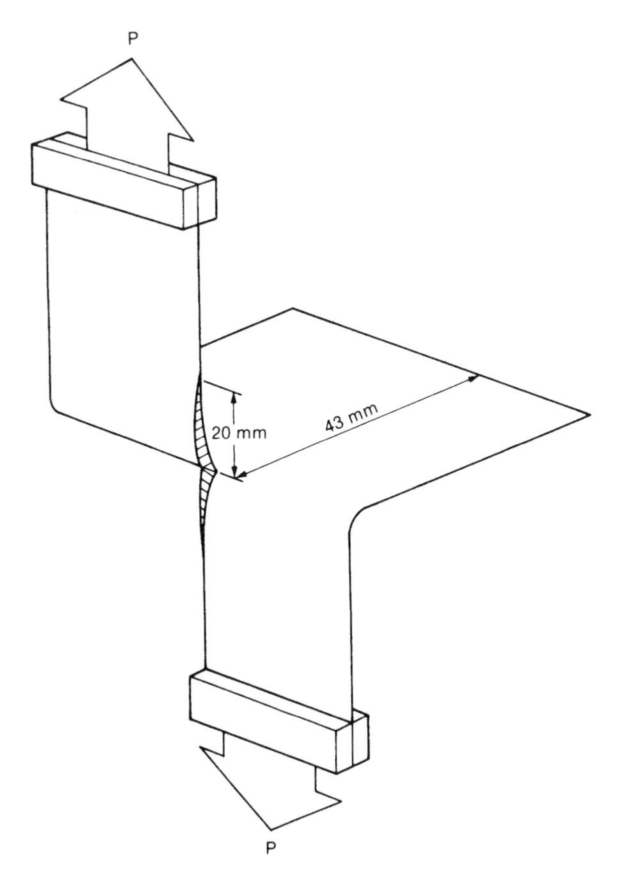

Fig. 71 Elmendorf tear.

A valid Elmendorf tear test is one in which the tear is complete and propagates directly across the sample, parallel to the direction of the initial slit. For the TAPPI test procedure, it is recommended for consistency that the number of sheets of a given test be varied until a scale reading of approximately 0.4 J/m is obtained. Of course, the scale readings are then modified for the actual number of sheets used. More detailed information is given in TAPPI Standard T414 ts-65 [177].

A sample 63 mm wide is used in the in-plane tear test. A 20 mm slit is made in one side of the sample, which is clamped in a testing machine by rotating the upper and lower jaws through a suitable misalignment angle α before clamping. The span between the jaws can be variable. Typically, a value of 50 mm is used in conjunction with a test speed of 50 mm/min [194]. The work done on the sample, i.e., the area under the load—elongation curve, to completely pull it apart is divided by 43 mm to obtain the length of an average energy consumption value for the in-plane tear test. The results of both tests can be divided by thickness to obtain values for energy consumption per unit crack area.

The Brecht–Imset tear tester employs a 60 mm wide specimen held against the tester face. Before the test, the specimen is initially cut and then torn through a short fixed distance by manually turning the pendulum. During the test, the pendulum is

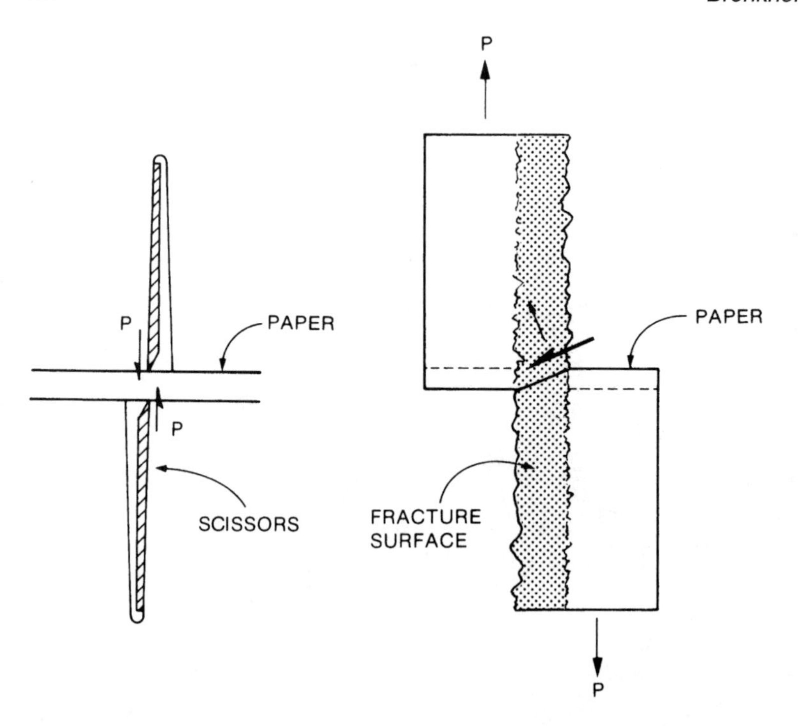

Fig. 72 Comparison between fracture tearing and mixed-mode tearing of paper.

released and a metal bar 31 mm wide and 3 mm thick emerges from a vertical slot in the instrument and causes the central zone of the specimen to be bent around the leading blunt edge of the bar and to be torn away from the rest of the specimen, which is held back by means of fixed metal fingers attached to the instrument face plate.

Unlike the Elmendorf tear tester, in which the initial cut is sharp and the tearing distance is fixed, the Brecht–Imset tester continues a tear already started. It produces a further total length of tear that is not fixed but depends on the instrument constants and the tearing resistance of the specimen. However, the method is such that the instrument measures the work expended to tear the specimen through a unit distance. The instrument readings of the Brecht–Imset tester are twice those of the Elmendorf tear tester and are therefore divided by 2 to get a tear force comparable with that given by the Elmendorf tear tester.

Analysis of Results For Elmendorf tear, it is evident that the orientation of the fracture surface varies during the test from almost 90° to the faces at the beginning to approximately 180° at the end. It might be expected that any number of factors could influence the results, particularly if these factors are activated as the test proceeds.

According to Van den Akker's tear strength theory [191], tearing work is composed of two different phenomena: the work associated with breaking fibers and frictional work associated with pulling fibers out of the fiber network. The balance between fiber fracture and fiber pull-out will depend, of course, on the degree of fiber bonding in the paper sheet, fiber length, and fiber strength.

Fig. 73 Elmendorf tearing tester.

Continued increase in fiber bonding will eventually lead to excessive fiber fracture, which will generally result in loss of tear strength, even though the tensile and bursting strengths continue to improve.

Helle [65] suggests that there are also other energy-consuming mechanisms in tearing, e.g., interaction between torn edges of the sample (bending and splitting the tearing sample). Elmendorf tear strengths for single and multiple sheets with identical total basis weights are shown in Table 3. It is seen that the use of single and multiple sheets does not give the same results. Because multiple sheets can slip relative to one another, less energy is consumed elsewhere in the sample compared with that in creating the fracture surface. The violent bending and steep peeling angle that prevail during Elmendorf tearing can cause stiffer specimens to consume more energy in other modes than is used in fracturing the specimen.

The fact that the tear strength in general depends on the number of plies torn simultaneously has been known for some time [70], and it has been confirmed over the years by several other researchers, e.g., Seth [153]. The mechanisms responsible for this difference are not completely understood. Therefore, at the present time, it is not possible to predict the discrepancy for specific pulps or for the degree of fiber bonding in the sheet. What is clear is that the discrepancy is different for different pulps and for different degrees of fiber bonding within the same pulp sheet.

Table 3 The Effect of Flexural Stiffness on Elmendorf Tear Results

Basis weight (g/m^2)	Elmendorf tear strength		
		Multiple sheets	
	Single sheet (J/m)	J/m	No. of sheets
60	1.96	1.57	1
120	5.20	2.45	2
180	9.02	5.00	3
240	13.14	6.67	4
300	16.08	8.04	5

Source: Data from Ref. 104.

Essentially a layered structure, paper tends to split when torn in the Elmendorf mode; tearing in this mode occurs under out-of-plane loading. The extent of splitting depends on sheet basis weight, the degree of fiber bonding present, and the number of plies being torn simultaneously. In the single-ply test, splitting of the sheet is allowed to proceed freely; however, it is arbitrarily restricted in the multi-ply test.

Seth [153] warns the papermaker who is selecting pulps from tear–tensile data that the plots of single-ply tear strength versus tensile strength will give a different perspective of a pulp's response to refining than will the multi-ply results. It is interesting to note that for the tear–tensile plots reported by Seth [153], the discrepancy in tear strength vs. tensile strength results between single-ply and multi-ply (four-ply) sheets decreases as sheet tensile strength is increased. For values of tensile strength equal to or greater than 6 km, there is little difference between single-ply and multi-ply tear results.

Single sheets are generally torn with the Brecht–Imset tester, whereas the Elmendorf tear tester tests four sheets simultaneously. A good correlation between the standard Elmendorf and Brecht–Imset tear strengths has been observed for many commercial papers; however, Seth and Blinco [155] indicate that such a correlation does not hold for weakly bonded handsheets with high tearing resistance. For these handsheets, the Elmendorf tear strength was found to be much higher than the Brecht–Imset tear strength. This discrepancy was attributed to the use of multiple plies for the Elmendorf tear test. When single plies were used for the Elmendorf tear test, results were similar to those for the Brecht–Imset test.

Results for the in-plane tear test are obviously a function of the misalignment angle α. The evidence can be seen in Table 4, where the angle has been varied over a wide range. The angle causes the sample to buckle or wrinkle, complicating the manner in which the paper microstructure resists fracture. From preliminary work of Van den Akker et al. [194], 6° has often been selected as the angle where minimum resistance to this particular kind of tear occurs. On the other hand, Helle [65] found that the minimum resistance is a variable related to other properties such as tensile elongation at rupture. Seth and Page [156,157] compare the in-plane test with the fundamental fracture resistance test method as well as Elmendorf tear and show that all three test methods yield different results. Based on the theory of fracture mechanics, however, only the critical crack extension force measurement has a solid theoretical foundation.

Table 4 In-Plane Tear Energy as a Function of the Misalignment Angle

Misalignment angle (deg)	In-plane tear energy					
	Machine direction[a]			Cross-machine direction[a]		
	1	2	3	1	2	3
1.6	0.86	7.60	18.82	1.36	6.51	—
4.0	0.85	4.57	13.82	1.35	5.04	20.29
6.8	0.81	4.00	16.56	1.35	4.86	20.97
9.7	0.90	4.17	19.01	1.70	4.98	—
	Handsheets[b]					
	1	2	3	4	5	
2	5.78	—	—	—	—	
3	5.00	8.78	—	—	—	
4	4.94	7.67	8.32	—	—	
6	4.88	6.76	6.96	9.42	9.94	
8	5.40	6.44	6.96	8.38	8.64	
12	6.24	6.56	6.96	8.00	8.19	
15	8.04	7.74	8.00	8.45	8.52	

[a] Three types of paper were used: 1, newsprint; 2, 100% rag bond; 3, kraft liner. Units of tear energy are J/m. Data from Ref. 194.
[b] Five levels of restrained shrinkage were used with handsheets, units of the tear energy are Jm/kg (basis weight = 0.65 kg/m^2). Data from Helle [65].

A large body of knowledge exists in the paper literature relating various raw material and processing variables to tear test results [29,115,127,159]. It is not the purpose of this chapter to review these relationships or to explore the appropriateness of using a particular test for a given situation It is difficult, however, to let this section end without offering a warning about the empirical nature of the traditional tear tests. Results of Elmendorf tear, in-plane tear, and tensile energy absorption (TEA) properties on sheets of different densities are shown in Fig. 74. A wide range of behavior is evident. This variation is simply a result of the fact that a wide range of behavior is being measured. Which behavior is the right behavior can be answered only by the investigator, who must decide on how the results should be related to end-use performance.

K. Transverse Compression of Combined Board

Background In our discussion thus far, we have concentrated on test methods that are used primarily to characterize failure of sheet material. In the next sections, we will consider two test methods for corrugated board and two for corrugated shipping containers. It is our purpose in using these selected test methods to illustrate how failure of an object is related to failure of its parts.

During fabrication, a transverse compressive load is applied through the liner facings to adhere them to corrugated medium. Once the corrugated structure has been formed, impact loading during converting and printing of the corrugated board

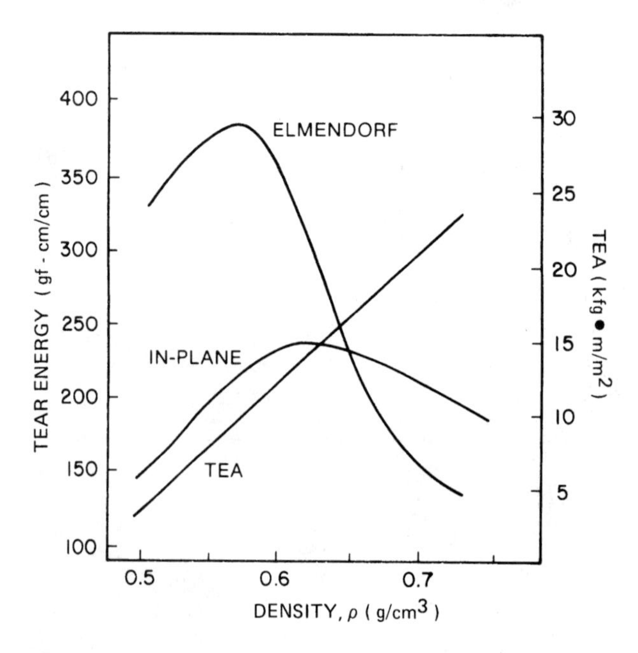

Fig. 74 Influence of sheet density on tensile energy absorption (TEA), Elmendorf tear, and in-plane tear. (From Ref. 194.)

or handling and shipping of the container can also be the source for damaging transverse compressive loading of the corrugated structure. If these transverse deformations are cumulatively extensive enough, then the corrugated structure will be compromised and the bending stiffness [36,123,167,204] as well as the compressive strength of the container will be adversely affected.

To characterize the transverse deformation failure of the corrugated structure, three TAPPI standard tests are used: (1) flat crush test of corrugated board (flexible beam method), T 808 om-97; (2) flat crush of corrugated medium, T 809 om-93; and (3) flat crush test of corrugated board (rigid support method), T 825 om-96. In all of these tests a compressive load is applied perpendicular to the plane containing the flute tips of the corrugated medium (Fig. 75) (with or without liners). Although a number of instability events occur during testing, only the maximum compressive load is generally used to characterize transverse deformation resistance to failure. Failure is defined as the complete collapse of the side walls of the corrugations. This collapse generally occurs when the original thickness is reduced by approximately one-half of its original value. A schematic illustration of the instability events that occur during testing is shown in Fig. 76. Note that a shearing instability can occur for irregular flute patterns. Also note that these test methods are sensitive to nonsymmetrical flute patterns and consequently are not recommended for evaluating transverse deformation behavior of double- or triple-wall corrugated board.

Procedure The compression specimen for a flat crush test of corrugated board should have a surface area of 3230 or 6450 mm^2. A circular shape is preferable. After moisture content conditioning, the sample is placed in a compression testing machine

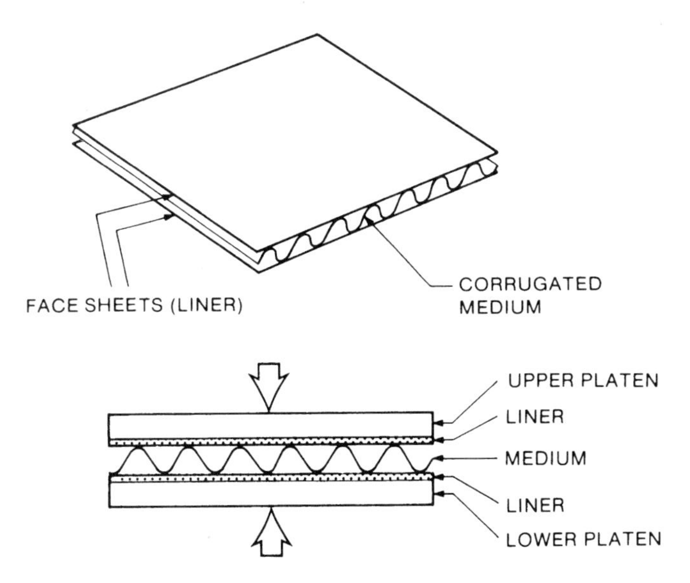

Fig. 75 Transverse compression deformation of corrugated board between parallel plates.

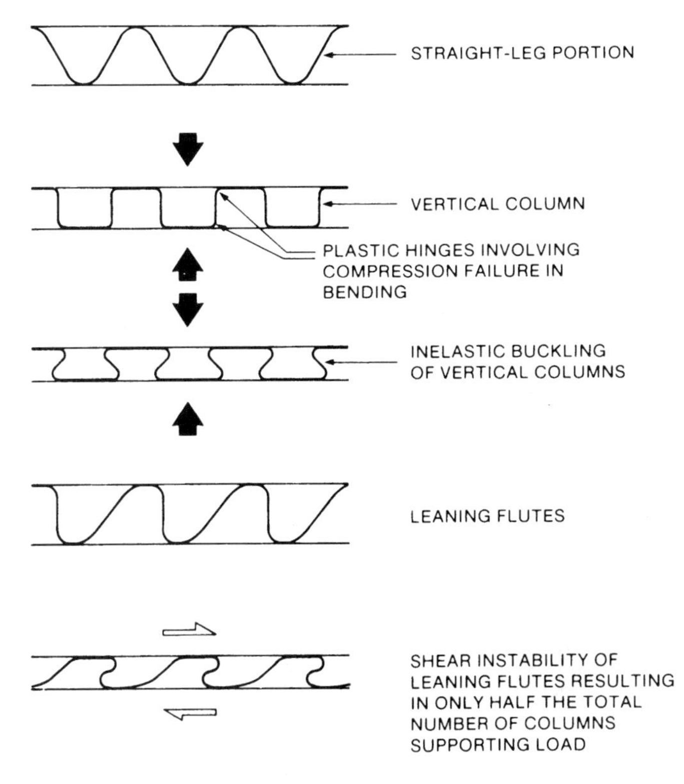

Fig. 76 Possible deformation modes of corrugated board during transverse compression loading.

that has platens that are held rigidly parallel and free from lateral movement. The samples should be compressed at a constant rate of 12.5 mm/min and the load recorded to the nearest 2.2 N. The flat crush test result should be reported as an ultimate stress (Pa or psi). A photograph of the type of mechanical system generally used for a standard test is shown in Fig. 77. Generally only the maximum load is measured and recorded. Load and displacement can be easily measured using conventional mechanical testing equipment that is equipped for compression testing.

Only an A-flute sample is specified for testing in determining the crushing resistance by TAPPI Standard T 809 om-93, corrugating medium test (CMT). A test strip is cut 13 × 150 mm with the larger dimension parallel to the machine direction of the material. The strip is corrugated, placed in a rack-and-comb device, and then adhered to a 130 mm strip of double-coated tape. The assembly is then placed in a compression testing machine where a compression load is applied by contact with the 10 flute tips. Only the maximum load is generally recorded.

Analysis of Results A typical load–deformation curve as reported by Crisp et al. [36] of a CMT test (A flute) is shown in Fig. 78. Initially, the load increases rapidly with imposed deformation to the point labeled *A*. From the origin to *A*, the curve represents the extent to which corrugated board behaves as an essentially elastic structure; that is, deformation is recovered on removal of applied load.

Fig. 77 The type of device used for standard flat crush testing of corrugated board.

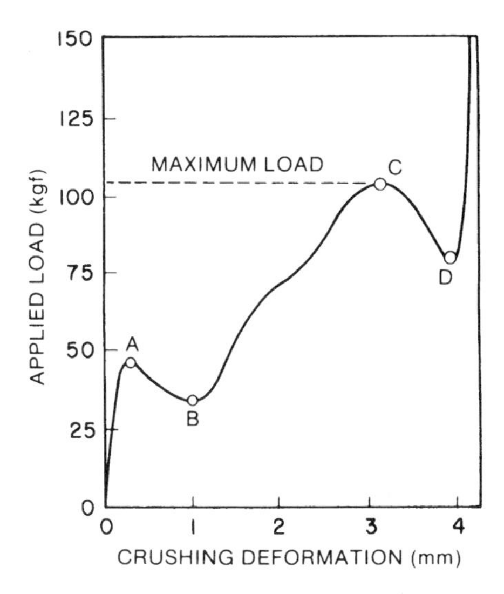

Fig. 78 A typical load versus deformation curve for transverse compression deformation of corrugated medium. (From Ref. 36.)

Eventually, inelastic bending takes place in the region *A–B* and hinges develop at the flute tips. Simultaneously, considerable rotation of the straight-leg portion of the flute occurs until it is positioned vertically (see Fig. 76). At this stage of loading the corrugated board may be thought of as a set of vertical columns with plastic hinge supports that carry compression load between *B* and *C* until inelastic buckling of the columns occurs at *C*. Collapse of these columns is associated with region *C–D*. Point *C* represents the maximum transverse load supported by the corrugated structure.

The sequential deformation of a corrugated board sample, sketched from photographs taken during deformation experiments, is shown in Fig. 79. The load–displacement curve and the locations on the curve corresponding to these sketches are shown in Fig. 80. The profile for the curve given in Fig. 80 is not uncommon for commercially manufactured corrugated board. Here, the radius of curvature of one set of flutes is greater than that of the other. Therefore, the broader tips will flatten and fail first, giving rise to the first peak (*3*). The flute tips on the opposite side will subsequently flatten and fail, giving rise to the second peak in the curve (*7*). The point of maximum load (*12*) is used to calculate the ultimate transverse deformation resistance of the corrugated structure.

Although the complete flat crush curve is of interest in evaluating its impact on subsequent box performance, only the maximum crushing load is measured and recorded for the conventional flat crush test. Experimental results have conclusively demonstrated that in actual practice, when board is run through converting machines, the perpendicular forces that act on the surface of the board can produce extensive board damage [167]. This information is generally not revealed by the flat crush test, as the maximum transverse load and residual board thickness are virtually unaltered by prior deformation events, especially if they were not severe enough to reach ultimate structural failure. Nevertheless, combined board stiffness and compression strength of containers constructed from the board are dramatically reduced. Because not much information is collected during standard measurement, care must

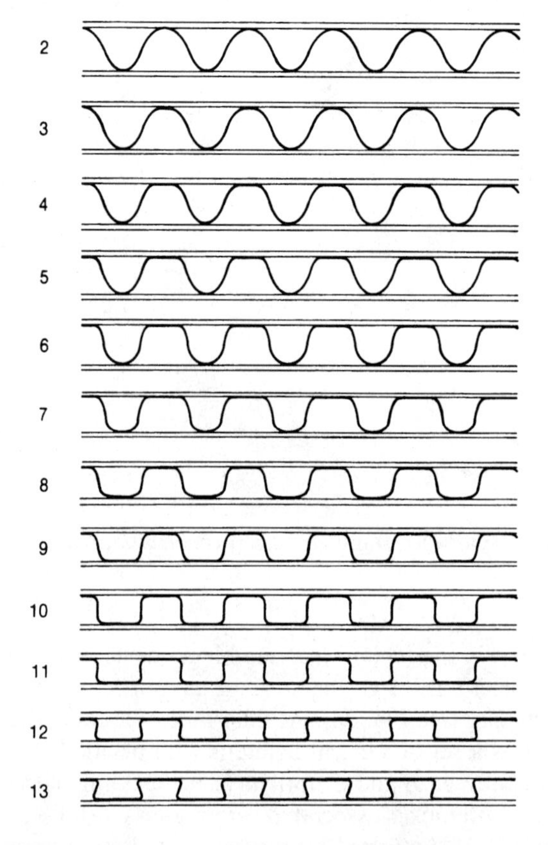

Fig. 79 Deformed geometry for corrugated board during transverse loading. Each of the configurations corresponds to a point on the load versus displacement curve given in Fig. 80. Note the difference in radii between the top and bottom flute tips.

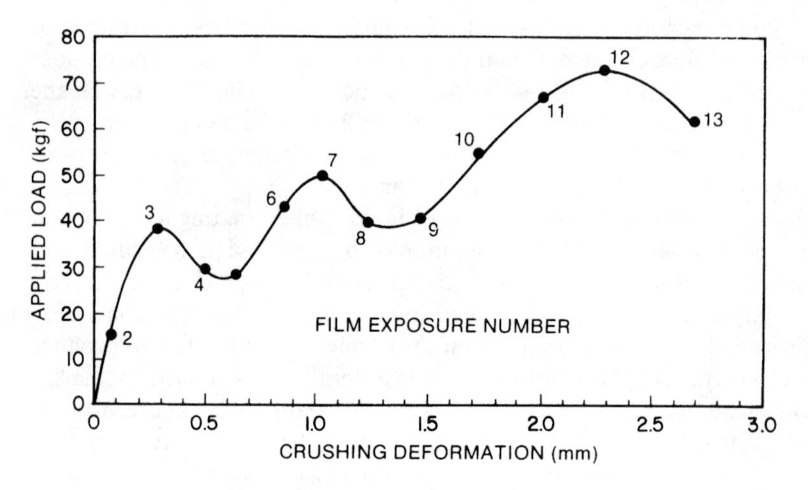

Fig. 80 A typical load versus deformation curve for single-wall corrugated board with an asymmetrical flute structure. (See Fig. 79 for corresponding deformed geometry.)

be taken in interpreting experimental results obtained from conventional flat crush tests.

From this brief discussion it should be quite clear that corrugated board may undergo significant irreversible damage before the maximum transverse compressive failure load is attained. Thus, even for a relatively small degree of transverse loading, the structure of the combined board can be damaged. Combined boards that have been intentionally predeformed to various levels prior to flat crush testing exhibit load–deflection curves in which a peak is replaced by an inflection point [123]. Typically, predeforming damages the structure to the point at which it feels soft to the touch and exhibits an inability to resist further thickness reduction.

L. Short-Column Crush Test

Background Another compression test method developed to evaluate the in-plane crushing resistance of corrugated board is the short-column crush (SCC) test. This test is also referred to as the edge crush test (ECT). The justification for the method stems from an attempt to simulate the compression load in a representative portion of the side panels of a corrugated container under a stacking load [112,114]. It has already been pointed out that failure modes of compression tests are dependent on the size of the test specimen. The short-column crush test is no exception. It is even more complex because the test specimen is more like a *structure* than a single *element* (such as a beam column). Consequently, overall failure is associated with failure of subelements such as buckling of linerboard miniplates (Fig. 81), buckling of the medium, or actual intrinsic compression failure of the corrugated medium or liner.

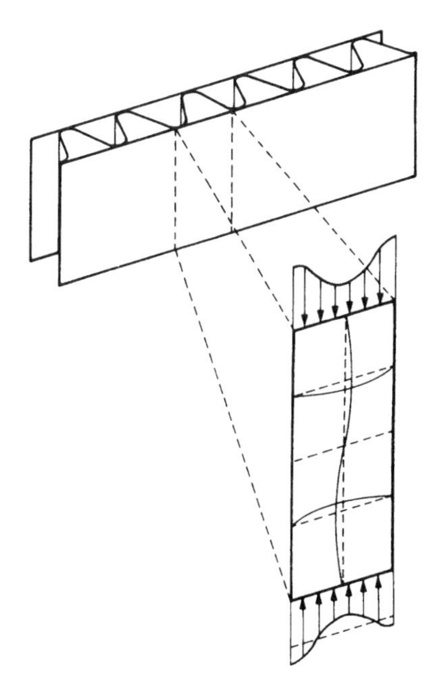

Fig. 81 Local buckling of miniature plates between flute tips of corrugated fiberboard. (From Ref. 74.)

Although the purpose of the short-column crush test is to duplicate service conditions, it is simply not possible to accomplish this goal exactly. The edges of the short-column crush specimen are free. This would not be true if the same material were part of the side panels; the edges would be restrained. Despite these differences, however, the results of edge crush tests for combined board can be used as a measure of stacking strength performance for corrugated containers.

Procedure A square test specimen, 50 mm on a side, has been found appropriate for the edgewise compression test of corrugated board. The details of sample preparation and testing are described in TAPPI Standard T811 or T823. A compressive force is applied parallel to the flute ridges as shown in Fig. 82. The bearing ends are waxed and placed in slotted guide blocks to ensure a fixed end-load condition. It is essential that the edges be cleanly cut and perfectly parallel. A parallel-plate testing machine is used to deliver the compression load at a rate of 125 N/s (Fig. 83). The wax is used to reinforce the loading edges of the specimen and to help prevent edge failure. Unfortunately, wax dipping of specimens is a laborious and time-consuming process. Urbanik et al. [189] investigated the possibility of reducing the reconditioning time of waxed samples from 24 h down to 2 h without significant change in strength values. Some attention has been given to the use of unwaxed necked-down specimens as an alternative to the waxed sample. However, experimental results of Koning [95] indicate that there is no significant difference in the edgewise compressive strength values obtained using necked-down specimens and those obtained using the TAPPI test method.

The maximum load per unit width (N/m, kg_f/cm, or lb_f/in.) is reported as short-column or edge crush strength. Division of maximum load by actual cross-sectional bearing area yields ultimate compression strength (Pa, kg_f/cm^2, or psi).

Analysis of Results A series of maximum column strengths for various column lengths is shown in Table 5 for A flute test strips 76 mm wide. Clearly, longer corrugated columns suffer buckling mode failures at low loads, whereas shorter

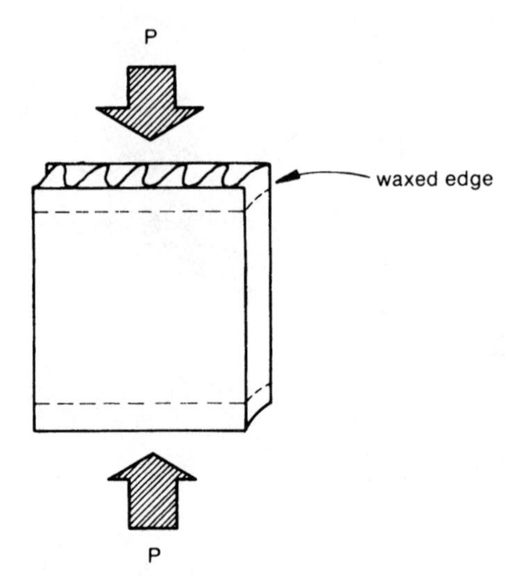

Fig. 82 Short-column crush sample.

Fig. 83 Short-column crush apparatus.

columns appear to reach a threshold strength indicative of a true compression strength. The length of the short-column crush test specimen, 50 mm, is appropriate for determining a form of the intrinsic strength rather than for characterizing the onset of an overall buckling mode. This does not mean, however, that the test specimens never undergo local instability.

Table 5 Column Strength as a Function of Column Height for a Corrugated Board

Slenderness ratio[a] λ	Column height (mm)	Column strength (kN/m)
35	50	6.57
69	100	6.47
104	150	5.69
139	200	4.71
208	300	2.45
277	400	3.73
346	500	0.69
416	600	0.39

[a] Flute thickness 5 mm; $\lambda = 2\sqrt{3}(\ell/h)$, where $\ell =$ column height and $h =$ thickness; load applied parallel to flute tip ridges.
Source: Data from Ref. 112.

A load–deformation curve of a short-column crush specimen is shown in Fig. 84, from IPC (Institute of Paper Chemistry) Report 1108-4 [74]. Results are also shown from edgewise compression tests of the two components of the combined board. A predicted load–deformation curve for the short-column crush specimen can be obtained from a rule-of-mixtures type of formula. This is shown as a dashed line in Fig. 84. The short-column crush strength (S_{SCC}) can, in principle, be approximated from the edgewise compressive strengths of the liner S_1 and corrugated medium S_2:

$$S_{SCC} = 2S_1 + t_2 S_2 \tag{51}$$

where t_2 is the take-up factor for the corrugated medium.

The agreement between experimental measurements of compressive strength of corrugated board as measured by the short-column crush test and predictions based on this simple rule of mixtures has been found to be less than satisfactory [152]. The measured combined board strength was found to be 70–80% of the predicted combined board strength. Several reasons for this discrepancy are discussed by Seth [152], including localized instability of the component liner or medium and premature failure of the loaded edge of the board in the short-column crush test. Despite these shortcomings, it is obvious that short-column crushing strength is related to the strength of the liner and corrugated medium.

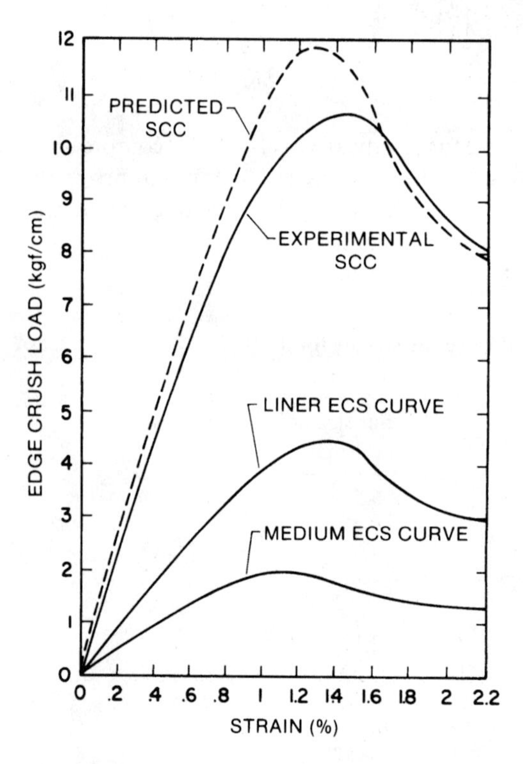

Fig. 84 Edge crush loading of corrugated fiberboard and component sheet material. (From Ref. 74.)

At the U.S. Forest Products Laboratory an attempt was also made to correlate the component compressive properties with the compressive properties of the combined board [75]. Equation (51) can be extended by adopting an empirical, two-parameter linear model determined by regression analysis for a given set of tests for combined board and component sheet material. Although this model gives a reasonable estimate of the corrugated board compressive strength, its disadvantages are that the regression analysis has to be repeated for each combination of liner and medium and that any inconsistencies in test methods are hidden in the regression analysis. This has been done by several authors including Seth [152] and Kainulainen and Toroi [81]. A theoretical approach using thin-plate theory was considered by Moody [118]. This work was extended by Johnson et al. [75] and later applied by Urbanik et al. [189] and Ince and Urbanik [71] to evaluate the relative importance of liner and medium in combined board strength.

Westerlind and Carlsson [200,201] present an experimental technique that enables an accurate measurement of the entire load–deformation curve of combined board subjected to edgewise loading. By suitable specimen and test design, end crushing, which can often lead to premature failure of the combined board specimen, is avoided. Special side supports are used to prevent global buckling of the sample and allow for the use of a long gauge length. This approach, along with a simple scaling law based on Weibull statistics, enables straightforward prediction of the compressive strength of corrugated board from edgewise compression testing of component sheet material.

Some work has been done on testing short-column crush specimens at angular orientations other than parallel to flute ridges. Results of applying compressive loads at various angles are shown in Fig. 85 for combined board specimens with a wide range of face densities [10]. The faces were handsheets; therefore, material anisotropy (MD/CD) effects are not embodied in the results. These results indicate that structural failure of corrugated containers might be the result of lower stresses in directions other than parallel to flute ridges.

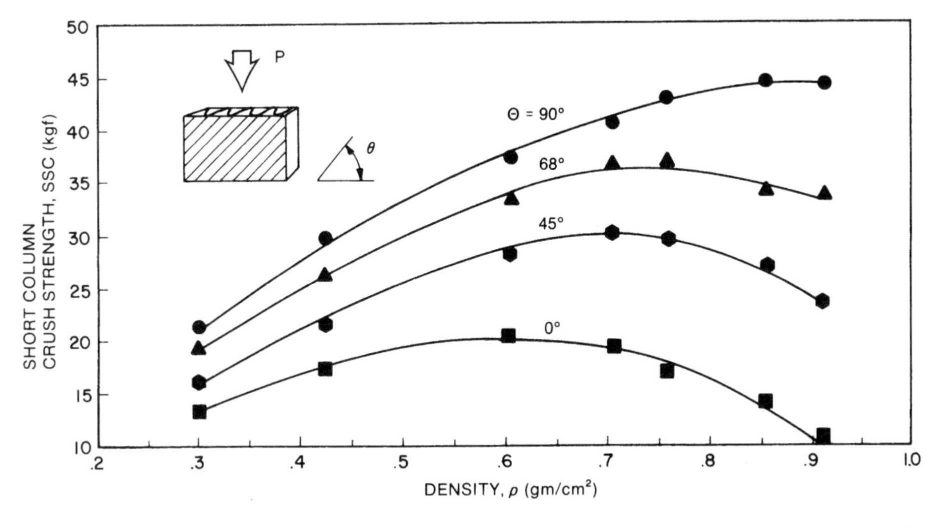

Fig. 85 Edge crush of short-column corrugated fiberboard at various orientations of the structural components. (From Ref. 10.)

M. Top-Load Compression of Corrugated Containers

Background The deformation of corrugated containers has been discussed in the context of four different size scale–dominated events [13]:

1. Material behavior—elastic and inelastic deformation behavior of paperboard
2. Small-scale structural behavior—out-of-plane deformation response of paperboard on the size scale of the flute and mechanical coupling between the paperboards
3. Large-scale structural behavior—out-of-plane response of the side panels
4. Pallet structural behavior—loading sharing between neighboring containers

Although seemingly simple in their function, the deformation behavior of corrugated containers is where several of the modes of paper deformation and failure discussed up to this point are coupled with the possibility of buckling failure at the size scale of both panel height and flute pitch. The most common measure of corrugated container performance is the top-to-bottom compression strength, which has been used as an overall measure of structural performance for several decades [111].

McKee and coworkers [111,113,114] recognized the importance of an engineering mechanics approach to an understanding of the deformation response of corrugated containers. They sought measurable quantities close to the material behavior. McKee et al. [114] developed a relationship between corrugated container compression strength and combined board properties. They began with a semiempirical relationship relating the ultimate edgewise compression strength of a plate to its buckling instability load,

$$P_z = cP_m^b P_{\mathrm{cr}}^{1-b} \tag{52}$$

where

P_z = ultimate edgewise compression strength of the plate, N/m
P_m = in-plane compression strength of the plate material, N/m
P_{cr} = critical plate buckling instability load, N/m
b, c = constants, dimensionless

The critical plate instability load was approximated by

$$P_{\mathrm{cr}} = 12k_{\mathrm{cr}}\frac{\sqrt{D_x D_y}}{W^2} \tag{53}$$

where

D_x = MD combined board bending stiffness, Nm
D_y = CD combined board bending stiffness, Nm
W = combined board panel width, m

The quantity k_{cr} is a buckling coefficient and was taken as

$$k_{\mathrm{cr}} = \frac{\pi^2}{12}\left[\left(\frac{r}{n}\right)^2 + \left(\frac{n}{r}\right)^2 + 2K\right] \tag{54}$$

where

$$r = (D_x D_y)^{1/4} \frac{d}{W} \tag{55}$$

and n = number of half-waves in the bucked plate, dimensionless
K = plate characteristic constant, dimensionless
d = combined board panel height, m

When Eqs. (52)–(55) are combined with three simplifying approximations*—(1)$K = 0.5$; (2) $(D_x D_y)^{1/4} = 7/6$; and (3) $W = Z/4$, where Z is the container perimeter—the following relationship for the top-to-bottom compression strength of a regular slotted container is obtained:

$$P = cP_m^b \left(16\pi^2 \frac{\sqrt{D_x D_y}}{Z^2} \left\{ \left[\frac{14}{3} \left(\frac{d}{nZ} \right) \right]^2 + \left[\frac{3}{14} \left(\frac{nZ}{d} \right) \right]^2 + 1 \right\} \right)^{1-b} \tag{56}$$

One additional simplifying assumption was made in that the constant c and the term in curly brackets were replaced by a new constant a. The resulting relationship is

$$P = aP_m^b (D_x D_y)^{\frac{(1-b)}{2}} Z^{2b-1}, \qquad \text{for } d/Z \geq 1/7 \tag{57}$$

The constants a and b must be evaluated experimentally. Note that this relationship contains the edgewise compression strength of the combined board, the geometric mean bending stiffness of the combined board, and the perimeter of the container as independent variables. McKee et al. [114] suggested that the edgewise compression strength term is more important than the bending stiffness term in Eq. (57). McKee et al. [114] evaluated constants a and b to be on the order of 2.0 and 0.75, respectively.

Relying on Eq. (57) for characterizing the container strength, Koning and Moody [90–93,118] worked to relate P_m, D_x, and D_y in Eq. (57) to the elastic and inelastic mechanical behavior of the linerboard and corrugating medium. Moody [118] viewed the combined board as a plate structure. The edgewise compression strength of the combined board (P_m) is determined by the local buckling stability of the linerboard and medium between the flute lines.

The same general approach as that of Moody was followed by Johnson and Urbanik [76–78], Johnson et al. [79], and Urbanik [186–188], although much more extensively. They focused on analytically quantifying the link between the edgewise compression strength of combined board and the failure of the component materials. In addition to taking a more rigorous theoretical approach, they recognized the importance of allowing for different behavior in the liner and medium and also allowed for both the strength and buckling failure of the paperboard (both liner and medium). Therefore their work allowed for edgewise compression failure to occur by in-plane compression failure or buckling instability failure of either the linerboard or corrugated medium. Mechanistically, this dual approach also suggests that the failure of corrugated containers is caused by either in-plane compression failure or buckling failure of the combined board component materials.

* The rationale for these simplifications can be found in McKee et al. [114].

The response of corrugated containers to internal loading was also studied extensively by Peterson [132–134] and Fox and coworkers [51-54], through both experiments and the development of a model of a square panel. The model allowed for quantification of the stress and displacement field in the plate during the loading process. The plate model also allowed prediction of failure location. Only material failure was considered in their evaluation of panel failure. Structural failure of the components of the combined board panels was not considered.

Most recently, the deformation and postbuckling response of loaded corrugated containers were examined computationally [15,124–126,129,130,137]. This approach allows for a more complete examination of corrugated container strength by allowing for nonlinear geometry and material behavior

Procedure In this test, a sealed empty container is compressed between the fixed platens of a suitable testing machine at a constant rate of platen displacement, 13 mm/min. The load and deflection are recorded continually until failure (maximum load) is reached. This load (N or lb_f) is reported as the quasi-static compression strength of the box. The test is performed under controlled conditions of temperature and humidity. More details are given in TAPPI Standard T804 om-97 [177]. A container deformed top to bottom in compression is shown in Fig. 86. Note that in this figure the top compression platen is allowed to swivel, whereas generally the test is done with a fixed upper compression platen.

Analysis of Results A typical load–deflection curve for the top-to-bottom compression deformation of a corrugated container is shown in Fig. 87 [111]. A load-deflection curve for a much simpler shape—a tube—is also shown for comparison. The top-to-bottom compression deformation response of the majority of corrugated containers generally progresses as follows [111,112]. As the load is applied vertically to the top of the container, the stress state in the side panels becomes increasingly compressive along the direction of loading. At some point during loading, the critical buckling load for the side panel (not to be confused with the critical buckling load of the material between the flutes as discussed above) is reached. When that point is reached, the walls of the container will begin to bow transversely. The maximum bowing will occur at the center of the panel, but bowing will be zero at the corners. Simultaneously with that event, the distribution of load around the perimeter of the container will become nonuniform (Fig. 88). From this point forward, a greater proportion of the applied load will be supported by the corners of the container. Loading will increase until failure occurs in the linerboard or corrugated medium due to intrinsic compression failure or buckling failure, which will likely initiate at or near the corners of the container [135].

A number of factors can affect the compression strength of boxes. The comparison of the behaviors of tube and box in Fig. 87 indicates how flaps and scoring can change the behavior of the structure [26,83,97,111]. The bearing surface through which the force is applied is also known to affect compressive strength. For example, a comparison of the strength of single boxes with that of boxes stacked three high, but aligned, showed a 30% reduction of strength for the stacked boxes. For stacked boxes that were not aligned, the reduction was as great as 50% [84]. Thus, how well a box supports a compressive load depends on the bearing surface as well as on how evenly the load is distributed around the perimeter.

Fig. 86 Corrugated container after failure during top-to-bottom compression testing. Note that the top loading platen is free to rotate. Generally the measurements are made with a top loading platen that is not free to rotate.

N. Top-Load Compression Creep of Corrugated Containers

Background Although not a test routinely made on corrugated containers (because of the time required), measurement of the dead-load compression creep response of corrugated containers is a way to quantify warehouse stacking performance. Such tests are also a good way for us to understand the events that lead to the ultimate failure of these paperboard structures under simulated warehouse conditions. As in other modes of deformation, failure of corrugated containers under a constant dead load is not a single event but rather a series of events. We define failure here as the complete collapse of the corrugated container.

The influence of moisture on the creep behavior of corrugated containers has been recognized for some time [85,171]. This understanding is generally compatible with the well-documented reaction of the mechanical behavior of paper to equilibrium moisture environments [e.g., 146]. As the moisture content of a corrugated container is increased, the time to ultimate failure decreases. More recently, it has been established that when corrugated containers under a top-to-bottom compressive load are exposed to a changing moisture environment, their lifetime under load is much shorter than if the container were exposed to a constant high moisture

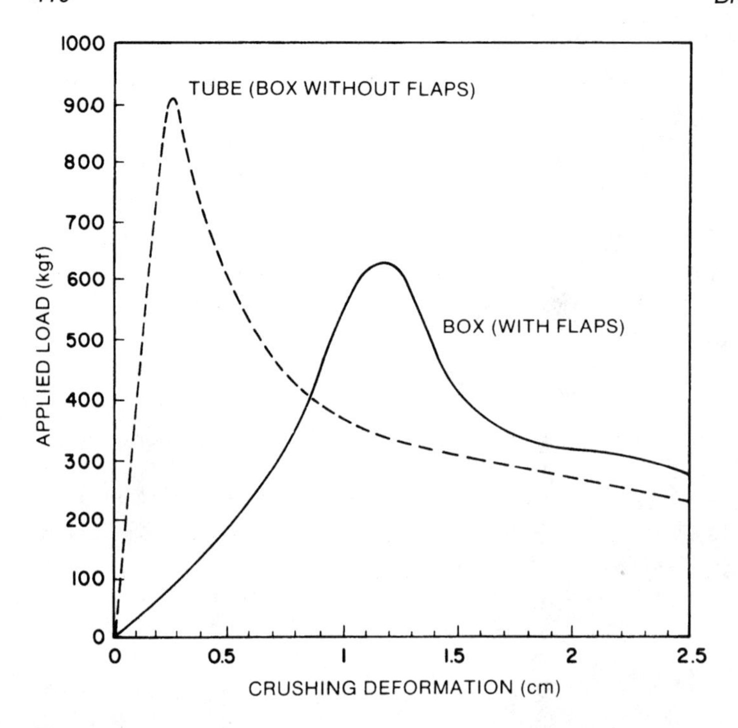

Fig. 87 Typical load–deflection response of corrugated containers with and without the top and bottom flaps. (From Ref. 111.)

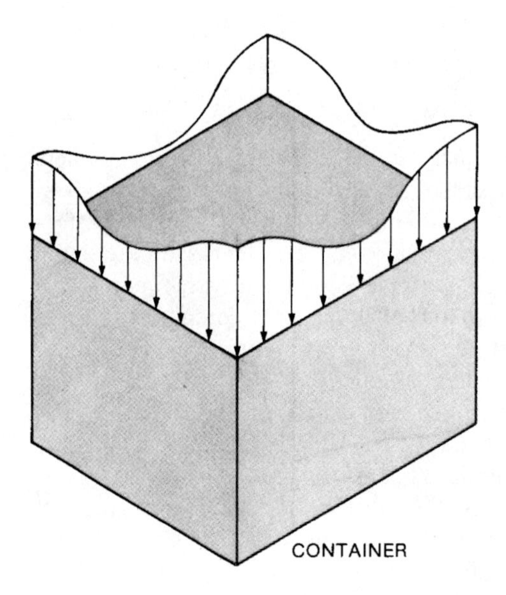

Fig. 88 Typical load distribution around the perimeter of a corrugated container after the panel buckling instability load has been exceeded.

environment [13,99,100]. It has also been shown that combined board and paperboard under a constant edgewise load (either tension or compression) will deform much more rapidly when exposed to a changing moisture environment than if exposed to a constant high moisture environment [18,19,21,23,31,58,166]. It is important to understand this behavior because most corrugated containers are used in environments in which the temperature and relative humidity are constantly changing. This section will focus on the behavior of corrugated containers exposed to cyclical moisture environments, because the failure events occur much more rapidly under these conditions.

Procedure The experiments are generally very simple. A constant magnitude load is applied to the top of a corrugated container, and the change of height of the container is measured on a regular basis until ultimate failure occurs. The container is usually placed on a flat, rigid loading platform, and a flat, rigid loading plate is then placed on top of the container. The load can be applied to the top loading plate as free weight or via dead weight loading frame (Fig. 89). The top loading plate is generally not forced to remain coplanar with the bottom loading platform. Change of height of the corrugated container as a function of time is usually taken as the change of distance between the loading platform beneath the container and the loading plate on top of the container. Because the two loading plates are usually not coplanar, the recorded container height is taken as the height at the container center (as viewed from the top).

The temperature and moisture environment substantially impacts the creep behavior of containers. Therefore, it is important to maintain good environmental control throughout the test. If the experiments are performed in a prescribed cyclical or changing environment, one should prescribe a cycle time that is appropriate for the moisture transfer characteristics of the container. If the container is not vented to the outside environment, it will take several hours for the side panels to absorb or desorb a significant amount of moisture. If the container is vented, the amount of time required for side panel moisture transfer will be reduced. If the containers are to be examined in a constant environment, be sure to allow enough time for moisture equilibration before applying a load. The amount of time required is a function of the container design, combined board design, and target environment. Moisture

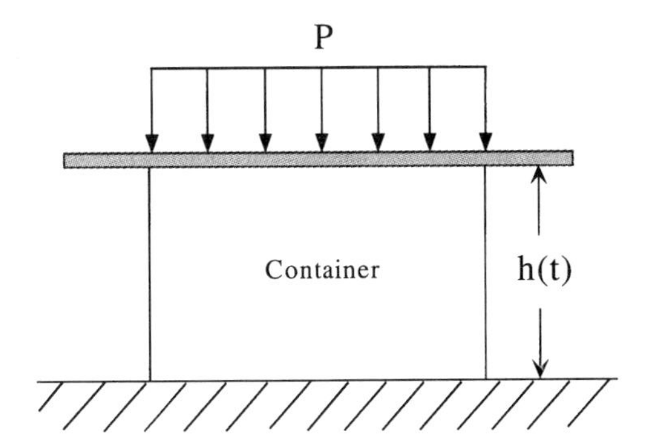

Fig. 89 Schematic representation of corrugated container creep tests for both free weight and loading frame methods of load application.

transfer rate is an important point to consider in the design of the experiment [13,101]. The reader is advised to consult the literature for details of corrugated container creep test procedures [13,94].

Analysis of Results The general creep deformation behavior of corrugated containers can be considered a combination of the following.

1. Material behavior: Elastic and creep deformation behavior of paperboard; transfer of moisture through the paperboard; mechanosorptive behavior; damage to the paperboard during the corrugation process
2. Small-scale structural behavior: Out-of-plane deformation response of paperboard on the size scale of the flute; mechanical coupling between the paperboards; moisture content variation through the combined board; combined board damage or defects created during the corrugation process
3. Large-scale structural behavior: Design and geometry of the container; out-of-plane response of the container side panels; deformation behavior of the top and bottom surfaces of the container; moisture content variations throughout the container; evolution of load distribution along the top and bottom perimeters of the container; container damage or defects created during the corrugation process or during use

One may observe that the size scale increases from category 1 to category 3. In addition, category 1 mechanisms are controlled by paperboard manufacture, category 2 mechanisms by combined board manufacture, and category 3 mechanisms by container manufacture.

The creep strain versus time curve is a map of the deformation response of corrugated containers to a constant compressive load and, in the case of cyclical moisture environments, a changing moisture content (Fig. 90). The shape and character of the curve are influenced by physical events that are characteristic of the way in which corrugated containers respond to load and the environment.

For the curve in Fig. 90, the load was applied to the container between the first and second data points. The total elapsed time between these points is very short (the time to place the load); therefore, the accumulated strain between these two points can be attributed primarily to the elastic response of the side panels and the top and bottom surfaces. As time progresses through the primary creep region, the additional strain can be attributed to the flattening of the top and bottom surfaces and the distribution of load directly to the side panels along the horizontal score lines. Just as McKee and Gander [111] observed during quasi-static compression deformation of corrugated containers, the top and bottom surfaces are actually being pushed into the container so that the load is more directly applied to the top and bottom edges of the side panels. Throughout the primary region, however, the side panels must necessarily be deforming in a steady-state way as dictated by the creep deformation response of the combined board. The end of the primary creep region is an indication that the deformation events at the top and bottom surfaces have diminished to the point of being insignificant compared with the steady-state deformation of the side panels.

The steady-state deformation of the side panels dominates the creep strain versus time curve in the secondary creep region. Early in the region, the side panels

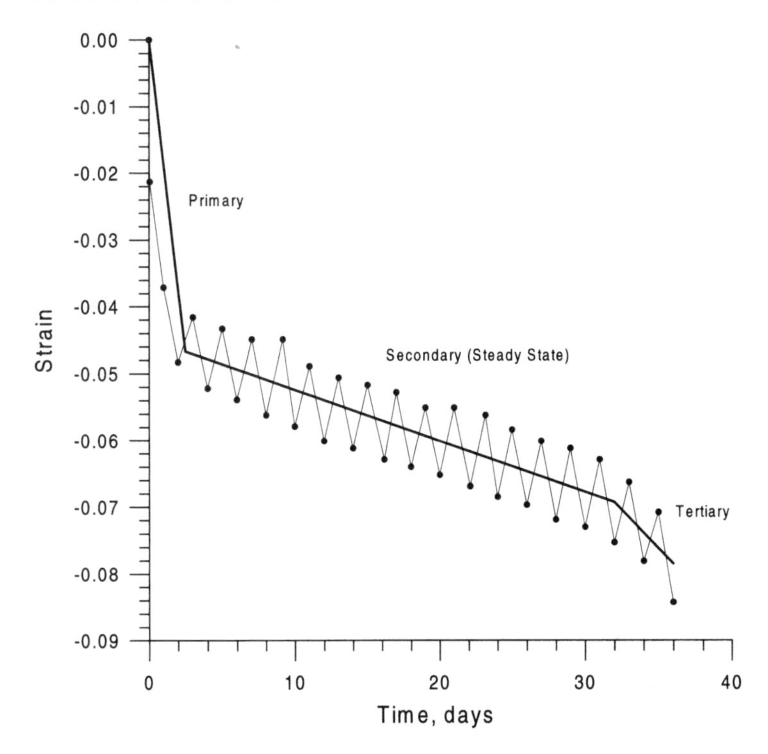

Fig. 90 Creep strain versus time curve from an actual corrugated container creep test performed in a cyclical relative humidity environment. Note that there are three regions to the curve (primary, secondary, tertiary). The oscillation in the creep curve is due to the natural hygroexpansion of the corrugated container as its moisture content changes in response to the cyclical relative humidity.

are nearly flat and vertical. As time passes and strain accumulates linearly, the center of the panels moves outward.* Because perimeter length of the container must be preserved (minus any net circumferential strain), the outward displacement of the panel centers must occur during the inward displacement of the corners. This deformation advances throughout the secondary region. The speed of deformation is believed to be dictated by both the steady-state creep response of the linerboard and structural evolution of the combined board. A photograph of a container tested well into the secondary region appears in Fig. 91. Whenever container panels bow outward under an edgewise compressive load, the linerboard on the concave side of the panel is subjected to compressive stress in the direction of loading. Since the majority of the linerboard surface area is laterally unsupported (the distance between flutes), eventually the linerboard will buckle in the direction of maximum compressive stress. This linerboard buckling between flutes usually occurs on the outside of the container in the neighborhood of each corner and on the entire inside surface of the container. This phenomenon will be called *interflute buckling* here. This structural deformation mechanism has been discussed by Byrd [22]. The buckling on the

* For the discussion here, only containers whose four side panels bow outward will be considered.

Fig. 91 Photograph of a corrugated container during creep testing in a cyclical relative humidity environment. Note that the side panels are bowing out and interflute buckling has formed in the corner region.

outside of the container becomes visible to the naked eye approximately one-third of the way through the secondary region (Fig. 91). Interflute buckling of the inner linerboard probably becomes noticeable at about the same time. Examination of failed containers reveals that the entire inside surface of the side panels is covered with interflute buckles. It is not necessary to have separation of the linerboard from the corrugated medium in order for interflute buckling to occur.

Near the end of the secondary region, the interflute buckling in the corner regions becomes so severe that the interflute buckles begin to cross the flute lines (Fig. 92). The result is that a single, continuous, nearly horizontal buckle line is formed at the corner, then grows in depth and length. This buckle line eventually penetrates through the thickness of the combined board and collapses the flutes along its length. The result is the formation of a hinge line and a severe reduction of the load-carrying capability of that corner (Fig. 92). This reduction translates into a significant increase in the slope of the creep curve and entry into the tertiary region. End of life for the container usually comes soon after the start of the tertiary creep region.

A three part straight-line representation of the creep strain versus time curve for a corrugated container is given in Fig. 93. Note that the line representing the secondary creep region is extended through the primary and tertiary creep regions. The steady-state deformation mechanisms that are active in the secondary creep region are also expected to be active in the primary and tertiary creep regions. Total strain to failure of the container can therefore be expressed as the sum of the three strain components:

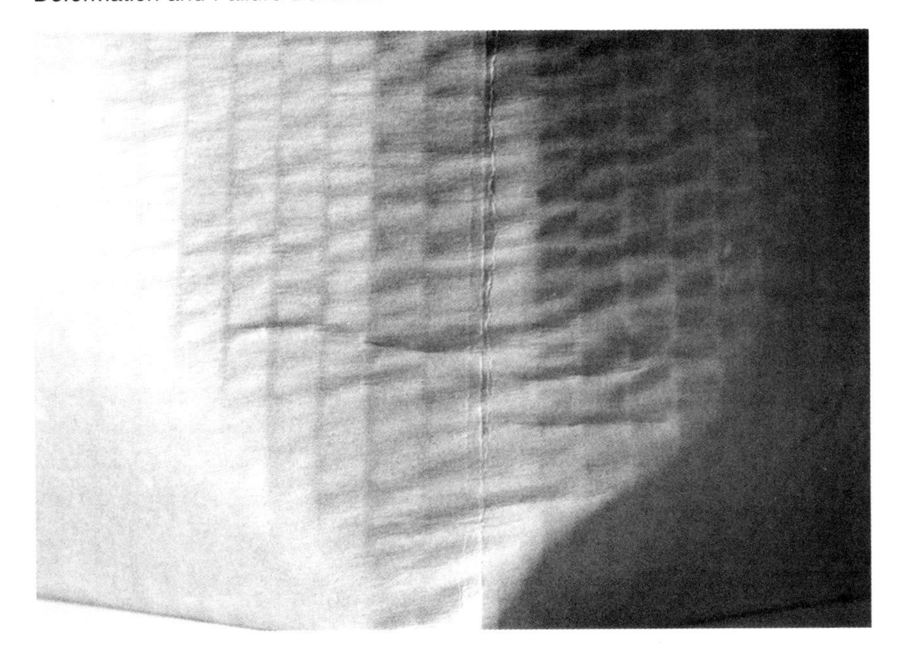

Fig. 92 Photograph of a portion of a corrugated container during creep testing in a cyclical relative humidity environment. The interflute buckles are rather prominent and have begun to coalesce along a line.

$$\varepsilon_f = \varepsilon_1 + \varepsilon_2 + \varepsilon_3 \tag{58}$$

If we assume that steady-state creep processes are active throughout the lifetime of the container, then

$$\varepsilon_2 = \dot{\varepsilon}_2 t_L \tag{59}$$

where t_L is the container lifetime.

After substitution of Eq. (58) into Eq. (59) we get

$$t_L = \left(\varepsilon_f - \varepsilon_1 - \varepsilon_3\right)\frac{1}{\dot{\varepsilon}_2} \tag{60}$$

which can be generalized to

$$t_L = A\frac{1}{\dot{\varepsilon}_2} \tag{61}$$

where the constant A is dimensionless and can be interpreted as a measure of steady-state compression creep ductility. The magnitude of A may be affected by such variables as container design, combined board design, material behavior, moisture and temperature environment, and load magnitude. This relationship has been used to represent corrugated container data found in the literature [13] and is reproduced in Fig. 94.

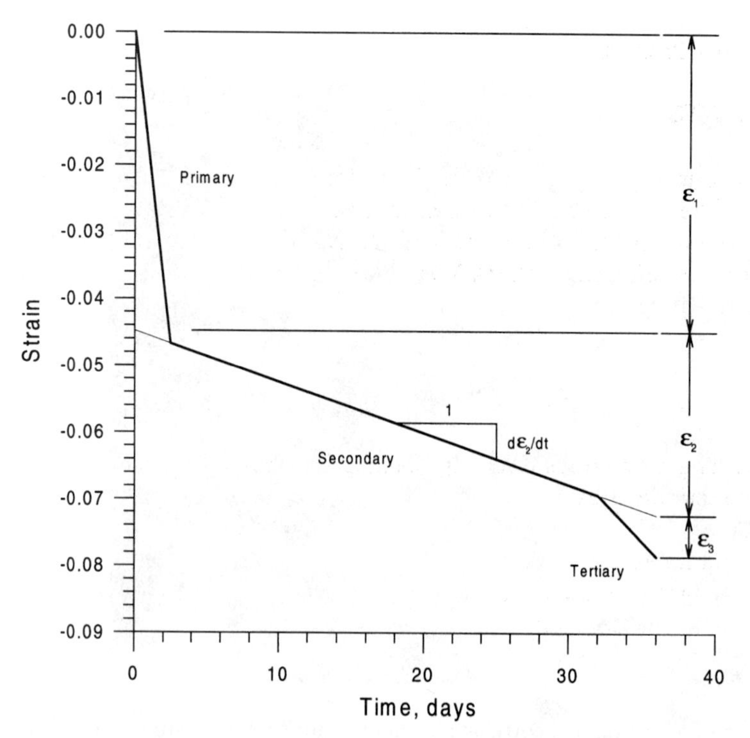

Fig. 93 Schematic representation of the corrugated container strain versus time curve showing the primary, secondary, and tertiary regions of deformation. Note that the secondary creep strain ε_2 is accumulated over the entire lifetime of the container and not just within the secondary creep region.

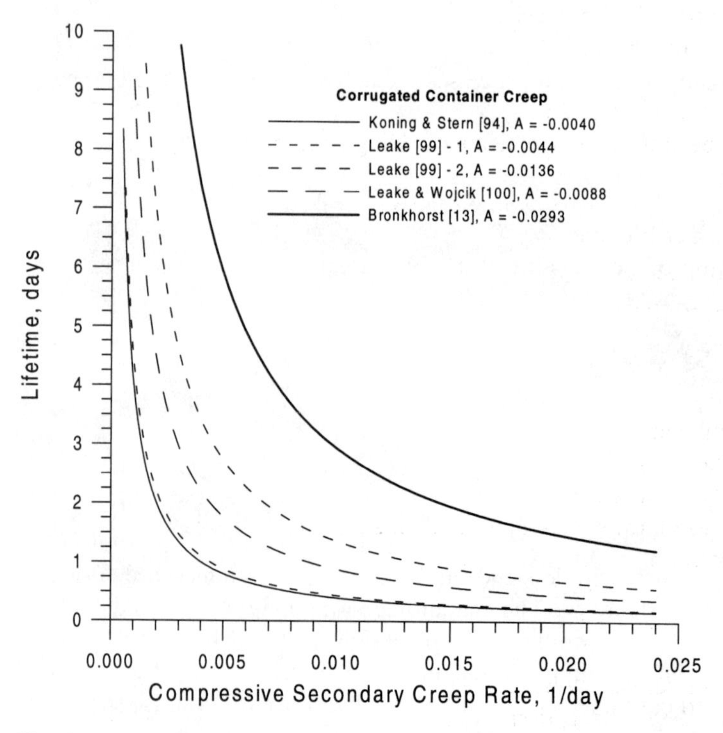

Fig. 94 Plot of Eq. (61) representing corrugated container creep data found in the literature. Values of the constant A for the data sets of Refs. 94, 99, 100, and 13 are listed in the inset table.

ACKNOWLEDGMENTS

We wish to express our appreciation to Drs. J. A. Johnson and H. M. Montrey, coauthors with K. A. Bennett of the first edition chapter "Failure Phenomena" from which this chapter was adapted. We also thank H. Nutwell for her expert assistance in document preparation. Finally, we thank Weyerhaeuser Company for their partial support of this work.

NOTATION

a	crack length or radius of clamp opening in the burst tester
A	cross-sectional area
b	width
BI	bonding index
BW	basis weight or grammage
CD	cross-machine direction
D	plate flexural stiffness
E	Young's modulus of elasticity
ECT	edge crush test
F	force or load
FCT	fluted edge crush test
FLI	fiber length index
FSI	fiber strength index
g	acceleration due to gravity
G	shear modulus
h	thickness
I	area moment of inertia
k	spring or stiffness constant or buckling coefficient
ℓ	length
L	breaking length or zero-span clamp width
m	mass
N	total number of fibers
OI	orientation index
p	pressure
P	force
r	radius
R	residual span
RCT	ring crush test
s	force per unit width
S	spring strength, short-column crush strength, or span
SCC	short column crush
SSC	short-span compression test
SSTC	short-span tensile curve
t	take-up factor
T	torque or tangent subscript
TEA	tensile energy absorption
u, v, w	displacements in x, y, z directions, respectively

U^*	strain energy density
V	volume
W	work
x, y, z	rectilinear coordinates (material reference frame)
X, Y, Z	rectilinear coordinates (spatial reference frame)
Y	geometric correction factor
Z	perimeter of a box

Greek

α	misalignment angle of tear test
β	ratio of spring constants
γ	shear strain or surface energy
δ	transverse compression deformation or in-plane shear deformation
ε	normal strain or mathematical strain
θ	polar coordinate or angle between coordinate systems
λ	slenderness ratio or deformation stretch ratio
ν	Poisson ratio
ρ	density
σ	normal stress
τ	shear stress
ψ	fiber orientation
Ψ	function of Poisson ratios

Subscripts

A	area
cr	critical
f	fiber
max	maximum
M, C, Z	machine, cross-machine, and through-the-thickness directions
$p\ell$	proportional limit
SCC	short-column crush
T	tangent
x, y, z	rectilinear direction components
u	ultimate
1, 2, 3	principal directions or system components
I, II	spring types or safe and damage zones, respectively
I, II, III	slenderness ratio regions for edgewise compression testing of paper and paperboard or three failure modes for tear
I, II, III, IV	Four normal stress quadrants

Superscript

*	strain energy density or specific strength

REFERENCES

1. Abrams, A. (1925). A study of the Mullen tester of paperboard. *Paper Trade J. 81*(6):54–63.
2. Aldrich, L. C. (1960). Standardization of Model C Mullen testers in a multimill company. *Tappi 43*(5):221A–225A.
3. Baum, G. A. (1993). Subfracture mechanical properties of paper and board. In: *Products of Papermaking*, Vol. 1. C. F. Baker, ed. Pira Int., Leatherhead, U.K., pp. 1–126.
4. Benson, R. E. (1971). Effects of relative humidity and temperature on tensile stress-strain properties of kraft linerboard. *Tappi 54*(5):699–703.
5. Berger, B. J. (1985). The effects of heat and pressure on the machine calendering of linerboard. Report A291. The Institute of Paper Chemistry, Appleton, WI, February.
6. Bernard, E., and Bouchayer, H. (1975). Manufacturing parameters affecting the characteristics of corrugating papers. *ATIP Rev. 29*(4):113–131 (in French).
7. Blockman, A. F., and Wikstrand, W. C. (1958). Interfiber bond strength of paper. *Tappi 41*(3):191A–194A.
8. Bodig, J., and France, R. C. (1974). A necked-down tensile test specimen for paper. *Tappi 57*(11):125–128.
9. Bohmer, E. (1962). The analogy between burst testing and conical shells. *Norsk Skog Ind. 16*(9):382–387.
10. Boitnott, R. L. (1979). The effect of various linerboard processing variables upon the edgewise compression strength of corrugated fiberboard. Master's Thesis, Virginia Polytechnic Institute and State University, Blacksburg, VA.
11. Boucai, E. (1971). Zero-span tensile test and fiber strength. Tech. Paper No. T313. *Pulp Paper Mag. Can. 72*(10):73–76.
12. Brezinski, J. P., and Hardacker, K. W. (1982). Poisson ratio values. *Tappi 65*(8):114.
13. Bronkhorst, C. A. (1997). Towards a more mechanistic understanding of corrugated container creep deformation behavior. *J. Pulp Paper Sci. 23*(4):J174–J181.
14. Bronkhorst, C. A., and Duong, T. (1997). Paper as an elastic-plastic fibrous composite. Sen-I Gakkai Preprints. *Society of Fiber Science and Technology*, Tokyo, Japan, p. 6–42.
15. Bronkhorst, C. A., and Riedemann, J. R. (1994). The creep deformation behavior of corrugated containers in a cyclic moisture environment. *Symposium on Moisture-Induced Creep Behavior of Paper and Board Proceedings*. C. Fellers and T. L. Laufenberg, eds. STFI, Stockholm.
16. Bronkhorst, C. A., and Riedemann, J. R. (2001). Numerical analysis of the time-dependent deformation response of a corrugated container to constant top-to-bottom loading (to be submitted).
17. Bronkhorst, C. A., Kalidindi, S. R., and Anand, L. (1992). Polycrystalline plasticity and the evolution of crystallographic texture in FCC metals. *Phil. Trans. Roy. Soc. Lond. A.341*:443–477.
18. Byrd, V. L. (1972). Effect of relative humidity changes during creep on handsheet paper properties. *Tappi 55*(2):247–252.
19. Byrd, V. L. (1972). Effect of relative humidity changes on compressive creep response of paper. *Tappi 55*(11):1612–1613.
20. Byrd, V. L. (1981). Fiber orientation and drying restraint. Effects on interlaminar shear and other paper properties. *Svensk Papperstidn. 84*(15):R105–R109.
21. Byrd, V. L. (1984). Edgewise compression creep of fiberboard components in a cyclic relative humidity environment. *Tappi 67*(7):86–90.
22. Byrd, V. L. (1986). Adhesives influence on edgewise compression creep in a cyclic relative humidity environment. *Tappi 69*(10):98–100.

23. Byrd, V. L., and Koning, J. W. (1978). Edgewise compression creep in cyclic relative humidity environments. *Tappi 61*(6):35–37.
24. Byrd, V. L., Setterholm, V. C., and Wichmann, J. F. (1975). Method for measuring the interlaminar shear properties of paper. *Tappi 58*(10):132–135.
25. Campbell, W. B. (1933). Relation of bursting to tensile tests. Forest Products Labs, Canada. *Pulp Paper Lab. Quart. Rev. 16*:1–4.
26. Carlson, T. A. (1941). Some factors affecting the compressive strength of fiber boxes. *Tappi 24*(1):473–476.
27. Carson, F. T., and Worthington, F. V. (1931). Critical study of the bursting strength test for paper. *Bur. Stand. J. Res. 6*(2):339–353.
28. Cavlin, S., and Fellers, C. (1975). A new method for measuring the edgewise compression properties of paper. *Svensk Papperstidn. 78*(9):329–332.
29. Chatterjee, A., Teki, S., Roy, D. N., and Whiting, P. (1991). Effect of recycling on tear and fracture behavior of paper. *Proc. Int. Paper Phys. Conf.* TAPPI Press, Atlanta, pp. 143–149.
30. Clark, J. d'A. (1944). The ultimate strength of pulp fibers and the zero-span tensile strength. *Paper Trade J. 118*(1):29–34.
31. Considine, J. M., Gunderson, D. E., Thelin, P., and Fellers, C. (1989). Compressive creep behavior of paperboard in a cyclic humidity environment. Exploratory experiment. *Tappi 72*(11):131–136.
32. Cowan, W. F. (1975). *Shot Span Tensile Analysis.* Pulmac Instruments Ltd., Montreal.
33. Cowan, W. F., and Cowdrey, E. J. K. (1974). Evaluation of paper strength components by short-span tensile analysis. *Tappi 57*(2):90–93.
34. Crandall, S. H., and Dahl, N. C. (1959). *Introduction to the Mechanics of Solids.* McGraw-Hill, New York, p. 444.
35. Crandall, S. H., Dahl, N. C., and Lardner, T. J. (1978). *An Introduction to the Mechanics of Solids.* 2nd ed. with SI Units. McGraw-Hill, New York.
36. Crisp, C. J., Stott, R. A., and Tomlinson, J. C. (1968). Resistance of corrugated board to flat crushing loads. *Tappi 51*(5):80A–83A.
37. Crosby, C. M., Eusufzai, A. R. K., Schaff, J., Choi, D., Furukawa, I., Ramasubramanian, M. K., Mark, R. E., and Perkins, R. W. (1991). Implications of the various test procedures in the evaluation of mechanical properties of paperboard. Empire State paper Res. Inst. Rep. No. 94, pp. 135–155.
38. de Ruvo, A., Carlsson, L., and Fellers, C. (1979). Biaxial strength of paper. *Proc. Int. Paper Phys. Conf.,* Tech. Sect. Can Pulp Paper Assoc., Montreal, pp. 23–30.
39. de Ruvo, A., Carlsson, L., and Fellers, C. (1980). The biaxial strength of paper. *Tappi 63*(5):133–136.
40. de Ruvo, A., Fellers, C., and Engman, C. (1978). The influence of raw material and design of the mechanical performance of boxboard. *Svensk Paperstidn. 81*(18):557–566.
41. Donner, B. C. (1989). The impact of structural formation on compression strength of paper. *Weyerhaeuser Tech. Rep.*, Weyerhaeuser Co., Federal Way, Washington.
42. Douty, D. E. (1910). Bursting strength. *Paper Trade J. 50*(6):271–276.
43. El-Hosseiny, F., and Anderson, D. (1999). Effect of fiber length and coarseness on burst strength of paper. *Tappi 82*(1):202–203.
44. El-Hosseiny, F., and Bennett, K. A. (1985). Analysis of zero-span tensile strength of paper. *J. Pulp Paper Sci. 11*(4):J121–J127.
45. Elias, T. C. (1967). Investigation of the compression response of ideal unbonded fibrous structures. *Tappi 50*(3):125–132.
46. Fellers, C. (1977). Procedure for measuring the interlaminar shear properties of paper. *Svensk Paperstidn. 80*(3):89–93.
47. Fellers, C., and Jonsson, P. (1975). Edgewise compression strength of liner and corrugating medium. An analysis of test methods. *Svensk Paperstidn. 78*(5):172–175.

48. Fellers, C., de Ruvo, A., Elfstrom, J., and Htun, M. (1980). Edgewise compression properties. A comparison of handsheets made from pulp of various yields. *Tappi 63*(6):109–112.
49. Fellers, C., Westerlind, B., and de Ruvo, A. (1983). An investigation of the biaxial failure envelope of paper: Experimental study and theoretical analysis. In: *The Role of Fundamental Research in Paper Making*, Vol. 1. J. Brander, ed. Mech. Eng. Pub., London, pp. 527–559.
50. Fong, J. T., Rehm, R. G., and Graminski, E. L. (1977). Weibull statistics and microscopic degradation model of paper. *Tappi 60*(1):156–159.
51. Fox, T. S. (1978). Shipping containers and cartons shown to fail only in compression when loaded internally. *Paperboard Packag.,* March, pp. 23–36.
52. Fox, T. S., Nelson, R. W., Watt, J. A., and Whitsitt, W. J. (1978). Shipping containers and cartons shown to fail only in compression when loaded internally: Part 2. *Paperboard Packag.,* April, pp. 28–36.
53. Fox, T. S., Nelson, R. W., Watt, J. A., and Whitsitt, W. J. (1978). Shipping containers and cartons shown to fail only in compression when loaded internally: Part 3. *Paperboard Packag.,* May, pp. 48–56.
54. Fox, T. S., Whitsitt, W. J., Nelson, R. W., and Watt, J. A. (1978–1979). Determination of the fundamental combined board and component properties which govern corrugated container performance. Reports 1 and 2, Project 3272. The Institute of Paper Chemistry, Appleton, WI.
55. Gibson, L. J., and Ashby, M. F. (1997). *Cellular Solids*. 2nd ed. Cambridge Univ. Press, Cambridge, UK.
56. Gunderson, D. (1983). Edgewise compression of paperboard: A new concept of lateral support. *Appita 37*(2):137–141.
57. Gunderson, D. E., and Rowlands, R. E. (1983). Determining paperboard strength-Biaxial tension, compression, and shear. *Proc. Int. Paper Phys. Conf.,* TAPPI Press, Atlanta, pp. 253–263.
58. Gunderson, D. E., and Tobey, W. E. (1990). Tensile creep in paperboard: Effect of humidity change rates. *Mater. Res. Soc. Symp. Proc. 197*:213–226.
59. Gunderson, D. E., Bendtsen, L. A., and Rowlands, R. E. (1984). A mechanistic perspective of the biaxial strength of paperboard. *Proc. Empire State Paper Res. Associates Meeting*, pp. 54–85.
60. Gunderson, D. E., Bendtsen, L. E., and Rowlands, R. E. (1984). Predicting the biaxial strength of paperboard: A mechanistic approach. *Progress in Paper Physics Seminar*, STFI, Stockholm, pp. 99–113.
61. Gurnagul, N., and Page, D. H. (1989). The difference between dry and rewetted zero-span tensile strength of paper. *Tappi 72*(12):164–167.
62. Gurtin, M. E. (1981). *An Introduction to Continuum Mechanics*. Academic Press, New York.
63. Habeger, C. C., and Whitsitt, W. J. (1983). A mathematical model of compressive strength in paperboard. *Fibre Sci. Tech. 19*(3):215–239.
64. Heckers, W., and Göttsching, L. (1980). A method on testing in-plane shearing strength of paper and board. *Das Papier 34*(1):1–5 (In German).
65. Helle, T. (1979). The tearing test: Its fundamentals and significance. *Proc. Int. Paper Phys. Conf.,* Tech. Sect. Can. Pulp Paper Assoc., Montreal, pp. 13–22.
66. Hill, R. (1948). A theory of the yielding and plastic flow of anisotropic metals. *Proc. Roy. Soc. Lond. A. 193*:281–297.
67. Hill, R. (1950). *The Mathematical Theory of Plasticity*. Oxford Univ. Press, London, p. 356.
68. Hoffman-Jacobsen, P. M. (1925). New method of determining the strength of chemical pulp. *Paper Trade J. 81*(22):52–53.

69. Hood, P. F. (1975). Practical evaluation of Mullen tests shows wide range of test results. *Paperboard Packag. 60*(2):44–49.
70. Houston, P. L. (1922). A supplementary study of commercial instruments for determining of the tearing strength of paper. *Paper Trade J. 74*(10):43–46.
71. Ince, P. J., and Urbanik, T. J. (1986). Economics of fiber cost and compressive strength of singlewall corrugated boxes. *Tappi 69*(10):102–105.
72. Institute of Paper Chemistry. (1938). The meaning of the bursting strength test. *Res. Bull. Inst. Paper Chem. 4*(3):46–51.
73. Ionides, G. N., and Mitchell, J. G. (1978). The capabilities and limitations of short span tensile characterization of pulp strengths. *Paperi ja Puu 60*(4a):233–238.
74. Institute of Paper Chemistry. (1966). *IPC Report 1108-4*. Effect of component properties and geometry on cross-direction edgewise compression strength of corrugated board. *Institute of Paper Chemistry*, Appleton, WI.
75. Johnson, M. V., Urbanik, T. J., and Denniston, W. E. (1979). Optimum fiber distribution in singlewall corrugated fiberboard. *Lab Rep. 348 U.S. Forest Products*, Madison, WI.
76. Johnson, M. W., and Urbanik, T. J. (1987). Buckling of axially loaded, long rectangular paperboard plates. *Wood Fiber Sci. 19*(2):135–146.
77. Johnson, M. W., and Urbanik, T. J. (1984). A nonlinear theory for elastic plates with application to characterizing paper properties. *ASME J. Appl. Mech.* (March):146–152.
78. Johnson, M. W., and Urbanik, T. J. (1989). Analysis of the localized buckling in composite plate structures with application to determining the strength of corrugated fiberboard. *J. Compos. Technol. Res. 11*(4):121–127.
79. Johnson, M. W., Urbanik, T. J., and Denniston, W. E. (1979). Optimum fiber distribution in singlewall corrugated fiberboard. *Rep. 348, U.S. Forest Products Lab,* Madison, WI.
80. Johnson, W., and Mellor, P. B. (1983). *Engineering Plasticity*. Ellis Horwood, West Sussex, England.
81. Kainulainen, M., and Toroi, M. (1986). Optimum composition of corrugated board with regard to the compression resistance of boxes. *Paperi Ja Puu 68*(9):666–668.
82. Kajanto, I., Laamanen, J., and Kainulainen, M. (1998). Paper bulk and surface. In: *Paper Physics*. K. Niskanen, ed. TAPPI Press, Atlanta, GA, pp. 88–115.
83. Kellicutt, K. Q. (1960). Compressive strength of corrugated containers. Part 1. Influence of scores. *Paperboard Packag.*, February, p. 74.
84. Kellicutt, K. Q. (1963). Effect of contents and load bearing surface on compressive strength and stacking life of corrugated containers. *Tappi 46*(1):151A–159A.
85. Kellicutt, K. Q., and Landt, E. F. (1951). Safe stacking life of corrugated boxes. *Fibre Containers 36*(9):28–38.
86. Kellogg, R. M., and Wangaard, F. F. (1964). Influence of fiber strength on sheet properties of hardwood pulps. *Tappi 47*(6):361–367.
87. Kill, E. N., and Maltenfort, G. G. (1974). The effect of specimen clamping pressure on corrugated burst strength. *Tappi 57*(11):67–70.
88. Kimura, M., and Mark, R. E. (1980). Mechanical properties in relation to network geometry for press-dried paper. Part II. *Empire State Paper Res. Inst.* Rep. No. 73, pp. 153–181.
89. Kimura, M., and Mark, R. E. (1987). Mechanical properties in relation to network geometry for press-dried paper. *Proc. Int. Paper Phys. Conf.,* Can. Pulp Paper Assoc. Tech. Sect., pp. 167–172.
90. Koning, J. W. (1975). Compressive properties of linerboard as related to corrugated fiberboard containers: A theoretical model. *Tappi 58*(12):105–108.
91. Koning, J. W. (1978). Compressive properties of linerboard as related to corrugated fiberboard containers: Theoretical model verification. *Tappi 61*(8):69–71.

92. Koning, J. W., and Moody, R. C. (1969). Effect of glue skips on compressive strength of corrugated fiberboard containers. *Tappi 52*(10):1910–1915.
93. Koning, J. W., and Moody, R. C. (1971). Predicting flexural stiffness of corrugated fiberboard. *Tappi 54*(11):1879–1881.
94. Koning, J. W., and Stern, R. K. (1977). Long-term creep in corrugated fiberboard containers. *Tappi 60*(12):128–131.
95. Koning, J. W., Jr. (1986). New rapid method for determining edgewise compressive strength of corrugated fiberboard. *Tappi 69*(1):74–76.
96. Koubaa, A., and Koran, Z. (1995). Measure of the internal bond strength of paper/board. *Tappi 78*(3):103–111.
97. Kutt, H., and Mithel, B. B. (1968). Studies on compressive strength of corrugated containers. *Tappi 51*(4):79A–81A.
98. Law, K. N., Koran, Z., and Barceau, J. J. (1979). On the short span tensile analysis of mechanical pulps. *65th Annual Meeting of CPPA*, Montreal, pp. A105–A110.
99. Leake, C. H. (1988). Measuring corrugated box performance. *Tappi 71*(10):71–75.
100. Leake, C. H., and Wojcik, R. (1989). Influence of the combining adhesive on box performance. *Tappi 72*(8):61–65.
101. Leake, C. H., and Wojcik, R. (1993). Humidity cycling rates: How they influence container life spans. *Tappi 76*(10):26–30.
102. Lindem, P. E. (1994). An instrument to assess the biaxial stress during paper drying. *Tappi 77*(5):169–173.
103. Lyne, M. B., and Bjelkhagen, H. (1979). The application of speckle interferometry to the analysis of elongation in paper and polymer sheets. *Proc. Int. Paper Physics Conf.*, Tech. Sect. Can. Pulp Paper Assoc., Montreal, pp. 97–104.
104. Lyne, M. B., Jackson, M. A., and Ranger, A. E. (1972). The in-plane, Elmendorf, and edge tear strength properties of mixed furnish papers. *Tappi 55*(6):924–932.
105. Malvern, L. E. (1969). *Introduction to the Mechanics of a Continuous Medium*. Prentice-Hall, Englewood Cliffs, NJ.
106. Mark, R. E. (1983). [Discussion] In: *The Role of Fundamental Research in Paper Making*. J. Brander, ed. Mech. Eng. Pub., London, p. 416.
107. Mark, R. E., Cardwell, R. D., Allerby, I. M., Perkins, R. W., and Uesaka, T. (1983). The relationship of edgewise compressive strength to interlaminar shear properties of commercial paperboards. *Proc. Int. Paper Phys. Conf.* TAPPI Press, Atlanta, pp. 121–129.
108. Mark, R. E., Perkins, R. W., Furukawa, I., Crosby, C. M., Lazo, N., Ramasubramanian, M. K., Allerby, I. M., Cardwell, R. D., and Eusufzai, A. R. K. (1984). The nature of compression failure in machine-made paperboard. *Empire State Paper Res. Inst. Rep.* 81, pp. 75–99.
109. Markström, H. (1995). Smart feature of the burst tester measures strain during the course of bursting. In: *paper Testing and Process Optimization Catalog*. Lorentzen & Wettre AB, pp. 122–127.
110. McClintock, F. A., and Argon, A. S. (1966). *Mechanical Behavior of Materials*. Addison-Wesley, Reading, MA.
111. McKee, R. C., and Gander, J. W. (1957). Top-load compression. *Tappi 40*(1):57–64.
112. McKee, R. C., Gander, J. W., and Wachuta, J. R. (1961). Edgewise compression strength of corrugated board. *Paperboard Packag. 46*(11):70–76.
113. McKee, R. C., Gander, J. W., and Wachuta, J. R. (1962). Flexural stiffness of corrugated board. *Paperboard Packag. 47*(12):111–118..
114. McKee, R. C., Gander, J. W., and Wachuta, J. R. (1963). Compression strength formula for corrugated boxes. *Paperboard Packag. 48*(8):149–159.
115. McKenzie, A. W. (1989). The tear-tensile relationships in softwood pulps. *Appita 42*(3):215–221.

116. Mohlin, U.-B. (1990). Fiber deformation and its implications in pulp characteristics. *Proceedings of 24th EUCEPA Conference—Pulp Technology Energy*. Swedish Assoc. Pulp Paper Eng., Neddelande, pp. 207–221.
117. Mohlin, U.-B., Dahlbom, J., and Hornatowska, J. (1996). Fiber deformation and sheet strength. *Tappi 79*(6):105–111.
118. Moody, R. C. (1965). Edgewise compressive strength of corrugated fiberboard as determined by local instability. *Rep. 46, U.S. Forest Products Lab*, Madison, WI.
119. Morgan, V. T. (1992). Measuring the thickness of stacked sheets of paper. *Tappi 75*(12):118–120.
120. Niskanen, K. (1993). Strength and fracture of paper. In: *Products of Papermaking*, Vol. 2. C. F. Baker, ed. Pira Int., Leatherhead, U.K., pp. 641–726.
121. Niskanen, K. (1998). *Paper Physics.* TAPPI Press, Atlanta.
122. Niskanen, K., and Kärenlampi, P. (1998). In-plane tensile properties. In: *Paper Physics.* K. Niskanen, ed. TAPPI Press, Atlanta, pp. 138–191.
123. Nordman, L., Kolhonen, E., and Toroi, M. (1978). Investigation of the compression of corrugated board. *Paperboard Packag. 63*(10):48–62.
124. Nordstrand, T. (1995). Parametric study of the post-buckling strength of structural core sandwich panels. *Compos. Struct. 30*(4):441–451.
125. Nordstrand, T. M., and Carlsson, L. A. (1997). Evaluation of transverse shear stiffness of structural core sandwich plates. *Compos. Struct. 37*(2):145–153.
126. Nordstrand, T., Carlsson, L. A., and Allen, H. G. (1994). Transverse shear stiffness of structural core sandwich. *Compos. Struct. 27(3)*:317–330.
127. Paavilainen, L. (1989). Effect of sulphate cooking parameters on the papermaking potential of pulp fibers. *Paperi ja Puu 71*(4):356–363.
128. Page, D. H. (1969). A theory for the tensile strength of paper. *Tappi 52*(4):674–681.
129. Patel, P. (1996). Biaxial failure of corrugated board. Licentiate Thesis, Lund University.
130. Patel, P., Nordstrand, T., and Carlsson, L. A. (1997). Local buckling and collapse of corrugated board under biaxial stress. *Compos. Struct. 39*(1–2):93–110.
131. Perkins, R. W., and McEvoy, R. P. (1981). The mechanics of the edgewise compressive strength of paper. *Tappi 64*(2):99–102.
132. Peterson, W. S. (1980). Minimum-cost design for corrugated containers under top-to-bottom compression. *Tappi 63*(2):143–146.
133. Peterson, W. S. (1980). Unified container performance and failure theory: Theoretical development of mathematical model. *Tappi 63*(10):75–79.
134. Peterson, W. S. (1980). Unified container performance and failure theory: Comparisons between experimental data and mathematical model. *Tappi 63*(11):115–120.
135. Peterson, W. S., and Fox, T. S. (1980). Workable theory proves how boxes fail in compression. *Paperboard Packag.*, October, pp. 136–144.
136. Pfeiffer, J. D. (1981). Measurement of the K_2 factor for paper. *Tappi 64*(4):105–106.
137. Pommier, J. C., Poustis, J., Fourcade, E., and Morlier, P. (1991). Determination of the critical load of a corrugated cardboard box submitted to vertical compression by finite element methods. *Proc. Int. Paper Phys. Conf.* TAPPI Press, Atlanta, pp. 437–448.
138. Qiu, Y. P., Millan, M., Lin, C. H., and Gerhardt, T. D. (1997). Nonlinear properties of high strength paperboards. In: *Mechanics of Cellulosic Materials*. R. W. Perkins, ed. AMD-Vol. 22 (MD-Vol. 77). Am. Soc. Mech. Engrs., New York, pp. 1–18.
139. Randall, E. B., Jr. and Lashof, T. W. (1970). Interlaboratory study of the measurement of the bursting strength of paper. *Tappi 53*(5):799–809.
140. Ranger, A. E., and Hopkins, L. F. (1962). A new theory of the tensile behavior of paper. In: *Formation and Structure of Paper*, Vol. 1. F. Bolam, ed. Tech. Sect. Brit. Paper Board Makers Assoc., London, pp. 277–310.

141. Reynolds, W. F. (1974). New Aspects of internal bonding strength of paper. *Tappi 57*(3):116–120.
142. Rodal, J. J. A. (1989). Soft-nip calendering of paper and paperboard. *Tappi 71*(5):177–186.
143. Rosenthal, M. R. (1977). Effective thickness of paper: Appraisal and further development. *Rep. 287. U.S. Forest Products Lab*, Madison, WI.
144. Rowlands, R. E., Gunderson, D. E., Suhlling, J. C., and Johnson, M. W. (1985). Biaxial strength of paperboard predicted by Hill-type theories. *J. Strain Anal. 20*(2):121–127.
145. Sachs, I. B., and Kuster, T. A. (1980). Edgewise compression failure mechanism of linerboard observed in a dynamic mode. *Tappi 63*(10):69–73.
146. Salmén, L. (1993). Responses of paper properties to changes in moisture content and temperature. In: *Products of Papermaking*, Vol. 1. C. F. Baker, ed. Pira Int., Leatherhead, U.K., pp. 369–430.
147. Salmén, N. L., and Back, E. L. (1978). Effect of temperature on stress–strain properties of dry papers. *Svensk Papperstidn. 81*(10):341–346.
148. Schaffrath, H. J., and Göttsching, L. (1991). The behavior of paper under compression in the Z-direction. *Proc. Int. Paper Phys. Conf.* TAPPI Press, Atlanta, pp. 489–510.
149. Seeley, F. B., and Smith, J. O. (1967). *Advanced Mechanics of Materials*. 2nd ed. Wiley, New York, p. 680.
150. Seo, Y. B., Chaves de Oliveira, R., and Mark, R. E. (1992). Tension buckling behavior of paper. *J. Pulp Paper Sci. 18*(2):J55–J59.
151. Seth, R. S. (1984). Edgewise compressive strength of paperboard and the ring crush test. *Tappi 67*(2):114–115.
152. Seth, R. S. (1985). Relationship between edgewise compressive strength of corrugated board and its components. *Tappi 68*(3):98–101.
153. Seth, R. S. (1991). Implications of the single-ply Elmendorf tear strength test for characterizing pulps. *Tappi 74*(8):109–113.
154. Seth, R. S. (1999). Zero-span tensile strength of papermaking fibres. 85th Annual Meeting, Pulp and Paper Technical Association of Canada Preprint A. pp. 161–173.
155. Seth, R. S., and Blinco, K. M. (1990). Comparison of Brecht-Imset and Elmendorf tear strengths. *Tappi 73*(1):139–142.
156. Seth, R. S., and Page, D. H. (1974). Fracture resistance of paper. *J. Mater. Sci. 9*(11):1745–1753.
157. Seth, R. S., and Page, D. H. (1975). Fracture resistance: A failure criterion for paper. *Tappi 58*(9):112–117.
158. Seth, R. S., and Soszynski, R. M. (1979). An evaluation of methods for measuring the intrinsic edgewise compressive strength of paper. *Tappi 62*(10):125–127.
159. Seth, R. S., Jantunen, J. T., and Moss, C. S. (1989). The effect of grammage on sheet properties. *Appita 42*(1):42–48.
160. Seth, R. S., Soszynski, R. M., and Page, D. H. (1979). Intrinsic edgewise compressive strength of paper: The effect of some papermarking variables. *Tappi 62*(12): 97–99.
161. Setterholm, V. C. (1974). A new concept in paper thickness measurement. *Tappi 57*(3):164.
162. Setterholm, V. C., and Gertjejansen, R. O. (1965). Method for measuring the edgewise compressive properties of paper. *Tappi 48*(5):308–313.
163. Setterholm, V. C., Benson, R., and Kuenzi, E. Q. (1968). Method for measuring edgewise shear properties of paper. *Tappi 51*(5):196–202.
164. Snyder, L. W. (1927). A study of the Mullen Paper tester. *Paper Trade J. 85*(5):47–49.
165. Sokolnikoff, I. S. (1956). *Mathematical Theory of Elasticity*. McGraw-Hill, New York.

166. Söremark, C., and Fellers, C. (1991). Mechano-sorptive creep and hygroexpansion of corrugated board in bending. *Proc. Int. Paper Phys. Conf.* TAPPI Press, Atlanta, pp. 549–559.
167. Staigle, V. H. (1967). Corrugated board: Is it crushed during fabrication? *Tappi 50*(1):45A–47A.
168. Stenitzer, F. (1967). Bursting strength (of paper). *Das Papier 21*(11):822–828 (in German).
169. Stockmann, V. E. (1976). Edgewise compressive strength of paper: A physical or structural property. *Proc. Tappi Annual Meeting*, Mar. 15–17, New York, pp. 257–264.
170. Stockmann, V. E. (1976). Measurement of intrinsic compressive strength of paper. *Tappi 59*(7):93–97.
171. Stott, R. A. (1959). Compression and stacking strength of corrugated fibreboard containers. *Appita 13*(2):84–89.
172. Strickwerda, J. C., and Considine, J. M. (1991). Analysis of the burst test geometry: A new approach. *Proc. Int. Paper Phys. Conf.* TAPPI Press, Atlanta, pp. 579–584.
173. Suhling, J. C. (1985). Constitutive relations and failure predictions for nonlinear orthotropic media. Ph.D. Thesis, Mechanics Department, Univ. Wisconsin—Madison.
174. Suhling, J. C. (1987). Analysis of paperboard deformations using nonlinear plate and membrane models. *Proc. Int. Paper Phys. Conf.,* Tech. Sect. Can. Pulp Paper Assoc., Montreal, pp. 139–144.
175. Suhling, J. C., Johnson, M. W., Rowlands, R. E., and Gunderson, D. E. (1989). Nonlinear elastic constitutive relations for cellulosic materials. In: *Mechanics of Cellulosic and Polymeric Materials.* AMD-Vol. 99 (MD-Vol. 13). R. W. Perkins, ed. ASME, New York, pp. 1–14.
176. Suhling, J. C., Rowlands, R. E., Johnson, M. W., and Gunderson, D. E. (1985). Tensorial strength analysis of paperboard. *Exp. Mech.*, March, pp. 75–84.
177. TAPPI Test Methods 1996–1997. TAPPI Press, Atlanta.
178. Taylor, D. L. (1964). Thickness and apparent density of paper. *Tappi 47*(7):165A–167A.
179. Timoshenko, S. P. (1957). *Strength of Materials*, Part 2. 3rd ed. Van Nostrand, New York, p. 572.
180. Ting, T. H. D., Chiu, W. K., and Johnston, R. E. (1997). Network changes in paper under compression in the Z-direction: The effect of loading rate and fiber wall thickness. *Appita 50*(3):223–229.
181. Tryding, J. (1994). A modification of the Tsai–Wu failure criterion for the biaxial strength of paper. *Tappi 77*(8):132–134.
182. Tsai, S. E., and Wu, E. M. (1971). A general theory of strength for anistropic materials. *J. Compos. Mater.* 5(1):58–80.
183. Uesaka, T., Murakami, K., and Imamura, R. (1979). Biaxial tensile behavior of paper. *Tappi 62*(8):111–114.
184. Uesaka, T., Perkins, R. W., and Mark, R. E. (1981). Determination of interlaminar shear modulus in paperboard by the torsion pendulum method. *Empire States Paper Res. Inst. Rep.* No. 74, pp. 49–66.
185. Uesaka, T., Perkins, R. W., and Mark, R. E. (1982). Determination of edgewise compressive strength of paperboard. *Empire State Paper Res. Inst. Rep.* No. 76, pp. 121–130.
186. Urbanik, T. J. (1981). Effect of paperboard stress-strain characteristics on strength of singlewall corrugated fibreboard: A theoretical approach. Report 401. U.S. Forest Products Lab, Madison, WI.
187. Urbanik, T. J. (1981). The principle of load-sharing in corrugated fiberboard. *Paperboard Packag.*, November, pp. 122–128.

188. Urbanik, T. J. (1990). Correcting for instrumentation with corrugated fiberboard edgewise crush test theory. *Tappi 73*(10):263–268.
189. Urbanik, T. J., Catlin, A. H., Friedman, D. R., and Lund, R. C. (1994). Edgewise crush test streamlined by shorter time after waxing. *Tappi 77*(1):83–86.
190. Urruty, J. P., Huchon, R., and Pouyet, J. (1996). Development of a biaxial tensile testing machine and a nondistributing displacement measurement method. *Tappi 79*(3):283–289.
191. Van den Akker, J. A. (1944). Instrumental studies. XLVI. *Paper Trade J. 118*(5):13–19.
192. Van den Akker, J. A. (1970). Structure and tensile characteristics of paper. *Tappi 53*(3):388–400.
193. Van den Akker, J. A., Lathrop, A. L., Voelker, M. H., and Dearth, L. R. (1958). Importance of fiber strength to sheet strength. *Tappi 41*(8):416–425.
194. Van den Akker, J. A., Wink, W. Å., and Van Eperen, R. H. (1967). Instrumentation studies. LXXXIX. Tearing strength of paper. III. Tearing resistance by the in-plane mode of tear. *Tappi 50*(9):466–470.
195. Van Liew, G. P. (1974). The Z-direction deformation of paper. *Tappi 57*(11):121–124.
196. Voelker, M. H. (1972). Role of base sheet properties in the development of coated and supercalendered sheet properties. *Tappi 55*(2):253–257.
197. von Mises, R. (1913). Mechanics of a solid body in plastic deformation. *Göttinger Nachrichten*, Math.-Phys. Kl. 4:582–592 (in German).
198. Waterhouse, J. F. (1983). Out-of-plane shear deformation behavior of paper and board. *Proc. Int. Paper Phys. Conf.* TAPPI Press, Atlanta, pp. 111–119.
199. Waterhouse, J. F. (1984). Out-of-plane shear deformation behavior of paper and board. *Tappi 67*(6):104–108.
200. Westerlind, B. S., and Carlsson, L. A. (1991). Compressive response of corrugated board. *Proc. Int. Paper Phys. Conf.* TAPPI Press, Atlanta, pp. 641–654.
201. Westerlind, B. S., and Carlsson, L. A. (1992). Compressive response of corrugated board. *Tappi 75*(7):145–154.
202. White, C. S., Bronkhorst, C. A., and Anand, L. (1990). An improved isotropic-kinematic hardening model for moderate deformation metal plasticity. *Mech. Mater. 10*:127–147.
203. Whitsitt, W. J. (1985). *Compressive Strength Relationships Factors* Institute of Paper Chemistry, IPC Tech. Ser. No. 163.
204. Whitsitt, W. J., and Sprague, C. H. (1987). Compressive strength retention during fluting of medium. *Tappi 70*(2):91–96.
205. Wink, W. A., and Van Eperen, R. H. (1962). The development of an improved zero-span tensile test. *Tappi 45*(1):10–24.
206. Wink, W. A., and Van Eperen, R. H. (1967). Evaluation of Z-direction tensile strength. *Tappi 50*(8):393–400.
207. Wink, W. A., Watt, J. A., Whitsitt, W. J., and Baum, G. A. (1984). Role of fiber axial modulus on compressive strength. *Fibre Sci. Technol. 20*(4):245–253.
208. Yamauchi, T. (1987). Measurement of paper thickness and density. *Appita 40*(5):359–366.
209. Yamauchi, T. (1989). Compressibility of paper measured by using a rubber platen thickness gauge. *Appita 42*(3):222–224.

8
FRACTURE OF PAPER

M. T. KORTSCHOT
University of Toronto
Toronto, Ontario, Canada

I. INTRODUCTION

Fracture mechanics is the field of applied mechanics that deals with the prediction of crack initiation and propagation in solid materials. Fracture mechanics provides for a quantitative prediction of fracture loads in cracked structures. It also provides a foundation for the design of fracture-resistant materials [1]. In its most general sense the term "fracture mechanics" applies to any study of material fracture, but over the past 40 years, it has come to refer more specifically to the study of crack propagation based on the fracture criterion derived by A. A. Griffith [2].

In the past 20 years, there have been many studies dealing with the application of fracture mechanics to paper and board. Unfortunately, much of the paper fracture literature has focused on the measurement of "fracture toughness" or some variation thereof, without properly presenting the case for the application of fracture mechanics to paper in the first place. For example, newsprint can fail in tension in a printing press, but the justification for using a fracture mechanics approach instead of a simple tensile strength measurement to predict this failure is not clear.

In this chapter, the phenomenon of paper fracture will be discussed and the basis of the fracture mechanics approach will be described. The application of this methodology in several practical situations will be outlined. One of the more important of these, pressroom runnability of newsprint, will be discussed in detail.

II. PAPER FAILURE

In engineering terms, the word *failure* means "failure to perform the intended function." Box failure occurs when a box collapses under compression loads, but a curled sheet that jams in a photocopier has also failed. *Fracture* is a particular mode of failure in which the material separates into two or more pieces. Fracture is often not the primary concern of end users of paper; in many grades, the need for a certain stiffness or opacity requires a furnish that has more than adequate strength. Nevertheless, in the ongoing struggle to reduce costs by reducing basis weights or increasing filler loading, sheet fracture will continue to be an important consideration in the design of paper.

There are a number of places in which fracture can occur:

- On the paper machine. Fracture can occur in the first open draw, in the press section, in the dryer section, and in the winding and rewinding operations. Failure at the wet end of a paper machine is one of the primary concerns of machine operators.

- •• During conversion. Paper and board can fracture in the printing press, in box-making operations, and in general operations of folding, slitting, etc. The ability of newsprint to pass through the printing press is known as its "press-room runnability" and is one of the principal selling points.
- ••• In service. Service fractures are important for many grades—directory, sack kraft, tissue, etc.

In order to predict and prevent fracture, papermakers and converters must find a correlation between the fracture and the material properties and applied loads and displacements. This might be done using a brute force statistical approach. In this case, a (typically large) set of quality control parameters is monitored and the correlation of the performance of the sheet to these parameters is calculated. This effort can certainly be successful and is often appropriate in a mill environment. Nevertheless, a study of the underlying physics of the fracture process can be used to substantially reduce the required experimental effort. This is the primary motivation for the extensive study of fracture mechanics found in the paper physics literature.

In order to predict the mechanical failure of a box, web, or indeed any material, a *failure criterion* must be established. A failure criterion normally takes the form of an equation in which an *applied quantity*—usually a stress or strain or some combination of these—is equated to a *material property*. Provided the material property is known, a failure criterion allows for the prediction of the stresses or strains at failure. Two examples are given in Eqs. (1). Each describes a different type of failure. The details of these equations are not important at this stage; it is the form of the equations that is of interest.

$$\tau_{max} = \tau_{yield} \tag{1a}$$

$$\sigma_{max} = \sigma_f \tag{1b}$$

where τ_{max} is the maximum shear stress, τ_{yield} is the shear yield strength, σ_{max} is the maximum tensile stress, and σ_f is the tensile strength.

These equations have a common format. On the left-hand side is an applied quantity, and on the right-hand side is the material property, the critical value of the applied quantity that is constant for a particular material. Equation (1a) is the Tresca criterion for the yielding of ductile materials such as pure aluminium [3]. It says that the material will yield when the maximum shear stress on any plane equals a critical value, the shear yield strength. Equation (1b) involves tensile rather than shear stresses and can be used to predict the failure of materials such as chalk or glass, which fail by fast fracture.

A failure criterion is said to be useful and correct if it always describes the onset of the particular failure mode in question. For example, the Tresca criterion [Eq. (1a)] is found to describe the yield of ductile metals reasonably well regardless of the other stresses and strains. A failure criterion is tested by comparing the stress and strain fields at the failure point in a variety of specimens. The researcher looks for some aspect of the stress or strain field that was constant at failure, regardless of the specimen geometry or applied loading conditions. In glass sheets containing large cracks, for example, it is the product of applied stress and the square root of the crack length that is found to be constant at the onset of fracture. This fact is at the heart of fracture mechanics and has a theoretical basis that will be explained in detail, but for now it may be treated as an empirical equation: Regardless of the size of the crack, $\sigma\sqrt{a}$ is found to be approximately constant at failure in glass. Other aspects of the stress field, such as σ_{max} or τ_{max}, are not found to be constant at failure in glass containing a large crack, and so Eqs. (1a) and (1b) have no significance for precracked glass sheets.

Papermakers and end users have two reasons to determine and use the appropriate failure criteria. First, papermakers need to know which material property governs failure so that they are monitoring the right quantity for quality control purposes. Secondly, both the papermakers and the end users need to be able to make predictions of failure in service conditions so that they can determine if a particular sheet will be adequate for its intended purpose. The second requirement is more demanding than the first; if a simple ranking of paper performance potential from day to day is all that is required, it is enough to measure a property that is merely proportional to the right-hand side of the appropriate failure criterion. However, if an actual prediction of failure is to be made, then nothing less than the exact failure criterion and associated property measurement will do.

The need for a rigorous physical understanding of paper fracture can be illustrated with two sets of data. In Fig. 1, three material properties are plotted as a function of breaking length: Elmendorf tear, in-plane tear, and fracture resistance [4]. Each of these properties has been used in the past as an indicator of the strength potential of paper, and yet not only are they poorly correlated but in addition the relationship between them depends on the type of pulp.

Figure 1 clearly illustrates the need for a fundamental understanding of the fracture process and the danger of relying on the measurement of individual properties in the absence of such understanding. Imagine three sheets with three different breaking lengths. A material ranking based on tear strength could produce an entirely different order than a ranking based on in-plane tear or fracture resistance. It must therefore be determined which of these properties can be used to predict service failure in order to determine which is the most suitable property for quality control purposes.

Davison [5] produced a second set of data that demonstrates the need for detailed understanding of the failure process. In Fig. 2, the relationship between the length of a specimen and its tensile strength is illustrated. As the specimen length increased, the average tensile strength was reduced. This observation suggests that it may be difficult to use the results of standard tensile tests to predict the strength of a running newsprint web where the effective specimen length might be many kilometers. The observed relationship between specimen size and tensile strength is characteristic of materials that fail from the largest of a population of internal flaws, because larger specimens are more likely to contain a large flaw [6,7].

The data in Fig. 2 indicate that a tensile strength criterion may not be adequate for predicting the failure of a sheet of paper. These data, and a hundred years of accumulated experience, tell us that paper is flaw-sensitive and that fracture often proceeds from a built-in crack or flaw such as a shive. It is not surprising, then, that paper physicists turned to fracture mechanics to analyze paper fracture. Fracture mechanics has been very successful in the analysis of the fracture of other materials. The evidence accumulated over the last 20 years (see Sections IV and V) clearly indicates that the fracture of paper can be effectively described by standard fracture mechanics analysis. The data of Fig. 1 suggest that a good understanding of the underlying principles is required, because different types of measurements may pro-

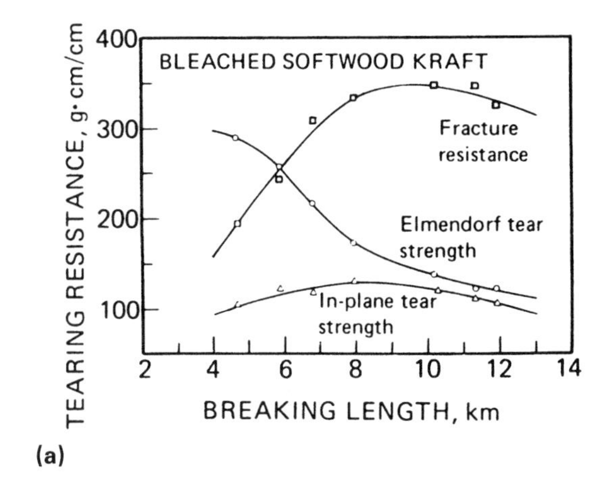

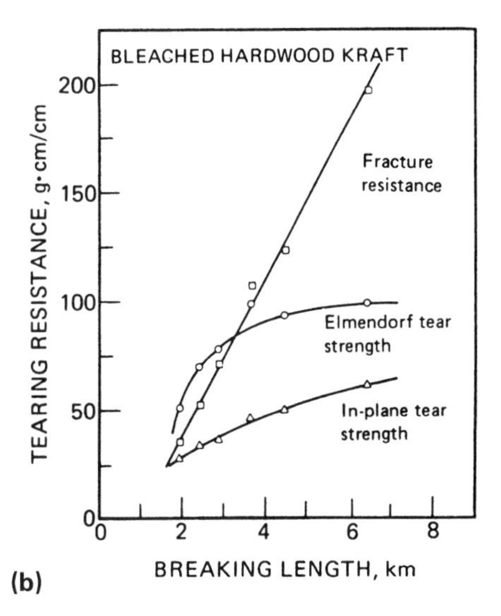

Fig. 1 Elmendorf tear, in-plane tear, and fracture resistance versus breaking length for (a) bleached softwood kraft and (b) bleached hardwood kraft. (From Ref. 4.)

duce different and even contradictory results. An explanation of these principles and their application to paper is provided in the remainder of this chapter.

III. FRACTURE MECHANICS

The study of the fracture of paper has followed, with some time lag, the concepts developed in the mainstream fracture literature, so it is instructive to first consider the historical development of the study of fracture of metals and other structural materials. The leading edge of fracture research tends to concentrate on metals because these materials are used in applications, such as airframes and nuclear reactors, where fracture can have devastating consequences. There are many good

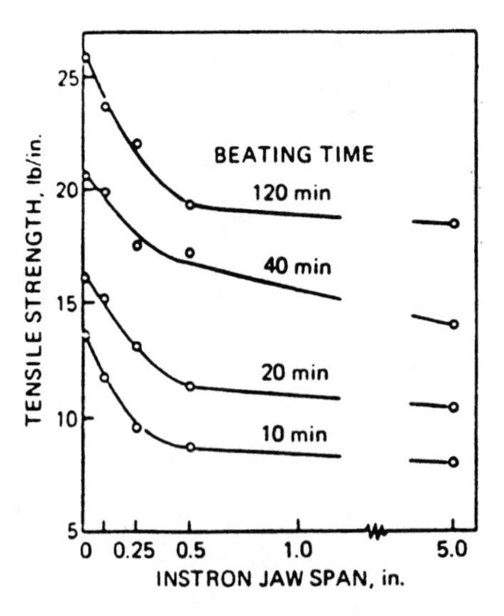

Fig. 2 Tensile strength depends on the length of the specimen in handsheets. (From Ref. 5.)

texts on fracture mechanics [1,8–12] and a number of review articles in the paper literature [13–16]. The American Society for Testing and Materials (ASTM) has published a fracture mechanics retrospective [17] that contains a collection of classic papers published before 1966.

In this section, the theories and techniques of fracture mechanics will be presented in a general form without any particular emphasis on paper. Section IV will focus on the application of these basic concepts to paper and will cite references from the paper physics literature.

Throughout the following development, plane stress rather than plane strain expressions will be used, because they are most relevant to paper. In the plane stress formulation of the equations, the material is assumed to freely contract in the thickness direction as tensile stress is applied in the plane, and therefore the through-thickness stress is assumed to be zero. A further discussion of plane stress versus plane strain will be presented in Section III.B.

A. Stress Concentrations

The *theoretical tensile strength* of a material is calculated by assuming that in order to fracture a specimen, every atomic bond on a given cross section (the fracture plane) is stretched to the breaking point simultaneously. In a simple crystalline material, the maximum point in the force–displacement curve for an individual bond can be related to the initial slope of the curve. Using this relationship, the theoretical strength of a crystalline material can be related to its stiffness. Very approximately, it is found that

$$\sigma_{\mathrm{th}} \cong E/15 \tag{2}$$

where σ_{th} is the theoretical strength and E is the tensile (Young's) modulus.

A cursory examination of a materials database [18] shows that for almost all materials the actual failure strength, σ_f, is much less than E/15 [3]. In fact, it is obvious that when real materials fail, the bonds are not all stretched to their breaking point simultaneously. In materials that fail by fast fracture, bonds are found to fail sequentially when the crack grows, and only the bonds at the tip of the crack are carrying their maximum load at any instant in time. To predict the failure of such materials, the conditions of crack initiation and growth must be understood.

It has been known for many years that the presence of sharp cracks or defects can lead to the failure of a structural component at unexpectedly low loads. In 1913 Inglis [19] noted that "The destructive influence of a crack is a matter of common knowledge, and is particularly pronounced in the case of brittle non-ductile materials." Inglis derived an equation for the general stress field around an elliptical hole in an infinite two-dimensional sheet of an istropic material [19]. (An istropic material has properties that do not depend on the orientation of the specimen.) The maximum concentrated stress is found at the edge of the hole and may be expressed as

$$\sigma_{\max} = \sigma_{\infty}\left(1 + 2\sqrt{\frac{a}{\rho}}\right) \tag{3}$$

where $\sigma_{\max}$ is the maximum concentrated stress at the edge of the hole, σ_{∞} is the remote stress applied to the specimen far from the hole, a is the axis of the ellipse perpendicular to the applied stress, and ρ is the radius of a circle coincident with the contour of the ellipse at the end of this axis, referred to as the notch root radius as shown in Fig. 3.

The general equations for stress around an ellipse are particularly useful because an ellipse can represent a range of hole geometries from a circle ($a = \rho$) to a sharp crack ($\rho \to 0$). For a circle, the maximum tensile stress is three times the remote stress, and therefore the *stress concentration factor* ($= \sigma_{\max}/\sigma_{\infty}$) is 3. However, for a sharp crack, Eq. (3) predicts that the maximum stress is very large. Stress concentration factors for many other notch shapes have been calculated and are tabulated in handbooks [20].

Cracks in real materials can have a minimum radius at the notch tip of the order of atomic dimensions. Equation (3) predicts that for a lattice spacing of 1 nm,

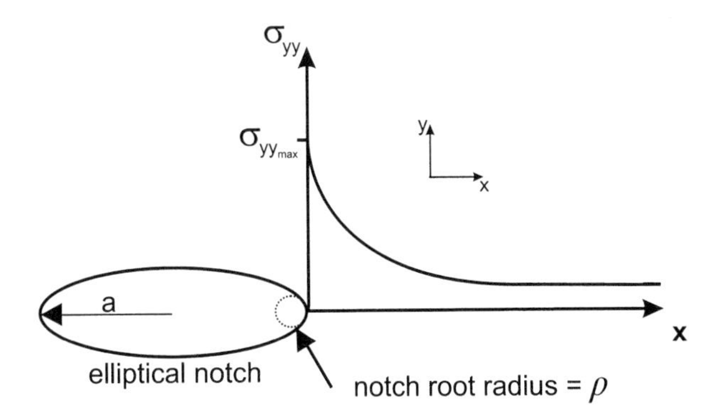

Fig. 3 Schematic of an elliptical hole in an infinite plate with the normal σ_{yy} stress plotted as a function of the distance from the notch tip, x. (Not to scale.)

even a crack just 100 μm long would produce a stress concentration of 633. If it is hypothesized that the material fractures when the stress at the tip of the crack reaches the theoretical strength, then a specimen of steel containing a 100 μm crack would have a strength of just 21 MPa. In reality, a specimen of steel containing a 100 μm crack would typically be much stronger than this. In most materials there is plastic flow at the crack tip, and this relieves the intense buildup of stress in this region, increasing the effective value of ρ in Eq. (3). Since the maximum stress at the tip of a crack with a yield zone is unknown, it would be difficult to use Eq. (3) as the basis for a fracture criterion. Hence Eq. (1b) would be of no use as a failure criterion in this case.

Equation (3) also states that the maximum stress is a function of the shape of the defect rather than its absolute size. Thus a circular hole in a ship's hull produces a stress concentration of 3, regardless of whether it is a rivet hole or a cargo hatch. Experience tells us that the larger hole is much more likely to cause early failure, and this provides further impetus for finding a more effective fracture criterion.

B. Linear Elastic Fracture Mechanics

In 1920, Griffith suggested a thermodynamic basis for fracture, saying that for an elastic solid "the equilibrium position, if equilibrium is possible, must be one in which the rupture of the solid has occurred, if the system can pass from the unbroken to the broken condition by a process involving a continuous decrease in potential energy" [2]. Griffith considered not the stresses at the crack tip but the overall energy of the system and went on to derive a fracture criterion in terms of a balance between the energy needed to create new crack surfaces and the release of elastic energy stored in the body as the crack progressed. Griffith's original work and the theory presented in this section are restricted to elastic materials with linear stress–strain curves.

Consider the geometry depicted in Fig. 4. A certain amount of work must be expended to extend the crack by a distance *da* in both directions. Griffith assumed

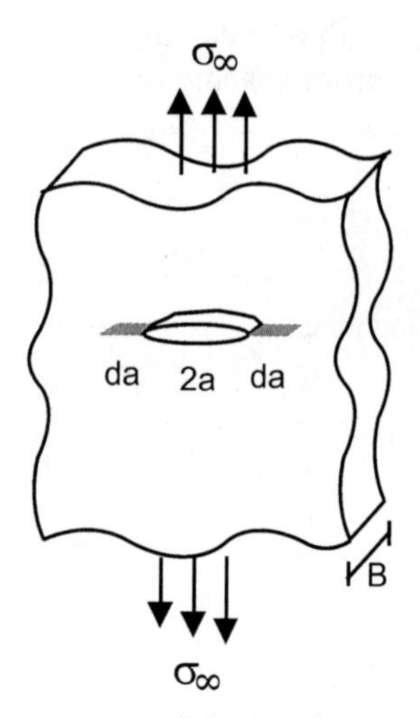

Fig. 4 Crack and specimen geometry in a sheet of infinite dimensions.

that the work required was equal to the new surface area of material created multiplied by its surface energy. If the work needed to create an incremental amount of new crack area is denoted dW, then

$$dW = 2\gamma(2\,da\,B) \tag{4}$$

where γ is the surface energy of the solid and B is the thickness of the specimen. Surface energy has dimensions of energy per unit area and can be measured independently by contact angle analysis. The additional factor of 2 in front of the γ term is needed because the new cracks have both top and bottom surfaces.

Griffith also calculated the strain energy that would be released if the crack grew. To picture this, imagine a rubber sheet that is stretched as depicted in Fig. 5b. The sheet contains stored elastic energy, which is equal to the work done in deforming it, and the amount of stored energy can be explicitly calculated from a knowledge of the stress field solution. For a volume V of material under constant stress, the strain energy is equal to $(1/2)\sigma\varepsilon V$, so in a sheet with a crack in it, the total stored energy is calculated as $\sigma\varepsilon/2$ integrated over the whole volume. If the positions of the specimen ends are fixed, then when the crack extends the stretched material that was in front of the crack prior to crack extension is unloaded as depicted in Fig. 5c, and the total strain energy in the system is reduced. To calculate the change in strain energy, the total energy in the sheet is simply computed before and after crack growth. Because the stress field solutions are quite complex, this calculation is not trivial, but the form of the answer turns out to be remarkably simple. If the change in the stored energy for an incremental amount of crack growth is denoted dU, then in a thin, elastic, isotropic sheet material,

$$dU = -\frac{\sigma_\infty^2 \pi a}{E} 2B\,da \tag{5}$$

Equation (5) has two significant characteristics. dU is proportional to σ_∞^2, so as the stress increases, more energy is released for a given increment of crack growth. dU is also proportional to the crack length, a, so an increment of crack growth, da, at the tip of a long crack produces a greater energy release than the same amount of crack growth at the tip of a short crack.

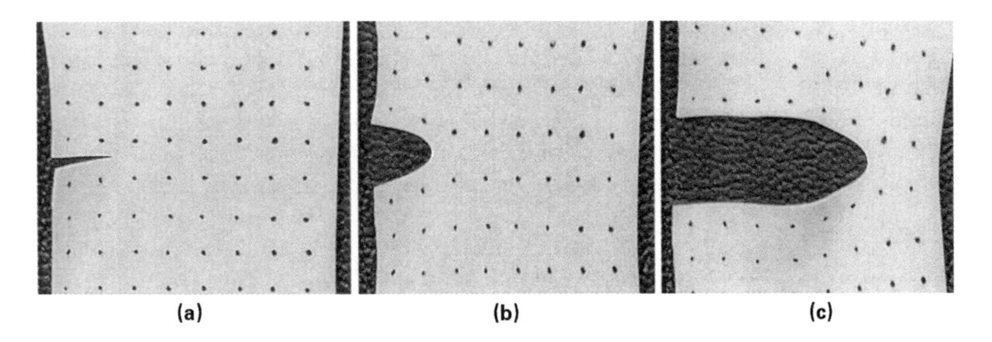

Fig. 5 (a) Slack rubber sheet with an inscribed grid. (b) Stretched rubber sheet with stored elastic energy. (c) Crack extension relieves the stress in part of the sheet; reducing the total stored elastic energy.

The discussion of Eq. (5) has been based on "fixed grip" conditions, in which the positions of the specimen ends (far from the crack) are fixed during crack growth, and the release of stored elastic energy is calculated. Of course, fracture can also occur under fixed load conditions, where the remote load is constant, as would be the case for a suspended specimen with a weight hanging from its lower end. In this case, the stored elastic energy in the specimen actually increases as the crack grows and the specimen becomes more compliant, but the potential energy of the applied load is reduced. When these two terms are summed, the result for the overall change in the system energy is also given by Eq. (5). Refer to any of the standard texts for a more detailed discussion of this point.

A fracture criterion can be derived by equating the released energy, dU, to the energy needed to create the new crack area, dW. As the applied stress increases, the amount of energy released, dU, during crack growth da increases. Fracture can occur as soon as

$$dU + dW = 0 \tag{6}$$

$$\frac{\sigma_c^2 \pi a}{E} 2B\,da = 4\gamma\,B\,da \tag{7}$$

$$\sigma_c = \left(\frac{2E\gamma}{\pi a}\right)^{1/2} \tag{8}$$

where σ_c is the remote stress at fracture, the critical stress. σ_c is not a material property, because it is dependent on the crack length.

Equation (8) relates the failure stress, σ_c, to both the crack length and the material properties E and γ. It states that a plot of σ_c versus $1/\sqrt{a}$ would produce a linear relationship, and this is found to be true for linear elastic materials that fail by fast fracture. However, in almost all materials, σ_c is found to be much larger than predicted by Eq. (8). This paradox was resolved by Irwin [21], who proposed that the work required to create new crack area is actually much greater than 2γ. Irwin suggested that in most materials a substantial amount of plastic flow accompanies crack growth and that the energy required to produce this flow would be much greater than the surface energy of the material. This is easily visualized by tearing a specimen of tracing paper: The whitened zoned that forms at the crack tip and tracks along with it is caused by a combination of fiber plasticity and debonding. In Fig. 6, images of growing cracks in tracing paper and a polymer film show stress-whitened zones, where energy is absorbed by plastic flow and other mechanisms. These processes cost energy and are always associated with crack growth in these materials. The amount of additional "non-surface energy" work associated with crack growth is said to be a property of the material itself, so the equation for energy absorption, Eq. (4), is modified by replacing the term 2γ with a new term, G_c.

$$dW = G_c(2\,da\,B) \tag{9}$$

where G_c is the total energy needed to extend the crack and has units of joules per square meter (J/m^2).

Equation (8) can now be rewritten using the new definition of dW.

$$\sigma_c = \left(\frac{EG_c}{\pi a}\right)^{1/2} \tag{10}$$

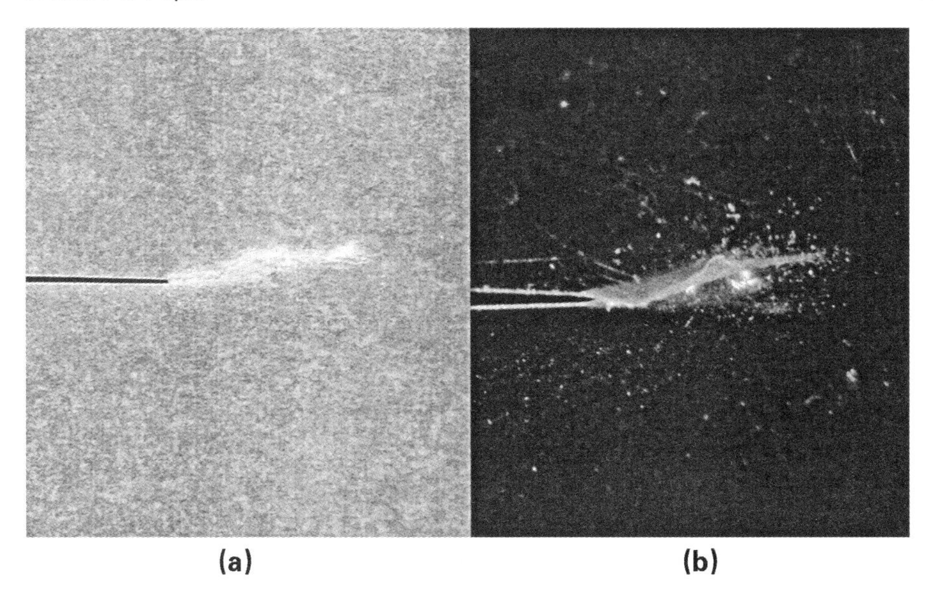

Fig. 6 Plastic and/or damage zone associated with crack growth in (a) tracing paper and (b) a polymer film.

This is the most important and most fundamental equation of fracture mechanics. It is a direct expression of Griffith's theory, modified to account for small-scale plasticity accompanying crack growth. Rewritten in the form $\sigma_c\sqrt{\pi a} = \sqrt{EG_c}$, it takes on the form of the failure criteria expressed by Eqs. (1a) and (1b). On the left-hand side is an applied quantity, and on the right-hand side are material properties.

If Eq. (1) is a valid fracture criterion for a material, then fracture mechanics, which was used to derive this equation, is also valid. Any failure criterion, such as Eq. (1) or (10) can be validated by testing a variety of specimens with differing configurations or loading schemes. To test Eq. (10), a set of very large specimens (from a single material) with different crack sizes are tested. Having measured Young's modulus for the material, E, separately, the data are fit using Eq. (10) with G_c as a fitting parameter. For many materials that fail by fast fracture, including many samples of paper [22], the stress at failure, σ_c, decreases with increasing crack size, in a manner well described by Eq. (10). A single value of G_c produces an adequate fit of all of the data, and hence it is concluded that the fracture criterion works. This is a sufficient validation of fracture mechanics for the material in question.

Materials that fail by fast fracture generally contain a characteristic population of internal flaws: tiny cracks, crystalline defects, voids, etc. They will fail by fast fracture from the most critical internal defect (generally the largest one) in the absence of a macroscopic crack. Because the population of flaws is the same in different specimens of the same material, the critical stress given by Eq. (10) is more or less a constant for these specimens, and it is used as a material property: the tensile strength, σ_f. In other words, although it is Eq. (10) that is really predictive of failure in such a material, Eq. (1b) is used together with the apparent tensile strength, σ_f, if all the cracks in the material are simply characteristic of that material.

When a prediction of failure is needed for the same material in a configuration where there is a large crack or cutout present, Eq. (10) must be used.

G_c is dependent on the structure of the material and the details of the energy-absorbing processes around the crack tip. G_c for tracing paper would be relatively low, and G_c for sack kraft would be relatively high. Equation (10) states that the failure stress, or strength, of the cracked sheet is not dependent on the tensile strength of the material but on the product EG_c. Although G_c is related to the commonly measured tear energy, there are some significant differences between these properties, and the nature of these will be discussed in Section IV.B.

It is often convenient to express the energy absorption and release differentials in terms of their derivative with respect to the crack area. Dividing dU by the change in crack area, $dA(= 2B\,da)$ produces a term $G \equiv -dU/dA$, which is referred to as the "strain energy release rate." Rearranging Eq. (9) yields $G_c \equiv dW/dA$, and thus G_c is referred to as the *critical strain energy release rate* although the term *work of fracture* is often used. The failure criterion, Eq. (10), can now be expressed in the classic form as

$$G = G_c \tag{11}$$

G_c is normally measured by measuring the remote stress at failure in a specimen with a known crack in it. Unfortunately, the crack must be located in an infinitely large sheet for Eq. (10) to be directly applicable. The amount of energy released will be slightly different than indicated by Eq. (5) when the sheet has a finite width or if some other aspect of the geometry is different than that depicted in Fig. 4. Provided the stress distribution in the specimen is known, however, an expression for dU can be derived and a modified version of Eq. (10) can be used to predict failure. However, it can be awkward to use complex equations for dU, and a more convenient (but equivalent) stress-based analysis is often used instead.

Stress-Based Analysis Westergaard [23] derived a solution for the stress field near a perfectly sharp crack tip in an infinite sheet of elastic material. Near the crack, all components of the stress field can be described by an equation of the form

$$\sigma_{ij} = \frac{\sigma_\infty \sqrt{\pi a}}{\sqrt{2\pi r}} f_{ij}(\theta) \tag{12}$$

where σ_{ij} is the ij component of stress at a point (r, θ) (see Fig. 7a) and $f_{ij}(\theta)$ is a function of θ that is different for each stress component. Additional terms must be added to Eq. (12) if it is to be used to predict stresses far from the crack, but these terms become relatively insignificant as r approaches zero (i.e., near the crack tip).

Equation (12) shows that as $r \to 0$, $\sigma \to \infty$ [for $f_{ij}(\theta) \neq 0$]. Of course, this is the solution for a completely elastic solid. Real materials cannot support infinite stress, and the material very near the crack tip generally yields, creating a plastic zone that elevates the value of G_c. Equation (12) is valid outside this plastic zone but sufficiently close to the crack tip that additional terms are not needed, as illustrated in Fig. 7.

It is the product of σ_∞ and $\sqrt{\pi a}$ that governs the crack tip stress field, and the combined term is given a special name—the *stress intensity factor*, K. (Although the names are similar, the stress intensity factor is quite distinct from the stress concen-

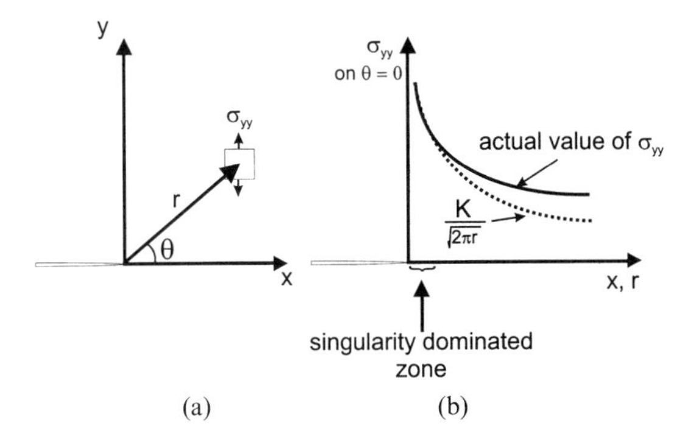

Fig. 7 (a) Polar coordinates used to describe stress field near the tip of a sharp crack. (b) Normal stress near the tip of a sharp crack along a line $\theta = 0$.

tration factor defined earlier.) In a sheet of infinite dimensions containing a central crack of length $2a$,

$$K \equiv \sigma_{\infty}\sqrt{\pi a} \tag{13}$$

In a finite-width sheet or a sheet with an angled crack or some other geometry, a more general expression for stress intensity is

$$K \equiv Y\sigma_{\text{ref}}\sqrt{\pi a} \tag{14}$$

where σ_{ref} is some reference stress in the stress distribution (often taken to be σ_{∞}) and Y is a geometric correction factor. Note that the stress intensity factor is an applied quantity, not a material property: It belongs on the left-hand side of any failure criterion. Solutions for K for some of the more common specimen configurations are listed in the Appendix. The geometric correction factor, Y, is determined using stress analysis to ensure that Eq. (15) will hold near the crack tip in an elastic solid.

$$\sigma_{ij} = \frac{K}{\sqrt{2\pi r}} f_{ij}(\theta) \tag{15}$$

The normal stress, σ_{yy}, is given by Eq. (15) with

$$f_{ij}(\theta) = \cos\left(\frac{\theta}{2}\right)\left[1 + \sin\left(\frac{\theta}{2}\right)\sin\left(\frac{3\theta}{2}\right)\right] \tag{16}$$

Now imagine that you are standing at the tip of a crack that is just about to propagate. When you look around, all the stresses, strains, and displacements are dependent on only the single parameter K, defined by Eq. (14). It makes sense, therefore, that if one specimen fails at a particular value of K, all other specimens of the same material must also fail at the value of K [8]. This is indeed found to be true in materials that fail by fast fracture, and the critical value of K is called the *critical stress intensity factor*, K_c. The other term commonly used for K_c is *fracture toughness*. K_c is a material property, and the fracture criterion based on this parameter is

$$K = K_c \tag{17}$$

Like other failure criteria previously discussed, the left-hand side of this equation is the applied quantity, and the right-hand side is the material property.

Previously, it was stated that Eq. (11), based on the thermodynamics introduced by Griffith, was a universal descriptor of fracture in elastic materials that fail by fast fracture. Now a second, quite different fracture criterion has been developed without any reference to the energy content of the system! There must be a relationship between these two equations, because they both describe the same phenomenon [9]. In fact, by comparing the equations $G \equiv -dU/dA = \sigma_\infty^2 \pi a/E$ and $K \equiv \sigma_\infty \sqrt{\pi a}$, which are both valid for a central crack in an infinite sheet, it is apparent that for thin sheet isotropic materials

$$K = \sqrt{EG} \tag{18a}$$

and thus

$$K_c = \sqrt{EG_c} \tag{18b}$$

For anisotropic materials such as machine-made paper, the relationship between K and G is somewhat more complex [see Eq. (23)]. Equation (18) can also be derived by using Eq. (15) to compute the crack closure energy [1] or by using dimensional analysis [9].

Equations (18a) and (18b) show that the energy formulation (the strain energy release rate approach) and the stress formulation (the stress intensity factor approach) really amount to the same thing, and either method may be chosen for a particular application. For composite material delamination and adhesive joint failure the energy approach is often adopted, but for most other analyses the stress intensity factor method [Eq. (17)] is generally found to be the most convenient. One reason for this is that the equations of the stress intensity factor of specimens with complex geometry are readily derived and are widely available in the literature [24]. If paper is to be treated as an elastic material, as suggested by some authors [22], then the stress intensity approach is more convenient than the original Griffith–Irwin approach. Expressions for the stress intensity factor in some common specimen configurations are listed in the Appendix.

It should be noted that for very small cracks, narrow specimens, or specimens with low yield strength and high K_c, Eq. (10), (11), or (17) might predict a failure load that is larger than the product of the tensile strength of the unnotched material and the cross-sectional area of the *ligament*. [The ligament is the band(s) of material between the crack tip and the edge of the specimen. Its area is calculated as (width of specimen − total length of cracks)$\times B$.] Clearly, the material in the ligament will fail if it is carrying a stress greater than its tensile strength, and this would result in specimen failure. There is a *competition* between two failure modes: crack propagation and general yield or failure. Whichever criterion is met first determines the failure load. Since Eqs. (10), (11), and (17) predict an increasing critical stress as the crack size, a, approaches zero, general failure of the ligament takes over as the failure mode when the crack gets small enough. In paper, this could be expected when the crack in the fracture specimen is reduced to less than the size of the internal flaws in the material itself (e.g., shives).

Plane Stress/Plane Strain Virtually all materials have a plastic zone (where there is permanent deformation) that forms at the tip of the crack and accompanies subsequent crack growth. This means that in almost all materials G_c is much larger than 2γ, generally by several orders of magnitude. Provided the plastic zone is relatively small compared to the overall specimen and crack dimensions, the analysis presented thus far can be used without modification. For metals, "small" is defined by relating the radius of the crack tip plastic zone to the various length dimensions in the material. For example, one of the standard methods used for fracture toughness testing of metals, ASTM E399, stipulates that the thickness of the specimen should be greater than $2.5(K_c/\sigma_y)^2$, where σ_y is the yield stress. This requirement, while somewhat arbitrary, is intended to ensure that most of the material near the crack tip will be in a state of plane strain.

The term *plane strain* applies when the highly stressed material at the crack tip is restrained from through-thickness Poisson contraction by the surrounding material. In metals, this means that plastic flow near the crack tip will be minimized and ensures that the measured toughness will have the minimum possible value. This criterion cannot be satisfied in paper, and therefore standard plane strain fracture toughness testing is impossible. Paper is typically so thin that it is in a state of *plane stress*, meaning that it freely contracts in the thickness direction as stress is applied in the plane, and the through-thickness stress is assumed to be zero. In plane stress, the measured toughness (G_c or K_c) of a metal is usually a function of the thickness of the specimen [1]. Although there is experimental evidence that the toughness is independent of specimen thickness in paper [25], care must be taken in using toughness measured on sheets of one thickness to make predictions of fracture in sheets of differing thickness.

There is a critical distinction between paper and metals that is often ignored in discussions of the fracture toughness of paper. The dimensional requirements specified by ASTM and the normal discussion of plane stress and plane strain have been formulated for metals, in which plane stress promotes high shear stresses, which enhance plastic flow. The ASTM thickness specification is designed to ensure a dominant flat fracture zone in the center of the specimen. This consideration is inappropriate for paper, as suggested by Helle [13]. Paper does not yield and flow like a metal but absorbs energy in a "deformation zone" by tensile and shear failure of the fibers and bonds. There is experimental evidence to suggest that the measured toughness of paper is independent of paper thickness, so the plane strain may not be required to get a measurement of toughness independent of the specimen thickness [25]. Indeed, a triaxial stress state in the center of a sheet caused by true plane strain conditions could enhance bond failure and might either increase or decrease the toughness of the material, depending on the microstructure.

Other dimensions—the specimen width, crack length, ligament width (specimen width minus crack length), and specimen length—should also be much larger than the singularity-dominated zone at the crack tip in order to apply the theory presented thus far. In other words, it must be the case that "the fracture events are limited to a region ahead of the crack which is so small, compared with the other specimen dimensions, that the single K-parameter describes the stress field in the critical region with almost completely accuracy" [1]. In practical terms, one method of determining the validity of the fracture toughness testing approach is simply to manufacture a series of specimens with differing crack sizes and dimensions to see if the measured values of K or G are indeed constant at failure.

Measuring G_c and K_c G_c is the energy required to create an increment of new crack area and has units of joules per square meter. Although it might be tempting to simply divide the work of fracture (the area under the load–displacement curve) by the new crack area, this approach will be valid only under very special circumstances [26] and is not generally used. One major problem with this approach is that crack growth is often unstable after the critical load is reached, so much of the energy stored in the specimen is converted into kinetic energy of the rapidly moving specimen halves. Because this "lost" energy was not used to drive the crack itself, it cannot be counted as part of G_c. Another problem is that the actual value of G_c can increase during the first bit of crack growth, leading to R-curve behavior. This occurs when the crack tip plastic zone expands as the crack grows or there is an accumulation of fibers bridging the crack surfaces just behind the tip. For a discussion of R-curve behavior and the analysis needed to predict fracture for materials displaying this behavior, refer to any of the standard texts [1,8–10].

For an unstable fracture proceeding from a central crack in an infinite sheet, Eq. (10) can be rearranged to express G_c in terms of the critical stress. In order to deal with finite-sized specimens or specimens with other configurations, the energy content calculations that led to Eq. (10) must be redone for the modified stress fields. Rather than doing this, an approach based on the stress intensity factor is normally used.

There is one specimen geometry that allows for the direct calculation of G without knowledge of the stress field. Composite materials and adhesives are often tested using a specimen referred to as the double cantilever beam, as illustrated in Fig. 8.

In the double cantilever beam method, the specimen can be unloaded after a small increment of crack growth, because the growth is usually stable if the test is being conducted in displacement control. The area under a load–displacement trace equals the work done on the specimen and hence the elastic energy stored in the specimen. G can be measured by calculating the difference in energy stored before

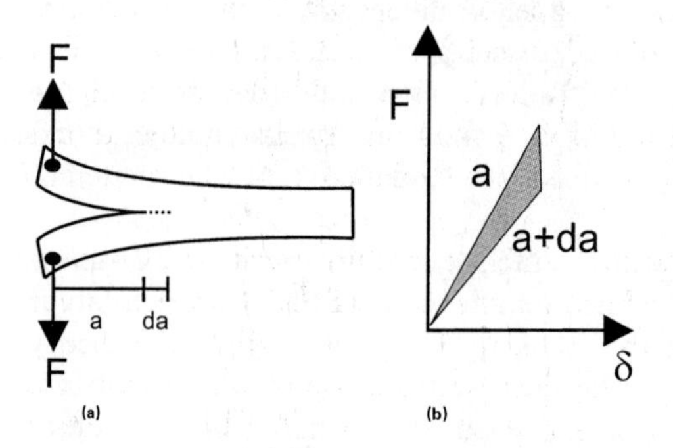

Fig. 8 (a) Double cantilever beam specimen for testing the G_c of composites or adhesive joints. (b) Force–displacement curve showing a small amount of stable crack growth. The shaded area represents the energy released upon crack extension.

and after a small increment of crack growth and then using the expression $G = -dU/dA$. $-dU$ is simply the shaded area in Fig. 8. Because crack growth was induced by the applied stress, the value of G computed in this way is equal to G_c. This method was applied to paper by Fellers and Steadman [27]. It cannot be used for specimen configurations where crack growth is unstable.

A more common way of measuring the toughness is to measure K_c using a specimen which there is a known relationship between σ and K. For an infinitely wide specimen containing a center crack of length $2a$, $K = \sigma_\infty\sqrt{\pi a}$, where σ_∞ is the remote stress far from the crack. For finite-width specimens or other specimen configurations, K may be obtained by stress analysis. Stress intensity factors have been derived and are tabulated for many different specimen configurations. A good reference is *Compendium of Stress Intensity Factors* [24], and a table of solutions for simple configurations is provided in the Appendix.

To determine the value of K_c experimentally, a cracked specimen is tested to failure, and the appropriate equation from the Appendix is used. A paper specimen can be prepared by using a razor blade to carefully create the notches. Obviously, great care should be taken to align the notches and not to damage the material near the notch tip, because it is this material that has the greatest influence on the failure load. The K_c measured with a specimen containing a crack prepared in this way will be different from that measured with a crack that has grown under the influence of external loads due to the crack blunting (R-curve) behavior described previously. The value of K_c measured in this way represents the critical stress intensity factor for crack initiation.

As discussed previously, one method of determining whether or not K_c is truly a material constant for paper is to test specimens with differing sizes and configurations. Some studies of paper have found K_c to be relatively constant [22,28], whereas others have not [29]. If K_c is not constant, then the plastic zone is probably too large in comparison to the specimen dimensions, and an elastic-plastic fracture characterization methodology is indicated. Swinehart and Broek [28] used a specimen 5 in. wide, 8 in. long, and containing a 1.5 in. central slit. For board samples, they suggested a specimen 4 in. wide, and they also suggested a length-to-width ratio of more than 1.5 for all specimens. They stated that "predictive capability, but not ranking, is lost with smaller samples" [28].

Making Failure Predictions with G_c and K_c The values of G_c and K_c measured using the methods described in the previous section can be used in two ways. They may simply be used as quality control parameters in much the same way that tear strength is commonly used in the paper industry today. This is undoubtedly the most common reason for measuring toughness in the paper industry. Since it was established in Fig. 1 that fracture resistance (K_c) is not necessarily correlated to tensile strength or tear, even this simple objective justifies a fracture mechanics approach provided it is believed that linear elastic fracture mechanics (LEFM) governs failure in service.

The toughness can also be used to predict failure loads using the equations $G = G_c$ and $K = K_c$. Having measured the material property G_c or K_c using a laboratory specimen, it remains only to determine the values of G or K as a function of applied load for the end use situation. The value of K for a small cut in a large sheet of newsprint is known, for example, even when the cut is at an angle or is close to the edge. K at the tip of a crack of a fixed length is roughly twice as large when it is at the edge of a sheet as when it is in the middle (see Appendix). This supports the common observation that failure in a printing press is more likely to be caused by a shive at the edge of the sheet. Eriksson and Viglund [29] made direct measurements of the effect of failure load on the proximity of a crack of fixed size to the edge of the sheet and found a strength reduction of approximately 30% as the crack approached the edge. In this study, they also examined the effect of crack orientation on fracture, and Ferahi et al. [30] have also studied the effect of crack orientation on fracture.

For complicated configurations that are difficult to analyze using closed-form solutions, a numerical technique known as finite element analysis (FEA) is often used to determine the relationship between K and the applied loads. For example, FEA could be used to determine the relationship between internal pressure and K in a cracked boiler or between end loading and K in a cracked diving board.

C. Elastic-Plastic Fracture Mechanics

Although linear elastic fracture mechanics can be applied to material with a limited amount of plasticity at the crack tip, it cannot be applied to very ductile materials where there is widespread yielding at the crack tip, because the elastic stress solutions no longer describe the real stress field adequately. In general, paper is both nonlinear and viscoelastic, and for more ductile grades a more complex form of fracture mechanics may be required to properly characterize fracture. An illustration of various forms of nonlinear stress–strain behavior is given in Fig. 9.

A number of approaches have been developed to deal with nonelastic materials. Many of the theories discuss plastic (nonlinear, strain rate independent) materials but not viscoelastic (strain rate dependent) materials, because they were developed primarily for metals. The fracture of viscoelastic polymers has attracted some attention however, and viscoelastic fracture has been specifically addressed in the literature [12].

Crack Tip Opening Displacement One of the first methods of dealing with extensive plasticity was developed independently by Cottrell [31] and Wells [32], who noticed that the crack tip opened appreciably when they were testing ductile materials. The crack tip opening displacement (CTOD) was postulated as a fracture parameter, because it characterized the conditions at the crack tip. In paper, however, the crack faces are hard to define because of the presence of pulled-out fibers. The CTOD can be calculated from a measurement of crack opening some distance

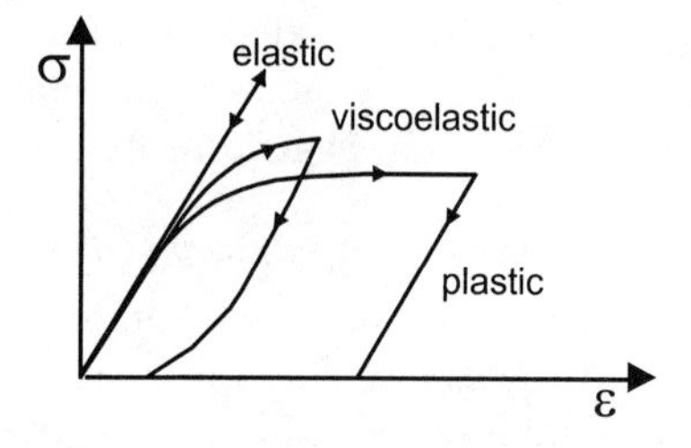

Fig. 9 A comparison of elastic, plastic, and viscoelastic behavior.

from the crack tip, but even so this method is not readily adapted for use with a thin film material and has not been used extensively in the paper literature. Tanaka and Yamauchi [33] made quantitative observations of the crack tip during paper fracture.

***J* Integral** For materials with a nonlinear stress–strain curve, the theory presented thus far cannot be applied. For these materials, a new quantity, the J integral, is commonly used to characterize fracture, and it has been extensively tested for paper over the past 15 years.

Just as K is computed using a knowledge of the stress fields around a crack in a linear elastic material, the J integral, or J, is computed using a particular path-independent integral in a nonlinear material. The details of this stress-based calculation are not needed, because the same calculation of energy changes in the system used to compute G in a linear elastic material can be used to compute J in the nonlinear elastic material.

For nonlinear elastic materials, the J integral may be interpreted as a crack tip energy release rate [34] and as a single parameter characterizing the crack tip stress fields. In a nonlinear elastic material, the same stress–strain curve is followed during loading and unloading. For this type of material, Rice showed that J, defined in Eq. (19), also represents the strain energy available for crack growth and is analogous to G.

$$J \equiv -\frac{dU}{dA} \tag{19}$$

In Fig. 10, the area between the two nonlinear curves represents the energy that would be released upon crack growth from a to $a + da$ in a nonlinear elastic material at fixed grip displacement. Nonlinear elastic materials are rare, however, and engineers are more interested in elastic-plastic and viscoelastic materials. In elastic-plastic materials, unloading follows a line parallel to the initial elastic portion of the loading curve. In a viscoelastic material such as paper, there are a wide range of possible unloading curves, depending on the loading and unloading rates, but the curve illustrated in Fig. 9 would be typical. In these materials, it is clear that J no longer represents the energy available for crack growth, and hence the J integral loses the elegant thermodynamic basis of the original Griffith approach. Nevertheless, in materials with monotonically increasing stress–strain curves such as paper, the J integral has been found to be relatively constant at failure even when

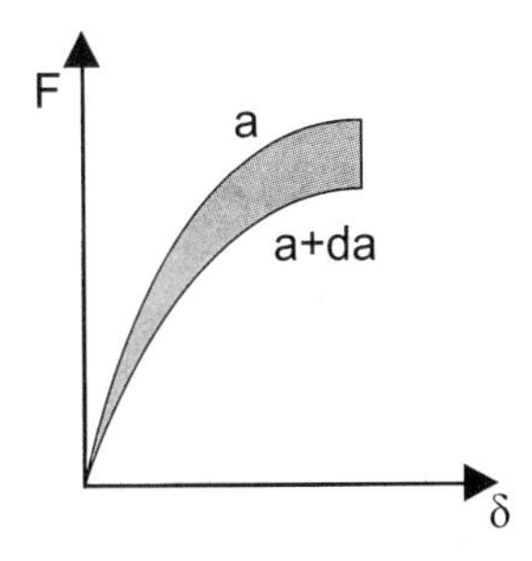

Fig. 10 Strain energy release rate for a nonlinear elastic material. The area between the curves represents energy available for crack growth if the material is nonlinear elastic.

there is permanent plastic deformation. For these materials, stress analysis indicates that the J integral uniquely determines all the stresses in the vicinity of the crack tip, and thus the stress-based arguments suggesting that the stress intensity factor should be a constant at fracture hold for the J integral in elastic-plastic materials.

The critical value of the J integral is known as J_c and can be used to characterize the toughness of elastic-plastic thin sheet materials such as polymer films and paper. The new failure criterion has the same form as all the others:

$$J = J_c \tag{20}$$

Note that if the stress–strain curve is more or less straight, as might be the case for a stiff machine direction newsprint, for example, then the value of J is simply equal to the value of G, because both are defined as $-dU/dA$.

D. Measuring J_c

The Multiple Specimen Method Paradoxically, although J_c cannot be interpreted as the critical strain energy release rate in plastic materials, the value of J at any load is still evaluated by treating a plastic material as if it were a nonlinear elastic material. This statement holds for materials with monotonically increasing stress–strain curves where there is no unloading. This is the basis of the multiple specimen that is illustrated in Fig. 11.

In the multiple specimen method, a set of specimens with identical overall dimensions but differing crack sizes are tested in tension, and the entire stress–strain curve is recorded [35]. For paper, the double-edge-notched (DEN) specimen illustrated in Fig. 12 would be suitable. The energy (area under the curve) at several different displacements is obtained (Fig. 11a) and then the data are cross-plotted as energy versus crack length for different displacements (Fig. 11b). Defining J as the energy release rate given in Eq. (19), J is obtained as the slope of any curve in Fig. 11b. The data are again cross-plotted as J versus displacement for each crack length, as illustrated in Fig. 11c. In principle, J should be constant at failure, regardless of crack length, and this value is labeled J_c.

The multiple specimen method requires at least three specimens to obtain a value of J_c, but because it is based directly on the definition of J, the results of this test are often used as the "true" value for comparison with the more convenient methods described in the following sections.

The Single Specimen Methods of Determining J The first single specimen method was derived by Rice, Paris, and Merkle [36] as a means of obtaining J_c with less experimental effort and is generally referred to as the RPM method [37]. A simpler approach for characterizing J with less experimental effort was developed by Liebowitz and Eftis [38] and applied to paper by Westerlind et al. [39]. This approach uses the Ramberg–Osgood model for a nonlinear load–displacement curve of the notched specimen:

$$\delta = \frac{P}{M} + k\left(\frac{P}{M}\right)^n \tag{21}$$

where δ is the displacement, P is the load, M is the initial stiffness (N/m) of the notched specimen, and k and n are material parameters fitted from the nonlinear

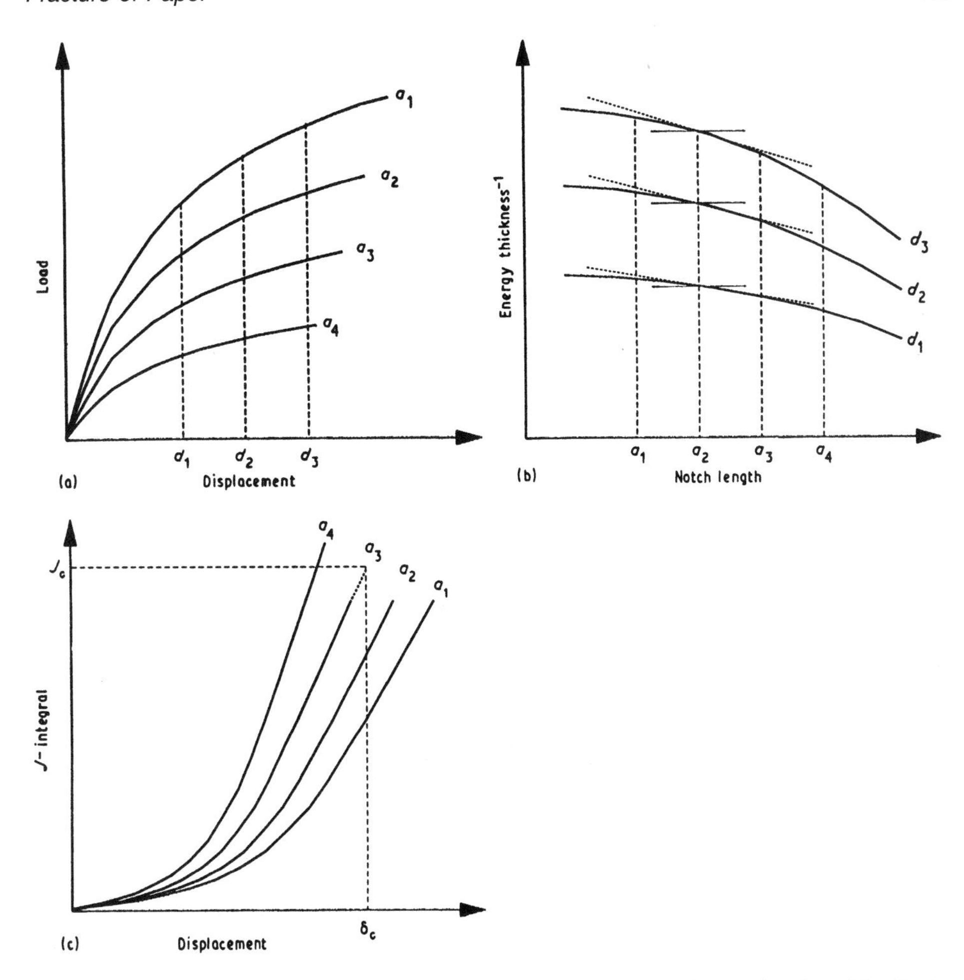

Fig. 11 Schematic of the multiple specimen method of evaluating J_c developed by Begley and Landis. (From Ref. 37.)

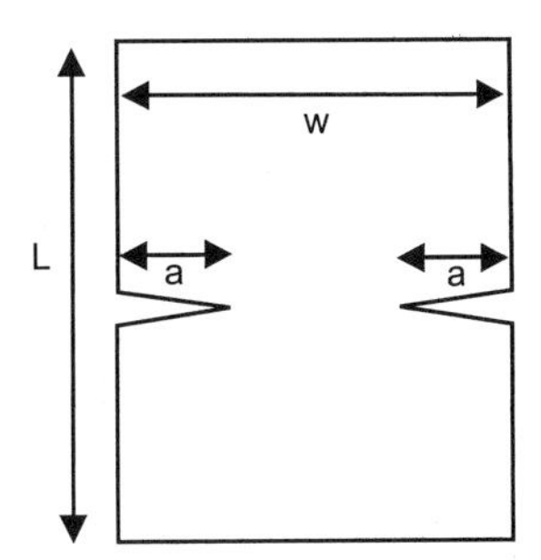

Fig. 12 Finite-width double-edge-notched specimen.

stress–strain curve. Liebowitz and Eftis [38] derived a nonlinear fracture parameter G^* that approximates J and is expressed as

$$G^* = \left[1 + \frac{2nk}{n+1}\left(\frac{P}{M}\right)^{n-1}\right] J_e \tag{22}$$

where J_e is the elastic part of J and is equal to G, which may be obtained by using the expression for K given in the Appendix and the general relationship between G and K.

$$J_e = G = K^2/E^* \tag{23}$$

where [37]

$$\frac{1}{E^*} = \frac{1}{(2E_x E_y)^{1/2}}\left[\left(\frac{E_x}{E_y}\right)^{1/2} + \frac{2E_x}{E_{45}} - \frac{E_x + E_y}{2E_y}\right]^{1/2} \tag{24}$$

for orthotropic materials such as machine-made papers, E_x is the modulus perpendicular to the loading direction, E_y is the modulus parallel to the loading direction, and E_{45} is the modulus at 45° to the loading direction. These moduli are the same for handsheets and isotropic papers, and $E^* = E$.

In practice, G^* is evaluated by testing a center-notched tension specimen, using three points of the load–displacement curve for evaluating M, n, and k, and using the appropriate formula to find J_e. By inserting the critical load, P_c into Eq. (22), the critical value of G^* is obtained. The value of G_c^* compares well with J_c, so this method will be referred to as the means of obtaining J_c in the rest of this chapter. As discussed later, this simple method of evaluating J_c has been applied to paper with good success. An alternative computation of J, also based on a Ramberg–Osgood material model, was presented by Swinehart and Broek [22].

E. Making Failure Predictions Using J_c

Once J_c has been measured for a particular material, then failure prediction is simply a matter of knowing the relationship between applied loads and J in the end use situation of interest and applying the failure criterion $J = J_c$. Unfortunately, the computation of the applied value of J for complex loading situations is generally difficult and often requires numerical analysis [40]. There are no widely available solutions for J except for rather simple geometries and loading situations. Most studies in the paper literature have focused on the measurement of J_c as an appropriate material property for ranking papers in terms of toughness and have ignored the prediction of failure for particular end use conditions such as pressroom runnability.

A discussion of the difficulties involved in predicting failure using J is given by Swinehart and Broek [22].

F. Essential Work of Fracture

The essential work of fracture (EWF) is another method used to characterize the toughness of ductile sheet materials such as polymers. The method was originally developed by Broberg [41] and was used more recently for polymers by Mai and Cotterell [42,43] and for paper by Seth and coworkers [44–46]. The method is based on the observation that the specific work of fracture, or area under the load–displacement curve divided by crack extension, is dependent on the amount of crack extension. It was postulated that the dissipation of energy could be divided into two components:

- The energy associated directly with crack tip processes
- The energy associated with remote plasticity

The crack tip energy dissipation divided by the amount of crack propagation is referred to as the "essential work of fracture" and denoted by the symbol w_e. w_e is said to be a material constant that characterizes the material's resistance to crack growth. The energy dissipated by remote plasticity divided by the amount of crack propagation is by implication the "nonessential work of fracture," w_p. By assuming that the size of the remote plastic zone varies as the square of the ligament length, a formula for the total normalized work of fracture can be derived:

$$w_f = w_e + \beta w_p l \tag{25}$$

where β is a parameter that accounts for the shape of the outer plastic zone and l is the ligament length. On a plot of the normalized work of fracture versus ligament length for specimens with a variety of ligament lengths, the intercept represents the essential work of fracture.

The linearity of the data in Fig. 13 is an indication that the form of Eq. (25) is correct and that the basis for the plot is correct. The principal objection to the EWF method is that the "nonessential" work of fracture, which is excluded from the measurement by definition, is bound to influence a service failure. It is not at all clear that it should be ignored.

G. Measuring w_e

The essential work of fracture is measured by preparing a series of specimens with various ligament lengths. The tensile energy absorption is recorded for each specimen, and a plot such as that presented in Fig. 13 is prepared. In order for this plot to be meaningful, all specimens must fail in a stable way so that the entire work of fracture is dissipated in the plastic zone. In addition, the essential work of fracture is intended to be a plane stress measurement, although this concept may be misleading when applied to paper as described previously. To ensure plane stress, the ligament length must be substantially greater than the sheet thickness; normally the criterion is $l \geq 3t$ [42]. In addition, the plastic zones from the two notch tips must merge to form a single zone of approximately circular shape so that w_p varies linearly with l. To satisfy this constraint, the notch length must be set so that $W \geq 3l$ [42]. Provided these three conditions are met, a set of experimental data as depicted in Fig. 13 will be obtained and a valid value for w_e can be determined.

H. Making Failure Predictions with w_e

The essential work of fracture has been related to the J integral in polymers [47], which suggests that J-integral prediction methodology described previously may be

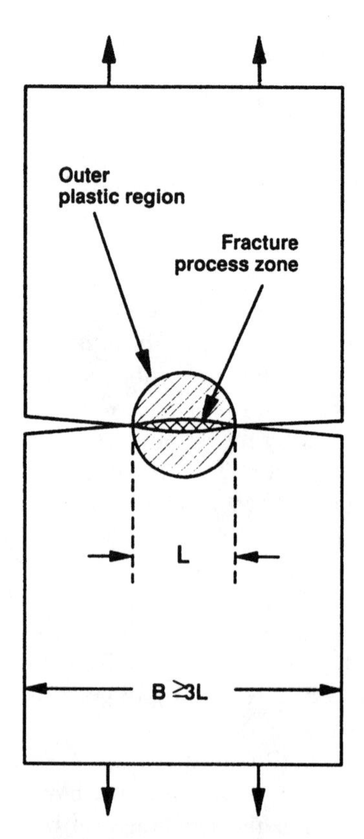

Fig. 13 Specimen geometry and data analysis for the essential work of fracture test. (From Ref. 44.)

used. Recent work on paper, however, has shown that J_c and w_e are not the same for this material [48]. It is more common to see w_e used as a simple ranking parameter because it is difficult to make predictions of failure based on J_c in the first place. w_e is most useful when the mechanisms of energy absorption at the crack tip, such as fiber debonding and pullout, are being studied.

I. Summary

In this section, the basic theory of fracture mechanics has been described, and several fracture parameters have been discussed in isolation. Although it is the value of K_c that relates directly to the critical load in a given end use situation, the value of G_c is more directly related to the processes of energy dissipation. Of course, these two are related by the modulus of the material, as given in Eq. (18) [or Eq. (23) for anisotropic materials]. It is possible for a material with lower G_c than another material to have a higher K_c if it also has a higher modulus. The material with the higher value of K_c is the one that will fail at higher stress, not the one with higher G_c, and on this basis it has been argued in the literature that K_c is a more appropriate measure of toughness [22]. In a printing press, however, the applied load is also dependent on the modulus in any fixed open draw situation. The fracture stress may not be as significant as the fracture strain in this case. Nevertheless, in order to rank materials based on the expected critical load in the presence of a flaw of a particular length, it is K_c that should be used. G_c is more closely related to the micromechanics of energy absorption and should be used to model the effect of furnish and process changes on toughness. At the same time, however, the effect of these changes on modulus, if any, should be accounted for. Clearly a good understanding of the service conditions is essential.

For nonlinear materials, such as more ductile grades of paper, a nonlinear fracture mechanics parameter such as the J integral is more likely to be appropriate. Unfortunately, failure load predictions using J are more complex, but they are certainly possible for simple configurations [22].

IV. FRACTURE OF DRY PAPER AND BOARD

A. Introduction

This section will focus on the study of fracture in paper, from its early roots to some of the more advanced concepts. Historically, the tensile test and tear tests were used in tandem to evaluate the strength of a sheet and the potential of reinforcement pulps. Other properties, such as the Mullen burst strength, are also used to evaluate the properties of paper. In fact, properties such as these often form part of the specifications provided by the converter or end user, so the papermaker has little choice but to base production decisions on measurements of these parameters. Nevertheless, as environmental and cost considerations drive the basis weights lower and lower, there is an ever-present desire to maximize the resistance of the sheet to fracture.

Research into the application of LEFM to paper began in the 1960s [49], and research in this area has been active ever since. In 1979, Uesaka et al. [50] pointed out that it was not practical to make paper specimens of sufficient size to satisfy the requirements of LEFM theory and proposed the use of the J integral to characterize the toughness of paper. Seth et al. [44] proposed the use of the essential work of fracture for paper in 1993. More recently, however, Swinehart and Broek [22] argued strongly in favor of the original LEFM approach for paper, based on the ease of testing and the relative simplicity of fracture predictions made using this approach. These studies present conflicting statements about the utility of the various approaches, leaving the newcomer to the field bewildered. In the discussion presented below, the various methods of characterizing "toughness" of paper are described in roughly chronological order. Comparison of these methods for various paper grades are used to draw conclusions about the most appropriate tests.

B. Elmendorf Tear

Even without the scientific foundation provided by Griffith, Irwin, and others, papermakers have used some measure of energy absorption capacity for many years. The Elmendorf test is most widely used in North America, whereas Europeans use the Brecht-Imset tear tester [51]. Both tests are based on a direct measurement of the energy consumed in driving a crack through a unit distance in a sheet. In the Elmendorf test, four plies of paper are mounted in a clamp and are torn in an out-of-plane action by a swinging pendulum. The potential energy loss of the pendulum is divided by the new crack length to produce value for tear energy in units of joules per meter or newtons. Note that the energy required per unit length is equivalent to the applied force, so the tear energy is often referred to as the "tear strength." The Brecht-Imset tester follows the same basic principles although the details of testing are different. The results of these two tests are not always well correlated [51].

The Elmendorf test has been used for more than 60 years because of its simplicity and reproducibility. It is often used as a specification by end users such as pressrooms; clearly, these users believe that the tear energy is a good indicator of performance. But in spite of its long history, the principles of the tear test have been called to question frequently over the years. Early work by Wink and Van Eperen [52] examined the effect of clamping conditions and number of plies on the measured tear energy. They discovered that differences in the method of clamping could yield differences of up to 10% in the results and that differences in the number of plies could yield differences of up to 100% in the results. The variations were attributed to differences in the extent of "splitting" or delamination resulting from different test parameters. They also noted that since splitting is not generally observed in end use situations, "the property measured by the Elmendorf is not generally the property of interest" [52]. It should be noted that the amount of variation they observed depended quite strongly on the degree of bonding in the sheet, with the tough, poorly bonded sulfite papers showing the greatest variations in tear energy.

Seth and Blinco [51] also observed that the Elmendorf tear energy is dependent on the number of plies being torn and discussed the possibility that the reduction in tearing speed associated with four-ply tests on very tough kraft fiber sheets may result in artificially high values of tear strength. They suggested single-ply tests, although in single-ply tests there is often extensive delamination, which is not at all representative of the fracture surfaces that would be produced in a pressroom break. Further experimental evidence calling the tear test into question was published by Lyne et al. [53] and by Seth and Page [4] in their original study of fracture toughness. Figure 1 (which is similar to Fig. 6 in Lyne et al.'s work) provided a comparison of tear strength, tensile strength, and in-plane tear strength for two different pulps as a function of beating. Because the results of the various tests are not proportional to one another, it is apparent that if one of these properties is proportional to an end use property such as runnability, the others will not be proportional. Lyne et al. [53] also observed a substantial difference in tear energy for a multiple sheet test and a single-sheet test even when the single sheet has a grammage equal to the total grammage of the sheets torn together in the multiple sheet test. This difference was attributed to the much higher flexural stiffness of the single, heavier sheet.

The Elmendorf tear test, like the essential work of fracture test, cannot really be used to predict failure load for a particular end use geometry and loading. Instead, it is intended to be used to rank the relative toughness of different products or to monitor the quality of a particular product on a hourly or daily basis. For it to be effective as a ranking test, however, it must correlate to the end use performance. Unfortunately, end use performance itself has been poorly characterized in the past; in particular, the pressroom runnability of a sheet is notoriously difficult to quantify. The arguments in favor of more rigorous fracture tests are therefore based largely on intuitive arguments about the mode and characteristics of fracture such as those made by Wink and Van Eperen [52]. Because paper in a printing press typically fails under the influence of in-plane stresses, it seems obvious that tensile strength, or in-plane tear energy (to be described in the next section), will be a better predictor of runnability. Many studies of paper fracture have made this statement, but in fact there are good reasons to question it in some instances

First, the difference between in-plane tests (such as the essential work of fracture test) and out-of-plane tests (the tear test) is said to be one of fracture mode. For thick specimens, three independent loading modes exist as illustrated in Fig. 14. All of the theory discussed earlier referred to mode I testing.

In fiber-reinforced polymers there are large differences in the fracture energies for the three different modes [54]. Although the tear test appears superficially to be a mode III test, in fact, recent work on composites shows that it is extremely difficult to generate true mode III conditions at the crack tip [55], and these conditions are certainly not created in the Elmendorf tear test. Nevertheless, the lack of proportionality between the Elmendorf tear strength and in-plane fracture results indicates that the details of crack propagation are somehow different. Good experimental evidence of this has been provided by Trakas [56] in a series of experiments involving the counting of broken fibers in both J_c (in-plane fracture) and tear tests. As the data in Fig. 15 show, more fibers are broken in tear testing, most likely because the frictional forces on a fiber being withdrawn from the network are higher as it is pulled "around a corner." It is possible that fibers that break contribute less to the overall fracture energy than fibers that pull out from the network, because the latter process involves larger displacements. Consequently, if the application of (nominal) mode III forces in the tearing test increases the percentage of fibers that break, it may reduce the total fracture energy. This simple mechanism can be used to explain all of the data in Fig. 1. A complete discussion of the origin of fracture toughness in paper is beyond the scope of this review. The elementary pull-out model can be found in several references: Kane [57], Shallhorn and Karnis [58], Yan and Kortschot [59], and Johnson [60]. It is also worthwhile to read an alternative view of the micromechanical origins of toughness presented by Page [61].

The primary argument against the use of tear strength comes from a consideration of data such as those presented in Fig. 1, where in-plane and out-of-plane toughness values are plotted against the level of beating. It is important to note that these curves are not particularly relevant in the day-to-day operations of a typical mill. In a newsprint mill, for example, tear strength may be controlled by kraft pulp

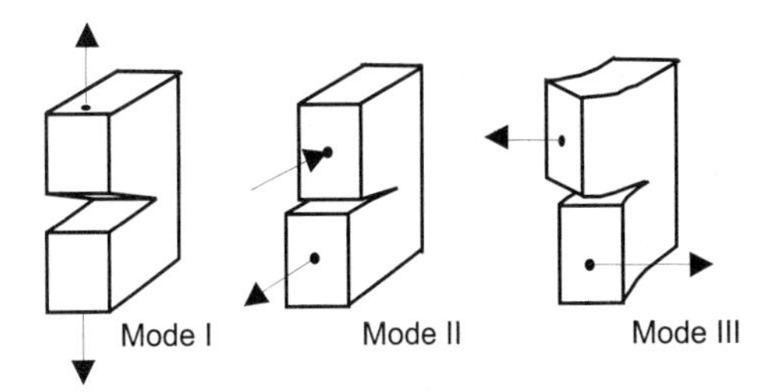

Fig. 14 Three modes of fracture for thick materials.

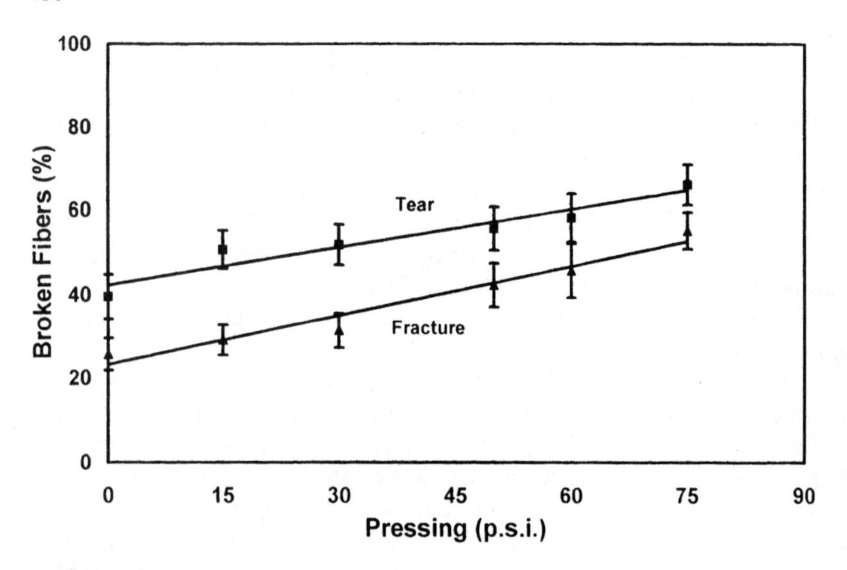

Fig. 15 Plot of broken fibers versus degree of wet pressing for both J_c (in-plane) and Elmendorf tear (out-of-plane) tests. (From Ref. 56 with permission of K. Trakas.)

additions The furnish varies only slightly from day to day, and for the range of furnishes possible, tear energy and fracture toughness may be well correlated. Work by Kazi and Kortschot [62] showed that this was more or less true for thermomechanical pulps (TMP) reinforced with kraft pulp, although for the particular furnish they tested there was an unexpected dip in the initial J_c measurements for 2.5% and 5% added kraft. If tear strength and J_c or G_c are well correlated for a limited range of furnishes, then tear strength can be used as a ranking parameter even if it cannot be used to predict fracture loads. The difficulty arises in comparing two pulps with very different furnishes. A press operator using the same tear strength specification for sheets made with northern spruce and sheets made with southern pine is probably not specifying sheets of equal press runnability.

Clearly, the tear strength should be interpreted with caution but may still be useful under certain circumstances. It cannot be used to predict fracture loads or strains directly, but it might be useful for ranking paper within some narrow range. Even so, it seems logical that any of the in-plane fracture tests described previously would be more suitable for ranking end use performance in situations where failure was dominated by in-plane stresses. The corollary to this is that Elmendorf tear is probably quite suitable for monitoring the resistance of a sheet to out-of-plane tearing.

Experimental Method The detailed experimental procedure for Elmendorf tear testing is contained in Tappi Standard T414 om-88 [63].

C. In-Plane Tear

In 1967, Van den Akker and colleagues described an in-plane tear test, in which a rectangular sheet of paper with a small edge crack is loaded in an offset manner to create stable crack growth as illustrated in Fig. 16 [64]. In principle, this test will replicate a typical end use fracture more closely than the Elmendorf tear test. The data in Fig. 1 clearly illustrate that fracture resistance is more closely proportional to in-plane tear than to Elmendorf tear [4].

The in-plane tear test is a direct measure of the crack growth resistance and should be reasonably well correlated to more fundamental fracture mechanics parameters. The arrangement of the specimen is such that the stress is very highly concentrated at the crack tip and the remainder of the specimen is virtually unstressed. The measured energy absorption will depend on the size of the highly stressed crack tip process zone, which in turn depends on the offset angle chosen. In the study from which the data of Fig. 1 were taken, Seth and Page [4] used a total offset angle of 12°, as illustrated in Fig. 16. The resulting in-plane energy absorption was much less than the fracture toughness. Note that the specimen used to measure fracture toughness is analogous to an in-plane tear specimen with an offset angle of 0°. As the offset angle is reduced, the measured energy absorption remains roughly constant to about 6°

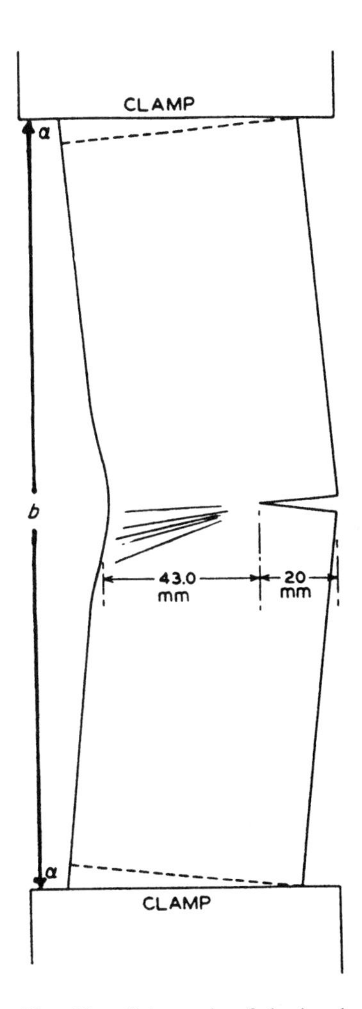

Fig. 16 Schematic of the in-plane tear test. (From Ref. 64.)

and then increases because of the increasing size of the process zone at the crack tip [13]. Because the choice of offset angle in the in-plane tear test is arbitrary, the absolute value of the measured toughness is also rather arbitrary. Furthermore, the data reported by Seth and Page (Fig. 1) indicate that the in-plane tear is not equal to or even proportional to fracture toughness even for a single furnish beaten to differing degrees. The ratio between the two varies from about 1.3 to about 3. Nevertheless, this test might be useful for internal comparisons and quality control within a single mill. It should be noted that fracture toughness (K_c) measurement is used as the reference here, not because of a demonstrated correlation with performance in service but because of its sound theoretical underpinning and its demonstrated ability to predict the failure load for cracked paper specimens with a variety of crack lengths.

Experimental Method The in-plane tear test is carried out on a standard tensile machine with rigid line clamps, which are required to maintain the prescribed tearing angle. (A pendulum-based tester capable of producing in-plane tearing was proposed by Lyne et al. [53]). The specimen is mounted as illustrated in Fig. 16. The specimen must be long enough that the grips do not impinge on the plastic zone at the crack tip, but this is quite localized in any case. The results are simply reported as the area under the load–displacement curve divided by the length of new crack created and normalized by grammage, if desired.

D. Tensile Energy Absorption

There are a number of methods of measuring "toughness" that do not involve precracking the specimen. In these cases, the energy is consumed in both initiation and propagation of the crack rather than in propagation only. Because the initiation energy may be a large fraction of the total, this type of test may not be appropriate for service failures that begin from a flaw such as a shive.

Tensile energy absorption (TEA) is the most commonly used measure of the energy-absorbing capacity of the sheet, because it is a trivial matter to extract this number from a normal tensile test. TEA is the area under the load–displacement curve for a standard tensile test in which a specimen is progressively strained in tension to failure. The energy is normally divided by the mass of the specimen between the grips, and the result is expressed in units of joules per gram. Sanborn and Diaz [65] investigated TEA in response to a need for new ways of measuring the properties of highly extensible papers. They studied the effect of specimen dimensions and strain rate on TEA. There was little change with specimen width or strain rate, but shorter specimens had higher values of TEA, and this was attributed to the fact that these specimens had fewer flaws.

The problem with TEA is that it measures the energy absorption of the whole specimen. In a service failure, high strain may be concentrated in a small zone near the tip of a preexisting flaw. Corte et al. [66] attempted to isolate the energy absorption in the rupture zone itself by measuring TEA for a series of specimens with differing gauge lengths and extrapolating these data to a gauge length of zero. This extrapolation is based on the premise that the TEA is composed of two parts, and the data obtained were fit extremely well by the equation

$$\text{TEA} = r + lp \quad (26)$$

where r is the energy directly associated with bond and fiber failure in the fracture zone, l is specimen length, and p is the plastic energy dissipated throughout the specimen.

As the data in Fig. 17 indicate, the energy consumed in the rupture zone may be only a small part of the total energy and can therefore be rather hard to isolate with any precision using this extrapolation.

Goldschmidt and Wahren [67] proposed a slightly different method that depended on incomplete fracture in unnotched specimens. After fracture, the residual work (area under the load–displacement curve) required to separate the specimen was added to the elastic energy stored in the specimen at the moment of fracture to produce a toughness value that was correlated to that produced by Corte et al.

In the methods proposed by Corte et al. and Goldschmidt and Wahren, the entire specimen (including the material along the final rupture path) was strained plastically prior to rupture. When the crack finally initiates and propagates, it is therefore propagating through prestrained material. Seth [68] pointed out that this leads to estimates of fracture resistance that are less than those obtained using methods based on LEFM, where the energy needed to drive a crack through unstrained material is measured. Seth's LEFM-based fracture toughness was reduced when the specimen was prestrained to 85–90% of its breaking load prior to notching and testing, and Seth [68] concluded that the method of Corte et al. and Goldschmidt and Wahren did not "measure the true fracture energy of paper."

Experimental Methods Tensile energy absorption (TEA) is simply the area under the load–displacement curve from a standard tensile test divided by the mass of the specimen between the grips. It is a useful number only if the load does not fall immediately to zero after crack initiation. Corte et al.'s method [66] involves the extrapolation of TEA for several different gauge lengths to find the intercept at zero gauge length. Goldschmidt and Wahren's method [67] is somewhat more complex and involves a subtraction from the TEA of the plastic work dissipated prior to

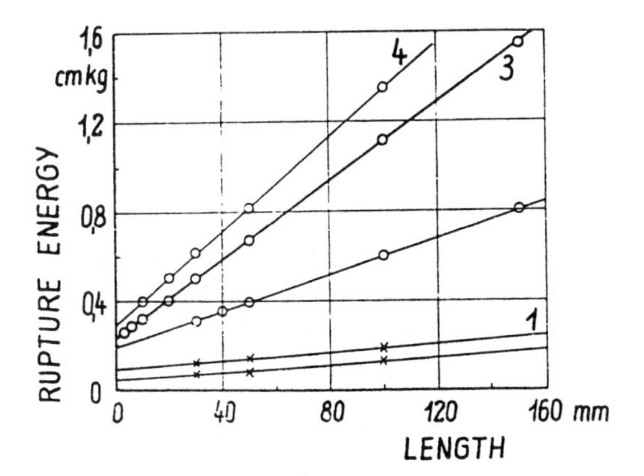

Fig. 17 Rupture energy for paper strips of various lengths. *1*, Unbeaten; *3*, 35° Schopper–Riegler; *4*, 45° Schopper–Riegler. (From Ref. 66.)

crack initiation The interested reader is referred to the original references for more detailed descriptions of the methods.

E. Linear Elastic Fracture Mechanics

Balodis [49] was the first researcher to apply LEFM to paper. He tested a range of eucalypt and *Pinus radiata* kraft handsheets with both center-notched and double-edge-notched geometries and applied the standard $K = K_c$ analysis (although the math was all done in terms of G and G_c). He found that G_c was geometry-dependent, first increasing and then decreasing with increasing precrack size. He also used a regression fit to model the relationship between critical stress and crack size. Recall that Eq. (10) states that if LEFM is correctly describing failure, then $\sigma \propto 1/\sqrt{a}$. Balodis found that $\sigma \propto 1/a^{0.78}$ for his materials and concluded that the plasticity of the sheets was responsible for this discrepancy. Eriksson and Viglund [29] tested much larger single-edge-notched sheets with crack lengths of up to 50 mm and found that $\sigma \propto 1/a^{0.6}$. Andersson and Falk [69], however, found that LEFM did describe failure in sulfite handsheets by confirming the $\sigma \propto 1/\sqrt{a}$ relationship, provided that they corrected for subcritical crack growth using a plastic zone size correction factor, which they used as a fitting parameter.

Seth and Page [26] studied the fracture toughness of paper using the quasi-static method developed by Gurney and Hunt [70]. The Gurney and Hunt test involves a precracked test specimen designed so that the crack growth is completely stable. For this to occur in a material with a constant value of G_c, the derivative of the strain energy release rate must be negative:

$$\frac{\partial G}{\partial A} = -\frac{\partial^2 U}{\partial A^2} \leq 0 \tag{27}$$

(It is assumed here that the critical strain energy release rate, G_c, is constant as the crack grows.) When the derivative of strain energy release rate is positive, as it is for a conventional fracture toughness test, then the crack accelerates as the imbalance between available and required energy becomes larger. In this case, the critical stress and the known solution for K (or G) is used to find the value of K_c (or G_c). However, if Eq. (27) holds, the crack growth is stable, and the value of G (and hence G_c) can be computed directly from the area under the load–displacement curve. This computation could be done by unloading the specimen after a known increment of crack growth, as illustrated for the double cantilever beam test described in Fig. 8. Seth and Page recommended a specimen geometry for which crack growth is stable for the entire duration of the test, and hence the area under the stress–strain curve can simply be divided by the total crack length for this specimen. This was accomplished using a very stiff tensile machine together with a short, wide specimen (see Fig. 18). The specimen geometry is also adjusted so that at failure the yield stress of the material is not exceeded outside the crack tip region.

Seth and Page also used a more conventional fracture toughness test with unstable crack growth and compared the results to the quasi-static test for a wide variety of furnishes, producing a one-to-one relationship with an r^2 of 0.917 [4]. This is surprising, because paper is a viscoelastic material and the crack tip plastic zone would be expected to be smaller for higher crack speeds. However, a change in strain rate had no pronounced effect on quasi-static crack growth resistance in Seth and

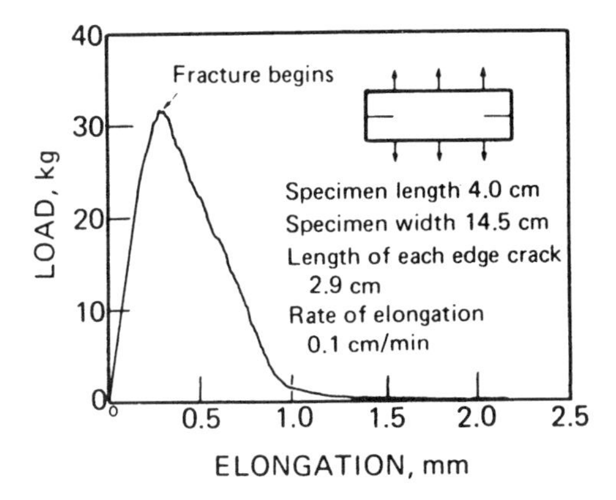

Fig. 18 Specimen geometry and load–elongation curve for a quasi-static fracture test. (From Ref. 4.)

Page's work [71]. Further data for a wider range of paper and processing conditions may be found in Ref. 4.

Seth and Page [26] also compared their results to those obtained by Balodis [49] and Andersson and Falk [69] for similar papers and found that their values were 3–10 times higher. Since both of the previous authors used 1.5 cm wide specimens, Seth and Page concluded that the difference could be attributed to specimen geometry effects and that the previous studies had not obtained a true value of G_c because the crack tip plastic zone was not small compared to the crack length and specimen width. Indeed, Seth and Page's data showed a stable value of G_c for specimens wider than 15 cm, with a significant decrease as the width was reduced below this value.

In the 1980s, most of the paper fracture literature was devoted to methods of measuring the J integral for paper, and there was very little additional work on LEFM. However, Swinehart and Broek advocated the use of a simple K_c-based LEFM approach in 1995, arguing that it was difficult to make failure predictions using the J_c approach and that in most cases MD paper was sufficiently elastic to justify the simpler method [22]. They justified this with an explicit comparison of the elastic and nonlinear components of the J integral that showed that the elastic component dominated for MD sheets, although not for cross-machine direction (CD) orientations. They also argued against methods based on a calculation of total work of fracture, such as the quasi-static test, because the work of fracture actually increases slightly over the first few millimeters of crack growth. This phenomenon is dependent on the micromechanics of crack advance and crack wake fiber bridging and is known as R-curve behavior. The interested reader is referred to any of the basic texts on fracture mechanics cited previously for a discussion of this behavior.

In 1996, Ferahi et al. [30] published a study of the effect of crack orientation and propagation direction on fracture toughness and failure prediction. They compared the known relationships between crack orientation and stress intensity, K, to measured relationships between crack growth direction and K_c to derive an "anisotropic failure criterion." They concluded that "the MD fracture toughness approach overestimates the fracture load and may result in non-conservative predictions of paper breaks" [30]. This paper presents a useful discussion of fracture mechanics in an anisotropic material such as paper.

Experimental Methods The two principal methods used in the fracture studies are the quasi-static method of Gurney and Hunt [70] and the traditional LEFM method. The advantage of the quasi-static method is that no measurements of the elastic properties are needed: A single fracture test yields a value for the work of fracture. This test does, however, require very wide and rigid grips. The traditional LEFM method does not need special apparatus.

A full description of the quasi-static method was provided by Seth [25], and a brief summary follows. The main requirements for the quasi-static test are

1. Slow stable crack growth,
2. Plastic deformation confined to the crack tip region

Seth and Page used specially developed wide hydraulic clamps with alignment rods used to maintain grip faces parallel. They used a specimen 5 cm long between the grips, 15 cm wide, and containing a single 5 cm edge crack along one edge. The crack was made with a sharp razor blade. The alignment rods are quite important for this geometry because the specimen is not symmetrical. In the original work [4,26], two symmetrical edge cracks were used, yielding a more balanced specimen and reducing the need for the alignment rods. In this case, the specimen width was 15 cm and the edge cracks were 3 cm long. In either case, the a/b ratio should be in the range 0.25–0.4. A rate of elongation of 1.0 mm/min was used, giving a test duration of approximately 2 min. The area under the curve was measured with a planimeter in the original work [25], but most modern computerized machines can now report this number with great accuracy. Rather than letting the test run until the specimen is completely fractured, it could be unloaded, and the enclosed area on the load–displacement diagram divided by the crack advance would represent the work of fracture. Using this method, the absence of remote yielding can be confirmed, because unloading should result in a return to the origin on the load–displacement diagram. Unfortunately, however, it is difficult to determine the extent of crack growth accurately because of the diffuse nature of the crack front, and therefore a complete fracture is recommended.

Traditional LEFM testing is conducted by making a center-notched or double-edge-notched specimen and testing it in tension to obtain the critical load. The disadvantage of center-notched panels is that they buckle, and a suitable antibuckling guide is required to ensure planar deformations. The guide usually consists of two sheets of plexiglass separated by a film two to three times as thick as the specimen as illustrated in Fig. 19. Buckling restraint is known to influence the toughness of thin metal sheets [1], but it is not clear what influence, if any, it should have on paper. Antibuckling guides are not required for double-edge-notched specimens, and since it is hard to conceive of a service failure in which buckling is restricted, it is probably best to use this configuration.

The specimen dimensions should be sufficient to ensure that the plastic zone is small with respect to the crack length and specimen width and length. Uesaka [72] provides a good discussion of the influence of various specimen dimensions. The size required for a valid test depends on the modulus and yield strength of the paper; the dimensions required for CD tests are much larger than those required for MD tests,

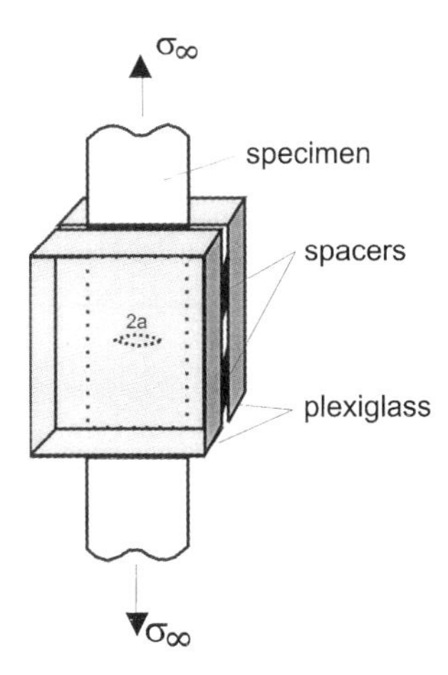

Fig. 19 Center-notched specimen with antibuckling guide.

for example. For various grades of paper loaded in the machine direction (i.e., with cracks growing in the cross-machine direction), a specimen width of 10 cm seems to be sufficient, with a total crack length-to-specimen width ratio of 0.25–0.4. If there is significant nonlinearity in the resulting load–displacement curve, then the dimensions chosen must be increased, because no significant yield is permitted in a valid LEFM toughness test. The specimen length should be sufficient to ensure a uniform remote stress field, because this is the assumption upon which the K solutions are based, and a length-to-width ratio of 6:1 is probably sufficient for this purpose [72]. A smaller length could be used if an appropriate stress intensity solution is also used to derive K. It is evident that a valid K_c test is difficult to achieve with handsheets.

It should be noted that in the open draw of a printing press, the specimen dimensions are much larger and the initial crack length might be much smaller.

Once the critical load is measured, the appropriate formula (see Appendix) is used to determine the value of K_c. For a finite-width double-edge-notched specimen, one form of the expression for K is given as [1]

$$K = \sigma \left\{ W \left[\tan\left(\frac{\pi a}{W}\right) + 0.1 \sin\left(\frac{2\pi a}{W}\right) \right] \right\}^{1/2} \tag{28}$$

Equation (28) replaces the expression $K = \sigma\sqrt{\pi a}$, which is valid only for a centrally notched infinite plate. An alternative, polynomial expression for K can be obtained from the Appendix.

If a value of G_c is required, Eqs. (23) and (24) are used, and hence E_x, E_y, and E_{45} must be measured in independent tensile tests. Of course, these three numbers are identical for isotropic handsheets.

F. *J*-Integral

Balodis, in the first application of LEFM to paper [49], concluded that in order "to obtain an unbiased estimate of fracture energy, it may be necessary to consider in detail the change in energy expended in plastic deformation and to separate the components which facilitate fracture from those not affecting the strength of the material." Hence from the very beginning it was known that for ductile grades of paper the fracture toughness parameters based on LEFM, K_c and G_c, might not be true material properties, independent of the specimen test geometry. For other materials, such as ductile metals and polymers, fracture mechanics was extended to include the CTOD and J-integral concepts, to provide this geometric independence Not surprisingly, then, these concepts were also applied to paper.

Uesaka et al. [50] were the first to suggest the use of the J integral for paper in 1979. In this study, both K_c and J_c were found to be roughly geometry-independent, but their plastic zone size calculations suggested that the handsheet specimens were too small for K_c testing. Uesaka [72] introduced the RPM method of J-integral evaluation in 1983 (see Section III.D). Both methods were used on highly ductility sack paper at the Swedish Pulp and Paper Research Institute (STFI) [73]. Whereas the results from the multiple specimen method were reasonably geometry-independent, the RPM method produced wide variations in J_c. Westerlind et al. [39] used the Liewbowitz nonlinear energy method and compared the results with the multiple specimen and RPM methods. The Liebowitz method compared well with the multiple specimen method and produced geometry independence; however, the RPM method once again produced different results. Yuhara and Kortschot [37] introduced a modified plasticity assumption and one extra parameter to the RPM method to improve the results. They also conducted a wide-ranging comparison of the various J-integral evaluation methods that included a computation of the essential work of fracture and some of the more advanced methods of crack growth detection available from the literature [14]. The comparison showed that in some paper grades the RPM method was not suitable for reliable toughness measurement. All methods showed some crack length dependence, except for the EWF method, which is based on an extrapolation through specimens with a variety of crack lengths and produces just a single value of toughness. The authors concluded that the Liebowitz technique was the most promising and simplest method for evaluating J_c.

Experimental Method The Liebowitz nonlinear method seems to be the most convenient way of measuring the J_c value for paper because the parameter G^* has been shown to be equivalent to J in practice. Westerlind et al. [39] used a centrally notched specimen, and Yuhara and Kortschot [14] used double-end-notched specimens, with both studies making use of antibuckling guides. Westerlind et al. used a gauge length of 15 cm, a specimen width of 5 cm, and crack lengths of 1, 2, and 3 cm. The crosshead speed was 1 cm/min, producing a test duration on the order of 10 s. Evaluation of the J_c parameter is a simple matter of curve fitting the load–displacement curve to obtain k, M, and n in Eq. (21) and using Eq. (22) to evaluate J. Typically the maximum load point is used to find J_c, although in principle the load at crack initiation should be used. Because it is impossible to determine the point of crack initiation optically in paper, a method based on the load–displacement trace known as the key curve method was implemented by Yuhara and Kortschot [14]. This is probably not required in most instances.

Yuhara and Kortschot [14] also used a very narrow specimen to evaluate toughness and compared the results to those obtained with wide specimens for a range of papers (see Fig. 20). Their reasoning was that most labs are not equipped with the wide rigid line clamps and that even if the resulting value of J_c was not

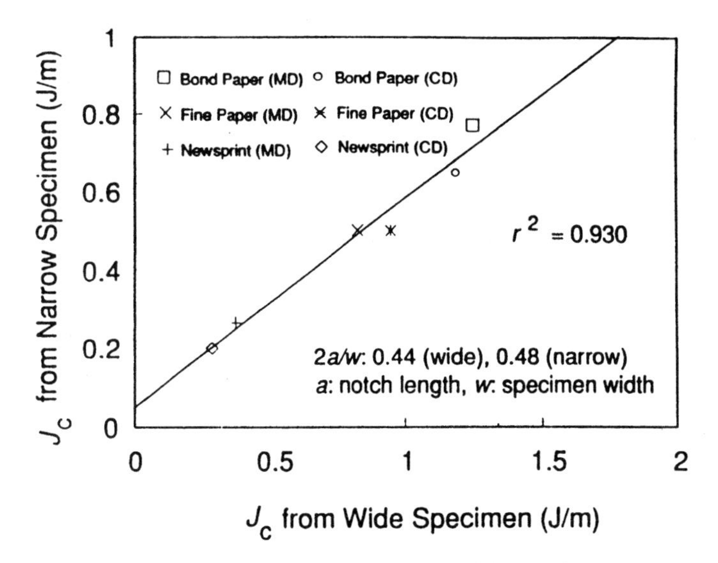

Fig. 20 J_c testing with narrow (25 mm wide) specimens compared to the results for wide (90 mm wide) specimens. (From Ref. 14 with permission of the Fundamental Research Society.)

correct it could still be reasonably well correlated to the correct value. For a wide variety of papers, they obtained a correlation coefficient $r^2 = 0.93$, indicating that the narrow specimen test would be suitable for ranking if not predictive purposes. Because it is difficult to make predictions based on J_c in any case, this test may be quite appropriate for mill quality control programs.

G. Essential Work of Fracture

In recent years, Seth and colleagues [44–46,74] have published a number of papers describing the application of the essential work of fracture (EWF) method to paper. Recall that for this method, plastic energy dissipation is divided into two components: energy dissipated in the fracture process zone, which is directly proportional to the ligament width, and energy dissipated in the outer plastic zone, which is proportional to the square of the ligament width. They found the plot of w_f versus L to be highly linear, which is an indication that the basic assumptions underlying this method are correct. Typical data are illustrated in Fig. 21.

The existence of the inner process zone can be confirmed by close inspection of the fractured edge of the specimen. In this zone, fiber breakage, debonding, pull-out, and plasticity are all occurring. In tracing paper, the debonding leads to a "stress-whitened" zone that is clearly visible and of constant width as the fracture progresses. The existence of an outer plastic zone has also been investigated directly. Mai et al. [74] saturated a light blue copy paper with wax, and the extent of the plastic zone was revealed visually by a marked change in opacity in this region. Tanaka et al. [75] used infrared thermography to measure the temperature change associated with plastic deformation. In both cases, a roughly circular outer plastic zone was observed, supporting the assumptions of the EWF test.

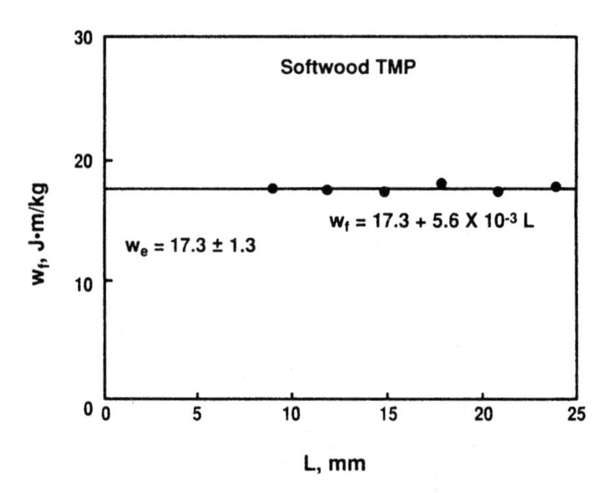

Fig. 21 Essential work of fracture test for TMP softwood. (From Ref. 44.)

The essential work of fracture method has several advantages over conventional fracture toughness testing. It is simple, although time-consuming. The underlying theory is easily understood, and the method clearly measures the effect of crack tip processes, so the effect of changes in sheet structure can be interpreted directly.

One additional feature of the essential work of fracture is that it provides a direct measure of the ratio between remote energy dissipation and crack tip process zone energy dissipation. If the slope of the plot used to obtain w_e is low, as is the case in Fig. 21, then the majority of the work is consumed right at the crack tip, and remote plasticity is quite restricted. This suggests than an LEFM method might be appropriate. Of course, the essential work of fracture is useful only for materials that display some remote ductility, where stable cracking is achieved. For more brittle grades of paper, such as MD newsprint, a method based on the quasi-static or critical stress based LEFM approaches are more suitable. Note that for brittle MD newsprint, the nonessential work of fracture is virtually zero [44], and therefore the EWF method produces the same result as the quasi-static G_c method used by Seth and Page [4], as expected.

Experimental Method Seth has provided a complete description of the measurement of the essential work of fracture for paper, and the following summary is taken from Ref. 45. Typically, double-edge-notched specimens are used without antibuckling guides. The specimen dimensions must be adjusted so that there is a fully developed yield zone in the ligament between the crack tips. This zone must not extend to the specimen edges. Seth suggested a set of self-similar specimens with widths in the range of 20–90 mm and ligament lengths fixed at 30% of the width (6–27 mm). The specimen length was 90 mm, and the symmetrical notches were made with razor blades mounted in a special jig. Four different ligament lengths were used, with 10 replicates of the smallest and at least six replicates for the others. Wide line clamps with alignment rods mounted in linear bearings were used in a standard tensile machine with a cross-head speed of 2 mm/min. The test is valid if the crack growth is stable for all ligament lengths, and the data fall on a straight line when plotted as illustrated in Fig. 21.

H. Conclusions

In concluding which of the various methods of characterizing toughness to choose, the two possible objectives of such experiments must be considered:

1. To rank materials in terms of their performance in a particular end use situation
2. To predict the failure stress or strain for a specific end use situation.

There is sufficient evidence to show that the response of paper containing a crack cannot be predicted by simple tensile strength, stretch, or TEA. For example, in Fig. 22 TEA is plotted against crack length for two different kraft sheets. Clearly, a ranking based on unnotched specimens (notch length = 0) would not represent the behavior of notched sheets such as sheets containing calender cuts or poorly bonded shives. Therefore, some sort of measurement of fracture toughness is required when fracture from a preexisting flaw is responsible for service failure. Based primarily on intuitive arguments about fracture mode, it has been widely suggested that the Elmendorf tear test, which involves out-of-plane deformations, is not the most suitable test when in-plane deformations are responsible for most service failures.

The fracture toughness tests based on fracture mechanics includes the G_c/K_c LEFM approach, the J integral, and the essential work of fracture. For relatively brittle MD sheets, including most newsprint sheets, a simple K_c approach is probably quite adequate. The suitability of the approach can be confirmed by testing several different specimen configurations—specimens with a variety of crack sizes, for example. If the value of K_c or G_c obtained is relatively constant, LEFM adequately describes fracture. The advantages of this approach are that it is simple and direct and that, if needed, K_c can be used to make predictions about service failures. Readers interested in straightforward LEFM testing should refer to the work of Swinehart and Broek [22].

For most CD-oriented sheets or for more ductile sheets such as sack kraft, LEFM may not be adequate to describe fracture. In these cases, the J-integral or essential work of fracture methods may be used. The J integral has a more complex

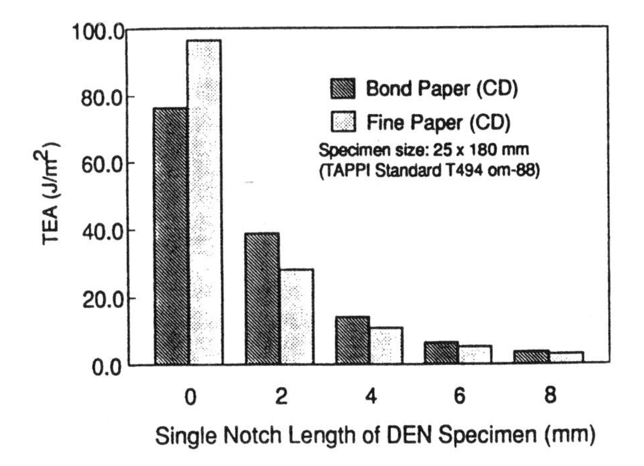

Fig. 22 Tensile energy absorption of double-edge-notched (DEN) and unnotched specimens (From Ref. 14 with permission of the Fundamental Research Society.)

theoretical basis but is easily measured using the Liebowitz method. The EWF is conceptually very straightforward but requires as many as 30 specimens to provide a reliable measurement of w_e. J_c also reduces to G_c if the specimen is elastic, whereas the EWF method is suitable only for sufficiently ductile materials. Karenlampi et al. [48] studied the relationship between w_e and J_c for sheets containing a variety of controlled structural changes including variable levels of beating and variable levels of reinforcement pulp. They found that the absolute values of the numbers were different, with J_c typically lower than w_e, and that the responses of the two measures of toughness to structural changes were not identical. They suggested that J_c may be more relevant for situations where the sheet is loaded monotonically to failure in service, whereas the essential work of fracture approach may be more relevant for situations involving multiple loading–unloading cycles [48].

For simple ranking purposes such as the day-to-day quality control requirement within a single mill, the requirement is only for a test result that is proportional to the actual material property governing end use failure. If the furnish changes little from day to day, Elmendorf tear may be correlated to any measure of in-plane fracture toughness. In general, however, because there are few specific studies of end use failure, a good principle is to have the quality control test emulate the service failure as closely as possible in terms of failure mode, test speed, etc. On this basis, an in-plane test would be preferable. A narrow K_c or J_c test, such as the one suggested by Yuhara and Kortschot [37], may be ideal when equipment is limited.

It is important to recognize that even for the purposes of ranking, an understanding of the underlying principles is important. In particular, the failure load is proportional to K_c, not G_c or J_c. Two sheets of equivalent K_c have the same failure load if they also contain precracks of the same size and location. However, two sheets of equivalent G_c but differing elastic modulus do not have the same failure load. On the other hand, neither parameter is directly proportional to the strain at failure, so neither can be used directly to rank the potential of a sheet to pass through a fixed draw in a printing press, because this will also depend on elastic modulus. Hence even for the purposes of ranking materials, it is important to choose an appropriate test or set of tests.

V. APPLICATIONS OF FRACTURE MECHANICS

Fracture toughness, strength, or any other property of paper is measured only because it is thought to be related to some aspect of the sheet "performance." Fracture toughness is measured in order to make predictions of fracture in typical converting or service conditions.

This section briefly summarizes three aspects of paper performance that have been studied using the tools of fracture toughness: pressroom runnability, bag drop performance, and crease cracking performance. These situations all depend on the properties of dry paper or board. Of course, there has also been extensive study of wet web runnability. Because the sheet is fully ductile at the wet end of the paper machine, LEFM cannot be used to make predictions of failure. Instead, the literature has focused on wet web strength, wet web stretch, and drainage characteristics of the pulp. This topic is beyond the scope of the present review.

A. Runnability of Newsprint

In a typical North American pressroom, two or three rolls out of every 100 may fracture in the press. In large pressrooms, this can translate to a number of breaks each night, with each break resulting in lost production time and wasted material. Because of tight production schedules, such failures can cause delivery delays, and ultimately circulation and advertising can suffer as a result. Pressrooms therefore care about the runnability of their newsprint, and they pass this concern on to their suppliers. Suppliers respond by modifying their furnish and papermaking conditions to produce stronger or tougher sheets. These modifications inevitably add to the cost base of the newsprint, and consequently suppliers are extremely interested in having a good means of predicting the runnability of their sheets so they can keep their costs to a minimum. The quest for a means of predicting newsprint runnability has been, in fact, one of the main driving forces for the study of fracture over the years. A comprehensive review of the runnability literature was published in 1995 by Farrell and McDonald [76]. A guide to troubleshooting runnability problems was published by Roisum [77].

Because it is known that fracture mechanics controls failure in laboratory tests of a variety of grades of paper, it is logical to assume that it will control failure in a press. In order to understand such failures, recall the central fracture criterion of linear elastic fracture mechanics:

$$\sigma\sqrt{\pi a} = K_c \tag{29}$$

This equation suggests that three factors control the failure of newsprint in the printing press:

- K_c, the fracture toughness. LEFM should be adequate for MD newsprint, so K_c is a suitable parameter for characterizing toughness.
- a, the flaw size. A variety of defects may be present in newsprint.
- σ, the applied stress. In a printing operation, this is a function of the web tension.

Fracture Toughness Fracture toughness has already been discussed in great detail. For MD newsprint, essential work of fracture tests show that there is little or no nonessential work of fracture, and therefore all plastic deformation is confined to a small region around the crack tip [44]. Seth and Page's original work on LEFM [26] also showed that for newsprint the K_c results were geometry-independent. On this basis, a simple LEFM fracture toughness test should be adequate to characterize MD newsprint.

Fracture toughness cannot be used to predict failure from round defects such as slime holes, because the stress distribution in front of a hole cannot be expressed in terms of a stress intensity factor. Recently, Kortschot and Trakas [78] adapted a simple fracture criterion from the fiber composites literature for characterizing the resistance of a web to fracture from a preexisting round hole. The method is based on the concept that the sheet will fail when the stress at some characteristic distance away from the crack tip reaches the unnotched tensile strength of the material. The method has been widely used to predict the failure of composite materials and can be used to predict the influence of large holes using tests of relatively small specimens.

Flaw Size A wide variety of flaws may be found in newsprint: pinholes, shives, slime holes, calender cuts, crepe wrinkles, etc. A general discussion of defects is presented by Snider [79]. A number of older studies identified shives, or thick fiber bundles, as the principal cause of newsprint failure during conversion. Shives are typically bonded quite poorly to the surrounding network and can debond completely at low stress, so they do act as cracks. In 1965 Sears et al. [80] studied paper breaks in a specially designed web strainer that imposed gradually increasing strain on the web as it passed through a series of rolls at high speed. They found that the web typically failed at strains as low as 20% of the measured stretch. All but five of the 3200 breaks they observed could reasonably be attributed to some type of defect:

21 were due to slime spots or holes
14 seemed to be due to hairs or hair-induced cuts
2 were caused by preexisting wrinkles
3 appeared to be due to roll damage
3155 were attributed to shives

The effect of shives was most significant for shives near or on the edge of the sheet. Note that LEFM tells us that a crack at the edge of the sheet produces the same stress intensity as a crack approximately twice as long in the center of the sheet (see Appendix). Further analysis by Macmillan et al. [81] identified a potentially harmful shive as one "greater than 3.5 mm long, greater than 0.12 mm wide, and approaching half the thickness of the paper." Recent theoretical work by Ferahi et al. [30] confirmed the importance of small cracks such as those that would result from a debonded shive and provided quantitative results of the effects of crack orientation and thickness on the stress intensity factor. The data in Fig. 23 demonstrate that a reduction in the thickness of shives would be far more effective than a reduction in the length of the shives.

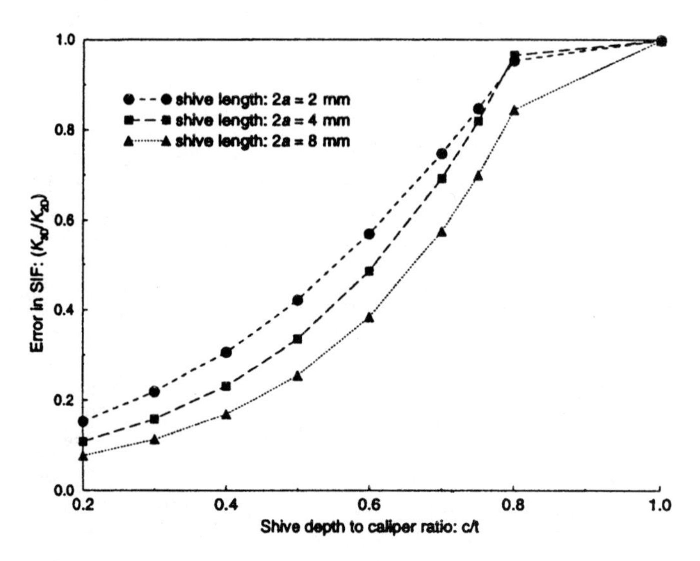

Fig. 23 The effect of shive depth on the stress intensity factor. Results are plotted as a ratio between the true value of K and the erroneous value obtained by assuming that the shive acts as a through-thickness crack. (From Ref. 30 with permission of PAPTAC.)

Two points are worth noting. First, as pulping, screening, and cleaning methods improve, the relative importance of shives is decreasing, and shives may no longer be important for many furnishes. Adams and Westlund [82] in a 1982 study similar to that performed by Sears et al., found that about 70% of breaks were caused by shives in a web strainer. Second, for typical press tension and fracture toughness values, a quick calculation shows that a crack of 25–50 mm would be needed to initiate fracture. This suggests that pressroom failures are often caused by a combination of a large flaw and a tension surge.

Low grammage in the vicinity of the shive might also contribute to increased tension and early failure. There have been a number of recent studies of the influence of formation on the strain distribution in paper [83–85]. Low grammage areas concentrate the strain and serve as failure initiation points.

Web Tension It is obvious that excessive tension in the printing press will lead to failure of the web. Typical web tensions in the press are comparable to the wound-in tensions in the roll and range from 0.09 to 0.26 kN/m [76]. The strength of a typical 48 grams per square meter (gsm) newsprint is on the order of 1.5–2.8 kN/m, and therefore under normal conditions the paper should never break. It is known that there are substantial CD and MD web tension fluctuations in the press, so much of the research has focused on these. It is important to note, however, that even for extreme fluctuations of tension, the expected break frequency obtained by comparing press tension to tensile strength is negligible. In an example given by Phillips [86], a press with a tension of 0.5 kN/m and a standard deviation of 0.25 kN/m running newsprint with a strength of 2.5 kN/m and a standard deviation of 0.25 kN/m would result in only one break in 120 million tonnes of paper. This suggests that defects must also be present for newsprint to fail in the press.

Machine and cross-machine direction web tension have been studied extensively by the printing community [e.g., 87–90]. In these studies, web tension is measured directly using a variety of techniques [91], and plots of MD and CD variations are produced. The studies deal with a range of topics, from practical situations such as tension variation during pasting to very sophisticated treatments of paper vibration.

Perhaps surprisingly, papermakers have substantial control over the stresses eventually experienced by the sheets they produce. Cross-machine direction roll structure and roll tension profiles affect runnability because the draws on the machine must be adjusted to prevent excessive flutter in the slackest part of the sheet. If the sheet is drier at the edges, shrinkage will have occurred, which together with the higher modulus produces greater web tension at the edge. Because fractures generally initiate at flaws on the edge of the sheet, this is an important point. It was once a common practice in some pressrooms to use a fine water mist on the roll edges to increase the moisture content and reduce the modulus in these regions.

Machine direction tension variations consist of a number of disturbances with different frequencies [89]. Out-of-round elements on the press create high frequency disturbances of relatively low amplitude. Out-of-round rolls created by uneven winding or handling problems can also cause these disturbances [76]. Larger tension fluctuations occur during roll changes [88,89] and speed changes [87]. Normal variations in wound-in tension through the thickness of the roll also affect the location of breaks [92].

It is generally agreed that tension fluctuations, and hence the maximum tension in the press, can be controlled by striving for roll and reel quality and uniformity. Moisture content, caliper, and basis weight must be consistent from roll to roll and within a roll. For some machines, particular attention must be paid to the reel edges and ends, because rolls from these regions are the most prone to fail in the press. A study by Frye [93] showed that the runnability of newsprint improved dramatically if edge rolls were not used.

Predicting Runnability There are effectively two ways to develop a laboratory test capable of predicting runnability. The first method is to devise a test that simulates, as closely as possible, the service conditions. The second method is to measure only true material properties and to deduce the correct combination of properties using fundamental physics. The application of fracture mechanics to paper is an attempt to follow this second route.

Because press breaks involve tensile failure of a nominally unnotched web, the first pass at a *simulation test* is a simple tensile test. Unfortunately, the tensile strength of paper is dependent on the volume of the sample, and there is little hope of predicting the strength of a specimen 100 km long using just five standard tensile specimens. Even lab-scale studies of strength versus volume cannot really be used, because they do not include specimens large enough to contain a representative sample of the flaws that cause failure in practice. A proper simulation of press failure must involve very large samples indeed, and for this reason a number of studies have been done with modified winders or web straining devices. Roisum [94] suggested that this is the only reliable way of predicting pressroom runnability.

The study by Sears et al. [80] was performed on a continuous web straining device with seven zones of incrementally increasing strain. This allowed the researchers to test large quantities of paper and to determine quite precisely the relationship between applied strain and fracture. A plot of the logarithm of the number of breaks per 10,000 ft versus strain was reasonably linear, so that an extrapolation to strains typical of a press provides a prediction of the expected break rate in the press. A similar study and semilog plotting technique was used by Adams and Westland [82], and their results are plotted in Fig. 24. Obviously, the extrapolation required reduces the confidence in the predicted runnability at low press tension, but the predicted break frequency of 5.32 breaks per 100 rolls was reasonable.

Lab-scale runnability tests are time-consuming and require huge amounts of paper. The resulting prediction of runnability depends on the extrapolation as illustrated in Fig. 24, and the uncertainty of this depends on the length of extrapolation. Furthermore, the conditions in a commercial press are much more complicated than those in a web strainer. Another objection to this method is that it provides little insight into the specifics of the failure process, although it does provide a means of identifying the critical flaws.

In response to these shortcomings, a number of authors have attempted to use fracture mechanics in conjunction with runnability data to come up with a predictive methodology. The first published attempt to do this was by Page and Seth in 1982 [95]. After a very labor-intensive study they were able to obtain the relatively weak correlation between measured fracture toughness (K_c) and runnability depicted in Fig. 25.

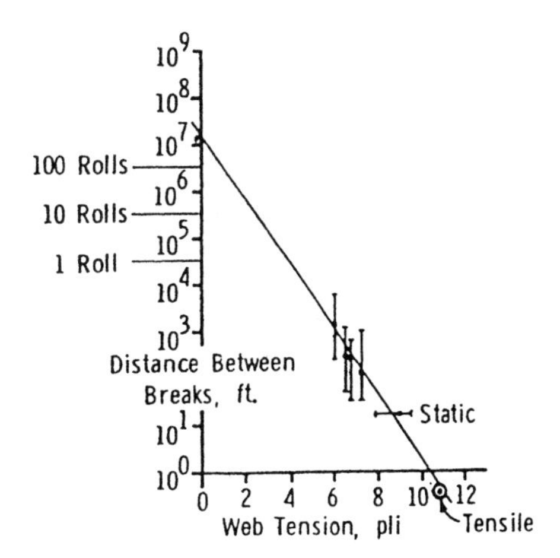

Fig. 24 Data from a lab scale runnability test. The extrapolation to typical web tensions of 2 pounds per linear inch would result in 5.32 breaks per 100 rolls. (From Ref. 82 with permission of PAPTAC.)

Page and Seth pointed out many of the pitfalls associated with this type of work. Statistical analysis showed that to distinguish between paper running at three breaks per 100 rolls and one running at 4.2 breaks per 100 rolls with any level of confidence would require a comparison of more than 1000 rolls of each kind of paper under similar conditions. Of course, within the lot of 1000 rolls, both the furnish and the humidity in the pressroom may change, introducing experimental noise. For example, their data showed that humidity had a major effect on the break frequency [95].

Weide et al. [96] published a study similar to that of Page and Seth in 1985. Their study involved three different pressrooms but was distinguished by the fact that they measured fracture toughness at a high displacement rate (5 m/s) to simulate the condition of press failures more precisely. Unfortunately, for a variety of reasons, including the logistics of running and sampling the required number of rolls in a controlled manner, they also obtained only poor correlation of fracture toughness to runnability.

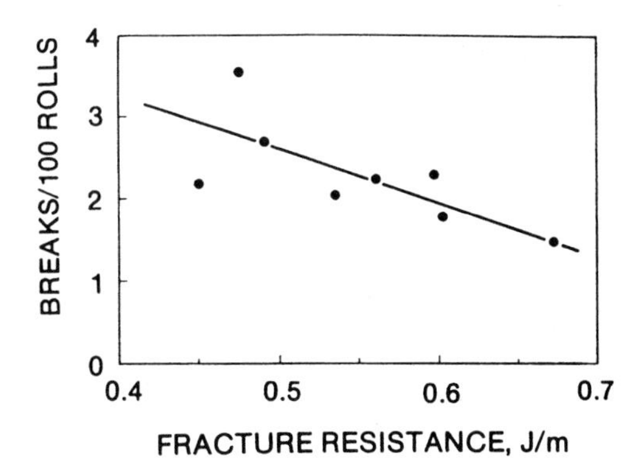

Fig. 25 Plot of runnability versus fracture toughness for papers from different suppliers. (From Ref. 95.)

One problem with these two studies is that they considered only one of the three factors identified in Eq. (29), K_c. The crack size and instantaneous press tension at failure were not measured. There may be a quite complex relationship between K_c, press tension, and flaw size, because each of these is related to the furnish and processing conditions on the paper machine. Although the press tension might have been relatively constant during the studies, small fluctuations in the moisture content at the edge of the sheet would lead to changes in the tension in this critical region and therefore a variation in the applied stress intensity factor. Without an associated program of flaw size sampling and tension monitoring, it is doubtful that a study of runnability based on measurements of toughness alone can succeed in predicting the relative performance of sheets from different mills.

Swinehart and Broek [28] modeled coater breaks by considering both the strength of the sheet and the flaw size population. They collected hole size data, measured the tensile strength (which was correlated to toughness for the furnishes studied), and made predictions of the length between breaks. They had to assume constant web tension but obtained good correlation between runnability and tensile strength. The correlation was slightly improved when they accounted for the hole size data in the prediction.

Of course, if a single mill that is adding variable amounts of kraft to an otherwise constant newsprint furnish plots the resulting runnability versus fracture toughness, tear, or some other strength property, a good correlation will probably result. This is far less satisfying than a truly universal predictive model. It is quite possible that the infrequency of breaks in a press leads to such enormous logistical problems that a valid study can never be done. Nevertheless, the basic form of Eq. (29) still provides for some guidance as to the relative merit of various furnish and process conditions when improvements in runnability are sought.

B. Bag Drop

Sanborn and Diaz [65] used the simple TEA measurement to characterize the "toughness" of sack kraft materials. Good correlation with the number of bag drops to failure was observed, even though this property was not well correlated to other strength measures such as tensile strength and tear. The number of bag drops to failure is, in fact, a *simulation test* but can be thought of in this case as being representative of the performance of the product in service.

C. Crease Cracking/Die Cutting

Fellers et al. conducted a study of the cracking during die cutting of corrugated board. In this case, the die blade creates a slit but also acts as a wedge, creating a stress intensity at the tip of the slit that can cause a ragged crack to extend beyond the edge of the die. They compared the total length of cracking created under fixed conditions to the values of J_c, K_c, and other conventional tests [97]. They obtained good correlation between the value of J_c and the total length of cracks caused by die cutting, whereas there was virtually no correlation when K_c was plotted (see Fig. 26). Because kraft linerboard is generally more ductile than newsprint, J_c would be

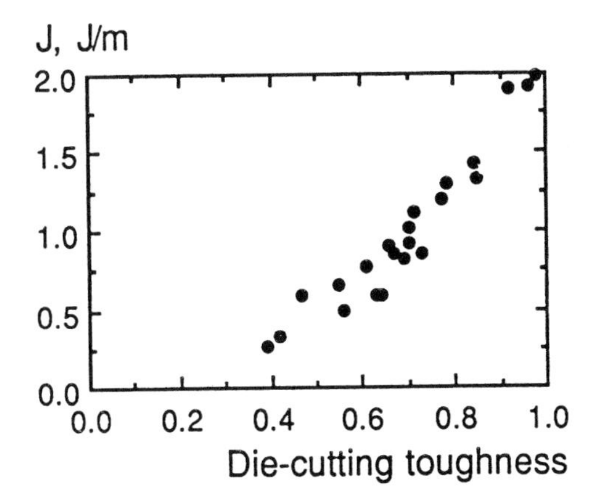

Fig. 26 Plot of J_c versus die cutting toughness. (From Ref. 97 with permission of PAPTAC.)

expected to be the better measure of toughness. In addition, fracture is not proceeding under constant load conditions in this case.

VI. CONCLUSIONS

The study of the fracture mechanics of paper has been quite intensive considering the limited experimental evidence tying this property to the performance of paper in converting and end use conditions. Although a number of methods for characterizing the resistance of a sheet to fracture have been developed, there are relatively few demonstrated applications of the fracture mechanics methodology to paper in practice. Nevertheless, the correctness of the fracture mechanics approach is amply demonstrated by the geometric independence of fracture toughness measured in laboratory studies, and hence the basic concepts can be used to formulate strategies to improve performance. Provided that there is a limited amount of plasticity during crack growth, the central equation of LEFM can be used for such guidance:

$$\sigma\sqrt{\pi a} = K_c$$

This equation [presented earlier in Eq. (29)] suggests that for a sheet containing a flaw, three approaches could be used to control the failure of a paper or board in service.

- K_c fracture toughness. This property is controlled by the microstructure of the sheet: the furnish, the fiber flexibility, the degree of bonding, etc. One method of reducing fracture is to modify the microstructure (e.g., by adding a reinforcing pulp) so as to increase this property. Care must be taken, however, that such modifications do not inadvertently increase the values of a or σ, the latter being affected by E in some converting operations. There have been many studies of the effect of microstructure on toughness, but they are beyond the scope of this review.
- a, flaw size. Fracture can also be suppressed by ensuring that the preexisting flaws in the sheet are minimized. Screening and cleaning the furnish is one method of reducing the number and size of flaws in the sheet.

- ●●● σ, the applied stress. Clearly, minimizing the applied stress can minimize fracture. Although the service loads may be beyond the control of the papermaker, reducing variability in the sheet can nevertheless minimize local stress.

Over the past two and a half decades, paper physicists have expended a great deal of effort to find the "correct" measurement of fracture toughness for paper and board. For many sheets displaying limited ductility, an approach based on LEFM as expressed by Eq. (29) is probably sufficient. For more ductile sheets, both the EWF and J-integral measurements have been used. If a simple ranking of paper performance potential from day to day is all that is required, it is enough to measure a property that is correlated to fracture toughness. However, if an actual prediction of failure is to be made, then nothing less than the exact fracture criterion and associated property measurement will do. Even so, when ranking sheets using an energy-based criterion G_c, J_c, or EWF, the user must keep in mind that the failure stress is related to the product of these quantities and the elastic modulus. The user must also keep in mind that the applied stress that the sheet experiences in service may be a complex function of material properties. Because of this, fracture mechanics is best suited to provide the general guidance described by the three points given above rather than specific predictions of sheet failure in complex service environments such as presses.

ACKNOWLEDGMENTS

I thank Dr. W. Ding and Mr. N. Woon-Fat for assistance in preparing the manuscript. Substantial financial support for research in paper physics was provided by the National Science and Engineering Research Council of Canada and the Mechanical Wood Pulps Network.

APPENDIX K_I SOLUTIONS FOR SIMPLE SPECIMEN CONFIGURATIONS[a,b]

Specimen configuration	Expression for Y in $K_I = Y\sigma_\infty\sqrt{\pi a}$
Center notch (length $2a$), infinite width	$Y = 1$
Center notch (length $2a$), finite width (width W)	$Y = \left(\sec\frac{\pi a}{W}\right)^{1/2}$
Single edge notch (length a), infinite width	$Y = 1.12$
Two symmetrical edge notches (each of length a), infinite width	$Y = 1.12$
Two symmetrical edge notches (each of length a), finite width (width W)	$Y = 1.12 + 0.43\frac{a}{W} - 4.79\left(\frac{a}{W}\right)^2 + 15.46\left(\frac{a}{W}\right)^3$

[a] σ_∞ is the tensile stress far from the crack.
[b] For more complex specimen configurations, see Ref. 24.
Source: Refs. 1, 8, and 19.

REFERENCES

1. Knott, J. F. (1973). *Fundamentals of Fracture Mechanics*. Butterworth, London.
2. Griffith, A. A. (1920). The phenomenon of rupture and flow in solids. *Phil. Trans. Ser. A 221*:163–198.
3. Ashby, M. F., and Jones, D. (1980). *Engineering Materials I*. Pergamon, Oxford, UK.
4. Seth, R. S., and Page, D. H. (1975). Fracture resistance: A failure criterion for paper. *Tappi J. 58*(9):112–117.
5. Davison, R. W. (1972). The weak link in paper dry strength. *Tappi J. 55*:567–573.
6. Weibull, W. (1951). A statistical distribution function of wide applicability. *J. Appl. Mech. 18*:293.
7. Creyke, W. E. C., Sainsbury, I. E. J., and Morrell, R. (1982). *Design with Non-Ductile Materials*. Applied Science, London.
8. Broek, D. (1986). *Elementary Engineering Fracture Mechanics*. 4th rev. ed., Kluwer, Dordrecht.
9. Broek, D. (1989). *The Practical Use of Fracture Mechanics*. Kluwer, Dordrecht.
10. Anderson, T. L. (1991). *Fracture Mechanics Fundamentals and Applications*. CRC Press, Boca Raton, FL.
11. Hellan, K. (1984). *Introduction to Fracture Mechanics*. McGraw-Hill, New York.
12. Williams, J. G. (1984). *Fracture Mechanics of Polymers*. Ellis Horwood/John Wiley, Chichester, UK.
13. Helle, T. (1987). Fracture mechanics of paper. In: *Design Criteria for Paper Performance, 1984 Seminar on Progress in Paper Physics*. P. Kolseth, C. Fellers, L. Salmen, and M. Rigdahl, eds. STFI, Sweden, pp. 93–125.
14. Yuhara, T., and Kortschot, M. T. (1993). The *J*-integral as a parameter for characterizing the fracture toughness of paper. In: *Products of Papermaking*. Trans. Tenth Fundamental Res. Symp., Vol. 2. C. F. Baker, ed. pp. 783–807.
15. Pouyet, J., and Voloizinskis, X., Poustis, J., and Lataillade, J. L. (1989). Investigation of the fracture toughness of paper. Third Joint ASCE/ASME Mechanics Conference, La Jolla, CA, pp. 133–140.
16. Niskannen, K. (1993). Strength and fracture of paper. In: *Products of Papermaking*. Trans. Tenth Fundamental Res. Symp., Vol. 2. C. F. Baker, ed. pp. 641–727.
17. Barsom, J. M., ed. (1987). *Fracture Mechanics Retrospective: Early Classic Papers (1913–1965)*. ASTM, Philadelphia.
18. Ashby, M. F. (1992). *Materials Selections in Mechanical Design*. Pergamon, Oxford.
19. Inglis, C. E. (1913). Stresses in a plate due to the presence of cracks and sharp corners. *Trans. Inst. Naval Architects 3*:219–230.
20. Peterson, R. E. (1953). *Stress Concentration Design Factors*. Wiley, New York.
21. Irwin, G. R. (1948). Fracture dynamics. *Trans. ASM 40*:147–166.
22. Swinehart, D., and Broek, D. (1995). Tenacity and fracture toughness of paper and board. *JPPS 21*(11):J389–J397.
23. Westergaard, H. M. (1939). Bearing pressures and cracks. *J. Appl. Mech. 6*:49–53.
24. Rooke, D. P., and Cartwright, D. J. (1976). *Compendium of Stress Intensity Factors*. Her Majesty's Stationery Office, London.
25. Seth, R. S. (1979). Measurement of fracture resistance of paper. *Tappi J 62*(7):92–95.
26. Seth, R. S., and Page, D. H. (1974). Fracture resistance of paper. *J. Mater. Sci. 9*:1745–1753.
27. Fellers, C., and Steadman, R. (1984). Fracture resistance of paper evaluated by a double cantilever beam. *Progress in Paper Physics*, STFI, pp. 143–151.
28. Swinehart, D. E., and Broek, D. (1996). Tenacity©, fracture mechanics, and unknown coater web breaks. *Tappi J. 79*(2):235–237.

29. Eriksson, L., and Viglund, J. (1984). Fracture properties and web break frequency of newsprint. *Progress in Paper Physics*. STFI, pp. 140–142.
30. Ferahi, M., Kortschot, M. T., and Dodson, C. T. J. (1996). Effect of anistropy on the fracture behavior of newsprint. *J. Pulp Paper Sci. 22*(11):J439–J445.
31. Cotrell, A. H. (1961). Iron and Steel Inst. Spec. Rep. 69.
32. Wells, A. A. (1961). Unstable crack propagation in metals: Cleavage and fast fracture. Proceedings of the Crack Propagation Symposium, Cranfield, U.K. 1:210.
33. Tanaka, A., and Yamauchi, T. (1997). Crack propagation of paper under fracture toughness testing. *J. Packag. Sci. Technol. 6*(6):324–332.
34. Rice, J. R. (1968). A path independent integral and the approximate analysis of strain concentration by notches and cracks. *J. Appl. Mech. 35*:379–386.
35. Begley, J. A., and Landes, J. D. (1972). The *J*-integral as a fracture criterion. ASTM STP 514. American Society for Testing and Materials, Philadelphia, pp. 1–20.
36. Rice, J. R., Paris, P. C., and Merkle, J. G. (1973). Some further results of *J*-integral analysis and estimates. ASTM STP 536. American Society for Testing and Materials, Philadelphia, pp. 231–245.
37. Yuhara, T., and Kortschot, M. T. (1993). A simplified determination of the *J*-integral for paper. *J. Mater. Sci. 28*:3571–3580.
38. Liebowitz, H., and Eftis, J. (1971). On non-linear effects in fracture mechanics. *Eng. Fract. Mech. 3*:267–281.
39. Westerlind, B. S., Carlsson, L. A., and Andersson, Y. M. (1991). Fracture toughness of liner board evaluated by the *J*-integral. *J. Mater. Sci. 26*:2630–2636.
40. Wellmar, P., Fellers, C., Nilsson, F., and Delhage, L. (1997). Crack-tip characterization in paper. *JPPS 23*(6):J269–J276.
41. Broberg, K. B. (1968). *Int. J. Fracture 4*:11–19.
42. Mai, Y.-W., and Cotterell, B. (1986). On the essential work of ductile fracture in polymers. *Int. J. Fract. 32*:105–125.
43. Mai, Y.-W., and Cotterell, B. (1985). "Effect of specimen geometry on the essential work of plane stress ductile fracture," *Eng. Fract. Mech. 21*(1), pp. 123–128.
44. Seth, R. S., Robertson, A. G., Mai, Y.-W., and Hoffmann, J. D. (1993). Plane stress fracture toughness of paper. *Tappi J 76*(2):109–116.
45. Seth, R. S. (1995). Measurement of in-plane fracture toughness of paper. *Tappi J 78*(10):177–183.
46. Mai, Y.-W., He, H. Leung, R., and Seth, R. S. (1995). In-plane fracture toughness measurement of paper. ASTM STP 1256. American Society for Testing and Materials, Philadelphia, pp. 587–599.
47. Mai. Y.-W., and Powell, P. (1991). Essential work of fracture and *J*-integral measurements for ductile polymers. *J. Polym. Phys. B: Polym. Phys. 29*:785–793.
48. Karenlampi, P., Cichoracki, T., Alava, M., Pylkko, J., Paulapuro, H., and Pylkko, J. (1998). A comparison of two test methods for estimating the fracture energy of paper. *Tappi J 81*(3):154–160.
49. Balodis, V. (1963). The structure and properties of paper. XV. Fracture energy. *Aust. J. Appl. Sci. 14*:284–304.
50. Uesaka, T., Okaniwa, H., Murakemi, K., and Imamura, R. (1979). Tearing resistance of paper and its characterization. *Japan Tappi 33*:403–409.
51. Seth, R. S., and Blinco, K. M. (1990). Comparison of Brecht-Imset and Elmendorf tear strengths. *Tappi J 73*(1):139–142.
52. Wink, W. A., and Van Eperen, R. H. (1963). Does the Elmendorf tester measure tearing strength? *Tappi J 46*(5):323–325.
53. Lyne, M. B., Jackson, M., and Ranger, A. E. (1972). The in-plane, Elmendorf, and edge tear strength properties of mixed furnish papers. *Tappi J 55*(6):924–932.

54. Trakas, K., and Kortschot, M. T. (1997). The relationship between critical strain energy release rate and fracture mode in multidirectional carbon-fibre/epoxy laminates. ASTM STP 1285, pp. 283–304.
55. Sharif, F., Kortschot, M. T., and Martin, R. H. (1995). Mode III delamination using a split cantilever beam. ASTM STP 1230, pp. 85–99.
56. Trakas, K. (1993). An investigation of bonding effects on paper fracture in the tear and tensile modes. B.A.Sc. Thesis, Dept. of Chemical Engineering and Applied Chemistry, University of Toronto.
57. Kane, M. W. (1960). Beating, fibre length distribution and tear. *Pulp Paper Mag. Can. March*:236.
58. Shallhorn, P., and Karnis, A. (1979). Tear and tensile strength of mechanical pulps. *Trans. Tech. Sect. CPPA* 5(4):TR92–TR99.
59. Yan, N. and Kortschot, M. T. (1997). Micromechanical modeling of tear strength in kraft and TMP papers. *Proceedings of the 11th Fundamental Research Symposium*, Cambridge, U.K., Sept. 22–26, Vol. 2, pp. 1249–1271.
60. Johnson, R. E. (1995). A simplified shear-lag model of paper as a "long" fibre reinforced material. *Proc. 49th Appita Annual General Conference*, pp. 405–409.
61. Page, D. H. (1992). A note on the mechanism of tearing strength. *Tappi J 77*(3):172–174.
62. Kazi, S. M., and Kortschot, M. T. (1996). The fracture toughness of TMP newsprint reinforced with kraft pulp. *Tappi J. 79*(5):197–202.
63. T414 om-88. *Tappi Test Methods, 1994–1995*. TAPPI Press, Atlanta.
64. Van den Akker, J. A., Wink, W. A., and Van Eperen, R. H. (1967). Instrumentation studies. LXXXIX. Tearing strength of paper. III. Tearing resistance by the in-plane mode of tear. *Tappi J 50*(9):466–470.
65. Sanborn, I. B., and Diaz, R. J. (1959). The stress–strain toughness test for paper and paperboard. *Tappi J 42*(7):588–597.
66. Corte, H., Schaschek, H., and Broens, O. (1957). The rupture energy of paper and its dependence on the stress-time characteristics during drying. *Tappi J 40*(6):441–447.
67. Goldschmidt, J., and Wahren, D. (1968). On the rupture mechanism of paper. *Svensk Papperstidn. 71*:477–481.
68. Seth, R. S. (1979). On the work of fracture in paper. *Tappi J 62*(3):105–106.
69. Andersson, O., and Falk, O. (1966). Spontaneous crack formation in paper. *Svensk Papperstidn 69*:91–99.
70. Gurney, C., and Hunt, J. (1967). Quasi-static crack propagation. *Proc. Roy. Soc. (Lond.) A299*:508–524.
71. Seth, R. S., and Page, D. H. (1975). Reply to "Comment on 'fracture resistance of paper'." *J. Mater. Sci. Lett. 10*:1273–1274.
72. Uesaka, T. (1983). Specimen design for mechanical testing of paper and paperboard. In: *Handbook of Physical and Mechanical Testing of Paper and Paperboard*, Vol. 1. R. E. Mark, ed. Marcel Dekker, New York.
73. Steadman, R., and Fellers, C. (1986). Measuring the fracture strength of tough papers. 1986 Progress in Paper Physics Seminar, Minnowbrook, NY.
74. Mai, Y.-W., He, H., Leung, R., and Seth, R. S. (1995). In-plane fracture toughness measurement of paper. In: *Fracture Mechanics*, Vol. 26. ASTM STP 1256. W. G. Reuter, J. G. Underwood, and J. C. Newman Jr. eds. ASTM, Philadelphia, pp. 587–599.
75. Tanaka, A., Otsuka, Y., and Yamauchi, T. (1997). In-plane fracture toughness testing of paper using thermography, *Tappi J., 80*(5):222–226.
76. Farrel, W. R., and McDonald, J. D. (1995). A papermaker's view of newsprint runnability: A literature review of the problem of pressroom breaks. Paprican Misc. Rep. MR 318. Paprican, Montreal, July.

77. Roisum, D. R. (1990). Runnability of paper, Part 2: Troubleshooting web breaks. *Tappi J 73*(2):101–106.
78. Trakas, K., and Kortschot, M. T. (1998). Predicting the strength of paper containing holes or cracks with the point stress criterion. *Tappi J 81*(1):254–259.
79. Snider, E. H. (1979). Why does the web ever break? *Pulp Paper Can. 80*(8):T238–T243.
80. Sears, G. R., Tyler, R. F., and Denzer, C. W. (1965). Shives in newsprint: The role of shives in paper web breaks. *Pulp Paper Mag. Can. 66*(7):T351–T360.
81. MacMillan, F. A., Farrell, W. R., and Booth, K. G. (1965). Shives in newsprint: Their detection, measurement and effects on paper quality. *Pulp Paper Mag. Can 66*(7):T361–T369.
82. Adams, R. J., and Westlund, K. B. (1982). Off-line testing for newsprint runnability. 1982 International Printing and Graphic Arts Conference, Quebec City, Sept. 28–Oct. 1, pp. 13–18.
83. Wong, L., Kortschot, M. T., and Dodson, C. T. J. (1996). Effect of formation on local strain fields and fracture of paper. *J. Pulp Paper Sci. 22*(6):J213–J219.
84. Thorpe, J. L. (1982). Simulation of tensile properties of newsprint. *Appita 36*:198–204.
85. Korteoja, M. J., Lukkarinen, A., Kaski, K., Gunderson, D. E., Dahlke, J. L., and Niskanen, K. J. (1996). Local strain fields in paper. *Tappi J 79*(4): 217–223.
86. Phillips, B. (1994). Defining the right performance criteria for newsprint, *Paper Tech. 35*(6):28–33.
87. Larsson, L. O. (1984). What happens in the press to cause web breaks? *Pulp Paper Can. 85*(9):T249–T253.
88. Eriksson, L. (1991). Measurement and control of the tension across the web in a newspaper printing press. First Int. Conf. on Web Handling, Oklahoma.
89. Linna, H., and Lindqvist, U. (1988). Web tension in a paper machine and runnability characteristics of web presses. *Pulp Paper Can. 89*(11):51–56.
90. Hellentin, P., Eriksson, L. G., Johnson, P., and Kilmister, G. T. F. (1993). Web tension profiles as measured in a rewinder and in a printing press. 2nd Int. Conf. on Web Handling, Oklahoma, June 6–9.
91. Linna, H., and Moilanen, P. (1988). Comparison of methods for measuring web tension. *Tappi J 71*(10):134–138.
92. Eriksson, L. (1987). What happens to a paper roll in the printing plant? *Advances and Trends in Winding Tech.*, Proc.1st Int. Conf. on Winding Tech., Stockholm, pp. 195–212.
93. Frye, K. G. (1994). Pressroom to mill feedback for problem solving and product development. Proc. Tappi Papermakers Conference, pp. 589–601.
94. Roisum, D. R. (1990). Runnability of paper, Part I: Predicting runnability. *Tappi J 73*(1):97–101.
95. Page, D. H., and Seth, R. S. (1982). The problem of pressroom runnability. *Tappi J 65*(8):92–95.
96. Weide, W., Ramaz, A., and Poujade, J. L. (1985). The fracture resistance of newsprint. 18th IARIGAI, June 2–8, Williamsburg, VA.
97. Fellers, C., Fredlund, M., and Wagberg, P. (1991). Die-cutting toughness and cracking of corrugated board. Proc. 1991 International Paper Physics Conference, Kona, HI, pp. 203–210.

9

EDGEWISE COMPRESSION STRENGTH OF PAPER

CHRISTER FELLERS
STFI, Swedish Pulp and Paper Research Institute
Stockholm, Sweden

BENJAMIN C. DONNER
Weyerhaeuser Pulp, Paper, Packaging R&D
Tacoma, Washington

I. INTRODUCTION

One fundamental end use property of a carton or corrugated box is its ability to protect the contents against compression forces during packing, storage, and distribution. A numerical link has been established between the properties of the material and the stacking strength of the final package. It is generally accepted that the two most important material properties for carton board and corrugated board are the bending stiffness and the compression strength [49,51]. These properties of the corrugated board can further be related to the properties of its individual plies, that is, the linerboard faces and the corrugated medium [34,35,39,40,44].

Two principally different uses for an edgewise compression test of single paper sheets can be visualized. The first use is for quality control in the mill and for quality acceptance tests in the box plant. In addition to being precise and correlative to the performance of the carton or of the corrugated container, a quality control test should be simple, rapid, and reproducible over a long period of time. The second use is as a research and development tool, where a test method is needed to establish design criteria and to permit optimization of carton board and corrugated board with regard to their constituent components, which means that the method should be based on fundamental principles of mechanics.

The theoretical considerations and practical experiments reported in this chapter compare the various principles and test methods in use, which in fact differ significantly. Different measures of compression strength can differ by 30% or more. Understanding the origins of these discrepancies has shed light on shortcomings of the various tests and has improved understanding of the meaning of compression strength itself.

The ideal test for purposes of this chapter measures compression strength as a material property while being simple and rapid to use. In the tests described, there is a trade-off between simplicity and accuracy. Quality assurance tests can deviate from the ideal, emphasizing speed and simplicity, whereas laboratory tests can be more rigorous at the expense of preparation time and difficulty. In the end, choosing a suitable test for a given purpose will benefit form coming to understand the *meaning* of compression strength in paper, understanding the artifacts arising during testing, and rationalizing the differences among the various test methods.

Principal sources of differences among the test methods are

- Sample size (weak link concept)
- Strain rate (including out-of-plane shear effects)
- Structural (not material) contribution to failure
- Sample preparation
- Sample shape (cylindrical vs. planar; aspect)
- Sample surface roughness

These influences on test results puts excessive demands on the precision of the testing equipment, on specimen preparation, and on the shape, size, and surface evenness of the specimen itself. For example, the relatively low out-of-plane shear stiffness of paper compared to the in-plane stiffness creates difficulties for the experimentalist trying to grip the material, leading to test artifacts that can make compression strength measurements misleading or erroneous.

A big portion of the strength difference among test methods is actually due to difference in sample size caused by the local variation in basis weight and fiber orientation and the weak-link effect. A large sample, failing at the weakest point, will always be lower in strength than a small sample. An equation governing the relationship between compression strength of small and large samples is provided later as Eq. (18).

Demonstrating that the strength differences measured by different test methods are traceable to specific causes is more than an academic exercise. These identical influences arise when paper and paperboard are put into service as structural materials. Strength of a material in service will almost always be lower than the "material strength" measured in the laboratory, traceable to the same specific causes that control the strength differences among test methods.

II. THEORETICAL CONSIDERATIONS

The basic challenge of all compression strength tests for paper is to introduce a compressive force into the plane of the sheet in a way that causes a pure in-plane compression deformation and failure (though failure itself will involve out-of-plane deformations). The compression of sheet materials in this way often leads to deformations out of plane, such as bending or buckling. Bending out of plane is a "structural" response, not a material response, and must be understood and eliminated before the material's compression strength can be determined.

Flat and cylindrical specimens are analyzed in this section, to understand the influence of specimen geometry on strength. For example, the relationship between the length/thickness ratio (slenderness) and the axial load at which buckling will occur is described, based on the mechanics of column buckling. This provides a means for evaluating the effectiveness of strip compression tests.

The cylinder is naturally more stable in compression, but the geometry of the sample must be carefully chosen. Buckling of the cylinder provides the understanding necessary to evaluate the ring crush test and other cylindrical compression tests.

Deviations from flatness in the strip or roundness of the cylinder ("imperfections") are also analyzed to determine their influence on buckling and strength. Some additional cylindrical specimen design considerations are given in Chapter 7 of this volume.

A. Flat Specimens

Perfect Columns Consider a slender bar or column with hinged ends subjected to an axial compression load P as in Fig. 1a. Buckling of slender columns was first discussed by Leonhard Euler in 1744. His works are cited in Ref. 76. A slender column has a length much greater than the cross-sectional dimensions. If the load P is less than a critical value, the column remains straight and undergoes only axial compression. This is a state of stable elastic equilibrium; that is, if a lateral force is applied and a small deflection produced, this deflection disappears when the lateral force is removed and the column becomes straight again. If P is gradually increased, a condition is reached in which the equilibrium becomes unstable. The *critical load* P_{cr} is then defined as the maximum axial load the column can resist while remaining

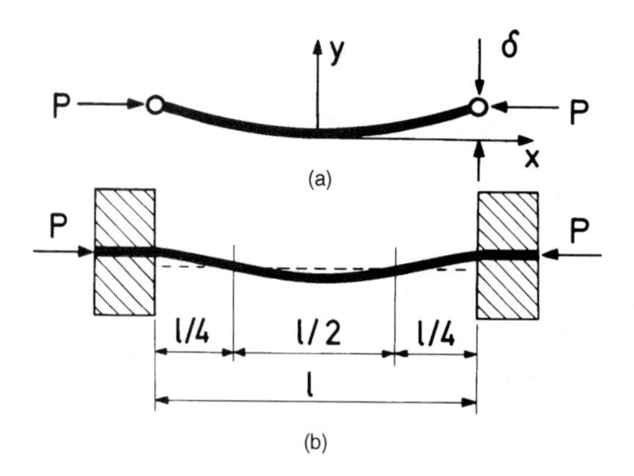

Fig. 1 (a) Column with hinged ends. (b) Column with fixed ends. P is the critical axial load to keep the column in a slightly bent shape.

straight; any increase in load results in a lateral deflection. This critical load can be calculated using the differential equation of the bent column and assuming small deflections. For a column of length ℓ, modulus E, and moment of inertia I, the smallest value of P is

$$P_{\mathrm{cr}} = \frac{\pi^2 EI}{\ell^2} \tag{1}$$

Equation (1) gives the *critical load* for the column with hinged ends (zero moment boundary condition) represented in Fig. 1a; that is, it is the axial load that represents the transition between stable and unstable loading. The energy of the buckled shape is lower than the energy of the straight, unbuckled shape when the load exceeds the critical load.

Equation (1) may be modified to also include the effect of shear [77]:

$$P_{\mathrm{cr}} = \frac{\pi^2 EI}{\ell^2}\left(\frac{1}{1 + (EI/GA)(\alpha\pi^2/\ell^2)}\right) \tag{2}$$

where $\alpha = 3/2$ for a rectangular cross section, G is the interlaminar shear modulus, and A is the cross-sectional area. It can be seen from Eq. (2) that shear always reduces the critical load.

Since A is the cross-sectional area, $r = (I/A)^{1/2}$ is the smallest radius of gyration and $\lambda = \ell/r$ is the slenderness ratio. When the slenderness ratio is large, the denominator of Eq. (2) is small, shear can be neglected, and the *critical stress* σ_{cr} in terms of the slenderness ratio is

$$\sigma_{\mathrm{cr}} = \frac{P_{\mathrm{cr}}}{A} = \frac{\pi^2 E}{\lambda^2} \tag{3}$$

Under these conditions, the only property of the material that influences the critical or elastic buckling stress is the stiffness of the material as measured by the elastic modulus E. The influence of column geometry on the buckling stress is embodied in the slenderness ratio λ.

Isotropic engineering materials such as steel have a ratio E/G of 2.6. For practical purposes, the role of shear becomes unimportant when the length/thickness ratio of the column is about 100. Computation using Eq. (2) shows the effect of shear has dropped to 0.38%. For paper, the elastic modulus is roughly 50 times the interlaminar shear modulus [16,48], making E/G about 20 times larger for a paper strip than for a steel strip. Consequently, the influence of shear doesn't disappear until the paper strip is 20 times longer than the steel strip. The behavior of a long column will be discussed below, including the role of nonideal loading and geometry; the role of shear will be discussed in a later section.

In the case of a column with rectangular cross section, the *slenderness ratio* becomes $\lambda = \sqrt{12}\ell/t \sim 3.46\ell/t$, where t is the thickness of the column. In the case of a column with fixed ends (Fig. 1b), there are reactive moments that keep the ends form rotating during buckling. There are *inflection points* where the dashed line in Fig. 1b intersects the deflection curve. At the points of inflection, the curvature of the column is zero because the bending moments at these points sum to zero. These points divide the column into three portions, where the center portion is in the same condition as the column represented in Fig. 1a. Hence, the *critical stress* for a column with fixed ends is found from Eq. (6) by substituting $\ell/2$ for ℓ, which gives

$$\sigma_{\text{cr}} = \frac{4\pi^2}{\lambda^2} \tag{4}$$

Equations (3) and (4) are called the Euler equations and are represented graphically as the long-column portion of Fig. 2b. It is important to note that the buckling curves do not represent the behavior of a single column but rather the behavior of an infinite number of ideal columns of different length. Figure 2 also shows the relationship between the compression stress–strain curve for a material when it is prevented from buckling and the column buckling curve [57]. The column buckling curve is the buckling stress for columns of different slenderness ratios.

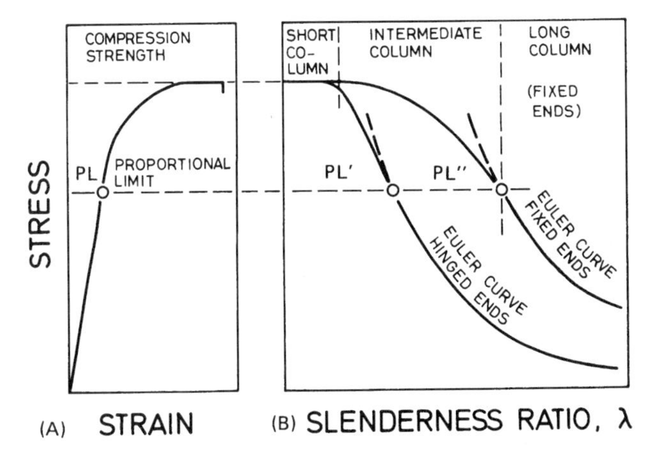

Fig. 2 Influence of inelastic behavior on buckling and failure. (a) The stress–strain curve in compression. (b) The stress at failure in compression versus the slenderness ratio. (Loosely based on Ref. 57.)

For the stress–strain curve in Fig. 2a, the material behaves elastically in the stress range from zero to the proportional limit PL. The portions below PL′ and PL″ of the Euler curves in Fig. 2b are applicable in such a case. The Euler curve beyond the elastic range is shown in the figure by dashed lines.

Columns that buckle elastically are sometimes referred to as long columns. For long columns, the load-bearing capacities are fixed ends and hinged ends are in a ratio of 4 : 1. The compression strength for intermediate columns when the material is subjected to stresses beyond the proportional limit PL in the stress–strain curve can be predicted by considering a column subjected to combined bending and compression [76]. For intermediate columns, the benefit derived from applying restraints to the ends is less than that achieved with long columns. Columns that have low slenderness ratios and exhibit no buckling phenomena are called short columns. For short columns, the curves for columns with fixed ends and those with hinged ends merge. It makes little difference whether a short column has hinged or fixed ends, because strength rather than elastic buckling determines the behavior.

Information from column buckling curves is normally used to design columns with certain safety factors. In our case it will be used in the reverse manner to find the conditions at which the true compression strength can be evaluated.

Imperfect Columns No column is perfectly straight, and applied forces are in general imperfectly centered, so the behavior of real columns is established by investigating the sensitivity to imperfections in the specimen or eccentricity in the application of load. Figure 3 represents results for structural steel with rectangular cross sections according to Timoshenko [76]. The results are based on elaborate calculations of the stress at failure as a function of the eccentricity of the applied load or the initial curvature of the specimen. The parameter in Fig. 3 is the eccentricity ratio e/t,

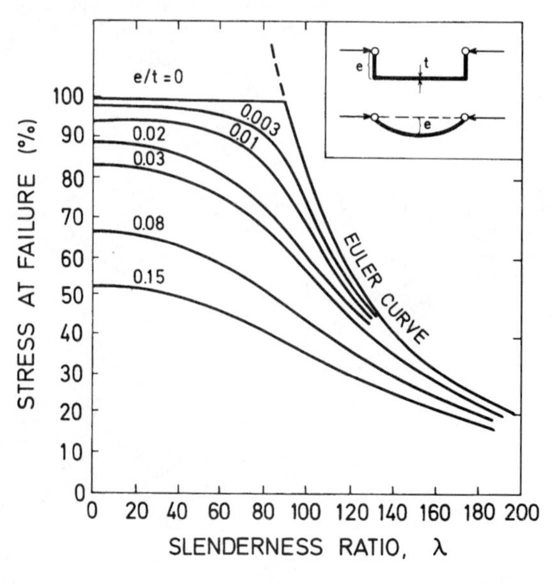

Fig. 3 Effects of eccentricity ratio e/t and slenderness ratio λ on the compressive failure stress for columns. (Loosely based on Ref. 76.)

where e is the eccentricity (defined pictorially in the inset diagram in Fig. 3) and t is the thickness, also defined in the figure.

As can be seen in Fig. 3, the effect of the eccentricity ratio on the stress at failure becomes more pronounced the lower the slenderness ratio; the true strength in compression can be evaluated only if the loading conditions and specimen straightness are close to ideal. Comparable experimental results have been obtained for unidirectional fiber-reinforced composites where imperfections in the form of bowed fibers, misaligned fibers, and unbonded fibers were found to have a significant influence on the microbuckling strength [9]. Microbuckling is the local buckling of fibers in the matrix.

Paper is indeed not a straight, homogeneous, or uniform material. An example of this nonuniformity is illustrated in Fig. 4, which shows the thickness variation along six lines in the cross direction (CD) of a $200\,g/m^2$ linerboard sheet. The integrated mean value is chosen as zero on the thickness scale, and, for clarity, except for the first (front) curve, only variations above the mean value are shown. As can be seen, the differences between peaks and valley are on the order of 5×10^{-5} m. Even though the thickness variations of paper are not in phase, which is a precondition for the calculations in Fig. 3, the testing of paper in compression requires special precautions in order to avoid buckling on different structural levels of the specimen. It is consequently not to be expected that the compression strength of such a thin and uneven material as paper can be as easily defined and evaluated as that of a smooth, straight engineering material.

B. Cylindrical Specimens

In order to make the evaluation of compression strength possible, the structural stability of the specimen must be maximized. One way to increase the stability of a material during compression is to form the sheet into a cylinder. In general, cylinder buckling theory is not as accurate as column buckling, and experimental buckling loads are frequently found to be substantially lower than calculated values. Nonetheless, the theory presented below for isotropic materials provides

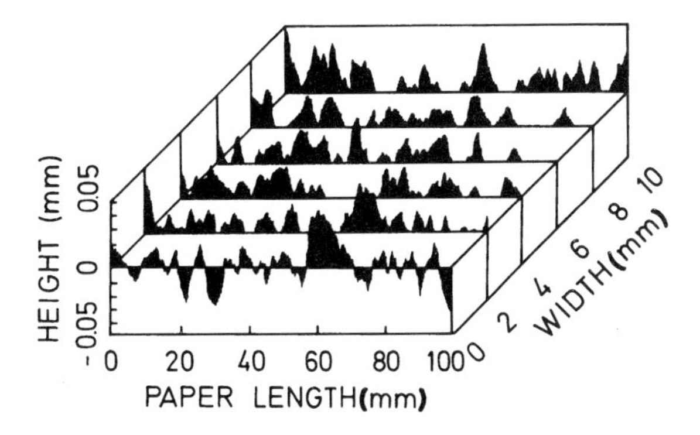

Fig. 4 Measured thickness variations of a $200\,g/m^2$ linerboard. The samples was scanned on opposing sheet surfaces simultaneously, using two spherically tipped needles. Tip diameter was 4.5 mm.

a qualitative basis for considering the effect of dimensions of the cylinder on the compression strength. The reader is also urged to consult Chapter 3 of this volume in this regard.

The following account of cylinder buckling is summarized from works by Gerard and Becker [22] and by the Institute of Paper Chemistry [29]. A circular cylinder subjected to an axial compression load may buckle in any of four modes as illustrated in Fig. 5, depending upon the height, radius, and wall thickness of the cylinder. In the following, the *buckling stress* σ_{cr} for each mode is given as a function of the elastic modulus E and physical dimensions of the cylinder, where $H =$ height, $R =$ radius, and $t =$ wall thickness and k_1, k_2, k_3 and k_4 are buckling coefficients [30].

Classical Euler Buckling may occur for very long cylinders. This is the familiar buckling of slender columns as discussed in the preceding passage; the entire cylinder bends away from its initial straight configuration:

$$\sigma_{cr} = k_1 \frac{ER^2}{H^2}, \qquad \frac{H^2}{Rt} \gg 100 \qquad \text{Euler column} \tag{5}$$

where k_1 is a function of end boundary conditions.

When height H is reduced, the cylinder will not buckle as an Euler column but will exhibit local buckling modes. The cylinder buckles into the classical long-cylinder mode with diamond-shaped buckles (reticulation) over the cylinder wall:

$$\sigma_{cr} = k_2 \frac{Et}{R}, \qquad \frac{H^2}{Rt} > 100 \qquad \text{Long cylinder} \tag{6}$$

where k_2 is a function of R/t.

A transition range is identified where the critical stress is defined by the expression

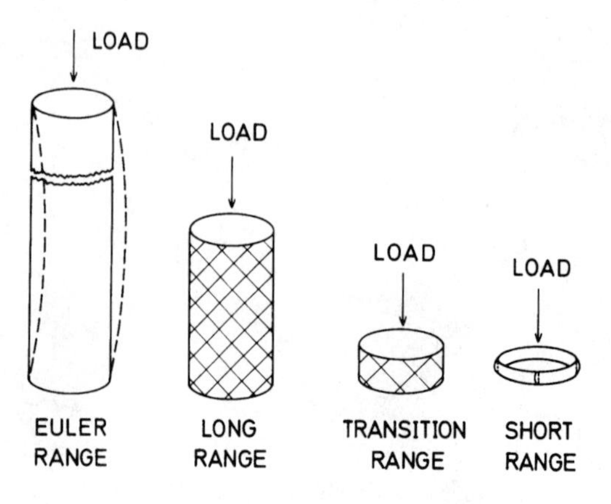

Fig. 5 The mode of buckling for a circular cylinder under axial compression depends on the height for a given radius of curvature and on sheet thickness. (From Ref. 30.)

$$\sigma_{cr} = k_3 \frac{ET^2}{H^2}, \qquad 1 < \frac{H^2}{Rt} < 100 \qquad \text{Transition range} \tag{7}$$

where k_3 is a function of H^2/Rt and end fixity.

For short cylinders, buckling appears as a simple bowing of the cylinder walls rather than as diamond-shaped buckles. The behavior of short-range cylinders is described by Eq. (8), analogous to the Euler equation, Eq. (5), for columns:

$$\sigma_{cr} = k_4 \frac{Et^2}{H^2}, \qquad \frac{H^2}{Rt} < 1 \qquad \text{Short range} \tag{8}$$

where k_4 is a function of the boundary conditions.

Following Eqs. (5)–(8), the effect of height on the stress at failure in compression of the cylinders may be shown schematically as in Fig. 6. This diagram reveals that the stress at failure essentially varies inversely as the square of the height, with the exception of the long-cylinder range, where the buckling strength is independent of height and depends instead on the radius of the cylinder. As the height of a cylinder of constant radius and thickness is progressively decreased, the stress at which buckling occurs progressively increases. Eventually a height will be found where the stress at which instability occurs is equal to the compression strength of the material. This is shown as point B in Fig. 6. With a further decrease in height, the cylinder is crushed before it buckles, and the maximum stress sustained by the cylinder remains constant at the compression strength of the material. This behavior is illustrated by the plateau A–B for the case where the crossover point lies in the so-called transition range where $1 < H^2/Rt \ll 100$. Even if the stability level is reached where the specimen is crushed before it buckles, the aforementioned strength dependence of eccentric loading and imperfect specimens still remains.

Furthermore, it should be noted that when a thin material such as paper is bent into a cylindrical shape, high stresses may occur perpendicular to the loading direction. Consider a cylinder with a wall thickness of 3×10^{-4} m and a radius of 0.024 m.

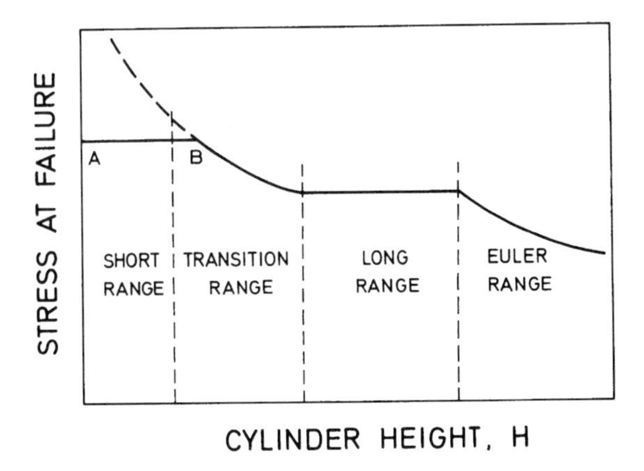

Fig. 6 Schematic representation of compressive failure stress versus cylinder height. (From Ref. 30.)

The familiar equation $\varepsilon = t/2R$ provides an approximate estimation of how the strain in the surface fibers, ε, is related to the sheet thickness, t, and radius of curvature, R [77]. With the previously mentioned thickness and radius, the maximum strain perpendicular to the loading direction is 0.6%, which is of the same order as the uniaxial strain to failure in compression of paper. This limits the maximum curvature (minimum radius), because decreasing the radius further leads to failure and weakening of the cylinder in compression.

It can be seen that thicker papers can definitely be crushed as result of being bent to a cylindrical specimen form. Consequently, there may be cases when there is conflict between stability demands and the desire to avoid failure caused by such bending of the paper. In fact, for steel, in order to avoid buckling a maximum radius-to-thickness ratio of 20 is needed, resulting in a maximum strain of 2.5% [21]. For paper and board, the optimum cylinder geometry was investigated by Uesaka. Uesaka's results are detailed in Chapter 7 of this volume.

C. Material Failure, Structural Buckling, and Bifurcation

A long steel column loaded axially as in Fig. 1 will buckle when the axial stress reaches the Euler critical stress. The assumption is that the stress remains below the proportional limit in Fig. 2b. Shorter columns will exhibit plastic deformation. The critical stress will be lower than the Euler critical stress, even though the column still deforms into the same sigmoidal shape.

The long column undergoes a shape change at the critical stress. At the critical load, in the absence of any transverse load, the column remains straight. As the load increases, the column deforms into the classic sigmoidal shape. The column does not collapse when the critical stress is reached, but the stiffness of the column is reduced, meaning that the slope of the load–deformation curve is reduced. The Euler critical stress is therefore a *bifurcation point* and represents the point at which the *shape* of the column changes as the load increases. For some engineering purposes, buckling of a column may represent failure that we can designate as a structural failure, but it is important to note that the load can still increase above the critical load without collapse of the column.

For the very short steel column, there is no bifurcation. Stress as a function of deformation arises purely from the material response. It should be clear that it is very difficult to infer material properties from a test that includes a macroscopic structural response such as column buckling. It is also worth noting that inhibiting this buckling does not eliminate changes in structure. However, the structural changes occur at the microstructural level, and the difference between a "material" response and a "structural" response is blurred.

Examination of the steel at a microstructural level reveals grain boundaries and atomic level slip plans that define the transition from elastic to plastic behavior as seen in the stress–strain response. In the same way, paper has a microstructure that can be considered in understanding the nature of and origins of failure, including material failure, structural buckling, and bifurcation. Current understanding of failure mechanisms in paper discussed in the next section arises from examining the mechanical behavior of other structured engineering materials, such as wood, polymers, and fibrous composites.

III. FAILURE MECHANISMS

A. Continuum Models and Bifurcation

The long column discussed in Section II is a continuum model. In such a model, the behavior of the material can be examined without detailed discussion or knowledge of the microstructure. In the continuum models discussed here, the analysis seeks to identify a transition in mechanical behavior analogous to the transition from elastic to plastic behavior in steel.

Perkins and McEvoy [56] used a theoretical approach to predict the critical instability load in terms of the incremental behavior of paper, using the stress–strain curve and the transverse shear modulus. Their analysis was based on work by Biot [5]. Biot identified a transition from homogeneous elastic compression to a wavy, shear-dominated compression. The transition is a *bifurcation*, meaning that he waviness develops smoothly as the stress increases, starting with zero amplitude at the bifurcation point.

The Biot analysis is for an orthotropic material. This type of material is elastic, but with elastic stiffness that depends on direction in the material. Orthotropic materials have three planes of symmetry, which simplifies analysis when loading is in one of the planes of symmetry. Perkins and McEvoy have modeled paper as an orthotropic solid, as have many others. The planes of symmetry correspond to the plane of the paper, a plane perpendicular to the plane of the sheet and in the direction of manufacture (the machine direction), and a plane perpendicular to both the plane of the sheet and the machine direction.

For an orthotropic solid column analogous to the Euler column analyzed above, the critical stress is given as a function of slenderness ratio in Fig. 7. At very low slenderness ratio (short columns), the theoretical critical stress is asymptotic to the shear modulus of the material. [This is somewhat higher than the stress predicted by Eq. (20).]

As observed by Perkins and McEvoy, this theoretical critical stress is higher than the compression strength of the material, suggesting that the material will never bifurcate into the shear-dominated mode identified by Biot. Other mechanisms have been proposed with the objective of overcoming this problem.

Uesaka [79] treated the failure as a kink band analogous to the behavior of oriented fibrous composites. Examination of compression failure zone morphology suggests that this mechanism can act in paper when the fibers are highly aligned in the direction of loading, but not in the orthogonal direction.

Habeger and Whitsitt [26] treated the influence of fibrous structure on the material stability. This brings together ideas of continuum and fibrous models of failure and yields some useful equations for predicting compression strength from ultrasonic properties.

Donner and Backer [12] treated the failure as a shear-dominated instability in a way similar to Perkins and McEvoy [56]. The Donner–Backer approach yields the same critical stress as that determined by Perkins and McEvoy and is therefore susceptible to the same criticism raised by Habeger and Whitsitt and by Uesaka. Donner and Backer examine the consequences of such a bifurcation on subsequent material response and argues that the bifurcation does take place.

Figure 8 shows a compressive stress–strain curve for a 336 g/m^2 commercial linerboard, plotted against its tensile counterpart. The initial part of the compression

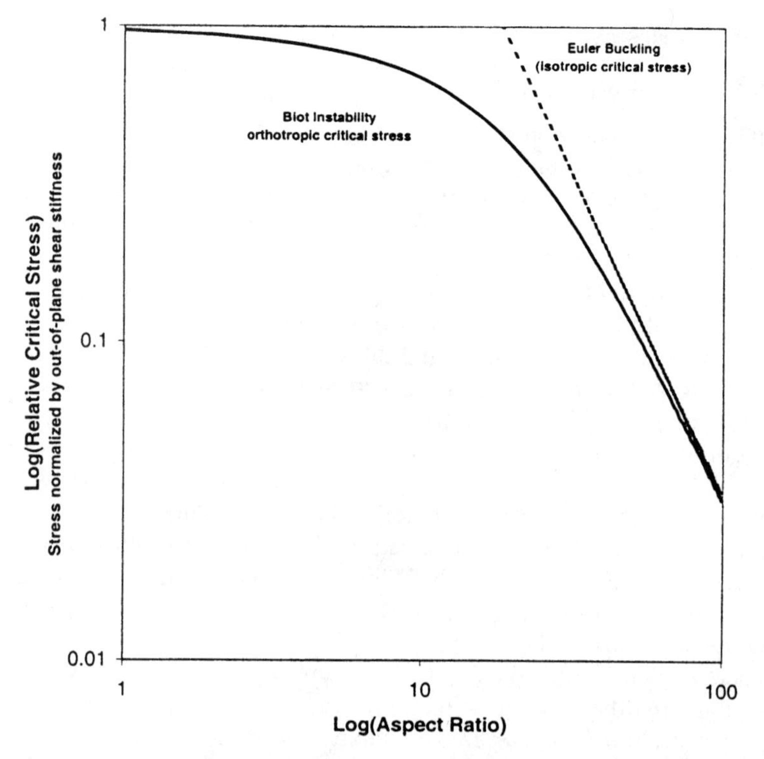

Fig. 7 Theoretical buckling load for an orthotropic solid, showing the essential role of shear in buckling.

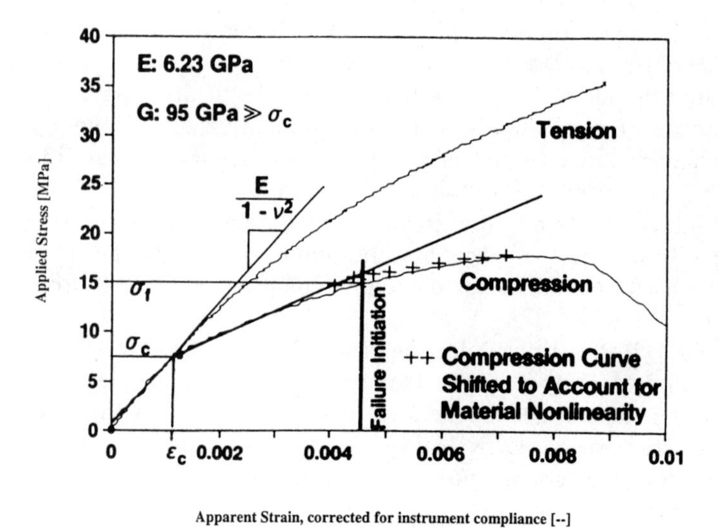

Fig. 8 Verification of theoretical result for compression behavior of paper.

stress–strain curve traces out the tensile curve but then bifurcates onto a lower stiffness path. This is a transition into the shear-dominated mode identified by Biot [5] and Perkins and McEvoy [56], but occurs at a much lower stress than predicted by theory. Because the continuum model has not ability to consider local stress variations due to microstructure, as does the Habeger and Whitsitt [26] model, this disparity between predicted and measured theoretical stress should not be a surprise—microstructure will have the same influence as imperfections in Fig. 2. Donner and Backer tested the theory in a different way: by examining the behavior of a continuum orthotropic material loaded beyond the critical stress.

Advanced analysis based on work by Koiter [37] and Budiansky [7] is able to predict the postbifurcation stiffness, that is, the slope of the curve in Fig. 8 following the departure from the tensile curve. As shown in Fig. 8, bifurcation is not the same as failure but rather a transition to a shear dominated mode that progressively translates the direct compressive stress applied by the experimental apparatus into a shear stress. Additional compression of the structure is required to create the shear stress, which in turn sets up the fiber cell wall dislocations, or local shear failure, and leads to the failure zone morphology observed in test specimens.

A common feature of the continuum approaches is the concept that there is a transition from one mode (homogeneous compression) to a second—kink band, shear-dominated waviness, or local lamina buckling. This gives the context for investigating fiber level origins of failure itself.

B. Composites, Wood, and Polymers

The compression characteristics of structural materials such as polymers, fiber-reinforced composite materials, and wood have been treated theoretically and studied experimentally. The compression failure of polymers is usually characterized by shear slip mechanisms of different natures. As an example, the extrudate of solid linear polyethylene with lamellar crystals and amorphoous layers stacked alternatively along the extrusion direction was investigated by Shigematsu et al. [69] As illustrated in Fig. 9, the typical compression failure is characterized by a kink band through the specimen, caused by slip between microfibrils and/or along crystal boundaries.

Many theoretical treatments of the compression failure in fiber-reinforced composites start from the suggestion that the problem is analogous to that of the buckling of a column of an elastic foundation. Figure 10 illustrates two microbuckling modes that have been analyzed, the extension mode and the shear mode [9,19,43]. The extension mode usually applies for fiber volume fractions less than 10%, whereas the shear mode applies for larger fiber volume fractions.

One assumption underlying the theoretical treatment is that compression failure in composites due to fiber buckling is a uniform process, which, according to De Ferran and Harris [19], seems to be unreasonable. On the basis of experimental observations, they suggest a mechanism started by local failure of a single fiber due to shear and partial splitting, which subsequently spreads throughout the sample much as dislocation-controlled yielding does through a crystalline solid.

In a review by Chamis [9], it is argued that a fiber composite material can fail by several different mechanisms, depending on the properties of the fibers and the matrix:

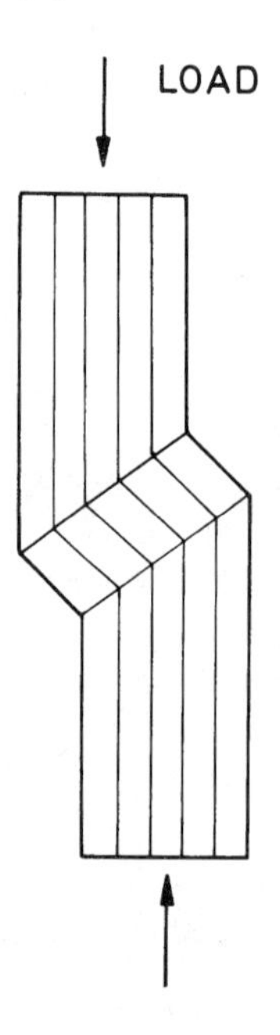

Fig. 9 Shear dislocation failure, called a kink band, in drawn, oriented polyethylene subjected to uniaxial compression.

- ● Microbuckling of the fibers in an elastic matrix
- ●● Matrix yielding followed by microbuckling of the fibers
- ●●● Debonding and interlaminar shear followed by microbuckling of the fiber
- ●●●● Shear failure
- ▬ Ply separation by transverse tension through the thickness

Variables such as specimen geometry, end condition, fiber diameter, fiber volume fraction, and properties of the fibers and the resin matrix are found to affect the failure mode and the compression strength.

In the case of wood, a theoretical model as suggested by Grossman and Wold [23] in which the wood sample is treated as an assembly of long hollow tubes. These

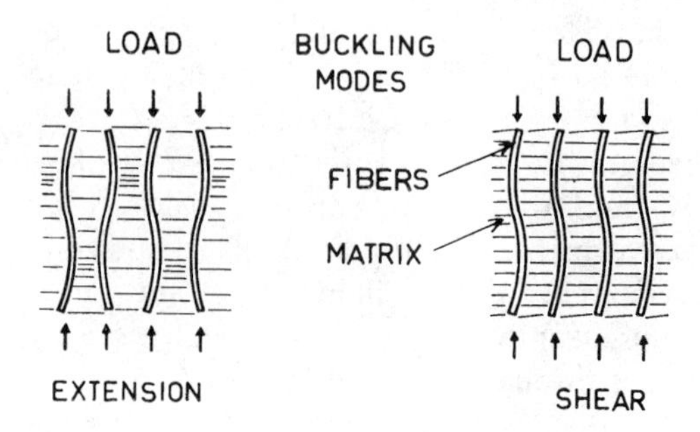

Fig. 10 Two microbuckling modes that have been treated mathematically for unidirectionally reinforced composite materials. (Loosely based on Ref. 43.)

tubes correspond to slip planes in the wood fibers. After local damage to the hollow tubes, the tubes are considered to be separated from their neighbors and to behave like unstable columns. The results show that the predicted stress at failure according to this theory is several times greater than the experimentally observed values. This discrepancy between theory and experiment arises from the treatment of the cells as tubes. Since the walls of tracheids and other cells are in reality wrinkled to some extent, they behave more like thin shells than like unstable columns.

Other studies on wood provide evidence of microstructural changes in the fibers [11,38]. These relate to the occurrence of slip planes. A slip plane is generally regarded as a dislocation of the fibrils that make up the cell wall. The number of slip planes has been found to increase with increasing compression until continued stressing results in gross buckling of the cell walls, at which state it is possible to observe the failure macroscopically, as shown in Fig. 11. Once initiated locally, a minute deformation may cause an overload of an adjacent fibril. Thus, slip lines propagate like a wave. This makes it very hard to predict the detailed phenomenon of fracture, because it is influenced by stochastically distributed features of the structure.

Fig. 11 Shear dislocation failure of wood subjected to compression loading. (Photo courtesy of Dr. Treiber, STFI.)

C. Fiber Level Mechanisms

For paper, with its very complex fibrous structure, no theory has yet been developed for the prediction of failure based on the properties of the constituent components; only descriptive interpretations of the failure mechanism have been proposed. Setterholm and Gertjejansen [68] consider fiber buckling to be the governing factor for compression failure.

Seth et al. [66] show that at low degrees of bonding compression strength depends strongly on bonding, but that as bonding is increased a limiting strength is reached. They conclude that the compression strength of paper becomes controlled by the fiber compression strength at higher degrees of bonding. This behavior is similar to that of tensile strength, according to a theory advanced by Page [54].

Sachs and Kuster [63] used a short-span method in connection with microscopic techniques to study the failure mechanism of linerboard in compression. This test method was similar to the STFI short-span test described below. Sachs and Kuster present evidence that the sequence of morphological changes leading to failure starts from dislocation of the cell wall tissues. An interfacial separation of the S1 and S2 cell wall layers leads to separation of fiber-to-fiber bonds and severing of fiber cell walls, promoting delamination and failure of the linerboard. Buckling of fibers follows failure.

Fellers and coworkers [8,17,18,60–62] view failure in compression as the result of an unstable state of loading with two complementary mechanisms. Failure is triggered either by buckling of fiber segments, predominant in low density sheets (Fig. 12a), or by shear dislocations in the fiber walls, predominant in high density sheets (Fig. 12b). In both cases, the number of fiber dislocations increases as the compression is continued until the whole sheet becomes unstable and collapses. The propagation of macroscopic dislocations often has the appearance of a shear slip dislocation as shown in Fig. 13, where delamination has also occurred.

D. Material Failure

Stockman [71] identifies the shear strength at an angle of about 45° through the sheet thickness as the factor governing compression failure. McEvoy [48] considers compression failure to be governed by a debonding of the fiber network that has the appearance of either a shear slip or a crushing type of failure. Shear slip failures were also observed by the Institute of Paper Chemistry [32] that were caused by either bond breakage or fiber buckling; they were also observed by Seth and Soszynski [65].

E. Structural Buckling and the Role of Shear

Paper has low out-of-plane shear stiffness relative to its in-plane stiffness. This low shear stiffness reduces the columnar buckling load below that of an isotropic material. At low slenderness ratios, the low out-of-plane shear results in out-of-plane deformation visible to the naked eye. In corrugated container side walls, such out-of-plane movement is observed prior to failure [80]. Figure 7 suggests that shear dominates the deformation of paper near failure. This same role of shear occurs at the fiber level, where fiber buckling and bond failures take place in a high shear environment, although at a local level rather than throughout the specimen.

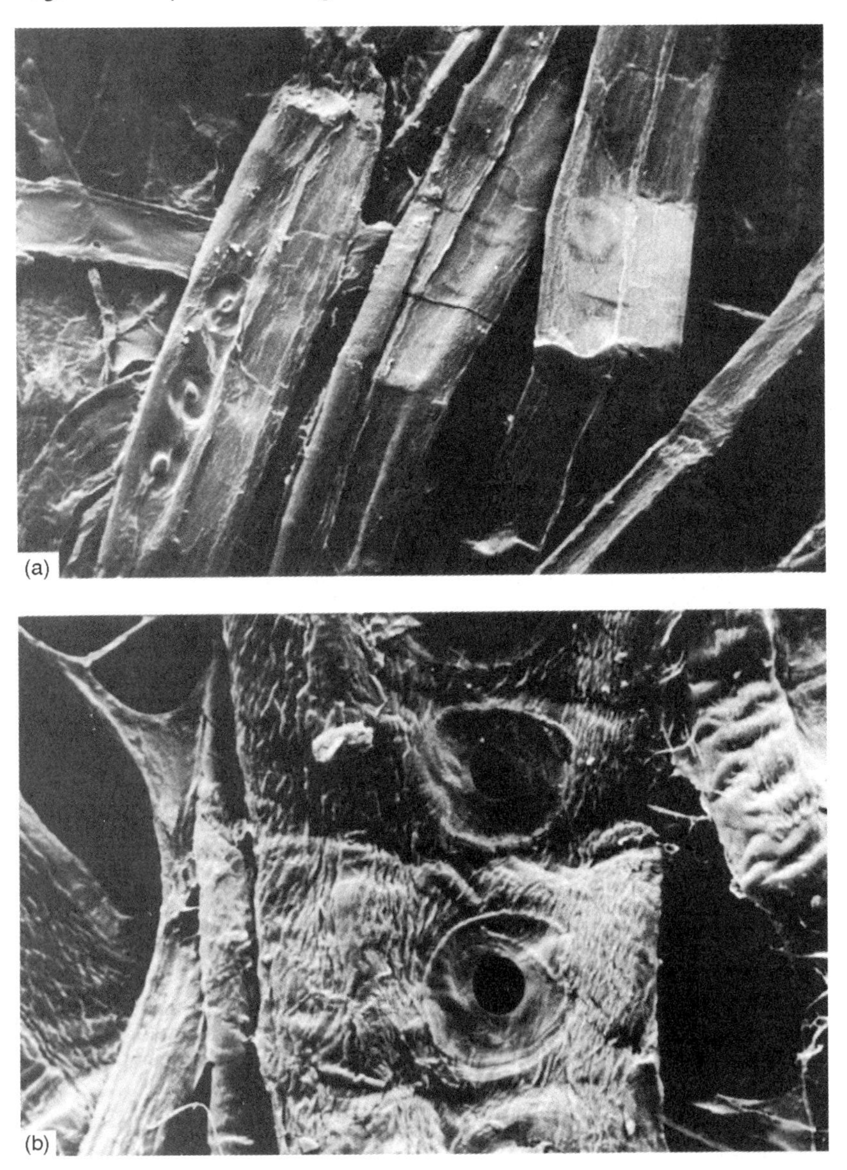

Fig. 12 Dislocations in the fiber cell walls in a sheet of paper subjected to compression loading. (a) Low density structure; (b) high density structure. (From Ref. 41.)

Equation (2) shows the role that shear plays in the buckling behavior short columns. The initial loading response of paper in compression is dominated by the elastic modulus E, but buckling is influenced by the ratio E/G (actually, $E_{MD}/G_{MD\text{-}ZD}$ or $E_{CD}/G_{CD\text{-}ZD}$, assuming that the paper is loaded in either the machine or cross-machine direction.) E/G begins to dominate as the slenderness decreases (column gets shorter). The unusually high E/G ratio in paper also means that clamping and loading specimens in compression require considerable care to suppress unwanted clamping and edge effects. Donner [13] examined the

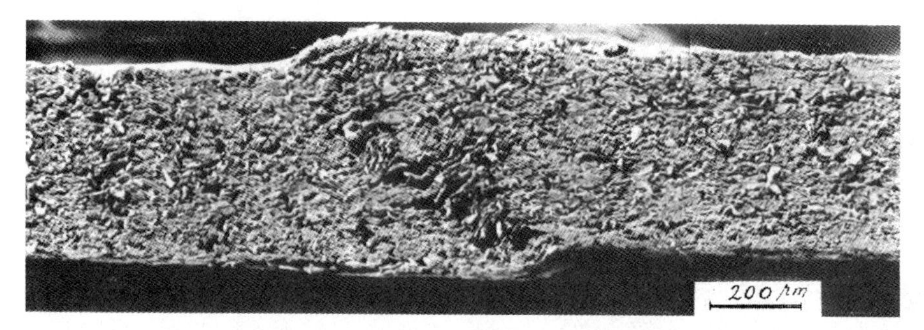

Fig. 13 Shear dislocation type of failure of a 300 g/m^2 linerboard subjected to compression loading.

role of shear in clamping of short-span compression specimens. Figure 14 represents the role of the clamp in compression testing, assuming that the load is transferred by shear and doesn't cause any thickness distortion. The shear loading of the material distorts the cross section as shown. The result clearly shows that shear arises as an experimental artifact that must be considered in modeling and understanding compression tests. Conversely, because shear arises as a natural, even essential, part of the compression behavior of paper, it is very difficult to separate the role of shear in clamping and testing from shear arising in compression. Shear-related artifacts often influence the deformation behavior and strength measurement in compression testing. After the different testing principles are introduced, the artifacts introduced by methods based on these principles will be examined.

IV. DIFFERENT TESTING PRINCIPLES

Based on the theoretical considerations and observations of the compression failure mechanism, the following criteria are considered necessary to define an adequate evaluation of the compression strength of paper.

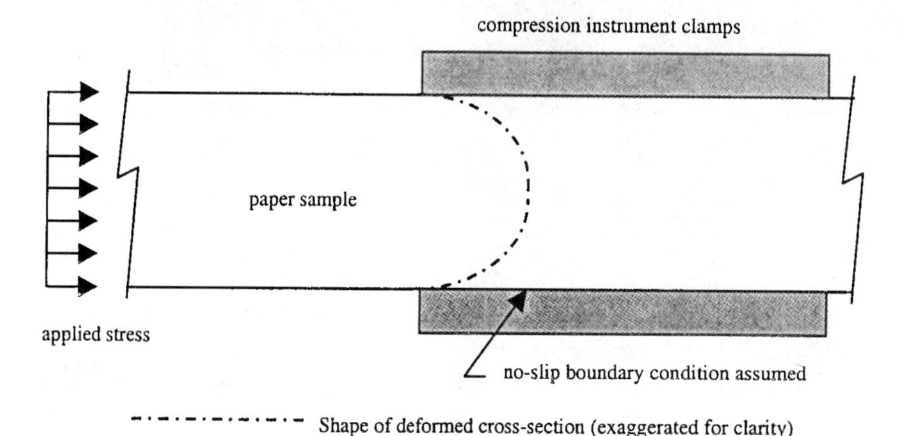

Fig. 14 Schematic of sample in compression, showing that clamp jaws apply compression load through shear.

- Strength degradation due to damaged edges or stress concentrations close to clamps should be avoided.
- •• Structural instability of the paper sheet should be avoided in order to be able to evaluate a material property, which should be independent of specimen size.
- ••• Instabilities of fibers within the sheet, which must be regarded as being associated with the true compression behavior of the material, should not be inhibited so that the sheet strength is artificially enhanced.

Three ways of evaluating the compression strength of sheet materials emerge from the preceding discussion; either by using a short span, by supporting the specimen, or by employing a tubular or corrugated specimen to gain geometric stabilization. One may also refer to pioneering work in compression testing that used the plug test [6] and the pack test [1,27], methods that are currently not used in the paper industry.

A. Aspects of Sheet Characterization

Paper may in many cases be treated as a homogeneous engineering material in spite of its fibrous, porous structure. Engineering measures of strength and modulus are given in units of force per unit cross-sectional area. However, any measure of the cross-sectional area poses special challenges because of the fibrous structure and rough surface. For this reason, normalizing strength and stiffness by area is generally considered unsuitable for evaluating process variables that affect the properties of paper [59]. Paper in its end use is often judged instead by its properties per unit width, which also has the advantage of making it easy to compute the load-carrying capacity of the paper. The engineering compressive failure stress (σ_c) and elastic modulus (E) are related to the compression strength and extensional stiffness, respectively, with the dimension of force per unit width (indicated by the superscript b):

$$F_c^b = \sigma_c t \tag{9a}$$

$$E^b = Et \tag{10a}$$

Since these expressions are dependent on sheet grammage (w), the measurement is typically divided by w. Expressions for the specific compression strength, i.e., compression index, and specific elastic modulus, i.e., tensile stiffness index, are then obtained. These are equivalent to dividing the compression stress at failure and elastic modulus by the density according to the equations

$$\sigma_c^w = \frac{\sigma_c}{\rho} = \sigma_c \frac{t}{w} = \frac{F_c^b}{w} \quad [\mathrm{N\,m/kg}] \qquad \text{compressive index} \tag{9b}$$

$$E^w = \frac{E}{\rho} = E\frac{t}{w} = \frac{E^b}{w} \quad [\mathrm{N\,m/kg}] \qquad \text{tensile stiffness index} \tag{10b}$$

The Euler equation, Eq. (3), for fixed ends is accordingly modified to be valid for specific properties, yielding

$$\sigma_c^w = \frac{4\pi^2 E^w}{\lambda^2} \quad [\mathrm{N\,m/kg}] \tag{11}$$

A separate evaluation of the thickness is still needed for the determination of the slenderness ratio (λ). Methods for the determination of thickness are discussed in Chapters 6 and 11 of Volume 2.

B. Short-Span Specimen Geometry

The basic principle behind the short-span specimen geometry is to use such a short span that crushing of the specimen occurs before any buckling of the specimen takes place. The Concora liner test (CLT) and the STFI short-span test, reported here, are examples of methods currently used in accordance with this principle. One may also refer to pioneering work by the Institute of Paper Chemistry [28] and by Ranger [59].

Concora Liner Test A frequently used method that employs a short flat specimen is the Concora liner test (CLT) designed by the Container Corporation of America [46,73]. As shown in Fig. 15, one edge of the 152 mm long specimen is clamped between two strips of metal so that a specimen free span of 6.3 mm protrudes above the clamp. A load is applied with a flat platen to the free edge of the specimen. In some cases an additional upper plate is used to reduce the influence of curl on the specimen [47,52]. The influence of height on the CLT value was investigated by the Institute of Paper Chemistry [30], and it was concluded that the CLT specimens always exhibit bending failure or edge crushing within the range of heights studied. The structural response reduces the test value below the material compression strength, limiting the effectiveness of the test.

STFI Short-Span Test A short-span method for the evaluation of the compression strength of paper was developed at the Swedish Pulp and Paper Research Institute (STFI) [8,15,42]. Figure 16 shows the apparatus with the same boundary conditions as those given in Fig. 1b, that is, a column with fixed ends. The free span is 0.7 mm. Two clamps, with a force on the sample applied by pneumatic cylinders, hold the specimen in a straight position. One clamp is attached to a load cell, and the other is moved by means of a motor. As pointed out earlier in Fig. 3, any eccentric loading on the specimen is likely to severely lower the strength of the material. The straight clamping configuration and precision-ground clamps ensure minimal eccentric loading.

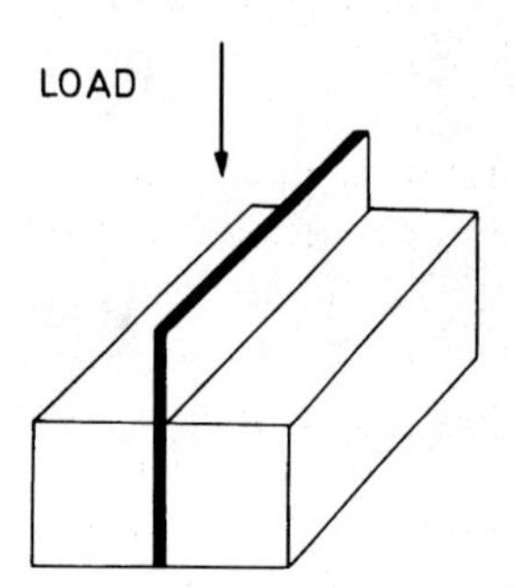

Fig. 15 Concora liner test (CLT).

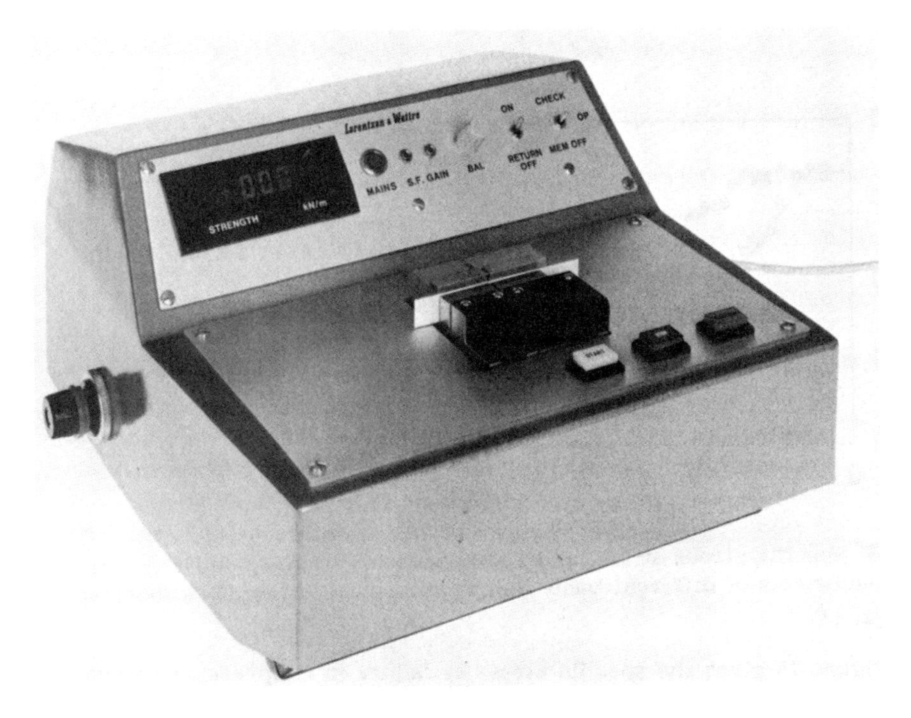

Fig. 16 Commercialized STFI short-span test instrument.

The effect of clamping force on strength has been investigated. The results show that the strength increases with lateral force at low forces up to approximately 1.5–3.5 kN, when a plateau is reached. If the force is further increased, the compression strength is reduced by stress concentration caused by the clamps. Details on clamping arrangement are given in ISO 9895: 1989(E).

By analogy with Figs. 2 and 3, Fig. 17 shows the specific stress at failure versus slenderness ratio for a commercial 125 g/m^2 linerboard. The test results are obtained with an STFI short-span compression test, which can be employed with different span lengths. The trend of the curves in Fig. 17 is representative of carton board, linerboard, and corrugated medium in general. Agreement with the Euler equation Eq. (9), is shown in the long-column range down to a slenderness ratio of 300, which corresponds to a slenderness ratio of 150 in Fig. 3. This indicates an elastic buckling failure. As the span decreases, the specimen shows deviations from the Euler curve. In the intermediate-column range, elastic buckling is gradually replaced by plastic failure until a well-defined plateau is reached at slenderness ratios smaller than 20, i.e., spans shorter than six times the thickness. This plateau was identified in Fig. 2 as the short-column range, representing the compression strength of the material. It should be noted that the short-column range is much shorter for paper than for steel. The reason for this is ascribed to the nonuniform thickness of paper, which has the effect that a shorter span is required to obtain sufficient stability in the sample.

Figure 18 gives the specific compressive failure stress versus span for two kraft pulp sheets of different grammage values. As in Fig. 17, the apparatus used is the STFI short-span method, but using different free span lengths. It is evident from

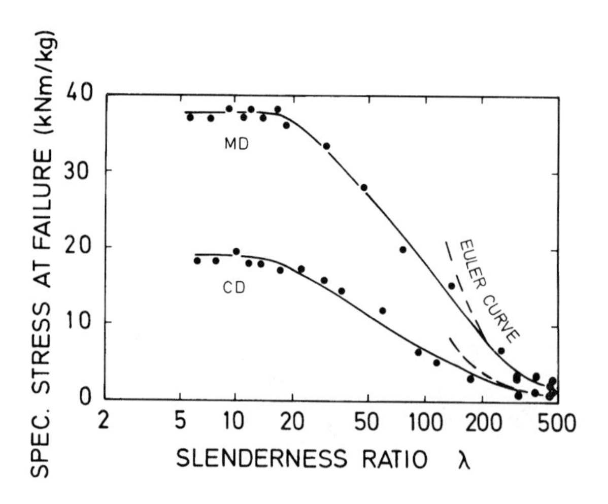

Fig. 17 Stress at failure in compression versus slenderness ratio for 125 g/m² linerboard according to the STFI short-span test.

these results that the plateau region represents a well-defined property of the material for the two specimens. The higher the grammage, the less dramatic is the decrease in stress at failure with increasing span.

In Fig. 19, strengths corresponding to the plateau at short spans in Fig. 18 show that the specific compression strength is independent of basis weight in the range of 100–350 g/m². This will be true for a given pulp, provided the manufacturing conditions for the laboratory sheets are the same.

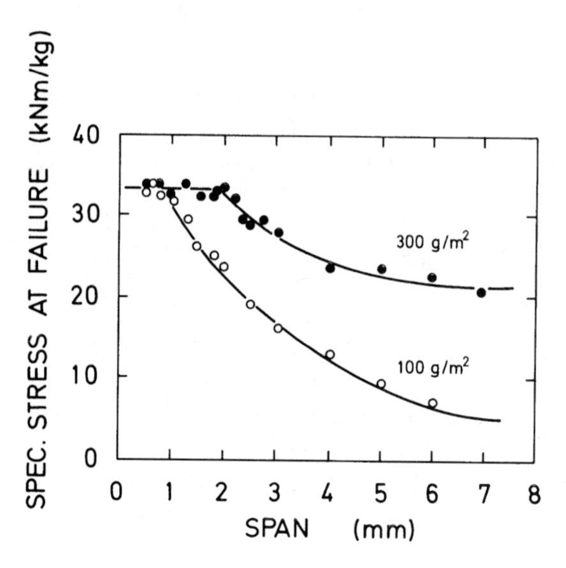

Fig. 18 Compressive failure stress versus span for two kraft pulp handsheets of different grammages according to the STFI short-span test.

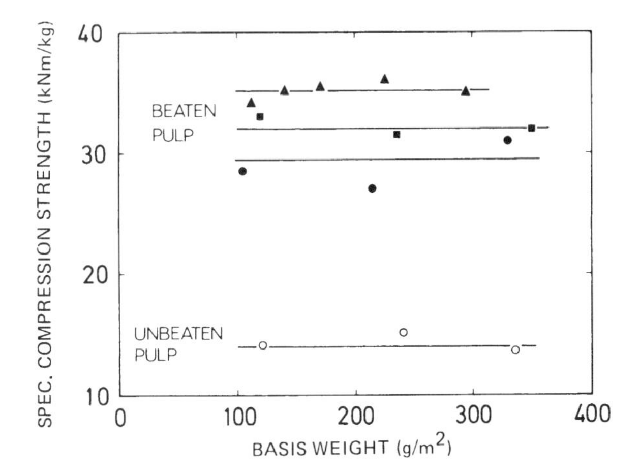

Fig. 19 Specific compression strength versus grammage for handsheets according to the STFI short-span test: (▲) Kraft; (■) bleached softwood; (●, ○) bleached hardwood.

This result can be obtained only at short spans on the order of 0.5–1 mm. At these spans, the structural deformation is inhibited and fiber compression failure controls the strength. Figures 11–13 show the compression failure zones for wood and paper, revealing that the compression failure occurs on a scale much smaller than fiber length. The mechanism of failure does not change with gauge length, showing that the strength plateau in Figs. 17–19 is in fact the material strength. This effect should not be compared to the so-called zero-span tensile value which is aimed at evaluating the fiber strength in tension. In the zero-span test the tensile strength monotonically increases with decreasing span all the way down to zero span, indicating a *change* in failure mechanism with gauge length [10].

If a longer span is used, the specific compression strength values obtained will increase as grammage increases, indicating that the measurement does not represent a material property.

The possible influence of an artificial strength increase at short spans has been investigated by the use of thick, smooth laboratory sheets [64] made from pulps at different beating levels [8,60,61]. Equal results were obtained for short-span values and values obtained by using 20 mm long specimens that were prevented from buckling by lateral supports. These experiments demonstrate that no artificial strengthening of the sheet occurs at short spans.

C. Plate-Supported Specimen

The short-span compression test is able to simply provide a measure of material compression strength, but the influence of the clamps prevents the measurement of stiffness properties. Stress–strain characteristics of different sheet materials in compression may be evaluated with the sheet restrained on each side by solid metal plates. A summary of different methods that adopt this principle is given in an ASTM standard [3] and by Ramberg and Miller [58]. When a single sheet is laterally supported, the predominant source of error, especially with thin samples, is the

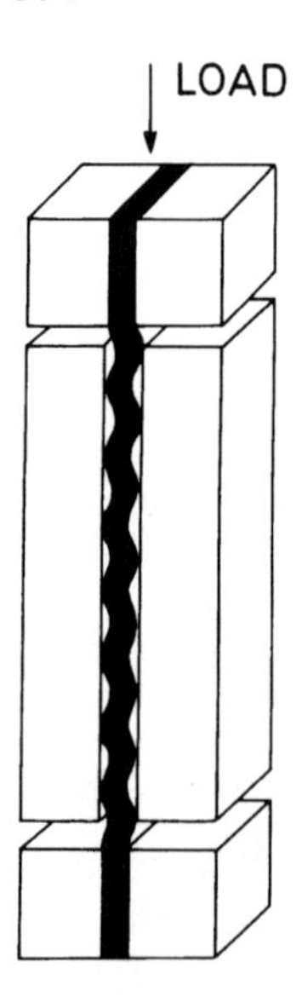

Fig. 20 Undulations may occur between lateral supports in the final part of a compression deformation if the lateral supports are not sufficiently stiff and not tightly pressed against the specimen.

development of small-amplitude waves as illustrated in Fig. 20. To avoid these waves in smooth steel samples, the clearance between the specimen and the supports must often be less than 0.02 times the thickness. The effect of waves on eccentricity of loading is demonstrated by Fig. 3. Errors due to direction may occur, on the other hand, if there is too close a fit between the specimen and the end supports. For this reason the specimen is often lubricated. Three different methods based on this principle have been developed for the determination of the stress–strain curve of paper in compression.

STFI Solid Support Test Fellers et al. [17] developed a test method based on the solid lateral support principle. The equipment is shown diagrammatically in Fig. 21. A 15 mm wide paper strip is fastened in two clamps and loaded horizontally. One clamp is attached to a load cell, while a motor moves the other. A sliding zone is created between the clamps, where 20 mm long steel blocks attached to each clamp support the specimen on each side to prevent buckling. The clearance between the paper and the supports can be regulated by turning micrometers. The free span is chosen to be 1 mm, which is short enough to ensure that buckling of the unsupported part of the specimen does not occur. Failure occurs randomly everywhere along the free span, showing that friction is not controlling failure behavior and is not subtracted from the failure load. If friction did have an important role, then the load would vary along the length of the free span, and the location of the failures would show a statistical clustering. This was not seen.

A strain gauge is used for evaluating strain and elastic modulus, to measure displacement in the free span independent of slippage or deformation in the clamps. As pointed out by Setterholm [67,68], the use of clamp movement for strain mea-

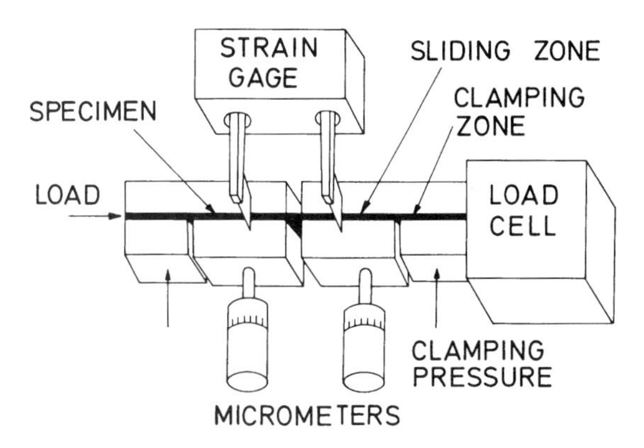

Fig. 21 STFI solid support test.

surement (as used by several research workers) may cause great errors due to slippage in the clamps, which will result in lower modulus and higher strain measurements.

It was pointed out earlier (in Fig. 4) that paper is a rather uneven material that contains a number of peaks and valleys, and because of this fact, local buckling of the sheet is highly likely to occur. For this reason, the clearance between the lateral supports and the sheet influences strength and stiffness measurements.

The effect of lateral clearance between the specimen and the supports on compression strength and the apparent elastic modulus in compression is illustrated in Fig. 22 for a 200 g/m^2 linerboard. Zero lateral clearance means that the lateral supports just touch the thickness peaks in Fig. 4, and a negative value means that he supports are pressed into the sample. As can be seen, the compression strength and elastic modulus in compression are in fact dependent on the clearance. If, however, the clearance is kept between +0.01 and −0.01 mm, the relative error in strength is less than 5%. Positive clearance should be avoided because of the tendency for the

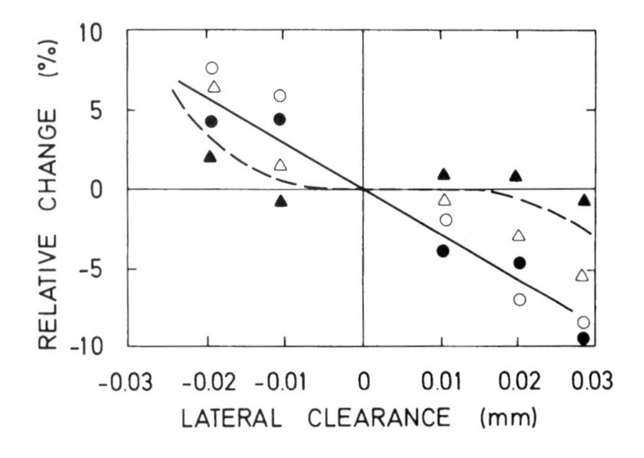

Fig. 22 Relative change in properties versus lateral clearance between the specimen and the support related to the properties at zero clearance for the STFI solid support test. Elastic modulus: (▲) MD; (△) CD. Compression strength: (●) MD; (○) CD.

specimen to bow, as shown in Fig. 20. Even though these results show the general trends of the compression behavior versus clearance, it should be noted that the sensitivity to clearance may differ, depending on basis weight, orientation, and thickness uniformity of the paper.

The specific stress–strain curve in compression as shown in Fig. 23 is a well-defined material characteristic. This is obtained with a clearance of ±0.01 mm, as described in the previous discussion. This makes the method useful for research purposes, especially when the compression creep and stress relaxation properties are of specific interest [17].

PPRIC Plate Support Test The compression apparatus shown in Fig. 24 was developed by Seth and Soszynski [65] at the Pulp and Paper Research Institute of Canada. Flat aluminum faces support the specimen in this device. The necked specimen is clamped pneumatically at the two ends. The lower clamp is fixed while the upper clamp moves vertically. One of the lateral supports is fixed to the base and backs the specimen. The other can move back and forth. The lateral force developed during the compression is determined by measuring the deformation of a beam that rests against the movable support. The beam deformation is 0.4 μm/N. Accordingly, a lateral deformation of 0.020 mm occurs at a lateral force of 50 N. The specimen, in fact, undulates between the supports as shown in Fig. 20. The coefficient of friction, which ranges from 0.1 to 0.2 between the specimen and the supports, was determined in situ by pushing the specimen against a known applied lateral force [70].

Figure 25 shows a typical compressive load–deformation curve for a paper specimen held between the flat plates, together with the lateral force developed during compression. The curve was obtained by subtracting the frictional component (static coefficient of friction × lateral force) from the observed compression loads. The compression deformation was calculated from the displacement of the upper clamp.

The application of the lateral force is required to eliminate lateral buckling or undulation of the specimen between the flat plates. The friction caused by slip along the plates can be removed by measurement and subtraction, at least on average.

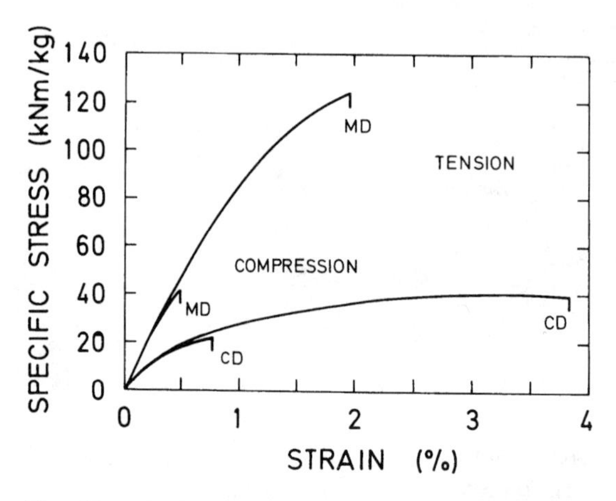

Fig. 23 Typical specific stress–strain curves in tension and compression for paper.

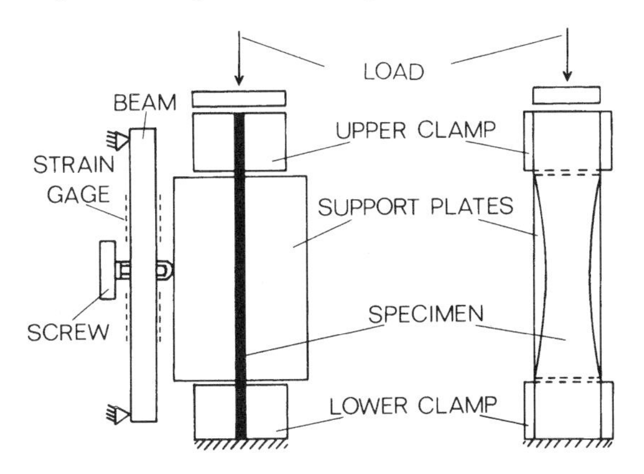

Fig. 24 PPRIC plate support test. (From Refs. 65 and 66.)

Deformation in the clamp can be compensated for, giving an improved strain measurement, as discussed by Setterholm [67] and Setterholm and Gertjejansen [68]. Therefore, the measured stress–strain curve can be demonstrated to be a valid material characteristic. On the other hand, compression failure is essentially the transition form in-plane deformation to out-of-plane deformation. This deformation is restricted by the plates. For this reason, the compression strength measured by this apparatus may be artificially high.

FPL Lateral Support Test A lateral support method for paperboard was reported by Jackson et al. [33] of the U.S. Forest Products Laboratory. Figure 26 shows their device, which is designed to provide almost complete lateral support to the paperboard specimen while the load is being applied. The inside faces of the device are ground parallel to the loading axis. In addition, material is removed from each face between the top and bottom clamping areas, to a depth of 0.025 mm, to allow the

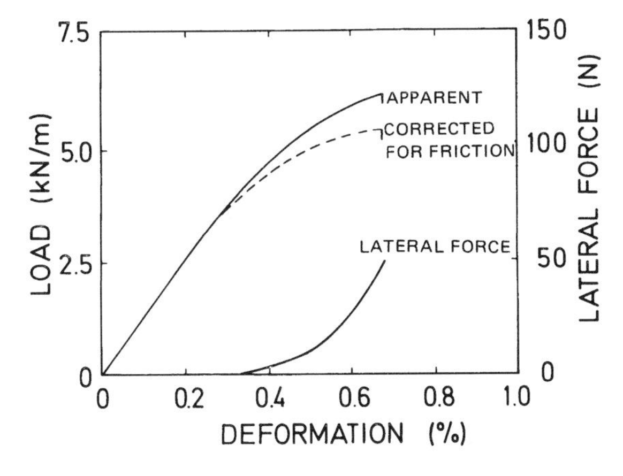

Fig. 25 Load–deformation curve in compression and the lateral forces developed during the compression according to the PPRIC plate support test. (From Ref. 65.)

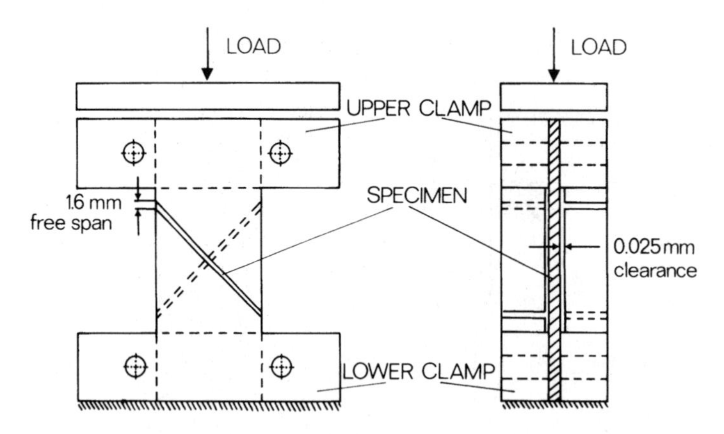

Fig. 26 FPL lateral support test. (From Ref. 65.)

specimen to slide along the effective gauge length of 34 mm. The top half of the device is approximately 1.6 mm above the lower half, allowing adequate room for compression of the specimen. A specimen is positioned in the center of the device. By progressive turning of the clamp screws, an essentially uniform clamping pressure on the specimen is achieved. The deformation is measured by recording the displacement between the upper and lower loading plates. Using this technique, Koning and Haskell [41] investigated the effect of various factors on the compression strength of linerboard.

The influence of lateral clearance between the specimen and the lateral supports on strength has not been specifically reported [31,33]. According to the work by Seth and Soszynski [65] using the PPRIC plate support test, the friction component should increase the apparent strength by 5–10%. Other influences such as the weak-link effect and clamping also need to be considered in future use of this test method.

D. Blade-Supported Specimen

It is possible to evaluate the stress–strain properties of sheet materials, such as veneer or plywood, by providing lateral support with slender steel blades. This is described in an ASTM standard [2]. These support blades provide a very stiff lateral restraint while allowing the specimen relative freedom of motion in the direction of loading. The distance between the blades must be short enough to prevent a buckling type of failure, and the supporting blades must be rigid enough to withstand lateral forces set up by the specimen.

STFI Blade Support Test The steel blade support principle, using a spacing of 1 mm between the blades was developed by Fellers and coworkers at the Swedish Pulp and Paper Research Institute (STFI) [8,60,61]. Agreement as to test results between the STFI blade support and the STFI short-span methods has been obtained for very smooth sheets with grammage above 380 g/m^2. The use of the method for more uneven materials, however, has not been successful.

Weyerhaeuser Lateral Support Test Another blade support test method was reported by Stockman [71] of Weyerhaeuser. The Weyerhaeuser tester, shown in Fig. 27, is loaded in a universal testing machine or other device capable of providing a load–deformation curve by recording the movement of the clamps. To prevent development of severe stress concentrations at the clamps, the test specimens have a continuous necking with a radius of 200 mm. The total test length between the clamps is 90 mm, and the minimum width of the specimen is 20 mm. Between each pair of supporting blades, the end conditions of the specimen correspond to the Euler column with hinged ends.

Figure 28 shows how the specific stress at failure in compression varies with the span between the supporting blades on a 200 g/m^2 linerboard [71]. A plateau value is found at slenderness ratios of 10–30, which corresponds to a span of 2.9–8.7 times the sheet thickness. Stockmann [71] defines the plateau value as the compression strength of the material. The shear dislocation, observed at failure, showed that it was the transition to out-of-plane deformations in the sheet that prevented a further increase in applied load. Below a slenderness ratio of 10, corresponding to a span of 0.7 mm, between the blades an increasing failure load is observed, which is interpreted by Stockmann as an artificial strengthening of the sheet by the prevention of internal dislocation.

The apparatus is difficult to use and requires careful experimental practice. Seth and Soszynski [65] noticed that slight misalignment between blades sometimes triggers the specimen to go into low-wavelength buckling as in Fig. 20, thus leading it to fail prematurely. In thesis work by McEvoy [48], the plateau region at small slenderness ratios could not be verified. The plateau is likely to depend on the thickness uniformity of the paper sample as well as on the condition and use of the experimental apparatus. These shortcomings led to the development of improved equipment based on the same principles, such as the vacuum restraint system described next.

Forest Products Laboratory Vacuum Restraint Gunderson's vacuum restraint method [24] uses air pressure to hold the specimen against vertical rods. The rods are

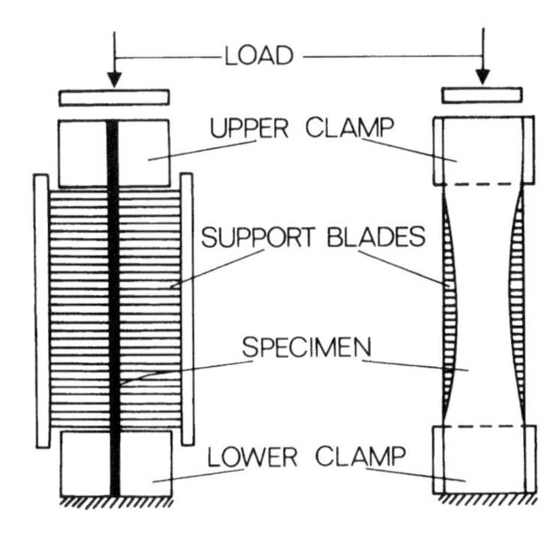

Fig. 27 Weyerhaeuser lateral support test. (From Ref. 65.)

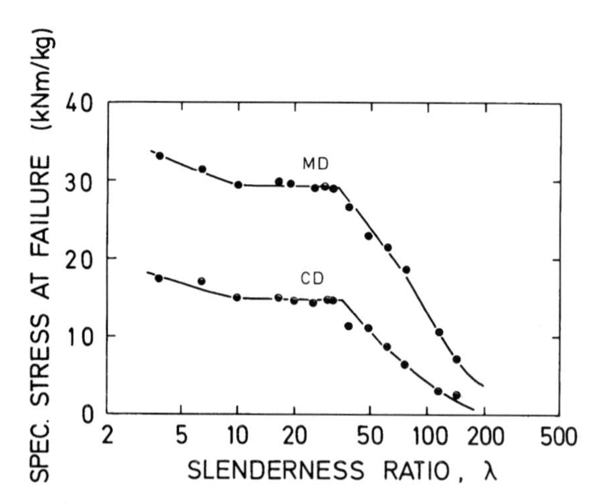

Fig. 28 Specific stress at failure versus slenderness for a 200 g/m^2 linerboard according to the Weyerhaeuser lateral support test. (From Ref. 71.)

embedded in an elastomer at open end and support the specimen at the other end. The space around the rods is partially evacuated to create the pressure different that holds the sample in place. As the sample deforms under the influence of the applied load, the rods provide the specimen relative freedom of motion in the plane of the sheet. The method of rod support gives the rod the ability to move biaxially. The apparatus has taken advantage of this to assess biaxial deformation of paper in compression. In this case, a cruciform sample shape is used, and the sample is loaded along both axes.

The sample is relatively large compared to short-span specimens and has an unobstructed surface, unlike the Weyerhaeuser method. Strain was directly measured with displacement gauges, similar to setups for direct measurement of tensile strain. This eliminated the influence of clamping on strain and stiffness measurements.

Gunderston [25] demonstrated that the restraint pressure has an influence on the compressive strength and strain to failure but that a pressure can be chosen that allows accurate measurement of the material's compression failure properties.

Failure locations in the gauge section of the specimen were distributed uniformly, giving further demonstration that strain was distributed uniformly throughout the sample. This uniform distribution allows accurate measurement of deformation and strain with the displacement gauges, which in turns gives accurate measurements of elastic modulus. Compressive modulus measurements are shown to be independent of restraint pressure as well.

E. Cylindrical Geometry

The cylinder test, successfully tested for metals, employs a sheet of the material rolled to form a circular cylinder that is held in that shape by bonding the vertical edges together [21]. One may also refer to pioneering work by Kellicutt [36], who used a specimen consisting of a number of turns of a paperboard strip rolled into the form of a hollow cylinder. In addition to the cylindrical specimen tests described here, the reader is urged to consult Chapter 7 in this volume in regard to specimen configuration.

All cylinder-type tests for planar materials have the purpose of giving structural stability to the specimen in a way that allows loading in the plane of the sheet. A key problem concerns the support of the edges as the load is applied. Because the radius of the cylinder tends to increase as a compression load is applied to the edge, the sample does not remain truly cylindrical unless special precautions are taken.

A second typical problem is that rolling a specimen into the cylindrical shape creates a seam. Different methods scarf and bond the seam, hold the seam under pressure without bonding, or use a lap joint with adhesive. The original ring crush test used a butt joint without any additional means of support, although one of the loaded ends was inserted into a circular groove. Both the unsupported load edge and the unsupported seam created difficulties for this test on lightweight materials even when the test was properly performed. Several modifications discussed below are motivated in part by the shortcomings of this standard test.

Ring Crush Test The ring crush test (RCT) was at one time the most common linerboard compression test and is still used for quality control in the paper industry. It is illustrated in Fig. 29. This test employs a paper strip 152 mm long and 25.4 mm wide that is rolled to form a cylinder in an annular holder with an accompanying island (47 mm in diameter and 6.3 mm high. The cylinder is subjected to a compression load on the upper edge of the sample, and the maximum load is registered [72]. A study of the effect of specimen height on the failure load in compression measured by the ring crush test was performed by the Institute of Paper Chemistry [30] using a sample of 205 g/m^2 linerboard. In Fig. 30, it can be seen that the maximum load increased almost linearly with decreasing height of the specimen. The line terminates at a specimen height of 6.3 mm, which corresponds to zero free span and is thus the limiting height for the RCT. There is consequently no evidence that there is a range of short heights over which crushing failure rather than buckling failure occurs.

These results clearly show that the ring crush test does not satisfy the criteria for an adequate compression test of linerboard. For medium weight and heavy

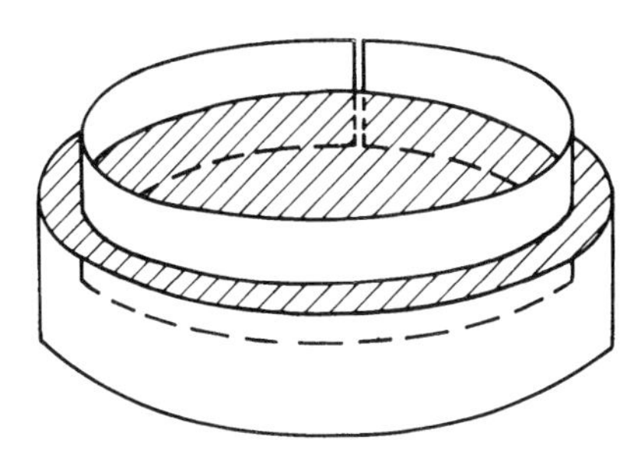

Fig. 29 The king crush test (RCT).

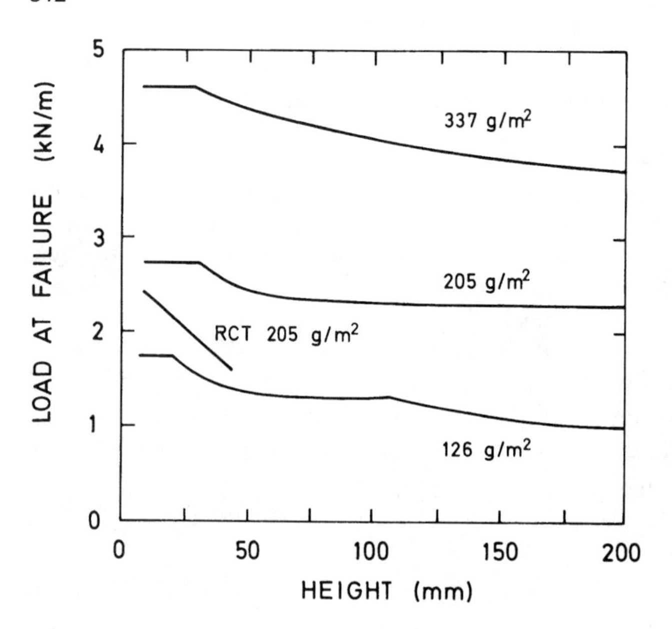

Fig. 30 Compression strength versus height according to the IPC-modified ring crush test (RCT). (From Ref. 30.)

weight linerboard, however, the test is a practical method for doing quality assurance testing. Testing of lighter weight materials for this purpose has more limited value, because structural as well as material aspects of the test sample influence the test result. Caveat emptor.

The RCT was further investigated by Travers [78], who suggested means for improving the reliability of the test. Furthermore, the Linerboard Technical Committee of the Japan Paper Association made an examination of the instrumental errors encountered in the RCT [45].

IPC-Modified Ring Crush Test A modified ring crush test was developed at the Institute of Paper Chemistry [29,30]. As illustrated in Fig. 31, the cylinder has the vertical edges sealed with contact cement and the loading edges reinforced with wax. For a suitable edgewise compression test, buckling of the specimen must be prevented. Therefore, consideration is given here to the buckling of paperboard cylinders, as discussed in Section II, this being the basic form of the modified ring crush specimens.

Some results for this test are given in Fig. 30 [30], which presents a graph of load at failure versus free height. The diameter of the cylinder was 47 mm for most papers. For the 126 g/m^2 paper, the cylinder diameter was 16 mm. It can be seen that the load at failure was virtually constant for free heights between 10 and 20 mm. It is also of interest to note that he 126 g/m^2 specimens appear to have experienced Euler buckling and passed through the long-cylinder range to the short-cylinder range as the height was decreased. The 337 g/m^2 specimens, on the other hand, were very sturdy, and it appears that the long-cylinder range was never entered with the heights employed in this study; certainly the Euler range was not entered. All 337 g/m^2

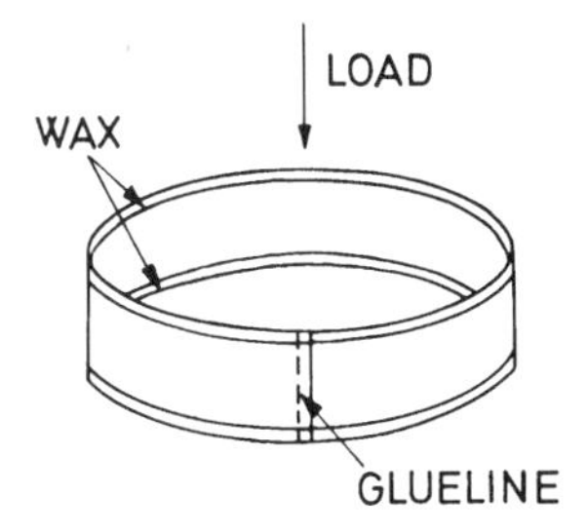

Fig. 31 IPC-modified ring crush test. (From Ref. 29.)

specimens exhibited a crush type of failure with no visual or audible evidence of long-cylinder buckling.

The load at failure is independent of height for the shorter specimens, meeting the requirement that structural buckling be eliminated. The failure involves crushing in the body of the specimen rather than the edges, showing that measured strength is not artificially low because of edge failure. However, as pointed out in Section II, when the specimen is formed into a cylindrical shape, high stresses may develop in the paper perpendicular to the loading direction. These radial stresses may seriously decrease the load at failure of the cylinder, but additional study is required to determine the magnitude of this effect.

FPL Supported Cylinder Test Figure 32 shows a cylinder tested with a method developed by Setterholm and Gertjejansen [68], which enables the stress–strain curve to be recorded by means of an attached strain gauge. This test uses a ring specimen 100 mm in height support from within by a split cylinder to prevent any inward buckling as compression loads are applied to the ends of the tube. Specimens are formed from flat sheets of paper rolled two turns in a tubular form on a mandrel with exactly the same 25 mm diameter as the inner, supporting split cylinder. Thicker paperboard can be tested using larger diameter cylinders. Tape is used to hold the outside wrap. The contribution to the total load of the tape used for joining the sheet ends has been found to be insignificant. Specimen ends have been reinforced against end crushing to ensure failure within the body of the specimen. Because a close fit of the sheet around the split cylinder is desirable to prevent buckling at low loads, frictional forces cannot be avoided, but they can be minimized by coating the inside of the specimen with powdered graphite or Teflon. The primary drawbacks of the method are the difficulty in sample preparation, which is very tedious, and the stresses that develop perpendicular to the loading direction due to the forming of the paper into a cylinder, as discussed in Section II. Using this technique, Mayers [50] investigated the effect of drying on the compression properties of paper.

F. Corrugated Specimen

A method was developed for use with corrugated specimens to evaluate the contribute of the medium to the compression strength of the combined board. It exists under various names, including the Concora fluted crush test (CFC-0) for corrugated medium [46,75], the H&D* stiffness test [53,74], and the corrugated crush test (CCT)

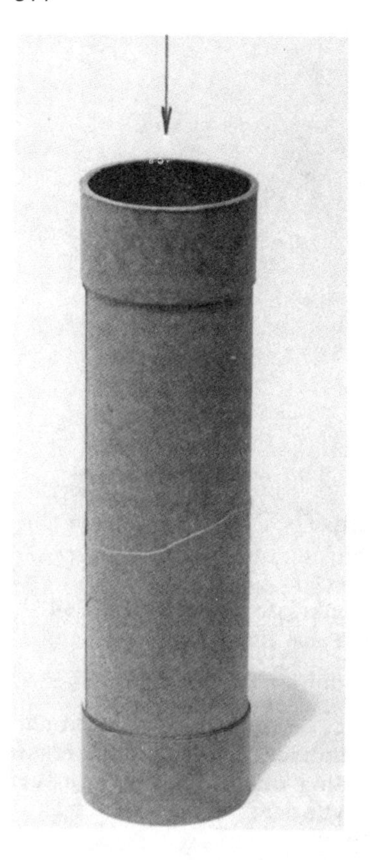

Fig. 32 FPL supported cylinder test (M 125 765). (Photo courtesy of Forest Products Laboratory, Madison, WI.)

[44]. This method is applicable to a corrugated specimen and employs a holder as shown in Fig. 33. The 152 mm long paper specimen is fluted on a Concora fluter, a device that uses heated meshed gears to simulate a standard corrugator. The fluted sample is fastened in jaws shaped to simulate an A-flute corrugation. The fluted configuration structurally hinders buckling of the specimen, but there is a risk that the method underestimates the true compression strength because of edge crushing of the specimen. The influence of the deformations during corrugation on the edgewise compression strength has not been specifically reported in the literature, but they are expected to affect the value in a way similar to that occurring in practice. Drawbacks to the use of this method are the time necessary for the conditioning the fluted specimen and the limitation in basis weight of the sheets that can be tested.

G. Comparison of Different Test Methods

An extensive study of the CD compression strength of linerboard measured by different test methods was performed at the Institute of Paper Chemistry [31]. Samples from this investigation have also been evaluated by the two STFI methods.

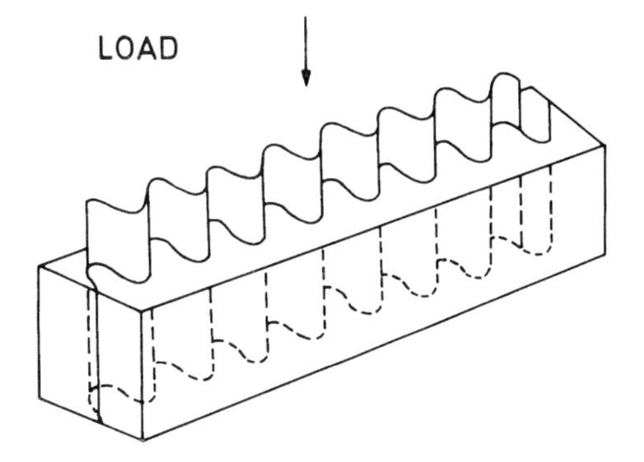

Fig. 33 Principle of the compression test of a corrugated medium specimen.

Figure 34 shows results obtained according to the different methods at equal strain rates. The compression strength obtained by the STFI short-span test is roughly 15% higher above 300 g/m^2 than that obtained by the three plate-support methods but virtually equal below 300 g/m^2. The plate-support methods in turn give values 20% higher than the Weyerhaeuser and modified ring methods. The lowest values are obtained with the RCT and Concora liner test (CLT) methods.

Differences in measured compression strength of the same material by different methods can be due to

Moisture content	Temperature
Deformation rate	Loading uniformity
Friction	Sample preparation
Clamping effects	Artificial constraints

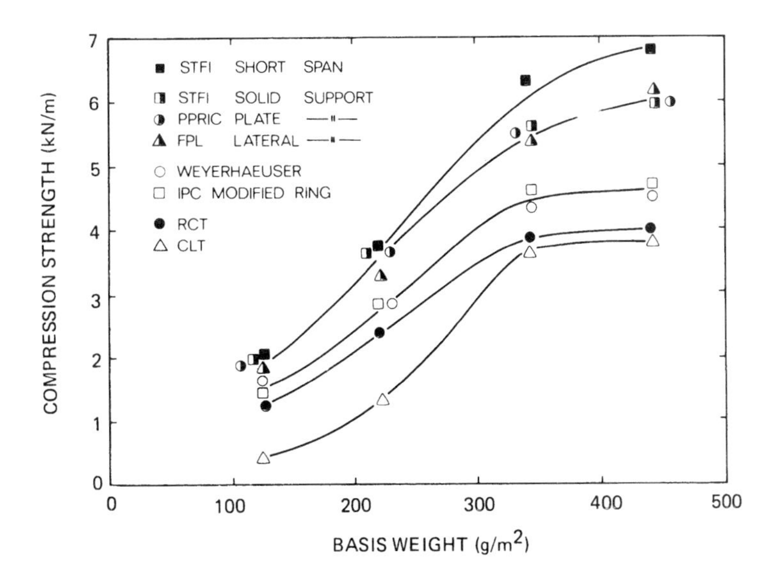

Fig. 34 Comparison between the CD compression strength of linerboard versus basis weight according to different test methods. (Courtesy of Christer Fellers, STFI, Stockholm, Sweden, and Raj Seth, PPRIC, and from Ref. 31.)

Assuming that these are properly handled, as in the above comparison, there still remain unaccounted discrepancies in measured compression strength. The two major themes of this review account for these discrepancies as being due to large-scale structural instability of the sample and the weak-link effect.

The first theme in this review is the prevention of large-scale structural instabilities during testing, which when they occur always reduce the measured strength to below the material strength. (If the material strength is reached first, one never gets to the point of structural instability.) This is the logic that underlies the search for the test that gives a maximum value for strength, friction aside.

Competing with this point of view is the idea that structural instability at the fiber level defines failure. Differences among test values may also reflect the level of transverse constraint, for example between plate-support tests and short-span tests. This is further complicated by surface roughness, which changes the relative constraint in the plate-support test. For smooth sheets, the short-span test gives values almost equal to those obtained using the plate-support methods. However, the plate-support methods give lower values, in spite of the virtually perfect lateral support of the specimen, when sheet thickness is not as regular, due to the surface irregularities on a larger scale of magnification (present for the sheets at higher basis weights).

The second theme in this review, developed in detail below, brings to light the weak-link effect in compression testing, which arises from differences in sample size among the different tests. This implies that larger specimen sizes will always have lower strength, even if shear instabilities are successfully inhibited and the test artifacts tabulated above are considered. This offers a qualitative understanding of the strength ranking in Fig. 34 but clouds the meaning of compression strength. It remains to develop a quantitative comparison, which will facilitate a clear understanding of the relationship between strength and sample size.

Figure 35a represents an area of paper under compression. Because of the fibrous structure of the material, there are variations in basis weight, density, bonding, fiber orientation, and other aspect-influencing strength, which we might call

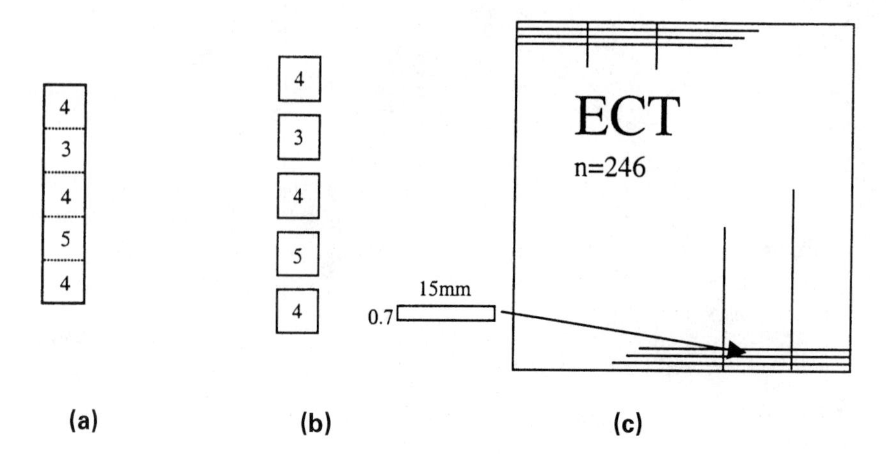

Fig. 35 Illustration of weak-link concept. (a) Composite test. (b) Five individual tests. (c) Schematic of edge crush test (ECT) conceptually composed of test areas equivalent in size to the STFI testing zone.

structural formation, to indicate that more than just variation is being considered. Structural formation gives rise to local variations in compression strength. This weak-link approach assumes that the strength of the larger sample is defined by the weakest area of the sample rather than the average. This is illustrated in Fig. 35b. Quantification of the strength of the larger sample means defining the *strength distribution* of the weakest area in the large sample. Strength of the larger sample can then be computed using order statistics.

Order statistics is a well-developed branch of statistics that investigates the marginal probability density functions for data that have been arranged by value. We will focus on the statistics—average, variance, and skewness—of the minimum value of a set of strength tests.

Peirce [55] successfully studied the strength of textile yarns of different lengths, assuming that he underlying strength was normally distributed. Fisher and Tippett [20] in earlier work exploited by Peirce, showed that the distribution had to be either normal or Weibull (although the name Weibull would not attached until a decade later). The definition of the Weibull probability density function is given below.

Because our objective is to interrelate results of different test methods, it is convenient to consider the larger area specimen in Fig. 35c to be composed of n small areas equal in size and shape to those of the STFI short-span test ($10.5\,mm^2$). The presumption is that the minimum strength in this set of n areas controls the strength of the total area. In the work described below, the "small areas" are tested using the STFI test, the probability density function of the test values is determined, and strength estimates are made of the total area.

Figure 36 shows the STFI strip method compression results for 500 tests on a single type of $275\,g/m^2$ linerboard. Superimposed on the histogram are the two- and three-parameter Weibull distributions and the normal distribution. The normal distribution permits negative strengths, the two-parameter Weibull permits zero strength, and the three-parameter Weibull uses the additional parameter to define the minimum positive strength of the distribution.

The different probability density functions to the test data can be fitted in several ways, but for clarity we can focus on parametric methods. We can represent the data in terms of the average, variance, and skewness. Variance represents the width of the distribution, whereas skewness represents the degree of asymmetry. A positive skewness in the strength data of Fig. 36 implies a long right-hand (strong-side) tail; a negative skewness would have a long left-hand (weak-side) tail.

Distributions with two parameters are able to represent the average and variance of the data but not the skewness. The normal distribution is symmetrical, whereas the two-parameter Weibull distribution has a large negative skewness. The three-parameter Weibull distribution adequately represents the average, variance, and skewness of strength data.

The distribution parameters are used to compute the minimum value statistics, meaning the average and variance of the minimum strength Both the two- and three-parameter Weibull distributions have closed-form solutions giving the difference in strength between the large- and small-area samples. These solutions are given below The normal distribution has the disadvantage of requiring numerical methods to determine the average strength of the distribution and therefore of the larger sample.

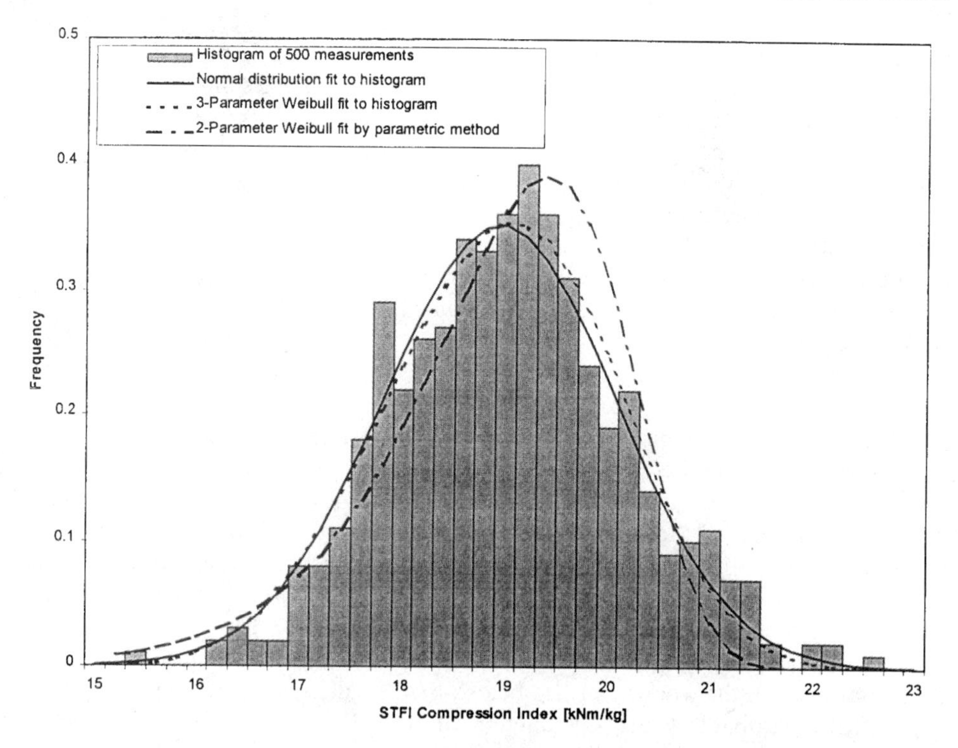

Fig. 36 Distribution of strength obtained by STFI short-span compression method, with normal distribution two- and three-parameter Weibull distributions.

Westerlind and Carlsson [81] used two-parameter Weibull distribution and the methodology of the composites literature to show that the strengths s_{large} and s_{small} of large and small samples are related to the volume ratio:

$$\frac{s_{\text{large}}}{s_{\text{small}}} = \left(\frac{V_{\text{large}}}{V_{\text{small}}}\right)^{\xi/\xi_0} \tag{12}$$

where V is the volume of material of either the large- or small-area specimen, ξ is the short-span coefficient of variation, and ξ_0 is the Weibull shape parameter defining the underlying distribution to be Weibull with zero minimum strength.

Actual distributions obtained using the STFI short-span method tend to be symmetrical or have slight positive skewness (the distribution in Fig. 36 is skewed slightly to the right) [14]. The two-parameter Weibull distribution underlying Eq. (12) corresponds to a substantial negative skewness; the ease of use may be offset by inaccuracy of the prediction if the volume ratio is large.

For a symmetrical distribution, Donner [14] used the three-parameter Weibull to show that

$$\frac{s_{\text{large}}}{s_{\text{small}}} = 1 - 3.243\xi\left[1 - \left(\frac{V_{\text{large}}}{V_{\text{small}}}\right)^{-0.2776}\right] \tag{13}$$

Although it isn't obvious, Eq. (13) transforms into Eq. (12) at the large, negative skewness of the two-parameter Weibull.

Before Eq. (13) can be applied to assess the role of structural formation and the weak link in controlling the difference among the various compression strength tests reported above, an additional correction is required. The proximity of the clamps in the short-span test constrains the material from transverse (widthwise) expansion during compression. If we presume that the material between the clamps is in plane strain, then the material will appear to be stiffer by $1/(1 - \nu^2)$, where ν is the geometric mean in-plane Poisson ratio, taken to be about 0.3 [5]. This makes the short-span strip in-plane strain about 9% stiffer than indicated by a uniaxial tensile test for measurement of modulus. Based on the assumption that the material will fail at the same strain whether or not the sample is in plane strain, the sample will be 9% stronger than the same material in uniaxial stress as well.

Figure 37 demonstrates the application of Eq. (13) including the Poisson correction. The horizontal axis represents the strength measured using a tube with a constrained seam. The tube is 25 mm in diameter and 25 mm high. The sample is epoxied into steel rings and uniformly loaded using an MTS servohydraulic system. The vertical axis of Fig. 37 represents the STFI short-span compression strength of the same material.

If the two methods were to give the same result, the data would plot on the solid line. Instead, the short-span data lie well above the line, showing that the short-span strength is considerably stronger than that of the tube.

The triangles in Fig. 37 are the tube strength calculated from the STFI test, using Eq. (13) and applying the correction for the Poisson effect. Because the predicted strength is similar to the measured strength (the data lie in close proximity to the unity slope line), the figure shows that the weak-link correction factor brings the small-area strip compression test into correspondence with the large-area tube sample. Previously, it was believed that the difference between the tube strength and the STFI strength showed structural buckling in the tube specimen, but application of Eq. (13) shows that this is unlikely.

H. Meaning of Compression Strength

Figures 34 and 37 clearly show that different test methods yield different results for compression strength. The key reasons for these differences seem to be a clamping effect in the STFI short-span instrument and a structural formation effect in the larger area tests. Together, these accounted for about 36% systematic difference in measured compression strength (Fig. 37). There always remains the possibility of structural buckling, especially in MD testing. The point is that even in the absence of structural buckling there exists an underlying reason why compression strength of different size samples will be different in general.

Figure 38 shows the results of modifying a standard STFI instrument to allow measurement of compression strength over a narrow range of gauge lengths. The small increase in strength with reduction of gauge length is qualitatively consistent with the weak-link argument. Clearly, the concept of a compression strength plateau requires some additional investigation, but for most quality assurance purposes this

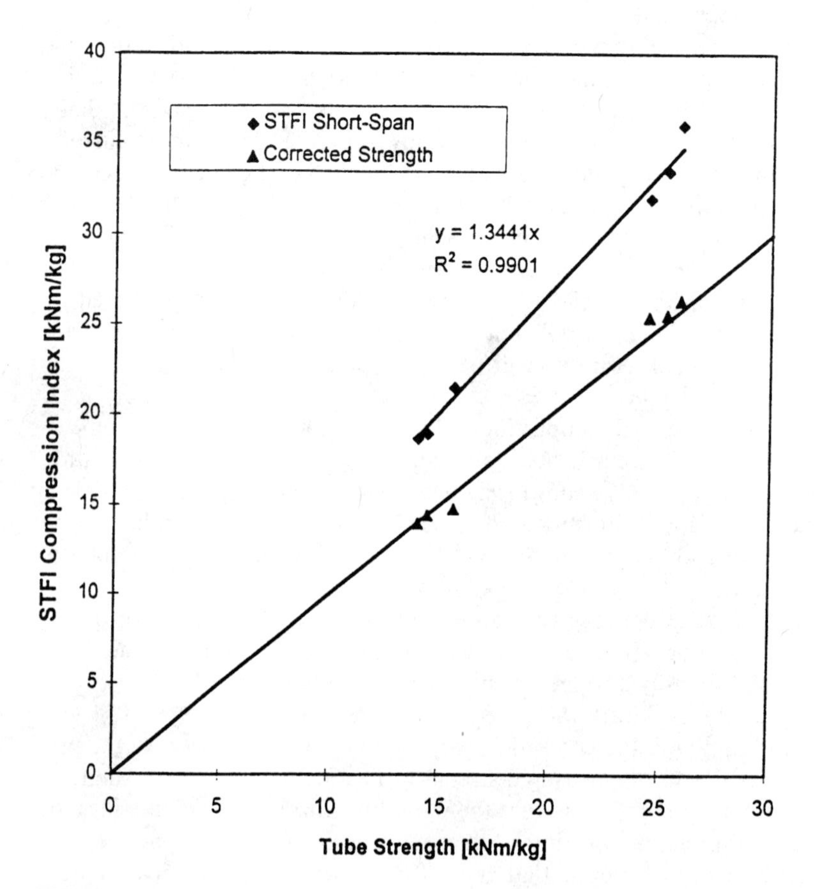

Fig. 37 Comparison of STFI short-span test with the Weyerhaeuser tube test. Area ratio is 193.

investigation is not warranted because the slenderness ratio is substantially constant for a given grade to be tested.

The commercial objective of compression strength measurement is often to determine the suitability of liner and medium components for use in the corrugated container side wall. The large area of the side wall means that Eq. (13) needs to be used to relate the STFI short-span test result to the strength of the side wall. This involves the coefficient of variation, a measure of the structural formation of the material.

Different headboxes, top formers, and table configurations lead to differences in structural formation, in turn leading to differences in compression strength variability. The STFI short-span method for measuring average compression strength seems to be the most reliable method in the manufacturing environment, but the average needs to be augmented by the coefficient of variation to provide complete information of component performance in the container.

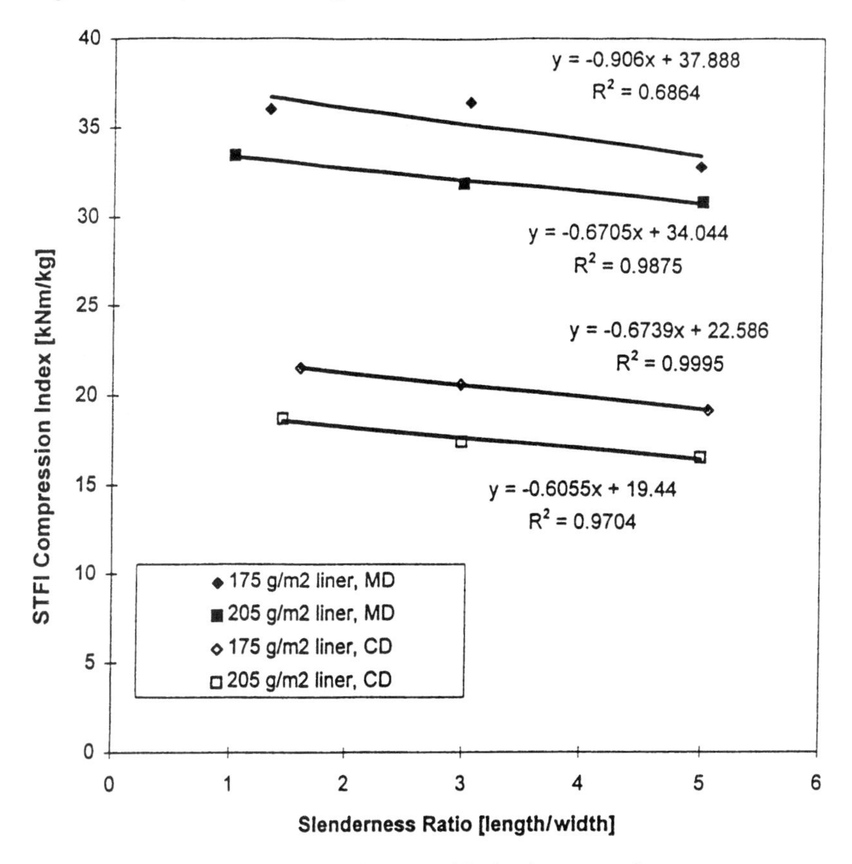

Fig. 38 Close-up of strength changes with slenderness ratio.

I. Commercially Available Instruments

For the testing methods described in this chapter, the following instruments are commercially available:

1. Concora liner test (CLT)
2. STFI short-span test
3. Ring crush test (RCT)
4. Corrugated crush test (CCT), H&D stiffness test, and Concora fluted crush test

In Europe these instruments may be bought through Lorentzen and Wettre, Box 4, S-16393, Stockholm, Sweden; in North America through Testing Machines, Inc., 400 Bayview Avenue, Amityville, NY 17701, USA.

REFERENCES

1. Aitchison, C. S., and Tuckerman, L. B. (1939). The "pack" method for compressive tests of thin specimens of materials used in thin-wall structures. National Advisory Committee for Aeronautics, Report No. 645.

2. American Society for Testing and Materials. (1974). Method for testing veneer, plywood and other glued veneer constructions. ASTM Design D 805.
3. American Society for Testing and Materials. (1977). Standard methods of compression testing of metallic materials at room temperature. ANSI/ASTM E9-77.
4. Baum, G. A. (1981). Orthotropic elastic constants of paper. *Tappi 64*(8):97–101.
5. Biot, M. A. (1963). Exact theory of buckling of a thick slab. *Appl. Sci. Res. Sec. A.* 12:182–193.
6. Barlowe, H. W., Stillwell, H. S., and Lu, H.-S. (1941). The "plug" method for obtaining the compression elastic properties of thin-walled sections. *J. Aeronaut. Sci. 8*(3):109–118.
7. Budiansky, B. (1974). Theory of buckling and post-buckling behavior of elastic structures. *Adv. Appl. Mech. 14*(2):1–65.
8. Cavlin, S., and Fellers, C. (1945). A new method for measuring the edgewise compression properties of paper. *Svensk Papperstidn. 78*(9):321–332.
9. Chamis, C. C. (1974). Micromechanics strength theories. In: *Composite Materials*, Vol. 5, *Fracture and Fatigue*. L. J. Broutman, ed. Academic Press, New York, pp. 93–151.
10. Cowan, W. F., and Cowdrey, E. J. K. (1974). Evaluation of paper strength components by short-span tensile analysis. *Tappi 57*(2):90–93.
11. Dinwoodie, J. M. (1968). Failure in timber. 1. Microscopic changes in cell-wall structure associated with compression failure. *J. Inst. Wood Sci. 21*(10):37–53.
12. Donner, B. C., and Backer, S. (1987). The role of continuum instabilities in compression and bending failure of paper and paperboard. *Proceedings* of the 1987 International Paper Physics Conference, Mont-Rolland, Quebec, Canada.
13. Donner, B. C. (1985). High-curvature bending of fibrous structures, with emphasis on paper and paperboard. Doctoral Thesis, Department of Mechanical Engineering, MIT, June 1985.
14. Donner, B. C. (1989). The Impact of Structural Formation on Compression Strength of Paper. Weyerhaeuser Internal Report, August 1989.
15. Fellers, C., and Jonsson, P. (1975). Compression strength of linerboard and corrugated medium: An analysis of testing methods. *Svensk Papperstidn. 78*(5):172–175 (in Swedish).
16. Fellers, C. (1977). Procedure for measuring the interlaminar shear properties of paper. *Svensk Papperstidn. 80*(3):89–93.
17. Fellers, C., de Ruvo, A., Elfstrom, J., and Htun, M. (1980). Edgewise compression properties. A comparison of handsheets made from pulps of various yields. *Tappi 63*(6):109–112.
18. Fellers, C. (1980). The significance of structure on the compression behavior of paper. Ph.D. Thesis, The Royal Institute of Technology, Stockholm, Sweden.
19. De Ferran, E. M., and Harris, B. (1970). Compression strength of polyester resin reinforced with steel wires. *J. Compos. Mater. 4*:62–71.
20. Fisher, R. A., and Tippett, L. H. C. (1928). Limiting forms of the frequency distribution of the largest or smallest member of a sample. *Proc. Camb. Phil. Soc. 24*:180–190.
21. Franks, R., and Binder, W. O. (1941). The stress-strain characteristics of cold-rolled austenitic stainless steels in compression as determined by the cylinder test method. *ASTM Proc. 41*:629–645.
22. Gerard, G., and Becker, H. (1957). *Handbook of Structural Stability*. 3. *Buckling of Curved Plates and Shells*. NACA TN 3783. Washington, DC.
23. Grossman, P. U. A., and Wold, M. B. (1971). Compression fracture of wood parallel to the grain. *Wood Sci. Technol. 5*:147.
24. Gunderson, D. E. (1983). Edgewise comparison of paperboard: A new concept of lateral support. *Appita 37*(2):137–141.
25. Gunderson, D. E. (1984). A comparison of three methods for determining the edgewise compressive properties of paperboard. *Appita 37*(4):307–313.

26. Habeger, C. C., and Whitsitt, W. J. (1983). A mathematical model of compressive strength in paperboard. *Fibre Sci. Technol. 19*:215–239.
27. The Institute of Paper Chemistry. (1956). Development of liner stiffness test. Part II. Correlation of pack strength with box-compression strength. Compression Report 62.
28. The Institute of Paper Chemistry. (1961). An investigation of test methods for determining the edgewise compression strength of the components of combined board. Part I. Description of a clamped-end column test and comparison of five methods of testing components in edgewise compression. Compression Report 74.
29. The Institute of Paper Chemistry. (1963). An investigation of test methods for determining the edgewise compression strength of the components of combined board. Part II. Development of a modified ring compression test and its relationship to combined board strength. Compression Report 78.
30. The Institute of Paper Chemistry. (1966). Effect of specimen dimensions on edgewise compression test of linerboard and corrugated medium. Part I. Effect of specimen height on modified ring compression and Concora liner tests. Compression Reports 82 and 83.
31. The Institute of Paper Chemistry. (1977). Comparative evaluation of methods of evaluating edgewise compression strength of containerboard. Project 2694-13.
32. The Institute of Paper Chemistry. (1978). Compression failure morphology. Project 2695-20.
33. Jackson, C. A., Koning, J. W., and Gatz, W. A. (1978). Edgewise compressive test of paperboard by a new method. *Pulp Paper Mag. Can. 77*(10):T180–T183.
34. Johnson, M. W., Urbanik, T. J., and Denniston, W. E. (1979). Optimum fiber distribution in single wall corrugated fiberboard. Forest Service Research Paper FPL 348.
35. Johnson, M. W., Urbanik, T. J., and Denniston, W. E. (1980). Maximizing top-to-bottom compression strength. *Paperboard Packag. 65*(4):98–108.
36. Kellicut, K. Q. (1959). Compressive strength of paperboard. *Package Eng. 4*(3):76–77.
37. Koiter, W. T. (1970). The Stability of Elastic Equilibrium Translation. Tech. Rep. AFFDL-TR-70-25. Air Force Flight Dynamics Laboratory, Air Force Flight Systems Command, Wright-Patterson Air Force Base, Ohio, February.
38. Kollmann, F. F. P. (1963). Phenomena of fracture in wood. *Holzforschung 17*(3):65–71.
39. Koning, J. W. (1975). Compressive properties of linerboard as related to corrugated fiberboard containers: A theoretical model. *Tappi 58*(12):105–108.
40. Koning, J. W. (1978). Compressive properties of linerboard as related to corrugated fiberboard containers: Theoretical model verification. *Tappi 61*(8):69–71.
41. Koning, J. W., and Haskell, J. H. (1979). Papermaking factors that influence the strength of linerboard weight handsheets. U. S. Forest Service Research Report FPL 323.
42. Kubát, J., and Rudström, L. (1968). Stiffness, edgewise compression strength and creasability of carton board (in German). 12th EUCEPA Conference, Berlin.
43. Lager, J. R., and June, R. R. (1969). Compressive strength of boron-epoxy composites. *J. Compos. Mater. 3*:48–56.
44. Langaard, Ö. (1968). Optimization of corrugated board construction with regard to compressive resistance of boxes. *Verpackungs-Folien/Papiere 11*:14–20 (in German).
45. Linerboard Technical Committee, Japan Paper Association. (1980). An examination of instrumental errors in the ring crush test of paperboard. *Jn. Tappi 34*(5):375–384.
46. Maltenfort, G. G. (1956). Compression strength of corrugated containers. Parts 1–4. *Fiber Containers 41*, Part 1 (7):44, 49–50, 52, 57; Part 2 (9):60–62, 67–68, 72, 74; Part 3 (10):52–54, 59; Part 4 (11):80.
47. Maltenfort, G. G. (1961). Improvements in Concora liner testing. *Paperboard Packag. 46*(5):76–77.
48. McEvoy, R. P. (1979). An experimental study of the compressive strength of paper. Masters Thesis, Syracuse University, Syracuse, NY.

49. McKee, R. C., Gander, J. W., and Wachuta, J. R. (1963). Compression strength formula for corrugated boxes. *Paperboard Packag. 48*(8):149–159.
50. Mayers, G. C. (1967). Effect of restraint before and during drying on edgewise compressive properties of handsheets. *Tappi 50*(3):97–100.
51. Moody, R. C. (1965). Edgewise compression strength of corrugated fiberboard as determined by local stability. U.S. Forest Service, Research Paper FPL 46.
52. Morris, R. M., and Van Liew, G. P. (1975). An improved edgewise compression test for linerboard. *Tappi 58*(11):110–113.
53. Ostrowski, H. J. (1960). The new H&D stiffness test for predicting the stacking strength of corrugated containers. *Pulp Paper Mag. Can. 61*(C):T130–T132 (convention issue).
54. Page, D. H. (1969). A theory for the tensile strength of paper. *Tappi 52*(4):674–681.
55. Peirce, F. T. (1926). 32X tensile tests for cotton yarns, v. "the weakest link": Theorems on the strength of long and of composite specimens. *J. Textile Inst.* T355–T368.
56. Perkins, R. W., and McEvoy, R. P. (1981). The mechanics of the edgewise compressive strength of paper. *Tappi 64*(2):99–102.
57. Popov, E. P. (1976). *Mechanics of Materials*. Prentice-Hall, Englewood Cliffs, NJ.
58. Ramberg, W., and Miller, J. A. (1946). Determination and presentation of compressive stress-strain data for thin sheet metal. *J. Aeronaut. Sci. 13*(11):569–580.
59. Ranger, A. (1967). Evaluation of fibrous materials for boardmaking and converting. *Paper Technol. 8*(3):245–250.
60. de Ruvo, A., Cavlin, S., Engman, C., Fellers, C., and Lundberg, R. (1975). Characterization of the mechanism of compressive failure in paper by viscoelastic methods. International Paper Physics Conference, Ellenville, NY.
61. de Ruvo, A., Cavlin, S., Engman, C., Fellers, C., and Lundberg, R. (1975). Viscoelastic characterization of the failure mechanism in compression of paper. *Das Papier 29*(7):280–288 (in German).
62. de Ruvo, A., Fellers, C., and Engman, C. (1978). The influence of raw material and design on the mechanical performance of boxboard. *Svensk Papperstidn. 81*(18):557–566.
63. Sachs, I. B., and Kuster, T. A. (1980). Edgewise compression failure mechanism of linerboard observed in a dynamic mode. *Tappi 63*(10):69–73.
64. Sauret, G., Trinh, H. J., and Lefebvre, G. (1969). Experiments to reproduce multi-ply boards in the laboratory. *Das Papier 23*(1):8–12 (in German).
65. Seth, R. S., and Soszynski, R. M. (1979). An evaluation of methods for measuring the intrinsic edgewise compressive strength of paper. *Tappi 62*(10):125–127.
66. Seth, R. S., Soszynski, R. M., and Page, D. H. (1979). Intrinsic edgewise compressive strength of paper: The effect of some papermaking variables. *Tappi 62*(12):97–99.
67. Setterholm, V. C. (1956). Method for determining tensile properties of paper. U.S. Dept. Agriculture, Forest Products Laboratory, Madison, WI. Report No. 2066.
68. Setterholm, V. C., and Gertjejansen, R. O. (1965). Method for measuring the edgewise compressive properties of paper. *Tappi 48*(5):308–312.
69. Shigematsu, K., Imada, K., and Takayanagi, M. (1975). Formation of kink bands by compression of the extrudate of solid linear polyethylene. *J. Polym. Sci.* (*Polym. Phys. Ed.*) *13*:73–86.
70. Soszynski, R. M. (1979). Personal communication.
71. Stockmann, V. (1976). Measurement of intrinsic compressive strength of paper. *Tappi 59*(7):93–97.
72. Tappi Standard T472 Su-68 (1968). Compression resistance of paperboard (ring crush test).
73. Tappi Useful Method 801. The Concora liner (edge) crush test (CLT).
74. Tappi Useful Method 805. Stiffness test for fluted corrugating medium.
75. Tappi Useful Method 811. The Concora fluted crush test for corrugating medium.
76. Timoshenko, S. (1936). *Theory of Elastic Stability*. McGraw-Hill, New York.

77. Timoshenko, S. (1976). *Strength of Materials*, Parts 1 and 2. Krieger, Huntington, NY.
78. Travers, R. (1976). Improving the reliability of the ring crush test. *Appita 30*(3):235–240.
79. Uesaka, T. (1983). Edgewise compressive strength of paper board as an instability phenomenon. *Svensk Papperstidn*. R191–R197.
80. Urbanik, T. J. (1981). The principle of load-sharing in corrugated fiberboard. *Paperboard Packag*. November:122–128.
81. Westerlind, B., and Carlsson, L. (1982). Compressive response of corrugated board. *Tappi J*., July:145–154.

10

RESIDUAL STRESSES IN PAPER AND BOARD

JOHN FREDERICK WATERHOUSE*
Institute of Paper Science and Technology
Atlanta, Georgia

*Retired.

I would like to dedicate this chapter to the memory of three paper scientists who encouraged and inspired me to seek a career in the discipline of paper physics. The first is Dr. Bertil Ivarsson, who was my first research director when I joined Westvaco Corporation in 1966. Bertil not only encouraged my involvement in paper physics through refining but, as I later discovered, had done landmark research into the nature of drying stresses. The second is Dr. Otto Kallmes, a great teacher and friend in my formative days of paper physics. The third is Dr. David Rutland, for many years a scientist with Xerox Corporation and well respected in the paper physics community, who contributed to an understanding of the nature and contribution of internal stresses to the dimensional stability of paper.

I. INTRODUCTION

This chapter is mainly concerned with residual stresses in paper and board, the term "residual stresses" being often used interchangeably with "internal stresses". Later, consideration is given to Kubat stresses, a term which is used to refer to "internal" or dried in stresses.

A. Definition of Residual Stresses

A body can be in a state of stress even though it has no external forces or constraints acting on its boundary. Such a body is said to have a residual stress distribution. The residual stress distribution may sometimes give rise to an undesirable distortion of the body.

Residual stresses can be produced in most materials through a variety of means and are usually employed to enhance material performance, e.g., to improve a material's strength. On the other hand, residual stresses may have an adverse effect on a material's performance, e.g., by leading to optical distortion in glass. Therefore a balance is usually sought between the positive and negative aspects of residual stresses.

In what follows we will use the term "paper" to describe both paper and board unless otherwise stated. It is surprising that little attention has been paid to the presence of residual stresses in paper, that is, to their importance, origin, measurement, and control. Dimensional stability and strength-related properties are areas where residual stresses should play an important role. The papermaking process is considered vital to the establishment of residual stresses, and converting processes such as calendering and coating may further modify them.

II. RESIDUAL STRESSES IN MATERIALS

Residual stresses can arise in most materials such as glass, wood, metals, ceramics, polymers, and composites and are important to their performance.

Thermal toughening (or strengthening) is a process used to create a residual stress distribution in glass that effectively enhances its strength [1]. When glass, a brittle material, is subjected to bending stresses, it will fail when the tensile stress in the outer layers reaches a critical level, giving rise to rapid crack growth and catastrophic failure. Therefore, if the outer layers of the glass can be put into compression, then this compressive stress level has to be overcome before the surface layer goes into tension, i.e., the glass is effectively strengthened. The outcome of the thermal toughening process is to have the outer layers of the glass in compression and, core in tension so that there are no resultant forces on the glass. The residual stress distribution is parabolic as shown in Fig. 1. The conditions for producing a parabolic stress distribution will be discussed later.

The thermal toughening process consists of heating the glass to a temperature in the range of 600–750°C, withdrawing it from the furnace, and rapidly quenching it by using jets of compressed air. The magnitude of the residual stresses is dependent on the initial glass temperature and the maximum temperature differential established between the outside and the center of the glass. During quenching the outside layers cool more rapidly and become more rigid while the more fluid inner layer continues to cool and shrink, thus putting the outside layers into compression. During the initial quenching phase the outer layers may initially be in tension, and if these stresses become too large then fracture occurs.

Residual stresses play an important role in the curing of wood [2]; incorrect drying procedures can lead to damage such as checking and cracking. Distortion of the ubiquitous 2×4 is familiar to everyone! Test prong samples shown in Fig. 2 illustrate the different stress states that can be produced. In a somewhat analogous situation to glass, tension stresses can arise both at the surface and in the interior of the wood depending on the drying history. The analysis of the development of residual stresses in wood is a complex nonlinear problem. Lewis et al. [3] used finite element analysis to calculate the development of drying stresses in wood. In an analysis that was both simplified and more realistic, the development of drying stresses was calculated and the phenomenon of stress reversal demonstrated, i.e., the outside layers of the beam that were originally in tension went into compression.

The manufacture of polymer parts by casting or injection molding can sometimes result in an asymmetrical residual stress distribution, producing an unacceptable distortion of the part as discussed earlier in the definition of residual stresses. Furthermore, the residual stress distribution has to be carefully controlled in the production of plastic pipes, because the inside surface layer, if not correctly cooled, may be in tension. As with glass, this is undesirable, because flaws may be more readily propagated.

Residual stresses are inevitable in composite materials due to differences in the shrinkage behavior of the matrix and reinforcing fiber [4]. Warpage and cracks may arise as a consequence of an unfavorable residual stress distribution. The wood fiber is viewed as a natural composite; however, the residual stress state of a wood fiber

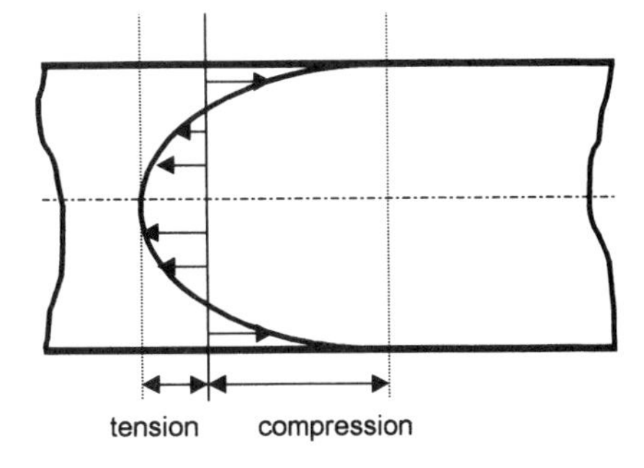

Fig. 1 Residual stress distribution.

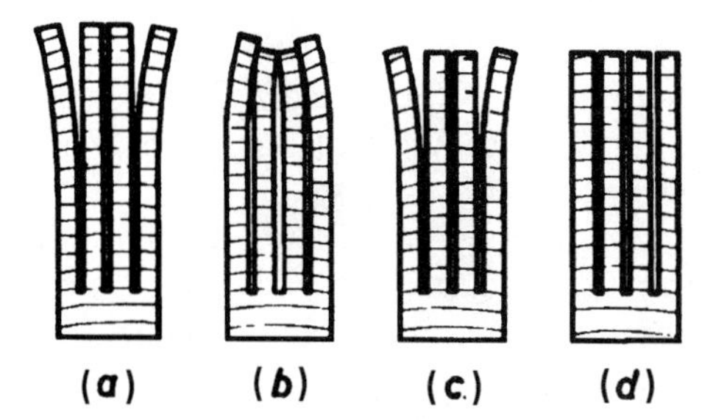

Fig. 2 Test prong samples that illustrate the presence of drying or residual stresses in timber. (a) Perpendicular stresses on the board surface at the beginning of drying. (b) Perpendicular stresses in the center of the board and compressive stresses on the surface (casehardening) for any further sharp drying. (c) Casehardening removed by steaming or conditioning. (d) Stress-free dried timber. (From Ref. 2.)

has yet to be determined. Birefringence is one manifestation of a residual stress distribution and may be used e.g., with glass, for the nondestructive determination of residual stresses.

Paper, basically a random network of self-bonding wood fibers, may have residual stress states at different levels of organization including both the fiber and network level. The origin of these stresses is not yet completely understood, but Salmen et al. [5] proposed a model involving the role of both intra- and interfiber regions as shown in Fig. 3. Furthermore, papermaking materials and their distribution in the thickness or z direction of paper, papermaking and drying strategies, calendering, and other mechanical treatment may create or modify the level and distribution of residual stresses in paper. Residual stress states in paper are expected to be important with respect to a number of performance areas including dimensional stability, strength, and fiber rising (induced surface roughness in coating and other aqueous finishing processes) [6–12].

In some instances residual stress states are undesirable because they can give rise to distortion that can adversely affect other properties such as the optical properties of glass used in lenses and mirrors. In this situation annealing processes are used to minimize residual stress levels. In the thermal toughening of glass a balance has to be sought between residual stress development for strength and induced distortion; which can negatively impact optical properties.

III. RESIDUAL STRESSES IN PAPER

Residual stress states in materials can originate in a variety of ways, including differences in crystal growth, viscoelastic behavior, differences in chemical composition, mechanically induced stresses, shrinkage, cooling, and drying or curing processes.

A. The Viscoelastic Nature of Paper

Paper, a highly anisotropic structure, can be produced from almost any source of fiber, although wood is by far the most common. Wood fibers can be described as

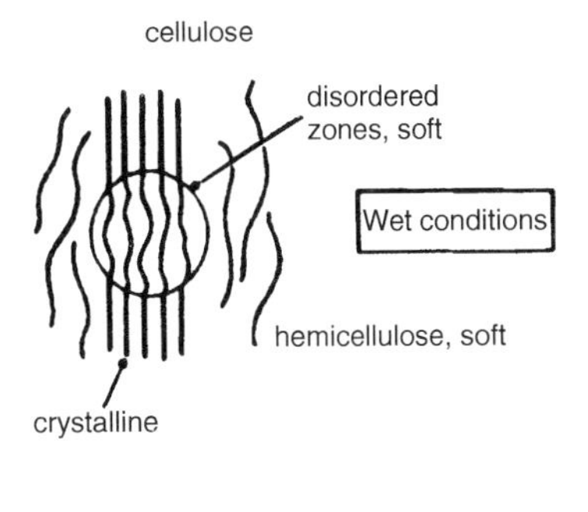

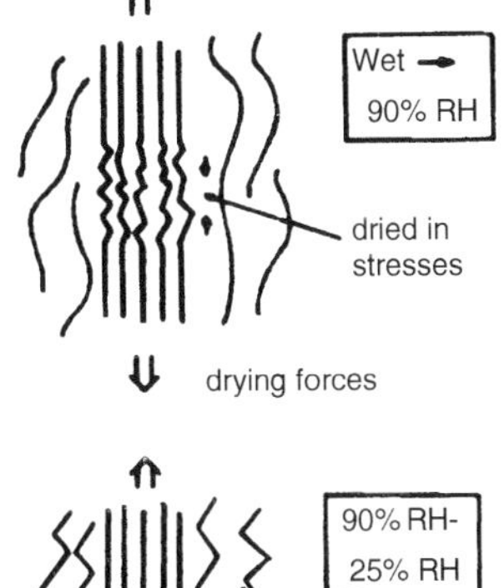

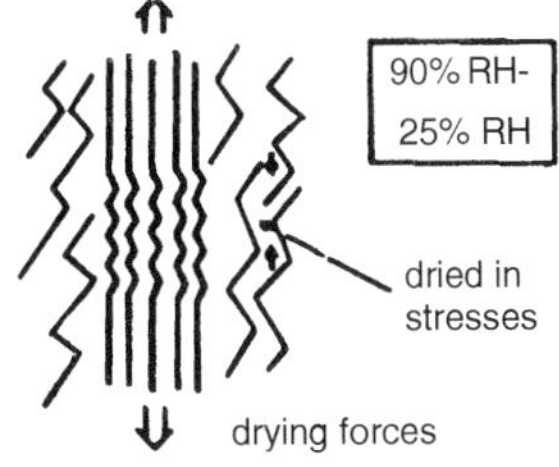

Fig. 3 Origins of dried-in stresses. (From Ref. 5.)

semicrystalline viscoelastic composites. The bulk of the wood fiber is contained in the middle secondary wall, which consists of cellulose microfibrils embedded in a matrix of lignin and hemicelluloses. The microfibrils are load-bearing elements that are aligned at an angle to the axis of the fiber known as the fibril angle. The lignin and hemicelluloses are amorphous polymers, which together with the less ordered regions of the cellulose microfibrils are responsible for the viscoelastic behavior of the fiber.

The cellulose fiber can be modeled as a composite consisting of highly ordered cellulose chains (crystallites) embedded in a matrix of amorphous cellulose and lignin. It is the "amorphous" components, i.e., nonordered cellulose regions, hemicellulose, and lignin, that are mainly responsible for the fiber's association with water, viscoelastic behavior, and internal stress state (see Fig. 3). Water acting as either a plasticizer or an antiplasticizer [13] modifies this behavior by changing the "softening" temperature. The viscoelastic behavior of paper is directly related to the viscoelastic behavior of its fibers but may be mediated by bonding and structural factors.

The moisture sensitivity, hygroexpansivity, and shrinkage behavior of paper are also dependent on the presence of amorphous polymers. The greater the amorphous content, the higher the moisture content will be at a given relative humidity. Dimensional changes with changes in moisture content are also strongly dependent on the level of amorphous polymers present as well as the reinforcing effect of the cellulose microfibrils. This is best illustrated by the shrinkage behavior of a wood fiber. It has been found that the fiber shrinks minimally, e.g., 1.5%, along its length, whereas the lateral shrinkage may be as much as 25% in going from the wet to the dried state. In the former case the cellulose microfibrils, which are closely parallel to the axis of the fiber, are effective in resisting the shrinkage of the amorphous polymers. By comparison there is relatively no resistance to lateral shrinkage, because the microfibrils are now in series with the amorphous polymers. As with composites, differences in shrinkage behavior between fiber (microfibrils) and matrix (hemicellulose) may produce significant levels of residual stress. However, stress relaxation may diminish the magnitude of these stresses.

Furthermore, by interactions at interfiber bond regions a residual stress state can be expected to develop in the fibrous network. Composition and structure may also vary from layer to layer in the *z* or thickness direction of paper; differences in fiber type, orientation, and fines distribution are not uncommon.

B. Development of Residual Stresses in Fibers and Paper

Opportunities exist in papermaking and converting for the development of residual stresses in fibers and papers, particularly in going from the wet to the dry state, and such stresses can also be induced mechanically.

As mentioned previously, wood fibers undergo considerable lateral shrinkage in going from the wet to the dry state, e.g., 25%, whereas shrinkage along the axis of the fiber is only about 1.5%. It is well documented that a sheet of paper can undergo considerable shrinkage during drying depending on the extent of bonding within the network. Furnish, refining, and wet pressing are the main factors controlling bonded area. It has also been shown that bonding is mainly established between fibers prior to the onset of major fiber shrinkage during drying. Thus if the network is unrestrained it will undergo significant shrinkage, e.g., 12%, during drying. This occurs because the fiber's transverse shrinkage results in axial compression of the fibers to which they are bonded if they are not restrained.

A sheet of paper dried without restraint will have a very uneven (cockled) appearance, presumably due to buckling and distortion of fiber segments. On the other hand, when shrinkage is partially or totally prevented, drying stresses will be developed in the network as demonstrated by Ivarsson [14], whose results are shown in Fig. 4. The overall level of (or average) drying stress is controlled by the final moisture content to which the sheet is dried, rather than the rate of drying, which might be expected to influence the residual stress distribution [10]. Furthermore, as a sheet of paper loses moisture, i.e., when moisture is lost from the cell wall of the fiber, a critical moisture content is reached at which significant shrinkage begins to occur. Any shrinkage that takes place prior to the sheet being restrained will also have an effect on the ultimate drying stress induced in the sheet. The ultimate level of drying stress will depend on many factors, including the extent of refining, the level of interfiber bonding, and the shrinkage potential of the fibers. As shown in Fig. 4, the establishment of drying stresses is monotonic, and there is no evidence of stress relaxation (although this has not been tested over very long times). Htun [10] also made very careful drying load measurements employing temperature-regulated clamps. In a related publication, Htun and de Ruvo [15] commenting on stress

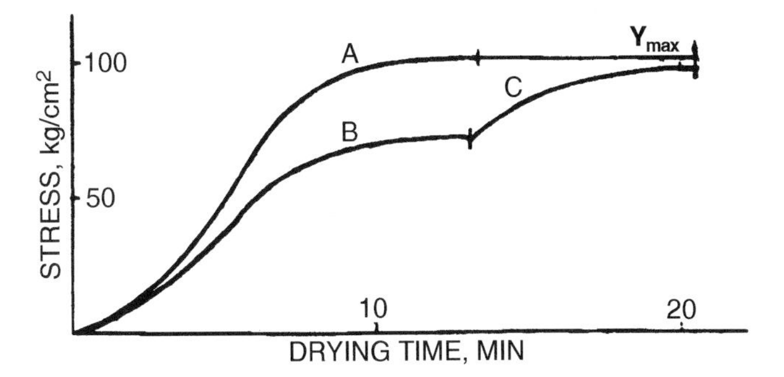

Fig. 4 Development of drying stresses as a function of drying time. Curve A: Drying with infrared heat (60°C) to $m = 0.04$. Curve B: air drying to $m = 0.09$, followed by infrared drying (curve C) to $m = 0.04$. (From Ref. 14.)

relaxation in the plateau region, stated that no relaxation is seen with chemical pulps but that there is a slight one with mechanical pulps.

If the sheet is fully restrained during the drying process, then it cannot change its in-plane dimensions as a result of lateral fiber shrinkage. Therefore, axial stresses will develop in the fiber and in turn contribute to the overall level of drying stress in the network, i.e., $\sigma_D(t) = E_s(t)\varepsilon_s(t)$, where $E_s(t)$ and $\varepsilon_s(t)$ are the effective modulus and drying strain, respectively, in the network at time t and the prevailing moisture content and temperature. This stress level will not be maintained in the early phases of drying because of stress relaxation. I am not aware of any attempts thus far to model this process. Clearly such a procedure could also be used to predict the residual stress distribution in paper as has been done for other materials such as wood [3].

Residual Stresses in Fibers In the single fiber, microstresses originate in the fibril hemicellulose–lignin matrix according to Salmen et al. [5], whereas Giertz [16] and Page [17] proposed that at the paper level (as illustrated in Fig. 5) macrostresses are developed at and between the interfiber bond regions in paper.

Jentzen [18], Spiegelberg [19], and Kim et al. [20] examined the effect of loads applied to single fibers during drying and their impact on the subsequent stress–

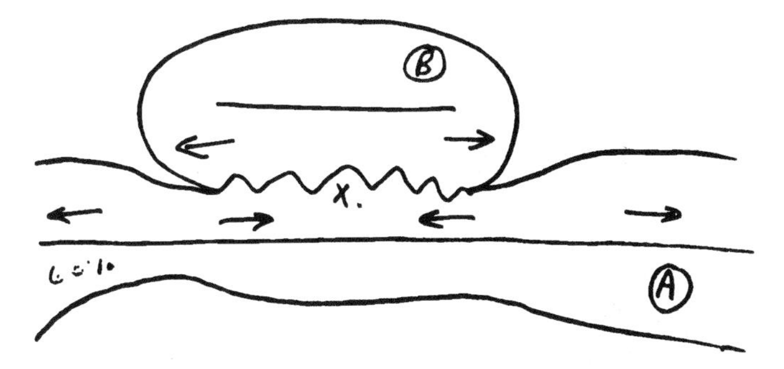

Fig. 5 Origin of drying stresses according to Page. (From Ref. 17.)

strain behavior of the fibers. Note that in these experiments it was the applied external loads rather than the stresses developed during drying that were measured. So we do not have any direct knowledge of the drying stresses included in fibers at different levels of restraint. In applying drying loads we have the possibility of creep during the drying and, in the as-yet-unexplored case, stress relaxation. The extent to which the results from one apply to the other remains to be determined. In the case of paper it is web strain (shrinkage or draw) that is controlled during drying.

Ignoring this difficulty for the moment, both Jentzen [18] and Spiegelberg [19] found that the stress–strain behavior of single fibers was significantly affected by drying load. Jentzen found that both the modulus and tensile strength of earlywood and latewood fibers increased as drying load increased, as shown in Table 1.

For summerwood fibers, Spiegelberg also found that as modulus and tensile strength decreased as hemicelluoses were extracted.

In a more extensive single-fiber studies, Kim et al. [20] investigated the effect of drying under load on single-fiber deformation behavior. Their drying loads covered a greater range than those of Jentzen [18] and Spiegelberg [19]. Their main conclusions were that single-fiber strength could be improved by drying under load if conditions were such as to allow shearing to take place between fibrils to remove cell wall defects such as local changes in fibril orientation. Thus the hemicellulose matrix must be sufficiently swollen, and therefore fibers that have been dried may not show an improvement in strength unless they can be reswollen by, say refining. This is stated to be the only mechanism possible at low fibril angle, whereas at high fibril angle both shearing and changes in fibril angle could be responsible for improvement of fiber strength.

The mechanisms proposed by Kim et al. [20] were also offered as an explanation of the results of Jentzen and Spiegelberg. However, it should be remarked that the fibers investigated by Kim et al. were subjected to two drying cycles before being rewet and subjected to a drying load.

The implications of Kim et al.'s findings are that if never-dried defect-free fibers of low fibril angle are dried under restraint, then any associated drying stress would not significantly increase their modulus or tensile stress. What might be the magnitude of the drying stresses induced in an ideal single fiber comprising fibrils at zero angle embedded in a hemicellulose matrix? If we assume that all the potential shrinkage comes from the hemicellulose matrix and that no slippage occurs, then the fiber can shrink laterally without hindrance by, say, about 25%. If the fibrils are fully restrained in the axial direction, then as the matrix shrinks the fibrils will be put into

Table 1 Selected Modulus and Tensile Strength Data of Jentzen

	Modulus E (dyn/cm^2)			Tensile strength σ (dyn/cm^2)		
	Load (9)			Load (9)		
	1	3	5	1	3	5
Spring wood	4750	5510	—	91.0	109.8	—
Summer wood	5370	6370	5630	129.3	145.1	135.4

Source: Ref. 18.

compression and the matrix will be in tension. This would give rise to a residual stress and a measurable drying stress in going from the wet to the dry state. How this might vary with fibril angle is also an unknown. The matrix modulus is considerably lower than that of the fibrils, and presumably the internal stress state developed would not be sufficient to distort or buckle the fibrils. Page et al. [21] stated that fiber strength is almost exclusively controlled by the cellulose fibrils and therefore the contribution from the internal stress state may be negligible, at least for fibers having a small fibril angle. The drying stresses that are developed in a network of fibers are probably dominated by the stresses that develop at the interfiber bonds. Page and Tydeman [22], Rance [23], and others demonstrated that the shrinkage of paper is potentially much greater than the axial shrinkage of isolated fibers, as discussed earlier in this section.

Gray [24] uses the term "chiral stresses" to refer to a phenomenon that supposedly occurs from the molecular to the continuum level in paper. Mark [25] has also demonstrated this winding and unwinding effect at the single-fiber level. One manifestation of this is the winding and untwisting of fibers as a result of residual stress changes with changes in moisture. The right-handed helical structure in the S2 layer of the wood fiber produces a right-handed twist in the fiber. This leads to a right-handed twist in paper strips the magnitude of which is dependent on the degree of fiber orientation.

Bell and Edie [26] calculated the internal stress distribution in melt-spun polymer fibers. However, I believe that predicting a residual stress distribution in wood fibers would be quite challenging.

Residual Stresses and Interfiber Bonds Salminen et al. [27] attempted to model the internal stress state induced by drying at the interfiber bond. The influence of this stress state on axial, shear, and z-direction bond strength has been modeled. The level of drying stress Δs is defined as the difference between the axial and transverse shrinkage of the fiber. The specific bond strength is greatest in the z direction, followed by out-of-plane shear and then axial loading of the fiber. Inferences about specific bond strength by Waterhouse [28] and Stratton [29] based on out-of-plane shear and z-direction tensile testing indicate that the z-direction strength of the bond is considerably lower than its shear strength. It does not appear that Salminen et al. accounted for the anisotropic nature of the fibers. Therefore, if one simply accounts for a lower modulus in the transverse direction of the fiber, there would be closer agreement with the findings of Waterhouse [28] and Stratton [29].

According to Salminen et al. [27], drying stresses are predicted to have an adverse effect on both z-direction and shear specific bond strngths. That is, freely dried sheets will have a higher specific bond strength than restrained dried sheets. They also offered an explanation—less shrinkage difference or a decrease in bonded area—for the greater specific bond strength of latewood fibers compared with earlywood fibers. Bither and Waterhouse [30] speculated that specific bond strength differences for earlywood and latewood fibers might be explained by the work of Button [31], who found, using linear fracture mechanics, that specific bond strength increased as the bond width-to-thickness ratio increased. These predictions were in good agreement with bond strength measurements made using cellophane model fibers.

Residual Stresses in Paper When paper is subjected to some level of restraint during drying, stresses are developed. Generally, the drying stress level increases with increased interfiber bonding resulting from refining, wet pressing, and a more favorable realignment of the fibrils. After drying and stress relaxation is essentially complete and restraints on the sheet are removed, the external stress level is reduced to zero.

As shown in Figs. 6a and 6b, when the drying stress is constant from layer to layer in the thickness direction of paper, then when the paper is no longer held under restraint, i.e., no external loads are being applied, there should be no residual stress distribution. However, when the drying stress is nonuniform through the thickness and the paper is no longer held under restraint, a residual stress distribution will be present as shown in Figs 6c and 6d.

When released from restraint, the paper will remain planar only if the residual stress distribution is symmetrical. Furthermore, if layers of the paper can be carefully removed without inducing residual stresses, then curvature will develop, as evidence of equilibrating an imbalance in the residual stress distribution as a result of layer removal. This is a somewhat simplistic view, because it is possible that residual stresses exist at other levels of organization and are not apparent in the layer removal procedure just described.

The drying process and stress relaxation appear to play a key role in the establishment of residual stresses in paper.

C. "Internal" or "Kubat" Stresses

Htun [10,32] reviewed the nature and measurement of internal stresses in paper and stated [32] that a number of terms are used to describe internal stresses, including microstress, macrostress, residual stress, thermal stress, and others, and are often used interchangeably. The term "internal stress" appears to be more frequently used for polymers [33], whereas "residual stress" is used almost exclusively in the metals

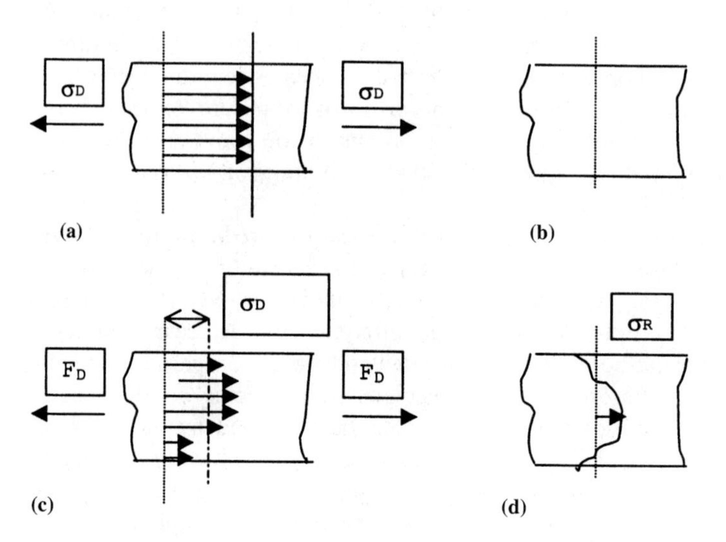

Fig. 6 (a,b) Uniform and (c,d) nonuniform internal stress distributions.

field [34], but it appears that these terms are used interchangeably. However, when discussing paper, and the avoid confusion with these terms, it is recommended that the term "internal" or more preferably "Kubat" stress [5] be used to refer to so-called internal stresses derived from stress relaxation procedures as discussed in more detail below.

Kubat [35], in his dissertation, examined similarities in the stress relaxation behavior of high polymers and metals. The relationship between the inflection slope F of the stress–log time curves $(d\sigma/d\ln t)_{\text{inflection}}$ and the total decrease in stress $\Delta\sigma$ was found to be $\mathbf{F \approx 0.1\Delta\sigma}$. This relationship was said to hold for a wide variety of materials in the absence of "internal" stresses, i.e., for carefully annealed materials. No specific discussion of the nature of the "internal" stresses was presented.

In the same thesis a contribution by Johansson et al. [36] dealing with the stress relaxation behavior of cellulose gel, films, and paper claimed that the above relationship between F and $\Delta\sigma$ still held. To quote from Johansson et al., "The difference observed between virgin and water-treated paper was typical of the influence of internal stresses on $F(\sigma_0)$" [36]. This is again a reference to internal stresses that are not defined but presumably relate to the drying stress to which the paper has been subjected. These authors also state that a greater deviatiaon in the F–$\Delta\sigma$ plot is produced when the "internal" stress is increased by strain hardening or drying under stress.

In a later paper, Johnasson et al. [7] defined what they meant by internal stresses: "In the present paper, the term dried in stress is replaced by the more general term internal stress." In this context, "internal" is not equivalent to residual stress.

The stress relaxation technique was later used by Johansson and Kubat [37] to determine the level of "internal" stress in paper and how it was influenced by papermaking variables. Htun [32] recognized that the meaning of "internal" stress in paper is not well understood; nevertheless, he has used the term synonymously with drying stress.

IV. MEASUREMENT OF RESIDUAL STRESSES IN PAPER

A variety of methods have been developed for determining the residual stress states in materials, including layer removal [38], sectioning techniques [39], hole drilling [40], X-ray diffraction [41], stress relaxation [37], neutron diffraction [42], ultrasonic [43], Raman [44], and photoelastic techniques [1]. Recently a technique referred to as successive grooving has been developed for polymer composites [45] the principle of which has been adapted for paper [46] and called diffusion-assisted stress relaxation.

A. Stress Relaxation Methods and "Kubat" Stresses

The stress relaxation method as described below is used to determine the "internal" or "Kubat" stress level in paper and other materials. In some instances it may strongly correlate with the drying stress to which the paper was subjected. However, it is *not* a measure of residual stress.

Craven [47], Johansson et al. [36], Johansson and Kubat [37], Kubat and Rigdahl [49], Robertson [50], Lepoutre and Skowronski [12], Htun [10], Back and Klinga [51], and Waterhouse et al. [52] have used stress relaxation techniques to determine the internal or Kubat stress levels in paper and other materials and how they are influenced by papermaking and converting operations.

Stress relaxation, a time-consuming method for determining Kubat stress, consists of generating for a given sample a series of stress relaxation curves at different applied stress levels σ_0. The maximum slope of stress σ vs. log time curve is then determined. A plot is made of the variation of this maximum slope with the initial applied stress σ_0. A linear relationship is generally obtained, and the intercept on the σ_0 axis is defined as the Kubat stress σ_i as shown in Fig. 7. This is the stress level at or below which no further stress-activated relaxation occurs provided the environment does not alter. It is claimed that if the sample is annealed or stress relieved there should be no Kubat stress and thus the line shown in Fig. 7 should go through the origin.

Htun and de Ruvo [48] demonstrated that the Kubat stress, measured by the above procedure, has a high degree of correlation with drying stress and proposed that they are equivalent, i.e., $\sigma_D = \sigma_i$ is illustrated in Fig. 8. Furthermore, this relationship is claimed to be independent of refining, level of wet pressing, and the fiber fraction from which the handsheets were made, i.e., whole pulp or fines-free pulp.

Kubat stress (as does drying stress) increased with refining and wet pressing and is dependent on the shrinkage allowed during drying.

Kubat [35] found for a wide range of materials that K, given by the equation below, is quite constant, i.e., 0.1 ± 0.01. Htun [10,32] found that K had a somewhat wider range for paper, i.e., 0.08–0.13.

$$\left(-\frac{d\sigma}{d\ln t}\right)_{\max} = K(\sigma_0 - \sigma_\infty)$$

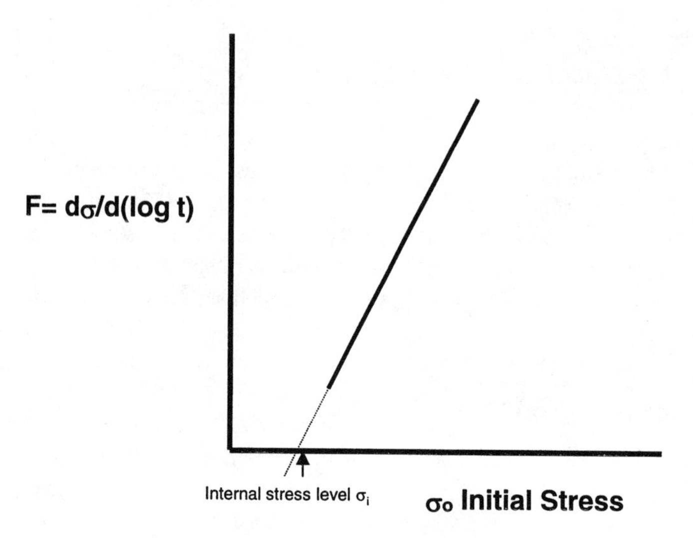

Fig. 7 Measurement of internal stress from stress relaxation measurements.

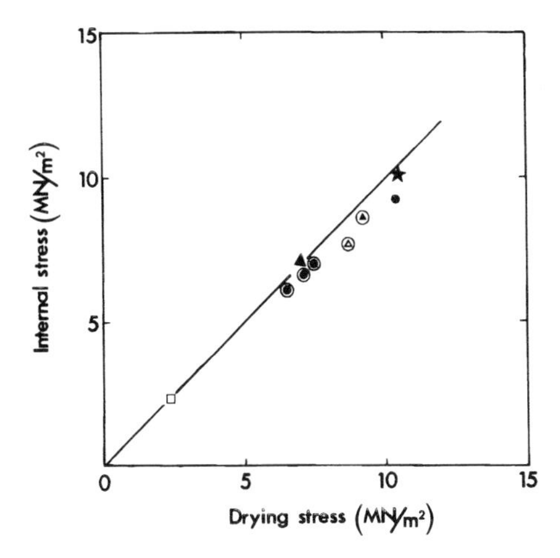

Fig. 8 Variation of internal stress with drying stresses. (*) Beaten to 50°SR; (•) beaten to 43°SR, whole pulp (◉) first beaten to 43° SR and later fractionated. Sheets containing different fiber fractions: (▲) beaten to 21° SR followed by wet pressing at 0.45 MPa; (△) beaten to 21°SR followed by wet pressing at 1 MPa, (▲) beaten to 21°SR followed by wet pressing at 2 MPa; (□) unbeaten pulp. (From Ref. 48.)

where σ_∞, the long time stress level, is equal to σ_i according to White's definition [53].

B. Sectioning, Layer Removal, and Successive Grooving Methods

Waterhouse et al. [52] used sectioning to determine the Kubat and residual stress distribution in paper, and Franke [54] and Waterhouse and Franke [55] used layer removal methods to determine the residual stress distribution in paper.

It is difficult to make a convincing argument that sectioning and layer removal techniques do not relax and/or induce additional residual stresses. This problem is not peculiar to paper. Layer removal techniques have been developed for polymers, and it has been demonstrated that if appropriate machining techniques are used, induced stresses are negligible. Isayev and Crouthamel [38] demonstrated by using annealed samples that curvature was not induced by their chosen machining method.

The grinding method used is one developed by Wink and reported by Beckman and Plucker [56] for determining z-direction filler distributions, etc., in paper. Waterhouse et al. [52] also demonstrated that by reducing the amount of material removed by grinding per pass, e.g., 0.004 in., induced stresses due to the grinding process were minimal. The grinding technique involves the use of a commercial flat bed surface grinder equipped with a suction chuck (replacing the magnetic chuck) to hold the paper samples in place during the grinding operation. A standard 78 grit grinding wheel was used. It is preferable that the grinding be conducted at low relative humidity, in the range, say, of 15–20% RH. However, if properties are then evaluated at standard conditions of 50% RH, some change in curvature and elastic properties may be expected due to moisture absorption.

Schwantes [57] made a thorough study of alternative methods of layer removal, including plasma etching, ion milling, and the excimer laser. Lasers are used commercially to "machine" paper—for greeting cards, for example—but have not been used for layer removal as far as I am aware. The excimer laser is now used routinely for changing the curvature of the eyeball. Interestingly, studies of the interaction of lasers with cellulose demonstrated that cellulose "melts" [58]. Schwantes [57] proposed that disturbances induced by layer removal, for example by an excimer laser with a suitable optical arrangement, should be very localized and therefore the induced stresses would be negligible, with no mechanical stresses and very limited thermal effects. Plasma etching was applied to paper by Sapieha et al. [59] and explored by Brodeur and Waterhouse [60] as a possible layer removal technique. Preliminary results indicated that layer removal was not uniform. We were also concerned about the environment, i.e., temperature and vacuum, and its impact on residual stresses.

One interesting and not unexpected effect seen with the grinding technique for layer removal is the relaxation of initial curvature, as shown in Table 2 for a sample of commercial linerboard.

Sunderland et al. [45] developed a successive grooving technique for measuring residual stresses in polymer composites. The method involves machining a series of progressively deeper grooves and measuring the bending response due to reequilibration of the stresses after each cut. A finite element procedure is used to evaluate the residual stress distribution. In adopting this principle to measure the residual stress distribution in paper, Sunderland et al. [46] claim that it is difficult to cut grooves in paper and therefore used local moisture diffusion to achieve the same effect. They called the technique diffusion-assisted stress relaxation. It may not be so difficult to machine grooves in paper and board using modern laser techniques.

Nevertheless, the method of Sunderland et al. [46] produces reasonable results, although one might suspect that the introduction of moisture into the sample, albeit locally, makes determination of the original residual stress distribution questionable. Relaxation effects, the precise zone of moisture diffusion, and the question of whether one can strictly assume that the irreversible stress component has been completely released are additional concerns. There is also the assumption of constancy of properties in the z direction, which is unlikely, particularly with commercial board products. The magnitude of the residual stresses is around +1 and −1 MPa, with a maximum differential of about +1 MPa tension and −1.8 MPa compression. These values are somewhat lower than those found by Waterhouse et al.

Table 2 Curvature Induced in Commercial Linerboard as a Result of Layer Removal by Surface Grinding

Sample	Initial (m^{-1})	After 1 hr (m^{-1})	After 24 h (m^{-1})	After 1 wk (m^{-1})
Control	5	5	5	5
2 mil removed	9	10	10	10
4 mil removed	11	14	14	14
6 mil removed	13	16	16	16
8 mil removed	17	22.5	25	27

[52]. The distribution is asymmetrical, with the compressive stress being higher on the wire side of the sheet, but appears to be approximately parabolic.

C. Evaluation of Residual Stress Distribution

Evaluation of the residual stress distribution from geometry, curvature, and modulus distributions is based on the method of Trueting and Read [62]. White [53] has summarized three cases as given below.

Case 1. Uniaxial Stress, $\sigma_{iy} = 0$

$$\sigma_{i,x}(z_1) = \frac{-E}{6}\left[(z_0 + z_1)^2 \frac{d\rho_x(z_1)}{dz_1} + 4(z_0 + z_1)\rho_x(z_1) - 2\int_{z_1}^{z_0} \rho_x(z)\, dz\right]$$

where σ, E, and ρ are the residual stress, elastic modulus, and radius of curvature, respectively. There is no loss of generality if specific values of modulus and stress are used.

Case 2. Negligible Curvature in One Direction, $\sigma_y = 0$.

$$\sigma_{i,x}(z_1) = \frac{-E}{6(1 - v^2)}\left[(z_0 + z_1)^2 \frac{d\rho_x(z_1)}{dz_1} + 4(z_0 + z_1)\rho_x(z_1) - 2\int_{z_1}^{z_0} \rho_x(z) dz\right]$$

and $\sigma_{iy}(Z_1) = \nu\sigma_{ix}(Z_1)$.

Case 3. Equi-biaxial Stresses $\rho_x(Z_1) = \rho_y(Z_1) = \rho$ and $\sigma_x(Z_1) = \sigma_y(Z_1) = \sigma$

$$\sigma_{i,x}(z_1) = \frac{-E}{6(1 - \nu)}\left[(z_0 + z_1)^2 \frac{d\rho}{dz_1} + 4(z_0 + z_1)\rho - 2\int_{z_1}^{z_0} \rho\, dz\right]$$

Paterson and White [62] also developed a numerical procedure for calculating the residual stress distribution when the modulus is varying from layer to layer in the thickness direction.

D. Residual Stress Distribution in Paper

Waterhouse et al. [52] initially used a sectioning method and the calculation scheme of Ribiki et al. [39] to make a crude estimate of the residual stress distribution in linerboard. The curvatures developed after grinding and the calculated residual stress distributions are shown in Fig. 9. An example of the properties obtained for the whole board felt, middle, and wire side sections are summarized in Table 3.

We also note from Table 3 that the Kubat stress σ_i varies from layer to layer in the z direction. This might imply, according to the results of Htun and de Ruvo [48], that there is a z-direction variation of drying stress and hence residual stresses present. Kubat and Rigdahl [49] developed a model for the variation of Kubat stress in

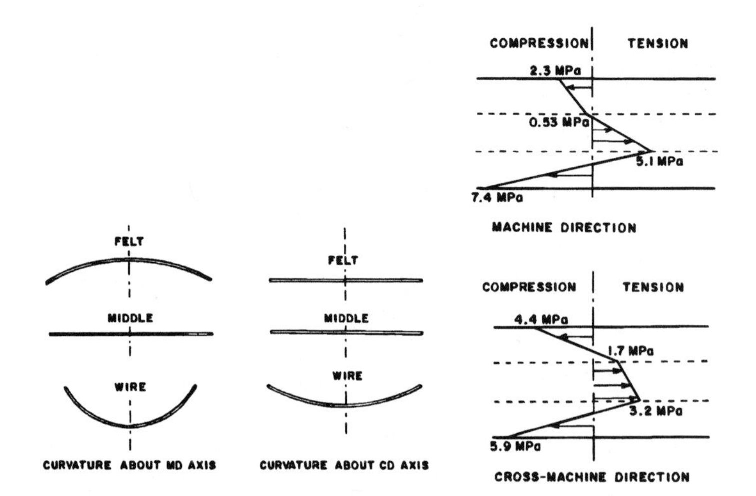

Fig. 9 (Left) Curvature and (right) residual stress measurements on 42lb/1000 ft^2 linerboard sections. (From Ref. 52.)

the thickness direction of injection-molded plastics. In this case the Kubat stress is equated with the residual thermal stress. The Kubat distribution is approximately parabolic, with the outside layers in compression suggesting that a residual stress distribution of the same form may also be present as suggested by White [53].

In later studies, Franke [54] and Waterhouse and Franke [55] used the layer removal procedure and the Trueting and Read analysis [61], as already outlined, to estimate the residual stress distribution developed in handsheets made on the Formette Dynamique and dried using different procedures.

Franke [54] investigated the effects of wet pressing and drying on residual stress and property variations from layer to layer in the thickness direction of paper. The pulp used was a fully bleached southern pine kraft market pulp that has been beaten to 600 mL Canadian Standard Freeness (CSF) in a Valley beater. Handsheets made on the Formett Dynamique had close to a random fiber orientation and a grammage of 255 g/m^2. After couching, the handsheets were stored between damp blotters in a cold room prior to wet pressing and drying. The sample size used for wet pressing and drying was $5\frac{1}{2} \times 5\frac{1}{2}$ in. The sheets were pressed and dried using a Carver heated flat platen press. The Formette sheets, which were at approximately 25% consistency, were sandwiched between polyester nonwoven milk filters to prevent sticking;

Table 3 Properties of Sectioned Commercial Linerboard

Sample	Basis weight (g/m^2)	Caliper (μm)	Density (g/cm^3)	E_{MD}/ρ (km/s)^2	E_{CD}/ρ^2 (km/s)^2	R (= E_{MD}/E_{CD})	σ_{iMD} (N m/g)	σ_{iCD} (N m/g)	Radius of cure, MD (cm^{-1})	Radius of curv, CD (cm^{-1})
Whole	207.5	287	0.723	13.1	6.23	2.10	7.72	1.56	—	—
Felt	94.1	121.9	0.772	12.4	5.19	2.39	6.65	2.06	67.1	8.08
Middle	98.7	135.8	0.727	11.4	5.04	2.27	3.95	3.12	142.2	28.2
Wire	86.9	119.1	0.729	12.1	3.81	3.17	5.61	1.10	8.46	3.68

then a blotter and a 200 mesh bronze screen were placed above and below the filters as shown schematically in Fig. 10. The sandwich was then placed in a heated Carver press, and the desired press load was quickly applied. Three press loads were used: low (23 kPa), medium (342 kPa), and high (1140 kPa).

Two drying strategies were employed. The first consisted of heating both platens to 350°F and is referred to as symmetrical pressing and drying. For asymmetrical pressing and drying, only the top platen was heated to 350°F; the lower platen was initially at room temperature but rose to 200°F before the sample was removed from the press. In both cases no shrinkage occurred until after the sheets were fully dry and released from the press.

After preconditioning and conditioning the samples were subjected to the layer removal technique described in Section III.B. The symmetrically dried samples were surface ground from the wire side of the sheet in 0.002 in. increments up to a maximum of 0.010 in. The asymmetrically dried samples were surface ground from both the wire side (which was next to the heated platen during drying and the felt side of the sheet in 0.001 in. increments up to a maximum of 0.007 in. The samples were again conditioned before testing, which included basis weight, caliper, in-plane and out-of-plane elastic constants, moisture content, and optical scattering coefficient measurements.

Table 4, which summarizes sheet properties prior to grinding for the sets of asymmetrical and symmetrical drying procedures, indicates that the properties were in reasonable agreement, although the symmetrical drying condition resulted in slightly less dense sheets and consequently lower overall elastic properties.

Results for the symmetric and asymmetric dried sheets are given in the Appendix and shown in Figs. 11a–11d.

Sheet density after grinding is expected to increase depending on the initial level of sheet roughness. In the symmetrical drying case (Fig. 11a and Tables A1 of the Appendix), there is virtually no z-direction density gradient for the low and medium wet pressing levels, whereas a density gradient is evident for the high level of wet pressing. Clearly there is a reduction in elastic properties as material is removed by surface grinding, particularly for the out-of-plane elastic modulus. In

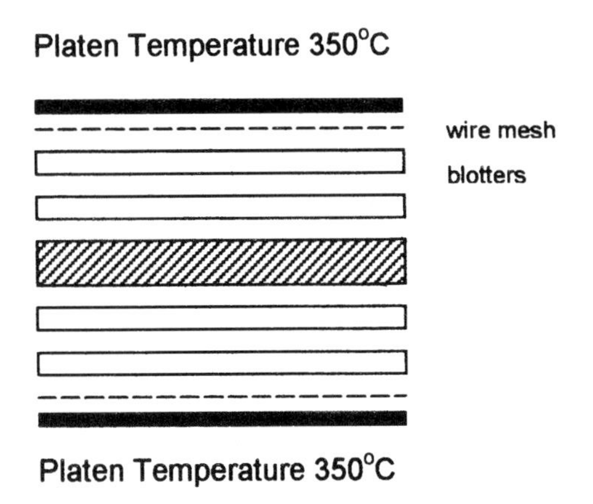

Fig. 10 Symmetrical drying arrangement.

Table 4 Summary of Press-Drying Experiments

Condition	Density (g/cm^3)	E/ρ $[(km/s)^2]$	$E_z/\rho[(km/s)^2]$	$R(=E_{md}/E_{cd})$
Dried symmetrically				
Low pressure	0.747	8.49	0.215	1.25
Medium pressure	0.826	—	0.353	—
High pressure	0.966	10.8	0.375	1.02
Dried asymmetrically, from heated to ambient				
Low pressure	0.668	8.82	0.174	1.29
Medium pressure	0.853	8.99	0.350	1.04
High pressure	1.065	11.8	0.559	1.20
Dried asymmetrically, from ambient to heated				
Low pressure	0.667	8.82	0.171	1.28
Medium pressure	0.852	9.44	0.355	1.05
High pressure	1.072	11.50	0.554	1.13

the case of the asymmetrically dried sheets (Figs. 11b and 11c and Tables A2 and A3), the z-direction variation in density is not significant at all levels of wet pressing. Again we see a significant variation in out-of-plane elastic properties, which increase in level with increased wet pressing. There do not appear to be any large differences in properties when layer removal is performed from either the heated or ambient side of the samples.

The variation in scattering coefficient with layer removal for the asymmetrical drying case (Fig. 11d) is relatively small for all levels of wet pressing. Table 5 summarizes regression analysis of the z-direction gradients of elastic properties with basis weight reduction for both the symmetrical and asymmetrical drying cases. The critical question is whether these gradients are real or an artifact of the layer removal process. The z-direction changes in elastic properties generally increase as wet pressing increases. This may be due to the release or relaxation of drying stresses with layer removal, the severity of which increases with increased wet pressing.

The unexpected finding of these experiments was that no significant curvature was measured for either of the two drying conditions. This implies that the z-direction variation in residual stress is negligible.

In a further series of experiments, Waterhouse and Franke [55] examined the properties of four sets of handsheets made on the Formette Dynamique. The overall properties of these handsheets are given in Table 6. The first set was made from a 100% softwood kaft furnish having a CSF of 600 mL. The three additional sets made included two where the sheet composition was varied in the thickness direction. The two pulps used had CSF values of 690 mL and 390 mL, respectively. These three sets comprised (1) a sheet having a uniform blend of the two pulps; (2) a three-ply sheet where the middle 50% was made of the pulp with a CSF of 390 mL and its outer layers (25%) of the pulp with a CSF of 690 mL; and (3) the same construction as set 2 but with the middle 50% comprising a pulp at 690 mL. For each of these sheet-making conditions, two levels of fiber orientation were used as shown in Table 6.

Table 5 Summary of Regression Analysis of Elastic Properties Versus Basis Weight

Symmetrical			Heated → Ambient			Ambient → Heated		
g/pass (g)	Slope $(km/s)^2m^2/g$	R^2	g/pass (g)	Slope $(km/s)^2m^2/g$	R^2	g/pass (g)	Slope $(km/s)^2m^2/g$	R^2
In-plane Slope, $\Delta(E/\rho)/\Delta(BW)$								
Low pressure								
21.0	−0.00829	0.395	6.03	−0.0305	0.919	8.23	−0.00275	0.194
Medium pressure								
24.5	−0.0120	0.916	16.8	−0.0054	0.126	15.9	−0.00097	0.013
High pressure								
23.1	−0.0240	0.733	16.3	−0.0177	0.688	18.3	−0.01512	0.989
Out-of-plane Slope, $\Delta(E_z/\rho)/\Delta(BW)$								
Low pressure								
21.0	−0.00023	0.848	6.03	−0.00088	0.863	8.23	−0.00050	0.839
Medium pressure								
24.5	−0.00136	0.990	16.8	−0.00192	0.950	15.9	−0.00157	0.991
High pressure								
23.1	−0.0012	0.990	16.3	−0.00378	0.938	18.3	−0.00197	0.953

Table 6 Formette Handsheet Properties

Sheet No.	$R(=E_{md}/E_{cd})$	Apparent density (g/cm^3)	Mean in-plane elastic constant $(km/s)^2$	Out-of-plane elastic constant $(km/s)^2$
	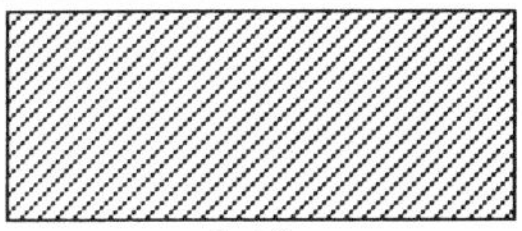			
60	1.19	0.830	9.21	0.273
61	1.74	0.834	9.29	0.264
	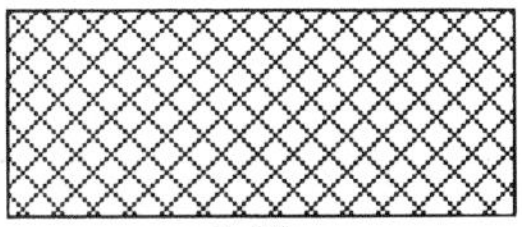			
48	1.12	0.806	9.34	0.244
50	1.73	0.790	8.77	0.244
	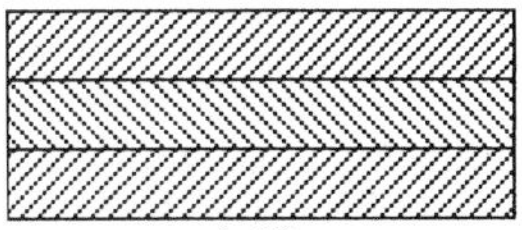			
51	1.19	0.801	9.30	0.266
53	1.85	0.809	9.42	0.276
	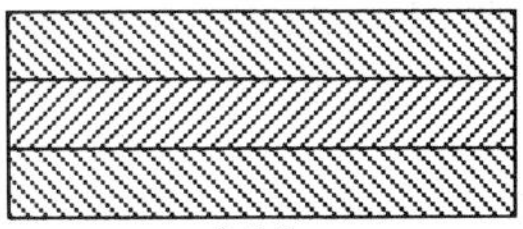			
56	1.15	0.790	9.57	0.253
58	1.90	0.804	9.80	0.267

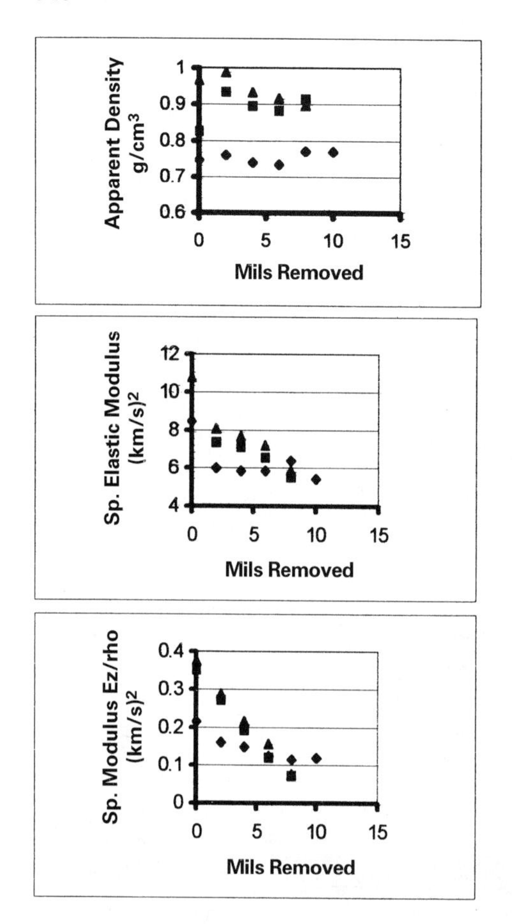

Fig. 11a Property variations with layer removal for symmetrical dried handsheets. (◆) low pressure; (■) medium pressure; (▲) high pressure.

The Formette handsheets were wet pressed and dried on a 32 in. diameter steam-heated drum dryer under full restraint; i.e., the shrinkage during drying was minimized using *z*-direction restraint in the form of a dryer felt. A schematic of the press–dryer combination is shown in Fig. 12. The level of felt tension required to achieve this condition had been previously determined.

The MD curvatures (curvature about the CD axis) developed as a result of layer removal for the four sheet structures examined are shown in Figs. 11a–11h. There was relatively little curvature about the MD axis. For the sheets of uniform composition, with reference to Table 6, sheets 60, 61, 48, and 50 (Figs. 13a–13d), the variation of curvature with mils removed is approximately linear, and as the MD/CD modulus ratio increases the degree of curvatures is reduced.

When the *z*-direction composition of the sheet is changed (with reference to Table 6, sheets 51, 53, 56, and 58), we see that the curvature is no longer linear with the thickness of material removed. Furthermore, there is a marked change in curvature close to the interface between the two layers. When the middle layer is

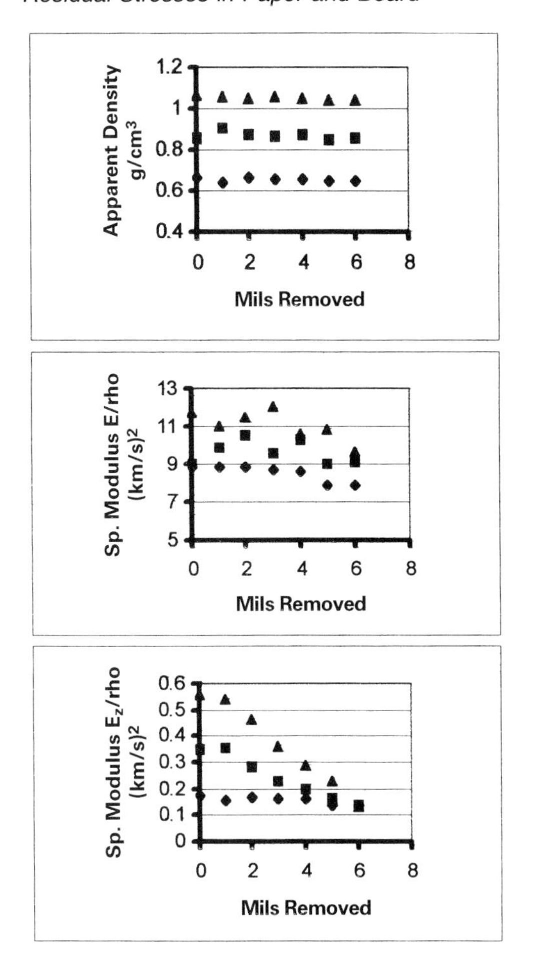

Fig. 11b Property variations with layer removal for asymmetrically, dried handsheets—heated to ambient. (◆) low pressure; (■) medium pressure; (▲) high pressure.

composed of the low CSF pulp (Figs. 13e and 13f) the curvature decreases, whereas when the middle layer is composed of the high CSF pulp (Figs. 13g and 13h) the curvature increases. As can be seen, the trends occur irrespective of the degree of fiber orientation, although their magnitude changes.

As noted by White [53], using the Trueting and Read analysis [61], it can be shown that substituting a linear variation of curvature with caliper in Case 1 of Section IV.C results in a parabolic residual stress distribution as illustrated in Fig. 14. The variation of residual stress with nondimensional caliper for sheet 60 is shown in Fig. 15. It should be noted that after drying the handsheet was close to being flat and did not retain the radius of curvature of the steam-heated drum upon which it was dried. Drying around a curved surface, where the sheet is restrained by a drying felt, puts the outside layers into tension and the inside layers into compression. Presumably persistence of this stress from the wet to the dry state is in part responsible for the sheet being approximately flat on exit from the

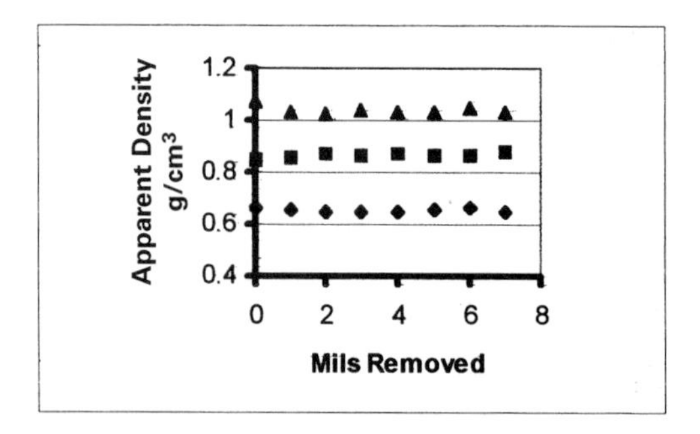

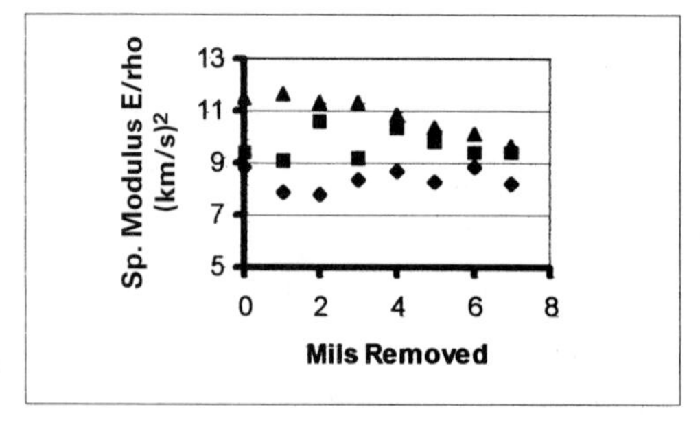

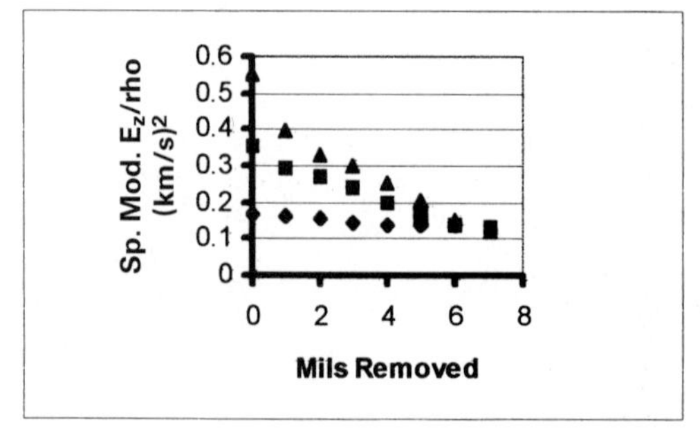

Fig. 11c Property variations with layer removal for asymmetrical dried handsheets—ambient to heated. (◆) low pressure; (■) medium pressure; (▲) high pressure.

dryer. It is also possible that some moisture redistribution and stress relaxation contributed to this result.

Why there should be a residual stress distribution in this case and not in the symmetrical and asymmetrical cases considered earlier is not yet understood. It is not believed to be an artifact of the layer removal procedure, although this is always a possibility. In the symmetrical and asymmetrical drying cases it is possible that very uniform drying stresses were developed from layer to layer in the thickness direction and that differences were further minimized by stress relaxation. This is illustrative of the case shown in Figs. 6a and 6b.

Waterhouse et al. [52] found significant curvature development when the layer removal procedure was applied to production lineboard samples. Large curvatures also developed when impulsed dried (a high intensity drying procedure) handsheets were subjected to this procedure (the conditions were such that the impulsed dried

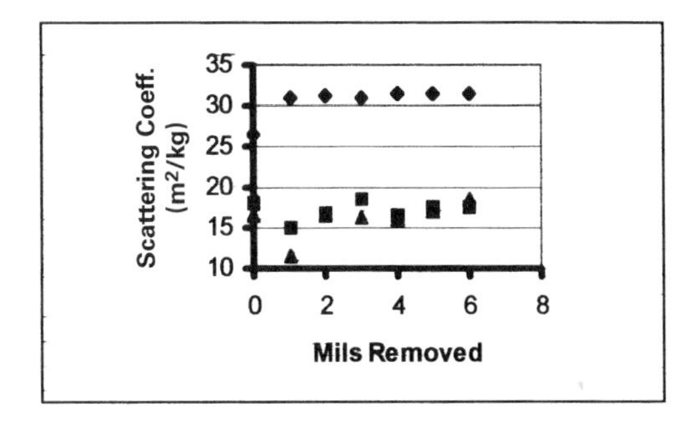

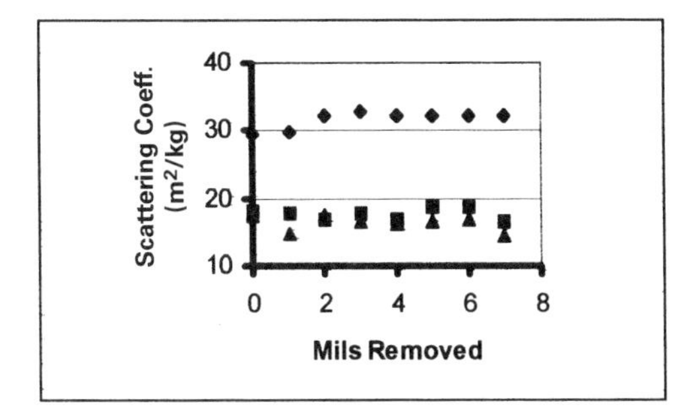

Fig. 11d Scattering coefficient variation with layer removal for asymmetrically dried handsheets—heated to ambient. (◆) low pressure; (■) medium pressure; (▲) high pressure.

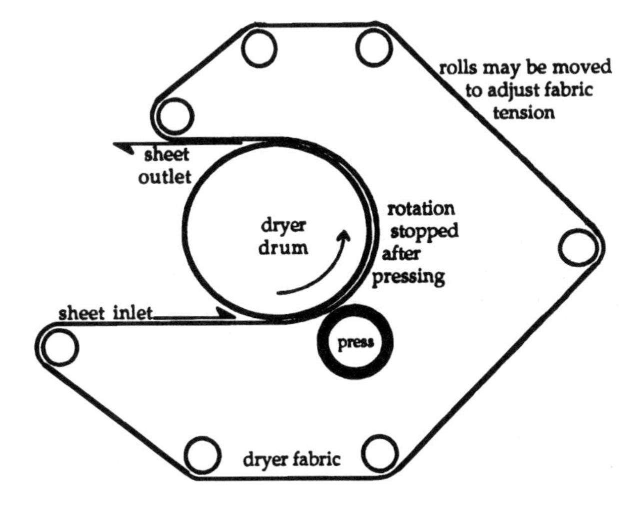

Fig. 12 Schematic of press and dryer combination.

(a) Sample 60

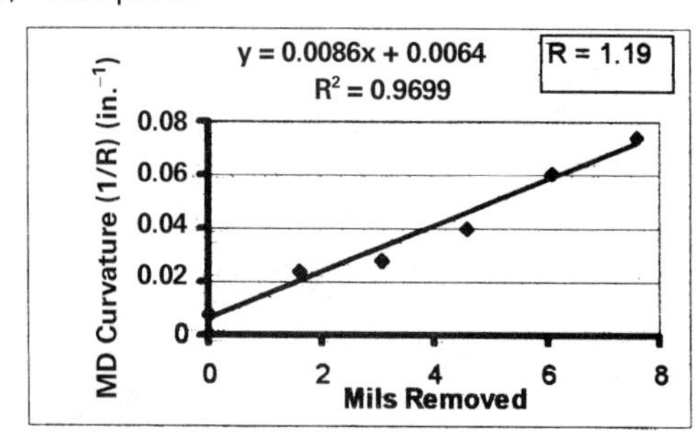

(b) Sample 61

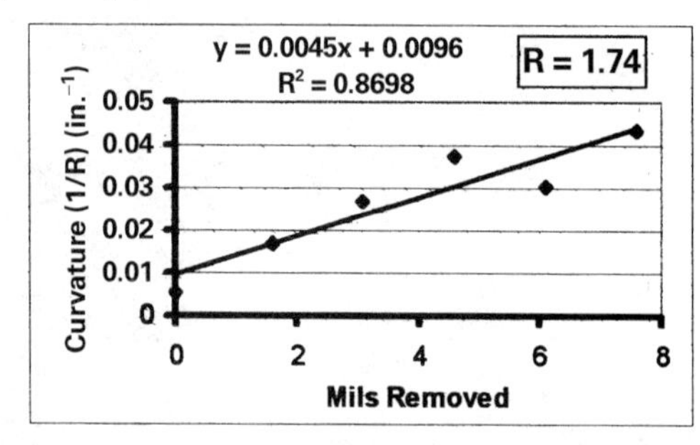

(c) Sample 48

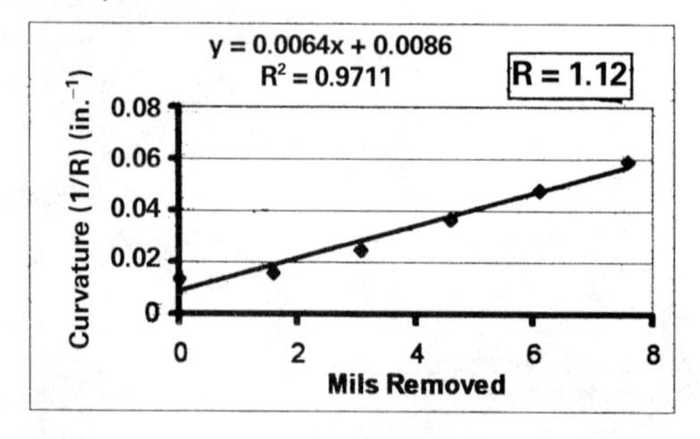

Fig. 13 (a,b, & c,d,e, & f,g,h) Curvature measurements on Formette handsheets wet pressed and dried on press–dryer combination.

sheets were not completely dry prior to removal from the dryer). Kubat stresses were also determined for the linerboard samples using the Kubat–Johansson technique. Properties of the whole board and the sections produced from it by surface grinding are given in Table 3.

Figure 16 shows the variation of the geometric mean specific modulus with geometric mean specific Kubat stress for both the whole board and the three sections for which data are given in Table 3. It suggests that the method of layer removal has resulted in the release of drying stress (Kubat stress). Htun has shown that Kubat stresses are highly correlated with drying stress, modulus,

(d) Sample 50

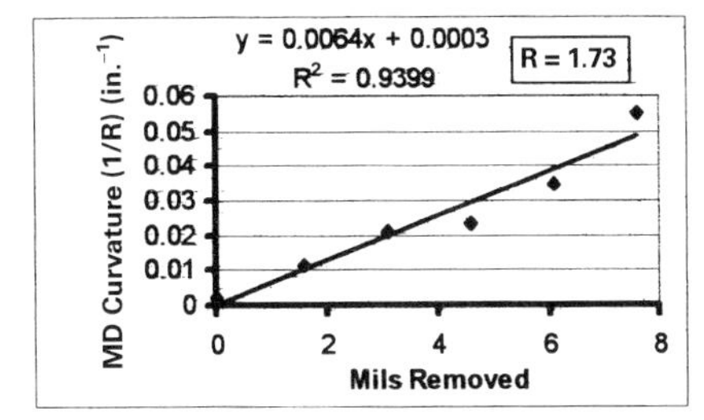

(e) Sample 51

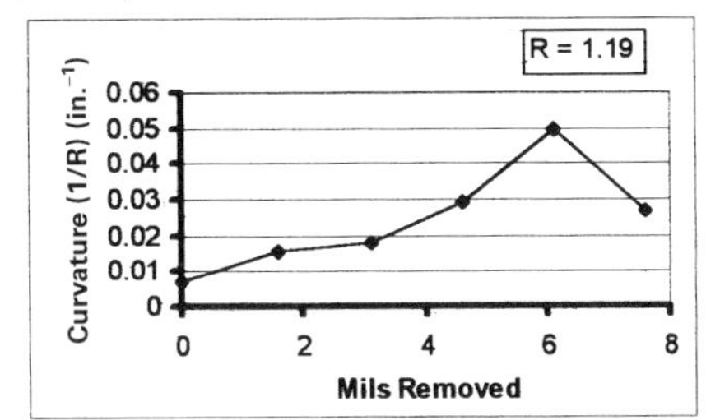

(f) Sample 53

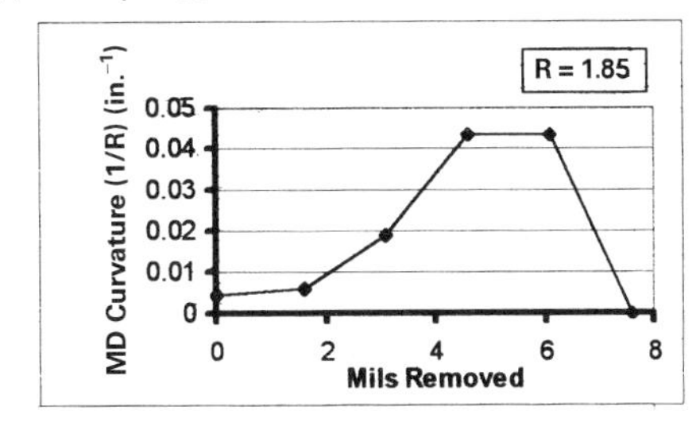

Fig. 13 *(continued)*

and other properties [48] as discussed earlier. Therefore, we tentatively conclude that the variation in in-plane specific modulus may be attributed in part to a variation in drying stress.

V. THE INFLUENCE OF RESIDUAL AND KUBAT STRESSES ON PAPER PROPERTIES

It can only be speculated that residual stresses are important for the performance of paper, because there have been virtually no studies directed toward their measurement and their impact on paper properties. Therefore, the bulk of the material in this section deals with the relationship of Kubat stress measurements to paper properties.

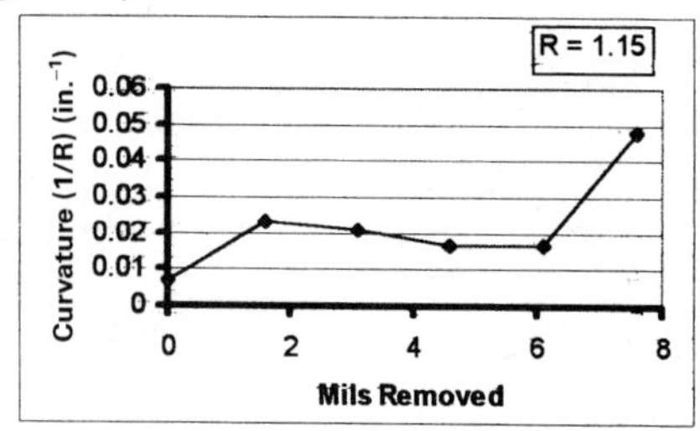

(h) Sample 58

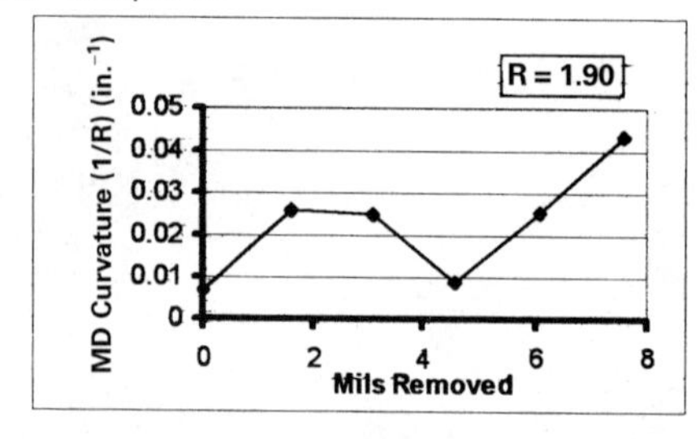

Fig. 13 *(continued)*

A. Elastic and Failure Properties of Paper

Htun [10] found a strong correlation between drying and Kubat stresses. Therefore, a strong correlation is also expected between modulus and other strength-related properties and Kubat stress, because these properties usually have a good correlation with drying stress. Htun and de Ruvo [48] also found this correlation.

Elastic and strength properties are controlled for a given furnish, refining, fiber orientation, and level of wet pressing by the restraints applied during drying, one

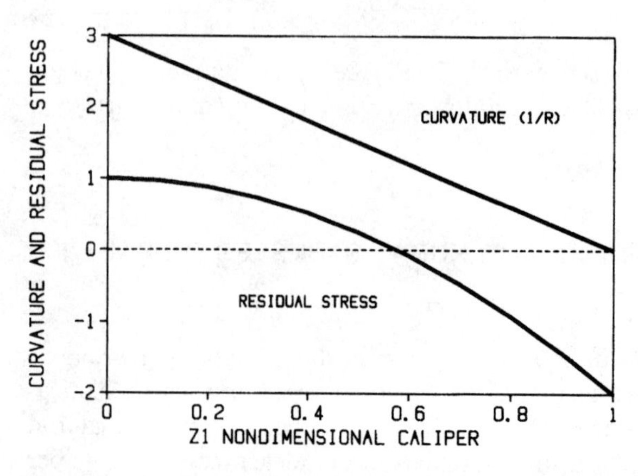

Fig. 14 Curvature and residual stress variations with thickness for a linear elastic solid.

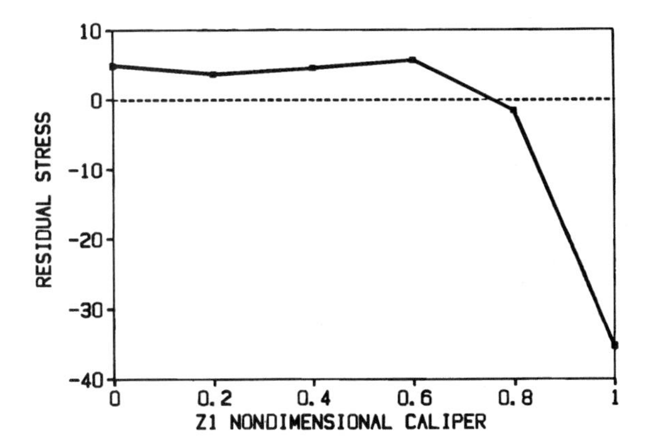

Fig. 15 Residual stress variation with nondimensional caliper for sample 60. Furnish: 100% softwood; $R = 1.19$.

manifestation of which is the level of drying stress induced. The Kubat stress level $\sigma_i = \sigma_D$ determined using the Johansson–Kubat [37] stress relaxation technique is one way of assessing this drying stress σ_D. It also follows that the Johansson–Kubat approach can also be used to assess the drying stress level on sections of board that have been produced by layer removal [52], with the caution that the layer removal process itself might modify both the residual and Kubat stress states. Annealing, work hardening, aging, calendering, and other mechanical processes may also modify the residual and Kubat stress states. In this situation we cannot directly determine the level of drying stress, but we might conclude that the level of Kubat stress measured is an altered drying stress.

B. Internal Variation of Paper Properties

It is evident that paper-related properties—apparent density, relative bonded area, elastic properties, strength, etc.—are not uniform from layer to layer in the thickness direction of paper [63–65]. *z*-Direction changes in composition, fiber orientation, and consolidation would appear to be, in part, responsible for this behavior, and as a consequence there is likely to be present a residual stress distribution. Waterhouse

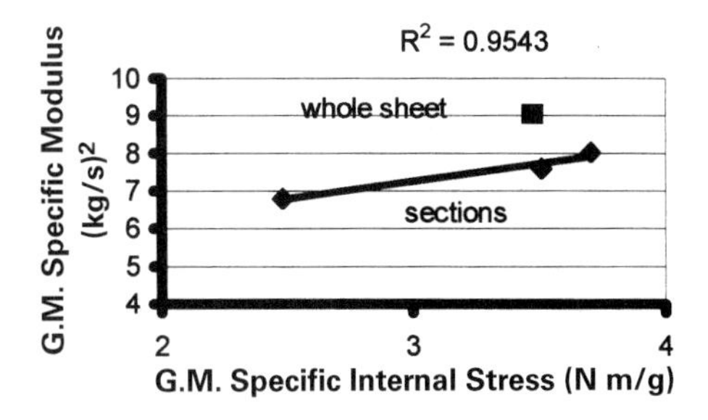

Fig. 16 Variation of geometric mean specific modulus with geometric mean specific internal stress.

[65] stated that property variations in the thickness direction might be attributed in part to variations in drying stress.

C. Dimensional Stability

Related dimensional stability problems include curl [8,9], fiber rising [11] and roughness induced by the application of aqueous coatings [12], e.g., coating base performance. Curl is often removed by changing the residual stress state using either mechanical means (decurlers) or chemical means (starch coatings). Curl has both reversible and irreversible components. Irreversible curl results when an asymmetrical residual stress distribution is produced by mechanical and/or environmental conditions.

Uesaka [9,66] examined the history or time dependence of curl using classical linear viscoelasticity and lamination theory. He states that it is usual to use a "stress-free" state as the reference condition. In the case he considers, Uesaka [66] uses the residual stress distribution at the end of drying as the reference condition. The residual stress distribution is assumed to be constant, and its possible change with moisture cycling is not accounted for. Nevertheless, the model that Uesaka [66] developed appears to simulate qualitatively the dimensional and curvature changes that paper undergoes during humidity cycling. In the model an asymmetrical residual stress distribution is used, and thus the paper has an initial curvature. The hygroexpansivity also varied in the thickness direction. The simulation shows an initially large component of irreversible curl with exposure to high moisture levels followed by a regime of reversible curl. Further exposure to a higher moisture level results in a smaller increment in irreversible curl. Uesaka [9] attributed the irreversible curl to axisymmetrical drying stresses (or residual stresses) and differences in the relaxation moduli between the two surfaces. Hygroexpansivity differences were the main factor controlling reversible curl.

D. Creep and Fatigue

According to recent work of Habeger and Coffin [67], residual stresses can also play an important role in creep and thus in accelerated creep. These authors have provided a simple explanation of accelerated creep and related phenomena. Accelerated creep is, according to their explanation, basically no more than creep, and no new mechanisms have to be invoked.

Residual stresses represent one type of inhomogeneity and result in stress concentration. Therefore, the creep rate or creep compliance increases. Habeger and Coffin [67] speculate that when high levels of drying stress occur in paper the residual stress level is small and thus the paper should be less compliant. We have seen in an earlier section that samples can be dried such that in one situation the residual stress level is negligible and in another a significant level of residual stress is produced. It is believed that in the former case the restraint conditions were such that residual stress differences were completely relaxed out. Craven [47] commented that relaxation effects are interfiber rather than intrafiber phenomena, i.e., a redistribution of stress between bonds. Furthermore, he states that sheets dried under heavy tension are inferred to have more homogeneous bond structures in the sense that frozen-in stresses are less. Craven does not define the term "frozen-in stresses."

Fatigue has been described as being no more than creep failure. This means that the cycles to failure can be equated with the time to failure during a simple creep test. One can therefore speculate that residual stresses play an important role in fatigue. Fold endurance is the procedure used to measure the fatigue characteristics of paper and board.

VI. IMPACT OF CONVERTING OPERATIONS ON RESIDUAL AND KUBAT STRESSES

Converting operations can be placed into one of the following general categories: coating, calendering and supercalendering, printing, forming and molding, laminating, impregnation, modification of deformation behavior, gluing, and size reduction. The role of Kubat or residual stresses in these operations is rarely mentioned, although I expect that these stresses are important with respect to their impact on the converting and end use performance of paper and board.

A. Calendering and Supercalendering

We have already noted that the Kubat and residual stresses present in paper can be modified by mechanical means. Stress relaxation can also be influenced by changes in temperature and moisture. Calendering and supercalendering are processes that potentially can induce such changes.

In a student project Ebert [68], using the Johansson–Kubat technique, measured changes in Kubat stress level produced by the soft calendering of an uncoated free sheet at two temperatures. The results, which are shown in Fig. 17, indicate that both MD and CD "internal stresses" are relaxed as a result of calendering.

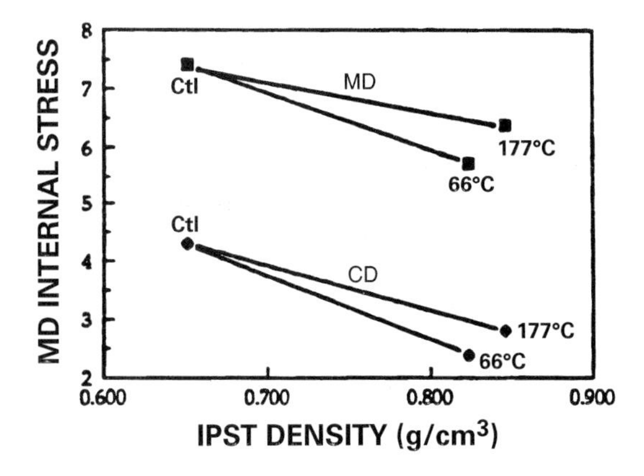

Fig. 17 Change in internal stress with calendering. Ctl = control. (From Ref. 68.)

B. Coating

Lepoutre and Skowronski [12] investigated the interaction between paper coating and base paper and its effect on stress relaxation. Stress relaxation in the base paper occurs because of the absorption of moisture from the coating. The Kubat stresses consisting of intrafiber and interfiber stresses were measured before and after coating. Three distinct slopes were obtained when $d\sigma/d\log t$ was plotted against the initial stress σ_0, these were designated α for network stress, β for sliding between unbonded fibers, and γ for intrafiber motions. Coating leads to roughening as a result of irreversible stress relief through debonding and swelling. It was also stated that calendering reduces Kubat stress, which is in agreement with the findings given in Section VI.A.

Clearly, coating can induce significant curl if the residual stress distribution is partially relaxed and becomes asymmetrical. Similarly, if correctly controlled, coating can be used to reduce curl.

VII. END USE PERFORMANCE OF PAPER

The satisfactory end use performance of paper is vital to its economic well-being, and we have briefly seen how some converting processes have the potential to modify the Kubat and residual stress distribution in paper and board. Dimensional stability may be an important consideration in maintaining register in printing processes, whereas curl and warp are important for sheet-fed converting and printing operations. Accelerated creep, as we briefly discussed, is also important for container performance. A practical solution to this problem will result in considerable raw materials savings.

Residual stresses may negatively impact paper and board performance, but in other cases performance may be enhanced. Generally the performance of such materials as paper, glass, polymers, and metals is greatly improved by significant levels of residual stress.

Almost without exception, paper should remain flat during its end use without developing curl. Agbezuge [69,70] and Rutland [8] examined the impact of Kubat stresses on xerographic papers. Agbezuge used a nonlinear (power law dashpot) viscoelastic model to describe stress relaxation. He found that stress relief achieved by rewetting the sheet and drying without restraint did not reduce the Kubat stress level to zero, as had been claimed [37]. Franke's findings [54] are in agreement with this. Agbezuge speculated that "internal stresses" might be important in paper transfer and jamming problems. In his second paper, Agbezuge [70] used the Halsey–White–Eyring nonlinear viscoelastic model to model the deformation behavior of paper. This model was then used to estimate the paper's stress relaxation behavior and hence predict its Kubat stress. These predictions differed significantly from the results he obtained using the Johansson–Rigdahl method [37]. Nevertheless, the method of Ref. 70 is claimed to be much less time-consuming.

ACKNOWLEDGMENTS

I wish to acknowledge the constructive reviews and criticism of Dr. Charles Habeger of IPST, Atlanta, and Dr. Leif Carlsson of Florida Atlantic University, Boca Raton, Florida.

APPENDIX

Table A1 Properties of Symmetrically Dried Formette Handsheets After Surface Grinding from Wire to Felt Side

Mils removed	Caliper (μm)	Grammage (g/m^2)	Density (g/cm^3)	E/ρ $[(km/s)^2]$	R $(= E_x/E_y)$	G/ρ $[km/s)^2]$	E_z/ρ $[(km/s)^2]$	S (m^2/kg)
Low pressure								
0	267.1	357.4	0.747	8.49	1.25	3.47	0.215	
2	254.4	334.4	0.761	6.02	1.59	1.95	0.159	
4	211.7	285.9	0.740	5.84	1.31	1.99	0.148	
6	173.8	236.9	0.734	5.82	1.35	1.94	0.124	
8	146.3	189.8	0.771	6.40	1.02	2.25	0.117	
10	108.9	141.5	0.770	5.40	1.13	1.87	0.120	
Medium pressure								
0	255.9	309.8	0.826	–	–	–	0.353	
2	244.4	261.7	0.934	7.33	1.29	2.58	0.273	
4	189.7	212.0	0.895	7.14	1.37	2.44	0.193	
6	143.0	162.2	0.882	6.53	1.19	2.20	0.122	
8	103.8	113.7	0.914	5.55	1.65	1.83	0.073	
High pressure								
0	262.4	273.3	0.966	10.8	1.02	4.14	0.375	
2	256.1	262.3	0.987	8.07	0.999	2.69	0.289	
4	200.9	215.1	0.932	7.71	1.48	2.72	0.217	
6	153.6	167.8	0.916	7.22	1.54	2.52	0.157	
8	78.3	87.9	0.896	5.86	1.18	2.09	0.077	

Table A2 Properties of Asymmetrically Dried Formette Handsheets After Surface Grinding from Heated to Ambient Side

Mils removed	Caliper (μm)	Grammage (g/m^2)	Density (g/cm^3)	E/ρ [(km/s)^2]	R ($= E_x/E_y$)	G/ρ [km/s)^2]	E_z/ρ [(km/s)^2]	S (m^2/kg)
Low pressure								
0	381.4	254.7	0.668	8.81	1.29	3.50	0.174	26.5
1	390.7	251.8	0.644	8.87	1.19	3.63	0.158	31.1
2	380.5	251.8	0.662	8.81	1.19	3.50	0.168	31.2
3	381.5	249.6	0.654	8.66	1.21	3.36	0.162	31.1
4	367.7	241.1	0.656	8.56	0.931	3.55	0.161	31.5
5	355.1	229.7	0.647	7.88	1.03	3.23	0.140	31.4
6	337.1	218.5	0.648	7.88	1.14	3.49	0.136	31.5
Medium pressure								
0	308.4	263.0	0.853	8.99	1.04	3.79	0.350	17.9
1	294.7	265.7	0.902	9.91	1.06	4.02	0.353	15.0
2	283.1	247.1	0.873	10.5	1.00	4.02	0.283	16.8
3	262.5	225.9	0.861	9.58	1.29	3.80	0.230	18.4
4	234.7	204.1	0.870	10.3	0.99	4.08	0.197	16.4
5	209.1	177.7	0.850	9.01	0.93	3.77	0.165	17.6
6	188.9	162.0	0.858	9.10	1.05	3.32	0.141	17.4
High pressure								
0	240.2	255.8	1.065	11.75	1.20	4.49	0.559	16.5
1	246.4	259.4	1.053	11.02	1.10	4.31	0.542	11.6
2	240.2	252.2	1.050	11.52	1.16	4.24	0.462	16.4
3	225.5	237.6	1.054	12.01	1.12	4.66	0.361	16.2
4	206.1	216.4	1.050	10.61	0.96	3.96	0.288	16.1
5	184.4	192.6	1.044	10.81	1.09	4.17	0.230	16.9
6	151.3	157.9	1.044	9.62	1.24	3.96	0.131	18.5

Table A3 Properties of Asymmetrically Dried Formette Handsheets After Surface Grinding from Ambient to Heated

Mils removed	Caliper (μm)	Grammage (g/m²)	Density (g/cm³)	E/ρ [(km/s)²]	R (= E_x/E_y)	G/ρ [km/s)²]	E_z/ρ [(km/s)²]	S (m²/kg)
Low pressure								
0	382.0	254.7	0.667	8.82	1.28	3.60	0.171	29.6
1	388.8	255.7	0.658	7.88	1.45	3.15	0.162	29.7
2	389.9	252.6	0.650	7.79	1.09	3.33	0.155	32.2
3	379.6	246.4	0.649	8.39	0.99	3.59	0.146	32.8
4	362.4	234.3	0.647	8.68	0.96	3.58	0.139	32.1
5	343.8	226.1	0.658	8.30	1.21	3.29	0.137	32.1
6	323.3	213.7	0.661	8.87	1.20	3.58	0.139	32.1
7	302.2	197.1	0.652	8.19	1.41	3.29	0.127	32.3
Medium pressure								
0	308.8	263.0	0.852	9.44	1.05	3.81	0.355	18.2
1	294.4	253.1	0.860	9.06	1.04	3.60	0.296	17.9
2	280.9	246.1	0.876	10.6	1.05	4.22	0.268	16.8
3	257.0	221.9	0.863	9.16	0.92	3.70	0.238	17.7
4	228.4	199.6	0.874	10.4	0.92	3.91	0.201	17.0
5	204.3	177.0	0.866	9.80	1.31	3.79	0.160	18.6
6	183.7	158.6	0.863	9.40	1.01	3.36	0.140	18.6
7	172.4	151.7	0.880	9.43	1.17	3.76	0.133	16.7
High pressure								
0	238.7	255.8	1.072	11.49	1.13	4.40	0.554	17.4
1	247.6	256.3	1.035	11.62	1.08	4.61	0.394	14.8
2	242.1	248.2	1.025	11.36	1.07	4.16	0.329	17.5
3	226.5	236.5	1.044	11.31	1.09	4.34	0.299	16.5
4	204.0	210.8	1.033	10.82	1.07	3.92	0.252	16.4
5	181.0	187.2	1.034	10.35	1.06	3.80	0.205	16.6
6	149.3	156.4	1.048	10.10	1.16	3.81	0.151	17.0
7	123.7	127.9	1.034	9.61	1.23	3.53	0.122	14.5

REFERENCES

1. Gardon, R. (Glass toughening) in Aben, H., and Guillemet, C. (1993). *Photoelasticity of Glass*. Springer-Verlag, New York.
2. Hildebrand, R. (1970). Drying process and drying stresses in kiln drying of sawn timber. Robert Hildebrand, Maschinenbau GmbH, Oberboihingen/Wertt.
3. Lewis, R. W., Srinatha, H. R., and Thomas, H. R., and Thomas, H. R. (1984). A finite element study of the drying stresses in timber using viscoelastic and elastoplastic rheological models. In: *Numerical Methods in Coupled Systems*. R. W. Lewis, P. Bettess, and E. Hinton, eds. Wiley, New York, Chap. 5.
4. Hahn, H. T. (1984). Effects of residual stresses in polymer matrix composites. *J. Astronaut. Sci. 32*(3):252–267.
5. Salmen, L., Fellers, C., and Htun, M. (1987). The development of dried-in stresses in paper. *Nordic Pulp Paper Res. J. 2*:44–48.
6. Fahey, D. J. and Chilson, W. A. (1963). Mechanical treatments for improving dimensional stability of paper. *Tappi 46*(7):393–399.

7. Johansson, F., Kubat, J., and Pattyranie, C. (1967). Internal stresses, dimensional stability and deformation of paper. *Svensk Paperstidn.* 10(May):333–338.
8. Rutland, D.F. (1992). Dimensional stability and curl. In: *Pulp and Paper Manufacture*, Vol. 9. Joint Textbook Committee of the Paper Industry, TAPPI/CPPA, Atlanta, GA, pp. 132–151.
9. Uesaka, T. (1991). Dimensional stability of paper: Upgrading paper performance in end use. *J. Pulp Paper Sci. 17*(2):J39–J46.
10. Htun, M. (1980). The influence of drying strategies on the mechanical properties of paper. Ph.D. Thesis, Department of Paper Technology, The Royal Institute of Technology, Stockholm, Sweden.
11. Aspler, J. S., and Beland, M-C. (1994). A review of fiber rising and surface roughening effects in paper. *J. Pulp Paper Sci. 20*(1):J27–J32.
12. Lepoutre, P., and Skowronski, J. (1986). Water-paper interactions during paper coating. Part II. Network and intrafiber stress relaxation. CPPA Annual Meeting (Montreal), Jan. 28–29. Preprints A:45–51.
13. Pankonin, B., and Habeger, C.C. (1988). A strip resonance technique for measuring the ultrasonic viscoelastic parameters of polymeric sheets with application to cellulose. *J. Polym. Sci. B: Polym. Phys. 26*:339.
14. Ivarsson, B.W. (1954). Introduction of stress into a paper sheet during drying. *Tappi 37*(12):634–639.
15. Htun, M., and de Ruvo, A. (1978). Correlation between the drying stress and the internal stress in paper. *Tappi 61*(6):75–77.
16. Giertz, H. W. (1972). Mechanism of paper strength development. *Svensk Paperstidn. 75*(9):252–253.
17. Page, D.H. (1977). In: *Fiber–Water Interactions in Papermaking*. Trans. Symp. held at Oxford, September 1977. FRC Tech. Div., ed. Brit. Paper and Board Industry Federation. London, pp. 489–490.
18. Jentzen, C.A. (1964). The effect of stress applied during drying on some of the properties of individual pulp fibers. *Tappi 47(7)*:412–418.
19. Spiegelberg, H. L. (1966). The effect of hemicelluloses on the mechanical properties of individual pulp fibers. *Tappi 49*(9):388–396.
20. Kim, C. Y., Page, D. H., El-Hosseiny, F., and Lancaster, A. P. S. (1975). The mechanical properties of single wood pulp fibers. III. The effect of drying stress on strength. *J. Appl. Polym. Sci 19*:1549–1562.
21. Page, D.H., Seth, R. S., and El-Hosseiny, F. (1985). Strength and chemical composition of wood pulp fibers. In: *Papermaking Raw Materials*, Vol. 1. Trans. 8th Fundam. Res. Symp., Oxford, September. V. Punton, ed. Mech. Eng. Publ., London, pp. 77–91.
22. Page, D. H., and Tydeman, P. A. (1962). A new theory of the shrinkage, structure and properties of paper. In: *Formation and Structure of Paper*, F. Bolam, ed. Tech. Sect. B.P. & B.M.A., London, pp 397–413.
23. Rance, H. F. (1954). Effect of water removal on sheet properties. The water phase evaporation. *Tappi 37*(12):640–653.
24. Gray, D. G. (1989). Chirality and curl of paper sheets. *J. Pulp Paper Sci. 15*(3): J105–J109.
25. Mark, R. E. and Murakami, K. eds. (1983). *Handbook of Physical and Mechanical Testing of Paper and Paperboard*, Vol. 1. Marcel Dekker, New York.
26. Bell, W. P., and Edie, D. D. (1987). Calculated internal stress distributions in melt-spun fibers. *J. Appl. Polym. Sci. 33*:1073–1088.
27. Salminen, L. I., Raisanen, V. I., Alava, M. J., and Niskanen, K. J. (1996). Drying-induced stress state of inter-fiber bonds. *J. Pulp Paper Sci. 22*(10):J402–J407.
28. Waterhouse, J. F. (1991). Failure envelope of paper when subjected to combined out-of-plane stresses. *TAPPI Conference Proceedings*, 1991 Int. Paper Physics Conf., Kona, HI, Sept. 22–26. *2*:629–640.

29. Stratton, R. (1993). Characterization of fiber-fiber bond strength from out-of-plane mechanical properties. *J. Pulp Paper Sci. 19*(1):J6–J12.
30. Bither, T. W., and Waterhouse, J. F. (1991). Strength development through refining and wet pressing. *Tappi J. 11*:201–208.
31. Button, A. F. (1979). Fiber-fiber bond strength: A study of a linear elastic model structure. Ph.D. Thesis, Institute of Paper Chemistry, Appleton, WI.
32. Htun, M. (1986). Internal stress in paper. In: *Paper Structure and Properties*. A. J. Bristow and P., Kolseth, eds. Marcel Dekker, New York, pp. 227–239.
33. Struik, L. C. E. (1990). *Internal Stresses, Dimensional Instabilities and Molecular Orientations in Plastics*. Wiley, New York.
34. Lu, J. (1996) (editor). *Handbook of Measurement of Residual Stresses*. Society for Experimental Mechanics.
35. Kubat, J. (1965). A similarity in the stress relaxation behavior of high polymers and metals. Swedish Forest Products Research Laboratory, Paper Technology Department, and Rheological Laboratory at the Royal Institute of Technology, Stockholm.
36. Johansson, F., Kubat, J., and Pattyranie, C. (1965). Experimental studies on stress relaxation in crystalline polymers. IV. Cellulose and paper. *Ark. Fys.* (*28*:317.
37. Johansson, F., and Kubat, J. (1964). Measurements of stress relaxation in paper. *Svensk Paperstidn. 20*:822–832.
38. Isayev, A. I., and Crouthamel, D. L. (1984). Residual stress development in the injection molding of polymers. *Polym.-Plast. Technol. Eng. 22*(2):177–232.
39. Rybicki, E. F., Shadley, J. R., and Shealy, W. S. (1983). A consistent-splitting model for experimental residual-stress analysis. *Exp. Mech.* December:438.
40. ASTM. (1982). A standard method for determining residual stresses by hole drilling method. *ASTM Annual Book of Standards*. 82nd ed. Part 10.
41. Noyan, I. C., and Cohen, J. B. (1987). *Residual Stress: Measurement by Diffraction and Interpretation*. MRE/Springer-Verlag, New York.
42. Pintschovius, I., Jung, V. Macherauch, E., and Vohringer, O. (1983). Residual stress measurements by means of neutron diffraction. *Mater. Sci. Eng. 61*:43–50.
43. Noronha, P. J. and Wert, J. J. (1975). An ultrasonic technique for the measurement of residual stress. *J. Testing Eval. 3*(2);147–52.
44. Hamad, W., and Eichhorn, S. (1997). Raman spectroscopic analysis of the micro-deformation in cellulose fibres. In: *The Fundamentals of Papermaking Raw Materials*. Trans. 11th Fundam. Res. Symp., Cambridge: September. C. F. Baker, ed. PIRA Int., Surrey, UK.
45. Sunderland, P., Yu, W. J., and Manson, J.-A. (1995). A technique for the measurement of process-induced internal stresses in polymers and polymer composites. *Proc. 10th Intl. Conf. Composite Material*, Whistler, BC, Canada *III*:125–132.
46. Sunderland, P., Sollander, A., Manson, J.-A., and Carlsson, L. A. (1999). Measurement of process-induced residual stresses in linerboard. *J. Pulp Paper Sci. 25*(4):130–136.
47. Craven, B. D. (1962). Stress relaxation and work hardening in paper. *Appita 16*(2):57–69.
48. Htun, M., and de Ruvo, (1977). Relation between drying stresses and internal stresses and the mechanical properties of paper. In: *Fiber-Water Interactions in Papermaking*. Trans. Symp. Oxford; September. Tech. Div., Brit. Paper Board Ind. Fed., London, pp. 477–487.
49. Kubat, J., and Rigdahl, M. (1975). A simple model for stress relaxation in injection molded plastics with an internal stress distribution. *Mater. Sci. Eng. 21*:63–70.
50. Robertson, A. A. (1973/1976). Modification of the mechanical properties of paper by the addition of synthetic polymers. In: *The Fundamental Properties of Paper Related to Its Uses*, 5th Fundam. Res. Symp., September, 1973 F. Bolam, ed. Brit. Paper and Board Ind. Fed., London, 1976, pp. 373–393.

51. Back, E. L., and Klinga, L. O. (1963). The effect of heat treatment on internal stresses and permanent dimensional changes of paper. *Tappi 46(5)*:284–288.
52. Waterhouse, J. F., Stera, S., and Brennan, D. (1987). z-Direction variation of internal stress and properties in paper. *J. Pulp Paper Sci. 13*(1):J33–J37.
53. White, J. R. (1984). Origins and measurement of internal stress in plastics. *Polym. Testing 4:165–191.*
54. Franke, M. (1987). Variation of residual stress and properties in the thickness direction of paper. A190 Independent Study Report, IPC, Appleton WI, March 13.
55. Waterhouse, J. F., and Franke, M. (1988). The layer removal technique for determining residual stresses in paper. Seminar abstracts Progress in Paper Physics, held at Miami University.
56. Beckman, N. J., and Plucker, E. I. (1973). z-Direction analysis of coated papers using large area sectioning techniques. TAPPI Coating Conference, April 30–May 3, pp. 85–102.
57. Schwantes, T. (1990). Internal investigation of elastic properties of paper using non-mechanical layer removal techniques. A390 Problem, IPST, June 10.
58. Nordin, S. B., Nyren, J. O., and Back, E. L. (1974). An indication of molten cellulose produced in a laser beam. *Textile Res. J.* February:152–154.
59. Sapieha, S., Wrobel, A. M., and Wertheimer, M. R. (1988). Plasma assisted etching of paper. *Plasma Chem. Plasma Proc. 8*(3):315.
60. Brodeur, P., and Waterhouse, J. F. (1990). Investigation of paper internal properties using microwave plasma etching. Unpublished exploratory work. Institute of Paper Science and Technology, December.
61. Trueting, R G., and Read, W. T., Jr. (1951). A mechanical determination of biaxial residual stress in sheet materials. *J. Appl. Phys. 22*(4):130–134.
62. Paterson, M. W. A., and White, J. R. (1989). Layer removal analysis of residual stress. Part 2. A new procedure for polymer moldings with depth-varying Young's modulus, *J. Mater. Sci. 24*(10):3521–3528.
63. Parker, J. D. (1972). *The Sheet Forming Process.* Special Technical Association Publication (STAP no 9). TAPPI, Atlanta, GA.
64. Wickes, L. (1982). The influence of pressing on sheet two sideness. *Tappi J. 65*(9):73–77.
65. Waterhouse, J. (1984). Out-of-plane shear deformation behavior of paper and board. *Tappi J. 67*(6):105–108.
66. Uesaka, T. (1989). History dependent dimensional stability of paper. *Rheol. Acta 28*:238-245.
67. Habeger, C. C., and Coffin, D. W. (2000). The role of stress concentration in accelerated creep and sorption-induced physical aging. *J. Pulp Paper Sci., 26*(4):145–157.
68. Ebert, R. (1990). The effect of internal stresses on the modification of paper properties by calendering. A190 Independent study Report. Inst. Paper Sci. Technol., Atlanta, GA.
69. Agbezuge, L. K. (1981). Numerical determination of internal stress and viscoelastic parameters for xerographic papers. *Polym. Eng. Sci. 21*(9):538–541.
70. Agbezuge, L. K. (1981). Internal stress levels in xerographic papers. *Polym. Eng. Sci. 21*(9):534–537.

PART 2. SPECIAL SITUATIONS — MECHANICAL PROPERTIES OF PRODUCTS

Emblem of Tlaximaloyan

Tarascans lived in the town of Tlaximaloyan, whose tribute to the Aztec rulers was in the form of hewn timbers. The emblem includes a tool, such as an axe or adze for hewing, and the trunk of a tree with some branches cut off. *Tlaximalli* is translated as hewn timber, *yan* as place, the combination giving a logical name for the tribute town.

11
CORRUGATED BOARD

ROB STEADMAN
Janus Research
Hamilton, New Zealand

I. INTRODUCTION

Corrugated board can be defined as "the structure formed by gluing one or more layers of fluted corrugating medium to one or more flat facings of linerboard" [45] (see Fig. 1). It has been used as a packaging material for more than 100 years, but it still maintains a healthy share of the total market despite increasing competition from plastics, in particular returnable plastic containers (RPCs). One reason for the continued demand for corrugated board packaging is its unique combination of attributes:

- It has a high stiffness-to-weight ratio. Compared to other rigid packaging materials, corrugated board delivers relatively high stiffness at a relatively low price. This was demonstrated by Steenberg et al. [163] in 1970, and the comparisons are still broadly valid (Fig. 2).
- It has reasonable shock resistance and cushioning properties, and the air space in the corrugated structure imparts good thermal insulation.
- It is made from a natural and renewable resource and is fully biodegradable and recyclable. In some European countries over 70% of all corrugated packaging is collected and recycled [28].
- It is versatile and can be used for the diverse packaging requirements of different industries. It can carry a high quality printed image for point-of-sale marketing and can be used on high speed automated packing lines.

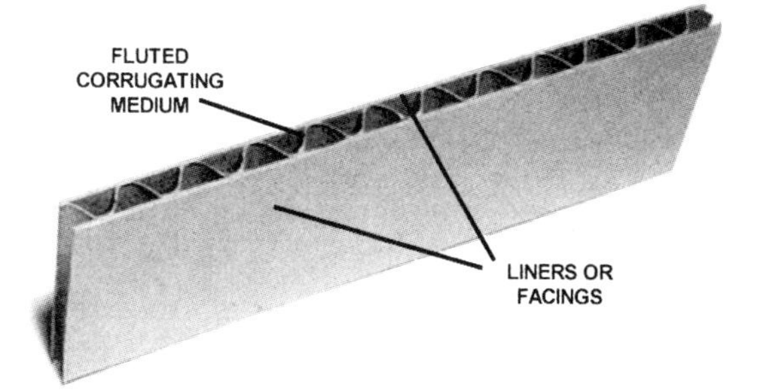

Fig. 1 Section through corrugated board showing the fluted structure.

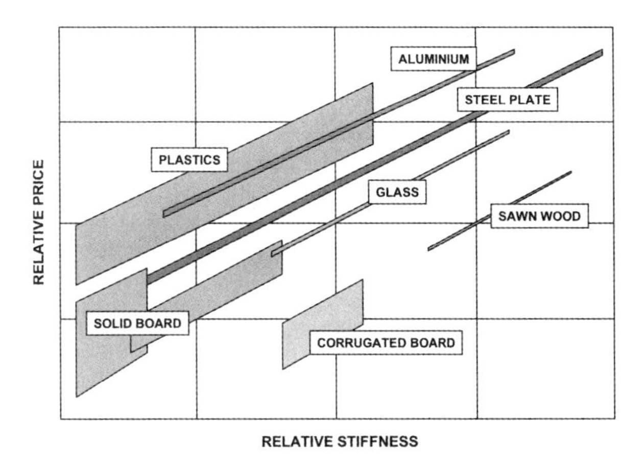

Fig. 2 Comparison of relative stiffness and price of various rigid packaging materials. (After Ref. 163.)

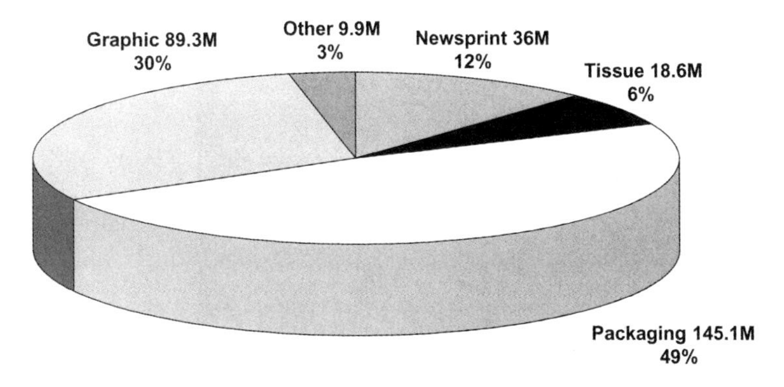

Fig. 3 World paper and board production in millions of tonnes by grade. (Data from Ref. 28.)

In 1997, total world paper and board production for all grades was 299 million tonnes [28]. Of the 145.1 million tonnes of packaging grades, almost 80 million tonnes was corrugating materials, which is second only to graphic papers in terms of volume (Fig. 3). Total world shipments of corrugated board in 1999 were 124.9 billion m^2 [64], or almost 125,000 km^2. By 2001, sufficient corrugated board will be produced annually in the world to totally cover a country the size of England in a continuous layer. Production increased by an average of 3.6% in the years 1995–1999 [64], and the industry is forecast to grow at a rate of about 1% faster than world GDP, or 3.1% per annum, up to 2012 [150].

II. BRIEF HISTORY

The development of corrugated board appears to be closely linked to the fashion industry in the late nineteenth century. At that time, it was fashionable to wear ruffled or pleated collars and cuffs (Fig. 4). Accordingly, the laundries of the day developed hand-cranked mangles with corrugated rolls to press the ruffles. The fashion in Victorian England was for tall hats to be lined with cylinders of paper to maintain their shape. In 1856, this prompted E. C. Healey and E. E. Allen to patent the use of corrugated paper to add stiffness to these hat liners. The corrugations were produced by wetting the paper and passing it through a modified heated ruffled collar press, as shown in Fig. 5.

The first patent for corrugated paper as a packaging material was awarded to A. L. Jones in 1871 [68]. The patent describes a "corrugated, crimped or bossed" sheeet that is wrapped around glass bottles or vials to provide superior cushioning during transportation. In 1874, Oliver Long [106] patented the concept of gluing a facing or liner to the corrugated material to prevent the stretching of the flutes. The machinery to make this material was developed over the next few years. During the 1880s there were several developments in machinery design that were assigned to Meech [133] and Thompson [171,172] as the process became more sophisticated. The first continuous paper corrugator was developed by J. T. Ferres for the Sefton Manufacturing Company in 1895. Langston [100] patented a machine with separate units to apply the first ("single facer") and second ("double facer") layers of liner to

Fig. 4 Ruffled lace collar from the nineteenth century.

Fig. 5 Hand-cranked modified collar press for producing early corrugated paper. (Courtesy of Fibre Box Association, Rolling Meadows, IL.)

the medium. The units operated independently but in tandem, a clearly recognizable feature of modern corrugators.

There are more historical details and illustrations in the review by Maltenfort [112], which includes reference to the comprehensive review by H. J. Bettendorf, now out of print.

III. THE MANUFACTURING PROCESS

The modern corrugator bears a strong likeness to the original Langston designs [100], with two independently controlled stations to apply the two facings (see Fig. 6).

The first facing, called the single face (SF) liner, is glued to the medium at the same station that forms the characteristic flutes (Fig. 7). The corrugating medium is fed from an unwind stand around a preheater roll, then into the labyrinth between two corrugated rolls (locations *2* and *3* in Fig. 7). These rolls are steam heated, and the nip pressure is controllable. The medium passes between the rolls, is compressed and thermally softened, and molds to the contours of the corrugated rolls. This process relies on a combination of paper properties such as hot coefficient of friction and ability to stretch and process variables such as web tension, as comprehensively reviewed in the work of Göttsching and Otto [53]–[58] and Whitsitt [183]. The medium is held in place on the corrugator roll by either vacuum or pressure (locations *4* and *6* in Fig. 7) or, on older machines, by the use of mechanical fingers. An adhesive is metered onto the tips of the medium corrugations by an applicator roll (location *5*), and the liner is bought into contact with the medium at the pressure roll (location *7*). The resulting structure is called single-face board, because it has only one liner glued to the medium. The bond between the two is still wet (or "green" as it is known) at this stage but has sufficient tack to hold the structure together. The factors affecting the development of green bond strength are explored in the articles by Batelka [2,3].

The board is transported to the next station via the bridge, which is a temporary holding facility. This is necessary because the two stations on the corrugator operate independently, and the process would be difficult to control without the provision of a buffer of single-face material.

The second layer of liner is added at the double facer, also known as the double backer station. Adhesive is applied to the exposed tips of the corrugating medium, and the second liner is bought into contact under light pressure. The corrugated board then passes over hot plates to cure the adhesive. The final stage is machine direction (MD) slitting and scoring at the triplex station, followed by

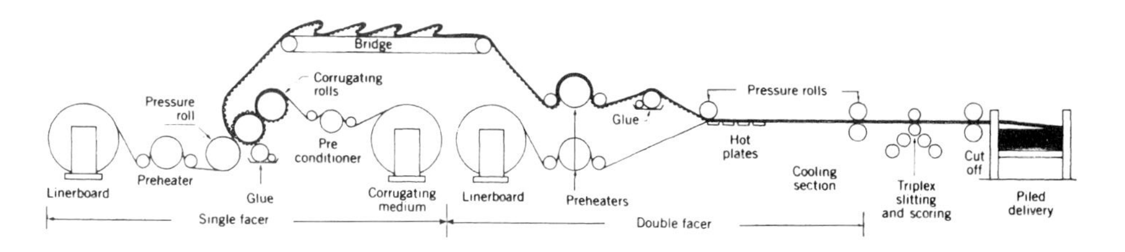

Fig. 6 Modern corrugator for manufacturing corrugated board. (From Ref. 167.)

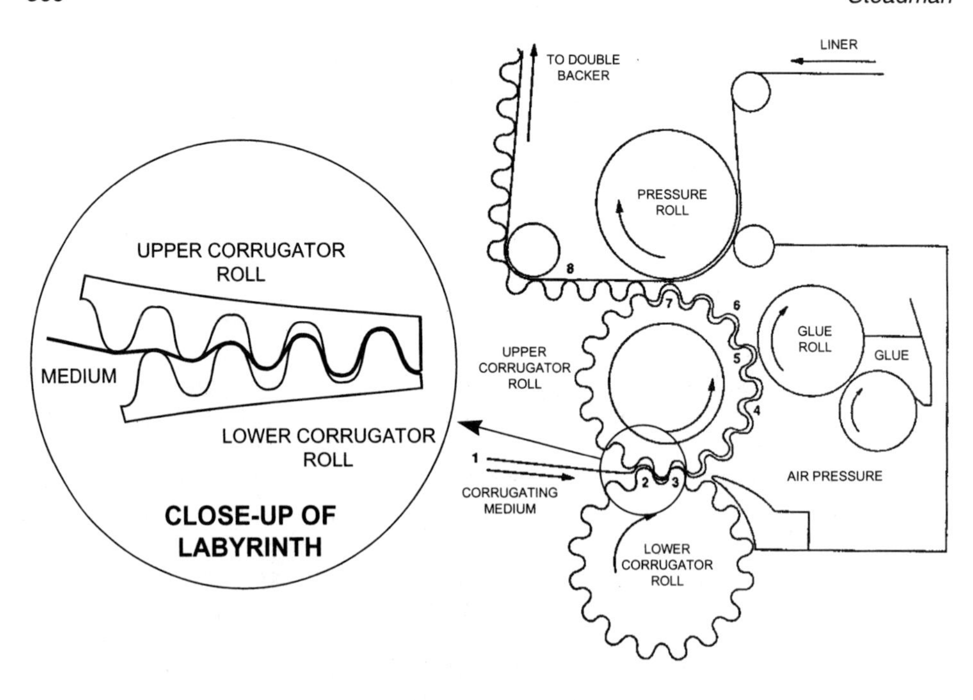

Location	Description	Residence Time, ms
1	Preconditioning before the corrugator roll nip	–
2	Labyrinth entering the corrugator roll nip	6.0
3	Labyrinth leaving the corrugator roll nip	1.5
4	Pressure compartment before the glue roll	94.5
5	Glue application	3.0
6	Pressure compartment before the pressure roll	60.0
7	Pressing the liner against the glued flute tips	1.15
8	Formation of the bond after the pressure roll	

Fig. 7 Schematic of the single facer station on a modern corrugator. The table shows approximate residence times at the various stages of the process. (Main diagram adapted from Ref. 62. Inset courtesy of Ian Chalmers, Papro, New Zealand.)

cross-direction (CD) cutting by a rotary cut-off knife. These operations produce a blank of the size required for the box-making operation. More details of the corrugating operation can be obtained in the book *The Corrugator* [6].

The corrugated board then passes through several other conversion processes, depending on the nature and complexity of the end use (Fig. 8). Details of these processes are beyond the scope of this book, but good overviews can be found in the literature, for example, Jönson [70]. More information is also available in the book by Schnell et al. [152] and the interactive CD ROM "How Corrugated Boxes Are Made" [10], both available from TAPPI Press.

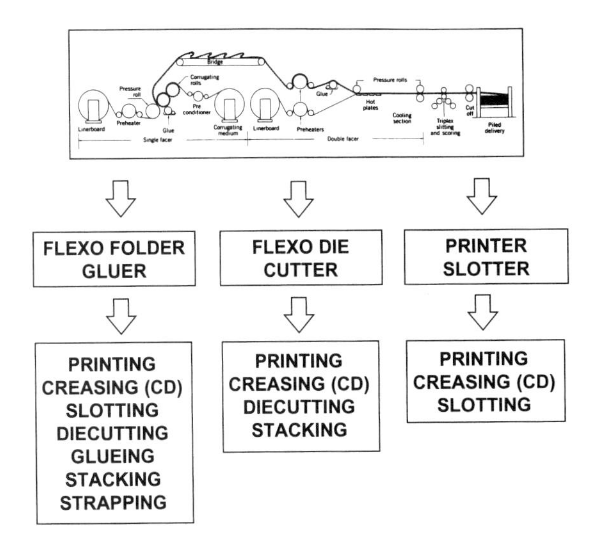

Fig. 8 Conversion operations on corrugated board.

A. Flexographic Printing of Corrugated Board

Most corrugated board is printed using a flexographic press. When printing is carried out on the linerboard before the corrugating operation, it is termed *preprint*. Printing after the corrugator is termed *postprint*. Flexography is a variation of letterpress printing, in which the image area in the press is raised above the nonprinting area. Flexographic printers use flexible elastomeric plates to carry the image (Fig. 9) and relatively low viscosity, fast-drying inks, which are generally water-based for corrugated board printing.

The press usually consists of four rollers. The fountain roll transfers the ink from the pan to the anilox roll, which is engraved with small cells—typically 100 per centimeter, with a depth of 20 μm. The anilox roll is commonly wiped with a reverse angle doctor blade and delivers a controlled volume of ink to the raised image on the plate cylinder. The image is transferred to the substrate in the nip between the plate

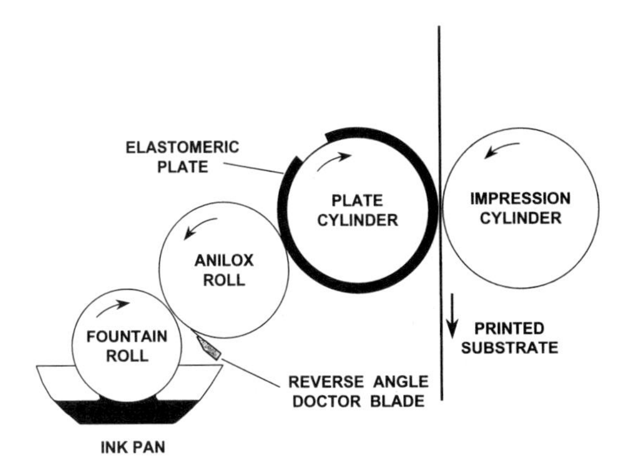

Fig. 9 Schematic of flexographic printing press.

and impression cylinders. More details of this process can be found in *Flexography Principles and Practices*, published by the Flexographic Technical Association. The wetting and penetration of aqueous liquids into paper is covered by Lyne (see volume 2, Chapter 7).

IV. THE STRUCTURE OF CORRUGATED BOARD

Corrugated board is a unique construction with a series of connected arches separating the two facings. This provides the structure with rigidity while maintaining an excellent strength-to-weight ratio. The structure has different properties in its three principal directions (Fig. 10).

The balance of properties in the three principal axes will depend on several characteristics of the materials and structure, including

1. The directional strength of the facings. The paper used for the facings or liners in corrugated board is generically known as linerboard. It is commonly a two-layered structure with greater strength in the direction of manufacture, or *machine direction*. This is due to the nature of the papermaking process, which tends to align fibers in the machine direction.
2. The ability of the medium to keep the facings apart. This will depend on the compressive and shear strength of the fluted structure, which will also be affected by the strength and integrity of the glue lines between the medium and the liners.
3. The weight per unit area of the facings and medium. Components with higher weight per unit area (*grammage*) will produce a structure with higher strength. The balance between the grammage of the liners and the medium will also affect structural strength.
4. The geometry of the flutes—the height and number of flutes per unit length (*pitch*).

Corrugated board can consist of combinations of layers with different liners and flute geometry, such as the triple-wall structure shown in Fig. 13. It is the ability to

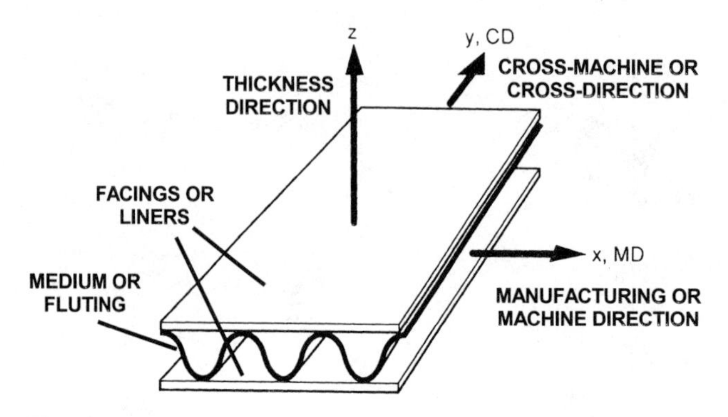

Fig. 10 Corrugated board structure showing the three principal axes. (Adapted from Ref. 69.)

change the material and structural attributes that makes corrugated board so versatile as a packaging material.

V. CHARACTERIZATION OF CORRUGATED BOARD

Corrugated board can be characterized by

- Flute geometry
- Linerboard and medium grammage
- Linerboard and medium furnish
- Number of layers

A. Flute Geometry

A primary characteristic of corrugated board is the height and pitch of the corrugations of the medium or fluting (Fig. 11). The four most common flute types are described in Table 1.

The take-up factor is a measure of the linear length of the medium per unit length of corrugated board. For example, one meter of A-flute board will contain 1.54 m of corrugating medium because of the fluted structure. The flutes are lettered in the order in which they were introduced.

A-flute board has the greatest thickness and consequently the highest bending stiffness (see Section X.B). Because the take-up factor is also high, the board has greater resistance to a compressive load applied in the direction of the flutes (see *edgewise compressive strength*, Section X.B). These two factors combine to give better box compression strength, as explained in more detail elsewhere in the text. However, the board is more susceptible to damage during manufacture and service life due to its poor resistance to a load applied perpendicular to the liner surface (see *flat crush strength*, Section X.B). The large spacing between the flutes can also lead to problems with flexographic printing.

B-flute board has relatively high resistance to perpendicular loading but lower thickness and bending stiffness and consequently produces boxes with lower compression strength. It is more resistant to perpendicular loading due to the greater number of flutes per unit length, which also yield a smoother surface with better flexographic printability. B-flute uses about 14% less corrugating medium than A- flute due to the lower take-up factor.

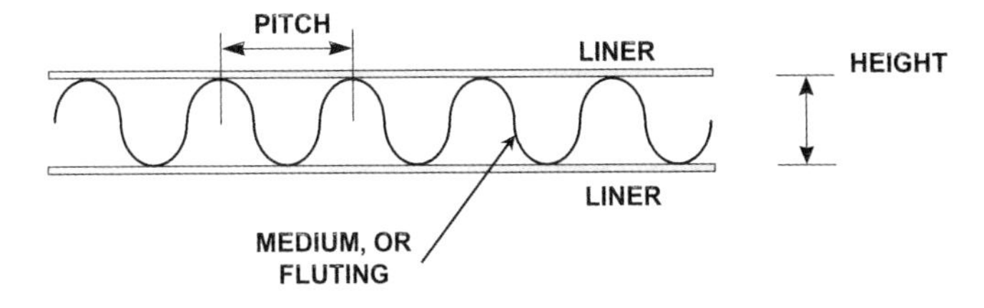

Fig. 11 Geometrical properties of medium corrugations.

Table 1 Geometrical Properties of the Four Common Flute Profiles

Flute type	Flute spacing (mm)	Flutes per meter	Flute height (mm)	Take-up factor
A	8.3–10	110 ± 10	4.67	1.54
C	7.1–8.3	130 ± 10	3.61	1.43
B	6.1–6.9	165 ± 10	2.46	1.32
E	3.2–3.6	295 ± 15	1.15	1.27

Conversion factor: in. = 25.4 mm

C-flute was introduced as a compromise between the A and B flutes and has a good balance of the critical board properties.

E-flute is not commonly used for shipping containers, but the large number of flutes per unit length impart a smooth surface for high quality printing. Consequently, it is used in display boxes where appearance is placed at a premium.

The D-flute designation is not assigned to a flute style.

B. Linerboard and Medium Grammage

The weight per unit area of paper or paperboard is a fundamental material property that is controlled in the manufacturing stage. *Grammage*, also known as *basis weight*, is defined as the weight of a square meter of paper or paperboard in grams (metric), or the weight of 1000 ft^2 in pounds (U.S. Customary System). The conversion factor for lb/1000 ft^2 to g/m^2 is 4.882.

Some common grammages used in the packaging industry are shown in Table 2. The list is by no means exhaustive, and other grammages are available from manufacturers.

Table 2 Some Common Grammages of Linerboard and Corrugating Medium

Linerboard			Medium		
U.S. basis weight (lb/1000 ft^2)	U.S. basis weight (g/m^2)	Metric grade (g/m^2)	U.S. basis weight (lb/1000 ft^2)	U.S. basis weight (g/m^2)	Metric grade (g/m^2)
26	127	125	26	127	125
33	161	150	28	137	140
38	186	175	30	146	150
42	205	200	33	161	160
47	229	225	36	176	175
—	—	250	40	195	200
—	—	275	42	205	200
—	—	300	50	244	250
69	337	330	52	254	250
90	439	—			

It is common practice to match the lower grammage liners with a lower grammage medium to obtain a balanced structure. Similarly, higher grammage linerboards are usually combined with high grammage medium. The effect of changing this balance is more thoroughly explored by Kellicut [84], Maltenfort [114], Johnson et al. [67], and Urbanik [176].

C. Linerboard and Medium Furnish

Linerboard is usually a two-layered sheet with a base layer providing strength and a top layer providing appearance and good printing characteristics. The properties of the linerboard will depend on the *furnish*, a term that encompasses the type of fiber, chemical additives, and fiber treatment. The furnish ranges from 100% virgin, or nonrecycled softwood kraft fiber, to 100% recycled fiber. Many kraft linerboards also contain a proportion of recycled fiber to reduce furnish cost. Corrugating medium furnish is dominated by 100% recycled fiber and semi-chemical hardwood and softwood pulps such as NSSC (neutral sulfite semi-chemical—a combined mechanical and chemical pulping process). The trend in recent years has been toward the use of recycled fiber for both linerboard and medium. Consumption of semi-chemical medium and natural kraft linerboard has been diminishing as board manufacturers move toward the recycled equivalent. In 1970, 31% of the world market was recycled products, compared to 35% in 1980, 42% in 1990, and 52% in 1997 [150]. The partition of the European market in 1999 is shown in Fig. 12.

There has also been a trend toward high performance linerboards, which deliver more compressive strength per unit weight, enabling the progressive reduction of corrugated board grammage. Figures from FEFCO [39] show that the average grammage of corrugated board in Europe decreased from 561 g/m^2 in 1992 to 551 g/m^2 in 1999.

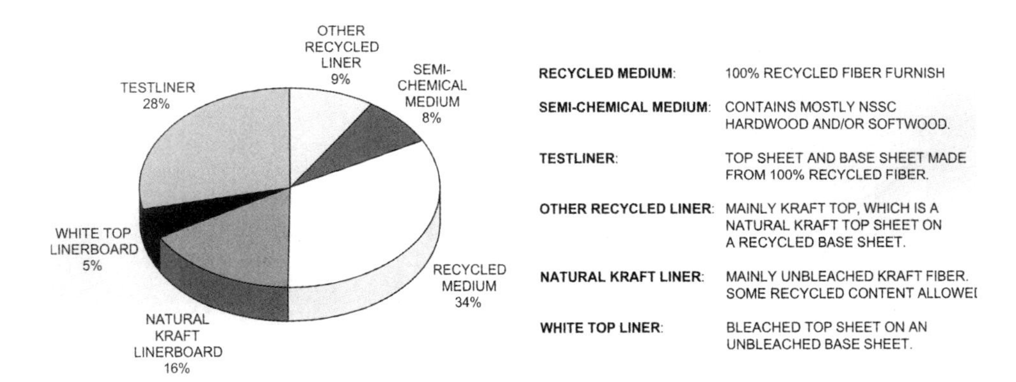

Fig. 12 Partition of the 1999 European market for linerboard and recycled components of corrugated board. (From Ref. 39.)

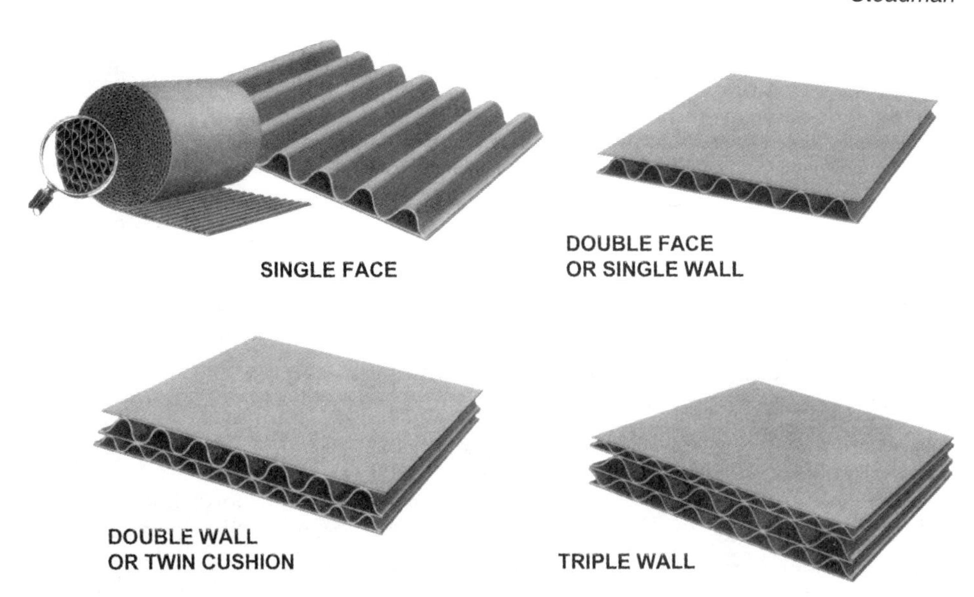

Fig. 13 Layered configurations of corrugated board. (Courtesy Fibre Box Association, Rolling Meadows, IL.)

D. Number of Layers

Most corrugators have more than one fluting station and can run them together to make structures with more than one layer (Fig. 13). Each of the layers, or walls, can have its own combination of flute type, component grammage, and furnish. In 1999, almost 89% of all manufactured corrugated board was single-wall, with 9.5% double-wall, about 1% single-face, and 0.6% triple-wall (source: Fibre Box Association).

VI. DEVELOPMENTS IN CORRUGATED BOARD

A. Mini-Flutes

There has been a trend toward smaller flutes to compete with established cartonboard grades, and the growth in this market was very strong toward the end of the 1990s. Mini-flutes are E-flute and smaller variants (Table 3). They are used for packaging where high quality printing is required and structural strength is of secondary importance. The surface of the board is very smooth due to the high number of flutes per unit length. In the late 1990s this market was growing by 7% per year and was expected to increase to 10% per year in the early twenty-first century [63].

B. X-Flute

AMCOR in Australia developed and hold the patent on a new board structure called Xitex or X-flute, which incorporates two corrugating media glued tip-to-tip [5] [156] (Fig. 14).

Table 3 Mini-flute Grades of Corrugated Board

Flute type	Flute spacing (mm)	Flutes per meter	Flute height (mm)	Take-up factor
E	3.2–3.6	295 ± 15	1.15	1.27
F	2.3–2.5	422 ± 15	0.76	1.25
G	1.6–1.8	584 ± 25	0.53	1.2
N	1.8	555 ± 25	0.40	1.2

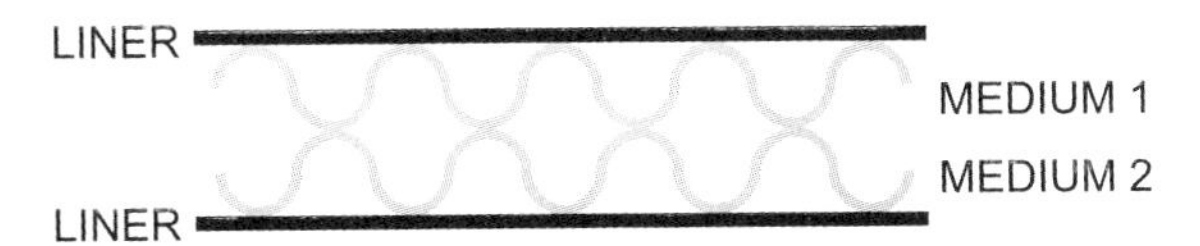

Fig. 14 X-flute board. (From Ref. 5.)

The board is manufactured using a highly modified single-facer unit (Fig. 15). The two media are corrugated between rolls AB and CD, respectively, and combine at the nip between B and C. The fluted carrier roll transports the combined medium to the point where the single-face liner is bonded to the structure using minimum pressure. The resulting single-face board is dried with an infrared heater before proceeding to the bridge and double backer.

X-flute board is essentially a symmetrical double-wall board without the middle liner. Because this liner is close to the neutral axis during bending, its omission will produce a fiber saving without unduly reducing the bending stiffness. The inventors claim that X-flute produces boxes that are up to 40% stronger at a given board weight, and the relatively small flute size yields board that will print well.

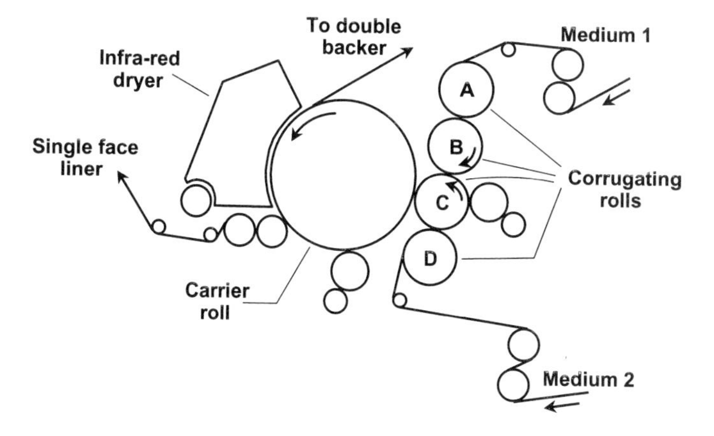

Fig. 15 Single facer unit for X-flute production. (From Ref. 156.)

C. Laminated Medium

Conventional corrugated board has a single layer of corrugating medium, but there is growing use of a laminated combination of two media (Fig. 16). This product is manufactured under a number of proprietary names, but the process is apparently not covered by existing patents. Two media are glued together—generally with a modified PVA adhesive—just before the corrugating rolls and are then combined on the single-facer unit of the corrugator.

The medium is rigid and is particularly useful in combination with the A-flute profile, because it corrects a deficiency of this structure, namely its resistance to a crushing load.

VII. RATIONALE FOR REVIEWING TEST PROCEDURES

Corrugated board is a product that is not usually used in the form in which it is manufactured—as flat sheets. It is made into boxes, and the performance of these boxes is the critical test of fitness for purpose. Consequently, it is logical to examine the important properties at each stage in the chain of manufacture, starting from the box and working back through the corrugated board to the board components (Fig. 17).

The major functions of corrugated boxes are

- To protect and contain the contents in the service environment
- To provide ease of handling
- To carry a marketing message about the contents

The dominant functions of the packaging will determine the balance between these requirements. For example, the main function of horticultural packaging is to protect sensitive fruit from damage in high and cyclical humidity environments, whereas an E-flute wine cask is expected to carry sophisticated graphics for consumer appeal.

Jönson [71] highlights the advantages and disadvantages of conducting tests in the service environment and in the laboratory (Table 4). The package engineer usually relies on a combination of past experience, laboratory testing, and limited field trials to establish the suitability of a package design for a particular application. This chapter concentrates on the laboratory methods that are available to test corrugated boxes and corrugated board and its components.

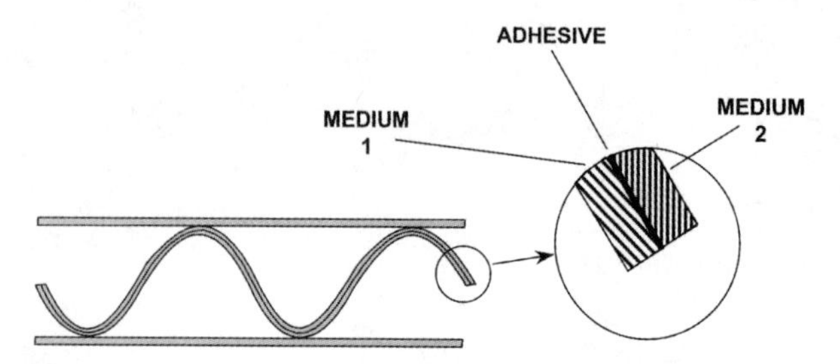

Fig. 16 Corrugated board with a laminated medium.

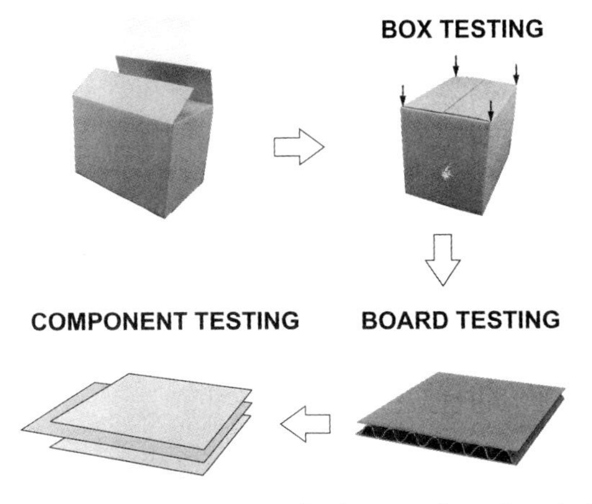

Fig. 17 Testing stages in the manufacturing chain of a corrugated box.

Table 4 Advantages and Disadvantages of Laboratory and Field Testing

Evaluation method	Advantages	Disadvantages
Field testing—practical tests in the service environment.	Packaging exposed to actual conditions that occur in the transportation and storage chain.	Testing is very time-consuming. Difficult to control and quantify the experimental variables. Difficult to evaluate the individual effect of experimental variables due to uncontrolled interactions. Difficult to typify a particular transportation and storage chain, requiring multiple tests that are prohibitively expensive.
Laboratory tests—simulation of the service environment.	Packaging exposed to a controlled environment, enabling factors to be selectively varied. Multiple sample testing possible. Cost relatively low. Package failure can be observed and understood. Testing can be completed in a relatively short time. Enables comparison of alternative package designs.	Cannot simulate all aspects and interactions in the service environment.

Source: Adapted from Ref. 71.

A. Standard Methods

Various countries have pulp and paper industry technical associations that publish standard methods for particular tests. The methods quoted in this chapter have been obtained from a number of sources:

- *Tappi Standard Methods (Tappi)*: Technical Association of the Pulp & Paper Industry, Inc., PO Box 105113, Atlanta GA 30348-5113.
  Telephone: 1-800-322-8686 (U.S.) +1-770-446-1400 (International)
  Internet Home Page: http://www.tappi.org/
  Internet Test Methods: http://www.tappi.org/public/test_methods
  e-mail: serviceline@tappi.org
- *ISO Methods (ISO)*: International Organization for Standardization, 1, rue de Varembé, Case postale 56, CH-1211 GENEVA 20, Switzerland.
  Telephone: +41 22 749 01 11
  Fax: +41 22 733 34 30
  Internet Home Page: http://www.iso.ch/
  Internet Test Methods: http://www.iso.ch/infoe/catinfo.html
- *SCAN-test Methods (SCAN):* SCAN-test, Box 5605, S-114 86 STOCKHOLM, Sweden.
  Telephone: +46 8 67 67 000
  Fax: +46 8 10 40 09
  Internet Home page: http://www.stfi.se/
  Internet Test Methods: http://www.stfi.se/contract/standard/scantest.htm
  e-mail: scantest@stfi.se
- *FEFCO Methods (FEFCO)*: Federation européenne des Fabricants de Carton Ondulé (European Federation of Manufacturers of Corrugated Board), 37 rue d'Amsterdam, 75008 PARIS, France.
  Telephone: +33 1 53 20 66 80
  Fax: + 33 1 42 82 97 07
  Internet Home Page: http://www.fefco.org/
  Internet Test Methods: http://www.fefco.org/contact/contact.htm
- *Appita Standard Methods (Appita)*: Australian and New Zealand Pulp and Paper Industry Technical Association, Suite 47, Level 1, 255 Drummond St, Carlton, VIC 3053, Australia.
  Telephone: +61 3 9347 2377
  Fax: +61 3 9348 1206
  Internet Home Page: http://www.appita.com.au/
  Internet Test Methods: Not available.
  e-mail: info@appita.com.au
- *JISC Methods (JIS)*: Japan Industrial Standards Committee, c/o Standards Department, 1-3-1, Kasumigaseki, Chiyoda-ku, TOKYO 100-8921, Japan.
  Internet Home Page: http://www.jisc.org/
  Internet Test Methods: http://www.jsa.or.jp/eng/catalog/frame.htm

In this text, the source of the standard method will be abbreviated by the term given within parentheses above. *Note*: Contact details were correct as of May 2001.

B. Equipment for Testing

Most of the test equipment described in the text is supplied by one manufacturer: Lorentzen & Wettre AB of Sweden. It must be emphasized that this does not constitute an endorsement of these particular instruments but reflects my familiarity with the equipment. There are a range of manufacturers who offer similar test instruments. An effective method of identifying alternative manufacturers of particular test instruments is to use the "equipment category search" function at http://www.paperloop.com/, specifying "testing (pulp & paper equipment & instruments.)" Alternatively, the compendium of TAPPI test methods [169] includes a list of equipment suppliers.

C. Nomenclature

There are variety of terms for describing the products and components covered by this chapter. To avoid confusion, the following terms will be used:

Liner:	The facing layers of corrugated board.
Linerboard:	The paperboard grade used for the liners in corrugated board.
Corrugating medium:	The fluted material in the center of corrugated board.
Corrugated board:	The combination of one or more liners with one or more corrugating media.
Box:	A container manufactured from corrugated board.

VIII. TESTING OF BOX PERFORMANCE

A. Some Notes on Box Construction

There are implications of the corrugated board conversion processes (see Fig. 8) that should be outlined before describing specific tests, because they can affect the physical properties of the box. Essentially, the corrugated board is cut to length and width on the corrugator using rotating knives and slitters to form a box blank. The blanks then undergo scoring and slotting, usually in combination with and after flexographic printing.

Scoring A score is a crease that is pressed into the board to facilitate the folding of the panels or flaps of the box. The crease can be impressed in the board by a rotating circular die on the top and underside of the board (Fig. 18) or by metal rules. The blanks are usually scored on the corrugator in the direction of manufacture.

The geometry, number, and shape of the scoring wheels will be determined by the type of fold required in the assembled box. Perforations may also be added along the crease to facilitate folding, particularly with heavier grades of board. The quality of the scores is important to the integrity of the final box and should be tested on a regular basis. Score quality can be measured following TAPPI T 829 om-96, which measures the force required to break the score (Fig. 19).

The apparatus and test piece dimensions are specified in the method. The load to break the specimen at the score is compared to the load at an unscored position:

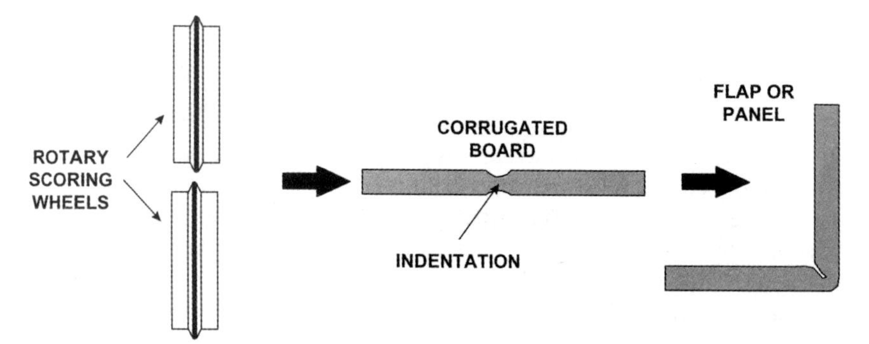

Fig. 18 Point-to-point crease formed by rotary scoring wheels.

$$\text{Score quality ratio} = \frac{\text{average force to break the score (Position A)}}{\text{average force to break unscored board (Position B)}}$$

Ten specimens are measured to obtain the average force values. This test is crude and yields little information on the folding properties of the crease. An instrument developed at the PTS Institute in Munich bends the crease and calculates 26 different properties (Fig. 20). The creased board specimen is inserted into pneumatic clamps, which are then rotated at a controllable speed to an adjustable angle. The specimen rests against a load cell that measures the force exerted as the clamps rotate. The force will depend on the quality of the crease, in particular the local bending stiffness. The device can produce plots of score bending moment versus bending angle, relaxation curves, and other information that can be used to optimize the scoring operation.

Slotting The box will also usually be slotted. This is because the external perimeter of the finished box will be greater than the internal perimeter, so a slot is needed at the corners to allow the proper folding of the flaps. The slotting operation is carried out by knives in flat-bed or rotary die cutters or on a flexo-folder-gluer.

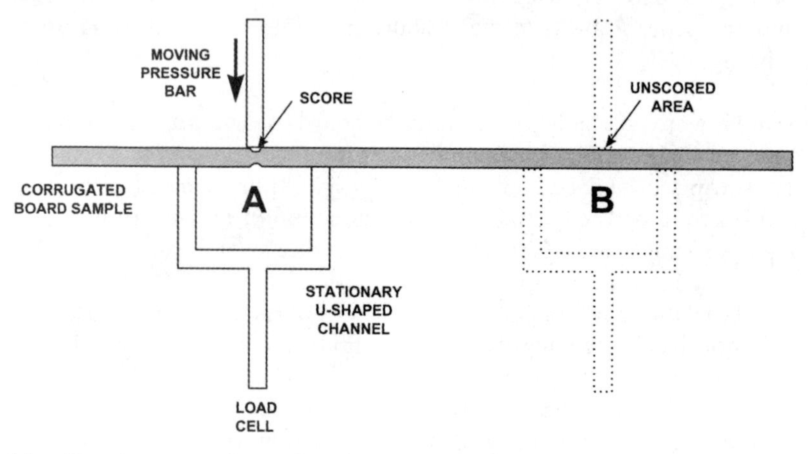

Fig. 19 Apparatus for testing the quality of scores in corrugated board.

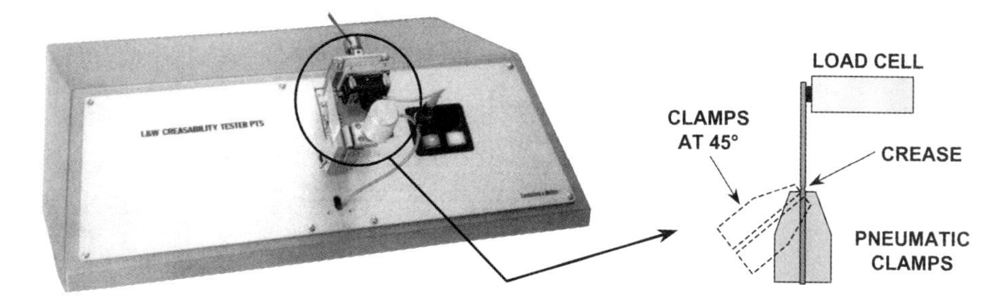

Fig. 20 Lorentzen & Wettre creasability tester PTS for testing the quality of scores.

Manufacturers Joint In order to form the box, the sides must be fastened together. This is achieved with the manufacturers joint, which usually consists of a flap of board that is glued, stitched, or stapled at a corner. The flap can be either external or internal to the box, depending on the requirements of the customer. The quality of the manufacturers joint is important to box integrity, because a poor joint will tend to fail under service and the box will lose its ability to contain the contents. All types of joints can be tested in tensile mode following TAPPI T 813 om-99. The test measures the tensile strength of a strip cut perpendicular to the joint (Fig. 21).

This mode of loading exerts a shear stress on the joint, and if failure occurs at the joint rather than in the corrugated board the full strength potential of the box will not be realized. However, it should be noted that the stresses acting on the joint are more complex, particularly when the box is under a compressive load. The joint will be subjected to out-of-plane stresses as the corner of the box deforms, and these stresses are not applied during the test procedure. If the manufacturers joint is glued, the strength of the bond can be evaluated following TAPPI Test Method T 837 pm-95, which does exert out-of-plane stresses similar to those experienced during the use of the box.

Complete Box Blank The final box blank will include scores, slots, and a flap for a joint (Fig. 22). The regular slotted container (RSC) is one of the simpler box designs, but there are a large number of other designs that are used for corrugated

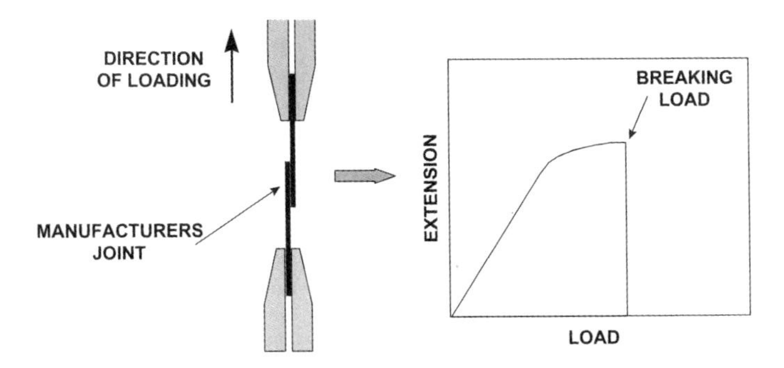

Fig. 21 Tensile testing of the manufacturers joint.

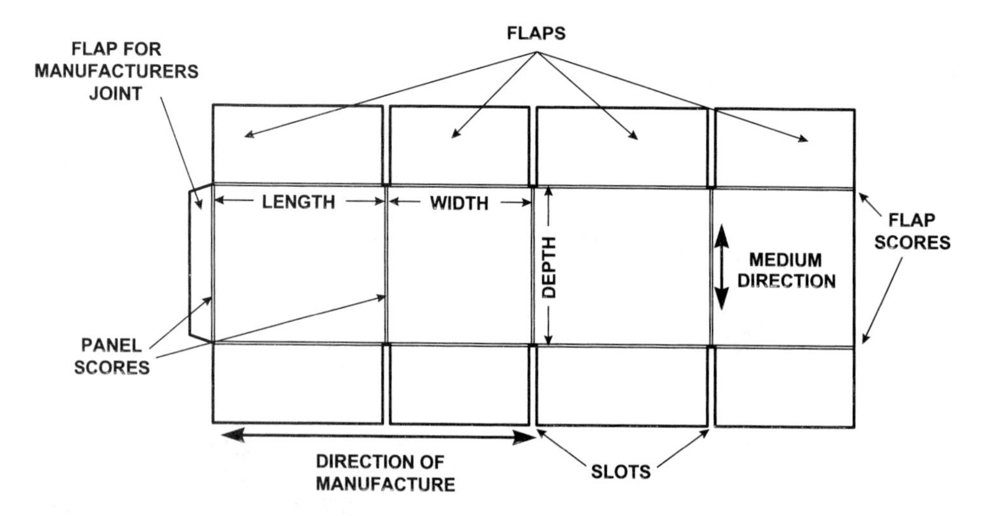

Fig. 22 Prepared blank for a regular slotted container (RSC) showing slots, scores, and joint.

boxes. Fortunately there is an international code that describes box construction, originally developed by FEFCO and adopted by the International Corrugated Case Association (see Ref. 43). The RSC is International Box Code 0201 (Fig. 23). Note that the flutes are vertical, which is the common configuration in boxes expected to resist compressive loads. The blank can also be cut so that the flutes are horizontal, which is more appropriate to resist side-to-side loading.

B. Box Compression Resistance

Most corrugated boxes are expected to withstand external compressive loading during their service life. This loading is usually the result of stacking the boxes and their contents on top of one another. The ability to withstand compressive stress is a complex property that depends on the environmental conditions and the magnitude and duration of the loading. The box compression test (BCT), also known as the top-to-bottom test (TTB) gives an indication of strength under a compressive load.

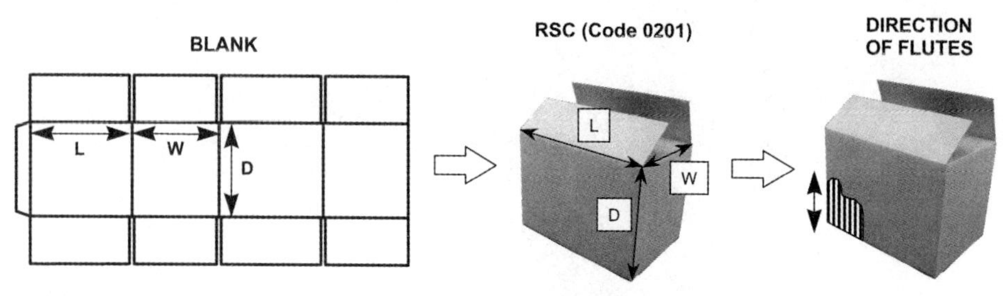

Fig. 23 Blank and assembled box type RSC, International Box Code 0201.

Standard Methods

ISO	TAPPI	JIS	SCAN	Appita	FEFCO
—	T 804 om-97	—	—	AS/NZS 1301 800s-87	No. 50

Procedure The apparatus consists of two platens, one of which is driven to compress the box (Fig. 24). The platens are flat and parallel to tolerances specified in the relevant standard. They may or may not be restrained from tilting, depending on the method.

The first stage in the test is to erect the box and seal the outer flaps to the inner flaps using tape, glue, clips, or other approved method. The box is then placed between the platens and subjected to a small preload to ensure a defined contact with the box and level off irregularities. This becomes the zero point for box deformation measurement. The box is then compressed at the specified rate until it fails. This rate can be specified as either deformation (in millimeters) or load (in newtons) per unit time. The deformation and load are recorded by a suitable device, and both the maximum load and the deformation at that load are reported. If the manufacturers joint fails during the test, this should be noted with the results. The procedure is repeated on at least five boxes.

Comments on the Box Compression Test

Sampling It is important that the tested boxes be representative of the batch from which they are taken. Failure to ensure this can result in sampling faults that will compromise accuracy and give misleading results. The types of sampling faults can be classified as *sampling error* and *sampling bias*, as defined in most statistical texts (for example [101]).

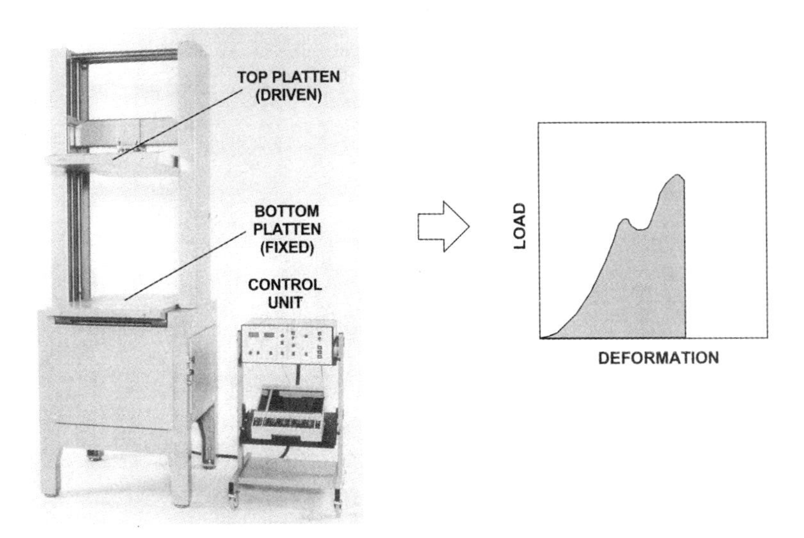

Fig. 24 Typical apparatus for testing box compression resistance.

Sampling error is due to the tendency to select samples that are not representative of the population as a result of the method of selection. Continuous manufacturing processes are particularly prone to this type of error. Such processes can produce periodic variations in the product, so it is important to randomize the timing of the sampling regime. It is also important to avoid taking batches of adjacent specimens, because there is no guarantee that they are representative of the whole run.

Sampling bias is the tendency to favor samples that have particular characteristics. For example, there may be a tendency to select well-made box blanks for testing and avoid those that have minor faults that would not normally be rejected by any quality control procedures.

It is important that the sampling regime be set up with due consideration to these factors. Further information on sampling can be found in statistical texts.

Sample Size TAPPI T 804 om-97 recommends that "at least five" specimens be tested for box compression resistance. In practice, a small sample size will be inappropriate due to the high variability of the corrugating and box-making processes. A more valid approach to determine sample size is to base it on expectations of accuracy and the intrinsic variability of the results. This can be accomplished by following a stepwise procedure [101].

Step 1 Set a tolerable error level e. This will depend on the requirements of the particular testing situation. For example, an audit test for a quality accreditation agency may have stringent requirements for accuracy, but an internal test may not be as demanding.

Step 2 Set a target for the acceptable level of confidence in the result, and find the value of the normal deviate z from statistical tables. This assumes that the population is normally distributed.

Step 3 Establish a value for the population standard deviation σ. This can be an estimate, or there may be historical data for particular box types that can be used to establish σ.

Step 4 Calculate the required sample size:

$$n = \frac{z^2 \sigma^2}{e^2} \tag{1}$$

Example A box maker wants to establish the sample size required to obtain a box compression resistance result with an accuracy of ± 200 N and a confidence level of 95% ($z = 1.96$). The estimate of σ is 500 N from historical data. How many boxes should be tested?

$$n = \frac{(1.96)^2 (500)^2}{(200)^2} = 24 \tag{2}$$

If σ is unknown, the box maker can estimate a value, test the required number of boxes, then recalculate n from the standard deviation of the results.

Flaps and Flap Scorelines This is arguably the most important aspect of box testing, because flaps and flap scorelines have a strong influence on the compression characteristics of the box. McKee and Gander [123] compared the compression resistance of RSC boxes and tubes made to the same dimensions but without flaps

(Fig. 25). They found that the tube exhibits a higher compression resistance and lower deflection at a given load. They concluded that box deformation is 90% due to the flaps and flap scorelines, which reduce the inherent stiffness of the tube structure. It is therefore important that the test method includes procedures to handle both factors in a reproducible manner.

- The outer and inner flaps should be secured to each other either by hot melt glue or by the use of clips, staples, or tape (see, e.g., Ref. 116). As the box is compressed, the flap will rotate into the body of the box because the scoreline acts like a hinge [141]. The amount of rotation will depend on the stiffness of the hinge, as will the contribution to box compression resistance. Securing the flaps to each other is a simple technique to reduce this source of variation.
- •• The flaps should be bent back 180° and then forward 270° to the closed position. This is to ensure that the scoreline receives a reproducible pretreatment prior to box compression.

The sensitivity of box compression resistance to flaps and scorelines also offers the box maker an opportunity to improve box performance by optimizing the scoring process.

Fixed vs. Floating Platens Some methods offer a choice between fixed and floating platens. As the names suggest, fixed platens are constrained from movement in the horizontal and vertical planes. Floating platens are attached to a central universal joint that allows freedom of movement. The choice of platen type has a contentious element. According to Maltenfort [108], a fixed platen test will not truly reflect the quality of the box, because failure will not necessarily occur at the weakest point. The TAPPI test method strongly recommends fixed platens for quality assurance and comparative testing of boxes but concedes that floating platens are useful for investigating weak points in a stacking pattern of several boxes. Singh et al. [157] tested five different RSC-type boxes with fixed and floating platens and concluded that the difference between the results was small in relation to normal levels of strength variation. Platen configuration should be reported as part of the test procedure. As long as a particular configuration is used consistently, the choice should not be an issue.

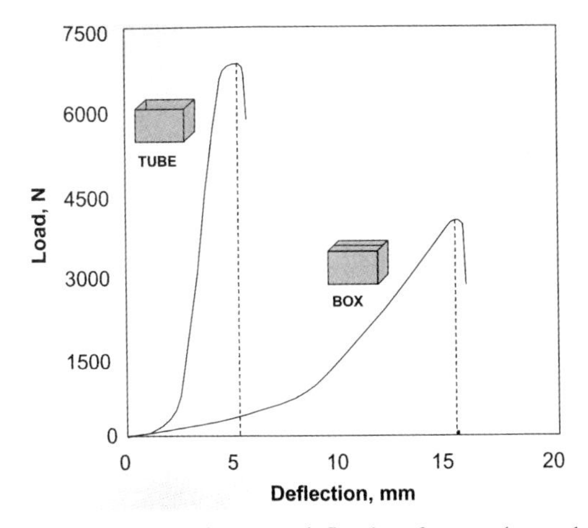

Fig. 25 Load versus deflection for a tube and equivalent box. (Adapted from Ref. 123.)

Use of a Dummy Load An empty box can fail by bulging either inward or outward, but a box filled with contents can normally bulge only outward. This can reduce the measured compression strength of the box by more than 20% [108]. If it is not possible or desirable to test the boxes with their contents, a dummy load should be used. The purpose of the dummy load is to prevent inward bulge, and this can be achieved in several ways. One method is to cut a solid block of polystyrene with a sliding fit into the box, but leave appropriate headroom to enable the box to deform under load.

Use of a Preload A preload is applied to the box to ensure proper contact with the top platen and to level off any irregularities due to construction. This preload either is fixed or varies according to the number of walls in the corrugated board. For example, the TAPPI test method stipulates 2230 N as a preload for triple-walled boxes, but the equivalent Appita method uses a standard preload of 220 N for all board constructions. Because box deformation is measured from the preload point, the two methods will give significantly different results.

Variability of the Results Several factors can contribute to the between-box variability, such as variations in flap score quality and corrugated board quality. A graph comparing load–deformation curves for 15 boxes illustrates this point (Fig. 26). These boxes were RSC type and made under carefully controlled conditions as part of a trial:

Mean box compression resistance	5764 N
Maximum	6420 N
Minimum	5100 N
Standard deviation	374 N
Coefficient of variation	6.5%

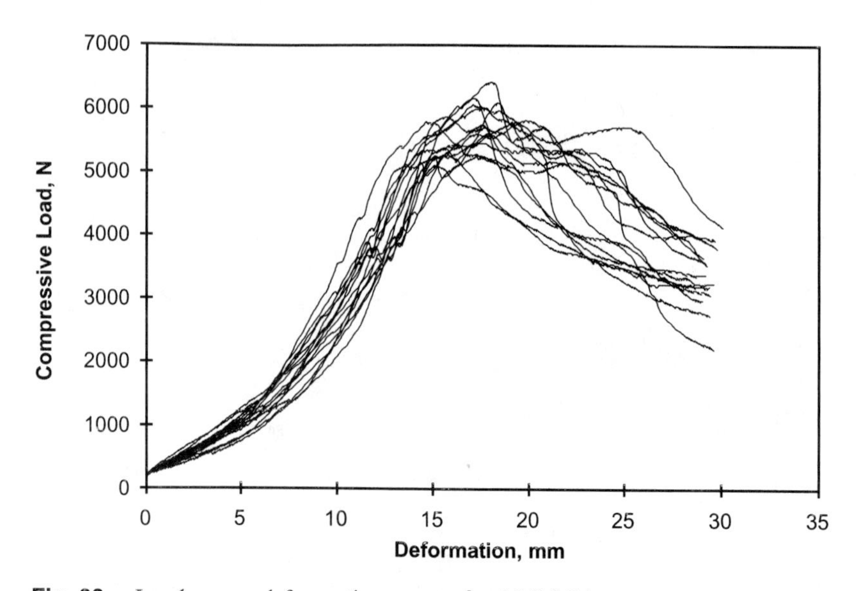

Fig. 26 Load versus deformation curves for 15 RSC boxes.

The range of values is quite high—over 1300 N—and the variability is also relatively high, considering that the boxes were made in a carefully controlled trial. To be 95% confident that the mean value was within 5% of its true value, 26 boxes would have to be tested.

Result variability can also be used as a quality control index. The value of this approach is that it recognizes the importance of a reasonable level of variability in the box compression results. For example, the mean value may be acceptable, but if the coefficient of variation (*standard deviation* σ divided by the *mean* expressed as a percentage) is excessive there may be a substantial proportion of the population that has unacceptably low box compression resistance. Excessive variability also points to manufacturing problems in the box-making process that warrant further investigation.

Interpretation of Load–Deformation Curves The load versus deformation curves contain useful information about the way the box has failed (Fig. 27). The curve in Fig. 27 can be broken into several segments:

A Any unevenness in the box is leveled out by the fixed platen. The top scorelines of the box begin to roll as the box starts to take load. The amount of deformation and the slope of this segment mainly depend on the quality of the score.

B The steepest corners of the box start to take load. This part of the curve has the steepest slope, because these corners are the stiffest part of the structure.

C A subpeak is caused by the small-scale yielding of one of the fold scorelines.

D The long panels begin to bow outward, and the load is taken up by the short panels.

E The maximum load. After this point, the box corners compress and the short panels begin to buckle. The box moves into a general type of failure.

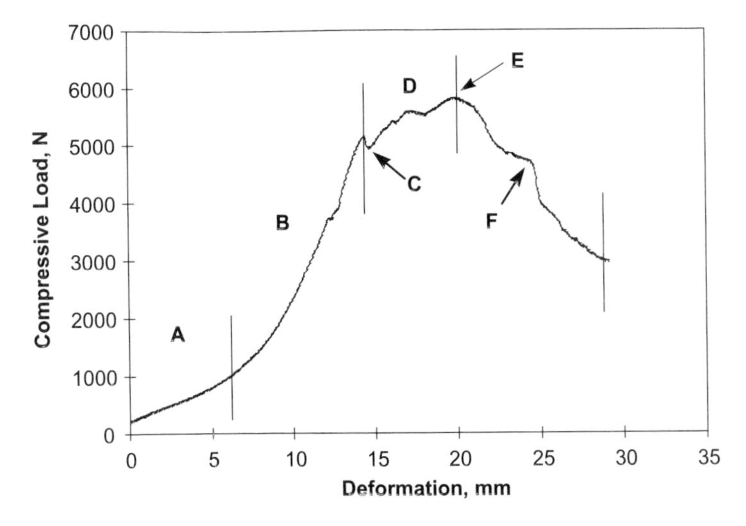

Fig. 27 Load versus deformation for an A-flute RSC box using fixed platens.

F A localized stability caused by the buckled structure temporarily assuming a more stable configuration.

This type of information can be used by the packaging engineer to optimize the contribution of the different box elements and to troubleshoot the box-making process.

Usefulness of Box Compression Resistance There is no doubt that the box compression test is highly useful as a quality control tool to monitor the box-making operation. It is also useful as a comparison between different box constructions and box types. However, its usefulness to predict how a box will perform in the service environment is limited for several reasons (see also Fig. 28):

1. Boxes are tested individually. If the boxes are stacked in patterns other than in columns, the full strength potential will not be realized [46,78,88].

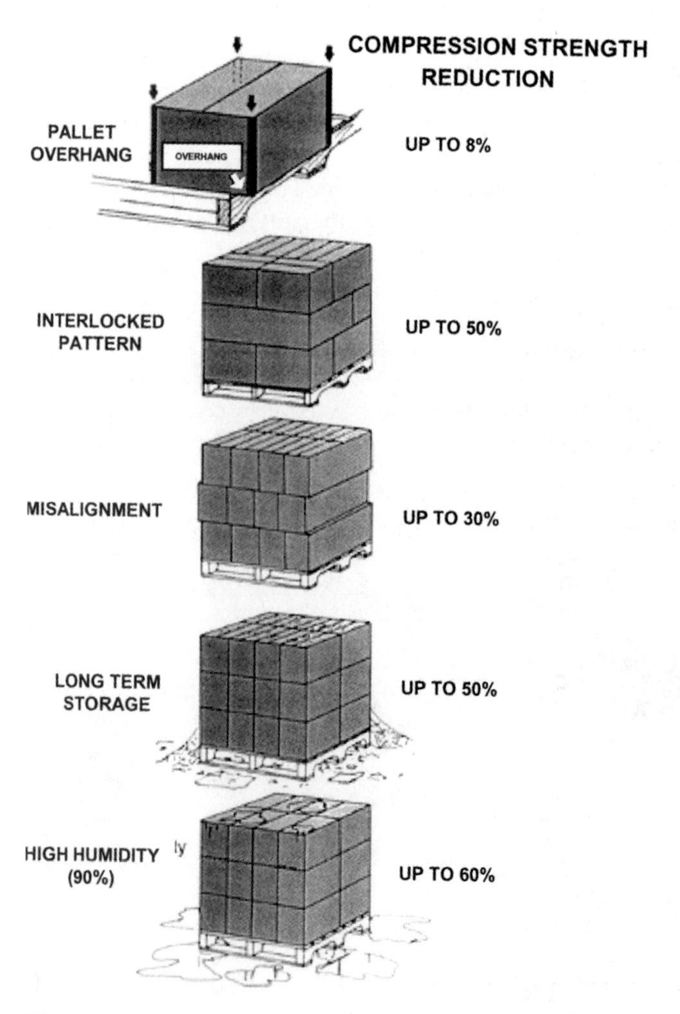

Fig. 28 Effect of various factors on realized compression resistance of stacked boxes.

2. The service environment will have climatic conditions or variations that may degrade box compression strength. This is discussed further in Section VIII.C.
3. The box will probably be subjected to static and sustained loads that invoke creep behavior that accelerates failure. Even in the simplest cases, box compression results do not relate well to stacking conditions. Kellicut and Landt [83] measured the compression resistance of boxes, then stacked the filled boxes. They found that the bottom box would fail in minutes if the load was only about 90% of the measured compression resistance.
4. The box may be subjected to dynamic loading such as vibration that will accelerate failure.

These limitations are usually compensated for by using a safety factor to ensure that a box will perform in a particular environment. These factors are purely empirical and rely on the experience of the box designer and past performance under similar service conditions. The subject of safety factors is discussed in more depth by Jönson [72].

C. Box Compression Resistance Under Static Loading—Creep

When a box is subjected to a constant load for a sustained period of time, it will deform in a manner that is related to the magnitude of the load and the length of time the load is applied. This behavior is termed *creep* and is discussed in detail in Chapter 2 of this volume. The importance of the creep behavior of boxes was reported in 1951 by Kellicut and Landt [83]. They noted that the time to failure of boxes under static loading decreased logarithmically as the dead load approached the laboratory-tested box compression strength (Fig. 29). The time to failure with a dead load of 95% of the compression strength was only 1.3 min. At 78%, the box failed within 7 h. Clearly, this has serious implications for the long-term stacking of corrugated boxes.

There is no standard method for creep testing of boxes, but the test is quite basic to conduct. The essential elements are a method to apply a static load to the

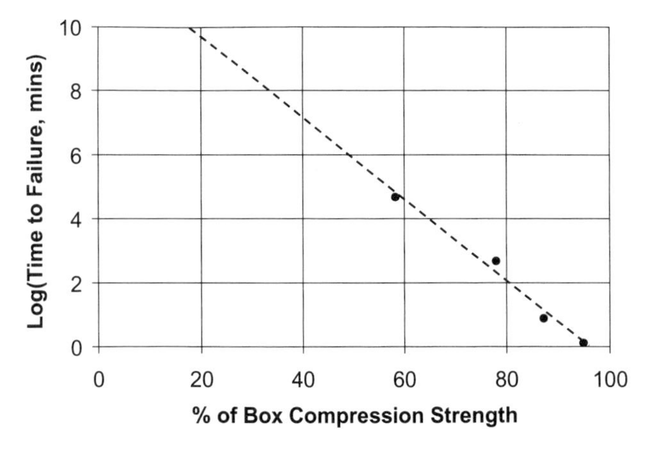

Fig. 29 Effect of load on time to failure. (Data from Ref. 83.)

box and a method to measure box deformation as a function of time. The simplest way to achieve this is by putting a deadweight on top of a rigid platen that covers the perimeter of the box. The deformation is monitored by periodically measuring the height of diagonally opposite corners of the platen with respect to a reference plane, usually the floor (Fig. 30).

Some recommendations for creep testing of boxes are the following.

- The floor should be sufficiently rigid that any deformation caused by the weight of the test rig is insignificant compared to the deformation of the box.
- The load can be applied either by a deadweight or by other means such as a pneumatically (Fig. 31) or hydraulically loaded platen. The load exerted on the box must be constant.
- The platen must also be rigid and not deform significantly under dead load.
- The deformation of the box should be measured at regular intervals or continuously if possible. The number of measurements should be sufficient to characterize the complete deformation curve to failure.
- The relative humidity and temperature of the environment should be held constant and continuously monitored.

A typical record of box deformation versus time should exhibit three distinct zones (Fig. 32) [11].

Primary creep is the response of the box to the initial period of loading. The deformation rate is rapid at the onset of load application but decreases as the box settles under the load.

Secondary creep is a stable phase of continuous deformation. The slope of this line is the secondary creep rate.

Tertiary creep is a zone of increasing deformation as the box begins to fail.

According to Bronkhorst [11], the time to failure of the box T_L can be predicted from the slope of the secondary creep curve:

$$T_L = A\left(\frac{1}{\dot{\varepsilon}_2}\right) \tag{3}$$

where $\dot{\varepsilon}_2$ is the secondary creep rate, expressed as strain per unit time (day).

Bronkhorst interprets the constant A as the "steady-state creep ductility." By adjusting the value of A, he was able to obtain good agreement with box creep results from the literature by Koning and Stern [90], Leake [102], and Leake and Wojcik [103] in different environmental conditions. Secondary creep rate is a useful criterion

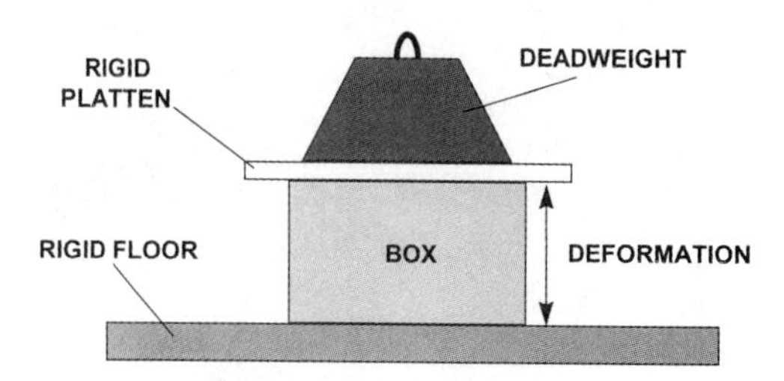

Fig. 30 Experimental setup for measuring the compressive creep response of boxes.

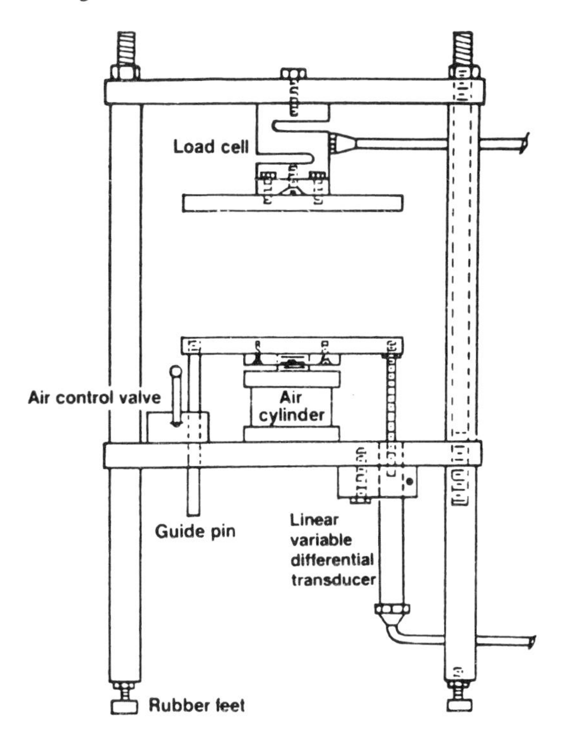

Fig. 31 Pneumatic static loading rig for studying box creep. (From Ref. 102.)

for comparing the performance of boxes with different components in the corrugated board or different box styles. However, there are experimental considerations that need some thought:

The load on the box. This must be chosen carefully, because a load too close to the compression strength (BCT) of the box will cause failure before secondary creep is established and a load that is too light will result in

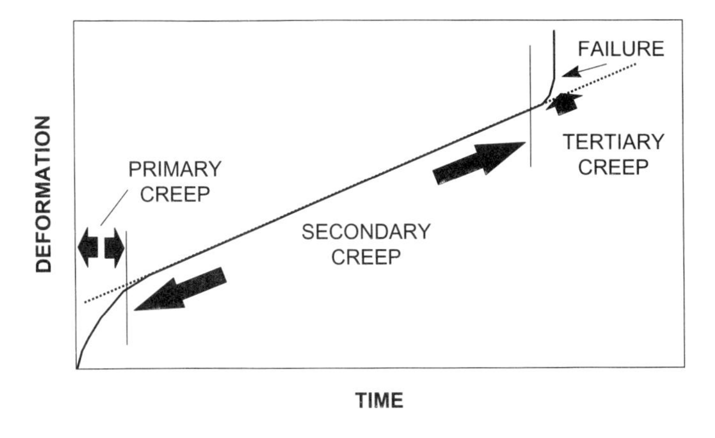

Fig. 32 The three zones of creep behavior. (From Ref. 11.)

impractically slow deformation. One useful technique is to carry out a preliminary experiment with several loads to determine a suitable creep load and then use this load in subsequent experiments. For standard laboratory conditions (23°C 50% RH), loads of 60–80% of BCT should be suitable.

The number of boxes to test. BCT results are variable, and creep results can be expected to be at least as variable, so a suitably large number of boxes should be tested. Equation (1) is useful to establish the required sample size. Note that testing a large number of boxes may exclude the use of relatively sophisticated and expensive rigs such as the one shown in Fig. 31, particularly if available testing time is limited.

The environmental conditions. Testing under laboratory conditions has limited value if the service conditions are different. This is due to the moisture content sensitivity of box creep rate, which is covered in the next section.

D. Effect of Environmental Conditions on Box Creep

One of the major disadvantages of packaging made from paper or paperboard is the moisture sensitivity of its strength properties. For example, increasing the moisture content of a box made from corrugated board from 7% to 16% can decrease box compression strength by about 55%, according to data published by Kellicut and Landt [83]. The equilibrium moisture content of paperboard is a function of the relative humidity of the environment (Fig. 33). Above 90% RH, the moisture content is increasing by more than 1% and box compression strength is decreasing by about 5% [83] for every percent increase in humidity.

The environmental conditions experienced by the boxes can be of either natural or artificial origin. For example, Leake and Wojik [103] reported the naturally occurring temperature and humidity variations in a warehouse in Bridgewater, NJ

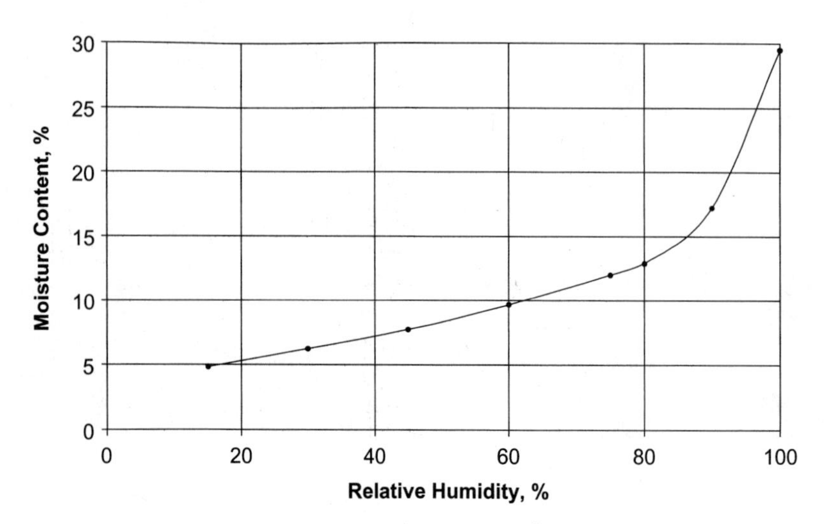

Fig. 33 Equilibrium moisture content of 240 g/m^2 bleached kraft paperboard as a function of relative humidity. (Data from Ref. 27.)

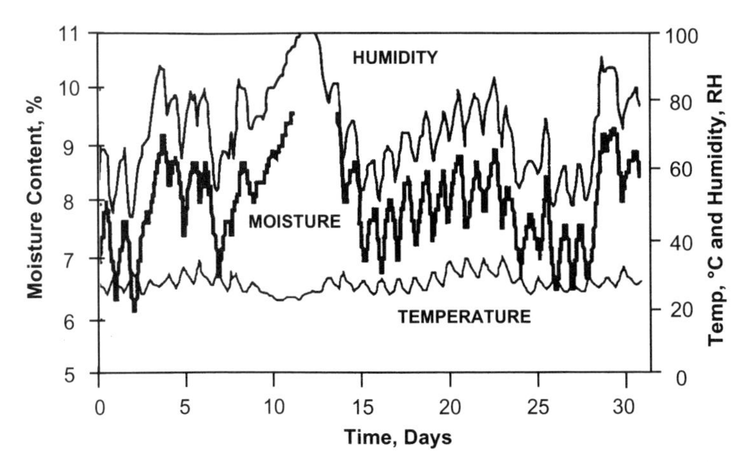

Fig. 34 Temperature and humidity profiles in Bridgewater, NJ for August 1992 and the effect on box moisture content. (From Ref. 103.)

(Fig. 34). Box moisture content responded quickly to changes in the humidity of the warehouse. There was a strong diurnal (24 h) cycle but also longer term trends, which were associated with changes in the weather.

The boxes may also be subjected to controlled changes in the environment. An example is the storage of fresh fruit, which requires low temperature and high humidity to minimize transpiration of moisture and reduce the rate of quality degradation. For example, the optimum environment for long-term apple storage is −1 to +2°C and 95% RH [33], which will severely compromise the stacking performance of the boxes.

Leake and Wojcik [103] investigated the effect of humidity cycling rate on the secondary creep rate of double-wall RSC boxes subjected to a static load of 91 kg (Fig. 35). They found that environmental cycling reduced time to failure by at least 70%. The faster cycles also increased time to failure from 3 days (1/2 cycle/day) to 14 days (2 cycles/day).

Clearly, the best method to determine box performance in a given environment is to simulate the environmental conditions as closely as possible, then test box stacking performance using a realistic load.

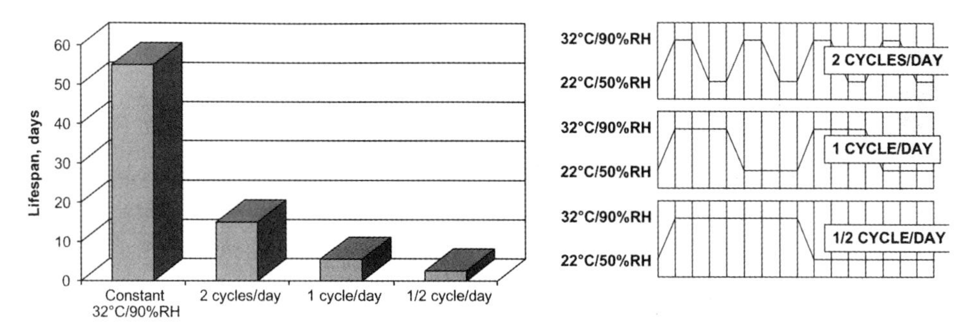

Fig. 35 Time to failure of boxes under a static load. A comparison of the effect of constant conditions and environmental cycling. (Data from Ref. 103.)

Moisture-induced creep of paper, paperboard, and boxes is incompletely understood. See Haslach (Chapter 4) for a review of the subject. A series of international symposia proceedings have also presented current and developing theories [144–147]. As mentioned previously, box creep tests should be carried out under service conditions wherever possible. However, there may be occasions when comparisons should be carried out under standardized conditions, such as for cooperative research between institutions. In these cases, it is prudent to adopt the conditions recommended at the 1997 International Symposium [146]:

- Cyclical humidity creep performance should be reported as *isochronous curves* of stress versus strain. The hyperbolic equation of the curve [61] should also be reported, so that values of the creep stiffness can be obtained if required. Isochronous curves are an alternative method of presenting creep data, as explained in the paper by Haraldsson et al. at the 1994 international symposium [145].
- •• The relative humidity cycle should be as shown in Fig. 36. There is no recommendation for the temperature during the cycle, but it is presumed to be 23°C.
- ••• Specimens should be preconditioned in the test environment before loading to release internal stresses in the materials. There are no further details in the recommendation, but two complete humidity cycles should suffice.

E. Drop Testing of Boxes

Another important performance parameter is the ability of the box to withstand a freefall impact and continue to protect the contents. This is particularly important for containers that are designed to carry dangerous goods such as flammable, toxic, or explosive substances. The method is described in FEFCO No. 51 and TAPPI Standard T 802 om-99. In the Tappi method the box and its contents are successively dropped onto a steel plate from a chosen height on one corner, three edges, and six faces for a total of 10 drops. The container and contents are then examined for damage and put through another drop cycle starting with a different corner. There are additional procedures to cover other handling and transportation conditions.

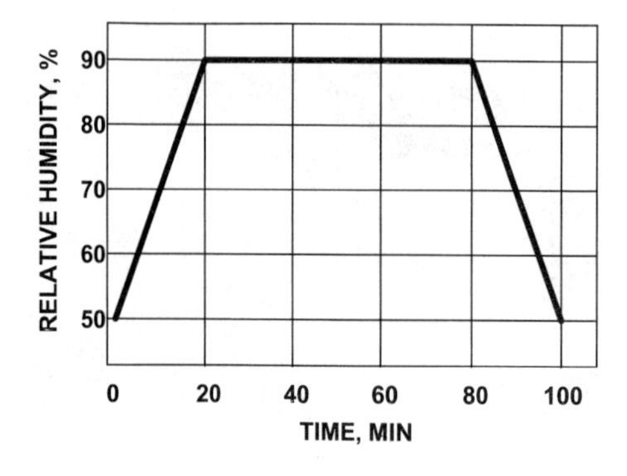

Fig. 36 Standardized cycle for cyclical humidity testing of paper, paperboard, and containers. (From Ref. 146.)

The drop test equipment must be able to control the height and orientation of the drop. Hoists fitted with remotely operated release hooks and dropping tables fulfill these requirements (Fig. 37).

The report should include details of the procedure, drops to failure, damage to the box and contents, and box moisture content if measured. The repeatability [168] of the method is poor unless there are several replications—the TAPPI test method quotes 27% for three tests.

F. Other Impact Tests for Boxes

Impact Resistance TAPPI T 801 om-94 describes an inclined plane test that is used to evaluate the impact resistance of large boxes with a gross weight in excess of 45 kg. The purpose of the test is to simulate shocks and impacts that may be encountered during transportation and handling. The basic apparatus—also known as the Conbar test—is an inclined plane with a dolly running on a track and a backstop (Fig. 38).

The dolly in Fig. 38 is shown fitted with an optional bulkhead. The programming material at the front of the bulkhead or on the backstop is to control the form of the shock pulse experienced on impact. A test involves pulling back the dolly to a position on the inclined plate that will produce a particular impact velocity. The container with contents is oriented with a corner, edge, or face at the front, and the dolly is released. The test is repeated until the container fails or until it has survived a predetermined number of impacts. There is no reported figure for the precision of this test due to variations in the design and construction of the apparatus. In particular, the rigidity of the backstop is a source of variability. However, the apparatus is useful for simulating such events as the shunting of railcars and their effect on container integrity. A related method is described in FEFCO No. 53.

Drum Test This is a TAPPI classical test method (TAPPI T 800 cm-96) that tests the resistance of a box to a range of shocks and impact stresses. The apparatus is a revolving hexagonal drum with baffles and protuberances on the inner faces. A test

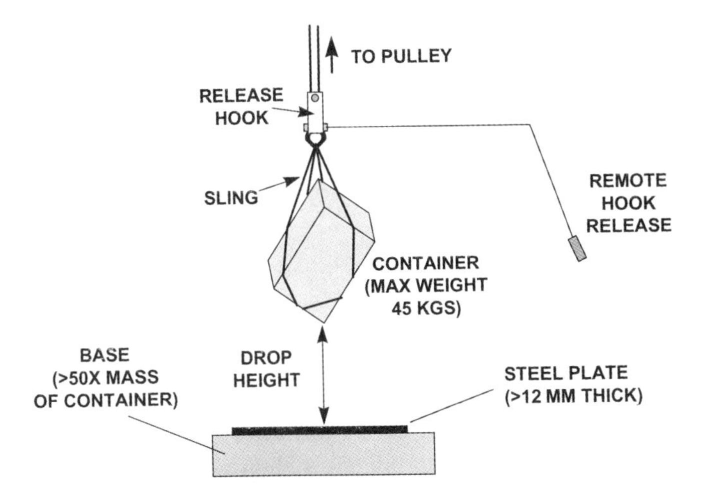

Fig. 37 Hoist equipped with remote hook released for box drop testing.

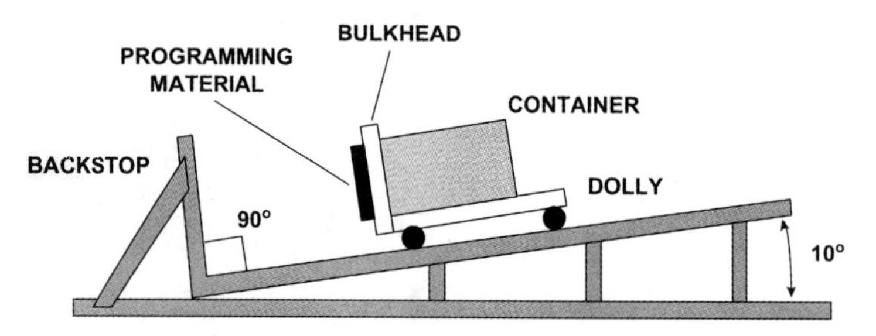

Fig. 38 Inclined plane apparatus for testing the resistance of containers to impacts (Conbar test).

involves putting a single filled box in the drum, starting it in motion, then counting the number of falls. After a prescribed number of falls, the drum is stopped and the container is examined for damage. The test is highly variable, and due to the random nature of the falls it is difficult to get reasonable precision even within a single laboratory. As such, it is rarely used for practical testing, and laboratories tend to favor more definable impact tests such as the drop test. However, it can be used to simulate the rough handling experienced during package handling by postal or courier services [89].

G. Note on Cushion Theory

The ability of boxes to survive impact tests depends on their ability to cushion the contents from shock damage. In turn, this depends on the ability of the corrugated board to absorb the energy of impact and spread the impact load [136]. The theory of cushioning is summarized by Jönson [73] and covered in more detail by Paine [136]. The performance of a cushioning material can be characterized by the cushioning factor C, which is the ratio of the stress exerted on the materal to the impact energy per unit volume. The relationship between C and impact energy is termed the *cushioning performance* of the material (Fig. 39).

The material will be most effective as a cushioning material at the minimum value C_{MIN}. The range considered as efficient cushioning is $\pm 1.2C_{MIN}$. Typical values of $1.2C_{MIN}$ for corrugated board range from 1.8 to 3.6, which compares poorly with materials such as bubble wrap (C_{MIN} 4.1–5.2) and polystyrene beads (C_{MIN} 4.0–4.4) [73]. In terms of corrugated board, A-flute has the best cushioning properties, followed by C-flute, then B-flute.

H. Other Box Tests

Several other container tests appear in the literature, and some of them are briefly reviewed here. Godshall [51,52] describes methods to test the response of containers to forced vibrations. These vibrations occur during freight transport and can produce compressive loading forces far greater than the load exerted by the stacked boxes [51]. Urbanik [173–175] extended Godshall's work to include pallets of boxes. Long and Penney [105] investigated the thermal insulation properties of boxes using thermocouples to measure rates of heat loss.

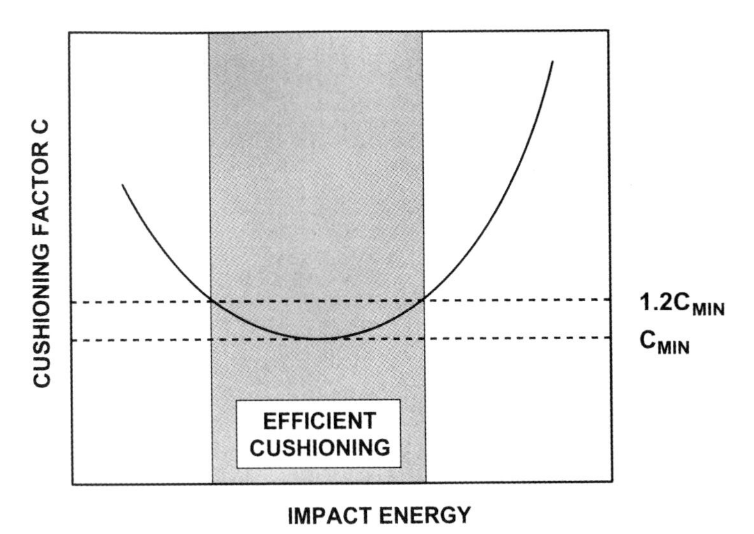

Fig. 39 The performance of a cushioning material. (Adapted from Ref. 73.)

IX. MODELING THE CORRUGATED BOX

The performance of a corrugated box is determined by the ability to protect its contents under service conditions. Because containers are usually stacked and subjected to high compressive loads, the box compressive strength is widely recognized as an important indicator of performance. A corrugated box is an engineered structure that is fabricated from corrugated board. Consequently, there have been several attempts to use engineering principles to predict the compressive strength of the box from the properties of the corrugated board.

A. Empirical and Semiempirical Models

It is generally accepted that the first researcher to treat corrugated board as an engineering material was T. A. Carlson of the Forest Products Laboratory, Madison, WI [20,22]. He developed and published work describing several combined board properties, including flexural rigidity, column compression, and shear strength, in terms of the behavior of the board components. However, he did not attempt to relate these parameters to box compression strength. The first to attempt to analyze box compression in terms of plate theory were McCready and Katz [119, 120]. Their paper and others on the subject are reviewed by McKee et al. [126]. The first widely used formula for predicting box compression strength is attributed to Kellicut and Landt [80,83]. They developed an expression derived from the theory of ultimate compression strength of orthotropic thin plates that included the ring crush value of the board components and box and flute factors. The box factor depends on flute size and the type of manufacturers joint, and the flute factor depends on flute size alone.

Ranger [149] treated a container as a series of struts with a load-bearing capacity that depended on height and distance from a vertical edge. He developed an empirical equation to predict box compression strength:

$$W = \frac{2L(5CD^2 + 2SL^2)}{5D^2 + 2L^2} + \frac{2W(5CD^2 + 2SW^2)}{5D^2 + 2W^2} \tag{4}$$

where

W = box compression strength
C = short-column compression strength of corrugated board
S = strut length $= \pi^2 F/D^2$
F = cross-direction flexural rigidity of corrugated board
L, W, D = length, width, and depth of box, respectively

Box compression strength is calculated by measuring L, W, and D and testing the board for C and S. Ranger claims a prediction accuracy that is generally within ±15%.

Maltenfort [110] described an edgewise compression test for linerboard. The test, called the Concora liner test (CLT), is a compression test that has similarities to the ring crush test (Section XI), but the sample is linear rather than ring-shaped. He developed three empirical relationships between CLT, the dimensions of the box, and box compression strength (BCT):

Flute style	Regression equation
A-flute	BCT = 5.8L + 12W − 2.1D + 6.5(CD CLT) + 365
B-flute	BCT = 5.8L + 12W − 2.1D + 5.4(CD CLT) + 212
C-flute	BCT = 5.8L + 12W − 2.1D + 6.5(CD CLT) + 350

A total of 190 types of boxes of different sizes and with different types of manufacturers joints and corrugated board components were tested to develop these equations.

Kawanishi [76] derived a statistical formula based on multivariate analysis for estimating the compression strength of a box. The variables in the study were the size and type of box, grade of corrugated board, printed area, and moisture content:

$$F = 3.79 \times 10^{-8} K^{0.379} W^{0.65} w^{1.2} d^{-4.15} y^{2.45} t^{3.43} Z^{0.565} k^{-0.315} p^{0.0602} S^{-1.1} \tag{5}$$

where

F = box compression strength, kgf
K = factor dependent on linerboard furnish
W = total grammage of the linerboards, g/m^2
w = grammage of corrugating medium, g/m^2
d = medium takeup factor (1.59 for A-flute, 1.36 for B-flute)
y = average corrugation count (34 for A-flute, 50 for B-flute)
t = thickness of corrugated board, mm
Z = box perimeter, cm
k = box style factor (1.0 for an RSC)
p = print ratio (1.0 for no print, 0.01 for solid print)
S = box moisture content, %

The number of box variations tested to develop Eq. (5) is not reported. However, the reported accuracy of box compression strength prediction averaged ±9% for a wide range of box specifications and moisture contents.

B. The McKee Model

The semiempirical McKee model [126] is still widely used in the industry almost 40 years after its first publication. It is interesting to revisit the derivation of the model, because it was developed at a time when desktop calculators were crude. Consequently, several assumptions and simplifications were made to facilitate calculations.

McKee and associates started from a combination of observation of box behavior under compression and strain measurements on the box panels. Motion picture analysis showed that box failure was initiated when the combined board failed at or near a panel corner. Because the vertical edges of the panel are essentially constrained to remain flat, the edgewise compressive strength of the combined board must play a major role in box compressive strength. The center regions of the panel also support the load, but here it is the panel dimensions and the bending characteristics of the combined board that determine load-bearing ability. As the load is applied to the box, the panels eventually begin to bulge and become unstable. The time from the onset of bulge to failure is called the postbuckling range. It is difficult to describe this behavior in theoretical terms, because the instability has a complex effect on the load-bearing capacity of the panels.

The postbuckling behavior of the box (Fig. 40) is treated using a semiempirical approach attributed to Cox [30] and developed to predict failure loads for isotropic flat plates by Gerard [50].

$$\frac{P_Z}{P_{CR}} = c\left[\frac{P_M}{P_{CR}}\right]^b \tag{6}$$

where

P_Z = ultimate strength of the plate, lb/in.
P_{CR} = instability load, lb/in.
P_M = edgewise compression strength of the plate materal, lb/in.
c, b = constants

This expression can be rearranged to solve for P_Z:

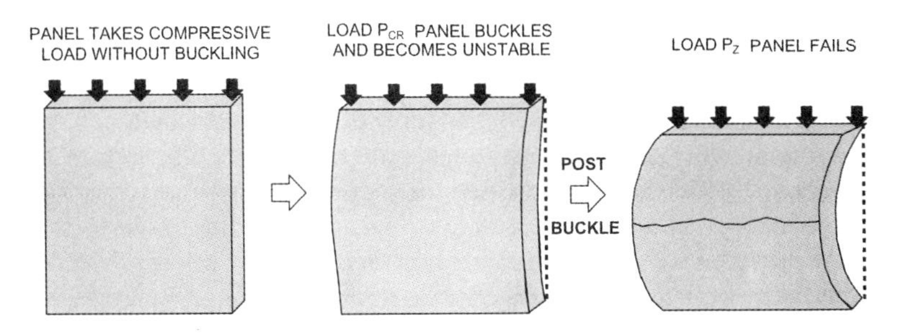

Fig. 40 Buckling behavior of isotropic, thin flat plates.

$$P_Z = cP_M^b P_{CR}^{1-b} \tag{7}$$

Once P_Z has been evaluated, the total compression strength of the box can be calculated by multiplying P_Z by the panel width and summing for the four panels. However, it is first necessary to evaluate the variables P_M and P_{CR} and the constants b and c. For corrugated board, P_M can be measured using a short-column test (Section X). An expression for the instability load P_{CR} for thin orthotropic flat sandwich panels was developed by March and Smith [115] and applied to corrugated board. The approach assumes that the stresses in the panel remain elastic up to the point of instability, which was deemed appropriate by McKee et al.

$$P_{CR} = 12k_{CR}\sqrt{\frac{D_x D_y}{W^2}} \tag{8}$$

where

k_{CR} = buckling coefficient
D_x = corrugated board MD flexural stiffness, lb-in.
D_y = corrugated board CD flexural stiffness, lb-in.
W = panel width, in.

The buckling coefficient is an expression involving a function of MD/CD flexural stiffness ratio and panel depth-to-width ratio and a dimensionless plate parameter K, which is itself dependent on the mechanical properties and cross-sectional geometry of the combined board:

$$k_{CR} = \frac{\pi^2}{12}\left(\frac{r^2}{n^2} + \frac{n^2}{r^2} + 2K\right) \tag{9}$$

where

$r = (D_x/D_y)^{1/4}(d/W)$
where D_x, D_y are the flexural stiffness of the corrugated board in the machine (x) and cross-machine (y) directions

d = panel depth, in.
W = panel width, in.
n = number of half-waves in the buckled panel in the direction of load.
$n = 1$ if $r \leqq \sqrt{2}$
$n = 2$ if $\sqrt{2} \leqq r \leqq \sqrt{6}$
$n = j$ if $\sqrt{j(j-1)} \leqq r \leqq \sqrt{j(j+1)}$

The treatment was considered too complex for practical application, and a series of approximations were developed:

1. The plate parameter K is a complex function of several corrugated board and liner properties. The derivation is not quoted in the McKee treatment, but the value $K = 0.5$ was adopted from practical evaluations on typical corrugated board.
2. $(D_x/D_y)^{1/4} = 1.17$ from practical measurements.
3. $W = Z/4$, assuming that the box is square with perimeter Z.

This led to the development of a simplified expression for total box load P:

$$P = c(4\pi)^{2-2b} P_M^b (\sqrt{D_x D_y})^{1-b} Z^{2b-1} k^{1-b} \quad (10)$$

where k = modified buckling coefficient (see Ref. 126.) The further simplification of $k^{1-b} = 1.33$ when $b = 0.76$ for depth-to-perimeter values ≥ 0.143 produced the general form of the McKee equation for box compression strength:

$$P = aP_M^b (\sqrt{D_x D_y})^{1-b} Z^{2b-1} \quad (11)$$

where P is the box compression strength and a, b are constants. The constants a and b were evaluated using 63 samples of A-, B-, and C-flute RSC boxes that were considered to be representative of the capabilities of the U.S. industry at that time:

$$P = 2.028 P_M^{0.746} (\sqrt{D_x D_y})^{0.254} Z^{0.492} \quad (12)$$

This is the industry-specific form of the McKee equation. Further simplification, based on an empirical relationship between board caliper, geometric mean flexural stiffness, and P_M, yielded the estimate:

$$P = 5.87 P_M \sqrt{h} \sqrt{Z} \quad (13)$$

where h = combined board thickness, in.

Equation (13) has been converted into SI units by Kainulainen and Toroi [75]:

$$P = 375\ ECT^{0.75}\ FR^{0.25}\ Z^{0.5} \quad (14)$$

where

P = box compression resistance, N
ECT = cross-direction edgewise compression strength of the board, kN/m
FR = $(D_x D_y)^{1/2}$ the geometric mean flexural rigidity of the board, N m
Z = box perimeter, m

McKinlay and Shaw [132] took the simplification of Eq. (13) one stage further:

$$P = K\sqrt{Z} \quad (15)$$

where K = board stiffness factor. They derived K for single- and double-walled corrugated board that was made using corrugated board components manufactured by Amcor in Australia. Interestingly, they also included a *panel type factor* to accommodate different box styles, in contrast to the RSC-type limitation of the McKee approach.

Comments on the McKee Model One emphasis of this model is the simplification of complex expressions to describe the postbuckling behavior of panels. It is a testament to the level of understanding of McKee and his coworkers that the approach yielded sensible numbers despite the compound application of approximations and simplifications. The model is still widely used within the packaging industry as a tool for predicting box performance from the properties of the corrugated board. However, it has limitations.

1. The constants a and b in Eq. (11) were evaluated on typical boxes from one country (United States) in the early 1960s. With the increasing use of high performance lineboards and advances in converting machinery, it is unlikely that these constants still apply. Box makers who wish to use this

equation should recalculate a and b using boxes that are representative of their range of production following the specified procedure [126]. For example, Nordman et al. [134] adapted the McKee equation to fit the performance of Finnish boxes:

$$P = 3.73 P_M^{0.14} (D_x D_y)^{0.22} (Z/2)^{0.47} \tag{16}$$

2. The McKee formula assumes that the boxes are square, but the length-to-width, or aspect, ratio will affect the load-carrying ability of the box [185]. Wolf [186] made an empirical modification to the McKee equation to account for the effect of aspect ratio.
3. The equation predicts maximum load but gives no information on the deformation of the box. Since the major function of the box is to protect the contents, the maximum load is of little use if the deformation at that load is excessive.
4. The influence of shear on the corrugated board is ignored. According to March and Smith [115], there is a transverse shear component that tends to decrease the buckling load. This was acknowledged by McKee et al. [126], but they claimed that the correction was of "minor importance," so it was removed from their considerations. Examining boxes during failure often reveals a pattern that suggests the presence of a shear stress near the vertical edges. Shear displacements are mainly resisted by the corrugating medium [22], and these may be significant, particularly at low medium grammages.

It would be interesting to review the assumptions and approximations in the McKee approach, and investigate the possibilities for improvement.

C. Mechanistic Models

Mechanistic models do not require the use of empirical factors and are based on structural analysis and engineering mechanics. Peterson [139,140] followed the work of Fox et al. [48] and developed a failure load analysis procedure to describe the large-deflection behavior of individual box panels. The mathematical model was based on a Rayleigh–Ritz displacement approach using Kirchhoff large-deflection strain relations. As such, the model is complex and is useful only to researchers with good mathematical skills. Pommier et al. [142] used a finite element method to model the buckling of a flapless corrugated box. There are also anecdotal reports that private companies have developed or are developing advanced models, but the results have not been published.

D. Comments—Box Models

Developing a deeper understanding of how boxes function as engineering structures is important for several reasons:

- It enables identification of the corrugated board properties that are critical to the performance of the box. The consensus of studies to date is that the CD edgewise compression strength of the corrugated board is the most important variable. This is followed by board thickness, then board bending stiffness in both the machine and cross-machine directions, with a smaller contribution from the in-plane shear stiffness.

- It allows the packaging engineer to optimize the design of the box. For example, Peterson [138] and Kainulainen and Toroi [75] developed approaches based on the McKee equation to calculate the minimum cost of a container for a given required performance.
- It facilitates comparison of different corrugated board constructions. For example, Koning [91] used the compressive stress–strain properties of the corrugated board components and concluded that it was more effective to add grammage to the medium for A-flute with liners heavier than 280 g/m^2. Johnson et al. [67] started from the premise that board edgewise compression strength is the important variable for box compression strength. They developed a model for short-column compression that yielded strength and weight contours that indicated when it was advantageous to add grammage either to the medium or to the liners.

It is unfortunate but understandable that much of the current research on box modeling may remain unpublished. The development of such models is time-consuming and expensive and can be justified in the commercial sector only if it yields a competitive advantage.

X. CORRUGATED BOARD TESTING

Corrugated board tests are split into three categories:

- *Physical characteristics* Nondestructive tests that measure attributes of the board that are important to end use performance.
- *Mechanical properties* Tests that measure structural integrity; can be either destructive or nondestructive. As a class of tests, they involve subjecting the board to a specified stress and measuring the response.
- *Cohesive tests* Corrugated board is held together by the bonds between the liners and the medium. These tests concentrate on the ability of the glue lines to hold the structure together under different loading conditions.

A. Physical Characteristics

Grammage A fundamental property of the corrugated board is its weight per unit area.

Standard Methods

ISO	TAPPI	JIS	SCAN	Appita	FEFCO
536:1995	T 410 om-98	P 8124:1998	P 6:75	AS/NZS 1301 405s-92	No. 2

Procedure Details can be found in the relevant test methods. Essentially, the technique involves cutting a representative area from the corrugated board with dimensions within a specified tolerance. The sample has first been pre-conditioned and then conditioned in the specified environments. It is then weighed on a suitably accurate balance, and the weight per unit area is calculated. In most of the world, the accepted term is *grammage*, the weight in grams of 1 m^2 of board. However, the use of nonmetric units still persists, particularly in the United States. The unit for the basis weight of corrugated board is $lb/1000\,ft^2$. The conversion is $g/m^2 = lb/1000\,ft^2 \times 4.882$.

Corrugated Board Components Measuring the grammage of the corrugated board components presents problems:

- The structure is glued together with a starch-based adhesive, and these bonds must be broken to separate the components.
- Some of the starch will be absorbed by the corrugating medium and linerboard, thus increasing the apparent grammage.
- The corrugating medium has flutes in it that have been heat-set during the manufacturing process.

FEFCO Testing Method No. 10 describes a procedure involving the soaking of the board in cold or hot water, and the flattening out and recutting of the corrugating medium. However, wetting and drying the specimens will cause an expansion and subsequent shrinkage of the components that will make it difficult to obtain an accurate test area. It is also difficult to flatten out the medium and obtain an accurate test area due to the high level of stretch in the machine direction, i.e., along the direction of the corrugations. In my experience, an accuracy of $\pm 5\,g/m^2$ is as good as can be expected from this method.

Corrugated Board Thickness The thickness of corrugated board exerts a strong influence on the compressive strength of boxes through its relationship with bending stiffness, as demonstrated in Eq. (15). It is also an indicator of the damage experienced by corrugated board during the converting processes.

Standard Methods

ISO	TAPPI	JIS	SCAN	Appita	FEFCO
3034:1975	T 411 om-97	P 8118:1998	P31:71	AS/NZS 1301 426s:94	No. 3

Procedure The specified methods all implement a deadweight micrometer to measure corrugated board thickness. The instrument may or may not be motorized (Fig. 41). The important features of the apparatus are stationary and moving anvils with flat and parallel faces, and a mechanism to provide the required measuring load on the moving anvil. The rate of descent of the anvil and the dwell time before taking a thickness reading are detailed in the relevent test methods. The TAPPI test method specifies a measuring pressure of 50 kPa, and the ISO standard uses 100 kPa. The SCAN standard is specific to corrugated board and quotes a measuring pressure of 20 kPa. Nordman et al. [134] claimed that this measuring pressure is inadequate. During manufacture, the corrugated board is subjected to compressive forces as it

Fig. 41 A motorized digital deadweight micrometer. (Courtesy Carter Holt Harvey Pulp & Paper Limited, Kinleith, New Zealand.)

passes through conversion machinery. The major part of the consequent thickness reduction is recovered when the load is removed, but the structure has been irreversibly damaged. The compressive load used in micrometer determinations of thickness are insufficient to register this damage. Nordman et al. define the *effective thickness* of the corrugated board as the thickness measured at a pressure of 80 kPa, or four times the pressure specified in the SCAN standard. They claim that effective thickness is a superior measure for corrugated board and present supporting results. Similar information can be extracted from the hardness test for corrugated board discussed in Section X.B.

Fig. 42 Device for determining the thickness of corrugated board components. (From Ref. 92.)

Thickness of Components The thickness of the corrugated board components can be measured in situ using a micrometer with modified anvils designed by Koning (Fig. 42) [92]. The device can also measure flute height and total board thickness, but the results will not be comparable to micrometer measurements due to the small anvils and the different—and undefined—compressive load.

Soft Platen Method The thickness of corrugated board can be measured by using a deadweight micrometer with soft rubber platens, as detailed in TAPPI Standard Method T 551 om-98. These platens conform to surface irregularities and yield thickness measurements that are lower than results obtained following TAPPI T 411 om-97. The method is particularly useful for corrugated board, because the surface of the board is not usually flat. This is due to the use of water-based adhesive such as starch to glue the tips of the fluted medium to the liners. As the starch dries, it will tend to pull the linerboard down the flanks of the medium, as illustrated in Fig. 43. Steel platens will rest on the peaks of the flutes and define the thickness T_1. The soft platens will conform to the surface irregularities, giving a smaller measured thickness T_2. This undulation is called *washboarding*. In severe cases it is classed as an appearance defect for corrugated board [21].

Coefficient of Friction There are two coefficients of friction (COF) that determine how one body will slide across another. The *static* COF is a measure of the force that is required to initiate motion between two surfaces. The *kinetic* COF is a measure of the force required to keep the body moving. It is important that corrugated board and boxes have a sufficiently high coefficient of friction to resist unwanted movement. This may be the movement of box blanks against one another during stacking at the end of the corrugator, or during transport within the converting facility. Filled and stacked boxes may also move during transport, causing damage to contents and handling difficulties. In some applications, the coefficient of friction must be sufficiently low to enable packages to slide down inclined chutes. Depending on the conditions during service, both the static and dynamic coefficients of friction can be important for corrugated board. For more coverage of paper friction, see Back (Volume 2, Chapter 12).

Standard Methods

ISO	TAPPI	JIS	SCAN	Appita	FEFCO
15359:1999	T 815 om-95 T 816 om-92	p 8147:1994	—	—	—

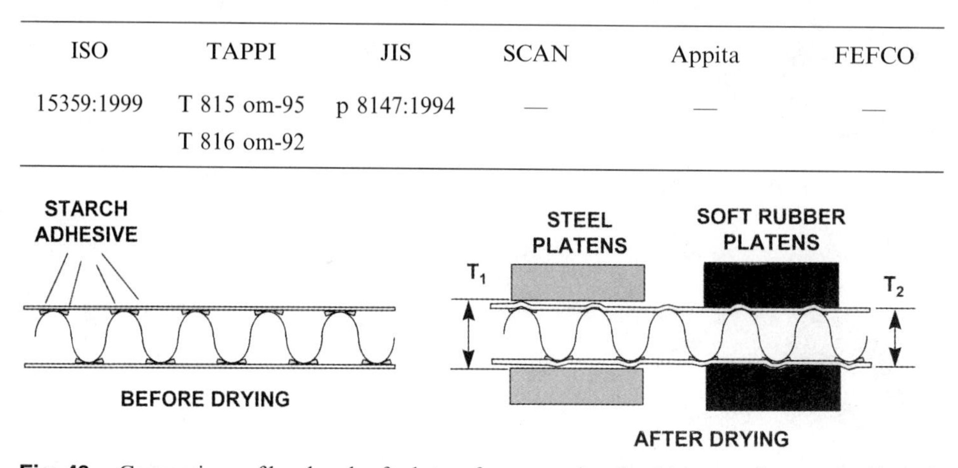

Fig. 43 Comparison of hard and soft platens for measuring the thickness of corrugated board.

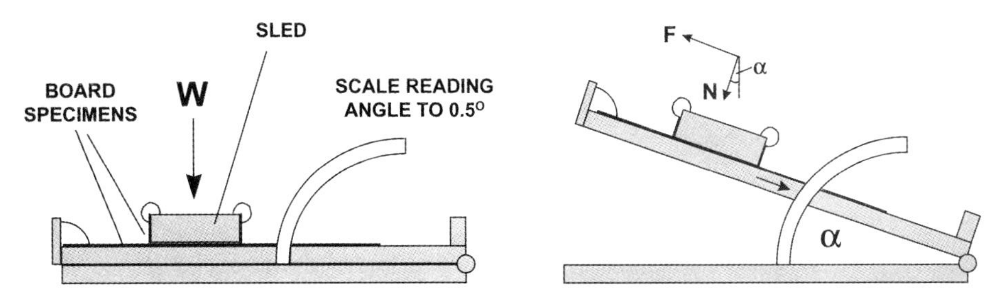

Fig. 44 Inclined plane method for determining the static coefficient of friction.

Procedure The two TAPPI test methods for static friction determination on solid fiberboard and corrugated board are different techniques that yield essentially equivalent results. T 815 om-95 is an inclined plane method (Fig. 44). The frictional force (F) opposes the motion of the sled. It is proportional to the normal force (N) acting on the sled. As the inclined plane begins to lift, the normal force will decrease as the inclination angle increases:

$$N = W \cos \alpha \tag{17}$$

where W is the weight of the sled.

The frictional force F will increase as the inclination angle increases:

$$F = W \sin \alpha \tag{18}$$

This produces a relationship between F and N:

$$F = N \tan \alpha \tag{19}$$

At some angle α_{crit} the gravitational component of the weight of the sled will be sufficient to overcome the frictional forces that oppose motion. The sled begins to move, and the inclined plane is stopped immediately. The angle defines the first static coefficient of friction, μ_{s1}:

$$\mu_{s1} = \tan \alpha_{\text{crit}} \tag{20}$$

TAPPI T 815 om-95 recommends repeating this procedure three times to get the third static coefficient of friction μ_{s3}. This is because friction measurements become constant only after repeated slides, as the surface is conditioned by the repeated action of the sled. The previous history of the specimen may be unknown, so this technique brings the surface to a more uniform and repeatable level. Five specimens are tested for each determination.

TAPPI T 816 om-92 is a horizontal plane method (Fig. 45). A corrugated board specimen is attached to the sled and to a movable table. The base is connected to a drive that can pull the table at 150 ± 25 mm/min. The sled is pulled over the specimen attached to the table twice; then the force F_{crit} required to initiate motion is read from the gauge at the beginning of the third pass. The ratio of this force to the weight of the sled W is the third static coefficient of friction:

$$\mu_{s3} = \frac{F_{\text{crit}}}{W} \tag{21}$$

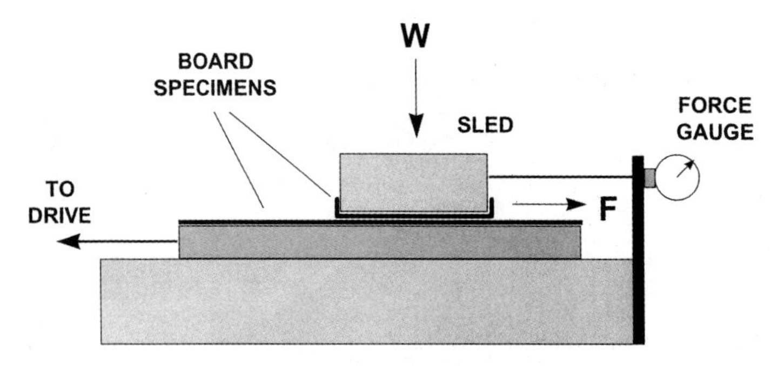

Fig. 45 Horizontal plane method for determining the static coefficient of friction.

Comments—Friction Testing It is important that both the slide angle and horizontal plane tests be set up to measure the appropriate frictional forces:

- Paper frictional forces are highly anisotropic, so the orientation of the samples must reflect the end use application. If there is no dominant orientation, then the machine direction should be tested against the cross-machine direction.
- ●● The test should reflect the service conditions, wherever possible. For example, if the behavior of boxes on a chute is of interest, then the board should be tested against a similar material.

Both standard methods describe the condition of the surface of the corrugated board at the time of the test. The presence of even trace amounts of fatty substances such as waxes can lower μ_{s3} dramatically (Fig. 46) [121].

The slightest contamination by handling can invalidate results. For this reason, measurements of friction are best carried out in the laboratory under carefully controlled conditions. The expectation that meaningful results will be obtained in a manufacturing environment is optimistic.

Kinetic Friction Neither TAPPI test method measures the kinetic coefficient of friction. A new device for measuring both static and kinetic coefficients was developed by STFI, Stockholm, in cooperation with the Forest Products Laboratory, Madison, WI [66]. Work on this instrument formed the basis for the ISO standard (15359:1999) for paper-to-paper friction.

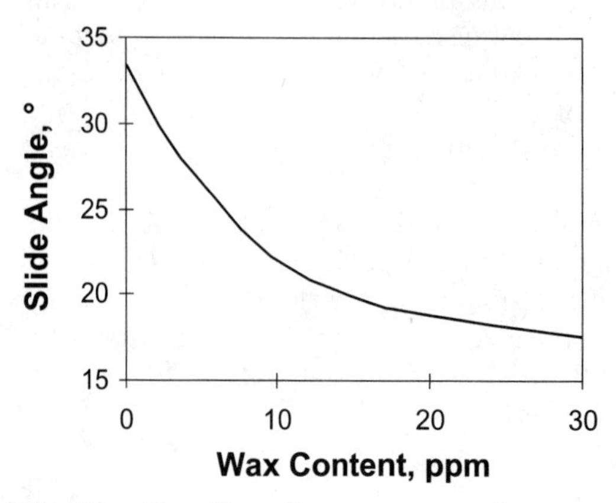

Fig. 46 The effect of trace amounts of wax on the coefficient of friction of paperboard.

The ISO standard was formulated after a review of available techniques and the observation that some of the critical variables that affect friction were unspecified [41]. The apparatus must have the following attributes, which control these variables and avoid sources of operator error:

1. The sled is lowered into position without any movement of the table surface.
2. The sled rests in position on the table for a defined time before measurement begins.
3. The sled speed is varied during measurement to ensure accurate determination of both static and dynamic coefficients of friction.
4. The sled is steered by guide rods to maintain a parallel orientation to the table and prevent wobble.
5. The apparatus is rigid to avoid the stick–slip phenomenon during kinetic measurements.
6. The test sequencing is computer-controlled to avoid operator error.

The method has been used to measure the coefficients of friction of linerboard and the effect of moisture content and paper surface energy [41]. A commercial instrument which conforms to ISO 15359:1999 is available from Mu Measurements Inc., Madison, WI (Fig. 47). The method tests paper–paper friction, but replacing one of the specimens with the appropriate material might enable service conditions to be more closely duplicated.

Thermal Resistance Corrugated board can be used as a thermal insulator. In some applications, such as cool storage of horticultural products, the thermal resistance will slow the rate of chilling of the contents. Ramaker [148] used a heat flowmeter designed to ASTM C 518 to determine the thermal resistance R_T of corrugated board and its components. He made the following observations:

1. Between 92% (A-flute) and 85% (B-flute) of the thermal resistance of corrugated board comes from the fluted section.
2. Increasing the grammage of the liners from 205 to 440 g/m^2 caused only an 8.5% increase in R_T for A-flute board.

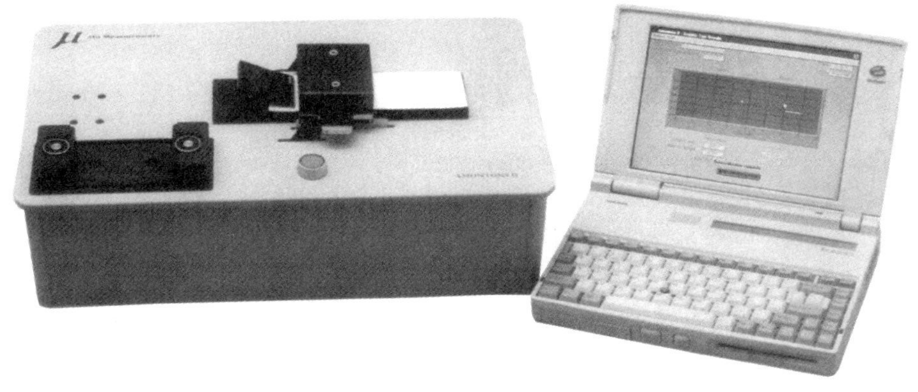

Fig. 47 Amontons II apparatus for measuring static and kinetic paper-to-paper friction. (Courtesy of Mu Measurements Inc., Madison, WI.)

3. Increasing medium grammage from 127 to 160 g/m^2 decreased R_T by 7% for an A-flute board.
4. Conditioning and testing at 0°C and 90% RH decreased R_T by 82% for A-flute board. Under these conditions, corrugated board loses its insulating properties.

Ramaker also observed that the thermal resistance of multiple layers of corrugated board is comparable to that mineral wool, vermiculite, and molded polystyrene as long as the board stays dry. See Bøhmer, Volume 2 Chapter 10, for more coverage of the thermal properties of paper.

Warp of Corrugated Board Warp is defined as a deviation from flatness in corrugated board [47]. The causes of warp can be quite complex and are beyond the scope of this chapter. However, good reviews of both simple and more complicated forms can be found in the literature [111,170,178,184]. Uesaka (Chapter 3) also covers paper curl, which is a related subject. Excessive warp will cause misfeeds in conversion equipment and automatic box erection equipment. In order to control the level of warp, it is first necessary to characterize it and measure it.

There are six basic forms of warp (Fig. 48). The most common forms of warp—normal or "up" warp and reverse or "down" warp—occur across the corrugator. End-to-end warp occurs along the direction of manufacture. There are also forms of warp that are combinations of these basic patterns, such as S-warp and twist warp, a serious type of warp where the board is twisted out of plane.

There are several ways to determine the degree of warp. The simplest method is to cut a square from the board and draw the curvature on a sheet of paper [135]. The degree of warp can be defined by the curvature of the board [178]:

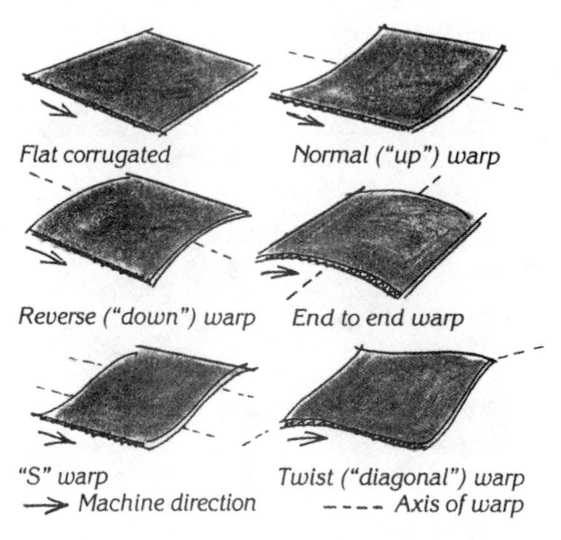

Fig. 48 Forms of warp in corrugated board. (Courtesy Amcor Kiwi Packaging Limited, Auckland, New Zealand.)

$$W = \frac{1}{R} \tag{22}$$

where W is the degree of warp per meter and R is the radius of curvature.

If the board is assumed to deform in an arc of a circle, R can be calculated from geometry (Fig. 49):

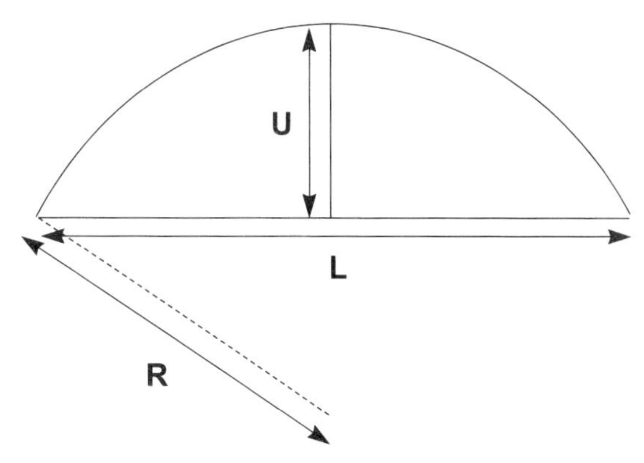

Fig. 49 Relation between radius of curvature and bulge for warp measurement.

$$R = \frac{L^2}{8U} \tag{23}$$

Where R = radius of curvature, m
L = sheet width, m
U = bulge, m.

Whitsitt and Hoerschelman [184] describe a method to measure twist warp by placing a 12 in. × 40 in (30.2 cm × 101.6 cm) board sample on a centrally placed support, then measuring the height of the corners and midpoints of the sides. This yields eight numbers, which is cumbersome. Eriksson et al. [36] describe a device that makes similar measurements using optical transducers. The test produces a series of numbers to characterize the form and extent of the warp, which may be useful for research purposes but is too complex for routine measurements. The Swedish Development Group for Corrugated Board [35] developed a simple device for measuring the simpler forms of warp (Fig. 50). The sample of corrugated board is placed between the stationary prongs of the tester, then a central probe advances into contact with the board. The readout is the degree of warp in inverse meters (m^{-1}). Billerud AB in Sweden also produced a simple tool for measuring warp [36,178]. These simple devices are good for measuring the less complex forms of warp with just one axis of curvature. However, Wennerblom [178] describes a technique that extends their application to twist warp (Fig. 51). In this technique, a square specimen is cut from the corrugated board with sides of about 20–25 cm. The board is carefully cut along the diagonals, and the warp is measured along each diagonal edge. Care must be taken to use the correct sign for the direction of warp; all up warp is positive and all down warp is negative. This technique, with reference to Fig. 51, yields the equation

$$W = \text{warp } (AC, \text{m}^{-1}) - \text{warp } (BD, \text{m}^{-1}) \tag{24}$$

Where AC and BD are the diagonals of the square.

Fig. 50 Device for measuring the warp of corrugated board. (Courtesy Lorentzen & Wettre, Kista, Sweden.)

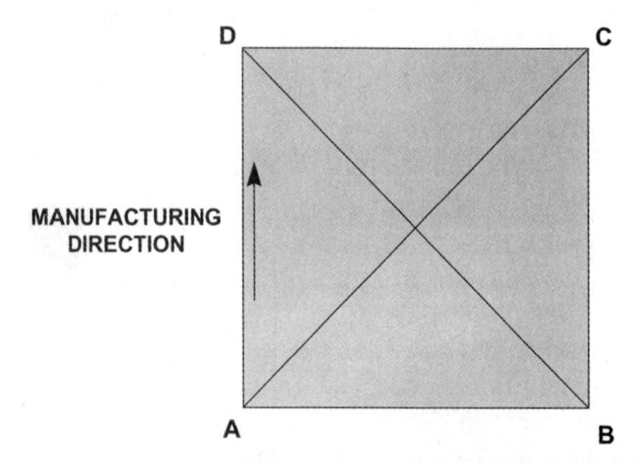

Fig. 51 Technique for measuring twist warp. (After Ref. 178.)

B. Mechanical Properties

The mechanical properties of corrugated board influence the ability of the boxes to perform in the service environment, provide a measurement of quality, or both. The tests described in this section involve exerting a defined stress on the specimen of corrugated board and measuring either the resistance to that stress or the strength of the structure.

Impact of Freight Carrier Regulations In the United States, the impact of freight carrier regulations on the development and testing of corrugated board packaging has been profound. Road and rail carriers are liable for in-transit damage to goods, so not surprisingly they developed specifications to ensure that packaging was fit for its purpose. Early specifications were based on board thickness and weight per unit area, but these were augmented by the addition of board burst testing (Section X.B) as a measure of packaging performance. However, there was widespread agreement that burst testing did not reflect the ability of the corrugated board to perform in service. In 1991, both the National Motor Freight Classification (Item 222) for road transport and the Uniform Freight Classification (Rule 41) for rail transport were modified to include an alternative criterion (Table 5). The alternative is based on the edgewise compression strength (ECS) of the corrugated board, as covered in the next section.

Kroeschell [97] gives an excellent overview of the history and development of this legislation. The alternative criterion has given box manufacturers more scope to use stronger and lighter components in corrugated board.

Edgewise Compression Strength Edgewise compression strength is defined as the strength of corrugated board when it is subjected to a compressive stress as a short column in the direction of the flutes. It is commonly known by two acronyms: ECS, for edgewise compression strength or edge crush strength, and ECT, for edgewise compression test or edge crush test. The terms short-column crush (SCC) and column crush test (CCT) also appear in the literature. Edgewise compression strength is arguably the most important corrugated board test due to its dominant influence on box compression strength [74,126]. Dagel and Jönson [32] carried out trials in Sweden with some 11,000 boxes and found that ECS is also the best indicator of the protective capability of corrugated board.

The historical development of edgewise compression tests for corrugated board is reviewed by Kroeschell [98]. The initial development work for ECS testing was carried out as early as 1943 [104]. Work at the Forest Products Laboratory started in 1947 and was continued by Kellicut [77]. McKee et al. [126] studied both necked-down and waxed specimens and recommended that the latter should become the standard method for ECS determination. This later became the TAPPI standard for ECS measurement and is still the only technique that is accepted by freight carriers.

Table 5 Original and Alternative Specifications for Corrugated Boxes According to Rule 41/Item 222

		Original specification		Alternative
Maximum weight of box and contents [lb (kg)]	Maximum external perimeter of the box [in. (cm)]	Minimum burst strength [lbf/in.^{2} (kPa)]	Minimum combined board grammage [lb/1000 ft^2 (g/m^2)]	Minimum edge crush strength [lbf/in. (kN/m)]
20 (9.1)	40 (102)	125 (862)	52 (254)	23 (4.0)
35 (15.9)	50 (127)	150 (1034)	66 (322)	26 (4.6)
50 (22.6)	60 (152)	175 (1206)	75 (366)	29 (5.1)
65 (29.4)	75 (191)	200 (1379)	84 (410)	32 (5.6)
80 (36.2)	85 (216)	250 (1724)	111 (542)	40 (7.0)
95 (43)	95 (241)	275 (1896)	138 (674)	44 (7.7)
100 (45.3)	105 (267)	350 (2413)	180 (879)	55 (9.6)

However, there is considerable ongoing debate about the measurement of ECS, and there is still no single internationally recognized method.

Standard Methods

ISO	TAPPI	JIS	SCAN	Appita	FEFCO
3037:1994	T 811 om-95 T 838 pm-95 T 839 pm-95 T 841-pm-95	—	P33:71	AS/NZS 1301 444s-92	No. 8

Procedure The various test methods are fundamentally similar. A test specimen of corrugated board is placed between the platens of a compression or crush tester with the flutes oriented vertically (Fig. 52). The specimen is subjected to an increasing compressive force until it fails. The maximum force that it can withstand, expressed per unit length of the specimen, is the edgewise compression strength. The methods differ by the dimensions and geometry of the test specimens, the cutting equipment used for specimen preparation, the rate of loading, and the use of edge reinforcement. Several techniques are available, and these can be broadly categorized into two groups:

1. Methods that avoid specimen edge failure by edge reinforcement, sample neck-down, or clamping.

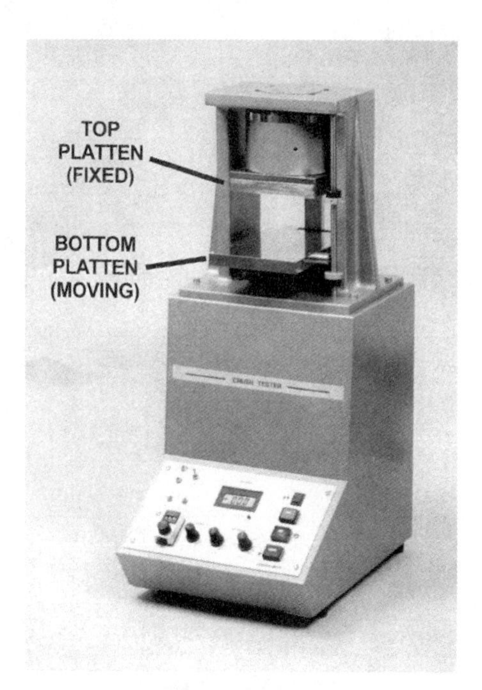

Fig. 52 Crush tester suitable for measuring the ECS of corrugated board. (Courtesy Lorentzen & Wettre, Kista, Sweden.)

2. Methods that use rectangular specimens and minimize the contribution of edge damage by careful sample preparation.

Edge Reinforcement, Neck-Down, and Clamping

TAPPI T 811 om-95 is the only test method that is approved for alternative Rule 41/Item 222 (Table 5). In this method, the sample is cut to a width of 50.8 ± 0.8 mm and a height that depends on the flute type and number of walls:

Specimen height (mm)	Application
31.8 ± 1.6	B-flute
38.1 ± 1.6	C-flute
50.8 ± 1.6	A-flute. All double- and triple-wall board

The different specimen heights are used to maintain the slenderness ratio λ at about the same value:

$$\lambda = 2\sqrt{3}\frac{H}{t} \tag{25}$$

where H is the specimen height (m) and t is the specimen thickness (m). This ensures that the specimen will fail in true compression and not by elastic or inelastic buckling (*see also Chapter 7*). The loaded edges of the specimen are reinforced by dipping in molten wax. It is then reconditioned for 24 h in a standard environment, as specified in TAPPI T 402 sp-98. The specimen is compression tested in either a rigid support (Fig. 52) or flexible beam tester at the specified compression rate. It is maintained in the vertical position during initial loading by metal blocks placed on either side. The rebate in the blocks is to accommodate the extra thickness of the waxed zone (Fig. 53).

The wax treatment will increase the stiffness of the specimen close to the loaded edges and retard failure. However, the waxing stage is both tedious and time-consuming and several alternatives have been proposed to make the test more convenient.

Specimen Neck-Down TAPPI T 838 pm-95 describes a necked-down specimen with no edge reinforcement (Fig. 54). The waist on the specimen will double the intensity of the compressive stress at the narrowest point and ensure that failure occurs away from the edges (Fig. 55). The testing procedure is essentially the same as for TAPPI T 811, and the same support blocks can be used in the initial part of the test.

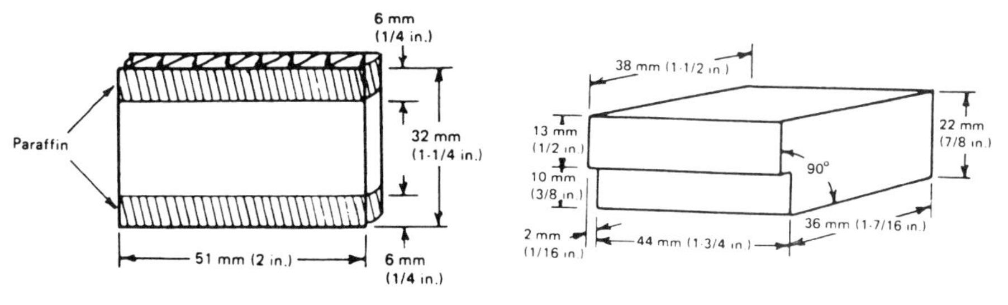

Fig. 53 ECS test specimen and metal guide blocks used in TAPPI T 811.

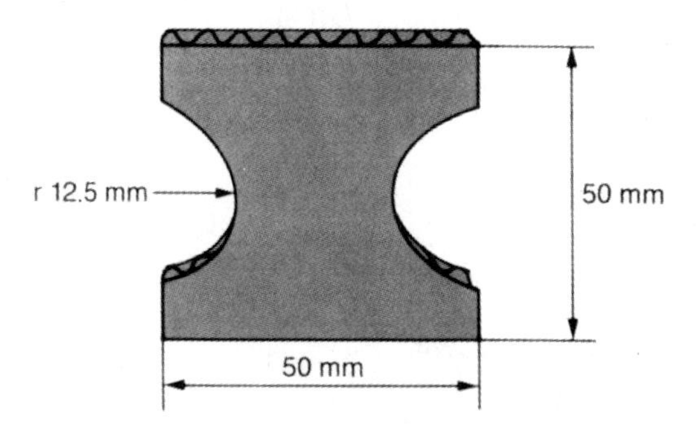

Fig. 54 Necked-down sample for ECS testing according to TAPPI T 838.

Edge Clamping TAPPI T 839 pm-95 circumvents the use of edge waxing by using a clamp to reinforce the edges (Fig. 56). The clamping device is Model D-105 made by the Sumitoto Corporation of Japan. It effectively clamps the upper and lower 20 mm of the specimen, leaving an unsupported span of approximately 10 mm in the center. Clamping pressure is between 55 and 83 kPa, which was determined by experimentation [153]. Specimen size is 50.8 mm width × 50.8 mm height and is the same for all flute types and for single-, double- and triple-wall corrugated board. The test rig is mounted centrally in a crush tester, and a compressive force is applied until the specimen fails. Results from specimens that fail at the clamping lines are discarded.

TAPPI T 841 pm-95 is another clamping method accredited to Morris (cited in Ref. 109) (see Fig. 57). The specimen holders are made of acylic sheet or metal. The critical feature is an adjustable groove with a sliding tapered plate on one side. The ECS specimen—again 50.8 mm × 50.8 mm for all types of board—is placed in the groove in the bottom holder, and the sliding plate is adjusted until the specimen is held. The groove in the top holder is adjusted to the same width. The distance between the top and bottom platens is adjusted so that the specimen can be positioned in the center of the groove without forcing. The specimen is then compressed in a suitable tester until it fails.

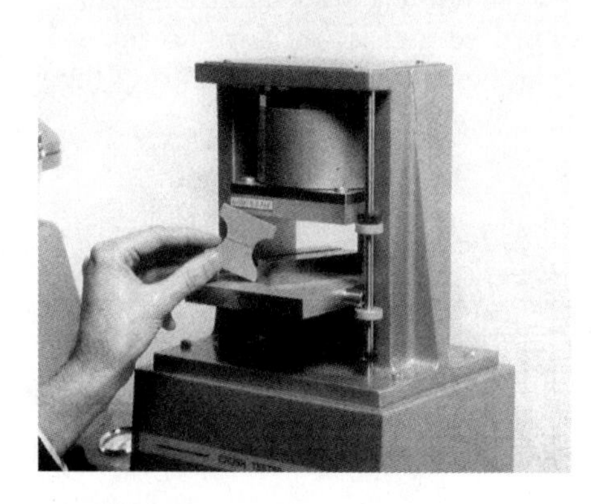

Fig. 55 Necked-down specimen after compression testing, showing failure at the waist. (Courtesy Lorentzen & Wettre, Kista, Sweden.)

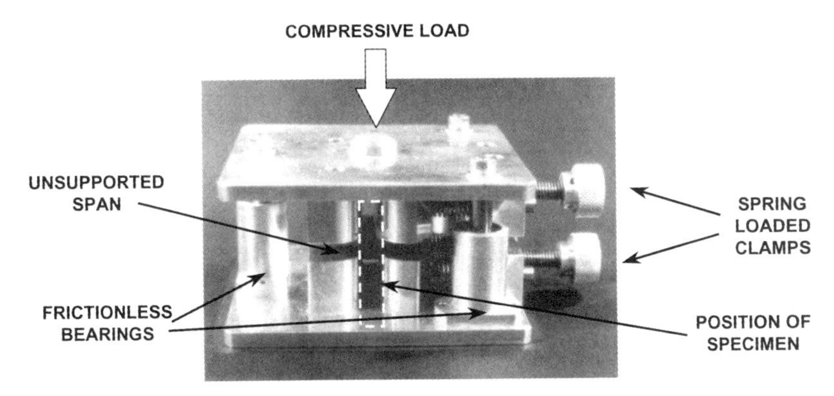

Fig. 56 Clamping device for ECS testing used in TAPPI T 839 pm-95. (From Ref. 153.)

Reinforced Liners Azens (cited by Eriksson [37,38]) stuck four pieces of liner onto the faces of a 50.8 mm × 50.8 mm specimen to provide edge reinforcement (Fig. 58). The liner is attached to the specimen using double-sided adhesive tape, and the compression test is carried out in the usual manner.

Long Span Method Westerlind and Carlsson [179] developed a technique to measure the complete load–deformation curve of corrugated board under compression (Fig. 59). The specimen is laser cut to 100 mm wide and 150 mm long with a 55 mm necked section of constant width. It is restrained from buckling during compression testing by two plates. The front plate is made from acrylic sheet to enable observation during testing, and the rear plate is aluminum. Displacement is measured by two gauges that are physically attached to the specimen with screws. These gauges fit into two horizontal slots in the front plate. The rig is fitted into a universal testing machine, and the compressive load is applied by the descending crosshead and measured by a load cell. Westerlind and Carlsson [179] used the technique to compare the compression load–deformation curves of corrugated board to the long-span compression curves of the board components. The latter were measured using an apparatus developed by Cavlin and Fellers [23].

Rectangular Specimens with No Edge Reinforcement The SCAN (P33:71), FEFCO (No. 8), and ISO (3037:1994) standard methods specify a different specimen size with no edge reinforcement (Fig. 60). The specimen is cut with the flutes oriented vertically as shown in Fig. 60, and the compression test is carried out in a crush

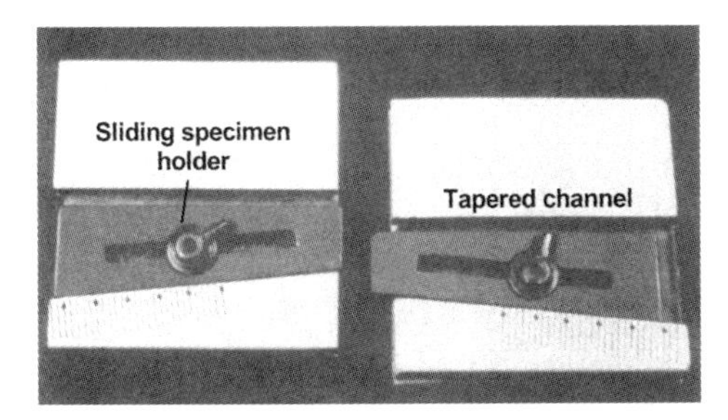

Fig. 57 Specimen clamps used in TAPPI T 341 pm-95 (the Morris method). (From TAPPI UM 814.)

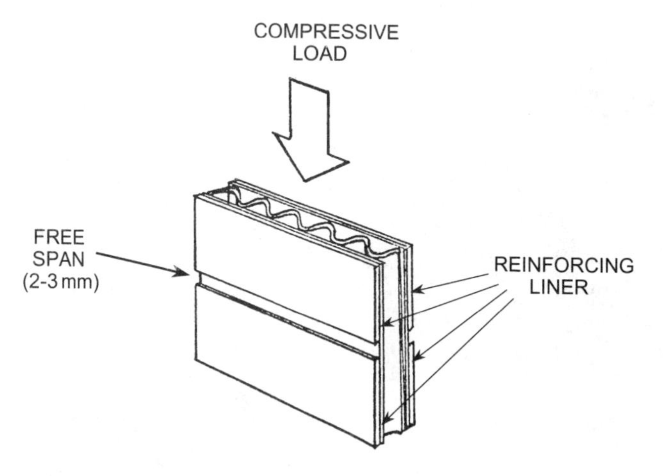

Fig. 58 Edge reinforcement using liners on the specimen facings. (From Ref. 38.)

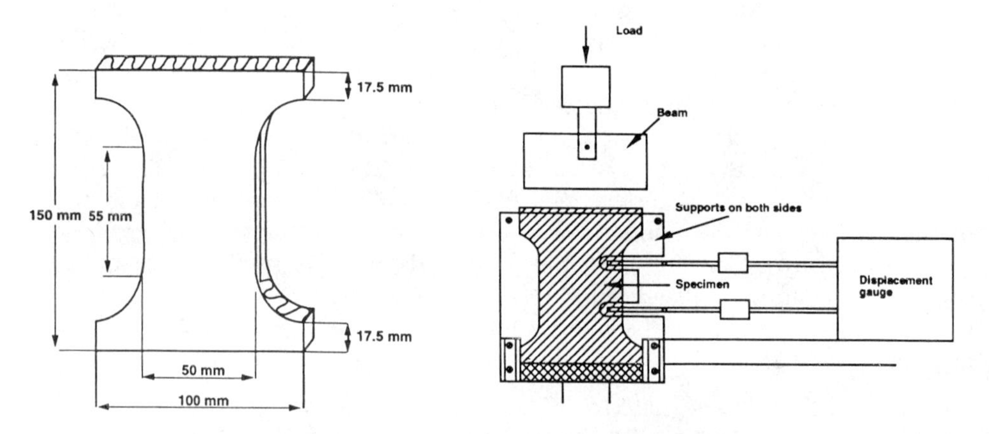

Fig. 59 Specimen geometry and test apparatus—Westerlind and Carlsson technique. (From Ref. 179.)

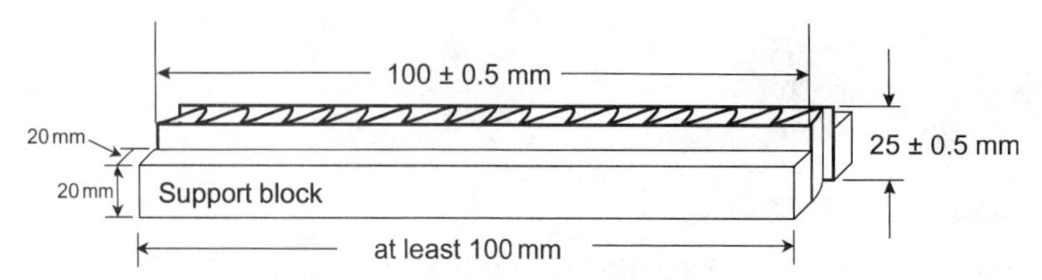

Fig. 60 Specimen size and support blocks for ECS testing according to SCAN P33:71.

tester. The specimen is supported between two blocks 20 mm × 20 mm in cross section and at least 100 mm long to keep it perpendicular to the platens of the compression tester during the initial loading. When the force reaches approximately 50 N, the blocks are removed, and the test is continued until the specimen fails. The maximum force sustained by the test piece is its edgewise compression strength, reported in newtons per meter.

Comments—Edgewise Compression Strength Testing The previous sections have demonstrated different approaches that are available to measure the edgewise compression strength of corrugated board. Not surprisingly, there is a great deal of controversy about which method is the most suitable for the purpose. Before carrying out a critique of the methods, it is useful to review the purpose of ECS testing. McKee et al. [125] studied the behavior of a corrugated box under the influence of an increasing compressive load. They noted that the compressive load is not evenly distributed around the box perimeter but is concentrated at the corners (Fig. 61) [125]. As the compressive load on the box increases, the side panels become unstable and begin to deflect outward. This affects the ability of the central portion of the panels to accept a further increase in load. Consequently, the load will concentrate at and in the proximity of the vertical edges of the box. However, the onset of panel buckling does not usually correspond with the maximum compressive load. The material near the vertical edges is in a more stable configuration due to the presence of a perpendicular adjacent panel and remains essentially flat. Box failure is usually precipitated by the compressive failure of the corrugated board at or near the vertical edges, which then propagates across the panels. McKee et al. [125] make the reasonable presumption that the corrugated board fails at a load intensity that is equal to or related to its intrinsic edgewise compression strength. Consequently, any ECS test method must measure this property with the exclusion of effects due to specimen geometry and

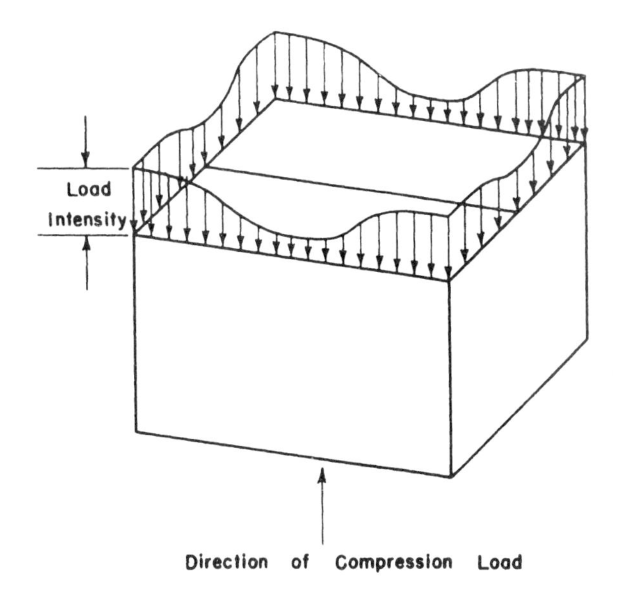

Fig. 61 Distribution of compressive load around the perimeter of a box. (From Ref. 125.)

edge-loading conditions. The type of failure exhibited by the specimen should also be similar to the failure experienced by the corrugated board in the box.

Following these criteria, the FEFCO, SCAN, and ISO methods can be excluded due to the influence of the specimen top and bottom edges. Eriksson [37,38] reported the sensitivity of test results to the quality of the loaded edges and the accuracy of the cut. In general, degradation of edge quality and deviation from a rectangular specimen with perpendicular loading edges will reduce the measured value of ECS. It is also acknowledged that even with a good quality edge, these methods will consistently give results that are on average 12% lower than TAPPI T 811 om-95, which includes edge reinforcement [99,165]. This is because the edge is a discontinuity in the structure of the liners and medium that is intrinsically weaker than the rest of the specimen [125]. Consequently, failure is usually the result of the crushing of one or both loaded edges (Fig. 62).

Clearly, these test methods do not measure the intrinsic compression strength of corrugated board.

Other test methods use different techniques to avoid edge failure. However, most of these techniques involve complications that detract from their usefulness or validity

- TAPPI T 841 pm-95 and TAPPI T 839 pm-95 use clamping to avoid premature edge failure. However, the clamping pressure is stated to be critical and must be adjusted for different grades of corrugated board [117]. Too much pressure will cause local crushing of the board. In itself, this will have a minor effect on ECS, causing a reduction of less than 5% for a reduction in board thickness of up to 40% [166]. However, the crushed zone will create another discontinuity in the structure, and the specimen will tend to fail at the clamping line. Too little pressure, and the specimen will tend to fail at the loaded edge.
- •• TAPPI T 811 om-95 specifies wax reinforcement of a 6 mm zone from the loaded edges of the specimen. The waxing procedure is complex and time-consuming and could be a source of operator error [160]. According to this method, specimens must also be reconditioned for 24 h after wax dipping, but Urbanik et al. [177] demonstrated that 2 h is sufficient. The complexity of the method has prevented its adoption as the international standard.
- ••• The long-span technique [179] uses a complex specimen that is cut by laser. The test apparatus is also too complex for a routine test method.

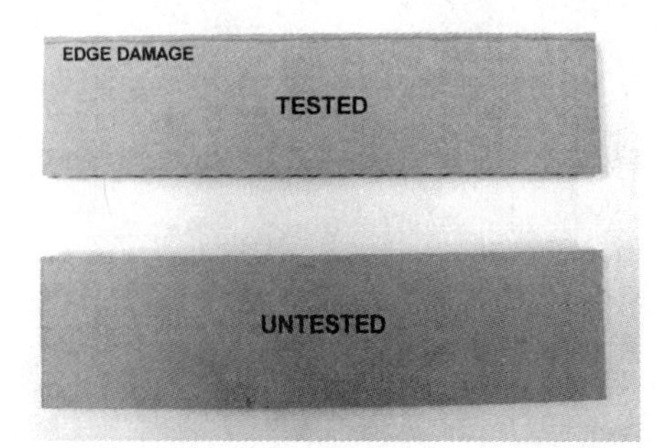

Fig. 62 Short-column specimen showing edge damage during compression testing.

TAPPI T 838 pm-95, which specifies a necked-down specimen, is clearly the superior test. This geometry was first suggested by McKee et al. [125] and further explored by Koning [86], who put it forward as the prospective international standard method [87]. The test fulfills the criteria for a test of the intrinsic edgewise compression strength of corrugated board:

- Loaded edge failure is avoided by significantly reducing the compressive stress on the specimen edges.
- The type of failure is very similar to that experienced by corrugated board in a box that has been compressed to failure.
- The test gives the highest load level before compressive failure.

Specimen preparation is greatly simplified by the use of an automatic cutter to prepare the 50.8 mm square blank (Fig. 63). The square specimen is obtained by first cutting a 50.8 mm strip from a piece of corrugated board along the direction of the flutes. The strip is rotated 90°, and a second cut is made perpendicular to the first one. The result is a 50.8 × 50.8 mm square with clean perpendicular edges. The next stage is to cut the waisted section by removing two semicircular segments. This can be achieved with a specifically designed cutting apparatus (Fig. 64), or by using an electric drill mounted on a vertical bench stand and fitted with a hole borer with the required diameter.

An Alternative to Edgewise Compression Strength Measurement There is a school of thought that questions whether ECS testing is necessary at all [99]. This is because there are several reports in the literature of a close correlation between the edgewise compression strength of the board components and the measured value of ECS. Seth [154] measured corrugated board ECS using TAPPI T811 and component intrinsic edgewise compression strength (IECS) using the STFI Short Span Compression Test (SCT) (see Section XI). He found that there was a strong correlation ($r^2 = 0.942$–0.978) between the measured ECS and an additive model:

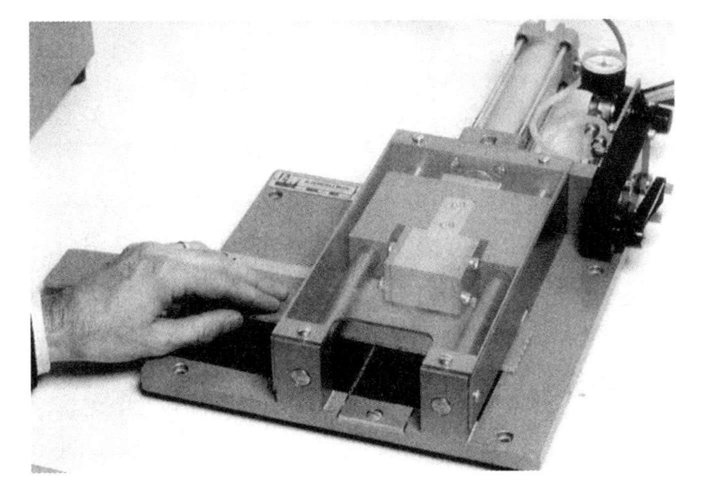

Fig. 63 Billerud-type automatic cutter for necked-down blank preparation. (Courtesy Lorentzen & Wettre, Kista, Sweden.)

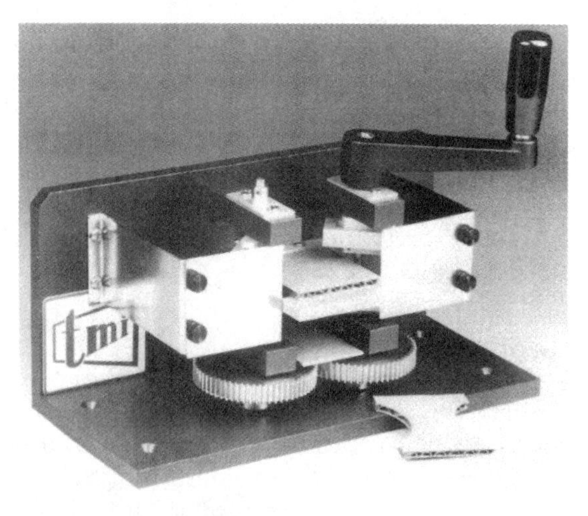

Fig. 64 Specialist apparatus for cutting a waisted specimen. (Courtesy Testing Machines Inc., Amityville, NY.)

$$S_p = L_1 + \alpha M + L_2 \tag{26}$$

where

S_p = cumulative IECS, kN/m
L_1 = IECS of linerboard 1, kN/m
L_2 = IECS of linerboard 2, kN/m
M = IECS of medium, kN/m
α = take-up factor for the particular flute

Koning [86] measured the ring crush values (Section XI) of the components and corrugated board ECS, and found

$$\text{Sum of ring crush values} = 0.768 \times \text{ECS} \tag{27}$$

Sandström and Titus [151] measured the tensile stiffness index (TSI) (see Section XI) of linerboard and medium using an ultrasonic device. They then correlated the TSI values with ring crush strength of the components and predicted ECS using the relationship:

$$\text{ECS} = k_1[C_{L1} + (aC_M) + C_{L2}] + k_2 \tag{28}$$

where k_1, k_2 are constants for a linear model; the $C's$ are short-span compression strength of the liners (L_1, L_2) and the medium (M); and a is the take-up factor for the medium.

It appears that the compression strength of the board components can be measured by a range of techniques, then fitted to simple models to predict the value of ECS. Howver, caution should be exercised when using this approach, because it assumes that there is no influence from the corrugated board manufacturing process. For example, if the glue-lines between the medium and the liners are excessively weak or discontinuous, the edgewise compression strength of the board will be reduced regardless of the strength of the components [95]. Nevertheless, ECS values calculated from the component properties should be a good indication of the strength potential of the board in compression.

Edgewise Compression Creep The effect of creep on box performance is covered in Section VIII.C. The box-modeling section also established that the edgewise compression strength of the corrugated board is a major factor that determines the compression strength of the box. It is therefore reasonable to expect that the edgewise compression creep of the board will strongly influence the creep behavior of the box. Byrd [17] found that creep tests carried out under constant humidity conditions were inadequate to predict the performance of corrugated board in a service environment with cyclical changes in humidity. He devised a cumbersome rig to measure the creep response of a corrugated board specimen subjected to a compressive load [18] (Fig. 65).

Byrd demonstrated that a cycling humidity environment will accelerate the edgewise compression creep rate [18]. Byrd and Koning [19] also found that corrugated board with linerboard manufactured from high yield pulps (pulps with a higher lignin content) or recycled pulp had a detrimental effect on cyclical humidity creep rate. Gunderson and Laufenberg [60] developed a device for measuring the edgewise compression creep of multiple specimens of corrugated board. The device consisted of up to 70 compression test load frames that were jointly loaded by a remote weight system. The specimens were edge-waxed and conformed to the requirements of TAPPI Standard T 811 om-95. The compression of each specimen was monitored by a non-contact linear voltage displacement transducer (LVDT) mounted on a gantry system over the array of load frames. This enabled the efficient

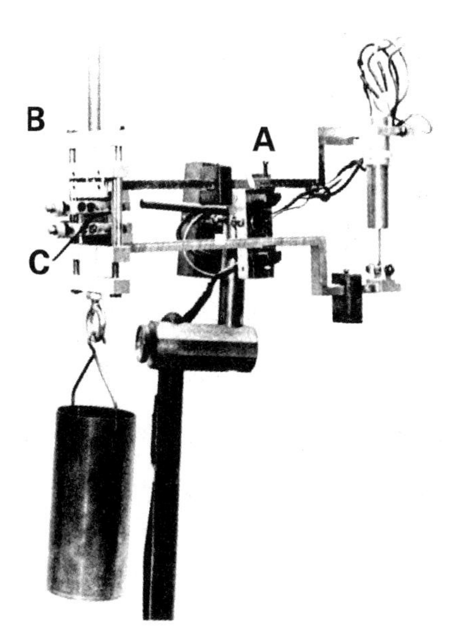

Fig. 65 The edgewise compression creep rig used by Byrd. A = electromechanical strain gauge; B = specimen cage to exert compressive load; C = corrugated board specimen. (From Ref. 18.)

collection of multiple sets of creep data to characterize the behavior of different board samples.

The experimental considerations that apply to edgewise creep studies of corrugated board are a development of the requirements for testing at 23°C and 50% humidity:

- The platens applying the compressive load must be flat, parallel, and free from excessive lateral movement. The specifications in TAPPI Standard T 811 om-95 will suffice.
- The specimen geometry must ensure that edge failure is avoided. The necked-down specimen described in TAPPI T 838 pm-95 is suitable.
- The corrugated board must be glued together with a water-resistant adhesive to ensure that slippage or failure at the glue lines does not contribute to creep displacement.
- The compressive creep load must be constant for the duration of the test.
- Sufficient displacement data must be taken to characterize the complete creep curve and allow evaluation of the secondary creep rate.
- The number of replicates should be sufficient to be statistically confident that the mean creep rate is within reasonable tolerance.

Bending Stiffness The ability of corrugated board to resist a bending moment is also important for box performance, because it controls the degree of bulge in the side panels. This in turn has an important influence on the compression strength of boxes. To recap, McKee et al. [126] developed a formula to predict the compression strength of boxes that includes the bending stiffness of the corrugated board:

$$P = 2.028 P_M^{0.746} \left(\sqrt{D_x D_y}\right)^{0.254} Z^{0.492} \tag{12}$$

The term in parentheses is the geometric mean of the corrugated board bending stiffness in (D_x) and across (D_y) the direction of manufacture, or machine direction (Fig. 66).

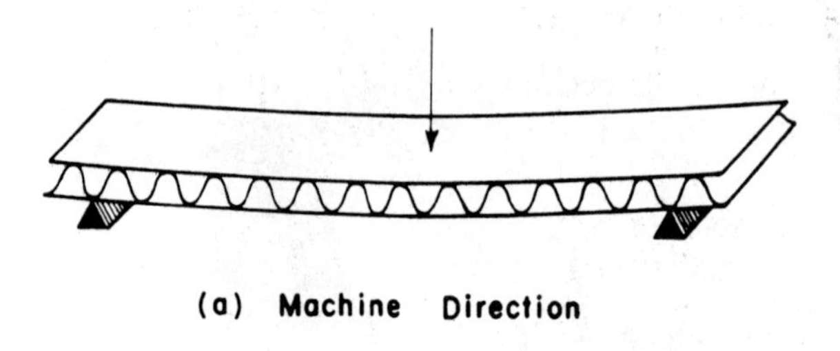

(a) Machine Direction

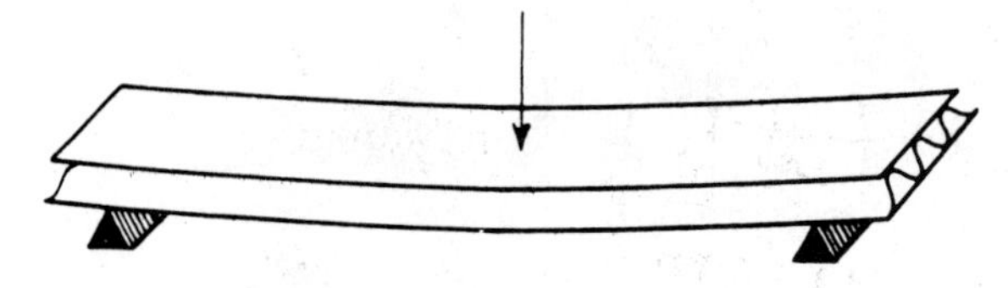

(b) Cross - Machine Direction

Fig. 66 Machine and cross-machine direction specimens of corrugated board for testing bending stiffness. (From Ref. 124.)

In the paperboard industry, bending stiffness is also known as flexural rigidity and flexural stiffness. See also Fellers and Carlsson (Chapter 5).

Standard Methods

ISO	TAPPI	JIS	SCAN	Appita	FEFCO
5628:1990	T 820 cm-00 T 835 pm-95	—	P65-91	AS/NZS 1301 453s-91	—

Three-Point Bending Test Early work on measuring the bending stiffness of corrugated board was carried out by Carlson [20,22] and Kellicut [79]. They used the three-point static bending test, so called because the specimen is supported at two points toward the opposite ends of the span and loaded at the midpoint. This led to the development of the American Society for Testing and Materials (ASTM) Standard D 1098-61 (discontinued 1986), Static Bending Test for Corrugated Fiberboard (Fig. 67). The specimen is supported centrally on the outer supports, which either have rounded edges or incorporate roller bearings. The bending load is supplied at midspan by a loading block with a rounded edge mounted on a load cell connected to the crosshead of a universal testing machine. The deflection is measured at midspan for several values of load using a dial gauge micrometer. The bending modulus of elasticity is then calculated from the linear portion of the load–deflection curve.

This type of test was criticized by McKee et al. [124] because the loading geometry introduces shear effects in the test specimen. Treating the specimen as a simply supported beam with constant cross section, the effect of a point load W at center span becomes clear when the shear force and bending moment diagrams are examined (Fig. 68). The shear force diagram is derived by considering the vertical equilibrium of the beam. The shear force is constant at $-W/2$ (negative by convention) over the left half of the beam and $+W/2$ over the right half. The bending moment diagram is derived by taking the moments at points A, B, and C:

Points A and C are free ends, so the bending moment is zero.

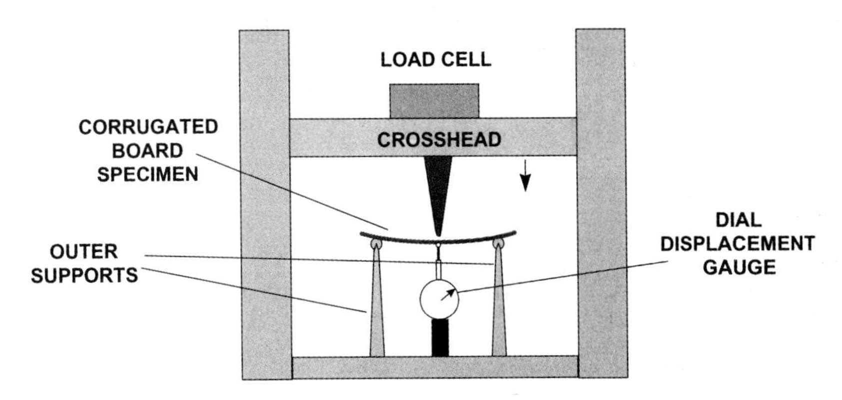

Fig. 67 Apparatus for the static bending test for corrugated board according to ASTM D 1098.

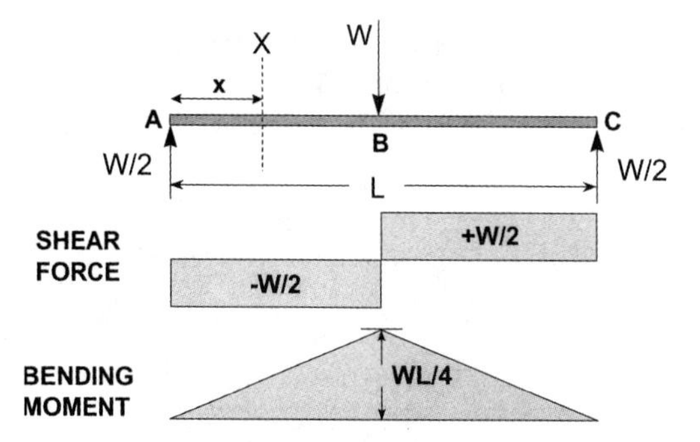

Fig. 68 Shear force and bending moment diagrams for a three-point bending beam test.

At point B, the bending moment is at its maximum value of $+(L/2)(W/2)$, or $+WL/4$ (positive by convention).

Since there are no loads imposed between A and B and between B and C, these portions are straight lines.

The deflection of the beam will be the resultant of the bending and shear stiffness. The bending moment will increase in proportion to the span length, but the shear force will remain constant. Consequently, this method will underestimate the true bending stiffness by a factor that diminishes with increasing span length (Fig. 69).

The apparent bending stiffness is significantly higher in the machine direction. This is because the two liners have a higher tensile stiffness in the machine direction, as detailed later. Note that the shear effects are also more pronounced in the machine direction. The shear stiffness in this direction is relatively low, owing to the poor ability of the fluted structure to resist shear forces. Consequently there are large deflections of the specimen, leading to a higher sensitivity to span length. From Fig. 69, the span lengths required for accurate determination of bending stiffness

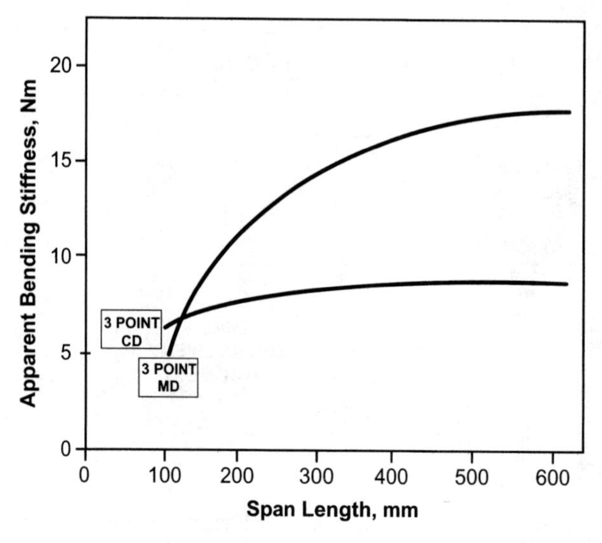

Fig. 69 Bending stiffness versus span length for three-point testing. (After Ref. 124.)

for this particular corrugated board are > 600 mm in the machine direction and > 300 mm in the cross-machine direction, which are impractically long. McKee et al. [124] reported a correction factor based on testing at two different span lengths [8,22]:

$$D = \frac{D_{a1} D_{a2} [L_1^2 - L_2^2]}{D_{a2}^2 L_1^2 - D_{a1}^2 L_2^2} \tag{29}$$

where

D = true flexural stiffness, N m
D_{a1} = apparent flexural stiffness (N m) at span L_1(m)
D_{a2} = apparent flexural stiffness (N m) at span L_2(m)

However, this would double the amount of testing required to obtain a true bending stiffness.

Four-Point Bending Test The influence of shear stress can be removed by using a loading geometry that applies a pure bending moment to the specimen. This can be achieved by the four-point bending test, as demonstrated by McKee et al. [124], Buchanan [15], and Koning and Moody [94]. The shear and bending moment diagrams (Fig. 70) show that the central span carries no shear stress and is consequently under a pure bending stress.

The work of McKee and others was the foundation for the TAPPI test method (T 820 cm-00) for measuring the bending stiffness of corrugated board by a four-point beam test. The method has been assigned "classical" status, which means that it is technically sound but has been superseded by more advanced technology. The specimen is placed on two supporting anvils, and the bending stress is applied by two loading anvils (Fig. 71). The top anvils are loaded with weights of equal increment, and the deflection at midspan is measured with a micrometer. The load applied by the micrometer is kept to a minimum by using an electrical contact method (detailed in the test method) to indicate when contact is made. The loading weight W is plotted against deflection Y to obtain the slope of the linear part of the curve. Bending stiffness is then calculated as

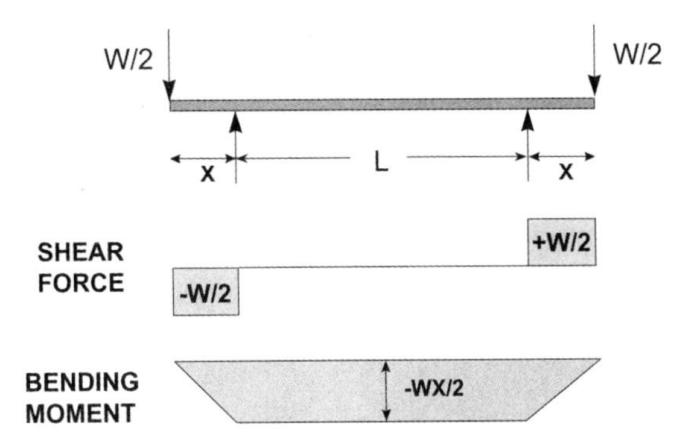

Fig. 70 Shear force and bending moment diagrams for a four-point beam test.

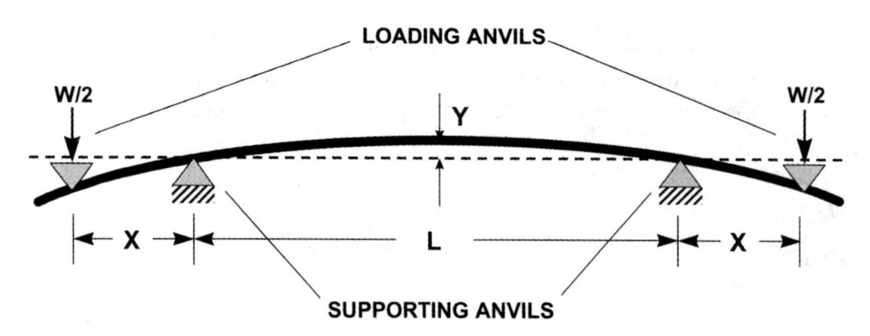

Fig. 71 Four-point beam loading method for bending stiffness determination according to TAPPI T 820 cm-00.

$$\text{Bending stiffness } = \frac{1}{16}\left(\frac{W}{Y}\right)\left(\frac{L^3}{w}\right)\left(\frac{X}{L}\right) \tag{30}$$

where

W = sum of the weights applied to the beam, N
Y = deflection at the center of the beam, m
L = distance between the support anvils, m
X = distance between the support anvil and loading anvil, m
w = sample width, m

The following points should be noted about four-point testing of bending stiffness:

1. Board warp either along or across the span will affect the result. The effect can be reduced by testing equal numbers of samples with the single facer side up and with that side down and by minimizing the width of the specimen. TAPPI T 820 specifies a width of 25–50 mm.
2. Care must be taken to position the anvils on the flute tips in MD specimens. Errors as high as 40% can result from the anvils deforming the linerboard between flute tips [124]. Carlson [20] recommends the use of a narrow plate between each anvil and the specimen as a stress distributor.
3. The ratio X/L should be between 0.25 and 1.0 [124], and the supporting anvils should be at least 10 cm apart.

TAPPI T 836 pm-95 and SCAN P65-91 describe another approach to four-point bending stiffness measurement developed by Fellers and Carlson [40]. The sample is clamped at both ends, and a pure bending moment is applied by rotating the clamps (Fig. 72). The clamping pressure P is specified as 14 ± 4 kPa with allowance for higher or lower clamping pressure depending on board construction. The bending stiffness (S^b) (N m) is calculated as

$$S^b = \frac{F}{d}\left(\frac{HL^2}{8w}\right) \tag{31}$$

where

F = bending force, N
H = length of moment arm, m

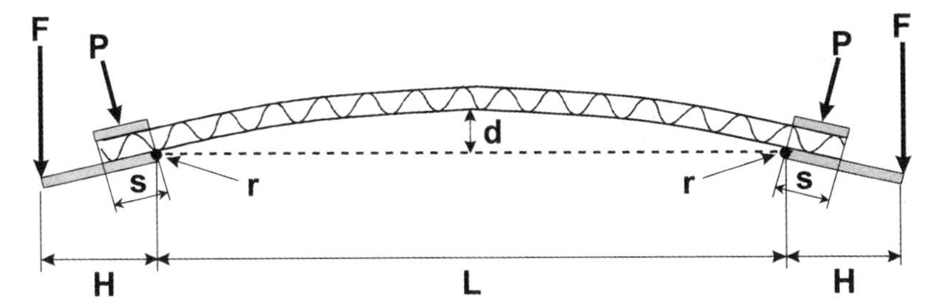

Fig. 72 Specimen loading for bending stiffness testing according to TAPPI T 836 pm-95.

L = free span between clamps, m
d = displacement at center of the free span, m
w = width of test piece, m

The method is quite complex and specifies a span length that depends on the flute type and board construction. The maximum allowable strain in the liners must also be in the range 0.03–0.05%, and this must be checked before testing for bending stiffness. The clamping pressure is also critical, because it is applied at the position of maximum bending moment.

An instrument that fulfills the requirements of T 836 pm-95 has been developed commercially (Fig. 73). A key element of this tester is the method of specimen clamping. The clamps are pneumatically operated and are movable to eliminate the influence of warp and twist in the specimen.

Properties of Corrugated Board Components That Affect Bending Stiffness The bending stiffness S^b of an elastic beam is defined as the ratio of the bending moment to the resulting curvature:

$$S^b = \frac{M}{k} = EI \tag{32}$$

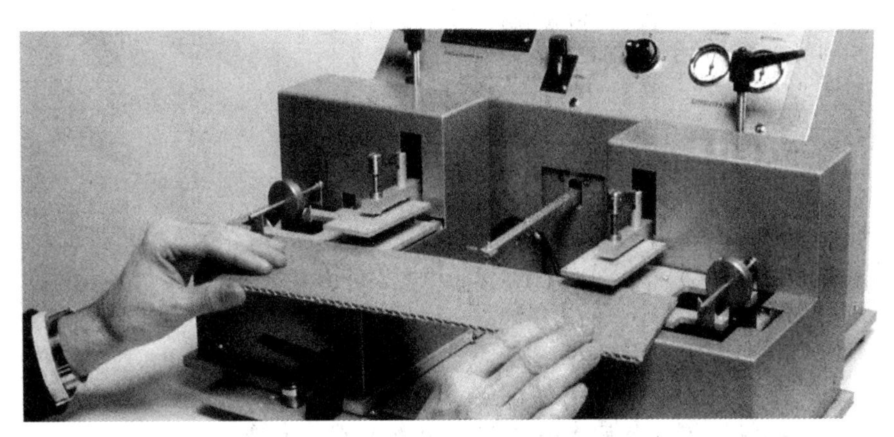

Fig. 73 Commercial instrument for testing four-point bending stiffness. (Courtesy Lorentzen & Wettre, Kista, Sweden.)

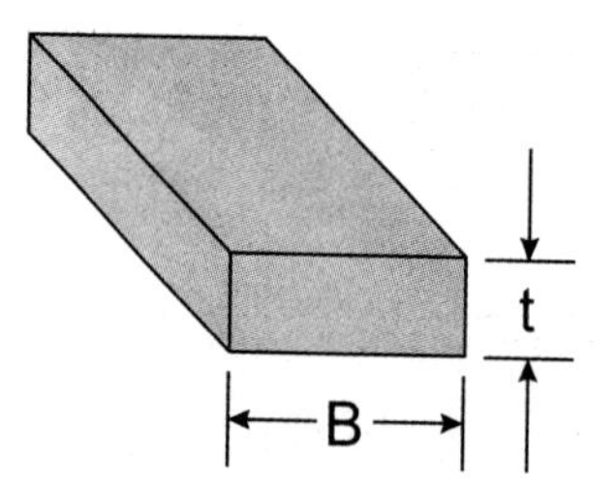

Fig. 74 Cross section of a uniform, solid rectangular beam.

where

$E =$ elastic modulus in bending, N/m^2
$I =$ second moment of area of the cross section, or *moment of inertia*
$M =$ bending moment, N m
$k =$ curvature $= 1/R$, $R =$ the radius of curvature

For a rectangular beam of uniform solid cross section, the calculation for the moment of inertia is (Fig. 74).

$$I = \frac{Bt^3}{12} \tag{33}$$

where

$B =$ width, m
$t =$ thickness, m

For corrugated board, the cross section is not continuous, and determination of the moment of inertia is more complex. It also depends on whether the flutes are oriented along or across the specimen (Fig. 75).

For a symmetrical board construction, the neutral plane runs through the central plane of the cross section (Fig. 76). The second moment of area about the neutral plane is calculated by summing all the elements dy across the thickness of the board. Clearly, this will be dominated by the contribution of the two liners, because their mass is furthest away from the neutral plane. To a first approximation [124,126]:

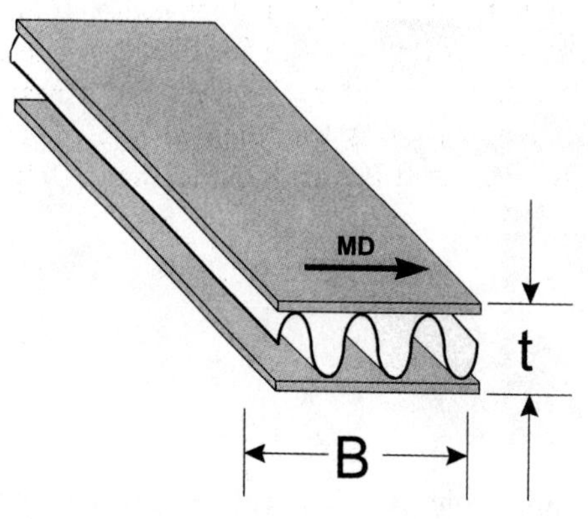

Fig. 75 Machine direction section through corrugated board.

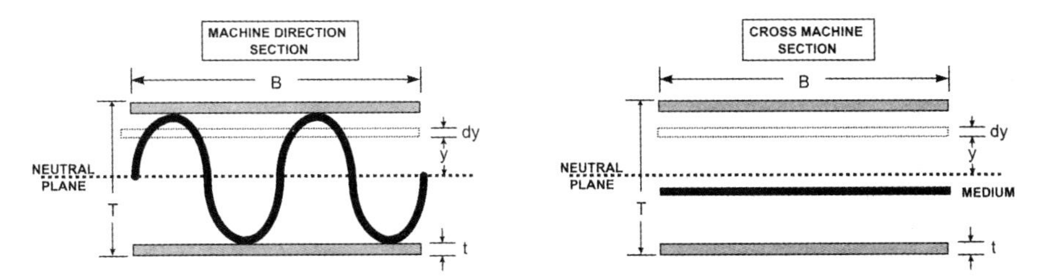

Fig. 76 Machine and cross-machine direction sections through corrugated board showing the neutral plane.

$$I = \frac{tT^2}{2} \tag{34}$$

where

T = thickness of the corrugated board, m
t = thickness of the linerboard, m

The bending stiffness (S^b) in either direction can now be approximated by multiplying the moment of inertia by the elastic modulus of the linerboard:

$$S^b = \frac{EtT^2}{2} \tag{35}$$

where E is the elastic modulus of the linerboard in the direction of the test. By definition, Et = tensile stiffness of the linerboard, so

$$S^b = \frac{ST^2}{2} \tag{36}$$

where S is the tensile stiffness of the linerboard in the direction of the test. This equation can be used to predict the bending stiffness of corrugated board. Bending stiffness will be at a maximum when the board thickness and linerboard tensile stiffness are as high as possible for a particular flute type.

The approximation used to develop Eq. (36) ignores any contribution of the corrugating medium. Clearly (Fig. 76), there will be a small but appreciable contribution from the medium when the board is bent in the cross machine direction. Koning and Moody [85] measured the contribution of the medium for B-flute single-walled board. They found that there was no measurable contribution to MD stiffness but a 13% contribution to CD stiffness. A more rigorous treatment of the cross section of corrugated board can be found in Kellicut and Peters [82] and Luey [107]:

$$D = E_{SF}I_{SF} + E_{DB}I_{DB} + E_{CM}I_{CM} \tag{37}$$

$$I_{SF} = I_{DB} = hx^2 + \frac{h^3}{12} \tag{38}$$

$$I_{CM} = \frac{\pi}{4}(a_1^3c_1 - a_2^3c_2) \tag{39}$$

where

D = bending stiffness in a particular test direction, N m
I = moment of inertia of the particular component, m^2
  $_{SF}$ = single-faced linerboard
  $_{DB}$ = double-backed linerboard
  $_{CM}$ = corrugating medium
E = modulus of elasticity of the particular component, $N\,m^2$
h = linerboard thickness, m
x = corrugated board thickness, m

and a_1, a_2, c_1, c_2 are geometric factors assuming that the profile of the flute is elliptical (Fig. 77).

Gartaganis [49] used this model to predict the bending stiffness of corrugated board made with two different grammages of medium. The average error between the predicted and measured results was 9%, which he attributed to the use of a crude four-point apparatus to test bending stiffness.

Peterson [138] developed a model that includes a complex mathematical treatment of the contribution of the medium to corrugated board bending stiffness in both directions:

$$D_x = (E_x I)_L + (E_x I)_M \tag{40}$$

$$D_y = (E_y I)_L + (E_y)_M \frac{t_M^3}{12} \tag{41}$$

where

D_x, D_y = MD and CD bending stiffness respectively
E_x, E_y = elastic modulus of liner (L) or medium (M) in MD and CD
I_L = liner bending moment of inertia
t_M = thickness of the medium
I_M = CD moment of inertia of medium, determined by numerical integration (see Ref. 138 for details).

It is clear from these three approaches that the thickness of the board and the tensile stiffness of the liners dominate the bending stiffness of corrugated board. However, the medium will also have an influence, and there are models available to predict this contribution.

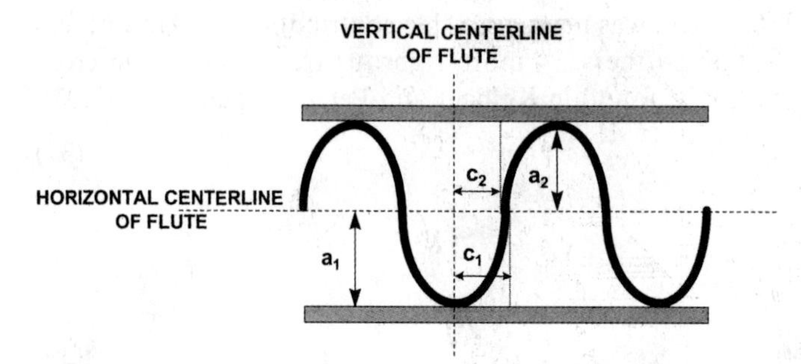

Fig. 77 Geometric factors a_1, a_2, c_1, c_2 for the Kellicut and Peters model of the bending stiffness of corrugated board. (From Ref. 82.)

Bending Creep The bending creep behavior of corrugated board will influence the long-term stacking strength of boxes and box performance in the service environment as discussed in Section VIII. Söremark and Fellers [159] used a commercially available four-point bending stiffness tester (Fig. 73) to evaluate the bending creep behavior of corrugated board. The specimens were 460 mm long and 100 mm wide, with the central 200 mm span under pure bending and the deflection was measured at the center of this span. A bending moment of 0.179 N m was used as the creep load. Gunderson and Laufenberg [60] used an array of four-point test rigs to obtain multiple measurements of bending creep (Fig. 78). The creep load was applied by the load frames, which were attached to a remote deadweight system to ensure that all specimens were loaded equally and simultaneously. The displacement of each specimen was measured at center span using a non-contact LVDT mounted on a gantry. McKenzie [127] modified this method and used a laser displacement sensor to monitor midspan displacement (Fig. 79). He found that the long-term creep behavior could be modeled by using an empirical curve-fitting model developed by Pecht [137].

Experiment considerations for bending creep measurement are as follows:

- The specimen should be loaded in pure bending. A four-point test rig is appropriate.
- The corrugated board must be glued together with a water-resistant adhesive to ensure that slippage or failure at the glue lines does not contribute to creep displacement.
- The bending creep load should be selected to ensure that the specimen creeps at a reasonable rate.
- The bending creep load must be constant for the duration of the test.
- Sufficient displacement data must be taken to characterize the creep curve and allow evaluation of the secondary creep rate. Note that for bending creep there is often no clear end point to the test.
- The number of replicates should be sufficient to yield statistical confidence that the mean creep rate is within a reasonable tolerance.

Flat Crush Resistance Flat crush resistance is defined as the ability of corrugated board to resist a compressive load applied perpendicular to its surface. During its manufacture and conversion, corrugated board passes through several processing

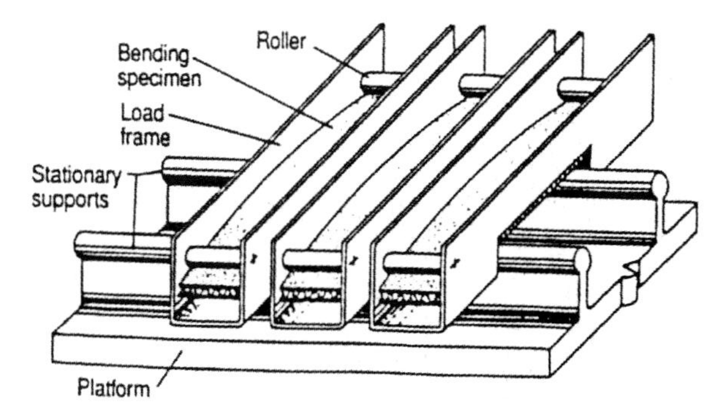

Fig. 78 Rigs for measuring the four-point bending creep of corrugated board. (From Ref. 60.)

Fig. 79 Rig used by McKenzie to study the bending creep of corrugated board, shown in a controlled environment chamber. A: Traversing head with laser displacement transducer, B: Specimen cradles, C: Overhead gantry.

stages that involve nips between rollers or belts. This processing can cause damage to the corrugated structure in the thickness direction. The flat crush test is purported to be a measure of the resistance of the board to this type of damage. The test involves cutting a specimen, which may be a circular or rectangular, and inserting it within the platens of a compression tester (Fig. 80). A compressive load is applied until a maximum is reached that corresponds with the collapse of the corrugations. This is the flat crush resistance of the specimen, reported in units of force per unit area, or pressure.

Standard Methods

ISO	TAPPI	JIS	SCAN	Appita	FEFCO
3035:1982	T 808 om-97 T 825 om-96	—	P32-71	AS/NZS 1301 429s-89	No.6

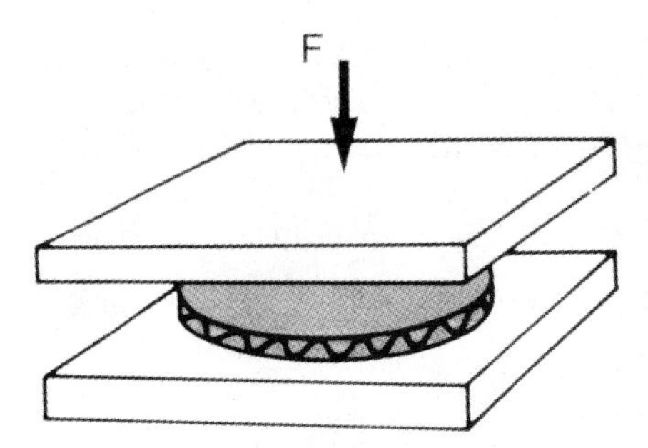

Fig. 80 Schematic of specimen loading during the flat crush test. (Courtesy Lorentzen & Wettre, Kista, Sweden.)

Procedure There are two TAPPI standard methods that describe this test. TAPPI T 808 om-97 specifies a flexible beam compression tester with a constant rate of load application in the range $111 \pm 22\,N/s$. TAPPI 825 om-96 is similar but specifies a rigid support compression tester with a constant rate of strain of $12.5 \pm 0.25\,mm/min$. The mechanical difference between flexible beam and rigid testers is explained in the literature [96]. In both standards, flat crush is defined as the maximum load per unit area sustained before the corrugations collapse completely. The SI units are kilopascals.

$$FCT = \frac{F}{A} \tag{42}$$

where

F = maximum load, kN
A = test piece area, m^2

The SCAN, FEFCO, and ISO test methods are basically similar but specify circular specimens of 65 or $100\,cm^2$. They do not make the distinction between constant rate of loading and constant rate of strain.

Comments—Flat Crush Test The two different TAPPI methods—constant rate of load application (T 808 om-97) and constant rate of strain (T825 om-96)—can give different results [96]. The flexible beam-type testers yield values that are 7–15% higher than those given by the rigid testers. This is because the rate of strain can be more than 27 times faster for the flexible testers [96], and the strength properties of corrugated board are rate-sensitive.

Flat crush values can be misleading and may not reflect the damage imparted to corrugated board during its manufacture. This was clearly demonstrated in the work of Crisp et al. [31], who deliberately crushed corrugated board between solid rollers set at various clearances. They found that the maxima on load–deflection curves for A-flute board were not significantly affected until the level of crushing exceeded 2.9 mm, which was more than 50% of the original board thickness (5.5 mm) (Fig. 81). With one exception the maximum load occurs at about the same level, which means that the maximum load is insensitive to high levels of crushing in the thickness direction. Increasing crushing also eliminates the first yield point and decreases the compressive rigidity, as measured by the slope of the first part of the load–deformation curve.

The flat crush method is susceptible to misinterpretation due to flute rolling during the test. This rolling—due to *leaning flutes*—is caused by lateral instability of the corrugations (Fig. 82). Maltenfort [109] claims that the incidence of learning flute failure due to an "undetectable" misalignment of the tester platens can be as high as 80% for B-flute board. This type of failure can reduce the test value by up to 35% [34]. It can be very difficult to decide whether leaning flutes have contributed to a low flat crush value [12]. This is because leaning flutes can be created during the manufacturing process due to excessive tension on the corrugating medium or as a result of the test. Leaning flute failure is reduced by using high friction surfaces such as emery paper on the tester platens [81] and by keeping the lateral play in the platens to tight tolerances [12]. However, a better approach is to prevent leaning flutes by constraining the lateral movement by using a special test rig [31] (Fig. 83). In this rig, a circular specimen fits into the outer housing without binding. The inner hous-

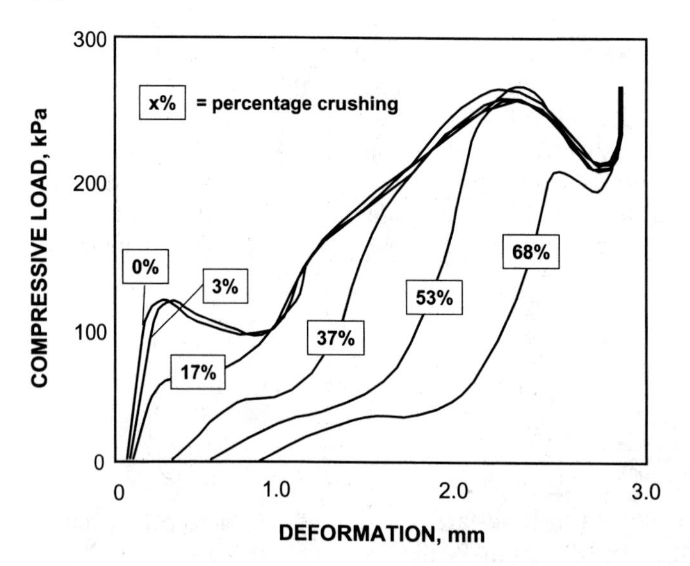

Fig. 81 Load versus deflection for different levels of board crushing. (Modified from Ref. 31.)

ing is machined to slide into the outer housing without lateral play, and the compressive load is applied to the center of this housing. The use of a suitably designed test piece holder to prevent lateral movement is part of the Appita test method for the flat crush resistance of corrugated board. Maltenfort [109] states that this type of rig should not be used for routine testing, because it will induce normal failure in board that has leaning flutes due to faulty manufacture. However, it is better to handle the lateral instability problem in a defined and reproducible manner, so using the rig is arguably preferable to omitting it.

Hardness Test The insensitivity of the flat crush test to severe levels of board damage led Crisp et al. [31] to suggest an alternative. They defined the maximum load at or before a deformation of 0.25 mm as the *hardness* of the board, and this

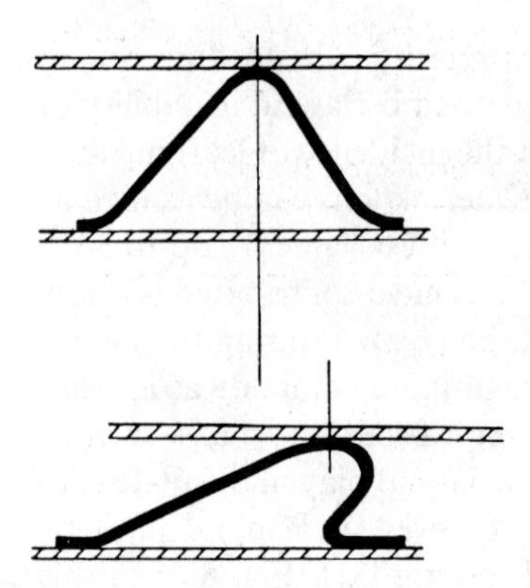

Fig. 82 Lateral instability of the corrugations during flat crush testing. (From Ref. 12.)

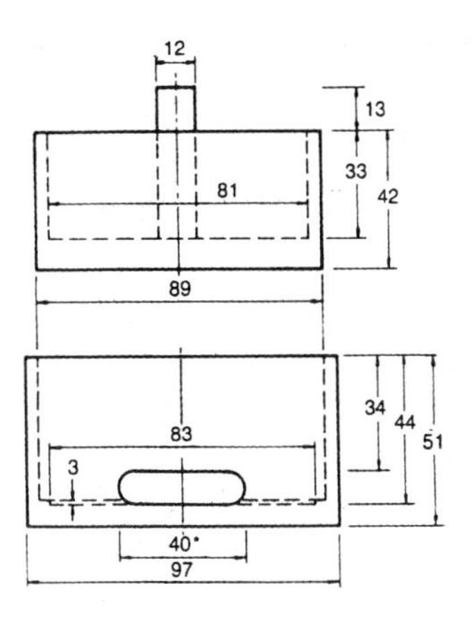

Fig. 83 Fixture for flat crush testing of corrugated board. (Drawing from Ref. 31.)

later became a separate Appita standard method (AS/NZS 1301.445s-89). Hardness is a sensitive indicator of board damage, as shown by comparison with the effect of crushing on a range of board properties (Table 6).

The flat crush values either remain constant or increase after a moderate degree of crushing. This confirms observations by Staigle [162]. Board thickness was measured using a deadweight micrometer. The thickness values are insensitive to crushing, probably due to an insufficient load on the specimen [134] as described in Section X.A. The hardness values decrease significantly and provide a superior measure of board damage.

In conclusion, the results from the flat crush test must be interpreted with care, and the test is of limited use as an indicator of board damage. Hardness, defined as the maximum load at or before 0.25 mm deflection and measured using a suitable test piece holder, is a sensitive indicator of flute damage.

Burst Strength Burst strength is defined as the resistance of the board to rupture in the thickness direction when subjected to increasing pressure exerted by an extend-

Table 6 Effect of Crushing on the Mechanical Properties of Corrugated Board

	A-flute	B-flute	C-flute
% Crushing	17%	20%	22%
Hardness	−50%	−46%	−62%
Flat crush	+1%	nil	+14%
MD flexural rigidity	−37%	−23%	−17%
CD flexural rigidity	−3%	−8%	−7%
Edge crush strength	−2%	+17%	+7%
Board thickness	−2%	−3%	−3%
Box compression strength	No data	−7%	−7%

Source: Adapted from Ref. 31.

ing, circular rubber diaphragm. The burst test was developed in the late nineteenth century by J. W. Mullen to simulate service conditions, and the term "Mullen" is still used interchangeably with "burst". The theory of the test is well covered in Chapter 5. In summary, burst strength is related in a complex manner to the strain at rupture, modulus of elasticity, and Poisson ratio in both the machine and cross-machine direction. For thick samples such as corrugated board, combined membrane and bending stresses exist through the thickness of the specimen.

Standard Methods

ISO	TAPPI	JIS	SCAN	Appita	FEFCO
2759:1983	T 810 om-98	P 8131:1995	P25-81	AS/NZS 1301 438s-97	No.4

Procedure The burst strength test involves the rupture of the board by an expanding spherical membrane (Fig. 84). A specimen is placed over the diaphragm and clamped around its periphery. The diaphragm is then bulged by pumping hydraulic fluid under it at a constant rate. The pressure of the fluid when the specimen ruptures is defined as the burst strength in kilopascals.

Comments—Burst Strength The principle of this test is simple, but the results are subject to serious errors if instrument maintenance is neglected or improper procedures are used. The effect of instrumental variables is comprehensively reviewed by Whitsitt [180–182]. The choice of appropriate clamping pressure is one of the more controversial issues. If the clamping pressure is too low the specimen will slip during loading, resulting in an erroneous reading. If the clamping pressure is too high, the corrugated structure under the platens will collapse. This will bring the two liners into close proximity and result in a significant decrease in the test result, as explained by Maltenfort [109]. McKenzie [128] also studied the effect of clamping pressure on the measured value of burst strength (Fig. 85). He found that burst strength was a monotonously decreasing function of clamp pressure between 200 and 700 kPa.

The TAPPI test method has a complex specification for clamping pressure that depends on the type of flute. The Appita test method specifies 800 ± 20 kPa. The

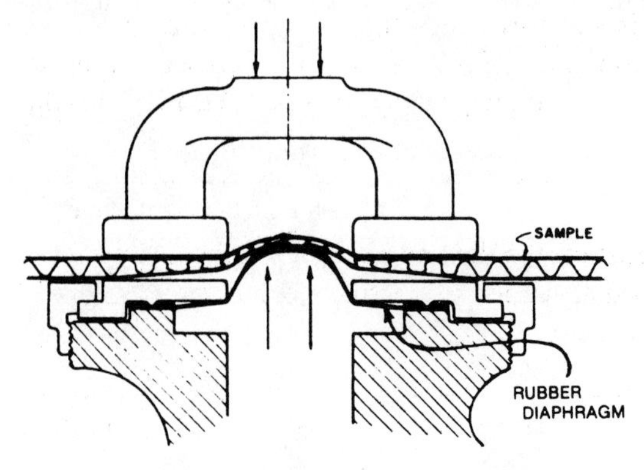

Fig. 84 The Mullen tester for measuring the burst strength of corrugated board. (From Ref. 109.)

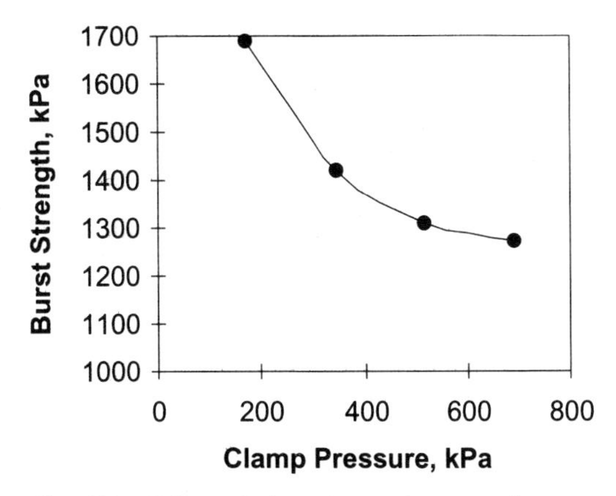

Fig. 85 Effect of clamping pressure on burst strength. (Adapted from Ref. 128.)

FEFCO method does not specify a set clamping pressure and permits flute crushing. Consequently, results obtained following the different standard methods will not be comparable.

The occurrence and handling of "double pops" is also a source of difference between the test methods. A "double pop" occurs when the two liners do not fail simultaneously. If the events are sufficiently separated in time, the failures of the liners can be heard as distinct sounds—hence the name. However, Maltenfort [109] reports that this effect can be inaudible if the two are very close together. A "double pop" will give a lower burst result. These results are included in the Australian test method, included but labeled in the TAPPI method, and rejected in the FEFCO method.

The relationship between burst strength and the compression strength of boxes is very poor [93]. This is not surprising, because there is no reason that a material with high burst strength should also have good compression strength and be able to perform in a corrugated box. For example, a textile materal can have high burst strength but would make a rather flimsy box, as remarked in the literature [93,116]. Remarkably, burst strength is still regarded as a sufficient performance requirement according to the Rule 41/Item 222 freight carrier regulations. On the positive side, there is some evidence that it is an indicator of the ability of boxes to contain their contents [13,14]. It is also a useful tool to determine whether the liners have been damaged by pressure cutting on the corrugator [109].

Puncture Resistance Puncture resistance is defined as the energy required for a pyramidal anvil to penetrate a clamped specimen of corrugated board.

Standard Methods

ISO	TAPPI	JIS	SCAN	Appita	FEFCO
3035:1975	T 803 om-99	P 8134:1998	P23:68	—	No. 5

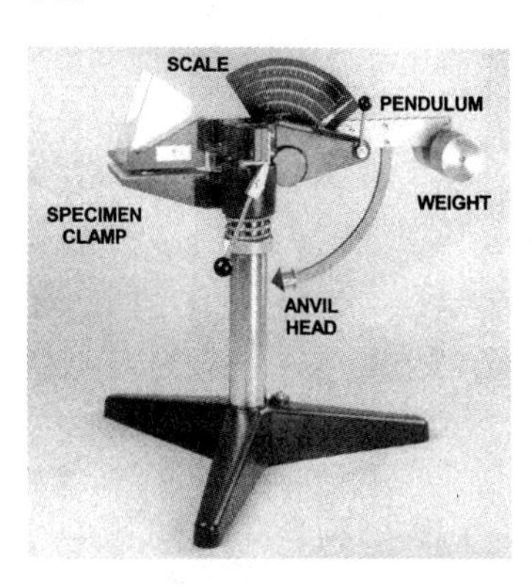

Fig. 86 Beach/GE puncture tester, showing a close-up of the anvil head. (Photograph courtesy Messmer Büchel, Gravesend, England.)

The puncture resistance tester was introduced by R. L. Beach in 1939 [4] as an instrument to quantify the suitability of corrugated board for box making. The instrument is still known as the Beach tester or as the GE (for General Electric Co.) tester after the original manufacturer (Fig. 86).

Procedure The anvil head has a pyramidal geometry and is 2.54 cm high to simulate the corner of a box. The head is attached to a curved arm, which is then attached to a pendulum fitted with an adjustable weight. The procedure involves clamping the corrugated board specimen, then releasing the pendulum and allowing the anvil head to swing into the corrugated board specimen. The energy required to puncture the board is read by the pointer on the scale, which is the "puncture test" in joules. The result is quoted as the average of four tests with the specimen in different orientations (face up, face down, flutes parallel, and flutes perpendicular).

Comments—Puncture Test The puncture test measures the energy expended during the initiation and propagation of three tears in the corrugated board and the energy required to bend the specimen as the puncture proceeds. It is not a particularly good indicator of board quality, as noted by McKee [122]:

1. The method has very low precision, which casts doubt on its usefulness as a criterion for board quality in terms of box performance.
2. It is not a fundamental test and depends on a number of physical properties of the board. Consequently, board with the same puncture value can produce boxes with very different performance.

Maltenfort [113] was even more dismissive: "In summary, G.E. puncture is suitable neither as a freight rule nor any other type of package specification since it tells us nothing that other tests cannot tell us better." However, the test does have validity as a specialist procedure for determining the resistance of corrugated board to a particular type of contact, such as impact from the corner of a wooden pallet. Its usefulness is perhaps indicated by the continued status of the test method as an official TAPPI standard.

Shear Stiffness Shear stiffness is defined as the ability of corrugated board to resist shear forces.

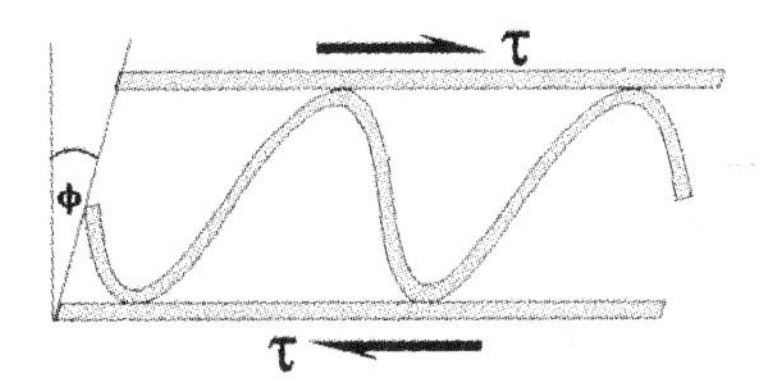

The diagram above shows a section through corrugated board in the machine direction. Shear stiffness, G (N/m^2) is defined as

$$G = \frac{\tau}{\phi} \tag{43}$$

where $\tau =$ shear stress, N/m^2, and $\phi =$ shear strain, rads.

The corrugated structure is intrinsically weak in shear in the machine direction and strong in the cross-machine direction. In terms of structural integrity, MD shear stiffness is therefore the critical property. Carlson [22] developed a method to test the shear stiffness of corrugated board in both directions. The method, based on ASTM D 1037, involves gluing the board specimen between two wooden blocks. The top block is then displaced horizontally, and the shear load and displacement are measured. McKinlay [129] points out the difficulties in applying a pure shear load to a corrugated specimen without introducing unwanted deflections in other loading modes. He developed a new instrument based on the relationship between the MD shear stiffness and shear twist of corrugated board [130], which he confirmed using finite element analysis. A device has been patented to measure the MD shear stiffness of thin corrugated board strips [131] (Fig. 87).

A specimen 25 mm wide and 220 mm long is cut in the machine direction (across the flutes) using a long-stroke Billerud-type cutter (similar to that of Fig. 63) to ensure clean and parallel edges. The strip is inserted into the self-centering

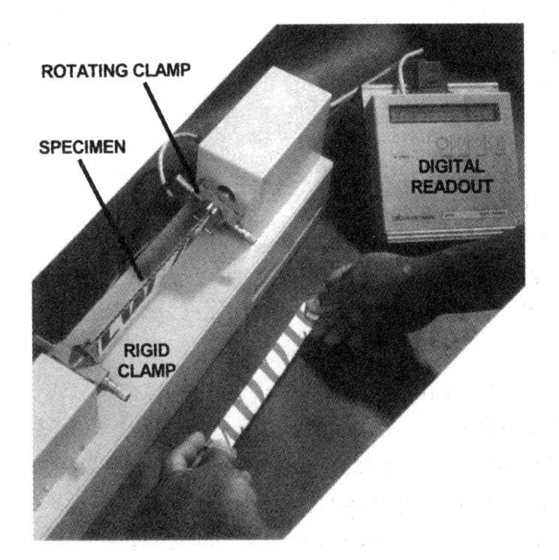

Fig. 87 Machine direction shear tester for corrugated board. (From Ref. 131.) (Courtesy Amcor Limited, Melbourne, Australia.)

clamps, and a set clamping force is applied by the spring-loaded micrometer screws. In a standard test, one end of the specimen is then rotated through 10° at a set rotational speed under microprocessor control. The instrument then calculates the MD shear stiffness (MDST) according to the equation

$$\text{MD shear stiffness} = \frac{6M}{kw^3} \tag{44}$$

where

M = twisting moment applied to the specimen, N mm
k = degree of twist $= \theta L$ rads/mm
L = specimen length, mm
w = specimen width, mm

Comments—MD Shear Test McKinlay [130] claims that the MD shear test is an excellent indicator of board quality at the corrugating plant and that it is very sensitive to board structural damage during manufacture and conversion. I have limited experience of this apparatus, but I can confirm that it has good potential as a quality control tool.

C. Structural Integrity

Corrugated board is a compound structure held together by adhesive. The ability of the adhesive to maintain the structural integrity of the board under service conditions is critical, and requires evaluation of the quality of the adhesive bond.

Pin Adhesion Pin adhesion measures the perpendicular force required to separate the liner from the medium in a standard atmosphere (23°C, 50% RH). Pin adhesion is the most widely used test for the strength of the adhesive bond.

Standard Methods

ISO	TAPPI	JIS	SCAN	Appita	FEFCO
—	T 821 om-96	—	—	AS/NZS 1301 430s-89	No.11

Procedure The various test methods are essentially the same and involve mounting a specimen of corrugated board on a test rig with elongated pins (Fig. 88), then pulling apart the bonds between one of the liners and the medium by applying a perpendicular force.

The corrugated board specimen is rectangular and either 150 mm × 50 mm for A-flute or C-flute or 101.6 mm × 31.8 mm for B-flute. The short dimension is in the direction of the flutes. The side of the board to be tested is placed face downward, and the pressure pin assembly (Fig. 88B) is carefully inserted into the spaces between corrugations (Fig. 89). The support pins (Fig. 88C) are then inserted as shown in Fig. 89. The assembly is completed by fitting the pins to the appropriate base plate (Fig. 88A or 88D). The rig is then placed in a compression tester and a load is applied. The

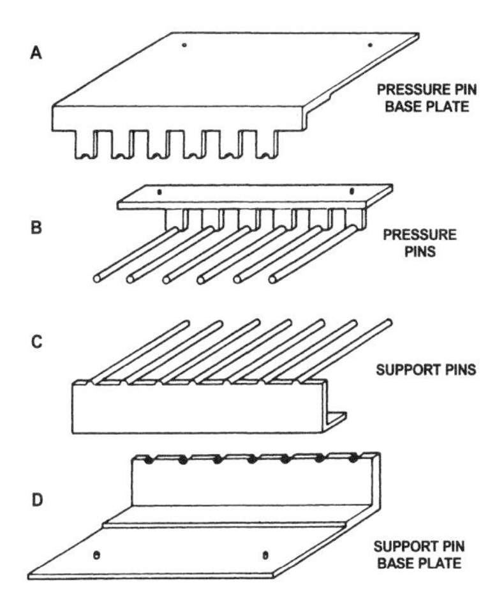

Fig. 88 Pin position in the pin adhesion test—single-walled board. (From TAPPI T 821 om-96).

force required for separation is the pin adhesion, reported in newtons per meter of bond length in the specimen (N/m).

Comments on Pin Adhesion Clearly, the stresses exerted by the pins are quite complex and are a combination of bending, tensile, and peeling. The out-of-plane loading is also questionable, because the bonds will not usually be subjected to this mode of loading in service conditions.

Glue Bond Shear Glue bond shear (GBS) measures the resistance of the glue bonds of a presoaked specimen to a shearing force.

Standard Methods

ISO	TAPPI	JIS	SCAN	Appita	FEFCO
—	T 842 pm-99	—	—	AS/NZS 1301 458rp-94	—

Procedure The glue bond test is a refinement of the Linke test for the shear resistance of glue bonds. It was developed by Allan et al. at the Amcor Research & Technology Center, Melbourne [1]. A specimen of board is positioned on a sample

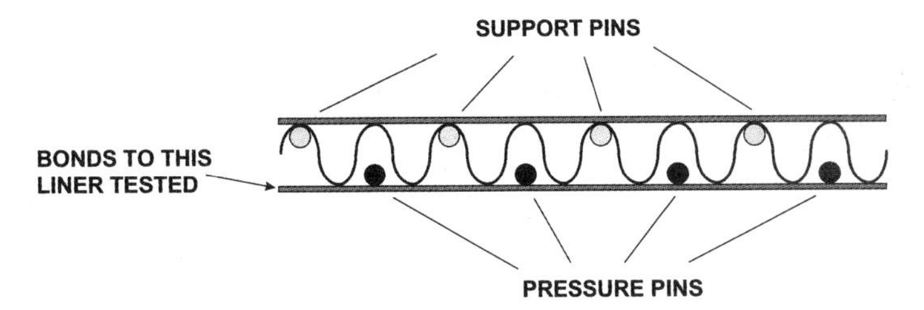

Fig. 89 Support and pressure pin positions for pin adhesion test.

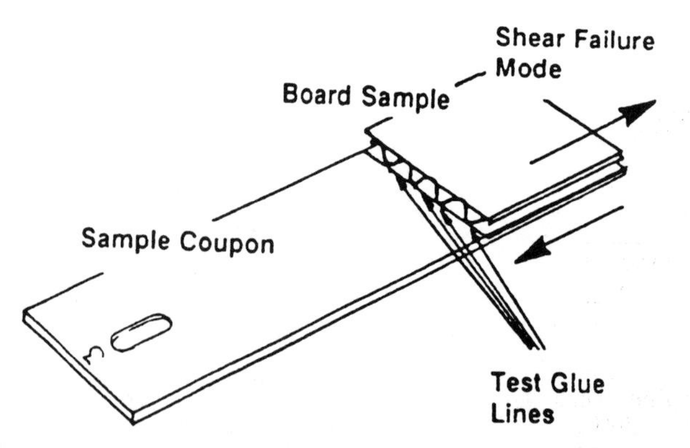

Fig. 90 Board specimen attached to a coupon in glue bond shear tester. (From Ref. 1.)

coupon with double-sided tape (Fig. 90) and immersed in water at 40°C for 15 min. The coupon is then attached to the base of a testing machine. A comb attachment is fitted into the spaces between the flutes, and the specimen is sheared to failure. The glue bond shear result is expressed as newtons per meter of flute length. The ability to measure the water resistance of the glue bonds in corrugated board is important for high humidity packaging applications. The glue bonds between the medium and liners must be sufficiently water resistant to maintain bond stability in the service environment. If the glue bond strength degrades with increasing humidity, the board will have a lower resistance to compression creep and box failure will be accelerated [16,90]. The test provides a relatively fast and quantifiable measure of the water resistance of glue bonds that can be used as a quality control tool in the board manufacturing process.

Other Wet Adhesion Tests

TAPPI T 812 om-97 This is a qualitative ply separation test that applies to both solid and corrugated board. The specimen is soaked in water for 24 h and then checked for adhesion between the components using the "thumb-flick test". As the term implies, this is a simple motion of the thumb to establish whether the bonding between the components is due to surface tension or the presence of a waterproof adhesive. This test is both subjective and time-consuming and does not provide any information on the relative strength of the bond—it either survives the soaking treatment or it does not.

FEFCO No. 9/ISO 3038:1975 These methods are technically identical and measure the water resistance of the glue bond by immersion (Fig. 91). Specimens are 200 ± 1 mm wide and 150 mm long, with flutes oriented as shown in Fig. 91. The top and bottom fixtures are inserted through reinforced holes. The two liners are cut with a sharp knife to isolate the five bonds to be tested. The assembly is placed in a water bath at 20 ± 3°C; then the weight is applied. The time taken for the weight to drop to the bottom of the bath—indicating shear failure of the five bonds—is the test value. At least five specimens are tested at once. The failure time for each specimen is reported in terms of a chosen test interval, which may be as long as 24 h. The test is both tedious and time-consuming. The *glue bond shear (GBS)* test described earlier is more convenient and reports the glue bond strength in engineering units.

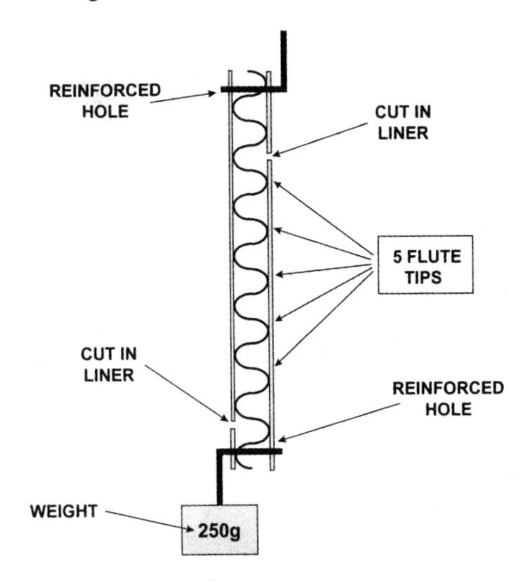

Fig. 91 FEFCO method for determining the water resistance of glue bonds. (From FEFCO No. 9.)

XI. COMPONENT PROPERTIES THAT AFFECT CORRUGATED BOARD PERFORMANCE

Table 7 presents a brief coverage of the important strength properties of the linerboard and corrugating medium that affect the performance of corrugated board. All strength properties increase with the grammage of the components, so this factor is omitted from the table. The synopsis of the test methods applies to the relevant TAPPI standard method. References to other standard methods are presented in Table 8.

The tensile stiffness of linerboard can be conveniently tested by measuring the speed of propagation of sound through the material. This principle was used to develop the Lorentzen & Wettre tensile stiffness orientation (TSO) tester (Fig. 92). This tester measures the velocity of propagation of ultrasonic sound through the sheet between six pairs of diametrically opposed transducers. This velocity depends on the elastic modulus and density of the linerboard:

$$\frac{E}{\rho} = kv^2 \tag{45}$$

where

E = elastic modulus, N/m^2
ρ = density, kg/m^3
k = dimensionless constant
v = velocity of ultrasonic pulse, m/s

The term E/ρ is the specific elastic modulus, or tensile stiffness index (TSI), of the linerboard. The tensile stiffness is then

$$S_t = kv^2 w \tag{46}$$

where S_t = tensile stiffness, N m/kg, and w = linerboard grammage, g/m^2.

Table 7 The Properties of Board Components Which Affect the Properties of Corrugated Board

Corrugated board property	Principal component properties affecting board property	Test methods
Edgewise compression strength F CD OF LINERS AND MEDIUM	CD compression strength of liners CD compression strength of medium	**SHORT-SPAN COMPRESSION TEST** See Chapter 9. A 15 mm wide specimen at least 70 mm in length is cut in the cross-machine direction of the liner or medium. It is clamped with a free span of 0.7 mm, then the clamps are driven together until the specimen fails in compression. **RING CRUSH TEST** See Chapter 9. A 12.7 mm wide specimen of length 152.4 mm is cut in the cross direction of the liner or medium. It is carefully inserted into a special jig that forms the specimen into a ring, placed between the platens of a compression tester, and compressed until it collapses.
Bending stiffness (a) Machine Direction (b) Cross - Machine Direction	MD and CD tensile stiffness of liners MD and CD compressive stiffness of liners	**TENSILE STIFFNESS** See Chapters 6 and 7. A 25 mm wide specimen of length 180 mm is cut in the test direction of the liner. It is clamped in the jaws of a tensile testing machine, and the jaws are driven apart at a speed of 25 mm/min until the specimen fails. The tensile stiffness is calculated from the slope of the initial part of the load vs elongation curve: $S_t = (\Delta f \bullet L)/(w \bullet \Delta L)$ where S_f = tensile stiffness, KN/m^2; Δf = incremental load, kN/m; L = initial test span, m; w = initial specimen width, m; ΔL = incremental length corresponding to Δf, m. **TEST OF COMPRESSION STIFFNESS** See Chapter 9. There is no standard method for testing the compression stiffness of the liners. However, the compression stiffness can be expected to increase in proportion to the tensile stiffness [42].

Burst strength

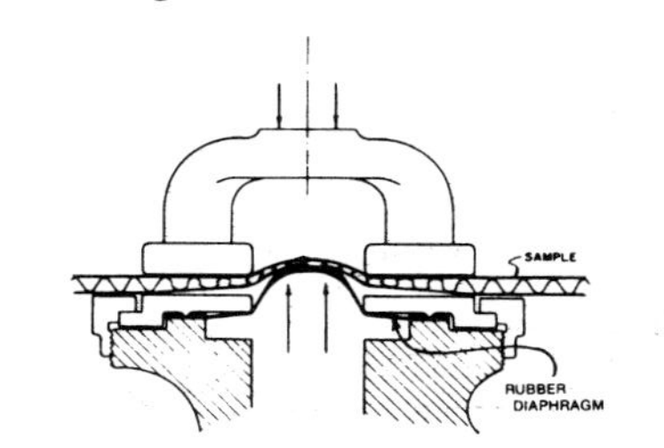

MD and CD tensile energy absorption of the liners

TEST OF TENSILE PROPERTIES A 25 mm wide specimen of length 180 mm is cut in the test direction of the liner. It is clamped in the jaws of a tensile testing machine, and the jaws are driven apart at a speed of 25 mm/min until the specimen fails. The tensile energy absorption (TEA) is proportional to the area under the load–elongation curve:

$$\text{TEA} = A/Lw$$

where A = area under the load–elongation curve, J; L = initial test span, m; w = initial specimen width, m. The units of TEA are J/m^2

Flat crush resistance

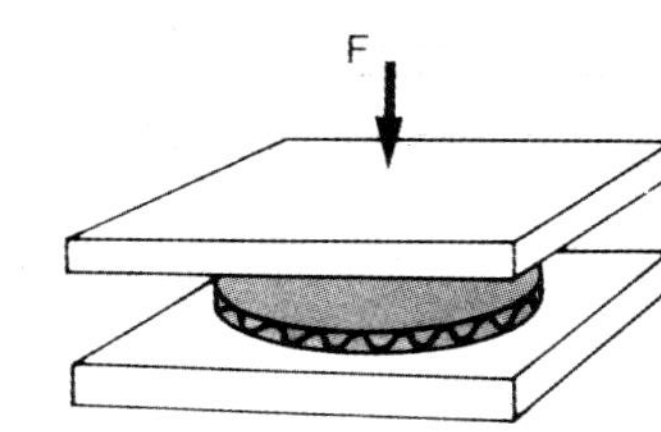

Flat crush of medium (CMT test)
MD compression strength of medium

CORRUGATING MEDIUM TEST (CMT) A strip of medium 152.4 mm by 12.7 mm is cut in the machine direction and fed through a laboratory fluter. The fluted strip is taped using a rack-and-comb rig, then placed in a compression tester and tested for flat crush. The result is expressed in units of force (N).
SHORT-SPAN COMPRESSION TEST As under Edgewise compression strength, but medium is tested in the machine direction.
RING CRUSH TEST As under Edgewise compression strength, but medium is tested in the machine direction.

Puncture test

ANVIL
HEAD

MD and CD tear resistance of liners
MD and CD tensile stiffness of liners

TEAR RESISTANCE see Chapter 8. A 52 mm long specimen with a width of 63 cm is cut in the test direction. The specimen is mounted in the clamps of the tester, and a central tear is initiated using a sharp blade. The energy to propagate the tear across the remaining specimen width is measured by a pendulum device.
TENSILE STIFFNESS As under Bending stiffness.

Table 8 Standard Methods for Mechanical Testing of Components Relevant to the Strength Properties of Corrugated Board[a]

Test	ISO	TAPPI	JIS	SCAN	Appita
Short-span compression	9895:1989	T 826 pm-92	—	P 46:83	AS/NZS 1301:450rp-89
Ring crush	—	T 822 om-93	P 8126:1994	P 34:71	AS/NZS 1301:407s-97
Tensile properties	1924-2:1994	T 494 om-96	P 8113:1998	P 38:80	AS/NZS 1301:404s-98
Corrugating medium (CMT)	7263:1994	T 809 om-99	—	P 27:69	AS/NZS 1301:434s-97
Tear resistance	1974:1990	T 414 om-98	P 8116:1994	P 11:96	AS/NZS 1301:400s-98

[a]There is no comparable FEFCO test.

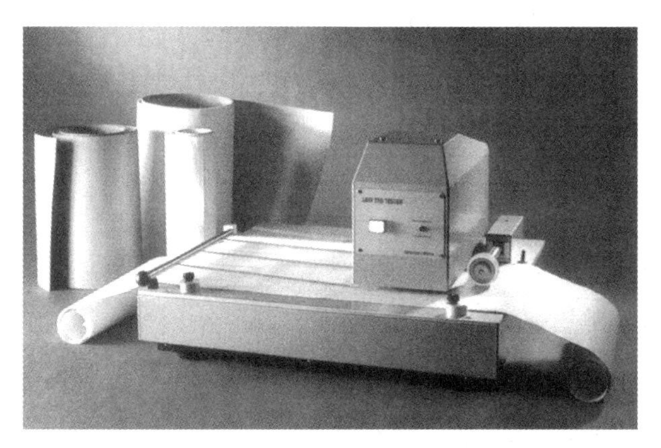

Fig. 92 Device for testing the tensile stiffness orientation of linerboard. (Courtesy Lorentzen & Wettre, Kista, Sweden.)

Note that the elastic modulus E in Eq. (45) is more correctly termed the sonic modulus (see Chapter 6). It is numerically greater than E determined from stress–strain measurements, but a linear relationship between the two variables has been reported for paper [26]. The device also produces a polar diagram of TSI at several positions across the width of the sheet, which can be used to evaluate the tensile stiffness orientation [118], which is important to control the propensity of the linerboard to produce warp in corrugated board.

A. Creep Testing of Corrugated Components

The creep behavior of corrugated boxes and board is influenced by the creep behavior of the components. Compression creep is difficult to measure, because the application of a compressive load will tend to buckle the specimen. Techniques have been developed that involve supporting the specimen between two rigid plates [42,65,155] and between opposed sets of closely spaced blades [23,164]. Neither type of support is ideal. The close-fitting plates will prevent the proper circulation of air and preclude the use of changing or cyclical humidities. The closely spaced leaves make it difficult to directly measure strain on the specimen. Researchers at the Forest Products Laboratory in Madison, WI have developed an apparatus that supports the specimen on an array or slender, flat-topped vertical rods [59] (Fig. 93). The specimen is held in place by masking the open area of the array and applying a vacuum. The specimen can be loaded in compression or tension or biaxially by using the appropriate clamps, and the top surface of the specimen is available for strain measurements in any planar direction. The device has been incorporated in a housing supplied with air temperature and humidity control [61] and marketed as the VCA 1000 (vacuum compression apparatus) by Isthmus Engineering of Madison, WI (Fig. 94). The apparatus is capable of testing compression and tensile creep properties of both corrugated components and corrugated board as well as dimensional stability and Poisson ratio [24,25].

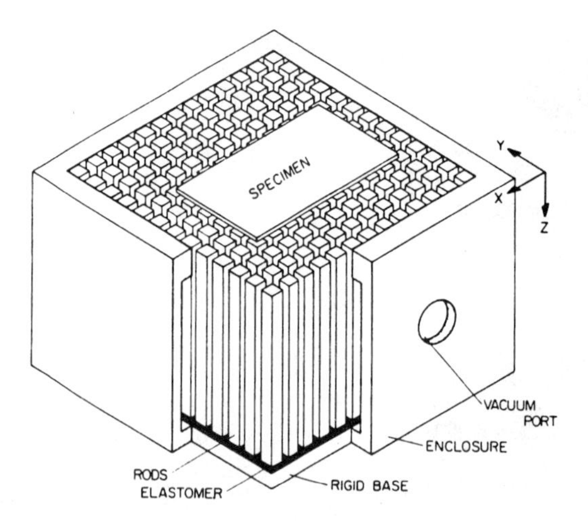

Fig. 93 Specimen support system developed by Gunderson. (From Ref. 59.)

XII. CLOSING REMARKS

A. The Service Environment

The ultimate test of a corrugated box is its ability to perform in the service environment. That environment may involve

- Long-term compressive loading, which will invoke a creep response
- Short-term impact loading, for example from a forklift truck
- Vibration
- Storage in a high and/or cycling humidity environment

The ideal test regime would simulate the service environment as closely as possible, thus enabling the boxes, board, and components to be engineered for specific applications. Although some box manufacturers are moving along that path, there are still many who rely on certification based on burst strength or ECS testing and use a

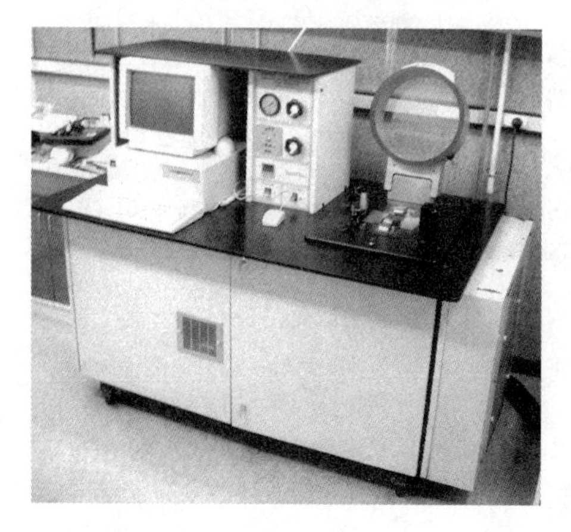

Fig. 94 The VCA 1000 vacuum compression apparatus. (Courtesy Papro, Rotorua, New Zealand.)

conservative safety factor to ensure performance in the field. This no doubt leads to overpackaging and the associated costs. Clearly, full service environment testing is technically challenging, expensive, and time-consuming. However, the potential savings in raw materals and the gain in market share by demonstrated technical competence may provide a sufficient incentive. There is certainly potential to move forward beyond the present bounds of Rule 41/Item 222.

B. High Performance Components

The change in Rule 41/Item 222 has led to a progressive decrease in the grammage of corrugated board components. This has been made possible by the development of high performance linerboards that are significantly denser and stronger than previous grades. Consequently, the corrugated board has the same edgewise compression strength (ECS), although the weight per unit area is lower. This is good news for the box manufacturer and end user, because box weight and associated freight costs are reduced. However, there may be some serious implications from this progressive densification of board components. There is anecdotal evidence that high performance linerboards do not perform well in high and cyclical humidity environments. For example, reports from Asian markets suggest that banana exporters tend to avoid high performance materials, because there is a perceived tendency for the incidence of box failures to increase during humid, cool storage.

This observation is supported by data presented by Boonyasarn et al. [9] They show that high performance boxes lost significantly more compression strength than equivalent "standard" boxes after exposure to an environment cycling between 30% and 85% relative humidity. In contrast, Considine et al. [29] compared the compression creep rate of high performance (HP) and standard linerboards in cyclical humidity. They found no consistent differences to suggest that HP linerboard exhibits a faster creep rate. Interestingly, the creep rate of the linerboard specimens was best predicted by hygroexpansive strain. Clearly, there is a need to develop a better understanding of the performance of high performance corrugated components in high and cycling humidity and the implications of box performance in the service environment.

C. Recycled Components

There is also an increasing tendency to replace kraft linerboard and semichemical corrugating medium with substitutes manufactured from recycled fiber. Advances in papermaking technology—particularly long-nip wet pressing of the sheet—have produced recycled sheets that can match the strength of virgin materials. There is a clear cost incentive, because the same nominal performance can be obtained with a cheaper raw material. However, care must be taken with recycled components, particularly corrugating medium.

Unpublished work at Carter Holt Harvey Pulp & Paper Limited in New Zealand has shown that increasing the recycled fiber content in an NSSC corrugating medium has a detrimental effect on box performance in a service environment (Fig. 95). The boxes were RSC style with external dimensions $L = 390\,\text{mm}$, $W = 252\,\text{mm}$, $H = 285\,\text{mm}$. The corrugated board was single-walled C-flute with two $250\,\text{g/m}^2$ unbleached linerboard facings and a $160\,\text{g/m}^2$ medium. The boxes were partially filled to prevent inward bulge and loaded to 90 kg with a deadweight. The storage

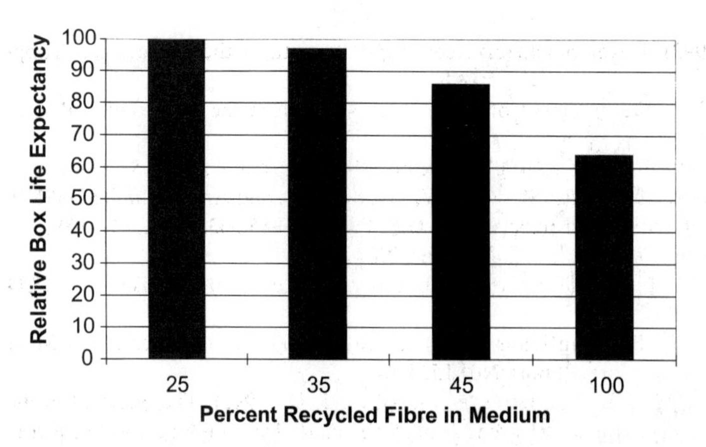

Fig. 95 Effect of recycled fiber content on the relative life expectancy of boxes in a service environment. (By kind permission of Carter Holt Harvey Pulp & Paper Limited.)

environment was a commercial coolstore running under conditions that are typical for fruit storage in New Zealand, namely 1°C/95% RH. Adding recycled fiber to the furnish of the NSSC medium had little effect up to 35% content, but relative life expectancy decreased by about 15% at 45% content. A medium with 100% recycled fiber decreased life expectancy by almost 40%.

Conversely, trials conducted by the Gaylord Container Corporation [158] showed that boxes made from 100% recycled liner and medium failed at the same time as boxes made from virgin components when exposed to creep loading in a cyclical humidity environment.

It is important to confirm the effect of recycled fiber content on box performance in a cyclical humidity environment. This will contribute to better understanding of the commercial and technical implications of the continuing trend toward recycled components for corrugated board.

ACKNOWLEDGMENTS

I thank my former employers Carter Holt Harvey Pulp and Paper Limited, particularly Jim Newfield, for providing time to work on the early stages of this chapter. I am also grateful to Ian Chalmers of Papro New Zealand and Dave McKenzie, Andrew McKenzie, and Tom Clark of Carter Holt Harvey Pulp & Paper Limited for peer reviewing the script and offering constructive suggestions. Thanks also to Chuck Habeger, my editor, for his patience and helpful suggestions.

REFERENCES

1. Allan, R., Kirkpatrick, J., Luke, A. McKinley, P., Plancon, A. E. and Tseglakoff, C. (1993). Development and automation of testing of water resistant bond quality. *Appita* *46*(5):371–374.

2. Batelka, J. J. (1992). Development of green bond strength in the single facer. Tappi 75(10):94–101
3. Batelka, J. J. (1994). Development of green bond strength in the single facer: Part 2. Tappi 77(7):71–73
4. Beach, R. L. (1939). Puncture testing of box board. Paper Trade J. 108(5):30–34
5. Bennett, P. G., McKinlay, P. R., Shaw, N. W., Scott, R. A., and Amcor Limited (1989). Method of making corrugated paperboard. U.S. Patent 4,886,563. (Dec. 12, 1989).
6. Bessen, A. H. (1999). *The Corrugator*. Jelmar, New York.
7. Billerud AB. (1987) Testing of corrugated board and its components. Billerud R&D Centre, Säffle, Sweden.
8. Boller, K. H. (1947). Buckling loads of flat sandwich panels in compression. USDA Forest Products Laboratory Report No. 1525-A.
9. Boonyasarn, A., Harte, B. R., Twede, D., and Lee, J. L. (1992). The effect of cyclic environments on the compression strength of boxes made from high performance corrugated fiberboard. *Tappi 75*(10):79–85.
10. Brittain, J., et al. (2000). How Corrugated Boxes Are Made (CD ROM). TAPPI Press, Atlanta, GA.
11. Bronkhorst, C. A. (1997). Towards a more mechanistic understanding of corrugated container creep deformation behaviour. *J. Pulp Paper Sci. 23*(4): J174–J181.
12. Brynhildsen, Av H. O., and Dagel, Y. (1959). Reproducibility and sources of error in the determination of the flat crush of corrugated board. *Svensk Papperstidn. 62*(18):631–639 (in Swedish).
13. Buchanan, J. A. (1965). Corrugated Case Performance Packaging *36*(425):31–34.
14. Buchanan, J. A. (1965). Corrugated Case Performance Packaging *36*(426):50–55.
15. Buchanan, J. S. (1968). Testing methods for the components of fiberboard cases. *Tappi 51*(2):65–72
16. Byrd, V. L. (1986). Adhesives influence on edgewise compression creep in a cyclic relative humidity environment. *Tappi 69*(10):98–100.
17. Byrd, V. L. (1984). Edgewise compression creep of fiberboard components in a cyclic relative humidity environment. *Tappi 67*(7):86–90.
18. Byrd, V. L. (1972). Effect of relative humidity changes on compressive creep response of paper. *Tappi 55*(11):1612–1613.
19. Byrd, V. L., and Koning, J. W., Jr. (1978). Corrugated fiberboards: Edgewise compression creep in cyclic relative humidity environments. *Tappi 61*(6):35–37.
20. Carlson, T. A. (1940). Bending tests of corrugated board and their significance. *Fiber Containers 25*(3):24–30.
21. Carbone, J. T. (1999). Corrugating Defect/Remedy Manual Tappi Press, Atlanta.
22. Carlson, T. A. (1939). A study of corrugated fiberboard and its component parts as an engineering material. *Fibre Containers 24*(7):22–35.
23. Cavlin, S., and Fellers, C. (1975). A new method for measuring the edgewise compression properties of paper. *Svensk Papperstidn. 78*(9):329–332
24. Chalmers, I. R. (1994). The performance of packaging grade paper in high and cycling relative humidity. In: *Proceedings, Moisture-Induced Creep Behaviour of Paper and Board*, (Fellers, C. and Laufenberg, T. L., eds.) STFI, Stockholm, Sweden December 5–7, pp 233–238.
25. Chalmers, I. R. (1994). Hygroexpansive theta angle versus ultrasonic TSO. In: *Proceedings, Moisture-Induced Creep Behaviour of Paper and Board*, (Fellers, C. and Laufenberg, T. L. eds.), STFI, Stockholm, Sweden December 5–7, pp. 239–247.
26. Chatterjee, P. K. (1969). Sonic pulse propagation in a paperlike structure. *Tappi 52*(4): 699–704.
27. Chatterjee, S. G. and Ramaro, B. V. Hysteresis in water vapour sorption equilibria of bleached kraft paperboard. In: *Proceedings, Moisture and Creep Effects on Paper Board*

and Containers. 3rd International Symposium, (Chalmers, I. R. ed.), Rotorua, New Zealand, February 20–21, pp. 121–134.
28. Confederation of European Paper Industries. (1998). 1998 Annual Statistics. *http://www.cepi.org/publications/index.htm*
29. Considine, J. M., Stoker, D. L., Laufenberg, T. L. and Evans, J. W. (1994). Compressive creep behavior of corrugated components affected by humid environment. *Tappi 77*(1):87–95.
30. Cox, H. L. (1954). Computation of initial buckling stress for sheet stiffener combinations. *J. Roy. Aeronaut Soc. 58*(6):634–638.
31. Crisp, C. J., Stott, R. A., and Tomlinson, J. C. (1968). Resistance of corrugated board to flat crushing loads. *Tappi 51*(5):80A–83A.
32. Dagel, Y., and Jönson, G. (1972). Transport trials in Europe and the USA. Final report. Proc. 1972 FEFCO Congress, Paris.
33. Debney, H. G., Blacker, K. J., Redding, B. J., and Watkins, J. B. (1980). Handling and storage practices for fresh fruit and vegetables: Product manual. Australian United Fresh Fruit and Vegetable Association.
34. Drewden, P. (1943). The flat crush test. *Paper Trade J. 116*(11): 119–122.
35. Eriksson, L.-E. (1979). Swedish group develops meter for rapid measurement of warp. *Boxboard Containers* 87(2): 62–64.
36. Eriksson, L.-E., Cavlin, S., Fellers, C., and Carlsson, L. (1987). Curl and twist of paperboard — theory and measurement. *Nordic Pulp Paper Res J. 2*(2):66–70.
37. Eriksson, L.-E. (1979). A review of the edge crush test. Part 1. *Boxboard Containers 86*(8):34–38.
38. Eriksson, L.-E. (1979). A review of the edge crush test. Part 2. *Boxboard Containers 86*(9):64–67.
39. Fédération Européenne des Fabricants de Carton Ondulé (FEFCO). (2000). Paper Consumption—Corrugated Board. http://www.fefco.org/pages/statistic/statisti.htm.
40. Fellers, C., and Carlsson, L. (1979). Measuring the pure bending properties of paper: A new method. *Tappi 62*(8):107–109.
41. Fellers, C., Bäckström, M., Htun, M, and Lindholm, G. (1998). Paper-to-paper friction. Paper structure and moisture. *Nordic Pulp Paper Res. J. 13*(3): 225–232.
42. Fellers, C., de Ruvo, A., Elfstrom, J., and Htun, M. (1980). Edgewise compression properties: A comparison of handsheets made from pulps of various yields. *Tappi 63*(6):109–112.
43. Fibre Box Association. (1992). In: Basics of boxes; Box styles. *Fibre Box Handbook*. 20th ed. Fibre Box Association, Rolling Meadows, IL, pp. 17–32.
44. Fibre Box Association. (1992). *Fibre Box Handbook*. 20th ed. Fibre Box Association, Rolling Meadows, IL, p. 9.
45. Fibre Box Association. (1992). *Fibre Box Handbook*. 20th ed. Fibre Box Association, Rolling Meadows, IL, p. 103.
46. Fibre Box Association. (1992). *Fibre Box Handbook*. 20th ed. Fibre Box Association, Rolling Meadows, IL, p. 53–54.
47. Fibre Box Association. (1992). *Fibre Box Handbook*. 20th ed. Fibre Box Association, Rolling Meadows, IL, p. 60.
48. Fox, T. S., Whitsitt, W. J., Nelson, R. W., and Watt, J. A. (1978). Container failure tests: Shipping containers and cartons shown to fail only in compression when loaded internally. *Paperboard Packag. 63*(5):48–56.
49. Gartaganis, P.A. (1975). Strength properties of corrugated containers. *Tappi 58*(11):102–108.
50. Gerard, G. (1957). Failure of plates and composite elements. Part IV. *Handbook of Structural Stability*. NACA TN 3784.

51. Godshall, W. D. (1968). Effects of vertical dynamic loading on corrugated fiberboard containers. Research Paper FPL 94. Forest Products Laboratory, Madison, WI.
52. Godshall, W. D. (1971). Frequency response, damping and transmissibility characteristics of top-loaded corrugated containers. Research Paper FPL 160, Forest Products Laboratory, Madison, WI.
53. Göttsching, L., and Otto, W. (1976). Running characteristics of corrugating mediums Part 1. *Papier 30*(6):221–228 (in German). Translation available from IPST, Atlanta, GA.
53a. Gottsching, L. and Otto, W. (1976). Running characteristics of corrugating mediums. Part 2a. *Papier 30*(10): 417–425 (in German). Translation available from IPST, Atlanta, GA.
54. Göttsching, L., and Otto, W. (1976). Running characteristics of corrugating mediums. Part 2b. *Papier 30*(11): 457–465 (in German). Translation available from IPST, Atlanta, GA.
55. Göttsching, L., and Otto, W. (1977). Running characteristics of corrugating mediums. Part 3. *Papier 31*(2): 45–53 (in German). Translation available from IPST, Atlanta, GA.
56. Göttsching, L., and Otto, W. (1977). Running characteristics of corrugating mediums. Part 4a. *Papier 31*(3): 85–94 (in German). Translation available from IPST, Atlanta, GA.
57. Göttsching, L., and Otto, W. (1977). Running characteristics of corrugating mediums. Part 4b. *Papier 31*(4): 129–136 (in German). Translation available from IPST, Atlanta, GA.
58. Göttsching, L., and Otto, W. (1977). Running characteristics of corrugating mediums. Part 5. *Papier 31*(5): 169–179 (in German). Translation available from IPST, Atlanta, GA.
59. Gunderson, D. E. (1983). Edgewise compression of paperboard: A new concept of lateral support. *Appita 37*(2):137–141.
60. Gunderson, D. E., and Laufenberg, T. L. (1994). Apparatus for evaluating stability of corrugated board under load in a cyclic humidity environment. *Exper. Techn. 18*(1): 27–31.
61. Gunderson, D. E., Considine, J. M., and Scott, C. T. (1988). The compressive load–strain curve of paperboard. Rate of load and humidity effects. *J. Pulp Paper Sci. 14*(2):J37–J41.
62. Highton, A. P. (1993). Physical aspects of single-facer operation. *Tappi 76*(7):130–139.
63. Huck, C. (1999). Customer demand drives mini-flute market. *Paperboard Packag. 84*:(4):30–40.
64. International Corrugated Case Association. (2000). Annual Statistics—Corrugated and Containerboard Production. http://www.iccanet.org/annual.htm
65. Jackson, C. A., Koning, J. W. Jr., and Gatz, W. A. (1976). Edgewise compressive test of paperboard by a new method. *Pulp Paper Mag. Can. 77*(10): T180–T183.
66. Johansson, A., Fellers, C., Gunderson, D., and Haugen, U. (1998). Paper friction. Influence of measurement conditions. *Tappi 81*(5):175–183.
67. Johnson, M. W. Jr., Urbanik, T. J., and Denniston, W. E. (1980). Maximising top-to-bottom compression strength. *Paperboard Packag. 65*(4):98–108.
68. Jones, A. L. (1871). Improvement in paper for packaging. U.S. Patent 122,023 (19 Dec. 1871).
69. Jönson, G. (1993). *Corrugated Board Packaging*. Pira Int., Leatherhead, England, Chap. 14.
70. Jönson, G. (1993). *Corrugated Board Packaging*. Pira Int., Leatherhead, England.
71. Jönson, G. (1993). *Corrugated Board Packaging*. Pira Int., Leatherhead, England, Chap. 17.
72. Jönson, G. (1993). *Corrugated Board Packaging*. Pira Int., Leatherhead, England, Chap. 13.

73. Jönson, G. (1993). *Corrugated Board Packaging*. Pira Int., Leatherhead, England, Chap. 16.
74. Jönson, G., and Toroi, M. (1969). Edge crush test for corrugated board provides better information for buyers. *Paperboard Packag*. *54*(7):35–38.
75. Kainulainen, M., and Toroi, M. (1986). Optimum composition of corrugated board with regard to the compression resistance of boxes. *Paperi ja Puu 68*(9): 666–668.
76. Kawanishi, K. (1989). Estimation of the compression strength of corrugated fiberboard boxes and its application to box design using a personal computer. *Packaging Tech. Sci.* 2(1):29–39.
77. Kellicut, K. Q. (1959). Structural design notes for corrugated containers. Note No.8. *Packag. Eng. 4*(9):92–94.
78. Kellicut, K. Q. (1963). Effect of contents and load bearing surface on the compressive strength and stacking life of corrugated containers. *Tappi 46*(1):151A–154A.
79. Kellicut, K. Q. (1959). Structural design notes for corrugated containers. Note No.6: Stiffness of corrugated board in relation to box rigidity. *Package Eng. 4*(7):78–79.
80. Kellicut, K. Q., and Landt, E. F. (1951). Basic design data for the use of fiberboard in shipping containers. *Fibre Containers 36*(12): 62–80.
81. Kellicut, K. Q., and Landt, E. F. (1951). Suggestions in making flat crush tests of corrugated. *Fibre Containers 36*(5):82.
82. Kellicut, K. Q., and Peters, C. C. (1959). Structural design notes for corrugated containers. Note No. 7. *Packag. Eng. 4*(8):84–86.
83. Kellicut, K. Q., and Landt, E. F. (1951). Safe stacking life of corrugated boxes. *Fibre Containers 36*(9):28–38.
84. Kellicutt, K.Q. (1972). How liner/medium weight relationships affect the strength of corrugated. *Boxboard Containers*, *79*(8): 51–56.
85. Koning, J. W. Jr., and Moody, R. C. (1971). Predicting the flexural stiffness of corrugated fiberboard. *Tappi 54*(11):1879–1881.
86. Koning, J. W. Jr (1964). A short column test of corrugated fiberboard. *Tappi 47*(3):134–137.
87. Koning, J. W. Jr (1988). Towards an international standard for the edgewise compression test of corrugated board. A second opinion. *Tappi 71*(10):62–64.
88. Koning, J. W. Jr., and Moody, R. C. (1966) Slip pads, vertical alignment increase stacking strength 65%. *Boxboard Containers 74*(4):56–59.
89. Koning, J. W., Jr (1995). *Corrugated Crossroads: A Reference Guide for the Corrugated Containers Industry*. TAPPI Press, Atlanta, GA, p. 240.
90. Koning, J. W., Jr and Stern, R. K. (1977). Long-term creep in corrugated fiberboard containers. *Tappi 60*(12):128–131.
91. Koning, J. W., Jr. (1978). Compressive properties of linerboard as related to corrugated fiberboard containers Theoretical model verification. *Tappi 61*(8):69–71.
92. Koning, J. W., Jr. (1971). Measuring linerboard thickness and flute height of corrugated fiberboard. *Tappi 54*(2):236–238.
93. Koning, J. W., Jr. (1983) Corrugated fiberboard. In: *Handbook of Physical and Mechanical Testing of Paper and Paperboard.* R. E. Mark, ed. Marcel Dekker, New York, Chap. 9, pp. 385–408.
94. Koning, J. W., Jr. and Moody, R. C. (1971). Predicting flexural stiffness of corrugated board. *Tappi 54*(11):1879–1881.
95. Koning, J. W., Jr. and Moody, R. C. (1969). Effect of glue skips on compressive strength of corrugated fiberboard containers. *Tappi 52*(10):1910–1915.
96. Koning, J. W., Jr., Kuenzi, E. W., Moody, R. C., and Godshall, W. D. (1972). Improving comparability of paperboard test results: Using flexible and rigid type test machines. *Tappi 55*(5):757–760.

97. Kroeschell, W. O. (1991). New carrier regulations for corrugated shipping containers are now in effect. *Tappi 74*(7):63–65.
98. Kroeschell, W. O. (1984). The edge crush test of corrugated board: Development of TAPPI test methods T 811 and T 823. *Tappi 67*(10):56–58.
99. Kroeschell, W. O. (1992). The edge crush test. *Tappi 75*(1):79–82.
100. Langston, S. M. (1908). Manufacture of cellular boards. U.S. Patent 878,403. (Feb. 4, 1908).
101. Lapin, L. (1980). *Statistics: Meaning and Method.* Harcourt Brace Jovanovich, New York.
102. Leake, C. H. (1988). Measuring corrugated box performance. *Tappi 71*(10):71–75.
103. Leake, C. H., and Wojcik, R. (1989). Influence of the combining adhesive on box performance. *Tappi 72*(8):61–65.
104. Little, J. R. (1943). A theory of box compressive resistance in relation to the structural properties of corrugated board. *Paper Trade J. 116*(24):31–34, 143–137
105. Long, D., and Penney, J. M. (1970). CCA investigates thermal insulation properties of corrugated flower boxes. *Boxboard Containers 78*(4):72–76.
106. Long, O. (1874). Packings for bottles, jars, etc. U.S. Patent 9,948 (Reissued 29 Nov. 1881).
107. Luey, A. T. (1963). Stiffness of multi-ply boxboard. *Tappi 46*(11):159A.
108. Maltenfort, G. G. (1988). Testing VI. Methods for shipping containers. In: *Corrugated Shipping Containers: An Engineering Approach.* Jelmar, Plainview, NY, Chap. 21
109. Maltenfort, G. G. (1988). Testing IV. Methods for combined board. In: *Corrugated Shipping Containers: An Engineering Approach.* Jelmar, Plainview, NY, Chap. 19.
110. Maltenfort, G. G. (1989). Compression strength of corrugated containers. In: *Performance and Evaluation of Shipping Containers.* Jelmar, Plainview, NY, Chap. 17.
111. Maltenfort, G. G. (1988).Testing V. Methods for paperboard related to manufacturing. In: *Corrugated Shipping Containers: An Engineering Approach.* Jelmar, Plainview, NY, Chap. 20.
112. Maltenfort, G. G. (1988). Understanding the industry. In: *Corrugated Shipping Containers: An Engineering Approach.* Jelmar, Plainview, NY, Chap. 1.
113. Maltenfort, G. G. (1989). Mullen vs puncture for corrugated. In: *Performance and Evaluation of Shipping Containers.* Jelmar, Plainview, NY, Chap. 47.
114. Maltenfort, G. G. (1989). Compression load distribution on corrugated boxes. In: *Performance and Evaluation of Shipping Containers.* Jelmar, Plainview, NY, Chap. 24.
115. March, H. W., and Smith, C. B. (1945). Buckling loads of flat sandwich panels in compression. Various types of edge conditions. USDA Forest Products Lab. Rep. No. 1525.H105.
116. Markström, H. (1988). *Testing Methods and Instruments for Corrugated Board.* Lorentzen & Wettre, Stockholm, Sweden, p. 15.
117. Markström, H. (1988). *Testing Methods and Instruments for Corrugated Board.* Lorentzen & Wettre, Stockholm, Sweden, pp. 17–25.
118. Markström, H. (1991). *The Elastic Properties of Paper-Test Methods and Measurement Instruments.* Lorentzen & Wettre, Stockholm, Sweden, pp. 13–24.
119. McCready, D. W. and Katz, D. L. (1939). A study of corrugated fiberboard: The effect of adhesive on the strength of corrugated board. Univ. Mich. Eng. Res. Bull. No. 28.
120. McCready, D. W. and Katz, D. L. (1939). Effect of the adhesive on strength characteristics of corrugated board. *Fibre Containers 24*(2):20–28.
121. McDonnell, W. T. (1993). Wax: A source of low surface friction in kraft recycled linerboard. *Tappi 76*(10): 31–36.
122. McKee, R. C. (1989). Puncture and box performance. In: *Performance and Evaluation of Shipping Containers.* G. G. Maltenfort, ed. Jelmar, Plainview, NY, pp. 389–397.
123. McKee, R. C., and Gander, J. (1957). Top load compression. *Tappi 40*(1):57–64.

124. McKee, R. C., Gander, J. W., and Wachuta, J. R. (1962). Flexural stiffness of corrugated board. *Paperboard Packag. 47*(12):111–118.
125. McKee, R. C., Gander, J. W., and Wachuta, J. R. (1963). Edgewise compression strength of corrugated board. *Paperboard Packag. 48*(11):70–76.
126. McKee, R. C., Gander, J. W., and Wachuta, J. R. (1963). Compression strength formula for corrugated board. *Paperboard Packag. 48*(8):149–159.
127. McKenzie, A. M. (1999). Bending creep of corrugated fibreboard in cycling relative humidity. M. Appl. Sci. Thesis, Massey University, New Zealand.
128. McKenzie, A. W. (1972). Influence of clamping pressure on the bursting strength test of corrugated fiberboard. *Appita 25*(4):278–282.
129. McKinlay, P. R. (1992). Machine-direction shear stiffness of corrugated board: New test to determine corrugated board quality. KCL Paper Physics Seminar, Otaniemi, Finland, June 8–11, 1992, pp. 92–95.
130. McKinlay, P. R. (1993). Analysis of the strain field in a twisted sandwich panel with application to determining the shear stiffness of corrugated fiberboard. In: *Products of Papermaking*, Vol 1. C. F. Baker, ed. Pira Int., Cambridge, U.K., pp. 575–599.
131. McKinlay, P. R. (1989). Shear stiffness tester. Australian Patent 603502 (8 Feb 1989).
132. McKinlay, P. R., and Shaw, E. Y. N. (1981). Technology for the non-technologist—new method of predicting compression strength of containers. *Appita 34*(5):418–420.
133. Meech, H. B. (1879). Carpet lining. U.S. Patent 215,648 (20 May 1879).
134. Nordman, L., Kolhonen, E., and Toroi, M. (1978). Investigation of the compression of corrugated board. *Paperboard Packag. 63*(10):48–62.
135. Ostlund, S. (1978). How moisture and heat affect warp on the corrugator. *Boxboard Containers, 85*(9):44–46.
136. Paine, F. A. (1981). *Fundamentals of Packaging*. Brookside Press, Leicester, England, pp. 81–96.
137. Pecht, M. G. (1985). Creep of regain: Rheologically simple hydrophilic polymers. *J. Strain Anal. 20*(3):179–181.
138. Peterson, W. S. (1980). Minimum-cost design for corrugated containers under top-to-bottom compression. *Tappi 63*(2):143–146.
139. Peterson, W. S. (1980). Unified container performance and failure theory I: Theoretical development of mathematical model. *Tappi 63*(10):75–79.
140. Peterson, W. S. (1980). Unified container performance and failure theory II: Comparisons between experimental data and mathematical model. *Tappi 63*(11):115–120.
141. Peterson, W. S., and Schimmelpfenning, W. J. (1982). Panel edge boundary conditions and compressive strengths of tubes and boxes. *Tappi 65*(8):108–110.
142. Pommier, J. C., Poustis, J., Fourcade, E., and Morlier, P. (1991). Determination of the critical load of a corrugated cardboard box submitted to vertical compression by finite elements method. Proc. Tappi Int. Paper Physics Conf., Kona Hawaii, Sept. 22–26, pp. 437–448.
143. Matussek, H., and Stefan, V., eds. (1997). *PPI International Fact & Price Book 1999*. Miller Freeman, San Francisco, CA.
144. Laufenberg, T. L. and Leake, C. H. eds. (1992). Proceedings: Cyclic Humidity Effects on Paperboard Packaging. Forest Products Laboratory, Madison, WI September 14–15.
145. Fellers, C. and Laufenberg, T. L. eds. (1994). Proceedings: Moisture-Induced Creep Behaviour of Paper and Board, STFI, Stockholm, Sweden December 5–7.
146. Chalmers, I. R. ed. (1997). Proceedings: Moisture and Creep Effects on Paper, Board and Containers. 3rd International Symposium, Rotorua, New Zealand 20–21 February.
147. Serra-Tosio, J-M. and Vullierme, I. eds. (1999). Proceedings: Moisure and Creep Effects on Paper, Board and Containers. 4th International Symposium, Grenoble, France 18–19 March.

148. Ramaker, T. J. (1974). Thermal resistance of corrugated fiberboard. *Tappi 57*(6):69–72.
149. Ranger, A. E. (1960). The compression strength of corrugated fibreboard cases and sleeves. *Paper Technol. 1*(5):531–541.
150. Resource Information Systems, Inc (RISI). (1998). *World Containerboard Study — Forecast & Written Analysis.*
151. Sandström, J. and Titus, M. (1993). Field experience in predicting corrugated board strength with ultrasonic testing. *Tappi 78*(10) Focus on Corrugated Containers: 19–22.
152. Schnell, P., Perkins, S., and Brittain, J. (2000). *The Corrugated Containers Manufacturing Process.* TAPPI Press, Atlanta, GA.
153. Schrampfer, K. E., and Whitsitt, W. J. (1988). Clamped specimen testing: A faster edgewise crush procedure. *Tappi 71*(10):65–69.
154. Seth, R. S. (1985). Relationship between edgewise compressive strength of corrugated board and its components. *Tappi 68*(3):98–101.
155. Setterholm, V.C., and Gertjejansen, R. O. (1965). Method for measuring the edgewise compressive properties of paper. *Tappi 48*(5): 308–312.
156. Shaw, N. W., Selway, J. W., McKinlay, P. R., Bennett, P. G., and Hill, M. (1998). A revolution in board design and manufacture. *Tappi 81*(10):27–34.
157. Singh, S. P., Brugess, G., and Langois, M. (1992). Compression of single wall corrugated shipping containers using fixed and floating test platens. *J. Test. Eval. 20*(4):318–320.
158. Soderberg, R. V. (1992). All fall down ... together. In: Proceedings, Cyclic Humidity Effects on Paperboard Packaging (Laufenberg, T. L. and Leake, C. H., ed.) Forest Products Laboratory, Madison, WI. September 14–15, pp. 117–132.
159. Söremark, C., and Fellers, C. (1993). Mechano-sorptive creep and hygroexpansion of corrugated board in bending. *J. Pulp Paper Sci. 19*(1):J19–J26.
160. Sprague, C. H. and Whitsitt, W. J. (1982). Compressive strength relationships and evaluation. *Tappi 65*(12):104–105.
161. Staigle, V. H. (1989). In: *Performance and Evaluation of Shipping Containers.* G. H. Maltenfort, ed. Jelmar, Plainview, NY, pp. 259–262.
162. Staigle, V. H. (1967). Corrugated board—Is it crushed during fabrication? *Tappi 50*(1):45A–47A.
163. Steenberg, B., Kubát, J. and Rudström, L. (1970). Competition in rigid packaging materials. *Svensk Papperstidn. 73*(4):77–92.
164. Stockmann, V. (1976). Measurement of intrinsic compressive strength of paper. *Tappi 59*(7):93–97.
165. Stott, R. (1988). Towards an international standard method for the edgewise compression test of corrugated board. *Tappi 71*(1):57–60.
166. Stott, R. (1975). A comparison of edgewise compression test methods. *Appita 29*(1):29–32.
167. Swec, L. (1986). Boxes, corrugated. In: *The Wiley Encyclopedia of Packaging Technology*. M. Bakker, ed. Wiley, New York, p. 73.
168. Tappi Test Method T 1206 sp-91. Precision statement for test methods.
169. TAPPI. (1999). Test equipment suppliers for Tappi test methods. In: *TAPPI Test Methods 1999–2000.* TAPPI Press, Atlanta, GA, pp. 25–50.
170. TAPPI. (1997). Warp in corrugated board. TAPPI TIP 0304-07.
171. Thomson, R. H. (1882). Machinery for manufacturing packing and lining fabrics. U.S. Patent 252,547 (17 Jan. 1882).
172. Thomson, R. H. (1890). Machine for making packing and lining fabrics. U.S. Patent 430,447 (17 June 1890).
173. Urbanik, T. J. (1978). Transportation vibration effects on unitised corrugated containers. Research Paper FPL 322, Forest Products Laboratory, Madison, WI.

174. Urbanik, T. J. (1981). Method for determining the effect of transportation vibration on unitised corrugated containers. *Shock and Vibration Bull. 51*(3):213–224.
175. Urbanik, T. J. (1990). Force plate for corrugated container vibration tests. *J. Testing and Eval. 18*(5):359–362.
176. Urbanik, T. J. (1981). Principle of load sharing in corrugated fibreboard. *Paperboard Packag. 66*(11):122–128.
177. Urbanik, T. J., Catlin, A. H., Friedman, D. R., and Lund, R. C. (1994). Edgewise crush test streamlined by shorter time after waxing. *Tappi 77*(1):83–86.
178. Wennerblom, A.B. (1992). Twist warp: Causes and remedies. *Tappi 75*(4): 97–101.
179. Westerlind, B. S., and Carlsson, L. A. (1992). Compressive response of corrugated board. *Tappi 75*(7):145–154.
180. Whitsitt, W. J. (1972). Theory, use and calibration of bursting strength testers. Part 1. *Boxboard Containers 80*(1):34–36.
181. Whitsitt, W. J. (1972). Theory, use and calibration of bursting strength testers. Part 2. *Boxboard Containers 80*(2):53–57.
182. Whitsitt, W. J. (1972). Theory, use and calibration of bursting strength testers. Part 3. *Boxboard Containers 80*(3):50–56.
183. Whitsitt, W. J. (1987). Runnability and corrugating medium properties. *Tappi 70*(10):99–103.
184. Whitsitt, W. J., and Hoerschelman, E. (1990). Liner orientation effects of combined-board warp. *Tappi 73*(1):89–95.
185. Wolf, M. (1971). How box length, width, affect its compression strength. *Package Eng. 16*(10):68–70.
186. Wolf, M. (1972). New equation helps pin down box specifications. *Package Eng. 17*(3):66–67.

12

PHYSICAL AND MECHANICAL PROPERTIES OF TOWEL AND TISSUE

M. K. RAMASUBRAMANIAN
North Carolina State University
Raleigh, North Carolina

I. INTRODUCTION

The primary distinguishing characteristic of the towel and tissue paper grades is their low density. From here on, we will use the terms tissue and towel grades and low density grades interchangeably. In addition to this characteristic, the low density grade is commonly low in basis weight, low in strength and stiffness, high in stretch, particularly in the machine direction (MD), and highly absorbent. The apparent density of tissue and towel grades is generally below 300 kg/m^3, and the basis weight is generally less than 35 g/m^2 for tissue and less than 50 g/m^2 for towel grades. Low density paper grades are used primarily in sanitary applications such as bathroom and facial tissue and in household applications such as kitchen towels and dinner napkins.

The flexibility and absorbency of tissue arises primarily from several factors: fiber morphological properties such as length, coarseness, and stiffness; large distance between the fiber–fiber bonds; intentionally poor fiber–fiber bonding; and fiber axial microcompressions due to creping. The large distance between bond centroids and poor fiber–fiber bonding give the bonds the ability to rotate and to fail gradually upon loading. The fiber axial microcompressions render the fibers less stiff and amplify the difference between the stiffness of the fibers in tension and that in compression. It has been shown by Lyne [1] that the sheet flexibility increases rapidly as the basis weight falls below 20 g/m^2. It is commonly observed that when a tissue sample is strained in a tensile tester, the fiber-to-fiber bonds begin to fail from the very early stages of the stress–strain curve and completely slip away before the fibers are substantially strained. Thus, the fiber–fiber bonds play a very important role in determining towel and tissue mechanical properties. A satisfactory theoretical approach to predicting the mechanical properties of paper that takes into account the fiber–fiber bond properties, the effect of finite fiber length, and the bending of the free segment has been presented by Perkins [2]. The elastic modulus of low density sheets has been shown to correlate well with the density of the sheet when finite fiber length is incorporated into the network theory [3]. Network theory has been extended to the inelastic regime by Ramasubramanian and Perkins [4].

II. TOWEL AND TISSUE MANUFACTURING PROCESSES

The physical and mechanical properties of low density grades are strongly dependent on their manufacturing process. It is important to have a basic understanding of the manufacturing process in order to understand the properties of low density grades. There are two primary classifications of tissue manufacturing processes today: the conventional wet press (CWP) process and the through air drying (TAD) process. In the CWP process, tissue is formed on a forming fabric, in either a suction breast roll or a twin-wire configuration, and the embryonic web is transferred to a papermaking felt and dewatered by pressing with the help of a pressure roll nip once or twice against the Yankee dryer. The pressing process also transfers the sheet to the Yankee dryer surface. An adhesive solution is sprayed on the dryer surface prior to sheet transfer in order to provide good bonding between the sheet and the dryer surface. The sheet is removed from the Yankee dryer surface by the creping process. In the TAD process, the sheet is formed on a fabric and transferred to another fabric, called the imprinting fabric, at about 25% consistency [5] and the sheet is dried in contact with this fabric by blowing hot air through the sheet and the fabric until about 40–80% sheet solids before transferring to the Yankee dryer for creping. By physically deflecting the embryonic web into a specially designed through air dryer (TAD) fabric, high and low basis weight regions and high and low density regions are interspersed in the fibrous network, dried to 40–80% solids, and subsequently creped to achieve both a high strength and a highly soft and absorbent structure [6]. To achieve extensibility when wet, wet foreshortening is practiced [7] in which there is a differential velocity transfer of a wet-laid embryonic web of relatively low fiber consistency from a forming fabric to a substantially slower moving, open mesh transfer fabric with a substantial void volume and thereafter the web is dried while substantial macroscopic rearrangement of the fibers in the plane of the web is prevented. Care is taken not to compact the web during differential transfer. At the extreme, creping is avoided and wet microcontraction is used to achieve extensibility by drying the paper on the fabric in a through air drying fashion for hand or wiper towel applications [8,9].

Recreping is the process of applying a pattern of a bonding material such as acrylate latex rubber emulsion through a rotogravure process and subsequently creping the sheet. This results in a grid pattern of strong bonding with very low density in between to achieve additional softness without a corresponding loss in strength [10–12]. Furthermore, low density grades may be formed from a homogeneous furnish, or it may be layered. Typically, a two-layered tissue is used to produce two-ply products and a three-layered tissue is used to produce single-ply products. The layering is done to achieve optimum consumer response in terms of "softness." Hardwood fibers (about 0.2–1.5 mm fiber length), such as eucalyptus, are used in the outer layers (Yankee dryer side of the sheet), and the sheet is creped while long softwood fibers (about 2 mm fiber length) make up the top layer, which provides the strength to the network [13].

Although uncreped through-dried technology [8,9] is useful in some applications, creping is the most commonly used foreshortening mechanism for most towel and tissue grades, especially the premium grades. Creping is a process in which the sheet density is decreased and the strength is increased. As a consequence, the softness and absorbency characteristics found in the towel and tissue grades are essentially created at the creping blade. After creping, the tissue and towel base sheet is obtained, which is subsequently converted into tissue and towel products. Embossing, laminating into two-ply products, calendering to increase the smoothness of the surface, and in some cases recreping are common towel and tissue converting processes.

III. THE CREPING PROCESS

To elucidate the mechanical behavior of towel and tissue paper, it is important to discuss in detail the fundamentals of the creping process. In this process, a wet low basis weight sheet, predried to 70% solids in the case of the TAD process or to about 35% solids in the case of the CWP process, is transferred to the Yankee dryer. The Yankee dryer is a steam-heated drum with a polished cast iron surface about 8–18 ft in diameter. A polymeric adhesive is sprayed onto the drum prior to the pressure roll nip. The sheet is transferred to the Yankee dryer surface, adhesively bonded, and scraped off with a creping blade. Creping reduces the stiffness and strength of the web substantially and increases the stretch. The sheet is reeled off at the other end. The surface speed at the reel is lower than that of the creping dryer. The difference between the Yankee dryer surface speed and the reel speed divided by the Yankee dryer surface speed is known as the degree of creping, expressed in percent.

There are three types of creping, distinguished by the moisture content at the time of creping. In the wet crepe process, the sheet is about 60–85% solids at the time of creping. In this case, the fibers still have the ability to form hydrogen bonds after creping during subsequent drying. As a consequence, the sheet is less soft, is stiff, and has lower stretch. In the semidry creping process, the sheet solids content is about 85–93%, and in the dry creping process it is about 93–97%. Of the three conditions, dry creping produces the greatest reduction in strength and increase in softness; this will be discussed in detail here.

The mechanism of the creping process has been studied experimentally by Hollmark [14]. He conducted experiments on an experimental tissue machine operating at 140 m/min. The primary variables were basis weight, creping angle, and degree of adhesion adjusted through percent resin content. The crepe wavelength was determined. Based on the experiments, Hollmark proposed the following mechanism for creping. The sheet is separated from the dryer so that a loop is formed. A second loop is formed on top of the first one. The loop formation continues until the pile collapses and the web follows the dryer surface to hit the doctor blade behind the first pile of loops.

Recently, a laboratory creping dryer capable of creping discrete samples up to 1100 ft/min was developed and used to understand the mechanics of the creping process [15].

From this study [15], a mechanistic description of the creping process can be developed as follows (Fig. 1):

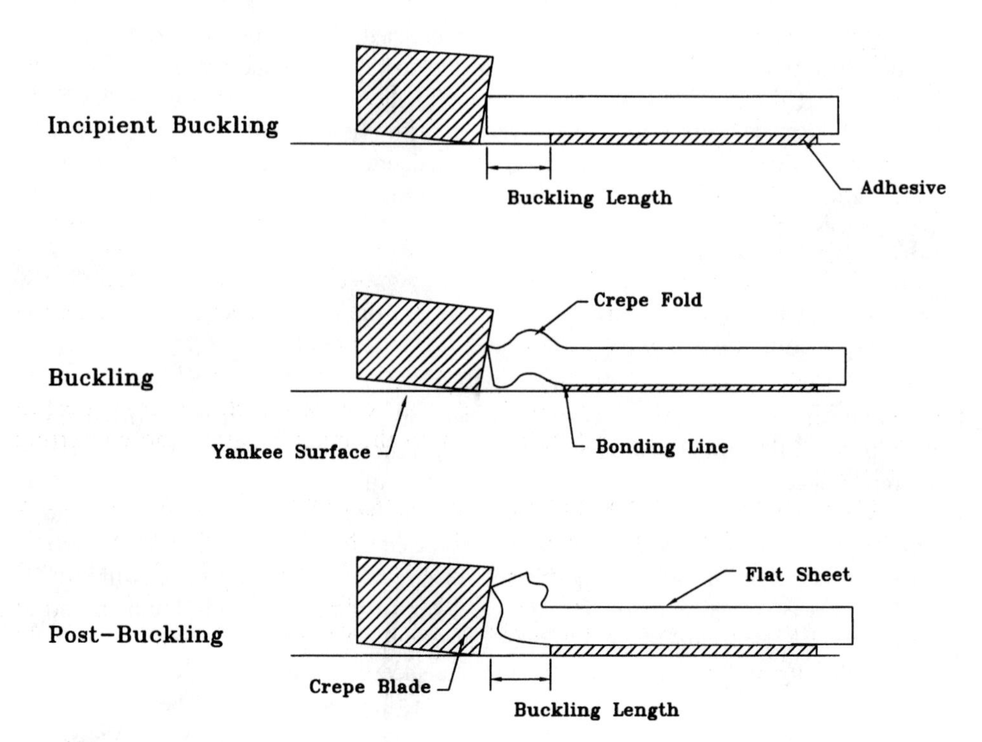

Fig. 1 Mechanistic description of the creping process.

1. As soon as the sheet hits the blade, stresses develop in the sheet and in the adhesive layer.
2. As the Yankee dryer rotates, the papers/adhesive interfacial shear strength is exceeded and a crack develops, releasing the energy due to web compression. This results in a portion of the sheet detaching from the dryer and simply lying on it.
3. The crack continues to propagate until the free segment of the web becomes unstable and buckles.
4. While the sheet buckles, crack propagation stops, and upon complete collapse of the free portion (folding) the leading edge of the bonded segment of the sheet hits the blade again and the process repeats itself.
5. The sheet is constantly pulled off at a relatively lower speed by a suitable reel mechanism, thus leaving the desired stretch in the sheet. This stretch is typically 10–20%.

The distance between consecutive peaks or valleys in the crepe pattern is defined as the crepe wavelength. The crepe wavelength is dependent upon the flat sheet characteristics prior to creping, adhesive characteristics, the crepe blade geometry, and process conditions such as the operating speed, moisture content at the time of creping, the use of release agents, and adhesive layer buildup. Shamgin's result [15] shows that as the creping angle decreases, the crepe wavelength increases or the crepe becomes coarse. The effect of creping angle on crepe wavelength is shown in Fig. 2; the results are consistent with that of Hollmark [14]. Creping angle is defined as the angle between the tangent to the Yankee dryer surface and the blade face (the surface that the sheet impacts against during creping).

Similarly, as the basis weight increases, the crepe wavelength increases as shown by Hollmark [14] (Fig. 3). Further, the effect of adhesive concentration on the crepe structure has been studied by Shamgin [15]. He observed that higher adhesive concentration leads to finer crepe, all other conditions remaining constant. The effect of adhesive concentration when polyvinyl alcohol was used is shown in Fig. 4. The crepe wavelength is an important characteristic of the web, because it influences critical properties. For very short wavelengths, the sheet undergoes short-column bucking in which severe internal damage occurs, and the structure of the crepe is known as the microcrepe (Fig. 5). When the crepe wavelength is large, the crepe is called macrocrepe. In the latter case, there is relatively less internal damage and the structure retains substantial strength and stiffness and is therefore not as soft. It is desirable to achieve microcrepe conditions while retaining the strength necessary for end use and web handling during the manufacturing and converting processes.

In summary, the creping process is primarily responsible for decreasing the density, stiffness, and strength and increasing the stretch and absorbency of the paper product. This is accomplished by causing internal fiber–fiber bond damage in the structure and creating void space within the structure. In the following section we discuss the measurement of physical and mechanical characteristics of towel and tissue fibrous structures.

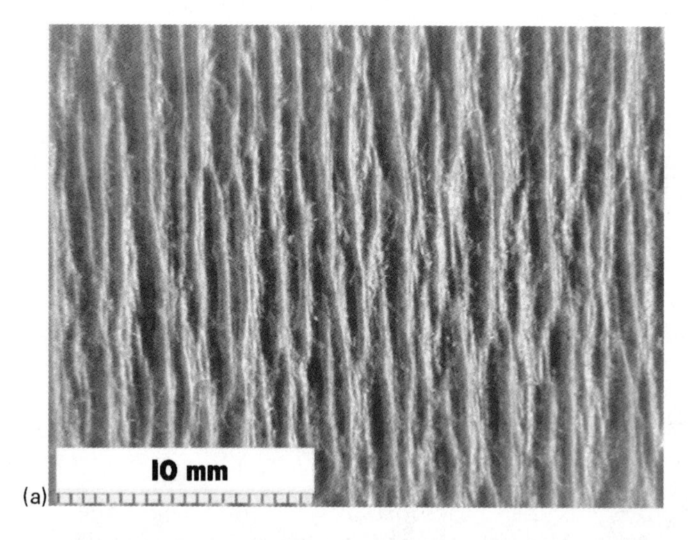

Fig. 2 Effect of creping angle on crepe structure. (a) 75°; (b) 80°; (c) 85°; (d) 90°. (From Ref. 15.)

IV. PHYSICAL AND MECHANICAL PROPERTIES

A. Density and Caliper

Density is perhaps the most important property that distinguishes tissue from other grades of paper. Perceived softness is directly proportional to bulk (inverse of density) [16]. Apparent density is defined as basis weight divided by caliper. The standard caliper measurement $50 \pm 2\,\text{kPa}$ ($7.3 \pm 0.3/\text{lb/in.}^2$) recommended by TAPPI T 411 om-89 is not suitable for measuring tissue paper thickness. Caliper should be measured using a soft platen under reduced loading. The influence of loading pressure on thickness of sanitary tissue was studied by Rowe and Volkerman [17]. They concluded that the TAPPI standard pressure (they report standard pressure to be

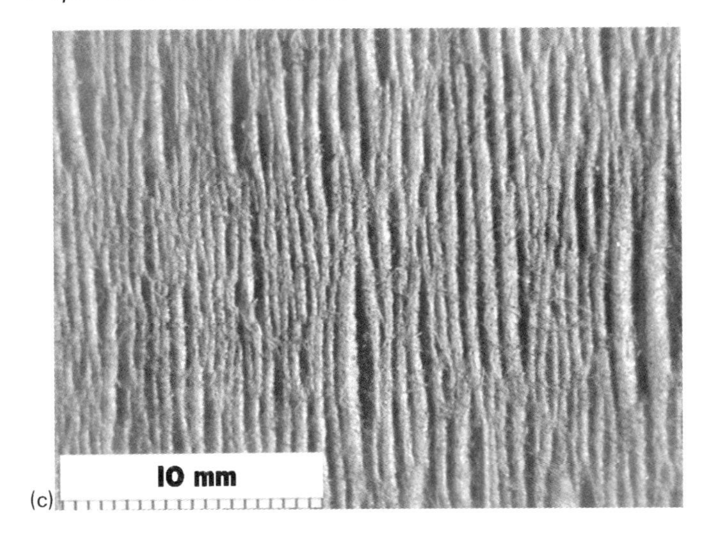

(c)

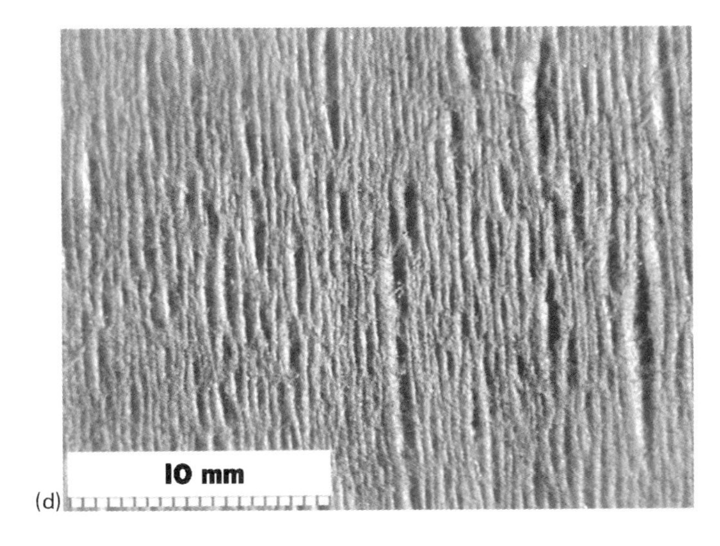

(d)

62 ± 8.6 kPa or 9 ± 1 lb/in.^{2}) is too high for testing tissue grades. They measure the thickness of tissue and towel products in an Instron tester using compression platens. In the pressure range of 0–10 lb/in.^{2} (68.95 kPa), uncreped papers showed, in the worst case, an increase of 15% in thickness at 0.05 lb/in.^{2} (0.34 kPa) as compared to 10 lb/in.^{2} (68.95 kPa). Dry creped toilet tissue and facial tissues showed substantial changes, with a 60–75% increase over the same pressure range. Embossed towels and napkins showed much larger increases (Fig. 6). Rowe and Volkerman [17] also studied the effect of loading rates by carrying out the measurements at 0.02, 0.1, and 0.2 in./min (0.51–5.1 mm/min). They observed a tendency for the fast tests to give slightly higher thickness values, the largest difference between measurements being about 4%. Based on their results, a measurement pressure in the range of 0.1–5 lb/in.^{2} (0.69–3.45 kPa) seems reasonable.

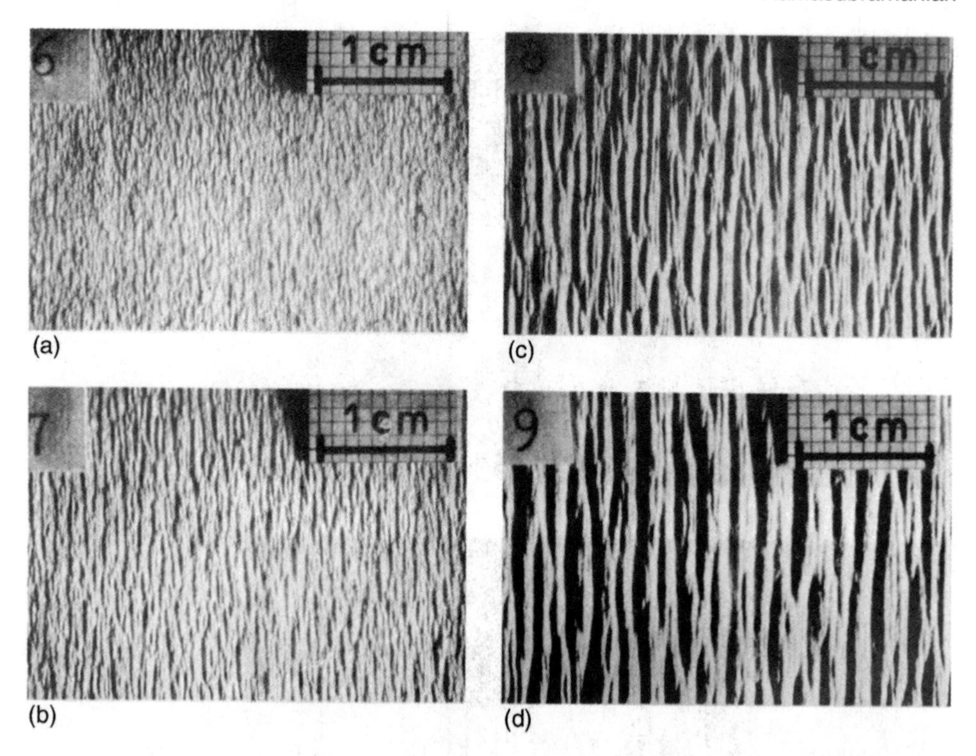

Fig. 3 Effect of basis weight on crepe structure (a) 11 g/m^2; (b) 15 g/m^2; (c) 30 g/m^2; (d) 60 g/m^2. (From Ref. 14.)

Morgan [18] reported using a pressure of 80 g/in.^{2} (0.17 lb/in.^{2} or 1.2 kPa) under a 54.8 mm (2 in.) diameter anvil to measure tissue caliper. Habeger et al. [19] reported a soft platen caliper measurement at 9 kPa (1.28 lb/in.^{2}). In this measurement, a tissue sample is introduced into the gap between a set of transducers intended for ultrasonic time-of-flight measurements. The apparatus consists of a set of two neoprene-faced polyvinylidene fluoride (PVDF) transducers. The transducer surfaces are allowed to conform to the surface of the samples, and the distance between them is measured using a linear variable displacement transformer (LVDT) to obtain the caliper. It appears that there is no standard pressure or loading rate for measuring tissue caliper. Furthermore, it is a common practice to determine the caliper of a stack of sheets, typically four plies, as opposed to measuring individual samples. In general, a soft platen caliper with a large-diameter (54.8 mm; 2 in.) foot at a reduced pressure can be used to measure tissue caliper. However, it is important to report the method used and the testing conditions with the results.

Caliper can also be measured using a digitizing technique, looking at cross sections at high magnifications. This method is tedious and highly variable. A large number of measurements are necessary to obtain statistically meaningful results. In this method [20], a sample 2.54 cm × 5.1 cm is stapled onto a rigid cardboard holder. The cardboard holder is placed in a silicon mold. Six parts Versamid resin, four parts Epcon 812 resin, and three parts 1,1,1-trichloroethane are mixed in a beaker. The resin mixture is placed in a low speed vacuum desiccator and the bubbles are removed. The mixture is then poured into the silicon mold with the cardboard sample holders so that the sample is thoroughly wet and immersed in the mixture. The sample is cured for at least 12 h until the resin mixture has hardened. The sample is removed from the mold, and the cardboard holder is removed from the sample. The sample is placed in a microtome and leveled. The edge of the sample is removed in slices by the microtome until a smooth surface is achieved. Slices have a thickness of about 100 μm are taken from the smooth surface. At least 10–20 slices, in series, are photographed [20]. Several vertical lines are drawn at random across the cross section, and the distance between the top surface of a fiber in the top of the sheet to the bottom surface of a fiber in the bottom of the sheet is measured. Several hundred measurements are made, and an average value of the thickness is calculated. By this method, the void space in the cross section, and hence a density, can also be calculated [5,21].

B. Crepe Frequency

Crepe frequency may be defined as the number of corrugations in the machine direction per unit length of the web [6]. Crepe frequency can be measured by a suitable stereomicroscope, a camera and a frame grabber card, and suitable image analysis software. In the image analysis approach, a magnification of 60×–70× has been used for crepe frequency measurements [20]. First the system is calibrated with a 10 mm stage micrometer at the desired magnification, and the magnification should not be changed after the calibration. A sample of cellulosic fibrous web is placed on the stage of the microscope and focused without changing the magnification. Using an image analysis program, the distance between two points of interest, such as peaks or valleys in the crepe, is measured. The reciprocal of the measurement is recorded as the crepe frequency. The measurement is repeated several times to ensure statistically significant results. At a sufficiently low magnification, it is possible to see the crepe peaks and valleys directly by viewing perpendicular to the sheet surface without having to prepare cross sections.

Alternatively, a profilometer may be used to measure the surface profile, and the resulting trace can be transformed into the frequency domain using the fast Fourier transformation algorithm (FFT). The dominant periodic component is usually caused by the crepe folds, and the information can be extracted. In this method [20] a sample of tissue is placed on a profilometer table. Rasch and Henser [20] used a diamond stylus of 2.54 μm (0.0001 in.) radius and a vertical force loading of 1.96 mN (200 mg). The sample was scanned at a rate of 60 mm/min. Data were digitized at a rate of 20 data points per millimeter. The sample was scanned 30 mm in one direction and indexed while it was in motion 0.1 mm in the transverse direction. This process was repeated until the entire square was scanned. Both sides of the sample were scanned. The data were analyzed using the FFT technique to identify periodic events such as the crepe frequency in the signal.

Confocal microscopy can also be used to get the out-of-plane thickness of the sample [20]. A sample of the fibrous structure measuring 2 cm × 6 cm is placed on top of a glass microscopic slide. The sample is viewed at a magnification of 40×. The magnification enlarges the field of view sufficiently to show several surface features. The sample is optically sectioned parallel to the *XY* plane at intervals of 10–40 μm, for a total depth of 800 μm normal to the plane of the sample. The computer in the

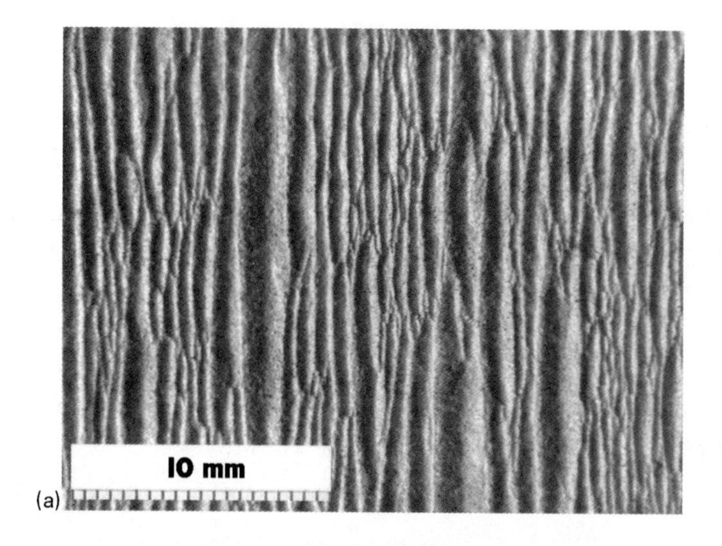

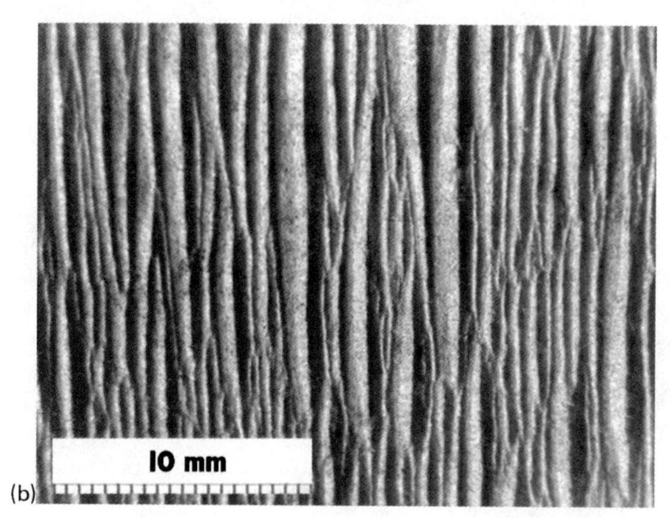

Fig. 4 Effect of adhesive concentration on crepe structure. (a) 0.25%; (b) 0.5%; (c) 0.75%; (d) 1.0%. (From Ref. 15.)

microscope acquires the desired number of XY slices at the desired interval. Each slice is viewed in the computer. A three-dimensional image is reconstructed and cross-sectional images are generated. Vertical lines are drawn across the cross section, and the thickness is calculated as described previously.

Microtome sectioning and image analysis techniques described earlier for the determination of caliper and density can also be used to determine the crepe frequency.

C. Basis Weight and Spatial Mass Distribution

As discussed in an earlier section, recent tissue-making technologies have evolved in such a direction that the spatial distributions of mass and density are controlled in a

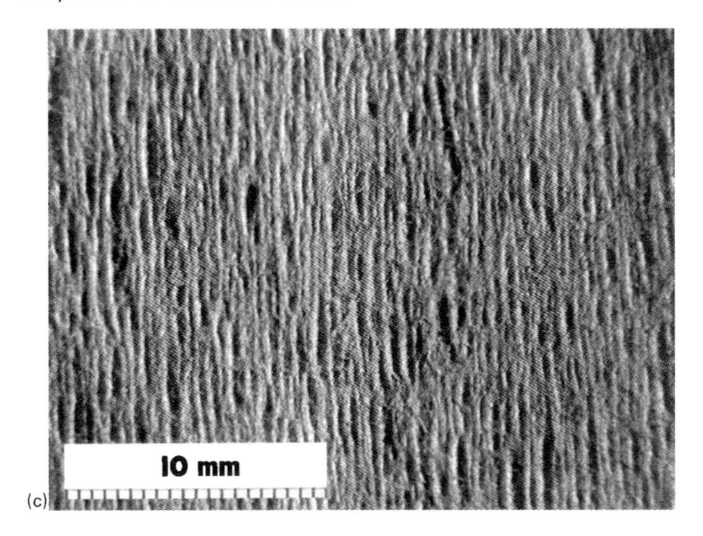

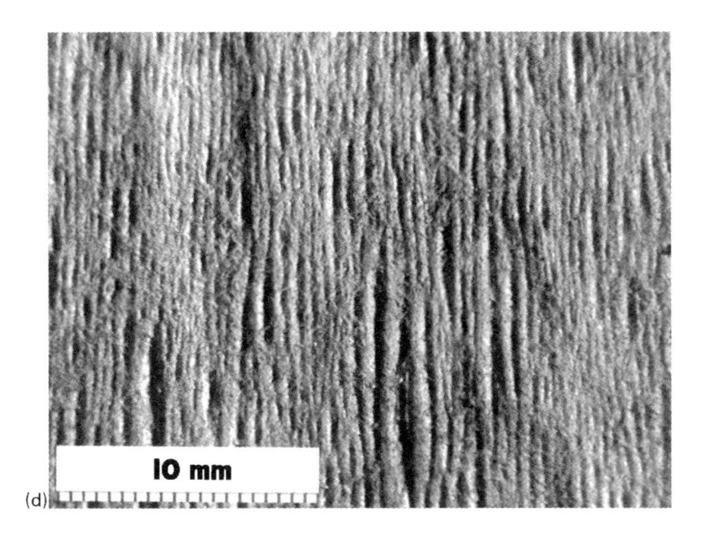

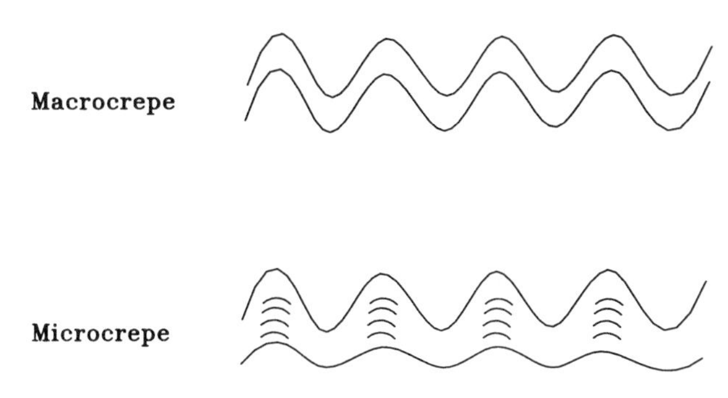

Fig. 5 Types of crepe structures.

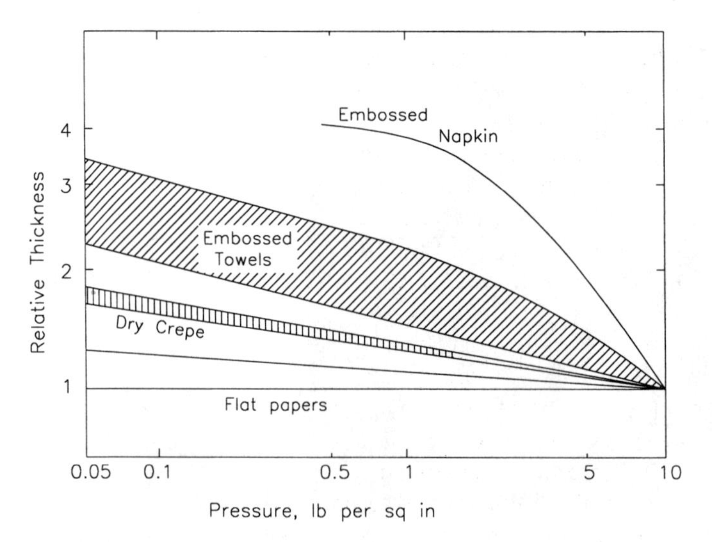

Fig. 6 Thickness vs. pressure. (From Ref. 17.)

desired pattern to obtain softness without a proportional loss in strength. Hence, it has become important to assess the spatial mass and density distribution in tissue grades.

The basis weight itself is measured according to TAPPI T 410 om-88. In low density grades, however, measurement of the spatial basis weight variation is important, because it influences the perception of softness and influences the strength properties. A sheet with greater small-scale grammage uniformity will usually be stronger and posses better optical properties [22]. Stronger tissue would help improve runnability. The variation of low basis weight is measured by the standard deviation $\sigma(w)$, where w is the local basis weight measured over a small area, or the coefficient of variation, which is defined as the standard deviation $\sigma(w)$ divided by the average basis weight $\bar{w}$, denoted by $V(w)$ or the symbol F, or the power spectrum [22]. There are four principal methods for the determination of small-scale basis weight variation, namely, weighing, absorption of light, absorption of beta rays [22], and soft X-ray imaging [23].

Weighing In the weighing method, the sample is cut into 1 mm squares and weighed. The weights are converted to basis weight units and spatially mapped. However, this method is difficult to carry out because of the high precision required in the preparation of samples. In fact, to achieve 4% accuracy in local basis weight, samples 1 mm square have to be cut to an accuracy of 20 μm edge length, i.e., the width of a single fiber [22]. Hence, this method is suitable only for larger scale basis weight variation assessment.

Light Absorption The light absorption technique can be used to determine the "look-through" characteristics of mass distribution, commonly called formation, as opposed to the true mass distribution. In this technique the variation of light transmitted through a sheet is measured [22,24]. The coefficient of variation of local basis weight is equal to the coefficient of variation of absorbance if the reflectivity and the scattering coefficient of the sheet can be assumed to be constants and the incident light is diffused, especially for lightweight sheets. The light source can be either a He-Ne laser that emits monochromatic light at 630 nm or an iodine incandescent lamp that emits white light. In the case of the incandescent lamp, the wavelength distribution can be modified with appropriate filters. The light is brought into a fiber-optic bundle (10 mm diameter) and then concentrated onto the measuring area by suitable optical means. A filled high density polyethylene (HDPE; 12 g/m^2) sheet is placed between the paper sample and the light source, which ensures that the light is well diffused [24].

Both the incident light and the transmitted light are converted to voltages by using semiconductor photodiodes. The light source is modulated at a suitable frequency (103 Hz in this case), which improves the accuracy and makes it insensitive to ambient light conditions. The stability of the optical system is excellent as measured by the coefficient of variation of repeat measurements. Measurements by Ebeling and Komppa [24] show a variation of 0.004% for 1000 measurements. The sample is mounted on a stage that is indexed using a suitable X-Y positioning mechanism capable of incrementing position precisely. Measurements are made when the sample is stationary after indexing. The indexing process is continued until the entire area is covered. Ebeling and Komppa used a sample area of $30\,mm \times 150\,mm$. Transmittance is measured in preselected increments covering the entire sheet area, and the coefficient of variation is calculated. The increments are in the millimeter to submillimeter range.

In the case of towel and tissue, the sheet is made of fibers alone and no filling materials are present. However, the tissue is often locally densified, calendered, and embossed. All of these operations are known to affect the scattering coefficient and hence influence the areal mass distribution measurement through optical means. They can provide only relative changes in mass distribution for sheets made under similar process conditions.

Beta-Ray Absorption In this method, beta radiation is used instead of diffuse light and its transmission through the sheet is measured. Beta rays are attenuated and absorbed in proportion to the mass they pass through. The wavelength of the beta radiation is so large that the rays are not scattered as they travel through a sheet [22]. The absorption of beta rays depends only on the total mass, not on the composition of this mass. The incident beta radiation is transmitted through the paper according to the formula [24]

$$T = e^{-\mu w}$$

where T is the transmission factor, μ is the absorption coefficient, and w is the basis weight. The absorption coefficient of beta radiation is known and is a constant for all known components of paper. There are two ways of carrying out the beta-ray absorption technique: the direct counting method and beta radiography.

Direct Counting Method In this technique, a beta-radiation source (1 mCi ^{147}Pm or ^{14}C of 1 mm diameter is placed behind the sample. A detector collimator with a 1 mm orifice is placed against the opposite face of the sheet. The measurement of beta transmittance is achieved by pulse counting. A constant number of pulses, usually 50,000, are counted at each point and the time needed is recorded. The longer the time, the more the mass the radiation has to go through. The sample is indexed spatially as described in the earlier section on light scattering measurements, and the pulse is counted at every location. Mylar samples of known basis weights are used to calibrate the transmittance of beta rays with the actual basis weight [24].

Beta Radiography The direct measurement technique is time-consuming and tedious. An alternative to the counting procedure is beta radiography [25]. In this technique, a beta radiograph is made instead of direct point-to-point measurements. A source with uniform strength over its area is placed in contact with the sheet to be tested, and transmitted radiation is recorded with an X-ray film. A ^{14}C plastic film (102 mm × 102 mm, 6.16 mCi) is the source. The sample is sandwiched between the source sheet and the X-ray film. Two foam pads press the source–sample–film sandwich to ensure close contact between all components, and it is placed in an aluminum chamber under vacuum. The beta rays that traverse the sample are absorbed by the photographic emulsion of the X-ray film to create the distribution of mass density (DMD) image. A wedge of Mylar film (basis weight up to 87 g/m^2) is always exposed with the sample. The densities of this wedge are used to fit a calibration curve. The curve is then used to convert optical densities into local basis weight variations. Exposure times can vary from 1 to 24 h or more, depending on the sensitivity of the film and the basis weight of the sample. After exposure, the radiograph is developed according to standard X-ray film development procedure. The absolute optical density values of the film can be measured either manually with a microdensitometer [22] or by using an image analyzer [25]. With the measurement of optical densities, gray levels, and known basis weight standards, a relationship between the gray level and the basis weight can be established. It is important to note that the beta-radiography method is influenced by the film type, exposure time, temperature, and developer concentration. The image analysis system also influences the results through the light intensity and camera aperture. The effects of these variables have been studied by Cresson et al. [25]. For reference, their experimental conditions were Kodak Industrex-R film, exposure times 8 h, developing time 5 min, developing temperature 22°C, developer concentration 21.3%, light intensity between 80% and 100% of the nominal intensity, and f-stop of the lens aperture near $f/22$.

Soft X-Ray Technique Soft X-rays are characterized by their low energy (1.2–12.0 keV, wavelength 1.0–10 Å) compared to beta radiation (^{14}C 150 keV). X-rays are attenuated at all energy levels according to the same equation that governs beta rays. Similar to beta radiography, the sheet samples are in intimate contact with the film. Kodak LP7 film was used with an exposure time of 60 min with an energy of 8 kV_p (peak), and developed with Ektaflow Type 1 developer with a development time of 2 min. Radiographs were taken with a Hewlett-Packard cabinet X-ray system (HP Faxitron Model 43855A) and developed with a JOBO film processor rotary tube unit [20]. The X-ray cabinet consist of a source that is about 0.5 mm in diameter, a 0.64 mm thick beryllium window, and a 3 mA continuous current source. The distance from the source to the sample is about 61 cm, and the voltage is about 8 kV_p. The only variable is the exposure time, which is adjusted to achieve maximum contrast. The sample is die cut to dimensions of about 2.5 cm by about 7.5 cm. The X-ray film (DuPont NDT 35, for example) is placed on the Hewlett-Packard Faxitron X-ray cabinet, emulsion side facing up and the paper sample is placed on the film. In addition to the sample, 15 mm × 15 mm calibration standards of known basis weight are placed on the X-ray film. Calibration samples are required to obtain the basis weight versus gray level relationship each time the image of the sample is exposed and developed. Helium is introduced into the chamber in order to purge the air and prevent absorption of X-rays by the air. Following the helium purge, the sample is exposed to soft X-rays for a predetermined amount of time. The procedure is repeated for different exposure times, and the exposed films are processed under standard conditions. The image made at each exposure time is then digitized by using a high resolution radioscope linescanner. The image is digitized with a spatial resolution of 1024 × 1024 discrete points representing an area of 8.9 cm × 8.9 cm of the radiograph. A histogram of the frequency of occurrence of each gray level value is plotted, and the standard deviation is recorded for each exposure time.

The radiograph with the exposure time that yields the maximum standard deviation is selected for subsequent analysis. If a maximum standard deviation is not apparent, the exposure time range should be expanded and the experiment should be repeated. The maximum standard deviation is used to maximize the contrast obtained by the scatter in the test data. The optimum radiograph is redigitized at higher resolution (12 bit mode). Next, the uniformity of exposure must be checked. This is done by examining the gray level values across the exposed areas of the radiographs not blocked by the sample or the calibration standards. The radiograph is judged acceptable if there are no gradients in gray levels from top to bottom or left to right. Next the image is processed by a suitable image processor. In the image processor, the gray level of the background is measured and subtracted from the gray level of the sample to obtain a new image. The values of this final gray level—the maximum, minimum, standard deviation, median—and pixel area of each image are recorded.

Similar measurements are made for the calibration standard images. A gray scale histogram is constructed. These histogram data are used to relate the gray level to the mass per pixel of the standard. With the calibration curve, sample gray levels are converted to mass. Basis weight is calculated by dividing the mass by the pixel area. The coefficient of variation of the data for the radiograph can be calculated and is a measure of the basis weight distribution. In tissue, the periodic variation component of mass distribution may be important. Hence, a simple coefficient of variation may not be sufficient. The pattern densified regions can be identified in the image by an operator and by using a light pen, digitize the periphery of these regions can be marked. By measuring the gray level in that region and dividing by the area, the local basis weight can be determined. Power spectrum analysis can also be used to determine the dominant frequencies in the gray scale data.

All these techniques in addition to electrography, which is the use of the transmission electron microscope, were compared by Tomimasu et al. [26] for characterizing paper formation. They concluded that in terms of spatial resolution, beta radiography and electrography were superior to light transmission but inferior to soft X-radiography. However, beta radiography and electrography are suitable for formation analysis. The light transmission technique is greatly affected by density and surface roughness, but it is a quick and easy method to implement. The soft X-ray technique is suitable for obtaining finer resolution, especially when studying fiber bonding and orientations and the mass distribution of sheets containing intense wire marks.

D. Tensile Characteristics

Standard Tensile Test Determination of the entire stress–strain curve for tissue and low density grades is important for several reasons. The tissue elastic properties are related to its softness. The stretch in the cross-machine direction is a critical parameter for successful converting operations, especially embossing. The strength characteristics influence runnability.

The tensile properties of low density papers can be determined using a standard tensile tester. Line contact grips are used to hold the sample as directed by TAPPI T 494 om-88. One-inch-wide samples are cut using a die cutter. Care should to be taken not to stretch the samples while handling. The standard gauge length for a tensile test is 180 ± 5 mm (7.1 ± 0.2 in.). The testing rate is typically 1 in./min (25 ± 5 mm/min). The sample is stretched to failure, and the maximum force to break the sheet is recorded. The load–elongation curve can be recorded by a suitable data acquisition system. The resulting load–elongation curve is plotted. Tests are conducted in the machine and cross-machine directions. Typical stress–strain curves for tissue and towel papers are shown in Figs. 7 and 8, respectively. The test is also used to determine the wet tensile strength, which will be discussed in a later section.

Sample Length Considerations While testing tissue products, it may not be possible to obtain a 203 mm (8 in.) sample to conduct the test with 180 mm (7.1 in.) gauge length. This is primarily due to the fact that tissue product perforations are typically spaced at about 4.5 in. (114 mm). So the maximum length that can be tested is 89 mm (3.5 in.), allowing for gripping. For paper towels and facial tissues it is

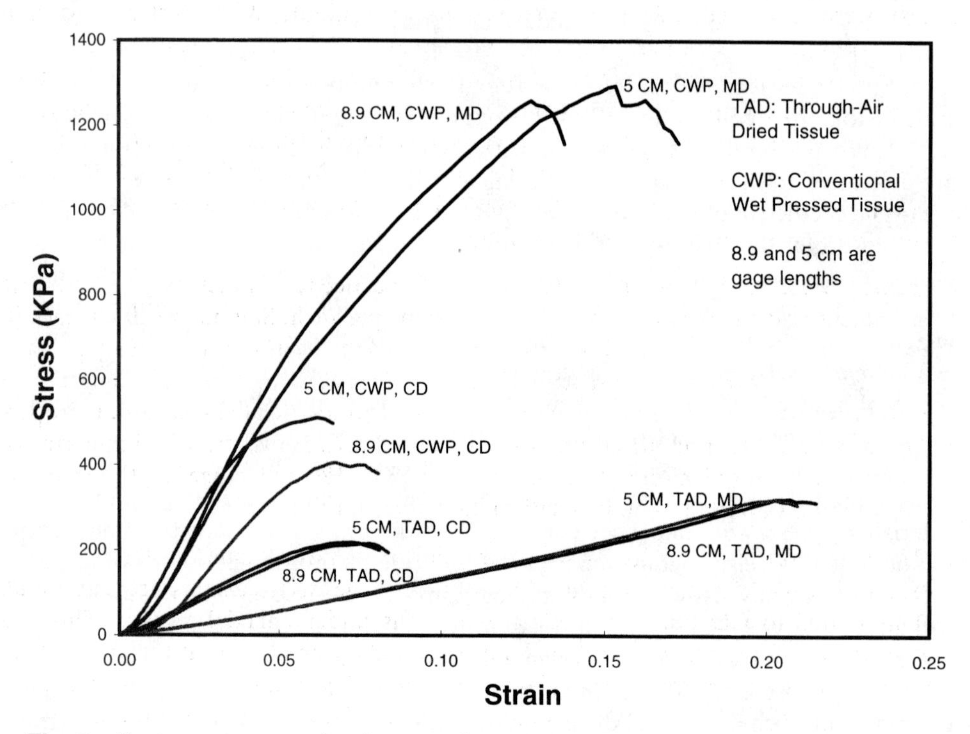

Fig. 7 Stress–strain curves for tissue products. 8, 9, and 5 cm are gauge lengths.

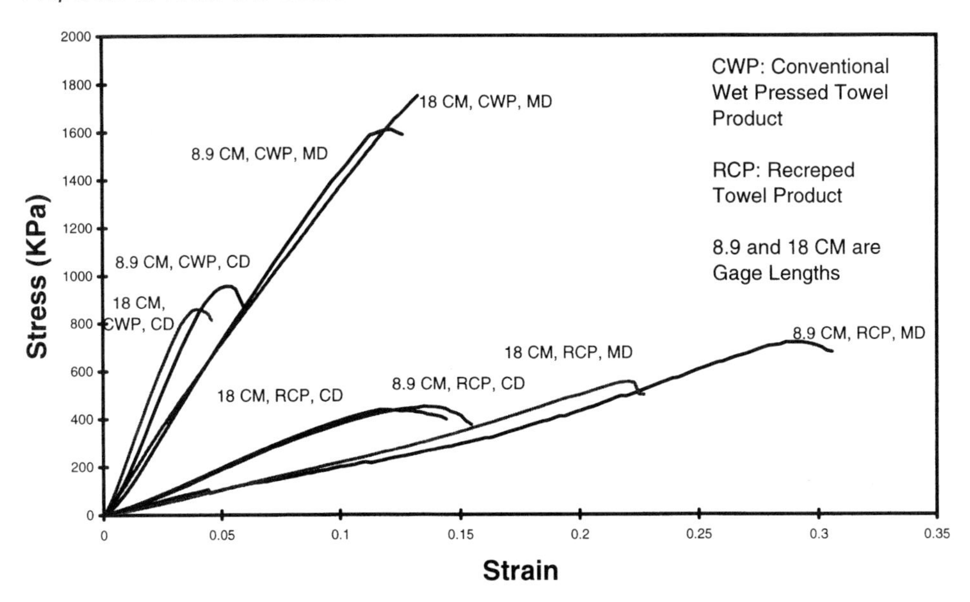

Fig. 8 Stress–strain curves for towel products. 8, 9, and 18 cm are gauge lengths.

possible to use 180 mm gauge length, conforming to the TAPPI standard procedure. For tissue samples with perforations, a gauge length of 89 mm (3.5 in.) or less is often used. A gauge length as small as 54.8 mm (2 in.) has been reported [18,27], and the tests were conducted at 25.4–101.6 mm (1-4 in.) per minute. With shorter spans, care must be taken to maintain the same strain rate as with the longer span. For instance, a 89 mm (3.5 in.) gauge length test should be conducted at 0.5 in./min (12.7 mm/min), and a 54.8 mm (2 in.) gauge length test should be tested at 7.1 mm/min (0.28 in./min). From Figs. 7 and 8, it can be concluded that tissue and towel grades exhibit higher stiffness in the cross direction than in the machine direction. The MD stretch is substantially larger than the CD stretch. It is important to specify the test conditions when reporting tensile testing results.

Sample Width Considerations It is a general practice to conduct tests on 3 in. wide specimens and then express the results as grams per inch. Sample width changes in general do not make much difference as long as the width is not too large in proportion to the gauge length. In the case of a 3 in. wide specimen, the load is divided by 3 to obtain load per inch. In one study to measure softness-related mechanical properties, a sample width of 101.6 mm (4 in.) was used [27]. However, a 25.4 mm sample width with a gauge length of 89 mm for tissue and a 180 mm gauge length for towels offer the best ability to conduct satisfactory uniaxial tests.

Tension Wrinkling Under uniaxial straining, the out-of-plane displacement that occurs is called tension wrinkling. The out-of-plane displacements are sinusoidal. This is due to CD buckling while the paper is under uniaxial tension. This CD compressive force arises from the clamped boundary conditions at the ends, which induce stresses other than pure uniaxial tension in the specimen. The resultant compressive stress may be very small compared to the axial tension but sufficient to cause tension buckling. The compressive stress is maximum at the center of the width and zero at the edges [28]. Due to this, it is reported that the tension buckling starts at the center of the width and propagates toward the edges. The mode of severity depends on the specimen geometry, specimen mechanical properties, grip boundary conditions, and applied tensile loading. For a constant width, thinner paper is much more prone to tension buckling than a thicker paper, other things being equal.

It has been shown that the use of a necked-down specimen provides a flat specimen at some distance away from the grips. For a thin rectangular specimen, on the other hand, tension buckling is found throughout the length of the sample. Seo et al. [28] tested 25 g/m^2 machine-made tissue samples at two different orientation distributions and recommend the use of necked-down specimens with a ratio of specimen width at the grips to width at the neck region of 2.5 of more to avoid wrinkles and induce a tensile stress in the specimen in both the elastic and inelastic range up to failure.

Number of Plies A special problem with tissue and towel samples is that some of them are two-ply and some are single-ply. It is general practice to test them as a product configuration, i.e., two-ply product is tested as two plies together. The reason is that the tissue samples are easily deformed and separating a two-ply product into two separate plies is likely to introduce stretching and deformation prior to the test. Sometimes more than four plies are tested together by combining several samples. In this case, however, it is possible to introduce different amounts of slack into each of the plies. As a result, different plies would be at different portions of their stress–strain curve and the combined result would be erroneous.

Kawabata Tensile Tester A series of four testers were developed by Kawabata to determine the mechanical properties of fabrics to predict "fabric hand." One of the testers is the KES-FB1 tensile and shear tester [29]. Some investigators have used this test to determine tensile properties to develop a correlation model for subjective softness of tissue [30]. A rectangular sample 5 cm long and 20 cm in width is clamped between the jaws in the tester, and the tensile load is applied in the length direction. Because of the large width of the sample, the transverse strain is assumed to be approximately zero, providing the conditions of "strip biaxial deformation." Strain rate is kept constant at 4.00×10^{-3} s. The strip is loaded to a maximum load of 500 g_f/cm (0.49 mN/m) and then unloaded. The loading–unloading curve is recorded. In the case of tissue testing, the maximum load has to be reduced substantially, perhaps by a factor of 5–10. A load of 50 g_f/cm has been used by Kim et al. [30]. Figure 9 shows a schematic load–deformation curve obtained by this test. From this curve the following quantities are calculated:

LT: linearity (nondimensional).
WT: tensile energy per unit area, measured in mN m/m^2 (g_f cm/cm^2)
RT: resilience (%)

LT is defined as WT/WOT, where

$$\mathrm{WT} = \int_0^{\epsilon_m} F\, d\epsilon$$

and WOT $= F_m \epsilon_m / 2$, where F_m = maximum force per unit width [0.49 mN/m (500 g_f/cm)], and ϵ_m = maximum strain at 0.49 mN/m (500 g_f/cm).

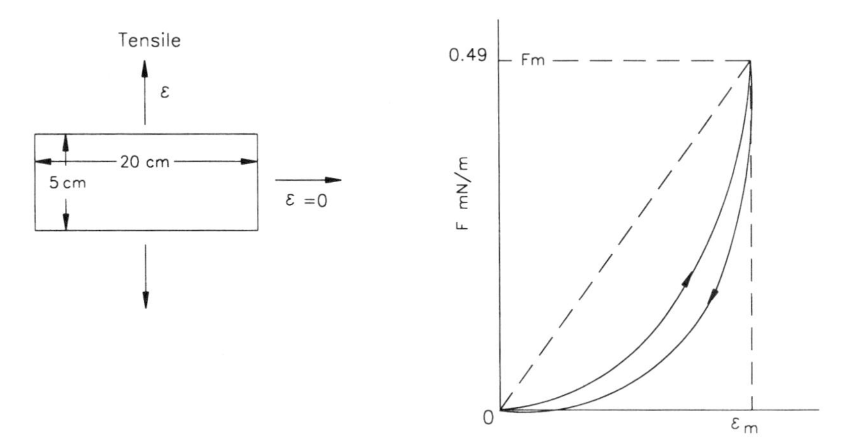

Fig. 9 Schematic cyclical tensile test curve from KES-FB1 tester. (From Ref. 29.)

$$RT = (WT'/WT) \times 100$$

where

$$WT' = \int_0^{\epsilon_m} F'\, d\epsilon$$

where F' is the force value during recovery and WT$'$ is the area under the recovery curve. A set of data for commonly available facial tissue was tested using all four Kawabata testers and was published by Brandt and Wagner [31].

Other Tests Ampulski et al. [27] reported a parameter called "total flexibility," which is actually a stiffness parameter that is useful in softness prediction. This parameter is calculated as follows. A tensile test with a sample 101.6 mm wide is conducted at a gauge length of 50.8 mm and a straining rate of 25.4 mm/min. The sample elongation is recorded when the load reaches 20 N/m, and the stiffness at this load is calculated by dividing 20 N/m by the strain at this load. The test is carried out in both the machine and cross-machine directions, and a geometric mean stiffness value is calculated.

In addition, the peak tensile loads in the MD and CD tests are recorded and a geometric mean tensile strength is computed. The geometric mean stiffness divided by the geometric mean tensile strength is another parameter that is recorded.

E. Lateral Compression Properties

Compressive Work Value "Sponginess" is part of the impression of total softness of a person who handles the towel and tissue products [18]. Compressive work value (CWV) is defined as the total work required to compress the surfaces of a single-ply sheet inward toward each other to a unit load of 125 g/in.^{2} (1.9 kPa). A standard tensile tester, Instron for example, can be used with compression platens. A 4 in. × 4 in. (101.6 mm × 101.6 mm) square sheet of sample is located between the compression plates and loaded at a rate of 2.54 mm/min (0.1 in./min) until the load reaches 1.2 N (125 g). The area under the compression load–deformation curve is the CWV value. It is claimed that this work is similar to the work done by a person who pinches the sample between the thumb and forefinger to get an impression of softness [10].

Compressive Modulus The compressive modulus is interpreted as the intrinsic resistance to compression of the material at a particular point on the stress–strain curve [18]. First, the differential change in deformation when the compressive load increases from 2.94 to 4.91 N (from 300 to 500 g) and the caliper at a load of 400 g (3.92 N) are measured. The compressive modulus is calculated by dividing 1.96 N (200 g differential load) by the product of differential compressive deformation and the sample surface area and multiplying by the sample caliper at 3.92 N load (400 g) measured from the load–deformation curve. Lower compressive modulus values are generally desirable in tissue and towel products.

Kawabata KES-FB3 Compression Tester In this Kawabata tester [29] a specimen 2.5 cm $\times$ 2.0 cm is compressed between two circular steel plates that each have an area of 2 cm^2. The loading rate is maintained at 20 μm/s until the compressive force reaches 4.91 kPa (50 g_f/cm^2), and the unloading part of the test is carried out at the same rate. For towel and tissue applications, the maximum load should be reduced, consistent with soft platen measurement discussed earlier [19]. Kim et al. [30] used 1.96 kPa (20 g_f/cm^2) satisfactorily for tissue samples. A schematic loading–unloading compressive curve is shown in Fig. 10. The following values are calculated:

LC: linearity (unitless)
WC: energy required for compression, Nm/m^2 ($g_f\,cm/cm^2$)
RC: resilience (%)

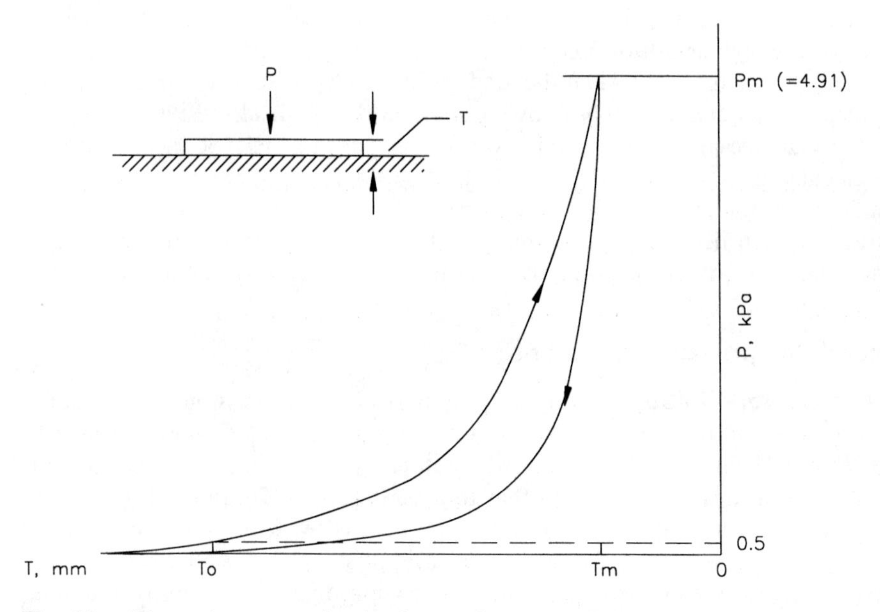

Fig. 10 Schematic compression load curve from a KES-FB3 tester. (From Ref. 29.)

Following the definitions of characteristic values calculated in the case of a tensile test, the definitions of the compression test are given below:

$$\mathrm{LC} = \frac{\mathrm{WC}}{\mathrm{WOC}}$$

$$\mathrm{WC} = \int_{T_m}^{T_0} P\, dT$$

and

$$\mathrm{RC} = \frac{\mathrm{WC}'}{\mathrm{WC}}$$

where

T = thickness of the specimen
T_0 = thickness of the specimen at a pressure of $0.5\,\mathrm{g_f/cm^2}$ (0.0491 kPa)
T_m = thickness of the specimen at the maximum pressure of $50\,\mathrm{g_f/cm^2}$ (4.91 kPa)

$$\mathrm{WOC} = P_m \frac{T_0 - T_m}{2}$$

$$\mathrm{WC}' = \int_{T_m}^{T_0} P'\, dT$$

where P' represents the force measured on the unloading curve.

As an additional measured value, the thickness of the sample is taken as the value of thickness measured when $P = 0.5\,\mathrm{g_f/cm^2}$. Although this thickness measurement may be applicable for textiles, a soft platen caliper is desirable for low density paper grades.

Out-of-Plane Ultrasonic Measurements The out-of-plane elastic modulus (C_{33}) has been measured for tissue samples [19], and a correlation between the subjective softness has been attempted. The method consists of placing a single ply of tissue cut into a rectangle 4.5 in. × 8 in. in a vertical gap between two ultrasonic transducers. The apparatus consists of two neoprene-faced polyvinylidene fluoride (PVDF) transducers to couple ultrasonic energy into one surface of the sample and detect it on the other side. The transducers are precisely aligned and mounted in a caliper gauge. This allows simultaneous measurement of time of flight and caliper at a repeatable time after the neoprene face comes into contact with the sample. The contact pressure is maintained at 20 kPa for tissue. After waiting for a predetermined amount of time, one transducer is excited with a 1.5 MHz single-cycle sine wave pulse. This produces disturbances that propagate through the sample into the other transducer and on the other side. The resulting electrical signal is amplified and digitized at a rate of 100 MHz, and the data are stored. The cross-correlation function between this signal and the one obtained when an aluminum foil is placed in the vertical gap is calculated.

With the known transit time through the aluminum foil, the time of flight through the sample is calculated. The caliper is determined from the output of a linear variable differential transformer (LVDT). Given the time of flight and the caliper, the velocity of sound through the sample can be computed [V_{ZD} = caliper/(time of flight)]. The out-of-plane bulk elastic stiffness (C_{33}) can be computed from $C_{33} = V_{ZD}^2\rho$ [32], where ρ is the density. The ZD (Z-direction) acoustic impedance Z is defined as ρV_{ZD} which is simply basis weight divided by time of flight. Another parameter called the *attenuation coefficient* A is calculated as follows: First, the ratios of the amplitudes of the Fourier components of the signal through the foil to those through the sample are calculated. These ratios are weighted according to the square of the amplitude through the sample and averaged. A is defined as 20 times the base 10 logarithm of the average. The attenuation coefficient has been hypothesized to relate to the interfacial effects for tissue at 20 kPa.

F. Flexural Rigidity and Bending Modulus

Flexural rigidity is an important property of towel and tissue products; it is a measure of the resistance to bending. Its value is the product of the elastic modulus and the moment of inertia of the cross section of the sheet under consideration. In a pure bending test, flexural rigidity is defined as the couple on either end of a strip of unit width bent into unit curvature in the absence of any tension [33].

The Shirley Stiffness Tester ASTM Standard Method 1388 is performed with the Shirley stiffness tester. In this method, a strip of fabric or the test material is slid parallel to its long dimension so that its edge projects from the edge of a horizontal surface. The length of overhang is measured when the tip of the test specimen is depressed under its own weight to the point where the line joining the tip to the edge of the platform makes an angle of 41.5° with the horizontal. One-half of this length is the bending length of the specimen. The cube of this quantity multiplied by the weight per unit area of the specimen is defined as the flexural rigidity. The apparatus consists of a horizontal platform with a minimum area of 1.5 in. × 6 in. (38 mm × 150 mm) and having a smooth, low friction, flat surface such as polished metal or plastic. A leveling bubble must be incorporated in the platform. It also includes an indicator inclined at an angle 41.5° below the plane of the platform surface and a weight consisting of a metal bar not less than 1 in. × 6 in. (25 mm × 150 mm) by about 1/8 in. (3 mm) thick. Both the MD and CD flexural rigidities are determined, and a geometric mean of the MD and CD values is reported as the overall flexural rigidity.

Kawabata Bending Tester This is the KES-FB2 tester in the Kawabata tester series [29]. In this test the sample is clamped and bent in such a way as to induce pure bending. This is accomplished by ensuring that the specimen has a constant radius of curvature throughout at any given instant during the test. One end of the sample is clamped, and the other end is rotated to produce pure bending deformations. The torque induced by the bending deformation is measured by an LVDT as a very small amount of rotation angle of the rod. A complete cyclical test of loading and unloading is carried out. A sample 2–20 cm long and 1 cm wide is bent into an arc of a circle in the direction of the 1 cm width. The sample is gradually bent between curvatures $K = 2.5$ and $+2.5\,\text{cm}^{-1}$ with a rate of change of curvature at $0.5\,\text{cm}^{-1}\,\text{s}^{-1}$. The torque induced by the bending deformation is measured. Figure 11 shows a schematic of the bending–curvature relationship obtained using the FB2 tester. The following characteristics are measured from the curve:

B	bending rigidity per unit length, mN m/m^2 (g$_f$ cm^2/cm)
2HB	moment hysteresis per unit length (g$_f$ cm/cm)

B is defined as the slope between $K = 0.5$ and 1.5, and 2HB is measured as shown in Fig. 11, as the difference in moment M at a curvature between 0.5 and 1.5 cm^{-1}, for the loading and unloading curves.

G. Surface Characterization

There are two primary characteristics of a surface: the topography, which is a purely geometrical attribute, and the nature of interaction of the surface with other surfaces, which is a physical parameter. The geometrical parameter can be measured in many different ways: A profilometer, which is a general-purpose surface roughness measuring instrument instrumented with data acquisition capability, can be used; a cross-sectional micrograph can be obtained; and subsequent digitizing of the position of the surface fibers relative to a datum can provide a surface profile data set. Results can be plotted as elevation versus position.

The interaction of one surface with others can be characterized by measuring the friction coefficient or by measuring the number of fibers projecting from the surface of the sheet to give a "velvet"-like feel, called the free fiber ends (FFE) [16] after a controlled interaction with another surface or an inferred parameter from the surface profile, called the human tactile response (HTR) [16]. In this section we will discuss methods for measuring the surface characteristics of tissue and towel grades.

Profile Measurement Surface profile or roughness is a commonly measured property of many engineering surfaces. Metal surfaces are machined to obtain

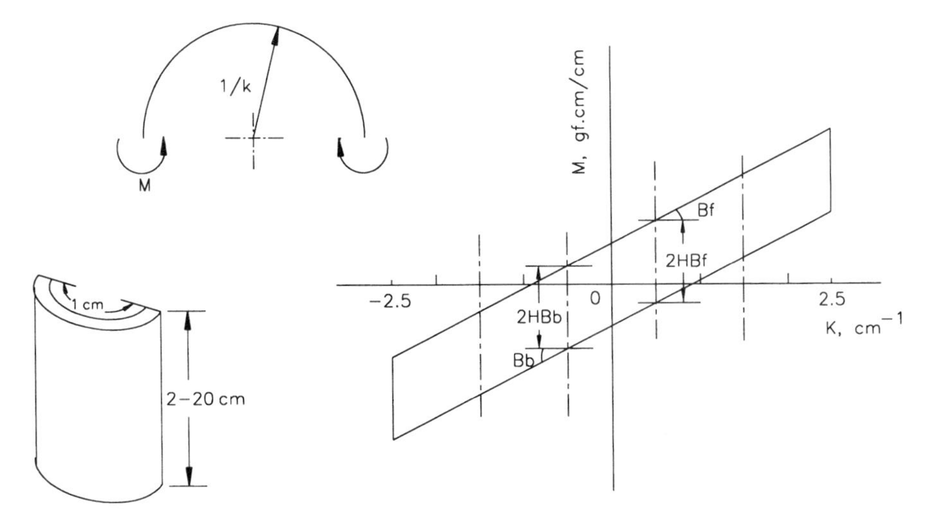

Fig. 11 Bending–curvature relationship from an FB2 tester. (From Ref. 29.)

desired levels of surface smoothness. A common method for measuring hard surfaces is to use a profilometer. The same basic arrangement can be used for tissue paper structure. The only difference is that the tissue surface is highly deformable and the stylus type and shape, the load on the stylus during the test, and the speed of traverse on the surface must be optimized for tissue. Ampulski et al. [27] used a Gould Surfanalyzer with a diamond stylus probe tip with a radius of 0.0127 mm and a normal force of 2.45 mN (0.25 g). A total length of 3.5 mm is scanned. During scanning, the displacement of the stylus with respect to a datum is recorded. An FFT algorithm is used to identify the periodic components in the signal. Quantities such as the RMS (root-mean-square) roughness value can easily be calculated. Another design that integrates the data acquisition and signal processing specifically for tissue and fabrics, called the mechanical stylus surface analysis (MSSA) [34], uses an interchangeable stylus (2.54–800 μm radius) with a normal contact force of 0.5 mN to increase the sensitivity of measurement. A raw data profile obtained by Allen et al. [34] with a 2.54 μm diamond stylus moving at 254 μm/s for a tissue sample is shown in Fig. 12. From the data, RMS values can be calculated and an FFT analysis can be carried out to separate the crepe-related surface profile component (and crepe frequency) and intrinsic surface roughness.

An interferometer non-contact surface profile measurement system was reported by Knapp and Lampman [35]. In this technique, an interference microscope with a white light source is used to measure the degree of fringe modulation, or coherence. The entire area of the sample is scanned along scan lines, and a three-dimensional image of the surface topography is reconstructed with the use of com-

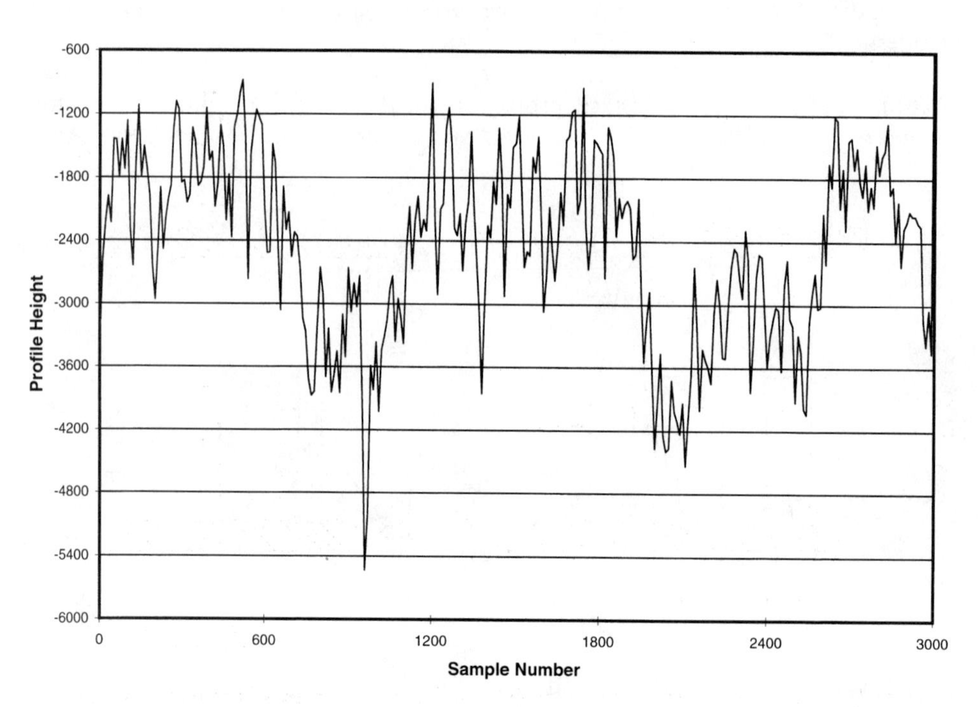

Fig. 12 Raw surface data profile for a tissue sample. (From Ref. 34.)

puter software. Standard roughness characterization parameters can be extracted from the data. The instrument is capable of measuring RMS roughness ranging from 0.1 nm to 500 μm in measurement steps of 100 μm. Although this method has been applied to determine the profile of coated papers, no data are available from a tissue application.

Kawabata Surface Profile Test The Kawabata surface profile measuring unit, called the KES-FB4 tester, is both a profile and surface friction measurement system. The sample is draped over a rectangular platform, one end is attached to a rotating drum and a weight is attached to the other. As the drum turns slowly, the sample moves from one side to the other. The stylus is made out of a piano wire, 0.5 mm in diameter, bent in the shape shown in Fig. 13 and is applied with a reduced pressure of 5 g. The vertical movement of the stylus is output to a chart recorder. A schematic representation of the results obtained is shown in Fig. 14. From the results, the mean deviation of surface roughness, SMD, is calculated as

$$\mathrm{SMD} = \frac{1}{X}\int_0^X \left|T - \bar{T}\right| dx$$

where

x = displacement of the contactor on the surface of the specimen
X = 2 cm is the standard length of measurement
T = thickness of the specimen at position x (a measured output)
$\bar{T}$ = mean value of T measured in the machine and cross-machine directions.

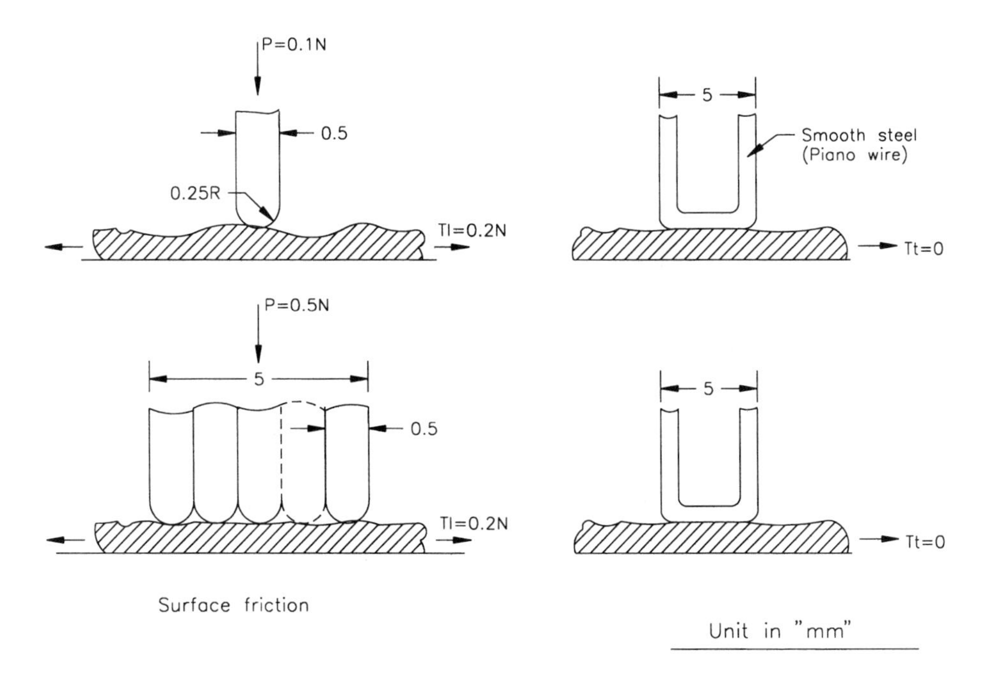

Fig. 13 Kawabata surface tester probes. Indicated dimensions is millimeters. (From Ref. 29.)

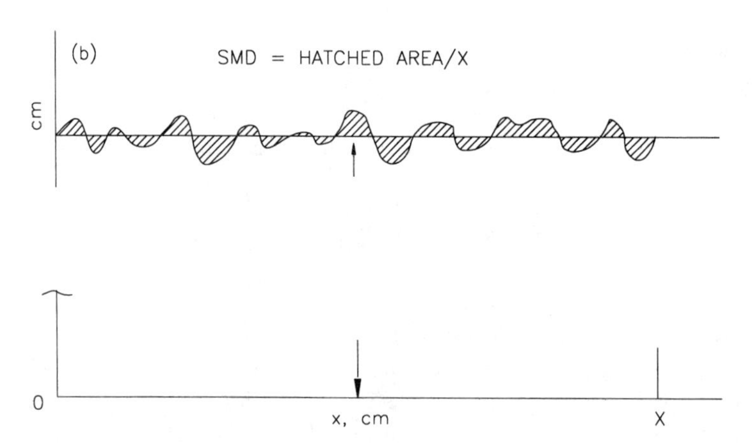

Fig. 14 Schematic Kawabata surface profile results from FB4 tester. (From Ref. 29.)

Kawabata Friction Test This test is carried out on the FB4 surface tester. The stylus for the friction test is made up of the same bent piano wire as in profile measurement, except that 10 such wires are bunched together (Fig. 13). The stylus is attached to the bottom of a long rectangular plate that is connected to a horizontal load cell. The stylus is placed on the sample, and additional compressive force of 0.5 N (50 g_f) is applied by a deadweight. For tissue property measurement, it has been found that the standard Kawabata friction probe is not suitable. A circular glass frit (40–60 μm), 2 cm in diameter, has been used successfully with a normal force of 12.5 g [27]. The sample moves past the probe at 0.1 cm/s for a total distance of 2 cm. The signals from both the surface and the friction tests are passed through a high-pass filter, and signal wavelengths less than 1 mm are allowed to pass. Figure 15 shows a schematic output from the friction instrument. From the results, the mean friction

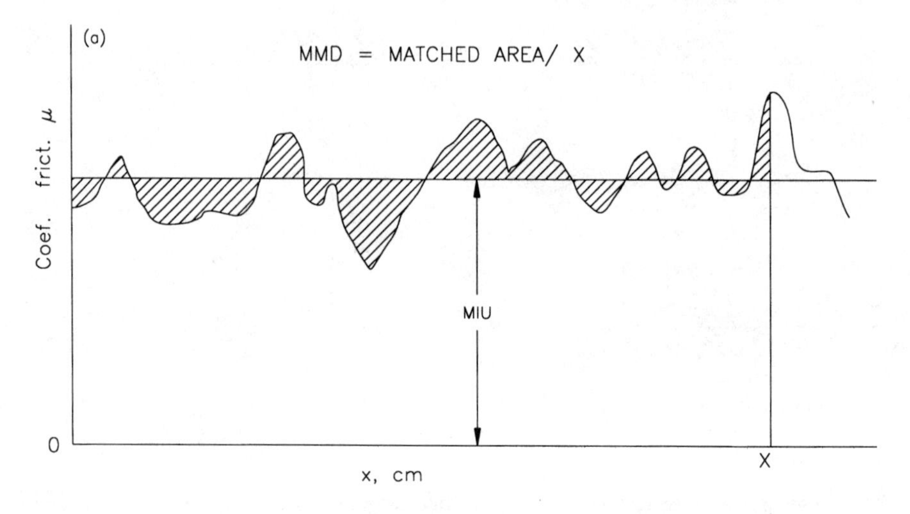

Fig. 15 Schematic friction coefficient output from the FB4 tester. (From Ref. 29.)

coefficient (MIU) and the mean deviation of the coefficient of friction (MMD) are calculated as follows:

$$\mathrm{MIU} = \frac{1}{X}\int_0^X \mu\, dx$$

$$\mathrm{MMD} = \frac{1}{X}\int_0^X |\mu - \bar{\mu}|\, dx$$

where

$x =$ displacement of the contactor on the surface of the specimen
$X = 2$ cm is the standard length of measurement
$\mu =$ friction coefficient determined by dividing the tangential force signal by the normal compressive force component
$\bar{\mu} =$ mean value of μ

The measurement is made in the machine and cross-machine directions. A modified version of the instrument in which the tissue samples are simply laid on a table and held down by a rectangular metal frame is marketed as the KES-SE tester. This tester is better suited for tissue and towel samples than the FB-4 tester.

Free Fiber Ends The free fiber ends (FFE) test method is described in a patent issued to Carstens [16] that was developed from a need to quantify the human tactile response of velvety smoothness and softness, especially for bathroom tissue and facial tissue products, as a means to justify the use of layered structure with hardwood fibers on the Yankee side of the structure. When creped, the several short hardwood fibers tend to be freed at one end and standing up from the surface, providing a "velvetlike" tactile perception when a consumer rubs a finger over the sheet, and the softness perception increases with increases in short papermaking fibers on the surface of the sheet. The test method consists of cutting a 1 in. wide sample at 45° to the machine direction, gently brushing the surface to raise the surface fibers with a force corresponding to what a human subject would apply to feel its softness, and counting the number of fibers that are projecting from the surface by more than 0.1 mm. After the sample is cut, it is placed on a stage. Either the stage or a sled that is set on the sheet will move, causing relative motion between the surfaces. The sled surface that contacts the paper surface is covered with a material like a corduroy fabric or chamois leather. The sled is dragged on the surface to cause brushing. The brushed sample is folded over the edge of No. 1.5 glass slide cover, and the assembly is sandwiched between microscope slides. The sample set up is shown in Fig. 16. The sample is illuminated from the bottom and viewed through an optical microscope, preferably one that has an adjustable focus. The free fibers that are over 0.1 mm in length are counted over a sample length of 1.27 cm. The resulting number of free fibers is called the free fiber ends index. Only visible free fiber ends are counted. The loading or the nature of the sled material, the brushing rate, and the influence of fiber loss due to brushing are not available. The pressure during pressing is maintained at 0.5 kPa (5 g/cm^2), which is less than half of the thumb-and-forefinger pressure applied by a human who is asked to feel a tissue for evaluating softness. The purpose of the brushing is to orient the fibers to facilitate counting and not to cause any additional free fibers to appear. FFE results can vary

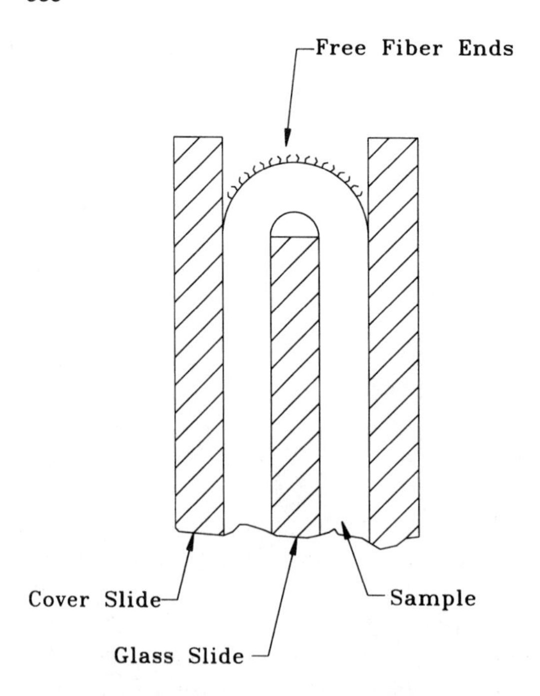

Fig. 16 Sample configuration for FFE index determination. (From Ref. 16.)

over a wide range, from about 30 to as high as 200. FFE index value increase with the increase in fiber consistency during creping, especially above 95%. The standard deviation of the test results is generally high. Hence, a very large number of samples have to be tested in order to get statistically meaningful results.

Human Tactile Response (HTR)—Texture This test is carried out in a profilometer with an additional feature of zigzag scanning across the surface of the sample [16]. A hemispherical stylus with a diameter of 508 μm (0.02 in.) counterbalanced to provide a normal loading of 1.2 kPa (12.4 g/cm^2), which is in the range of the pressure applied by humans who grab the sample between their fingers to evaluate softness, is moved back and forth on the sample while the sample holder is moved perpendicular to the movement of the stylus, thus causing a zigzag motion on the sample. The stylus is moved at a range of 2.54 mm/s (0.1 in./s) while the orthogonal motion is carried out at 0.0025 in./s for a test period of about 200 s, giving a total traversed distance of about 51 cm (20 in.). The data are converted to the frequency domain and plotted as stylus amplitude measured in mils versus texture measured in cycles per inch. The area under the curve between frequencies of 10 and 50 cycles/in. and above the datum amplitude of 0.1 mils (25.4 μm), expressed in mils-cycles per inch, is referred to as the HTR index. A frequency of 10–50 irregularities per inch was chosen because it was found to be the most significant range of frequencies for human tactile response [16]. A sufficiently large number of samples must be tested to get statistically meaningful results. Samples are cut off-axis at 45° to the machine direction. In the case of single-ply tissue, both sides are measured and the results are averaged. In the case of two-ply sheets, the Yankee sides of each ply is individually tested and the results are averaged. As the moisture at the time of creping decreases, the HTR value also decreases. A small HTR index indicates a softer tissue; values range from 3.0 to 0.1 or less.

H. Other Physical Characteristics

In addition to the properties discussed so far, there are some other properties specific to the towel and tissue grades that are also important and therefore also measured. Some of these properties are discussed in this section.

Lint Resistance Lint resistance is the ability of the fibrous product and its constituent webs to bind together under end use conditions, such as in mirror wipe, when they are either wet or dry. The higher the lint resistance, the lower the propensity of the web to lint [13].

Dry Lint Dry lint is measured using a Sutherland rub tester, a piece of black felt, a 1.81 kg (4 lb) weight, and a Hunter color meter. The Sutherland tester is a motor-driven instrument that can stroke a weighted sample back and forth across a stationary sample. The piece of black felt is attached to the weight, and the tester rubs or moves the weighted felt over a stationary tissue sample for five strokes. The Hunter color L value of the black felt is determined before and after the rubbing. The difference in the two Hunter color readings constitutes a measurement of dry linting.

Wet Lint The general wet lint methodology consists of passing a tissue sample into a nip where one of the rolls is moistened by water. Lint from the sample is transfered to the steel roll, which is continuously moistened by a water bath. The continuous rotation of the steel roll deposits the lint into the water bath. The lint is recovered and then counted by a suitable method. The GFL Fluff Tester Model No. SE 56, manufactured by Lorentzen & Wettre, Stockholm, Sweden, has been used for wet lint testing in facial tissues [36]. The sheets are passed into a nip formed by a steel roll and a rubber roll with a nip loading of 490 N/m. The rubber roll is 40 mm in diameter and 180 mm in length, and the bottom steel roll is 92 mm in diameter and 190 mm in length. The steel roll is partially submerged in 250 mL of water. The rolls are turned at a peripheral speed of 0.4 m/s. As the samples are passed through the nip, lint from the sample is transferred to the steel roll and washed off by the water bath. A doctor blade controls the amount of water on the steel roll. Two 5 mL samples are analyzed using a Kajaani FS 100 fiber analyzer. The entire procedure is repeated 20 times to get a statistically valid result, and an average lint count per sample or per unit area of the sheet is computed. Kajaani offers the additional ability to determine the fiber length distribution along with the count.

Wet Strength Measurement High wet strength is desirable in towel grade paper due to the nature of its end use, which is typically to wipe up spills or wipe wet surfaces such as windows and mirrors. There are at least two known approaches to determining the wet strength of towel and tissue grades. The first uses the standard tensile tester, and the second employs a specialized form of burst test, called the ball burst test.

Sponge Wet Tensile Test In this test, a towel or tissue sample is mounted between the jaws of a tensile tester as in the standard dry tensile test described earlier.

The sample is grabbed by a pair of wet sponges in the middle of the testing span with the help of a pair of tongs to which the sponges are attached. The sample is allowed to sit for a predetermined amount of time, and a tensile strength test is conducted in the machine and cross-machine directions.

Ball Burst Test This test is a modification of the burst test method. In this test a steel ball, 15.87 mm (5/8 in.) in diameter, is attached to the load cell in a tensile tester by means of an extension rod. The sample of paper is clamped between two concentric annuli by means of pneumatic pressure. The inner diameter of the clamp is 89 mm (3.5 in.). The circular disk of exposed sample is pushed against the ball at its center, and the load–deflection response is recorded [37]. Analysis shows that the ball burst strength can be satisfactorily predicted by the product of the geometric mean of the MD and CD strengths multiplied by the CD stretch. In the case of tissue, a temporary wet strength is desirable. Whereas wet strength is necessary for functionality, fugacity is essential for disposal. Hence, the duration of wetting is a critical parameter of the test that must be measured and controlled.

Smear Index The smear index is another aspect of lint resistance, where the ability to wipe a glass surface without leaving lint is the desired property to be measured [36]. First, a glass slide 75 mm × 50 mm is thoroughly cleaned and dried in a dust-free environment. The dried glass slide is placed on a horizontal surface, and a metal template, 50.8 mm × 76.2 mm (2 in. × 3 in.), with a 2.54 mm (1 in.) hole in the middle is placed over the glass slide. A 50.8 mm × 50.8 mm (2 in. × 2 in.) tissue sample is placed over the hole in the template. The narrow end of a No. $5\frac{1}{2}$ rubber stopper is placed over the hole in the template and resting on the tissue sample. A 5 lb weight is placed over the rubber stopper and allowed to stand for 3 min. After 3 min the 5 lb weight and the rubber stopper are turned 180° clockwise followed by 180° counterclockwise. Care must be taken not to exert any additional normal load. The weight, sample, and template are lifted off, resulting in a smeared glass slide. The slide is illuminated with a slide projector, with all other room lighting off, and the reflected light is detected by a video camera. The slide is held at 20° to the horizontal, and the light beam hits the slide at a 20° angle to the horizontal. The video camera is directly above the slide. The video image is processed by an image analysis system to calculate the smear index, which is a brightness value between 0 and 63 (a 64-value gray scale).

I. Softness Properties

Softness is perhaps the most important property governing consumer acceptance of tissue grades, especially in sanitary applications. We may begin by defining softness as a tactile sensory response to a texture that is pleasing to touch and handle or a feeling of delicate texture lacking firmness. This can be tested by crumpling a tissue by hand and stroking the surface with the finger. The combined sensory response is summarized as the psychophysical measure of softness. Softness has been traditionally measured using human panelists and consumers. The softness determined in this manner is called subjective softness. Later in this section we will discuss attempts to relate subjective softness to measurable physical properties. Softness determined by using a combination of physical properties is called objective softness. The relationship between subjective softness and objective softness is dependent upon several factors, especially the method used for obtaining subjective softness. These studies can be grouped for discussion with the idea of developing a softness model for towel and tissue paper products.

Subjective Softness This subjective property is measured in any of several ways: by a sensory panel consisting of mill experts, a consumer panel consisting of a selected group of consumers brought to a central location for conducting the tests, or by consumers carrying out a home use test where products are given to them to use and their perceptions are recorded by means of a questionnaire. There are two major approaches used in conducting subjective softness experiments.

Magnitude Estimation This method is based on the idea that a panelist can translate the intensity of perception into a number. Each panelist judges the softness or its components in relation to a standard and assigns numbers on a scale. One of the softest in the sample set of tissues can be chosen as the standard and assigned the number 100, and the softness of test samples can be estimated as a percentage of the softness of the standard [38]. For instance, if a test sample is perceived to be half as soft as the standard, it would be given a rating of 50%. Panelists are prevented from seeing the sample during the test to avoid visual cues.

Another approach to magnitude estimation is described by Winakor and Kim [39] in evaluating "fabric hand," which is the equivalent of softness in textiles. In this case judges are asked to handle one sample at a time behind a screen so that they can not see it but can handle it freely. Adjective pairs such as limp–crisp, scratch–silky, and flexible–stiff are developed, and one adjective is chosen as the left with an assigned value of 1 and the other is the right adjective with a value of 99. After handling, the panelists indicate on a 99-point certainty scale which of the two polar adjectives better described the fabric hand by assigning a whole number between 1 and 99, 1 for strong agreement with the left adjective, 99 for strong agreement the right adjective, and 50 for uncertainty as to whether the left or the right adjective better describes the hand of the test fabric. Scores closer to the extremes, 1 or 99, reflect greater certainty about the choice than do those close to 50.

The data are transformed to normal deviates by the method of Wolins and Dickinson [40] to spread out the tails of the scale and push together the middle scores. Thus, scores at the extreme ends of the scale receive higher weighting, because the panelists were more certain of their choices; scores near the middle receive less weight, reflecting a lower degree of confidence. Transformed response ranges from −2.33 to +2.33, with 0 representing the midpoint of the scale using the PROBIT procedure in the SAS (Statistical Analysis Systems) software [30].

Paired Comparison Method In this method, a pair of samples A and B are presented to panelists who are asked to judge their relative softness. In each pair, one is the reference and the other is the test sample. This is done in a round-robin fashion so that the reference sample in one test will have its turn as test sample in another pair. If a grade of 0 is assigned, then the two samples are judged to be equally soft; a grade of +1 indicates that sample A may be a little softer than sample B, and a grade of −1 is given if sample B may be a little softer than sample A; a grade of +2 is given if sample A is judged to be surely softer than sample B, and a grade of −2 is given if sample B is judged to be surely softer than sample A; a grade of +3 is given if sample A is a lot softer than sample B, and a grade of −3 is given if sample B is a lot softer than sample A; a grade of +4 is assigned to sample A if it is a great deal softer than sample B, and −4 is assigned if sample B is softer than sample A. The resulting data from all judges and all sample pairs are then pair-averaged and rank ordered according to their grades. An arbitrary softness value of 0 is assigned to a sample chosen as standard. Other samples then have plus or minus values as determined by their relative grades with respect to the zero base standard. The resulting values are called the samples' panel softness units (PSUs) as a measure of softness [16]. A PSU change of 0.2 represents a significant difference in subjectively perceived softness.

Another scheme, described by Back [41] uses a ratio scaling procedure where several standards are prepared to represent a scale from 20 to 86. Test samples are compared with selected standard samples in a pairwise comparison fashion. The results are treated as if they belong to a ratio scale and are converted to numerical values by interpolation.

Objective Softness Measurements The subjective softness tests can provide relative ranking of tissue samples between different products over a wide range of softness. However, in a production environment, for quality control, it is required to measure small changes in softness and make process adjustments to control the property. The panel ranking may be insensitive to small changes. Even if the differences are measured, subjective softness does not offer insight into what process parameters need adjustment. To overcome this difficulty, there have been several attempts to relate softness to measurable physical quantities, which in turn can be related to process parameters. These attempts to develop an objective measure of softness can be grouped into two categories: the direct softness measurements, in which a single test method is used in an attempt to measure a parameter that can be correlated to softness, and the softness component approach, in which components of softness are measured and a model is developed to relate these physical properties to the subjective softness.

Direct Softness Measurements There have been several attempts to invent a single "softness-measuring apparatus." Some of the inventions are discussed in this section, but none of these tests has withstood the test of time as a softness predictor.

Handle-O-Meter (TAPPI T 498) This instrument measures the force necessary to push a sample into a 6.35 mm (0.25 in.) slot. It is used to conduct both MD and CD tests. The instrument is sensitive to surface characteristics such as crepe undulations, embossing, and friction between the sheet and the metal surface as well as bending characteristics of the sample and gives an indication of the handle, softness, and drape [42]. A 114.3 mm × 114.3 mm (4.5 in. × 4.5 in.) sample is cut for each test. MD tests are conducted by aligning the sample machine direction parallel to the Handle-O-Meter blade, and CD tests are conducted with the cross-machine direction parallel to the blade. The sample is placed across the slot, and the force required to push the sample into the slot is measured. Results are expressed in grams [18]. Lower Handle-O-Meter values indicate less stiff, smoother samples and hence point toward higher softness and drape. Studies on the performance of the Handle-O-Meter have been reported [43,44]. It was concluded that the Handle-O-Meter "instrument error" is no more than 4% and the coefficient of variation was not significant. However, roll-to-roll variation and position along roll variabilities are significant. Further, the result is sensitive to slot width. This test is not widely used in the industry, because it does not offer any more information than that of a panel test for process parameter changes.

Clark's Softness Tester In this tester [45], a strip of paper is gripped in a steel-to-steel nip at one end and is free at the other. The two rollers forming the nip are mounted on a frame that can be rotated about a horizontal axis, parallel to the length of the nip. In a given position where the nip is pointing upward, the paper sample falls over in one direction due to its weight and flexibility. The nip is rotated through 90°, and the sample is observed. If the sample still continues to fall over in the same direction as before, the length for the overhang of the sample is reduced until, when the nip is rotated through 90° the sample would fall over equally on either side of the nip. Under these conditions, the overhang length is measured from the nip to the tip of the sample. In addition to this length L, the caliper and basis weight are used to calculate a softness value as

$$\text{Softness} = 10 \times 10^6 \frac{\log(t+1)}{L^3 W}$$

where W is the basis weight, t is the caliper, and L is the overhang length from the tester. The test does not consider the surface effects on softness and is purely a stiffness-based parameter.

Brown Softness Tester This test method [46] consists of a glass beaker and a graduate that fits into the beaker, like a cylinder and a piston. Two sheets of paper 0.228 m × 0.3048 m (9 in. × 12 in.) are separately crumpled loosely in the hand just enough to be stuffed into the beaker. The graduate is then rested on the paper in the beaker and weighted so as to crush, within 1 min, the wad of paper sufficiently to bring the base of the graduate to within 25–35 mm from the bottom of the beaker. The softness is expressed in arbitrary units as one-thousandth of the height in millimeters of the was of paper after being loaded for 1 min, multiplied by the weight in grams required to bring the height within the stated range of 25–35 mm. This instrument was tested by Hollmark [38] and was found to achieve a correlation coefficient of 0.97 when compared with subjective values. However, the variability induced in sample preparation (crumpling) was too high, and the agreement between users was very poor.

Other Methods Several attempts have been made to predict softness by mimicking the subjective evaluation methods. In one method [47], a sample of tissue is placed over a knoblike head containing a sensitive microphone, and another sample is placed on a similar head facing the first. The samples are rubbed against each other with a predetermined and known degree of force while the "amount of sound" produced is measured on a meter. The sound emitted is amplified and recorded. The intensity of sound level is correlated with softness perception.

Drape measurements is very important in woven fabrics. Drape is the property of textile materials to conform to a shape when the sample is spread over an object and subjected to its own weight. However, drape alone does not satisfactorily include all the critical tissue softness-related attributes. Hence, force drape methods were studied for applicability in tissue softness measurements. Two of the force drape testers are the C. H. Dexter softness tester and the Fabricometer [38]. In the Dexter test method, the specimen is draped around the top of a rod and forced through her annulus created between the rod and a circular opening. The softness is related to the work done in forcing the specimen through the annulus. An instrument similar to the

Fabricometer was studied by Wahren and Nilsson [48]. A circular sample 240 mm in diameter is held in the center in between two circular plates that are 40 mm in diameter. The plates are attached to a rod mounted on a load cell. Another plate with a 55 mm diameter hole in the middle is moved downward over the 40 mm plates and the sample at a rate of 250 mm/min. The maximum force transmitted is recorded. Correlation with subjective bulk softness was found to be 0.97.

Measurements of Softness Components Another approach to objective softness measurement is to identify the components of softness, measure them individually, and construct a softness model that combined these parameters to predict the overall subjective softness. This approach has been successful in predicting the softness of tissue. It has been widely accepted that softness can be separated into two primary components: bulk softness and surface softness [49,50]. Watching a jury of panelists, Bates [51] concluded that the panelists' fingers and hands were measuring the amount of force necessary to cause the material they were feeling to deform within elastic limits, which is a measure of bulk softness. Stroking the surface with the finger contributes to the feeling of surface texture. The bulk softness and flexibility are closely related [50], and softness should be inversely related to bending stiffness or flexural rigidity. The surface softness consists of three components, namely, contour, friction, and surface impression [52]. Physical properties that are important are compressibility, extensibility, resiliency, density, pliancy, and surface characteristics.

In the field of psychophysics, a generalization can be made regarding all human senses. Perception is related to the stimulus by the power function [49]

$$I\ (\text{perceived}) = kS^n$$

where I is the intensity of sensation, S is the intensity of the stimulus, and k and n are numerical coefficients. The coefficient n will vary, depending on the human sense being employed and the type of sensation being perceived. This generalization can be used to develop a softness model in terms of its components.

Bulk Softness Measurement Bulk softness of tissue is the perception of softness obtained when a sheet is crumpled between the hands [53]. This property is dependent upon the flexural rigidity (or bending stiffness) that can be determined by the test methods discussed earlier. However, the fact that high bulk materials are perceived to be soft indicates that the thickness may not strongly influence the bulk softness perception, although bending stiffness will be influenced. Hence, an attempt was made to measure the tensile stiffness (per unit width, defined at Et, where E is the elastic modulus and t is the thickness) and evaluate the correlation to bulk softness perception [53]. When evaluating softness the MD and CD stiffness should be measured and a geometric mean of the stiffness ($S_T = \sqrt{S_{\text{MD}} S_{\text{CD}}}$) should be calculated. This quantity was shown to correlate well with the bulk softness perception through a power law model:

$$R_1\ (\text{perceived}) = kaS_T^m$$

where R_1 is the perceived bulk softness. For the set of products used in the study, it was found [50] that $a = 99$ and $m = 0.36$ gave a coefficient of correlation of 0.86.

Surface Softness Measurement Surface softness is the softness perception obtained when the sample is touched or the fingers gently brush over it. This perception is related to two components, the surface topographical smoothness and the surface friction characteristics. A velvetlike material with short, smooth, and flexible fibers on the surface will be perceived as soft. The surface profile can be measured by several methods and instruments discussed earlier. The surface friction properties can be assessed by the Kawabata surface analyzer [29]. A modified phonograph stylus was used by Hollmark to obtain a surface softness parameter [53]. The HTR index discussed earlier is another surface profile measurement parameter.

Softness Model Development With the ability to measure the two main components of softness, bulk softness and surface softness, it is possible to obtain a predictive model for the overall softness by using the power law model [53]

$$\text{Softness} = aS_T^m L^n$$

where L is the surface softness measurement obtained from the surface softness analyzer. Hollmark showed that for $a = 61$, $m = -0.22$, and $n = 0.31$, overall softness can be predicted with a correlation coefficient of 0.98 when the surface softness was evaluated using the synthetic finger [50]. Results from the Kawabata testers for tissue samples have been used to establish a model for softness of tissue in terms of combination of individual physical parameter measurements. Kim et al. [30] reported a model involving the Kawabata parameters, ϵ_m (extensibility parameter), and SMT (surface roughness parameter). Habeger et al. [19] developed a regression model for softness perception with ultrasonic out-of-plane measurements. The model involved the ZD acoustic impedance, the attenuation coefficient A, and the basis weight BW.

It is clear from Hollmark's work [53] that the bulk softness component can be predicted by the geometric mean stiffness, determined through tensile tests. The surface component, however, needs reexamination. A surface profile alone cannot predict the surface softness component because the profile itself does not indicate how the tissue sample will interact with the fingers during softness evaluation. The measurement of friction properties offers the ability to quantify the interaction between surfaces and possibly relate it to softness. Surface softness evaluation is a result of dynamic interaction between the fingers and the tissue sample. The finger senses not only the friction force but also the variation in the force. Hence, the standard deviation of the friction parameter MMD is expected to be significant. On the basis of these arguments, it can be hypothesized that the surface component in Hollmark's power law model could be replaced by a friction measurement parameter to take into account the physical nature of the interaction between the human skin and the tissue sample. Thus a power law softness model would be of the general form

$$\text{Softness} = a[\text{GM(stiffness)}]^m[\text{GM(friction parameter)}]^n$$

The parameters a, m, and n can be determined by obtaining a wide range of tissue samples, obtaining a softness assessment through a suitable human panel, measuring the physical parameters indicated and determining the constants from a regression procedure. GM indicates the geometric means of the MD and CD values for the parameters indicated. Surface perception is a dynamic property. It is dependent on the variation or frequency of stimulation of the senses rather than the absolute value. Hence, the deviation of the friction coefficient, MMD, should be of significance in evaluating the surface softness component.

V. SUMMARY AND CONCLUDING REMARKS

In this chapter the physical and mechanical properties of towel and tissue grades of paper have been discussed. Published literature is scarce in the area of towel and tissue relative to other papermaking areas due to the competitive nature of the products. However, abundant information is available in the patent literature, as can be seen from the number of patent citations in this chapter. The only drawback with the patent literature is that it is not peer-reviewed like a journal article and the test methods may be designed to support a particular process or product claimed in that patent or by that company. I have attempted to discuss the essential principles of the tests. These ideas may be taken as the starting point in conducting a systematic laboratory study to modify the test methods to serve particular needs. A fundamental analysis of the mechanisms and implications of test conditions may lead to industry-wide standard test conditions for various towel and tissue property measurements.

REFERENCES

1. Lyne, L. M. The mechanism of softness in paper. *Pulp Paper Mag. Can. 51*(8):80–82.
2. Perkins, R. W. (1986). Fiber networks: Models for predicting mechanical behavior of paper. In: *Encyclopedia of Materials Science and Engineering*. M. B. Bever, editor-in-chief. Pergamon Press, New York, pp. 1711–1720.
3. Hollmark, H., Anderson, H., and Perkins, R. W. (1978). Mechanical properties of low density sheets. *TAPPI 61*(9).
4. Ramasubramanian, M. K., and Perkins, R. W. (1987). Computer simulation of the uniaxial elastic-plastic behavior of paper. *ASME J. Eng. Mater. Technol.*
5. Sanford, L. H., and Sisson, J. B. (1967). Process for forming absorbent paper by imprinting a fabric knuckle pattern theoreon prior to drying and paper thereof. U.S. Patent 3,301,746. (Jan. 31).
6. Trokhan, P. D. (1987). Tissue paper. U.S. Patent 4,637,859 (Jan. 20).
7. Wells, E. R. (1984). Wet-microcontracted paper and concomitant process. U.S. Patent 4,440,597 (Apr. 3).
8. Cook, R. F., and Westbrook, D. S. (1991). Non-creped hand or wipe-towel. U.S. Patent 5,048,589 (Sep. 17).
9. Sudall, S. J., and Engel, S. A. (1995). Uncreped throughdried towels and wipers having high strength and absorbency. U.S. Patent 5,399,412 (Mar. 21).
10. Becker, H. E., McConnel, A. L., and Schutte, R. W. (1979). Bonded differentially creped, fibrous webs and method and apparatus for making the same. U.S. Patent 4,158,594 (June 19).
11. Salvucci, J. T., Jr., and Yiannos, P. N. (1974). Soft, absorbent, fibrous sheet material formed by avoiding mechanical compression of the elastomer containing fiber furnished until the sheet is at least 80% dry. U.S. Patent 3,812,000 (May 21).
12. Gentile, V. R., Hepford, R. R., Jappen, N. A., Roberts, C. J., Jr., and Steward, G. E. (1975). Absorbent unitary laminate-like fibrous webs and method for producing them. U.S. Patent 3,879,257 (Apr. 22).
13 Van Phan, D., Trokhan, P. D., Kelly, S. R., Ostendorf, W. W., and Hersko, B. S. (1995). Multi-ply facial tissue paper product comprising biodegradable chemical softening compositions and binder materials. U.S. Patent 5,437,766 (Aug. 1).

14. Hollmark, H. (1972). Study of the creping process on an experimental paper machine. *STFI-Meddelande, Ser B*, No. 144 (in Swedish).
15. Shmagin, D. L. (1997). Mechanics of the creping process. M.S. Thesis, Mechanical and Aerospace Engineering, North Carolina State University.
16. Carstens, J. E. (1981). Layered paper having a soft and smooth velutinous surface, and method of making such paper. U.S. Patent 4,300,981 (Nov. 17).
17. Rowe, S., and Volkerman, R. J. (1965). Thickness measurement of sanitary tissues in relation to softness. *Tappi 48*(4):54A–56A.
18. Morgan, G., and Rich, T. F. (1976) Process for forming a layered paper web having improved bulk, tactile impression and absorbency and paper thereof. U.S. Patent 3,994, 771 (Nov. 30).
19. Habeger, C., Pan, Y., and Biasca, J. (1989). Empirical relationships between tissue softness and out-of-plane ultrasonic measurements. *Tappi*, November 1989, pp. 95–100.
20. Rasch, D. M., and Hensler, T. A. (1995). A papermaking belt. U.S. Patent 5,431,786 (July 11).
21. Crosby, C. M., Eusufzai, A. R. K., Mark, R. E., Perkins, R. W., Chang, J. S., and Uplekar, N. V. (1981). A digitizing system for quantitative measurement of structural parameters in paper. *Tappi 64*(3):103–106.
22. Wahren, D., and Norman, B. (1976). Mass distribution and sheet properties. In: *The Fundamental Properties of Paper Related to Its Uses*. F. M. Bolam, Tech. Sect., Brit. Paper and Board Makers Assoc., London, pp. 7–70.
23. Farrington, T. E., Jr. (1988). Soft X-ray imaging can be used to assess sheet formation and quality. *Tappi J.* May 1988, pp. 140–144.
24. Ebeling, K., and Komppa, A. (1983). Correlation between the areal and optical densities in paper. In: *The Role of Fundamental Research in Papermaking*. Transactions of the symposium held at Cambridge, September 1981. J. Brander, ed. Mech. Eng. Pub., London, pp. 603–631.
25. Cresson, T. M., Tomimasu, H., and Luner, P. (1990). Characterization of paper formation. Part 1: Sensing paper formation. *Tappi* July 1990, pp. 153–159.
26. Tomimasu, H., Kim, D., Suk, M., and Luner, P. (1991). Comparison of four paper imaging techniques: β-Radiography, electrography, light transmission, and soft X-radiography. *Tappi*, July, pp. 165–176.
27. Ampulski, R. S., Sawdai, A. H., Spendel, W., and Weinstein, B. (1991). Methods for the measurement of the mechanical properties of the tissue paper. *Proceedings of the International Paper Physics Conference*, pp. 19–30.
28. Seo, Y. B., Chaves de Oliveira, R., and Mark, R. E. (1992). Tension buckling behavior of paper. *J. Pulp Paper Sci. 18*(2):55–59.
29. Kawabata, S. (1980). *The Standardization and Analysis of Hand Evaluation*. 2nd ed. The Hand Evaluation and Standardization Committee, The Textile Machine Society of Japan.
30. Kim, J. J., Shalev, I., and Barker, R. (1994). Softness properties of paper towels. *Tappi 77*(10):83–89.
31. Brandt, H., and Wagner, J. R. (1986). Establishment of a tissue ranking index using the Kawabata instruments and a panel of judges. *Proceedings, TAPPI 1986 International Process and Materials Quality Evaluation Conference*, pp. 75–81.
32. Craver, J. K., and Taylor, D. L. (1965). Nondestructive sonic measurement of paper elasticity. *Tappi 48*(3):142–147.
33. ASTM. (1975). ASTM Test Method D 1388-64. Standard test methods for stiffness of fabrics.
34. Allen, D. B., Rust, J. P., Shalev, I., and Barker, R. L. (1954). Development of a mechanical stylus based surface analysis system for soft paper products. *Proceedings of the TAPPI 1994 Nonwovens Conference*, pp. 133–138.

35. Knapp, J. K., and Lampman, R. D. (1995). Characterization of calender roll surface features using interferometric profiling. *Proceedings of the 1995 Tappi Finishing and Converting Conference*, pp. 15–24.
36. Walter, R. S., and Rosch, P. M. (1920). Multi-functional facial tissue. U.S. Patent 4,950,545 (Aug. 21).
37. Ramasubramanian, M. K., and Ko, Y. C. (1989). Relationship between the in-plane paper properties and the ball burst strength for paper. Presented at Symposium on Mechanics of Cellulosic and Polymeric Materials, ASMD/ASCE Joint Mechanics Conference, La Jolla, CA, July.
38. Hollmark, H. (1983). Mechanical properties of tissue. In: *Mechanical Properties of Paper and Paperboard*. R. E. Mark, ed. Marcel Dekker, New York, pp. 497–524.
39. Winakor, G., and Kim, C. J. (1980). Fabric hand: Tactile sensory assessment. *Textile Res. Inst. J.* October:601–610.
40. Wolins, L., and Dickinson, T. L. (1973). Transformations to improve reliability and/or validity for affective scales. *Educ. Psych. Meas. 33*:711–713.
41. Back, S. (1996). Production of soft paper products from old newspapers. U.S. Patent 5,582,681 (Dec. 10).
42. Schwartz, M. A., and Nothnagle, M. A. (1955). Softness tester for sheet material. U.S. Patent 2,718,142 (Sept. 20).
43. Lashof, T. W. (1960). Note on the performance of the Handle-O-Meter as a physical test instrument for measuring the softness of paper. *Tappi 43*(5):175–178.
44. Wardwell, F. B. (1965). Papermaking variables and the Handle-O-Meter test. *Tappi 48*(4):60–61.
45. Clark, J. D. A. (1935). Determining the rigidity, stiffness and softness of paper. *Paper Trade J. Tech. Sect.* Mar. 28, pp. 169–175.
46. Brown, T. M. (1939). A method for determining softness of soft papers. *Paper Mill News 62*(23):19–21.
47. Pearlman, J. (1962). Testing paper tissues and the like. U.S. Patent 3,060,719 (Oct. 30).
48. Wahren, D., and Nilsson, K. Softness and smoothness of tissue paper. *STFI-Meddelande Ser. B 32* (in Swedish).
49. Gallay, W. (1973). Textural properties of paper: Measurements and fundamental relationships. In: *The Fundamental Properties of Paper Related to Its Uses*. Transactions of the Symposium Held at Cambridge. F. Bolam, ed. pp. 684–695.
50. Hollmark, H. (1983). Evaluation of tissue paper softness. *Tappi* February: 97–99.
51. Bates, J. D. (1965). Softness index: Fact or mirage? *Tappi 48*(4):63–64.
52. Ray, J. E. (1965). Control of softness in sanitary tissues. *Tappi 48*(4):57–58.
53. Hollmark, H. (1976). The softness of household paper products and related products. In: *The Fundamental Properties of Paper Related to Its End Uses*. F. M. Bolam, ed. Technical Division, The British Paper and Board Industry Federation, London, pp. 696–703.

PART 3. STRUCTURAL PARAMETERS — FIBERS, BONDS, AND PAPER

Emblem of Huaxtepec

Located in the "Hill of Huaxin Trees," this tribute town provided a fruit similar to carob derived from the Huaxin (*Leucaena esculenta* Benth.). The emblem clearly shows a green hill surmounted by a stylized *Leucaena* tree with bananalike red flowers. The Aztec word for hill was *tepec*. The modern city of Oaxaca is also thought to derive its name from the leguminous Huaxin tree.

13
FIBER STRUCTURE

LEENA PAAVILAINEN
Finnish Forest Cluster Research Programme—Wood Wisdom
Helsinki, Finland

I. INTRODUCTION

This chapter gives an overview of wood and plant fiber properties, their alteration in different process stages, and the role of fiber characteristics in the formation of end product properties. It has been seen worthwhile also to provide the readers of a paper testing reference book, especially those coming from outside the paper industry, with some fundamentals of fiber structure.

When measuring fiber and paper properties it should be kept in mind that both the physical and chemical properties of wood and plant fibers are altered in processing. The pulp fiber properties are a result of raw material properties (structural and chemical), pulping method, process conditions, and process design (Fig. 1). However, the raw material properties determine the papermaking potential of fibers produced by a certain pulping method. This is especially the case with chemical pulp fibers.

Customer friendliness (quality, price, environmental aspects) and the functional properties of each end product set the requirements for the properties and proportions of different fiber components in the paper furnish. In practice, the required fiber properties can be achieved by optimization of raw material quality, cooking, bleaching, and beating/refining.

Wood, with a 94% share of all papermaking fiber used, is and will be the dominant papermaking raw material. Thus, this chapter is devoted to fibers of wood origin. Until the 1880s, non-wood raw materials were the only source of papermaking fiber, and although wood pulp has replaced non-wood fibers in many paper grades, non-wood fibers are still an important papermaking raw material, especially in regions suffering a shortage of wood.

All plant fibers used in papermaking are seed plants that belong to the botanical division *Spermatophyta*, and structural similarities can be found among them. There are two major groups of seed-bearing plants, *Gymnospermae* and *Angiospermae* (Table 1) [39]. *Coniferae* (called softwoods) is a class of the gymnosperms. The angiosperms are divided into two classes, *Monocotyledonae* and *Dicotyledonae* that include broadleaf trees (called hardwoods) and non-wood fibers. According to their end use requirements, the fiber sources are classified into long (typical length 2–7 mm) or short (typical length 0.7–1.7 mm) fibers. Softwoods and

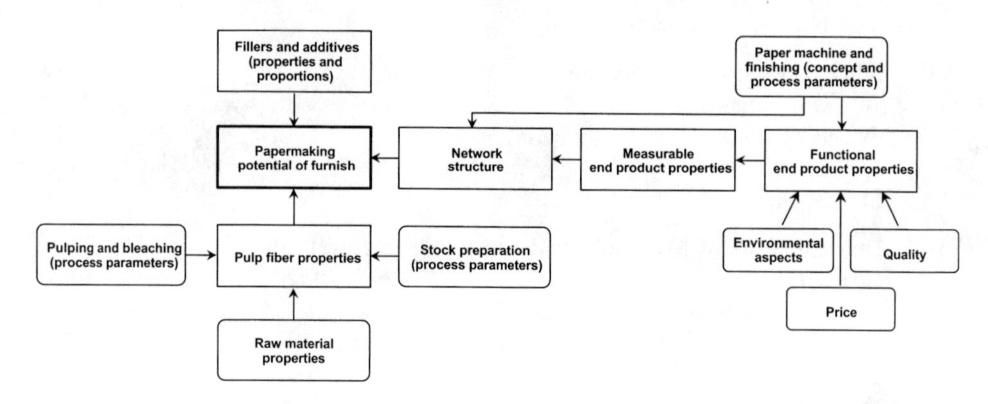

Fig. 1 Factors affecting functional properties of paper.

Table 1 Classification of Papermaking Fibers

Subdivisions	Classes	Plants	Species	Usability
Gymnospermae	*Coniferae*	Coniferous trees	Softwood species	Long fibers
Angiospermae	*Monocotyledonae*	Leaf fiber plants	Abaca (*Musa textilis*) Sisal (*Agave sisalana*)	Long fibers
		Palms (*Palmae*)	Oil palm (*Elaeis guineensis*)	Long fibers
		Grasses (*Gramineae*)	Sabai (*Eulaliopsis binata*) Bamboo (*Dendrocalamus strictus*)	Long fibers
			Bamboo (*Dendrocalamus strictus*) Esparto (*Stipa tenacissima*) Reed (*Phragmites communis*) Sugar cane (*Saccharum officinarum*) Straws (*Triticum, Zea, Oryza*)	Short fibers
	Dicotyledonae	Broadleaf trees	Hardwood species	Short fibers
		Bast fiber plants	Flax *Linum usitatissimum*) Hemp (*Cannabis sativa*) Jute (*Corchorus capsularis*) Ramie (*Boehmeria nivea*) Kenaf (*Hibiscus cannabinus*)	Long fibers
		Fruit fiber plants	Cotton (*Gossypium* spp.) Kapok (*Ceiba pentandra*)	Long fibers

some non-wood plants produce long fibers, whereas short grass fibers can replace hardwood fibers in many end products.

II. WOOD STRUCTURE

There are two major fiber types serving as fiber sources for paper and board. Long softwood fibers are used in the furnish to give the strength required, whereas short hardwood fibers are used in papermaking to provide good printability and stiffness to the end product.

The most commonly used softwoods include Nordic firs (*Abies* spp.), hemlocks (*Tsuga* spp.), larches (*Larix* spp.), Douglas fir (*Pseudotsuga*), spruces (*Picea* spp.), and pines (*Pinus* spp.). In the future, more and more pulpwood is expected to come from the fast-growing pine (*Pinus radiata, Pinus patula*) plantations of the southern hemisphere. A typical cross section of Scots pine is shown in Fig. 2.

Hardwood pulpwood includes aspens and poplars (*Populus* spp.), maples (*Acer* spp.), birches (*Betula* spp.), and oaks (*Quercus*) in the northern hemisphere and eucalypts (*Eucalyptus globulus, Eucalyptus grandis*) and, in the future, acacia (*Acacia mangium*) in the southern hemisphere. Eucalyptus and acacia pulpwood is principally grown in plantations. A typical cross section of birch wood is presented in Fig. 3.

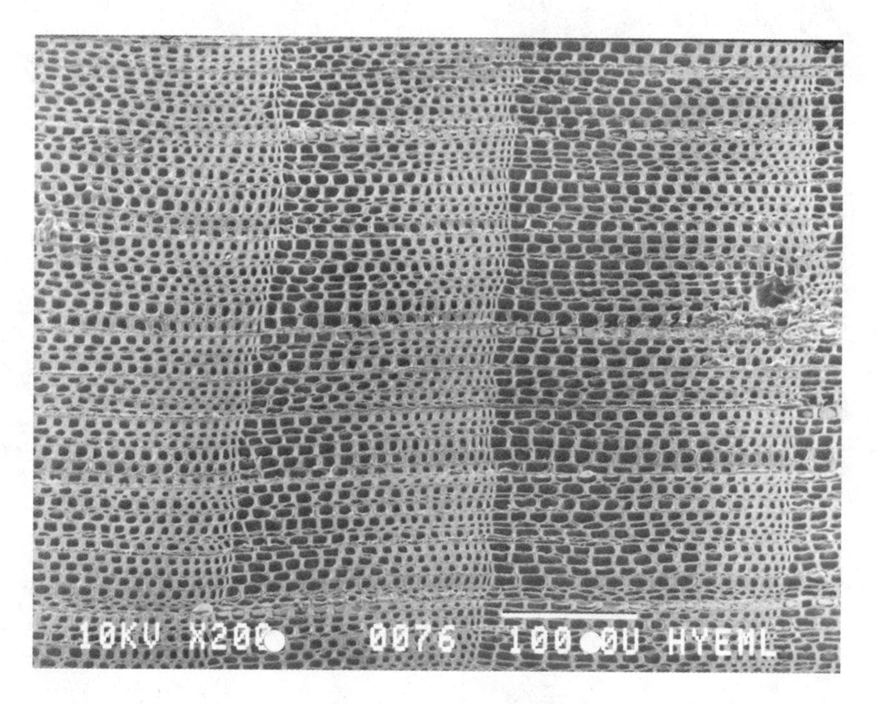

Fig. 2 Cross section of Scots pine, *Pinus sylvestris*. (Courtesy of P. Saranpää, Finnish Forest Research Institute.)

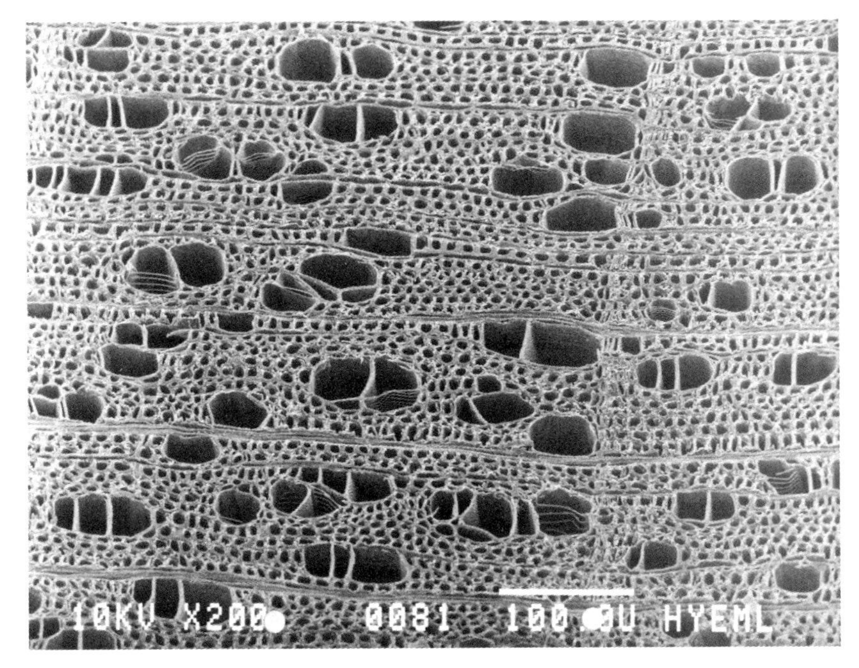

Fig. 3 Cross section of birch, *Betula verrucosa*. (Courtesy of P. Saranpää, Finnish Forest Research Institute.)

A. Macrostructure of Wood

In wood characterization three planes are recognized: the transverse plane (cross section), radial plane, and tangential plane. The center of the stem, seen in cross section, is the pith formed during the first year of growth. The pith is surrounded by xylem, which consists in the mature tree of the inner heartwood part (difficult to pulp) and the outer part, called sapwood. As the diameter of the stem grows, the central core begins to change to nonliving heartwood. Surrounding the sapwood is the cambium, a layer of living and dividing cells. The outer part of the tree is the bark, composed of the living inner bark (phloem) and the dead outer bark (rhytidome).

Wood consists of axially oriented cells and radially oriented cells that are aggregated in rays (see Table 2). In temperate zones, trees form annual rings that are visible in cross section as concentric rings. The thin-walled fibers formed at the beginning of the growing season are called earlywood or springwood fibers, and the thick-walled fibers formed later during the growing period are known as latewood or summerwood fibers. The fibers are interconnected through openings called pits. Wood species, especially softwoods, can be identified according to differences in pit characteristics. In hardwood fiber identification, other features in vessel elements have to be considered [39].

Softwood Fibers The main cell types in conifers are longitudinal tracheids, generally referred to as fibers, which make up over 90% of the wood. The rest are bricklike, short parenchyma cells and ray tracheids (Table 2).

Table 2 Softwood and Hardwood Cell Types

Cell type	Softwoods	Hardwoods	Function
Axial cells	Tracheids	Fibers	Support/conduction
		Libriform fibers	
		Fiber tracheids	
		Vessel elements	Conduction
	Short tracheids	Tracheids	Conduction
	Strand tracheids	Vascular	
		Vasicentric	
Radial cells	Ray tracheids		Conduction
Axial cells	Parenchyma cells	Parenchyma cells	
	Longitudinal	Longitudinal	Storage
	Epithelial	Epithelial	Excretion of resin
Radial cells	Ray parenchyma	Ray parenchyma	Storage, etc.

In commercial pulpwood, the average length of tracheids varies between 2 and 6 mm and their width between 20 and 60 μm [39,75]. Cell wall thickness varies from 2 to 10 μm. The length of thin-walled parenchyma and ray cells varies between 0.02 and 0.2 mm and their width between 2 and 50 μm.

Hardwood Fibers Hardwoods may contain several cells types: libriform fibers, fiber tracheids, vessel elements, and parenchyma cells. Libriform and fiber tracheids (the hardwood fibers) constitute 30–70% of the stem volume, vessels 15–40%, and parenchyma cells 5–30%.

The dimensions of hardwood fibers used in papermaking are, on average, length 0.7–1.7 mm, width 14–40 μm, and cell wall thickness 3–4 μm. The vessel elements are wide (30–130 μm) [75] and rather short (0.3–0.6 mm). In addition to the usual vessels, some hardwoods contain cells resembling softwood tracheids or small vessels.

B. Ultrastructure of Wood

The cell wall of plant fibers is a composite material. It consists of cellulosic fibrils surrounded by a matrix of hemicelluloses and lignin. The structure of the native cell wall, together with the changes occurring in various process stages (pulping, refining, drying) determines the strength and the swelling tendency of fibers and thus their conformability.

The smallest building element of the cellulose skeleton is often considered to be an elementary fibril consisting of about 36 parallel cellulose molecules [88]. The way in which cellulose molecules are arranged in the elementary fibrils and microfibrils is still debated. It has been suggested that cellulose molecules form crystalline regions that change into amorphous regions without any distinct boundary, then back into crystalline regions. However, all the electron microscopic, electron diffraction, and atomic force microscopic (AFM) studies so far have failed to show evidence of these putative amorphous regions.

Elementary fibrils are organized into larger unit microfibrils which are visible in an electron microscope. Microfibrils, embedded in hemicelluloses and lignin, form fibrils and lamellae. After delamination and fibrillation, fibrils and pieces of lamellae are visible in pulp as a fine material. Fiber microstructure is discussed in detail in Refs. 7, 29, and 88.

The cell wall consists of concentric layers, namely, the middle lamella, primary wall, outer layer of the secondary wall (S1), secondary wall (S2), inner layer of the secondary wall (S3), and in some species a warty layer (see Fig. 4). These layers have differences in structure and chemical composition. The microfibrils can wind alternately around the cell axis either to the right or to the left (crossed fibril angles). The S2 layer is a right-handed helix.

The middle lamella consists mainly of lignin which binds the fibers in wood. The primary wall is a thin layer in which the microfibrils form an irregular network. The middle lamella and primary wall are removed in cooking. The middle layer of the secondary wall (S2 layer) forms the main part of the cell wall and thus controls the mechanical properties of the fiber. The microfibril angle in the S2 layer varies between 5° and 10° in the latewood part of softwood fibers and between 20° and 30° in the earlywood part. Some recent measurements indicate that the microfibril angle in earlywood and latewood differs by about 10° [10]. The thickness and fibril angle of the S2 layer principally determine the stiffness and strength of the fibers.

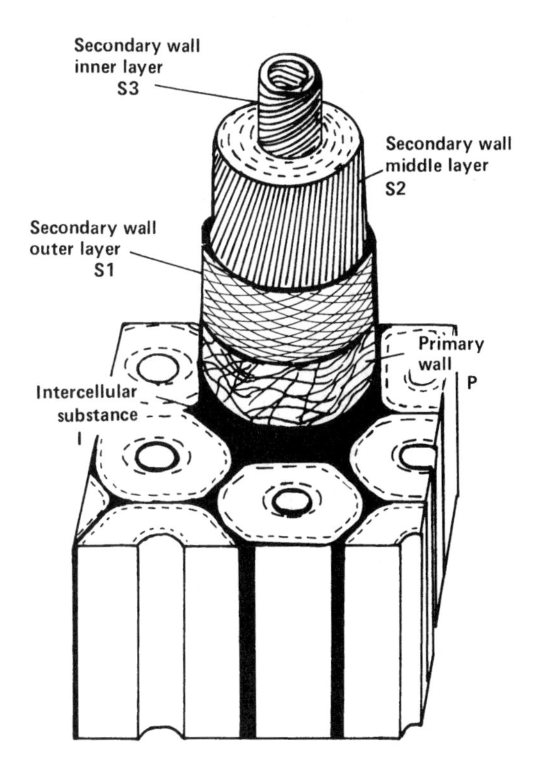

Fig. 4 Lamellar structure of a cell wall.

III. CHEMICAL COMPOSITION OF WOOD AND PULP FIBERS

The three main building components of fibers are cellulose, hemicelluloses, and lignins. Wood may also contain small amounts of resins.

Cellulose is the main component of wood, located for the most part in the S2 cell wall. Cellulose is a long-chain unbranched polymer composed of β-D-glucopyranose units linked together by (1-4)-β-glycosidic bonds. The polymerization degree of wood cellulose is around 10,000 glucose units. Cellulose molecules have a high tendency to form intra- and intermolecular hydrogen bonds and thus aggregate together into microfibrils [29,88]. Because of its high crystallinity and unbranched form, cellulose is the least accessible of the wood components to solvents.

Whereas cellulose is a homopolysaccharide, hemicelluloses are heteropolysaccharides consisting of two or more different monosaccharide units. The molecules also contain side chains. The degree of polymerization is in the range of 50–300. Amorphous hemicelluloses are easily hydrolyzed to their monomer units, namely to D-glucose, D-mannose, D-galactose, D-xylose, L-arabinose (L-rhamnose), and glucuronic as well as uronic acids.

The main softwood hemicelluloses are galactoglucomannan, arabinoglucuronoxylan, and arabinogalactan, in that order. Hardwood hemicelluloses consist of glucuronoxylan and glucomannan. As hydrophilic polymers, hemicelluloses have a high ability to absorb water and swell. Thus, a high hemicellulose content improves fiber flexibility.

The third component of wood fibers is lignin. Lignins are amorphous cross-linked three-dimensional polymers (DP 20,000) formed of phenylpropane units [31]. The phenylpropane units are joined together with stable covalent bonds (ether and C—C linkages) and less stable hydrogen bonds. Lignin also contains functional groups such as methoxyl, phenolic hydroxyl, benzyl alcohol, and carbonyl groups. Functional groups are of major influence in the reactivity of lignin. The typical softwood lignins are predominantly guaiacyl type, whereas guaiacyl and syringyl groups are represented in hardwoods.

IV. FIBER CHARACTERIZATION

The characterization of the papermaking potential of pulp fibers should be based on the functional properties of the end product and the quality requirements set by them for pulp fiber and network properties. If the papermaking potential of fibers is characterized only by measurement of handsheet properties (as often happens), it is almost an impossible task for a papermaker to

- • Select the best possible market pulp for a certain end product
- •• Solve paper quality problems
- ••• Develop the pulp quality by woodyard rearrangement, modifying the cooking and/or bleaching by fractionating pulp or by improving the processes and process control.

Clark has presented that the papermaking potential of fibers can be characterized by using so-called basic fiber properties (fiber length, coarseness, intrinsic fiber strength, cohesiveness, and wet fiber compactability) [12]. This approach was further developed by Paavilainen [66]. In this chapter the papermaking potential of wood and pulp fibers is characterized in terms of fiber (wood fiber and pulp fiber) and network properties (Fig. 5). The effects of wood fiber properties and process parameters on the papermaking potential of fibers are evaluated by studying first their influence on pulp fiber properties and then the effect of pulp fiber properties on network and handsheet properties [66]. A similar approach is suitable for the characterization of mechanical pulp quality [62]. Examples of the usefulness of the structural hierarchy concept can be found in Kortschot's review [53].

Wood fiber properties are divided into morphological properties (cell wall thickness, fiber width, and fiber length), chemical properties (chemical composition and structure), and fine structural properties (S2 fibril angle, crystallinity degree, and weak points in fiber wall). Pulp fiber properties, on the other hand, consist of coarseness, fiber length (length-weighted), fines, intrinsic fiber strength, conformability (wet fiber flexibility and collapsibility), and degree of external fibrillation.

Network properties affected by pulp fiber properties and papermaking operations (web formation, wet pressing, and drying) are defined in terms of the bonding ability of fibers (bonding strength and bonded area) and network structure (geometrical properties).

V. VARIATIONS IN PULP WOOD FIBER STRUCTURE AND COMPOSITION

The morphological properties (fiber length, width, cell wall thickness), fine structural properties (S2 fibril angle, crystallinity), and chemical properties (chemical composition, degree of polymerization of hemicelluloses on the fiber surface) of wood fibers may vary between species, within an annual ring, between different parts of the stem, and due to the effects of different growing conditions.

A. Softwoods

A general assumption can be made that, within a species, fiber property variations follow the earlywood/latewood ratio. Long softwood fibers have thicker walls than shorter softwood fibers.

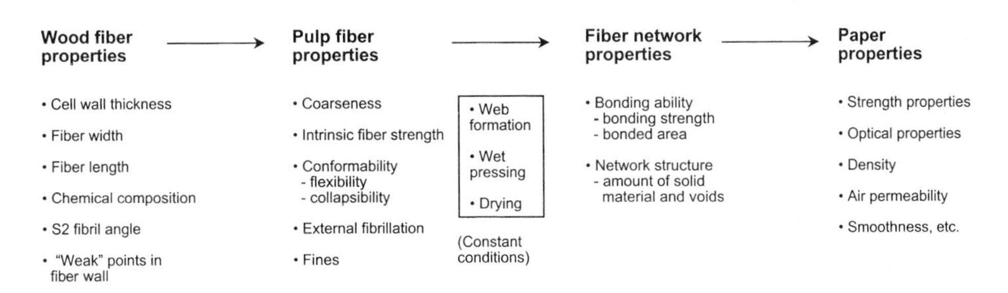

Fig. 5 Method for characterization of pulp quality. The boxed items are process variables. (From Ref. 66.)

Variation in Fiber Morphology Fiber morphology includes fiber length, cell wall thickness, and fiber width. Softwood fibers have a wide *fiber length* distribution. Fiber length consists of two different elements: fibers (tracheids) and primary fines (ray cells and axial parenchyma cells). Fiber length is a highly genetically controlled property, more so than cell wall thickness or fiber width. Thus it is possible to improve fiber length with genetic manipulation [100].

Fiber length varies between different wood species (3–6 mm) [39]. Fibers of the Pacific Northwest of North America (Douglas fir, Sitka spruce, and hemlock) have longer fibers than most of the softwoods that are used in papermaking.

Latewood fibers are longer than earlywood fibers. The length variation within the annual ring is 12–25% [75]. It has been found [37] that Douglas fir latewood contains more extra long fibers than earlywood, although the amounts of short fibers are approximately the same. Fiber length varies at different heights of the trunk. In general, softwoods used in papermaking have longer fibers in the base than in the top, and the maximum fiber length is reached with all probability at the medium height of the trunk.

Older trees form longer fibers [103]. At any height of the trunk, fiber length increases from the pith to the cambium until the tree attains a certain age. The length of both earlywood and latewood fibers increases, and thus the latewood fibers are also longer in sapwood [75]. Unfavorable growing conditions tend to increase fiber length [102], whereas fertilization decreases it [25]. Thus the longest fibers can be found in the medium height and outer parts of the trunk grown under unfavorable conditions and in the latewood part of the annual ring.

The cross-dimensional fiber properties are *cell wall thickness* and *fiber width*. Wood density is related to cell wall thickness and latewood content; basic wood density increases with increase in cell wall thickness and/or latewood content. Latewood fibers have greater cell wall thickness than earlywood fibers and are also radially narrower. The cell wall thickness of latewood fibers measured from wood varies between 4 and 8 μm, and that of earlywood fibers between 2 and 4 μm. Fiber tangential width varies from 30 to 60 μm [39].

The average cell wall thickness also varies between wood species. Southern pines (loblolly and slash pine) have thick-walled fibers, and spruces (Norway spruce, white and black spruce) have thin-walled fibers. The latewood content in southern pines is 50–75%, whereas in softwoods growing in more northerly climes it is only 15–30%.

Cell wall thickness and fiber width increase from the pith to the cambium. The cell wall thickness of latewood fibers is 15–30% higher near the cambium than near the pith, and the wall thickness of latewood fibers is 15% higher than that of earlywood fibers [75]. One may presume that the latewood content and cell wall thickness vary in the axial direction as does fiber length, being higher in the lower bole than in the top of the trunk.

Unfavorable growing conditions can cause the formation of thin-walled fibers. On the other hand, it has been shown that fertilization may decrease the latewood content and thus also cell wall thickness [25,75]. Cross-dimensional properties can be affected by genetic manipulation. In practice, genetic manipulation is used to increase wood basic density, which usually also means an increase in cell wall thickness.

Fine Structural Properties The S2 fibril angle is dependent on wood species and varies according to the position in the tree from 5° to 55° and depending on the measurement technique used [10]. Spruce fibers have a smaller S2 fibril angle than pine fibers [57]. Latewood fibers have a smaller S2 fibril angle than earlywood fibers. For most species, the fibril angle of latewood fibers is on average 10–28° and that of springwood fibers 20–38° [10,88].

Wood age affects the fibril angle. Fibril angle decreases from the pith to the cambium, and the decrease is most significant during the juvenile period [18]. Fibril angle increases in a given annual ring in the stem from the base of the tree to the top [36]. Acceleration of tree growth increases the fibril angle [26,103]. There seems to be a relationship among fiber length, cell wall thickness, and fibril angle within a species; longer and thicker walled fibers tend to have a smaller fibril angle than shorter, thin-walled fibers.

Chemical Composition The differences in chemical composition affect not only pulp yield and delignification rate but also the papermaking potential. The lignin content of softwoods lies generally in the range of 24–33%, hemicellulose content 22–32%, and cellulose content 41–47% [24]. There are also differences in resin content, which limits the use of some pines, especially in mechanical pulping. The extractives content varies from 2% up to 14% in different species [88].

Cellulose and lignin contents are slightly lower in latewood than in earlywood [28]. Earlywood contains more xylan and less glucomannan than latewood [55]. Xylan tends to condense onto the latewood fibers [30].

Cellulose content decreases from the base to the top of the stem (like latewood content) [96]. Also, sapwood and heartwood differ from each other, especially in resin content. Pits in heartwood are closed and thus prevent impregnation by cooking liquids. Lignin and hemicellulose contents as well as the proportion of certain hemicelluloses decrease from the pith to the cambium [26,98]. Genetic manipulation of chemical composition and lignin structure has recently given promising prospects.

B. Hardwoods

With hardwood fibers the main structural and chemical variation occurs between different wood species, between sapwood and heartwood, and according to the age of the tree. Although large variations in fiber properties and chemical composition are typical for natural tropical hardwoods, large variability has also been observed in wood density, fiber characteristics, and pentosan content of different eucalyptus species and hybrids of the same age [99].

Variation in Fiber Morphology Compared to softwoods, hardwood fibers have a narrow fiber length distribution. Fiber length varies between 0.7 and 1.7 mm. Cell wall thickness varies from 2.5 to 5 μm, and fiber width from 15 to 40 μm [39]. Narrow hardwood fibers, such as the eucalyptus fibers (e.g., *Eucalyptus globulus*, *Eucalyptus grandis*) used in papermaking can be quite thick-walled. The cell wall thickness of birch fiber is comparable to that of eucalyptus fibers, but birch fibers are much wider. Oak fibers are wide and thick-walled. Cell wall thickness increases and the number of fibers per unit weight decreases with increasing wood density.

The vessel elements in northern hardwoods (e.g., birch, aspen) are long. Some papermaking hardwoods, such as ash, elm, and oak have short and wide earlywood vessel elements. In the heartwood of such hardwood, the vessel elements may be filled by tyloses and are thus impenetrable to liquids. The vessel elements in tropical hardwoods are in general short and wide [39].

Chemical Composition The biggest difference in chemical composition of hardwoods is in hemicellulose and extractive content. The cellulose content of hardwoods is at the level of 44% (40–48%), similar to that of softwoods. However, hardwoods generally contain less lignin (17–25%) and more hemicelluloses (28–42%) than softwoods [24].

Chemical composition varies between species and between varieties, which affects pulping characteristics (pulp yield, delignification rate). Birch and acacia have a lower lignin content (21–29%) than eucalyptus (19–21%). Eucalyptus has the highest and acacia the lowest cellulose content, whereas birch has more hemicelluloses (pentosan content 23–25%) than the other hardwoods (pentosan content 15–22%) compared here. Acacia has a higher extractive content (around 5%) than eucalyptus or birch (1.5–4% in ethanol/benzene solubles).

VI. INFLUENCE OF FIBER STRUCTURE ON PULP FIBER AND PAPER PROPERTIES

A considerable amount of work has been carried out concerning the relationship between fiber morphology and paper properties (5,6,13,19–21,23,33,34,37,38,43,50, 51,58,66,86,87,92,99,101]. The subject was popular in the 1950s and 1960s and became important again during the 1990s with the tougher pulp quality requirements set by the end products. The earlier studies were based on regression analysis of experimental data or on generally accepted hypotheses about the effect of morphological properties on paper properties. In the more recent studies, the role of fiber morphology on paper properties has been studied through certain fiber characteristics such as coarseness, fiber length, number of fibers per unit weight, fiber strength, flexibility and collapsibility, and bonding ability [20,66,86,87].

The approach presented in Fig. 5 can be used to evaluate the effect of morphological properties as well as the effect of the cooking process and the cooking parameters on the papermaking potential of fibers. Examples are given in this section on how to study, by utilizing pulp fiber and network properties, the influence of wood characteristics on the papermaking properties of softwood fibers in Fig. 6 and hardwood fibers in Fig. 7. The effect of cooking and bleaching on the papermaking properties can be demonstrated by using the same model of presentation (Fig. 8). According to Kortschot [53], paper is a material with high structural hierarchy and should therefore be characterized with a set of properties describing the structure at an appropriate scale of interest; one may ignore the structure at smaller size levels as has been done in this chapter.

A. Softwood Chemical Pulp

The role of softwood fibers in the furnish is to ensure the required strength for the fibrous network. In addition to excellent strength potential (consisting of fiber strength, length, and bonding potential), reinforcement fibers should also fulfill certain printing quality requirements. For example, they should possess high conformability.

Each end product sets its own requirements for the pulp fibers. Based on the requirements set by the functional properties of paper for pulp, it can be stated, for example, that an excellent lightweight coated (LWC) reinforcement fiber is long, strong, and flexible, with good bonding potential. Fiber strength is not as critical for supercalendered (SC) reinforcement fiber as for LWC reinforcement fiber, but high bonding potential is required. For fine paper reinforcement fibers, the most important properties are adequate fiber stiffness and bonding potential.

Morphological properties have been shown to be responsible for 70–90% of the paper property variations [6,21,66]. The most important wood fiber properties with regard to softwood pulp fiber properties are cell wall thickness [65], fiber length [63], and S2 fibril angle [64]. The influence of these wood fiber properties on pulp fiber, network, and paper properties is shown in Fig. 6 and explained in the following.

Pulp Fiber Properties Intrinsic fiber strength increases [64] and fiber conformability (flexibility and collapsibility) [67] and number of fibers in a paper of specific basis weight decrease with increase in cell wall thickness. A small S2 fibril angle is beneficial for fiber strength [64]. Cell wall thickness also controls the development of pulp fiber properties in beating [67].

Network Properties Bonding strength, bonded area, and network structure determine the bonding potential of softwood pulp fibers. The most important of these for reinforcement fiber seem to be the bonded area and network structure. Bonding strength and bonded area are discussed in detail in Chapters 15 and 16.

The increase in the bonded area of reinforcement pulp fibers during beating is mostly attributable to the increase in wet fiber flexibility and not to fiber collapse [64]. Flexible fibers respond better to surface tension forces and get closer to each other, forming fiber–fiber bonds. Intense external fibrillation and fines formation

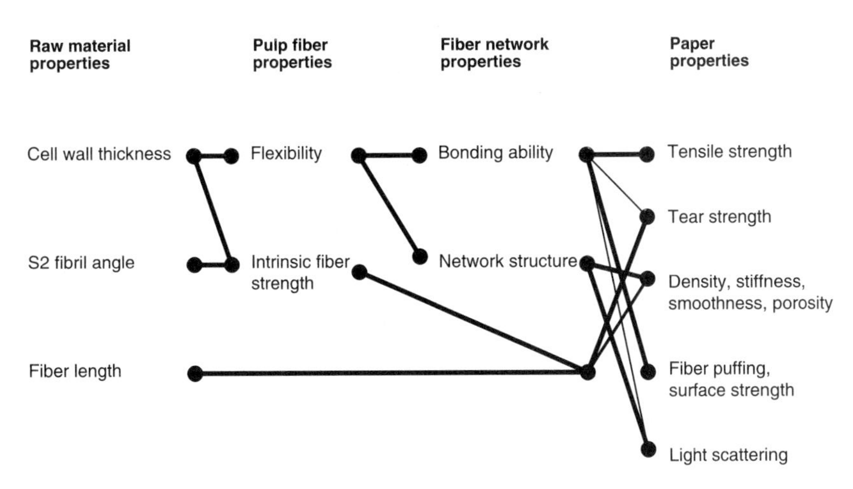

Fig. 6 Influence of softwood fiber properties on pulp fiber and network properties and through them on paper properties. (The main effects are presented as thick lines.)

increase surface tension forces and thus promote bonding. A high number of fibers in paper, on the other hand, increases the bonded area only if the fibers are so flexible that they can get close together and form interfiber bonds [68]. In sheet formation, thin-walled, flexible fibers can get close together and form numerous fiber–fiber bonds. Such fibers have excellent bonding potential.

Paper Properties Bonding properties, especially fiber flexibility, mainly determine the tensile strength of reinforcement fibers. Together with the number of fibers in a sheet, fiber flexibility is also largely responsible for the light-scattering coefficient. Tear strength is a product of intrinsic fiber strength, length, and stiffness. Flexibility also strongly influences sheet density, air resistance, and surface smoothness. Thus, variation in cell wall thickness explains over 80% of the variations in the strength and printing properties of long fibers [66]. This conclusion is reached because fiber conformability and intrinsic fiber strength are related to cell wall thickness, and cell wall thickness controls the development of these properties during beating.

The pulp quality variations caused by cell wall thickness variations are often erroneously explained by fiber length. Generally, however, long softwood fibers are also thick-walled [12,65].

Pulp with high tear strength can be produced from strong thick-walled fibers or fibers with a small S2 angle such as sawmill chip fibers originating from sapwood. If it is, however, important to gain both excellent strength (tear and tensile strength) and printing properties, it would be wise to use flexible, strong, long fibers such as spruce fibers. If the papermaking potential of reinforcement pulps made from different species is compared, it can be concluded that, for example, the fir (*Abies*) and hemlock (*Tsuga*) in northwestern North America produce excellent LWC reinforcement pulp. The thick-walled southern pines (*Pinus elliottii* and *Pinus taeda*), on the other hand, give low quality reinforcement pulp.

B. Hardwood Chemical Pulp

Morphological properties have been shown [33,101] to be responsible for over 80% of the papermaking potential of hardwood fibers. According to Demuner et al. [20], the basic density of wood, the number of fibers per gram, and the pentosan content explain 80% of the papermaking properties of the eucalypts. Cell wall thickness increases with increasing wood density. A popular way to characterize hardwood has been to use the Runkel ratio, i.e., the ratio of twice the cell wall thickness to the lumen diameter [12]. The rationale for using the Runkel ratio for pulp evaluation is given in Chapter 16.

In summary, it can be stated that the main raw material characteristics that affect the papermaking properties of short-fiber chemical pulp are the morphological properties: the ratio of fiber width to cell wall thickness, fiber length, fiber coarseness, and hemicellulose content [71] (see Fig. 7). These factors apply to both short wood fibers and non-wood fibers [70]. Fiber strength is not usually a critical property for short-fiber pulp, because paper can be reinforced by using long fibers.

Pulp Fiber Properties The role of short fibers is to give the paper and board good printability and stiffness. Thus, the critical fiber properties of short-fiber pulp are

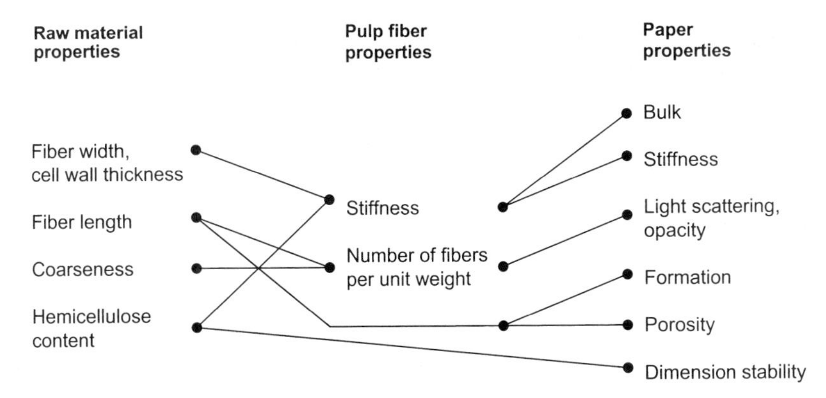

Fig. 7 Influence of hardwood fiber properties on pulp fiber and paper properties. (From Ref. 71.)

stiffness and the number of fibers per unit weight. Thick-walled narrow fibers that have a low hemicellulose content are especially stiff. Short, thin-walled wood fibers with low coarseness, on the other hand, give pulp with a high number of fibers per unit mass.

Eucalyptus (*Eucalyptus globulus* and *Eucalyptus grandis*) fibers represent stiff hardwood fibers. The ratio of cell wall thickness to fiber width is small, and the fibers have low hemicellulose content. Short, thin-walled acacia (*Acacia mangium*) gives pulp with the highest number of fibers per unit weight.

Paper Properties Pulp that contains a high number of fairly stiff fibers per unit weight, such as acacia, has fewer air/fiber interfaces reflecting light than other hardwood pulps. As examples, eucalyptus and especially acacia fibers impart an excellent light-scattering property to paper.

Stiff, uncollapsed fibers form an open network structure, which results in high bulk and good compressibility. Thus, mixed tropical hardwood pulps give high bulk to paper, whereas Scandinavian birch pulp gives low bulk.

The smaller the fibers, the better they can fill the pores in the fiber surface and fiber network. Fibers that have a narrow fiber length distribution, such as eucalyptus and acacia, produce paper with a homogeneous pore size distribution, which favors ink penetration. They also give good surface smoothness to paper. Short fibers and a narrow fiber length distribution are also prerequisites for good formation.

Fibers have good dimensional stability if they have a low tendency to swell (low hemicellulose content), as is characteristic of commercial eucalyptus pulp. Short, stiff fibers tend to have low picking resistance and surface strength. High hemicellulose content, as in birch, and low cell wall thickness guarantee good bonding ability for the pulp. Thus, Scandinavian birch has a high and Indonesian mixed tropical hardwoods a low bonding ability. Stiff and long fibers, as with hardwoods of the southern United States (e.g., *Quercus, Cornaceae*), give the highest tear strength. Stiff eucalyptus fibers that can form a porous network structure impart excellent drainage properties.

VII. FIBER ALTERATION IN PROCESSING

In practice, the papermaking potential of wood fibers is not fully utilized; part is lost in pulping (wood handling, pulping/refining, bleaching, beating). This loss is greatest in the intrinsic fiber strength. Raw material quality variations cannot be compensated for by any means in cooking and bleaching.

A. Cooking

Cooking and bleaching affect the chemical state of the cell wall (chemical composition of cell wall and fiber surface and length of hemicellulose and cellulose chains) and cell wall structure (dislocations, etc.) (see Fig. 8).

Cooking and bleaching conditions should be optimized so that the intrinsic fiber strength is preserved (by preventing the degradation of cellulose) and the development of weak points in the fiber wall is prevented. High hemicellulose, especially glucomannan content, and degree of polymerization of hemicelluloses on the fiber surface are, on the other hand, beneficial for gaining good bonding properties. The most effective way to affect fiber flexibility in cooking and bleaching is to keep the hemicellulose content as high as possible.

Intrinsic fiber strength decreases in processing because of chip damage, inhomogeneous cooking, and poorly controlled bleaching. Both damaged and stiff fibers are susceptible to fiber shortening in beating. For example, a reduction of 30% in tear strength (intrinsic fiber strength) has been shown to be possible in the hot blow [84]. The strength reduction can be decreased by improving the uniformity of cooking for example, by using thin chips [32] or by improving impregnation. An even chip quality (uniform chip size, density, moisture content) promotes good impregnation.

In conventional sulfate cooking, fiber strength and flexibility can be affected by optimizing the alkali charge and the sulfidity level. Intrinsic fiber strength can better be preserved in cooking if the sulfidity level can be kept high [61]. Lowering the alkali charge increases the hemicellulose content in pulp. Fibers become more flexible and

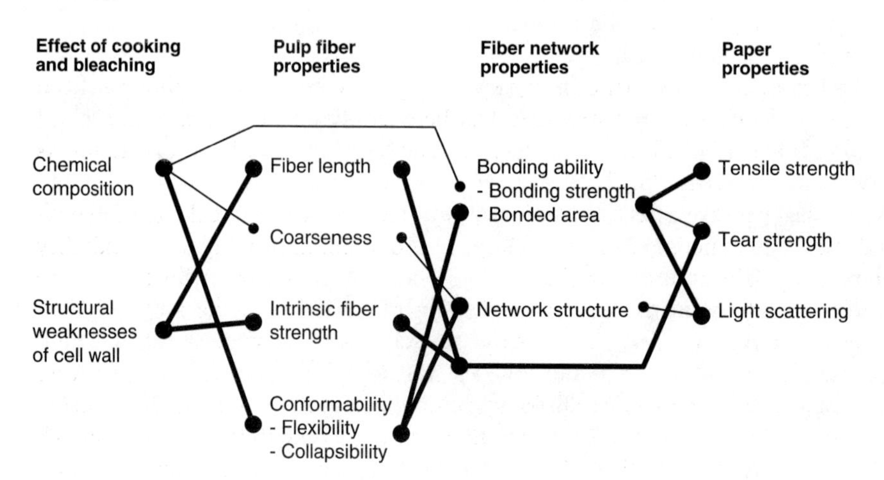

Fig. 8 Influence of cooking and bleaching on pulp fiber and network properties and through them on paper properties. (The main effects are presented as thick lines). (From Ref. 69.)

thus undergo less shortening during beating. However, the variations in pulp fiber properties caused by raw material variations [69] cannot be compensated for with the cooking parameters.

The latest trends in pulping—low kappa targets, and especially totally chlorine-free (TCF) bleaching—have not necessarily been favorable for reinforcement fiber properties. Fiber structure plays an important role with regard to intrinsic fiber strength when reinforcement pulp is cooked to a kappa level below 20 with either modified batch or continuous cooking [93]. Strong fibers such as spruce fibers and thick-walled pine fibers seem to preserve their strength potential better than thin-walled fibers. In extended delignification, hemicelluloses are also more prone to dissolution than in conventional cooking. The swelling ability and flexibility of fibers decrease and the amount of beating has to be increased in order to reach the specific bonding ability that is needed.

B. Bleaching

The swelling tendency of fibers is reduced by bleaching. Ozone and dioxide reduce the total charge of fibers and decrease fiber swellability and thus fiber flexibility [54]. Elemental chlorine-free (ECF) pulp fibers swell more than TCF pulp fibers and require less refining energy to reach the bonding potential needed [80]. Hardwood fibers, on the other hand, swell more, owing to a higher total charge and higher hemicellulose content than softwood fibers [54].

The production of fully bleached TCF reinforcement pulp requires an ozone stage. Ozone is an unselective and intense oxidizing agent that can cause a severe reduction in fiber strength. TCF softwood pulps have been shown to have a higher number of dislocations and lower intrinsic strength than ECF pulps [76]. The damaged fibers cannot resist the intensive beating used for conventional bleached sulfate pulp. It seems that the strength potential of fully bleached TCF fibers is lost in beating; the damaged fibers break.

C. Beating

Beating is a key operation in papermaking with regard to the papermaking potential of fibers. Its influence on fiber structure has been studied widely, and many good review papers are available [12,22,73]. Beating changes the cell wall microstructure as well as fiber dimensions.

During beating, the fibers are subjected to forces that cause breakage of hydrogen bonds between lamellae. Fibers delaminate and swell, improving fiber flexibility and collapsibility. This internal fibrillation, together with the external fibrillation, are the two main beneficial effects of beating. External fibrillation changes the physical and chemical nature of the fiber surfaces by loosening lamellar structures from the P and S1 layers and fibrillar bundles from the S2 layer. If the shear forces are strong enough, they peel fragments of fibrils from the fiber surfaces, forming fines. External fibrillation is responsible for bonding strength and, together with fines formation, for bonded area. External fibrillation increases surface tension forces. In web consolidation, flexible fibers respond better to these forces than do stiff fibers; fiber–fiber bonds will be formed. Beating also causes some undesirable effects, such as the formation of weak points in the fiber wall that make it susceptible to fiber shortening as well as swelling [73]. Illustrations of fiber damage and other irregularities appear in Chapter 14.

Cell wall thickness controls the development of pulp fiber properties in beating [67]. In particular, the flexibility of thin-walled fibers increases during beating. On the other hand, over 70% of both thin-walled and thick-walled beaten fibers collapse during drying. Stiff, thick-walled fibers are more prone to fiber shortening during beating. Stiff fibers are also more susceptible to external fibrillation and fines formation, whereas thin-walled curly fibers develop more dislocations during beating [64,63]. The bonding strength of fibers increases with latewood content and with beating [68].

D. Mechanical Pulping

Mechanical pulping always causes fiber damage (see Figs. 42 and 43 of Chap. 14). Not only do fibers separate from each other in the middle lamella but, depending on wood structure, pulping method, and conditions, some of the fibers are cut and thus shortened, some remain attached to each other, forming shives and fiber bundles, and cell wall fragments and fibrils are removed from the fiber wall. High quality mechanical pulp has a high proportion of long, flexible fibers, a large specific surface area (fibrillated fibers and fines) and fines that have good bonding ability.

Cell wall thickness, together with the viscoelastic characteristics of lignin, controls the fiber separation phenomena. Pulp produced from thick-walled fibers contains a considerable amount of fiber fragments and shives. The thin-walled fibers break in refining and form a large amount of fines, whereas thick-walled, stiff fibers fibrillate; the fines formed in this way tend to have high bonding ability. Fiber development in refining includes two competing mechanisms: fiber breakage into small pieces and removal of material from fiber surfaces by peeling [48,91].

The raw material quality requirements of high quality, high yield pulping are (1) low density, (2) low resin content, (3) long, thin-walled fibers, and (4) light color. Thus, sapwood is more suitable for mechanical pulp than heartwood. There are also differences between different softwood species. Long, thin-walled fibers meet the raw material requirements best. Dense wood with coarse fibers, on the other hand, produces pulp with low strength, light-scattering ability, and smoothness but with high bulk and porosity.

The quality of hardwood mechanical and chemimechanical pulps is dependent on wood density. Light woods, such as aspen (*Populus* spp.), can be used in high yield pulping. Hardwood pulps are not comparable with high yield softwood pulps in terms of pulp strength but may have an extremely high number of fibers per unit weight and therefore a high light-scattering coefficient.

VIII. DETERMINATION OF MORPHOLOGICAL AND FINE STRUCTURAL FIBER CHARACTERISTICS

In this chapter some methods to determine particular wood fiber properties (fiber length, cross-dimensional properties, and S2 fibril angle) are reviewed. Other structural fiber characteristics are presented in Chapters 14 and 16. In practice, cell wall thickness can be characterized with fiber coarseness (defined as the average weight per unit length of fibers, mg/m), and fiber length with length-weighted fiber length [62,66]. Fibril angle, on the other hand, determines the Young's modulus of fibers and therefore plays a key role in the mechanical properties of paper and board.

A. Fiber Length

Fiber length can be determined by either direct measurement (e.g., microscopic counting, image analysis, measurements from slurry) or indirect fractional methods.

In many mills, especially in those producing mechanical pulp, fiber length is measured with the Bauer-McNett classifier (TAPPI Standard T 233) [9]. In the older literature such classifiers as the Clark classifier [81] and the Hillbom apparatus [4] can be found. In classifiers the weight-weighted average fiber length is determined by weighing the pad from each compartment and using the assumption that the fiber length of each fraction is reasonably constant for a certain pulp type [14,94].

Classifiers thus give an indication of fiber length, but the real fiber length value can be determined directly only by measuring single fibers, e.g., with a microscope. Microscopic measurements are presented in Chapter 16. Conventional measurements using a micrometer eyepiece in a microscope are tedious and time-consuming. Fiber length measurement became faster when performed manually from a projected image with a wheeled probe (TAPPI Standard T 232) [40,90,97]. Further advances have led to image analyzers—first manual ones [1,16] and then the sophisticated image analyzers that calculate fiber length automatically from a screen image of dyed fibers [45].

Fiber length can also be measured directly from the pulp slurry by using optical devices. Optical on-line fiber length analyzers that are commercially available include the BGT fiber analyzer (TP1), Sunds Defibrator pulp quality monitor (PQM), and Kajaani fiber analyzer. In the first two devices, three different fiber length classes can be obtained by measuring light intensity variations originating from light scattering and absorption in a suspension [83].

Kajaani fiber measurements are based on the ability of individual fibers to change the level of light polarization as they pass through a narrow capillary. The resolution is 50 μm over the entire measurement range, and the analyzer divides fibers into 144 length classes (TAPPI T 271 pm-91). The trend in fiber analysis is toward simultaneous measurement of other basic fiber properties such as kinks and curl. One such laboratory apparatus is the commercial OpTest fiber quality analyzer (FQA) [82] developed from a prototype imaging fiber analyzer (IFA) [60], in which polarized light is used to measure the images of oriented fibers in a slurry with a two-dimensional CCD camera.

Light scattering and absorption measurements can be disturbed by slurry characteristics such as consistency, color of fibers and liquid, beating degree, and entrained air. These factors do not affect polarization measurements. Consequently, the accuracy of fiber length measurements by light polarization surpasses that of the optical methods (see Table 3).

B. Cross-Dimensional Properties and Coarseness

The measurement of cell wall thickness and fiber width on individual fibers or their cross sections [52], even if image processing is used, is laborious and time-consuming.

Table 3 Comparison of Fiber Length Measurement Methods for Softwood and Hardwood Pulps

	Fiber length (mm)			
	Numerical[a]		Length-weighted[a]	
Method	Av.	Std. dev.	Av.	Std. dev.
Microscope [16]				
HW, northern kraft			0.79	0.04
SW, western (cedar) kraft			2.80	0.10
Bauer-McNett [94]				
SW 14 mesh	3.74	0.39		
28 mesh	1.87	0.19		
48 mesh	1.21	0.11		
100 mesh	0.69	0.06		
Image analyzer [16]				
HW, northern kraft			0.87	0.04
SW, western (cedar) kraft			2.62	0.16
TP1 (±4%); e.g., SW/83/			2.00	0.08
Kajaani FS-200 [63]				
HW, birch	0.585	0.003	0.863	0.006
SW, Scots pine	0.889	0.009	2.140	0.017

HW = hardwood; SW = softwood.
[a] Av. = average; Std. dev. = standard deviation.

Confocal microscopy has made the measurement of transverse fiber dimensions easier [42]. Microscopic measurement methods for cell wall thickness and fiber width are presented in Chapter 16. Cell wall thickness can be measured from wood cross sections or from single fibers or paper sheet cross sections. The results obtained using different methods are not comparable on account of the swelling tendency of pulp fibers.

Although cell wall thickness is mainly responsible for the papermaking potential of reinforcement pulp, there are no suitable cell wall thickness measurement methods for process control purposes. Thus coarseness determined by cross-sectional area and cell wall density is the best practical measures for characterizing the papermaking potential of different softwood fibers. However, one should remember that thick-walled, narrow, stiff, and strong fibers can give the same coarseness value as thin-walled, wide fibers with a large lumen and excellent bonding properties.

Except for the Kajaani and OpTest FQA methods, all coarseness measurement methods are based on microscopic counting. In TAPPI Standard T 234, coarseness is measured by calculating the number of crossings of a known weight of dried fibers per unit on a prepared slide over lines of known length. Britt [8] determined coarseness by calculating the number of fibers per unit area of wood cross sections, measuring wood density, and multiplying the result by pulp yield. Sastry et al. [85] determined coarseness distribution by weighing single fibers with a sensitive microbalance and measuring their lengths with a microscope.

Jordan et al. [46] took the first step toward the practical application of coarseness measurement by image analysis. They determined coarseness by measuring the chord length of the void and applying a statistical network theory. The void area, on the other hand, is related to cell wall thickness [44]. Clark et al. [11] also used image processing in proposing a new definition for coarseness: the weight per unit projected area of fibers, in which the shape of the fibers is also included.

Although image processing facilitates the measurement itself, sample preparation is still rather complicated. A coarseness measurement method for routine testing has therefore been developed in which fiber coarseness is measured from a pulp slurry. When the weight of the sample is known, the coarseness can be determined by using a Kajaani analyzer or OpTest FQA, which measure the number of fibers and the numerical average fiber length value as fibers pass through a narrow capillary tube [63,82]. In its latest version the Kajaani device (FiberLab) is able to measure fiber length, fiber width, and cell wall thickness index (resolution of 1 μm), based on gray level differences, which are measured with a CCD camera system [49]. The STFI Fiber Master [47] also measures the image of oriented fibers passing through a capillary with a CCD camera. The device gives fiber length, width, and curl index (Table 4).

C. Fibril Angle

The S2 microfibrillar angle has been measured by using light or electron microscopy [15,95]. In light microscopy special techniques have to be used, such as metal shadowing, phase contrast, polarized light, and ultraviolet light with or without fluorochrome stains, to enhance the appearance or render the striations visible. A near-ultraviolet method that uses a combination of phase contrast with filtered mercury burner illumination has also been developed [17]. Although it does not

Table 4 Comparison of Coarseness Measurements for Softwood Kraft Pulps

	Coarseness		Repeatability (%)
Method	(mg/m)	(Std. dev.)	(R) at 95% confidence level
Kajaani FS-100 [65]			
Scots pine[a]	0.159	0.003	4.3
Modified Kajaani [65]			
S-200[a]			
Southern pine	0.295	0.006	4.6
Fiber Quality Analyzer (FQA) [82]	0.140	—	11.0
TAPPI T234 [85]			
Western hemlock[b]	0.207	0.050	47.0
Britt's method [8]			
Western hemlock	0.187	—	—
Jordan et al. [45]			
Softwood, 14 mesh fraction	0.213	0.002	2.1

[a] $R = 2.26$ std. dev. ($n = 10$).
[b] $R = 1.96$ std. dev. ($n > 20$).

require any special sample preparation, the method is seldom used, because it is expensive, metallographic plates are hard to find, and UV protection is necessary for the eyes. The latest technique developed uses the elongated cavities formed by soft rot decay to measure microfibril angle and microfibrillar orientation at the tips of fibers [2]. It also brings new information on the trajectories of microfibrils near the pits and implications for better understanding of mechanical properties of fibers [3]. The cavities are easy to distinguish and, although it is formed primarily in the S2 layer, the fibril angle can be determined in both thin-walled earlywood and thick-walled latewood fibers.

Techniques based on polarized light have been widely used to measure the fibril angle. The Sénarmont method measures the birefringence of secondary cell wall layers. It is tedious and not suitable for single fibers [18,78]. Several sample preparation techniques [56,59,78] have been developed to determine the mean fibril angle with polarized light transmitted through a single cell wall. Page et al. [72,74] used reflected polarized light to determine the fibril angle in a single cell wall after mercury was inserted into the fiber lumen. This method is not suitable for thick-walled fibers, because the S1 and S3 layers can affect the polarization and thus the results.

Recent developments in microfibril angle determination include improvements of the X-ray diffraction technique [27,89] and the use of new methods, such as micro-Raman spectroscopy [77] and microscopic transmission ellipsometry [104,105], all of which may be prone to errors from other wall layers. In addition, the new confocal microscopic technique has been used for fibril angle determination. Butterfield [10] provides a good overview on the determination of microfibril angle.

In the X-ray diffraction method [35,79], a measurement is made of the mean crystallite angle of a large number of cells that are scanned simultaneously. Although it is a rapid method, its weakness is that it is suitable for fibril angle determination only from wood samples. Thus it does not give information on between-tracheid variability. The development of microfibril angle measurement with X-ray scattering is ongoing.

In the confocal microscopic technique the fibril angle can be determined from the change in fluorescent intensity as the polarization angle of the excitation light is varied [41]. This rapid and simple measurement technique is applicable to pulp and wood fibers in either the dry or wet state. The advantage of confocal laser scanning microscopy is its ability to analyze thin layers of a translucent specimen with minimal interference from adjacent layers above or below. It correlates well with the Page method, but only if the thickness of the S2 layer lies between 1.3 and 3.4 μm.

REFERENCES

1. Abbott, J. C., Jaycox, L. B., and Ault, G. M. (1979). A simple rapid method for determining fiber dimensions. *Proc. Printing Reprography Testing Conf.*, TAPPI Press, Atlanta, GA, pp. 119–132.
2. Anagnost, S. E., Mark, R. E., and Hanna, R. B. (1998). Utilization of soft rot cavity orientation for the determination of microfibril angle. *Empire State Paper Res. Inst. Rep. No. 108*, pp. 70–79.
3. Anagnost, S. E., Mark, R. E., and Hanna, R. B. (1999). Utilization of soft rot cavity formation as a tool for understanding the relation between microfibril angle and

mechanical properties of cellulosic fibers. In: *Mechanics of Cellulosic Materials 1999*. AMD Vol 231 (MD-Vol. 85). R. W. Perkins, ed. Am. Soc. Mech. Eng., New York, pp. 43–63.

4. Andersson, O. (1953). An investigation of the Hillbom and Bauer-McNett fiber classifiers. *Svensk Papperstidn. 56*(18):704–709.
5. Annergren, G., Rydholm, S., and Vardheim, S. (1962). Influence of raw material and pulping process on the chemical composition and physical properties of paper pulps. *Proc. 6th Eucepa Symposium*. Eucepa, Stockholm, pp. 4:1–4:15.
6. Barefoot, A. C., Hitchings, R. G., and Ellwood, E. L. (1964). Wood characteristics and kraft paper properties of few selected loblolly pines. I. Effect of fiber morphology under identical cooking conditions. *Tappi 47*(6):343–356.
7. Bristow, J. A., and Kolseth, P. (1986). *Paper Structure and Properties*. Marcel Dekker, New York.
8. Britt, K. W. (1965). Determination of fiber coarseness in wood samples. *Tappi 48*(1): 7–11.
9. Butler, W. T. (1948). The use of the Bauer-McNett fiber classifier. *Pulp Paper Mag. Can. 49*(3):133–136.
10. Butterfield, B. G. (1998). Microfibril angle in wood. Proc. IAWA/IUFRO International Workshop on the Significance of Microfibril Angle to Wood Quality. University of Canterbury, Christchurch, New Zealand.
11. Clark, B., Ebeling, K. J., and Kropholler, H. W. (1985). Fiber coarseness: A new method for its characterization. *Paperi Puu 77*(9):490–499.
12. Clark, J. d'A. (1985). *Pulp Technology and Treatment for Paper*. 2nd ed. Miller Freeman, San Francisco.
13. Clark, J. d'A. (1962). Effects of fiber coarseness and length. I. Bulk, burst, tear, fold and tensile tests. *Tappi 45*(8):628–634.
14. Corson, S. R., and Uprichard, J. M. (1972). Fiber length of Clark screen fractions. *Tappi 55*(11):1620.
15. Côté, W. A., Jr. and Day, A. C. (1965). Anatomy and ultrastructure of reaction wood. In: *Cellular Ultrastructure of Woody Plants*. W. A. Côté, Jr., ed. Syracuse Univ. Press, Syracuse, NY, pp. 391–418.
16. Crosby, C. M. (1981). A digitizing system for quantitative measurement of structural parameters in paper. *Proc. TAPPI Annual Meeting*. TAPPI Press, Atlanta, pp. 83–90.
17. Crosby, C. M., and Mark, R. E. (1974). Precise S2 angle determination in pulp fibers. *Svensk Papperstidn. 77*(17):636–642.
18. Crosby, C. M., de Zeeuw, C., and Marton, R. (1972). Fibrillar angle variation in red pine determined by Sénarmont compensation. *Wood Sci. Technol. 6*(3):185–195.
19. Dadswell, H. E., and Watson, A. J. (1962). Influence of the morphology of wood pulp fibers on paper properties. In: *formation and Structure of Paper*, Vol. 2. F. Bolam, ed. British Paper and Board Makers' Assoc., London, pp. 537–572.
20. Demuner, D. B., Vianna Doria, E. L., Claudio da-Silva, E. Jr., and Manfredi, V. (1991). The influence of eucalyptus fiber characteristics on paper properties. *Proc. International Paper Physics Conf.* TAPPI Press, Atlanta, pp. 185–196.
21. Dinwoodie, J. M. (1966). The influence of anatomical and chemical characteristics of softwood fibers on the properties of sulfate pulp. *Tappi 49*(2):57–66.
22. Ebeling, K. (1980). A critical review of current theories for the refining of chemical pulps. *Proc. Symposium on Fundamental Concepts of Refining*. Institute of Paper Chemistry, pp. 1–33.
23. Einspar, D. W. (1964). Correlations between fiber dimensions and fiber and handsheet strength properties. *Tappi 47*(4):180–183.
24. Emerton, H. W. (1982). The fibrous raw materials in papermaking. In: *Handbook of Paper Science*, Vol. VI. H. F. Rance, ed. Elsevier, Amsterdam.

25. Ericson, B., and Harrison, A. Th. (1974). Douglas fir wood quality studies. Part I. Effects of age and stimulated growth on wood density and anatomy. *Wood Sci. Technol. 8*(3):207–226.
26. Ericson, H. D., and Arima, T. (1974). Douglas-fir wood quality studies. Part II. Effects of age and simulated growth on fibril angle and chemical constituents. *Wood Sci. Technol. 8*(4):255–265.
27. Evans, R., Hughes, M., and Menz, D. (1999). Microfibril angle variation by scanning X-ray diffraction. *Appita 52*(5):363–367.
28. Fengel, D. (1969). The ultrastructure of cellulose from wood. Part I. Wood as the basic material for isolation of cellulose. *Wood Sci. Technol. 3*(3):203–217.
29. Fengel, D., and Wegner, G. (1989). *Wood: Chemistry, Ultrastructure, Reactions.* Walter de Gruyter, Berlin.
30. Galdstone, W. T., and Ifju, G. (1975). Nonuniformity in the kraft pulping of loblolly pine. *Tappi 58*(4):126–129.
31. Glasser, W. G., and Sarkanen, K. V. (1989). Lignin properties and materials. ACS Symp. Ser. 397. American Chemical Society, Washington, DC.
32. Gullichsen, J., Kolehmainen, H., and Sundqvist, H. (1992). On the nonuniformity of the kraft cook. *Paperi Puu 74*(6):486–490.
33. Gurnagul, N., Page, D. H., and Seth, R. S. (1990). Dry sheet properties of Canadian hardwood pulps. *J. Pulp Paper Sci. 16*(1):J36–J41.
34. Hatton, J. V., and Cook, J. (1990). Managed Douglas fir forests: IV. Relationships between wood, fiber, pulp and handsheet properties. *Proc. CPPA Annual Meeting. Tech. Sect.* CPPA, Montreal, pp. A51–A66.
35. Hermans, P. H. (1949). *Physics and Chemistry of Cellulose Fibres.* Elsevier, New York, p. 248.
36. Hiller, C. H. (1964). Variations of fibril angle with wall thickness of tracheids in summerwood of slash and loblolly pine. *Tappi 47*(2):125–128.
37. Holger, W. F., and Levis, H. F. (1950). The characteristics of unbleached kraft pulps form western hemlock, Douglas fir, western red cedar, loblolly pine and black spruce. VII. Comparison of springwood and latewood fibers of Douglas fir. *Tappi 33*(2):110–112.
38. Ifju, G., and McLain, T. E. (1982). Quantitative wood anatomy based on stereological methods and its use for predicting paper properties. *Proc. TAPPI Research and Development, Div. Conf.* TAPPI Press, Atlanta, pp. 15–28.
39. Ilvessalo-Pfäffli, M.-S. (1995). *Fiber Atlas. Identification of Papermaking Fibers.* Springer Ser. Wood Sci. Springer-Verlag, Berlin.
40. Ilvessalo-Pfäffli, M.-S., and Alftan, G. (1957). The measurement of fiber length with a semiautomatic recorder. *Paperi Puu 39*(11):509–516.
41. Jang, H. F. (1998). Measurement of fibril angle in wood fibers with polarization confocal microscopy. *J. Pulp Paper Sci. 24*(7):224–230.
42. Jang, H. F., Robertson, A., and Seth, R. S. (1992). Transverse dimensions in wood pulp fibers by confocal laser scanning microscopy and image analysis. *J. Mater. Sci. 27*:6391–6400.
43. Jayme, J., and Gessler, H. (1963). Contribution to the study of the biological properties of sulfate pulp: Morphology and swellability. *Das Papier 17*(1):5–14 (in German).
44. Jordan, B. D. (1988). A simple image analysis procedure for fiber wall thickness. *J. Pulp Paper Sci. 14*(2):J44–J45.
45. Jordan, B. D., and Page, D. H. (1979). Application of image analysis to pulp fiber characterization. Part 1. *Proc. International Paper Physics Conf. Tech. Sect.* CPPA, Montreal, pp. 105–114.
46. Jordan, B. D., Nguyen, N. G., and Page, D. H. (1982). An image analysis procedure for pulp coarseness determinations. *Paperi Puu 64*(11):691–701.

47. Karlsson, H., and Fransson, I.-P. (1994). STFI Fiber Master gives the papermakers new muscles. *Svensk Papperstidn. 97*(10):26–28 (in Swedish).
48. Karnis, A. (1994). The mechanism of fiber development in mechanical pulping. *J. Pulp Paper Sci. 20*(10):J280–J288.
49. Kauppinen, M. (1988). Prediction and control of paper properties by fiber width and cell wall thickness measurement with fast image analysis. *Proc. PTS Symposium Image Analysis for Quality Assurance and Enhanced Productivity*. PTS, Munich, pp. 1–8.
50. Kellogg, R. U., Thykeson, E., and Warren, W. G. (1977). The influence of wood and fiber properties on kraft converting. Paper quality. *Tappi 58*(12):113–116.
51. Kibblewhite, R. P., (1982). The quality of Radiata pine papermaking fibers. *Appita 35*(4):289–298.
52. Kibblewhite, R. P., and Bailey, D. G. (1987). Measurement of fiber cross-section dimensions using image processing. *Proc. International Paper Physics Conf., Tech. Sect.* CPPA, Montreal, pp. 25–35.
53. Kortschot, M. T. (1997). The role of the fiber in the structural hierarchy of paper. In: *Fundamentals of Papermaking*. Proc. 11th Fundamental Research Symposium, Vol. 1. C. F. Baker, ed. Pira Int., Leatherhead, U.K., pp. 351–399.
54. Laine, J. (1996), The effect of cooking and bleaching on the surface chemistry and charge properties of kraft pulp fibers. D. Tech. Thesis, Helsinki Univ. of Technology, Lab. Forest Products Chemistry.
55. Larson, R. R. (1966). Changes in chemical composition of wood cell walls associated with age in Pinus resinosa. *Forest Prod. J. 16*(4):37–45.
56. Leney, L. (1981). A technique for measuring fibril angle using polarized light. *Wood Fiber 13*(1):13–16.
57. Leopold, B. (1966). Effect of pulp processing on individual fiber strength. *Tappi 49*(7):315–317.
58. Matolcsy, G. A. (1975). Correlation of fiber dimensions and wood properties with the physical properties of kraft pulp of Abies balsamea L. (Mill). *Tappi 58*(4): 136–141.
59. Meylan, B. A. (1967). Measurement of microfibril angle by X-ray diffraction. *Forest Prod. J. 17*(5):51–58.
60. Olsson, J. A., Robertson, A. G., Finnigan, T. D., and Turner, R. R. H. (1995). An analyzer for fiber shape and length. *J. Pulp Paper Sci. 21*(11):J367–J373.
61. Paavilainen, L. (1989). Effect of sulfate cooking parameters on the papermaking potential of pulp fibers. *Paperi Puu 71*(4):356–363.
62. Paavilainen, L. (1989). Importance of coarseness and fiber length in papermaking. In: *Quality Assurance: Its Management and Technology*. TAPPI Press, Atlanta, pp. 269–279.
63. Paavilainen, L. (1990). Importance of particle size—fiber length and fines—for the characterization of softwood kraft pulp. *Paperi Puu 72*(5):516–526.
64. Paavilainen, L. (1991). Influence of morphological properties of wood fibers on sulfate pulp fiber and paper properties. *Proc. International Paper Physics Conf.* TAPPI Press, Atlanta, pp. 383–395.
65. Paavilainen, L. (1993). Importance of cross-dimensional fiber properties and coarseness for the characterization of softwood sulfate pulp. *Paperi Puu 75*(5):324–351.
66. Paavilainen, L. (1993). Influence of fiber morphology and processing on the softwood sulfate pulp fiber and paper properties. D.Tech. Thesis, Helsinki Univ. of Technology, Lab. Pulping Technology.
67. Paavilainen, L. (1993). Wet fiber flexibility and collapsibility of softwood sulfate pulp fibers. *Paperi Puu 75*(9/10):689–702.
68. Paavilainen, L. (1994). Bonding properties of softwood sulfate pulp fibers. *Paperi Puu 76*(3):162–173.

69. Paavilainen, L. (1994). Influence of fiber morphology and processing on the papermaking potential of softwood sulfate pulp fibers. *Proc. TAPPI Pulping Conference*. TAPPI Press, Atlanta, pp. 857–868.
70. Paavilainen, L. (1997). Non-wood fibers in paper and board grades. European perspective. TAPPI Short Course: Non-Wood Fiber for Paper in North America. TAPPI, Atlanta, GA.
71. Paavilainen, L. (1998). Quality-competitiveness of Asian short-fiber raw materials in different paper grades. *Proc. Asian Paper '98*, Applied Technology Conf., Session 1. Miller Freeman, Singapore.
72. Page, D. H. (1969). A method for determining the fibrillar angle in wood tracheids. *J. Microsc. 90*(2):137–142.
73. Page, D. H. (1989). The beating of chemical pulps: The action and the effects. In: *Fundamentals of Papermaking*. Proc. 9th Fundamental Research Symposium, Vol. 1. C. F. Baker and V. W. Punton, eds. Mech. Eng. Pub., London, pp. 1–38.
74. Page, D. H., El-Hosseiny, F., Winkler, K., and Lancaster, A. P. S. (1977). Elastic modulus of single wood pulp fibers. *Tappi 60*(4):114–117.
75. Panshin, A. J., and de Zeeuw, C. (1981). *Textbook of Wood Technology*, Vol. 1. *Structure Identification, Uses and Properties of Commercial Woods of United States and Canada*. 3rd ed. McGraw-Hill, New York.
76. Pihlava, M. (1998). Fiber deformation and strength loss in kraft pulping and softwood. Licentiate Thesis, Helsinki Univ. Technology, Lab. Pulping Technology.
77. Pleasants, S., Batchelor, W., and Parker, I. H. (1997). Measuring the fibril angle of bleached fibers using micro-Raman spectroscopy. Proc. 51st Appita Annual General Conf. Appita, Melbourne, Vol. 1, pp. 545–549.
78. Preston, R. D. (1974). *The Physical Biology of Plant Cell Walls*. Chapman and Hall, London.
79. Prud'homme, R. E., and Noah, J. (1975). Determination of fibril angle distribution in wood fibers: A comparison between the X-ray diffraction and the polarized microscope methods. *Wood Fiber 6*(4):282–289.
80. Rautonen, R., Rantanen, T., Toikkanen, L., and Malinen, R. (1996). TCF bleaching to high brightness: Bleaching sequences and pulp properties. *J. Pulp Paper Sci. 22*(8):J306–J314.
81. Reed, A. E., and Clark, J. d'A. (1950). An instrument for rapid fractionation of pulp. *Tappi 33*(6):292–298.
82. Robertson, G., Olson, J., Allen, P., Chan, B., and Seth, R. (1999). Measurement of fiber length, coarseness, and shape with the fiber quality analyser. *Tappi J 82*(10):93–98.
83. Rydefalk, S., Pettersson, T., Jung, E., and Lundqvist, I. (1981). The STFI optical fiber classifier. *Proc. International Mechanical Pulping Conf.*, Session II:4. Eucepa, Oslo.
84. Saarela, S. (1990). Influence of raw material quality, handling and cooking on the fiber properties of reinforcement fibers used in printing papers. M.Sc. Thesis, Helsinki Univ. Technology, Lab. Pulping Technology (in Finnish).
85. Sastry, C. B. R., Kellogg, R. M., and Wellwood, R. W. (1973). Measurement and prediction of fiber coarseness in western hemlock. *Tappi 59*(4):158–161.
86. Seth, R. S. (1990). Fiber quality factors in papermaking. I. The importance of fiber length and strength. *Proc. Mater. Res. Soc. Symp.*, Vol. 197. Materials Research Society, pp. 125–141.
87. Seth, R. S. (1990). Fiber quality factors in papermaking. II. The importance of fiber coarseness. *Proc. Mater. Res. Soc. Symp.*, Vol. 197. Materials Research Society, pp. 143–161.
88. Sjöström, E. (1992). *Wood Chemistry: Fundamentals and Applications*. 2nd ed. Academic Press, New York.

89. Stuart, S.-A., and Evans, R. (1995). X-ray diffraction estimation of the microfibril angle variation in eucalyptus wood. *Appita 48*(3):197–200.
90. Sugden, E. A. N. (1968). A semi-automated method for the determination of average fiber length by projection. *Pulp Paper Mag. Can. 69*(15):71–76.
91. Sundholm, J. (1999). *Mechanical Pulping. Papermaking Science and Technology*. Fapet Oy, Helsinki.
92. Surma-Slusarka, B., and Surewicz, W. (1981). The effect of chemical composition and anatomical structure of wood on the yield and properties of kraft pulps. *Cellulose Chem. Technol. 15*(1):77–97.
93. Svedman, M., Luhtanen, M., and Tikka, P. (1996). Effect of softwood morphology and chip thickness on pulping with displacement kraft batch cooking. *Proc. TAPPI Pulping Conf.* TAPPI Press, Atlanta, pp. 767–778.
94. Tasman, J. E. (1972). The fiber length of Bauer-McNett screen fractions. *Tappi 55*(1):136–138.
95. Timell, T. E. (1986). *Compression Wood in Gymnosperms*. Springer-Verlag, Heidelberg.
96. Tsoumis, G. (1968). *Wood as Raw Material*. Pergamon Press, Oxford.
97. Unger, E. W., and Freund, F. (1975). New developments of fiber length analysis by projected image. *Zellstoff Papier 24*(5):143–160 (in German).
98. Uprichard, J. M., and Lloyd, J. A. (1980). Influence of tree age on the chemical composition of Radiata pine. *N.Z. J. Forestry Sci. 10*(3):551–557.
99. Vasconcelos Dias, R. L., and Claudio da-Silva, E., Jr. (1985). Pulp and paper properties influenced by wood density. *Fundamentals of Papermaking*. Proc. 8th Fundamental Research Symposium, Vol. 2. V. Punton, ed. BPBSIF, Mech. Eng. Pub., London, pp. 7–35.
100. Velling, P. (1974). Phenotypic and genetic variation in wood. Basic density of Scots pine (Pinus silvestris L.). *Folia Forestalia 188* (in Finnish).
101. Wangaard, F. F. (1962). Contributions of hardwood fibers to the properties of kraft pulps. *Tappi 45*(7):548–556.
102. Wegelius, Th. (1946). Properties of Finnish spruce. *Svensk Papperstidn. 49*(3):51–61 (in Swedish).
103. Wellwood, R. W. (1957). Pulp quality and the forester. *Pulp Paper Mag. Can. 58*(2): 89–91.
104. Ye, C., and Sundström, M. O. (1997). Determination of S_2 fibril angle and fibril-wall thickness by microscopic transmission ellipsometry. *Tappi 80*(6):181–190.
105. Ye, C., Sundström, M. O., and Remes, K. (1994). Microscopic transmission ellipsometry: Measurement of the fibril angle and the relative phase retardation of single intact wood pulp fibers. *Appl. Opt. 28*(33):6626–6637.

14
MECHANICAL PROPERTIES OF FIBERS

RICHARD E. MARK*
Empire State Paper Research Institute
State University of New York
College of Environmental Science and Forestry
Syracuse, New York

*Retired.
Section IV.C is authored by José Iribarne, Engineering and Technical Services, Solvay Paperboard, Syracuse, New York.

I. INTRODUCTION

The technical uses of fibers in paper and paperboard, textiles, cordage, and fiber-reinforced composite materials have inevitably created a need for better characterization of fiber mechanical properties. Much of the original work in fiber testing was carried out by researchers who were primarily interested in textile fibers. Considerable use has been made of these pioneering studies by investigators in the paper field, who have made many modifications in textile fiber testing in order to provide information that is more specifically applicable to paper and board. Inherently, some mechanical characteristics (e.g., fiber friction properties) are of greater importance in the textile field, whereas others (e.g., fiber collapse) are of greater importance to papermakers. Fiber twisting properties are of interest to both textile and paper investigators, but for entirely different reasons.

Cellulosic (plant) fibers are overwhelmingly the substance from which paper and board are made. Wood fibers greatly predominate as the furnish for modern paper, and these fibers are structurally quite similar to other natural cellulosic fibers, that is, bark or bast fibers, leaf and stalk fibers, and (to a lesser extent) the fibers of seeds and fruits. The use of other types of fibers in papermaking has shown great promise and development. However, the relatively low cost and availability of the plant fiber resources available to the paper industry have kept the mineral and animal fibers and synthetics, including petroleum-based, metal, and ceramic fibers, relatively restricted to specialty papers and nonwoven materials.

This chapter is organized into seven principal sections. Sections II–V consist of a discussion of the organization of a cellulosic fiber from the standpoint of the mechanical functions of the structural components and how these components act together as a composite material. Section VI discusses some important physical properties of papermaking fibers. The seventh and largest section of the chapter is concerned with specific mechanical tests of fibers (and some indirect approaches).

II. STRUCTURE AND FUNCTION IN THE PLANT FIBER

A. Structure

In most native plant fibers, the composition of a small element of any layer within the fiber wall consists of a variety of materials.

A crystalline, filamentous material (the microfibril), which is composed nearly exclusively of cellulose I and is impenetrable to water, accounts for the largest fraction of the solid phase. The cellulose content of both hardwoods and softwoods is normally in the range of $42 \pm 2\%$ but is generally lower in compression wood and higher in tension wood and in bark and leaf fibers. For example, the cellulose contents of jute, flax, and sisal are approximately 63%, 71%, and 69%, respectively. Stalk fibers (e.g., bamboos) may have cellulose contents higher or lower than those of wood, depending on whether they are more woody or grasslike in nature. As for the seed hair fibers, cotton in the raw state has by far the highest cellulose content of any large commercial source. The content ranges from 88% to 96% of the dry weight. Kapok, on the other hand, contains 50–65%. Each of these percentages is with respect to the dry solid mass of the fiber. However, under normal ambient conditions, cellulosic microfibrils occupy a lower proportion of the total volume because air and water are present within the fiber cell wall.

In wood and most other plant fibers, two interpenetrating solid-phase systems—one composed of an extensively branched, three-dimensional, amorphous polymer (lignin) and one composed of a complex of relatively linear (but nevertheless partially branched), partially paracrystalline polymers of a variety of molecular sizes and solubilities (hemicelluloses)—account for most of the remaining solid fraction and create a matrix of polymer materials in which the filamentous microfibrils are embedded. Some, perhaps almost all, of the hemicellulose polymers exhibit little or no crystallinity. The fibers of cotton contain no lignin; flax fibers contain only a very small amount. On the other hand, compression wood and certain bark fibers such as those of redwood (*Sequoia sempervirens*) contain over 35%. All plant fibers contain at least some hemicellulose. The chemical pulping of wood principally removes lignin constituents; thus, the chemical pulp fiber has an increased proportion of cellulose and other polysaccharides.

A fourth, very delicate and tenuous, interpenetrating system of solid inorganic matter exists, in addition to the lignin and hemicelluloses, that remains as a recognizable structure when wood is carefully ashed [339]. It comprises only a small proportion, on the order of 1%, of the dry mass of wood. The same proportion of ash is normally found in other types of plant fibers. It should be noted that there are substantial variations. Most conifer woods tend to be low in ash content (less than 0.5%), whereas eucalyptus and certain other tropical hardwoods contain several percent. Also high in ash content are the fibers of bamboo and grasses, running as high as 5%, and certain cereal straws such as rice, which are high in silica and may have over 13% ash content [33].

Two more interpenetrating systems can be ascribed to the presence of water and air within the cell wall structure. Both water and air are always present, but their proportions vary with the dryness of the ambient environment. If little water is present, as in the case of wood chips or other fiber sources stored in a low humidity environment, the water will exist only in the form of (1) water of constitution and (2) adsorbed water, wherein the H_2O molecules are hydrogen-bound to the surfaces of the carbohydrate and protein molecules that are present. This binding may be monomolecular or polymolecular. If the fibers are wet or subject to extreme conditions of humidity, there may also be free water present, which penetrates and swells the spaces between microfibrils. Air, or in general gas, occupies space not occupied by solid or liquid substances.

Substances such as pectic substances, waxes, fats, starches, resins, gums, organic crystals and other extractives, and proteins are usually not regular components of the spaces or unit volumes of fibers but exist in specific locations. Except for the proteins and certain of the extractives, they are typically found only in the fiber lumens or on the fiber surface as inclusions or depositions and are thus not considered intrinsic molecular components.

B. Function

From the foregoing, it is evident that any given layer in a plant fiber can be described as a filamentary composite. Because the fiber is composed of several such layers (of different orientations), the fiber itself may be described as a laminated composite.

Based on experimental evidence (direct and indirect) and deduction from experience with analogous industrial materials and structures, we can assign structural functions to each of the components found in a typical unit volume of fiber wall. As with industrial multiphase materials, certain of the components in fiber cell walls act as the bulking and stiffening matrix in which the reinforcement is embedded. In fibers and other plant cell walls, the microfibril provides the structural reinforcement. The microfibril is principally if not exclusively cellulose I, so the mechanical properties of that crystalline substance are used identically for the properties of the microfibril. In some fibers, especially those of coniferous woods, some of the glucomannan hemicellulose is intimately associated with native cellulose in the microfibril and must therefore also be considered part of the reinforcing phase, or framework, of the wall. However, up until now, most computations for the elastic constants and other physical and mechanical properties of the solid-state microfibrillar entity have dealt with it as if it were composed exclusively of crystalline cellulose I, without any defects.

Matrix and Framework Components The matrix phase in a cellulosic fiber cell wall consists of the relatively unoriented or amorphous short-chained or branched polymers of various molecular species, that is, lignin and hemicellulose and any other polymers such as pectopolyuronides that might be present in a given layer, plus all of the tiny voids and the gases and adsorbed water associated with them—in other words, everything else that surrounds the microfibrils in their local environment. An up-top-date (as of 1999) source for characterization of the matrix components is Ref. 10.

The microfibrils, as noted earlier, constitute the structural framework and consist essentially of cellulose I. The last decade and a half have witnessed very dramatic developments in our understanding of the structure of this material.

Despite the fact that cellulose is the world's most abundant biopolymer, that it has a myriad of industrial applications, and that its basic structure has been known since the 1830s [291], its structural details have been the subject of intense scientific scrutiny and controversy since that time.

Native cellulose is also called cellulose I to distinguish it from modified or regenerated celluloses (cellulose II, III, etc.). It has long been known that cellulose I is a long-chain polymer of β-D-glucose that occurs in a largely crystalline state [291] in the microfibrils of all higher plants as well as many algae and some bacteria and tunicates. However, the polymer is never "pure," because other polymers and segments of polymers (such as hexose or pentose residues) are typically intimately associated with it. Various species of wood are found to contain (on a dry basis) 40–55% cellulose, 15–35% lignin, and 25–40% hemicelluloses. As stated earlier, the cellulose is organized within the microfibrils, which can be seen readily in the scanning electron microscope (SEM). See Figs. 1 and 2.

A microfibril can be broken down (for example, ultrasonically—see Fig. 3) into *elementary fibrils*, which are considered to be basic cellulose crystalline structures. Hanna and Côté [122] showed that elementary fibrils vary in size, with the smallest (1.0–1.5 nm width) occurring in the cambial regions. The sizes become larger during the maturation of the fiber cell wall, by either surface addition or aggregation or both. Newman [261] recently tabulated the range of widths of elementary fibrils in conifers from five references containing measurements obtained by electron microscopy and eight references with estimates of the widths by X-ray diffraction line broadening. The mean of these tabulated values is 2.9 nm. Microfibrils are aggregates of elementary fibrils and contain hundreds or more of cellulose chains in crystalline arrays [46,136,329].

Figure 4 shows two schematic drawings of elementary fibrils in which the cellulose chains are represented as rods. Thus, Fig. 4a illustrates an elementary fibril with 36 chains, which has an approximately rectangular cross section with dimensions of 3.6 × 3.2 nm. This model is supported by the electron microscopic evidence for marine algae such as *Valonia*. However, the above-noted studies also indicate that for conifer woods, bundles of fewer than 36 cellulose chain molecules (of unspecified shape) exist. Newman [261] suggests that the hexagonal shape shown in Fig. 4b may be an appropriate model for the elementary fibril of cellulose in wood, citing nuclear magnetic resonance (NMR) and other evidence:

> NMR spectra of wood show signals assigned to cellulose chains with hydroxyl groups exposed on surfaces of elementary fibrils, clearly distinguished from signals assigned to chains buried within the elementary fibril. The spectra do not show signals with the characteristics expected for noncrystalline cellulose.

These conclusions have great significance for our understanding of cellulose I.

Some of the more significant advances in understanding cellulose structure from the standpoint of its physical and mechanical properties are briefly mentioned below. There are other aspects of research on native cellulose, such as the modes of biosynthesis in various species, that are also actively studied but lie outside the purview of this book.

Considerable effort has been made in prior years to determine the length or molecular weight of the cellulose chain molecule. But in its natural state in the cell wall of a fiber from a tree or other woody plant, or from an alga or cotton seed hair, the length of a cellulose chain molecule is impossible to prove, because it is part of a solid-state crystalline structure. What can be measured is the degree of polymerization (DP) of cellulose molecules that have been extracted from this solid state and put into solution. The DP represents the number of glucose residues per chain molecule. Accordingly, the estimates of DP have risen from a few hundred or less in the 1920s [235,291,325] to "averages" of 14,000 in wood, flax, and ramie and 18,000 in the alga *Valonia* by the 1980s [113] as techniques for extraction became gentler. A DP of 14,000 is equivalent to a length slightly greater than 7 μm. But there is no reason to disallow much longer chain lengths, because the natural ends of microfibrils are rarely if ever seen in electron micrographs of plant cell walls. Reports of DP $\approx$ 26,500 for *Valonia* exist [362]. A DP of 36,000 [325] was found for cellulose isolated from flax (*Linum usitatissimum*).

Both acids and alkalies are capable of breaking glycosidic linkages in cellulose chains (and thus reducing DP). Mechanical damage to fiber cell walls, however, usually has little effect on DP [260]. Mechanical properties of fibers are relatively insensitive to DP unless the reductions are severe [141].

An important line of investigation in cellulose that has many implications concerning physical and mechanical properties has involved the related questions of whether chains are extended or folded, whether chains are all aligned in one direction within a given microfibril, and the crystal unit cell type and configuration in the lattices within the microfibril.

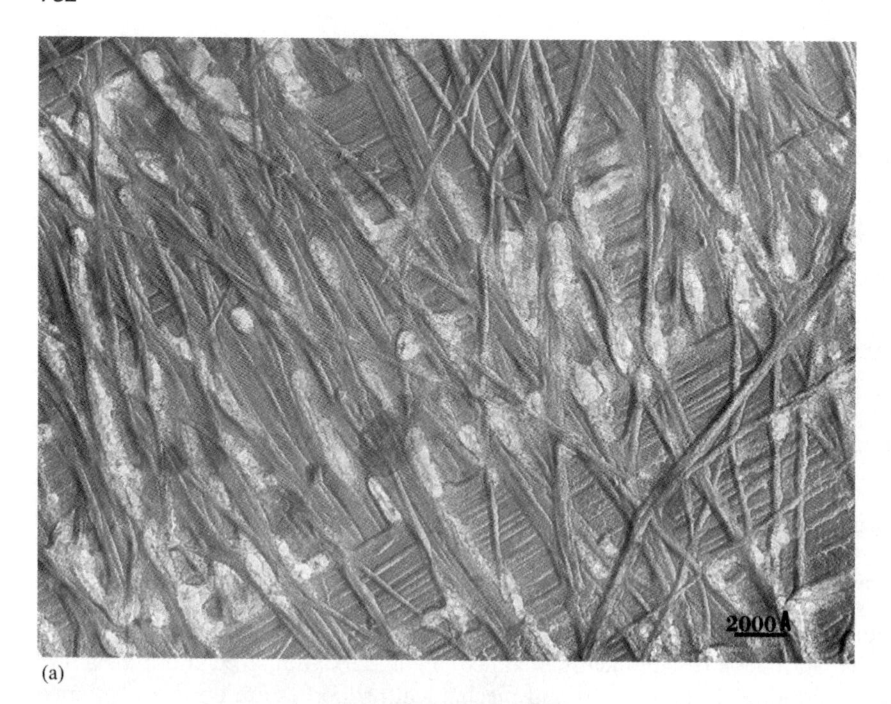

(a)

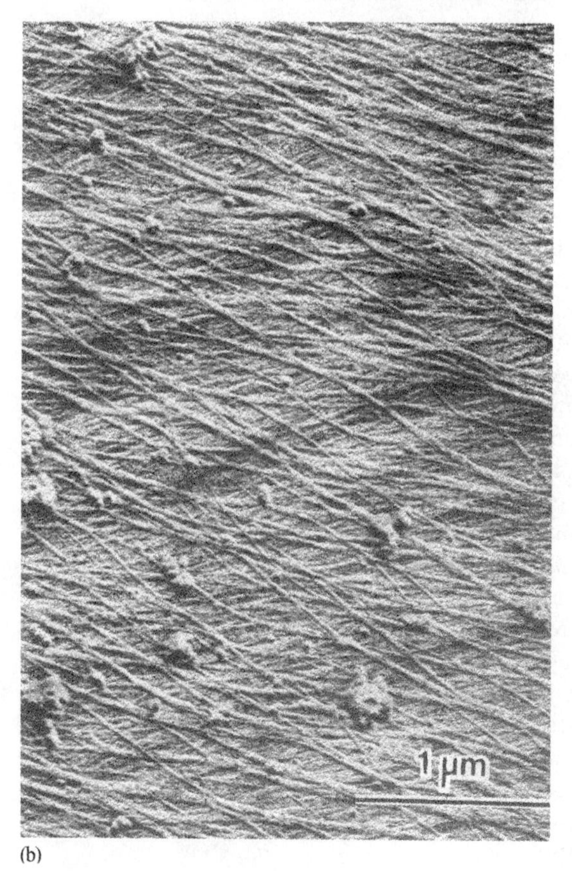

(b)

Fig. 2 Replica of microfibrils in a *Picea jezoensis* tracheid secondary wall, showing orientation of S1 and S2 in the region surrounding a pit. Bar = 1 μm. (From Ref. 124.)

Various models have been proposed for the spatial structure and packing of cellulose I chains in the crystallite. Accurate measurements of the crystalline unit cell have been published for at least two species [91,367]. Because cellulose I can be converted to cellulose II (which is considered to be bipolar) as a result of mercerization or solubilization and regeneration, many of the early proposals assumed that the chains in cellulose I also have bipolarity, i.e., they alternate directionally in the crystallographic unit cells so that half the chains run in one direction, half in the other [235,236]. This so-called antiparallel configuration model was largely based on X-ray diffraction studies of fibers. However, it was soon recognized that the evidence for antiparallelism was ambiguous, because various microfibrils in a typical fiber or other cell wall could be aligned up or down even if all the chains within one microfibril were aligned in one direction only. Obviously, any possibility of folded chains (often proposed in the literature) within microfibrillar structures would depend on positive determination of antiparallelism therein.

←

Fig. 1 Plant cell wall microfibrils. (a) Crossed parallel microfibrils in two successive laminae of the secondary cell wall of the marine alga *Valonia*. Note that the two laminae tend to align at right angles to each other. (Electron micrograph courtesy of Dr. R. B. Hanna [121].) (b) Microfibrils forming in a differentiating tracheid wall of *Pinus densiflora*. Note that the orientation of the newly deposited microfibrils does not lie at a right angle with respect to the underlying S1 layer. (From Ref. 98.)

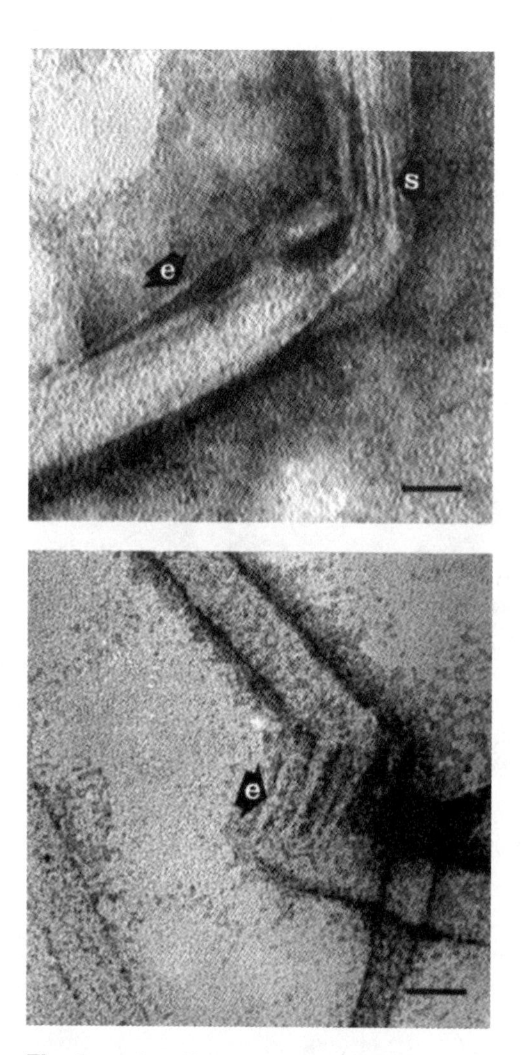

Fig. 3 Microfibrils of *Valonia* subjected to ultrasonication. The structure has broken into subunits of two widths: $e = 3.5\,\text{nm}$; $s < 3.5\,\text{nm}$. (Transmission electron micrographs courtesy of Dr. R. B. Hanna [121].)

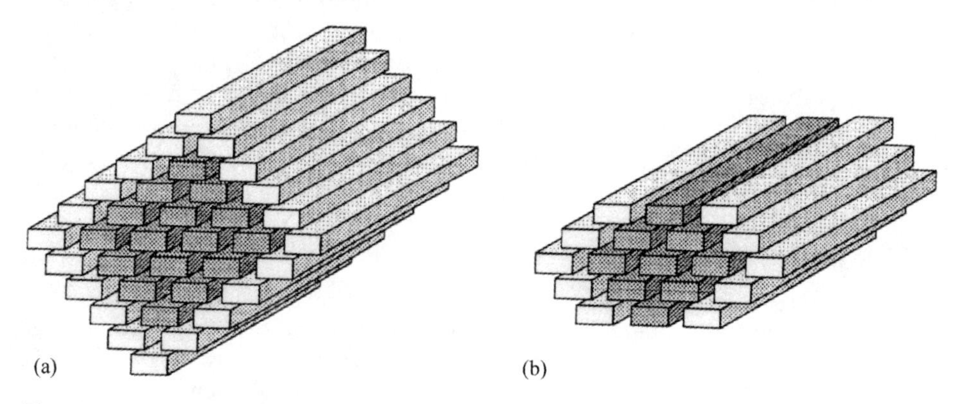

Fig. 4 Portions of elementary fibrils containing (a) 36 and (b) 19 chains. Pale shading indicates chains with hydroxyl groups exposed on surfaces. (From Ref. 261.)

The problem was examined in the 1960–1980 period from several different perspectives that all pointed to parallel chain packing of native cellulose (cellulose I). Two papers by Muggli and associates [248,249] showed that when fibers are sectioned transversely at various intervals, the molecular length (DP) varies in accordance with expectations based on extended chains in the principal layer (S2 layer) of the fiber wall. Murphey [251] and Jentzen [161] demonstrated that crystallinity increases if fibers are strained in tension, which indicates that the chain molecules align in more perfect lattices. Again, these results would meet expectations if the chains were extended but not if they were folded. A series of studies [107,158,214, 222] from the perspectives of crystal physics and solid mechanics concluded that the experimentally determined mechanical properties of wood fibers could not be achieved if the chains were folded and refuted some folded-chain proposals specifically. Ultrasonic studies [96,102,121] showed that a rodlike substructure of elementary fibrils was visible in electron micrographs when crystalline microfibrils were shattered by ultrasound (see Fig. 3); the elementary fibrils had diameters of only 3.5 nm—too small to permit folding of chains within yet still retain a repetitive lattice structure. From the spectroscopic side, it was concluded by Gardner and Blackwell [103] and Sarko and coworkers [312,367] that stereochemical packing analysis of the X-ray structure showed that crystallites of cellulose I are statistically much more likely to be based on parallel chains. Nevertheless, discussion, disputes, and proposals concerning chain folding and parallel versus antiparallel chain packing have continued [195,239,322].

Since 1980, six major technical improvements have permitted further clarification of cellulose structure questions: electron diffraction analysis of very small structures (e.g., a single microfibril), Fourier transform intrafred (FT-IR) spectroscopy, computer-based stereochemical (conformational) studies of enhanced diffraction patterns, molecular dynamics simulation, atomic force microscopy (AFM), and solid-state cross polarization/magic angle spinning methods in nuclear magnetic resonance spectroscopy (CP/MAS NMR). The latter method enhances the spectroscopic signal from carbon 13, eliminates dipolar line broadening from protons, and eliminates chemical shift anistropy. The result is to achieve spectra of exceptionally high resolution for crystalline cellulose samples. The ^{13}C CP/MAS NMR technique enabled Atalla and VanderHart [15,360] to conclude that all native celluloses are really a mixture of two crystalline modifications, which they designated cellulose Iα and Iβ. The alpha form is dominant in bacterial and algal celluloses, whereas the beta form is dominant in the celluloses of higher plants, according to these reports. In 1991, Sugiyama et al. [331] reported the results of an electron diffraction study on microfibrils of the alga *Microdictyon* that confirmed the existence of these two phases and concluded (1) that both cellulose Iα and Iβ are exclusively composed of extended chains in "rigorous alignment" with the microfibril axis; (2) that one form will dominate, then the other in sequential contiguous crystalline blocks as a *Microdictyon* microfibril is examined at 50 nm intervals (the periodicity is about 100 nm); (3) that cellulose Iα has a triclinic unit cell with a single chain in the repeat unit and cellulose Iβ has two chains and is monoclinic; and (4) that the Iα form is metastable and will convert irreversibly to Iβ if the material is annealed in 0.01 N NaOH at 260°C. (See also Refs. 330 and 371.) The latter extends the variation possible in cellulose Iα. At least four types of microfibrils (corresponding to different modes of synthesis) are known for algae.

Reducing-end staining trials showed that the cellulose microfibril has a parallel-chain structure for both forms Iα and Iβ [194].

Lennholm et al. [202] studied bleached pine kraft pulp using the ^{13}C CP/MAS NMR spectroscopy approach. They found small but systematic differences in the proportions of cellulose Iα and Iβ when kraft cooking temperatures of 160°C and 180°C were compared. The latter had a higher proportion of cellulose Iβ. When made into unbeaten pulp handsheets, it also demonstrated higher tensile stiffness indices.

Evans et al. [88] used several methods to show that the degree of crystallinity increases as kraft pulping proceeds. The changes are ascribed to the preferential removal of less ordered carbohydrate materials while the amount of crystalline cellulose that remains in the fiber stays about constant unless the pulping method reduces the yield to 57% or less, in which case damage to the cellulose structures occurs.

The type and proportion of cellulose in the tree appears to be different from the condition in pulp of the same species. Newman [262] obtained resolution-enhanced ^{13}C CP/MAS NMR spectra for the woods of seven different tree species—three conifers and four hardwoods. An experiment of this type is complicated by the large presence of noncellulose components in wood, which results in other, weak signals (e.g., from glucomannan) and background noise. However, signal-to-noise ratios can be improved by lengthy experiments from which the data can be averaged over several days. Newman found that cellulose in the woods of the three softwoods average about 1.8 : 1 in the ratio of cellulose Iα to Iβ (similar to the cellulose in algae). The Iα : Iβ ratio of the four hardwoods studied by Newman averaged 0.8 : 1, which puts these woods in the same cellulose group as cotton and ramie. He suggests that the fact that both hardwood and softwood sulfate pulps have predominantly Iβ-type cellulose is influenced by the temperature (typically 170°C) reached during kraft pulping; this was also indicated by other studies such as Ref. 202.

The enzymatic action of brown rot fungi on wood is the opposite of a desirable chemical pulping action—that is, pulping preferentially removes lignin, whereas the brown rots preferentially remove cellulose and other carbohydrates. Kim and Newman [180] used NMR in a study to observe the changes in cellulose and other pine wood components that occurred at weight losses of 0%, 8.6%, and 32% caused by a brown rot fungus. The crystalline structure at the surfaces of the cellulose microfibrils is degraded extensively before any damage occurs to the crystal interior. Thus, the surface-to-interior ratio of crystalline cellulose in this wood dropped from 0.8 to 0.5 during the change from 0% (original wood) to 32% decay loss. Kim and Newman conclude that their results provide evidence for "selective degradation of the triclinic Iα form" of the cellulose, which seems to be concentrated in the surface areas of the microfibrils. They also concluded that "solid state NMR spectra show little evidence for amorphous or noncrystalline cellulose in wood."

Baker et al. [17] showed that direct imaging by atomic force microscopy can be used to observe triclinic crystal structure of cellulose I on the surface of *Valonia* microcrystals (segments of microfibrils) with a resolution of approximately 0.5 nm. Detection of the 1.04 nm periodicity of the cellobiose units (see Fig. 5) is evident in the surface images; less clearly,the 0.52 nm glucose monomer intervals and intermolecular spacings of the same order of magnitude are observed. Analysis of the images of different planes shows that it is possible to detect regions with triclinic

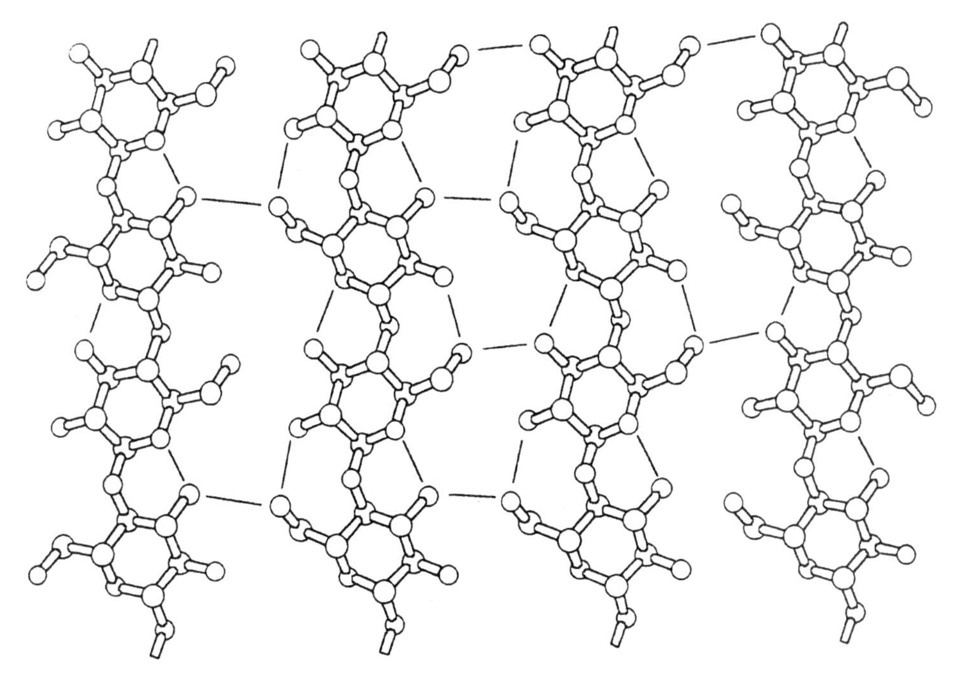

Fig. 5 Four anhydroglucosyl structural units (two cellobiose units) form each of four chains, represented by a horizontal row of four rods in either part of Fig. 4. Hydrogen atoms have been omitted, but H bonds are indicated by thin lines. (From Ref. 261.)

characteristics, indicating that a distinction between triclinic and monoclinic phases is possible with AFM. The distinction between these two phases is discerned by noting a different shift pattern of the cellulose chains, each by one-fourth of the cellobiose period. The monoclinic array will exhibit a staggered pattern, whereas the triclinic will shift diagonally in successive layers of chain molecules, as shown in Fig. 6.

Solid Mechanics of Cell Wall Components Among all properties, the set of elastic constants applicable to a given structural solid material is generally the most important. These elastic constants can be measured physically in laboratory tests on materials that are available in the form of large specimens. For microcrystalline materials such as cellulose I, these elastic constants can be derived on the basis of known crystal structure and the energetics of bond stretching and bond angle change within that structure. In the case of the cellulose I unit cell (the repeat unit in the space lattice), we equate the sum of the energies associated with deforming each bond length and bond angle within the unit cell to the overall elastic strain energy of the unit cell itself. In its simplest form, we can make an assumption that a unit cell of the cellulose I crystal undergoes a change in length in the direction of the chain axis so that the repeating unit length $L = 1.036\,\text{nm}$ experiences an incremental displacement

$$dL = Le_3 \tag{1}$$

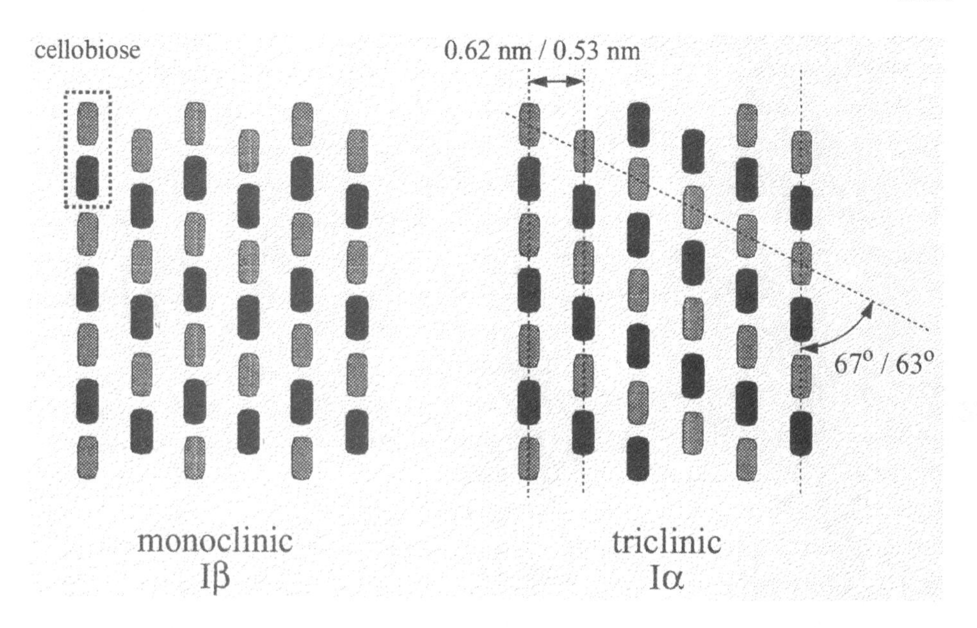

Fig. 6 Schematic diagrams showing the differences between monoclinic and triclinic forms of cellulose I. Each rectangle represents a single glucose unit, with a pair of glucose units (cellobiose) constituting the true repeat unit in the cellulose I crystal. (From Ref. 17.)

where e_3 is the strain in the direction of the chain axis. The axial deformation Le_3 is related to the deformations of all the primary and secondary bond lengths and angles within the unit cell. An apportionment of this displacement is made (vectorially) on the basis of the orientations of each bond length and angle. For a prescribed dL it is assumed that the bond deformation energies are minimized. The method yields an inverse gross spring force constant $1/K$ for the 1.036 nm repeating unit as the sum of terms containing $1/K_i$ values for all the bonds therein. Then, the energy in straining the unit cell volume is equated to the overall bond deformational energy by

$$W = \frac{1}{2} A_{33} e_3^2 \Delta V = \frac{1}{2} K (dL)^2 \tag{2}$$

where

$A_{33} =$ elastic stiffness constant in the chain axis direction
$\Delta V =$ unit cell volume

Solution of Eqs. (1) and (2) yields

$$A_{33} = \frac{KL^2}{\Delta V} \tag{3}$$

It is possible to derive a full set of elastic constants by this method [158,216,218]. Although a four-index notation is sometimes employed for elastic constants, a simpler operation is to abbreviate to a two-index system.

The simplifications are made as follows for stresses:

$$\begin{aligned} \sigma_{11} &= \sigma_1 & \sigma_{23} &= \sigma_4 \\ \sigma_{22} &= \sigma_2 & \sigma_{13} &= \sigma_5 \\ \sigma_{33} &= \sigma_3 & \sigma_{12} &= \sigma_6 \end{aligned} \tag{4}$$

and similarly for strains:

$$\begin{aligned} \epsilon_{11} &= \epsilon_1 & 2\epsilon_{23} &= \epsilon_4 \\ \epsilon_{22} &= \epsilon_2 & 2\epsilon_{13} &= \epsilon_5 \\ \epsilon_{33} &= \epsilon_3 & 2\epsilon_{12} &= \epsilon_6 \end{aligned} \tag{5}$$

Thus, Hooke's law becomes

$$\sigma_i = \sum_{j=1}^{6} S_{ij}\epsilon_j \tag{6}$$

where the S_{ij} are the compliance moduli. Accordingly, the matrix of stiffness moduli A_{ij} would be (for a triclinic crystal)

$$\begin{vmatrix} A_{11} & A_{12} & A_{13} & A_{14} & A_{15} & A_{16} \\ & A_{22} & A_{23} & A_{24} & A_{25} & A_{26} \\ & & A_{33} & A_{34} & A_{35} & A_{36} \\ & & & A_{44} & A_{45} & A_{46} \\ & & & & A_{55} & A_{56} \\ & & & & & A_{66} \end{vmatrix}$$

In the triclinic case, there are 21 independent constants (with reference to orthogonal axes). The matrix of stiffness moduli for a monoclinic crystal becomes

$$\begin{vmatrix} A_{11} & A_{12} & A_{13} & 0 & 0 & A_{16} \\ & A_{22} & A_{23} & 0 & 0 & A_{26} \\ & & A_{33} & 0 & 0 & A_{36} \\ & & & A_{44} & A_{45} & 0 \\ & & & & A_{55} & 0 \\ & & & & & A_{66} \end{vmatrix}$$

The above matrix yields 13 independent constants (with reference to orthogonal axes). For an (assumed) orthorhombic unit cell, nine constants will remain:

$$\begin{vmatrix} A_{11} & A_{12} & A_{13} & 0 & 0 & 0 \\ & A_{22} & A_{23} & 0 & 0 & 0 \\ & & A_{33} & 0 & 0 & 0 \\ & & & A_{44} & 0 & 0 \\ & & & & A_{55} & 0 \\ & & & & & A_{66} \end{vmatrix}$$

Table 1 summarizes the many attempts to calculate the theoretical modulus of elasticity for native cellulose in the chain axis direction. Only a few of the published

Table 1 Theoretical Moduli of Elasticity for Cellulose I

Authors and publication dates	Calc. on molecule or space lattice	Elastic moduli (GPa) in chain direction	Chain configuration in crystal	Full set of elastic constants?	Ref.
Meyer and Lotmar (1936)	—	77–121	Antiparallel [95,236]	No	234
Meyer (1942)	—	117	do.	No	233
Meredith (1956, 1959)	—	89	do.	No	231,232
Lyons (1959)	—	180	do.	No	209
Treloar (1960)	Molecule	57	do.	No	355
Jaswon, Gillis, and Mark (1968)	Molecule	57	do.	Yes	158
Jaswon, Gillis, and Mark (1968)	Lattice (est.)	> 150	do.	No	158
Gillis (1969)	Lattice	246	do.	Yes	104
Gillis (1969)	Lattice	319	do.	Yes	104
Tashiro and Kobayashi (1985)	Molecule	71	Parallel [103,367]	No	347
Tashiro and Kobayashi (1985)	Lattice	173	do.	No	347
Kroon-Batenburg, Kroon, and Northolt (1986)	Molecule	64	do.	No	196
Kroon-Batenburg, Kroon, and Northolt (1986)	Lattice	137	do.	No	196
Tashiro and Kobayashi (1991)	Molecule	~ 70	do.	Yes	348
Tashiro and Kobayashi (1991)	Lattice	168	do.	Yes	348
Reiling and Brickmann (1995)	Lattice	148	Iβ [331]	No	294
Reiling and Brickmann (1995)	Lattice	128	Iα [331]	No	294
Marhöfer, Reiling, and Brickmann (1995)	Lattice	134	Iβ (35%) and Iα (65%) [15,331,360]	No	212

papers provide complete sets of elastic constants—Jaswon et al. [158], Gillis [104], and Tashiro and Kobayashi [348]. Table 2 contains the results of experiments on single ramie fibers subjected to tensile testing (loads in fiber axial direction) and crystallite distortions as measured by X-ray diffraction, wherein some of the tests were on bundles of 100 fibers. Although many inconsistencies appear when a comparison is made of theoretical versus experimental results, one fact emerges clearly: Calculations for cellulose I made on the basis of the primary bonds alone, or the primary and intrachain hydrogen bonds alone (the "molecule" calculations in Table 1), uniformly produce values for chain direction modulus that are smaller than the experimental results. This cannot be.

Table 2 Experimental Moduli of Elasticity (MOE) for Crystalline Native Cellulose[a]

Authors and publication dates	MOE (GPa)	Ref.
Sakurada, Nukushina, Ito (1962)	134	305
Matsuo, Sawatari, Iwai, Ozaki (1990)	120–135	225
Nishino, Takano, Nakamae (1995)	138	263
Ishikawa, Okano, Sugiyama (1997)	90–140	147

[a] Values are based on X-ray diffraction measurements of crystallite distortion.

An examination of a wood pulp or other plant cellulosic fiber wall shows that the orientation of the microfibrils, layer to layer, differs from the fiber axis, often by a large angle. In the primary and S1 layers of wood fibers, the microfibrils lie at large angles to the fiber axis. In the main (S2) layer, the predominant orientation of the microfibrils is often 15–30° or more from the fiber axis and only rarely lies at less than 7° off-axis. S3 is less well aligned than S2. The fiber walls also contain pits and other pore spaces. The matrix materials that remain in the wall after processing are generally much weaker and softer (more compliant) than cellulose. Yet in spite of all the chemical, structural, and geometrical factors that would tend to diminish the cell wall properties, a very linear stress–strain relationship is exhibited by these fibers. Cellulose I is a high modulus material that vastly predominates over the effects of cell wall geometry and nonstructural molecular components [215]. Data or computations that show the cellulose to be weaker and softer than the fibers of which it is a part must be in error. One should accordingly look to the calculations that are based on the actual pattern of primary and secondary bonds in the cellulose I space lattice.

Table 1 reflects three chronological periods of discovery of the nature of the abundant and useful polymer we call cellulose. Up until the end of the decade of the 1970s, the prevalent concepts were that cellulose I possessed an antiparallel chain structure and that interchain hydrogen bonds existed in the diagonal planes of the unit cell. The cited papers in this period were based on these models. The next group of papers on this subject appeared in the 1980s and in 1990–1991. During this period it was generally accepted that the chain configuration was parallel and that the interchain hydrogen bonds were probably confined to the plane of the a axis of the crystal unit cell so as to form a sheetlike structure. The sheets are oriented roughly parallel to the fiber wall and are stacked together by van der Waals forces along the b axis. By the mid-1990s, the Atalla–VanderHart proposals [15,360] for a two-phase (triclinic α and monoclinic β) system in cellulose I had been experimentally confirmed by Sugiyama et al. [331] as previously noted. The beginnings of new theoretical work (utilizing molecular dynamics simulations) on the two-phase structure [212,294] appeared in the literature.

The physical and mechanical behavior of the matrix is dependent on the distribution of its molecular components and the presence or absence of moisture. Because the distribution of the hemicelluloses, lignin, void space, and so forth is nonuniform, there is point-to-point variation in the matrix elastic constants. Regions of dense molecular aggregation will have high local moduli of elasticity and rigidity, and the more open regions where air and/or water is present in the cell wall voids will have small, perhaps insignificant, moduli. The moduli overall will take on some sort of average values, but deformations will take place preferentially in the regions of lower aggregate macromolecular packing density.

As properties of the noncellulose molecular species and/or water content change, the matrix constants will change, whereas undegraded microfibrillar cellulose I is considered invariable, at least within constant temperature conditions. The matrix is collectively taken to be isotropic. Thus, a single value each is calculated for the modulus of elasticity, shear modulus, and Poisson ratio for each physical state condition.

It is important to note that work is required to deform the matrix molecular structure even if the matrix components are not bonded to each other or to the microfibrils. The physical location of the constituent polymer molecules between microfibrils requires that these molecules deform along with the microfibrils.

We can assume that the matrix polymers are subject to deformation via small rotations about various C—C and C—O bonds (see, for example, Fig. 7). Such rotations require less energy than the stretching or bending of primary bonds. Matrix moduli can be anticipated to be low, because deformations of this kind (in a noncrystalline arrangement) can occur so easily. In Fig. 7, a three-dimensional side group is rotated about a carbon–carbon bond whose direction is taken as the X axis. The dimensions in the Y direction will vary considerably with the angle of rotation. The same can be said of the Z direction, normal to the plane of the paper. However, the X dimension will stay constant. If additional rotation occurs in the carbon–oxygen bond, all three dimensions will change in a manner fixed by the relative rotational movement of the C—C and C—O bonds. In a polymer, there are many possible conformations because of the large number of rotatable bonds in each molecule. The energy barriers to intramolecular bond rotation arise from

1. Steric hindrance between substituted groups on the backbone atoms
2. Hydrogen bonding within the large molecule
3. Other attractions between the kinked segments of nonpolar molecules

If hindered bond rotation is the controlling factor, it is possible to make an estimate of the appropriate Young's modulus for an amorphous polymeric matrix. With reference to Fig. 7, let us assume that a C—C bond in a matrix molecule is rotated as shown through a small angle $\Delta\theta$. Furthermore, let us imagine that other molecular segments in the matrix are deforming in such a manner that the spatial domain occupied by this side chain is unchanged in the Z direction. Because the domain in the X direction will not change by rotation, the area A occupied by the

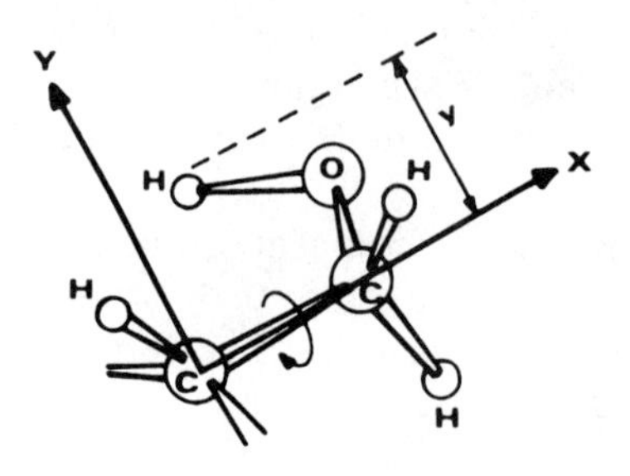

Fig. 7 Rotation of a CH_2OH (hydroxymethyl) group about the C—O axis. (From Ref. 216.)

CH_2OH group in theXZ plane will not change. However, a positive change in the Y direction, Δy, will occur, and a unit of work ΔU is accomplished by the applied force F:

$$\Delta U = F\Delta y \cong Fy\Delta\theta \tag{7}$$

where y is the domain occupied by this group in the Y direction. Accordingly,

$$F = \frac{1}{y}\frac{dU}{d\theta} \tag{8}$$

The stress acting on the area A will be F/A, and thus the matrix Young's modulus E_m can be found by taking the differential of stress, $d(F/A)$, to strain, dy/y. That is,

$$E_m = \frac{d(F/A)}{dy/y} = d\left(\frac{F}{A}\right)d\theta \tag{9}$$

The energy of rotation is estimated from

$$U = P\sin^2\phi \tag{10}$$

where

$U =$ energy of rotation
$P =$ height of energy barrier to C—C rotation ($\cong 2.2 \times 10^{-20}$ J)
$\phi =$ $(3/2)\theta$, as U must be zero when θ is rotated by 120° (a rotation from one minimum bond energy position to the next)

Combination of Eqs. (8) and (9) yields

$$E_m = \frac{d^2U/d\theta^2}{Ay} \tag{11}$$

The value of dU is obtained by differentiating the term $P\sin^2\phi$, and from this

$$\frac{d^2U}{d\theta^2} = \frac{9}{2}P\left(2\cos^2\phi - 1\right) \tag{12}$$

For small rotations, the term in parentheses becomes unity and $E_m = 4.5P/Ay$. Substituting $P = 2.2 \times 10^{-20}$ J, $A = 8.4 \times 10^{-20}\ \text{m}^2$, and $y = 0.15$ nm, we obtain 1.8 GPa. A value of 1.8 GPa for the Young's modulus E_m can be substituted into the appropriate equation for determining the shear modulus G_m of an elastically isotropic material,

$$G_m = \frac{E_m}{2(1+\nu_m)} \tag{13}$$

where ν_m is the Poisson ratio, here taken as 0.30. Solution of Eq. (13) yields $G_m = 0.7$ GPa.

The actual values for E_m and G_m vary substantially with the amount of water present per unit volume of cell wall ΔV. Water acts as a plasticizer for the non-cellulosic (matrix) phase. Water expands the distances between microfibrils and reduces the physical constraints to matrix molecular deformations. Heat also reduces E_m and G_m.

III. PROPERTIES OF THE COMPOSITE

Once the structural functions of the various chemical constituents in the plant cell wall are determined, a variety of modeling systems can be employed to determine the physical and mechanical properties of the composite unit volume element of the fiber wall. In general, the microfibrils are not aligned with the fiber axis. Therefore, a small element of a cell wall layer, such as is shown in Fig. 8, has a filament alignment that differs by some angle α from the axial direction of the fiber.

The appearance in Figs. 8 and 9 of a squarish or slightly rectangular shape of the microfibrils may well represent the ultrastructural arrangement in marine algae and some other plants. In wood fibers, however, there is a tendency for the microfibrils to aggregate in the tilted tangential-longitudinal or XZ plane in the cell wall. The XZ plane corresponds to the *ac* plane in the cellulose I crystal; it contains the interchain hydrogen bonds. The microfibrillar aggregates may therefore create exten-

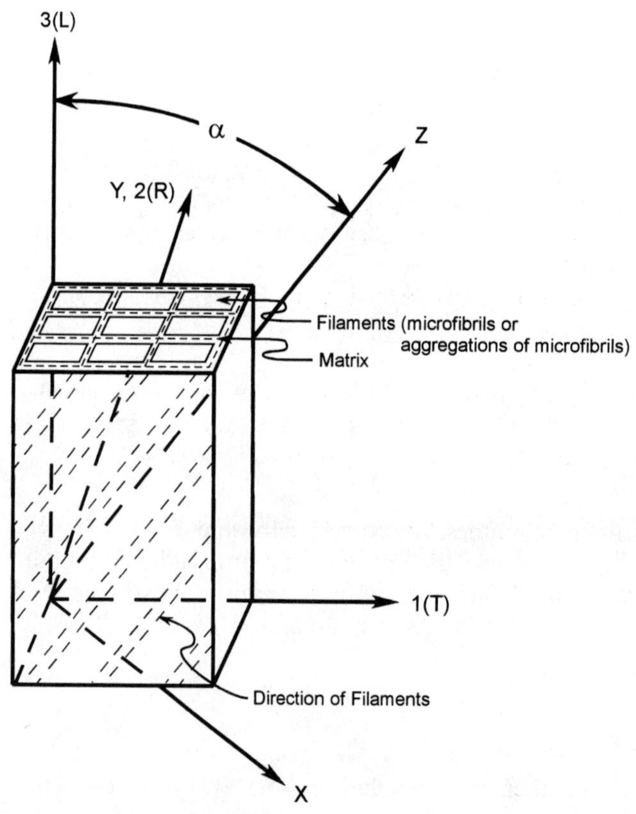

Fig. 8 Model of a submicroscopic element in a layer of a fiber wall. Coordinate axes 1, 2, 3 represent the directions transverse-tangential (T) to the wall, transverse-radial (R) to the wall (i.e., in the wall thickness direction), and longitudinal (L) or along the fiber axis, respectively. Coordinate axes X, Y, Z represent the directions normal to the microfibrils and parallel to the layer surfaces, normal to the microfibrils and aligned with layer thickness, and aligned with the microfibrillar axes, respectively. (Drawing courtesy of Dr. D. S. Keller.)

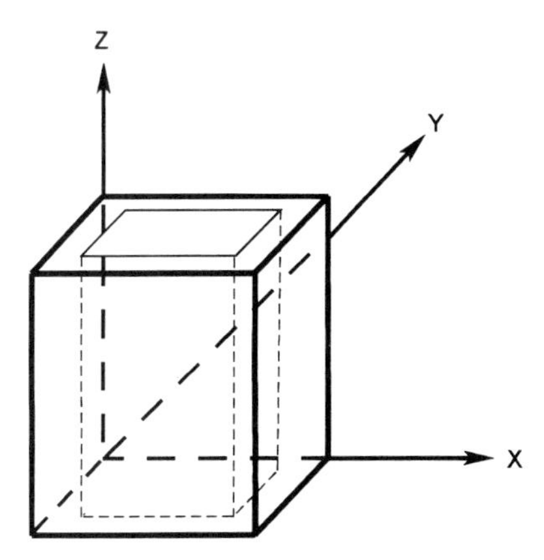

Fig. 9 Model of a basic unit of fiber or other plant cell wall. The unit shown consists of a single anisotropic filament of cellulose I surrounded by a noncellulose matrix. The filament may consist of one or more microfibrils or aggregates of microfibrils, creating concentric sheetlike laminae (see Figs. 10 and 11). (Drawing courtesy of Dr. D. S. Keller.)

sive concentric lamellae within the fiber walls. The evidence that such structures exist is demonstrated by Figs. 10 and 11. Figure 10 shows a cross section of a chemical pulp fiber that has been beaten and then embedded in methacrylate, which makes the wall swell and reveal the concentric laminar structure within a wood fiber. Another example (Fig. 11) is the concentric delamination observed (by cryogenic scanning electron microscopy) in a bleached sulfite softwood dry lap pulp beaten in a Lampen ball mill [244].

Regardless of the actual cross-sectional shape of the microfibril itself—squarish (Fig. 9) or more straplike—there are general constitutive laws in materials science that pertain to two-phase or multiphase composites [30,163]. Also, there is good evidence that cellulose I in the microfibril obeys Hooke's law very closely [215,216,221]. For many purposes, it is reasonable to assume that the matrix and the framework can both be treated as linear elastic materials, especially in low humidity environments.

If one wishes to determine the material constants that flow from elasticity theory using a given model such as those shown in Figs. 8 and 9, some variation of the rule of mixtures [137,163,274] is employed. The following set of equations might be used, for example, for the properties of a lamina of microfibrils embedded in a matrix:

$$
\begin{aligned}
E_Z &= fE^f_Z + mE_m & G_{XY} &= \frac{G^f_{XY}G^m}{mG^f_{XY} + fG^m} \\
E_X &= \frac{E^f_X E^m}{mE^f_X + fE^m} & G_{ZX} &= fG^f_{ZX} + mG^m \\
E_Y &= \frac{E^f_Y E^m}{mE^f_Y + fE^m} & G_{ZY} &= fG^f_{ZY} + mG^m
\end{aligned}
\qquad (14)
$$

where (with reference to Figs. 8 and 9)

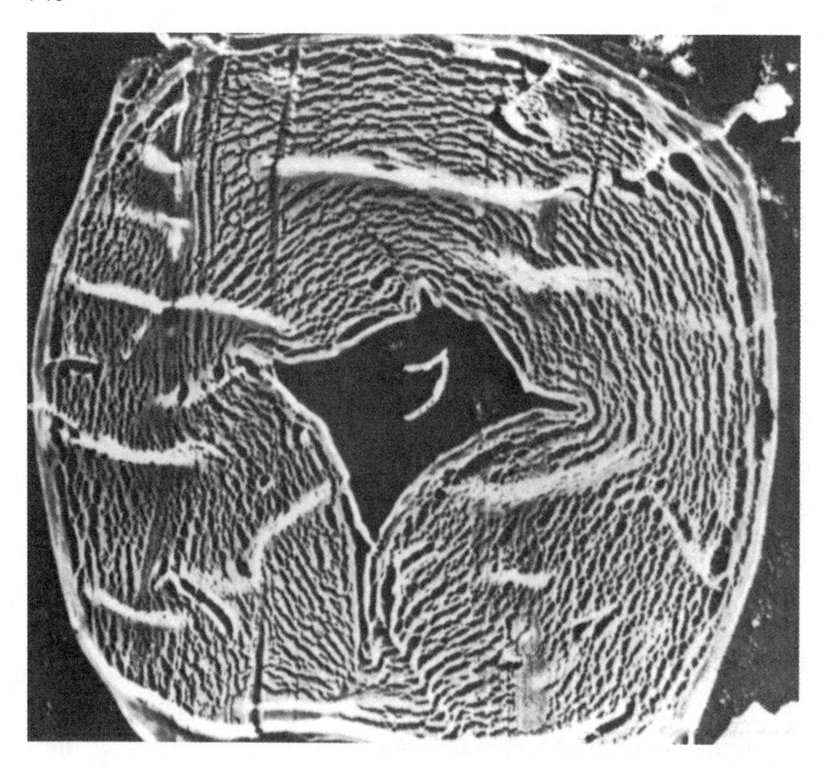

Fig. 10 Cross section of wood pulp fiber after beating and embedment in methacrylate. (From Ref. 226.)

$E_X, E_Y, E_Z =$ moduli of elasticity of a cell wall lamina in the tangential, radial, and longitudinal directions of the tilted cell wall element (Fig. 8), respectively

$G_{XY}, G_{ZX}, G_{ZY} =$ shear moduli of rigidity of the lamina in the transverse, tangential-longitudinal, and radial-longitudinal planes of the element, respectively.

$f =$ volume fraction of microfibril (framework)

$m = 1 - f =$ volume fraction of matrix

$E^f_Z =$ modulus of elasticity of microfibrils in the chain axis direction

$E^f_X =$ modulus of elasticity of microfibrils in the tangential direction (*a* axis of crystallite [348])

$E^f_Y =$ modulus of elasticity of microfibrils in the radial direction (*b* axis of crystallite [348])

$E^m =$ matrix modulus of elasticity

$G^f_{XY} =$ shear modulus of microfibrils in the transverse (*ab*) crystal plane

$G^f_{ZX} =$ shear modulus of microfibrils in the tangential-longitudinal (*ac*) plane

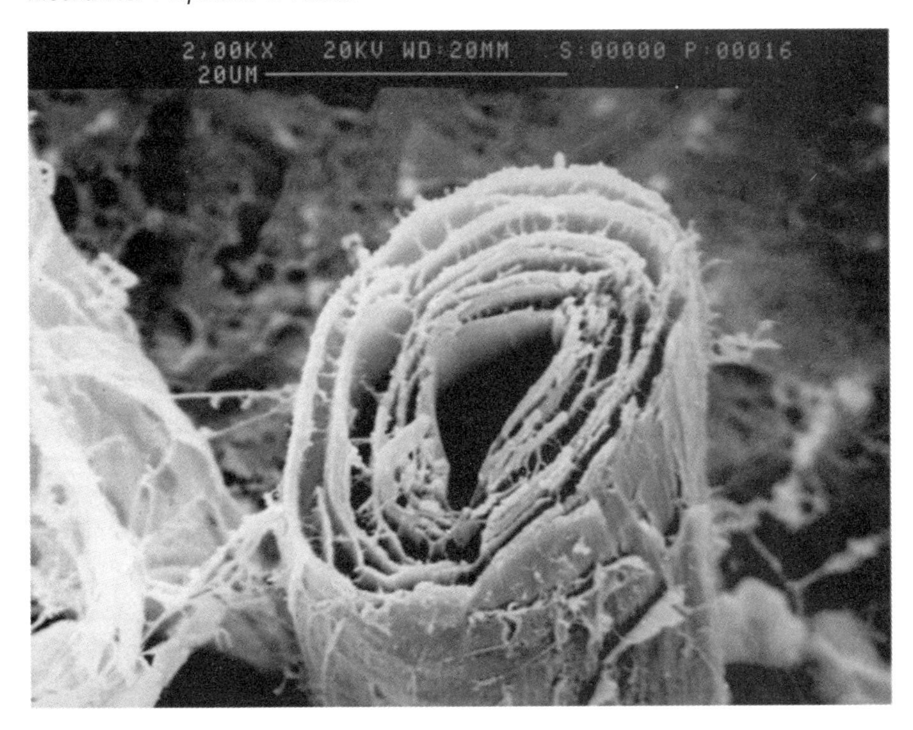

Fig. 11 Unembedded bleached sulfite softwood fiber that had undergone beating for 120 min in a Lampen ball mill. Observation by low temperature scanning electron microscopy after freezing in liquid N_2 and fracturing while frozen. Bar = 20 μm. (Cryo-SEM micrograph courtesy of Dr. Patricia A. Moss. See also Ref. 244.)

G^f_{ZY} = shear modulus of microfibrils in radial-longitudinal (bc) plane

G^m = shear modulus of rigidity of the matrix

There are also six Poisson ratios (three independent) that can be derived theoretically by considering the force–displacement relations for the unit cell of cellulose [Eqs. (1)–(3)] and its interaction with the mathematical matrix [Eqs. (14)] in the laminated composite; these calculations assume an orthorhombic unit cell for cellulose I. Because celluloses Iα and Iβ are actually triclinic and monoclinic, respectively, there are more stiffness constants in the A_{ij} areas shown sequentially after Eqs. (4)–(6). These additional constants modify the engineering elastic constants derived from the A_{ij}'s.

In the most reliable computation of the cellulose I crystal space lattice [348], Tashiro and Kobayashi derive monoclinic three-dimensional stiffness and compliance constant matrices. For these matrices, they use a two-index system, where 11 corresponds to the X direction, 22 to the Y direction, and 33 to the Z (chain-axis) direction. This system has been employed in Eqs. (1)–(5). To distinguish the notation used for various properties of the layer composite structure [X, Y, Z axes, as in Eqs. (14) and Fig. 45] from properties pertaining to the whole fiber (1, 2, 3 axes in Figs. 8 and 45), the latter will henceforth bear unprimed symbols with two indices whereas those of the composite will be primed. For example, A'_{33} is the stiffness constant for the microfibril axis (Z) direction of one of the layers or part thereof, while A_{33} is the stiffness constant for the whole fiber in its axial direction.

The Tashiro–Kobayashi X, Y, Z coordinate labeling system for cellulose has been used in connection with Eqs. (14). In their paper, the radial and tangential axes are shown as being along the crystallographic b axis (2 direction in their notation) and perpendicular to the b axis, respectively. The latter is given the 1 direction index notation. By substitution of matrix-inverted A_{ij} values from Ref. 348 into Eqs. (14), together with an assumption that $f = 0.7$ (reasonable for a high-cellulose fiber such as flax or sisal or for a chemical wood pulp fiber) and the matrix constants that arise from Eqs. (12) and (13), one obtains the set of values for elastic and shear moduli shown in Table 3. Use of non-inverted A_{ij}'s yields slightly higher moduli.

Table 3 illustrates that the cellulosic microfibrils are the dominant elements in the structural properties of the composite wherever the strains in those framework elements are essentially equal to the strains in the matrix, that is, E_Z, G_{ZX}, and G_{ZY}. If strains are equal, a parallel rule of mixtures applies. This condition cannot be expected for the other two elastic moduli E_X and E_Y or the transverse shear modulus G_{XY}. In these cases, strains are unequal (although stresses may be equal), the series rule is used, and the microfibrillar framework has less influence on the elastic constants. This lack of influence is best seen in a comparison of E_X and E_Y, where the stiffness of the framework increases by 3.6 times (owing to the presence of the interchain hydrogen bonds along the X axis), yet the laminar stiffness is only 29% greater in E_X. But the same hydrogen-bond-rich plane creates the largest (G_{ZX}) shear constant; the laminar shear stiffness in that plane is reflected in the resistance of the laminae to beating exhibited in Fig. 11.

Torsional tests of fibers impose shear strains in each of the planes for which we have computed moduli; however, the ZX plane probably experiences the highest stresses, other things being equal. Results of torsional testing of single wood pulp fibers typically show values consistent with the theoretical results in Table 3 [184–186].

Equations (14) allow for adjustments to be made for fibers with changed characteristics such as a larger proportion of matrix material (for example, in a TMP fiber) or softening of the matrix on account of heat, moisture, or exposure to chemicals [79,80,186,271].

Table 3 Elastic Constants for a Fiber Cell Wall Lamina [from Eqs. (14)] Compared with the A_{ij} Term for Cellulose I and the Applicable Rule of Mixtures

Laminar elastic constant	Rule of mixtures applied	A_{ij} term (GPa) from Ref. 348	Modulus value (GPa)
E_Z	Parallel	167.8	118.0
E_Y	Series	15.16	4.30
E_X	Series	54.55	5.54
G_{XY}	Series	4.53	1.51
G_{ZX}	Parallel	8.08	5.46
G_{ZY}	Parallel	3.53	2.50

IV. THE ARCHITECTURE OF A PAPERMAKING FIBER AT VARIOUS STAGES

A. The Layered Structure of Fibers

As mentioned in Section III and illustrated in Fig. 8, microfibrils do not generally align with the fiber axis. Also, it is evident in Fig. 12 that wood (and other) fibers of plant origin are layered composites; each layer has a number of laminae that consist of the bundles or "sheets" of microfibrils and matrix materials just described. The predominant angular orientation of the microfibrils is different in each layer.

Wood fibers normally possess three distinct layers that are readily visible in the polarizing microscope. The layers are also detected by a number of chemical, X-ray, and optical and electron microscopical techniques [20,289]. Botanists have designated these layers as S1, S2, and S3, which signifies that they are the three layers formed in the secondary wall of the wood fiber. There is also a primary (P) wall layer that envelops the other layers. All plant cells have a primary wall, but in the woody tissue of trees, only those cells that undergo extensive thickening with lignification are considered to possess secondary walls. The secondary wall layers are very thick in proportion to the primary wall, because the latter has a thickness of about 0.06 μm whereas the aggregate secondary wall thickness may be two orders of magnitude (or more) greater. The outermost (first-formed) secondary layer is S1, which is followed ontogenetically by S2 and S3, as shown in Fig. 12. The completion of the S3 layer is

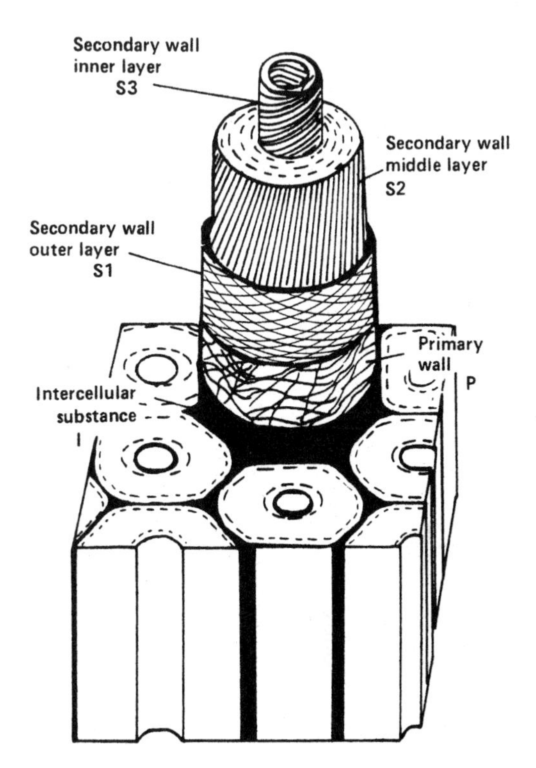

Fig. 12 Diagrammatic representation of dominant helical patterns in layers of a wood fiber. The thin P layer tends to be more disperse in its organization. (From Ref. 218.)

normally followed by the rapid death of the living cell contents. The lignin-containing intercellular substance (I in Fig. 12), also called the *middle lamella*, provides the intercellular bond that joins the fibers together in wood and in most other fibrous plant tissues. Leaf fibers, stalk fibers, bark fibers, and so forth contain lignin in and between the fibers, because they usually do not occur singly but in bundles. Cotton, kapok, and some of the other seed hairs, however, consist of single cells not bonded to each other; these fibers are generally devoid of lignin. Some phloem (bark) fibers tend to occur singly, at least in part; an example is the sugarberry or hackberry (*Celtis* sp.). The more usual arrangement is for fibers to be part of fiber bundles that are distributed in the bark according to a pattern that is typical for the species [254]. When chemical pulping methods are employed to separate the fibers in wood or fiber bundles from one another, partial or complete dissolution of the middle lamella substances occurs. This process is illustrated in Figs. 13 and 14.

Generally speaking, all wood fibers have the three-layered secondary wall structure illustrated in Fig. 12. The exceptions to this rule are usually found in the reaction wood tissue that forms in the stems of leaning trees and in branches (see

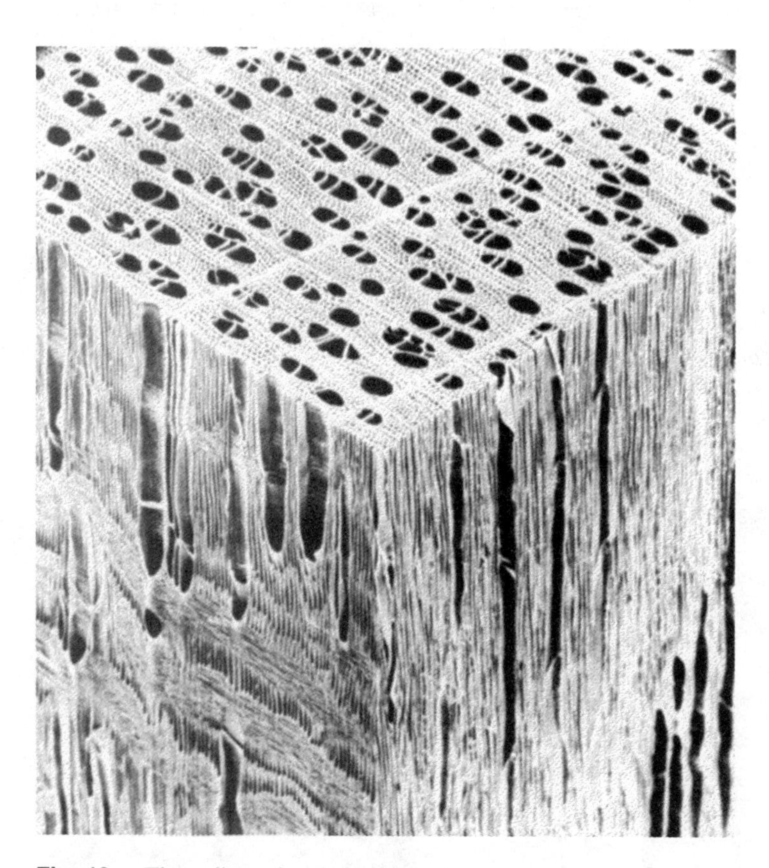

Fig. 13 Three-dimensional scanning electron micrograph view of a small block of birch (*Betula alleghaniensis*) wood before pulping. Magnification ~ 30×. (Courtesy of Dr. W. A. Côté, Jr. and A. C. Day and J. J. McKeon, SUNY College of Environmental Science and Forestry, Syracuse, NY.)

Fig. 14 Fibers and ray parenchyma cells of birch (*Betula*) separated from the wood as a result of pulping treatment. Magnification ~ 120×. (Courtesy of Dr. W. A. Côté, Jr. and A. C. Day and J. J. McKeon, SUNY College of Environmental Science and Forestry, Syracuse, NY.)

Ref. 237 and pp. 1–27 of Ref. 354). Another salient feature of wood fibers is that the S2 layer is normally much thicker than the other layers. Typically, S2 might make up about 73% of the solid volume in an aggregation of fibers in wood or about 78% of the volume of a wood pulp fiber. Saiki [303] determined the thicknesses of the layers and the proportions of each layer in the whole wall of the tracheids (fibers) in five conifer woods. The proportions of S2 varied from a low of 50.8% in the springwood of oriental white cedar (*Thuja orientalis*) to a high of 81.5% in the summerwood of Hinoki cedar (*Chamaecyparis obtusa*). Table 4 gives his thickness values and Table 5 the layer percentages for Japanese red pine. Similar values for hardwoods can be found in the literature; see for example Ref. 123. If one were computing the percentages for a chemical pulp fiber, the value for I + P would be set equal to zero and the remaining percentages adjusted accordingly.

A third feature of significance is that it has been shown [238] that the S2 helix is always right-handed (as shown in Fig. 12) in timber trees. It is common to refer to the right-handed helix as a Z helix, because the microfibrils always travel between upper right and lower left on the side of the fiber closest to the viewer. Thus, the diagonal bar of the Z is in front. When microfibrils are oriented between upper left and lower right on the closest side, the helix is referred to as S or left-handed. A

Table 4 Thickness of Layers in the Tracheid Walls of Akamatsu Red Pine (*Pinus densiflora*)

	Springwood		Summerwood	
	Radial walls, av. (min.-max.) (μm)	Tangential walls, av. (min.-max.) (μm)	Radial walls, av. (min.-max.) (μm)	Tangential walls, av. (min.-max.) (μm)
I + P	0.20 (0.11–0.35)	0.10 (0.05–0.17)	0.38 (0.20–0.50)	0.22 (0.10–0.30)
S1	0.29 (0.20–0.40)	0.30 (0.20–0.40)	0.62 (0.42–1.00)	0.46 (0.30–0.60)
S2	1.66 (0.90–2.25)	1.86 (1.10–2.90)	6.94 (4.90–8.78)	4.82 (3.20–7.00)
S3	0.10 (0.05–0.20)	0.10 (0.05–0.20)	0.14 (0.10–0.20)	0.14 (0.10–0.20)

Source: Ref. 303.

Table 5 Area Proportions of the Layers in the Tracheid Walls of Akamatsu Red Pine (*Pinus densiflora*)

	Springwood			Summerwood		
	Radial walls (%)	Tangential walls (%)	Total (%)	Radial walls (%)	Tangential walls (%)	Total (%)
I + P	7.6	4.1	11.7	4.2	3.7	7.9
S1	8.1	6.2	14.3	4.7	4.9	9.6
S2	36.4	33.6	70.0	40.2	40.2	80.4
S3	2.2	1.8	4.0	1.2	0.9	2.1
			100.0			100.0

Source: Ref. 303.

fourth notable feature is that the S1 layer has both S and Z helical orientations of microfibrils. It has been demonstrated [214,289] that the S and Z ply thickness totals are approximately equal in S1. The S and Z counterrotating helices are evident in S1 in Fig. 12.

Not all plant fibers have this organization, although the differences are usually not major. Bamboo fibers tend to have five or more secondary wall layers [281,290]. A model of a thick-walled, multilayered bamboo fiber is shown in Fig. 15. Cotton fibers are somewhat unique in that the outer (primary) wall layer has a crisscrossed microfibrillar organization, but the orientation of the microfibrils in the interior layers alternates many times along the length of the fiber, changing from an S to a Z helix periodically (see Fig. 16 and Refs. 68 and 160). As shown in Fig. 16, the helical orientation in the S1 layer, or *winding* layer as it is often called, tends to run at an angle of between 20° and 35° to the fiber axis, alternating with S and Z configurations. The reversals in the thickest layer (S2) are opposite in sense to those of the S1. However, because the filament winding angle in the S2 tends to be somewhat more axially oriented, there are areas in each fiber where the two come into phase. The S3 in cotton is extremely thin and has relatively little influence on the fiber structural properties. In the area of the reversals, the microfibrillar structure may be somewhat disturbed.

In general, normal phloem fibers and fiber-sclereids of bark can be classified into two major categories, models of which are shown in Fig. 17. The more common type possesses only two secondary wall layers, S1 and S2, as shown in Fig. 17a. According to Nanko and Côté [254], the two layers have opposed helices, one S and one Z. The other type includes, in addition to the usual primary wall and S1 and S2 layers, an inner G (for gelatinous) layer. The term *gelatinous* is applied because the layer often appears gel-like and swollen because of its extremely low or absent lignification; it is almost exclusively cellulose. This type of fiber, which is also typical of the type of reaction wood found in the wood of leaning hardwood trees [304,354],

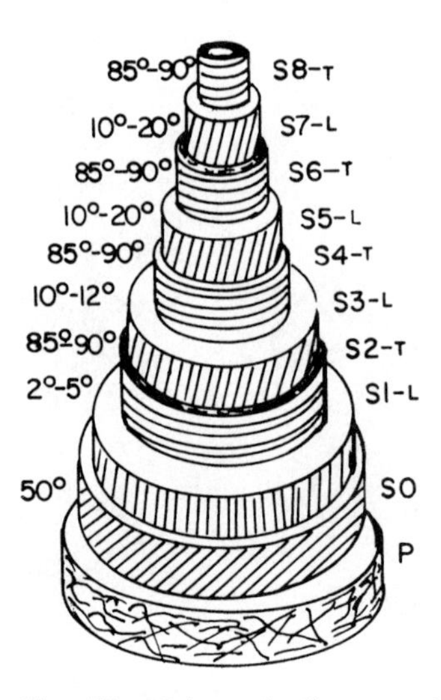

Fig. 15 Schematic diagram of the wall layer organization of a bamboo (*Bambuseae*) fiber. (Adapted from Ref. 281.)

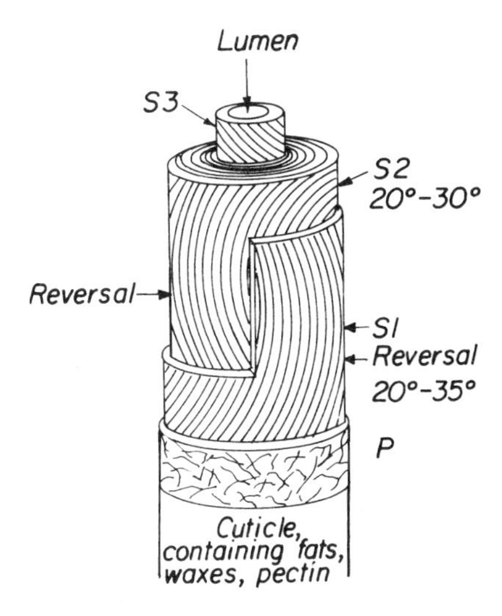

Fig. 16 Schematic diagram of the wall layer organization of a cotton (*Gossypium*) fiber. (Adapted from Ref. 160.)

is shown in Fig. 17b. Microfibrillar orientations are shown. Note that the microfibrils in G are approximately or nearly aligned with the fiber axis. Conifers exhibit a different kind of reaction wood, in response to tilting of the stem, characterized by a very thick S1 layer, no S3 layer, and an S2 that is checked or fissured, with microfibrils oriented at a large angle to the fiber axis [237,354] (see Fig. 18). Thus, phloem fibers have some of the characteristics of abnormal reaction wood fibers even though they are not necessarily associated with leaning stems or branches.

Finally, in regard to fiber layering, it may be noted that certain wood or other fibers with papermaking potential have helical thickenings. Examples include the tracheids of Douglas-fir (*Pseudotsuga*) and yew (*Taxus*) among the conifers and of

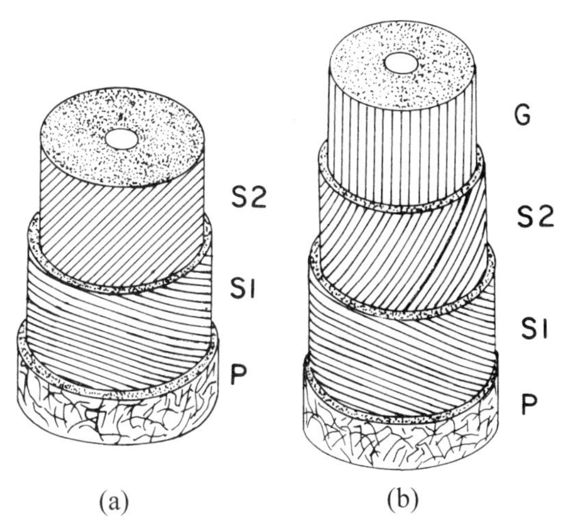

Fig. 17 Schematic diagrams of the wall layer organization of typical fibers found in hardwood barks. The type shown in (a) is more common. (Adapted from Ref. 254.)

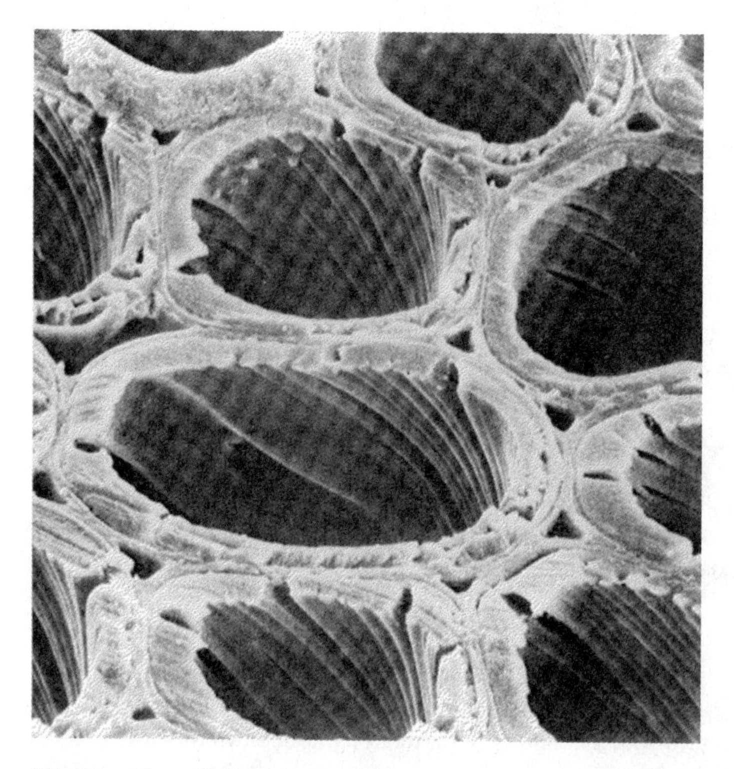

Fig. 18 Three-dimensional scanning electron micrograph of compression wood tracheids in *Pinus radiata*. (From Ref. 215. Photomicrograph courtesy Brian A. Meylan.)

rice (*Oryza*) among the cereal straws [54,142,282]. These are not to be confused with the helical checks shown in Fig. 18. Also, the electron microscope has revealed that there are often very thin transition layers of intermediate orientation between principal layers such as S1 and S2.

The nonfiber components of wood, which often find their way into pulp, have their own unique secondary wall structures [97,142].

B. Filament Winding Angles (Microfibril Angles)

In Fig. 12, one sees the typical pattern of microfibril alignment in the wood fiber cell wall layers. There are, however, many deviations from the "typical" arrangement, some of which were discussed in the preceding section. Even within a single species, there are substantial variations. Because these winding angles are of tremendous importance in understanding the development of good mechanical properties in wood and pulp fibers, this section has been included. The reader may find it helpful to read Chapter 13 in conjunction with this section, to gain more than one perspective.

The S2 layer typically dominates the fiber wall, accounting for an average of between 75% and 80% of the total volume of a pulp fiber (see preceding subsection). Because of this dominance, the term "microfibril angle" or "fibril angle" usually refers to the S2 angle, and in fact most of the methods used either measure just S2 and ignore the other layers or measure some integrated average for the whole fiber. Depending on the characteristics of the fiber, the latter type of measurement may be quite erroneous in some cases [84]. Even though measurements of the S2 angle may show the same trend, the actual angle measurements are often substantially different [76], as can be seen in Fig. 19. Huang et al. [140] employed nine different measurement techniques for comparison on very thin sections of U.S. southern pine wood; their tabulated results show correlation coefficients ranging from 0.50 to 0.98 when mean values by different methods were paired. The highest correlation was between an iodine staining procedure and a method that uses ultrasonic checking.

The Importance of the S2 Angle The literature contains several dozen papers in which the axial and occasionally transverse (circumferential and radial) elastic constants of the fiber cell wall are calculated for wood fibers, pulp fibers, cordage fibers, etc. under various assumptions. In Fig. 20, for example, the following parameters were used in a two-dimensional model:

1. Proportions (by volume) of cell wall in each layer: S1, 20%; S2, 74%; S3, 6%
2. Microfibril angle in each layer: S1, ±80°; S2, varies as shown from 0° to 50°; S3, 70°
3. Proportions (by volume) of matrix substances in all three layers: 30%
4. Proportion (by volume) of cellulose in all three layers: 70%

Fig. 19 Measurement of microfibril angle across a growth ring boundary by two methods. Overall averages indicate that pit aperture orientation measurements (solid line) are 10° higher than those obtained by polarized light (dashed line) [76]. The pit aperture method putatively measures the microfibril angle of the S2 layer, whereas the polarized light method used measures the maximum extinction position of single cell walls containing S1, S2, and S3.

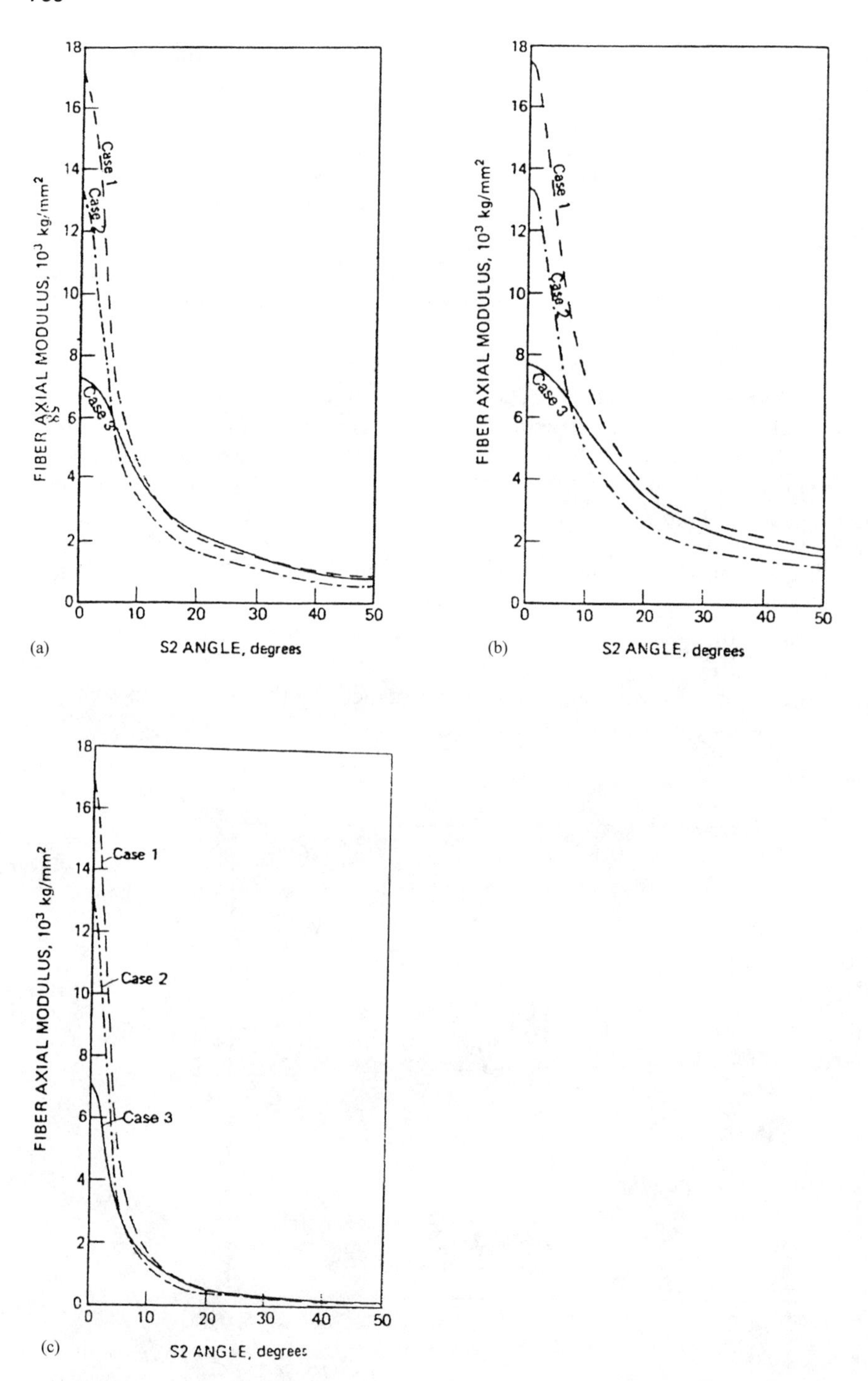

Fig. 20 Curves of theoretical variation of fiber modulus in the direction of the fiber axis versus S2 angle. The three sets of curves represent extreme variations in the theoretical elastic constants employed for the structural polysaccharide (cellulose) of the fiber cell wall, as explained in the text. The matrix constants are varied according to (a) best estimate, (b) very stiff matrix, (c) very compliant matrix. (From Ref. 221.)

5. Elastic constants of matrix (isotropic)
   Young's modulus:
   a. Best estimate: 204 kg/mm^2 (2 GPa)
   b. Very stiff: 703 kg/mm^2 (6.9 GPa)
   c. Very compliant: 20.4 kg/mm^2 (0.2 GPa)
   Shear modulus of rigidity: 38.4% of the Young's modulus for each of the three cases
   Poisson ratio = 0.30
6. Elastic constants of cellulose (anisotropic)
   a. Uniform dashed line (---) "Case 1"
      Axial Young's modulus: 32,540 kg/mm^2 (319 GPa)
      Transverse Young's modulus: 3,800 kg/mm^2 (37.3 GPa)
      Shear modulus of rigidity: 39.9 kg/mm^2 (0.385 GPa)
      Poisson ratios: 0.041, 0.0048
   b. Dot-dash (-·-) "Case 2"
      Axial Young's modulus: 25,130 kg/mm^2 (246 GPa)
      Transverse Young's modulus: 1,677 kg/mm^2 (16.4 GPa)
      Shear modulus of rigidity: 17.3 kg/mm^2 (0.17 GPa)
      Poisson ratios: 0.041, 0.0038
   c. Solid line (—) "Case 3"
      Axial Young's modulus: 13,700 kg/mm^2 (134 GPa)
      Transverse Young's modulus: 2770 kg/mm^2 (27.2 GPa)
      Shear modulus of rigidity: 449 kg/mm^2 (4.40 GPa)
      Poisson ratios: 0.10, 0.011

The above parameters are drawn from Refs. 105, 213, 214, and 216.

The computational basis for the generation of the curves in Fig. 20 is the "two-wall" analysis for the mechanics of fiber walls described by Mark and Gillis [219] with a provision of zero net twisting moment to satisfy the strain determination requirements for tests of isolated individual fibers. By contrast, the "two-wall" analysis applied to a fiber surrounded by other fibers in a piece of wood would assume complete shear restraint of the fiber wall.

Some interesting relationships emerge when Fig. 20 is examined. The most obvious is the effect of the S2 angle. When S2 angles are very small ($< 10^\circ$), the matrix properties have virtually no effect on the curves, but when S2 angles are greater than 10°, the levels of all curves rise substantially with an assumption of a stiff matrix and drop drastically when matrix moduli are low. For a given matrix condition, the three curves are fairly congruent in the range 10–50° for S2 also.

The theoretical fibers whose properties are plotted in Fig. 20 would be analogous to delignified fibers, because the cellulose constitutes 70% of the material of composition. In Fig. 20, the assumptions for Case 3 (solid line) yield curves that intersect the 0° axis between 7000 and 8000 kg/mm^2 (68.6–78.5 GPa).

Similar curves for wood, where the percentage of matrix is nearly 50%, intersect the 0° axis at around 4700 kg/mm^2 (46.1 GPa) [216] when the microfibril angles are constant within a layer. Cave [42] modeled a wood cell wall in which the S2 angles had a statistical distribution with a somewhat greater mean value; thus, the 0° axis intersect diminished to about 4200 kg/mm^2 (41.2 GPa). This theoretical curve is shown in Fig. 21. Note that the majority of experimental points are well above the

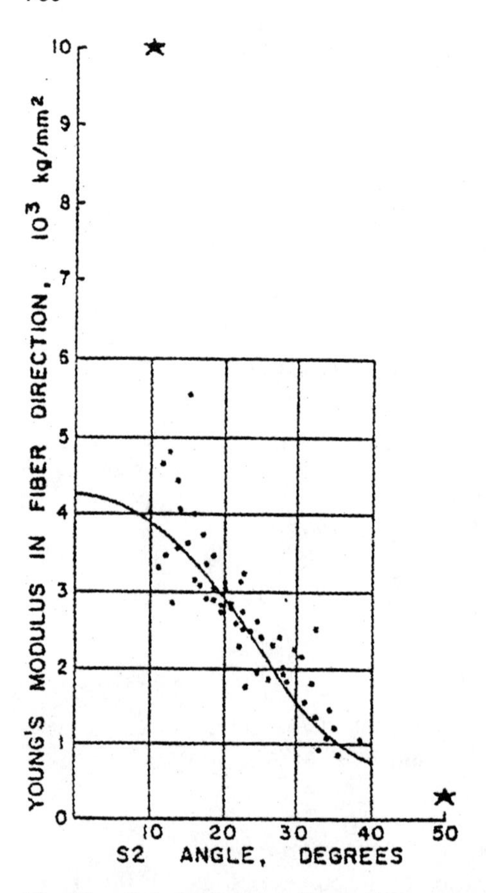

Fig. 21 Relation between "mean" S2 angle and fiber axial modulus for *Pinus radiata*, where the solid line gives the theoretical relation. Circular dots show experimental points. Added to the graph are experimental points for sisal (*Agave sisalana*). (From Ref. 221.)

theoretical curve when the S2 angle is less than approximately 15° and that there are no experimental points below 10°. The experimental points for sisal (stars) in Fig. 21 seem to be better fitted to the curve for Case 1 in Fig. 20a than to the theoretical curve in Fig. 21. A paper by Cowdrey and Preston [65] relating the Young's modulus of Sitka spruce (*Picea sitchensis*) to the S2 angle also showed no tendency to level or curve to 0° and no data for S2 angles smaller than 6.5°.

In contrast with the above observations, Page et al. [279] presented results of experiments of single wood pulp fibers of *Picea* spp. over a range of S2 angles from 0° to 50°. The technique used to measure the S2 angle used reflected polarized light. The reflective surface is created by introducing mercury into the fiber lumen. Thus, the polarized light beam passes through S1, S2, and S3 and is reflected back through these three layers. In this method, the "S2" angle that is recorded derives from the major extinction position when the fiber being examined is rotated on the microscope stage. A more recently developed method, polarization confocal microscopy, also integrates fiber layers, and S2 angles as low as 0° have been reported by Jang [150]. He notes that "the S1 layer can affect polarization of the excitation light, but its effect on the fluorescence is insignificant."

The data of Page et al. [279] did not show a very consistent relationship between the elastic modulus of a fiber and the S2 angle, contrary to expectations from theory. For example, fibers with fibril angles around 21–22° had moduli ranging from 24 to 50 GPa, as can be observed in Fig. 22. The disparities between experiment and theory have been examined by several researchers, who reach different conclusions. Page et al. [279] stated that the explanation lies in defects of the fibers—that fibers with low moduli contain microcompressions, dislocations, and other irregularities of structure. They also concluded that a curve-fitted "upper bound" of the experimental values represented "the experimental relationship between modulus and fibril angle for uniform, undamaged fibers" and further that "it is to be expected that this upper bound should fit the theoretical equations derived from orthotropic elasticity theory."

Salmén and de Ruvo [307] compared the results of their model fiber with the Page et al. data. The Salmén–de Ruvo model assumes a collapsed fiber (an assumption also made by Page et al.); however, they considered the full layered structure rather than a fiber wall consisting only of S2, as Page et al. did. Salmén and de Ruvo also assumed that the cellulosic reinforcement had a discrete length, even though microfibrillar ends are seldom *if ever* observed experimentally in the SEM except where the fibers have been intentionally severed. Instead of curve fitting to the upper bound of the experimental data, they produced a series of theoretical curves in which various parameters were changed. Figure 22 contains two theoretical curves and Page et al.'s data from Ref. 279. These curves both purport to predict the modulus of elasticity along the fiber axis as the S2 angle varies from 0° to 50°. The upper curve

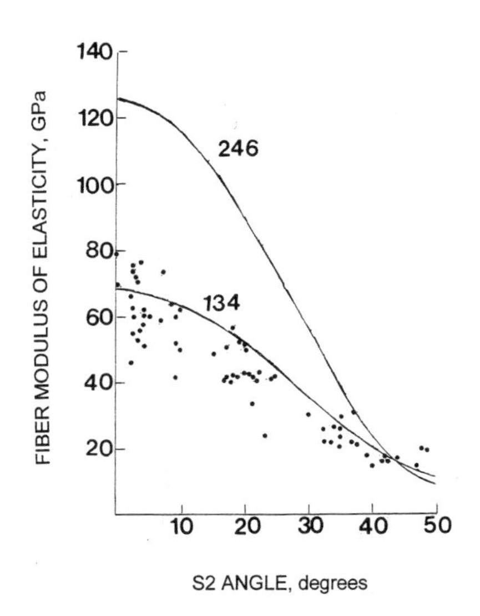

Fig. 22 Experimental data of Page et al. [279] plotted in relation to two theoretical curves as developed by Salmén and de Ruvo [307]. The numbers associated with the curves are values in GPa of the Young's moduli of crystalline native cellulose in the chain axis direction as determined theoretically by Gillis [104] and experimentally by Sakurada et al. [305]. (From Ref. 7.)

uses cellulose I elastic constants calculated by Gillis [104]; the lower curve uses an experimental value for the crystalline part of a cellulosic fiber [305] for the axial elastic modulus for cellulose and theoretical values developed by Mark [214] for other needed constants in the Salmén–de Ruvo model. Because the data points match the latter curve much more closely, the authors concluded that the constants employed for that curve were probably closer to the real values for cellulose in the microfibril.

More recently, Anagnost et al. [7] offered evidence that the large number (40%) of data points in Fig. 22 for the 0–10° range of S2 angles is probably erroneous and state, "We think it is likely that most, perhaps all, of these data points belong to the right of the positions they now occupy, which would place many of them between the theoretical curves." They noted, for example, that if the theoretical results of Tashiro and Kobayashi [348] were used in the same manner as the two curves in Fig. 22, they would yield a curve that lies between the two that are shown—at least in the low S2 angle range—and would intercept the 0° axis at about 84 GPa, just slightly greater than the highest experimental value.

It may well turn out that each of these three analyses identifies a factor in the results of experiment vs. theory.

One Fiber, Many Angles The experiments of Anagnost et al. [7] were based on soft-rot cavity formation in the S2 layer. This is a classical method for determining the S2 angle. It has been demonstrated many times that the soft-rot cavities lie parallel to the cellulose microfibrils, and measurements of their orientation accordingly provide a reliable direct method [9,16,66,115,204]. The special advantages of the soft-rot method include the following.

- • Soft-rot cavities form primarily in the S2 layer, so they are not confused with the other layers. Angles can be determined easily for both thin-walled, large-diameter fibers (e.g., earlywood or springwood) and thick-walled, narrow fibers (e.g., latewood or summerwood).
- •• The cavities reveal fine structural details of the cellulosic cell walls such as orientation of microfibrils near and at the tips of fibers, differential orientations in the vicinity of pits (including the encircling bands of microfibrils), and differential orientations in radial versus tangential walls.
- ••• Cavity shape is dependent on the cellulose structure (not the decay organism); it seems to follow the hydrolysis planes in the cellulose I crystal lattice. Cavities can be induced in both lignified and delignified fibers, although the time frame is somewhat longer for the delignified case.

Figures 23–26 illustrate the point that a fiber does not have just one S2 angle. This series of light micrographs was taken with Nomarski differential interference contrast (DIC) illumination. Characteristic of this illumination are large depth-of-field focus and some contour inversion, so that some cavities appear convex whereas they are actually biconical-tipped concavities. An example is shown in the fiber in Fig. 23. The front radial wall has S2 cavities that form right-handed (Z) helical patterns. The back radial wall has cavities in the S2 layer that appear to form left-handed (S) helical patterns. There is also a ring cavity in the thickened edge of a bordered pit (arrow).

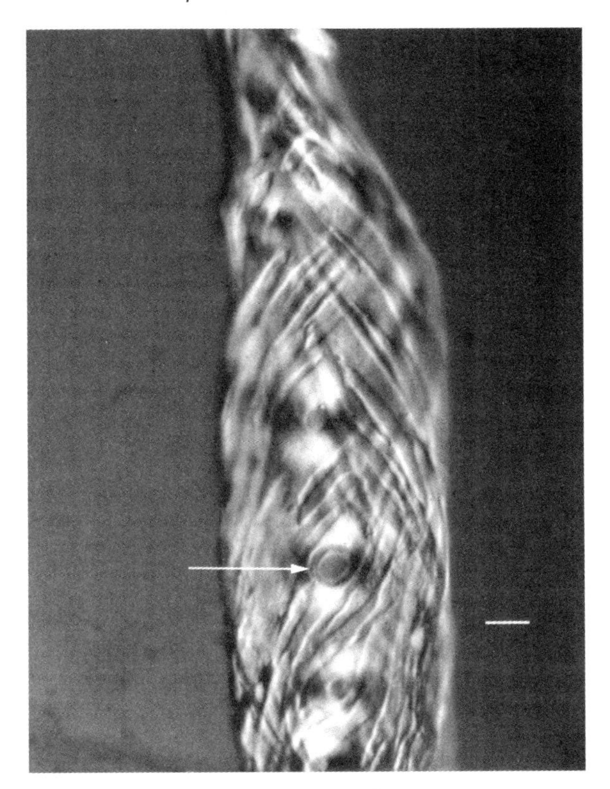

Fig. 23 Soft-rot cavities observed by Nomarski differential interference contrast (DIC) light microscopy in an earlywood tracheid of U.S. southern pine. The soft-rot cavities follow the helical orientation of the cellulosic microfibrils. A cavity is also visible in the border of an intertracheid pit. Bar = 10 μm. (Photomicrograph courtesy of Dr. Susan E. Anagnost.)

In Fig. 24, cavities in the front radial wall deviate around the intertracheid pits. Microfibrils do not hold a constant helical pitch in pit field areas. Any measurements in areas such as this must be confined to the central area of the radial wall; the additional curvature at the edge of the fiber as the cavity runs from the radial to the tangential wall is an optical artifact.

In Fig. 25, the soft-rot cavities at a fiber tip reveal a continuous helical trajectory for the microfibrils as they ascend the back wall, loop around the end of the fiber, and reenter the Z-helical pattern in a return trajectory to the middle of the fiber. This is an example of the "twisted skein of yarn" model for the microfibrils proposed by Mark in 1980 [218]. The continuous skein effect is one of the reasons Anagnost et al. [7] felt that very small ($\sim 7^\circ$ or less) S2 angles should be rare, especially at or near 0°.

The soft-rot cavity orientation measurement method is rather straightforward in pit-free areas of the fiber such as the one shown in Fig. 26, in which two biconical-tipped cavities are used to swing an angle of 27° from the fiber axial direction. Customarily, the measurement is made at the midpoint of the fiber. A case can be made for taking angular measurements at many points to determine the extent of within-fiber variability of the S2 angle; such variability was the subject of a recent

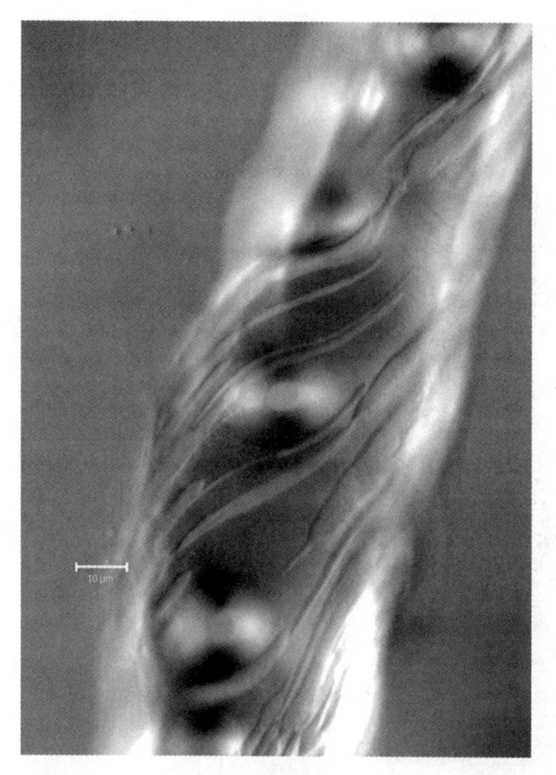

Fig. 24 Soft-rot cavities in a southern pine tracheid follow the microfibrils as they deviate around the intertracheid pits. Bar = 10 μm. [Photomicrograph (DIC) courtesy of Dr. Susan E. Anagnost.]

study [8]. On a larger scale, it is well known that there is systematic variability of the S2 angle in the wood of a single tree. Typically, the largest angle is found in the juvenile wood at the center of the base of the tree; the angle decreases with increased height and increased girth [139]. When all of the variabilities of tree position, ring position, and position in the fiber are taken into account, it is not surprising that different methods for determining "the" microfibril angle of a certain sample often yield quite disparate results. It is certainly advisable to compare methods, especially direct vs. indirect methods, but it is also important to get structurally consistent sample material.

C. Natural Irregularities and Processing Damage in Fibers (Contributed by José Iribarne)

When wood is pulped and the separated fibers are further processed, changes occur in the fiber structure, some beneficial and some harmful. An example of a generally beneficial change is illustrated in Figs. 10 and 11. The concentric delamination that occurs as a result of pulping, bleaching, and/or beating flexibilizes the fiber, facilitating collapse and creating greater sheet density and more opportunities for good

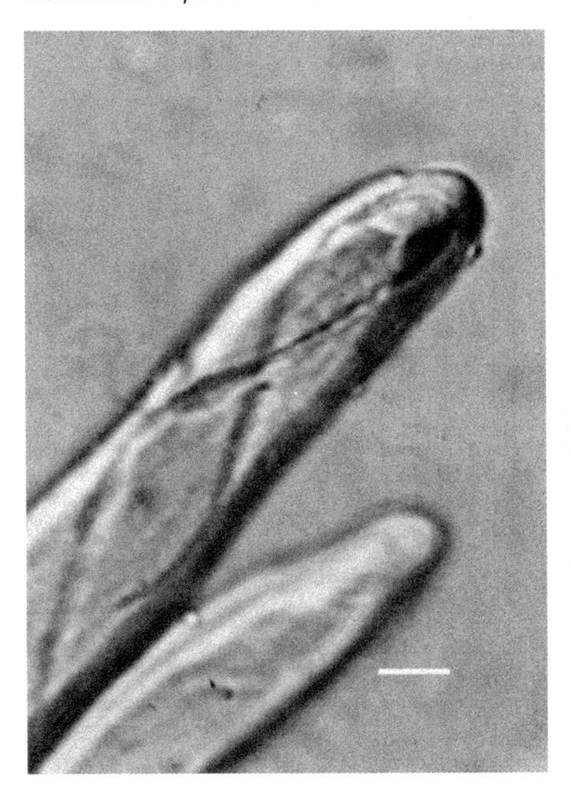

Fig. 25 A light micrograph of the tip of a tracheid of southern pine. Soft-rot cavities form a helical trajectory that comes from below, extends over the tip, and returns to below the tip on the opposite wall. Bar = 10 μm. (Photomicrograph courtesy of Dr. Susan E. Anagnost. From Ref. 7.)

fiber bonding [226], as illustrated in Fig. 27. During sheet consolidation, hydrogen bonds re-form internally and externally. External bonding is enhanced by surface fibrillation (see Chapter 16). Lightly beaten fibers may have just a minor surface roughening of the S1 layer, but as the refining proceeds there may be holes torn in the S1, exposing part of that layer (see Fig. 28a). Heavy refining may strip the S1 away entirely (Fig. 28b). Optimal refining varies with the furnish and with the use of the raw material in paper and other fiber products.

The pits that occur in tracheids, hardwood fibers, and other cellular elements in wood have often been considered detrimental points of weakness in the fiber walls, but it appears that, in most cases, fracture lines in wood, fibers, and paper are not associated particularly with pits [55,100,144,214,275]. It is evident in Fig. 29 that the fracture lines circumvent the bordered pits in this specimen of southern pine, for example. The reason for this path is that there is heavy microfibrillar reinforcement around the bordered pit of a conifer, as was shown in Fig. 2. Stress is relieved around the pit aperture. Thus, pits can be considered natural irregularities in wood fibers—irregular because they are holes and their distribution along the length of the fiber is nonuniform—but they should not be thought of as defects under most circumstances. The above references do cite evidence that natural

Fig. 26 Microfibril angle (S2 angle) measured with soft-rot technique. By measuring the angle of the cavity in relation to the longitudinal axis of the fiber, an accurate determination can be made. Customarily, one measurement is made at the midpoint of the fiber. Here the microfibril angle is 27°. The cavities lie on the back wall of the fiber. Bar = 10 μm. (Photomicrograph courtesy of Dr. Susan E. Anagnost.)

Fig. 27 Scanning electron micrograph of paper sheet top surface containing a large percentage of collapsed, flattened fibers, pulped by a soda-oxygen process. (Courtesy of Dr. R. Marton and A. C. Day, SUNY College of Environmental Science and Forestry, Syracuse, NY.)

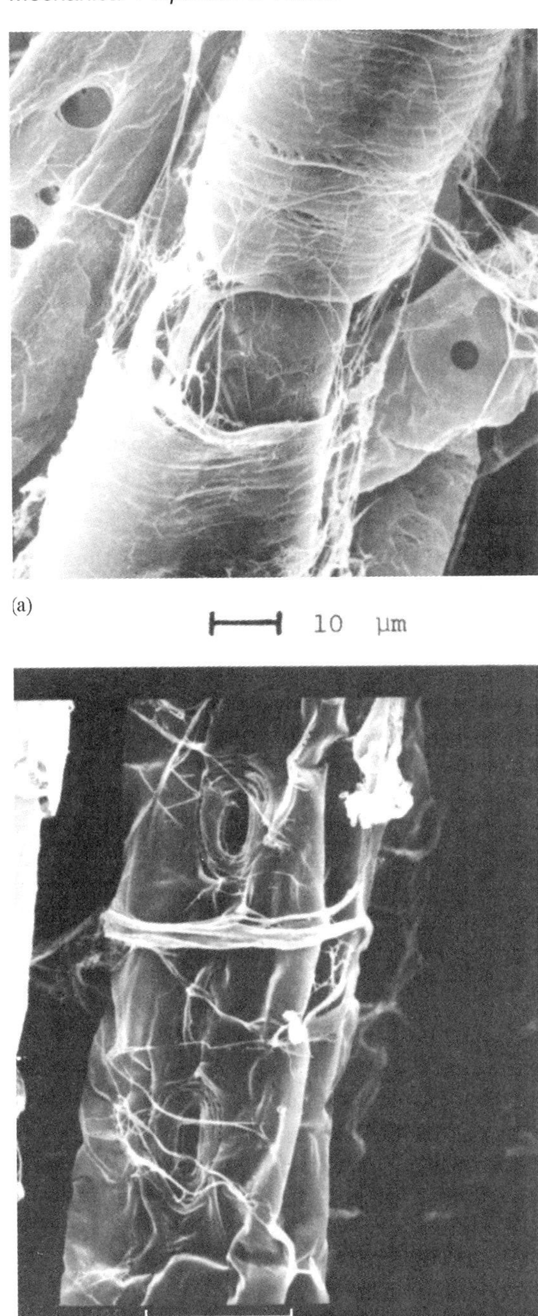

Fig. 28 Scanning electron micrographs of refined fibers. (a) Moderately heavily refined. Bar scale = 10 μm. (Courtesy of T. H. Hsu, SUNY College of Environmental Science and Forestry, Syracuse, NY.) (b) Very heavily refined. Bar scale = 10 μm.

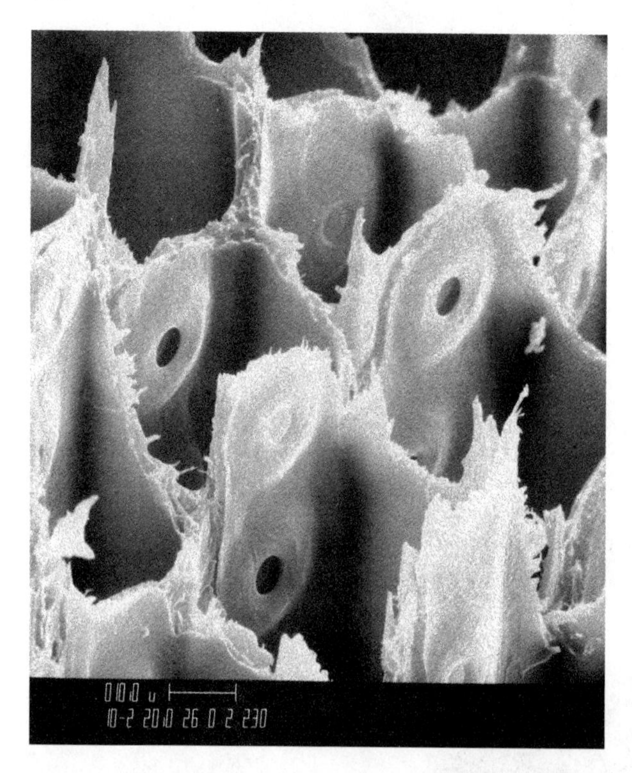

Fig. 29 Scanning electron micrograph of tensile failure in U.S. southern pine wood. The fractures circumvent the bordered pit areas; their trajectories skirt around the pit annulus and/or pass through areas between pits. Bar scale = 10 μm. (Courtesy of Dr. R. B. Hanna.)

irregularity of another type—the areas of fibers that were located adjacent to ray tissue in the wood—do tend to be weak spots, although perhaps not severely so. These "ray crossings" (Fig. 30) often contain dense clusters of cross-field pits that led to ray parenchyma when formed.

Processing damage to fibers has been given significant attention by, among others, Page et al. [280], Laamanen [199], Mohlin and Alfredsson [241], Hakanen and Hartler [114], and Iribarne [144]. The types of damage may be referred to by various authors as structural or morphological changes, "deformations" (deformities), singularities, and processing irregularities or defects. Illustrations of some of these processing defects are shown in Figs. 31–43. Irregularities caused by processing of chemical pulp include kinks, twists, folds, breaks, curl, microcompressions, and dislocations. Microcompressions are small transverse dislocations or slip planes in the alignment of fibrils. Large regions of deformation were designated as "nodes" by Forgacs [93], but the term more often used in the literature is "kinks." Typically, large dislocations (slip planes) extend across the fiber at each kink and are revealed distinctly by polarized light.

Eleven examples of processing irregularities are shown in Figs. 31–41. All the images correspond to never-dried fibers from unbleached kraft pulp made from U.S. southern yellow pine (a mixture of *Pinus taeda, P. elliotti*, and other species). The fibers were stained with Safranin O (Fisher Scientific, Pittsburgh, PA), mounted with

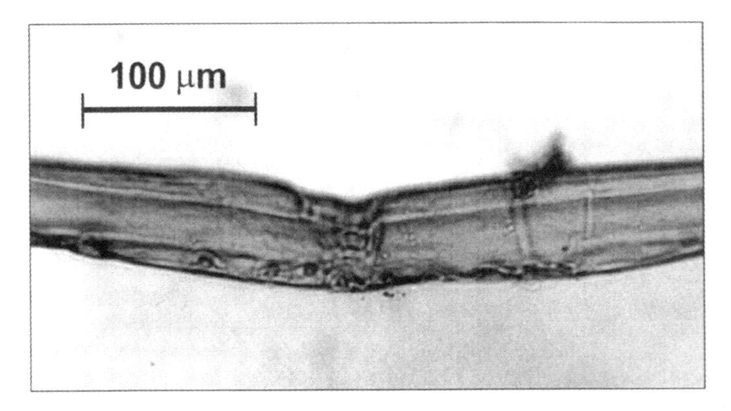

Fig. 30 Light micrograph of a U.S. southern pine earlywood pulp fiber containing a pit field typical of ray crossing locations at center. (This figure and Figs. 31–41 are courtesy of Dr. José Iribarne, who prepared them as part of dissertation research at SUNY College of Environmental Science and Forestry, Syracuse, NY.)

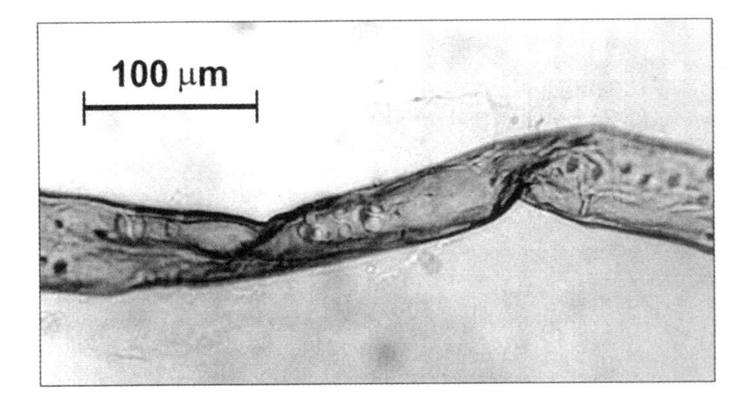

Fig. 31 Double-twist, bordered, and pinoid pits in an earlywood kraft pulp fiber segment.

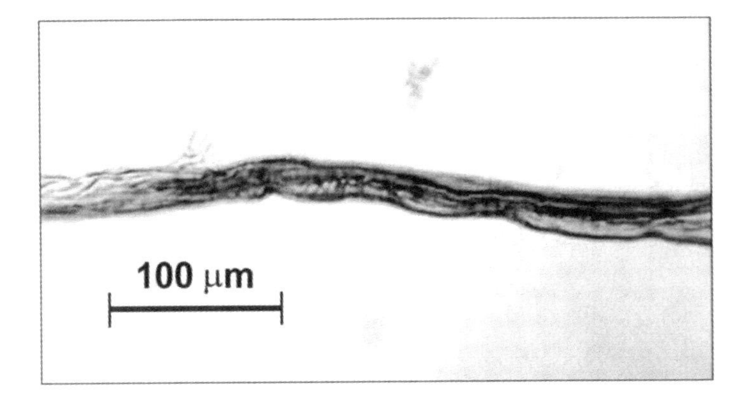

Fig. 32 Collapse and fine fibrillation in an earlywood kraft pulp fiber.

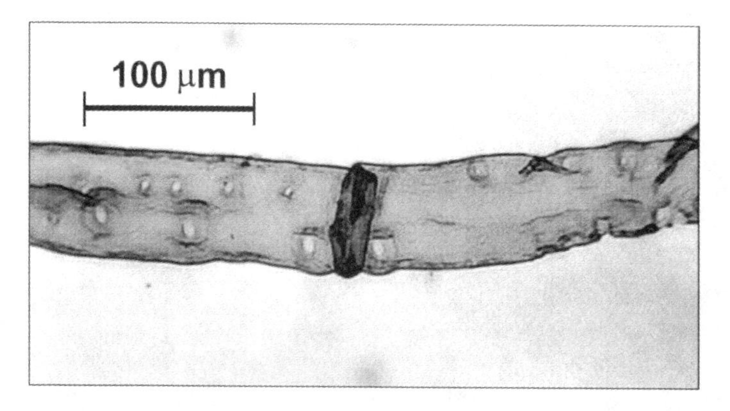

Fig. 33 Double fold and partial twist in an earlywood kraft pulp fiber.

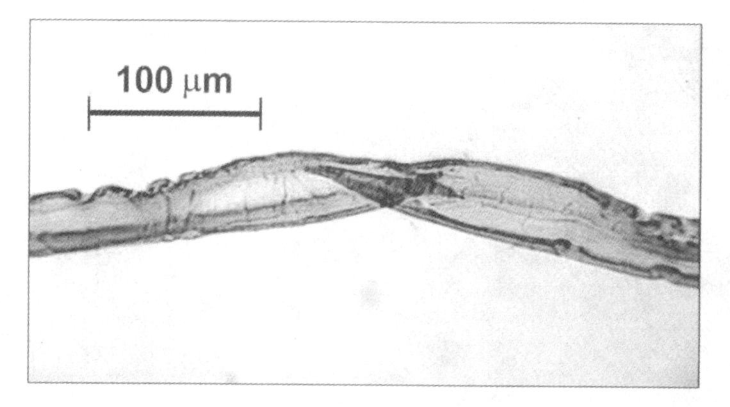

Fig. 34 Twisting and collapse of an earlywood kraft pulp fiber.

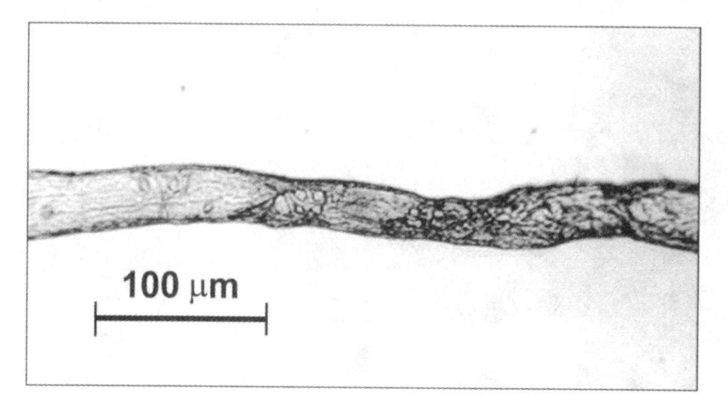

Fig. 35 Microcompressed segment of an earlywood kraft pulp fiber. Clusters of cross-field pits that were adjacent to rays also appear.

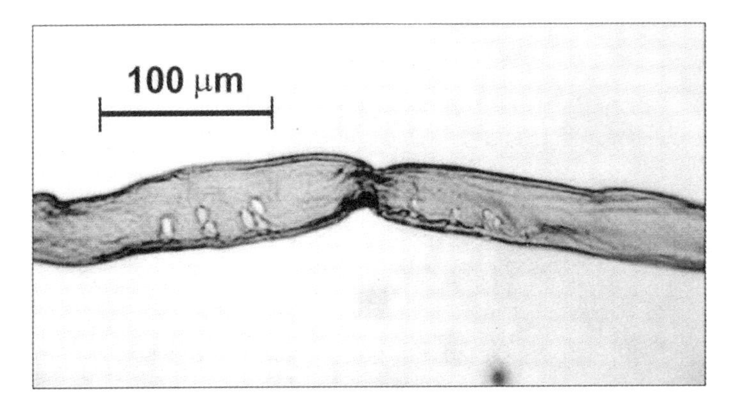

Fig. 36 Local collapse of a latewood kraft pulp fiber at a ray crossing.

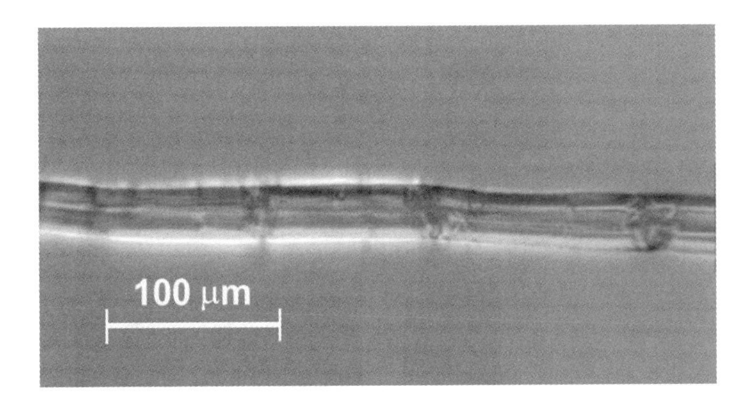

Fig. 37 Dislocations in a latewood kraft pulp fiber. Partially polarized light.

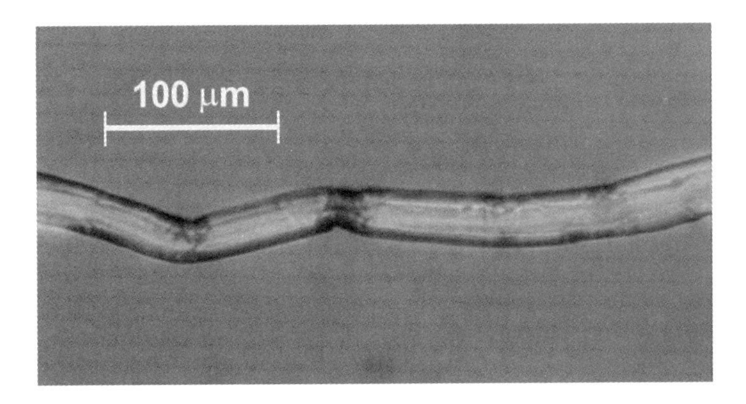

Fig. 38 Kinks and dislocations in a latewood kraft pulp fiber. Partially polarized light.

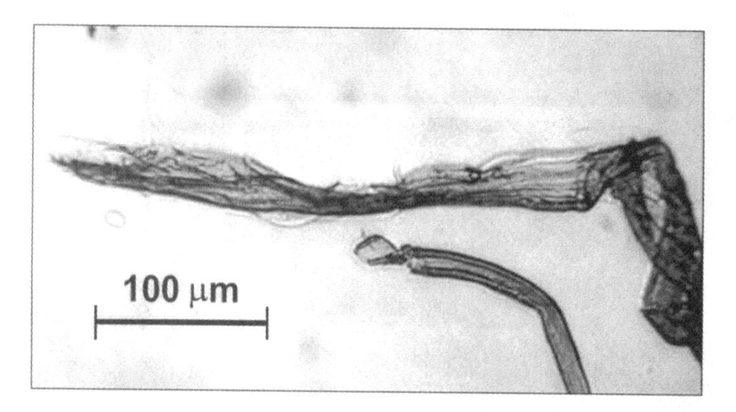

Fig. 39 Broken end of a fibrillated latewood kraft pulp fiber.

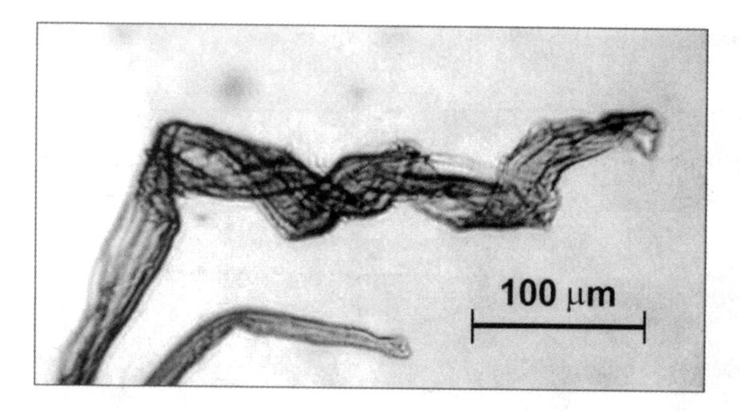

Fig. 40 Multiple twists, delamination and fibrillation at the other end of the fiber shown in Fig. 39.

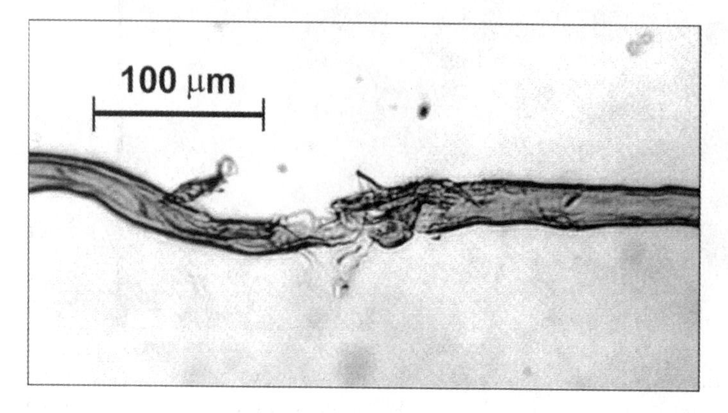

Fig. 41 Severe disruption and fibrillation of a latewood kraft fiber.

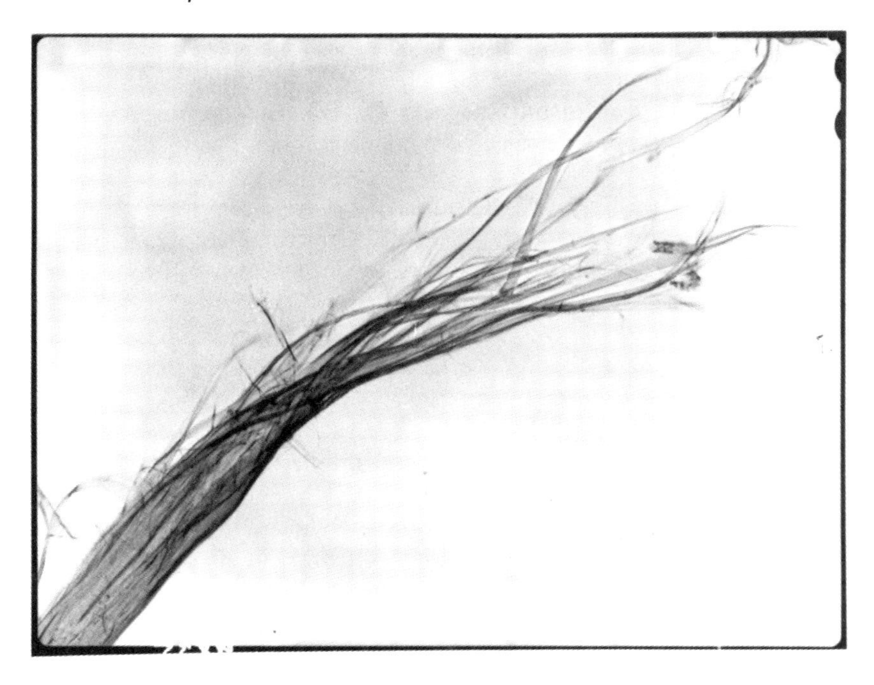

Fig. 42 Coarsely fibrillated fiber end of a stone groundwood mechanical pulp fiber. (From Ref. 199. Photomicrograph courtesy of Jouko Laamanen, KCL, Espoo, Finland.)

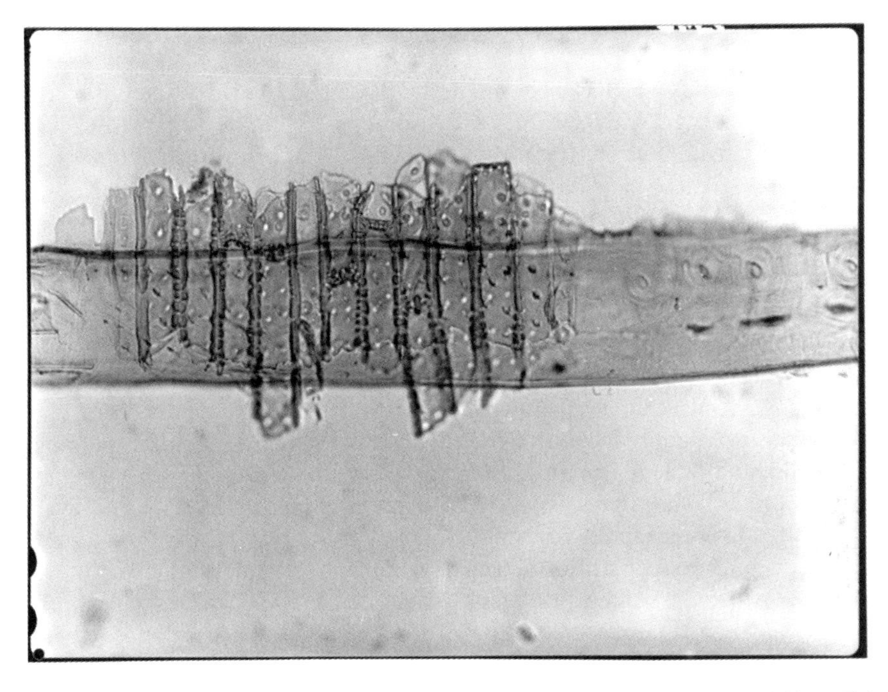

Fig. 43 Retention of ray cells and middle lamella on a stone groundwood fiber. This may be considered a form of minishive. (From Ref. 199. Photomicrograph courtesy of Jouko Laamanen, KCL, Espoo, Finland.)

water on glass slides, and imaged in transmission mode. Figures 42 and 43 show processing features often found in mechanical pulps [199].

There have been efforts to quantify some of the more important processing defects, especially curl and kink. One can find a more thorough treatment of these parameters in Chapter 16.

Most measurements of fiber irregularities that have been reported were obtained by visual microscopy of relatively few fibers. Approximately 40 microcompressions per millimeter of fiber length may be present in unbeaten softwood kraft pulp fibers [4,86,178,272]. However, not all studies support a consistent presence of microcompressions. Nanko and Ohsawa [255], who studied bond formation of *Fagus crenata* fibers, noted wrinkling of the fiber surfaces, not transverse to the fiber axis but more or less parallel to it. Also, these wrinkles appeared mostly on the free surfaces of the fibers compared with their scant presence on the bonded side. It may be that microcompressions are scarce or absent in fibers with relatively thick S2 layers (e.g., conifer latewood, hardwood fibers).

According to Forgacs [93], nodes (kinks) in softwood fibers tend to be regularly spaced at 0.2–0.4 mm intervals, corresponding to ray crossings in the original wood. This is equal to 2.5–5 major irregularities per millimeter of fiber length. Kibblewhite [178] counted the number of "zones of dislocation" in 210 pine kraft pulp fibers and found that they decreased upon PFI mill refining from 4.3 ± 0.3 to 2.1 ± 0.2 per millimeter. Hakanen and Hartler [114] counted between 3.1 and 8.6 major irregularities ("compressions and dislocations") per fiber in five mill pulps and between 1.6 and 5.5 major irregularities per fiber in the corresponding laboratory pulps. Using their reported fiber lengths for each sample, the counts correspond to 2.0 ± 0.8 irregularities per millimeter for mill pulps and 1.4 ± 0.7 irregularities per millimeter for laboratory pulps.

Although great progress has been made in the automated determination of distributions of fiber length [452], width and thickness [353], and shape [169,270], fiber irregularities remain elusive. Recently, Iribarne developed a prototype instrument to count a class of fiber irregularities that he designated as "major singularities" [144]. The number of major singularities per millimeter was well correlated ($R^2 = 0.91$) with the number of microcompressions per millimeter determined at PAPRO (New Zealand) according to the method of Kibblewhite [178]. The two determinations were completely independent, so the measurements provided by the prototype instrument appear to be valid. The result also suggests that different types of fiber features are cross-correlated, i.e., fibers that have large numbers of microcompressions may also have large numbers of twists, folds, or collapsed ray crossings.

Measurements of major singularities for fibers obtained using different pulping methods are shown in Table 6. Both mill pulp samples had more major irregularities per millimeter of fiber length than their laboratory counterparts. The difference was significant within confidence limits of 89% or higher. In contrast, there were minor differences between the irregularities detected in mill pulps and the holocellulose samples prepared from the same wood chips. These observations suggest that both the hot discharge of kraft pulp in the mills and the dispersion of holocellulose fibers in a refiner introduce more fiber irregularities than the mild dispersion of laboratory kraft pulps.

Ciné micrography of 90 spruce kraft fibers under tensile load by Page et al. [278] indicated that failures generally initiate in regions of fiber wall disruption. Some of

Table 6 Major Singularities Detected in Pulp Fibers from the Same Chip Source

Chip source	Pulp sample	Fiber count	Singularities $\pm 0.008^a$ (mm^{-1})
A	Mill kraft	3344	0.078
	Laboratory kraft	3544	0.048
	Holocellulose	2252	0.085
B	Mill kraft	2731	0.077
	Laboratory kraft	3744	0.056
	Holocellulose	1875	0.077

[a] Indicated precision limit is the standard deviation of four replicate determinations.
Source: Ref. 144.

their results are reproduced in Fig. 22. Notwithstanding technical limitations in the determination of fibril angles—resulting, for example, in possibly false zero-angle and low-angle results [7]—differences by a factor of 2 or 3 in the strength of fibers were apparent at similar fibril angles and could be explained by the extent and severity of defects along the fibers, as noted in the previous section. Fibers with gross defects were also the weakest [278]. Forgacs also reported that fibers swell and rupture preferentially at "nodes" [93]. It follows that both the number of irregularities in fibers and the severity of the associated disruptions must be important factors of fiber strength. This conclusion is brought out in the study by Laamanen [199], who found that fiber mechanical properties were influenced measurably by external fibrillation (Fig. 42), kink index and content of minishives (two or more fibers not separated, which occurs significantly in mechanical pulps; see Fig. 43). Other factors affected include wet fiber stiffness or flexibility (See Section VII.C) and fiber length.

The strength potential of conventional mill-made softwood kraft pulps, measured as the tear index at a given tensile index, is about 25% lower than that of laboratory pulps prepared using the same materials and process conditions [210]. This is a serious problem, because strength is the key commercial characteristic of those pulps. One of the hypotheses to account for the strength difference is the possible presence of more irregularities along the length of a mill pulp fiber. This hypothesis was suggested by Knutsson and Stockman [182], based on fiber cuts after mild hydrolysis; by MacLeod and Pelletier [211], based on qualitative microscopic observations; and by Hakanen and Hartler [114], who counted more irregularities in mill pulps than in the corresponding laboratory pulps (see above). The hypothesis was confirmed statistically by Iribarne [144] with the data presented in Table 6.

V. FIBER MECHANICS MODELING

There are many books and papers devoted to the mechanics of laminated composite materials [2,30,39,40,163,356]. Analytical models that have been employed for fiber-reinforced plastics, filament-wound structures, angle-ply laminates of wood veneer, etc. can be suitably modified to simulate the cell wall properties of a wood fiber, or pulp fiber, in various environments; the basic mechanical principles are the same or very similar. The outstanding difference is the diminutive size scale, which requires some evaluation and use of putative basic component properties originating from diffraction data and other indirect experimental methods, or pure theory. Salmén and associates have provided a good informational starting point for the researcher who is entering this specialty for the first time [40,187,306,308].

Figure 12 shows the laminated nature of the wood fiber. Figure 44a illustrates that the fiber in wood is effectively prevented from twisting by the bonding of adjacent fibers through the compound middle lamella (ML). There is accordingly restraint of the shear strain that would otherwise result from a tensile or compressive stress along the fiber axis. The restraint arises mainly because the S2 angles in adjacent cells are approximately equal and opposite; two adjacent walls of similar thickness must deform as a unit, as shown by Schniewind [313] and Schniewind and Barrett [314]. In contrast, isolated fibers are free to twist, a behavior that has been examined experimentally and theoretically [106,184–186,223]; see also Section VII.D. However, the collapse and internal S3–S3 bonding of the fiber (Fig. 44b) creates a partially restrained double-walled antisymmetrical laminate. If similar fibers are formed into a well-bonded sheet (Fig. 27), a very high proportion of the fiber surfaces are bonded, which regains much of the shear restraint and consequently resistance to twist.*

The construction of appropriate mechanical models for fibers has as its primary objectives the theoretical mechanical properties and stress distributions in a fiber as it responds to the forces and displacements applied to it. Theoretical predic-

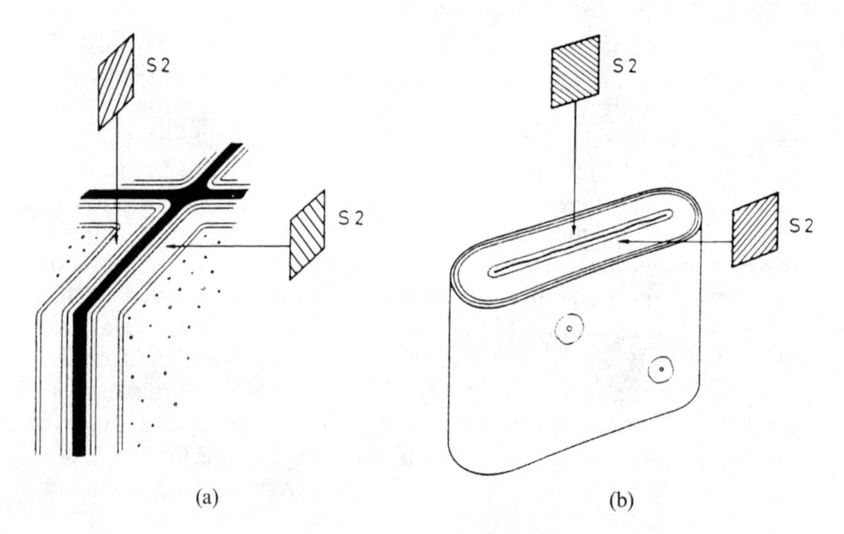

Fig. 44 Micromechanics elements in wood and pulp fibers: (a) Wood cell wall element consisting of a fiber wall, a middle lamella (ML), and an adjacent fiber wall. The element includes the layers S3, S2, S1, P, ML, P, S1, S2, S3. (b) Element of a collapsed pulp fiber, consisting of the layers P, S1, S2, S3, S3, S2, S1, P. (From Ref. 306.)

* Warping of paper and paperboard can be influenced by fiber twisting strains, but these are usually small in comparison with the effects of fiber orientation distribution throughout the thickness.

tions can be compared with actual experiments in which tensile or other loading is applied. Such tests may include the effects of damage, or time, temperature, and moisture conditions (applied cyclically or otherwise), which can be incorporated into the model.

The first step in building the model uses *micromechanics* to construct the directional properties of a single fiber cell wall lamina (or layer, if it can be assumed that layer properties are the same as those of the constituent laminae) from a knowledge of the properties of its components. An example of this step was carried out in Section III on the basis of reported theoretical and experimental work on the mechanical properties of native cellulose and matrix materials. Micromechanics principles are discussed in references such as 163, 274, and 306.

The second step considers the whole fiber wall, with each of the layers having a set of directional properties. When these layers are combined, the principles of lamination theory [30,40,163,357] are used, for example in Ref. 25, which will be discussed further along as an example of the process. For intact fibers, it is generally assumed that the fiber deforms as a unit. That is, all the layers will deform together without relative slippage under applied load. This assumption can be made with reasonable assurance for a dry pulp fiber subjected to small deformations, but it is highly unlikely that the assumption holds for delignified wet fibers or fibers subjected to large deformations.

For each wall layer in the absence of swelling, Hooke's law for the microfibrillar filaments can be written as

$$\sigma^f = A'^f \epsilon^f \tag{15}$$

Here, σ denotes the stress tensor, ϵ the strain tensor, A' the elastic stiffness tensor, and the superscript f refers to the cellulosic microfibrils or filaments. For the matrix material, we use superscript m and note that arbitrary swelling strains must be provided for in Hooke's law. By analogy with thermal strains, this can be written as

$$\sigma^m = A'^m(\epsilon^m - \epsilon^*) \tag{16}$$

where ϵ^* are the swelling strains. By convention, positive strains are taken to be tensile, i.e., swelling strains. Although we may refer to strains as either shrinkage or swelling, it is understood that negative strains are required to describe shrinkage.

Because $A'^m \epsilon^*$ defines a stress tensor, denoted as σ^*, the preceding equation can be written equivalently as

$$\sigma^m = A'^m \epsilon^m - \sigma^* \tag{17}$$

The overall behavior of the composite is related to its constituents through the principle of volume-average properties. Let f be the volume fraction of the filament and $m = 1 - f$ that of the matrix. Then the overall stress tensor σ is assumed to be the volume average $\sigma = f\sigma^f + m\sigma^m$ and therefore

$$\sigma = fA'^f \epsilon^f + mA'^m \epsilon^m - m\sigma^* \tag{18}$$

Thus, if the composite stiffness tensor A' is also defined by volume averaging, $A'\epsilon = A'(f\epsilon^f + m\epsilon^m)$, Hooke's law for the composite layer must be of the form

$$\sigma = A'\epsilon - m\sigma^* \tag{19}$$

The fiber as a whole can be treated as an axially symmetrical tube consisting of several layers. The contribution of each layer to the properties of the fiber is proportional to the relative volume it occupies in the fiber wall. In brief, the procedure is to use an expanded form of Eq. (19) for each layer, yielding stiffness constants that are orthotropic with respect to the reference frame of the layer, i.e., parallel and perpendicular to the predominant microfibril direction of that layer (one perpendicular axis lies in the plane of the wall layer; the other is oriented radially—in the thickness direction of the layer). The transformation equations are used to obtain stiffness constants A_{ij} with respect to *fiber* coordinates (1, 2, 3 axes) in Fig. 45 from the stiffness constants A'_{ij} with respect to layer coordinates (X, Y, Z, axes) in Fig. 45. These transformations are performed for each layer by accounting for the angle α between the mean microfibrillar direction of the layer and the fiber axis.

Within a layer, the normal stiffness constant A_{33} associated with the 3 direction will be reduced in comparison with that for the Z direction in accordance with the formula

$$A_{33} = A'_{33}\cos^4\alpha + A'_{22}\sin^4\alpha + \left(2A'_{23} + 4A'_{44}\right)\sin^2\alpha\cos^2\alpha \qquad (20)$$

Here A'_{33} is by far the largest coefficient, so the $\cos^4\alpha$ term dominates the relation. As examples, if the angle α in Fig. 45 is 15°, A_{33} for the layer will be about 15% less than A'_{33}. If $\alpha = 30°$, the reduction will amount to approximately 45%. The properties of the composite fiber, composed of several layers, are thus influenced heavily by the degree of microfibrillar orientation and the proportions of the cell wall in each layer.

In Fig. 46 is shown the calculated relationship between the S2 filament winding angle (the mean microfibrillar angle of the S2 layer) and the axial and transverse

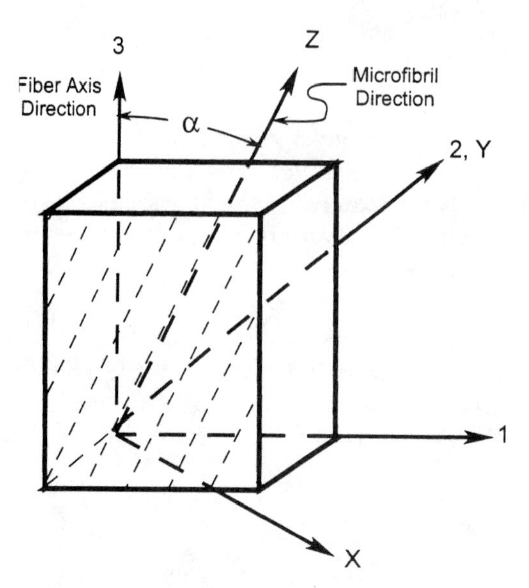

Fig. 45 Model element of a layer of the fiber wall, showing the reference frames with respect to microfibrillar orientation (X, Y, Z axes) and the fiber coordinates (1, 2, 3 axes). (Courtesy of Dr. D. Steven Keller.)

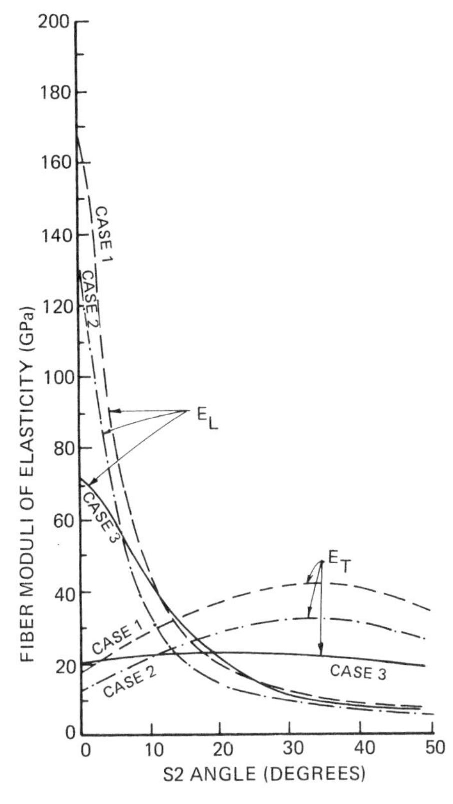

Fig. 46 Theoretical curves for axial (E_L) and transverse (E_T) elastic moduli of a fiber versus assumed S2 angle. (From Ref. 220.)

moduli of elasticity for a pulp fiber consisting of three layers (S1, S2, and S3). The three parametric curves represent three different sets of values for the elastic constants of cellulose I.

As in the discussion regarding Fig. 20, the Case 1 and 2 curves derive mainly from the cellulose I elastic constants shown in Table 1 for Gillis [104] in 1969; 319 GPa and 246 GPa, respectively. The Case 3 curve is based on an experimental value of 134 GPa by Sakurada et al. [305], shown in Table 2. The values in these tables are the A'_{33} values in Eq. (20). Derivations by Gillis [104] and Mark [214] were used for the other A'_{ij} values needed for Eq. (20). The other assumptions regarding fiber cell wall structure, matrix elastic constants, etc. are the same as those listed in Section IV.B in connection with Fig. 20.

One may compare assumptions for matrix elastic constants with the experimental E^m values provided by Cousins [56–58], which are moisture-related. An interesting result can be noted if values near the maximum observed by Cousins are employed for the matrix constants. Specifically, if E^m is set equal to 7.0 GPa and G^m is set at 2.7 GPa ($\nu^m = 0.3$ as before), the values for E_L in Fig. 46 (all cases) are approximately doubled for the range of S2 angles greater than 20°, show only moderate increases in the range between 10° and 20°, and show virtually no change when the S2 angle is $< 10°$. Figure 20 demonstrates this matrix–framework–S2 angle relationship clearly. For additional insight into the relationship of matrix and fiber softening induced by moisture, see Refs. 309 and 310.

Fiber axial stiffness is defined by the product of fiber cross-sectional area and the modulus of elasticity in the fiber axis direction. Thus, for fibers of equal cross-sectional area, the axial stiffness of fibers with large S2 angles is largely insensitive to the properties of the cellulosic framework and very dependent on the properties of the matrix; conversely, the stiffness of fibers with small S2 angles ($< 10^\circ$) is very insensitive to matrix properties but extremely dependent upon the properties of cellulose, as noted in Section IV.

Although Fig. 46 shows that the microfibrillar laminae that make up the layers of the fiber wall are extremely anisotropic, the overall elastic properties of the wood fiber are not likely to differ greatly in the axial and transverse (circumferential) directions as a consequence of the off-axis orientation of S2 and the relatively circumferential orientations of S1 and S3. There are several ramifications as to strength and other properties that arise from this cell wall arrangement [216,217]. However, because the principal reason for the similarity in properties when the S2 angle is in the 12–20° range is the presence of the S1 layer superposed on an off-axis S2 helix, any processing variable, e.g., beating, recycling, that tends to disrupt or break the S1 layer has as one consequence a greater level of fiber anisotropy. Similarly, fibers such as those of flax (*Linum*) may tend to be more anisotropic because of a characteristically small S2 filament winding angle.

Variations in Fiber Modeling The groundwork for today's developments in fiber properties was laid by many able researchers, including Barkas [19], whose thoughts on cellular elasticity were ahead of his time.

There is now a substantial body of literature devoted to the mechanics of fiber cell walls. The first two- and three-dimensional theoretical models that applied the orthotropic Hooke's law to the case of fibers, with compliance and stiffness matrices such as those introduced in Eqs. (4)–(6) and the A_{ij} arrays following them, were published in the second half of the decade of the 1960s. The years since then have seen this work expand considerably. This literature has found many useful applications; however, there are probably not a large number of readers of this book who need to have detailed descriptions of the many important research works in this specialized area. What is provided is a short overview of some of the main applications, a chronological listing of reference sources in the field by decade, some cautionary comments with appropriate explanations, and a discussion of one recent paper that will serve as an example of important current work.

Subject Areas in Fiber Modeling Application areas for fiber modeling include the cells walls as elemental units in wood, fiber bundles (in bark, leaves, stalks, etc.), shives, thin slices of woody tissue, collapsed and uncollapsed single fibers, and paper and board and as analogs for other biological tissues (e.g., bone). Boundary conditions that are typically considered involve loading modes, e.g., comparison of shear vs. tensile loading; complete, partial, or absent shear restraint; and effects of time, temperature, moisture, damage, and cycling. Many of the tests described in Section VII are used to evaluate the usefulness of fiber mechanical models.

Chronology The following references are fairly complete from 1965 through 1999. They are arranged by decade in (as best as can be determined) chronological order:

Decade	References
1960s	213, 214, 41, 42, 314, 222
1970s	219, 105, 223, 216, 313, 220, 335, 217, 106, 336, 21, 221, 85, 43, 268, 324, 269, 279, 45, 44, 373
1980s	218, 157, 276, 350, 309, 307, 206, 190
1990s	191, 158, 185, 128, 14, 257, 13, 25, 7

Cautionary Comments Each of the research reports listed above has contributed in some way, large or small, to the advancement of our understanding of fiber mechanics. However, there are also several areas of controversy that need to be mentioned, primarily to identify potential problem areas in methodology that could produce errors or important inaccuracies.

An obvious first step in avoiding error is to consider the fundamental constitutive relations. The boundary conditions and compatibility equations must be realistic. For example, most of the 3-D published work assumes that shear strains are equal at a layer interface (say, at a boundary between S2 and either S1 or S3 in a fiber subjected to axial tension or compression or twisting). Other work stipulates that shear stresses are equal at an interface, a condition that is difficult to justify.

A related equation concerns 2-D versus 3-D models. Barrett and Schniewind [21] demonstrated that there is rather good agreement between the results obtained from two- and three-dimensional models as to internal stress distributions among the layers and the prediction of angular twist of fibers in response to axial loads, as long as the assumption of equal shear strains at interfaces is maintained. Thus, the 2-D analysis techniques are satisfactory for many problems. The principal advantage of the much more complicated 3-D models is their ability to allow stress distribution determinations, including radial stresses, to be made for thick-walled fibers that can be reasonably modeled as layered cylinders.

In some reports, the author(s) have assumed that microfibrils within a layer such as S2 do not lie parallel to one another but instead are statistically distributed around some mean value. The main effect of this assumption is to diminish the composite Young's modulus value (E_Z) in the "microfibril direction" (see Fig. 45). In turn, reductions in E_Z values result in a reduction in the stiffness constant A_{33} for the fiber as a whole. There seems to be no evidence for disperse orientations of microfibrils throughout a given layer. On the contrary, there are at least two types of evidence for a generally *uniform* alignment: (1) a high degree of control of microfibril orientation(s) by the rosettes on the plasmalemma and (2) the pattern of cavity formation by soft rot fungi.

The high degree of control at the time of cellulose I biosynthesis, shown diagrammatically in Fig. 47 [261], is best evidenced by Figs. 1 and 2. When microfibrils are being created, they are aligned for maximum effectiveness in the locations and situations that exist at the instant of formation. No better example could be found than in the S1 layer, where measurements by polarized light [214] yield values of 50% Z helix at 80° and 50% S helix at −80°. Alignment dispersal would diminish the functionality of these layers. As for the action of soft rot fungi, it has

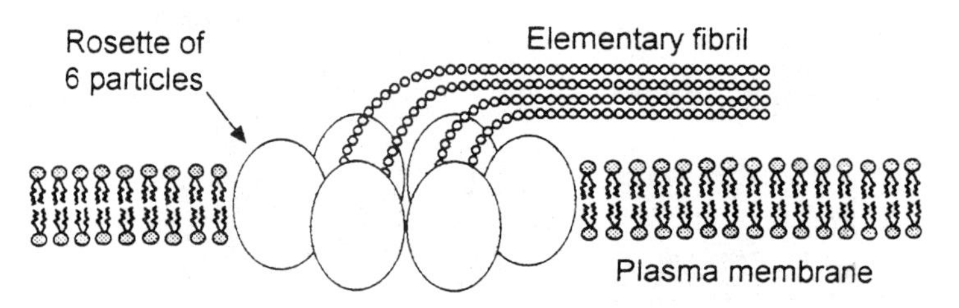

Fig. 47 Simplified representation of a cellulose I biosynthesis site. Four chains are schematically shown emanating from the six-particle rosette. The actual number is on the order of 100^+ to 200^+ chains. (From Ref. 261.)

been noted that the biconical-tipped cylindrical cavities in S2 are rigorously aligned with the microfibrillar orientation [7–9]. The cavities are often almost as wide as the entire S2 layer thickness. The geometrical cavity shape (Fig. 26) is therefore not an excavation from one or a few microfibrils but from hundreds or even thousands of packed microfibrils. Cell wall computations based on dispersal statistics or "homogenization" should be looked at from the standpoints of formation and functionality in the fiber.

Another type of reduced modulus computation occurs when the microfibrils are considered to be finite-length reinforcement rods. This concept implies that microfibrils are known to be of certain lengths or that they are so weak at intervals that they do not function as intact crystallites beyond a given length. But over 40 years of electron microscopy has not revealed discrete microfibrillar lengths in completed fiber walls. The visible structure (Fig. 1) is endless, a fact that led to the "twisted skein of yarn" concept of fiber wall formation—see Refs. 218 and 7 and Fig. 25. As for the idea that microfibrils possess alternating structural (crystalline) and nonstructural (amorphous) segments, Kim and Newman [180] noted that solid-state NMR spectral studies discount the idea of amorphous or noncrystalline cellulose in wood, and Sugiyama et al. [331] showed in their classical electron diffraction study that both cellulose Iα and Iβ are *exclusively* composed of extended chains that are "rigorously" aligned in the axial direction within the microfibril.

Of greater consequence is the concept that the behavior of the S2 layer can be considered akin to the behavior of the whole fiber. In most cases this is not a justifiable assumption. Figure 48 demonstrates the differences that occur in fiber moduli of elasticity (E_{33}, E_{11}) and shear rigidity (G_{13}) when a model comparison of a collapsed, flattened fiber (Fig. 27) composed of the usual three layers S1, S2, and S3 (Fig. 48a), is made with an otherwise identical fiber composed of S2 only (Fig. 48b).

The layer elastic constants used for the calculations whose results are displayed in Fig. 48 are based on the cellulose I elastic constant values derived by Tashiro and Kobayashi [348]. The layer (or lamina) properties are derived from Eqs. 14 when appropriate matrix constants (see Sections II. B. and IV. B) are combined with those for cellulose. The derived 2-D set in this case has the following values:

$$E_Z = 118\,\text{GPa} \qquad E_X = 5.54\,\text{GPa}$$
$$G_{ZX} = 5.46\,\text{GPa} \qquad \nu_{ZX} = 0.0554$$

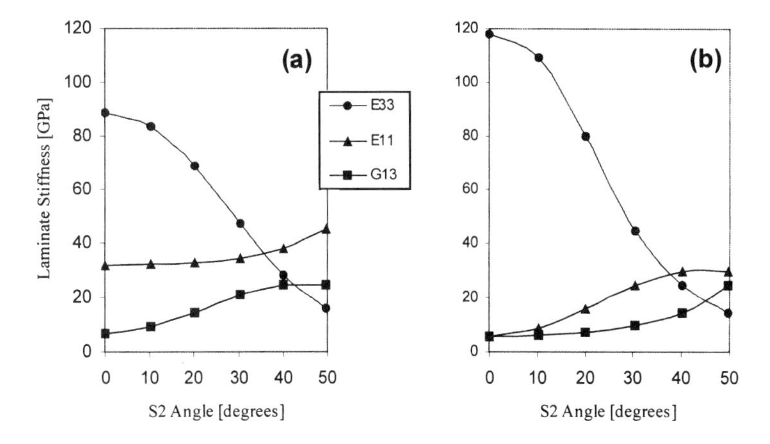

Fig. 48 Graphical representation of the fiber cell wall laminate calculations that appear in Table 7. (a) Elastic constant calculations of fiber wall when S1, S2, and S3 layers are all taken into account (Table 7A). (b) Comparable calculations resulting from an assumption that only the S2 layer exists in the laminate (Table 7B). The fiber axis direction is 33; the transverse (circumferential) direction is 11.

Results shown in Fig. 48b for the single-layered (S1 = S3 = 0%, S2 = 100%) collapsed fiber, which is a two-ply laminate, demonstrate that for an S2 angle of 0° the fiber axial modulus is identical with the above E_Z value. For the three-layered case (a six-ply laminate) (S1 = 20%, S2 = 74%, S3 = 6%), the comparable value is 88.8 GPa, a reduction of 25% ascribable to the real presence of S1 and S3. The transverse modulus is very small in Fig. 48b at S2 = 0°, resulting in an axial-to-transverse modulus ratio of over 21 (see Table 7). When three layers are considered, the ratio is found to be less than 3. The points of convergence for axial and transverse curves also differ greatly.

Table 8 contains the same type of information for Case 2 in Fig. 46. These curves are based on the layer elastic constants given below, derived from the cellulose I elastic constants of Gillis [104] and Eqs. (14). Other parameters are similar to those used for Figs. 20 and 48 and Table 7.

$$E_Z = 177\,\text{GPa} \qquad E_X = 6.8\,\text{GPa}$$

$$G_{ZX} = 2.6\,\text{GPa} \qquad \nu_{ZX} = 0.10$$

It was noted previously that in a collapsed three-layered fiber there are angular orientations of S1, S2, and S3 on one wall that are balanced by the same orientations of opposite sign on the other wall. However, in wood pulp fibers, the layers that "balance" each other (as a result of complete collapse of the lumen) are neither contiguous (except for S3) nor truly symmetrical through the thickness (although they are symmetrical with respect to the fiber axis). This antisymmetry factor is serious even in thin-walled laminates and may be accentuated in the case of thick walls. It has been shown theoretically that there is a coupling between shear strain and axial tension and between bending and tension in anisotropic laminates even when the governing equations are linearized [295,306,364,365]. This coupling phenomenon has been demonstrated experimentally [12] for orthotropic laminates and

Table 7 Illustration of Effect of Dropping Thinner Fiber Layers in Modeling, Based on Tashiro–Kobayashi Computations for Cellulose I [348] (See Fig. 48)

S2 angle (deg)	E_{33} (GPa)	E_{11} (GPa)	$G_{13}(=G_{31})$ (GPa)	E_{33}/E_{11} ratio
A. Calculated elastic constants for a collapsed three-layered pulp fiber, considered as a six-ply antisymmetrical angle-ply laminate				
0	88.8	32.0	6.67	2.78
10	83.8	32.2	8.86	2.60
20	68.9	32.9	14.4	2.09
30	47.9	34.4	20.7	1.39
40	28.5	37.7	24.8	0.756
50	15.9	45.2	24.8	0.352
B. Calculated elastic constants for a collapsed single-layered (S2) pulp fiber, considered as a two-ply antisymmetrical angle-ply laminate				
0	118.0	5.54	5.46	21.3
10	109.5	5.88	8.41	18.6
20	81.6	6.97	15.5	11.7
30	45.7	9.47	24.4	4.83
40	24.7	14.4	30.0	1.72
50	14.4	24.6	30.0	0.585

Note: Most of the computational work for Table 7 and Fig. 48 is courtesy of Dr. Richard W. Perkins, Jr.

Table 8 Illustration of Effect of Dropping Thinner Fiber Layers in Modeling, Based on Gillis "Case 2" Computations for Cellulose I [104] (See Fig. 46)[a]

S2 angle (deg)	E_{33} (GPa)	E_{11} (GPa)	G_{13} (GPa)	E_{33}/E_{11} ratio
A. Calculated elastic constants for a collapsed three-layered pulp fiber, considered as a six-ply antisymmetrical angle-ply laminate				
0	133	18.7	2.7	7.1
10	46.7	18.5	2.8	2.5
20	18.7	18.3	3.3	1.0
30	11.4	18.1	4.3	0.6
40	8.6	18.4	5.4	0.5
B. Calculated elastic constants for a collapsed single-layered (S2) pulp fiber, considered as a two-ply antisymmetrical angle-ply laminate				
0	177	6.8	2.6	26.0
10	60.0	6.7	2.8	9.0
20	21.7	6.5	3.5	3.3
30	11.9	6.5	4.7	1.8
40	8.3	6.9	6.2	1.2

[a] Exception: Value (experimental) chosen for G_{ZX} was 2.6 GPa [185].

will have to be taken into account as further refinements are made in the mechanical models for cell walls.

Figure 49 shows a two-layered strip fabricated of orthotropic layers of nylon-reinforced rubber under a tensile load. This tensile specimen is free to rotate, and thus only the axial stress resultant N_x is nonzero, whereas the transverse and shear stress resultants N_y and N_{xy} and all of the moment resultants are zero. The two layers are balanced with respect to the axial direction of the strip and therefore would be expected to extend without twisting or bending on the basis of simple orthotropic theory. As shown in Fig. 49, however, the strip twists under the tensile load. This effect is caused by the tendency for the two layers to exhibit equal and opposite shear deformations because of their different orientations with respect to the direction of loading. The following relation from Ashton et al. [12] shows the shear coupling term in addition to the normal stress resultants:

$$N_x = A_{11}e_x + A_{12}e_y + B_{16}K_{xy} \tag{21}$$

where

$A_{11}, A_{12}, B_{16} =$ elastic stiffness constants obtained through a transformation of the generalized Hooke's law for orthotropic materials

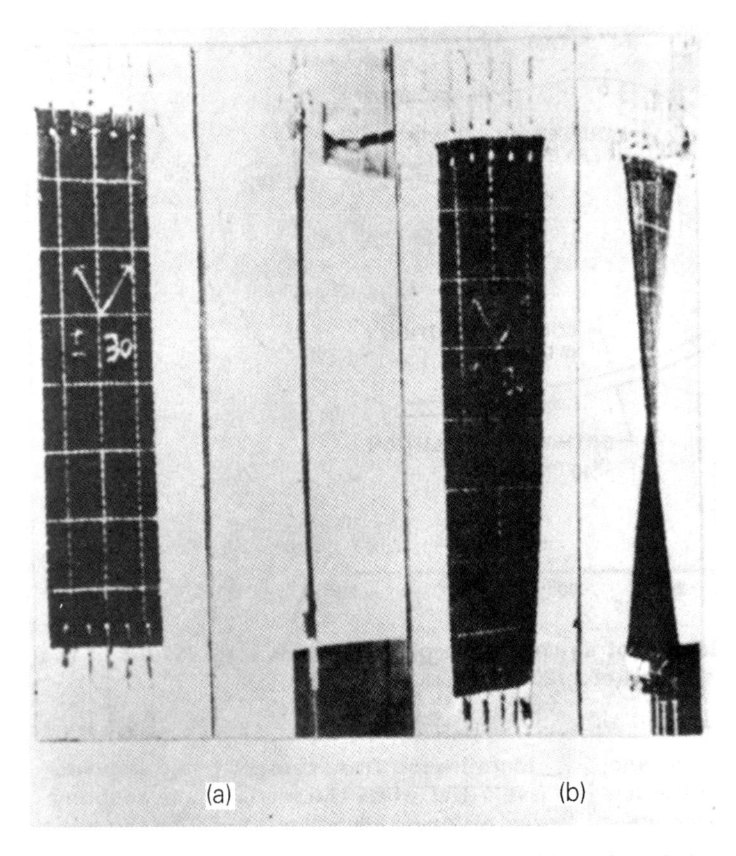

Fig. 49 Coupling between bending and stretching of a "balanced" two-ply laminate. (a) Before test. (b) Under uniaxial tension. (From Ref. 12.)

$e_x, e_y =$ axial and transverse strains, respectively
$K_{xy} =$ the inverse rate of change of the slope for deflection in the laminate thickness direction under the condition that displacements are small.

Similar phenomena have been demonstrated in flexure studies on composite plates. As Fig. 50 shows, the orthotropic elasticity solution for a plate composed of a two-ply balanced laminate loaded transversely would indicate deflection according to the lower curve. But when the interlaminar coupling terms are taken into account, the very different behavior shown by the upper curve is predicted. Careful experimental work has confirmed that this type of predicted behavior does in fact occur. A two-ply plate is not sufficiently symmetrical about its own middle plane to allow neglect of the coupling phenomenon. This effect diminishes when four-ply, six-ply, etc., antisymmetrical laminates are employed, and eventually the orthotropic solution is approximated when more than six plies are present in the balanced laminate (for example, eight-ply laminates, where there are two sets of balanced laminates on *either side* of the bisector axis running in the middle plane of the plate).

A Look at a Recent Paper Rather than specify a preference among the 40-odd references listed earlier, I decided to take one very recent publication to summarize briefly in the hope that this will give readers a sense of the state of advancement in the field.

Berg and Gradin's paper [25] describes a model for a wood fiber subjected to processing damage (e.g., in mechanical pulping) and loaded in tension or shear. Because the fibers are not pulped chemically, the middle lamella and primary wall

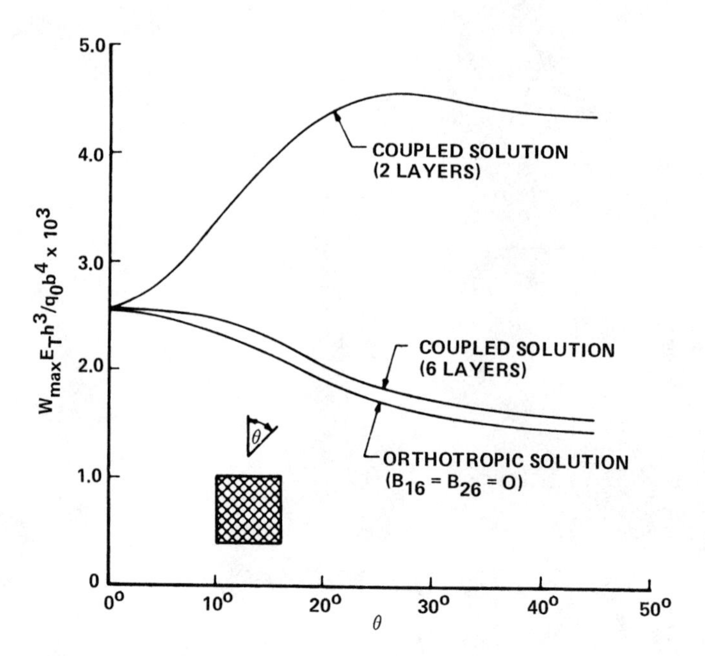

Fig. 50 Maximum deflection of square angle-ply composite plate loaded in the thickness direction. (From Ref. 365.)

are present. The authors develop different composite in-plane elastic properties for each of five layers—ML, P, S1, S2, and S3—under four different conditions of moisture: dry at 20°C, 12% MC at 20°C, wet at 20°C, and wet at 95°C. Consequently there exist 20 different sets of in-plane constants. Lignin is taken to be isotropic, whereas cellulose and hemicellulose are considered to be transversely isotropic. Berg and Gradin then construct a laminate model consisting of two combined layers: an outer layer consisting of ML and P plus the two orientations that are characteristic of S1, and an inner layer consisting of S2 and S3. The S2 angle is set at 15°. This writer would challenge several of the parametric values that are used, for example for S1. The coupling terms between shear strains and normal stresses are ignored on the basis that these terms will be small with the choice of a 15° S2 angle. The analyses of the model are based on an assumed displacement field together with the minimum total potential energy theorem For the damage development, it is assumed that during loading of the fiber, equally spaced cracks will appear in the inner layer, running parallel to the S2 angle and controlled by an energy criterion. Stiffness coefficients are calculated as a function of the damage state. Results indicate that stiffness will degrade under a tensile load to a greater extent than under a shear load.

VI. SOME IMPORTANT PHYSICAL PROPERTIES OF PAPERMAKING FIBERS

We introduce here some property measurements that are or may become essential in testing for fiber mechanical properties, especially in the light of new efforts to unify the tests and reporting units [90,342].

Fiber-using industries used to have their own individual systems for defining such basic properties as mass per unit length, now known as *linear density* or *coarseness*. Among the now obsolete terms for this property was *spyndle*, but there were others such as *linen lea number* and *cotton hank number*. Later, *denier* and *grex* and, relatively recently, *tex* units were employed for measuring this property, primarily in the textile industries but also occasionally in the pulp and paper industry. See Table 9.

A. Wood Density and Fiber Density

Different measures of density are used in the evaluation of various woods and fibers for properties of interest. The *basic density of wood* is the oven-dry mass (kg) of the

Table 9 Conversion Factors to Tex Equivalent

Tex units	= 0.111111 × denier units
	= 0.100000 × grex units
	= 1000.000 × kilotex units
	= 0.100000 × decitex units
	= 0.001000 × millitex units
	= 34.4482 × spyndle units
	= 0.590541 × grains per 120 yd
	= 0.708650 × grains per 100 yd
	= 0.496055 × lb/1,000,000 yd

wood substance contained in the original green volume in cubic meters. "Green" refers to the presence of moisture within the tree, so the measurement of "green volume" is done before the wood moisture drops below the fiber saturation point and the wood begins to shrink. If the basic density of wood is divided by the density of water at 4°C, which is 1000 kg/m^3, the units cancel and the result is called the *relative density* or *specific gravity*. Another measure sometimes used is the *mass density*, wherein both the mass and volume of wood are measured at the same moisture content. This measurement may take place at the point of arrival (in the laboratory, in the woodyard, etc.) and is inherently somewhat arbitrary.

Timell (see Ref. 354, pp. 470–488) describes the five common methods used for density determination of wood. They are (1) water displacement, (2) maximum moisture content, (3) mercury immersion, (4) beta-ray densitometry, and (5) X-ray densitometry. The latter method is especially convenient.

The density of wood substance itself is fairly constant among woods that are largely free of extractives. However, wood cell wall substance is often heavier in heartwood (which contains most of the extractives) than in sapwood. Berlyn [27] has shown by microspectrophotometry that the dry cell wall in wood has only a few or no voids. The technique and calculations are also given in Ref. 28, pp. 272–273. The "true" density determined for the wood fiber cell wall in normal wood typically has a value in the 1430–1500, kg/m^3 range when measured pycnometrically in toluene. A slightly higher range is found when water is the displacement liquid, because water will penetrate (and perhaps force open) void space that is inaccessible to toluene. Compression wood cell walls, which contain more lignin (specific gravity $\cong$ 1.38) and less cellulose (specific gravity $\cong$ 1.58) than normal wood, tend to have lower densities.

Fiber densities change upon removal from the original wood according to what proportions of cellulose, hemicelluloses (specific gravity $\cong$ 1.50), and lignin remain. If the pulping method is a purely groundwood operation, the true fiber density will be the same as in the original wood. On the other hand, most of the lignin is removed in a typical chemical pulp, and this removal may result in internal voids. Thus, two types of fiber density measurements are made: *bulk density* and *true density*. Bulk density (also called *apparent density*) may take into account all the voids within the fiber structure or be based on "bulk volume" (volume of fiber wall plus lumen). True density (also called *packing density*) is based on the solid matter present, at least to the best of one's ability to measure it. For papermaking fibers, true density measurements are generally specified, although there are many exceptions. For instance, Li et al. [206] showed that morphological properties that characterize the apparent densities of fibers (either collapsed or uncollapsed) have important practical applications to mill problems such as hydrocyclone fractionation of chemical pulp. Their analysis quantifies the extent of fiber collapse with changes in apparent density that affect the accepts/rejects ratio.

Evaluations of true fiber density can be made on the basis of measurements of fiber dimensions (length and cross-sectional area) by techniques discussed in the next section and data on constituent masses. The main constituents within the fiber wall structure are cellulose, hemicellulose, lignin, and inorganic matter. If the volumetric proportions and densities of these constituents are known or can be assumed, the true density can be calculated directly. Measurements of cell wall density can also be done by obtaining the mass of a small known amount of fiber material and either

(1) determining the volume of the mass [327,354] or (2) determining the total number and arithmetic mean length of the fibers in the mass by use of a fiber length analyzer [319] to determine coarseness, and applying the coarseness–density relation for fibers (see next section).

The measurement of density is sensitive to the manner in which the test is conducted. For example, is the dry pulp mass obtained by oven drying or with moisture-free air? Density data should be complemented with additional information such as the temperature at which the determination was made, especially if the fibers are immersed in a liquid. Reference to any use of a standard measurement procedure such as TAPPI Useful Method 18 [341] should be stated. Care should be taken to ensure that appropriate units are employed. The usual SI reporting unit for volumetric density (true or apparent) is kilograms per cubic meter (kg/m^3).

B. Coarseness (Linear Density)

The mass per unit length of a fiber is conventionally referred to as the *linear density* in the textile industry and as *coarseness* in the paper industry.

Accurate determination of coarseness of papermaking fibers is usually accomplished today by instruments that can rapidly count fibers that pass through flow fields and measure their lengths. This counting and measuring information, combined with an accurate determination of fiber dry mass, yields a coarseness value C when used in the equation

$$C = \frac{m}{n\ell_n} \tag{22}$$

where

m = a small mass of fibers with free and bound water removed

n = total number of fibers in the mass

ℓ_n = arithmetic mean length of the fibers

Two types of instruments are increasingly employed for coarseness determinations. One is based on passage of a known mass of fibers through a narrow orifice past a beam of polarized light, which renders the cellulosic fiber birefringent. The instrument has a detector capable of recording and measuring the length of each fiber, based on detection of the birefringence. Several models of instruments of this type are manufactured by Kajaani Electronics, Ltd., Kajaani, Finland.

In the second type of instrument, a dilute suspension of pulp fibers flows past optical and image-processing systems in a tapered sheath cell that orients the fibers into a thin, nearly two-dimensional (2-D) plane. A manufacturer of this system is OpTest Equipment Inc., Hawkesbury, Ontario, Canada. Their instrument is marketed as a fiber quality analyzer (FQA) that has an image detection system consisting of a two-dimensional 256×256 pixel charge-coupled device (CCD) camera.

Robertson et al. [299] compared coarseness measurements made by the FQA method with results from a Kajaani FS-100 instrument and with the standard manual method given in TAPPI 234 [337]. The three methods yield results that are quite similar for a range of chemical and mechanical pulps as well as rayon fibers. T 234 is based on observation and counting of a fiber sample in a microscope or projected image. Results are reported in units of milligrams per 100 meters (mg/100 m), which is also called decigrex (dg) in the TAPPI method. However, in most of the newer variations of coarseness tests, the reporting unit is milligrams per meter (mg/m) or micrograms per meter (μg/m). The largest or coarsest softwood pulp fibers have values approaching 0.5 mg/m [171]; most plant fibers have *much* smaller linear densities, especially in those hardwoods that possess thin-walled ("fine") fibers. A typical fiber might have a coarseness close to 0.2 mg/m.

From the previous section, it can be seen readily that the relation between true fiber density ρ and coarseness C is simply

$$C = A_w \rho \tag{23}$$

where A_w = cross-sectional area of the fiber. Therefore, if either coarseness or true density is known or can be determined, the other can be found by accurate measurement of A_w and the use of Eq. (23), provided that measurements are all at the same moisture-free conditions. Improvements in the measurement of coarseness have been accompanied by steady improvements in the determination of fiber cross-sectional area, aspect ratio, collapse behavior, wall thickness, etc. Until about 20 years ago the methods used for these parameters involved optical microscopes with ancillary equipment such as a 35 mm camera mounted on the monocular tube of a light microscope or a camera lucida or other light projection apparatus. Measurements were then made on the projected image or the film after calibration. For these methods, the fibers must be dehydrated and embedded (usually after staining) in a suitable medium such as paraffin or epoxide plastic resin and sectioned by microtome [28]. For measurements of cell wall cross-sectional area, an alternative was developed by Hardacker [126]. His "fiber lateral compaction device" compressed the fiber between sapphire anvils to close the lumen. When fibers are fully collapsed, the thickness of the fiber can be measured with a linear variable differential transducer (LVDT) connected to a data acquisition system. The width is measured under a microscope equipped with a graduated eyepiece. If the shape of the compacted fiber is considered to be a flat ribbon with rounded edges, the cross-sectional area A_w is found as

$$A_w = wt - \frac{4 - \pi}{4} t^2 \tag{24}$$

where w = width and t = thickness of the fiber.

Development of the Kajaani instrument in the early 1980s provided a more rapid way to determine the denominator terms in Eq. (22). Improvements in measurement of A_w were enhanced by improvements in image analysis at about the same time [164,165,179]. Major advances in the following decade arose from the use of length analyzers in combination with the aforementioned techniques for obtaining the data needed for Eq. (22) or (23) for example, in Ref. 273). Advances in image processing [50,170,179], especially the introduction of confocal laser scanning microscopy (CLSM) revolutionized the measurement of dimensional properties of fibers such as wall thickness, cross-sectional area, and collapse characterization, improving both speed and accuracy [109,151–153,245,370]. Figure 51 from Ref. 152 shows several features of interest in a partially collapsed fiber that are readily measurable with the use of confocal microscopy and image analysis.

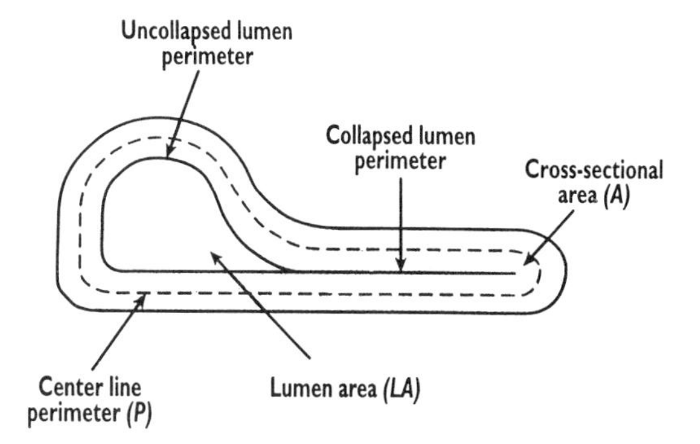

Fig. 51 Fiber collapse parameters as determined by the image analysis method of Jang and Seth [152]. Mean wall thickness $T = A/P$. Lumen perimeter LP is the sum of the perimeters of both collapsed and uncollapsed sectors.

Determination of the coarseness of a blended furnish was studied by Seth and Chan [319], who used different proportions of fibers of high (Douglas-fir: *Pseudotsuga menziesii*) and low (western redcedar: *Thuja plicata*) coarseness. This study demonstrated that the presence of debris in the pulp results in incorrect assessment of coarseness. The problem of debris, which can adversely affect the accuracy of coarseness measurements of single-species pulps as well as blended pulps, is related to the uncertainty of the number and length measurement values [as needed for Eq. (22)] when debris is present.

For long fibers (greater than 20 mm in length) a standard method is available (ASTM D 157-96) [6] that is based on the principle that the mass per unit length of a perfectly flexible fiber can be calculated when the fundamental resonant frequency of transverse vibration is measured under known conditions of length and tension. It is commonly known as the vibroscope method. In actuality, a fiber is not perfectly flexible; it has a finite bending stiffness. For very precise work, correction factors may have to be applied to account for cross-sectional shape, dimensions, and initial Young's modulus of the fiber. The vibroscope is standardized by a direct weighing method involving known numbers and dimensions of fibers as described previously. Where it can be used, the vibroscope provides a relatively rapid test procedure, and automatic features have been developed. However, papermaking fibers are normally 0.5–5 mm in length; the test is therefore generally employed where long fibers are more common—for example, textiles, reinforced plastics, tire cord, and ablative materials.

It was mentioned earlier that the textile industry generally uses the term "linear density" in place of coarseness. Different units are also employed. Linear density has been expressed in many units over industry history. In recent years there has been a movement toward the *tex* unit, which equals the mass in grams per kilometer of fiber length, although there is still considerable use of older unit systems, especially *denier*, which equals the gram mass per 9 km, and *grex*, which equals the gram mass per 10 km. One may encounter test results of fibers used in making paper and board expressed in these or other (older) units. Also, the textile industry uses the term *tenacity* or *breaking tenacity*. Breaking tenacity expressed in grams force per tex is identical to breaking length expressed in kilometers. Some conversion factors are given in Table 9.

C. Fiber Wetness Indexes

Çöpür and Makkonen [53] recently reported on a correlation between predicted and measured pulp yield from wood based on the hygroscopicity characteristic of the fibers. The concept behind measurement and use of this parameter is based on the differences between the various polymeric components of wood. Hemicelluloses are more hygroscopic than the cellulose component (and much more so than the lignin). Therefore, an indirect yield determinant is created by using fiber hygroscopicity as an indicator of the ultimate yield.

The wetness indexes reported by these investigators are based on coarseness and grammage units by expressing the amount of water in the fiber after humidity exposure as mass per unit length (μg/m) (same unit as for coarseness) or mass per unit area of the fiber ($\mu g/m^2$).

VII. MECHANICAL TESTING OF FIBERS

The most frequently performed laboratory tests for the mechanical properties of fibers may actually be fiber flexibility tests and zero-span or short-span tensile tests of (wet or dry) sheets. This section includes a subsection (VII.C) devoted to flexibility tests. Part of Section VII.A, entitled "Other Approaches to Evaluation of Fiber Tensile Properties," is focused on alternatives to single-fiber tensile testing, especially zero-span and/or short-span tests. The purpose of calling attention to these popular tests is to distinguish tests that can be run relatively quickly, some with commercially available instrumentation, as opposed to tests for basic research purposes that are designed to enhance our fundamental understanding of structure–properties relationships. The latter category of fiber tests often requires special instruments and systems.

Nevertheless, "fiber testing" is often used as a synonym for tensile testing in which the force is exerted in the fiber axial direction, one fiber at a time, at a rate of strain that may or may not satisfy the conditions of "static" testing from start to failure. Section VII.A covers static axial tensile tests, some new work in transverse tension, and the above-mentioned "other approaches." The remaining subsections explore other types of fiber testing—compression, bending (flexibility), torsion—as well as some of the effects of time, temperature, moisture, and machine-induced cycling and tests that relate fiber properties to fiber network properties.

A. Tensile Testing

Axial Tension The tensile properties of single fibers have been studied for over a century, starting with the experiments of O'Neill [316]. Some of the earliest tests of the mechanical strength of chemical pulp fibers were conducted by Rühlemann [300,301], who determined elongations and breaking loads of spruce tracheids. Since then, many investigators have performed such tests for a variety of purposes related to product and process development in order to understand the relationship between fiber properties and paper properties and also to elucidate the effects of fiber variables. Some of these variables are drying sequences, deinking, recycling, effects of drying under load, cross-sectional area, S2 microfibril angle, temperature, moisture, structural irregularities, and defects.

A discussion of the historical development of textile fiber testing instruments has been presented by Schwarz [316], who describes various devices classified according to the means by which load can be applied to the fibers: the hydraulic type, balance type, pendulum type, spring type, and chain-feed (chainomatic) type. Today, most fiber (and other tensile) testing is performed in machines in which the rate of strain (more precisely, rate of extension) is the controlling variable and load is recorded via load cells equipped with strain gauge transducers. The strain may be set at constant, variable, or cyclical rates through a programmable displacement of the crosshead to which the lower grip or clamp is attached. The upper grip or clamp is attached to the load cell. Test data generally are available in electronic form for direct use in computer programs that can convert load, extension, and physical data to material properties and also plot relationships of these properties with other parameters related to materials or processes.

It should be borne in mind that tensile tests of fibers, as with other materials, can be conducted under greatly different conditions, which will yield different material constants, different fracture energies, and so forth. The particular test conditions should be selected with the end use of the data in mind. Too often, breaking loads have been determined when design or acceptance criteria were sought in relation to performance of the material under low or moderate load or strain conditions, for example. On the other hand, the use of fiber elastic constants may be inappropriate in many cases because the material *in service* may be strained well beyond the elastic region (if it exists) or because changes in the material properties occur as a result of the service environment of the paper. The fiber properties in the sheet may be substantially different form those of pulp fibers [167,168,368,369]; see also the discussion in Ref. 127. Also when paper is subjected to load, the individual fibers and fiber segments will transmit forces of greatly different magnitudes and directions than the forces applied to the whole sheet, box, or sack. The latter caution is a general one; the others are for more specific cases. The important point is that the mechanisms of failure are in general very different from the mechanisms of small-strain deformation. Also, it has long been known that wood pulp fibers exhibit mildly viscoelastic behavior [127]. However, there are certainly cases in which analysis based on elasticity theory may be used for problems where the assumptions of elasticity may not truly hold yet the level of inaccuracy is small enough to be quite acceptable for an engineering solution, just as there are cases in which failure may be used to evaluate or predict satisfactory service that does not involve any kind of failure. One must always be ready to try methods that will prove or disprove the acceptability of the test data for the intended use.

When do the assumptions of an elastic stress–strain relation hold well enough for results from a static tensile test to be used in engineering calculations? The answers to that question again relates to the intended use of the data, because differences appear in stress-strain curves from tests conducted at different rates of strain (see Figs. 29 and 30 in Chapter 2), all other variables held constant. The same statement may be made when the rate of application of strain is varied. As with other materials testing, the rate of load (or strain) application is selected by the researcher; if an initial linear portion of the stress–strain curve is observed, it is generally possible to calculate a value for the elastic modulus from that linear relation. As long as the researcher recognizes that the same calculation of elastic modulus may not obtain at a different strain rate (or load rate), the results may be very usefully applied. Hardacker [127] has presented data that indicate a 20% rise in the calculated modulus as the rate of fiber loading increases from 0.005 to 0.5 N/s.

Fiber Tensile Test Theory and Practice Axial tensile tests of single fibers can be variously reported in units of force (e.g., as in breaking load, N), in units of stress (e.g., $N/m^2 = Pa$, where m^2 represents the units of the cross-sectional area of the fiber), or as *breaking length*, which is the calculated length of a fiber that would break under its own hanging weight. Breaking length has been a traditional reporting unit in the paper and other fiber-using industries, usually stated in units of meters or kilometers.

Nevertheless, the preferred reporting unit today is tensile index [90,342]. The relationship between tensile index and breaking length is demonstrated in four systems of measurement in Table 1 of Chapter 7. Tensile index is usually expressed in SI units of newton-meters per kilogram or per gram (N m/kg or N m/g). Since $1\,N = 1\,kg \times 1\,m/s^2$, a unit equivalent for tensile index (TI or σ_T^w) is

$$\frac{(\text{kg m/s}^2)\ \text{m}}{\text{kg}}$$

When the kilogram units are cancelled out, the residual expression divided by gravitational acceleration, 9,80665 m/s^2, yields the breaking length in meters. Data at prefailure stress levels are obtained by load–elongation tests, from which one can obtain stress–strain relations.

In Fig. 52 eight types of load–elongation curves observed in the testing of Douglas-fir (*Pseudotsuga*) summerwood fibers by Kersavage [174] are shown. For single chlorite-delignified fibers tested at 6% moisture content, he found that all specimen curves were of three types: type 1 (82%); type III (9%), and type IV

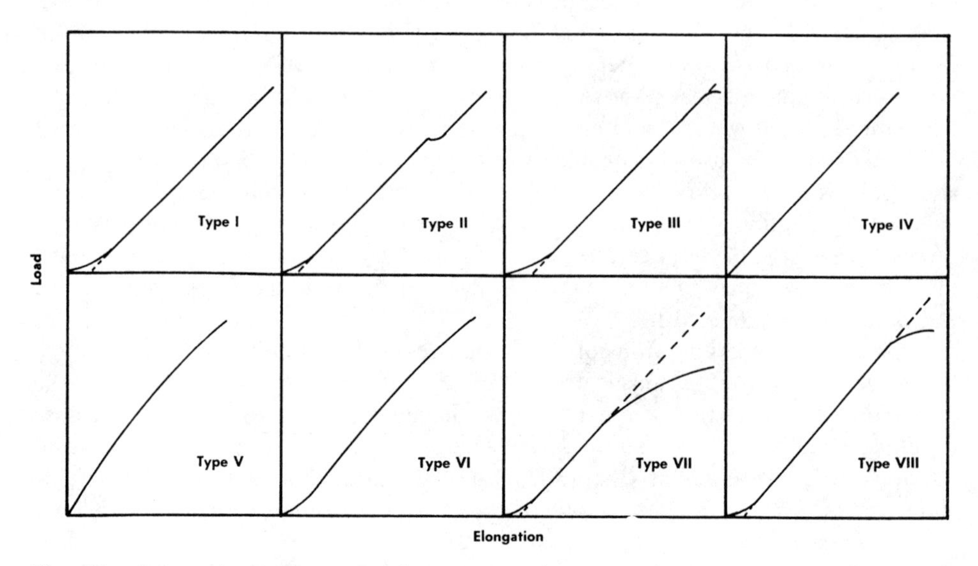

Fig. 52 Schematic drawings of eight representative types of stress–strain curves resulting from fiber tensile tests by Kersavage [174].

(9%). Higher moisture contents tended to raise the test proportion of type III and incur a scattered few instances of other types. The curves in Fig. 52 are unquantified, but if converted to stress–strain curves by taking load/cross-sectional area = stress and by taking elongation/free-span length = strain, they appear similar to results from other fiber studies (e.g., Refs. 159, 161, 167).

Figure 53 shows a curve for a hypothetical synthetic fiber that includes substantial plastic deformation after an initial phase of linear elastic behavior. Although this curve is useful for a general understanding of tensile test data, it should be borne in mind that synthetic pulp fibers show a much higher extension to break and a lower breaking stress than do conventional wood pulp fibers. The stress–strain curves of plant-derived fibers are also generally much more linear.

In materials science, "elasticity" refers to complete and rapid recovery of strain upon removal of stress. Thus, in Fig. 53, if the filament is loaded to 0.01 N, resulting in a stress in the filament equal to $50\,\mathrm{N/mm^2} = 50\,\mathrm{MPa}$ and a strain of 0.01 (= 1% extension), then removal of half the load would result in a return to an extension of 0.5%, and removal of all the load would return the filament to its original unstrained state. The elasticity may be linear, which is the case in the example just cited, or nonlinear if the path of unloading follows the same curve as the loading. On the other hand, plastic strain is nonrecoverable.

The slope of the initial linear portion of the stress–strain curve defines the Young's modulus, or modulus of elasticity E, observed in the test, i.e.,

$$E = \tan\phi = \frac{\Delta\sigma}{\Delta\epsilon} \tag{25}$$

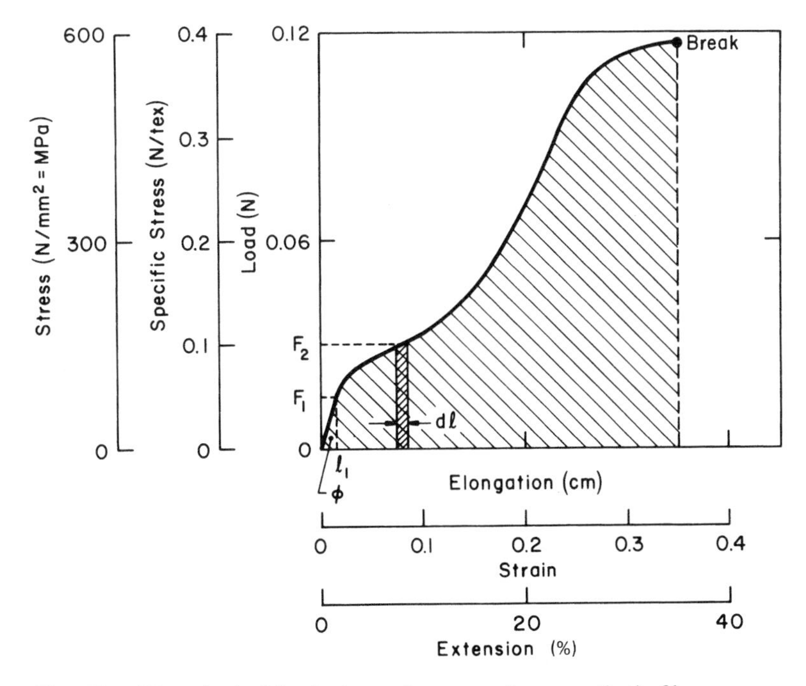

Fig. 53 Hypothetical load–elongation curve for a synthetic fiber.

which for the slope shown in Fig. 53 would equal 50 MPa/0.01 = 5 GPa.

If the deformation of the fiber is characterized by a time-dependent nonrecoverable strain superposed on an initial elastic strain, then its behavior can be modeled as an assembly of ideal elastic elements whose stress–strain characteristics are those of Eq. (25), together with viscous Newtonian dashpots that follow the relationship $\sigma = \eta\dot{\epsilon}$ (where η is the viscous coefficient of the dashpot and $\dot{\epsilon}$ is the rate of strain). One may refer to Chapter 2 and Ref. 243 for excellent treatments of viscoelasticity in paper and fibers.

In filamentary materials testing, one is often interested in the specific stress σ^w, which is defined as the load F in newtons divided by linear density (coarseness) ρ_c in kg/m, i.e.,

$$\sigma^w = \frac{F}{\rho_c} \tag{26}$$

so that units are expressed in N m/kg (or alternatively, Pa m^3/kg or N/tex). The first of these units is recommended for reporting tensile index (tensile strength divided by volumetric density) for paper [90] and is also used for fibers.

Similarly, the tensile stiffness index may be reported in N m/kg. For the example given just after Eq. (25) and a fiber density of 1500 kg/m^3, the tensile stiffness index is 5 GPa/1500 kg/m^3 = 3.3 MN m/kg.

One other aspect of the load–elongation curve for a fiber is shown in Fig. 53. Two shaded areas appear under the curve, the smaller of which shows an increment of work done as the fiber is stretched by a small amount $d\ell$. If the mean load on the fiber during the elongation $d\ell$ is F, the work of elongation W is

$$W = F\,d\ell \tag{27}$$

from which it follows that the total amount of work needed to break the fiber is the total shaded area under the curve (= the integral of all the increments of work such as $F\,d\ell$). That is, the work of rupture W_R is

$$W_R = \int_0^{\delta_T} F\,d\ell \tag{28}$$

where δ_T is the elongation at the point of rupture. Work of rupture is also known as toughness.

The SI unit of work and energy is the joule (J); 1 J equals the work done when a force of one newton experiences a displacement of one meter in the direction of the force, i.e., J = N m. One can determine the tensile energy per unit volume that the fiber has assimilated up to the point of rupture, referred to as tensile energy absorption (TEA), from

$$W_T = \text{TEA} = \frac{W_R}{V} \tag{29}$$

where V = fiber volume. The units will, of course, be J/m^3. The TEA is also commonly reported for tensile tests of paper, usually on the basis of sheet *area* (units: J/m^2).

In paper tensile testing, a quantity that is often determined is the tensile energy absorption index, which is equal to TEA divided by the basis weight. In the fiber testing field, the analogous quantity is often known as the specific work of rupture, which may be expressed (in both tex and SI systems) in units of kilojoules per gram; such a unit arises from dividing the work of rupture by the product of fiber coarseness and initial length, i.e., it is a measure of energy absorbed per unit mass. It is recommended [90] that units of J/kg be employed for paper, and again it seems most consistent to use an identical reporting unit for the fiber. High toughness, reflected by large TEA and TEA index values, is an important mechanical property for the fibers employed in products subject to shock loading such as occur in many converting operations, shipping sacks in service, kites, and so forth.

It is sometimes important to know the amount of energy that can be applied without causing irreversible strain, that is, in the fiber, nonrecoverable stretch. This quantity is known as the elastic strain energy or modulus of resilience. In Fig. 53, it would be equal to the area under the linear part of the stress–strain curve, i.e.,

$$\text{Max } W_E = \frac{1}{2} F_1 \ell_1 \tag{30}$$

where W_E stands for elastic strain energy. Up to the maximum W_E value, the energy that has gone into fiber deformation is recovered to the testing machine and/or the environment as the fiber is unloaded and its dimensions return to their preloading state.

In many of the reports of fiber tensile testing, it is apparent that no attempt has been made to obtain prerupture data. In such cases, the following results are usually reported (SI units are stated at right).

1	Elongation at break	δ_T (m or mm)
2	Strain at break	$\epsilon_T = \delta_T/\ell$ (m/m or mm/mm)
3	Breaking force	F_T (N)
4	Breaking length	BL or σ_T^ℓ (m or km)
5	Tensile index*	TI or σ_T^w (kN m/kg)
6	Tensile strength	σ_T (N/m^2 = Pa)

Breaking length was mentioned earlier as the traditional reporting unit in the paper industry. Also, a conversion from tensile index (specific tensile strength) to breaking length was shown. Tensile strength σ_T is reported only if a value for fiber cross-sectional area can be established.

Numerous investigators have tested papermaking and other fibers in tension [23,81,117,127,159,161,167,168,243,246,247,276,279,300,359,368,369]. The purposes and conditions under which these tests have been conducted vary greatly, yet the order of magnitude for tensile strength, modulus, etc. of papermaking fibers is generally found to be comparable to those for plant textile fibers (see Chapter 2 of Ref. 214). Some stress–strain curves for summerwood holocellulose fibers are shown in Fig. 54 that illustrate the influence of tension applied on the fiber during drying; the pre-tension has stiffened the fiber internally. The curves in Fig. 54 are for relatively undamaged fibers. In commercial pulps, damage to the fibers may occur during stock preparation. For examples of damage, see Section IV.C. Curves for such market pulp fibers may show considerably less linearity.

* Also called specific tensile strength (STS).

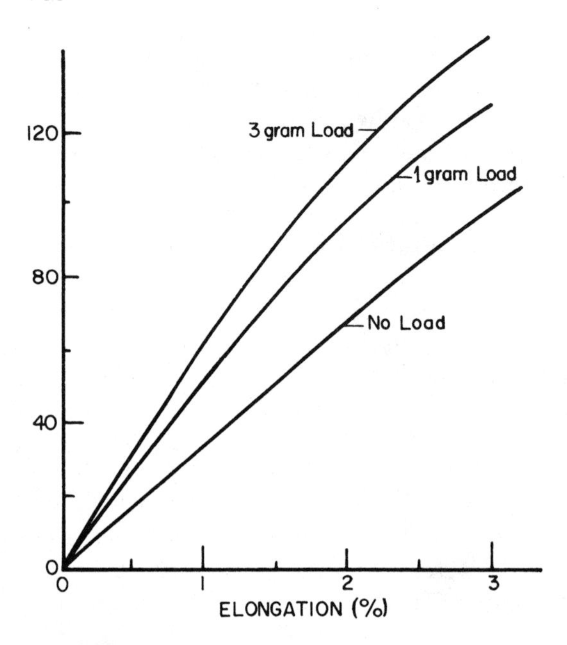

Fig. 54 Typical curves of stress versus elongation for summerwood fibers that had been pretensioned during drying at the three load levels indicated. (From Ref. 359.)

Experimental Determination of Elongation and Tensile Strain In many paper testing laboratories around the world, the measurement of tensile deformation (elongation) is taken as the crosshead movement or load (grip) displacement. However, the assumption that the deformation is the same as crosshead or grip displacement is often false, because the strain within the fiber in the vicinity of the gripping mechanism can be quite different from the strain in the rest of the specimen. There may also be softness, slack, or slippage in the mechanisms that grip the fiber or cause displacement. The shorter the length of the specimen, the more potentially serious is the error introduced by failing to measure deformation independently on the body of the specimen, so this precaution applies especially to fiber testing. Stated another way, it is always desirable to measure strain by recording deformation and gauge length directly on the fiber if at all possible. One cannot attach anything that would influence the intrinsic fiber properties; consequently, one must measure the distance between some natural mark (such as a bordered pit on a tracheid) or applied marked (such as a xerographic particle) and another similar fiducial mark if strain is indeed measured on the specimen. Fiber testing systems have now been devised in which a comparison can be made between measurements made directly on the fiber and measurements based on crosshead or grip displacement. An example will be given in this section. If both methods closely agree for a sufficiently large sample, then crosshead or grip displacement may be acceptably used. In such a case, the gauge length L becomes the observable span length between the edges of the grips, clamps, or jaws or droplets of adhesive that grip the fiber specimen. Fiber strain ϵ_f is then calculated from crosshead or grip displacement δ according to

$$\epsilon_f = \delta/L \tag{31}$$

The fiber tensile tests reported by Kappil [167] and Kappil et al. [168] serve as an example of the successful use of two such measurement systems. We provide here some experimental details that include, but are not limited to, the determination of elongation and tensile strain. The purpose of doing so is to place various aspects of the Kappil testing protocols in one location, which will be referenced elsewhere in the chapter.

In these experiments, the objective was to study changes in fiber properties in machine-made paper related to recycling, orientation in the sheet, and drying tension. An unbleached kraft paperboard furnish (70% hardwood, 30% softwood) was formed on a Fourdrinier and then dried (1) on the machine, (2) in frames that fully restrained the sheet in the machine direction during drying, or (3) while subjected to a 2% stretch in the machine direction during drying. Single fibers were then excised microsurgically from both the virgin and recycled paperboard. Recycled fibers showed reductions in tensile strength and modulus of elasticity while the strain to failure increased. But the tensile strength and elastic modulus of the excised fibers increased when the paperboard was restrained from shrinking during drying. Also, fibers that had been aligned in the machine direction were stiffer and stronger than those aligned in the cross-machine direction.

Viewed from the perspective of the drying forces imposed on the fibers as the web consolidates, these results can be expected. Indeed, fibers aligned in the machine direction that had greater stiffness and strength were a principal finding of Wuu and coworkers [368,369], whose work inspired the Kappil study. Wuu et al. also found that fibers aligned at an intermediate orientation such as 45° exhibit properties in between those of the MD-aligned and CD-aligned fibers.

In the system used by Kappil, 0.2% of the softwood pulp in the machine furnish was dyed with methyl violet prior to the paper machine run. The excision of the fibers involved the following steps:

1. Under a stereomicroscope, the target fibers (dyed, straight, and long) were identified.
2. With the help of microsurgical knives and blades, two cut lines were made, one on each side of a target fiber.
3. The fibers were pulled out from the paper using fine-edged tweezers, with minimal disturbance to the fiber system.

After mounting, the testing assembly was stored in a room conditioned to TAPPI standard conditions. The test fixture is described in Ref. 167.

To get an accurate system of reference points for the strain measurement, the fibers were carefully dusted with xerographic toner particles. A single-hair brush was used for this purpose. The fiber–particle assembly was heated to approximately 70°C for 5 min, which resulted in many of the particles sticking to the fiber surface.

The test setup included a microscope and a high resolution video camera that was connected to a Super VHS videocassette recorder and a high resolution television monitor. This system,* described more fully in Refs. 167 and 168, permitted full observation and recording during mounting and testing. The rate of travel for the

* The system was located at the Pulp and Paper Research Institute of Canada (PAPRICAN), Pointe Claire, Quebec. PAPRICAN also provided important technical assistance

crosshead was 0.02 cm/min. The loading force in the test was monitored by an Instron load cell (load cell A, maximum range 50 gf ≅ 0.5 N). A Fotonic sensor was used to monitor the crosshead motion. For monitoring the fiber deformation, two methods were employed:

1. An X-Y plotter was used to record the data, *based on crosshead movement*, as load–deformation curves.
2. A high resolution video system was used to record the fiber deformation *based on particle displacement.*

The cross-sectional areas of the individual fibers were determined by Hardacker's fiber lateral compaction method [126].

Comparative Analysis of Stress–Strain Data At the conclusion of mechanical testing, the first step in the analysis of strain data from the video images was to capture these images digitally to a computer with a high resolution monitor (in this case, Adobe Photoshop software was used with a Macintosh Quadra 700 computer). The original Super VHS tape was played back on an S-VHS editing player/recorder until the desired frame was reached. Then the player was stopped momentarily for each image (640 × 480 pixels) that was captured. (For each fiber test, 12–15 frames were captured.)

From the captured images, two significant points on each fiber were selected. These points (fused xerographic particles) had to be visible in all the frames of the test, which in the case of wood pulp fibers is often not achievable owing to the helical configuration of the cellulosic microfibrils. The S2 helix is a factor that causes fibers to twist as they dry (and untwist when axial loads are imposed [106,223], as in tensile testing). However, Kappil [167,168] was able to conduct sufficient tests with a complete set of images in which all frames contained the two reference xerographic particles to compare with crosshead displacement data. "NIH Image 1.47" software [originally designed to record gene positions on deoxyribonucleic acid (DNA) strands] was used to quantify displacements of the particle markers on the fibers.* The load–displacement curves were converted to stress–strain curves by dividing load by cross-sectional area [126] and the displacements of grips and particles by their respective gauge lengths.

The result of these efforts is the creation of two stress–strain curves for each fiber tested. These sets of representative stress–strain curves are shown in Figs. 55–57. Slopes of the linear positions of each curve (after the initial nonlinearity at low strain) are calculated as fiber moduli of elasticity.

A paired variate t test was performed to compare particle and crosshead modulus values. From these results it was concluded that there is no significant difference between the moduli of elasticity determined by these two methods. Accordingly, the data obtained by jaw movement was processed for 62 other successful tests, giving a total of 80.

For the 18 fiber tests in which there were both successful measurements of strain by particle displacement and crosshead movement, observation of the stress–strain results shows that the plots are essentially congruent for a particular sample.

* Equipment, software, and important technical assistance were provided by the Department of Education Communications, SUNY Upstate Medical University, Syracuse, NY.

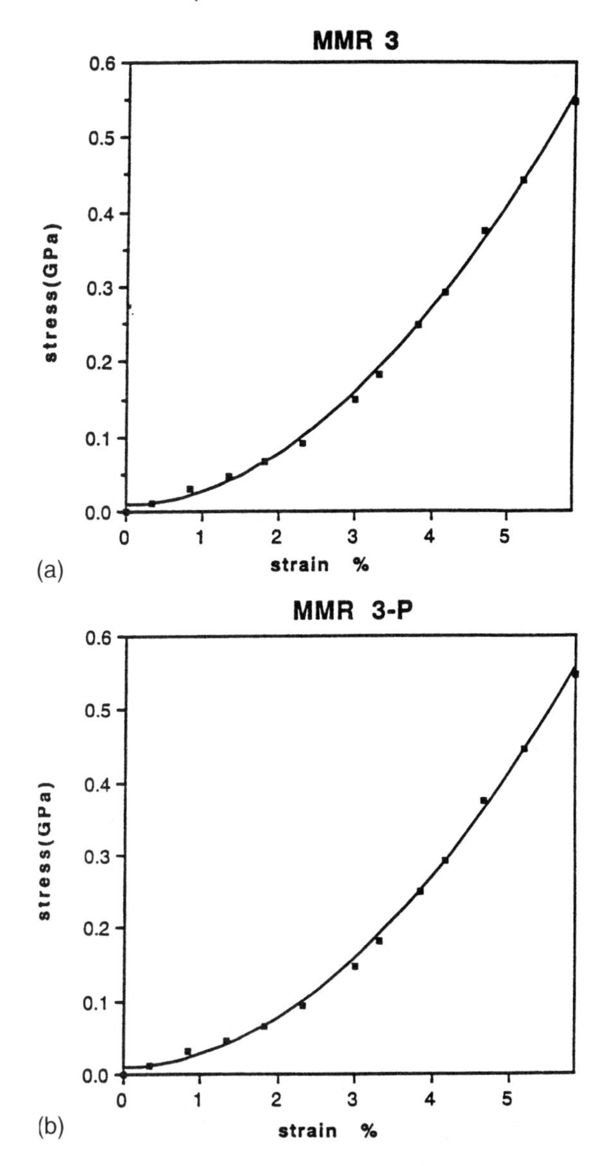

Fig. 55 Stress–strain curves of a cut-out fiber specimen. This specimen was a recycled fiber aligned with the machine direction in the sheet, dried on the machine. (a) Grip displacement $E = 13.76$ GPa. (b) Particle displacement $E = 13.64$ GPa. Curve is Type I in the Kersavage system [174]. (From Ref. 167.)

Marker Particles for Fiber Gauge Lengths The use of fused or semifused xerographic particles as strain gauge markers on cellulosic fibers dates back to at least the late 1960s, when it was employed by Perez [284]. Other particle markers can be used for this purpose, as for example in the work of Hamad [118] and Hamad and Rokbaa [120], who used 2–3 μm polyvinyltoluene (latex) spheres that stuck to the fiber surfaces. Sputtered gold was employed as a marker by Retulainen and Ebeling

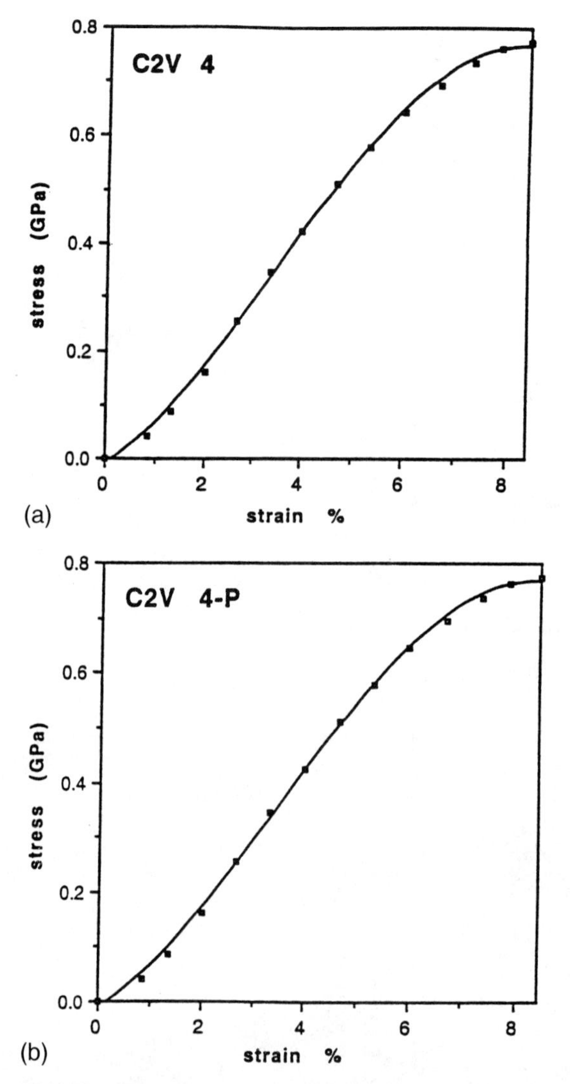

Fig. 56 Stress–strain curves of a cut-out fiber specimen. This specimen was a virgin fiber aligned with the cross-machine direction in the sheet, dried at 2% MD stretch on a frame. (a) Grip displacement $E = 12.21$ GPa. (b) Particle displacement $E = 11.76$ GPa. Curve is Type VII in the Kersavage system [174]. (From Ref. 167.)

[297[. Bergander and Salmén used a random pattern created by deposition of MgO particles [26]. Nanko and Wu [256] developed a silver grain method in combination with a reflection-type scanning laser microscope (SLM). In this technique, silver powder grains (1.4 μm diameter) are dispersed in water. The suspension is poured over a blotter, which is in turn pressed on a wet sheet (or wet fibers). A portion of the silver grains are thus transferred to the specimen. Silver grains have the following advantages:

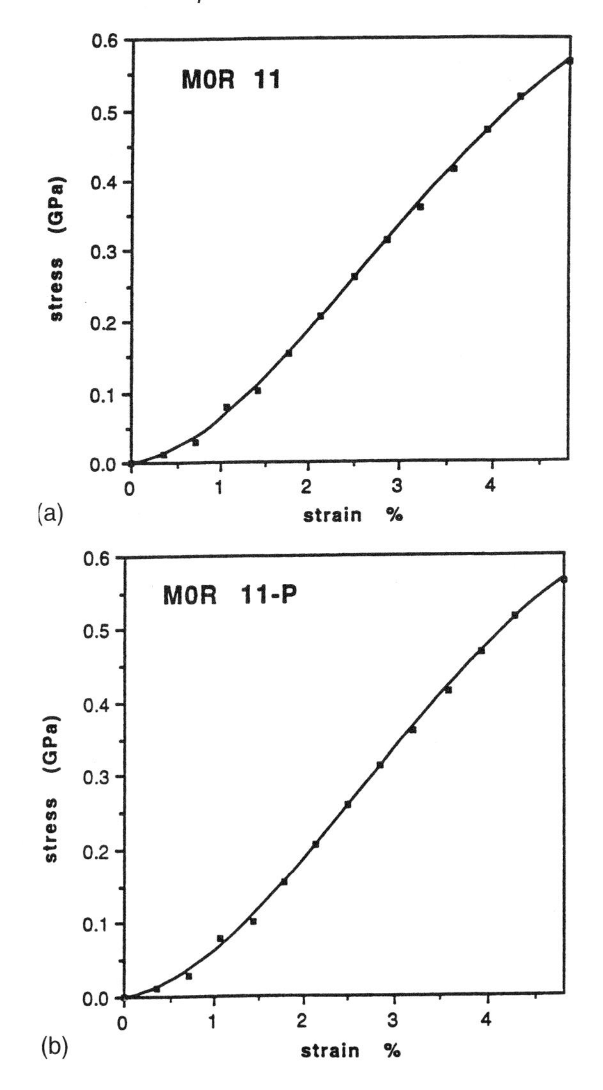

Fig. 57 Stress–strain curves of a cut-out fiber specimen. This specimen was a recycled fiber aligned with the machine direction in the sheet, dried under MD restraint (0% shrinkage) on a frame. (a) Grip displacement $E = 13.61$ GPa. (b) Particle displacement $E = 13.69$ GPa. Curve is Type VIII in the Kersavage system [174]. (From Ref. 167.)

- Small size
- Strongly reflect light (easy to detect under SLM)
- Readily dispersible in water
- Stay on the fiber surface during drying
- Do not require warming of the specimen

Stress Determination Accurate determination of stress carries with it a separate set of difficulties. Normally it is assumed (1) that the fiber has a constant cross-sectional area, (2) that it is aligned perfectly in the grips—that is, centered perfectly and at an angle of 0° from the axis of applied force, and (3) that the load transmitted at any cross section can be as large as that transmitted by any other cross section. If these assumptions are valid and the gripping arrangement is ideal, the tensile stress field in the fiber is uniform over the gauge length.

The first assumption is not technically valid in the case of a wood pulp fiber. However, the tapered ends of the fiber are not within the free-span distance, so the lack of uniformity in the cross section probably has little consequence in the test setup. As for the second assumption, Kersavage [174] developed a system of alignment that can be used when a high degree of coaxiality between the fiber and applied load is required. Epoxy droplets are applied to the ends of the fiber test specimen. After the epoxy has cured, the specimen is placed on two gripping forks that are coaxial with the machine load cell, as shown later in Fig. 87. This system was also applied very successfully by Mott et al. [109,246,247]. The third assumption does not hold if the free-span area of the fiber is damaged (e.g., during processing) or has natural irregularities (Section IV.C). Damage is always more serious, because natural fiber structures such as pits are reinforced with microfibrils that curve around the pit border, reducing stress concentrations in the fiber wall. Mill furnishes acquire more damage in processing than laboratory pulps. As might be expected, fiber stress–strain relationships are different for fibers with different free spans or gauge lengths, as one can note from the results of single-fiber tests and short-span testing [83,127,143].

Two further difficulties in plant fiber tensile testing should be borne in mind. One is that the gripping system may inhibit the Poisson effect (lateral contraction as the fiber is axially stretched). Because of the long, slender shape of a fiber, this effect is probably quite minor in most cases. However, it is also a good reason to maximize span length and to try to use a gauge length that is less than the full span. The second difficulty is more serious. When a plant fiber is loaded in tension by a free-hanging weight, some twist will occur because the axis of loading (the fiber axis) is not coincident with the axis of symmetry in its S2 layer, which dominates fiber behavior; the microfibrils do not line up with the fiber axis. The effect is analogous to that shown in Fig. 58. Mark and Gillis [106,219] reported that delignified Virginia pine (*Pinus virginiana*) tracheids showed twisting of 1.08–3.03 rad/mm under free-hanging

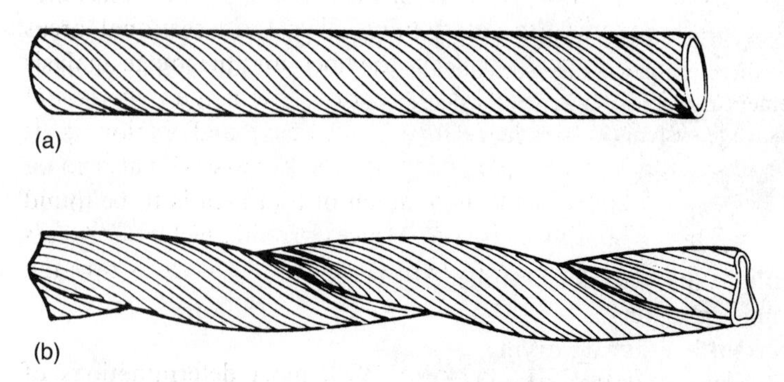

Fig. 58 A helically wound tube (a) subjected to large axial tensile elongation in fixed grips will deform by twisting and tension buckling to a shape such as shown in (b). The effect has been shown both experimentally and theoretically. Single fibers will show similar deformations because of the dominance of the S2 layer helix. (From Ref. 277.)

loads of approximately 0.03 N and much greater twisting if the fibers were wet. When fibers under tension are not free to twist, a buckling of the fiber occurs, which can be observed in high resolution ciné photography such as that obtained in the experiments of Page et al. [277,279] or in high resolution video such as that obtained in the Kappil et al. experiments [167,168].

Some researchers try to prevent fibers from twisting upon drying, which has both advantages and disadvantages. One the one hand, drying the fibers so that they cannot twist, as Mott [246] did, has results somewhat analogous to the behavior of fibers in paper, which form hydrogen bonds with adjacent fibers during consolidation of the web. These bonds prevent most of the twisting that occurs if fibers are freely dried. On the other hand, fibers in paper have irregularities (some natural, some induced during processing). Selection for fibers that lie perfectly flat and are essentially free of defects does not represent the fibers in a commercial mill furnish. Mott's restraint procedure was to select straight fibers without apparent defects and dry them slowly ($\sim$ 24 h) between moist glass slides.

Determination of the cross-sectional area of the tensile-tested fiber used to be more tedious and uncertain than today. Typically, fibers were embedded, sectioned, and mounted on glass slides for measurement optically or by planimeter [3,28,132, 203], taking care to use nonswelling media for the embedment. For high accuracy, a cut-out weighing method was used. In that procedure, a photomicrograph of a fiber cross section (400–7000× magnification) is cut from the photographic print and weighed in comparison with a geometric figure such as a rectangle cut from the same material, based on dimensions of a stage micrometer photo taken at the same magnification (see, for example, Refs. 28,146,214). Kersavage [174] used direct microscopic observation of the unembedded broken sections of fibers after rupture followed by slicing and mounting of cross sections, photographing the dry cut surface and determining the area planimetrically. Another method involved dividing the coarseness by the pycnometric density of the fiber [161]. The fiber lateral compaction method of Hardacker [124] is somewhat faster than the other classical methods, but it requires special equipment and considerable micromanipulative skill.

The advent of confocal laser scanning microscopy (CLSM) combined with advanced image analysis equipment and methods has provided the fiber researcher with a fast, highly accurate system for determining fiber cross-sectional areas [109,154–156,206,245]. The CLSM optically sections the fiber, and the image analysis system scans the images of the cross sections as calibrated for magnification and calculates and stores the area data automatically. Embedment and sectioning of fibers still has use if, for example, one needs not only the cross-sectional area of the whole fiber but also the breakdown as to how much of that area is to be found within each layer of the fiber. The above-cited references provide technical details about the microscopes and image processing systems used by the various research teams. See also Chapter 5 in Vol. 2.

Systems for Single-Fiber Tensile Tests Prior to 1970, most determinations of stress versus strain of wood pulp fibers were conducted, for example by Jayne [159], by using freely dried (and therefore twisted) fibers mounted in modified jewelers' pin vises attached to the crosshead and load cell of a tensile tester. A modification of this gripping system that enables the testing of wet specimens is shown in Fig. 59. The jaws of the pin vises are lined with very fine (No. 400A) silicon carbide paper. Elongation and load are measured from crosshead displacement and load cell transducers, respectively.

As pulp fiber testing increased in importance, efforts were made to improve the gripping, load application, and measurement methods. The Institute of Paper Chemistry did a series of studies that demonstrated how critical the nature of the fiber clamping or gluing is to the results obtained. A machine with a constant rate of loading was developed that permitted rotation of the upper grip. It also featured an alignment method that used pins and forceps designed to prevent prestressing of the fiber [125].

Other investigators sought better ways to observe the process of deformation and fracture by constructing manipulative tensile stages or operation on either light [11] or scanning electron microscopes [101]. In the work of Furukawa et al. [101], a video scanning device with a 30 frames/s speed was used for observation of the fiber under test, which was recorded on videotape. In such systems at that time, the fibers had to be coated to withstand the electron beam, and charging effects would obscure observation if the coating cracked during a test.

In the 1970s, substantial improvements in fiber testing were devised at several institutions. Figures 60 and 61 show schematically the microtensile testing system developed by Kersavage at the Pennsylvania State University [174]. Tests could be performed in a controlled humidity environment (RH 1–100%, providing nominal equilibrium fiber moisture contents of up to 30%). Another feature was the introduction of a clamp-free gripping system that also provided excellent coaxial alignment of the fiber and the load cell axis. The gripping system employed the principle of the ball-and-socket joint. Small droplets of an epoxy adhesive were affixed to the fiber specimen so as to leave a free span of at least 1 mm. The droplets (≈ 0.7 mm long and 0.5 mm wide) were allowed to cure and were then inserted in a gripping

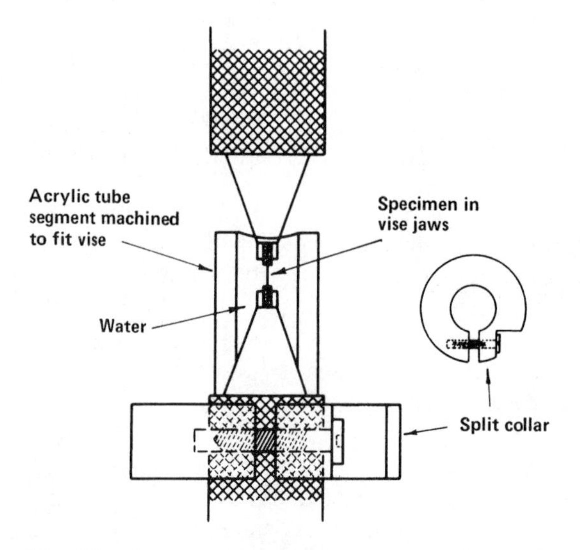

Fig. 59 Pin vise grips with water reservoir attachment for testing of wet fibers and other small specimens. (From Ref. 214.)

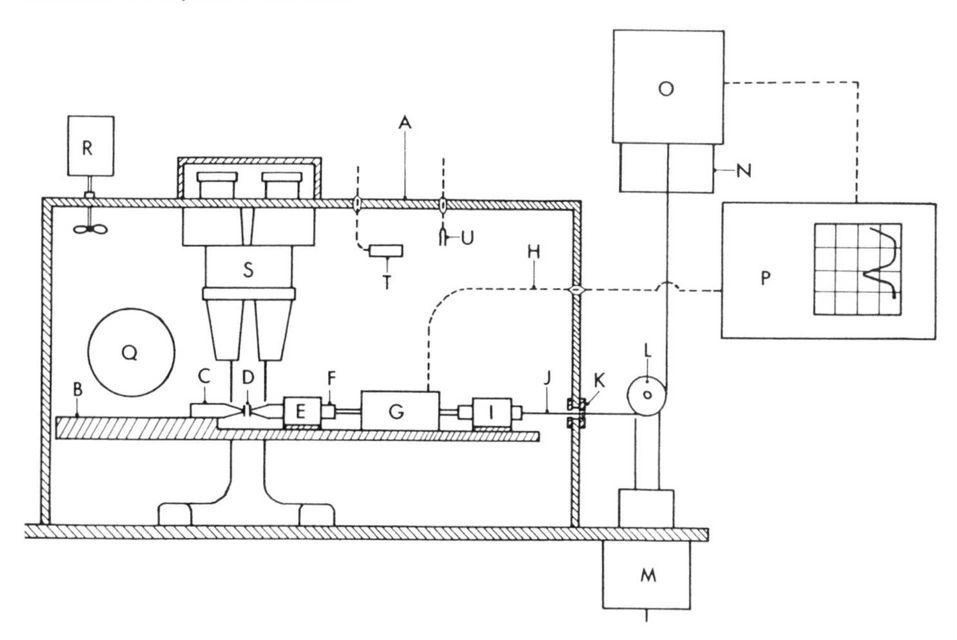

Fig. 60 Schematic diagram of microtensile testing system of Kersavage [174]. Components include conditioning chamber (*A*), plastic base (*B*), fixed tensile grip (*C*), fiber grips (*D*), ball bushings (*E*, *I*), movable tensile grip (*F*), linear transducer (*G*), displacement (transducer) wiring to plotter (H), tensile thread (*J*), mercury seal (*K*), pulley (*L*), movable crosshead (*M*), stationary crosshead (*N*), load cell (*O*), X-Y recorder (*P*), chamber entry port (*Q*), fan (*R*), stereoscopic microscope (*S*), humidity sensor (*T*), and thermocouple (*U*).

assembly consisting of slotted recessed stainless steel disks designed to retain the hardened droplet and self-align in the microtensile tester. Mott (University of Maine) [246] and Wild et al. (University of Victoria) [366] used Kersavage's idea, employing miniature forks to align and restrain the epoxy droplets (see Fig. 87). The humidity control chamber is now part of fiber testing systems at many laboratories, e.g., STFI in Sweden (see Section VII.E, Figs. 87–90).

Clamp-free mounting was also incorporated into the fiber testing system at PAPRICAN in Quebec, Canada [277,278]. To avoid the clamp effect as much as

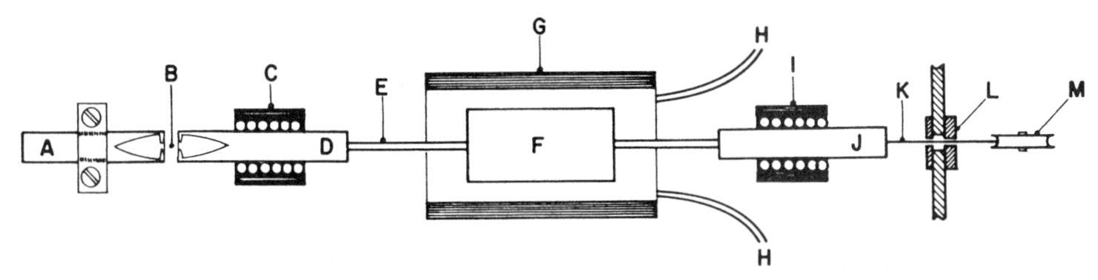

Fig. 61 Cut-away schematic illustrating top view of elongation-measuring system of Kersavage [174]. Components include stationary tensile grip (*A*), fiber grips (*B*), ball bushings (*C*, *I*), movable tensile grip (*D*), transducer rod (*E*), transducer armature (*F*), linear transducer (*G*), transducer wiring (*H*), brass tubing support (*J*), tensile thread (*K*), mercury seal (*L*), and pulley (*M*).

possible, test span lengths of at least 1 mm are used. Single fibers are transferred from a slurry of wet pulp to a drop of water on a glass slide. A small sliver of glass cover slip, about 1.5 mm wide and 4 mm long, is placed across the middle of the fiber so that the fiber ends protrude. Light pressure is applied to the glass while the fiber is drying, thus compacting it and preventing twisting. After drying, the cover slip sliver (with fiber attached by surface forces) is placed across two glass tabs that are held in position by virtue of the fact that they are both bonded to a stiff card. The fiber ends are bonded to the glass tabs with a well-defined epoxy edge line. The cover slip is removed, the glass tabs are mounted in the clamps of a tensile test machine, and the card is burned through with a hot wire, leaving the fiber as the load transfer link between crosshead and load cell. The fact that the ends of the fiber are embedded in the adhesive rather than held in metal clamps ensures a more even shear stress distribution on the fiber surface where it is held. The PAPRICAN research team also innovated real-time film recording of fiber tests and significant improvements in the measurement of elongation.

Variations of clamp-free mounting have been used elsewhere. In Fig. 62 is shown a fiber–shive assembly mounted on small pieces of glass that are bonded to manila card tabs. The tabs are held together by transparent tape prior to insertion into a microtensile stage apparatus. Before load is applied, the tape is burned through, leaving the fiber–shive assembly to transmit the pulling force.

Microtensile and Macrotensile Testing of Fibers Relative to the testing of a sheet of paper, all testing of individual fibers could be considered "micro" testing. A 1 mm long fiber typically has a mass of about 0.2 μg; a 40 g/m^2 basis weight sheet 1 m square has a mass 2×10^8 times as great. Mott and coworkers [109,246,247] introduced the term "microtensile" testing to denote a type of test that considers very small displacements in localized areas of a fiber cell wall such as a pit border or an area near a natural or induced defect. Their method for doing such measurements is to perform digital image correlations on high quality images of these localized areas. Tests that are directed toward evaluation of the whole fiber are referred to as

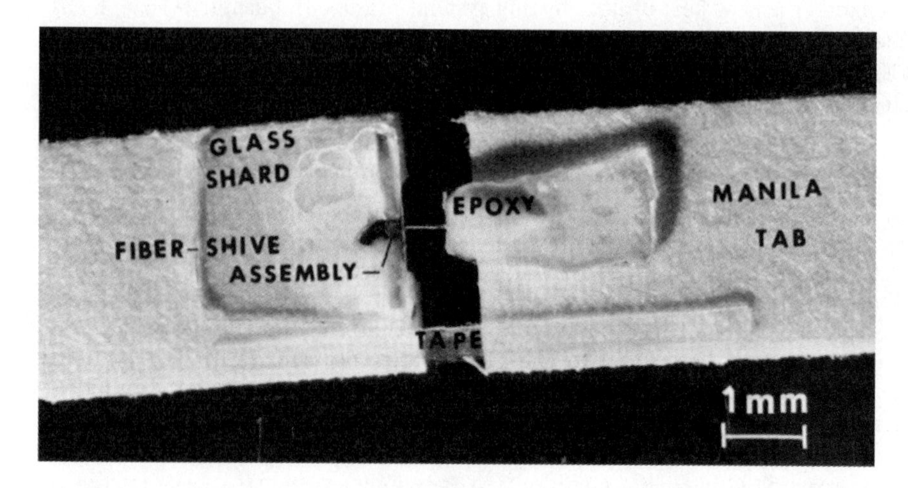

Fig. 62 Clamp-free mounting of a fiber–shive assembly. (Courtesy of A. Eusufzai, SUNY College of Environmental Science and Forestry, Syracuse, NY.)

"macrotensile" by contrast. Accordingly, all the tests described thus far fall into the latter category.

Mott et al. did macrotensile testing also, but the new work is what is described here. They expanded the focus of fiber mechanical properties studies by developing a methodology to measure strains in specific localized areas of the fiber cell wall, such as natural or induced irregularities. An example of an induced irregularity would be a crack, kink, or microcompression. The micromechanical behavior of these localized areas was ascertained by placing the fiber specimens in a dual-crosshead tensile stage designed to fit in an environmental scanning electron microscope (ESEM). High magnification (2500×) digital images obtained during tensile testing were then subjected to a digital image correlation (DIC) routine to construct full-field strain maps of localized regions of the fiber cell wall such as the above-mentioned irregular areas.

Environmental scanning electron microscopy (ESEM) differs from standard scanning electron microscopy (SEM) in that the specimen is not viewed under high vacuum. Instead, a gas (usually water vapor) is introduced into the chamber as air is expelled. Also, there are a series of different pressure zones in the ESEM, separated by pressure-limiting apertures. Thus, the electron source remains under high vacuum even though the sample is in a chamber containing low (ambient) pressure water vapor or other gas. The gas in the specimen chamber performs two important functions: (1) It prevents buildup of charge on specimen surfaces, eliminating the need for a conductive coating on the fiber or other type of sample; and (2) moist specimens can be observed when the gas used is water vapor. More details can be obtained by consulting Chapter 5 of Volume 2.

The tensile testing procedures that were employed in the Mott et al. study are well covered in Refs. 246 and 247. The procedures for ascertaining microstrains involved several steps. All tests were recorded on Super VHS video tape, which was then processed using a DIC system that has had application in measurements of strain in paper and wood [47,49,166,207]. Readers are referred to Chapter 3 for a brief but useful introduction to digital correlation techniques

Another recent development, by Wild et al. [366], is a single-fiber fatigue instrument that has the capability of imposing static, dynamic, and cyclic loads on fiber specimens. A brief discussion of this development (with two diagrams) is given in Section VII.E.

Transverse Tension Based on the properties of the three main polymers in wood, layer proportions, and fibril angles, various theoretical models have been applied to the development of material constants for the wood fiber cell wall and the consequent stress distribution in the cell wall and its constituent layers when the fiber is deformed. (See Section V.) The approaches to these studies have included two-dimensional and three-dimensional analyses, some with finite element simulations.

Attempts to prove that the theoretical approaches to fiber wall properties are well founded have been based on experimental work with fibers and very small (on the order of one fiber thick) wood sections. The experiments have included axial tension (discussed in the preceding subsection), axial compression (Section VII.B), flexure (Section VII.C), torsion (Section VII.D) and various environmental influences (Sections VII.E and VII.F). Indirect tests that provide insight into the transverse compressive properties of fibers—in particular as regards the radial (thickness) direction within the fiber wall—are also covered in Section VII.B. However, a dearth of experimental work on the transverse tensile properties of wood fibers has made it difficult to evaluate the predictions of fiber cell wall behavior in the tangential (circumferential) direction.

Bergander and Salmén [26] recently presented the results of an experimental study on the transverse tensile stress–strain behavior in the wood fiber cell wall. Specimens were cut from spruce (*Picea abies*) wood blocks in the manner illustrated in Figs. 63 and 64. Their technique uses micromechanical measurements of MgO particle displacements by cross-correlation of images obtained with electronic speckle photography (ESP). Subsequent differentiation of the displacement field enables calculations of strains in the radial (with respect to orientation in the tree) cell walls.

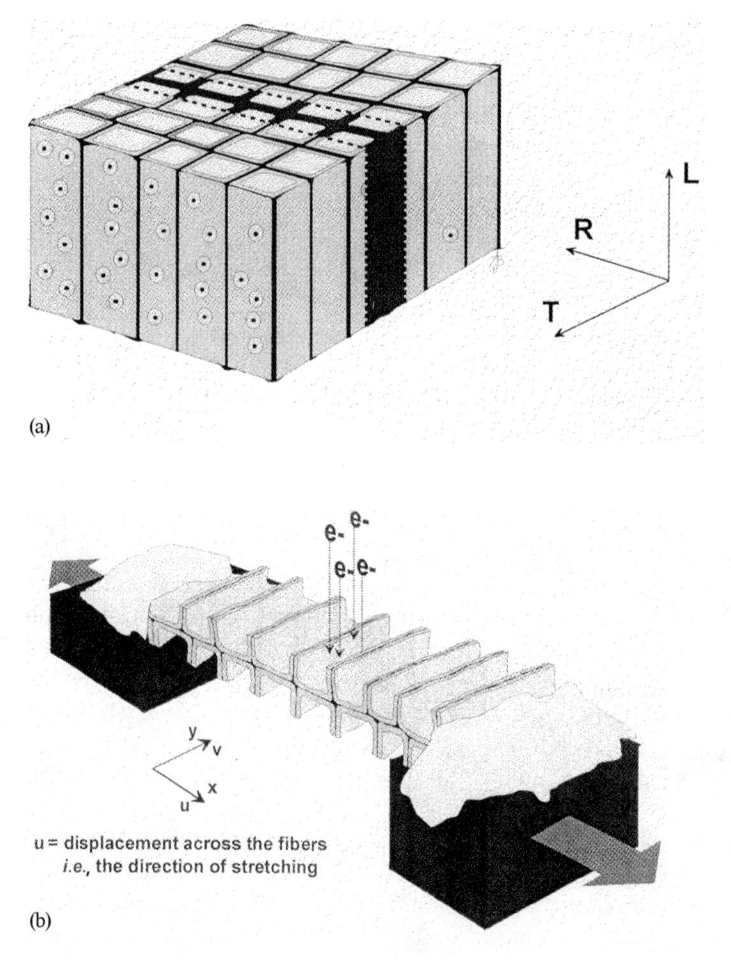

Fig. 63 Schematic illustrations showing procedures used by Bergander and Salmén [26] to obtain and mount transverse tensile specimens for testing. (a) Drawing shows position of specimen (black) within the wood block. (b) Drawing demonstrates positioning and attachment of the specimen (gripped by adhesive).

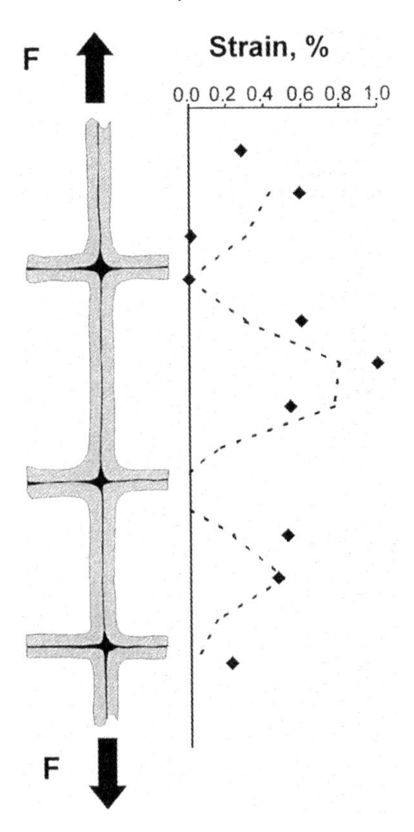

Fig. 64 Illustration of variation in strain across the fiber walls (over a specimen segment of three double radial walls) in experimental setup of Bergander and Salmén [26].

These tests were carried out on a tensile stage in an environmental scanning electron microscope. Accordingly, humidity conditions inside the test chamber could be varied, and the MgO particles could be applied directly to the natural cell wall surface. Because the strains are calculated from particle displacements across the width of each double wall in the specimen, the stress–strain and elastic modulus data obtained are for the tangential (or circumferential) direction with respect to the fiber wall. That is, the experimental moduli from the Bergander–Salmén tests are for the double radial wall of two adjacent fibers in the transverse direction that is parallel to the middle lamella that joins them.

The moduli obtained by Bergander and Salmén ranged from about 700 to 3000 MPa. Values calculated from a theoretical model used by Koponen et al. [189] tended to be about 8000 MPa for an earlywood (springwood) fiber wall with an S2 angle of 10° and a moisture content of 30% by weight. The reason for the large difference could possibly be related to the presence of pit fields, which are concentrated in the radial walls of springwood in temperature zone conifers. As shown in Fig. 24, there are great variations in local S2 fibril angle in the vicinity of pit fields as the microfibrils curve around the pit structures. Perhaps future research will provide a comparison between pitted and pit-free cell walls in this regard.

Other Approaches to Evaluation of Fiber Tensile Properties Generally speaking, single-fiber tensile tests are slow and laborious, require great care and precision in execution, and still leave open questions such as the correct rate of strain. Additionally, fibers are seldom used independently in industrial applications. Various alternatives to single-fiber testing have been designed. In the textile field, much of the fiber testing is actually done by testing bundles of fibers, such as ASTM D 1445 for cotton [5]. In the field of filament-reinforced composite materials, there are similar strand tests [48].

At times there is a lack of differentiation between single-fiber data and data obtained from tests on strands, yarns, flat bundles, and so forth. This can cause problems; more often than not, results from single-fiber and multiple-fiber tests are disparate. For example, the strand strength of glass fiber is only about 80% of the tensile strength σ_R for the single filament [283]. The bundle theory, based on Weibull statistics [283], has been used to explain such discrepancies.

Although papermaking fibers are seldom found in any strand or yarn form, it is equally as true in paper as in textiles and reinforced composites that they do not transmit forces independently In most cases, it cannot even be said that wood pulp fibers in paper undergo the same type of deformation that is observed in the same fibers tested singly. Page et al. [277] noted that single tracheids undergo twisting and tension buckling when subjected to axial tension (see Fig. 58). These are manifestations of the shear stresses that arise as a result of the inherent axial asymmetry of the layered fiber wall. Nissan [266] points out that the tensile stress–strain curves for fibers that are prevented from buckling are quite different from those of free fibers. Mark et al. [223] observed that the twisting that is observed in a single fiber during free drying is prevented (by bonding to other fibres) as the paper sheet is dried; they presented experimental and theoretical evidence of the energy changes that occur. Because it has been well demonstrated that fibers in most real papers are bonded at such frequent intervals that there is hardly any free fiber length [266,277,372], the following conclusion of Page et al. seems to be well justified: "while data obtained from isolated fibers are relevant to the behavior of loosely bonded structures such as nonwoven fabrics, they are not directly applicable to the behavior of tightly bonded structures such as paper and board" [277].

The experimental difficulty in conducting single-fiber tensile tests has led to widespread use of the so-called zero-span test as an acceptable measure of fiber tensile strength.

Zero-Span and Short-Span Tests Sheet tensile properties are developed by (1) the use of adequately strong fibers and (2) the enhancement of the fiber–fiber bonding properties. Good (or poor) fiber–fiber bonding is achieved primarily by the pulping, bleaching, refining, and drying systems. If one could eliminate the influence of bonding in sheet tests, then presumably the fiber mechanical properties could be deduced without resorting to tests on individual fibers. In theory, shortening the span between the two tensile grips to the point where the fiber bonds are insignificant—the ideal "zero" span—means that all forces between grips are transmitted by those fibers, lying at various orientations, that are held within both grips simultaneously. For an in-plane statistically isotropic sheet, the mean fiber strength, expressed as breaking force, is related to the zero-span tensile strength of the sheet by [358]

$$F_{zs} = \frac{8}{3} A \rho a \sigma_{zs}^{b} \tag{32}$$

where (in currently preferred SI notation)

F_{zs} = fiber mean axial breaking force, N
A = average fiber cross-sectional area (net), m^2
ρ = fiber density (often assumed equal to the density of cellulose, $\sim 1550\,kg/m^3$)
a = acceleration due to gravity, $\approx 9.81\,m/s^2$
σ_{zs}^{b} = zero-span tensile strength of paper sample, km

With the proper conversions (see p. 794 and Table 1 of Chapter 7), the units balance. It will be noted that a is a constant and little variation is usually assumed for ρ. Breaking force can be converted to ultimate tensile strength (Pa) by dividing F_{zs} by A.

The zero-span test has become widely used, and several standards organizations have published test procedures for it; for example, TAPPI issued T 231 cm-96 for the dry zero-span test [345] and T 273 pm-95 for the wet zero-span test [344]. Equipment is readily available as a fixture for the universal testing machine or as a stand-alone system (several models). As Bronkhorst and Bennett point out in Chapter 7, the zero-span tensile test was introduced to "distinguish the difference between the strength of the network and the strength of the fibers themselves." For softwood pulps, zero-span sheet strengths are typically 1.5–3 times as great as the strength values achieved in a conventional tensile test such as T 494 om-88 [340]. For hardwood pulps, the ratios tend to be greater.

Figure 65 illustrates the principle of the zero-span test. Prior to the start of the test, the gap is as close to zero dimension as machining permits (Fig. 65a). As the test proceeds, a finite jaw displacement occurs (Fig. 65b) so that some of the fibers no longer bridge the gap; there is now an uneven distribution of tensile stress in the specimen in the vicinity of the free span (Fig. 65c), and there are uneven shear stresses along the jaw/paper and anvil/paper interfaces (not shown). These stresses extend back from the free span for an indeterminate distance, not necessarily limited to those fibers that still bridge the gap. The concept that fibers lying at various angles φ_i and bridging the gap will transmit a force of $P_i = (\cos\varphi_i)\,\sigma_{fi}A_i$ (see Fig. 22 of Chapter 7) is based on some unlikely conditions, namely, that the fibers are perfectly straight and elastic and that fibers in the gap do not bend, twist, or rotate during the test.

Acknowledgement that the zero-span test conditions were at variance with the theoretical basis of the test led to the concept of short-span testing. Boucai [32] recognized that the actual span is not zero and that fiber failure occurs in a region extending beneath the edges of the grips and anvils. This extended region is referred to as the *residual span*. A test instrument was developed that can conduct a series of wet and dry tests at various short spans [64]. Tensile test data for several short-span tests are plotted in the manner shown in Fig. 24 of Chapter 7. When the curves are extrapolated to "true" zero span, an intercept with the ordinate establishes the *fiber strength index* (FSI). With reference to the same illustration, the *fiber length index* (FLI) is the distance from the "true" span (at the distance R to the left of "nominal"

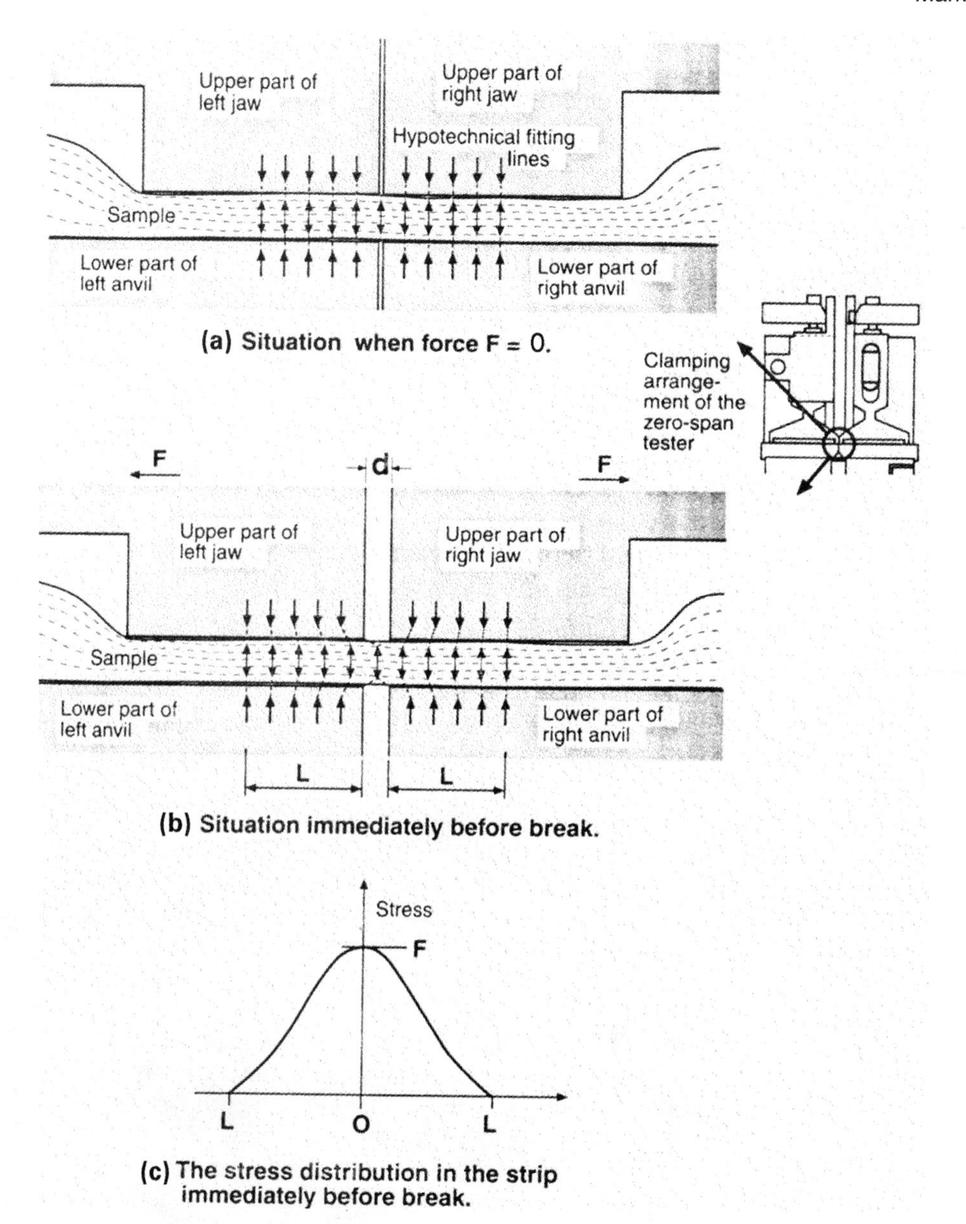

Fig. 65 Grip displacements and fiber alignments at different stages of a zero-span test. (From Ref. 205.)

zero span) to the position on the wet strength curve where it is equal to 0.5 FSI. The authors also developed two other indexes, a *bonding index* and an *orientation index*, which are explained in Section III.B of Chapter 7. The fiber strength index is an adjusted breaking length and is specified in the same units. In Chapter 7, Fig. 24, the extrapolation of the three curves to "true" zero span yields FSI = 14.2 km for a softwood kraft pulp handsheet. For machine-made paper, the FSI is the average intercept value for MD and CD data.

Various investigators have examined the zero-span/short-span technique. Ionides and Mitchell [143] found that it characterizes long-fibered pulps very well but is rather insensitive to groundwood and thermomechanical pulps (FSI varied only 10% while sheet breaking length (BL) varied 47% in TMP).

Cowan [59,60] suggested that because the dry zero-span tensile strength changes with interfiber bonding and is accordingly affected by sheet structure [32], the wet zero-span test using rewetted sheets would provide more useful results, because virtually all interfiber bonds would be broken and thus bonding would have little or no effect on the test. Also, it has been suggested that the ratio of the wet short-span and wet zero-span tensile strengths is a valid measure of the extent of fiber–fiber bonding in the sheet [59]. Cowan published results for evaluation of pulp quality using wet short- and zero-span methods [61–63] compared with TAPPI T 220 om-88 [339].

Various challenges have been raised to these procedures and suggestions. The concepts that wet or rewet sheets are essentially bond-free and that wetting test sheets has no effect on fiber properties have been subjected to experimental and theoretical review. Tests with 38 pulps by Gurnagul and Page [110] showed that wet sheets have lower zero-span tensile strengths than the same material tested when dry. However, loss of bonding is ruled out as the cause, because the specimens showing the greatest degree of bonding (as indicated by finite-span tensile strength values) show no pattern of strength loss going from dry to wet condition. Instead, there is a clear pattern of loss in strength related to the effects of pulping, bleaching, and refining on the noncellulosic matrix components of the fiber cell walls. Specifically, it was found that standard unbleached kraft pulps of both softwoods and hardwoods scarcely change in zero-span strength between dry and wet, but unbleached sulfite pulps are substantially weakened. Bleaching resulted in lower wet zero-span strengths for all kraft and sulfite pulps than for similar unbleached pulps. Beaten pulps and never-dried pulps showed larger strength reductions than corresponding pulps that were unbeaten or rewetted, respectively. The authors concluded that wetting has very significant and selective effects on fiber properties that were also found in individual fiber axial tensile tests.

On the other hand, Iribarne and Schroeder [145] consider the zero-span tensile test the only practical alternative to the conventional laboratory tests or tests on individual fibers. They reported on three experiments that were part of thesis research by Iribarne [144]. In the first experiment, never-dried unbleached kraft pulps of southern yellow pines were obtained from two different mills. The samples were PFI mill refined at five revolution levels and then zero-span tested [344,345] before and after decrilling (removal of secondary fines). Whole-pulp samples showed gradual increases in dry zero-span strength (DZS) up to 5000 rev, then leveled off. Wet zero-span strength (WZS) increased slightly up to 2500 rev, then decreased slowly. Therefore, there was an increase in the difference between DZS and WZS upon beating, as was found by Gurnagul and Page [110]. Data for decrilled samples indicated no significant changes between DZS and WZS after 2500 rev, which has been attributed to fiber straightening in the PFI mill [241,320]. The authors concluded that the increase in the difference between DZS and WZS upon beating was attributable to the secondary fines, which are a component of interfiber bonding.

In the second experiment, restraint-dried and free-dried handsheets were tested over two cycles of rewetting and drying and compared with never-dried handsheets.

The reduction from the DZS to the WZS was "completely reversible" for given drying conditions, indicating that the difference results from the breakage of accessible hydrogen bonds. Both DZS and WZS increased with drying restraint, which has also been observed by others [110,224]. These increases may be the result of fiber straightening [22,241,320], or due to an increase in fiber strength and stiffness upon drying under restraint [161], or both. However, El-Hosseiny and Bennett [83] concluded that the fiber modulus of elasticity has an insignificant influence on short-span tensile strength.

In the third experiment, handsheets were prepared in water with different concentrations of either cationic surfacant (a debonding agent) or cationic starch (which promotes bonding). The DZS changed in both cases, whereas the WZS remained approximately constant. These results, together with changes in the rates of fiber breakage to fiber pullout, indicated a strong influence of bond strength on DZS.

Fiber straightness, or the lack of it, seems to play an important role in the development of zero- and short-span test results [241,242,320]. As Mohlin et al. (who tested never-dried samples) state [242], "If zero-span tensile index is to be used as a measure of fiber strength, the measurement should be made on pulps containing only straight fibers," because fiber deformities, such as kink, curl, twist, and angular fold will influence the results. "If a laboratory pulp is carefully handled so that it consists only of straight fibers, the zero-span tensile index does not change with beating. If the pulp is a commercial pulp in which the fibers are deformed, the zero-span tensile index increases initially with beating to reach a plateau after about 3000–4000 revolutions" [242]. Mohlin and Alfredsson [241] had earlier demonstrated that when rewetted sheets were tested in the zero-span apparatus, the tensile index values showed a direct correlation with degree of fiber curl, independently of the type of beating. (See Fig. 66.) The tensile index values correlated positively with the curl index* (or negatively with shape factor, its inverse). It was concluded [242] that defects that change the direction of the fiber axis have the greatest (negative) effect on zero-span results.

Seth and Chan [320] conducted dry zero-span tests as a matter of conviction that wet fibers and wet sheets do not provide any measure of *fiber* strength. Their point is that what breaks in wet conditions are the bonds; fibers do not. Thus, DZS may be a reliable indicator of fiber strength if the fibers are straight. Because pulp fibers are usually kinked and curled, they must be straightened. One way to do this is to beat them at 10% consistency in a PFI mill for 1500 rev or more. Market pulp that has not been dried will need 2000–5000 rev, and if dried, it may need 4000–10,000 rev to effectively straighten the fibers for dry zero-span testing.

Although this discussion has touched on some of the factors that can influence zero-span and short-span testing, there are many other factors that have been investigated, such as fiber length, relative bonded area, and effect of wet pressing. The researcher may well find DZS or WZS testing advantageous but should also be aware of the pitfalls in interpreting results and know that there has been a great amount of study and trial related to this controversial approach to fiber evaluation.

* Mohlin and Alfredsson's "curl index" is identical to the fiber curl factor identified as "curl III" in Chapter 16.

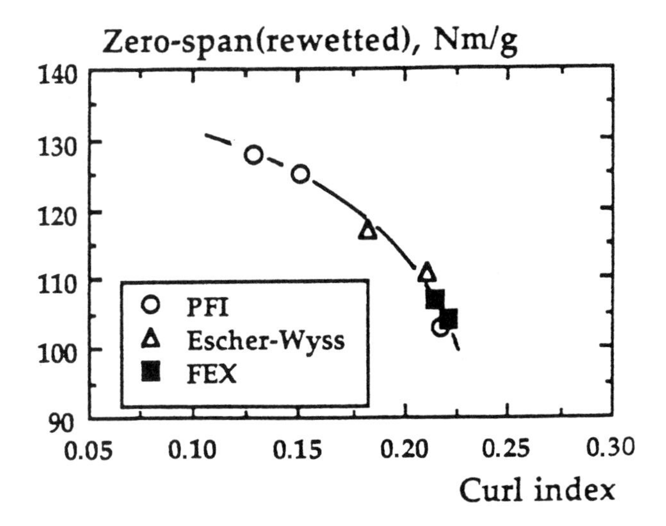

Fig. 66 Correlation between curl index and zero-span strength of rewetted sheets. Pulp stock was a commercial bleached softwood kraft. Beating was done in a PFI mill, in an Escher-Wyss Kleinrefiner, and in an industrial refiner (Beloit FEX). (From Ref. 241.)

Sonic Method on Thin Wood Sections to Determine Fiber Elastic Modulus Yiannos and Taylor [374] presented a method for determining the dynamic elastic modulus of fibers by measuring the velocity of sound wave propagation through thin wood sections before and after delignification by various processes. The method arose from earlier work by Taylor and Craver [349], who developed the sonic method for determining moduli of paper sheets at various orientations to the machine direction. The method uses the (approximate) relation between sonic velocity v and elastic modulus Q_{us} in the direction of sound propagation, that is,

$$Q_{us} = \rho v^2 \tag{33}$$

where ρ is the density of the material. The units for v are m/s, and for ρ, kg/m^3. Because $1\,N = 1\,kg \times 1\,m/s^2$, the units for Q_{us} are $N/m^2 = Pa$.

In the Yiannos and Taylor experiments, thin sections of wood were cut in the thickness range of paper. The basis weight used was about 100 g/m^2. Radial-longitudinal and tangential-longitudinal sections were prepared. The sections were then delignified to various yields. In their method, a short pulse of sound at about 10 kHz frequency is transmitted into the sample by a ceramic piezoelectric element; the pulse is received by a second transducer in contact with the specimen a measured distance away. The time of travel of the sound pulse is recorded; from the time and distance data, the velocity is determined.

Because the fibers in the thin wood sections are aligned parallel with each other, the sound propagation should be an unambiguous function of the fibers themselves without influence from the bonds that exist in fiber networks. The fiber moduli as determined by Yiannos and Taylor for nine species fell into the normal range of results obtained from single-fiber tensile testing. Koashi et al. [183] also conducted some velocity tests, using natural cellulosic fibers, as a means of determining the elastic modulus of microfibrillar cellulose. Long fibers were required.

Correlations with Solution Viscosity TAPPI T 254 cm-85 [338] states the significance of the viscosity test as follows:

> The purpose of measuring solution viscosity of a pulp is to determine the average cellulose molecular weight after processing and bleaching, hence, to measure the degree of degradation. In general, the lower the viscosity, the lower the physical strengths, such as burst and tear factor, breaking length, fold and rupture energy.

There is, accordingly, a rationale for evaluating mechanical properties of cellulosic fiber sheets—and, by extension, cellulosic fibers—that has been widely applied through the use of T 254 (the preferred procedure for pulps containing more than 4% lignin) and its counterpart T 230 cm-94 (recommended for bleached cotton and wood pulps as well as conventional kraft pulps with up to 4% lignin [343]) and similar standards worldwide. The correlations—viscosity versus tensile strength, tear strength, modulus of elasticity, zero-span strength, burst strength, stretch to failure, and shrinkage and other physical properties—are all empirical but can play and have played a significant role in decision-making in the mills because results are quickly obtained and often quite definitive.

Viscosity tests have, of course, come under scrutiny by scientists who would like to have a good justification for using these tests. This short discussion is included because results are often presented as "fiber strength" values without a clear connection between the properties of the fiber and the sheet or the pulp for a particular physical test.

Seth and Chan [320] examined the relationship between pulp viscosity and dry-sheet zero-span testing of dried, bleached commercial softwood kraft pulps beaten to 6000 rev in a PFI mill. Whether measured with the Pulmac Z-span 3000 system or the Pulmac Trouble Shooter benchtop zero-span tester, there was no apparent correlation between dry zero-span strength and pulp viscosities ranging from 11 to 24 mPa/s; the curves and standard deviation lines were quite flat. The authors concluded that "pulp viscosity is a poor predictor of fiber strength."

Gurnagul et al. [111] ran a series of experiments with cellulose-degrading treatments and reached the conclusion that there are two mechanisms that cause strength loss in fibers: *homogeneous degradation* and *localized degradation*.

Homogeneous degradation is random and causes little strength loss above a degree of polymerization (DP) of 1000. This type of degradation is exemplified by kraft cooking and vapor-phase acid hydrolysis. When the DP has been reduced to ~ 300, about half the strength is lost. Localized degradation, which occurs with stages of acid sulfite cooks, treatments with enzymes containing cellulases, treatments with vapor-phase HCl, etc. are processes that weaken the fibers beyond recovery. Some damage occurs even at DP 1600. The authors caution that measurements of pulp viscosity may be misleading when assessing strength loss by localized degradation. The nature of the degradation controls the extent of the strength loss.

B. Compressive Testing

Axial Compression In 1986 Sachs [302] reported on a testing procedure for individual fibers in axial compression and observations in the scanning electron microscope at various load levels to failure. He observed three stages in the process of compression failure. The apparatus is shown in Fig. 67. Sachs selected loblolly

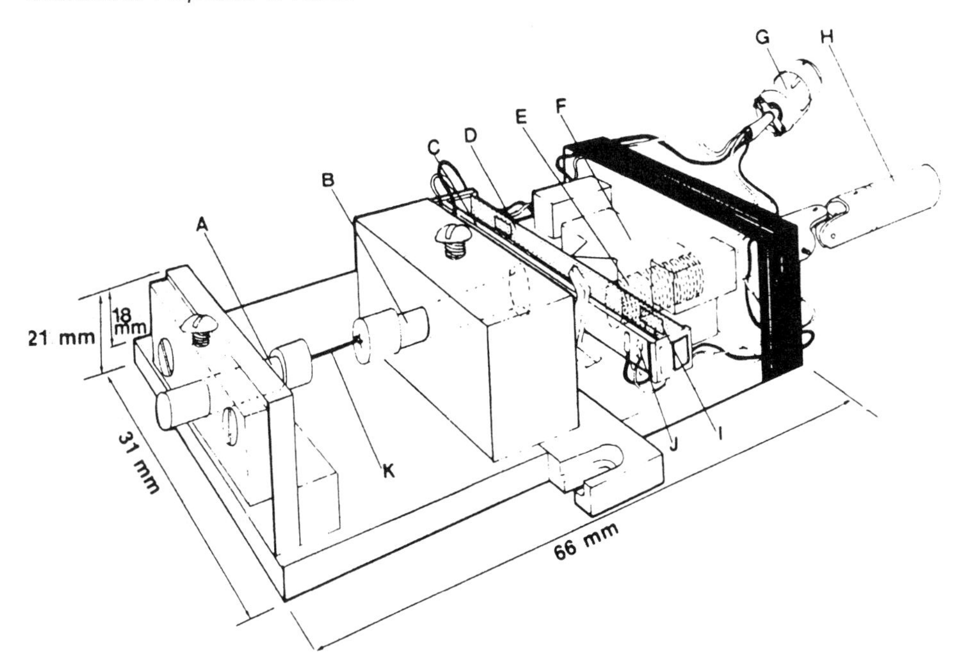

Fig. 67 Apparatus for longitudinal compression loading of individual pulp fibers. Components include (*A*) fixed rod, (*B*) sliding rod, (*C*) T bar, (*D*) load cell, (*E*) lead screw, (*F*) brass bar, (*G*) lead wires from strain gauges, (*H*) microscope rotation knob, (*I*) strain gauges, (*J*) terminal points, and (*K*) fiber specimen. (From Ref. 302.)

pine (*Pinus taeda*) summerwood tracheids that appeared to be free of twists, kinks, or other irregularities. These fibers were carefully cut into segments that averaged 1.5–2.0 mm in length (original fiber length was 3.0–4.6 mm). A segment was then cemented into two fixtures (one fixed, one movable) axisymmetrically. The free span between the fixed ends of the fiber specimens varied from 0.1 to 0.8 mm (estimated). Because the fibers were 32.0–45.0 μm in diameter, the length-to-width ratio was on the order of 12 : 1, thus creating an effective short column with fixed ends.

Prior to Sachs's work, axial compression tests had been performed on blocks or slices of wood, which were subsequently pulped [75]. Another approach was used by Dumbleton [77], who compressed individual longleaf pine (*Pinus palustris*) holocellulose fibers by aligning them between two prestretched porous rubber sheets, holding the rubber sheets firmly together with air pressure, and releasing the prestretched rubber by amounts corresponding to compressive strains of 0–20%. The fibers were allowed to dry while so held. The actual longitudinal compressive strain was determined from photomicrographs of the fibers before and after compression.

Compressive failure, meaning the inability to sustain compressive force, does not usually result in complete rupture in either fibers or sheets (see Chapter 7, Fig. 34). The specimen can still bear some tensile load. The results of such experiments generally show decreases in fiber tensile strength and modulus and increases in ultimate elongation and work to rupture. The load–elongation behavior of fibers given the compressive treatment is quite similar to that exhibited by sheets formed from pulp containing high proportions of longitudinally compressed fibers [77].

Transverse Compression The mechanical properties of a fiber in axial tension and short-column compression are not influenced (in theory) by such morphological factors as the ratio of wall thickness to total thickness, at least in the elastic region. In contrast, the force required to compress a fiber laterally is very much a function of fiber geometry. Luce [208] made a series of tests on large models (Fig. 68) that had round tubular shapes to simulate fibers of various kinds. These simulated fibers each have a shape factor S defined by

$$S = \frac{d_o^2 - d_i^2}{d_o^2 + d_i^2} \tag{34}$$

where d_o and d_i are the outside and inside diameters of the model. It was found that the collapsing force to close the lumen is linearly proportional to the shape factor. Experimental techniques for evaluating fiber collapse (see Fig. 51) are discussed in Chapter 16.

Once the fiber has collapsed, the wall material undergoes compression in the transverse (thickness) direction. Thus, a curve of load versus deformation (Fig. 69) has two distinct slopes. Balodis et al. [18] and Harrington [129] obtained similar results. If the fiber has collapsed on drying before the start of the test, only the steeper slope will be evident. Using an apparatus of the type shown in Fig. 70, Nyrén [267] and Hartler and Nyrén [131] measured the force on the fiber by means of a piezo force transducer. An LVDT displacement device was used to measure deformation. The lateral dimensions of the fiber were observable in the apparatus, thus enabling an approximation of the transverse modulus of elasticity

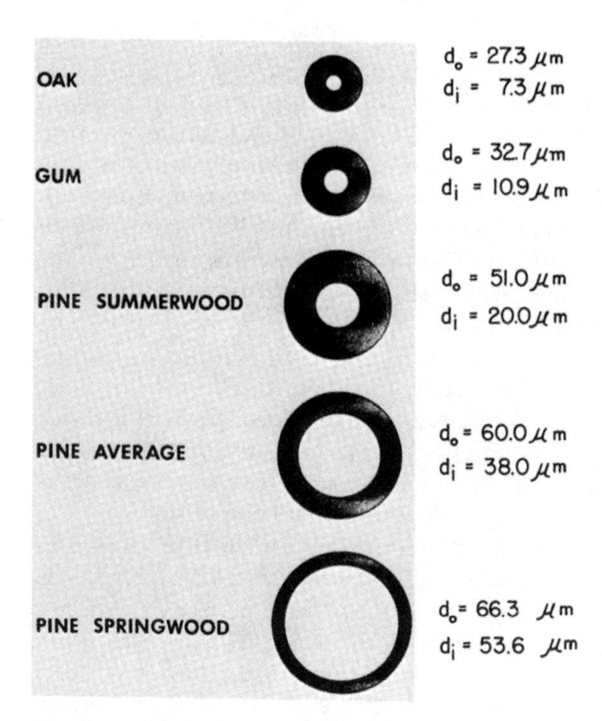

Fig. 68 Idealized cross sections of pulp fibers. (From Ref. 208.)

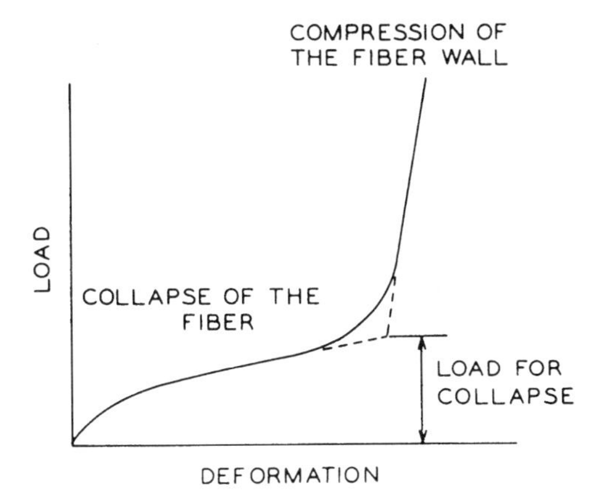

Fig. 69 Load–deformation curve of a fiber in lateral compression. (From Ref. 130.)

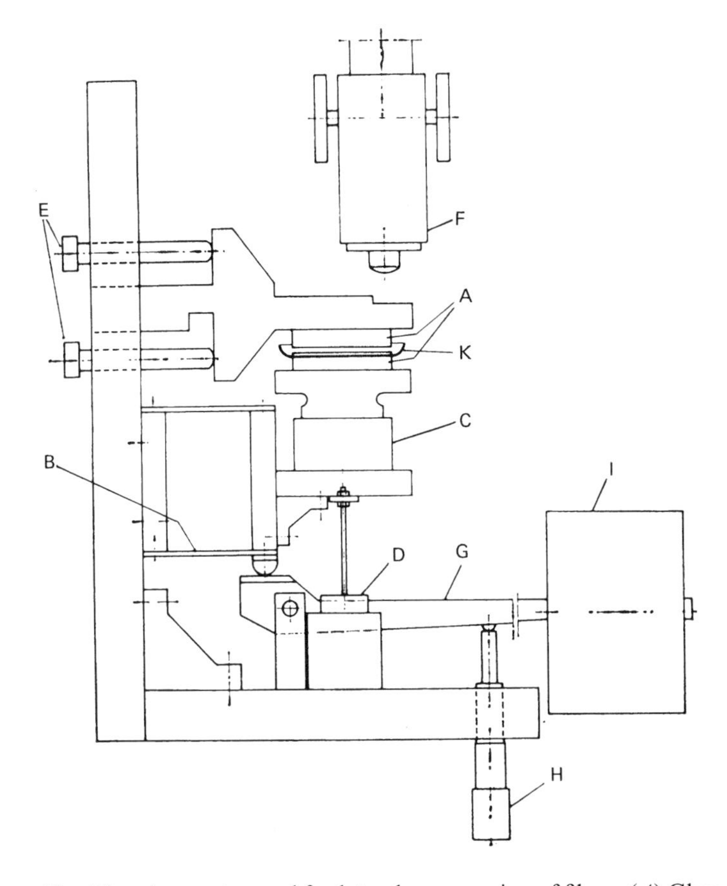

Fig. 70 Apparatus used for lateral compression of fibers. (*A*) Glass plates; (*B*) spring in box; (*C*) force transducer; (*D*) differential transformer; (*E*) adjusting screws; (*F*) measuring microscope and microscope equipped with Dyson eyepiece (interchangeable); (*G*) movable lever; (*H*) micrometer screw connected to gear box; (*I*) weight; and (*K*) rubber collar. (From Ref. 267.)

in the thickness direction (inexact because the deformation is not fully elastic). The reported values are one order of magnitude lower than the axial modulus. The apparatus recently develop by Wild et al. [366] shown later in Fig. 88 includes the capabilities for transverse compression testing of a single fiber plus other features such as cyclical loading and state-of-the-art data acquisition.

Although the thickness direction modulus can be estimated from the area of one surface of a fully collapsed fiber, Hartler and Nyrén [130] felt that the initial force of closure could not be precisely ascertained because the load-bearing area is poorly defined and varies during the collapse of the fiber. Accordingly, they expressed collapse force as a per-unit length (of fiber) quantity. Typical values for the force to collapse a latewood and earlywood fiber from a wet kraft pulp were 350 and 175 N/m, respectively. Fiber compressibility has been linked to wet mat compressibility by Binotto and Nicolls [29] and Hasuike [133].

Other Approaches to Evaluation of Fiber Compressive Properties Scientific progress in understanding the mechanical properties of fibers has been advanced to a large extent by the importance of these properties to successful design, production, and use of fibers in high performance composites, textiles, etc. The need for mechanical properties information has led to innovations that, at least in some cases, might be modified to yield advances in the evaluation of papermaking fibers.

Specifically, in the case of compressive modulus of elasticity, one finds considerable use of unidirectional composites. That is, specimens are fabricated with a known volume fraction of fibers, V_f, all of which are aligned in one direction in an isotropic matrix. Specimens of this type are subjected to compressive loading in the direction of the aligned fibers; other specimens, lacking fiber reinforcement, are given similar testing. Consider the relation [mathematically identical to the first of Eqs. (14)].

$$E_c = V_f E_f + V_m E_m \tag{35}$$

where

E_c = compressive modulus of elasticity of the composite
V_f, V_m = volume fractions of fiber reinforcement and matrix, respectively
E_f, E_m = compressive moduli of fiber reinforcement in fiber axial direction and of matrix, respectively.

It can be observed that the tests provide data for all terms other than E_f, which is the value sought. A composite of the type shown later in Fig. 93 possibly could be used for such an analysis, although in that case a thermoplastic matrix is used. Most fiber-reinforced composites use thermosetting polymer matrices such as epoxies.

Two recent papers [181,200] have made use of other means to evaluate the properties of synthetic fibers in compression. A test known as the *elastica test*, which is a derivative of a 50-year-old loop test for glass fibers [323], makes use of beam theory as applied to a fiber. The filament is wound into a single loop, which is then gradually contracted (see Fig. 71). The long (vertical) and short axes of the loop are measured during contraction. As long as the fiber deformation is elastic, the ratio of long to short axis lengths is constant, but when the elastic range is exceeded the ratio increases. The compressive strength can be calculated on the assumption that the

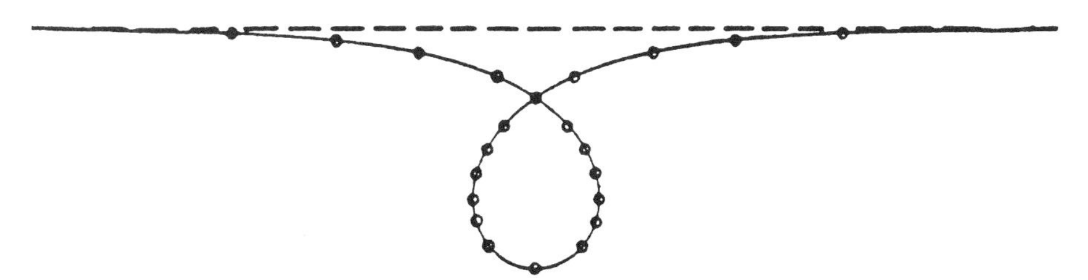

Fig. 71 Loop test of a glass fiber. Solid line: Tracing from test photograph. Circles: Theoretical points. (From Ref. 323.)

inelastic behavior is the result of compressive failure on the inside of the loop, according to the relation

$$\sigma_c = \frac{kE_c r}{(\text{v.a.})_{\text{crit}}} \tag{36}$$

where

σ_c = fiber compressive strength
k = constant (value 2.85 for test conditions of Ref. 200)
E_c = compressive modulus of elasticity
r = minimum radius of fiber in loop
$(\text{v.a.})_{\text{crit}}$ = dimension of vertical loop axis at the point where axis ratio starts to increase

Equation (36) requires the assumptions that (1) compressive and tensile moduli are equal and (2) the loop behaves like an elastic beam until compressive failure starts to occur. There is evidence to support both assumptions for many fibers. Synthetic fibers are made in lengths of 2 cm (the recommended loop test minimum) and longer. It seems possible that fiber bundles, naturally occuring very long fibers such as ramie (*Boehmeria*) (see pp. 346–347 of Ref. 142), and miniaturization could be investigated as possibilities for adaptation of the elastica test to studies on wood pulp fibers.

De Teresa et al. [71] conducted tests on a number of fibers and found the following experimental relationship between compressive strength σ_c and the torsional modulus G:

$$\sigma_c = 0.3G \tag{37}$$

For fibers that are aligned perfectly, the torsional modulus equals the internal shear modulus. Torsional testing of fibers is discussed in Section VII.D.

C. Fiber Flexibility and Related Phenomena

In recent years, considerable effort has been devoted to the testing of fibers from the standpoint of flexural rigidity (bending stiffness) or its reciprocal, flexibility. Fiber flexibility has been widely accepted as a property that is highly correlated with the papermaking process (especially flocculation) and with the achievement of good bonding and high density in the sheet, although the correlations are often poorer for mechanical pulps than for chemical pulps [199]. Fiber flexibility has also been studied in relation to energy efficiency in beating and refining [332]. Because mechanical forces are transmitted from fiber to fiber by surface bonding, a very effective way to achieve good force transfer is to use fibers that are sufficiently flexible in the wet state to maximize the available fiber–fiber surface contact area.

Bending stiffness S is defined for an object such as a beam, rod, or fiber as

$$S = EI \tag{38}$$

where

$E =$ modulus of elasticity of the material (an intrinsic property)
$I =$ cross-sectional moment of inertia (a geometric property)

Since the applicable SI units for E are $\mathrm{N/m^2}$ and the units for I are $\mathrm{m^4}$, the units for S are $\mathrm{N\,m^2}$ and the units for flexibility $F = 1/S$ are $\mathrm{N^{-1}m^{-2}}$.

Some confusion may arise from literature references to *flexibility ratio* or *coefficient of flexibility* or some similar term to describe purely morphological features of fibers that may correlate with paper properties. See, for example, Refs. 74, 92, 296. Because the transverse, or cross-sectional, fiber morphology does determine the moment of inertia, I, there may be a discernible relationship between fiber morphology alone and paper properties. However, the modulus of elasticity in bending, or flexural modulus, will have a major influence on fiber flexibility in almost all cases. Thus, flexibility measurements are generally more useful when correlations are sought between the properties of a given paper and the properties of its constituent pulp fibers, because these measurements reflect both E and I components.

Fiber *bendability* is a term that has recently been introduced by the makers of instruments designed to measure various fiber quality parameters by image analysis. It is defined [170] as "the difference in fiber *shape* between two measurements" on pulp fibers moving in flowing water. "Shape" in this case refers to the ratio of the diameter of the smallest circle that can contain the image of the fiber divided by the total length (contour length) of the fiber. This definition of shape is identical to the "curl II" measurement of fiber curl (see Chapter 16). Bendability is not to be confused with *bending factor* (a term Isenberg [146] applied to the reciprocal of curl I).

Another term that has appeared frequently in the literature is *collapse* or *collapsibility*. Collapse represents change of cross-sectional shape in a fiber from its "original" state (for example, in the stock chest) to its generally more flattened appearance in the sheet. (See Figs. 27 and 57.) It results from a combination of surface tension, which causes the opposing fiber walls to draw together as water leaves the fiber, and external forces of compression on the sheet, principally those developed in the press and drying sections [131,292].

Criteria that have been suggested for evaluating the degree of collapse of a fiber and measuring its collapsibility (the ease with which the lumen becomes partly or wholly collapsed as the paper sheet is formed) are found in Chapter 16 and in Ref. 152. Experimental methods include a variety of techniques such as evaluation of embedded fiber cross sections by light microscopy [24], stylus profilometry [108], and confocal laser scanning microscopy (CLSM) [151]. See also Chapter 5 of Volume 2.

Closely related to the concept of collapse is the *aspect ratio* of the fiber, which is the ratio of the major axis to the minor axis of a fiber that has been transversely sectioned. In fact, Bawden and Kibblewhite [24] define this ratio as "collapse." Aspect ratios can also be determined by a variety of microscopical and/or image analysis techniques (see, e.g., Refs. 151, 153, and 372). In the Jang et al. papers, measurements of aspect ratio on most types of mechanical pulp fibers were quite comparable with those for *collapse index*, which they defined by the relation

$$\text{C.I.} = 1 - \frac{A_c}{A_u} \tag{39}$$

where A_c, A_u are the collapsed and uncollapsed ("original") lumen cross-sectional areas, respectively, obtained (usually by image analysis) from the CLSM images. It is evident that collapse index and aspect ratio are both dimensionless.

Finally, the concept of conformability should be mentioned. As the word implies, *conformability* is the property that enables the fibers in a furnish to conform well to each other during the sheet consolidation process. Evaluation of conformability takes into account such factors as fiber length, length-to-diameter ratio (slenderness), coarseness, and number of fibers per gram of pulp in addition to fiber stiffness or flexibility [51]. Inherent in studies of conformability is the recognition that blended pulps (e.g., virgin–recycled, hardwood–softwood, and chemical–mechanical combinations) will have conformability characteristics different from those of more uniform furnishes. Conformability is also discussed in Chapters 15 and 16.

Testing Principles When a fiber is bent by some force, a moment (force × distance) is exerted upon it, the magnitude of which is determined by the location of the force. For example, if a fiber is held at one end and bent by a force F at the other end, the applied moment M_A at any distance x along the fiber from the force is $M_A = Fx$ and the maximum applied moment is exerted at the support. That is, for such a specimen (cantilever loading),

$$M_A = FL \tag{40}$$

where L is the full free length between the fixed support and the applied load.

If, on the other hand, the fiber is simply supported at two points and the force is applied at the midpoint, the maximum applied moment (at the midpoint) is

$$M_A = \frac{FL}{4} \tag{41}$$

In general, the applied moment, whether it is exerted by a cantilever, midspan, asymmetrical, distributed, or any other type of load, must be resisted by some internal distribution of forces that act about the neutral plane of the fiber acting as a miniature beam. The neutral plane in a beam of homogeneous material whose cross-sectional beam shape is vertically symmetrical is a horizontal plane through the centroid of the beam. However, for beams composed of layers or regions that have different material properties, whose cross sections may or may not be vertically symmetrical, the neutral plane for beams loaded elastically is found as a function of the area of reach region A_i, its elastic modulus E_i, and its centroidal distance z_i from a reference plane according to the relation

$$z_n = \frac{\sum E_i A_i z_i}{\sum E_i A_i} \tag{42}$$

If a reasonably uniform cross section along its length can be assumed, the bending stiffness S for a fiber can be found through use of the relation

$$S = EI = \sum E_i I_i \tag{43}$$

where I is the moment of inertia of the fiber cross section and I_i is the moment of inertia of a given small element of the cross section. It should be noted that the right-hand side of this equation will yield the precise value for Eq. (38) upon solution. A typical small element is shown by the shaded area in Fig. 72. The moment of inertia of this small square element A_i about the X axis is

$$I_x = \frac{s^4}{12} + A_i z^2 \tag{44}$$

where s = the side length of the square. The moment about the Z axis is

$$I_z = \frac{s^4}{12} + A_i x^2 \tag{45}$$

Suppose that the element A_i has dimensions of 1 μm on a side and the distance $z = 5\,\mu$m. From Eq. (44), the moment I_x about the horizontal axis thus becomes $0.083 + 25 = 25.083\,\mu\text{m}^4$. It can be seen that the magnitude of the moment of inertia of the fiber and consequently its bending stiffness are overwhelmingly dependent upon the *distribution of area and therefore the mass* and that a large-diameter, coarse, uncollapsed fiber will accordingly have a stiffness far greater than that of a small, thin-walled, collapsed one. Fiber shape and size, especially as regard thickness, are the most important factors controlling fiber stiffness, with the axial moduli of the layers also exerting an important effect.

In the study done by Claudio da Silva [51], the determination of values for E_i was made on the basis of microfibrillar orientation in the various cell wall layers, the layer area proportions, and the mechanical properties of the molecular constituents,

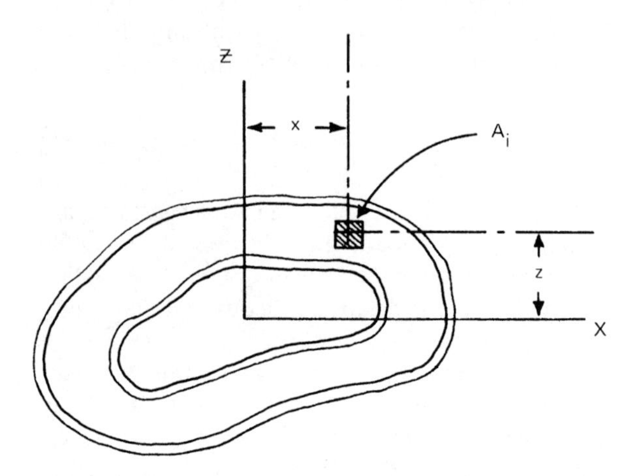

Fig. 72 Determination of the moments of inertia of a fiber cross section.

similar to the method shown in Sections III–V. When the *EI* calculations are done theoretically, as in Ref. 51, the *minimum* value is taken because the wide axis of a typical collapsed fiber cross section will usually lie in the plane of the sheet and contribute less to the sheet stiffness than it would if it were "on edge." As will be seen, most of the methods actually used for determination of bending stiffness or flexibility will also tend to yield the minimum stiffness (= maximum flexibility) because beamlike objects subjected to bending forces tend to align as much as possible in the orientation of least resistance, i.e., with minimum *EI* in the direction of the bending force.

Experimental Methods With the notable exception of the fundamental approach used by Claudio da Silva [51], most determinations of fiber flexibility have fallen into the following categories: (1) static cantilever beam tests, (2) dynamic cantilever beam tests, (3) other types of beam tests, (4) conformability tests, and (5) classification methods.

Static Cantilever Beam Tests Tests in this category represent the classical beginning of investigation into the nature of bending properties of papermaking fibers.

Seborg and Simmonds [317] built an apparatus in which the hooked end of a quartz spring engaged the fiber end, after which the fiber was moved against the spring tension. The spring extensions were measured by cathetometer. The apparatus had the capability of rotating the fiber about its long axis to obtain readings about several centroidal axes but did not have the capability of determining the flexural rigidity *EI*.

Schniewind et al. [315] improved on the foregoing method. In their experiments, loads are applied near the free end by means of a helical quartz spring and are measured by determination of its extension. The fiber span and deflection are measured microscopically. A series of readings for load and deflection are taken and plotted as shown in Fig. 73. The flexural stiffness *EI* is determined from the slope of the linear portion of the load–deflection diagram by taking matching load and deflection intervals along that straight line segment.

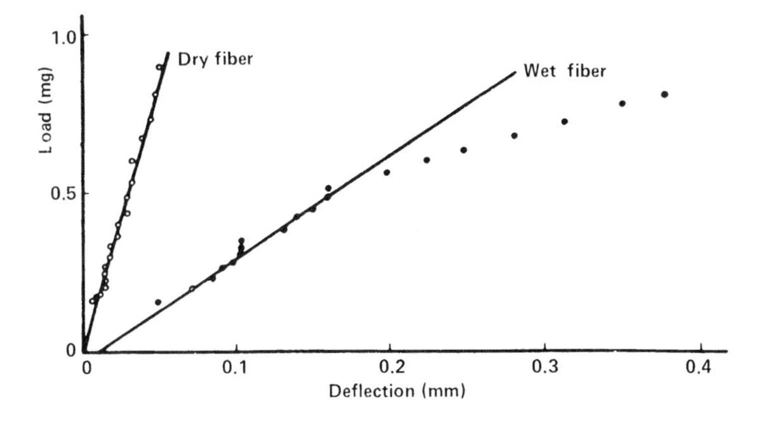

Fig. 73 Load–deflection plots for dry and wet beaten summerwood fibers. (From Ref. 315.)

The expression for deflection Δ at a free span distance L for a cantilever beam under a concentrated load P is (ignoring shear deflections)

$$\Delta = \frac{PL^3}{2EI} \tag{46}$$

which yields the desired value for EI for a given test fiber.

Haugen and Young [135] mounted a cantilever testing system on the stage of a research microscope. Loading of the fiber (as a cantilever) is accomplished by means of a torsion wire with a lever arm fixed to it that presses on the fiber end. The design of the apparatus permits variation of the rate of loading, maximum load, and total time under load. As with the Seborg and Simmonds device, the fiber can be rotated about its axis to any desired position. Fiber dimensions and free-end deflections are measured at 500× magnification.

Experimenters who are conducting fiber flexibility tests should determine if the tests are deflecting the fibers beyond the elastic range. If the fibers are deflected elastically, the determination of EI for a cantilever test will derive from Eq. (46). What needs to be measured is the vertical deflection Δ, the deflecting force P, and the free-span length (distance from support to point of application of force) L. Release of the deflecting force will result in elastic recovery of the fiber to its starting position at $\Delta = 0$. If the fiber does not recover its original position, the elastic limit has been exceeded and the EI measurement is suspect. Haugen and Young used the criterion of a maximum deflection of 10% of the span length, but variability in their tabulated results suggests that this deflection might have been excessive. Note that the value of EI from Fig. 73 is determined from the *linear* (elastic) portion of the load–deflection curve.

A newer version of the static cantilever test was used by Eissa et al. [82]. Schematically shown in Fig. 74, the apparatus has a clamp that holds the fiber in a horizontal position. The free end of the fiber is connected to the arm of an

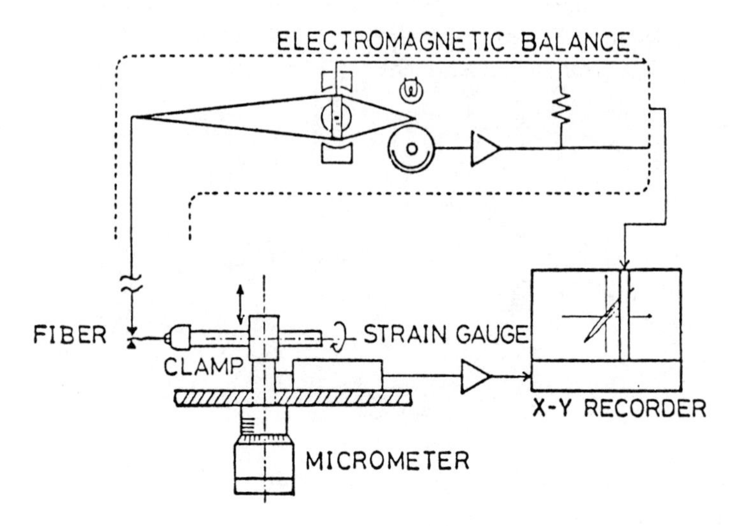

Fig. 74 Schematic drawing of apparatus for measuring flexural rigidity of single fibers. (From Ref. 82.)

electromagnetic balance by means of a very fine metal wire. The fiber clamp is mounted along with a strain gauge on the head of a micrometer. Small deflections of the fiber are measured by the strain gauge, and the corresponding deflection forces are measured with the electromagnetic balance. Thus, a bending load–deflection curve can be recorded. Haugen and Young [135] carefully selected their fibers for uniformity so as to avoid the tendency of most fibers to twist and thus present the lowest resisting moment to the applied force, but Eissa et al. accepted that the cross sections of pulp fibers are usually quite irregular and determined *EI* at four directions with respect to the fiber axis. They then averaged these four measurements for the purpose of plotting a graphical relationship between flexural and torsional rigidity.

Dynamic Cantilever Beam Tests Several approaches have been made to the determination of fiber flexibility by gripping the fiber at one end and deflecting it by means of a liquid flow or induced vibration.

Samuelsson and his associates [78,311] developed an apparatus in which fibers are fixed as cantilevers in a special holder that forms part of the wall of a flow channel (Fig. 75). The fiber is loaded by the force of the water stream, which is directed at a right angle to the fiber axis. The deflection of the fiber, which is microscopically observed through a port in the wall of the flow channel, is measured on an ocular micrometer.

The force of the flowing water on a small length of fiber (Fig. 75) is given by the equation

$$dF = c\frac{\rho w^2}{2}(2r)\,dx \tag{47}$$

where

c = friction coefficient
ρ = density of fluid
w = linear flow velocity
r = fiber radius (average)

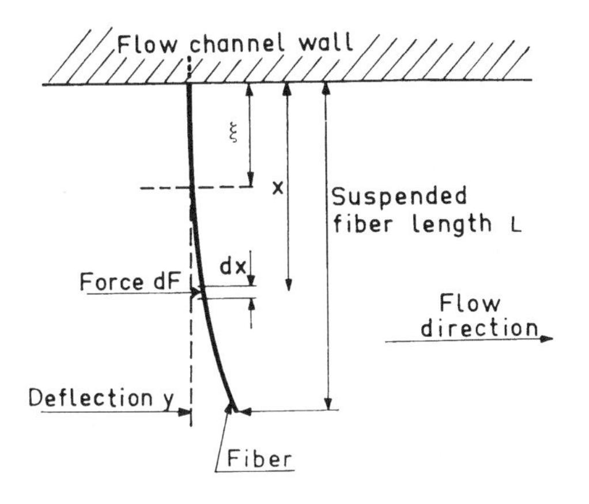

Fig. 75 Schematic of fiber deflected by water flow. (From Ref. 311.)

The bending moment at any point ξ on the fiber for the distributed force of the liquid is found from

$$M_\xi = \int_{x=\xi}^{L} (x - \xi)\, dF(x) \tag{48}$$

The elastic deflection of the fiber is governed by

$$\frac{d^2y}{d\xi^2} = \frac{M}{EI} \tag{49}$$

and if appropriate values for deflection y, suspended fiber length L, and flow rate can be determined, the stiffness EI and its reciprocal, flexibility, can be calculated.

Nethercut [259] and James [149] devised apparatuses that were also based on cantilever bending of the fiber. However, the methods for determining bending stiffness are based on a vibration principle. To carry out a vibration method in practice requires relatively straight fibers, which James obtained somewhat artificially by holding pulp fibers at each end as they dried so that some tension was exerted on each fiber; this procedure probably enhances the mechanical properties of the fiber to some extent.

Finished specimens consist of a single fiber or a segment of a fiber glued coaxially to a short piece of either stainless steel wire [149] or, as shown in the Weyerhaeuser modifications in Fig. 76, to a glass sliver. A small stainless steel ball is glued to the free end of the fiber. James used four ball sizes: 0.38, 0.64, 0.79, and 1.19 mm diameter. Experiments with the various ball sizes did not reveal any effect of size on logarithmic decrement, although there was a predictable effect on frequency. For this reason, James used only specimens with the smallest (0.38 mm) ball for experiments in which temperature and/or humidity was varied. However, as the magnetic spheres are large in relation to fiber diameter and the specimen and ball hang in a vertical position, there can be some pendulum effect.

The assembled fiber with the ball and its supporting clamp, a driver to start the fiber–ball cantilever vibrating, and a detector for measuring the vibration amplitude are all enclosed in a small, tight, environmentally controlled plastic chamber. The apparatus also incorporates an observing microscope, an audio generator to initiate the vibration, a preamplifier and amplitude discriminator to provide damping, a variable bandpass audio frequency filter to screen out signal noise, and an electronic counter to access the frequency and damping data.

The fiber is vibrated by a small electromagnetic coil wound on a ferrite core. The core is tapered to a blunt point and offset slightly to permit easy positioning of the ball on the fiber about 1 mm to the side of the tip of the ferrite core.

Vibration of the fiber is detected optically. A shadow of the ball on the end of the fiber is caused by a penlight bulb. The bulb is affixed to the end of a small, movable fiber-optic light guide that terminates in a transparent epoxy glue bond to the window of a small photoresistor whose output is transmitted to the preamplifier and amplitude discriminator.

Each specimen is oriented by rotation to give the position of minimum resonant frequency (maximum resonant frequency can also be ascertained). The frequency of vibration for a given position is varied until resonance, at which point the amplitude of vibration of the specimen is at a maximum. If a stroboscope is

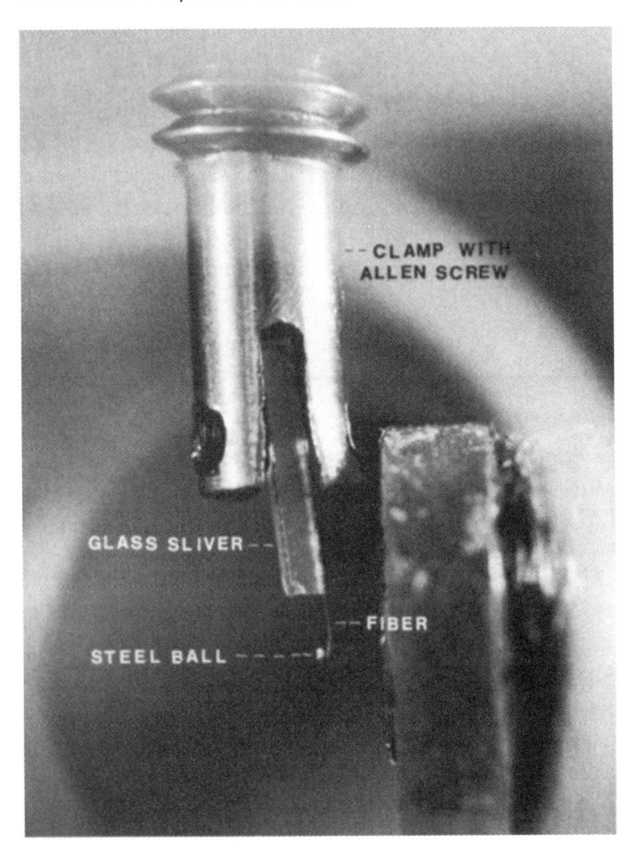

Fig. 76 Fiber with steel ball attached in position for vibration test. (Courtesy of Dr. K. A. Bennett, Weyerhaeuser Company Technical Center, Tacoma, WA.)

employed, the frequency of light flashes can be varied until they are in resonance with the oscillation of the specimen. In similar work with textile fibers, frequencies between 20 Hz and 10 kHz have been used [243]; James found that resonance for Douglas-fir (*Pseudotsuga*) fibers occurred in the range 135–210 Hz and for ramie (*Boehmeria*) at 30–50 Hz. The dynamic flexural rigidity, neglecting air friction, is

$$EI = \frac{4\pi^2 A\rho L^2 f^2}{m^4} \tag{50}$$

where

$A =$ cross-sectional area of fiber
$\rho =$ fiber density
$L =$ cantilever bending span length
$f =$ resonant frequency
$m =$ a constant that depends on the harmonic being excited and is derived by the relation $\cos m \ \cosh m = -1$ (for the fundamental harmonic, $m = 1.8751$)

Other Types of Beam Tests The main advantage of bending beam test methods is that if the test is conducted well the fiber behaves similarly to a homogeneous, isotropic, linearly elastic beam. When cantilever methods are used, however, there is a danger that the fiber may be crushed in the fixed grip if excessive force is applied. To overcome such clamping problems, Tam Doo and Kerekes [333,334] developed a simply supported beam method in which a fiber is supported in notches on opposite sides of a capillary tube. Deflection of the fiber is achieved by water flowing into the end of the tube. The fiber is modeled as a circular bar of constant cross-sectional area and modulus of elasticity. For these conditions, the parameters used to calculate flexibility involve the bulk Reynolds number, the dynamic viscosity of water, flow rate, and midspan deflection of the fiber. There is a disadvantage in that each fiber must be individually lifted from the pulp and set in place for the test, but this is still quicker and safer than clamping the fiber as a cantilever. The Tam Doo–Kerekes method has been used rather widely, because the apparatus is relatively easy to construct and the test is not complex [134,173]. There is no commercial version of the apparatus, however [172].

Kuhn et al. [198] developed a device (planned for commercial development) that also supports the fiber at two points as illustrated in Fig. 77. A fiber is temporarily detained as it exits from a capillary tube (vertical in illustration) and enters the main channel flow at a T-junction. The fiber deflects as it impinges on the capillary at points A and B. The position and deformed shape of the fiber are digitized by a charge-coupled device (CCD) camera. In contrast to the other beam methods, there is no individual handling of fibers by the operator, because fibers are being delivered continuously by water flow in the capillary. Fibers that bend at the T-junction are assumed to have deflections of 5% or less of fiber length (deflection measured as the perpendicular distance from the straight line BC to point C' in Fig. 77).

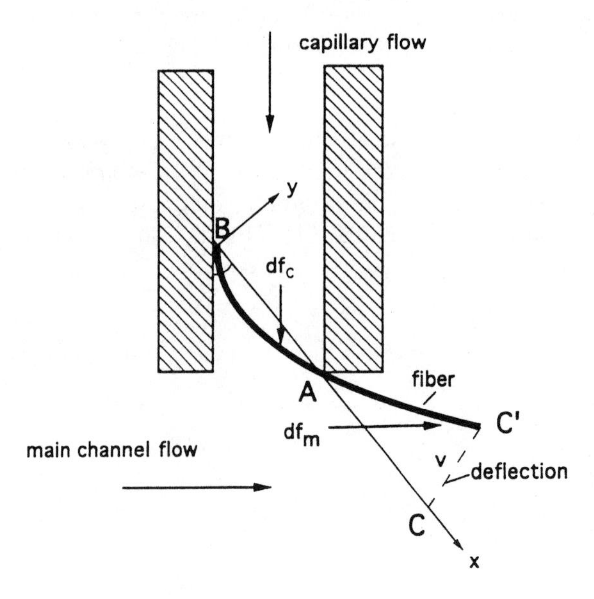

Fig. 77 Schematic of fiber transiently deformed by main channel flow as it starts to exit from a side capillary. At this point, the fiber can be analyzed as a simply supported beam with an overhanging end. (From Kuhn et al. [198].)

When these conditions are met, small deflection theory for beams is used. A schematic of the image capture system is shown in Fig. 78. Lawryshyn and Kuhn [240[performed an analysis that compares large deflection and small deflection theory as applied to test results of four kinds of fiber flexibility tests; these results will be discussed further on in this chapter.

Conformability Tests (Fibers and Fiber Networks Over Wires) The conformability method was first introduced by Mohlin [240], who noted that "conformability, stiffness, flexibility and collapsibility ... are interrelated," because they all depend at least partly on the Young's modulus of the fiber wall. Because the fiber conforms to other fibers in the sheet during the consolidation stage of papermaking, it seemed appropriate to determine this property experimentally by placing a wet pulp fiber over a glass rod that rests on a glass plate and allowing it to dry. She selected 0.06 mm for the diameter of the glass rod. As the fiber dries, it bridges over the glass rod and makes contact with the plate, bonding to it. The shorter the distance between the points of contact on the glass plate, the greater the conformability; that is, the measure of conformability is the reciprocal of the distance between fiber contacts in Fig. 79 in units of mm^{-1}.

Steadman and Luner [326] combined the conformability method of Mohlin just described with the contact ratio test for fiber bendability to devise a measure of "wet fiber flexibility" (WFF). Fiber stiffness and flexibility values are obtained by treating the fiber as a statically indeterminate cantilever beam and measuring the span length (one-half the distance shown by the arrow in Fig. 79) of fibers in

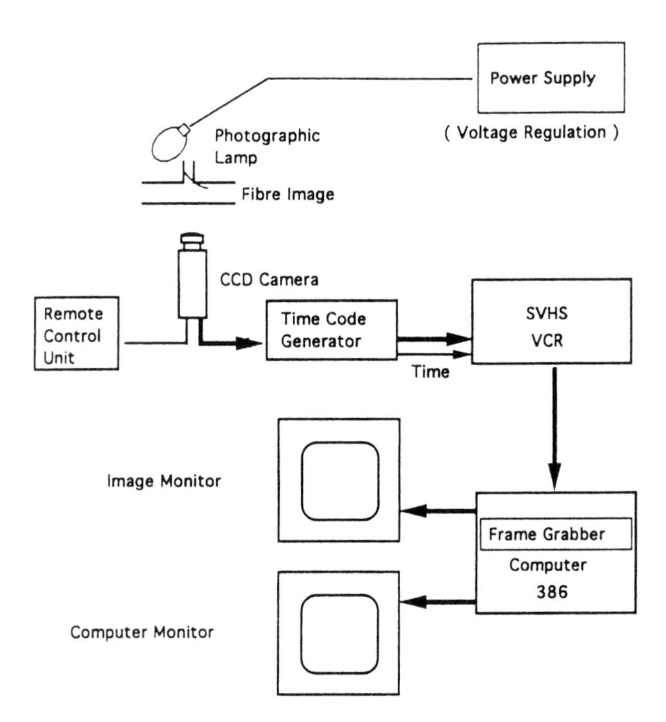

Fig. 78 A schematic of the image capturing system used for tests such as the one illustrated in Fig. 77. (From Kuhn et al. [198].)

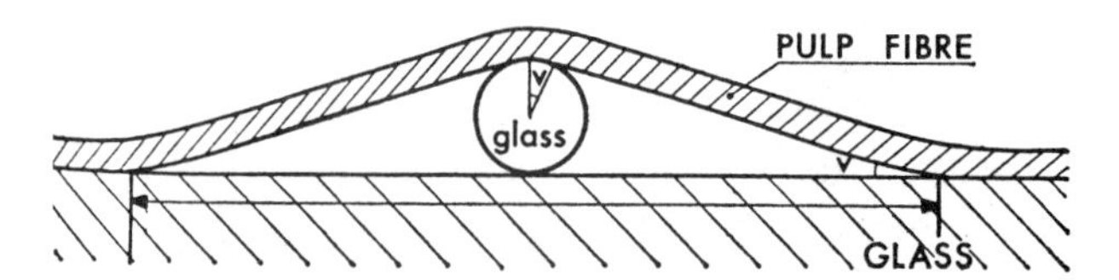

Fig. 79 Pulp fiber draped over glass rod in a conformability test by Mohlin [240].

a thin network pressed over 25 μm stainless steel wires with a known constant pressure. Their method provides EI values in units of $\mathrm{N\,m^2}$ (and accordingly, flexibilities in $\mathrm{N^{-1}\,m^{-2}}$). However, their theoretical treatment assumes that deflections are small, whereas the actual deflections may be quite large and the span length can be uncertain, as Fig. 80 illustrates. The ratio of beam deflection (in this case equivalent to the wire thickness) to span length can reflect significant shearing forces on the fiber as the ratio increases. A good discussion of the relationship between shear loading and beam deflection is given by Boresi et al. [31]. The examples they give apply particularly to the case of curved beams, especially those for which the radius of curvature is less than five times the beam depth. However, the same principles apply to straight beams, where the effects can be significant if the beam span length is less than five times the cross-sectional depth of the beam.

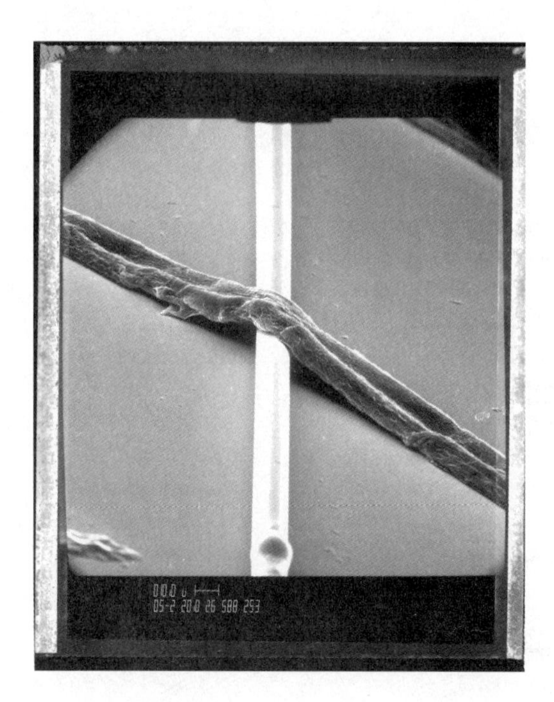

Fig. 80 Fiber positioned over small-diameter wire to determine wet fiber flexibility. There is some uncertainty as to span length even when the fiber is positioned at a right angle to the wire, as would be the normal test position. (Photomicrograph courtesy of Dr. Philip Luner, SUNY College of Environmental Science and Forestry, Syracuse, NY.)

Abitz and Luner [1] continued and refined the use of the WFF test. Automated commercial instrumentation is now available (CyberMetrics, Inc., Alpharetta, GA) to measure WFF as well as wet fiber conformability (WFC) and relative bonded area (RBA). This test is widely used, owing largely to the instrument availability.

Classification Methods Various methods have been developed that relate the flexibilities of fibers to their deformation behavior in well-defined flow fields such as in classifiers. These can be described as classification methods. One of the earliest examples is the orbital rotation method of Robertson et al. [298]. Using fluids such as 50 poise silicone (for dry fibers) and corn syrup diluted below 70 poise (for wet fibers), they observed the deformations and buckling behavior of fibers in relation to the characteristics of the flow fields.

Shallhorn and Karnis [321] developed a method for evaluating fiber flexibility based on the principle that fibers of equal length but different wet fiber flexibilities will pass at different rates through a square mesh screen. It had previously been shown that pass rates depend on fiber length [94]. Shallhorn and Karnis obtained fibers in different length classifications from a cascade classifier, then passed these groups separately through a screen subjected to a pressure differential. In this test, the ratio of passed to feed consistency, which is proportional to the corresponding fiber flow ratio, is taken as a (nondimensional) flexibility index.

Petit-Conil et al. [286] used a classifier in conjunction with a polarized light fiber length analyzer. The method, which was particularly developed for use with thermomechanical (TMP) and chemithermomechanical (CTMP) pulp fibers (but can be used for other types), yields a "dynamic flexibility coefficient" and a "static flexibility coefficient." The dynamic coefficient is defined as the average of three slopes of the straight line regression of classifier length versus weighted average length as determined by the Kajaani FS 100 fiber length analyzer. The static coefficient is defined as the average of the three classifier slopes versus the arithmetic average Kajaani length. Because these are slope measurements, the flexibility coefficients are unitless. In principle, the concept simply involves a calculation of the ratio of flexible length (the more flexible the fiber, the more likely it is that a given length will pass through a given screen size) to the actual length of the fiber. However, the details are more complex, and the reader is referred to the original paper. For a discussion of arithmetic average fiber length and weighted average fiber length, see Chapter 16.

Comparison of the Experimental Methods In connection with the discussion of Eq. (38) at the start of this section, it was noted that fiber bending stiffness and its reciprocal, flexibility, are determined from the modulus of elasticity (E) of the material of which the fiber is composed and the cross-sectional moment of inertia of the fiber (I), which is governed by the geometry (that is, the distribution of mass) of the cross section. In principle, it should be possible to compare any experimental results based on E and I with each other and with theoretical solid mechanics predictions. On the other hand, experimental results that are reported in other units or dimensionless units, such as the slope of a curve, may be difficult to compare because one can observe only the trends of empirical data.

It is also important that the determination of the modulus of elasticity E is based on Hookean behavior of the material, which is simply that stress (force per unit area) and strain are linearly proportional to each other. For E to be valid, this linearity must hold for both loading and unloading. In virtually any material of structural importance, there is a range of strains for which the linear relationship holds rather well. This range is often referred to as the small deflection range, and its extent varies with the type of material. When a beamlike object such as a fiber is bent, there will be a small deflection range in which the fiber bends and recovers elastically. Then, as the imposed forces are increased and the fiber undergoes sharper bending, some of the strains are nonrecoverable or only partially recoverable; frictional forces at supports and shear strains can become important relative to bending. In this condition, the fiber is undergoing large deflection, and a different theoretical analysis should be used.

Lawryshyn and Kuhn [201] analyzed four types of wet flexibility test results, using both small deflection and large defection analyses. This study is important because some test methods are likely to result in very large deflections. Up until the Lawryshyn–Kuhn paper, most authors had simply assumed that strains were small and the concept of constant elastic modulus (E) and constant moment of inertia (I) were givens.

The four methods that were analyzed were

1. The cantilevered fiber in a T-junction [198]
2. The clamped cantilevered fiber in a channel flow [311]
3. The simply supported fiber in a capillary [333,334]
4. The conforming fiber on a wire [326]

For reasons explained in their paper, Lawryshyn and Kuhn assumed that the results obtained with the large deflection analyses represent the actual state of fiber deflections, to which they compared the results from small deflection analysis as is usually assumed or implied by the experimenters. They found that for deflections of less than 20% of the span length, the small deflection analysis introduced an error of less than 10% in all four methods. However, the large deflection analysis showed that underestimating the coefficient of friction at the fiber supports leads to an overestimate of fiber stiffness in method (3), and cross-sectional deflection of a fiber in method (4) can lead to an underestimate of stiffness.

Large deflections lead to nonrecoverable strains. What actually happens in a fiber that undergoes large deflections varies with such factors as the thickness of the wall relative to the diameter of the fiber; the proportions of S1, S2, and S3 in the wall; the filament winding angles of the layers (particularly the S2 microfibril angle); damage to the fiber in processing; and the specific conditions of the test. Examples of what may occur in a test in which large deflections occur are

- • Fibers can deform inelastically, e.g., viscoelastically or plastically.
- •• Local separations may take place between layers, especially at the S1/S2 interface.
- ••• A sharp bending deflection may cause kinking or local buckling, thus changing the cross-sectional moment of inertia locally.

Such factors usually change the stress–strain relation from linear (or nearly so, within our normal capabilities of measurement) to nonlinear. If the strain (ϵ) in the linear portion of the stress–strain curve is designated as

$$\epsilon = \frac{\sigma}{E} \tag{51}$$

where σ = stress, then one of the ways to describe the cumulative strain beyond the proportional limit (say, when microcracking occurs between layers of the fiber), would be

$$\epsilon = \frac{\sigma}{E} + \frac{1}{k}\ln\sigma \tag{52}$$

where k is a function of one or more parameters that depend on the microcracking process. Other models are employed for other types of nonelastic behavior (see Chapter 1).

D. Twisting Tests

The properties of fibers that relate to twisting—torsional strength, breaking twist angle, cell wall shear modulus, torsional rigidity, and specific torsional rigidity—have been relatively little studied in papermaking fibers, although considerable attention has been given to them in the textile sciences because of their great importance in braiding, weaving, and other textile processes.

In 1969 and 1974, McMillin published two seminal papers [228,229] showing that thick-wall tracheids can be unravelled into ribbonlike strips of high bonding potential during the mechanical pulping process. Up to that time, few facts were known about the mechanical behavior of fibers subjected to torsional moment, either in refining or in sheet formation or in the paper sheet as used in converting, or to other service operations. It was later to be demonstrated exquisitely by de Ruvo and coworkers [70] that the measure of twist is an elegant way to relate the hygroexpansion of single fibers and further that the effects of the humidity changes on the torsional rigidity of the fibers are fully analogous to the effects of humidity on the sheet.

McMillin [229] devised a micromanipulative substage for applying torque to single fibers in the specimen chamber of a scanning electron microscope (SEM). The SEM was operated in a fast-scan mode, and the processes of deformation and fiber unwinding were observed on a television monitor. Videotape recordings were also made for subsequent analysis. McMillin noted that stresses are generally greatest at the surface of a shaft subjected to torsional forces. For a hollow circular cylinder of isotropic material, the maximum shear stress τ on the outer surface is related to the applied twisting moment, or torque, T, by the relation

$$\tau = \frac{16Td_o}{\pi(d_o^4 - d_i^4)} \tag{53}$$

where d_o and d_i are the outer and inner shaft diameters, respectively. McMillin reasoned that the shear strength of the shaft (fiber) might be limiting, particularly if the twist of the fiber tended to open (unravel) the S2 helix, and found this to be true in many actual cases.

In the analysis of twisting data for papermaking fibers, some simplifications have to be made. A circular shaft segment of length L (see Fig. 81) subjected to a torque T will undergo rotation at one end through the angle ϕ. If the shaft is solid

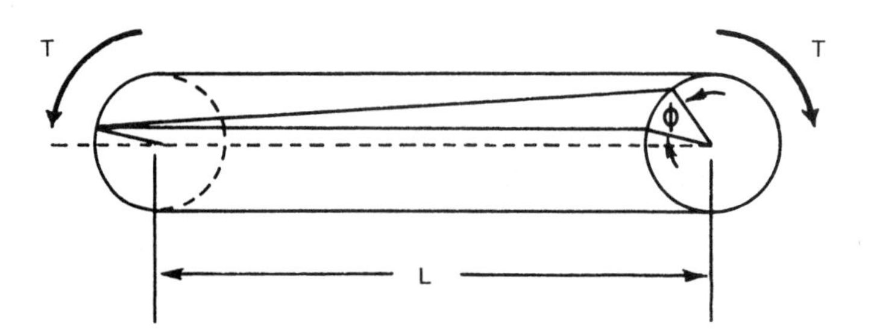

Fig. 81 Circular shaft subjected to torque.

and the material of which it is composed is homogeneous, isotropic, and elastic, the relationship between ϕ and the dimensions and properties of the shaft is expressed as

$$\phi = \frac{TL}{GJ} \tag{54}$$

where

$\phi =$ the end twist angle in radians
$G =$ the modulus of rigidity in shear in units of force per length squared, e.g., N/m^2
$T =$ torque or twisting moment, in units of force $\times$ length, e.g., N m
$J =$ polar moment of inertia (moment of inertia with respect to the axis of the shaft)

The units chosen for J and L will cancel out those for T and G.
For the solid circular cross section, $J = \pi d^4/32$, where d is the shaft diameter.
For a hollow circular shaft, $J = \pi(d_o^4 - d_i^4)/32$.
In general, the polar moment of inertia of any plane area with respect to an axis perpendicular to the plane of the figure where the axis intersects the plane is defined as the integral

$$J = \int_A r^2\, dA = \int_A (x^2 + z^2)\, dA \tag{55}$$

where r is the distance from the axis to the centroid of the element dA, and x and z are defined as in Fig. 72. For nonsymmetrical thin-walled tubes of irregular wall thickness (Fig. 82), an approximation is often used:

$$J = \frac{4A^2}{\int_0^s (1/t)\, ds} \tag{56}$$

in which s is the perimeter distance along the dashed middle line in Fig. 82. The wall thickness t can vary along s.

All of these approximations are used and have value in interpreting the measured torsional properties of fibers, but four factors should be borne in mind by those evaluating the results of the experiments:

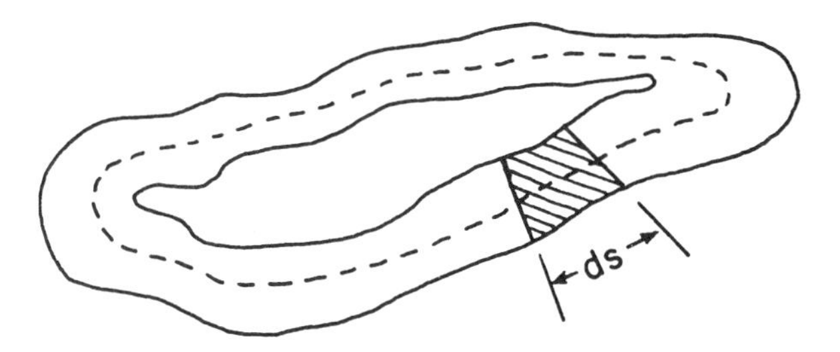

Fig. 82 Fiber cross section modeled as a nonsymmetrical tube.

- • A fiber is not homogeneous and isotropic. It is heterogeneous and anisotropic because of its oriented, layered structure.
- •• The cross-sectional area is not constant.
- ••• The axis through the (approximate) center of the fiber is not an axis of symmetry, for the reasons discussed in Section V.
- •••• In noncircular cross sections such as those exhibited by wood pulp fibers (Fig. 83), cotton fibers [142,287], etc., torsional rigidity R, which in the circular cross section cases cited previously would be equal to GJ, is also affected by out-of-plane warping. For these fibers, $R \neq GJ$, and a correction for shape factor [243] is often desirable.

Of the properties mentioned at the beginning of this section, we have defined only torsional rigidity. The shear modulus of the cell wall is an overall G value determined either experimentally or theoretically (cf. Refs. 184 and 187).

The torsional strength would normally be reckoned as the magnitude of τ in Eq. (53) at failure. The breaking twist angle is the angle of twist per unit length, ϕ/L, at which rupture initiates, usually in the outer fiber layers. It is given by

$$\tan\frac{\phi}{L} = \frac{\pi d}{n_b} \tag{57}$$

Fig. 83 Cross sections of kraft pulp fibers tested by Kolseth and de Ruvo. Top row: Microtomed after testing. Bottom row: Fracture surface. (From Ref. 185.) (See also Ref. 184.)

where n_b is the breaking twist in turns per unit length. Specific torsional rigidity is equal to the torsional rigidity divided by the linear density squared; it is a value used primarily in the textile field and is customarily expressed in units of mN mm/tex^2.

About 20 years ago, two research groups, one in Japan and one in Sweden, constructed specially designed testing devices to measure the torsional strength and elastic properties of wood pulp fibers and the logarithmic decrements and resonant frequencies of fiber–inertial mass systems [70,80,185,188,252,253]. Prior to that time, twisting and untwisting experiments and analysis had been performed by using simple free-hanging masses suspended from wood pulp fibers under different moisture regimes [106,223].

A plan view of a torsion apparatus built by Kolseth and de Ruvo [185] is shown in Fig. 84. Its basic elements are schematically illustrated. The sample is

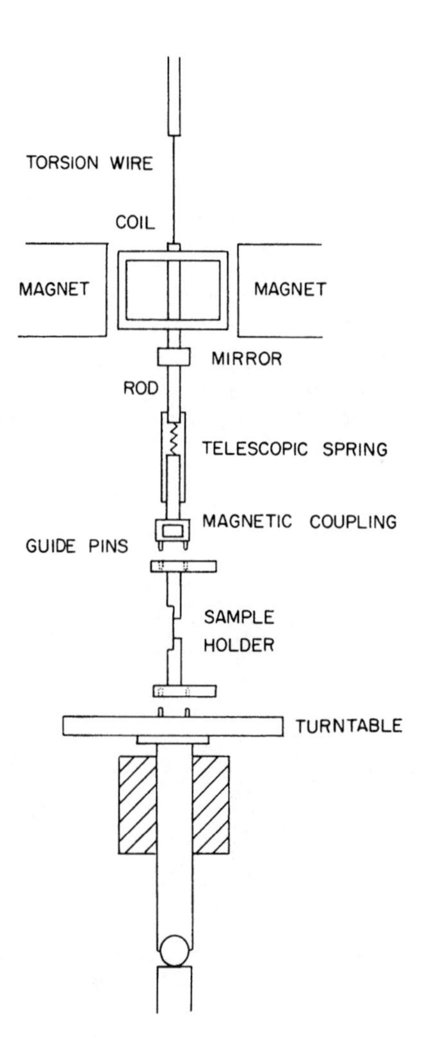

Fig. 84 Apparatus for torque-twist measurements constructed by Kolseth and de Ruvo. (From Refs. 184, 185.)

attached to a sufficiently thin torsion wire through a special coupling device. To measure the torque, an electric coil is placed axially on a rod connecting the torsion wire and the sample holder. In addition, the coil is placed in a magnetic field supplied by two magnets. A torsional deflection of the wire is sensed by a mirror placed on the rod that reflects a light beam on two light-sensitive resistors placed parallel to each other (see Fig. 85). A deflection leads to an increased light intensity on one of these resistors. To counteract this motion, the current is changed in the coil by a simple electronic circuit including the two resistors and the coil until the balanced position of the beam is reestablished. Thus, the system acts as a servo that enables continuous monitoring of the torque. According to the following formula, the torque is proportional to the current in the coil:

$$M = IBS \sin\theta \tag{58}$$

where

$M =$ torque
$I =$ current
$B =$ magnetic field strength
$S =$ surface area of the coil
$\theta =$ angle between the magnetic field and the normal of the surface of the coil

The angle θ is kept approximately constant throughout the experiments. The recording of the current is made directly on a potentiometric recorder over a 100 ohm resistance in series.

Because of the principle of the measuring device, it is necessary to have a specially designed sample holder. The fiber is mounted with a free span of about 1 mm on a brass pin and a steel pin; these pins are inserted in two triangular baseplates and are held firmly in place by a spring and a clamp. The ends of the pins are ground to allow the fibers to be glued on a horizontal surface and to be placed on the rotational axis of the pin. Each baseplate has two holes that fit into guide pins in the coupling arrangement of the measuring device. These precautions are taken to facilitate alignment of the fiber axis with the center of rotation of the instrument.

To attach the sample holder to the torsional device, it is placed on the turntable in the bottom of the apparatus. The guide pins ensure a nonslip contact and proper alignment. After that, the upper contact is made by pushing the turntable upward. When the upper plate makes contact with the triangular baseplate, coupling is

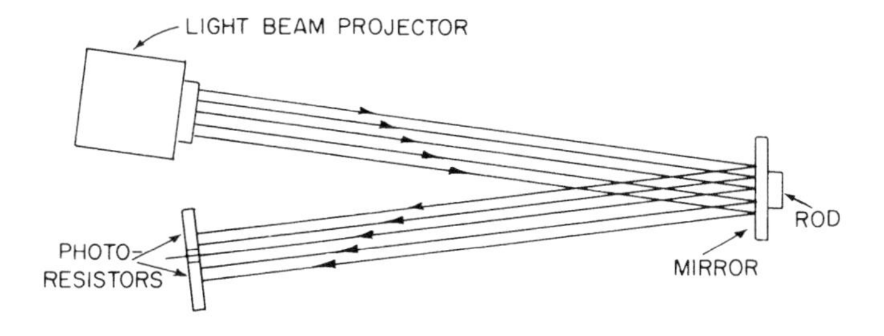

Fig. 85 Reflection of light beam by mirror on the rod shown in Fig. 84. Any rotation of the rod results in an increased light intensity on one of the two photoresistors [184,185].

ensured by means of a small magnet and the steel pin. The sample is released from the holder by removing the clamp for the baseplates. The fiber is now hanging freely from the torsion wire on which the measuring device for the torque is attached. A telescopic spring is also included to keep the longitudinal force on the fiber low. The measurements are performed by twisting the turntable with a driving unit operating at a constant speed of 1 rpm. The torsional rigidity may be obtained from the torque–twist measurements or by allowing the sample to perform free-resonance oscillations in torsion, using the lower baseplate to provide a mass moment of inertia. In this case the torsional rigidity can be derived (approximately) from the relation

$$R = \frac{4\pi^2 JL}{\eta^2} \tag{59}$$

where η is the period of the oscillations and J is the mass moment of inertia of the baseplate.

Evaluation of fiber torsional rigidity by Eq. (59) was the method used by Naito et al. [252]. Kolseth and de Ruvo [185] used this relation but also calculated the elastic and strength properties of the fiber in an approximate way by modeling the fiber as a hollow circular tube of isotropic material, the relations for which were discussed earlier. Similar torsional pendulum apparatuses have been built to test natural and synthetic textile fibers and ceramic and other high performance fibers for use in composites, etc. [72,99,230,288,293,361]. Because these fibers are usually much longer than wood fibers, more flexibility is available in choosing the sample length and method of attachment to the gripping fixture.

E. Time, Temperature, and Moisture-Dependent Fiber Properties

This handbook contains extensive coverage of analysis, tests, and procedures for determining time, temperature, and moisture-dependent properties of paper (see Chapters 1–4). The mechanical behavior of paper and other cellulosic materials is especially subject to *creep*, which is defined as "the time-dependent deformation of a (material) sample held under a constant load" [112]. The specific conditions of the test may require certain conditions of relative humidity or temperature in the ambient air within the test chamber or a specified moisture content in the sample. Typically, the deformation is nonlinearly progressive with time, and at least part of the deformation is nonrecoverable if the load is removed.

Many materials, including paper, undergo more creep deformation under a cyclical load than under a constant load, even when the cycling conditions are less severe (compared with the constant load conditions) with respect to the temperature and/or humidity environment [37,38]. This type of creep deformation is known as *accelerated creep* and will be discussed (as it applies to fibers) shortly.

If a fixed level of strain or deformation is applied to a fiber or other specimen, a reduction in the force required to maintain that strain level will be observed over time. This reduction results from internal molecular readjustments and is referred to as *relaxation*.

Another form of response to cyclical loading is known as *fatigue*. Classically, fatigue is defined as failure under cyclical straining at a load level that would not result in failure if the strain level did not cycle. Morton and Hearle [243] point out that for fibers subjected to extension cycling, the results are often confounded by the fact that the load steadily diminishes on account of stress relaxation. Thus, the failure may occur at or near the usual breaking deformation. For heavily loaded nylon fibers, failure is observed to occur in the same time and the same mode as in a creep test, and the results are consistent with creep theory except for the morphology of the fracture surface, which differs considerably between creep and fatigue specimen failures.

Both constant rate of load and constant rate of extension machines can be used to perform creep and relaxation tests on fibers. In tensile creep tests of individual pulp fibers, for example, Hill [138] used the Institute of Paper Chemistry's fiber load–elongation recorder (FLER) [125], with load applied at the rate of 2.0 gf/s ($\sim$ 0.02 N/s) initially until a constant load (10, 15, 20, or 25 gf) was achieved; the constant load was then maintained for periods of up to 2 days. Deformations were recorded photographically with a slit camera. Byrd's single-fiber tests, involving several relative humidity (RH) regimes, were made in a special apparatus in which deformations were observed on a strip chart recorder receiving a signal from a vertical displacement transducer [36]; the loads (about 45% of maximum fiber stress) were applied by attached weights. Ehrnrooth et al. [81] also measured fiber deformation with a displacement transducer in the apparatus drawn in Fig. 86. A traveling microscope was used to calibrate the signal from the transducer (LVDT) in order to be able to record fiber lengths. In their experiments, the fibers were preloaded with a force of 1 mN to balance the system and straighten the fiber. Various ambient conditions were combined with a load–unload regime to determine fiber creep characteristics. El-osta and Wellwood [87] used a tensile testing machine of the constant rate of extension type for creep tests of small wood specimens. Deformation was recorded by a strain gauge extensometer.

We return to the question of accelerated creep and fatigue behavior in wood pulp and other fibers. An intriguing aspect of recent research results is that moisture-

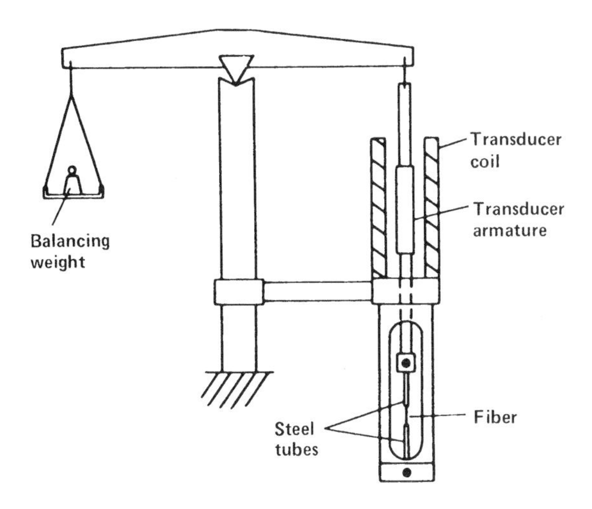

Fig. 86 Single-fiber creep test apparatus with environmental control [79,80].

induced accelerated creep is usually *not* observed in single-fiber tensile creep tests, even though paper made from the same fibers generally does experience the phenomenon [52,69,318]. An explanation for this difference, based on the creation of stress gradients in humidity cycling of paper that do not develop in the humidification–dehumidification time cycles of single fibers is part of a theory for accelerated creep that was postulated by Habeger and Coffin [112]. Their paper considers the phenomenon across a wide variety of cellulosic and noncellulosic materials, including man-made fibers, where test results differ (tests with nylon 6,6 and three types of rayon fibers also failed to show accelerated creep, whereas aramid fibers (Kevlar) do show the increased creep on cycling [112,148]). The cited papers contain some details about the equipment and procedures used in the above-cited tests (see also Ref. 119).

Considerable improvements in equipment designed to study creep and fatigue behavior have been made recently. Wild and associates [366] have developed a single-fiber fatigue test instrument that can apply cyclical loads to individual wood pulp or other fibers in either the axial or transverse direction. The instrument has found use in the study of fiber hysteresis, fiber stiffening, and fiber creep in response to various test parameters, including frequency (up to 1 kHz), temperature (up to 100°C), relative humidity (up to 100%) , and total work done on the fiber. Force–displacement data for each axis are collected as loading proceeds.

In the axial mode, as shown in Fig. 87, the fiber *5* has epoxy droplets *4* applied to each end (see also Refs. 174 and 247). The fiber is then placed on two small forks *6* that are mounted to a moving jaw *7* and a stationary jaw *3*. The jaws reside within an environmental chamber *17* that contains water and a heating element *16* so that tests can be conducted at elevated temperature and humidity. The position of the stationary jaw *3* can be adjusted by using the micrometer head *1* in conjunction with the linear slide *18* and carriage *2*. The moving jaw *7* is fixed to the load cell *9*. The load cell consists of a block *15* that is suspended on two strain-gauged metal strips *8*. Displacement is applied to the fiber, via the load cell and jaw, by the piezo actuator *10*, whose motion is indicated by the dashed lines. Fiber elongation is measured by an eddy current transducer *14*, which detects motion of the load cell block *15* relative to the base of the piezo actuator. The position of the actuator–load cell assembly can be adjusted manually with the micrometer head *12*, linear slide *13*, and carriage *11*.

When the loading is applied transversely to the fiber axis (Fig. 88), the lumen will compress so that the static load–deformation behavior would typically appear similar to that shown in Fig. 69. After fiber collapse, the slope of the upper part of the curve approximates the stress–strain behavior of the fiber wall in its radial direction. The fiber *5*, shown in exaggerated cross section, is, again, supported in the forks, not shown in Fig. 88. The stationary jaw or "anvil" *4* is fixed to a piezo-electric load cell *3*. The position of the stationary jaw *4* can be adjusted by using the micrometer head *1* in conjunction with the linear slide *12* and carriage *2*. The moving jaw or "hammer" *6* is fixed to the piezo actuator *8*. The motion of the hammer is measured with an L DT *7* mounted, via a bracket, to the base of the piezo actuator. The position of the actuator can be adjusted manually with the micrometer head *10*, linear slide *11*, and carriage *9*.

The developers foresee many applications for this instrument in research on the mechanical behavior of wood pulp fibers subjected to various processing environmental conditions such as exist in mechanical pulping, refining, and papermaking. For earlier work in this general area, consult Refs. 81, 116, 117, 119, and 138.

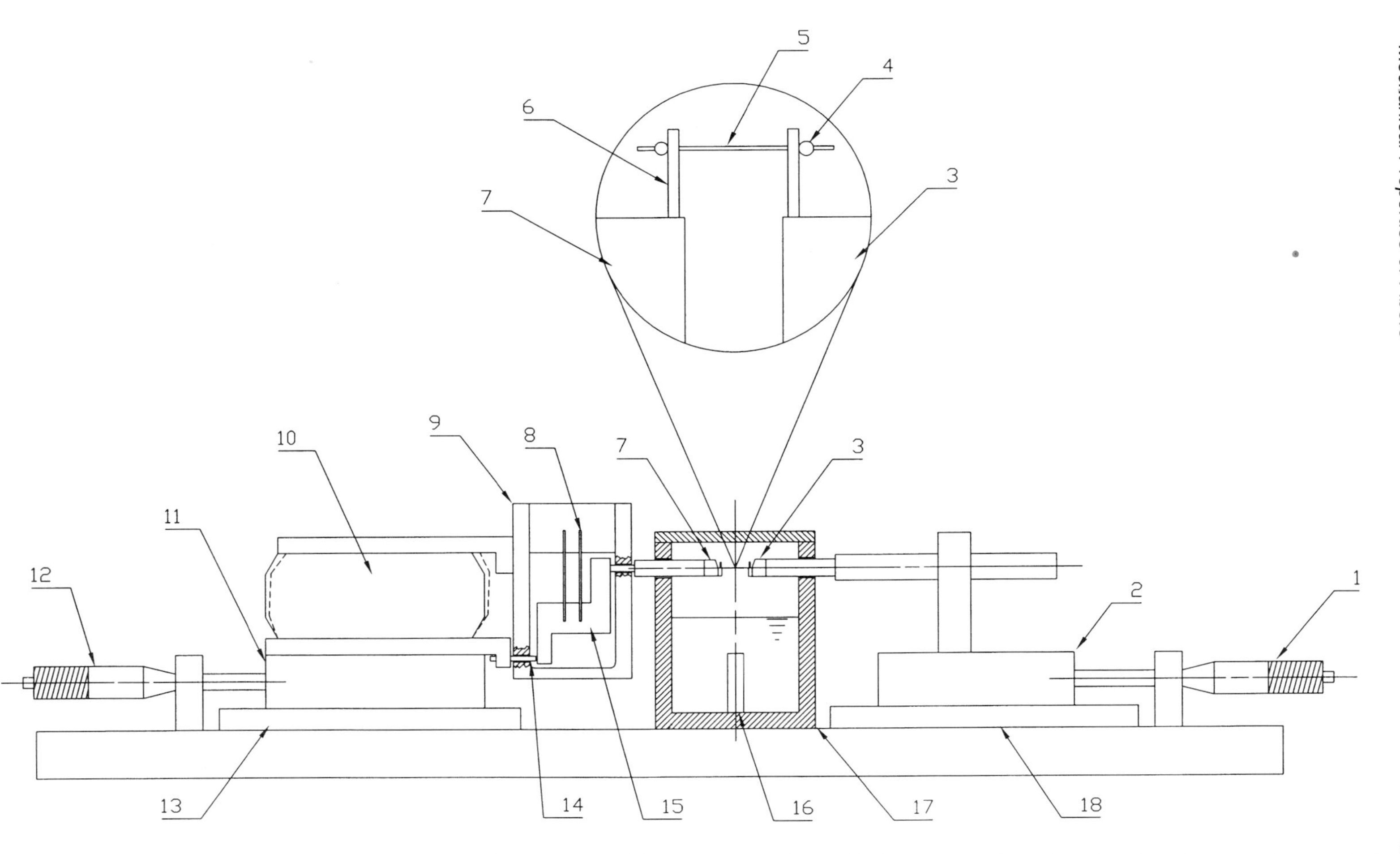

Fig. 87 Single-fiber fatigue–creep test instrument (axial mode). (Courtesy of Dr. Peter M. Wild, Queen's University Department of Mechanical Engineering, Kingston, Ontario, Canada.) (See also Ref. 366.)

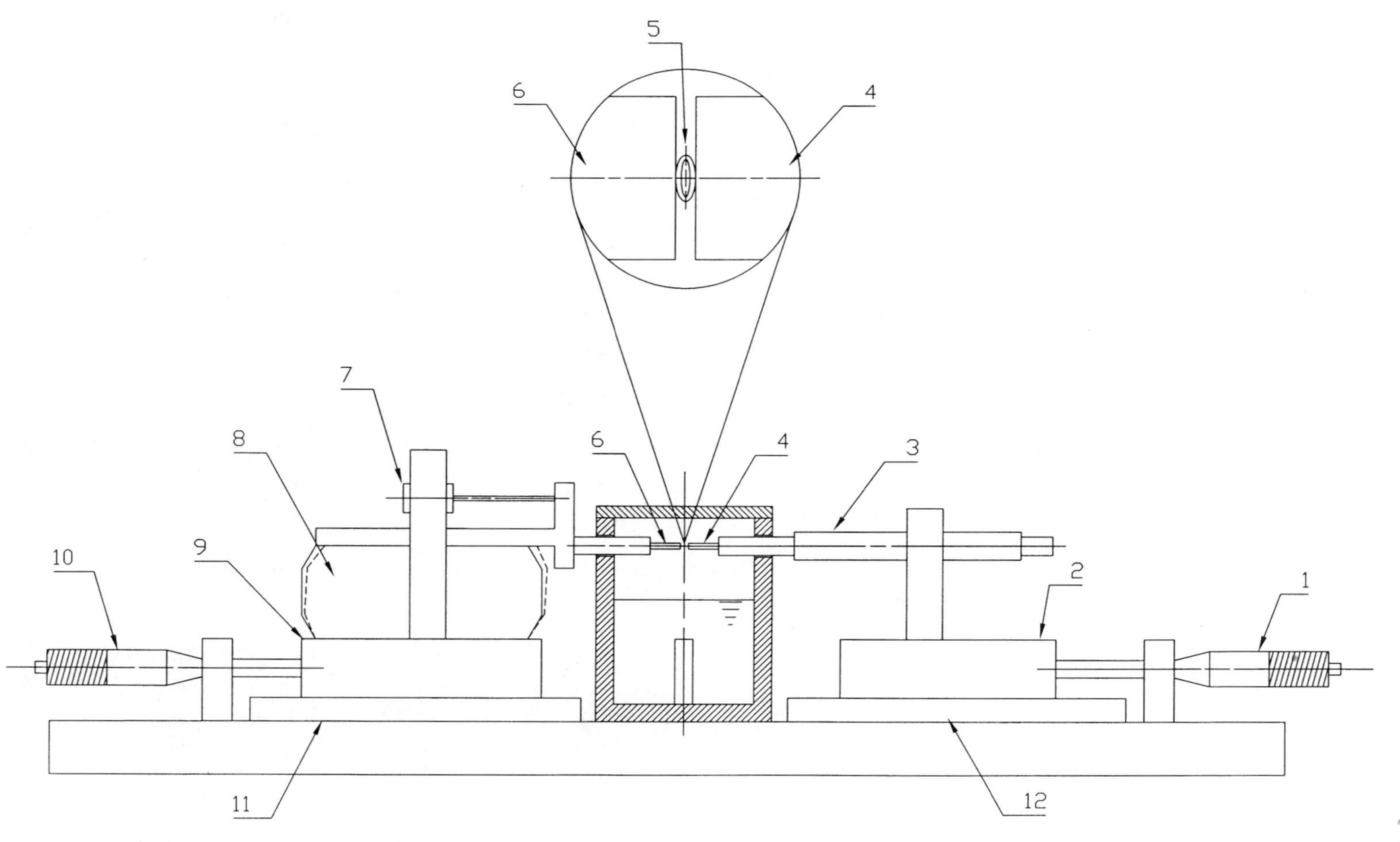

Fig. 88 Single-fiber fatigue instrument (transverse mode). (Courtesy of Dr. P. M. Wild.) (See also Ref. 366.)

Humidity-dependent torsional rigidity has been measured by Naito et al. [252] for both earlywood and latewood pulp fibers in a torsional pendulum apparatus. Their apparatus, similar to that shown in Fig. 89, is encased in a glass chamber; the humidity of the inside air can be varied between 10% and 98% RH via air recirculation through saturated salt solutions. As shown in Fig. 90, the Kolseth et al. [188] apparatus has been set up to provide variation in ambient temperature and humidity in two different experimental arrangements, one for temperature scans and the other for scanning RH at a constant temperature. In the temperature scanning arrangement of Fig. 90, the pendulum is situated inside an electric oven controlled by a temperature programmer. Two thermocouples are used. One measures the temperature at the fiber, and the other is connected to the temperature programmer. The oven is in the shape of an open cylinder with its upper end held close to a rigidly mounted plate carrying the pendulum. The lower end is covered with a glass disk, allowing for the motion of the pendulum to be measured optically from the detector housing described later. The oven is also equipped with an inlet and an outlet for environmental gas. Normally nitrogen is flushed past the pendulum, but other gases such as humid air can be used. The

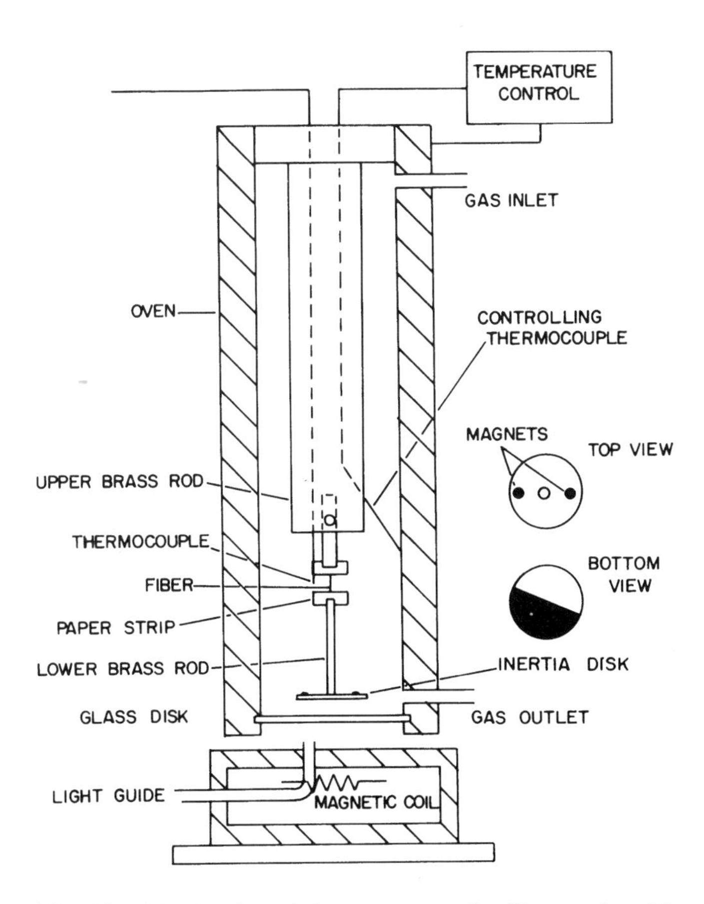

Fig. 89 Torsional pendulum apparatus for fibers enclosed in an environmental chamber. See also Ref. 188.

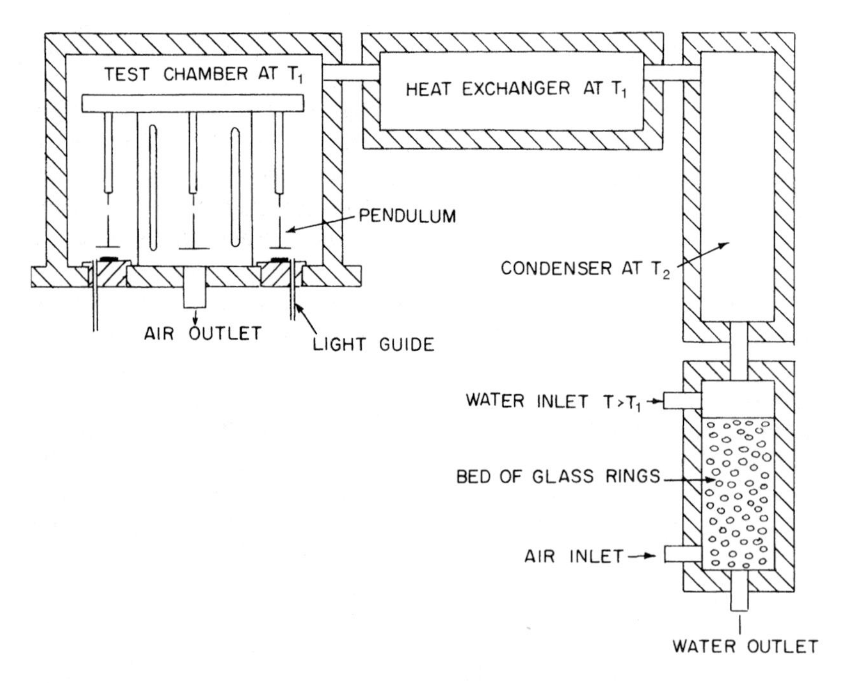

Fig. 90 Arrangement for control of temperature and humidity used with torsional pendulums in single-fiber testing. See also Ref. 188.

detector housing can be moved sideways, making it possible to lower the oven and make the pendulum accessible for change of fiber.

The humidity-scanning arrangement consists of a thermostated test chamber containing four pendulums and three heat exchangers (Fig. 90). The test chamber is kept at a constant RH by feeding it with air that has been passed through the system of heat exchangers. Fresh air is pumped into a column filled with glass rings meeting a countercurrent of heated water. The warm, humid air is then fed through the other two heat exchangers, the first a condenser held at the dew point wanted, and the second at a temperature equal to that of the test chamber. The rate of flow is enough to exchange the air in the test chamber every 2 min. Two thermocouples are used for measuring the air temperatures in the test chamber and in the condenser, respectively.

To record the torsional oscillations, which are induced by an electromagnet, a light beam is focused by fiber optics onto the borderline between reflecting and nonreflecting (black) zones on the inertia disk (Fig. 89. When the disk oscillates, the reflected light intensity varies with the oscillations; this intensity is measured by photodiode, and the output signal is recorded on a potentiometric recorder. The light source is mounted on a turntable, thus permitting restoration of the original intensity by rotation of the optical system. Thus, an absolute value of the full angular rotation of the disk can be obtained. Curves of the exponential damping of rotation are used to determine the logarithmic decrement. Also, as shown in Eq. (59), the torsional rigidity can be obtained, because it is proportional to the square of the frequency of the free oscillations. These experiments were performed in an atmosphere of either air or N_2, with temperature and humidity in the test chamber adjustable within the ranges 10–90°C and 5–59% RH.

Fatigue and relaxation tests have been performed on various textile and other polymeric fibers because of the importance of that property in composites as well as fabric structures (see Refs. 243 and 250). Several means of automatically recording the stress relaxation of fibers have been developed. In Fig. 91, a system is shown in which the changing load on the fiber is reflected as a reduced voltage output of the electrical strain gauge; the voltage is converted to a frequency, displayed on a counter calibrated so that the load can be read directly. A scanner is programmed to accept the output of the strain gauge at preset time intervals and positions. Kubát [197], working with a wide variety of metallic and nonmetallic materials, including paper, found that there is a basic similarity of shape in graphs relating stress σ to the logarithm of time according to the relation.

$$F \sim 0.1/\Delta\sigma \tag{60}$$

where $F = d\sigma/d\,\ell\mathrm{n}\,t$ at the experimental inflection point.

An alternative means of studying time dependence is to subject the specimen to an oscillating load such as in the torsion pendulum device shown in Fig. 89 [70,89].

Morton and Hearle [243] classify the most convenient time scales for fiber testing into four categories:

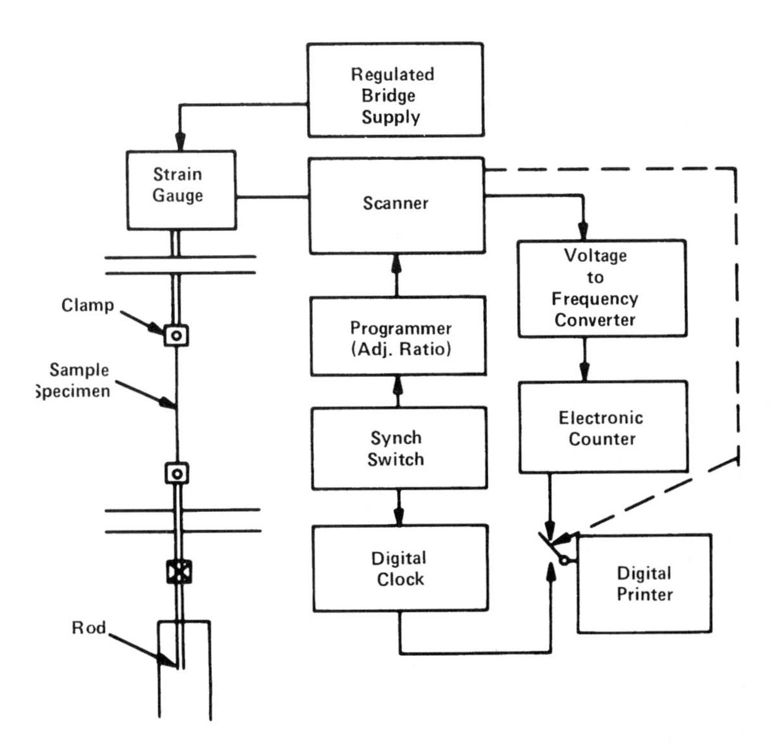

Fig. 91 Schematic diagram of a stress relaxation apparatus for fibers. (From Ref. 250.)

- • Creep: long times, from 1 min to 1 month
- •• Stress relaxation: medium to long times, from 1 s to 1 h
- ••• Stress–strain curves, including impact methods: short to medium times, from 0.01 s to 10 min
- •••• Dynamic testing: short times, from 0.1 ms to 1 s

These time ranges can be modified, but there are normally experimental difficulties in doing so.

In much of the foregoing discussion, references were cited that include some of the significant experimental testing of fibers under various conditions of relative humidity. Brunnberg et al. [34] developed an all-glass tensile testing device to enable the study of single filaments and paper and film strips in controlled temperature environments between −100° and +150°. The principles of this device are explained in Fig. 92.

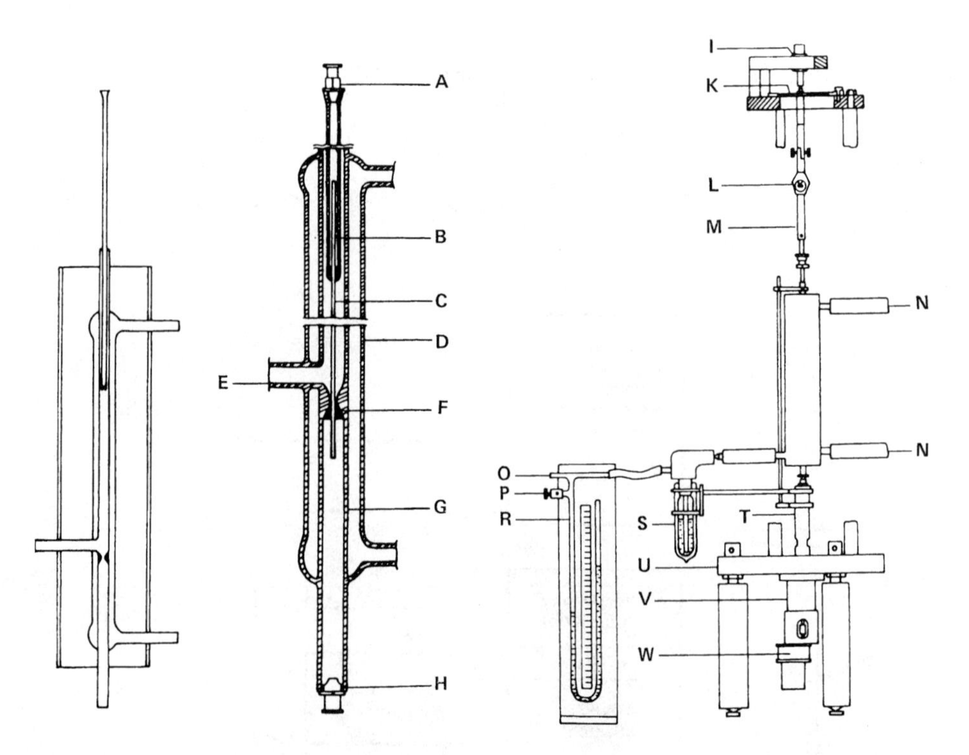

Fig. 92 Schematic diagrams of tensile testing device used for various fiber materials while immersed in a liquid at various temperatures. The all-glass sample holder with heat-insulating mantle is shown at left. In the center the details of the sample holder are given, and the complete apparatus with measuring devices is shown at right. (*A*, *H*) Bayonet catch; (*B*) glass tube; (*C*) filament sample; (*D*) jacket; (*E*) inlet tube; (*F*) constriction; (*G*) inner glass tube; (*I*) differential transformer; (*K*) stress-sensing steel membrane; (*L*) knife edge; (*M*) adjustable rod; (*N*) connections to thermostat; (*O*) connection to compressed nitrogen cylinder; (*P*) needle valve; (*R*) open manometer; (*S*) Dewar flask; (*T*) attachment to the straining device; (*U*) support plate for straining device; (*V*) straining device; (*W*) synchronous motor. (From Ref. 34.)

F. Tests That Relate the Mechanical Properties of the Fiber to Those of the Network

It has been shown in several studies that the mechanical properties of the fibers within a sheet of paper differ very substantially from the same fibers isolated from the pulp [167,168,368,369]. The same research demonstrated that these fiber properties vary with orientation as to MD, CD, or intermediate alignment in machine-made papers. The tests were conducted on fibers that had been excised microsurgically from the sheet.

Other studies have made use of pull-out or extraction tests to elucidate the mechanical properties of the fibers as they exist in paper [67,171]. Niskanen [264] notes that a "surprisingly large number" of fibers do not break despite being bonded to other fibers by numerous bonds. In the Davison experiments [67], the average peak force to extract a fiber was 50–80 mN, which is about 10 times the strength of a fiber–fiber bond. Niskanen [264] concludes that the bonded fiber segments—not the bonds per se—are the relevant structural units that control the mechanical properties of paper. However, he does add that "it is probably impossible to change bond strength properties without a related change in the mechanical properties of the bonded fiber segment."

Certainly the study of bond properties has been extensive over the years (see Chapter 15 and references such as 35, 227, 328, and 351). These investigations have usually involved a microtensile testing device, including capabilities in load and deformation measurement, and a magnification system for easier observation.

Niskanen and associates have taken a new conceptual and technical approach [175–177,265,375]. Their contributions have been to relate the fiber and bond studies to fracture energy and fracture toughness of the sheet. Fiber pull-out or bond rupture on the one hand and fiber breakage on the other have different effects on the fracture energy of paper. These factors are examined theoretically and experimentally. The research team has also developed some parameters for describing microscopic damage in two zones of the fracture surface, the pull-out width and the damage width. *Pull-out width* refers to the length of the fiber ends that can be extracted from paper along a fracture line, whereas *damage width* measures the area in which fiber debonding takes place. The pull-out width is at and near the fracture surface; the damage zone is behind the pull-out zone. Niskanen et al. use conventional procedures for mechanical testing (zero-span and in-plane tear tests) but have adopted a special method of silicone impregnation used by Korteoja et al. [192,193] and an image analysis procedure [176] to characterize the two parallel zones. The silicone is applied to the microscopic damage area that exists within a few millimeters of the fracture line. It reveals the microscopic fiber level fracture zones in the failed specimens as differences in gray level by image analysis. The profile of gray levels distinguishes the zones as a function of distance from the fracture line. Table 10 shows selected results from these experiments.

It has also been noted (Chapter 1) that fiber networks can be modeled as continuous media for many purposes. The ultimate example of this is when the network is impregnated with a nonfibrous matrix or when fibers are dispersed in a polymeric or other matrix for the purpose of reinforcement. Such systems were examined theoretically and experimentally by Johnson and Nearn [162], who obtained photostress patterns in polyethylene reinforced with Douglas-fir

Table 10 Selected Fiber and Sheet Properties

Sample set	Mean fiber length (mm)	Coarseness (mg/m)	Beating level	Apparent density (kg/m^3)	In-plane tear work (J m/kg)	Pull-out width (mm)	Damage width (mm)	Tensile index (N m/g)
Coarse abaca–TMP								
100% abaca	3.2	0.136	SR 40	609	51.8	2.8	4.8	70.9
65%				595	41.5	2.4	4.4	62.6
50%				566	31.5	2.3	3.1	60.8
25%				549	21.4	1.6	2.6	59.3
10%				538	14.6	1.0	2.0	46.8
100% TMP	0.8	0.220	CSF 40 mL	555	8.9	0.8	1.8	52.0
Fine abaca–TMP								
100% abaca	2.5	0.072	SR 27	641	34.7	2.7	3.8	61.7
65%				597	25.0	1.8	2.8	62.3
50%				590	22.2	1.5	2.4	59.4
25%				552	14.8	1.0	2.0	59.4
10%				553	11.0	1.1	1.8	54.2
100% TMP	0.8	0.220	CSF 40 mL	555	8.9	0.8	1.8	52.0

TMP = thermomechanical pulp.
Source: Ref. 175.

(*Pseudotsuga*) bark fibers. Figure 93 shows sequential photostress patterns around a fiber at various load levels. Such tests are very helpful in analyzing the load transfer mechanisms that take place as stresses are increased [73]. In the case shown in Fig. 93, the shear transfer zone steadily shifts from the fiber ends toward the center of the fiber, thus reducing the mechanically effective fiber length in the sample.

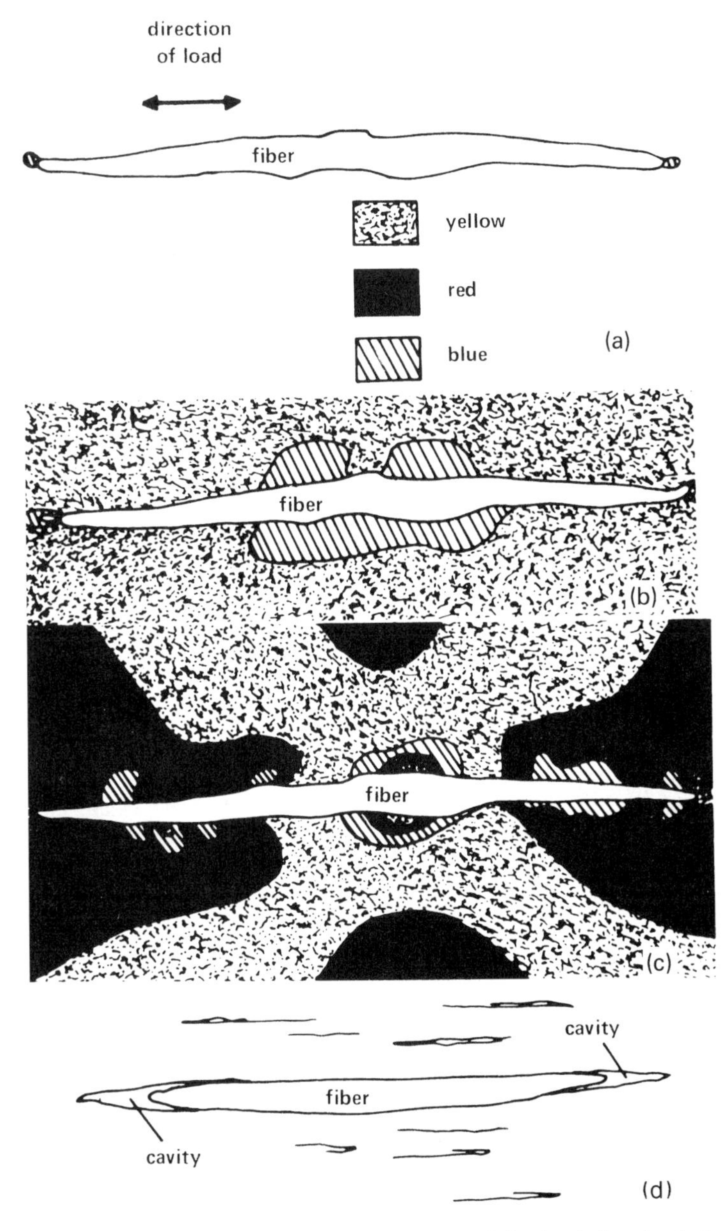

Fig. 93 Photostress patterns of a sequentially loaded polyethylene sample containing a single bark fiber. Stress levels are (a) zero stress, (b) 3.17 MPa, (c) 4.62 MPa, (d) 11.03 MPa. (From Ref. 162.)

REFERENCES

1. Abitz, P., and Luner, P. (1989). The effect of refining on wet fiber flexibility and its relationship to sheet properties. In: *Fundamentals of Papermaking* (Trans. 1989 Symp.), Vol. 1. C. F. Baker and V. Punton, eds. Mech. Eng. Pub., London, pp. 67–86.
2. Agarwal, B. D., and Broutman, L. J. (1980). *Analysis and Performance of Fiber Composites*. Wiley, New York.
3. Alexander, S. D., Marton, R., and McGovern, S. D. (1968). Effects of beating and wet pressing on fiber and sheet properties. 1. Individual fiber properties. *Tappi 51*(6):277–283.
4. Allison, R. W., Ellis, M. J., and Wrathall, S. H. (1998). Interaction of mechanical and chemical treatments on pulp strength during kraft pulp bleaching. *Appita J. 51*(2):107–113.
5. Am. Soc. for Testing and Materials. (1995). Standard D1445-95. Standard test for breaking strength and elongation of cotton fibers (flat bundle method).
6. Am. Soc. for Testing and Materials. (1998). Standard D 1577-96. Linear density of textile fibers.
7. Anagnost, S. E., Mark, R. E., and Hanna, R. B. (1999). Utilization of soft rot cavity formation as a tool for understanding the relation between microfibril angle and mechanical properties of cellulosic fibers. In: Proc. Symp. Mechanics of Cellulosic Materials 1999. AMD Vol. 231/MD Vol. 85. Am. Soc. Mech. Eng. New York, pp. 43–63. (Also, ESPRA Report 111, Chap. 11, 1999.)
8. Anagnost, S. E., Mark, R. E., and Hanna, R. B. (2000). Variation of microfibril angle within individual fibers. Empire State Paper Res. Inst. Report 112, Syracuse, NY, Chap. 5, pp. 53–68.
9. Anagnost, S. E., Mark, R. E., and Hanna, R. B. (2000). Utilization of soft-rot cavity orientation for the determination of microfibril angle. Part I. *Wood Fiber Sci. 32*(1):81–87.
10. Argyropoulos, D. S. (1999). *Advances in Lignocellulosics Characterization*. TAPPI Press, Atlanta.
11. Armstrong, J. P., Kyanka, G. H., and Thorpe, J. L. (1977). S2 fibril angle–elastic modulus relationship of TMP Scotch pine fibers. *Wood Sci. 10*(2):72–80.
12. Ashton, J. E., Halpin, J. C., and Petit, P. H. (1969). *Primer on Composite Materials: Analysis*. Technomic, Stamford, CT.
13. Astley, R. J., Harrington, J. J., Tang, S., and Neumann, J. (1998). Modelling the influence of microfibril angle on stiffness and shrinkage in radiata pine. In: *Microfibril Angle in Wood*. B. G. Butterfield, ed. Univ. Canterbury, Christchurch, N.Z., pp. 272–295.
14. Astley, R. J., Stol, K. A., and Harrington, J. J. (1998). Modelling the elastic properties of wood. Part 2: The cellular microstructure. *Holz Roh-Werkst. 56*(1):43–50.
15. Atalla, R. H., and VanderHart, D. L. (1984). Native cellulose: A composite of two distinct crystalline forms. *Science 223*:283–285.
16. Bailey, I. W., and Vestal, M. R. (1937). The significance of certain wood-destroying fungi in the study of enzymatic hydrolysis of cellulose. *J. Arnold Arbor. 18*: 196–205.
17. Baker, A. A., Helbert, W., Sugiyama, J., and Miles, M. J. (1997). High resolution atomic force microscopy of native Valonia cellulose I microcrystals. *J. Struct. Biol. 119*:129–138.
18. Balodis, V., McKenzie, A. W., Harrington, K. J., and Higgins, H. G. (1966). Effects of hydrophilic colloids and other non-fibrous materials on fibre flocculation and network consolidation. In: *Consolidation of the Paper Web*, Vol. 2. F. Bolam, ed. Tech. Sect. Brit. Paper and Board Makers' Assoc., London, pp. 637–691.

19. Barkas, W. W. (1953). Elasticity of the model cell. In: *Mechanical Properties of Wood and Paper*. R. Meredith, ed. North-Holland, Amsterdam, pp. 79–81.
20. Barnett, J. R., Chaffey, N. J., and Barlow, P. W. (1998). Cortical microtubules and microfibril angle. In: *Microfibril Angle in Wood*. B. G. Butterfield, ed. Univ. Canterbury, Christchurch, N.Z., pp. 253–271.
21. Barrett, J. D., and Schniewind, A. P. (1973). Three-dimensional finite-element models of cylindrical wood fibers. *Wood Fiber 5*(3):215–225.
22. Batchelor, W. (1999). Determination of load-bearing element length in paper using zero/short span tensile testing. Proc. 1999 Int. Paper Physics Conf. TAPPI Press, Atlanta, pp. 247–258.
23. Batra, S. K. (1998). Other long vegetable fibers: abaca, banana, sisal, henequen, flax, ramie, hemp, sunn and coir. In: *Handbook of Fiber Chemistry*. M. Lewin and E. M. Pearce, eds. Marcel Dekker, New York, pp. 505–575.
24. Bawden, A. D., and Kibblewhite, R. P. (1997). Effects of multiple drying treatments on kraft fibre walls. *J. Pulp Paper Sci. 23*(7):J340—J346.
25. Berg, J.-E., and Gradin, P. A. (1999). A micromechanical model of the deterioration of a wood fibre. *J. Pulp Paper Sci. 25*(2):66–71.
26. Bergander, A., and Salmén, L. (2000). The transverse elastic modulus of the native wood fibre wall. *J. Pulp Paper Sci. 26*(6):234–238 and erratum *26*(9):316.
27. Berlyn, G. P. (1969). Microspectrophotometric investigation of free space in plant cell walls. *Am. J. Bot. 56*:498–506.
28. Berlyn, G. P., and Miksche, J. P. (1976). *Botanical Microtechnique and Cytochemistry*. Iowa State Univ. Press, Ames.
29. Binotto, A. P., and Nicolls, G. A. (1977). Correlation of fiber morphological variation and wet mat compressibility of loblolly pine bleached kraft pulp. *Tappi 60*(6):91–94.
30. Bodig, J., and Jayne, B. A. (1993). *Mechanics of Wood and Wood Composites*. Reprint ed. Krieger, Melbourne, FL.
31. Boresi, A. P., Schmidt, R. J., and Sidebottom, O. M. (1993). *Advanced Mechanics of Materials*. 5th ed. Wiley, New York, pp. 362–391.
32. Boucai, E. (1971). Zero span tensile test and fibre strength. *Pulp Paper Mag. Can. 72*(10):73–76.
33. Browning, B. L. (1977). *Analysis of Paper*. Marcel Dekker, New York.
34. Brunnberg, I., Kubát, J., and Soderlund, G. (1969). Tensile testing device for measurements of polymeric materials between -100 and $+150^{\circ}$C. Mechanical behavior of cellulose in liquid media. *J. Appl. Polym. Sci. 13*:571–576.
35. Button, A. F. (1979). Fiber-fiber bond strength: A study of a linear elastic model structure. Ph.D. Thesis. Inst. Paper Chem., Appleton, WI.
36. Byrd, V. L. (1972). Effect of relative humidity changes during creep on handsheet paper properties. *Tappi 55*(2):247–252.
37. Byrd, V. L. (1972). Effect of relative humidity changes on compressive creep response of paper. *Tappi 55*(11):1612–1613.
38. Byrd, V. L., and Koning, J. W. (1978). Corrugated fiberboards: Edgewise compression creep in cyclic relative humidity environments. *Tappi 61*(6):35–37.
39. Carlsson, L. (1986). The layered structure of paper. In: *Paper: Structure and Properties*. J. A. Bristow and P. Kolseth, eds. Marcel Dekker, New York, pp. 347–363.
40. Carlsson, L., and Salmén, L. (1986). Basic relations for laminated orthtropic plates. In: *Paper: Structure and Properties*. J. A. Bristow and P. Kolseth, eds. Marcel Dekker, New York, pp. 369–375.
41. Cave, I. D. (1968). The anisotropic elasticity of the plant cell wall. *Wood Sci. Tech. 2*:268–278.
42. Cave, I. D. (1969). The longitudinal Young's modulus of Pinus radiata. *Wood Sci. Tech. 3*:40–48.

43. Cave, I. D. (1976). Modelling the structure of the softwood cell wall for computation of mechanical properties. *Wood Sci. Tech. 10*:19–28.
44. Cave, I. D. (1978). Modelling moisture-related mechanical properties of wood. 2. Computation of properties of a model of wood and a comparison with experimental data. *Wood Sci. Tech. 12*:127–139.
45. Cave, I. D. (1978). Modelling moisture-related mechanical properties of wood. 1. Properties of the wood constituents. *Wood Sci. Tech. 12*:75–86.
46. Chanzy, H., and Henrissat, B. (1985). Unidirectional degradation of Valonia cellulose microcrystals subjected to cellulase action. *FEBS Lett. 184*(2):285–288. (Original not seen.)
47. Chao, Y. J., and Sutton, M. A. (1990). Computer vision in engineering mechanics. In 1986 N.S.F. Workshop on *Solid Mechanics Advances in Paper Related Industries*. R. W. Perkins, R. E. Mark, and J. L. Thorpe, eds. Mech. Aerospace Eng. Dept., Syracuse Univ., pp. 286–323.
48. Chiao, T. T., and Moore, R. L. (1970). A tensile test method for fibers. *J. Compos. Mater. 4*:118–123.
49. Choi, D., Thorpe, J. L., and Hanna, R. B. (1991). Image analysis to measure strain in wood and paper. *Wood Sci. Tech. 25*(4):251–262.
50. Clarke, B., Ebeling, K. J., and Kropholler, H. W. (1985). Fibre coarseness: A new method for its characterization. *Paperi Puu 67*(9):490–499.
51. Claudio da Silva, E. (1983). The flexibility of pulp fibers: A structural approach. *Proc. 1983 Int. Paper Phys. Conf.* TAPPI Press, Atlanta, pp. 13–25.
52. Coffin, D. W., and Boese, S. B. (1997). Tensile creep behavior of single fibers and paper in a cyclic humidity environment. *Symp. on Moisture and Creep Effects on Paper and Containers.* PAPRO, Rotorua, N.Z. pp. 39–52.
53. Çöpür, Y., and Makkonen, H. (2000). Prediction of pulping yield from fiber dimensions and properties. Empire State Paper Res. Inst. Rep. 112, pp. 13–21.
54. Côté, W. A. (1980). *Papermaking Fibers: A Photomicrographic Atlas.* Renewable Materials Inst., SUNY College Environ. Sci. and Forestry and Syracuse Univ. Press, Syracuse, NY.
55. Côté, W. A., and Hanna, R. B. (1983). Ultrastructural characteristics of wood fracture surfaces. *Wood Fiber Sci. 25*(2):135–163.
56. Cousins, W. J. (1976). Elastic modulus of lignin as related to moisture content. *Wood Sci. Tech. 10*:9–17.
57. Cousins, W. J. (1977). Elasticity of isolated lignin; Young's modulus by a continuous indentation method. *N.Z. J. Forest Sci. 7*:107–112.
58. Cousins, W. J. (1978). Young's modulus of hemicellulose as related to moisture content. *Wood Sci. Tech. 12*:161–167.
59. Cowan, W. F. (1986). Zero/short span tensile testing can determine basic paper properties. *Pulp Paper 60*(5):84–86.
60. Cowan, W. F. (1988). Preprints, CPPA Annual Meeting. Tech. Sect. CPPA, Montreal, p. A149.
61. Cowan, W. F. (1994). Testing pulp quality: An alternative to conventional laboratory evaluation. *Tappi J. 77*(10):77–81.
62. Cowan, W. F. (1995). High-shear laboratory beating and fiber-quality testing offer new insights into pulp evaluation. *Tappi J. 78*(3):133–137.
63. Cowan, W. F. (1995). Explaining handsheet tensile and tear in terms of fiber-quality numbers. *Tappi J. 78*(1):101–106.
64. Cowan, W. F., and Cowdrey, E. J. K. (1974). Evaluation of paper strength components by short-span tensile analysis. *Tappi 57*(2):90–93.
65. Cowdrey, D. R., and Preston, R. D. (1966). Elasticity and microfibrillar angle in the wood of Sitka spruce. *Proc. Roy. Soc. (Lond.) B 166*:245–272.

66. Crossley, A., and Levy, J. F. (1977). Proboscic hyphae in soft rot cavity formation. *J. Inst. Wood Sci. 7*:30–33.
67. Davison, R. W. (1972). The weak link in paper dry strength. *Tappi 55*(4):567–573.
68. de Gruy, I. V., Carra, J. H., and Goynes, W. R. (1973). In: *The Fine Structure of Cotton: An Atlas of Cotton Microscopy*, Vol. 6. Fiber Sci. Ser. R. O'Connor, ed. Marcel Dekker, New York, pp. 171–225.
69. Denis, E. S., and Parker, I. H. (1995). Creep and dynamic testing of eucalypt paper and fibres in changing R. H. *Proc. Int. Paper Physics Conf., Niagara-on-the-Lake, Ontario.* Tech. Sect. Can. Pulp Paper Assoc., Montreal, pp. 143–147.
70. de Ruvo, A., Lundberg, R., Martin-Löf, S., and Söremark, C. (1976). Influence of temperature and humidity on the elastic and expansional properties of paper and the constituent fibre. In: *The Fundamental Properties of Paper Related to Its Uses*, Vol. 2. F. Bolam, ed. Tech. Div. Brit. Paper Board Ind. Fed., London, pp. 785–810.
71. De Teresa, S. J., Porter, R. S., and Farris, R. J. (1985). A model for the compressive buckling of extended-chain polymers. *J. Mater. Sci. 20*(5):1645–1659.
72. DeTeresa, S. J., Allen, S. R., Farris, R. J., and Porter, R. S. (1984). Compressive and torsional behavior of Kevlar 49 fibre. *J. Mater. Sci. 19*(1):57–72.
73. DiBenedetto, A. T. (1981). Evaluation of fiber adhesion in composites. In: *Adhesion in Cellulosic and Wood-Based Composites*. J. F. Oliver, ed. Plenum Press, New York, pp. 113–125.
74. Dinwoodie, J. M. (1965). The relationship between fiber morphology and paper properties: A review of literature. *Tappi 48*(8):440–447.
75. Dinwoodie, J. M. (1978). Failure in timber. 3. The effect of longitudinal compression on some mechanical properties. *Wood Sci. Tech. 12*:271–285.
76. Donaldson, L. A. (1998). Between-tracheid variability of microfibril angles in radiata pine. In: *Microfibril Angle in Wood.* B. G. Butterfield, ed. Univ. Canterbury, Christchurch, N.Z., pp. 206–224.
77. Dumbleton, D. P. (1972). Longitudinal compression of individual pulp fibers. *Tappi 55*(1):127–135.
78. Duncker, B., Hartler, N., and Samuelsson, L. G. (1966). Effect of drying on the mechanical properties of pulp fibers. In: *Consolidation of the Paper Web*, Vol. 1. F. Bolam, ed. Tech. Sect. Brit. Paper Board Makers' Assoc., London, pp. 529–537.
79. Ehrnrooth, E. M. L. (1982). Softening and mechanical behaviors of single wood pulp fibres: The influence of matrix composition and chemical and physical characteristics. Ph.D. Thesis, Univ. Helsinki.
80. Ehrnrooth, E. M. L., Kolseth, P., and de Ruvo, A. (1978). The influence of matrix composition and softening on the mechanical behavior of cellulosic fibers. In: *Fiber-Water Interactions in Paper-Making*, Vol. 2. F. Bolam, ed. Trans. Symp. British Paper and Board Ind. Fed., London, pp. 715–738.
81. Ehrnrooth, E. M. L., and Kolseth, P. (1984). The tensile testing of single wood pulp fibers in air and in water. *Wood Fiber Sci. 16*(4):549–566.
82. Eissa, Y. Z., Naito, T., Usuda, M., and Kadoya, T. (1983). Flow resistance of rotating dilute fibre suspensions. In: *The Role of Fundamental Research in Papermaking* (Trans. 1981 Symp.), Vol. 1. J. Brander, ed. Mech. Eng. Pub., London, pp. 309–324.
83. El-Hosseiny, F., and Bennett, K. (1985). Analysis of the zero-span tensile strength of paper. *J. Pulp Paper Sci. 11*(4):J121–J127.
84. El-Hosseiny, F., and Page, D. H. (1973). The measurement of fibril angle of wood fibers using polarized light. *Wood Fiber 5*(3):208–214.
85. El-Hosseiny, F., and Page, D. H. (1975). The mechanical properties of single wood pulp fibres: Theories of strength. *Fibre Sci. Tech. 8*:21–31.
86. Ellis, M. J., Duffy, G. G., Allison, R. W., and Kibblewhite, R. P. (1998). Fibre deformation during medium consistency mixing; role of residence time and impeller geometry. *Appita J. 51*(1):29–34.

87. El-osta, M. L., and Wellwood, R. W. (1972). Short-term creep as related to microfibril angle. *Wood Fiber 4*(1):26–32.
88. Evans, R., Newman, R. H., Roick, U. C., Suckling, I. D., and Wallis, A. F. A. (1995). Changes in cellulose crystallinity during kraft pulping. Comparison of infrared, x-ray diffraction and solid state NMR results. *Holzforschung 49*(6):498–504.
89 Falk, O., Kubát, J., and Rigdahl, M. (1978). Damping behavior of polymeric materials subjected to longitudinal loads. *J. Mater. Sci. 13*(11):2328–2332.
90. Fellers, C., Wellmar, P., and Kolseth, P. (1995). Unified symbols for expressing properties. Paper. Part 1. Mechanical properties. STFI Rep. P002.
91. Finkenstadt, V. L., and Millane, R. P. (1998). Crystal structure of Valonia cellulose Iβ. *Macromolecules 31*:7776–7783.
92. Food and Agriculture Organization, United Nations (1975). Pulping and Papermaking Properties of Fast-Growing Plantation Wood Species. FO MISC/75/31, Rome, Italy.
93. Forgacs, O. L. (1961). Structural weaknesses in softwood pulp tracheids. *Tappi 44*(2):112–119.
94 Forgacs, O. L., and Mason, S. G. (1958). The flexibility of wood-pulp fibers. *Tappi 41*(11):695–704..
95. Frey-Wyssling, A. (1955). On the crystal structure of cellulose I. *Biochim. Biophys. Acta 18*:166–168.
96. Frey-Wyssling, A., Mühlethaler, K., and Muggli, R. (1966). Elementary fibrils as structural elements of native cellulose. *Holz Roh Werkst. 24*:443–444 (in German).
97. Fuji, T., Harada, H., and Saiki, H. (1980). The layered structure of secondary walls in axial parenchyma of the wood of 51 Japanese angiosperm species. *J. Jpn. Wood Res. Soc. (Mokuzai Gakkaishi) 26*(6):373–380.
98. Fujita, M., and Harada, H. (1991). Ultrastructure and formation of wood cell wall. In: *Wood and Cellulosic Chemistry*. D. N.-S. Hon and N. Shiraishi, eds. Marcel Dekker, New York, pp. 3–57.
99. Fukunaga, H., and Goda, K. (1991). Tensile strength of Nicalon SiC fibers subjected to torsional strain. *J. Mater. Sci. Lett. 10*(3):179–180.
100. Furukawa, I., Mark, R. E., and Perkins, R. W. (1985). Inelastic behavior of a machine-made paper in relation to internal structural changes. Empire State Paper Res. Inst. Rep. 82, Syracuse, NY, pp. 79–117.
101. Furukawa, I., Saiki, H., and Harada, H. (1973). Continuous observation of tensile fracture process of single tracheid by scanning electron microscope. *J. Jpn. Wood Res. Soc. (Mokuzai Gakkaishi) 19*(8):399–402.
102. Gardner, K. H., and Blackwell, J. (1971). The substructure of crystalline cellulose and chitin microfibrils. *J. Polym. Sci. C 36*:327–340.
103. Gardner, K. H., and Blackwell, J. (1974). The structure of native cellulose. *Biopolymers 13*:1975–2001.
104. Gillis, P. P. (1969). Effect of hydrogen bonds on the axial stiffness of crystalline native cellulose. *J. Polym. Sci. A-2 7*:783–794.
105. Gillis, P. P. (1970). Elastic moduli for plane stress analyses of unidirectional composites with anisotropic rectangular reinforcement. *Fibre Sci. Tech. 2*:193–210.
106. Gillis, P. P., and Mark, R. E. (1973). Analysis of shrinkage, swelling and twisting of pulp fibers. *Cellulose Chem. Tech. 7*(2):209–234.
107. Gillis, P. P., Mark, R. E., and Tang, R.-C. (1969). Elastic stiffness of crystalline cellulose in the folded-chain solid state. *J. Mater. Sci. 4*:1003–1007.
108. Görres, J., Amiri, R., Grondin, M., and Wood, J. R. (1993). Fibre collapse and sheet structure. In.: *Products of Papermaking*. C. F. Baker, ed. Pira Int., Leatherhead, Surrey, UK, pp. 285–310.

109. Groom, L. H., Shaler, S. M., and Mott, L. (1995). Characterizing micro- and macromechanical properties of single wood fibers. *Proc. 1995 Int. Paper Physics Conf.* Tech. Sect., CPPA, Montreal, pp. 13–22.
110. Gurnagul, N., and Page, D. H. (1989). The difference between dry and rewetted zero-span tensile strength of paper. *Tappi J. 72*(12):164–167.
111. Gurnagul, N., Page, D. H., and Paice, M. G. (1992). The effect of cellulose degradation on the strength of wood pulp fibres. *Nordic Pulp Paper Res. J. 7*(3):152–154.
112. Habeger, C. C., and Coffin, D. W. (2000). The role of stress concentrations in accelerated creep and sorption-induced physical aging. *J. Pulp Paper Sci. 26*(4):145–157.
113. Haigler, C. H. (1985). The functions and biogenesis of native cellulose. In: *Cellulose Chemistry and Its Applications.* T. P. Nevell and S. H. Zeronian, eds. Halsted/Wiley, New York, pp. 30–83.
114. Hakanen, A., and Hartler, N. (1995). Fiber deformations and strength potential of kraft pulps. *Paperi Puu 77*(5):339–344.
115. Hale, M. D. C., and Eaton, R. A. (1983). Soft rot decay of wood: The infection and cavity-forming process of Phialophora hoffmannii (van Beyma) Schol-Schwarz. In: *Biodeterioration 5.* T. A. Oxley and S. Barry, eds. Wiley-Interscience, Chichester, UK, pp. 54–63.
116. Hamad, W. Y. (1997). Some microrheological aspects of wood-pulp fibers subjected to fatigue loading. *Cellulose 4*:51–56.
117. Hamad, W. Y. (1998). On the mechanisms of cumulative damage and fracture in native cellulose fibres. *J. Mater. Sci. Lett. 17*(5):433–436.
118. Hamad, W. Y. (2000). A non-contact microscopic technique for fibre strain measurement in non-woven materials. (Submitted for publication.)
119. Hamad, W. Y., and Provan, J. W. (1995). Microstructural cumulative material degradation and fatigue-failure micromechanisms in wood-pulp fibres. *Cellulose 2*:159–177.
120. Hamad, W. Y., and Rokbaa, H. (1998). Mechanics of deformation in wood-pulp fibres and fibrous networks. Paper A7. Presented at 1998 Progress in Paper Physics Seminar, Univ. British Columbia, Vancouver, B.C.
121. Hanna, R. B. (1973). Cellulose sub-elementary fibrils. Ph.D. Thesis, State Univ. New York, College of Environ. Sci. and Forestry, Syracuse, NY.
122. Hanna, R. B., and Côté, W. A., Jr. (1974). The sub-elementary fibril of plant cell wall cellulose. *Cytobiologie 10*:102–116.
123. Harada, H. (1962). Electron microscopy of ultrathin sections of beech wood (Fagus crenata Blume). *J. Jpn. Wood Res. Soc. (Mokuzai Gakkaishi 8*(6):253–358.
124. Harada, H. (1965). Ultrastructure and organization of gymnosperm cell walls. In: *Cellular Ultrastructure of Woody Plants.* W. A. Côté, Jr., ed. Syracuse Univ. Press, Syracuse, NY, pp. 215–234.
125. Hardacker, K. W. (1962). The automatic recording of the load-elongation characteristic of single papermaking fibers. IPC fiber load-elongation recorder. *Tappi 45*(3):237–246.
126. Hardacker, K. W. (1969). Cross-sectional area measurement of individual wood pulp fibers by lateral compaction. *Tappi 52*(9):1742–1746.
127. Hardacker, K. W. (1970). Effects of loading rate, span and beating on individual wood fiber tensile properties. In: *The Physics and Chemistry of Wood Pulp Fibers.* D. H. Page, ed. TAPPI STAP No. 8, pp. 201–216.
128. Harrington, J., Booker, R., and Astley, R. J. (1998). Modelling the elastic properties of softwood. Part 1. The cell-wall lamellae. *Holz Roh-u-Werkst. 56*(1):37–41.
129. Harrington, K. J. (1970). Measurement of the lateral compression of single wood pulp fibres. *Appita 23*(5):365–367.
130. Hartler, N., and Nyrén, J. (1970). Transverse compressibility of pulp fibres. 2. Influence of cooking method, yield, beating, and drying. *Tappi 53*(5):820–823.

131. Hartler, N., and Nyrén, J. (1970). Influence of pulp type and post-treatments on the compressive force required for collapse. In: *The Physics and Chemistry of Wood Pulp Fibers*. STAP No. 8. D. H. Page, ed. Tech. Assoc. Pulp Paper Ind., pp. 265–277.
132. Hartler, N., Kull, G., and Stockman, L. (1963). Determination of fibre strength through measurement of individual fibers. *Svensk Papperstidn. 66*(8):301–308.
133. Hasuike, M. (1973). On the physical properties and macrostructure of earlywood and latewood pulp sheets. *J. Jpn. Wood Res. Soc. (Mokuzai Gakkaishi) 19*(11):547–553 (in Japanese; Engl. Sum.)
134. Hattula, T., and Niemi, H. (1988). Sulphate pulp fibre flexibility and its effect on sheet strength. *Paperi Puu 70*(4):356–361.
135. Haugen, P., and Young, J. H. (1970). Rheology of single wood pulp fibres under static bending stresses. Effects of chemical processing, composition and physical structure. In: *The Physics and Chemistry of Wood Pulp Fibers*. STAP No. 8. D. H. Page, ed. Tech. Assoc. Pulp Paper Ind., pp. 242–264.
136. Hieta, K., Kuga, S., and Usada, M. (1984). Electron staining of reducing ends evidences a parallel-chain structure in Valonia cellulose. *Biopolymers 23*:1807–1810.
137. Hill, R. (1964). Theory of mechanical properties of fibre-strengthened materials. 1. Elastic behavior *J. Mech. Phys. Solids 12*:199–212.
138. Hill, R. L. (1967). The creep behavior of individual pulp fibers under tensile stress. *Tappi 50*(8):432–440.
139. Hirakawa, Y., Yamashita, K., Fujisawa, Y., Nakada, R., and Kijidani, Y. (1998). The effects of S2 microfibril angles and density on MOE in sugi tree logs. In: *Microfibril Angle in Wood*. B. G. Butterfield, ed. Univ. Canterbury, Christchurch, N.Z., pp. 312–322.
140. Huang, C.-L., Kutscha, N. P., Leaf, G. J., and Megraw, R. A. (1998). Comparison of microfibril angle measurement techniques. In: *Microfibril Angle in Wood*. B. G. Butterfield, ed. Univ. Canterbury, Christchurch, N.Z., pp. 177–205.
141. Ifju, G. (1964). Tensile strength behavior as a function of cellulose in wood. *Forest Prod. J. 14*(8):366–372.
142. Ilvessalo-Pfäffli, M.-S. (1995). *Fiber Atlas: Identification of Papermaking Fibers*. Springer-Verlag, Berlin.
143. Ionides, G. N., and Mitchell, J. G. (1978). The capabilities and limitations of short-span tensile characterization of pulp strengths. *Paperi Puu 60*(4a):233–238.
144. Iribarne, J. (1999). Strength loss in kraft pulping. Ph.D. Thesis, SUNY College Environ. Sci. and Forestry, Syracuse, NY. (Available from UMI Dissertation Services, Ann Arbor, MI).
145. Iribarne, J., and Schroeder, L. R. (1999). Investigations on the zero-span tensile test. Rep. 111. Empire State Paper Res. Inst., Syracuse, NY, pp. 63–79.
146. Isenberg, I. H. (1967). *Pulp and Paper Microscopy*. 3rd ed. Institute of Paper Chemistry, Appleton, WI.
147. Ishikawa, A., Okano, T., and Sugiyama, J. (1997). Fine structure and tensile properties of ramie fibres in the crystalline form of cellulose I, II, III_I and IV_I. *Polymer 38*:463–468.
148. Jackson, T., and Parker, I. (1997). Accelerated creep in rayon fibres? *Symp. on Moisture and Creep Effects on Paper and Containers*. PAPRO, Rotorua, N.Z., pp. 53–67.
149. James, W. L. (1973). A method for studying the stiffness and internal friction of individual fibers. *Wood Sci 6*(1):30–38.
150. Jang, H. F. (1998). Measurement of fibril angle in wood fibers with polarization confocal microscopy. *J. Pulp Paper Sci. 24*(7):224–230. Presented at 1998 Progress in Paper Physics Seminar.
151. Jang, H. F., Amiri, R., Seth, R. S., and Karnis, A. (1996). Fiber characterization using confocal microscopy: Collapse behavior of mechanical pulp fibers. *Tappi J 79*(4):203–210.

152. Jang, H. F., and Seth, R. S. (1998). Using confocal microscopy to characterize the collapse behavior of fibers. *Tappi J. 81*(5):167–174.
153. Jang, H. F., Howard, R. C., and Seth, R. S. (1995). Fiber characterization using confocal microscopy: The effects of recycling. *Tappi J. 78*(12):131–137.
154. Jang, H. F., Robertson, A. G., and Seth, R. S. (1991). Optical sectioning of pulp fibers using confocal scanning laser microscopy. *Tappi J. 74*(10):217–219.
155. Jang, H. F., Robertson, A. G., and Seth, R. S. (1991). Optical sectioning of pulp fibers with confocal scanning laser microscopy. *Proc. 1991 Int. Paper Phys. Conf.* TAPPI Press, Atlanta, GA, pp. 277–280.
156. Jang, H. F., Robertson, A. G., and Seth, R. S. (1992). Transverse dimensions of wood pulp fibres by confocal laser scanning microscopy and image analysis. *J. Mater. Sci. 27*:6391–6400.
157. Janssen, J. J. A. (1981). Bamboo in building structures. D. Tech. Sci. Thesis, Eindhoven Univ. Printed by Dissertatie Drukkerij, Helmond, Netherlands.
158. Jaswon, M. A., Gillis, P. P., and Mark, R. E. (1968). The elastic constants of crystalline native cellulose. *Proc. Roy. Soc. (Lond.) A 306*:389–412.
159. Jayne, B. A. (1959). Mechanical properties of wood fibers. *Tappi 42*(6):461–467.
160. Jeffries, R., Jones, D. M., Roberts, J. G., Selby, K., Simmens, S. C., and Warwicker, J. O. (1969). Current ideas on the structure of cotton. *Cellulose Chem. Tech. 3*: 255–274.
161. Jentzen, C. A. (1964). The effect of stress applied during drying on some of the properties of individual fibers. *Tappi 47*(7):412–418.
162. Johnson, J. A., and Nearn, W. T. (1972). Reinforcement of polymeric systems with Douglas-fir bark fibers. In: *Theory and Design of Wood and Fiber Composite Materials.* B. A. Jayne, ed. Syracuse Univ. Press, Syracuse, NY, pp. 371–400.
163. Jones, R. M. (1999). *Mechanics of Composite Materials.* 2nd ed. Taylor & Francis, Philadelphia.
164. Jordan, B. D., and Page, D. H. (1979). Application of image analysis to pulp fibre characterization: Part I. *Proc. Int. Paper Physics Conf.* Tech. Sect. Can. Pulp Paper Assoc., Montreal, pp. 105–114.
165. Jordan, B. D., Nguyen, N. G., and Page, D. H. (1982). An image analysis procedure for pulp coarseness determinations. *Paperi Puu 64*(11):691–694, 701.
166. Kajanto, I. M. (1996). Optimal measurement of dimensional stability. *Progress in Paper Physics Seminar.* STFI, Stockholm, pp. 75–77.
167. Kappil, M. O. (1994). Effect of drying restraint and recycling on mechanical properties of fibers in machine-made paper. M.S. Thesis, SUNY, College of Environ. Sci. and Forestry, Syracuse, NY.
168. Kappil, M. O., Mark, R. E., Perkins, R. W., and Holtzman, W. (1995). Fiber properties in machine-made paper related to recycling and drying tension. In: *Proc. Symp. Mechanics of Cellulosic Materials.* AMD Vol. 209, MD Vol. 60. Am. Soc. Mech. Eng., New York, pp. 177–194.
169. Karlsson, H., and Fransson, P.-I. (1994). STFI FiberMaster gives papermakers new muscles. *Svensk Papperstidn. 97*(10):26–28 (in Swedish).
170. Karlsson, H., Fransson, P.-I., and Mohlin, U.-B. (1999). STFI FiberMaster. Presented at SPCI 99. *6th Int. Conf. on New Available Technologies.* Swedish Assoc. of Pulp and Paper Eng., Stockholm, pp. 367–374.
171. Katsura, T., Murakami, K., and Imamura, R. (1978). Single fiber extraction test using long fiber sheets, *J. Jpn. Wood Res. Soc. (Mokuzai Gakkaishi) 24*(8):552–557 (in Japanese).
172. Kerekes, R. J. (1998). Personal communication.
173. Kerekes, R. J., and Tam Doo, P. A. (1985). Wet fibre flexibility of some major softwood species pulped by various processes. *J. Pulp Paper Sci. 11*(2):J60–J61.

174. Kersavage, P. C. (1973). A System for Automatically Recording the Load-Elongation Characteristics of Single Wood Fibers Under Controlled Relative Humidity Conditions. Bulletin 790. Pennsylvania State Univ. Agr. Exp. Sta.
175. Kettunen, H., and Niskanen, K. (1999). On the relationship of fracture energy to fiber debonding. *Proc. Intl. Paper Physics Conf.* TAPPI Press, Atlanta, pp. 125–143.
176. Kettunen, H., and Niskanen, K. (2000). Microscopic damage in paper. Part 1. Method of analysis. *J. Pulp Paper Sci. 26*(1):35–40.
177. Kettunen, H., Yu, Y., and Niskanen, K. (2000). Microscopic damage in paper. Part 2. Effect of fibre properties. *J. Pulp Paper Sci. 26*(7):250–265.
178. Kibblewhite, R. P. (1976). Fractures and dislocations in the walls of kraft and bisulphate pulp fibres. *Cellulose Chem. Technol. 10*(4):497–503.
179. Kibblewhite, R. P., and Bailey, D. G. (1987). Measurement of fibre cross-section dimensions using image processing. *Proc. Int. Paper Physics Conf.* Tech. Sect. Can. Pulp Paper Assoc., Montreal, pp. 25–34.
180. Kim, Y. S., and Newman, R. H. (1995). Solid state ^{13}C NMR study of wood degraded by the brown rot fungus Gloephyllum trabeum. *Holzforschung 49*(2):109–114.
181. Klop, E. A., and Lammers, M. (1998). XRD study of the new rigid-rod polymer fibre PIPD. *Polymer 39*(24):5987–5998.
182. Knutsson, T., and Stockman, L. (1958). The effect of mechanical treatment during the final stages of the cook on the beating properties and strength of sulphate pulp. *Tappi J. 41*(11):704–709.
183. Koashi, K., Mori, H., and Kyogoku, Y. (1974). Properties and structure of natural cellulosic fibers. *Sen-I Gakkaishi, J. Soc. Fiber Sci. Tech. (Jpn) 49*(10):92 (in Japanese).
184. Kolseth, P. (1983). Torsional properties of single wood pulp fibers. D.Tech. Thesis, The Royal Institute of Technology, Stockholm.
185. Kolseth, P., and de Ruvo, A. (1978). An attempt to measure the torsional strength of single wood pulp fibers. *General Constitutive Relations for Wood and Wood-Based Materials.* NSF Workshop Report. Syracuse Univ. Syracuse, NY, pp 57–79.
186. Kolseth, P., and Ehrnrooth, E. M. L. (1986). Mechanical softening of single wood pulp fibers. In: *Paper Structure and Properties.* J. A. Bristow and P. Kolseth, eds. Marcel Dekker, New York, pp. 27–50.
187. Kolseth, P., and Salmén, L. (1986). Relations involving the torsional rigidity of single fibers. In: *Paper, Structure and Properties.* J. A. Bristow and P. Kolseth, eds. Marcel Dekker, New York, pp. 367–368.
188. Kolseth, P., de Ruvo, A., and Tulonen, J. (1983). A torsion pendulum for single wood pulp fibers. (24-page paper bound into *Torsional Properties of Single Wood Pulp Fibers*, D. Tech. Thesis of P. Kolseth. Royal Inst. of Technology, Stockholm.
189. Koponen, S., Toratti, T., and Kanerva, P. (1988). Modeling mechanical properties of wood based on cell structure. (In Finnish). Report No. 5, Laboratory of Structural Engineering and Building Physics, Helsinki Univ. Tech. (Original not seen).
190. Koponen, S., Toratti, T., and Kanerva, P. (1989). Modelling longitudinal elastic and shrinkage properties of wood. *Wood Sci. Tech. 23*:55–63.
191. Koponen, S., Toratti, T., and Kanerva, P. (1991). Modelling elastic and shrinkage properties of wood based on cell structure. *Wood Sci. Tech. 25*:25–32.
192. Korteoja, M. J., Lukkarinen, A., Kaski, K., Gunderson, D. E., Dahlke, J. L., and Niskanen, K. J. (1996). Local strains in paper. *Tappi J. 79*(4):218–223.
193. Korteoja, M. J., Niskanen, K. J., Kortschot, M. T., and Kaski, K. (1998). Progressive damage in paper. *Paperi Puu 80*(5):364–372.
194. Koyama, M., Sugiyama, J., and Itoh, T. (1997). Systematic survey on crystalline features of algal celluloses. *Cellulose 4*:147–150.

195. Kroon-Batenburg, L. M. J., Bouma, B., and Kroon, J. (1996). Stability of cellulose structures studied by MD simulations. Could mercerized cellulose II be parallel? *Macromolecules 29*:5695–5699.
196. Kroon-Batenburg, L. M. J., Kroon, J., and Northolt, M. G. (1986). Chain modulus and intramolecular hydrogen bonding in native and regenerated cellulose fibres. *Polym. Commun. 27*(10):290–292.
197. Kubát, J. (1965). Stress relaxation in solids. *Nature 205*:378–379.
198. Kuhn, D. C. S., Lu, X., Olson, J. A., and Robertson, A. G. (1995). A dynamic wet fibre flexibility measurement device. *J. Pulp Paper Sci. 21*(10):J337–J342.
199. Laamanen, J. S. (1983). Morphological fiber characteristics in mechanical pulps and their relation to paper properties. *Proc. 1983 Int. Paper Phys. Conf.* TAPPI Press, Atlanta, pp. 1–11.
200. Lammers, M., Klop, E. A., Northolt, M. G., and Sikkema, D. J. (1998). Mechanical properties and structural transitions in the new rigid-rod polymer fibre PIPD (M5) during the manufacturing process. *Polymer 39*(24):5999–6005.
201. Lawryshyn, Y. A., and Kuhn, D. C. S. (1996). Large deflection analysis of wet fiber flexibility measurement techniques. *J. Pulp Paper Sci. 22*(11):J423–J431.
202. Lennholm, H., Wallbacks, L., and Iverson, T. (1995). A ^{13}C-CP/MAS-NMR-spectroscopic study of the effect of laboratory kraft cooking on cellulose structure. *Nordic Pulp Paper Res. J. 10*(1):46–50.
203. Leopold, B., and McIntosh, D. C. (1961). Chemical composition and physical properties of wood fibers. 3. Tensile strength of individual fibers from alkali extracted loblolly pine holocellulose. *Tappi 44*(3):235–240.
204. Levi, M. P., and Preston, R. D. (1965). A chemical and microscope examination of the action of the soft-rot fungus Chaetomium globosum on beechwood (Fagus sylvatica). *Holzforschung 19*:183–190.
205. Levlin, J.-E., and Söderhjelm, L. (1999). *Pulp and Paper Testing.* Book 17 of Papermaking Sci. Tech. Series. Fapet Oy, Helsinki.
206. Li, M., Johnston, R., Xu, L., Filonenko, Y., and Parker, I. (1999). Characterization of hydrocyclone-separated eucalypt fibre fractions. *J. Pulp Paper Sci. 25*(8):299–304.
207. Lif, J. O., Fellers, C., Söremark, C., and Sjödahl, M. (1995). Characterizing the in-plane hygroexpansivity of paper by electronic speckle photography. *J. Pulp Paper Sci. 21*(9):J302–J309.
208. Luce, J. E. (1970). Transverse collapse of wood pulp fibers: Fiber models. In: *The Physics and Chemistry of Wood Pulp Fibers.* STAP No. 8. D. H. Page, ed. Tech. Assoc. Pulp Paper Ind., pp. 278–281.
209. Lyons, W. (1959). Theoretical value of the dynamic stretch modulus of cellulose. *J. Appl. Phys. 30*(5):796–797.
210. MacLeod, J. M. (1986). Kraft pulps from Canadian wood species. *Pulp Paper Can. 87*(1):T25–T30.
211. MacLeod, J. M., and Pelletier, L. J. (1987). Basket cases: Kraft pulps inside digesters. *Tappi J. 70*(11):47–53.
212. Marhöfer, R. J., Reiling, S., and Brickmann, J. (1995). Computer simulations of crystal structures and elastic properties of cellulose. *Ber. Bunsen. Gesell. Phys. Chem. 100*(8):1350–1354.
213. Mark, R. E. (1965). Treatise on the tensile strength of tracheids. D.For. Dissert., Yale Univ., New Haven, CT.
214. Mark, R. E. (1967). *Cell Wall Mechanics of Tracheids.* Yale Univ. Press, New Haven, CT.
215. Mark, R. E. (1971). Mechanical behavior of cellulose in relation to cell wall theories. *J. Polym. Sci. C. 36*:391–406.

216. Mark, R. E. (1972). Mechanical behavior of the molecular components of fibers. In: *Theory and Design of Wood and Fiber Composite Materials*. B. A. Jayne, ed. Syracuse Univ. Press, Syracuse, NY, pp. 49–82.
217. Mark, R. E. (1973). On the transverse tangential strength of wood cell walls. *Wood Fiber 4*:347–349.
218. Mark, R. E. (1980). Molecular and cell wall structure of wood. *J. Educ. Modules Mater. Sci. Eng. 3*(2):251–308.
219. Mark, R. E., and Gillis, P. P. (1970). New models in cell wall mechanics. *Wood Fiber 2*(2):79–95.
220. Mark, R. E., and Gillis, P. P. (1972). The relationship between fiber moduli and S2 angle. Rep. 56. Empire State Paper Res. Inst. Syracuse, NY, pp. 55–68.
221. Mark, R. E., and Gillis, P. P. (1973). The relationship between fiber modulus and S2 angle. *Tappi 56*(4):164–167.
222. Mark, R. E., Kaloni, P. N., Tang, R.-C., and Gillis, P. P. (1969). Solid mechanics and crystal physics as tools for cellulose structure investigation. *Textile Res. J. 39*:203–212. See also *Science 164*:72–73.
223. Mark, R. E., Thorpe, J. L., Angello, A. J., Perkins, R. W., and Gillis, P. P. (1971). Twisting energy of holocellulose fibers. *J. Polym. Sci. C 36*:177–195.
224. Matolcsy, G. (1975). Drying under tension increases zero-span tensile strength of pulp handsheets. *Pulp Paper Mag. Can. 76*(5):80–85.
225. Matsuo, M., Sawatari, C., Iwai, Y., and Ozaki, F. (1990). Effect of orientation distributions and crystallinity on the measurement by X-ray diffraction of the crystal lattice moduli of cellulose I and II. *Macromolecules 23*:3266–3275.
226. McIntosh, D. C. (1967). The effect of refining on the structure of the fiber wall. *Tappi 50*(10):482–488.
227. McIntosh, D. C., and Leopold, B. (1962). Bonding strength of individual fibers. In: *Formation and Structure of Paper*, Vol. I. F. Bolam, ed. Tech. Sect. Brit. Paper Board Makers Assoc., London, pp. 265–270.
228. McMillin, C. W. (1969). Aspects of fiber morphology affecting properties of handsheets made from loblolly pine refiner groundwood. *Wood Sci. Tech. 3*:139–149.
229. McMillin, C. W. (1974). Dynamic torsional unwinding of southern pine tracheids as observed in the scanning electron microscope. *Svensk. Papperstidn. 77*(9):319–324.
230. Mehta, V. R., and Kumar, S. (1994). Temperature dependent torsional properties of high performance fibres and their relevance to compressive strength. *J. Mater. Sci. 39*(14):3658–3664.
231. Meredith, R. (1956). Dynamic mechanical properties. In: *The Mechanical Properties of Textile Fibres*. R. Meredith, ed. Interscience, New York, pp. 106–128.
232. Meredith, R. (1959). Mechanical properties of cellulose and cellulose derivatives. In: *Recent Advances in the Chemistry of Cellulose and Starch*. J. Honeyman, ed. Interscience, New York, pp. 213–239.
233. Meyer, K. H. (1942). *Natural and Synthetic High Polymers*. Interscience, New York.
234. Meyer, K. H., and Lotmar, W. (1936). On the elasticity of cellulose. *Helv. Chim. Acta. 19*:68–86 (in French).
235. Meyer, K. H., and Mark, H. (1930). *The Structure of High Polymer Organic Natural Substances*. Akademische Verlagsges, Leipzig, pp. 152–157 (in German).
236. Meyer, K. H., and Misch, L. (1937). Atomic positions in the new spatial model for cellulose. *Helv. Chim. Acta 20*:232–244 (in French).
237. Meylan, B. A., and Butterfield, B. G. (1972). *Three-Dimensional Structure of Wood*. Syracuse Univ. Press, Syracuse, NY.
238. Meylan, B. A., and Butterfield, B. G. (1978). Helical orientation of microfibrils in tracheids, fibres and vessels. *Wood Sci. Tech. 12*:219–222.

239. Mikelsaar, R.-H., and Aabloo, A. (1993). Antiparallel molecular models of cystalline cellulose. In: *Cellulosics: Chemical, Biological and Material Aspects*. J. F. Kennedy, G. C. Phillips, and P. A. Williams, eds. Ellis Horwood, London, pp. 57–60.
240. Mohlin, U.-B. (1975). Cellulose fibre bonding. Part 5. Conformability of pulp fibres. *Svensk Papperstidn. 78*(11):413–416.
241. Mohlin, U.-B., and Alfredsson, C. (1990). Fibre deformation and its implications in pulp characterization. *Nordic Pulp Paper Res. J. 5*(4):172–179.
242. Mohlin, U.-B., Dahlbom, J., and Hornatowska, J. (1996). Fiber deformation and sheet strength. *Tappi J. 79*(6):105–111.
243. Morton, W. E., and Hearle, J. W. S. (1993). *Physical Properties of Textile Fibres*. 3d ed. The Textile Institute, Manchester, U.K., pp. 341–398.
244. Moss, P. A., Kropholler, H. W., and Sheffield, E. (1989). Ltsem: Great potential for pulp evaluation *Paper Tech. 30*(9):12–14.
245. Moss, P. A., Retulainen, E., Paulapuro, H., and Aaltonen, P. (1993). Taking a new look at pulp and paper: Applications of confocal laser scanning microscopy to pulp and paper research. *Paperi Puu 75*(1/2):74–79.
246. Mott, L. (1995). Micromechanical properties and fracture mechanisms of single wood pulp fibers. Ph.D Thesis, Univ. Maine, Orono, ME.
247. Mott, L., Shaler, S. M., Groom, L. H., and Liang, B.-H. (1995). The tensile testing of individual wood fibers using environmental scanning electron microscopy and video image analysis. *Tappi J. 78*(5):143–148.
248. Muggli, R. (1968). Fine structure of cellulosic elementary fibrils. *Cellulose Chem. Tech. 2*:547–567 (in German).
249. Muggli, R., Elias, H.-G., and Mühlethaler, K. (1969). Fine structure of the elementary fibrils of cellulose. *Makromol. Chem. 121*:290–294 (in German).
250. Murayama, T., Dumbleton, J. H., and Williams, M. L. (1967). The viscoelastic properties of oriented nylon 66 fibers, 3. Stress relaxation and dynamic mechanical properties. *J. Macromol. Sci. B1*(1):1–14.
251. Murphey, W. K. (1963). Cell-wall crystallinity as a function of tensile strain. *Forest Prod. J. 13*(4):151–155.
252. Naito, T., Usuda, M., and Kadoya, T. (1980). Torsional property of single pulp fibers. *Tappi 63*(7):115–118.
253. Naito, T., Usuda, M., and Kadoya, T. (1983). Fatigue phenomena of single pulp fibers under repeated torsional stress. *Proc. 1983 Int. Paper Physics Conf.* TAPPI Press, Atlanta, pp. 197–201.
254. Nanko, H., and Côté, W. A. (1980). *Bark Structure of Hardwoods Grown on Southern Pine Sites*. Syracuse Univ. Press, Syracuse, NY.
255. Nanko, H., and Ohsawa, J. (1989). Mechanisms of fibre bond formation. In: *Fundamentals of Papermaking*, Vol. 2. C. F. Baker, ed. Mech. Eng. Pub., London, pp. 783–830.
256. Nanko, H., and Wu, J. (1995). Mechanisms of paper shrinkage during drying. *Proc. Int. Paper Physics Conf.*, Niagara-on-the-Lake, Ontario; Tech. Sec., Can. Pulp Paper Assoc., pp. 103–113.
257. Navi, P. (1998). The influence of microfibril angle on wood cell and wood mechanical properties, experimental and numerical study. In: *Microfibril Angle in Wood*. B. G. Butterfield, ed. Univ. Canterbury, Christchurch, N.Z., pp. 62–80.
258. Navi, P., Rastogi, K., Gresse, V., and Toulou, A. (1995). Micromechanics of wood subjected to axial tension. *Wood Sci. Tech. 29*:411–429.
259. Nethercut, P. E. (1957). A fundamental study of the softening mechanism of paper. *Tappio 40*(1):39–45.

260. Nevell, T. P. (1985). Degradation of cellulose by acids, alkalis and mechanical means. In: *Cellulose Chemistry and Its Applications*. T. P. Nevell and S. H. Zeronian, eds. Halstead/Wiley, New York, pp. 223–242.
261. Newman, R. (1998). How stiff is an individual cellulose microfibril? In: *Microfibril Angle in Wood*. B. G. Butterfield, ed. Univ. Canterbury, Christchurch, N.Z., pp. 81–93.
262. Newman, R. H. (1994). Crystalline forms of cellulose in softwoods and hardwoods. *Wood Chem. Tech. 14*(3):451–466.
263. Nishino, T., Takano, K., and Nakamae, K. (1995). Elastic modulus of the crystalline regions of cellulose polymorphs. *J. Polym. Sci. B 33:*1647–1651.
264. Niskanen, K. (1998). *Paper Physics*. Book 16 of Paper Sci. Technol. Ser., Fapet Oy, Helsinki (pub. in cooperation with Finnish Paper Eng. Assoc. and TAPPI).
265. Niskanen, K. J., Alava, M. J., Sepälä, E. T., and Aström, J. (1999). Fracture energy in fibre and bond failure. *J. Pulp Paper Sci. 25*(5):167–169.
266. Nissan, A. H. (1977). *Lectures on Fiber Science in Paper*. Pulp Paper Technol. Ser. No. 4. Joint CPPA-TAPPI Textbook Committee.
267. Nyrén, J. (1971). The transverse compressibility of pulp fibers. CPPA Tech. Sect. Paper T326. *Pulp Paper Mag. Can. 72*(10):81–83.
268. Ohgama, T. (1976). Viscoelasticity of wood as porous material. Ph.D. Thesis, Kyoto Univ. (in Japanese).
269. Ohgama, T., Masuda, M., and Yamada, T. (1977). Stress distribution within cell wall of wood subjected to tensile force in transverse direction. *J. Soc. Mater. Sci. (Jpn.) 26*:433–438.
270. Olson, J. A., Robertson, A. G., Finnigan, T. D., and Turner, R. R. H. (1995). An analyzer for fiber shape and length. *J. Pulp Paper Sci. 21*(11):J367–J373.
271. Olsson, A.-M., and Salmén, L. (1997). The effect of lignin composition on the visco-elastic properties of wood. *Nordic Pulp Paper Res. J. 12*(3):140–144.
272. Paavilainen, L. (1991). Influence of morphological properties of wood fibres on sulphate pulp fibre and paper properties. *Int. Paper Phys. Conf. (Kona, Hawaii)*. TAPPI Press, Atlanta, Proc. Bk. 2, pp. 383–395.
273. Paavilainen, L. (1993). Importance of cross-dimensional fibre properties and coarseness for the characterization of softwood sulphate pulp. *Paperi Puu 75*(5):343–351.
274. Pagano, N. J., and Tsai, S. W. (1968). Micromechanics of composite media. In: *Composite Materials Workshop*. S. W. Tsai, J. C. Halpin, and N. J. Pagano, eds. Technomic, Stamford, CT, pp. 1–8.
275. Page, D. H., and El-Hosseiny, F. (1976). The mechanical properties of single wood pulp fibers. 5. The influence of defects. *Svensk Papperstidn. 79*(14):471–474.
276. Page, D. H., and El-Hosseiny, F. (1983). The mechanical properties of single wood pulp fibres. Part IV. Fibril angle and the shape of the stress-strain curve. *J. Pulp Paper Sci. 9*(4):1–2.
277. Page, D. H., El-Hosseiny, F., and Winkler, K. (1971). Behavior of single wood fibres under axial tensile strain. *Nature 229*:252–253.
278. Page, D. H., El-Hosseiny, F., Winkler, K., and Bain, R. (1972). The mechanical properties of single wood-pulp fibres. Chapter 2: A new approach. *Pulp Paper Mag. Can. 73*(8):72–77.
279. Page, D. H., El-Hosseiny, F., Winkler, K., and Lancaster, A. P. S. (1977). Elastic modulus of single wood pulp fibers. *Tappi 60*(4):114–117.
280. Page, D. H., Seth, D. S., Jordan, B. D., and Barbe, M. C. (1985). Curl, crimps, kinks and microcompressions in pulp fibres. Their origin, measurement and significance. In: *Papermaking Raw Materials*, Vol. 1. V. W. Punton, ed. Mech. Eng. Pub., London, pp. 183–227.
281. Parameswaran, N., and Liese, W. (1976). On the fine structure of bamboo fibres. *Wood Sci. Tech. 10*:231–246.

282. Parham, R. A., and Kaustinen, H. M. (1974). *Papermaking Materials: An Atlas of Electron Micrographs*. Institute of Paper Chemistry, Appleton, WI.
283. Patrick, A. J., Jr., and Hood, J. H. (1965). Glass reinforcements for filament wound composites. *Proc. 20th Tech. Conf. Soc. Plastics Ind., Reinf. Plastics Divis.*, Sect. 9-D.
284. Perez, M. (1967). Research on the Behavior of Fibers in Paper Under Stress. St. Regis Tech. Center Rep. Presented at New York Metropolitan TAPPI meeting, January 1968.
285. Persson, K. (1997). Modelling of wood properties by a micromechanical approach. Licentiate Thesis, Rep. TVSM-3020, Lund Univ., Div. Structural Mechanics, Lund, Sweden (original not seen).
286. Petit-Conil, M., Cochaux, A., and de Choudens, C. (1994). Mechanical pulp characterization: A new and rapid method to evaluate fibre flexibility. *Paperi Puu 76*(10):657–662.
287. Petkar, B. M., Oka, P. G., and Sundaram, V. (1980). The cross-sectional shapes of a cotton fiber along its length. *Textile Res. J. 50*:541–543.
288. Phillips, D. G. (1987). Effects of humidity, aging, annealing and tensile loads on the torsional damping of wool fibers. *Textile Res. J. 57*(7):415–420.
289. Preston, R. D. (1974). *The Physical Biology of Plant Cell Walls*. Chapman and Hall, London.
290. Preston, R. D., and Singh, K. (1950). The fine structures of bamboo fibres. *J. Exp. Bot. 1*:214–230.
291. Purves, C. B. (1954). Chemical nature of cellulose and its derivatives. A historical survey. In: *Cellulose and Cellulose Derivatives*. 2nd ed. Part 1. E. Ott, H. M. Spurlin, and M. W. Grafflin, eds. Interscience, New York, pp. 29–53.
292. Pye, I. T., Washburn, O. V., and Buchanan, J. G. (1966). Structural changes in paper on pressing and drying. In: *Consolidation of the Paper Web*. F. Bolam, ed. Brit. Paper and Board Makers Assoc., London, pp. 353–370.
293. Rao, D. R. (1984). Torsional studies of wool fibers. *J. Textile Inst. 75*(5):336–341.
294. Reiling, S., and Brickmann, J. (1995). Theoretical investigations on the structure and physical properties of cellulose. *Macromol. Theory Simul. 4*:725–743.
295. Reissner, E., and Stavsky, Y. (1961). Bending and stretching of certain types of heterogeneous aeolotropic elastic plates. *J. Appl. Mech. 28*:402–408.
296. Retulainen, E. (1996). Fibre properties as control variables in papermaking? Part 1. Fibre properties of key importance in the network. *Paperi Puu 78*(4):187–194.
297. Retulainen, E., and Ebeling, K. (1985). Effect of paper on the load-elongation behaviour of fibre-to-fibre bonds. In: *Papermaking Raw Materials*, Vol. 1. V. Punton, ed. Mech. Eng. Pub., London, pp. 230–263.
298. Robertson, A. A., Meindersma, E., and Mason, S. G. (1961). The measurement of fibre flexibility. *Pulp Paper Mag. Can. 62*(1):T3–T10.
299. Robertson, G., Olson, J., Allen, P., Chan, B., and Seth, R. (1999). Measurement of fiber length, coarseness, and shape with the fiber quality analyzer. *Tappi J. 82*(10):93–98.
300. Rühlemann, F. (1925). On the strength determination of pulp: Influence of bleaching on technical properties of paper pulp fibers. (In German). Dissert. Tech. Hochschule Dresden (original not seen).
301. Rühlemann, F. (1926). On the determination of the strength of cellulose. *Paper Trade J. 82*(13):T168–T171.
302. Sachs, I. B. (1986). Microscopic observations during longitudinal compression loading of single pulp fibers. *Tappi J. 69*(7):98–102.
303. Saiki, H. (1970). Proportion of component layers in tracheid wall of early wood and late wood of some conifers. *J. Jpn. Wood Res. Soc. (Mokuzai Gakkaishi) 16*(5):244–249.
304. Saiki, H., and Ono, K. (1971). Cell wall organization of gelatinous fibers in tension wood. *Bull. Kyoto Univ. Forests 42*:210–220 (in Japanese).

305. Sakurada, I., Nukushina, Y., and Ito, T. (1962). Experimental determination of the elastic modulus of crystalline regions in oriented polymers. *J. Polym. Sci. 57*:651–660.
306. Salmén, L. (1986). The cell wall as a composite structure. In: *Paper, Structure and Properties*. J. A. Bristow and P. Kolseth, eds. Marcel Dekker, New York, pp. 51–73.
307. Salmén, L., and de Ruvo, A. (1985). A model for the prediction of fiber elasticity. *Wood Fiber Sci. 17*(3):336–350.
308. Salmén, L., and Kolseth, P. (1986). A mechanical model of softwood. In: *Paper, Structure and Properties*. J. A. Bristow and P. Kolseth, eds. Marcel Dekker, New York, pp. 377–380.
309. Salmén, L., Kolseth, P., and de Ruvo, A. (1985). Modeling the softening behavior of wood fibers. *J. Pulp Paper Sci. 11*(4):J102–J107.
310. Salmén, L., Kolseth, P., and Rigdahl, M. (1986). Modeling of small-strain properties and environmental effects on paper and cellulosic fibers. In: *Composite Systems of Natural and Synthetic Polymers*. L. Salmén, A. de Ruvo, J. C. Seferis, and E. B. Stark, eds. Elsevier, Amsterdam, pp. 211–223.
311. Samuelsson, L. G. (1963). Measurement of the stiffness of fibres. *Svensk Papperstidn. 66*(15):541–546.
312. Sarko, A., and Muggli, R. (1974). Packing analysis of carbohydrates and polysaccharides. 3. Valonia cellulose and cellulose II[1a]. *Macromolecules 7*:486–494.
313. Schniewind, A. P. (1972). Elastic behavior of the wood fiber. In: *Theory and Design of Wood and Fiber Composite Materials*. B. A. Jayne, ed. Syracuse Univ. Press, Syracuse, NY, pp. 83–95.
314. Schniewind, A. P., and Barrett, J. D. (1969). Cell wall model with complete shear restraint. *Wood Fiber 1*(3):205–214.
315. Schniewind, A. P., Ifju, G., and Brink, D. L. (1966). Effect of drying on the flexural rigidity of single fibres. In: *Consolidation of the Paper Web*, Vol. 1. F. Bolam, ed. Tech Sect. Brit. Paper Board Makers' Assoc., London, pp. 538–543.
316. Schwarz, E. R. (1954). Fiber testing methods. In: *Matthews' Textile Fibers*. H. R. Mauersberger, ed. Wiley, New York, pp. 1214–1251.
317. Seborg, C. D., and Simmonds, F. A. (1941). Measurement of the stiffness in bending of single fibers. *Paper Trade J (TAPPI Sect.) 113*:225–226.
318. Sedlachek, K. M., and Ellis, R. L. (1994). The effect of cyclic humidity on the creep of single fibers of Southern pine. *Conf. Moisture-Induced Creep Behavior of Paper and Board*. STFI, Stockholm, pp. 29–49.
319. Seth, R. S., and Chan, B. K. (1997). Measurement of fiber coarseness with optical fiber length analyzers. *Tappi J. 80*(5):217–221.
320. Seth, R. S., and Chan, B. K. (1999). Measuring fiber strength of papermaking pulps. *Tappi J. 82*(11):115–120.
321. Shallhorn, P. M., and Karnis, A. (1981). A rapid method for measuring wet fiber flexibility. *Pulp Paper Can. 82*(12):TR69–TR74.
322. Simon, I., Glasser, L., Scheraga, H. A., and Manley, R. S. J. (1988). Structure of cellulose. 1. Low-energy conformations of single chains. *Macromolecules 21*:990–998.
323. Sinclair, D. (1950). A bending method for measurement of the tensile strength and Young's modulus of glass fibers. *J. Appl. Phys. 21*(5):380–386.
324. Sobue, N., and Asano, I. (1976). Studies on the fine structure and mechanical properties of wood. On the longitudinal Young's modulus and shear modulus of rigidity of cell wall. *J. Jpn. Wood Res. Soc. (Mokuzai Gakkaishi) 22*:211–216 (in Japanese).
325. Stamm, A. J. (1964). *Wood and Cellulose Science*. Ronald Press, New York, pp. 78–121.
326. Steadman, R., and Luner, P. (1985). The effect of wet fibre flexibility of sheet apparent density. In: *Papermaking Raw Materials* (Trans. 1985 Symp.), Vol. 1. V. Punton, ed. Mech. Eng. Pub., London, pp. 311–337.

327. Stone, J. E., and Scallan, A. M. (1966) Influence of drying on the pore structures of the cell wall. In: *Consolidation of the Paper Web*, Vol. 1. F. Bolam, ed. Tech. Sect. Brit. Paper Board Makers Assoc., London, pp. 145–174.
328. Stratton, R .A., and Colson, N. L. (1990). Dependence of fiber/fiber bonding on some papermaking variables. In: *Materials Interactions Relevant to the Pulp, Paper and Wood Industries*. D. F. Caulfield, J. D. Passaretti, and S. F. Sobczynski, eds. Materials Res. Soc., Pittsburgh, pp. 173–181.
329. Sugiyama, J., Harada, H., Fujiyoshi, Y., and Uyeda, N. (1985). Lattice images from ultrathin sections of cellulose microfibrils in the cell wall of Valonia macrophysa Kütz. *Planta 166*:161–168.
330. Sugiyama, J., Okano, T., Yamamoto, H., and Horii, F. (1990). Transformation of Valonia cellulose crystals by an alkaline hydrothermal treatment. *Macromolecules 23*:3196–3198.
331. Sugiyama, J., Vuong, R., and Chanzy, H. (1991). Electron diffraction study on the two crystalline phases occurring in native cellulose from an algal cell wall. *Macromolecules 24*:4168–4175.
332. Tam Doo, P. A., and Kerekes, R. J. (1989). The effect of beating and low-amplitude flexing on pulp fibre flexibility. *J. Pulp Paper Sci. 15*(1):J36–J42.
333. Tam Doo, P. A., and Kerekes, R. J. (1982). The flexibility of wet pulp fibres. *Pulp Paper Can. 83*(2):T37–T41.
334. Tam Doo, P. A., and Kerekes, R. J. (1981). A method to measure wet fiber flexibility. *Tappi 64*(3):113–116.
335. Tang, R. C. (1972). Three-dimensional analysis of elastic behavior of wood fiber. *Wood Fiber 3*(4):210–219.
336. Tang, R. C., and Hsu, N. N. (1973). Analysis of the relationship between microstructure and elastic properties of the cell wall. *Wood Fiber 5*(10):139–151.
337. TAPPI. (1984). Test Method T 234 cm-84. Coarseness of pulp fibers.
338. TAPPI. (1985). Test Method T 254 cm-85. Cupriethylenediamine disperse viscosity of pulp (falling ball method).
339. TAPPI. (1993). Test Method T 211 om-93. Ash in wood, pulp, paper and paperboard: Combustion at 525°C.
340. TAPPI. (1988). Test Method T 494 om-88. Tensile properties of paper and paperboard (using constant rate of elongation apparatus).
341. TAPPI. (1991). Useful Method 18. Density or volume of small specimen (Hg pycnometer). In *TAPPI Useful Methods* 0104-UM91.
342. TAPPI. (1993). Units of measurement and conversion factors. TIP 0800-01.
343. TAPPI. (1994). Test Method T 230 om-94. Viscosity of pulp (capillary viscometer method).
344. TAPPI. (1995). Test Method T 273 pm-95. Wet zero-span tensile strength of pulp.
345. TAPPI. (1996). Test Method T 231 cm-96. Zero-span breaking strength of pulp (dry zero-span tensile).
346. TAPPI. (1998). Test Method T 220 om-88. Physical testing of pulp handsheets.
347. Tashiro, K., and Kobayashi, M. (1985). Calculation of crystallite modulus of native cellulose. *Polym. Bull 14*(3/4):213–218.
348. Tashiro, K., and Kobayashi, M. (1991). Theoretical evaluation of three-dimensional elastic constants of native and regenerated celluloses: Role of hydrogen bonds. *Polymer 32*(8):1516–1526.
349. Taylor, D. L., and Craver, J. K. (1966). Anisotropic elasticity of paper from sonic velocity measurements. In: *Consolidation of the Paper Web*, Vol. 2. F. Bolam, ed. Tech. Sect. Brit. Paper Board Makers' Assoc., London, pp. 852–874.
350. Thorpe, J., and McLean, D. (1983). Simulation of the effects of compression and shear on single fibers. *Proc. 1983 Int. Paper Phys. Conf.* TAPPI Press, Atlanta, pp. 51–58.

351. Thorpe, J. L., Mark, R. E., Eusufzai, A. R. K., and Perkins, R. W. (1976). Mechanical properties of fiber bonds. *Tappi 59*(5):96–100.
352. Tiikkaja, E. (1993). Experiences with a new on-line fibre length analyzer. *79th Annual Meeting Tech. Section. Preprints*. CPPA, Montreal, pp. A7–A10.
353. Tiikkaja, E., and Kauppinen, M. (1998). Fibre dimensions, their effect on paper properties and required measuring accuracy. *84th Annual Meeting Techn. Section, Preprints*. CPPA, Montreal. A369–A372.
354. Timell, T. E. (1986). *Compression Wood in Gymnosperms*, Vol. 1. Springer-Verlag, New York.
355. Treloar, L. R. G. (1960). Calculations of elastic moduli of polymer crystals. III. Cellulose. *Polymer 1*(3):290–303.
356 Tsai, S. W., and Hahn, H. T. (1980). *Introduction to Composite Materials*. Technomic, Westport, CT.
357. Tsai, S. W., and Pagano, N. J. (1968). Invariant properties of composite materials. In: *Composite Materials Workshop*. S. W. Tsai, J. C. Halpin, and N. J. Pagano, eds. Technomic, Stamford, CT, pp. 233–253.
358. Van den Akker, J. A. (1962). Some theoretical considerations on the mechanical properties of fibrous structures. In: *Formation and Structure of Paper*. Vol. 1. F. Bolam, ed. Tech. Sect. Brit. Paper Board Makers' Assoc., London, pp. 205–241.
359. Van den Akker, J. A., Jentzen, C. A., and Spiegelberg, H. L. (1966). Effects on individual fibres of drying under tension. In: *Consolidation of the Paper Web*, Vol. 1. F. Bolam, ed. Tech. Sect. Brit. Paper Board Makers' Assoc., London, pp. 477–506.
360. VanderHart, D. L., and Atalla, R. H. (1984). Studies of microstructure in native celluloses using solid-state ^{13}C NMR. *Macromolecules 17*:1465–1472.
361. Villeneuve, J. F., and Naslain, R. (1993). Shear moduli of carbon, SiCTiO and alumina single ceramic fibers as assessed by torsional pendulum tests. *Compos. Sci. Tech. 49*(2):191–203.
362. Wang, J. (1996). Ultimate mechanical properties of polyethylene and cellulose fibers: Breaking strength, breaking strain, and modulus. Ph.D. Thesis, SUNY College Environ. Sci. and Forestry, Syracuse, NY.
363. Weibull, W. (1939). *A Statistical Theory of the Strength of Materials*. Handbook 151. Royal Swedish Inst. Eng. Res., Stockholm.
364. Whitney, J. M. (1969). Bending-extensional coupling in laminated plates under transverse loading. *J. Compos. Mater. 3*:20–28.
365. Whitney, J. M., and Leissa, A. W. (1969). Analysis of heterogeneous anisotropic plates. *J. Appl. Mech. 36*:261–266.
366. Wild, P. M., Provan, J. W., Guin, R., and Pop, S. (1999). The effects of cyclic axial loading of single wood pulp fibers at elevated temperature and humidity. *Tappi J. 82*(4):209–215.
367. Woodcock, C., and Sarko, A. (1980). Packing analysis of carbohydrates and polysaccharides. 11. Molecular and crystal structure of native ramie cellulose. *Macromolecules 13*:1183–1187.
368. Wuu, F. (1993). Mechanical properties and structural changes in recycled paper. Ph.D. dissertation. SUNY College of Environ. Sci. and Forestry, Syracuse, NY.
369. Wuu, F., Mark, R. E., and Perkins, R. W. (1991). Mechanical properties of "cut-out" fibers in recycling. *Proc. Int. Paper Phys. Conf.* TAPPI Press, Atlanta, GA. Vol. 2, pp. 663–680.
370. Xu, L., Filonenko, Y., Li, M., and Parker, I. (1997). Measurement of cell wall thickness of fully collapsed fibers by confocal microscopy and image analysis. *Appita J. 50*(6):501–504.

371. Yamamoto, H., and Horii, F. (1993). CP/MAS ^{13}C NMR analysis of the crystal transformation induced for Valonia cellulose by annealing at high temperatures. *Macromolecules 26*:1313–1317.
372. Yang, C. F., Eusufzai, A. R. K., Sankar, R., Mark, R. E., and Perkins, R. W., Jr. (1978). Measurements of geometrical parameters of fiber networks. 1. Bonded surfaces, aspect ratios, fiber moments of inertia, bonding state probabilities. *Svensk Papperstidn. 81*(13):426–433.
373. Yang, C. F., Perkins, R. W., and Mark, R. E. (1978). The effect of lateral force on wood fiber deformation. I. Linear elastic behavior. Rep. 69. Empire State Paper Res. Inst., Syracuse, NY, pp. 23–46.
374. Yiannos, P. N., and Taylor, D. L. (1967). Dynamic modulus of thin wood sections. *Tappi 50*(1):40–47.
375. Yu, Y., Kettunen, H., Hiltunen, E., and Niskanen, K. (1999). Comparison of abaca and spruce as reinforcement fiber. *Proc. Int. Paper Phys. Conf.* TAPPI Press, Atlanta, pp. 161–170.

15

DETERMINATION OF FIBER–FIBER BOND PROPERTIES

TETSU UESAKA
Pulp and Paper Research Institute of Canada
Pointe-Claire, Quebec, Canada

ELIAS RETULAINEN
Asian Institute of Technology
Pathumthani, Thailand

LEENA PAAVILAINEN
Finnish Forest Cluster Research Programme—Wood Wisdom
Helsinki, Finland

RICHARD E. MARK* **and D. STEVEN KELLER**
Empire State Paper Research Institute
State University of New York
College of Environmental Science and Forestry
Syracuse, New York

*Retired.

I. INTRODUCTION

Bonding ability can be considered one of the most important fiber properties, especially for reinforcement fibers that are used to ensure the required strength in paper. In addition to providing necessary strength properties, the structure and properties of bonded regions also affect all mechanical, optical, thermal, and electrical properties of the paper sheet. The bonding ability of fibers is determined by bonding strength and bonded area.

- Bonding strength (specific bond strength) is a function of the strength of elementary bonds and their number per unit bonded area. Bonding strength is determined by bond type, surface topography, and the inherent strength of material within the bonded region.
- Bonded area is determined by fiber conformability and the total surface area of the fiber network. Fiber conformability is a function of fiber collapsibility and wet fiber flexibility. Total surface area is affected by the number of fibers in a sheet with a certain basis weight and by fiber width, external fibrillation, and fines.

Fibers can form fiber–fiber bonds if the following requirements are satisfied:

- Fiber surfaces are brought close enough together to form interfiber bonds. This depends on wet fiber conformability, surface tension forces, and external forces such as those introduced by wet pressing.
- Fiber surfaces possess physical and chemical properties that are conducive to bonding.

Figure 1 illustrates a classical model of fiber–fiber bond formation as presented by Van den Akker [79]. In water, beaten fiber cell walls tend to swell and delaminate to a certain extent. The fiber surfaces are covered by microfibrils and microfibril bundles. During consolidation on a paper machine, surface tension forces (the so-called Campbell effect) [5] contribute to three important events:

- Fibers and fibrils are brought into close contact.
- Fibrils readhere to the surfaces of the base fibers.
- Fibrils adhere to adjacent fibers.

Hydrogen and other molecular bonds may then form between the entangled fibrils and fiber surfaces.

Interfiber bonding is a principal interest in paper physics. Although the theoretical background is available, the link between theory and practice is frequently absent. Papermakers would, however, benefit from an improved understanding of interfiber bonding, for example when trying to minimize the proportion of reinforcement fiber in printing and writing grades and in striving to develop products with a lower basis weight. Strong bonding is also a prerequisite for increased filler content and for developing speciality papers that have superior strength.

The purpose of this chapter is to clarify the nature of fiber–fiber bonds, to define essential terms related to bonding, and to review the existing methods for determining bonding strength and bonded area of the fibers. Experimental approaches to the measurement of bonding ability are also discussed.

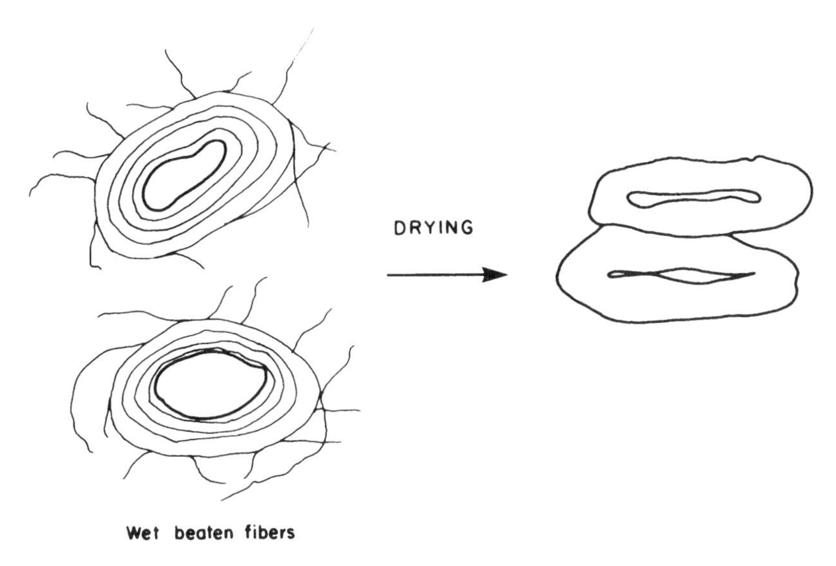

Fig. 1 Fiber–fiber bond formation.

II. MORPHOLOGY OF FIBER BONDING

A. Three-Dimensional Aspects of Bonds

The mechanical functionality of paper, like that of other types of fiber networks, depends on the transfer of tensile, compressive, and shear stresses from fiber to fiber. Often the fiber bond referred to in the literature gives an impression of being an essentially two-dimensional array of hydrogen bonds that connect one rigid fiber surface to another. In fact, the nature of the stress transfer zone, or *bonding region*, between adjacent fibers can be considered a three-dimensional array with surface contact of cell walls, entanglement of fibrils and aggregates, and the chemical bonds themselves.

The concept of two surfaces connected by hydrogen bonds has mechanical implications that can be illustrated by some simple calculations. For example, we can think of an O—H --- O bond as forming a local bridge across two surfaces; the distance that is bridged is less than 0.3 nm (< 3 Å), because direct distances of 2.5 Å or less between oxygen atoms are usually involved [25] and most nonlinear O—H --- O distances are in the 2.4–2.9 Å range. Now if this concept is extended to a local field or array containing hundreds or thousands of such bonds, it becomes apparent that there are spatial constraints to the deformations that could possibly exist in that field. An interlaminar shear strain of 0.5% would yield a displacement of less than 0.0015 nm between the two surfaces.

But delicate experiments on the shear properties of fiber–shive bonds show disablements in the bonded area of about 1.5 μm when bond strain is approximately 0.5% (at or near ultimate strain) as shown in Table 1 [78]. The data in the table are derived from high resolution photomicrographs of fiber–shive assemblies marked with well-positioned semifused xerographic particles; the test specimen material is pine thermomechanical pulp bonded at elevated temperatures. However, results for chlorite holocellulose and autoclaved thermomechanical pulps were quite similar

Table 1 Bond Properties of 160°C Thermomechanical Pulp[a]

Test No.	Max load (N)	Free fiber length[b] (μm)	Bond length[b] (μm)	Max. free fiber elong.[b] (μm)	Max. free fiber strain (%)	Max. bond elong.[b] (μm)	Max bond strain (%)	Gross area overlap[c] (μm^2)	Shear stress at max. load (MPa)
1	0.208	333	299	2.01	0.60	3.42	1.15	17,300	12.01
2	0.142	410	241	8.84	2.16	1.69	0.56	17,360	8.19
3	0.102	313	269	1.81	0.58	1.74	0.65	11,865	8.60
4	0.125	279	282	1.48	0.53	1.10	0.39	15,400	8.09
5	0.095	308	234	2.30	0.75	1.23	0.53	9,695	9.82
6	0.042	348	321	4.71	1.53	1.16	0.36	10,420	4.05
7	0.092	123	357	5.44	4.40	3.33	0.93	11,365	8.11
8	0.040	332	290	3.16	0.95	0.49	0.17	12,570	3.20
9	0.104	247	252	4.94	2.00	1.40	0.56	7,790	13.35
10	0.097	292	271	3.67	1.26	1.00	0.37	11,315	8.58
11	0.049	212	283	3.35	1.58	1.49	0.53	9,270	5.29
Avg.	0.100	291	282	3.79	1.49	1.64	0.56	12,214	8.12

[a] Bond formed by clamping fiber–shive assembly between glass plates and holding for 15 min in an autoclave at 210°C (steam pressure ~ 3.15 MPa).
[b] Measured from photomicrographs.
[c] Assumed to be identical with bond area. (*Note*: 1 MPa = 1 MN/m^2.)
Source: Ref. 78.

[11]. The significant point is that the experiments show a difference of six orders of magnitude compared with the limits imposed by the shear displacement capabilities of a planar array of hydrogen bonds.

Another way to view this *bonding region* from a spatial perspective is to consider the dimensions of the fibrillar material created by beating or refining pulp fibers. The loose fibrils create surface irregularities that are two to four orders of magnitude greater than hydrogen bond distances. Thus, the surfaces to be bonded are not smooth; they have profiles analogous to that of a geographical area with diverse features of topography and even fine-scale features such as trees and small plants rather than an open flat plain. For additional perspective on this subject, consult Salminen et al. [60].

The stress transfer zone between two fibers is a three-dimensional structure that most likely will extend within the secondary walls at least between the S1/S2 interfaces of bonded fibers (see Chapter 1). It has been demonstrated many times [6,12,34,38, 43,44,69] that the S1/S2 interface in the wood fiber, regardless of pulping or beating history, is weak and deformable relative to the rest of the fiber structure. In response to mechanical or drying forces, the S1/S2 interface typically delaminates at points and readjusts its surface conformation to accommodate the stress changes. In beaten pulp fibers, the S1 layer is typically a loose, fibrillated sheath that covers the S2 layer and acts to enlarge the contact surface between fibers [49,50,55]. As the web dries and the shrinkage of the cell wall causes an increase in internal stress, the S1 layers of adjacent fibers act as the adhesive to draw together the residual bulk of the adhered fibers, which is composed of the S2 and S3 layers. Layer thicknesses vary with species, position in tree or tree ring, pulping method, degree of beating and other factors. Based on data for wood, a double S1 thickness for a typical conifer species might be 0.3–0.8 μm [59]. Within such a thickness, there are many modes of deformation as the bond is subjected to mechanical forces. The opportunities for internal deformation increase greatly as a result of pulping followed by beating, which fibrillates the surface and partially delaminates and flexibilizes the layers of the fiber wall.

B. Concepts of Bonded Area

As pointed out by Retulainen and Ebeling [57], the number and area of fiber bonds largely determine the internal cohesion of paper. These geometrical/structural effects are critically important for all the physical and mechanical properties that are observed and measured. These authors compared eight methods commonly used to characterize the bond strength of paper but noted that the methods yielded different and sometimes contradictory results. They found three main sources of discrepancy, caused by differences in

- The mode of loading (in-plane or *z*-directional)
- The measurement principle employed (strength, force, or energy)
- The measurement of the bonded area

It is the third source of discrepancy that is considered in this section.

At first glance, it might appear to be a simple matter to measure the area of one bond or multiple bonds within a sheet of paper. One can, for example, see and measure the overlap area of two crossing fibers microscopically. Yet immediately it becomes evident, because fibers are somewhat rounded (even when they collapse during sheet consolidation), that the total overlap is larger than the actual contact area. "Optical contact" is measurable by microscopy, and Page et al. [53] concluded that the optically bonded area approximates the "true" bonded area. On the other hand, because bonds are three-dimensional, the internal topography of the bond inherently implies that the bond density within any designated or assumed bonded area will be highly variable. That variability in local conditions of hydrogen bond formation (or lack of same) negates the meaning of "bonded" to some extent.

Kallmes and Corte [30,33] developed a model for relative bonded area (RBA) based on a two-dimensional idealization of the structure of the layers of fibers in a sheet. In this model, fibers can exist in three states at any cross-sectional plane: (1) bonded on one side, (2) bonded on two sides, or (3) bonded on neither side. Thus, RBA is a percentage of the total potential bonded area in an idealized sheet of n layers with defined fiber characteristics.

The most common way to measure RBA experimentally is based on the idea that there is often a good correlation between sheet tensile strength (which is dependent on bonding up to the stress level required to start significant fiber ruptures) and light scattering coefficient. This method was originally developed by Ingmanson and Thode [21]. Standard methods such as TAPPI 425 [76], based on light scattering, are commonly employed. See also TAPPI TIS 0804-03 [77]. A good treatment of light scattering can be found in Chapter 4 of Volume 2. Another way to measure the ratio of unbonded fiber surface is the nitrogen gas adsorption (BET) technique [16,17]. Several other experimental techniques are discussed by Paavilainen [52]. Görres et al. [14] have presented a new analysis, which is described in the next section.

C. Bonding States of Fibers

The most basic process of papermaking may be the one in which fibers dispersed in water are deposited on a wire screen. This unique process yields variation in spatial distribution of fibers (formation) and variation of fiber–fiber bonding states in the paper sheet because of the random nature of fiber deposition. The fact of randomness indicates the need for some statistical methods for describing bonding states.

A simple example of a fiber crossing is illustrated in Fig. 2. Because the crossing angle can vary according to the statistical orientation distribution (Chapter 16), the overlapped area, which may be larger than the actual contact area between the fibers, can be described in only a statistical manner. For the case of random orientation (the case wherein the value of the distribution function of the crossing angle is constant over the range), the mean overlapped area $\langle S \rangle$ is given by

$$\langle S \rangle = \frac{\pi \omega^2}{2} \qquad \text{when } \lambda \gg \omega \tag{1}$$

where λ and ω are fiber length and width, respectively. The effective length of the overlapped area can be calculated as $\langle S \rangle / w = \pi\omega/2$ $(= 1.57\omega)$.

This length can be considered an average length of the possible bonding area. In this area, various chemical, physical, and mechanical interactions between fibers take place, such as the formation of hydrogen or other bonds and the setting up of residual stresses caused by different shrinkages in the crossing fibers. This simple example indicates that the possible bonding length is statistically about 60% larger

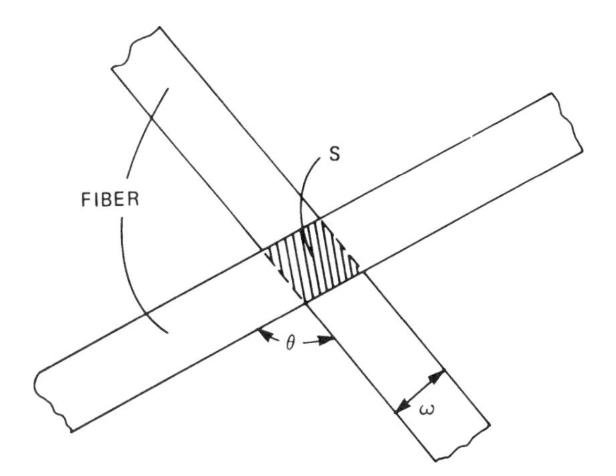

Fig. 2 Simple fiber crossing. θ = crossing angle; ω = fiber width; S = area of overlap.

than the fiber width and that the width of a fiber is a critical parameter in determining the bonding area.

Multiplanar Model The statistical description of the bonding states in paper began with a series of papers on the structure of paper by Kallmes and coworkers [27–32]. Assuming that the process of fiber deposition on a wire screen is a Poisson process, they developed a two-dimensional fiber network model (2-D sheets) and further extended their model to the three-dimensional case, calling it the *multiplanar model* (MPM). Although it is difficult to directly compare the prediction from the Kallmes et al. model with the actual system because of difficulties in determining the parameters involved and the assumptions employed, this model presents some important statistical features of paper structure.

The Poisson distribution [33] most successfully describes the following process. When N_f fibers are deposited randomly into an area A that is sufficiently large compared with fiber length, the probability that k fibers pile up on a point in the area A is given by

$$P(k) = \frac{e^{-H} H^k}{k!} \tag{2}$$

where H is the mean number of overlapping fibers given by

$$H = \frac{N_f \lambda \omega}{A} \tag{3}$$

The probability $P(k)$ can also be regarded as the fraction of the sheet area that is covered by k fibers in the plane. One of the most important parameters of bonding states, relative bonded area (RBA), which is defined as the ratio of the total bonded area to one-half the total external surface area of fibers, can be derived on the basis of the above distribution function.

The maximum relative bonded area RBA_{max}, the upper bound of RBA, can be calculated by assuming that k fibers piled upon a point are completely bonded to each other [33]. That is,

$$\mathrm{RBA}_{\max} = 1 - \frac{1}{H}\left(1 - e^{-H}\right) \tag{4}$$

The parameter H can be expressed in terms of the basis weight (BW) and the fiber weight w_f by the formula $H = (\mathrm{BW}/w_f)\lambda\omega$ [see Eq. (3)]. As basis weight increases, $\mathrm{RBA}_{\max}$ increases and approaches unity as $\mathrm{BW} \to \infty$ or $H \to \infty$. Results of sample calculations of $\mathrm{RBA}_{\max}$ versus basis weight are shown in Fig. 3 [28].* The potential of fiber–fiber bonding (i.e., $\mathrm{RBA}_{\max}$) is basically determined by three parameters: basis weight BW, fiber weight per unit length w_f/λ (coarseness), and fiber width ω. Because for actual fiber networks there is always a strong geometric interaction between fibers, the relative bonded area is lower than the $\mathrm{RBA}_{\max}$.

Kallmes and Bernier [27] obtained an expression of RBA for a multiplanar fiber network consisting of N_L layers of two-dimensional sheets ($N_L \gg 3$).

$$\mathrm{RBA} = \frac{1}{H_{2\mathrm{D}}}\Bigg(\beta_1[H_{2\mathrm{D}} - 1 + P(0)] + [1 - P(0)]^2 \times \left\{\left(1 - \frac{1}{N_L}\right)\left(\beta_2 + \frac{\mathrm{D}P(0)^2}{1 - P(0)^2}\right) - \frac{\mathrm{D}P(0)^2\left[1 - P(0)^{2(N_L-1)}\right]}{N_L\left[1 - P(0)^2\right]^2}\right\}\Bigg) \tag{5}$$

In this equation, $H_{2\mathrm{D}}$ is the mean number of overlapping fibers for 2-D sheets and $P(0) = \exp\{-H_{2\mathrm{D}}\}$. The parameters β_1, β_2, and D represent the completeness of fiber–fiber bonds within a 2-D sheet, between two contiguous layers, and between two layers separated by other layers, respectively. These parameters are therefore dependent on fiber flexibility and the shape of the fiber cross section. Kallmes and Bernier [27] determined β_1, β_2, and D experimentally by examining the crossings of a three-layer sheet, each layer of which was dyed differently. Figure 4 shows the

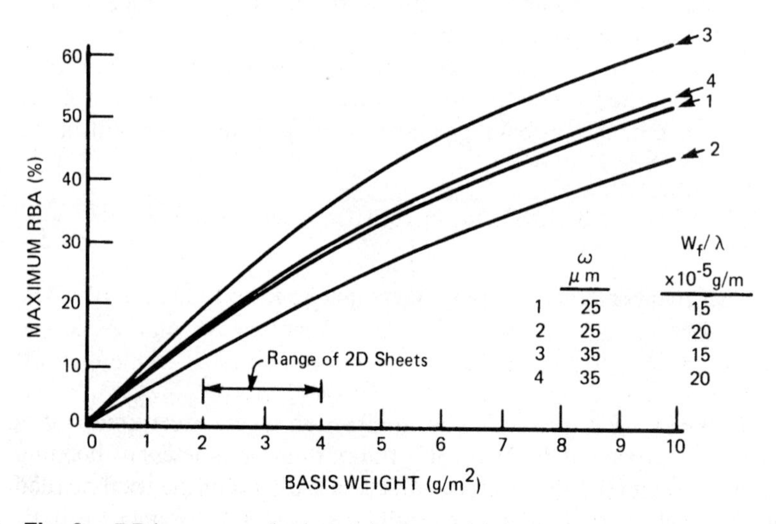

Fig. 3 $\mathrm{RBA}_{\max}$ versus basis weight for multiplanar sheet model. (From Ref. 27.)

* In this model the effect of fiber thickness is not taken into account explicitly (for two-dimensional fibers); the predicted $\mathrm{RBA}_{\max}$ and RBA are always greater than actual values.

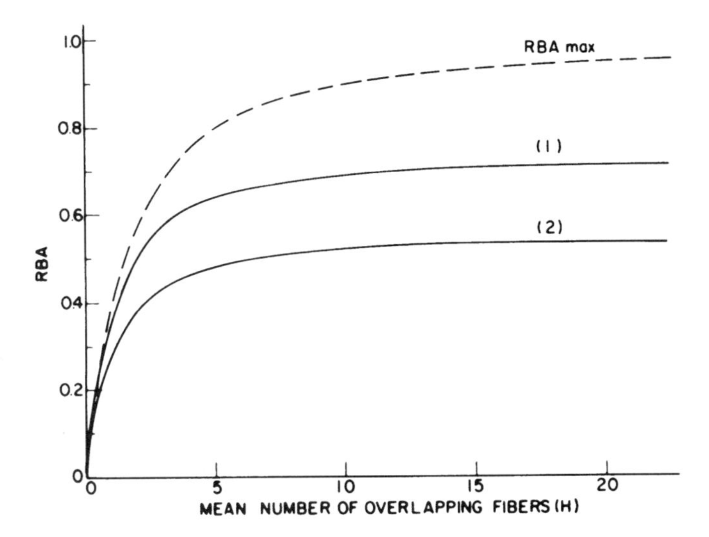

Fig. 4 RBA as a function of mean number of overlapping fibers (H) for a multiplanar sheet model. (1) Mildly beaten kraft ($\beta_1 = 0.98$, $\beta_2 = 0.82$, $D = 1.54$). (2) Unbeaten sulfite ($\beta_1 = 0.84$, $\beta_2 = 0.60$, $D = 1.08$). (Data for β_1, β_2, and D from Ref. 27.)

relative bonded area as a function of H calculated for different β_1, β_2, and D values (mildly beaten kraft and unbeaten sulfite pulps). As expected, the RBA values are highly influenced by fiber flexibility through the parameters β_1, β_2, and D. The effect of the mean number of overlapping fibers H (or basis weight) on RBA is more significant when H is less than 5 (i.e., the mean number of overlapping fibers is less than 5). Beyond this value, RBA approaches its limiting value very slowly as H (or basis weight) increases. The value of 100% for RBA is attainable only for sheets with infinitesimally smaller fiber thickness and an infinite number of fibers per unit area (infinite basis weight). These results suggest that

1. The basis weight for paper is not simply a scaling parameter such as thickness for a homogeneous continuum but is indeed one of the important structural parameters.
2. The degree of fiber–fiber bonding in a paper sheet generally varies with basis weight.

The concept of RBA refers only to the overall condition of the bonding states in the sheet. If one looks at a typical fiber in the paper sheet (see Fig. 5), one can see three different bonding states along the fiber: free on both surfaces of the fiber (0 state), bonded on one side (1 state), and bonded on two sides (2 state). The fiber segments that correspond to the above three bonding states in Fig. 5 may have different mechanical, optical, and electrical properties. An analytical relation between RBA and the three bonding states was developed by Kallmes et al. based on the *multiplanar fiber network model* (see Appendix 1 in Ref. 32). The relationships between the bonding states and RBA are illustrated in Fig. 6. As Kallmes indicated [33], the basic feature of the relations does not change very much, even for the case of complete interfiber bonding. It should be noted that the value of each fraction shown in Fig. 6 is the mean value for the entire sheet. The bonding states of fibers, however,

Fig. 5 Idealized bonding states of a typical fiber. (From Ref. 29.)

generally vary through the sheet thickness; that is, the layers near the sheet surface have different bonding states than the inner layers [28,32]. This inhomogeneity of bonding states through the thickness suggests the possibility of basis weight dependence of some properties.

Although the statistical model mentioned above describes the basic structure of randomly deposited fiber networks, there are some limitations inherent in the basic assumptions of the model. For example, it is assumed that the fiber is a flattened ribbon and that bonding is possible only on the upper and lower surfaces of the fiber, which is not the situation observed experimentally [84]. Development of a general three-dimensional statistical model of paper structure would be highly helpful in analyzing experimental results obtained by increasingly sophisticated image analysis techniques.

Interactive Multiplanar Model Görres et al. [15] introduced an "*interactive*" *multiplanar model* (IMPM) that they used to derive a sheet structural parameter similar to RBA from four measurable fiber properties (thickness, width, coarseness, and flexibility) and a sheet densification factor based on stylus profilometry measure-

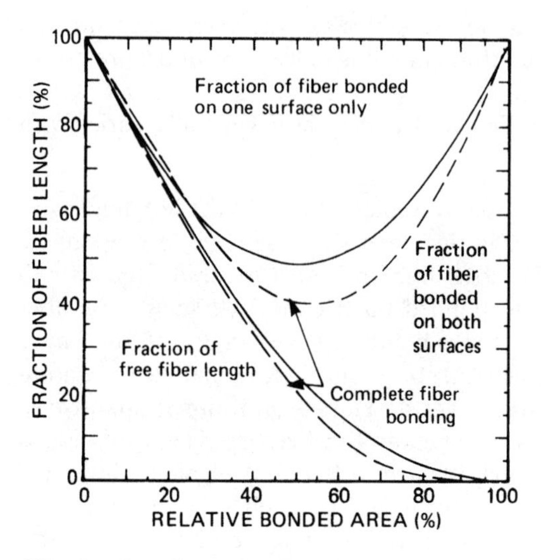

Fig. 6 Bonding state diagram for multiplanar sheet model. (From Ref. 32.)

ments of fiber collapse during wet pressing [13]. This model was used in the prediction of sheet apparent density as well as the shear bond strength [14]. This section outlines the theoretical basis for the model, from which bond strength predictions will be developed later in this chapter.

The IMPM is an extension of the MPM and shares the assumptions that

1. The structure of each layer within the sheet is adequately described by network statistics [30].
2. In the absence of pressing, the sheet layers are parallel and there are no interactions of fibers between layers.
3. Upon wet pressing, the out-of-plane deflection of fibers causes the interaction of fibers between layers.

The IMPM further assumes that fiber layers are one fiber thickness thick and that fibers deflected to a bonded state between layers remain deformed after pressing.

The fraction of bonded crossings between two fiber layers separated by m layers, $R_b(m)$, was represented by Görres et al. [15] as

$$R_b(m) = \exp\left[-4S_{xy}(m)\frac{\mathrm{BW}}{\pi C}\right] \tag{6}$$

where C is fiber coarseness and BW is basis weight of the network [15]. $S_{xy}(m)$ is the flexibility function derived by Steadman and Luner [70], who considered the deflection of a fiber as a deflecting cantilever beam $S_{xy}(m)$ is effectively a measure of how well fibers can be deflected in the z direction during pressing. $S_{xy}(m)$ is determined from fiber thickness to fiber width w, wet fiber flexibility F, and applied pressure P by the relationship

$$S_{xy}(m) = \left(\frac{72tm}{2PwF}\right)^{1/4} \tag{7}$$

The bond of fibers separated by m layers is limited by the amount of projected open area, or void space contained within the separating layers, having a basis weight $(\mathrm{BW})_1$. This void space can be determined from network statistics by the relationship

$$P(0, m) = \exp\left[\frac{-mw(\mathrm{BW})_1}{C}\right] \tag{8}$$

For central layers not bounded by the web surface, there are $2N$ layers contributing to increased weight as fibers from adjacent layers are deflected into the central plane. The total layer weight after wet pressing, Λ, can therefore be represented as

$$\Lambda = 1 + \sum R_b(m)P(0, m) \tag{9}$$

The bonded fiber area per mass of fiber, α, can be derived from the number of fiber crossings in a layer, N_C, the number of layers in the sheet, N_L, and the total layer weight after pressing, Λ, by the relationship

$$\alpha = \frac{2N_L N_C \Lambda}{A(\mathrm{BW})}\langle S\rangle \tag{10}$$

where A is the area of the sheet and $\langle S \rangle$ and N_L are the mean overlap area and number of layers in the sheet respectively, as defined above. The number of crossings, N_C, in a layer may be determined from the total fiber length λ_t^2 by using the equation

$$N_C = \frac{\lambda_t^2}{\pi A} \tag{11}$$

This parameter was used with the aid of the Page strength equation [54] to calculate shear bond strengths for sheets made from various chemical and mechanical pulps that resulted in good correlations with experimental values. This is discussed in details in Section III.B.2.

D. Experimental Determination of RBA

Direct Methods The direct determination of RBA requires measurement of both total surface area of fibers and bonded areas of fibers in the sheet structure. Yang and coworkers [84] and Eusufzai [10] pioneered the determination of RBA using image analysis of paper cross sections. This method essentially consists of image analysis of photomicrographs of sheet cross sections by the use of a digitizer interfaced with a computer. The perimeter length of fiber cross sections and the bonded length, as shown in Fig. 7, are measured to determine RBA. Today, with the use of a CCD camera, a computer, and powerful image analysis software, most of the time-consuming processes of data acquisition and image analysis are automated. An example can be found in this handbook in Chapter 16.

Although the principle of this technique is relatively straightforward, and image analysis software provides various image-enhancing routines, the accuracy of the measurement still resides in the quality of the images of the paper cross sections. First, the micro cross section must be very thin ($< 1\,\mu$m) to obtain a

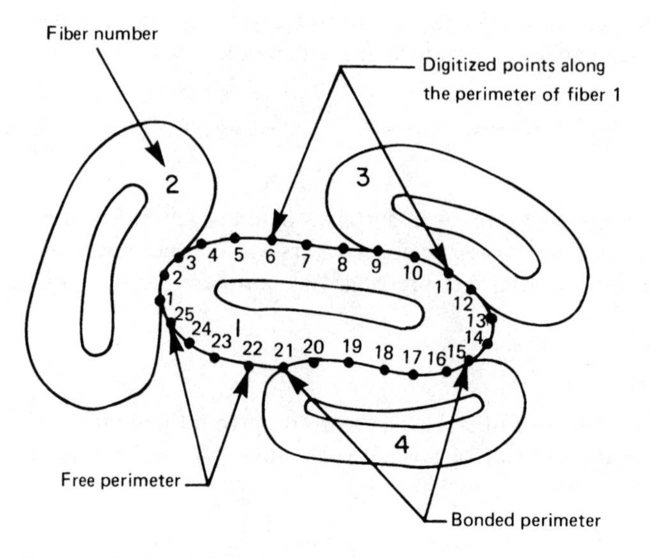

Fig. 7 Point mode loci of fiber cross sections. (From Ref. 84.)

sharp image of the bonded interface. Hasuike [18] demonstrated that the thickness of the cross section has an enormous effect on the accuracy of the fiber geometry measurements: Thicker cross sections consistently give artificially higher estimates of actual fiber perimeter and cross-sectional area. In addition, the requirement for thinner cross sections poses the challenge of preparing the sample with minimum damage through the entire microsectioning process.

In order to avoid the difficulties associated with thin microsectioning, Williams and Drummond [81] developed a method that involves the observation of a sectioned block, instead of a thin cross section, using scanning electron microscopy (SEM). The sample is first embedded in resin, and then the resin block containing the sample is sectioned to expose the cross section of the sheet structure. The sectioned surface of the block is polished and then slightly eroded with a solvent to enhance the contrast of the SEM images. This technique gives a high contrast image, which is particularly suited for image analysis. Another advantage is that it is effective in viewing fine structures, such as bonded area, at the higher magnifications available in SEM. At the same time, it allows observations of a large (long) cross section which is required for obtaining statistically significant analysis data [81].

In the techniques that use image analysis of sheet cross sections, RBA may be redefined as

$$\mathrm{RBA} = \frac{1}{N}\sum_{i}^{N}\mathrm{RBA}_i = \frac{1}{N}\sum_{i}^{N}\frac{\text{number of bonded sites at the } i\text{th position}}{\text{total number of available sites at the } i\text{th position}} \tag{12}$$

RBA can be measured at each position in the cross section, as shown in Fig. 8. This definition gives RBA = 1 when the fibers are completely bonded in the sheet structure, whereas the traditional definition still gives RBA values of less than unity because of the presence of the paper surface area [40].

Indirect Methods Current techniques for indirectly measuring RBA are based on either a gas adsorption method or optical scattering methods. Swanson [74] reviewed the early literature on these techniques.

Gas Adsorption Method The gas adsorption technique was employed by Haselton [16,17] to determine the bonded area with the use of nitrogen adsorption and the

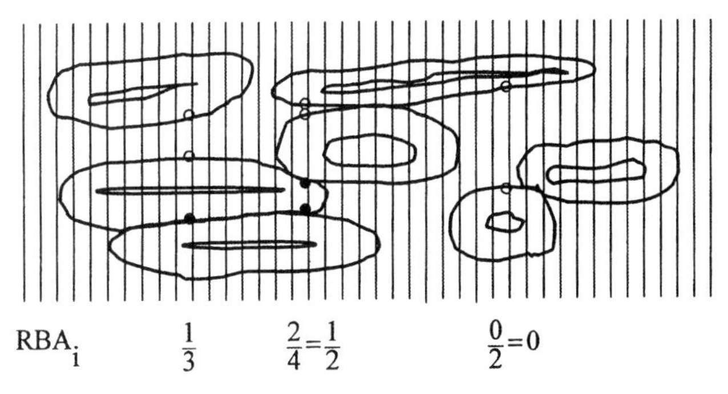

Fig. 8 Determination of RBA from line crossing. (○) Sites available for bonding; (●) bonded sites.

Brunauer–Emmett–Teller (BET) equation (see Chapter 6 of Volume 2). This technique measures both internal and external surface areas. These areas include the outer surface area of fibers, surface area of lumen, and any surface area in the cell wall accessible to the gas. The bonded area is defined as the difference between the specific surface area of the sheet and the specific surface area of the "unbonded" sheet.

Measurement of the "unbonded" area is, however, not straightforward. Haselton used two drying techniques to obtain an unbonded sheet. One is solvent-exchange drying from water to benzene (or butanol) to prevent hydrogen bonding between fibers and fibrils. The other technique is spray drying, where 0.2% consistency pulp slurry was sprayed onto a horizontal sheet of polyethylene and dried so that drying conditions were close to those of the usual water-dried sheet. Comparisons of these techniques showed that the solvent-exchange drying technique consistently gave higher values of specific surface area than those for the spray-dried fibers. This difference was explained by an electron micrograph showing that surfaces of solvent-exchange dried fibers still have pronounced surface fibrillation, whereas surfaces of spray-dried fibers area almost free of projecting fibrils.

These results clearly show the importance of subjecting fibers to a realistic drying process (or, in general, to a realistic papermaking process) to obtain a fiber surface that closely resembles the one that is actually present in the sheet.

Optical Scattering Method The most commonly used technique is the optical scattering method [21,23,35,39,40,42,75], which utilizes the specific scattering coefficient from application of the Kubelka–Munk theory (see Chapter 4 of Volume 2). The basic assumptions involved in this method are as follows:

1. The scattering coefficient S is proportional to the "external" surface area A of fibers in the sheet:

$$S = kA \tag{13}$$

2. The scattering coefficient of an unbonded sheet, S_{total}, is determined by the extrapolation of a tensile strength versus S curve to zero tensile strength. (Elastic modulus is also used for the same purpose.)

Therefore, RBA is determined as

$$\mathrm{RBA} = \frac{A_{bonded}}{A_{total}} = \frac{A_{total} - A}{A_{total}} = \frac{S_{total} - S}{S_{total}} \tag{14}$$

Evaluation of Indirect Methods There have been numerous discussions on the validity of these assumptions. First, it is known that the optical contact area is not necessarily the area where actual bonding exists. Therefore, the change in S is not necessarily a reflection of the change in the actual contact area (or actual non-contact area). Second, the fiber surface areas responsible for light scattering are not only the external surface area of fibers but also the surface areas in the lumen and, to a much lesser extent, the areas within the cell wall (such as large cracks in the fiber cell walls in heavily calendered sheets). Therefore, if some lumens are collapsed, for example, S also can change *regardless of the change in bonded area.*

Haselton [16,17] and Swanson and Steber [75] showed an excellent linear relationship between scattering coefficient and surface area measured by the BET nitrogen adsorption technique. This correlation, however, does not necessarily imply the validity of assumption 1. As mentioned earlier, the gas adsorption technique also measures both internal and external surface areas; therefore, when fibers and sheets are subjected to refining and wet pressing, both measurements are expected to yield similar changes in the surface areas.

Third, extrapolated values (S_{total}) have been known to be dependent on whether refining or wet pressing is used to change tensile strength. When refining is used, the degree of surface fibrillation varies among different samples, and thus the proportional constant k varies, violating assumption 1 [23]. On the other hand, when wet pressing is used, the degree of collapse of the lumens may vary among the samples [40], and as a result the one-to-one relationship between scattering coefficient and external surface area may not hold. In the case of using wet pressing to obtain the extrapolated value (S_{total}), the magnitude of the error depends on the degree of lumen collapse of the fiber used [24,40].

Therefore, in using these indirect methods and in interpreting the results obtained from these methods, the errors associated with the following effects must be assumed:

1. The structural scale that the technique can probe
2. The changes in surface areas other than external surface areas, such as lumen and cracks in the cell wall
3. The method to change the bonded area to obtain "unbonded" sheets

III. MECHANICAL TESTING AND EVALUATION OF FIBER BONDS

A. Determination of Bonding Strength

Bonding strength (specific bond strength), i.e., the bond strength per unit bonded area, is measured either by using single fiber crossing or from the network. The value obtained for bonding strength depends greatly on which measurement method is used. If laborious single-fiber measurements are used, the effect of sheet structure can be avoided. In addition, the bonding strength distribution would characterize average bonding strength better than measurements made from a sheet, because network breakage is a chain reaction that starts from a weak fiber–fiber bond or from a weak region in the fiber.

Direct Methods In measurements on single fibers, the shearing force needed to break a bond is in general measured per unit bonded area using a tensile strength apparatus. McIntosh [46], Schniewind et al. [61], Yang et al. [84], Eusufzai [10], and Stratton [72] measured the shear strength of contacts between a fiber and a shive, and Mohlin [47] determined the shear strength of single fiber–cellophane crossings. Mayhood et al. [45] used a modified chainomatic balance to measure the shear strength of individual fiber–fiber bonds. Bonded area is usually determined in these methods as the area of optical contact at fiber interfaces [53].

Indirect Methods Nordman et al. [51] assumed that the energy absorbed by a sheet during tensile straining is consumed in rupturing interfiber bonds. Thus, the transformation of bonded areas into unbonded areas can be determined by measuring the light scattering coefficient or by using the BET gas adsorption technique [71].

Bonding strength is given in terms of the energy required to reduce the bonded area by one unit of light scattering coefficient. However, breakage of intrafiber bonds and sheet straining seem to interfere with the measurement. Smith and Graminski [68] determined bonding strength from the load–elongation curve of 2-D sheets, where each jag presents the breakage of an interfiber bond. Skowronski [64,66] measured the force needed to delaminate a paper sheet by pulling on plastic tape fastened to opposite surfaces and obtained the bonding strength by dividing the bond-breaking energy by the increase in light scattering coefficient. Retulainen and Ebeling have presented a good evaluation of the indirect bonding strength measurement methods and the factors affecting them [56,57].

B. Estimation of Shear Bond Strength

Page Equation A semiempirical equation for tensile strength was proposed by Page [54] and later applied in the evaluation of unbleached kraft pulps by Jones [26]. This equation is expressed in terms of the tensile breaking length T as

$$\frac{1}{T} = \frac{9}{8Z} + \frac{1}{B} \tag{15}$$

where the first term uses zero-span tensile strength Z as an indication of fiber strength. The second term gives the contribution of interfiber bonding B as

$$\frac{1}{B} = \frac{12g}{b\lambda}\left(\frac{C}{P(\mathrm{RBA})}\right) \tag{16}$$

where

$b =$ shear bonding strength
$g =$ gravitational constant
$C = \rho A =$ coareseness
$\rho =$ cell wall density
$A =$ cross-sectional area of the fibers
$\lambda =$ fiber length
$P =$ perimeter of fiber cross section
RBA= relative bonded area

The shear bonding strength can be calculated from Eqs. (15) and (16) as

$$b = \frac{12g}{(1/T - 9/8Z)\lambda}\left(\frac{C}{P(\mathrm{RBA})}\right) \tag{17}$$

The determination of fiber cross-sectional parameters may be experimentally tedious because it involves image analysis of sheet cross sections [36] and/or measurement with confocal microscopy [63]. Methods for the determination of RBA were discussed earlier in this chapter.

In using this method to calculate the bonding strength of beaten fibers, one should remember that lumen collapse, external fibrillation, and fines formation may affect the RBA, although according to Ingmanson and Thode [21] the relationship between tensile strength and light scattering ability is the same irrespective of the beating degree. However, these factors do not necessarily increase bonded area by the same magnitude.

Modified Page Equation Görres et al. [14] suggested an alternative approach for estimating shear bond strength form the Page equation that avoids the difficulties in obtaining accurate values for sheet structural parameters. The method uses the bonded fiber area per mass of fiber, α, derived from the results of the *interactive multiplanar model* (IMPM) [15] as discussed in Section II.C.2. Using Eq. (10), the second term of Eq. (16) may be rewritten as

$$\frac{1}{\alpha} = \frac{C}{P(\mathrm{RBA})} = \frac{A(\mathrm{BW})}{2N_L N_C \Lambda}\left(\frac{1}{\langle S \rangle}\right) \tag{18}$$

Equation (18) can be solved by inserting experimental parameters calculated using Eqs. (6)–(9). Görres et al. [14] calculated the bond strength B by substituting the IMPM derived term for the second term in Eq. (16), which resulted in the relationship

$$B = \frac{b\lambda N_C N_L \Lambda \pi w^2}{12g(\mathrm{BW})A} \tag{19}$$

where g is the gravitational constant and the units are in breaking length (km). They found good agreement between calculated and measured bond strength for chemical and mechanical pulp fibers as illustrated in Fig. 9.

C. Determination of *z*-Directional Strength

In addition to machine and cross-machine directions, the *z*-directional (thickness) strength of paper is important as a functional property for several paper and board grades. It is also an important measure for the thickness or *z*-directional internal cohesion of paper, which is strongly affected by interfiber bonding.

In the layered structure of a handsheet, the *z*-directional strength of paper arises mainly from interfiber bonds. In this case, a *z*-directional test can be seen as

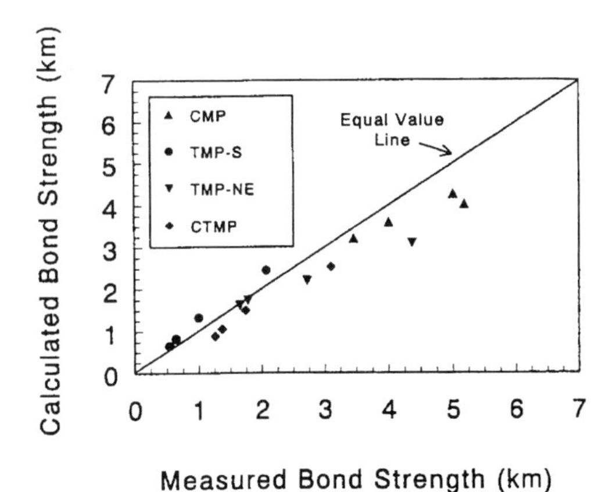

Fig. 9 Comparison of calculated and measured bond strength values B in units of breaking length for pulps and blends of kraft fibers with TMP, CTMP, and CMP. (From Görres et al. [14].)

an ideal method for measuring the strength of interfiber bonds. However, all commercial sheets, especially those formed at high consistency, contain fiber entanglement and z-directional orientation (see Chapter 16), so that fibers extend from one layer to another. Because fibers are much stronger axially than they are laterally and are stronger than the interfiber bonds, this kind of entanglement and z orientation can strongly contribute to the z-directional strength of paper. Although there are several methods for measuring the z strength of paper, they are all affected—in addition to pulp properties—by the forming process and the structure of the fiber network and its bond structure. Some of the methods are designed for measuring primarily the delamination energy, others for measuring the detachment force. Tests range from near static to dynamic, including impact-type tests. Force- or stress-based methods are sensitive to uneven loading and to local stress concentrations in the sample. Energy-based values tend to include also energy dissipated in plastic deformations that are not related to the actual rupture process.

Strength per actually debonded (actually bonded before test, then ruptured in the test process) fiber area, often called specific bond strength, may also be measured if the change in the nonbonded fiber surface area is estimated, for example, by light scattering measurements.

z-Tensile Strength The standard z-directional tensile test can be carried out using a conventional tensile tester and a special testing assembly. The paper sample is attached on both sides to metal blocks with two-sided pressure-sensitive tapes (Fig. 10). The sandwich thus obtained is pressed to get good adhesion between the tape and the sample and the tape and the block. The whole system is then placed into a tensile testing apparatus. The system is loaded at a constant speed until the paper sample is delaminated. The maximum force is measured and divided by the sample area to give the z-strength. In addition to the failure force, the elongation and failure energy can be recorded. Several standards are available for the measurement of z strength: TAPPI T 541 om-89, PAPTAC D.37P, and SCAN-P 80:98.

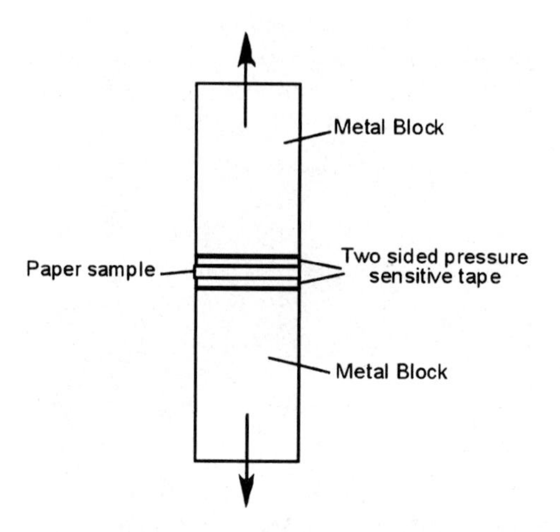

Fig. 10 Arrangement for measuring z-directional tensile strength of paper.

Dependent on the variation of the method, the cross section of the blocks (and thus also the shape of the samples used) may be circular or rectangular. The blocks can be made of metallic or polymeric material as long as their dimensional changes are negligible compared with the sample and their faces can be maintained clean and smooth. In some cases, glue has been used for fixing the samples to the blocks, but the results are somewhat different. It is important that the tape be fresh and of a constant quality. The failure should occur within the specimen, not between the sample and tape.

The *z*-directional deformation of paper was studied by Wink and Van Eperen [83], and Van Liew [80]. The *z*-directional modulus is typically about one-tenth of that in-plane. The fracture strain is normally 0.25–0.7% and is mostly recoverable but accompanied by a large hysteresis. Significant nonrecoverable deformations occur only at strains close to failure. Sample deformations can be explained by *z*-directional intrafiber deformation and fiber debonding. Measurable debonding takes place only at stresses exceeding about 85% of the breaking load. However, some debonding most likely takes place at lower load levels. Because debonding concentrates within narrow planes within the sample, it cannot be easily detected [80].

Debonding on the fiber level is assumed to take place by a peeling mechanism, which concentrates the stresses at the periphery of fiber bonds. Typically, delamination does not take place in parallel planes, indicating that there are stress concentrations within the specimen area. The sample does not necessary delaminate in the weakest zone of the sheet.

Andersson [1–3] made an extensive study of *z* strength for the characterization of different pulps. The results are independent of fiber strength and fiber length but are affected by the pressure applied during sample preparation, the adhesive used, and the straining rate. However, the result depends on the basis weight. In the low basis weight range, an increase in basis weight causes a decrease in the *z* strength [83], whereas the drying mode (free vs. restrained shrinkage) and fiber flocculation have very little effect. Further information on this method is given in Chapter 7.

Modified *z*-Tensile Strength Methods Two main modifications of the *z*-tensile strength tests are the Brecht–Knittweis method [19] (DIN 54516) and the modified Mullen method [7] (TAPPI UM 522, PAPTAC Useful Method D.8U). The Brecht–Knittweis method applies the force nonuniformly on the sample area (Fig. 11). The sample is fixed on a block that is loaded from one side of the sandwich. The modified Mullen method (Fig. 12) uses a burst tester with annular specimens. The sample is attached to an annular metal ring and a metal disk with double-sided tapes. The expansion of the diaphragm against the solid disk ruptures the specimen. The pressure reading of the tester is taken as the result. However, because the contact area of the diaphragm and the disk changes during the test, there is no means to convert the test result to a true force per sample area value.

Specific bond strength values have been also determined based on *z* tensile tests. Hieta et al. [20] used a *z*-directional tensile test for estimating the specific bond strength between a cellophane film and a handsheet that were dried together. An estimate for the bonded area between the sheet and the film was obtained by using a special staining technique and image analysis.

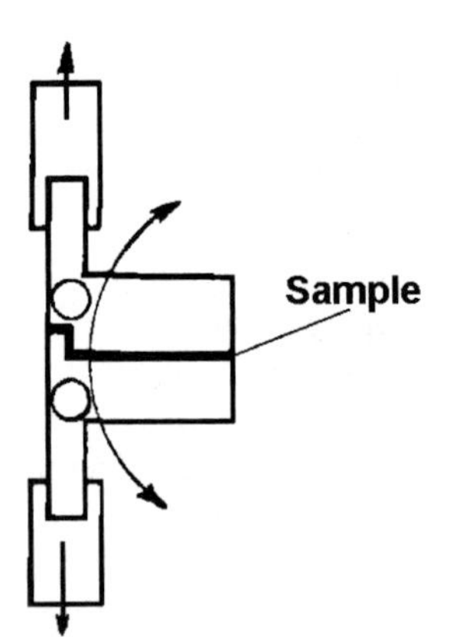

Fig. 11 Measurement of z-strength according to the Brecht–Knittweiss method. (From Ref. 19.)

Peeling Tests In peeling tests, a sample stripe is delaminated so that the delamination front progresses gradually through the specimen. Usually, strong pressure-sensitive tape adhered to both sides of the specimen is used to transmit the external force into the sample. A prerequisite for this test, as with others that use tapes, is that the adhesion between the tape and the sample is higher than the internal cohesion of the sample. The results are affected, for example, by the delamination speed, the support system, and the peeling angle. The main peeling test principles are illustrated in Fig. 13.

The peeling angle θ affects the results in that the force of peeling is smallest when the angle is between 120° and 140° [8,9]. The calculated work of peeling, however, increases with the angle of peeling.

In addition to the energy consumed in the breaking of the interfiber bonds, some energy is dissipated in plastic and viscoelastic deformation related to bending of the specimen. This energy increases with increasing peeling angle. The energy consumption is also dependent on the thickness of the top layer detached from the sample. There is always some variation in layer thickness or peeling angle, which causes variation in the measured values. Those methods where the peeling angle can be kept constant (i.e., Figs. 13c and 13d) can be considered most reproducible.

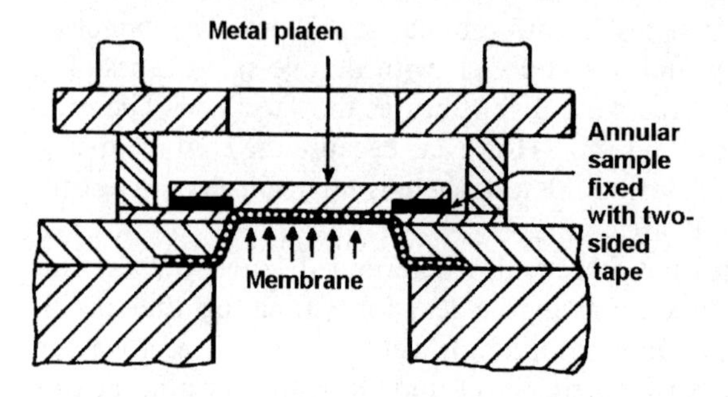

Fig. 12 Burst tester application in z-strength measurements. (From Ref. 3.)

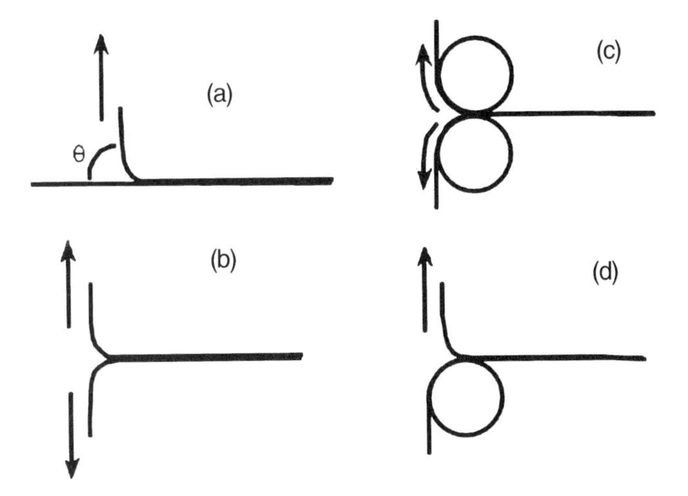

Fig. 13 Different peeling test modifications. (See text for details.)

Usually, pressure-sensitive tapes are used as the medium to transmit the delamination force to the sample. Tacky, highly viscous ink has been used for delaminating the sheet in a nip (see Fig. 13c). With the type of ink held constant, the speed needed to delaminate the sample is used as a measure for the *z*-directional strength [4,82]. In addition, the product of the ink viscosity and the delamination velocity is used as a measure of the internal strength of the sample [22,82]. However, with this ink-based method no unambiguous measure for the force or energy needed to delaminate the sheet is obtained. In any event, this test closely simulates conditions found in a printing nip and can give valuable information of practical importance. The high loading speed used makes this test more dynamic than the other methods in this group.

Although the basic principles of the tests discussed above are very similar, one should be careful when comparing results made with different methods. Testing conditions such as the delamination speed, peel angle, and the geometry and symmetry of the testing arrangement affect the results. The sheet structure and basis weight and its local variation also contribute to the result. For example, in the test method of Fig. 13c the delamination energy has been found to first decrease with the basis weight and then level off, whereas with that of Fig. 13d the delamination energy, after the initial decline, has been found to increase again [48,65]. With machine-made samples, results from the peeling methods may also be affected by the peeling direction. When the peeling takes place "downstream" in the machine direction, smaller values may be obtained than when peeling takes place upstream in the machine direction.

An advantage of the peel-type test is that when transparent tapes are used the change of light scattering coefficient that occurs in the peeling test can be measured and used as an estimate of the debonded area [65]. The specific bond strength of the paper can also be estimated.

Internal Bond Strength Test The internal bond strength method, also called the Scott bond test (TAPPI T 833 pm-94), is probably one of the most commonly used methods for estimating the z-directional strength of paper. A special apparatus is commercially available for this purpose. Double-sided adhesive tapes are used to attach the sample to two metal pieces, the upper one of which is an angle piece made of aluminum. A pendulum hits the angle pieces as illustrated in Fig. 14 and causes the sample to break in cleavage mode rather than in a shear mode [58]. The rupture of the sample begins at the front side of the specimen. Because of the high loading rate, fractures in the fiber wall occur [67]. The energy used in the impact phase is obtained from the change in the kinetic energy of the pendulum. Further analysis of this method appears in Chapter 7.

z-Toughness In analogy with the in-plane fracture toughness (see Chapter 7), the z-directional toughness can also be estimated. By definition, fracture toughness is the energy dissipated in the crack tip area during stable crack growth per unit crack area formed. The peeling methods are close to this concept. However, the question is how well the plastic energy dissipation can be limited to the crack tip area; also, are the deformations small enough to use elastic theory? Attempts have been made to solve this problem by using two stiff beams for transmitting the z-directional force as shown in Fig. 15 [41,62].

z-Modulus The in-plane modulus of paper is closely correlated with the extent of interfiber bonding. Because bonds play a greater role in the z direction than in the sheet plane, the z-directional modulus, or out-of-plane elastic stiffness C_{33}, should be expected to reflect the degree of bonding in paper rather closely. This has been found to be the case except when strength additives are used in the furnish [73]. This method is sensitive to the area of interfiber contacts but does not take into account the strength of the bonds. In addition to interfiber bonding, the transverse modulus of fibers affects the results [73].

Comparison of the Methods Several factors affect the mechanical properties of bonds, among them pulp type, beating degree, wet pressing pressure, basis weight, fines amount, sheet formation, drying shrinkage, wet strength chemicals, defoaming agents, and alum [2,19]. The effect of these variables, however, may be reflected in different ways with different test methods. No comprehensive attempts have been made to compare the different methods of measuring the z strength. However, several authors have compared two or three methods with some in-plane strength tests.

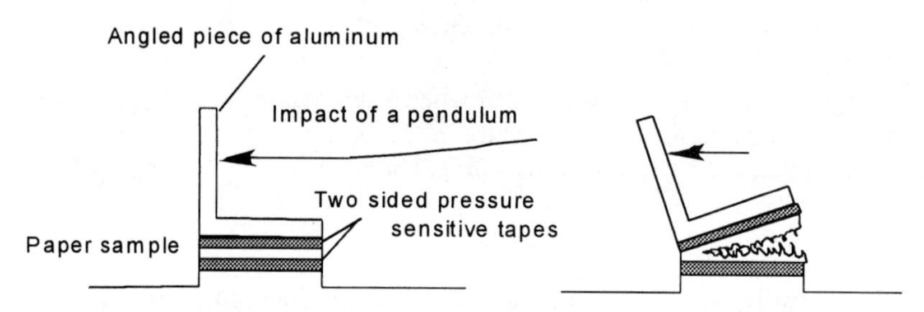

Fig. 14 Internal bond strength test measures the energy needed to split the paper sample.

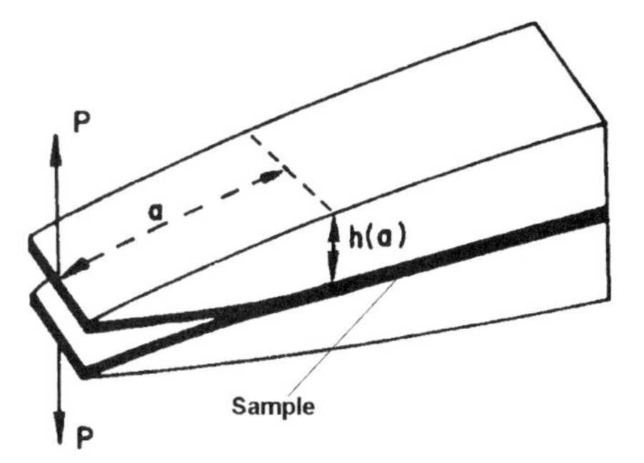

Fig. 15 Contoured cantilever beam arrangement for measurement of z toughness. (From Ref. 62.)

The internal bond strength and results from the Skowronski peeling method (see Fig. 13d) show increases with increasing basis weight, while the z tensile strength and z tensile energy do not. In thicker sheets, ruptures that are more partial can be observed. Fiber ruptures probably occur in the methods of high loading rate because of stress concentrations that do not have time to disperse or even out. Fiber ruptures have been observed in the internal bond strength test, but not in peel tests conducted at low delamination speed [67].

An important factor with the method where energy is measured is that energy is consumed not only in the breakage of interfiber bonds but also in other deformation processes [37,57]. Moreover, the amount of energy consumed in these other processes varies with the method and sample properties. Force-based test values correlate better with each other. Koubaaa and Koran [37] found that the z tensile test provides more consistent results than the internal bond strength method or Skowronski peel tests (type in Fig. 13d).

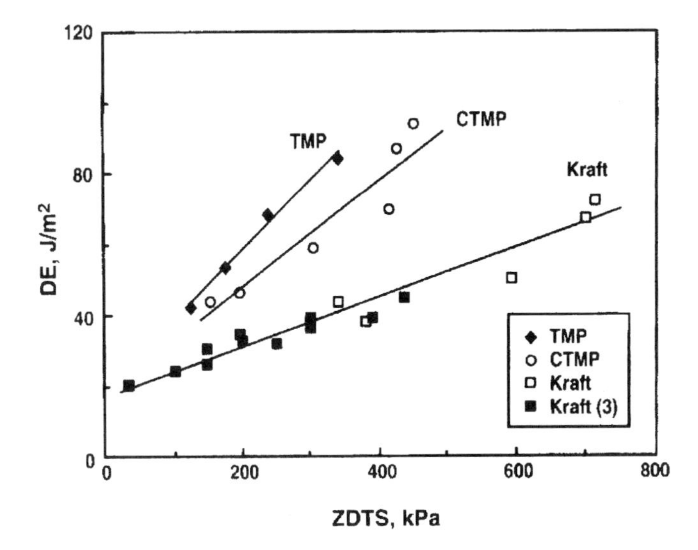

Fig. 16 The relationship between the delamination energy (DE) from Skowronski peel tests and z directional tensile strength (ZDTS) depends on the furnish. (From Ref. 37.)

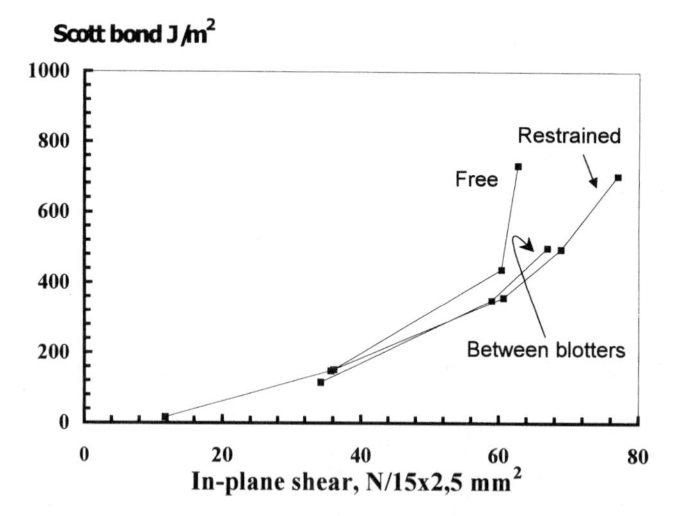

Fig. 17 The out-of-plane strength (Scott internal bond strength) at given in-plane shear strength values depends on the drying mode used. (From Ref. 57.)

Also, Retulainen and Ebeling [57] compared several methods. Usually the methods correlate rather well when only one sheet structure parameter is changed in a controllable way. However, if sheet structure or the furnish composition is changed, the situation becomes rather problematic. Certain fiber properties and sheet structures are related to certain bond structures. Different methods emphasize different structural aspects of the loading behavior of paper and fiber bonds. For example, a clear difference between in-plane and out-of-plane methods can be found, especially when drying shrinkage of the paper is increased (Fig. 17). The fiber bond is three-dimensional and anisotropic. The strength of a bond is different in the z direction than in the in-plane direction. There is no simple relationship between these two strengths.

REFERENCES

1. Andersson, M. (1981). Aspects of z-strength in pulp characterization. *Svensk Papperstidn. 84*(6):R34–R42.
2. Andersson, M. (1981). *z*-Directional strength to characterize semi-chemical pulp: Measurement techniques and the influence of sheet forming parameters. *Papier 35*(2):49–61 (in German).
3. Andersson, M. (1979). Z-tensile strength. In: *Pulp characterization. Mechanical properties*, P.O. Bethge, ed. SCAN Forsk Rapport 206. STFI, Stockholm, pp. 141–159 (in Swedish).
4. Attwood, B. W. (1969). A method of measuring the plybond strength of paperboard. *Paper Technol. 10*(2):125–128.
5. Campbell, W. B. (1933). Forest Service Bull. 84. Department of Interior, Canada.
6. Cisnersos, H. A., Williams, G. J., and Hatton, J. V. (1995). Fibre surface characteristics of hard wood refiner pulps. *J. Pulp Paper Sci. 21*(5):J178–J184.

7. Donahue, J. F., and Verseput, H. W. (1957). A method for routine measurement of the ply-bond strength of paperboard. *Tappi 40*(5):311–313.
8. El Maachi, A., Sapieha, S., and Yelon, A. (1995). Angle dependent delamination of paper. Part 1. Inelastic contribution. *J. Pulp Paper Sci. 21*(10):J362–J366.
9. El Maachi, A., Sapieha, S., and Yelon, A. (1995). Angle dependent delamination of paper. Part 2. Determination of deformation and detachment work in paper peeling. *J. Pulp Paper Sci. 21*(12):J401–J407.
10. Eusufzai, A. R. K. (1982). Sheet structure in relation to internal network geometry and fiber orientation distribution. M.Sci. Thesis, SUNY, College of Environmental Science and Forestry, Syracuse, NY.
11. Eusufzai, A. R. K., Thorpe, J. L., and Mark, R. E. (1975). Further studies on the mechanical properties of fiber-fiber bonds. Empire State Paper Res. Inst. Rep. 63, Chap VI, pp. 93–107.
12. Garland, H. (1939). A microscopic study of coniferous wood in relation to its strength properties. *Ann. Missouri Bot. Gard. 26*:1–95.
13. Görres, J., Amiri, R., Grondin, M., and Wood, J. R. (1993). Fibre collapse and sheet structure. In: *Products of Papermaking*, C. F. Baker, ed. Pira Int., Leatherhead, U.K., pp. 285–310.
14. Görres, J., Amiri, R., Wood, J. R., and Karnis, A. (1995). The shear bond strength of mechanical pulp fibres. *J. Pulp Paper Sci. 21*(5):J161–J164.
15. Görres, J., Sinclair, C. S., and Talentire, A. (1989). An interactive multi-planar model of paper structure. *Paperi Puu 71*(1):54.
16. Haselton, W. R. (1954). Gas adsorption of wood, pulp and paper. 1. The low temperature adsorption of nitrogen, butane and carbon dioxide by spruce wood and its components. *Tappi 37*(9):404–412.
17. Haselton, W. R. (1955). Gas adsorption of wood, pulp, and paper. 2. The application of gas adsorption techniques to the study of the area and structure of pulps and the unbonded and bonded area of paper. *Tappi 38*(12):716–723.
18. Hasuike, M. (1989). Morphological evaluation of the structure of paper. *Jpn. J. Paper Technol. 4*:1–7 (in Japanese).
19. Heckers, W., Baumgarten, H. L., and Gottsching, L. (1979). Determination of split-resistance according to the Brecht-Knittweis method. *Das Papier 33*(10):457–465 (in German).
20. Hieta, K., Nanko, H., Mukoyoshi, S., and Ohsawa, J. (1990). Bonding ability of pulp fibers. *1990 Papermakers Conference. TAPPI Proc.* pp. 123–131.
21. Ingmanson, W. L., and Thode, F. T. (1959). Factors contributing to the strength of the sheet of paper. II. Relative bonded area. *Tappi 42*(1):83–93.
22. Institute of Paper Chemistry Staff (1946). Instrumentation Studies. 55. Determination of the bonding strength of paper. *Paper Trade J. 123* (18):24,26,28–29; (19):24,26,28,30,32,34.
23. Jacobs, R. S., Cole, B. J. W., and Genco, J. M. (1994). The relative bonded area (RBA) of high-yield pulps. *1994 TAPPI Pulping Conference*, San Diego, CA, pp. 427–433.
24. Jang, H. F., and Seth, R. S. (1998). Using confocal microscopy to characterize the collapse behavior of fibers. *Tappi 81*(5):167–174.
25. Jayme, G., and Hunger, G. (1962). Electron microscope 2- and 3-dimensional classification of fibre bonding. In: *Formation and Structure of Paper*. F. Bolam, ed. British Paper and Board Makers Assoc., London, pp. 135–170.
26. Jones, A. R. (1972). Strength evaluation of unbleached kraft pulps. *Tappi 55*(10):1522–1527.
27. Kallmes, O., and Bernier, G. (1962). The structure of paper. 3. The absolute, relative, and maximum bonded areas of random fiber networks. *Tappi 45*(11):867–872.
28. Kallmes, O., and Bernier, G. (1963). The structure of paper. 4. The free fiber length of a multiplanar sheet. *Tappi 46*(2):108–114.

29. Kallmes, O., and Eckert, C. (1964). The structure of paper. 7. The application of the relative bonded area concept to paper evaluation. *Tappi 47*(9):540–548.
30. Kallmes, O., and Corte, H. (1960). The structure of paper. 1. The statistical geometry of an ideal two-dimensional fiber network *Tappi 43*(9):737–752.
31. Kallmes, O., Corte, H., and Bernier, G. (1961). The structure of paper. 2. The statistical geometry of a multiplanar fiber network. *Tappi 44*(7):519–528.
32. Kallmes, O., Corte, H., and Bernier, G. (1963). The structure of paper. 5. The bonding states of fibers in randomly formed paper. *Tappi 46*(8):493–502.
33. Kallmes, O. J. (1972). A comprehensive view of the structure of paper. In: *Theory and Design of Wood and Fiber Composite Materials*. B. A. Jayne, ed. Syracuse Univ. Press, Syracuse, NY, pp. 157–175.
34. Keith, C. T., and Côté, W. A., Jr. (1968). Microscopic characterizations of slip lines and compression failures in wood cell walls. *Forest Prod. J. 18*(3):67–74.
35. Kenney, F. C. (1952). Physical properties of slash pine semichemical kraft pulp and of its fully chlorinated component. *Tappi 35*(12):555–563.
36. Kibblewhite, R. P., and Hamilton, K. A. (1984). Fibre cross-section dimensions of undried and dried Pinus radiata kraft pulps. *N.Z. J. Forestry Sci. 14*(3):319.
37. Koubaa, A., and Koran, Z. (1995). Measure of the internal bond strength of paper/board. *Tappi 78*(3):103–111.
38. Lagergren, S., Rydholm, S., and Stockman, L. (1957). Studies on the interfibre bonds of wood. *Svensk Papperstidn. 60*:632–644, 664–670.
39. Leech, H. J. (1954). An investigation of the reasons for increase in paper strength when locust bean gum is used as a beater adhesive. *Tappi 37*(8):343–349.
40. Leskelä, M., and Luner, P. (1993). Light scattering and relative bonded area: Simulation of the effect of fiber collapse. *Paperi Puu 75*(8):601–605.
41. Lundh, A., and Fellers, C. (1998). Delamination resistance of paper. *1998 Progress in Paper Physics: A Seminar*. CPPA Tech. Sect., Montreal.
42. Luner, P., Kärnä, A. E. U., and Donofrio, C. P. (1961). Studies in interfiber bonding of paper: The use of optically bonded areas with high yield pulps. *Tappi 44*(6):409–414.
43. Mark, R. E. (1967). *Cell Wall Mechanics of Tracheids*. Yale Univ. Press, New Haven, CT, pp. 40–53.
44. Mark, R. E., and Gillis, P. P. (1970). New models in cell wall mechanics. *Wood Fiber 2*(2):79–95.
45. Mayhood, C. H., Kallmes, O. J., and Gauley, M. M. (1962). The mechanical properties of paper. Part II: Measured shear strength of individual fiber to fiber contacts. *Tappi 45*(1):69–73.
46. McIntosh, D. C. (1963). Tensile and bonding strengths of loblolly pine krafts fibres cooked to different yields. *Tappi 46*(5):273–277.
47. Mohlin, U.-B. (1974). Cellulose fiber bonding. Determination of interfibre bond strength. *Svensk Papperstidn. 77*(4):131–137.
48. Naito, T., Nishi, K., and Kawano, Y. (1995). Delamination resistance of paper. *Proc. 1995 International Paper Physics Conference*. Tech. Sect. CPPA, Montreal, pp. 125–130.
49. Nanko, H., and Ohsawa, J. (1989). Mechanisms of fibre bond formation. In: *Fundamentals of Papermaking*, Vol. 2. C. F. Baker, ed. Trans. Symp. (Cambridge). Mech. Eng. Pub., London, pp. 783–830.
50. Nanko, H., Ohsawa, J., and Okagawa, A. (1989). How to see interfibre bonding in paper sheets. *J. Pulp Paper Sci. 15*(1):J17–J23.
51. Nordman, L., Gustafsson, Ch., and Olofsson, G. (1954). The strength of bonding in paper II. *Paperi Puu 36*(8):315–320.
52. Paavilainen, L. (1994). Bonding potential of softwood sulphate pulp fibres. *Paperi Puu 76*(3):162–173.

53. Page, D. H., Tydeman, P. A., and Hunt, M. (1962). A study of fibre to fibre bonding by direct observation. In: *Formation and Structure of Paper*. F. Bolam, ed. British Paper Board Makers Assoc., London, pp. 171–193.
54. Page, D. H. (1969). A theory for the tensile strength of paper. *Tappi 57*(4):678–681.
55. Pang, L., and Gray, D. G. (1998). Heterogeneous fibrillation of kraft pulp fibre surfaces observed by atomic force microscopy. *J. Pulp Paper Sci. 24*(11):369–372.
56. Retulainen, E., and Ebeling, K. (1992). Fiber-fiber bonding and ways of characterizing bond strength. *46th Appita Conf.*, Tasmania, pp. 459–468.
57. Retulainen, E., and Ebeling, K. (1993). Fibre-fibre bonding and ways of characterizing bond strength. *Appita 46*(4):282–288.
58. Reynolds, W. F. (1974). New aspects of internal bonding strength of paper. *Tappi 57*(3):116–120.
59. Saiki, H. (1970). Proportion of component layers tracheid wall of early wood and late wood of some conifers. *J. Jpn. Wood Res. Soc. (Mokuzai Gakkaishi) 16*(5):244–249.
60. Salminen, L. I., Räisänen, V. I., Alava, M. J., and Niskanen, K. J. (1996). Drying-induced stress state of inter-fibre bonds. *J. Pulp Paper Sci. 22*(10):J402–J407.
61. Schniewind, A. P., Nemeth, L. J., and Brink, D. L. (1964). Fiber and pulp properties. I. Shear strength of single-fiber crossings. *Tappi 47*(4):244–248.
62. Schultz-Eklund, O., Fellers, C., and Olofsson, G. (1987). *z*-Toughness: A new method for the delamination resistance of paper. *Proc. 1987 Int. Paper Phys. Conf.* Tech. Sect. CPPA, Montreal, pp. 189–191.
63. Seth, R. S. (1995). The effect of fibre length and coarseness on the tensile strength of wet webs: A statistical geometry explanation. *Tappi 78*(3):99–102.
64. Skowronski, J. (1991). Fiber-to-fiber bonds in paper: Part II. *Proc. 1991 Int. Paper Phys. Conf.* TAPPI Press, Atlanta, pp. 539–547.
65. Skowronski, J. (1991). Fibre-to-fibre bonds in paper. Part II: Measurement of the breaking energy of fibre-to-fibre bonds. *J. Pulp Paper Sci. 17*(6):J217–J222.
66. Skowronski, J., and Bichard, W. (1987). Fiber-to-fiber bonds in paper: Part I. *Proc. 1987 Int. Paper Phys. Conf.* Tech. Sect. CPPA, Montreal, pp. 65–70.
67. Skowronski, J., and Bichard, W. (1987). Fibre-to-fibre bonds in paper. Part I. Measurements of bond strength and specific bond strength. *J. Pulp Paper Sci. 13*(5):J165–J169.
68. Smith, J. C., and Graminski, E. L. (1977). Characterizing the interfibre bond strengths of paper pulps in terms of breaking energy. *TAPPI Annual Meeting*, Atlanta, pp. 169–175.
69. Stone, J. E. (1955). The rheology of cooked wood. II. Effect of temperature. *Tappi 38*:452–459.
70. Steadman, R. K., and Luner, P. (1985). The effect of wet fibre flexibility on sheet apparent density. In: *Paper Making Raw Materials*, Vol. I. V. Punton, ed. Trans. Eighth Fundamental Res. Symp. Mech. Eng. Publ., London, pp. 311–337.
71. Stone, J. K. (1963). Bond strength in paper. *Pulp Paper Mag. Can. 64*(12):T528–T532.
72. Stratton, R. A. (1991). Characterization of fiber-fiber bond strength from paper mechanical properties. *Proc. 1991 Int. Paper Phys. Conf.* TAPPI Press, Atlanta, pp. 561–577.
73. Stratton, R. A. (1993). Characterization of fibre-fibre bond strength from out-of-plane paper mechanical properties. *J. Pulp Paper Sci. 19*(1):J6–J12.
74. Swanson, J. W. (1956). Beater additives and fiber bonding: The need for further research. A review of the literature on beater or wet-end additives. *Tappi 39*(5):257–270.
75. Swanson, J. W., and Steber, A. J. (1959). Fibre surface area and bonded area. *Tappi 42*(12):986–994.
76. TAPPI. (1991) TAPPI T 425 om-91. Opacity of paper (15% diffuse illuminant A, 89% reflectance backing and paper backing).
77. TAPPI. (1992) TAPPI TIS 0804-03. Interrelation of reflectance, R_0; reflectivity, R_∞; TAPPI opacity, $C_{0.89}$ scattering, s; and absorption, k.

78. Thorpe, J. L., Mark, R. E., Eusufzai, A. R. K., and Perkins, R. W. (1976). Mechanical properties of fiber bonds. *Tappi 59*(5):96–100.
79. Van den Akker, J. A. (1959). Structural properties of bonding. *Tappi 42*(12):940–947.
80. Van Liew, G. P. (1974). The z-direction deformation of paper. *Tappi 57*(11):121–124.
81. Williams, G. J., and Drummond, J. G. (1994). Preparation of large sections for the microscopical study of paper structure. *1994 TAPPI Papermakers Conference*, San Francisco, CA, pp. 517–523.
82. Wink, W. A., Schillcox, W. A., and Van Eperen, R. H. (1957). Instrumentation studies. 80. Determination of the bonding strength of paper. *Tappi 40*(7):189A–199A.
83. Wink, W. A., and Van Eperen, R. H. (1967). Evaluation of z-direction tensile strength. *Tappi 50*(8):393–400.
84. Yang, C. F., Eusufzai, A. R. K., Sankar, R., Mark, R. E., and Perkins, R. W. (1978). Measurements of geometrical parameters of fiber networks. 1. Bonded surfaces, aspect ratios, fiber moments of inertia, and bonding state probabilities. *Svensk. Papperstidn. 81*(13):426–433.

16

STRUCTURE AND STRUCTURAL ANISOTROPY

TSUTOMU NAITO
Nippon Paper Industries
Kumamoto, Japan

Section III is authored by Dr. Makio Hasuike, Hiroshima Research and Development Center, Mitsubishi Heavy Industries Ltd., Hiroshima, Japan.

I. INTRODUCTION

A quantitative knowledge of the geometric structure of paper and board is fundamental to the understanding of why these materials possess the properties they exhibit and how those properties are changed when the structure changes. In this chapter, some of the methods used to determine several important structural parameters are described and the principles behind these testing methods are discussed. The internal structural parameters that are the most important depend on whether one considers a relatively dense, highly bonded material such as linerboard or a low density material such as tissue, whether one is primarily interested in mechanical or other physical properties, and, within either of these last two categories, what type of physical (e.g., opacity or absorbency) or mechanical (e.g., elastic properties or properties at failure) characteristics are of particular interest. Perkins [152,153] identified the network parameters that are considered in theories relating to mechanical properties as follows:

1. Fiber length (including length/width and length/thickness ratios) and its distribution
2. Fiber curliness and kink
3. Fiber segment length (distance between bond centroids along a fiber)
4. Fiber cross-sectional shape (includes area, width, wall thickness, degree of collapse, aspect ratio); characteristics of the distribution of these quantities
5. Mechanical properties of fibers
6. Size and mechanical nature of the fiber–fiber bond area
7. Percentage of the fiber surface bonded to other fibers (relative bonded area)
8. Fiber orientation distribution
9. Sheet density and uniformity; distribution of mass
10. Sheet shrinkage strains

For testing related to other (physical or mechanical) properties, one may need information regarding some or all of the above parameters plus additional information on other properties such as

11. Basic weight (grammage)
12. Moisture content
13. Apparent fiber density

14. Proportions of cell wall layers (compound middle lamella, S1, S2, S3)
15. Filament winding angles (microfibril angles)
16. Conformation of molecular components
17. Surface charges
18. Internal surface volume (including pore size distribution)
19. Electrical conductivity and dielectric constant
20. Acoustical conductivity
21. Indices of refraction

The above list, by no means complete, includes many items that are considered in detail in other chapters. In this chapter we concentrate on methods for ascertaining some important parameters related to internal structural geometry. Special attention will be given to item 8, fiber orientation distribution, because the amount of previously published information available in the paper testing literature is quite limited despite the importance of this parameter.

II. FIBER ORIENTATION DISTRIBUTION

The influence of in-plane fiber orientation on the physical and mechanical properties of paper sheets has been well documented in the literature [30,50,83,86,90,91,153, 159,164,172,180,206,223]. Whether one is evaluating the effects of fiber alignment or processes that affect it or attempting to predict properties by analytical methods, it is important to obtain accurate data regarding the distribution of these alignments or orientations in the sheet.

In this section we consider both direct and indirect methods of making such determinations. Both have advantages and drawbacks. Both require some statistical manipulation to give the researcher a complete picture of the orientation parameters, which may or may not be coupled to other parameters such as fiber length distribution. In addition, it is important for the investigator to consider such questions as what actually needs to be measured and how representative samples are to be obtained, which are illustrated in this chapter for the case of a direct (digitizer or image analyzer) method. Certainly it can be well demonstrated (Chapter 1) that the choice of distribution function (as discussed in this chapter) used to represent fiber orientation has some serious implications relative to our ability to understand and predict elastic constants and other mechanical properties for sheets of different fiber alignment configurations.

A. Methods for Fiber Orientation Distribution

The initial work was done in this field by Danielsen and Steenberg [34], who developed a sort of rotatable protractor for directly measuring the orientations of dyed fibers that are added to the furnish when the sheet is made. The dyed fibers, amounting to less than 1%, are typical of the pulp stock used and are measured as a representative sample of the whole. Subsequently, other workers [50,86,87,159,180, 206] modified the Danielsen–Steenberg method. For example, Glynn et al. [50] constructed a circular turntable fitted with 48 equal sheet-metal bins on its periphery, each bin thus comprising 7.5° or arc. The paper sample, containing dyed fibers, was placed at the center of the turntable under fixed hairlines and a magnifying glass. A hopper at the edge of the turntable dispensed small metal balls, one at a time, into the bin whose arc corresponded with the orientation of a given fiber as observed by the hairlines. It was determined that 2000 measurements should be made to obtain reproducible results; thus, 2000 balls were required for one set of measurements. The number of fibers in each 7.5° of arc was then calculated from the weight of the balls in each bin. It required about 2.5 h for an operator to make 2000 counts on both sides of a sample in this way.

The alternative (indirect) methods that have been developed include mechanical testing [50,88,180,202], light diffraction [16,17,169,170,224], small-angle light scattering [15,112,115,127], X-ray diffraction [158,168], ultrasonic propagation [11,64], microwave attenuation [5,56,137], line intersection methods, and image analysis [123,206,222,227]. In the line intersection methods, used in various forms by Corte and Kallmes [30], Forgacs and Strelis [47], and Silvy [182], counts are made of the numbers of dyed fibers falling on straight reference lines. These counts provide indirect data that are converted to orientation distribution functions via mathematical treatment of the data.

The indirect method most often used is by calculation of the strength anisotropy of the sheet as determined from zero-span tensile tests. This method is embodied in the former TAPPI Suggested Method T 481 [202] and has been extensively discussed by Kallmes [88]. It has the virtues of being relatively simple, inexpensive, and rapid, and it employs equipment often found in paper testing facilities. On the other hand, the agreement between this method and the classic manual direct method (dyed fiber orientations) is not very good (Fig. 1), and great care must be employed to obtain accurate, reproducible results [31,61].

The most serious drawback of the zero-span method is that it is a measure not of orientation but of mechanical anisotropy. The mechanical anisotropy is related strongly to orientation, but it is also related to the draw or stretch imposed by the paper machine [182] and other aspects of drying, such as the solids content of the paper web at the time of application of drying stress [70]. Another mechanical method for determining fiber orientation is acoustic measurement. Because this method measures mechanical anisotropy, the same drawback that affects the zero-span test is involved in it. Despite this drawback, acoustic measurement is still a powerful tool for the diagnostics of machine variation, especially on orientation angle [12].

The methods of determining fiber orientation by electromagnetic wave (visible light, infrared, X-ray) scattering and microwave attenuation have been examined by several workers. Testers based on some of these methods have been devised. Although some show considerable promise for the future, further studies are required to determine if these indirect methods determine "real" fiber orientation.

Fortunately, devices designed to extract quantitative information from slides, photographs, and other images have been developed and enable much more rapid collection of data on geometric parameters. These *image analysis* devices are discussed in Section IV.A. One type of semiautomatic device is the graphic digitizer, which is basically an instrument for the rapid and accurate reading of coordinates. It is especially well adapted for accumulation of fiber orientation information in paper.

The digitizing system used by Mark and colleagues is shown in Fig. 2. A description of its operation is given in Refs. 33 and 154. They used the system

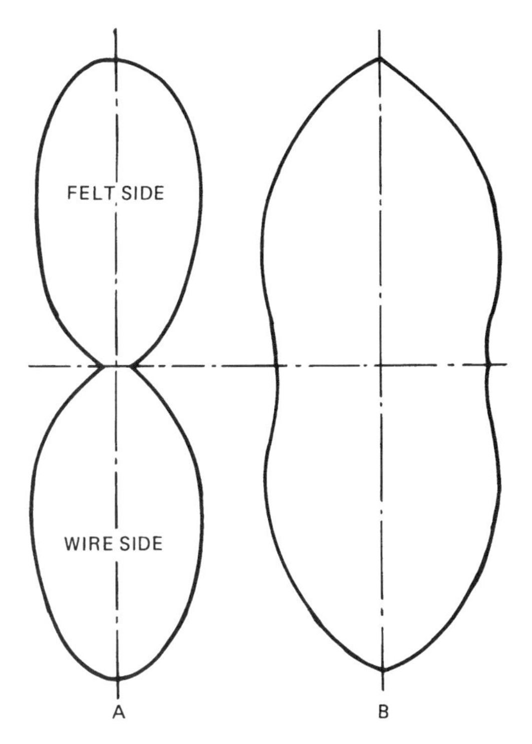

Fig. 1 Polar diagrams of fiber orientation (A) determined directly by protractor with turnable pointer and (B) a strength versus orientation relation determined from tensile tests in sheets conducted in various directions. The machine direction is vertical in these diagrams. (From Ref. 207.)

Fig. 2 Operation of graphic digitizer. (From Ref. 154.)

with a mainframe computer via device coupler and terminal. Nowadays one can apply personal computer (PC)-based systems such as that shown in Fig. 3 to the same task. Niskanen [130,131] used an automatic image analyzer with a TV camera system for this purpose. Some uses of the image analyzer to obtain structural information will be discussed in this chapter. A typical image-analyzing process for a sheet containing dyed fibers is shown in Fig. 4. These methods measure orientation one fiber at a time. Recent developments in the image analysis technique have made it possible to determine fiber orientation in a network image directly [72–74,123,206].

B. Direct Methods

Determination of the orientation of fibers by any direct method requires some preliminary decisions as to what types of information are needed.

Measurements Needed Because fibers in a sheet of paper or paperboard are of finite (short) length and usually contain bent, curled, or broken sections, one has to make a decision as to what constitutes a fiber for purposes of determining fiber orientation. The same statement can be made regarding determination of fiber length, and it should be emphasized that there is no basis a priori for assuming that fiber length and orientation are independent of each other. These parameters may, in fact, be highly correlated in some cases [24].

With reference to Figs. 5 and 6, one may observe several possibilities for defining the fiber in terms of orientation. Each possibility has different implications for length categorization. Figure 5 shows a sheet in which a small fraction (0.25%) of black dyed fibers show contrast against the remaining 99.75%, which are bleached. The paper samples are impregnated with silicone oil (with or without the assistance

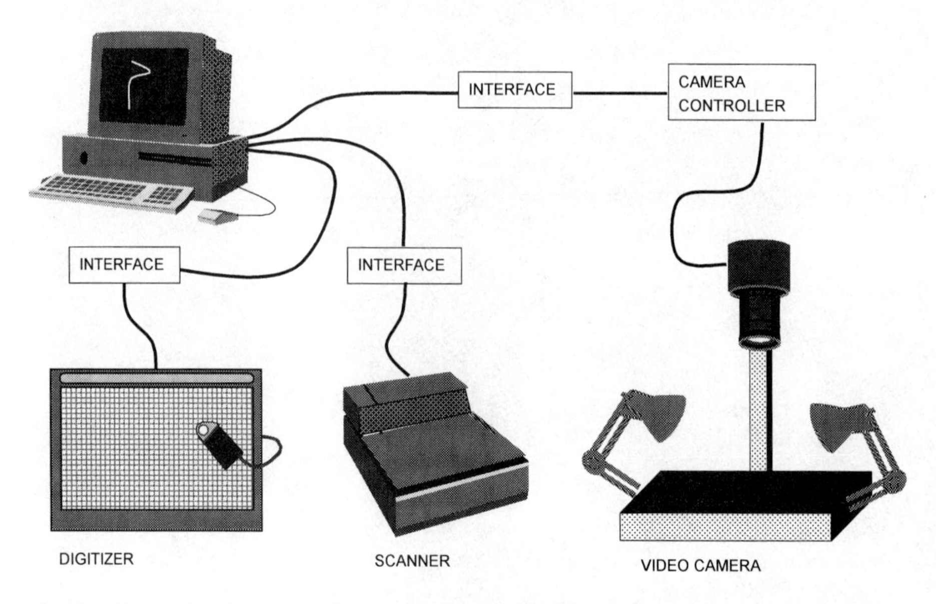

Fig. 3 Example of a personal computer based digitizing system for the direct measurement of fiber orientation.

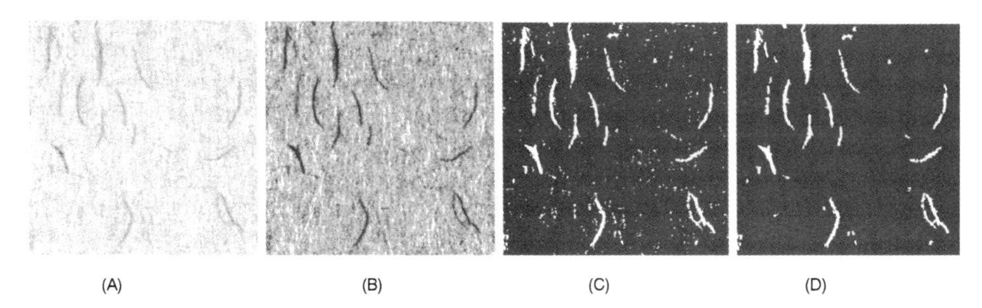

Fig. 4 Typical image analysis process for a sheet containing dyed fibers. The original image was obtained by a CCD TV camera with a microscope. (A) The loaded image in frame grabber; (B) the contrasted image; (C) the binary image; (D) the refined image. (Photo by L. Mignot. From Ref. 123.)

of vacuum) to enhance visibility of the dyed fibers by making the rest of the sheet almost transparent. The orientation of the fiber in Fig. 5 labeled *F*, for example, can be described in any of the following ways:

- With respect to a fixed axis (e.g., the machine direction, MD), the orientation of a straight line joining the ends of the fiber (Fig. 6A).
- •• With respect to a fixed axis, the orientation of the best fit of all the segments of the fiber together. Essentially, a regression line is drawn that best fits a set of points taken to lie at the midpoint of each segment of the fiber (Fig. 6B).

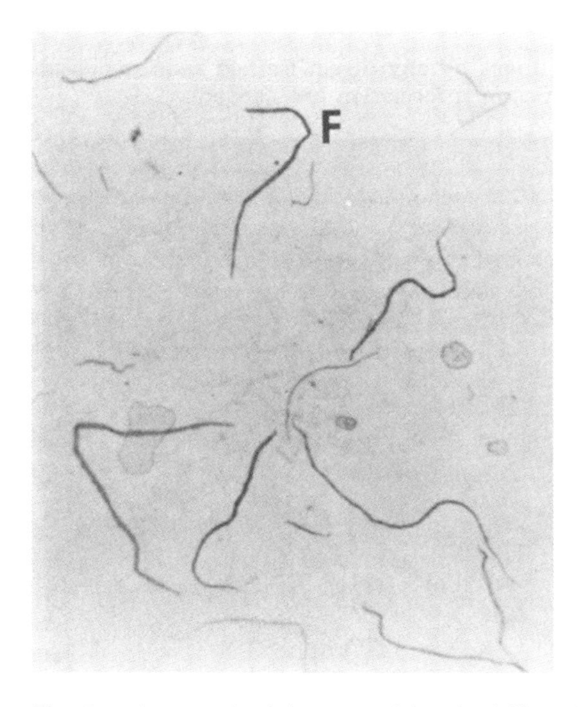

Fig. 5 Photograph of sheet containing dyed fibers. The sheet is impregnated with silicone oil. Its image, projected by the lamp of a photographic enlarger, clearly delineates the black fibers regardless of their depth in the sheet. (Photo by A. Eusufzai.)

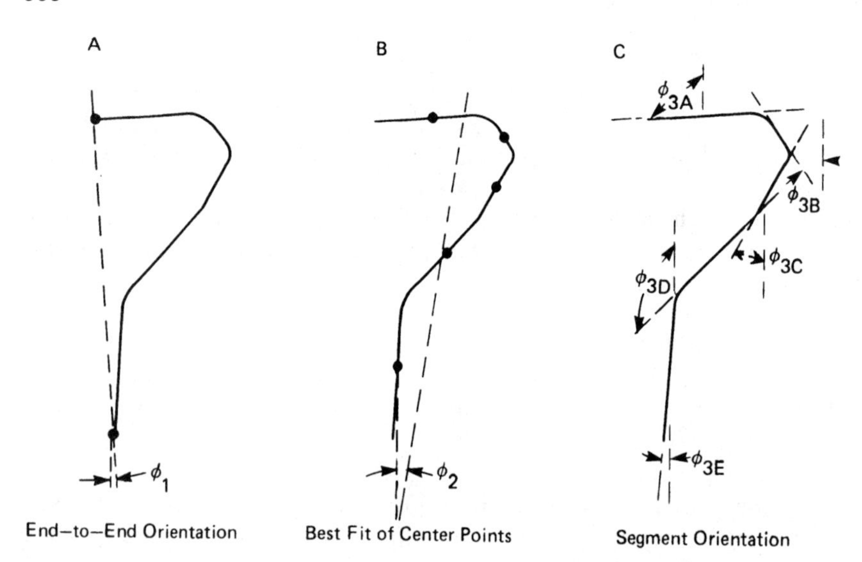

Fig. 6 Diagrammatic representation of dyed fiber F in 5. (A) End-to-end orientation. Black circles represent end points; dashed line represents end-to-end distance; ϕ_1 equals orientation of fiber in sheet. (B) Best fit of center points. For computational purposes, fiber is divided into five segments Black circles are segment midpoints; dashed line represents regression line of midpoints; ϕ_2 equals orientation of fiber in sheet by best fit of center lines. (C) Orientation ϕ_{3i} determined for each segment.

●●● With respect to a fixed axis, the orientation of each segment of the fiber taken individually (Fig. 6C).

●●●● With respect to a fixed axis, the orientations of segments according to length category.

In such cases, criteria have to be set as to how many segment length categories are needed. For example, a preliminary study can be made to determine the most probable (statistically) segment length one encounters in a given type of sheet material. One can establish the length categories according to some system based on most probable length (mpl). For example,

Segment group No. 1: $0 < \ell \leq \ell_{\mathrm{mpl}}$

Segment group No. 2: $\ell_{\mathrm{mpl}} < \ell \leq 2\ell_{\mathrm{mpl}}$

Segment group No. 3: $2\ell_{\mathrm{mpl}} < \ell \leq 3\ell_{\mathrm{mpl}}$

Segment group No. 4: $3\ell_{\mathrm{mpl}} < \ell$

The four modes of description itemized above are listed in increasing order of reliability as to the results obtained, assuming an adequate sample size. However, the difficulty of measurement also increases in the same order.

Fiber (or fiber segment) length data can be obtained at the same time as data on orientation. The acquisition of data on length distribution will usually be highly desirable, and plans should be made to record them at the start. As noted previously, length and orientation may or may not be independent variables. The probability of finding certain lengths in conjunction with certain orientations may be large for some materials. Another use for length data is to make possible the determination of fiber curl (difference between total fiber length and straight-line end-to-end length divided by total length*), which varies greatly according to the fiber processing methods used in making the sheet.

In some cases, such as when sheet curl is being studied (see Chapter 3), it may be necessary or desirable to examine the variation of fiber orientation (and/or length) distribution with respect to the Z direction of the sheet. To accomplish this, one can employ a sheet-splitting technique, using either a cryostatic-type apparatus or one that employs pressure-sensitive tape, to examine the fiber pattern at various depths (see Ref. 118 and Section III). If sheets from a Fourdrinier machine are examined, an apparatus that uses pressure-sensitive tape is generally able to split a sheet more finely. Alternatively, one can adopt a simplified (and much less accurate) assumption that there are only two orientations—those corresponding to the patterns observed on the top- and wire-side surfaces. However, if only the surfaces are to be examined the sheet should not be impregnated with oil. Oil impregnation makes all the dyed fibers stand out with about an equal intensity at all depths in the sheet, so an observer cannot tell if a dyed fiber lies near the surface or near the center.

Choice of Sector Size The frequency with which fibers or fiber segments are found with alignments falling in a given range (radians or degrees) is usually the immediate objective of the experiment. Preselection of an angular interval that makes it possible to obtain a comprehensive, accurate picture of fiber orientation distribution is essential. A very commonly selected interval is 5°.

Obtaining Representative Samples If possible, the incorporation of a small percentage (0.10–0.25%) of dyed fibers into the sheet is desirable. For many purposes, a chlorazol black E dye is excellent. These dyed fibers stand out among the other fibers, as illustrated in Fig. 5. The use of dyed fibers makes it possible to accurately cross check against other methods of determination. However, the presence of scattered dyed fibers may be unacceptable. In such cases, it is sometimes possible to incorporate dyed fibers into a sheet only during the last few minutes of a machine run. Alternatively, a colorless additive that fluoresces in light of an appropriate wavelength may be acceptable. A third possibility, developed by Glynn et al. [50], is to add tannic acid as a mordanting agent to a small fraction of the pulp stock. The mordant is then reacted with a metallic salt such as lead acetate. With approximately 1% of these fibers in the stock, machine-made paper is unaltered in appearance. Samples are cut from the roll and dipped in a bath containing basic red dye, which is taken up preferentially by the treated fibers. The contrast may be enhanced by dipping in chlorinated water, then washing and dipping in a yellow dye bath. After a final washing and drying, the mordanted fibers show up dark red on a buff yellow background. Samples should be taken across the width of the sheet, with sufficient replication to provide reproducibility of results. The dyed fibers should be representative of all the fibers in the sheet.

* This is one definition of fiber curl; other definitions are discussed in Section IV.B.

If it is not possible to incorporate any type of individually identifiable fiber into the sheets to be tested, one of the indirect methods will have to be used. However, it is essential that when any indirect method is considered for use on a particular material it is independently verified that the indirect method will yield results in agreement with a direct method using similar test material that does contain dyed fibers.

Compilation of Data

Angular Distribution Frequency Usually, fiber orientation data have to be compared or fitted to a generally bimodal (symmetrical with respect to the machine direction) mathematical function of some form to be useful. Distribution functions that have been used or suggested for use with fiber networks include the cosine [30], elliptical [47], wrapped Cauchy [181], and von Mises [116]. The goodness with which any of these functions will fit a set of experimental data for fiber orientation depends on the degree of anisotropy of the sheet, the scatter in the experimental data, and the shape parameter(s) for the function. Some explanation of these functions and parameters will illustrate this point.

Elliptical The elliptical distribution function has the form

$$f(\theta) = \frac{\lambda}{\pi}\left(\frac{1}{\cos^2\theta + \lambda^2 \sin^2\theta}\right) \tag{1}$$

The degree of ellipticity of this function is controlled by the shape parameter λ, which is equal to the ratio of the major and minor semiaxes. Selection or determination of an appropriate value for λ enables one to fit the curve of the probability density of finding a fiber within a given (say, 5°) sector of orientation. The determination of λ is usually done by the least squares error method (see Section II.D) when the elliptical function is fitted to the observed data. Allowable values of λ are never less than unity. The symbol θ refers to the angle of orientation. It varies from $-90°$ to $90°$ when the machine direction is taken to be 0°. Plotted in Cartesian form, the elliptical function will generate a smooth, symmetrical, rounded-peak curve. When plotted in polar coordinates, the function generates an ellipse, of course.

Cosine As shown by Cox [32], the cosine function is a useful distribution function consisting of a series expansion of cosine terms of the form

$$f(\theta) = \frac{1}{\pi}(1 + \eta_1 \cos 2\theta + \eta_2 \cos 4\theta + \cdots + \eta_n \cos\ 2n\theta) \tag{2}$$

In this expression there are a series of shape parameters η that modify the basic curve form. Again, this makes it possible to fit the function to the experimental probability distributions (i.e., data points) sector by sector. In the work of Corte and Kallmes [30], the series expansion was truncated to

$$f(\theta) = \frac{1}{\pi}(1 + \eta \cos 2\theta), \qquad 0 \le \eta \le 1 \tag{3}$$

for ease of mathematical manipulation. The form of Eq. (3) is designated as a single cosine term in this chapter. Plotted in polar coordinates, the form is bimodal cardioid. The truncation to only one cosine term may have undesirable consequences from the standpoint of determining or predicting certain mechanical properties. It is better to retain a second cosine term; that is, the expression

$$f(\theta) = \frac{1}{\pi}(1 + \eta_1 \cos 2\theta + \eta_2 \cos 4\theta) \tag{4}$$

is generally superior to Eq. (3) for purposes of analyzing experimental data. When plotted in polar coordinates, the two cosine term form is cusped differently than the single cosine term. The use of Eqs. (3) and (4) also generates rounded-peak curves when plotted in Cartesian coordinates. The shape parameters η_i in Eqs. (2) and (4) can also be determined by using the least squares error method. In Eqs. (2) and (4), these shape parameters are subject to limits in allowable values that ensure that no negative probabilities are generated. In Eq. (3) the limits are $0 \leq \eta \leq 1$.

Von Mises Another powerful function suited to the handling of fiber orientation distribution data is known as the von Mises distribution. More specifically, the function used here is a multimodal distribution of the von Mises type (see Ref. 116, pp. 36–45). Here, the probability density function is given by

$$f(\theta) = \frac{1}{\pi I_0(\kappa)} \exp[\kappa \cos 2(\theta - \mu_0)] \tag{5}$$

where $I_0(\kappa)$ is a modified Bessel function of the first kind and order zero, that is,

$$I_0(\kappa) = \sum_{n=0}^{\infty} \frac{1}{(n!)^2} \left(\frac{\kappa}{2}\right)^{2n} \tag{6}$$

Two parameters are present in Eq. (5). The parameter μ_0, which is discussed further in the ensuing paragraphs, establishes the mean direction, whereas κ is a shape (concentration) parameter. The determination of parameters κ and μ_0 is discussed by Mardia and Jupp [116]. Allowable values of κ must always be nonnegative. Because an increase in the anisotropy of the test material will result in a set of experimental points with a relatively high, narrow peak and, conversely, a more random sheet will have less variation between the MD and other directions, it can be inferred from Fig. 7 that the value of the concentration parameter κ will be larger for the distribution curve that approximates the points generated from the more oriented material.

Establishment of Mean Directionality It can often be assumed that the mean fiber orientation* will lie in a sector around the machine direction, for example, in the interval from -2.5° to 2.5° when MD lies at 0°. If this is a valid assumption, then the mean direction parameter that is included in Eq. (5), μ_0, has the value of 0°, as in Fig. 7. If the fibers are more oriented in the cross-machine direction (CD), then data plotted in the same way will approximate a curve whose peak lies at 90° (see Fig. 8) and $\mu_0 = 90^\circ$. In such cases θ will be taken for the interval 0–180°. For sheets formed under conditions wherein the greatest fiber alignment is neither MD nor CD, μ_0 will have to be determined according to the method described by Mardia and Jupp [116].

It should not be too readily assumed that $\mu_0 = 0^\circ$ or that $\mu_0 = 90^\circ$ (see Fig. 9). Some paper machines, in fact, operate in a manner that results in an offset mean direction [50]. One of the advantages in fitting Eq. (5) (for the von Mises distribution) to the experimental data is that the determination of u_0 can serve as an

* This refers to the interval -90° to 90° [116]. The term is also (differently) used with reference to a single quadrant [182].

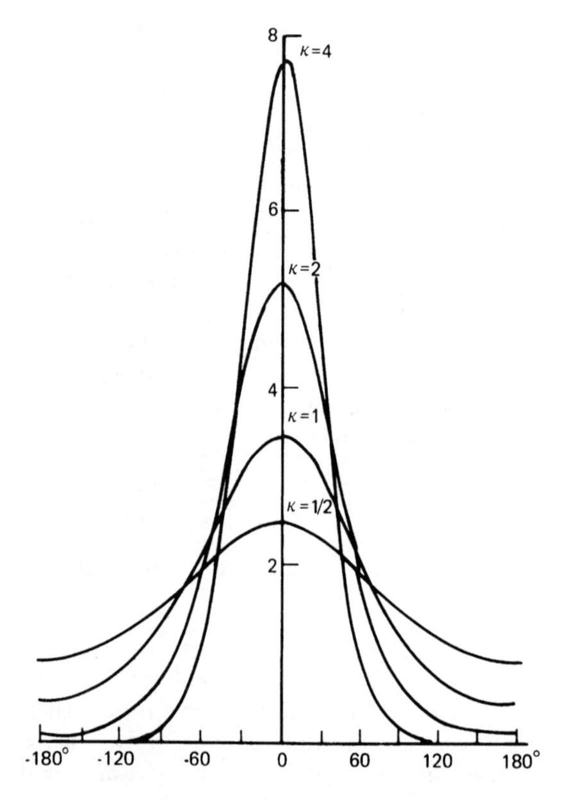

Fig. 7 Cartesian density of the von Mises distribution for $\mu_0 = 0°$ and $\kappa = 1/2$, 1, 2, 4. (From Ref. 116 with permission from K. V. Mardia, Statistics of Directional Data, 1972. Copyright Academic Press Inc. Ltd., London.)

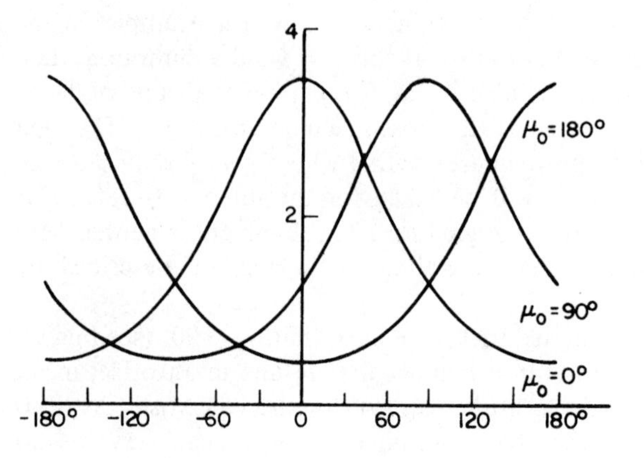

Fig. 8 Cartesian density of the von Mises distribution for $\kappa = 1$ and $\mu_0 = 0°$, 90°, 180°. (From Ref. 116 with permission from K. V. Mardia, Statistics of Directional Data, 1972. Copyright Academic Press Inc. Ltd., London.)

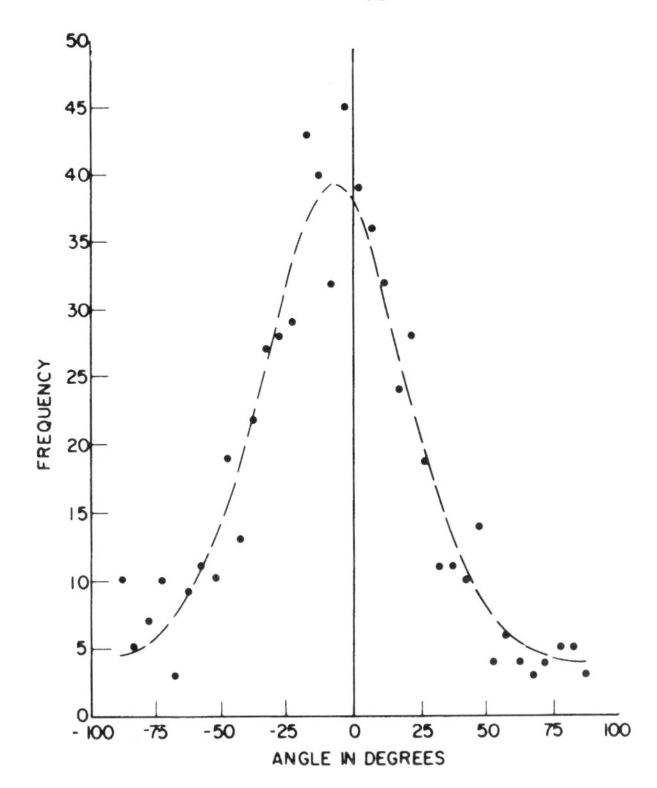

Fig. 9 Frequency data for fiber end-to-end angular orientation measurements in 5° increments compared with a von Mises distribution. This sheet has a high degree of fiber orientation. Note that the distribution is not centered around 0° but is displaced by −7.1° (−0.12 radians); therefore, $\mu_0 = 0.12$. The shape factor κ is 1.12. With these parametric values, P (probability) = 64.2%.

indication of whether the sample edges are, in fact, aligned with the machine and cross-machine directions in cases where that is the intention. The elliptical and cosine functions can be similarly modified by replacing θ with $\theta - \mu_0$ in Eqs. (1)–(4); this was done, for example, by Corte and Kallmes for the single cosine function [30]. The determination of μ_0 can at times provide insights that make it possible to correct faulty experimental or analytical procedures. The use of μ_0 is not limited to orientation distributions; other functions with angular dependence can be shifted by use of such a parameter.

C. Selection of a Representative Function

Once the true axes of orientation have been determined, the appropriate distribution function(s) (elliptical, cosine, or von Mises) can be used as analytical tools to develop structure–property relationships with respect to different directions in the sheet. One important example is the precise determination and prediction of anisotropic elastic constants. Given the true orientation axes, the term μ_0 is set equal to zero from that point on. Accordingly, the Fourier expansion forms for the above-mentioned distributions are very helpful. A Fourier expansion for the von Mises distribution has been given by Mardia and Jupp [116]:

$$f(\theta) = \frac{1}{\pi I_0(\kappa)} e^{\kappa \cos 2\theta} = \frac{1}{\pi}\left[1 + 2\sum_{n=1}^{\infty} \frac{I_n(\kappa)}{I_o(\kappa)} \cos\ 2n\theta\right] \tag{7}$$

where $I_n(\kappa)$ is a modified Bessel function of the first kind of order n. A polar diagram of the von Mises type of distribution shows that it is bimodal when plotted in that manner.

For the elliptical distribution, the Fourier series can be written as

$$f(\theta) = \frac{\lambda}{\pi}\left(\frac{1}{\cos^2\theta + \lambda^2 \sin^2\theta}\right) = \frac{1}{\pi}\left[1 + 2\sum_{n=1}^{\infty}\left(\frac{\lambda - 1}{\lambda + 1}\right)^n \cos\ 2n\theta\right] \tag{8}$$

A comparison of Eqs. (2), (7), and (8) shows that the n-term cosine function has n degrees of freedom, whereas the von Mises and elliptical functions each have only a single degree of freedom. The fact that the last two functions each have only one shape parameter makes them especially suitable in evaluating fiber orientation data. When the von Mises distribution is displayed in Cartesian coordinates it is a highly useful function for fitting the curve of a plot of probability density versus orientation angle, as shown in Fig. 9. A cumulative frequency distribution for the same data as are used in Fig. 9 is shown in Fig. 10. For other possible distribution functions, refer to Mardia and Jupp [116].

In general, angular frequency distribution functions can be used with results generated by indirect as well as direct methods. For example, use was made of the single cosine term function to interpret results obtained by optical diffraction in the study described in Section II.H.

D. Curve-Fitting Method and Goodness-of-Fit Test

Given a need to match a curve to a set of observed data, one inevitably demands that the assumed mathematical model fit the data as closely as possible. Therefore, some measure of goodness of fit is needed. Probably the most popular evaluation of fit involves application of the least squares principle.

According to the least squares method [67,157], the parameters of the assumed model are adjusted so as to minimize the sum of the squares of the differences between the observed and assumed distributions. It is usually necessary to solve a set of nonlinear simultaneous equations with the parameters as unknowns; however, when there is only one parameter involved [e.g., λ in Eq. (1), η in Eq. (3)] one nonlinear equation suffices. A numerical method is required to solve these equations [67,157]. An important method for making statistical inferences is known as the chi-square test. It is often used to justify whether a given distribution model is sufficiently acceptable in the sense of goodness of fit [2,68,157]. The principle of the chi-square test is based on the following rationale.

It can be shown that the variable chi-square

$$\chi^2 = \sum_i \frac{(\text{observed} - \text{expected})^2}{\text{expected}} = \sum_i \frac{(\chi_i - f_i)^2}{f_i} \tag{9}$$

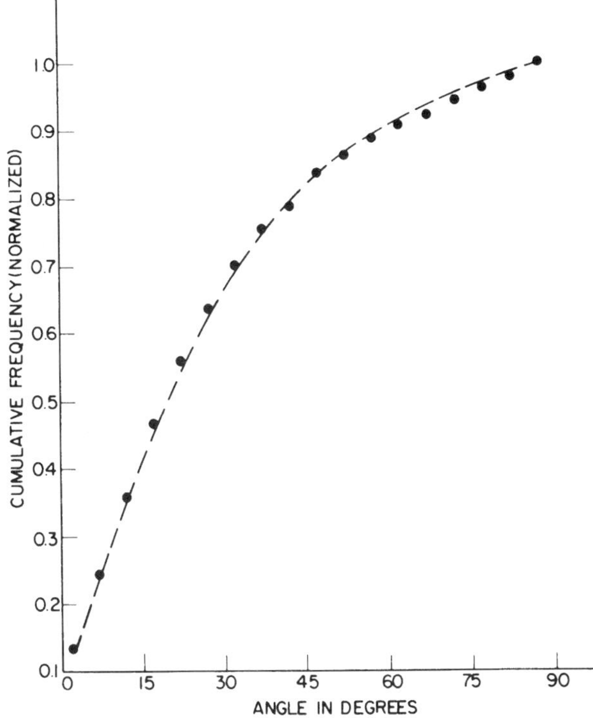

Fig. 10 Cumulative frequency data for the end-to-end angular orientation measurements used to plot Fig. 9. The cumulative von Mises distribution frequency probability in this case exceeds 99.9%.

obeys the chi-square distribution

$$f(\chi^2) = \frac{e^{-\chi^2}(\chi^2)^{(\nu/2)-1}}{2^{\nu/2}\Gamma(\nu/2)} \tag{10}$$

Here χ_i and f_i represent the observed and expected frequency, respectively, ν is the degree of freedom (number of classes subtracted from number of restrictions on expected frequency), and

$$\Gamma\left(\frac{\nu}{2}\right) = \int_0^\infty e^{-y} y^{(\nu/2)-1}\, dy \tag{11}$$

For a particular set of data, the chi-square χ_p^2 is calculated according to Eq. (9). One can then compute the probability P in accordance with the relation

$$P = 1 - \int_0^{\chi_p^2} f(\chi^2)\, d\chi^2 \tag{12}$$

P can be explained as the percent probability that the expected distribution (that is, the assumed mathematical distribution model) can describe the observed distribution. If the value of P is found to be equal to 1 (100%), then the assumed distribution model is actually a true representation of the observed distribution. For the other extreme case, the assumed distribution cannot describe the observed distribution at all if the probability is found to be zero. In general, one accepts that the assumed model is a good representation of the observed frequency when the probability P is greater than 0.05, that is, greater than 5%. In the examples shown in Figs 9–11 this requirement is satisfied for appropriate distribution functions. An important application of the chi-square test comes from the requirement that P be large to consider that a curve is well fitted to a set of data points. However, one can continue to adjust the shape parameters of an assumed model obtained by the least squares method until the probability is the largest one possible for that particular model.

The least squares method is a well-established method for selecting parameters in a way that adequately fits the data, but the following alternative procedure could be suggested:

1. Use the least squares method to find parameters (e.g., η_1, η_2) that will match the function to the data reasonably well.
2. Find P by the chi-square test.
3. Adjust parameters η_i so as to maximize P.

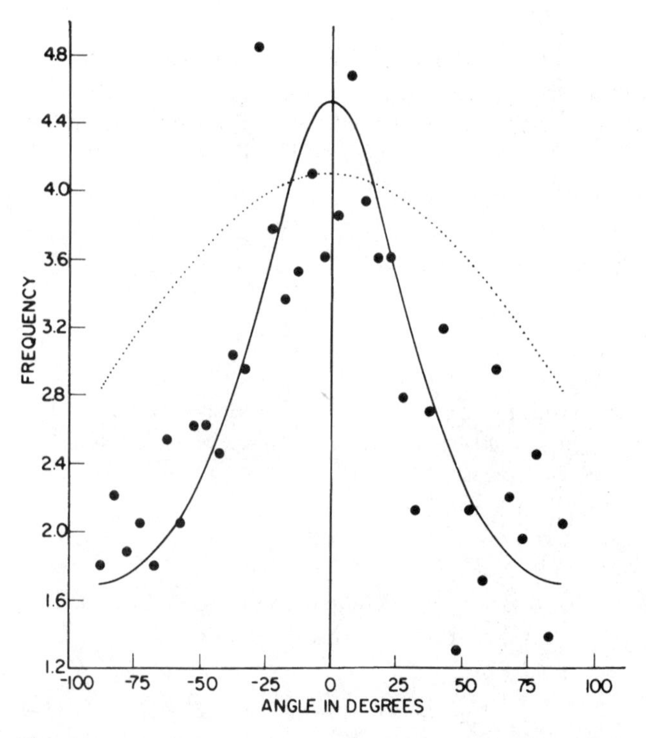

Fig. 11 Frequency data for fiber segment angular orientations measured in 5° increments compared with elliptical (solid line) and single cosine term (dotted line) distributions with shape parameters $\lambda = 1.459$ and $\eta = 0.368$, respectively. This sheet has a fairly low degree of fiber orientation. The mean direction parameter μ_0 has been assumed to equal 0°, so the distributions are symmetrical about 0°. The elliptical function shows the better fit in this case, with $P = 32.5\%$.

E. Relationships Between Orientation Functions

In Fig. 12, a graphic comparison is shown of the four distributions that have been discussed in this chapter. The graph may enable the reader to visualize the mathematical interrelationships between the functions. In this figure, a plot is made of a von Mises distribution for $\mu_0 = 0$ and arbitrarily assigned $\kappa = 1.5$ (corresponding to a highly anisotropic sheet). An attempt is made to adjust the other three functions to fit the von Mises curve, using the least squares method. It is seen that no congruence occurs; the best-fitting elliptical function generates a curve between -50°C and 50° that is narrower and higher, whereas the cosine function curves are wider and lower. In the 50–90° range, other disparities occur. This illustration is included to emphasize that the choice of distribution function may be different for each data set; there is neither a "right" function nor a universal correspondence between data and function or between function and function (cf. Refs. 156 and 181).

F. Silvy's Orientation and Segment Length Distribution Method: (The "Equivalent Pore")

Silvy [182] notes that the structure of the fibrous network in a sheet of paper may be characterized in such a way as to take into account the organization of pores in the sheet.

A floc in the headbox is probably spherical for symmetry reasons and to minimize the energy dispersed by turbulence in the suspension. Under the effect of shear force between the suspension and the outlet lips of the headbox or between the

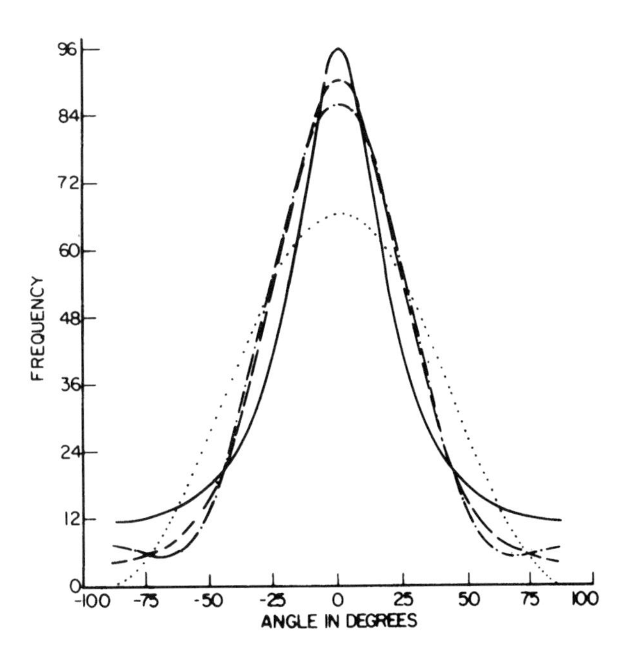

Fig. 12 Elliptical (solid line), single cosine term (dotted line), and two cosine term (dot-dashed line) distributions adjusted for best fit to a von Mises distribution (dashed line) with shape parameter $\kappa = 1.5$.

suspension and the forming wire, this floc is progressively distorted and becomes ellipsoidal: It has three main axes in space, resulting in two main axes by projection on the formation wire or on the plane of the sheet. With these considerations, the angular distribution function of the line segments of the fibers making up the flocs or pores in sheet is of an elliptical nature, and the directions of the two plane axes of its projection are generally the machine direction for the major axis and the cross-machine direction for the minor axis. In real paper, the pores are polygons that are enveloped by fibers in different directions.

The length-weighted orientation distribution of the segments of a fibrous network $n_{\ell\theta}$ is the length fraction of the segments whose angle with a reference direction lies between θ and $\theta + d\theta$.

$$n_{\ell\theta} = \frac{1}{L}\frac{dL_\theta}{d\theta} \tag{13}$$

where

L_θ = cumulative length of the line segments of the network whose angle lies between 0 and θ

L = cumulative length of the line segments of the network in all directions

Next, $R(\theta)$ is defined as the radius of curvature at point M, where the tangent is θ-oriented, along a centrosymmetric closed curve referred to as the equivalent pore [182] (see Fig. 13):

$$R(\theta) = \lim_{\Delta\theta \to 0} \frac{\Delta s}{\Delta\theta} \tag{14}$$

where Δs is the element of arc length over angle $\Delta\theta$ on the contour of the equivalent pore. Thus, Δs may be considered the sum of the lengths of the θ-oriented line segments in the analyzed fibrous network. Under these conditions, $n_{\ell\theta}$ varies with respect to θ in the same way as the length of arc $ds = dL_\theta$ along the equivalent pore. So

$$L = \mathbf{L}/2$$

$\mathbf{L}$ being the perimeter of the equivalent pore, and

$$R(\theta) = dL_\theta/d\theta$$

Consequently, using Eq. (13),

$$n_{\ell\theta} = R(\theta)/L \tag{15}$$

where

n_θ = orientation density in number of line segments

ℓ_θ = average length of the line segments in the θ direction

N = total number of line segments of the network, all directions included

λ = average length of the line segments over all directions

We can write

$$ds = dL_\theta = N\bar{\ell}_\theta n_\theta \, d\theta = R(\theta)\, d\theta$$

and

$$\mathbf{L}/2 = L = N\lambda$$

Therefore, using the concept of the equivalent pore [182], we can define the length-weighted orientation distribution of the segments of a fibrous network. Then $n_{\ell\theta}$ can be computed form the radius of curvature $R(\theta)$ of the equivalent pore at the point where the tangent is θ-oriented as shown in Fig. 13.

Silvy [182] also showed that the equivalent pore is homothetic to the average pore; the diameter of the equivalent pore in a given direction is proportional to the average directional secant measured in the pores of the fibrous network. Consequently, the equivalent pore can be derived simply from statistical measurements of the average directional secant (free fiber length) of the pores found in the fibrous network.

The shape of the diffused spot seen when a laser beam passes through a sheet of paper (see "Light Diffusion," Section II.H) may give a practical illustration of this theoretical concept [51,122,183].

When the equivalent pore is an ellipse, the anisotropy of the orientation density distribution of the fibers as measured by $n_{\ell\theta}$ is characterized by one parameter only—the ellipticity a/b—i.e., the ratio of large semiaxis a and small semiaxis b.

$$\frac{a}{b} = \left(\frac{R(\theta_m)}{R(\theta_c)}\right)^{1/3} = \left(\frac{n_{\ell\theta_m}}{n_{\ell\theta_c}}\right)^{1/3} \tag{16}$$

The mean angle of orientation θ^* is defined (with respect to the cross-machine direction) as

$$\theta^* = \frac{\int_0^{\pi/2} n_{\ell\theta}\theta \, d\theta}{\int_0^{\pi/2} n_{\ell\theta} \, d\theta} = 2\int_0^{\pi/2} n_{\ell\theta}\theta \, d\theta = 2\int_0^{\pi/2} \theta \, ds \tag{17}$$

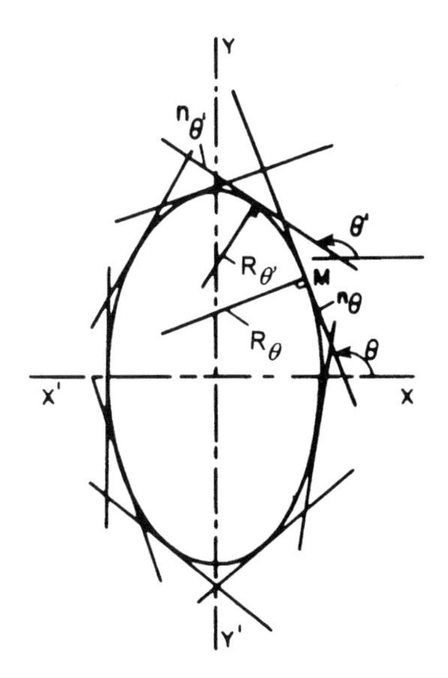

Fig. 13 Diagram of equivalent pore, showing that radius of curvature is directly proportional to orientation density. (From Ref. 182.)

Note that the integration interval is 0 to $\pi/2$. Calculated over the length of the perimeter of the pore, this integral has the value $\theta^* \cong \tan^{-1}(a/b)$. Thus, for the assumption of an elliptical shape for the equivalent pore, the mean orientation angle θ^* is sufficient to describe the ellipticity of the pore and accordingly the fiber orientation distribution in the plane of the sheet for a wide range of papers.

Several workers have verified that this shape applies for most machine-made papers. Thus, Dodson and coworkers [38,39,175] showed that $n_{\ell\theta}$ was expressed by using elliptical functions or their development in trigonometric series, the ellipticity a/b being the only parameter used. They also noted that length-weighted expressions of $n_{\ell\theta}$ are more precise and more realistic than other number-weighted orientation density distributions of the fibers n_θ.

Koran et al. [106] also compared the results of the mean angle of orientation θ^* from the equivalent pore, X-ray, and zero-span tensile strength methods.

G. Mean Orientation Angle

Several investigators have developed orientation indices or other orientation parameters for the purpose of simplifying the relationship between mechanical properties and structural anisotropy in the plane of the sheet. Such an index is usually based on the mean angle of orientation θ^* in the first quadrant, defined as

$$\theta^* = \frac{\sum_{i=1}^{\tilde{n}} n_i \bar{\theta}_i}{\sum_{i=1}^{\tilde{n}} n_i} \tag{18}$$

where $\bar{\theta}_i$ is the mean angle of the fibers in an angular increment (sector) containing n_i fibers and $\tilde{n}$ equals the number of sectors lying between orientations 0° and 90°.

The above expression, similar to the first form of Eq. (17), is limited in that

1. Either the angular increment θ_i must be small in order to justify an assumption that $\bar{\theta}_i$ lies at the center of the increment, or else $\bar{\theta}_i$ must be computed for each sector.
2. No weight is given to such factors as length of fibers ℓ or linear density of fibers ω.

To account for item 2, Eq. (18) would have to be modified by changing n_i terms to $n_i\ell_i\omega_i$ terms. It may be noted that Silvy's method [182], as given in Eq. (17), does take into account ℓ_i as well as n_i terms in the development of the equivalent pore.

The grouping of fiber orientation data in increments of θ_i lends itself to weighting according to the number n_i or frequency f_i of observations in the ith sector of one or more quadrants. Taking fiber mass $\ell\omega$ as an example of a weighted grouping, one can select M class intervals representing different $\ell\omega$ fractions and determine how frequently the various $\ell\omega$ fractions occur in each increment. Now each $\bar{\theta}_i$ is the angle of M vectors whose directions all lie at the center of the ith increment but whose lengths vary according to the magnitude of $\ell\omega$. The mean orientation θ^* is now defined as the direction of the resultant of these vectors, whose magnitude $\bar{R}$ is determined by

$$\bar{C} = \frac{1}{N}\sum_{i=1}^{N}\sum_{J=1}^{M}(\ell\omega)_j f_i \cos\theta_i \tag{19a}$$

$$\bar{S} = \frac{1}{N}\sum_{i=1}^{N}\sum_{J=1}^{M}(\ell\omega)_j f_i \sin\bar{\theta}_i \tag{19b}$$

$$\bar{R} = \left(\bar{C}^2 + \bar{S}^2\right)^{1/2} \tag{19c}$$

where

N = sum of the frequencies or numbers of observations
$\bar{R}$ = resultant magnitude of vectors $\bar{S}$ and $\bar{C}$

The mean orientation is found as

$$\theta^* = \tan^{-1}\left(\frac{\bar{S}}{\bar{C}}\right) \tag{20}$$

The mean orientation angle can be related to physical or mechanical properties (e.g., curl, MD versus CD breaking length) when other processing variables are held constant; experimental reports often show linear relationships [47,50,159,168,182]. In general, the elastic and inelastic (e.g., strength) mechanical properties of paper in the plane of the sheet should vary with mean fiber orientation angle in a nonlinear manner [180]. This variation follows the exponential laws in good agreement with theoretical prediction.

H. Fiber Orientation by Scattering

In addition to the indirect methods that were discussed earlier, efforts have been made to apply visible light and X-ray scattering techniques to the quantitative determination of fiber orientation in paper sheets [25,112,158,169,224]. Because scattering techniques are nondestructive and in their mode of transmission provide integrated average values of fiber orientation through the thickness of the sheet, there is potential application in routine and on-line control of commercially made sheet structures.

Light Diffraction The multiple scattering and absorption processes of visible light in paper structures create their opaque or translucent light-diffusing appearance, as described in Chapter 4 of vol. 2. The pattern of visible light transmitted through a sheet provides information about fiber orientation on the basis of Fraunhofer diffraction. In order to examine scattering details from individual fibers, it is necessary to isolate individual fibers and apply a suitable experimental setup. The laser light source is ideally suited for illuminating single fibers of small (1–100 μm) diameter as well as providing coherent illumination of larger objects through beam expansion. Low intensity helium-neon (He-Ne) and ruby lasers have been used for both light diffraction and small-angle light scattering (SALS) analysis (Fig. 14) [14,16,17,48,49,169,185,224].

The diffraction characteristics of single pulp fibers are similar to those of textile fibers where the scattered light intensity is broken up into a series of periodic maxima in the plane perpendicular to the fiber direction [114]. Shape distortions and internal

Fig. 14 Laser scan setup for light diffraction analysis. (From Ref. 224.)

structural heterogenities create imperfections in the symmetry and periodicity of the intensity peaks (Fig. 15).

The possibility of measuring fiber orientation by light diffraction depends on the spatial distribution of scattering streaks from individual fibers. The scattered light is collected in the back focal plane of a suitable lens systems and analyzed through a rotating aperture in a photomultiplier [169] or photographically via microdensitometry [224]. The individual scattering streaks are lost, but the diffraction

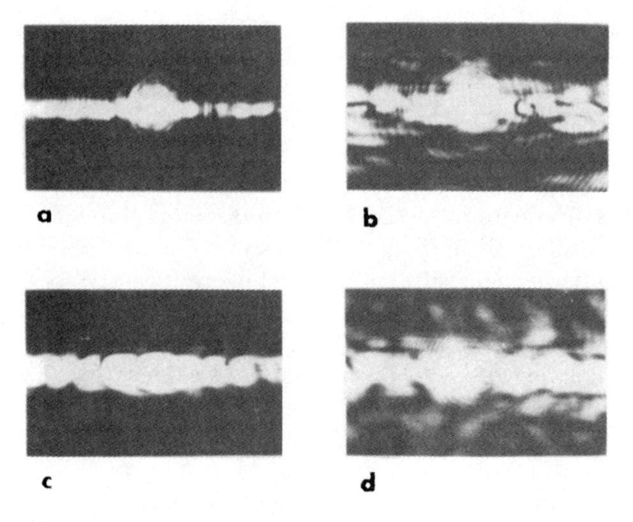

Fig. 15 Scattering from single paper fibers viewed from the forward direction (each fiber *a*–*d* aligned perpendicularly to scattering streak).

patterns (power spectra) of sufficiently well oriented fiber assemblies still indicate preferential orientation direction, as shown in Fig. 16. The data may be further treated by scanning the angular intensity distribution in the plane of the photograph and relating the variation to the probability that the constituent fiber is inclined to a preferential orientation direction. By including a parameter D_1 that depends on the number of (equal) sectors selected for scanning over a specified circular arc and an orientation distribution parameter D_2 determined by least squares fitting of the intensity data, the dropoff in probability density $F(\theta)$ can be expressed by the relation

$$F(\theta) = D_1\left(\pi^{-1} - D_2 \cos 2\theta\right) \tag{21}$$

as shown in Fig. 17 [224]. Note that Eq. (21) is a variant form of Eq. (3).

So far, this technique has been most successful in analyzing a thin sheet ($< 50\,\mu$m), split paper sheets, and synthetic nonwoven sheets due to the opacity of most paper structures made of pulp fibers. Niskanen and Sadowski [132,133] presented the experimental data and a theoretical argument suggesting that the light diffraction measurement is quite sensitive to basis weight. Thus, the results from samples with different basis weights cannot be directly compared.

Boulay and coworkers [14,16,17] and Gagnon et al. [48,49] developed another light diffraction method using a submillimeter (SMM) or far infrared (FIR) laser. The energy emitted from these lasers is not identical to the incident energy on the sheet, because the wave is absorbed differently by fiber and water in the sheet. The ratio of transmitted to incident energy is dependent on the orientation and the plane of polarization of the laser beam. Assuming that the alignment of fibers in the sheet is a linear polarizer to an electromagnetic wave, an optical setup such as the one shown in Fig. 18 provides information about the fiber orientation by rotating the beam plane of polarization. Because the usual diameter of the laser beam is small, this technique can be applied to the measurement of local variations of fiber orientation.

Light Diffusion When laser light is focused on a sheet, machine-made paper shows a simple anisotropic transmittance pattern. The major axis of the ellipse in this case is aligned in the machine direction. The ratio of major to minor axes may provide an index of fiber orientation anisotropy. Kohl and Hartig [105] explained that there can be two mechanisms for this pattern of anisotropy: (1) light scattering preferentially along fiber axes and (2) less attenuation of the scattered light along the direction of strongest orientation than perpendicular to it. Thus the measurement depends on the light scattering properties of paper as well as fiber orientation. An on-line instrument using this technique has been developed by Honeywell Lippke [7]. Adequate light transmittance can be achieved on sheets of up to $250\,\mathrm{g/m^2}$ [56].

Small-Angle Light Scattering Experimentally, the small-angle light scattering (SALS) technique is only slightly more complex than the visible light diffraction method.

One procedure is to embed the anisotropic scattering object in a liquid or resin of closely matching refractive index and filter the scattered light through a polarizer with its optic axis oriented as shown in Fig. 19 (H_v scattering) [15,25,112,115,127,

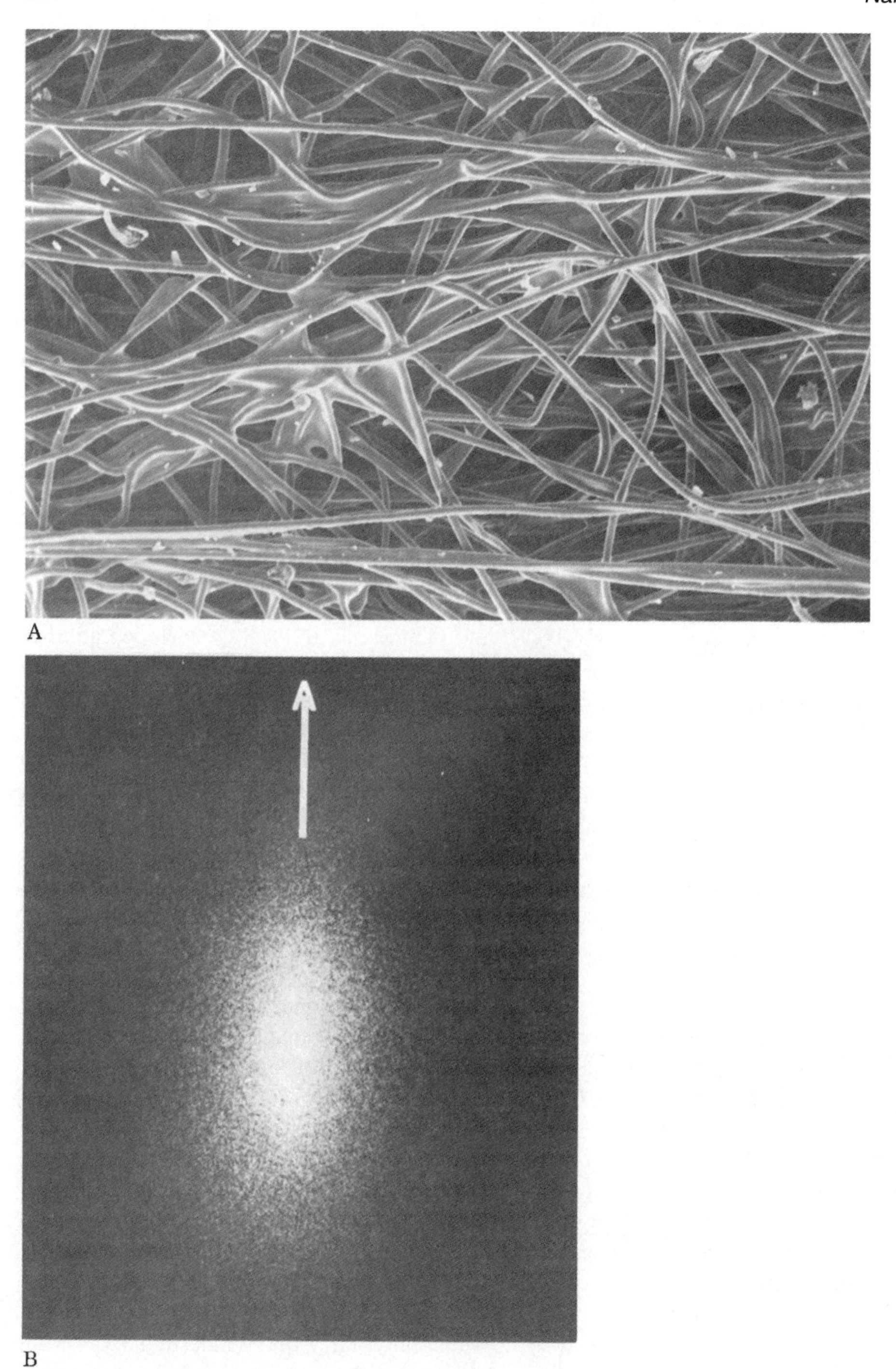

Fig. 16 A highly oriented synthetic fiber network (A) photographed at 60× magnification and (B) its diffraction pattern (Fourier transform). The arrow in (B) indicates the plane of maximum scattered light perpendicular to orientation direction. (From Ref. 224.)

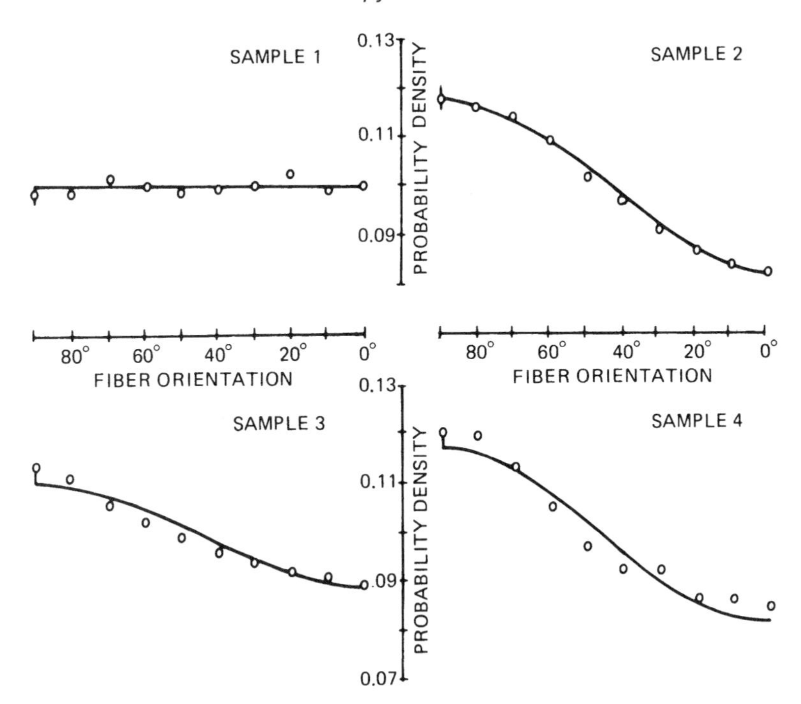

Fig. 17 Fiber orientation distributions of fiber assemblies of varying orientations. Values for D_2 [Eq. (21)] varied from 0 to 0.055, while $D_1 = 0.314$ for all cases. (From Ref. 224.)

158,185,215]. The scattered light intensity that is recorded is the result of optical anisotropy (birefringence) rather than of structural anisotropy. The interpretation of scattering patterns relies heavily on theories developed for visible light scattering from anisotropic polymer films [185]. For "rodlike" scattering (Fig. 19), the rod is visualized as consisting of scattering elements situated along the rod axis, for which scattered intensity will depend on the angular inclination of the rod axis toward the *XZ* plane. Consequently, a rodlike assembly oriented around a preferred angular direction will induce asymmetry to the distribution of scattered light (scattering envelope), as discussed by Charrier and Marchessault [25]. The H_v scattering envelope of condenser paper shows the MD orientation created during manufacture [15,25,112,115]. X-shaped scattering streaks are flattened around the horizontal

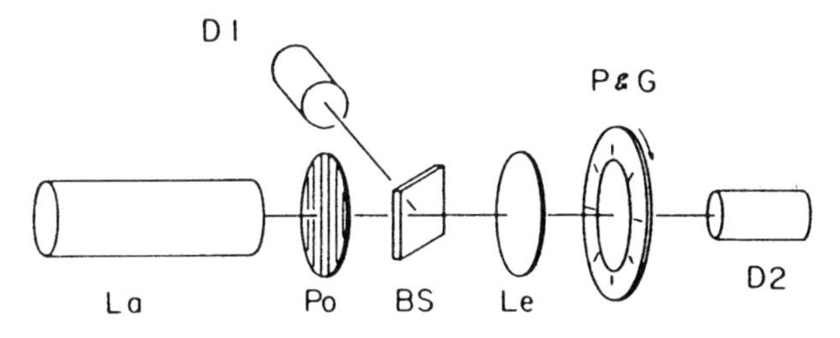

Fig. 18 Experimental setup for determining fiber orientation with a submillimeter laser (SMM): SMM laser (La), linear polarizer (Po), lens (Le), paper and goniometer (P&G), detectors (D1,D2). (From Ref. 17.)

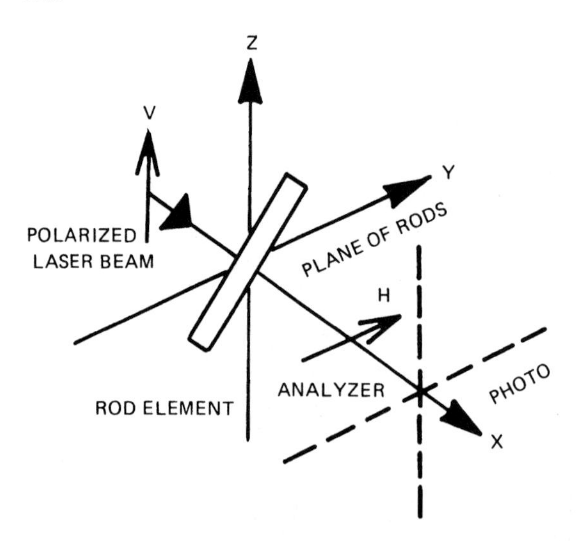

Fig. 19 Light scattering geometry. The incident laser beam is vertically polarized. The scattered light from elements of a rod assembly is filtered through a horizontally oriented polar (the analyzer).

axis when rod orientation (machine direction) is vertical (Fig. 20a). When rod orientation is 45° to the horizontal and vertical directions, the scattering streaks are orthogonal but of unequal intensity variation (Fig. 20b). The intensity variation in both patterns can be computer-calculated and related to orientation parameters and rod dimensions [25,112]. Additional V_v scattering envelopes (vertically oriented polarizer) provide additional structural information at the expense of added intensity contributions from imperfections in the sample and optical components of the scattering fixture (density fluctuations) [25,185].

The SALS technique provides a powerful method of analyzing cellulosic material that is strongly broken up (condenser paper) [15,25,112,115], regenerated (solvent cast films) [15], or elucidated for internal structure (single fibers) [158,215]. The most serious obstacles preventing the more universal use of this technique in paper science are those that pertain to fiber size and distribution and to the thickness of most commercially made papers. The radiation wavelengths of the visible light spectrum tend to favor the asymmetry of micrometer-size substructures rather than complete paper fibers in the intensity distributions, even at small scattering angles. A significant sample thickness requires corrections for multiple scattering effects similar to those needed for multiple scattering corrections in light diffraction analysis [25].

X-Ray Diffraction The orientation of crystalline cellulosic microfibrils in the middle secondary layer of the pulp fiber (S2 layer) affords a means of measuring fiber orientation by X-ray scattering [158]. The method is indirect in that the short-wavelength X-ray radiation interacts with the fibrillar subunits inside the collapsed fiber that are oriented at angles $\pm\beta$ to the fiber axis, itself inclined at angle ϵ to the YZ plane (Fig. 21). Consequently, it is necessary to measure S2 fibrillar orientation using one of the techniques discussed in Ref. 22, Chapters 13 and 14.

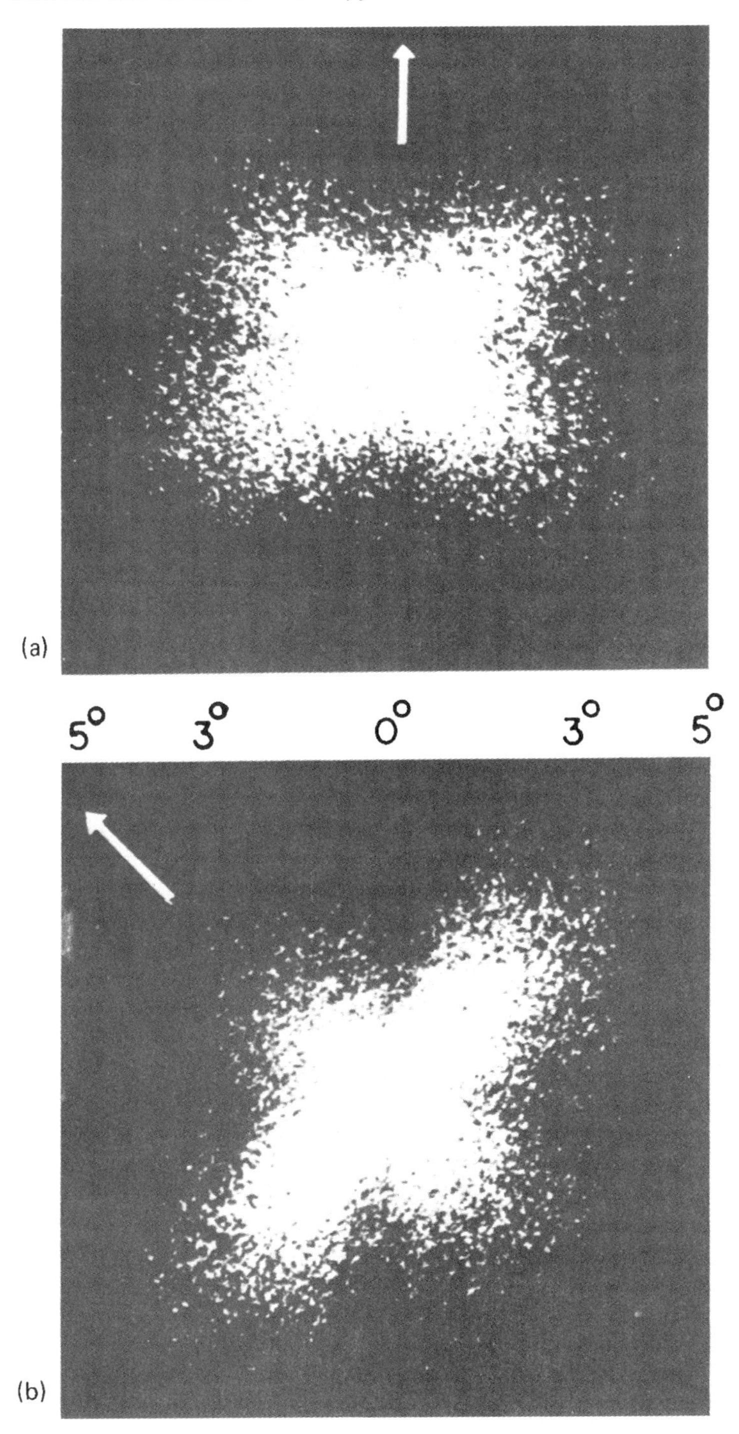

Fig. 20 Photographic H_v scattering patterns of condenser paper. Machine direction is indicated by arrows. (From Ref. 115.)

Both the SALS technique [25] and X-ray diffraction analysis [158] have been treated through orientation distribution functions where ϵ and C, an orientation parameter, provide calculation parameters for matching theoretically assumed distributions with experimentally measured intensity variations. For X-ray diffraction analysis, an elliptical-type orientation distribution has been employed, as in the equation

$$N(\epsilon) = \frac{C}{C^2 \sin^2 \epsilon + \cos^2 \epsilon} \tag{22}$$

The intensity distribution is both a function of fiber orientation distribution $N(\epsilon)$ and the fibril angle distribution $D(\beta)$, as described by Prud'homme et al. [158]. Experimentally, the X-ray method relies on the azimuthal intensity drop off of equatorial reflections in the X-ray pattern (Figs. 22A and 22B). Better mutual fiber orientation produces a stronger intensity decrease (Fig. 22A compared with Fig. 22B). The decrease is measured quantitatively and matched to a best fit value of the orientation parameter C for the fibril orientation in question (Fig. 23) [158].

Compared with the methods that employ visible light, this method is less ambiguous as to what creates the scattering. However, the added complexity in instrumentation and the need for knowledge of the fibril angle distribution create a procedure that is generally better suited for research and development than for routine and on-line control. On the other hand, the X-ray procedure may provide the only method available for paper embedded in plastics. Other X-ray methods suffer from limitations in evaluating scattering patterns—with associated complicated analysis procedures—as discussed by Rudström and Sjölin [169].

I. Fiber Orientation as a Component of Ultrasonic Measurement of Elastic Properties

The specific elastic stiffnesses for any material describe its stress–strain response. Acoustic methods are routinely used to measure elastic stiffness in sheetlike materials

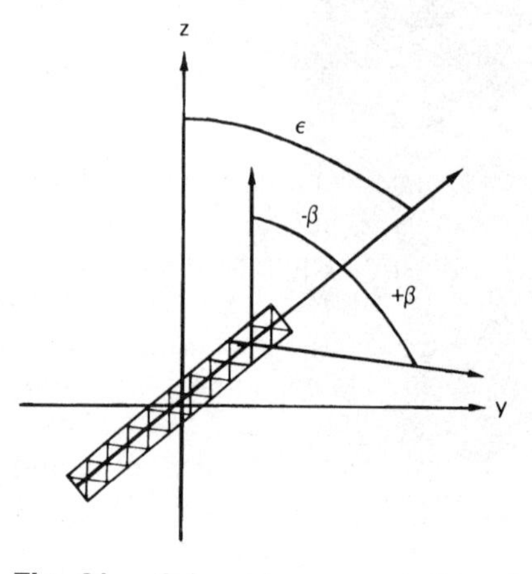

Fig. 21 Geometric representation of fiber in the YZ plane tilted at an angle ϵ. (From Ref. 158.)

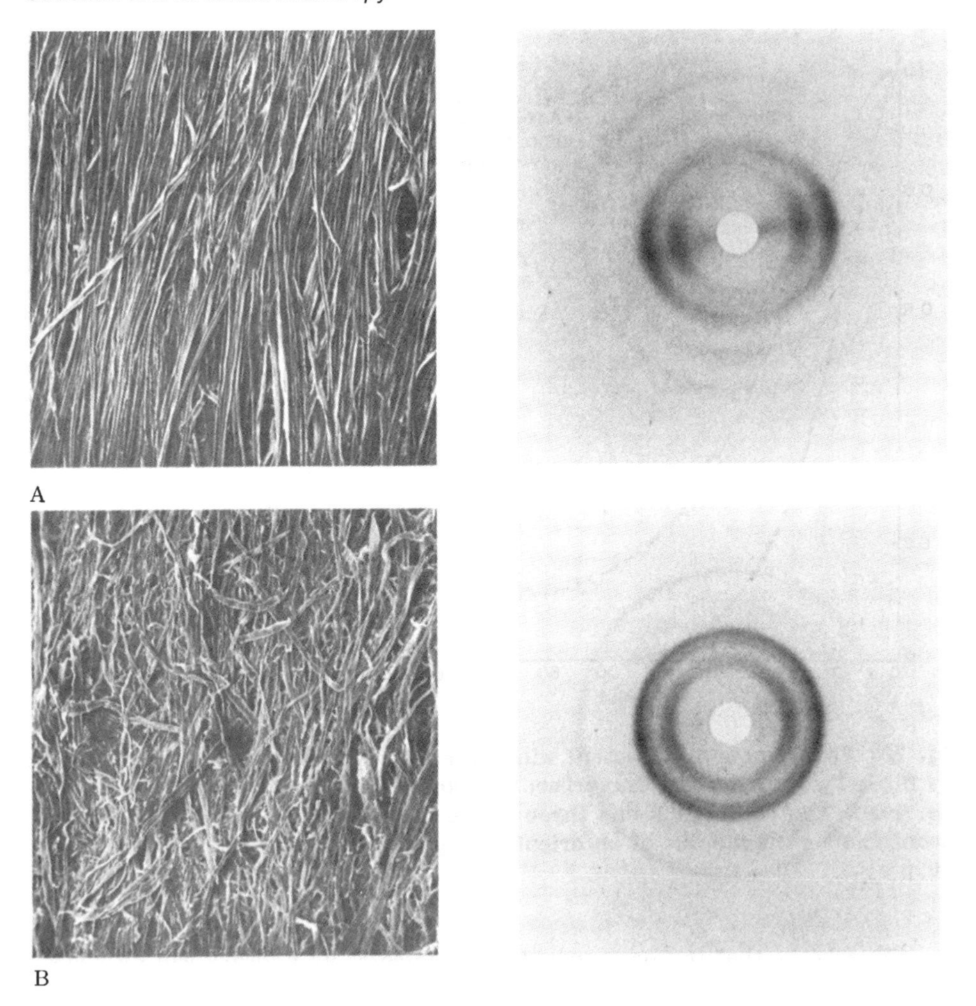

Fig. 22 Scanning electron micrographs of paper samples with different levels of orientation in the fiber structure at $\sim 30\times$. Better mutual orientation [(A) compared with (B)] produces stronger azimuthal intensity drop-off of equatorial X-ray reflections. (From Ref. 158.)

nondestructively. The elastic stiffnesses can be determined by measuring the velocity of ultrasound in the paper. The theory is described in detail in Chapter 6. The longitudinal specific elastic stiffness is determined by propagating longitudinal waves in the MD–CD plane. The acoustic measurements provide a mechanical MD/CD ratio more easily than most other methods. Several commercial testing instruments automatically measure specific elastic stiffnesses as a function of angle from the machine direction, mostly to predict the "fiber orientation" of the sheet (see Fig. 24). When the specific elastic stiffnesses determined by these instruments are plotted against angle from the machine direction, as a polar graph, the results typically look like those shown in Fig. 25. The major axis of the sheet can be computed from this plot by fitting the data points to an elliptical or other distribution function as discussed in Section II.C. Two definitions of the MD/CD ratio can be considered if there is a

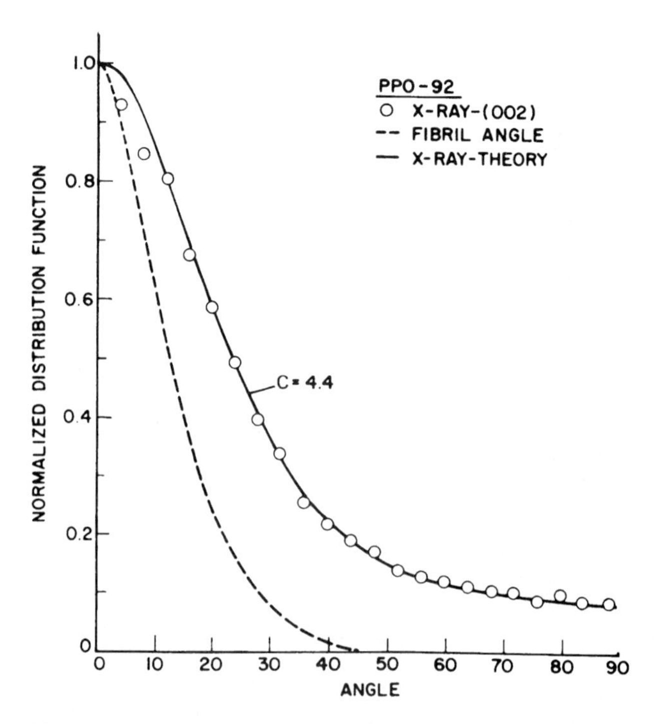

Fig. 23 Fibril angle (S2 filament winding angle) distribution function and X-ray intensity distribution (experimental points) for the sample shown in Fig. 22A. The continuous line through the experimental points is obtained theoretically with the use of an orientation parameter $C = 4.4$. (From Ref. 158.)

Fig. 24 An ultrasonic tester with an automatic turntable for measuring elastic stiffnesses as a function of angle. (Photo courtesy of Nomura Shoji, Tokyo, Japan.)

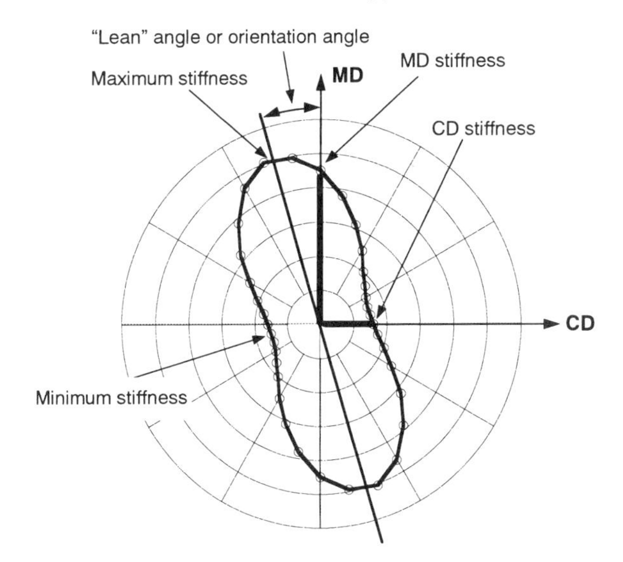

Fig. 25 Typical in-plane (MD–CD plane) polar diagram of specific elastic stiffness by an ultrasonic tester. The "lean" of the ellipse away from the MD results from the maximum in the fiber orientation distribution not lying along the machine direction.

"lean angle" between the original machine direction and the major axis of the ellipse; one is an elastic stiffness ratio for original machine and cross-machine directions, and another is that ratio for the maximum and minimum values. The lean angle typically varies across the width of a paper machine. Figure 26 shows eight diagrams taken at regular intervals across a 5 m width on a newsprint machine, from the front side to the drive side of the machine. As has been mentioned, these ultrasonic measurements should not generally be used alone to predict fiber orientation because they are also sensitive to wet straining and drying effects. Hess and Brodeur reported that the ultrasonic stiffness orientations and fiber orientations by image analysis are clearly different properties for bleached softwood kraft sheets formed on a Formette Dynamique machine and subsequently dried under various restraint conditions [64]. They concluded that the use of ultrasonic analysis of CD strips to diagnose headbox performance can be misleading. On the other hand, experiments by Baum [12] on commercial papers showed that the lean angle did not change as a result of rewetting and drying. He concluded that the relatively high lean angles at the edges of the web are related to the inclination of fiber orientation, perhaps caused by transverse stock flows at the slice.

Because acoustic techniques are nondestructive and quick, there has been rapid development of instrumentation. For instance, the SST (sonic sheet tester) of Nomura Shoji and TSO (tensile stiffness orientation) tester of Lorentzen & Wettre are routinely used.

J. Fiber Orientation by Microwave Attenuation

The interaction of microwave radiation with paper is sensitive to moisture content. Because water molecules in a sheet that has moderate moisture content are the main

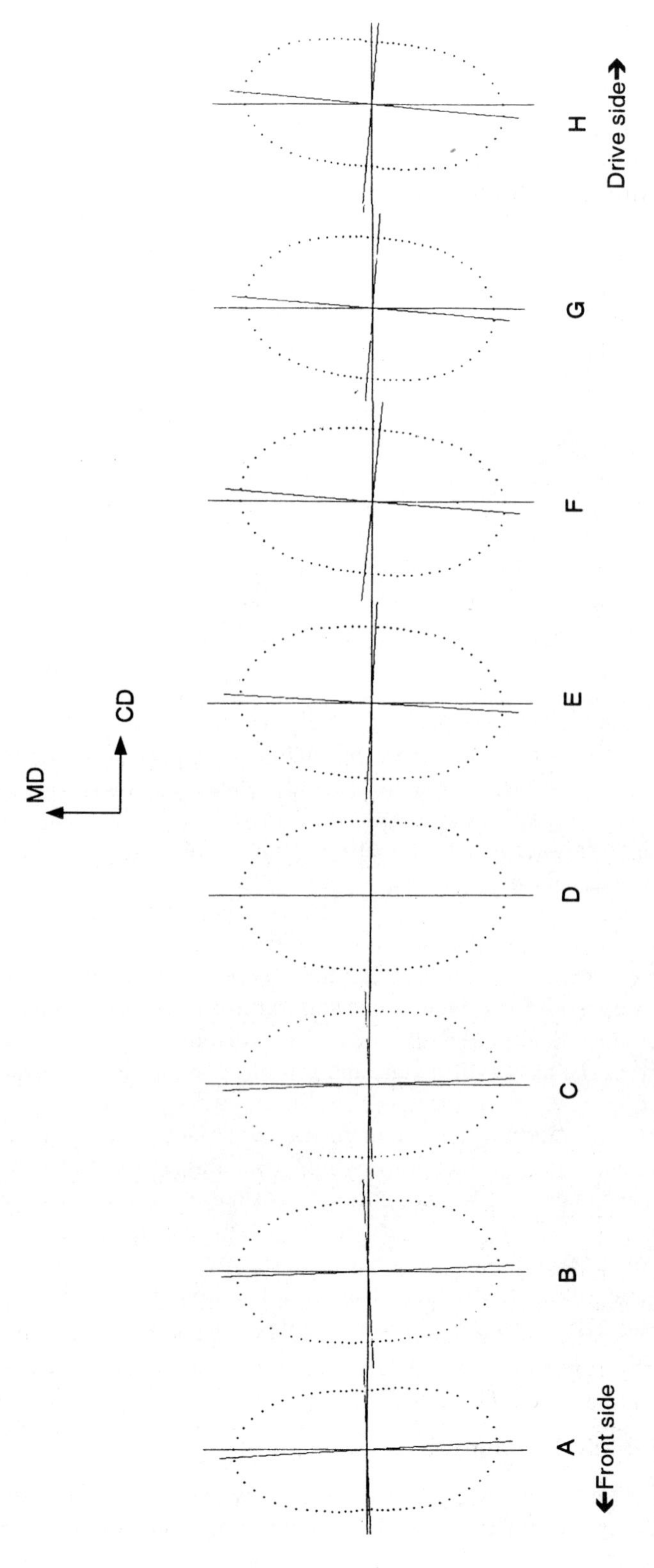

Fig. 26 Eight polar diagrams of ultrasonic measurements taken across a 5 m web. *A* is at front side of the paper machine, *H* at drive side.

source of labile dipoles, the dielectric constant (especially the imaginary part) depends mainly on the amount of water present. Microwave techniques for measuring moisture content, which are widely used in paper mills, are based on this strong interaction between water and microwave radiation.

The effective dielectric constant of paper, ε_p, is affected by the orientation of the electric field with respect to the principal axes of the paper. The absolute value of ε_p is greater for an in-plane than for an out-of-plane electric field [56,57]. Also, ε_p is larger for an electric field parallel to the machine direction than for one parallel to the cross-machine direction. There are two explanations of this dielectric anisotropy [56,57]. The fibers that principally make up paper lie in the plane of the paper and are preferentially aligned in the machine direction. The fiber-induced anisotropy in the geometry leads to dielectric anisotropy. The fiber fraction, which has a higher dielectric constant, is also more connected (bonded) along the machine direction, and ε_p is thereby larger in that direction. Roughly speaking, the structure is analogous to a parallel alignment of capacitors in the machine direction. Although the dielectric anisotropy can be explained in terms of fiber geometry, a secondary mechanism may also contribute. The bond between the absorbed water and the fiber restricts the lability of the water dipoles. On average, the water lability is different along the fiber axis than it is transverse to the axis. This variation, along with the dielectric anisotropy of the dry fiber, can result in a fiber matrix that is not dielectrically isotropic. The anisotropy in the fiber is reflected in the effective sheet properties.

The change in in-plane dielectric constant with the angle of the electric field to the machine direction is small; however, ε_p can be determined with sufficient repeatability to distinguish the machine and cross-machine directions and to detect changes in fiber orientation. Unlike mechanical anisotropy measurements (such as the MD/CD modulus ratio), microwave dielectric anisotropy measurements are insensitive to drying conditions (as was shown by Fleishman et al. [44]) and are thus a good nondestructive indicator of in-plane fiber orientation.

Instrumentation for this technique was developed by Habeger and Baum [56], Osaki [132], and Vepsäläinen et al. [214]. These methods measure the attenuation of a microwave signal passing through a sample. The attenuation depends on the amount of energy reflected at the boundary between sample and air and the amount of energy dissipated in the sample. These quantities, in turn, depend on the real and imaginary parts of the dielectric constant and the sample thickness. The effective dielectric constant is greater when the electric field is aligned in the machine direction; therefore, the ratio of attenuation of the electric field in the machine direction to that in the cross-machine direction (RM) is a measure of fiber orientation anisotropy. Briefly, the technique is to measure attenuations for a sample placed in a microwave waveguide at different orientations to the electric field. This is a simpler approach than finding the complex dielectric constant.

Figure 27 is a schematic diagram of the instrumentation employed by Habeger and Baum [56]. The microwave signal is produced by a microwave signal generator within a frequency range of 8.2–12.4 GHz. Habeger and Baum used a signal frequency of 9.25 GHz, whereas Osaki used a frequency of 3.9 GHz and Vepsäläinen et al. used a frequency of 2.2 GHz. The microwave signal is pulsed on and off with a 50% duty cycle at a 1 kHz rate. This amplitude-modulated microwave signal goes through a signal filter to eliminate spurious low frequency signals. The signal is next carried through a rigid coaxial cable to a rectangular X-band waveguide in order to

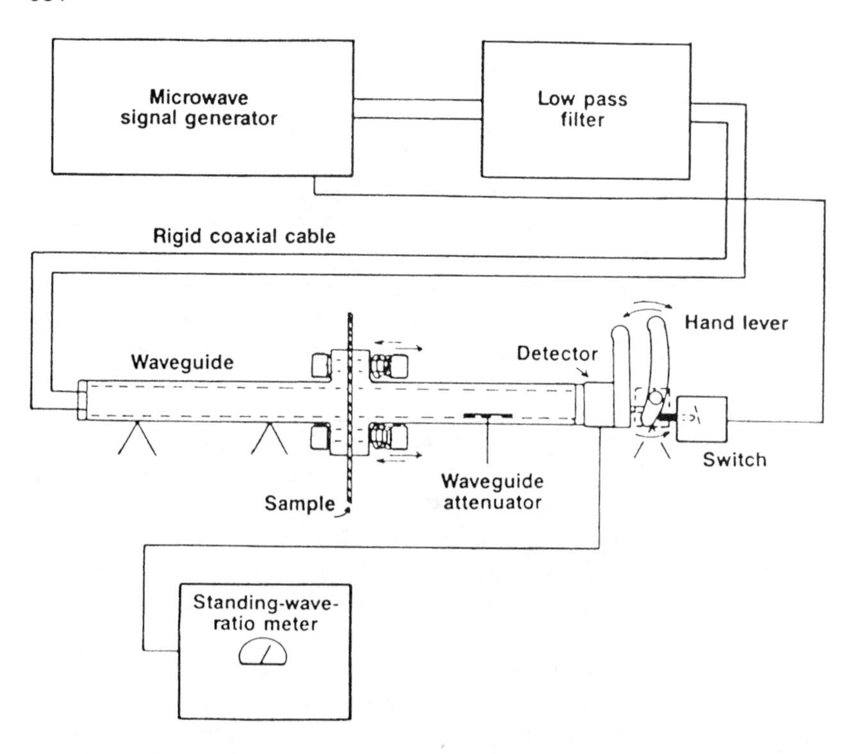

Fig. 27 Schematic diagram of a microwave fiber orientation gauge. (From Ref. 56.)

orient the electric field vertically. The end of the waveguide is attached to a variable microwave attenuator with spring-loaded bolts. The attenuator connects to a detector mount, which contains a low-barrier Schottky diode. The attenuator and detector mount are attached to a custom-made carrier that can be manually translated to open a gap between the waveguide and the attenuator. A switch shuts off the signal generator when the gap is open. Samples can be inserted in the gap with the machine direction vertical or horizontal. The signal, after passing through the sample and the attenuator, is rectified by the diode. The resulting 1 kHz square wave is applied to a standing wave ratio meter. This is a narrowband amplifier centered at 1 kHz. It registers—in decibels—the incoming signal on a needle dial.

Habeger and Baum discussed the sensitivity of this instrument to fiber orientation, basis weight, furnish, moisture content, and density. They concluded that the measurement of microwave attenuation anisotropy ratio is a good laboratory indicator of fiber orientation if the testing is done in a humidity-controlled room to avoid changes in moisture content; also, caution must be exercised when sheets whose densities have very little difference, are compared.

K. Fiber Orientation by Image Analysis

In addition to the analysis of tracer fibers, some image analysis approaches have been adopted to measure the fiber orientation. Yuhara et al. [227] developed a method using a soft X-ray microradiograph for determination of the mass distribution of sheet. The microradiographs obtained are analyzed with image analysis including two-dimensional Fourier power spectrum and edge detection. They found that fiber orientation distributions are approximated numerically by the angular distribution of the normalized power spectrum density rotated by $\pi/2$. For thin sheets or split sheets, fibers (or pores described by adjoining fibers) can be easily recognized on the fiber network image. Fiber orientation can be characterized by measuring the angle of each fiber [123] or by using transform functions [206] (applications to split sheets will be discussed in Section III). For example, the Fotocomp Orientation by Fotocomp Oy, Finland, is commercial software for analyzing fiber orientation in split sheets with their digital images.

III. STRUCTURAL VARIATIONS IN THE THICKNESS (*Z*) DIRECTION (CONTRIBUTED BY MAKIO HASUIKE)

Currently, paper is produced on Fourdrinier formers, twin-wire formers, and hybrid formers designed to combine the respective advantages of both Fourdrinier and twin-wire formers. High consistency forming and multilayer forming technologies are also under development in the paper machinery industry. Today's papermaker needs to have sophisticated knowledge in order to tailor the end product by optimizing the interior structure of the sheet on the available machines.

Quite often, in order to maintain desirable properties in certain grades, it is necessary to (1) enhance the degree of fiber dispersion in the fibrous layers and (2) reduce the structural and constituent differences between the top and bottom layers, as, for example, in the distribution of fiber orientation and/or fines and filler contents. The characteristics are closely related to the mode of aggregation of fibers in the Z direction. The mode of aggregation is influenced by the dewatering process, in which fibers are immobilized to form fibrous layers.

Techniques for paper structural analysis that make it possible to perform a three-dimensional analysis of fiber orientations, entanglement, and coherent flocs that aggregate in fibrous layers are described in the following pages.

A. Variation of In-Plane Fiber Orientation Ratio Through the Sheet Thickness

Fiber orientation differences in the sheet surface layers are dominant influences in the anisotropy of physical properties in machine-made paper. Paper two-sidedness is related to several types of out-of-plane deformation such as curl induced by changes in the ambient environmental conditions of the paper. Hence, various methods have been proposed to evaluate fiber orientation. These can be roughly classified into (1) methods for obtaining the orientation distribution indirectly from the directional physical properties of paper and (2) methods to directly detect the orientation distribution of fibers and molecular chains of cellulose (fibril orientation).

For evaluation of orientation in two-sided split sheets, which is the subject of this section, the indirect zero-span tensile test method has often been used. However, this method often shows a large variance in its results. Sonic measurements, on the other hand, are related to restraint conditions during the drying of paper as well as changes in elastic properties, etc. caused by the splitting. Direct methods also have limitations. The presence of small amounts of filler or other inorganic matter can affect automatic image analysis results. A small measuring range can be a problem, but by using adhesive tape for splitting, a very thin paper layer can be easily prepared. Then, by direct microscopic analysis of the split layer, the orientation distribution of fibers and/or fiber segments can be determined by image processing.

Measurement of orientation in a microscopically magnified image of a split layer proceeds as follows.

1. *Preparation of sample.* In any microscopic method, sufficient contrast to distinguish the individual fibers is required. Accordingly, the sample (4.5 × 7.0 cm) is given a preliminary staining. After drying, the sample is split into about eight layers for a sheet grammage of 64 g/m^2, as illustrated in Fig. 28. Because the thickness of each split layer is not constant, additional splitting is needed in some cases, until there are about equal proportions of fiber and interfiber space or a slight preponderance of the former in each specimen that is to be used for orientation analysis.

Next, the split sheet layer (with its attached transparent tape) is inserted between two glass plates (5.0 × 7.5 cm), and thus a specimen for orientation analysis with suppressed scattering of light from the fiber and adhesive surface is obtained, as shown in Fig. 29. By counting the fiber crossing points on the test lines when the split sheet is superimposed on the grid (Fig. 28), the geometrical orientation ratio is determined.

2. *Orientation evaluation by line crossing analysis.* The theoretical background described below refers to the textbook of Weibel [220]. For evaluation of the orientation of line elements on a plane, test line patterns are overlaid on a fiber network diagram and the number of crossing points per unit length of test lines is counted. The index of orientation η, defined by the orientation distribution function $f(\theta)$ in Eq. (23), is determined by the number of crossings $N(\theta = 0)$ of the test lines that lie in the direction parallel to the orientation direction of interest, the number of fiber segments in a unit length, and the number of crossings $N(\theta = \pi/2)$ of the test lines that lie in the direction perpendicular to the orientation direction of interest in the unit length.

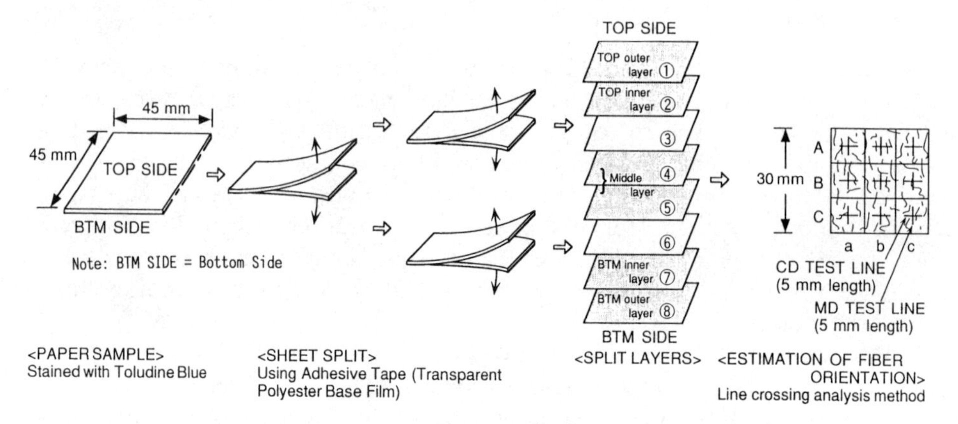

Fig. 28 Preparation of split sheets for fiber orientation analysis.

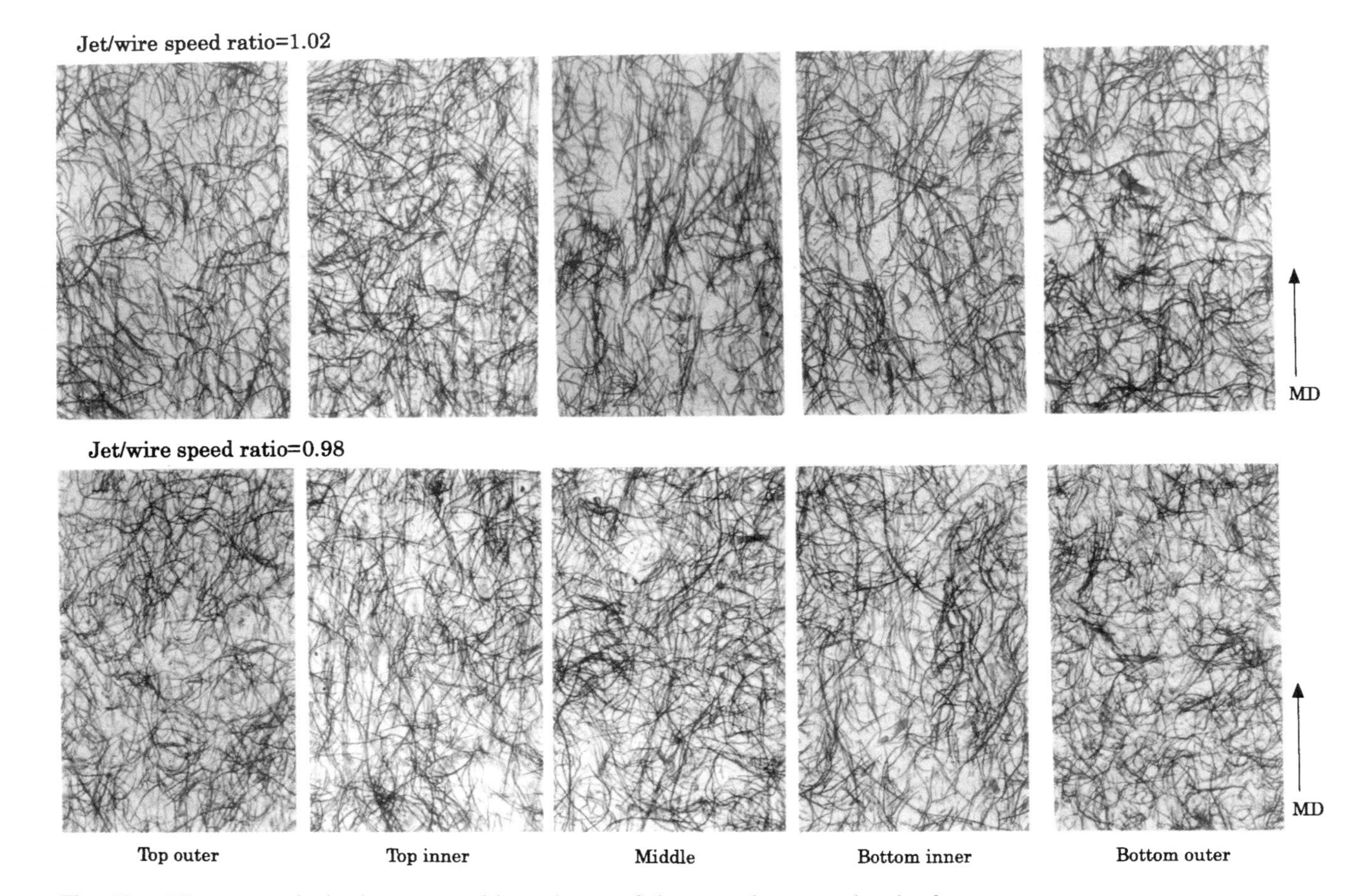

Fig. 29 Fiber networks in the outer and inner layers of sheets made on a twin-wire former.

In an oriented fiber network such as paper, the number of crossings $N(\theta = 0)$ of the fibers that fall within length L on test lines in the direction of the major axis of orientation is smaller than the number of crossings $N(\theta = \pi/2)$ in the direction perpendicular to the direction of the major axis of orientation. The angular distribution of the number of crossings $N(\theta)$ can be expressed by the relations

$$f(\theta) = \frac{1}{\pi}(1 + \eta \cos 2\theta), \qquad 0 \leq \eta \leq 1 \tag{23}$$

$$N(\theta) = a + b \cos 2\theta \tag{24}$$

where a and b in Eq. (24) are

$$a = \frac{2N_f \lambda L}{\pi A} \tag{25}$$

$$b = \frac{a\eta}{3} = \frac{2N_f \lambda L}{3\pi A}\eta \tag{26}$$

Equation (24) can now be written

$$N(\theta) = \frac{2N_f \lambda L}{\pi A}\left(1 - \frac{1}{3}\eta \cos 2\theta\right) \tag{27}$$

Therefore,

$$\eta = 3\left[\frac{N(\theta = \pi/2) - N(\theta = 0)}{N(\theta = \pi/2) + N(\theta = 0)}\right] \tag{28}$$

By using this index of orientation η and the probability density function $f(\theta)$ of orientation given by Eq. (23), the orientation ratio R can be calculated as

$$R = \frac{1 + \eta}{1 - \eta} \tag{29}$$

Estimation of the MD/CD ratio of fiber orientation may be made difficult by fluctuations in the count of crossings by the test line method. The relationship between these fluctuations and the length of the test line has been investigated with respect to the wire-side paper layer of machine-made sheet samples. It was necessary to count crossings of test lines of 18 mm or more on such samples. The number of measuring points could be approximated by a Poisson distribution, and there were fluctuations of about 400 ± 20 measuring points in the examined area of 20×20 mm containing approximately 12,500 fibers.

A system for analyzing the fiber orientation distribution in split layers—Fotocomp Orientation, produced by Fotocomp Oy, Jyväskylä, Finland—is now available commercially. For imaging of the split sheets, two systems can be used, one camera-based and the other scanner-based. Alternatively, externally generated images can be analyzed. The Orientation software includes automatic graphical presentation of the measurement results.

3. *Evaluation of fiber orientation in the middle layer of a machine-made sheet.* Thin-layer fibrous networks from sheets, as shown in Fig. 29, were made on a twin-wire former at different jet/wire speed ratios. Variations of the MD/CD ratio of fiber orientation of the outer and inner layers were observed as the wire speed ratios were

varied. Differences in the MD/CD orientation ratio were also significant in the networks of middle layers.

Another influence on the MD/CD orientation ratio of the inner layer of the sheet was the migration effect that is caused by gravity. Gravitation changes pulp slurry speed on the wire when the vertical or horizontal alignments of the dewatering devices of the twin-wire former are varied. In Fig. 30, calculated profiles of pulp slurry speed at various distances from the impinging pulp jet along the wire are shown. The cumulative shear factor was estimated by Eq. (30) for a range of vertical and horizontal alignments of the dewatering devices.

$$\tau_c = \int_0^L |U - V_m| dL \tag{30}$$

where

τ_c = cumulative shear factor
L = distance from the jet impingement
U = velocity of mobile pulp layer
V_m = velocity of immobilized pulp layer

A plot of the MD/CD orientation ratio versus cumulative shear factor demonstrates a consistent linear trend as shown in Fig. 31. The orientation ratio of the middle layer of a sheet made on the twin-wire former can now be interpreted in terms of the cumulative shear factor acting on the pulp slurry in the forming zone as estimated from the theoretical analysis of the dewatering.

Z-Directional Orientation It is possible to apply three-dimensional (3-D) reconstruction techniques from serial or sequential images of cross sections of the sheet to obtain an analysis of its Z-directional fiber arrangement. Reconstruction is a first step in developing a general method for the evaluation of the 3-D arrangement of fibers in machine-made paper. Computer graphics allow one to visually evaluate three-dimensionally reconstructed results from serial sections. One can also record photographs of serial thin sections of a specimen under a microscope as part of another method of investigation of 3-D structure.

Quantitative information on Z-directional fiber orientation in the sheet has been derived from the data of central points for the fibers and construction of a 3-D skeletonized image of the fiber segments. The orientation distribution of skele-

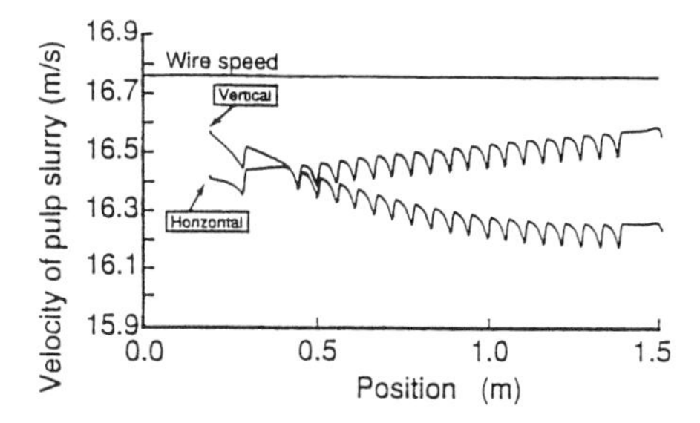

Fig. 30 Velocity profiles of pulp slurry between top and bottom wires.

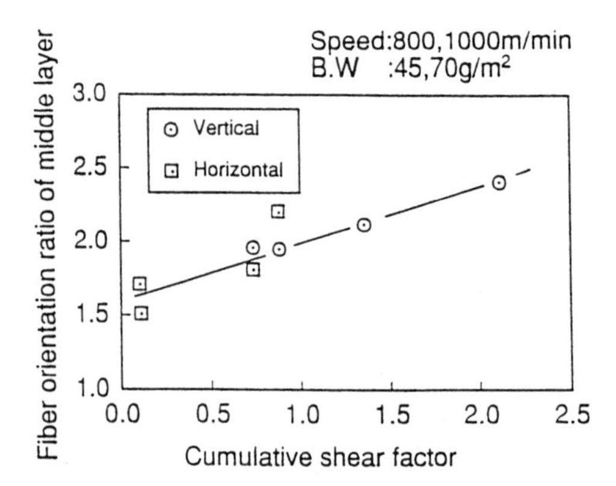

Fig. 31 Relationship between cumulative shear factor and fiber orientation ratio of middle layer. (Sheets of 45 and 70 g/m^2 basis weight made at 1000 and 800 m/min in wire speed, respectively.)

tonized fiber segments in the Z direction is shown in Fig. 32. The distribution curve projected on the XZ plane shows variations in penetration, but that on the YZ plane does not. The demonstrated difference of the component of orientation in the Z direction between the XZ and YZ planes contributes to the anisotropic structure of the pore system in the sheet.

Concerning the 3-D fiber orientation, Q_i [161] formulated a theoretical description that introduces ordering parameters Q_{20} and Q_{40} for the quantitative description of the Z-directional orientation of fibers. In so doing, he demonstrated fairly good fitting between 3-D angular distribution functions (ADFs) derived from the experimental data plotted in Fig. 32 and those obtained by using Q_{20} and Q_{40} estimated mathematically for the experimental results shown in Fig. 32. His mathematical description may be applicable for prediction of mechanical properties in the thickness direction of paper and paperboard.

B. Fiber Entanglement

Preparation of Serial Cross Sections and Visualization of Three-Dimensional Sheet Structure To visualize the three-dimensional arrangement of fibers in a sheet, it is indispensable to obtain precise three-dimensional coordinates of each fiber. It is therefore necessary to section a sheet serially and track the coordinates of the fibers in each section. As shown in Fig. 33, a sheet embedded in epoxy resin can be sectioned serially with an ultramicrotome to gain as many good sections as possible. Two threads of nylon fiber are embedded in the same resin block so that the images of their cross sections can be used as marks for aligning each sample. Although a minimum thickness of 0.5 μm is possible with this method, a somewhat thicker (4 μm) section will yield greater replicability over a slide sample size ($\sim$ 150 sections). The 4 μm thick sections will usually be 2 mm in width, taking the convenience of three-dimensional image processing, described later, into consideration. Figure 34 shows optical microscopic images of serial sections.

The optical microscopic images of serial sections are processed to extract the contour of the cross section of each individual fiber. Then these fibers are recon-

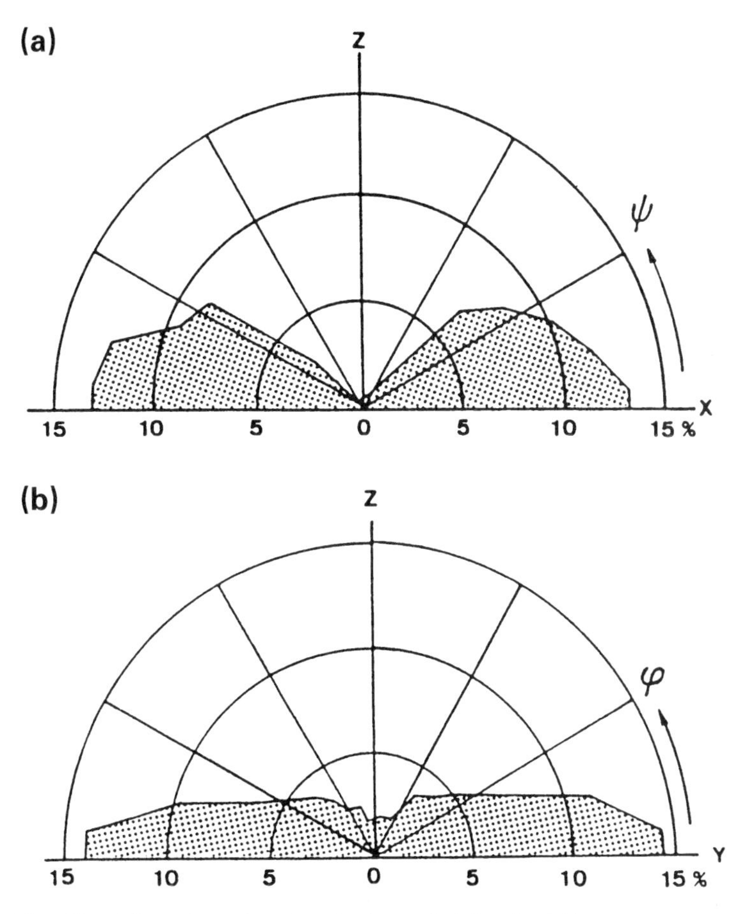

Fig. 32 The orientation distributions of skeltonized fiber segments projected on the (a) *XZ* and (b) *YZ* planes. (*X* axis: cross-machine direction; *Y* axis: machine direction.)

structed by connecting the contours of the individual cross sections of the fiber. In this way the three-dimensional sheet structure is visualized, as shown in Fig. 35.

When one looks at a cross section of paper sheet such as that shown in Fig. 33, the salient structural feature in the *Z* direction is the fact that the sheet is essentially "layered." Such an impression contrasts with the notion of "interwoven" or "felted" fibers in the sheets that is still widely found among papermakers.

The 3-D reconstructed image of the fibers located in the middle layer of the sheet, which consists of three marked fibers (numerical symbols 1, 3, 4) together with unmarked local fibers, is shown in Fig. 35A. Three fibrous segments that appear in the image of Fig. 35A are eliminated in Fig. 35B. From these figures, part of the 3-D fiber arrangement in the sheet can be visualized by computer graphics; fiber segments that are partially layered and partially interwoven in the sheet can be accounted for.

Analysis of Fiber Entanglement This section focuses on an individual fiber and describes a method of evaluating the degree of entanglement from the coordinates in three dimensions.

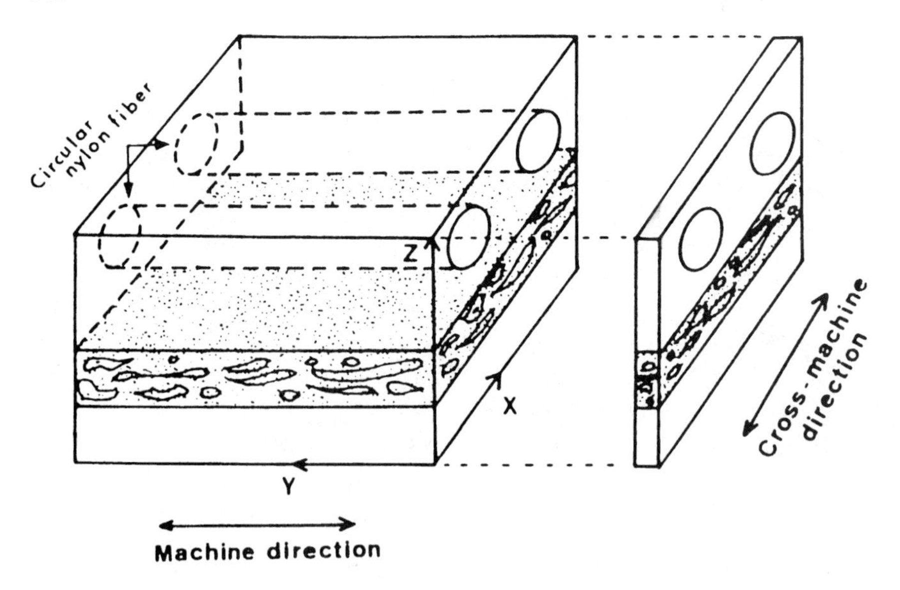

Fig. 33 Sectioning of a specimen block.

The state of overlapping of fibers can be observed when the cross section of a sheet is examined. If we assign an identification number to each fiber on each cross section—for example, No. 1 for a fiber on the top surface and larger numbers towards the wire side—it is to be expected that the order of numbers on a given cross section may be more varied than that on another cross section if the degree of fiber entanglement is higher.

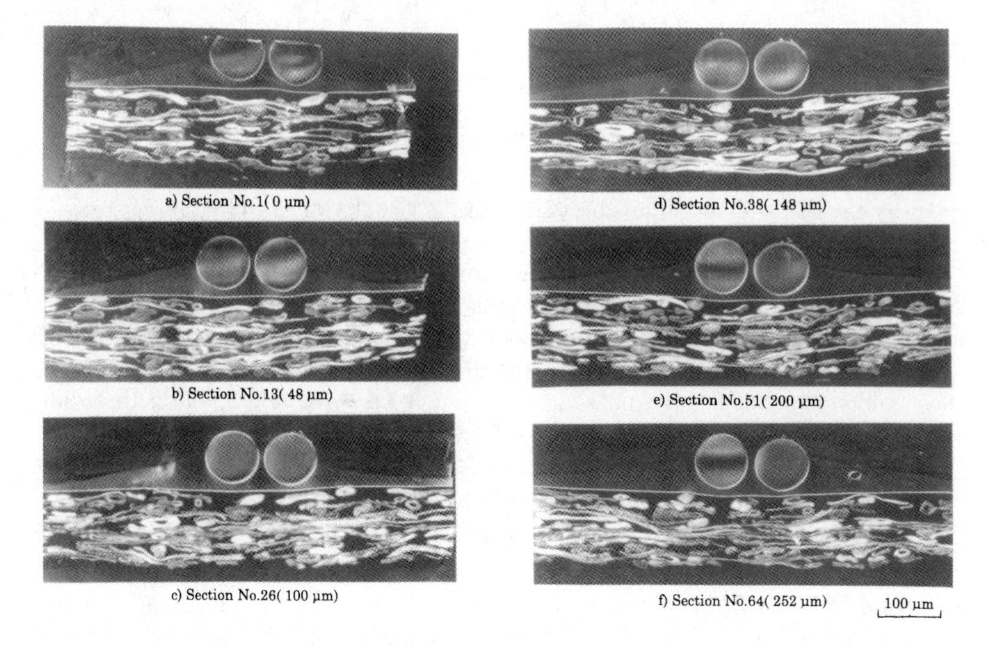

Fig. 34 Optical microscopic images of serial sections (thickness of a section = 4 μm)

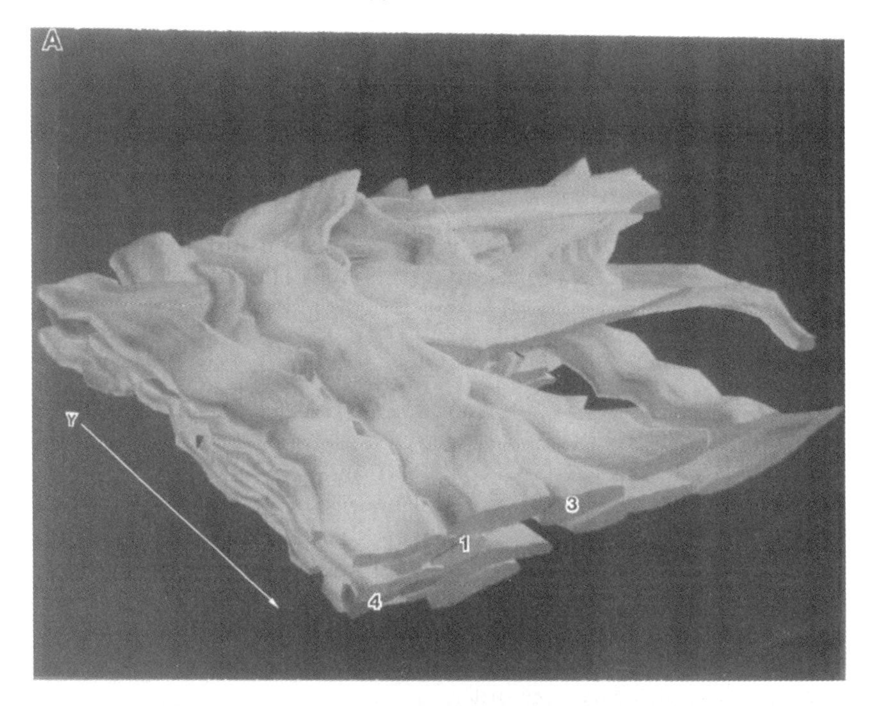

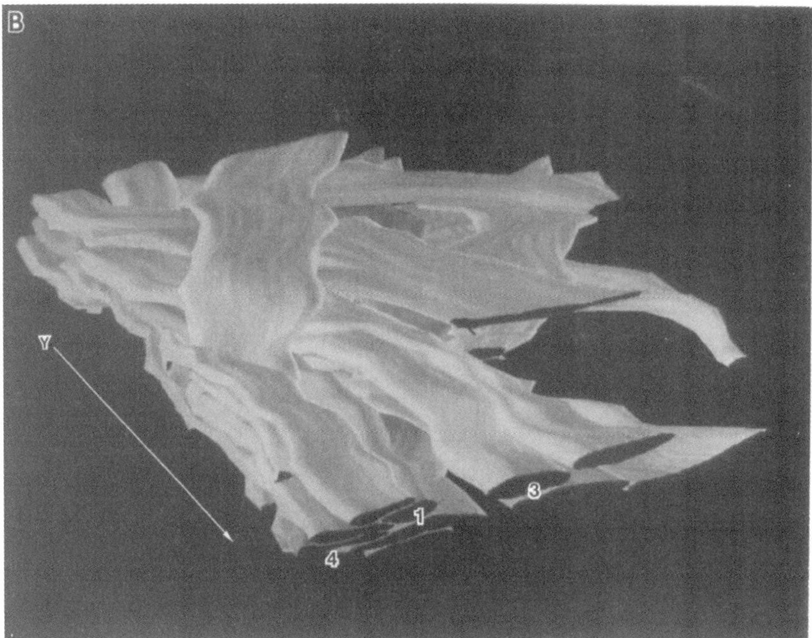

Fig. 35 Three-dimensional images of fiber segments located in the middle part of a sheet.

If S is defined as the identification number of a fiber on a given cross section, the normalized number is expressed as $0 \leq S \leq 1$. If a projected fiber length is expressed as l, the average order h of a fiber is expressed by

$$h = \frac{1}{L}\int_0^L S\, dl \tag{31}$$

If the probability density G is defined as the average order distribution of every fiber within the range of analysis, the following equation is obtained:

$$\int_0^L G\, dh = 1.0 \tag{32}$$

Frequency distributions of two contrasting sheet structures are shown in Fig. 36. They indicate that the distribution of a sheet with a layered structure shows a straight line (constant frequency), as shown in Fig. 36a, and that the distribution of a felt-like structure shows a curve with the center of the entanglement being at maximum, as shown in Fig. 36b [163]. If fibers are arranged like a felt, then some fibers on a given cross section may be located toward the top surface of the sheet whereas the same fibers on another cross section may be found toward the back side of the sheet. The average order will therefore be in the middle stratum. It is therefore possible to estimate the degree of entanglement in fibrous layers from the shape of the frequency distribution of the average order.

The number of fibers that cross over or under a given fiber on a certain cross section can be counted as illustrated in Fig. 37. The average order h of the fibers is estimated with the equation [63].

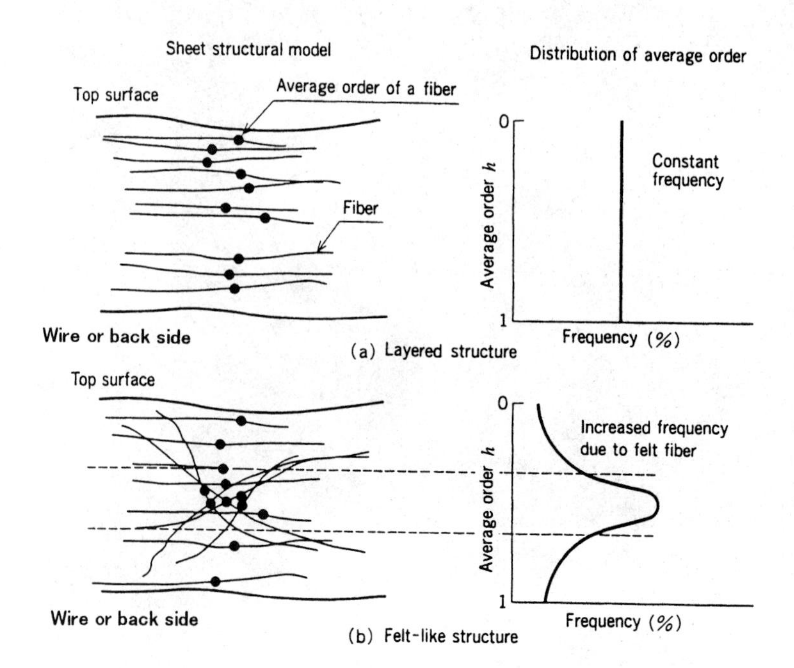

Fig. 36 Sheet structure model (left) and frequency distribution of the average order (right) for (a) layered structure and (b) felt-like structure.

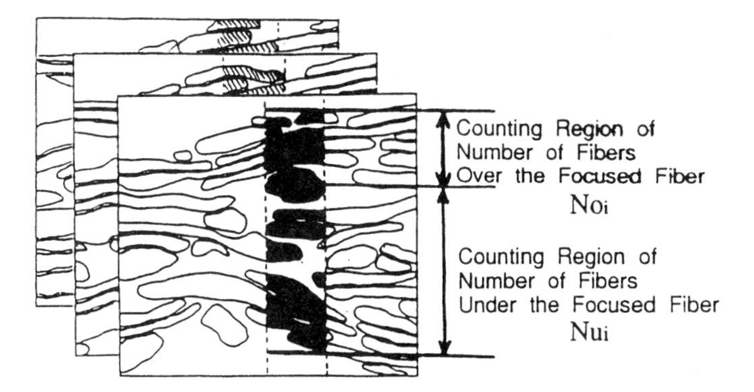

Fig. 37 Estimation of the average order by counting fibers that cross over and under a designated fiber.

$$h_i = \sum_{i=1}^{n} \left[\frac{N_{oi}}{(N_{oi} + N_{ui})} \right] \left(\frac{1}{n} \right) \tag{33}$$

where the subscript i represents the identification number of a cross section. N_{oi} and N_{ui} are defined as the numbers of fibers crossing over and under a given fiber on a certain cross section, respectively.

The frequency distribution of the average order can be estimated if the above process is executed for every fiber that is within the range of analysis

In the dewatering process, an $\sim 0.8\%$ concentration of fibers in suspension is enhanced to 20% during formation of the basic sheet structure. Paper qualities that depend on sheet structure are largely determined by this dewatering process. Figure 38 illustrates models of three different dewatering methods. In Fourdrinier-type

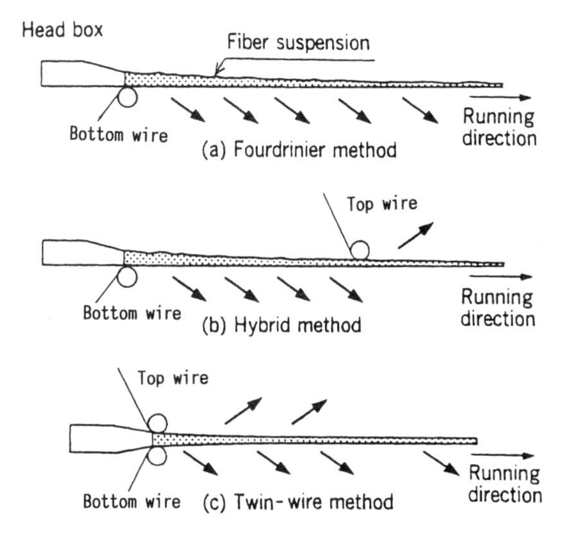

Fig. 38 Drainage models of the forming parts of three paper machine systems. (a) Fourdrinier method; (b) hybrid method; (c) twin-wire method.

dewatering, water is drained through the bottom wire only. In hybrid-type dewatering, water is drained through the bottom wire first, and subsequent dewatering is done through the top wire with the use of dewatering devices on top. In twin-wire-type dewatering, the initial dewatering is done through both top and bottom wires simultaneously, then water is additionally drained through the bottom wire. The analytical method herein described applies to sheets made using all three of these dewatering methods.

Distribution of Fiber Entanglement The frequency distributions of average order and estimated models of fiber arrangement in relation to each dewatering method are shown in Fig. 39. The frequency of average order increases if the degree of fiber entanglement becomes greater. The fiber arrangement is influenced by the dewatering direction in the sheet-forming process. In the Fourdrinier method, the areas with comparatively high frequency exist at a few locations; the highest frequency is found in the middle layer slightly closer to the back wire side. This arrangement indicates that most of the fibers there are entangled like a felt.

With hybrid machines, there is an area with a comparatively low frequency in the middle layer, lying between areas with comparatively high frequency. This type of frequency distribution shows that a flocculated area exists on both the top surface and the back side. In the twin-wire method, the frequency distribution is comparatively uniform, which suggests a layered structure with relatively few entanglements.

C. Extension of Flocs in the *Z*-Direction

Multiple Layer Interactions Described by the Nature of Stochastic Structure [38] Two layers, P and Q, with the same mean grammage may be independent when the variance of the sum is simply the sum of the variances.

The layers may reinforce one another if heavy regions fall onto heavy regions and the net variance is four times the variance of the individual layers. Or the layers may interact to deposit more material on previously lighter regions, to give a smoothing effect. Let p be the variable representing local grammage in P, and q the corresponding variable for Q, at some chosen scale, say 1 mm squares. Then the variance of the composite sheet is by definition

$$\mathrm{Var}(p+q) = \mathrm{Var}(p) + \mathrm{Var}(q) + 2\ \mathrm{Cov}(p, q) \tag{34}$$

For independent layers, $\mathrm{Cov}(p, q) = 0$. Therefore, $\mathrm{Var}(p+q) = \mathrm{Var}(p) + \mathrm{Var}(q)$, and hence, for statistically similar layers we conclude that

$$\mathrm{Var}(p+q) = 2\mathrm{Var}(p) \qquad \text{and} \qquad n_f(p+q) = n_f(p) \tag{35}$$

For reinforcing layers, $q = p$, so $\mathrm{Cov}(p, q) = \mathrm{Var}(p)$. Hence,

$$\mathrm{Var}(p+q) = \mathrm{Var}(2p) = 4\mathrm{Var}(p) \qquad \text{and} \qquad n_f(p+q) = 2n_f(p) \tag{36}$$

For smoothing layers, $\mathrm{Cov}(p, q) < 0$. Say, $\mathrm{Cov}(p, q) = -s\ \mathrm{Var}(p)$, for $0 \le s < 1$. Then

$$\mathrm{Var}(p+q) = 2(1-s)\ \mathrm{Var}(p) \qquad \text{and} \qquad n_f(p+q) = (1-s)n_f(p) \tag{37}$$

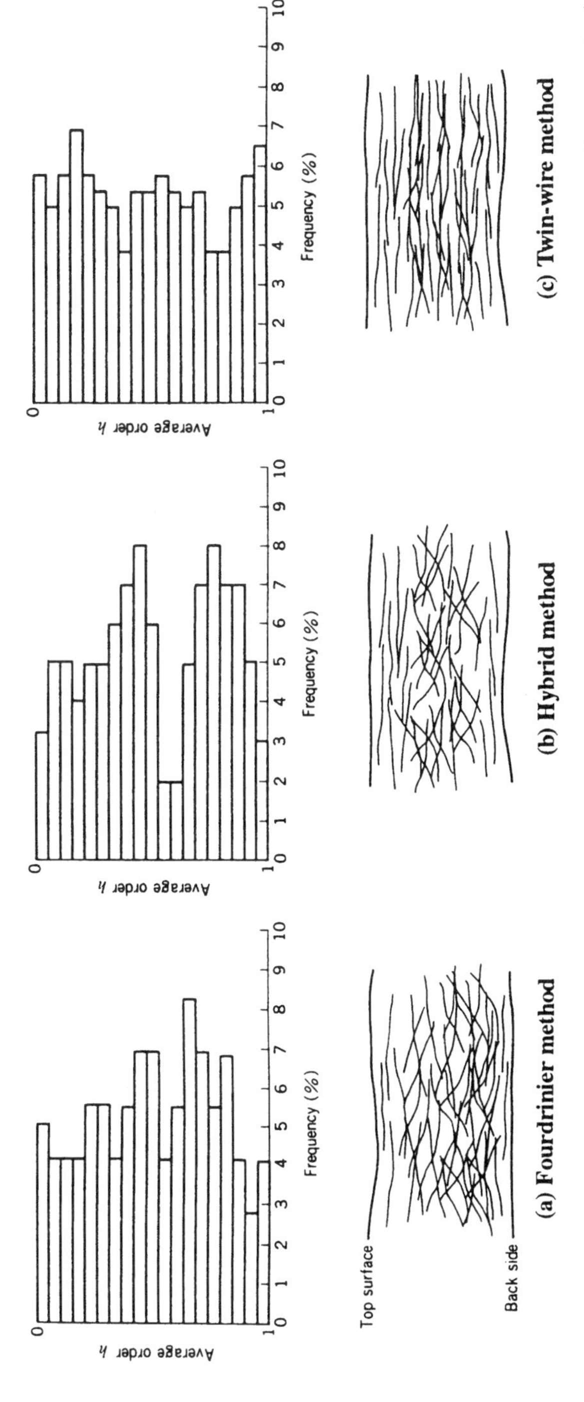

Fig. 39 Frequency distribution of the average order and models of fiber arrangements in paper sheets formed by different methods.

The smoothing layer case is the most realistic and provides us with an experimental method to investigate the phenomenon by measurement of the smoothing parameter s, which is the negative of the correlation coefficient between the two layers at the chosen scale.

From Eq. (37),

$$s = 1 - \frac{\mathrm{Var}(p+q)}{2\,\mathrm{Var}(p)} \tag{38}$$

The smoothing parameter s in Eq. (38) helps us understand the Z-direction structural changes under different forming conditions and for different formers. For more than two layers, the form of Eq. (38) remains similar, and the parameter s just measures the mean negative correlation between pairs of layers.

Cross Correlation Analysis of Split Layers When a sheet is split into many layers, a given floc is likely to exist through several of these layers. To determine the existence of split-layer flocs, two neighboring layers can be compared and the cross correlation coefficient estimated. This coefficient represents the degree of correlation between layers. With the use of image processing, the cross correlation coefficient is expressed by Eq. (39) [10].

The image of a split layer is composed of pixels having 0–255 degrees of gray scale. The image is represented by $f(i,j)$ and the neighboring one by $t(i,j)$. If a split layer is dislocated by a distance τ_1 in a certain direction, the cross correlation coefficient $C(m,n)$, where $m \times n$ represents the number of pixels in the overlapped area, is defined as follows.

$$C(m,n) = \frac{\sum_{(i,j)} \sum_{\in D} f(i,j)t(i-\tau_1,j)}{\left[\sum_{(i,j)} \sum_{\in D} f^2(i,j)\right]^{1/2} \left[\sum_{(i,j)} \sum_{\in D} t^2(i-\tau_1,j)\right]^{1/2}} \leq 1 \tag{39}$$

If the two split layers are identical, excluding the cases where one is simply an integer multiple of the other, the maximum value of 1 is obtained.

Because the correlation between each pair of split layers can be estimated from the dependence of the cross correlation coefficient upon τ_1, the ratio of the cross correlation coefficient for $\tau = \tau_i$ to that for $\tau = 0$ shown in Eq. (40) is used for the purpose of simplified analytical evaluation.

$$\begin{aligned}\frac{C_{\tau=\tau_1}(m,n)}{C_{\tau=0}(m,n)} &= \frac{\sum_{(i,j)} \sum_{\in D} f(i,j)t(i-\tau_1,j)}{\left[\sum_{(i,j)} \sum_{\in D} f^2(ij)\right]^{1/2} \left[\sum_{(i,j)} \sum_{\in D} t^2(i-\tau_1,j)\right]^{1/2}} \\ &\quad \times \left(\frac{\sum_{(i,j)} \sum_{\in D} f(i,j)t(i-\tau_1,j)}{\left[\sum_{(i,j)} \sum_{\in D} f^2(i)\right]^{1/2} \left[\sum_{(i,j)} \sum_{\in D} t^2(i-\tau_1,j)\right]^{1/2}} \right)^{-1} \\ &= \frac{\sum_{(i,j)} \sum_{\in D} f(i,j)t(i-\tau_1,j)}{\left[\sum_{(i,j)} \sum_{\in D} f^2(i,j)\right]^{1/2} \left[\sum_{(i,j)} \sum_{\in D} t^2(i-\tau_1,j)\right]^{1/2}}\end{aligned} \tag{40}$$

In Eq. (40), if $\tau_1 = 0$, the cross correlation coefficient is 1. If $\tau_1 = x$, one of the following applies:

1. If it is greater than 1, the split-layer flocs are of different origins.
2. If it is smaller than 1, the split-layer flocs are of the same origin, which shows that the same floc exists through different split layers, as shown in Fig. 40.

When split-layer flocs are determined to be of the same origin, those split layers can be grouped into a single floc layer, and the extension of the floc in the Z direction can be estimated by summing the thickness or fiber mass of each split layer that makes up the floc.

To estimate the floc distribution characteristics in the XY plane, the gray level of each pixel is converted into the value of mass by use of a calibration curve. The floc distribution characteristics in the XY plane are expressed by the coefficient of variation CV (CV = standard deviation SD/average W), which refers to the deviation of each pixel with respect to the average mass of paper. The distribution variance of a grouped floc layer $\mathrm{Var_{all}}(=\mathrm{SD}^2)$ can be estimated from Eq. (41) because the additive law exists for the variance Var_i of each split layer and the average W_i. The subscripts "all" and "i" represent a grouped floc and the number of split layers in the floc, respectively.

$$\mathrm{CV_{all}} = \frac{\mathrm{SD_{all}}}{\mathrm{W_{all}}} = \frac{\sqrt{\mathrm{Var_{all}}}}{\mathrm{W_{all}}} = \sqrt{\frac{\sum \mathrm{Var}_i\, \mathrm{W}_i}{\sum W_i}} \tag{41}$$

Degree of *Z*-Direction Extension of Flocs Figure 41 shows the cross correlation coefficient for split layers, those that are affected from the distance τ, and the floc unit layer models obtained by discrimination according to the value of cross correlation coefficients of split layers for $\tau > 5$ mm. The variation coefficient of mass for the floc unit layer is calculated from the data of split layers, Var_i and W_i, using Eq. (41).

Different numbers of increasing positive > 1 curves, which are indicative of flocs of different origins, correlate with different types of dewatering methods. In Fig. 41, for example, there are one for the Fourdrinier method, two for the hybrid method, and four for the twin-wire method. All the other curves for decreasing (< 1) curves, showing that a floc exists through several split layers in that particular range. In the Fourdrinier method, an interface of floc layers exists slightly closer to the back side than to the top surface, and the distinction between the surface layers and

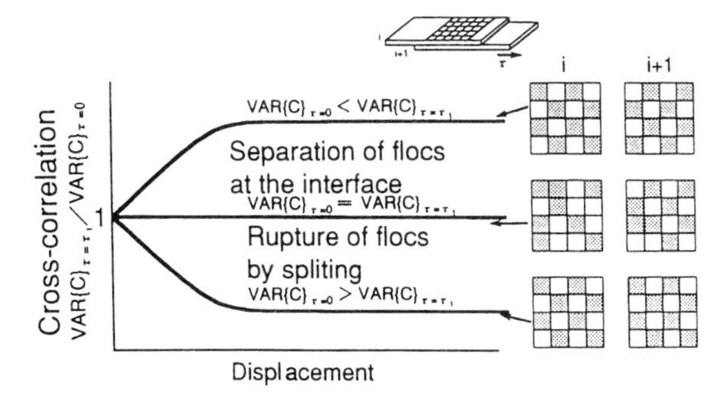

Fig. 40 Estimation of the location of split-layer flocs.

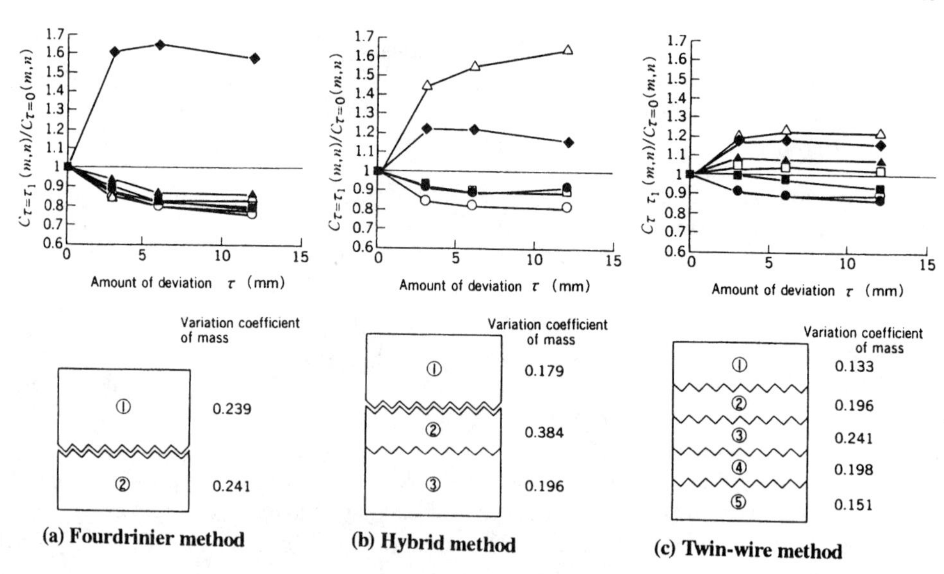

Fig. 41 Cross correlation characteristics and floc unit layer models for paper sheets formed by different methods.

middle layers is not clear. On the other hand, the middle layers are distinct in the hybrid method. The variation coefficient of mass of the surface layers in the hybrid method is smaller than that in the Fourdrinier method; this result is considered to be characteristic of a layered structure. However, the variation coefficient of mass of middle layers is large and the degree of flocculation is high. The floc unit layer model of the twin-wire method is composed of five floc units, and the extension of a floc in the Z direction is small. In addition, the variation coefficients of mass of the surface and back-side layers are smaller than with the other two dewatering methods. In particular, the degree of layering of both top and back layers is distinctive.

IV. DIMENSIONAL CHARACTERIZATION OF FIBERS

The term "dimensional characterization" includes such parameters as the length, width, cross-sectional area, surface area, degree of flattening (collapse), and aspect ratio of fibers and fines, as well as the kinking (segmentation),* curving, and twisting of fibers and the resultant curl that these deformed fibers exhibit. Fiber curl may be considered from the standpoint of in-plane curvature or combined in-plane and out-of-plane curvature and is quantitatively assessed as a ratio of the total convoluted length to the straight-line length between the ends of the fiber [99] or by several other criteria.

A great deal of work has been done on dimensional characterization by data-gathering observations and measurements with light and electron microscopes.

* That is, segmentation by abrupt bends (changes in direction). Segmentation in this sense does not mean a physical separation of the segments.

These excellent instruments have been joined by powerful electronic allies—the graphic digitizer and the automatic image analyzer—with data transmitted directly to computers for rapid processing. There are also indirect methods for quantitatively assessing certain structural parameters, some of which have become standard test procedures.

A. Fiber Length

The length of a typical fiber that is ultimately incorporated into a sheet of paper or paperboard undergoes two kinds of changes from the time it is part of a tree or plant stem until it is part of the sheet. First, the chipping and grinding and subsequent chemical, physical, and mechanical processes shorten it, and second, these same processes change the rather straight, spindle-shaped structure that is the original fiber to a generally convoluted (curled), roughened, and sometimes bent or twisted object. Thus, length is changed both absolutely and in the longest dimension (Fig. 42). For some purposes, it is important to ascertain fiber lengths at various stages.

In the living stem, fiber length is often regarded as a prime indicator of quality in evaluation of tree improvement trials and other genetic programs.

Fiber length, fiber strength, and fiber flexibility are generally highly correlated to the development of good mechanical properties in paper, with tear strength and folding endurance being especially dependent on length* [4,26,28,35,36,108,174, 217–219].

Fiber length is important in the development of strength, because although fibers transmit load axially, the load transmission from fiber to fiber takes place in the fiber–fiber bonds, which are stressed in shear. The actual shear transfer length is not great; thus concentrated stress transfer by shear occurs over a shorter percentage

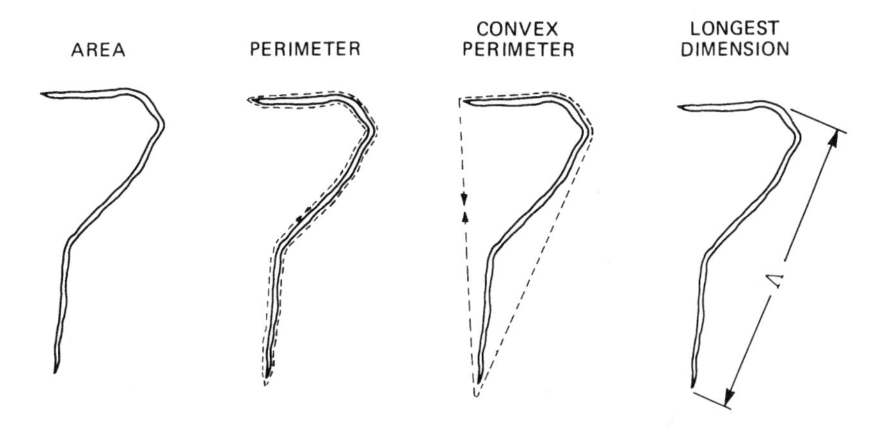

Fig. 42 Four of the basic measurements that can be made using fast logic circuitry in image analysis.

---

* Unlike most mechanical properties assays, tensile tests of paper and board often show little or no correlation or even a negative correlation with fiber length [95,119,145,169, 173,189,213,218] except in the case of wet web tensile strength [81,100].

of the total fiber length as undamaged fiber length increases and bond density remains constant. Reduced fiber length, reduced fiber moduli, and reduced bonding result in lower paper elastic moduli [141,143].

These concepts are embodied in the literature of paper physics and will not be expanded here. Interested readers can refer to Chapter 1 and the outstanding series of fundamental research symposium proceedings of the British Paper and Board Industry Federation Technical Division and the paper physics conferences organized by TAPPI and the Canadian Pulp and Paper Association (CPPA) Technical Section.

Classical Methods of Length Determination

Microscopy The determination of fiber length in the stem or other tissue of a plant is generally accomplished with the instruments available in a good light microscopy laboratory equipped to kill, fix, embed, section, stain, and mount tissues. In sections of wood, the longitudinal–tangential (L–T) plane is normally used to measure fiber length, because there is less ambiguity as to the locations of the fiber ends than in any other orientation of a section. Detailed descriptions of the procedures for obtaining satisfactory sections for observation are available in a good reference on botanical microtechnique, such as that of Berlyn and Miksche [13].

The light microscope has also been the classic tool for determination of fiber length in pulp. A well-known reference source is Isenberg's book [75], which includes 18 pages on methods of making length and width determinations microscopically.

Projection Most of the early microscopic procedures proved to be too time-consuming. A substantial developmental effort over approximately 25 years [26,27,53] produced projection methods such as those described in the book by Isenberg [75] and TAPPI Suggested Method T 232 [193]. In these methods, a slide containing a dispersion of pulp fibers is placed on the stage of a projection microscope. An image of the fibers is projected onto a screen that incorporates a ground plate glass grid so that the line spacing on the grid bears a known quantitative relation to the actual length dimensions on the microscope stage as determined by a transparent micrometer scale.

Measurements of fiber dimensions and coarseness can be made in the projection apparatus. For additional details, the reader is referred to the TAPPI suggested method for coarseness T 234 [195] and the description and references given in Chapters 13 and 14 (this volume) and Chapter 5 of Volume 2.

Screen Classification During roughly the same time period that witnessed the development of projection methods, screening methods were also developed for pulp fiber length determination [26]. The devices that have been developed for separation or fractionation by length depend on a series of screens of different mesh sizes through which the fibers pass, at a dilute consistency of about 0.15–0.4%, starting with the coarsest screen. Most commonly, four screen opening sizes are used, although more screen gradations can provide greater fractionation of the sample. Preferably, the series of screens used should be selected so that approximately one-fourth of the fibers are held on the coarsest screen. A given mesh size generally retains fibers whose lengths are over twice the mesh opening size, except in the case of very flexible fibers.

Two of the most popular types of these screening devices (classifiers) are (1) a horizontal type wherein the stock passes through a series of compartments separated by removable screens and (2) a series of vertical tanks arranged in a cascade series so that the stock not retained on the first tank's screen overflows to the next tank, and so on. These classifiers form the basis of such industry standards as JIS P-8207 [81], TAPPI T 233 [194], and SCAN-M6 [173].

Number-Average Length and Weighted Average Length There are basically three kinds of length calculations made using fiber length data. The number average (or arithmetic average) can be found for ungrouped fibers by dividing the total length of the fibers measured by the number of fibers measured,

$$\langle L \rangle = \frac{\sum_{i=1}^{n} L_i}{n} \tag{42}$$

where

L_i = actual measured length of the ith fiber
n = number of fiber length measurements

The same calculation made for fibers grouped into certain class intervals according to length has the form

$$\langle L \rangle = \frac{\sum_{i=1}^{n} F_i L_i}{\sum_{i=1}^{n} F_i} \tag{43}$$

where

F_i = frequency (number) of fibers in a particular length group or class interval
L_i = specified or measured length of the individual fibers in the ith class interval
n = number of fiber length groups or class intervals

There is little practical advantage in ascertaining $\langle L \rangle$ as in Eq. (42) or (43), for in such a calculation the smallest fiber fragments or other fines count for as much numerically as full-length fibers in the denominator even though they contribute very little to the sum in the numerator. Also, the lower size limit of the fines counted is inherently arbitrary or subjective.

A second, more useful value is obtained by weighting the fiber lengths to reflect the fact that longer fibers and thicker (coarser) fibers exert a disproportionate influence on most sheet properties. The weighted average length by length is found for ungrouped fibers as

$$\langle L \rangle_\ell = \frac{\sum_{i=1}^{n} L_i^2}{\sum_{i=1}^{n} L_i} \tag{44}$$

and for grouped fibers as

$$\langle L \rangle_\ell = \frac{\sum_{i=1}^{n} F_i L_i^2}{\sum_{i=1}^{n} F_i L_i} \tag{45}$$

The third kind of calculation refers to the distribution of mass and is referred to as the weighted average fiber length by weight. Because fiber mass represents a fundamental property of pulp and paper sheets, the calculation of weighted average fiber length by weight is generally the most useful of length calculations. One could theoretically obtain this value by summing the products of individual fiber lengths and linear densities multiplying by lengths, and dividing by the total weight of the sample. There has been no easy way to ascertain coarseness (linear density) of an individual fiber,* accordingly, the weighted average length by weight computation is generally restricted to grouped or classified fibers. The following formula applies:

$$\langle L \rangle_W = \frac{\sum_{i=1}^{n} F_i (Lc)_i L_i}{W} \tag{46}$$

where

$(Lc)_i$ = length of fiber times fiber coarseness for the ith fraction
W = total mass of sample

L_i is defined as for either Eq. (42) or (usually) Eq. (43). For the particular case of fiber length by classification [194], Eq. (46) takes the form

$$\langle L \rangle_W = \frac{W_1 L_1 + W_2 L_2 + W_3 L_3 + W_4 L_4 + W_5 L_5}{W} \tag{47}$$

where

W_i = oven-dry mass of the fiber fraction retained on each of four screens in the classifier for $i = 1$, 2, 3, and 4
W_5 = oven-dry mass of the pulp lost through the finest screen
L_i = average length of each fraction i in the pulp sample
W = oven-dry mass of the material supplied to the classifier

It is seen that in Eq. (47) the number of F_i terms is 5 and the W_i terms are the equivalent of the $(Lc)_i$ terms in Eq. (46). An alternative way to write Eq. (46) is therefore

$$\langle L \rangle_W = \frac{\sum_{i=1}^{n} F_i (Lc)_i L_i}{\sum_{i=1}^{n} F_i (Lc)_i} \tag{48}$$

Quite commonly, an assumption is made that the mass of a given fiber is proportional to its length. When this *constant coarseness* approximation is applied to Eq. (48) for grouped fibers, the result is

$$\langle L \rangle_W = \frac{\sum_{i=1}^{n} F_i L_i^3}{\sum_{i=1}^{n} F_i L_i^2} \tag{49}$$

and the corresponding relation for ungrouped fibers is expressed at times in the form

$$\langle L \rangle_W = \frac{\sum_{i=1}^{n} L_i^3}{\sum_{i=1}^{n} L_i^2} \tag{50}$$

The investigator should carefully consider the purposes for which fiber length data are to be acquired, because the different methods of weighting the fiber length data will yield results that differ from each other as well as from the number average. Tasman [203] observed that the average fiber lengths of fractions retained on the respective screens of a cascade-type classifier are relatively constant† and that the

* Readers are referred to Chapter 14.
† Corson and Uprichard [29] reported similar results for a horizontal classifier.

number-average distributions of the fractions are approximately normal (Gaussian). This cannot be said for the weighted averages, which tend to have a log-normal distribution. Yan [222] provided a method for transforming number distributions into weight distributions from classifier data and a procedure for determining the variance or standard deviation of the curve fit when the log-normal distribution is plotted. The microscopic and projection techniques yield both averages and standard deviations from data in a more direct but tedious procedure [75].

The constant coarseness approximation can be usefully exploited for a number of operations involving rapid means of data acquisition, provided it is valid for the particular pulp in question. The approximation may be quite reasonable for a kraft pulp of a species having fairly uniform fibers but invalid for a blend of kraft and groundwood, for example.

A manual trial should always be made to provide baseline data against which the assumptions can be checked. Coarseness in pulp should be determined by one of the methods described in TAPPI T 234 [195] or Chapter 14. A method for determining fiber coarseness in wood was described by Britt [18].

The fiber lengths should be determined, along with widths, for at least 800 fibers, fiber fragments, or other cells at a magnification of approximately 75×, which will permit measurements accurate to the nearest 0.01 mm. It should be verified that 800 fibers is a sufficient sample size for reproducible data.

From the coarseness data and the percentage frequency by which fibers, fragments, and other cells (e.g., ray cells) are classified, one can determine the relative amount and frequency distribution by number and by weight of each component of the pulp. The overall percentage of probable error for 800 measurements to determine weighted average fiber length should be around 3%, although the frequency distribution results will generally show larger deviations for the individual points used to construct a curve or histogram. Other things being equal, the number-average fiber length is subject to a higher percentage of probable error [75].

Newer Methods of Length Determination A series of improvements have been made as a result of the development of new instruments, adaptations of older instruments to improve speed and accuracy, and utilization of technology developed for other branches of science. Although some of the descriptions that follow discuss such developments principally as they apply to the measurement of fiber length and other fiber dimensions in pulp, the material on digitizing emphasizes technological advances that permit these measurements to be made in the paper sheet.

Coulter Counter and Other Particle Size Analyzers An unmodified Coulter particle counter is an electronic device that measures the sizes of particles in the diameter range of 1–1000 μm by displacement of electrolyte as particles pass through an aperture. At any moment there is a voltage drop between the two electrodes in the counter, and this drop is increased by the interruption of ion flow as approximately spherical particles pass through the aperture. The magnitude of the interruption is primarily governed by the relation of particle volume to aperture size. The counter has analysis circuits that generate a cumulative distribution curve of particle size.

Valley and Morse [212] modified such an instrument to accept slender objects such as fibers by changing the counter from a volume-sensing to a length-of-time-sensing device. Although the fiber lengths can be determined to within ±0.1 mm, the instrument is not well adapted for extremely curled or bent fibers or for determining other fiber dimensions.

The most recent developments in this field are the Kajaani fiber size analyzer FS-200 and the OpTest fiber quality analyzer (FQA). The FS-200 senses the length of fibers drawn in dilute (0.01%) solution through a narrow capillary under suction. Fibers moving singly through the capillary pass through a beam of polarized light. Because cellulosic fibers are birefringent in polarized light, the passage of the fiber creates a discrete interval in which the birefringence occurs. In this manner the passage of several thousand fibers can be recorded in a short time span. The optical fiber length data are then treated statistically in ancillary computing equipment to provide a frequency distribution for 35 length categories in increments of 50 μm. The method used with this apparatus is described in TAPPI T271 pm-91 [190]. The FQA is also an optical fiber analyzer that can measure not only fiber length but also the curl/kink using an image analyzing technique. The principle of this tester is described in another section of this chapter.

Planimetric and Related Techniques Projected fiber images can be measured by electromechanical instruments semiautomatically. The measuring gauges, based on the principle of a planimeter or map curvimeter (map reader), consist essentially of a hand-held probe with a measuring wheel at the lower end [135,186]. The operator traces the fibers. The measuring wheel movement along fiber lengths is usually transferred to electrical pulses through a photoelectric device. Some of these gauges have a computer interface, and the operator can process the data needed to make fiber length calculations of the type described earlier.

Automatic Image Analysis The electromechanical instruments for photo-optical planimetry and data processing described above collectively represented a major advance in our ability to acquire and process dimensional and related data on fibers. The next generation is represented by the semiautomatic image analyzer or digitizer, for which some description has already been given in this chapter, and the automatic image analyzer in its several configurations.

Analysis of an image can mean many things. The eye analyzes images constantly; it is most adept at pattern recognition (e.g., a signature, the facial features of an acquaintance, the cover of a familiar book) and spatial perception (e.g., distances and relative positions of objects). It is not well suited to the production of numerical information, particularly when large counts must be made. It suffers from fatigue, low productivity, and inaccuracy.

A digitizer interfaced to a computer or calculator may be thought of as a semiautomatic image analyzer. It requires manual operation of a light pen, cursor, or mouse to define the boundaries of the features to be measured on a map, photograph, or projected image, but the X, Y coordinate positions of these features are recorded and can be processed into useful data automatically by the computer (Fig. 43). Digitizing systems have the advantages of accurate and flexible boundary detection under the guidance of the human operator, detection and separation of true versus spurious elements in the image, and low cost. Their major disadvantage is that they are considerably slower than the fully automatic image analyzers. Some commercially available systems combine digitizing and automatic image analysis equipment.

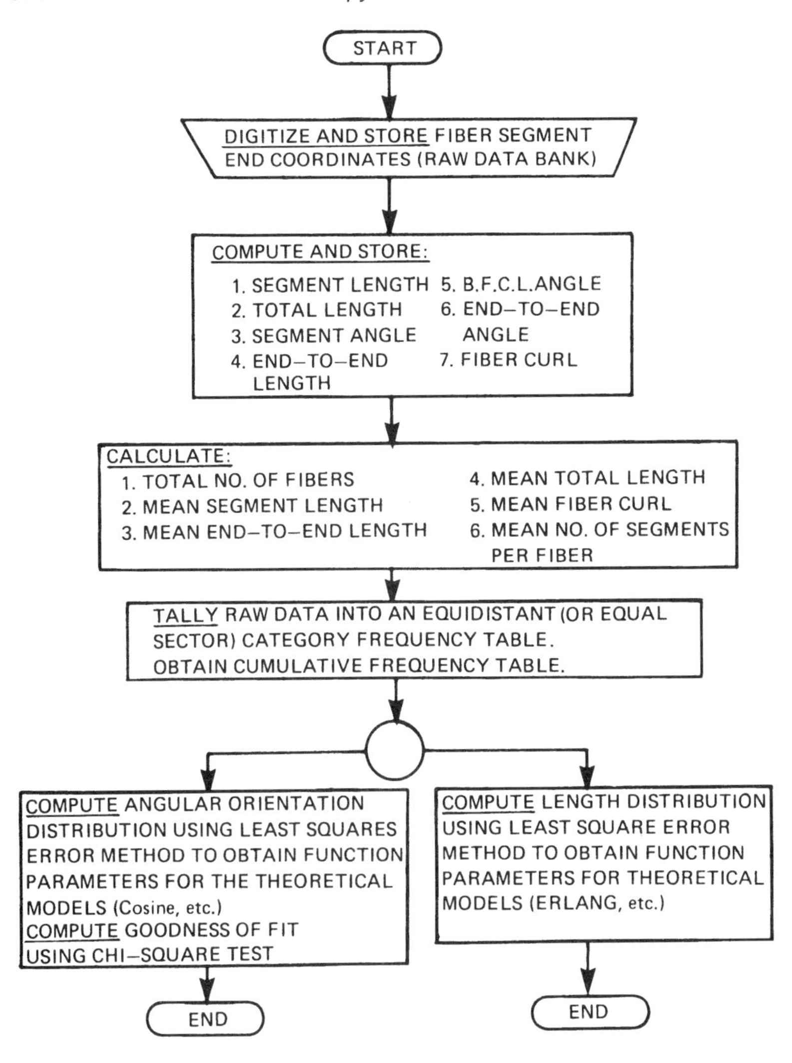

Fig. 43 Flowchart for acquisition and processing of fiber orientation and length data using a digitizing system.

Instruments designed to make rapid, fully automatic measurements on images have been developed to provide data that would otherwise require vast expenditures of human effort. In its typical configuration, an automatic image analysis system comprises a personal computer or minicomputer interfaced to a producer of images such as a microscope or high resolution camcorder. The advantages include speed, convenience, and accuracy. There are also some ancillary benefits. Because of the speed, a larger area of sample can be analyzed than would be practical using manual methods. A greater variety of information can often be obtained. With the computer interface, results can be automatically formatted and processed, because the direct connection between the measurement section and the data processing section eliminates the need to manually copy and key in the data. Some image analysis systems include refinements such as automatic microscope stage motion or focusing.

Types of Automatic Image Analyzers The major types of automatic image analyzers can be categorized as (1) television (TV) based and (2) scanning electron microscope (SEM) based.

Television-Based Image Analyzers. Television-based image analysis systems are equipped with a TV (video) camera to transform the image into electrical information from which measurement information can be obtained. Therefore, the instrument itself has the responsibility for defining the boundaries of the objects to be measured as well as for producing the desired counts and measurements.

Television-based systems can provide measurements of area, perimeter, longest dimension, breadth, and convex perimeter (see Figs. 42 and 44) automatically. Furthermore, these systems have the capability to make the measurements on each feature individually as well as cumulatively, so that the individual results are available for later data processing.

Because a large amount of data can be acquired in such a system, a computer is normally included to handle processing and reduction of the data. Reduction operations include

1. Sorting results into a histogram or cumulative size distribution
2. Extracting statistics concerning the measurement of features
3. Determining those features that exceed predetermined sizes
4. Determining the shape factors of features in the field of view
5. Selecting features based on specific shape factors

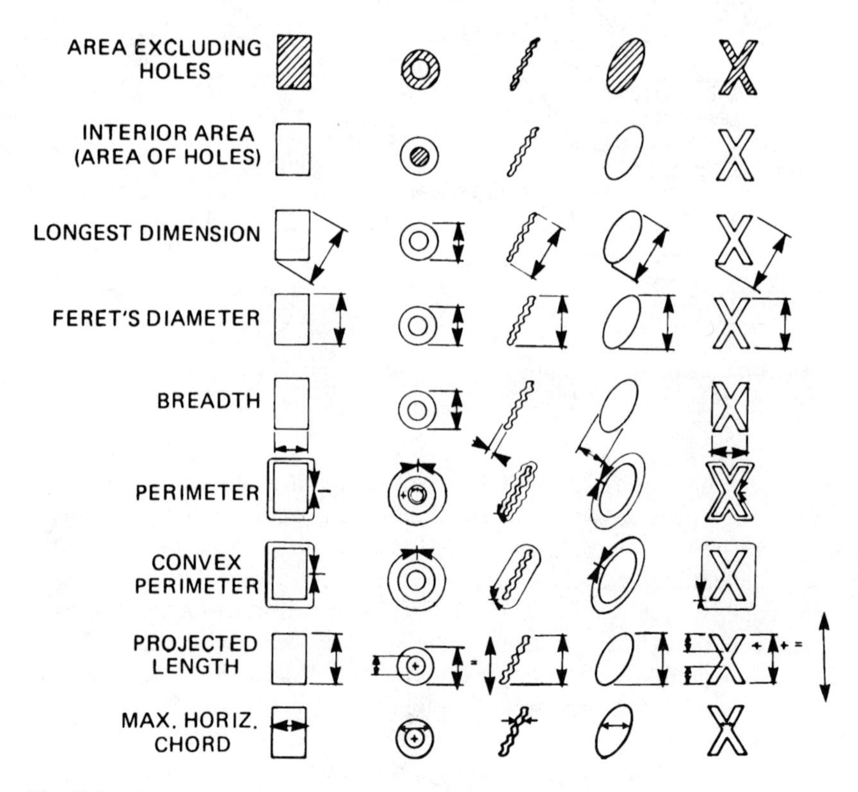

Fig. 44 Examples of commonly used measurements in image analysis produced by algorithms executed in the digital analysis function. (From Ref. 124.)

The data processing capabilities of such systems often include packaged software or operator-generated programs written in a suitable language such as C or Fortran. Usually, the standard equipment includes a display and hard copy (with plotting) features (Fig. 45).

Some systems perform size and shape identification. In these systems, algorithms are employed that subdivide the features in the image according to shape and size either by selection of a size parameter such as longest dimension (Figs. 42 and 44) or a shape factor derived from a computation. Examples of the latter type are (1) longest dimension divided by breadth (a measure of elongation) and (2) convex perimeter squared divided by area, to differentiate between two classes of features.

Image analysis systems in this type are frequently interactive, and the operator is given a choice as to which measurements best fit the selection requirements. He or she can change the criteria when the need arises. These systems invariably include a computer to perform the selection operations. In addition to the programming capabilities for sequencing and control, this type of TV-based system may have a special storage facility such as a disk or extended memory.

SEM-Based Image Analyzers. In these systems, the electronic video signals that represent the image are generated from a scanning electron microscope rather than a video camera. Except for a generally slower scanning rate, appropriate to the SEM, the configuration of image analyzers of this type is the same as that required for the operation with a TV scanner.

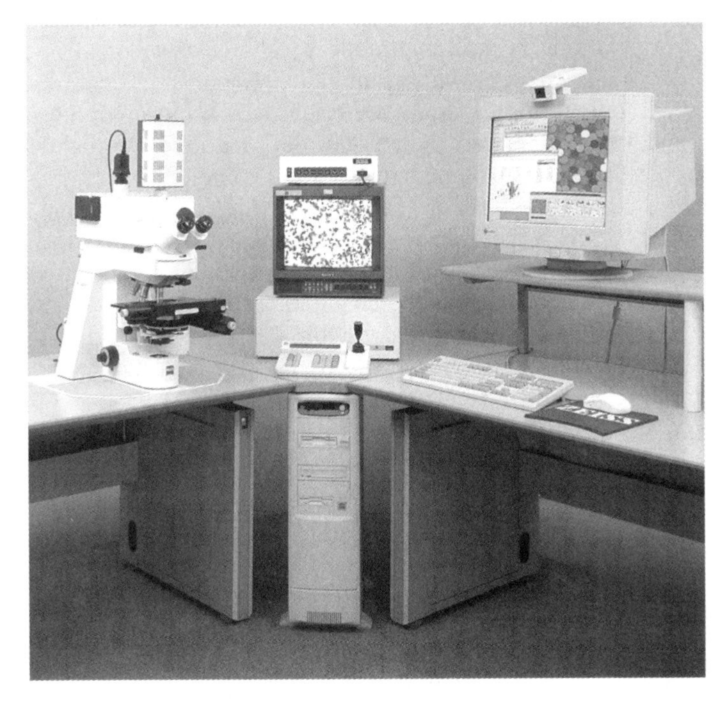

Fig. 45 An automatic image analyzer with (from left) a light microscope and television camera, a video processor and display, an output terminal and dedicated computer. (Photo courtesy of Carl Zeiss, Germany.)

Systems are now available that can produce from the SEM the same range of measurements and operations as would be obtained from an optical microscope but that provide the additional capability of locating the positions of inorganic and other particles in the field of view. These systems can switch the SEM to X-ray mode and direct the electron beam sequentially to the centroid of each of the particles of interest. When the centroid is targeted, the X-ray analyzer is activated and structural data for the particle are produced. The subsequent output from such systems provides not only size and shape information about the particles but also X-ray data concerning their physicochemical makeup. A related development is the energy-dispersive X-ray analysis (EDXA) system. The EDXA system in combination with scanning electron microscopy has the capability of providing information on the topochemical distribution of fiber cell wall constituents, on contaminants in pulp or in the air, and in many other areas of interest in the field of pulp and paper research.

Basic Functions of Image Analyzers An understanding of how an image analyzer derives numerical information from an image can perhaps be best explained in conjunction with the block diagram shown in Fig. 46.

The image source for a typical system is a microscope of good laboratory quality. Other specialized optical image sources include Petri dish viewers to view macerated fibers, holographic viewers to view reconstructed holograms, and high resolution monitors for analog-to-digital (A/D) video sources. Another type of image source is the transmission electron microscope (TEM), which usually requires a special type of interfacing. Also, when a TEM is used, it is employed either to produce micrographs or to operate in the scanning mode.

The illumination of the field is extremely important in image analysis. The most commonly used microscope illumination techniques are bright field transmitted illumination and bright field incident illumination. The use of dark field incident light, which normally requires a powerful external light source, is less common. As shown in Fig. 47, the direction of the illumination largely determines the type of imaging that is achieved. When illumination is from above the sample and lies within the collection cone of the objective, the light is reflected from the specimen directly into the objective; this is bright field incident light. When illumination is outside the collection cone of the objective, it is dark field incident light; and when illumination comes from below and passes through the specimen directly into the collection cone of the objective, it is called bright field transmitted light.

The purpose of the video camera is to pick up the image and convert the visual information to video signals. Video cameras specifically designed for image analysis are used; they are generally referred to as scanners. Most scanners accommodate several types of pickup tubes or semiconductor sensors. The most common type is the charge-coupled device (CCD), which is relatively inexpensive, can tolerate a wide range of light levels, and provides good sensitivity and resolution. However, other specialized pickup devices are available for more demanding uses. These scanners or pickup devices are usually based on an analog video signal system. Recent technologies have provided digital image scanners, digital still cameras, and digital video cameras. Because they have a good digital interface such as a small computer system interface (SCSI), the image data can be taken easily and directly into the personal computer.

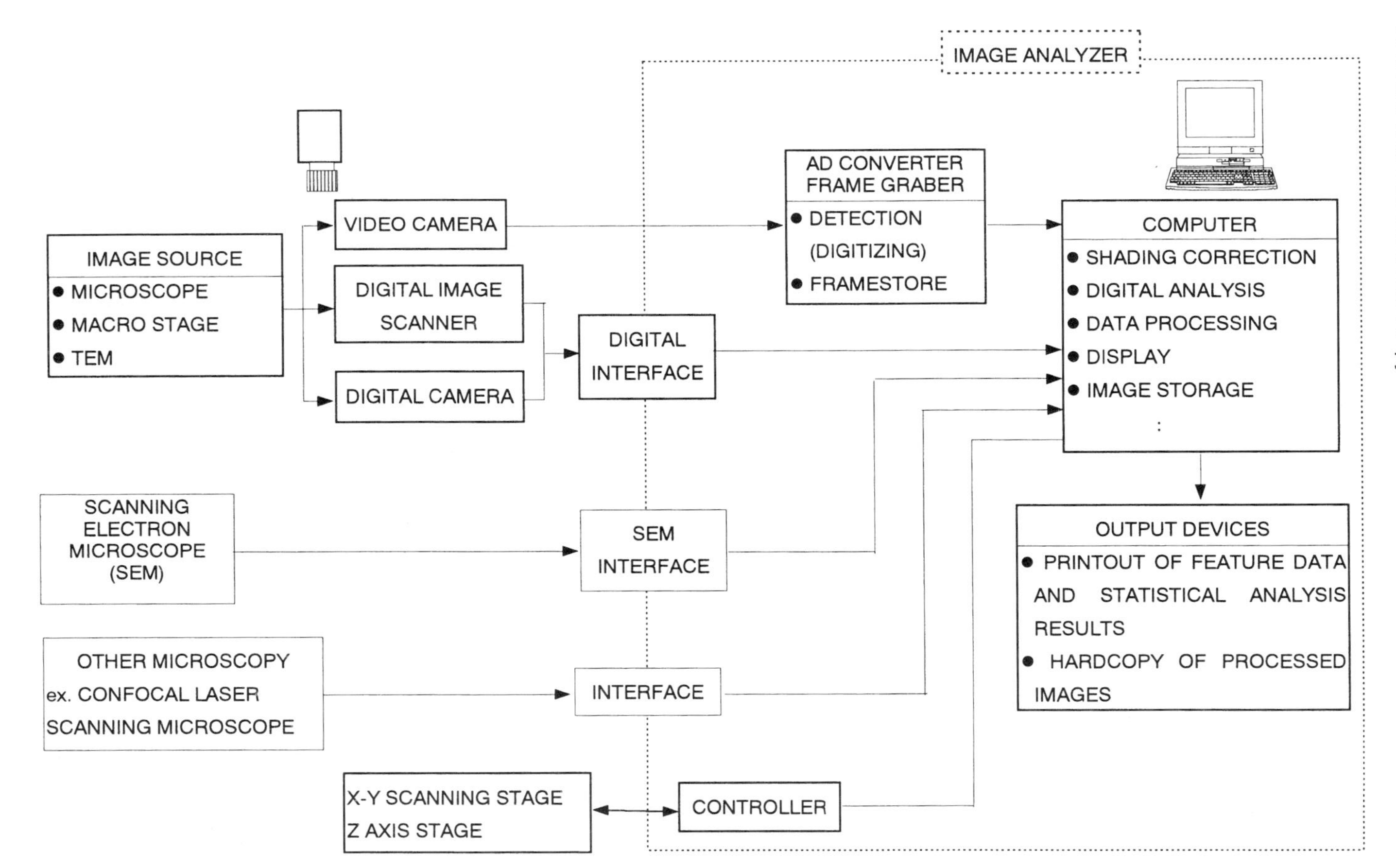

Fig. 46 Block diagram of a typical image analysis system.

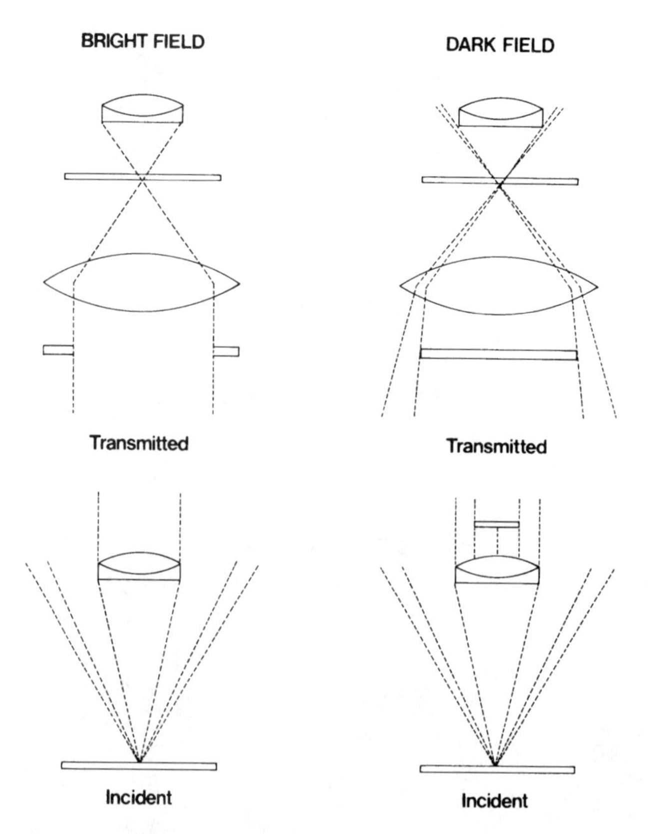

Fig. 47 Light paths for incident and transmitted illumination. (From Ref. 125.)

Although shading correction is an option in image processing functions, it is very desirable for fiber measurements. It ensures uniformity of contrast throughout the field of view.

Detection. The process of electronically determining the boundaries of the features of interest from the video signal is a critical step, because the objects are represented quantitatively by the list of position coordinates at which the edges of each object of interest intersect the scan lines (rasters) generated in the analog scanner. There are many detection techniques.

In older systems, the boundary (of a fiber, for instance) is defined as the point where the video signal crosses some threshold as determined by the operator or semiautomatically. In newer systems, all the processes, including shading correction, detection digital analysis, and data processing, can be programmed; it is thus possible to supersede gray level thresholding by performing different measurements on and around the detected image by program changes. In such systems, the hard-wired detectors of earlier systems are replaced with an analog-to-digital converter (ADC) or frame grabber device. During the scanning of the image, the intensity distribution of the selected subject is digitized in two ways. The first is a quantification in the X and Y axes in the image surface through the frame measurements; the second is a quantification of the intensity values through the ADC or frame grabber device that can digitize the analog signal.

The result is a rectangular matrix of screen pixels. The image area size is determined by the video camera and usually consists of 600–800 columns and 480–600 lines. The ADC converts the brightness of each individual pixel into a number—usually an eight-bit number—that results in assigned brightness values from 0 to 255. The gray values represent the intensity of the image over the entire area of possible X and Y coordinates. In the case of true color images, a combination of three such values—red, green, and blue—per pixel is needed to display the image. In addition to the gray value, every pixel can have a graphics descriptor or a graphics plane assigned to it, which can be used to designate a particular characteristic. Typical examples are regions that belong to several different classes depending on their gray values or characteristics. Membership in a class can be indicated in the graphics plane by different colors.

Images can also be enhanced with a range of image processing functions such as shading correction, contrast enhancement or normalization, smoothing, edge improvement, and image arithmetic. This process is labeled *gray image processing*.

Segmentation. Regions or interesting phases can be detected and separated from their environment on the basis of their gray values. This process, called "segmentation," creates a binary image from a gray value or true color image.

Binary images are a special kind of gray scale image. They contain only two gray values, normally 0 and 1 or 0 and 255. Binary images are a necessary transition step on the road from the original to the quantitative data. They define the areas of the regions that are to be analyzed. Segmentation results can be improved with binary image processing. This includes, for example, arithmetic operations, filling of holes, or filtering on the basis of size.

Measurement. A variety of algorithms are used to extract quantitative measurements such as those shown in Fig. 44 and others such as the X, Y coordinates of the centroid of an object, the moment of inertia of the object, and so on. The digital analysis function also corrects for features that partially intercept the border of the frame or field of view.

The need for data processing arises because many measurements are not in their most useful form. These data are generally stored in database files. We can therefore modify the definition of image analysis and refer to it as the process that crates database files from images; the data processing function extracts the length or other size distribution from the list of numbers. A database file is a sequence of data records. Every record contains the measurement values of the previously defined measurement parameters. The measurement values are extracted from the image and its components during the measurement process. Other data processing operations include calibration for magnification, performing statistical calculations such as the mean and the standard deviation, testing for homogeneity in the raw database, and so forth.

Evaluation. The measurement results can be listed, represented in various graphics, and evaluated statistically. When measurements are being taken, the images are stored in the memory of the computer. They can also be stored on the hard disk and downloaded from there. In the latter case the images should be stored in the standard image formats, for example, bitmap (bmp), tagged image file format (tiff), joint photography expert group (jpeg; only for true color images), CompuServe (gif), Mac Point (mc), and Zsoh Paintbrush (pcx).

The purpose of the TV display is to present the various images generated within the system: the image scanned by the scanner, the detected image, the results of measurements, and the outputs of the digital analysis function. The operator selectively interacts with this part of the image analysis system to obtain the desired visual presentation. Look-up tables allow us to determine the way in which pixels are displayed on the monitor. Normally a gray value of 0 is black, a gray value of 255 is white, and the gray values in between are displayed in gray increments. The 256 gray values can generally be mapped to any color combination available on the monitor or graphics card of the system.

Image Analyzer Operations in Fiber Length Determinations In Fig. 44, some examples of image analysis measurements are given. Examples of derived size measurements from automatic image analysis are shown in Fig. 48.

When fibers are to be measured, it is advantageous to dye them intensely and use very dilute suspensions in the slide mounts on the microscope stage. The intense staining enhances accuracy in locating the fiber boundaries on each raster by providing a strong contrast between neighboring pixels in the detection function. The dilution will minimize occurrences of crossed fibers, which the automatic image analyzer would, in general, recognize only as an X-shaped single object as in the right-hand column of Fig. 44.

Jordan and Page [84,85] used four basic measurements—fiber area, perimeter, convex perimeter, and longest dimension (Fig. 42)—and derived from them the length, width, coarseness, and curl of the fibers. From fiber boundary coordinates

Derived Measurements	Significance
Area excluding holes + area of holes	Total area of feature 
Convex perimeter/π	Average Feret's diameter 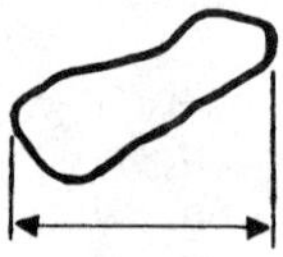
Area/longest dimension	Average feature width at right angles to the longest dimension 
π × area/perimeter	Average chord length 
$\dfrac{\pi \times (\text{Area})}{4 \times \text{longest dimension}}$	Equivalent cylindrical volume 

Fig. 48 Examples of derived size measurements from automatic image analysis. (From Ref. 125.)

sent to the digital analysis function by the detection function, their image analyzer generates the area of the fiber, its perimeter including concavities, its perimeter excluding concavities (the convex perimeter), and the length of the fiber, with its typical kinks and twists, projected onto a line. The orientation of the line is scanned at 2° intervals over 180°. The maximum projected length is specified as the longest dimension (see Fig. 42).

From these basic measurements, fiber length can be taken equal to one-half the perimeter (approximately). For greater accuracy, especially if the fibers tend to be short or square-ended, length may be computed as half the perimeter minus the width. Mean fiber width can be taken as the area divided by the length of the fiber. Because the method of preparation used by Jordan and Page transfers known weights of fibers to each slide, they are able to calculate coarseness from the total length of fiber divided by the fiber weight on each slide. Their measurement of fiber curl will be discussed in Section IV.B.

According to Jordan and Page [84,85], excellent agreement can be obtained between fiber lengths measured manually with a map reader from enlarged micrographs and lengths obtained by automatic image analysis. Figure 49 shows the length distribution obtained by image analysis for a commercial bleached kraft pulp that had previously been manually characterized. Very close agreement between the two methods was reported. From statistical considerations, they also found an empirical relationship between weighted average fiber length and number-average length, specifically,

$$\langle L \rangle_W = (1 + a^2)\langle L \rangle \tag{51}$$

where a is a constant whose value, for their (Bauer–McNett classified) samples and from results reported in the literature, is in the range of 0.32–0.40.

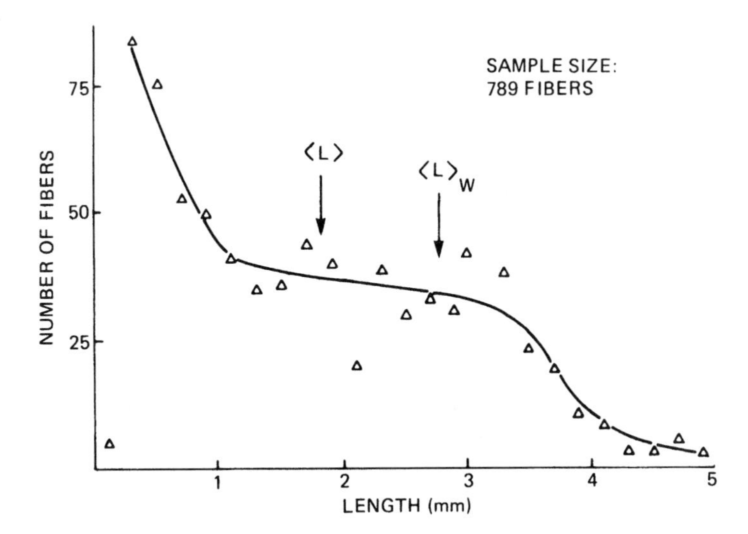

Fig. 49 The length distribution for a commercial bleached kraft pulp. The weighted average $\langle L \rangle_W$ and the number average $\langle L \rangle$ are shown by arrows. (From Refs. 6 and 88.)

The problem of crossed fibers has received special attention. A method called skeletonization was employed by Taylor and Dixon [204]. A computational algorithm reduces the detected widths of the overlapped fibers progressively until they are represented only by lines of unit width. At this point the crossed structure is decomposed into four segments, each with two free ends. A reconstruction of the segments that belong together is accomplished by matching these segments to form best fits on geometric grounds. Once the path of each individual fiber has been located, fiber length can be defined as simply the distance traced by a path along its successive constituent segments. Skeletonization does introduce some artifacts in measurement that have to be accounted for when other parameters—width, for example—are to be determined.

The measurement of fiber width, or average diameter, which is discussed further in Sec. V, is generally performed in automatic image analysis systems by either a diameter analysis or a grid analysis method. The software programs that are used in the two methods differ significantly. In diameter analysis, the program instructs the scanner to find a particle, then finds the center of gravity of the particle and performs a set of diameter determinations from the center. This method works fairly well for straight fibers and other more or less regularly shaped particles. However, a detection problem may arise in the case of kinked or highly curled fibers.

The grid analysis method is based upon the construction of a grid or network of lines that cover all solid parts of an object regardless of shape or of whether internal holes lie within it. The grid encounters boundaries of the object at points and uses an approximation to the Pythagorean theorem to calculate distances. A large amount of information is generated in such a system, so mass storage of approximately 40,000 bytes is needed. The types of output that can be obtained include length, width, length/width ratio, surface area, surface/volume ratio, area and volume equivalent diameters, and estimated ratio of perimeter to area.

Advantages and Limitations of Image Analyzers In their present (evolving) stage, automatic image analyzers are useful when large numbers of pulp fibers must be detected and measured. Several stringent criteria must be met. Among these are the following:

- Fibers are well defined, by appropriate staining and illumination, to provide high optical density and a large signal-to-noise ratio.
- The instrument is calibrated so that light intensity and other factors are fully reproducible.
- The instrument is not required to detect and measure at or below the limits of resolution of any of its optical components.

The automatic image analyzer can perform some functions—for example, width and coarseness determinations—with greater speed and accuracy than other available methods. Length measurements can certainly be made faster, but there is an out-of-plane dimensional component that may be significant in pulp suspensions as a three-dimensional system is reduced to two-dimensional analysis. The third dimension may, in some cases, create detection as well as measurement difficulties. Detection of fibers in the paper sheet is extremely difficult if not impossible for an automatic image analyzer, because the embedded fiber—even though stained intensely—appears as an object (or objects) of greatly varying optical density as crossing fibers partly obscure it. Furthermore, the appearance of identically stained fibers is vastly different at different depths in the sheet. For related reasons (of poor contrast between objects of interest in the field), automatic image analysis does not lend itself well to quantitative fiber bonding studies.

In summary, automatic image analysis lends itself best to situations in which fibers and other objects are clearly identifiable, as in pulp suspension. There has already been progress in developing these systems for on-line analysis (and feedback) of fiber measurements as part of paper pulp production control [54]. Automatic image analysis is currently deficient in making such measurements in paper sheets, where the detection function is best done with the human eye. In such cases, a graphic digitizer interfaced with a high speed programmable computer offers the fastest system for accurate, reproducible results.

Digitizing There has already been a short discussion outlining differences between the fully automatic image analysis systems and a semiautomatic system such as digitizing. A digitizing system has been shown photographically and schematically in Figs. 2 and 3, and an example of flowcharting for fiber length and orientation measurements has been presented in Fig. 43. Digitizing systems, with or without associated storage and computing facilities, graphics display screens, printers, video interfaces to allow tracing directly from video images, and so on, are commercially available from several manufacturers. Also available is interactive computer software to provide users with necessary data acquisitions, analysis, and storage capabilities.

Many early digitizers operated only in point mode, which required the operator to locate discrete points on maps, photographs, and other images with the pen or cursor. Newer systems can operate in either point or continuous line mode. Both have advantageous uses. The point mode type enables (with appropriate electronics) the calculation and presentation of (1) coordinate locations, (2) distances between points, (3) the area enclosed by discrete points, and (4) the total number of points entered.

In continuous line mode, which is in reality a dense spacing of discrete points, the cursor or pen is moved steadily along lines or other features of interest in the image, with the result that continuous coordinate information is acquired in the data base. This information enables the calculation of (1) individual point coordinates, (2) line length, (3) the area and perimeter of closed curves, (4) form factors of closed curves, (5) centroids and moments of inertia of plane figures, and (6) other parameters of the types shown in Figs. 44 and 48.

It must be borne in mind, however, that these operations are slower (generally much slower) than automatic image analysis. On the other hand, the detection function is under the operator's personal control, thus reducing spurious or ambiguous detection and ensuring that no features of interest are missed. The relative merits, in regard to accuracy and precision, of digitizers and automatic image analyzers will vary greatly according to the type of image source and the nature of the features to be measured.

A graphic digitizer lends itself well to the detection and measurement of dyed fibers in a sheet of paper, as illustrated in Fig. 5 and discussed earlier. The measurement of fiber or fiber segment length in the sheet—which is difficult, tedious, or impossible with various other methods—can be efficiently accomplished even in fairly thick sheets. For light-to-medium basis weight sheets composed of mostly bleached fibers, the method of oil impregnation described in Section II.B works very well to highlight the scattered dyed fibers to be measured. For sheets composed of unbleached or colored fibers, sheets with coatings or fillers, sheets of high basis weight, and any other type of sheet that does not permit observation by transmitted light after oil impregnation, it is necessary to employ one of the sheet-splitting techniques [118,150,201] in order to provide thin enough layers for observation. Refer to Section III for additional information on sheet splitting.

The fiber F in Fig. 5, shown diagrammatically in Fig. 6, can be measured in several ways. Based on the digitized coordinates of its end points and the intermediate points acquired in continuous line mode operation, it is possible to calculate fiber straight-line end-to-end length, fiber segment length, and total length based on summing of segment lengths. Fiber curl can also be calculated from the data thus accumulated [154].

Typically, the type of length distribution data that are generated are in the form of a skewed curve when lengths are plotted versus frequency of occurrence. The data will generally match quite well to a type I or type III frequency curve in the Pearson system. Fiber length distributions can in most cases be fitted readily to an Erlang distribution function [43], for example, using the least squares error method.

The Erlang probability density function is

$$f(x) = \frac{(x/b)^{c-1} \exp(-x/b)}{b[(c-1)!]} \tag{52}$$

where

$b =$ = scale parameter, always > 0

$c =$ shape parameter, an integer > 0

$bc =$ mean value of x

The range of this function is determined by $0 \leq x \leq +\infty$.

Figures 50 and 51 show the data points for total length distribution and cumulative length distribution, respectively, for a sample size of approximately 600 dyed fibers in a bleached softwood kraft sheet of 80 g/m^2 basis weight. The Erlang function (solid line) fits the data according to the chi-square test [Eqs. (9)–(11)] with a probability of 87% (Fig. 50); for the cumulative frequency, the probability that the curve fits the data is 94.7% by a chi-square test. Other probability functions, for example, Rayleigh [53], do not generally match fiber length distributions as well as the Erlang function.

The above results indicate that the sample size should be increased somewhat, in this case probably to approximately 800 fibers, to yield a better fit of data to function. In many cases, a sample size of 600 fibers will be sufficient with this method.

The procedure used for such samples by Mark and colleagues is to cut samples 32 mm × 19 mm (1.25 in. × 0.75 in.), impregnate them with silicone oil, and insert them in the film holder of a standard 35 mm photographic enlarger. The illuminating lamp of the enlarger projects the image of the dyed fibers directly onto the digitizer tablet (Fig. 2) for coordinate data acquisition by cursor. It should be emphasized that although a film holder is used, no film is involved. What is projected onto the digitizer tablet is the image of all the dyed fibers in the sheet itself, because the oil

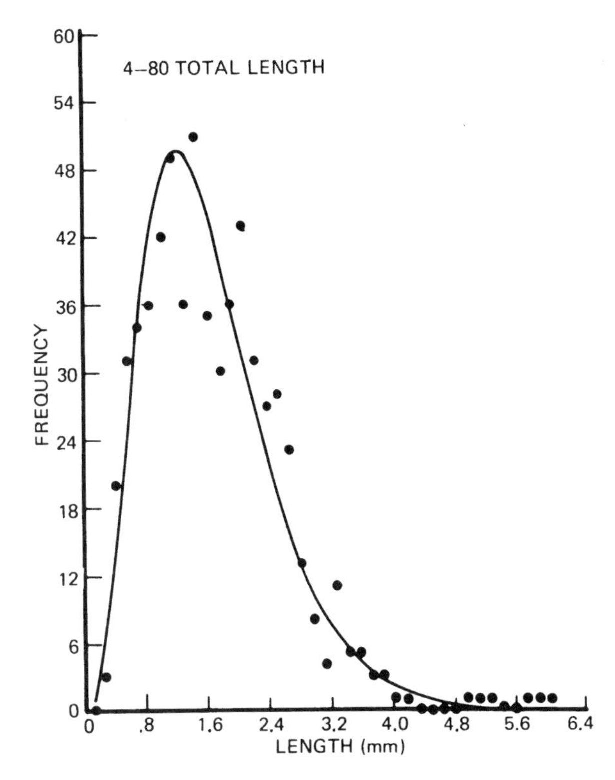

Fig. 50 Frequency data for fiber total length measurements in 0.2 mm increments compared with an Erlang distribution.

impregnant renders them visible regardless of the depth at which they are located within the sheet, as shown in Fig. 5. A rear projection digitizer may be useful for measurement of slide-mounted pulp fibers or fibers in thin sheets [76,211].

B. Fiber Curl and Kink

Fiber Curl Fiber curl, also called curliness, may exhibit itself by directional changes in the plane of the sheet, as in the case of the dyed fibers shown in Fig. 5.

Any out-of-plane deviation observed as the fiber passes over and under crossing fibers or as it folds and kinks in drying or mechanical straining may also be considered a component of fiber curl. Out-of-plane curl is quite evident in the synthetic fiber network shown in Fig. 16. The role of fiber curvature and curl in paper properties has been examined theoretically and experimentally by several investigators with somewhat divergent conclusions [84,92,107,142,151,154]. Its importance is highly variable, as an examination of the electron micrographs in the atlas of Parham and Kaustinen [149] will readily reveal. Fiber curl in a well-refined softwood kraft formed into a dense sheet is far less prominent than in nonwoven lens tissue made essentially of highly curlated rayon fiber, for example. These and other paper structures of wood fibers, other plant fibers, synthetic fibers, mineral fibers, and blends are all illustrated in the aforementioned reference.

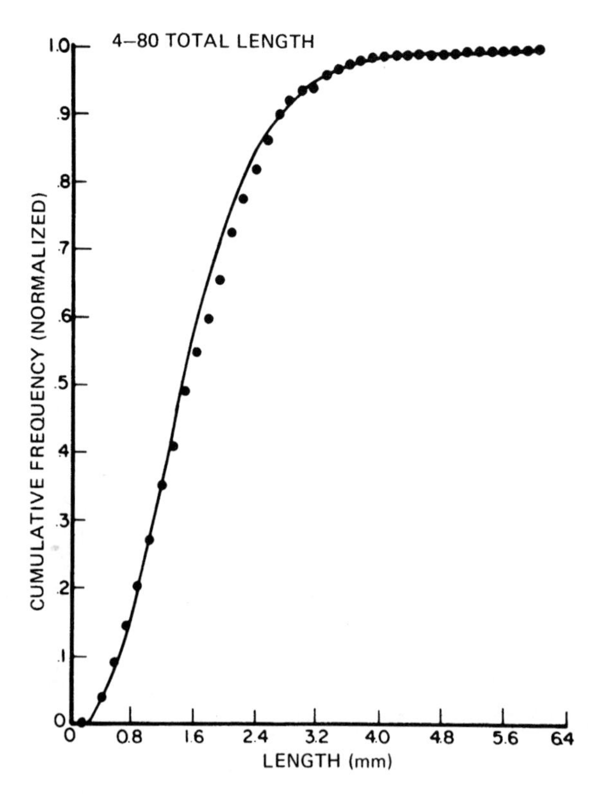

Fig. 51 Cumulative frequency data for fiber total length measurements compared with cumulative Erlang distribution.

In-plane curl has been given much greater quantitative assessment in the recent literature; considerable future effort needs to be applied to the measurement of out-of-plane curl in order to provide better data for the evaluation of Z-direction properties and other paper structural parameters.

In-Plane Curl, Bending Factor, and Curvature If one examines fiber F in Fig. 5, it is possible to measure its in-plane curl by the calculation mentioned in Section II.B or according to several other criteria. With reference to Fig. 52, if the straight-line end-to-end distance AB is referred to as L_1, the longest straight-line dimension AC is referred to as L_2, and the total fiber length (contour length) is designated by L, then curl can be variously and arbitrarily defined by relations such as the following:

$$\text{Curl I} = \frac{L_1}{L}, \qquad \text{Curl II} = \frac{L_2}{L}$$
$$\text{Curl III} = \frac{L}{L_2} - 1 \qquad \text{Curl IV} = 1 - \frac{L_1}{L} \tag{53}$$

It can be noted that numerical curl factors that employ definitions I, II, and IV will always be ≤ 1. Isenberg [75], Kibblewhite [99], Kallmes and Corte [92], and Perez and Kallmes [151] used the reciprocal of curl I, similar to the usage employed in polymer chain terminology (ratio of total length to linear end-to-end distance), and thus their curl factors are always ≥ 1. Isenberg refers to this reciprocal as the

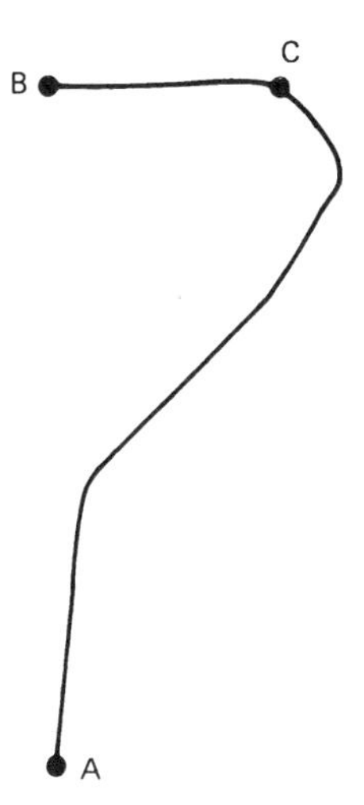

Fig. 52 In-plane length measurements made on a bent fiber and used for calculation of fiber curl.

bending factor. Curl III can be any number from zero up. Curl III is a measure of the fractional increase in the straight-line end-to-end length of a fiber that would result if the fiber were completely uncurled but not stretched. This definition of curl is widely employed in the textile field and is known as the *crimp ratio*. Page and Jordan also measured fiber contour length and longest projected dimension using an image analysis technique and called the curl of this definition with these dimensions "Jordan and Page's curl index" [85].

As fibers become straighter, curl factors I and II increase and factors III and IV decrease. It is evident that in evaluating the literature one must be careful to know clearly how the investigator has defined curl. When experiments are designed, thorough consideration should be given to the feasibility of measurements and the information that needs to be extracted from the data. For example, if a manual or digitizer method is to be employed, any of the four curl factors can be ascertained directly on the image, but the operation will proceed faster if curl I or curl IV is selected because the fiber ends are so much easier for the operator to identify than the longest dimension (L_2).

On the other hand, fiber end recognition to obtain L_1 is difficult and more error-prone in the case of automatic image analysis, whereas the longest dimension L_2 (Λ in Fig. 42, distance AC in Fig. 52) is a relatively easy measurement, executed in the digital analysis (or ADC with memory) function.* There are ways to approximate curl I or curl IV using fast logic circuitry; however, Jordan and Page used the relation

$$\text{Curl I} = 2 \times \frac{\text{convex perimeter} - 1}{\text{simple perimeter}} \tag{54}$$

* It should be borne in mind that automatic image analysis is not generally suitable for quantitative measurements of fibers in sheets. The comments pertaining to fiber curl measurements in the image analyzer refer to measurements on pulp fibers.

where the (simple) perimeter and convex perimeters are as shown in Fig. 42, as such an approximation. Determination of curl I on the image analyzer was used by them as an aid in detecting crossed fibers [84,85]. They used curl III as an evaluator of different refining and moisture regimes in pulp, and felt it might well be related to wet web extensibility.

Fiber curvature, like curl, can be measured (to some extent) by the manual and image analysis techniques that have been described, although its definition is arbitrary as well. Few fibers assume the shape of a smooth circular arc, for which the radius of curvature is defined as the perimeter length divided by the subtending angle. The normally irregular shape of a fiber can be defined by the inverse of the "radius," where the subtended arc length can be the end-to-end length, an identifiable segment of length, or the contour length. Curvature thus defined is usually measured in units of radians per millimeter; this measurement tends to give too much weight to waviness in the fibers. Perez and Kallmes [151] discussed the relationship between their curl factor (inverse of curl I) and fiber curvature.

The most recent development in this field is the OpTest Fiber Quality Analyzer (FQA), which was developed jointly by Paprican and the University of British Columbia. The FQA measures the length and shape of fibers with a two-dimensional (2-D) imaging system. Fibers are singly transported through a flow cell by flowing water and are oriented hydrodynamically. The image of the fiber under the diffuse illumination is acquired with a CCD video camera. The image data are then processed and analyzed by a personal computer to determine the fiber length, fiber curl factor (curl III), and kink index (see subsection titled "Fiber Kink"). These data are also treated statistically to provide their frequency distribution. Olson et al. [136] observed a linear correlation for results of the average fiber length between the Kajaani FS-200 and FQA instruments.

Out-of-Plane Curl The measurement of out-of-plane curl is restricted to examination of fibers within sheets that have been embedded and sectioned for microscopic examination or photomicrography or electronmicrography. Although any of the manual or semiautomatic techniques that are available could be used, the most powerful tools are the image analyzers or graphic digitizers that have been described in this chapter. Further information on digitizing methods is given in Section V.

In a cross section of a sheet such as that shown in Fig. 53, one can see what Page et al. [140] referred to as the undulatory structure of fibers in paper—out-of-plane deformations that can be considered in terms of wavelengths. These wavelengths can range in size from several fiber thicknesses down to molecular dimensions. The fiber labeled L in Fig. 53 exhibits a deformation of relatively long wavelength, whereas the fiber labeled S has a deformation whose wavelength is shorter than the fiber is wide. Short-wavelength (crimped) out-of-plane deformations are considered especially significant from the standpoint of paper shrinkage [144], and Page et al. offer polarized light microscopic evidence that the fibrillar orientation in the fiber walls is disarranged around these locations [140].

Long wave-length undulations may significantly affect the flexural properties of paper and board, especially at or near failure [23]. A single undulation can be approximated as a segment of a circle; the ratio R of its circumferential length to the distance between its ends is

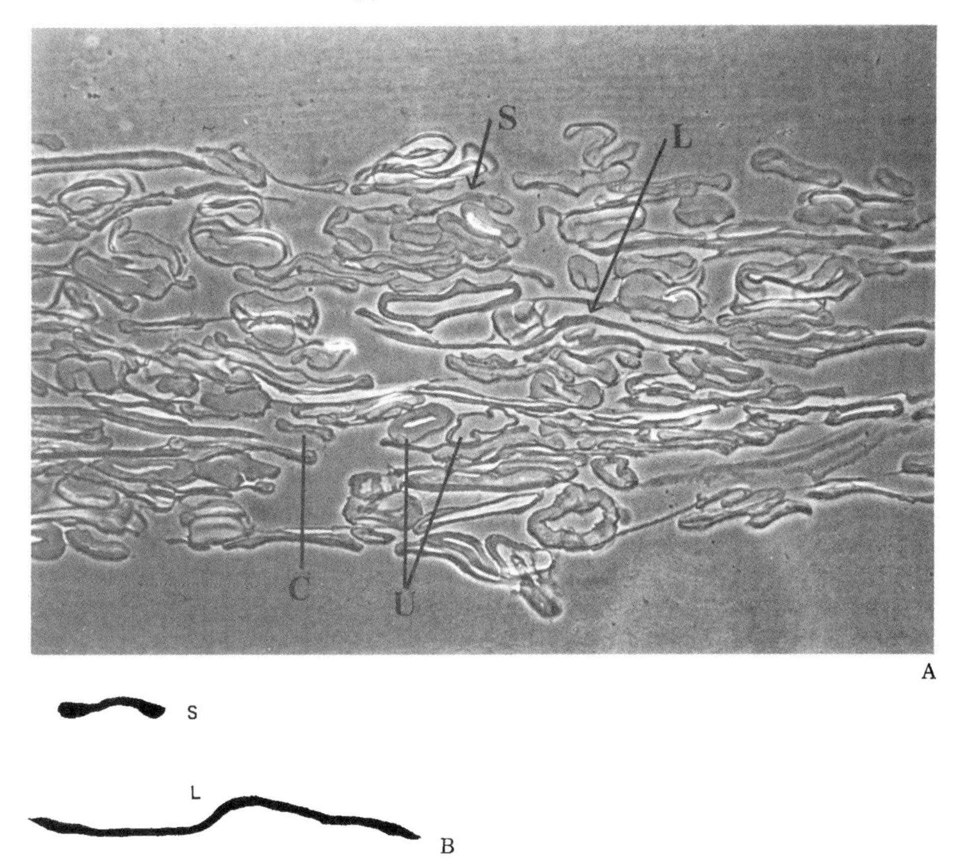

Fig. 53 (A) Cross section of 80 g/m^2 bleached softwood kraft embedded in epoxy resin and sectioned at 1.5 μm. *C*, collapsed fiber; *U*, uncollapsed fibers. (Photo by A. Eusufzai.) (B) Ink drawing of designated fibers of short (*S*) and long (*L*) wavelength undulations visible.

$$R = \frac{r\theta}{2z[\tan(\theta/2)]} \tag{55}$$

where

r = radius of curvature
θ = subtended angle
z = maximum out-of-plane local displacement of the fiber

One should take parenthetical note of the fact that whereas the flexibilities of two fibers of the same width and curvature might be quite similar in the plane of the sheet, those same two fibers might have very dissimilar flexibilities in the out-of-plane or Z direction. As noted in Chapter 14, particularly, bending stiffness is a function of the cube of the thickness.

Because out-of-plane curl is often of interest from the standpoint of bending deformations of the sheet, it is important that fiber thicknesses as well as curvatures be ascertained in a manner that does not result in statistical bias toward the more readily visible, easy-to-measure thick fibers.

Fiber Kink "Kink" refers to an abrupt change (sharp bend) in the direction of a fiber, usually caused by beating or other mechanical action on pulp fibers. In Figs. 6b and 6c, four kinks are recognized, dividing the fiber into five segments. Forgacs [46], who used the term "nodes," did an extensive microscopic study on their cause and occurrence; he concluded that their frequency of occurrence was related to the spacing of rays in the pulpwood, and their cause was the relative weakness of the S2 layer in the vicinity of the rays (see Fig. 54). In his description of the nodes, Forgacs included zones of dislocation that could serve as sites for kink formation.

Kibblewhite [99,100] discriminated between abrupt kink and gradual curl in his works. He calculated a kink index by allotting arbitrary weights of 1–4 to sharp bends in the categories 10–20°, 21–45°, 46–90°, and 91–180°, respectively. Kibblewhite's kink is the weighted sum of the number of kinks, divided by the total fiber length:

$$\text{Kink index} = \frac{N_{10\text{-}20^\circ} + 2N_{21\text{-}45^\circ} + 3N_{46\text{-}90^\circ} + 4N_{91\text{-}180^\circ}}{L} \tag{56}$$

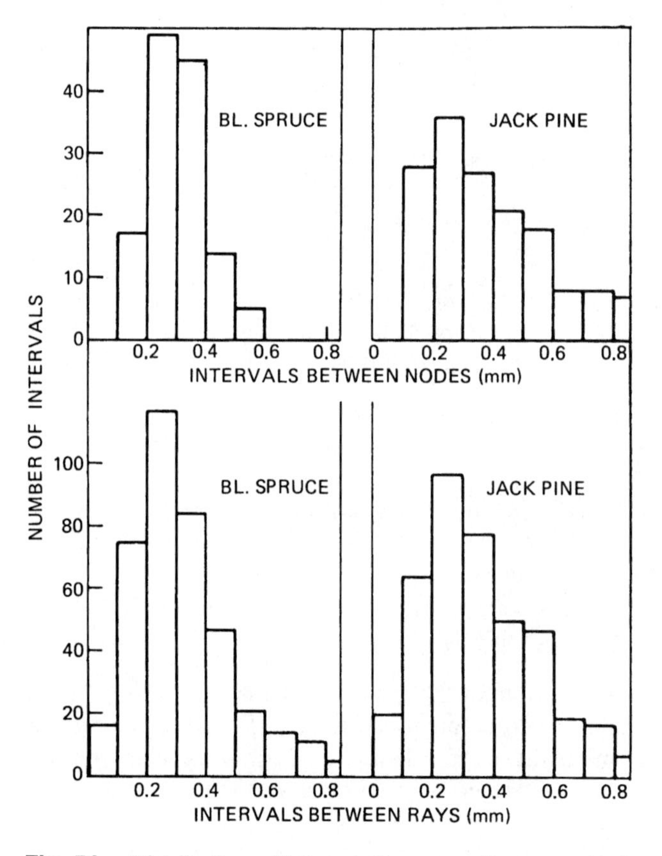

Fig. 54 Distribution of intervals between adjacent nodes in tracheids of 48% yield spruce and 48% yield jack pine sulfite pulps compared with the distribution of intervals between ray crossings in the corresponding woods. (From Ref. 46. Used with permission. Copies available from TAPPI, Technology Park/Atlanta, PO Box 105113, Atlanta, GA 30348.)

where $N_{(\alpha\text{-}\beta^\circ)}$ is the number of kinks in the angular range from α° to β°.

As described earlier, the OpTest FQA can also measure the kink index with the angular range above 21°.

See also "Microcompression" in Section V.

V. OTHER DIMENSIONAL PARAMETERS USED TO CHARACTERIZE FIBER RAW MATERIALS AND FIBER NETWORKS

The relationship of fiber coarseness, fiber shape, and fiber dimensions to tensile and other strength indices, folding endurance, porosity, density (and its reciprocal, bulk), opacity, and other sheet properties has been the subject of many studies and reports [35,66,143]. There have also been many studies and reported related to the influence of the fiber fragment and nonfiber components in paper [9,19–21,104,148,177,188]. Although the most useful data concerning fiber parameters normally come from measurements made on sheets, the expenditure of technician time in the dehydration, embedding, sectioning, mounting, photography or projection, identification of features, and manual recording of fiber measurements is generally very substantial for these operations. The advent of the graphic digitizer and other semiautomatic measurements by projection techniques has made it possible to reduce the time requirements for the recording of measurements.

Although these instruments often provide the capability for huge savings in time and increased reliability of data, the proper preparation of the specimens still represents a substantial effort with good equipment by skilled personnel. Much of the actual data, therefore, on fiber dimensional parameters have been obtained (1) from fibers in wood or other source material, (2) from fibers isolated from pulp, or (3) from fibers that have been removed from sheets by hydropulping or some other mechanical action. In the case of the fiber fragment and nonfiber components of paper, our ability to gather quantitative dimensional data has been greatly enhanced by all of the recent developments cited in Section IV, but especially by scanning electronic microscopy [149,216], SEM combined with image analysis, and the particle counter combined with a special drainage apparatus [209].

For a discussion of fiber coarseness, the reader is referred to Section VI.B of Chapter 14 and the paper by Britt [18]. Other dimension-related parameters, especially those related to fiber bonding and fiber surface area, have been treated in other chapters, such as Chapter 15 (this volume) and Chapter 4 of Volume 2.

For those parameters not covered elsewhere, we feel that this book requires some explanation of the most significant terms encountered in the dimensional properties testing field, combined with some brief exposition, either by discussion or by reference, that will help the reader locate or develop the specific technique or information needed. We call this "Notes on Procedures for Various Structural Parameter Determinations" and hope it will serve its intended purpose.

Notes on Procedures for Various Structural Parameter Determinations

Area, Contact (See Contact area, Crossing area, and Bond area.) Area, Gross cross-sectional; Area, Lumen; Area, Net cross-sectional These three terms refer to the space occupied by a fiber that is viewed in a transversely cut section. The area of fiber cell wall material is the net area after the area occupied by its hollow center (lumen), if any, is deducted from the gross area. The ratio of the net to the gross cross-sectional area is termed the *Mühlsteph ratio* (which see) in some works, especially those dealing with wood. In wood, these areas are often measured on slide-mounted sections by automatic image analysis [121], by photometric densitometry [55], or by a grid intersection or dot-counting method [109,117].

When individual fibers are measured, a planimeter, a cutout weighing method [117], or an image analysis is often used. Hardacker [60] determined net cross-sectional area of pulp fibers by compaction. The compacted shape is that of a flat ribbon with rounded edges and cross-sectional area equal to $ab - [(4 - \pi)/4]^2$, where a is width and b is thickness.

When either wood or individual fibers are embedded and sectioned for mounting for purposes of cell wall area determination, it is important that the sections be thin (never over 5 μm and preferably less than 1 μm) and even more important that embedding media such as methacrylates not be used, because they cause swelling of the specimen. A comparative study of 35 embedding–mounting combinations by Crosby and Mark [33] concluded that the epoxies are superior for embedding because expansion artifacts are virtually undetectable with these materials. Studies by Berlyn and Miksche [13] and Isenberg [75] support this conclusion.

The embedding technique developed by Quackenbush [162] is advantageous in that it eliminates the need for solvent dehydration of the specimen. Thus, possible chemical or physical alteration of the fibers is avoided. Also, the hardness of the resin medium can be varied to suit the particular material to be sectioned. Diamond or glass knives should be used for the sectioning. Clear epoxy resins are recommended for both embedding and mounting if the slide is to be permanent; if not, the choice of mounting medium is principally related to the desirability of matching the refractive indices of the embedding and mounting media as closely as possible. There are a number of commercially available mountants suitable for this purpose that can be subsequently dissolved to permit reuse of the slide.

Newer Methods. The confocal laser scanning microscope (CLSM) provides a nondestructive method of sectioning fibers and paper sheets in either a wet or dry state [77,126]. In the CLSM system, only structures in the focal plane are imaged. The microscope stage moves vertically through a set distance, one step at a time, so that a series of images, called "optical sections," can be collected from a fiber. These images are superimposed with a computer to make cross-sectional images of a fiber in the *XY* or *XZ* planes. Moss et al. [126] state that cross-sectional images of wet pulp fibers obtained by the CLSM are as close to the natural state as those from a low temperature SEM. Several researchers have developed image analysis procedures for obtaining fiber transverse dimensions and other features. Seth et al. [179] confirmed good accuracy of cross-sectional area measurement for various wood species by comparing their method with other methods such as an automatic fiber length analyzer. They also applied this technique to their studies on recycled fibers and mechanical pulp [78,79]. Although the primary advantage is simplicity in specimen preparation compared with other methods, technical skills are required to obtain images with good contrast that enhance the transverse size measurements. This issue has been well discussed by Donaldson and Lausberg [40] for wood cell dimensions.

Area, projected This is the total area that is occupied by all the fibers contained in a known mass of pulp if they are laid out singly in a field of view. The projected area is obtained by adding the individual values of area for each fiber sampled, that is, length times width. It is not correct to calculate this area by multiplying total length by mean width or total width by mean length. The objective of ascertaining projected area is to find the maximum surface coverage that would be achieved with a known quantity of a given fiber stock. It will, of course, be different for different conditions of moisture, temperature, and pressure.

Aspect ratio This is the ratio of the major to the minor axis of a fiber or other feature of interest (shive, particle, etc.) in a transversely sectioned sheet. Aspect ratios are easily determined by CLSM and image analysis [78–80] or by digitizing photomicrographs of serial or sequential sections of paper [165,225]. Aspect ratio determination provides the investigator with information on fiber flexibility and collapse (which see) and orientation; thus, aspect ratio values will, in general, be different for sections cut normal to the machine and cross-machine directions. Table 1 illustrates the way Rothenberg and Fernandez [165] used the aspect ratio and other sheet geometry parameters to evaluate the flexibilizing (and therefore enhanced bonding) action of various oxidizing agents on TMP handsheets as part of a larger study on the effects of bleaching, refining, and so forth, on different furnishes. It can be seen that aspect ratio correlates positively with tensile index and negatively with tear index. Aspect ratio in the sheets can usually be determined more quickly and reproducibly than degree of collapse (see **Collapse**) because it is less judgmental and therefore less operator-dependent. However, the algorithms used with CLSM make either task easier [79,80]. See also **Roundness**.

Bond area When two fibers are in optical contact (see **Contact area**), it is assumed that the forces of adhesion between the fibers are distributed within the optical contact area. The forces, principally those of hydrogen bonds in the case of wood pulp fibers, may or may not be uniformly distributed in that area; however, the problem of hydrogen bond distribution within a fiber bond area has so far been intractable. For further understanding of this problem, the reader is referred to Refs. 65, 144, 147, and 225 and Chapter 15.

Bond distance; Bond region These terms are used with reference to the Z direction in the sheet, or normal to the fiber axis. When assemblages of fibers bonded to shives, fibers bonded to fibers, or fibers bonded to other materials are tested so as to pull the fiber from the substrate in sliding shear, the resultant degree of elongation at the contact area is of the substantial. For example, Thorpe et al. [205] found an elongation of about 1 μm, equivalent to a bond strain of about 0.5%. Such deformations are far above the magnitude of hydrogen bond distances. For this reason, the "bond region" between fibers may be more appropriately thought of as some greater distance. The distances between S1/S2 interfaces of wood pulp fibers in contact with each other may be thought of as a bond region, or adhesive layer, whose properties will differ from those of the bulk of the fiber wall. These concepts are further discussed in Chapter 15.

Nordman et al. [134] pointed out that increases in bonding strength related to beating time may reflect transitions from a state of bonding involving only S1

Table 1 Comparison of Thermomechanical Pulps Modified with Oxidizing Agents

Pulp	Bonded surface (contact) area of fibers (%)	Aspect ratio	Moments of inertia (mm^4) I_x	Moments of inertia (mm^4) I_y	Bonding state probability	Apparent density (kg/m^3)	Tensile index (N m/g)	Tear index ($MN\,m^2/g$)
TMP No. 1 modified with								
(Untreated)	26.2 ± 11.1	3.0	6.2×10^{-9}	3.9×10^{-8}	2.33	393	41.5	4.54
3.0%O_3	27.6 ± 12.9	3.1	4.6×10^{-9}	3.3×10^{-8}	2.27	452	48.9	7.37
5.8%O_3	29.4 ± 13.8	3.1	4.6×10^{-9}	2.9×10^{-8}	2.61	526	62.4	6.52
7.5%O_3	31.1 ± 10.9	3.7	3.7×10^{-9}	2.0×10^{-8}	3.01	531	62.9	6.38
TMP No. 3 Modified with								
(Untreated)	21.2 ± 14.8	2.6	8.3×10^{-9}	3.1×10^{-8}	1.41	317	32.3	11.73
3.0%O_3	18.7 ± 12.9	2.5	7.1×10^{-9}	2.2×10^{-8}	1.51	337	38.5	12.24
4.2%O_3	29.2 ± 19.9	2.8	5.0×10^{-9}	2.4×10^{-8}	1.81	395	45.9	9.94
5.8%O_3	29.5 ± 17.2	2.8	6.2×10^{-9}	2.7×10^{-8}	2.11	402	48.7	9.48
CH_3COO_2H								
0.44 mol/100 g OD pulp	18.0 ± 13.9	2.7	5.4×10^{-9}	2.5×10^{-8}	1.57	386	43.6	10.14
0.50 mol/100 g OD pulp	23.7 ± 14.0	3.1	3.9×10^{-9}	2.3×10^{-8}	1.90	475	54.8	8.14
0.56 mol/100 g OD pulp	29.0 ± 17.7	3.4	3.8×10^{-9}	2.7×10^{-8}	2.24	480	65.0	8.35
1.8% ClO_2	21.6 ± 15.1	2.4	5.4×10^{-9}	2.1×10^{-8}	1.48	342	37.5	11.82
3.0% ClO_2	22.0 ± 15.5	2.8	8.7×10^{-9}	3.2×10^{-8}	1.59	367	39.8	9.81
7.0% ClO_2	27.8 ± 17.2	2.9	4.2×10^{-9}	2.3×10^{-8}	1.94	405	43.6	8.93

OD = oven-dried.
Source: From Refs. 165–167.

surfaces to those involving more contacts between S2 and S2 or S2 and S2 as the beating fibrillates and disrupts the fiber S1. Work by Nanko and associates [65,128,129] using transmission electron microscopy greatly elucidated the three-dimensional nature of the bond region.

Bonded area, Percent; Bonded area, Relative; Bonded area, Total As defined in Chapter 15, the relative bonded area (RBA) is the ratio of the total bonded area to one-half the total external surface area of component fibers. The concept of RBA is thus tied to the idea that sheets are assemblages of two-dimensional fiber planes. As shown in Fig. 6 of Chapter 15, the possible states of bondedness of any given fiber at a point along its length are (1) that it is bonded to another fiber on one side, (2) that it is bonded to other fibers on both sides, or (3) that it is unbonded. The total bonded area is, of course, the sum of the bonded areas on the fiber surface. Bonded area may also be expressed as a percent of the total surface area of the fiber.

According to the RBA concept developed by Kallmes and coworkers [89,93], RBA has a maximum potential (RBA_{max}) that is constrained by fiber width, fiber coarseness, and sheet basis weight. Uesaka et al. discuss the limitations of this concept in Chapter 15. Figures 7 and 8 of that chapter demonstrate that bonds vary in nature and kind and that they have a component in the third dimension.

Yang et al. [225,226] showed by a digitizer technique that fibers in paper sheets are actually bonded to more than two fibers at a point in a large number of cases, and this is shown in Figs. 55A–55D. By their technique, they also determined the bonded surface area of fibers directly on the same sheet cross sections, as shown in Fig. 56. Yang et al. determined the percent bonded area of fibers by calculating the ratio of the sum of that portion of the fiber perimeter in contact with other fibers in a section to the entire perimeter of the fiber, then summing up those ratios for all the sections for each fiber measured.

It is assumed that the bonded portion of the fiber perimeter is constant through the section thickness, which should be as thin as possible, preferably less than 1 μm. The calculation is expressed by

$$B = \frac{1}{N_f} \sum_{i=1}^{N_f} \frac{\sum_{j=1}^{N_s} S_b}{\sum_{j=1}^{N_s} S_t} \tag{57}$$

where

$B =$ percent average bonded surface area of fiber
$N_f =$ number of fibers whose trajectories are followed through successive sections
$N_s =$ number of sections used for measurements
$S_b =$ perimeter length of fiber within area of contact with other fibers
$S_t =$ total perimeter of fiber

If one compares Figs. 55 and 56, one can see that when the bonded surface area of kraft fibers reaches 50%, there is essentially no free fiber length (which see) left; at virtually every point along the fiber length, at least one other fiber is in contact (bond area is assumed equal to contact area for such measurements). In contrast, an RBA of 50% in Fig. 6 of Chapter 15 results in a theoretical free fiber length of over 20%. The

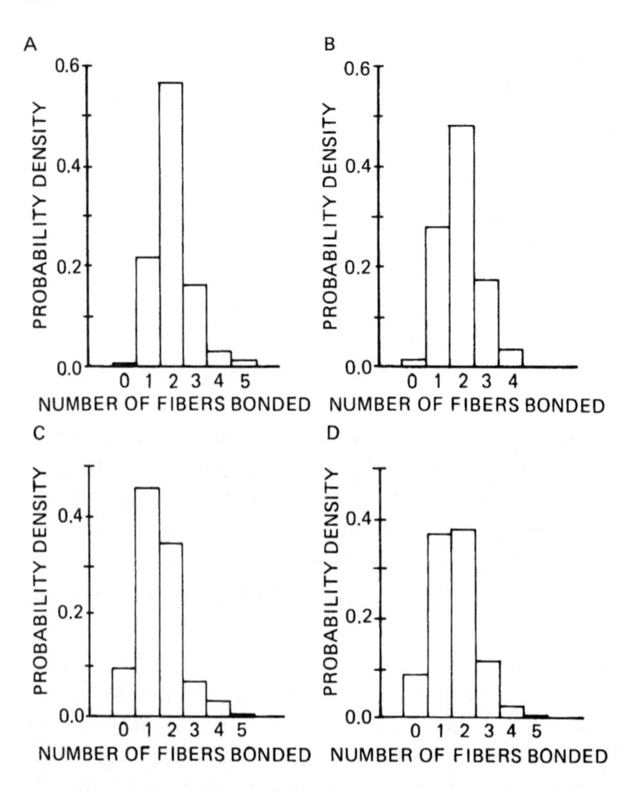

Fig. 55 Bonding state probability (number of fibers bonded to a crossing fiber within a 4 μm section of the crossing fiber). (A) kraft CD; (B) kraft MD; (C) bond paper; (D) newsprint.

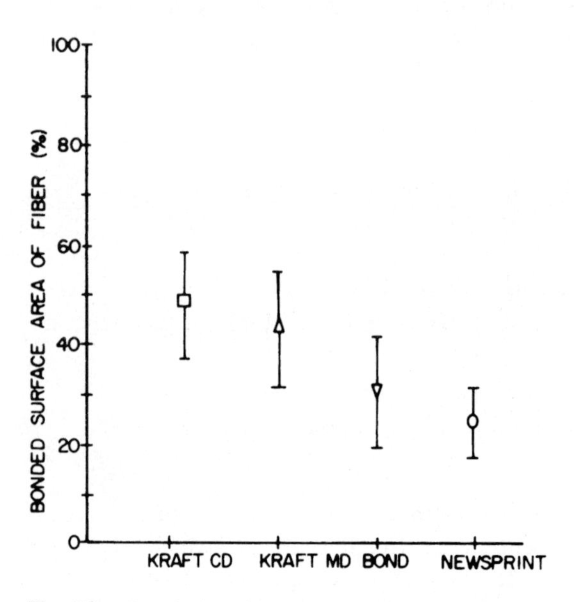

Fig. 56 Bonded surface areas of fibers in four types of paper (mean values ± one standard deviation).

digitizer results of Yang et al. merely confirm what was reported much earlier by Page et al. [146]—that when pulp is well beaten and fibers conform well, then at virtually any point along the fiber length the fiber is bonded to other fibers at one or more points on its surface. It is important for the investigator to bear in mind that the results of many tests such as those described in Chapter 15 are often interpreted on the basis of the multiplanar Kallmes model despite the disagreement of that model with structural measurements such as those that have been mentioned. For additional discussion of planar sheet bonding models, see Chapter 15 and Görres et al. [52].

Bonded fiber surface area This term is best expressed as a percentage of the total surface area of all fibers measured. In Table 1 the results of some digitizer measurements by Rothenberg and Fernandez [165–167] given in the first column of data include the mean values plus or minus one standard deviation. See also Refs. 102 and 103. The deviation indicates that a larger sample size might be advantageous in this direct method. The direct method of microscopic examination and indirect methods—optical, gas adsorption, and so on—are described in Chapter 15.

Bonding state This is the term used to describe the bondedness of a fiber at any point along its length. Conceptually, one must visualize a fiber with a series of circumferentially drawn lines on it and infer from data whether one or more crossing fibers are bonded at that imaginary line. From digitized data, Yang et al. [225] prepared the histogram of probability distribution for different types of sheets shown in Fig. 55. Similar histograms were developed by Rothenberg and Fernandez [165] and Kimura and Mark [102,103]; mean values for several types of thermomechanical pulp sheets are given in Table 1. Figure 55 may be considered an experimental bonding state diagram; Fig. 6 of Chapter 15 is a type of theoretical bonding state diagram, as discussed under **Bonded area**. For experimental work, it is critical that sections of sheets be made as thin as practicable, never over 5 μm; otherwise, depth-of-focus limitations will prevent good resolution of the contact surfaces between fibers and introduce spurious data.

Brushing out (see **Fibrillation.**)

Centroid; Centroidal axis The location of the center of gravity of a line, a plane area, or a solid is called the centroid. An axis through the centroid parallel to a coordinate axis is called a centroidal axis. Algorithms are available for automatic image analysis that locate the centroid of a feature (such as the fiber shown in Fig. 72, of Chapter 14) with respect to either coordinate axis. The object can be of any shape that will fit in the scanned image.

Circularity Used primarily to evaluate cross-sectional shapes of textile fibers (see, e.g., Ref. 156, the circularity or fullness is determined by the ratio $4\pi A : P^2$, where A is the cross-sectional area and P is the fiber perimeter distance. See also **Roundness**.

Collapse; Conformability Collapse represents change of cross-sectional shape in a fiber from its "original" state, for example, in the stock chest, so its generally more flattened appearance in the sheet. It results from a combination of surface tension, which causes the opposing fiber walls to draw together as water leaves the fiber, and external forces of compression on the sheet, principally those developed in the press and drying sections [160]. Page et al. [140] examined fibers in sheet cross sections and arbitrarily classified as collapsed any fiber whose inner walls touched along more than half the length of the lumen.

Kallmes and Eckert [94] used a 50% criterion in an entirely different way. They viewed individual fibers in extremely thin sheets of the type shown in Fig. 57 under dark field illumination microscopy. Those fibers that were generally dark across more than half of a scan line traversing the fiber were considered collapsed; the brighter fibers—less than 50% dark—were considered uncollapsed. High percentages of dark, collapsed fibers were found in sheets pressed at high pressure. Page [138] investigated the collapse behavior of individual pulp fibers by covering them with a nonpenetrating medium (cellophane tape adhesive) that had a refractive index similar to that of the fiber surface. In this way, those fibers (or portions thereof) that were collapsed appeared invisible in transmitted light; open lumina in the uncollapsed areas were revealed by reflection and refraction at the cellulose/air interface. With such a procedure, fibers can be classified as to their susceptibility to collapse upon drying.

Higgins et al. [66] adopted a criterion for collapse based on the relationship of lateral force F (the combined forces of surface tension and press application) to conformability, which has been defined [3] as the ability of the fiber to conform to the shape of a crossing fiber (for example, as in Fig. 5 of Chapter 15). The conformability of a fiber by the Higgins et al. definition is the reciprocal of the load C required for collapse; they define the degree of collapse as the proportion of fibers in which $1/C > 1/F$. Collapse in this view is an all-or-nothing phenomenon, and the value of C is correlated directly to the Luce shape factor [113] for the fibers in the

Fig. 57 Very low grammage "two-dimensional" sheet showing many collapsed (flattened) fibers and significant free fiber segment lengths. (Courtesy of Empire State Paper Research Institute, Syracuse, New York.)

original wood. It is my opinion that except for the pulp fiber experiment of Page [138] (which did not involve fibers in sheets), these definitions impose arbitrary structural appearance or shape criteria that are unnecessary in the light of today's data-gathering equipment. The definition of Page et al. [140] is too dependent on operator judgment and requires a great expenditure of time for sufficient data acquisition. The Kallmes and Eckert procedure, which is dubious on theoretical grounds, is restricted to extremely thin sheets or sheets that have been peeled apart, and peeling certainly causes some internal disruption. It is also very time-consuming if a sufficient sample is to be obtained.

The **Luce shape factor** (which see) does not adequately predict fiber collapse, because it is based on circular shapes for both fiber wall and lumen—features seldom possessed by real fibers. Whether a fiber collapses or not may be influenced by its location within the sheet—for example, by its proximity to a thick-walled fiber. The two uncollapsed fibers marked *U* in Fig. 53 have markedly different wall thicknesses. The one on the right would certainly have a much lower original Luce shape factor than the collapsed fiber marked *C* close to it.

Jang et al. [78–80] discussed the collapse behavior of fibers using confocal microscopy. They defined the "collapse index (C.I.)" as the fractional loss of lumen volume that results from fiber processing such as recycling or refining. For studies on sheets, their work should be consulted.

Contact area Although the concept of two fibers making contact with one another seems simple enough, the contact area is very difficult to measure satisfactorily. It is possible to observe the formation of optical contacts in paper under controlled moisture conditions by vertical (incident) polarization microscopy with the use of a porous plate apparatus (Fig. 58); Page and Tydeman [145] carried this experiment out very elegantly. They considered that bonding between fibers occurred when dark areas (the optical contact areas) appeared in the fiber crossing when the fibers were viewed normal to the sheet plane (see **Bond area**). For most purposes, bonding in sheets is more easily measured by the surface contact that appears when sheets are embedded and sectioned transversely, as in Fig. 53A cut at 1 μm with a glass knife. A semiautomatic method such as digitizing will produce information on bonded (contact) surface area and bonding state probability quite rapidly and efficiently, but the direction of the future would appear to be with confocal laser scanning microscopy and image analysis (see Chapter 5 of volume 2). The relationship between the area of optical contact by light and SEM microscopy and the actual bond area is discussed by Algar [3].

Count This is the number of fibers present in a unit mass, usually of pulp. A standard procedure for its determination is to disperse 0.5 g of dyed pulp in 1 L of water, then further subdivide and dilute until the suspension contains 10 μg (if softwood) or 5 μg (if hardwood). Known amounts of the suspension are pipetted onto slides, where the dyed fibers are counted under the microscope. No fibers or fiber segments less than 0.1 mm are counted. By proportion, the total count is usually given as number of fibers per gram.

Typical counts might be 5,000,000 for a softwood such as pine and 25,000,000 for a hardwood such as oak. Dependent on the degree of processing, the count may be increased very slightly or more than doubled, for example, for highly beaten,

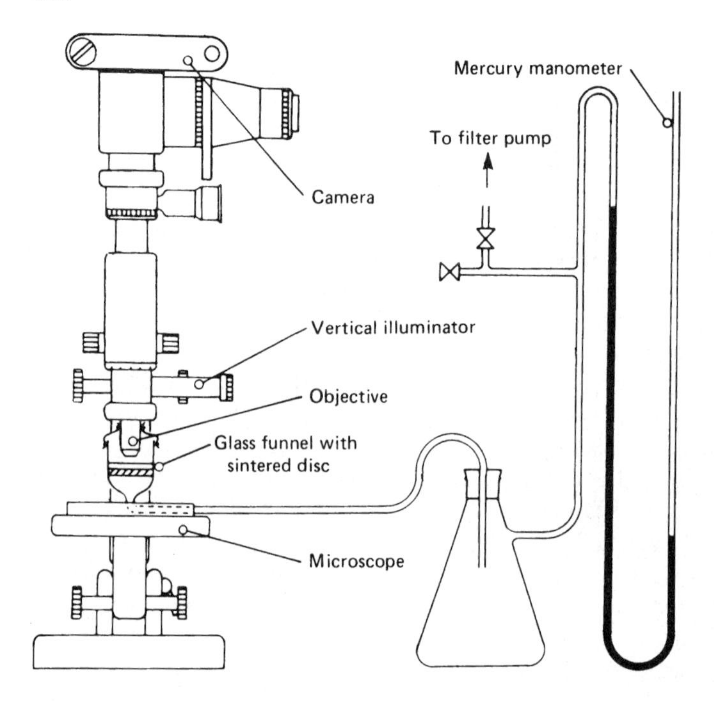

Fig. 58 A porous plate apparatus modified to follow bond formation in a drying web. (From Ref. 145.)

bleached, low yield, or degraded fibers. With improved procedures, commercial fiber analyzers such as OpTest FQA or image analyzers should be able to perform these measurements.

Crill (see **Fines.)**

Crimp ratio This is the fractional increase of the linear extent of a fiber as it undergoes a change from a curled to a fully straightened configuration without stretching. It is expressed mathematically as curl III in Eq. (53).

Crossing area If one considers Figs. 1 and 2 of Chapter 15, it is evident that the crossing area S in Fig. 2 is not equivalent to the contact area (which see), because fiber contact is often partly obstructed and the projected overhang of the curved fiber edge on the neighboring fiber in plan view results in a larger overlap than the actual contact area. In fact, fibers that cross may not be in contact at all. Thus the ease of making this measurement (on fibers lying near the sheet surface) bears little relation to the usefulness of the data.

Crushing A state of fiber compaction in which the fiber is pressed beyond **collapse** (which see), with resultant mechanical disruption of the fiber wall layers and the microfibrillar structure and plastic deformation [160]. Occurrences of crushing can be observed and counted by light or confocal microscopy and image analysis [221].

Diameter (see **Width.**)

Dislocations These are areas of the fiber wall that show evidence of abrupt changes in direction of the cellulosic microfibrils in the S2 layer (see Fig. 59). Dislocations, also called slip planes, may not be evident unless the fiber is viewed under polarized light. They tend to be formed at a characteristic angle [96]. They are the result of excessive axial (compressive) forces and are formed because of shear instability (sliding delamination) between the sheets of microfibrils. Chipping, refining, sectioning, and other mechanical action on fibers increases their frequency of occurrence [46,96,99]; however, sometimes they are present in the standing tree [38,96]. Dislocations may be counted (1) per fiber, (2) per unit length of fiber, or (3) per unit mass of fiber basis under the polarizing microscope. Dislocations often occur concentrated together. Kibblewhite [99] calls these concentrations "zones of dislocation" and believes that they are added to or extended through the development of local compression failures at sites of bond formation as the fibers shrink and mutually restrain each other during drying. Such zones he considers identical with the **microcompressions** (which see) described and photographed by Page and Tydeman [144].

Felting coefficient The ratio of fiber length to diameter or width (which see).

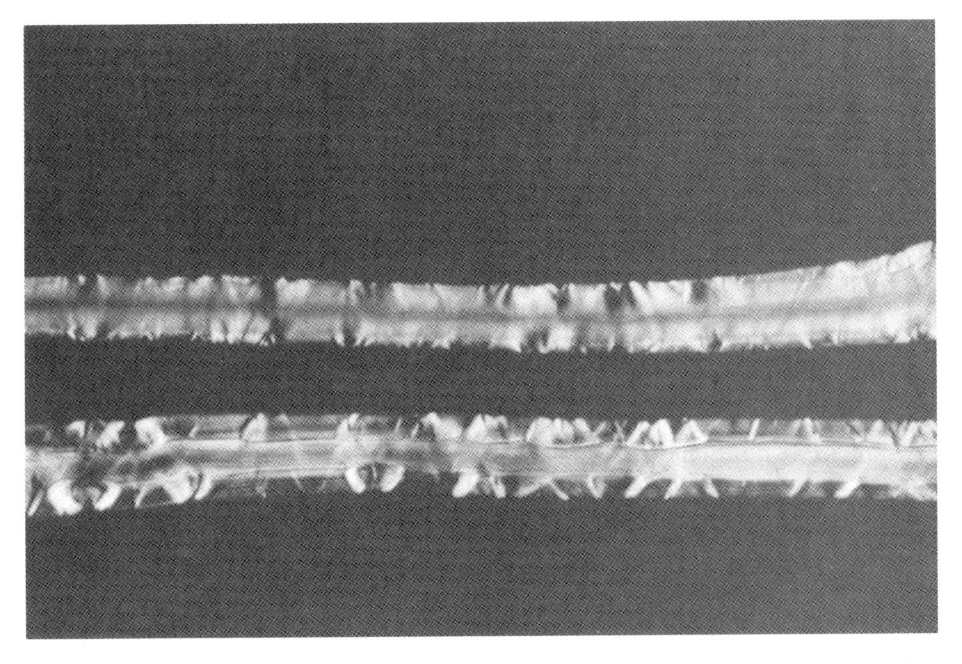

Fig. 59 Intense concentrations of dislocations in the fiber walls of a softwood, *Pinus radiata*. These fibers have undergone sufficient refining to create dislocation zones but not enough to cause visible fibrillation. The upper fiber is a thin-walled springwood tracheid; the lower summerwood tracheid exhibits some evidence of possible delamination in the microfibrillar structure. (Courtesy of R. P. Kibblewhite, Forestry Research Institute, Rotorua, New Zealand.)

Feret diameter A statistical diameter defined as the mean length of the distance between two tangents on opposite sides of the image of a particle (see Figs. 44 and 48). This parameter is readily obtainable by automatic image analysis. (See also **Statistical diameter**.)

Fiber composition The different types of fibers present in a sheet of paper or fiber stock can be expressed in terms of fractional proportion. A count (which see) is made of each fiber type present, determined typically by the length or other anatomical characteristics of the fibers and their staining reactions; each type is then expressed as a percent of the total.

Fiber segment This item can have more than one meaning and therefore more than one measure. As seen in Figs. 6B and 6C, the fiber marked *F* in Fig. 5 can be subdivided into five segments of markedly different orientations in the sheet. Measurements of length can easily be made on such fiber segments by image analysis or via a projected image or photograph on a digitizer tablet. Another operator, however, might distinguish additional short segments and thus divide the fiber further. From the standpoint of network mechanics (Chapter 1), it is the ability of the fiber to transmit axial load that determines segment length. Thus, one or more dislocations (which see) in an unbonded length might increase the tabulation of segments; conversely, fiber curvature through a densely bonded region may be treated satisfactorily as a single structural element, thus decreasing segment tabulation. Fiber segment length has also been used as a synonym for free fiber length (which see).

Fibril angle The predominant helical angular orientation of the microfibrils in the fiber wall. The term is usually used with reference to the S2 layer. For further details on fibril angle and its measurements see Ref. 22 and Chapter 13 and Section IV of Chapter 14.

Fibrillation The mechanical actions to which papermaking fibers are subjected, particularly those in the refining stage, result in the loosening of microfibrillar elements from the fiber wall and provision for greater fiber–fiber bonding surface as a result of the fibrillation. Fibrillation of fibers can be observed clearly in the scanning electron microscope, although the highest level of detail is achieved with transmission electron microscopy [127,128]. In light microscopy, fibrillation is easier to see in mechanical than in chemical pulp fibers (see Chapter 14, Section IV.C).

Filament winding angle (see **Fibril angle** and **Micellar spiral angle.)**

Fines The *Pulp and Paper Dictionary* [110] describes fines as "small particles of fiber that are shorter than normal wood pulp fibers. Sometimes called *wood flour*." A closely related term is *crill* [184], which is ordinarily used to designate the extremely fine particles (microfibrils and wall fragments) that are abraded from the surface of cellulosic fibers during refining. Thus, the term "fines" is somewhat more inclusive. Fines or crill can be quantitatively determined by fractionation through some type of filtering system or plug flow separation combined with weighing or microscopy. Further definition of fines fractions and a method for their determination is given in Section VI. Because commercial fiber analyzers such as Kajaani or OpTest FQA count fibers in a length range, the fines fraction may be measured as the length-weighted frequency of fibers in a particular length group (e.g., fibers less than 0.2 mm). The retention of fines in the sheet has been shown to affect many of its properties [19,71].

Free fiber length; Free fiber length aspect ratio; Mean free fiber length Considerable attention has been given in the paper physics literature to the concept that fibers in a sheet have segments of their length that are bonded to other fibers and segments that are not bonded. In extremely thin sheets (e.g., Fig. 57) or low density materials such as tissue and nonwovens, this is a reasonable distinction. For typical sheets of moderate and high density, it was first shown microscopically by Page et al. [146] and subsequently by image digitizing by others [165,225,226] that virtually the whole of the fiber length is bonded at some point on its periphery. Thus the concept of "free" or unbonded fiber segment length does not have much relevance to most grades of paper and board. A distinction should be made between the actual unbonded distance (the free length) and the distance between midpoints or centroids of adjacent bonds, especially as the frequency of contacts (see **Contact area**) increases. Algar [3] and others term the midpoint distance the *fiber segment length*, but the free fiber length has also been described by that term. Kallmes and Bernier [90] discuss several definitions of free fiber length. *Fiber segment* (which see) also has several other meanings.

Finally, it should be noted that, for different purposes, free fiber length could refer to the maximum, minimum, or average length between fiber contacts (Fig. 60A). Kallmes and Bernier [90] define the mean free fiber length as the mean total free fiber length per fiber divided by the mean number of free fiber lengths per fiber. Thus, the mean is affected by the defined choice of free fiber length. In Fig. 60B, the distance D bears a statistical relation to the free fiber length distribution. In another paper, Kallmes and Bernier [91] use the term *free fiber length aspect ratio*, which they mean to be the ratio of free fiber length to fiber width in their simulation study of idealized fiber networks. In this glossary, the term *aspect ratio* (which see) refers to actual measurement of major and minor axes of fiber cross sections.

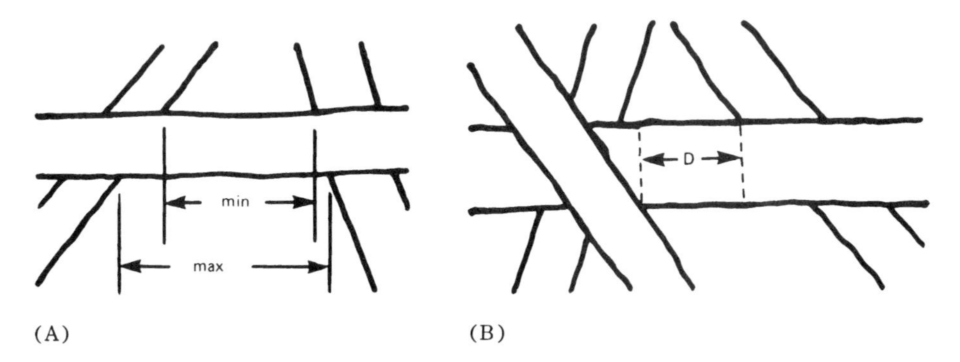

Fig. 60 Schematic of fibers crossing. (A) The minimum and maximum free fiber lengths are shown. (B) A free minimum distance D; the mean of such distances equals one-half the mean free fiber length by derivation in Ref. 90.

Free fiber surface area The percentage of the total surface area of a group of measured fibers that is not bonded to any other fiber; 100% less the bonded fiber surface area (which see).

Fullness (see **Circularity.**)

Inertia (see **Moment of inertia.**)

Kink (see **Section IV.B.**)

L/D ratio; L/W ratio The ratio of fiber length to width. It is one of the eight basic wood fiber characteristics considered significant for papermaking by the Food and Agriculture Organization of the United Nations (FAO) [45]. The determination of length and of width is discussed in Section IV.A. More information is found under *Width* in this section.

L/T ratio This is the ratio of fiber length to fiber cell wall thickness.

Luce shape factor Luce [113] made a series of tests on tubes of various shapes (illustrated in Fig. 68 of Chapter 14) and found that the lateral force required to collapse the tubes was proportional to $(d_o^2 - d_i^2)/(d_o^2 + d_i^2)$, where d_o and d_i are the outside and inside diameters of the model. This shape factor has been used by others such as Higgins et al. [66] to evaluate hardwood fibers for collapse (which see) in papermaking.

Major axis; Minor axis As defined by Yang et al. [225], a fiber viewed in a sheet cross section has as its major axis the maximum straight-line distance between any coordinates lying on the perimeter of the fiber; its minor axis is taken as the distance across the fiber normal to the major axis. These distances are most easily obtained by CLSM and image analysis [79,80]. The information is used to calculate the aspect ratio (which see) for the fiber in the section.

Martin diameter A statistical diameter defined as the mean length of a line that intercepts the boundary or perimeter (Fig. 42) of the image of a fiber, particle, or other object and divides the image into two portions of equal area. It is a dimension that can be obtained with an appropriate algorithm via automatic image analysis. The bisecting line is parallel to or coincident with the scan lines, irrespective of the orientation of the subject in the image.

Mean bond area The determination of the bond area (which see) for various pulps yields characteristic means for such measurements. For example, Page et al. [146] found microscopically that mean bond area of a bleached spruce sulfite increased by over 60% when beaten to 310 mL CSF.

Micellar spiral angle This is an archaic and erroneous term for the predominant helical orientation of the microfibrils in the S2 layer of the plant fiber wall. More acceptable terms are fibril angle, S2 angle, and filament winding angle. The determination of this parameter is discussed in Chapter 13. See also Ref. 22.

Microcompression A localized compressive deformation or wrinkling of a fiber such as that shown diagrammatically in Fig. 61 [60]. Microcompressions can be introduced into fiber structure by the processes of refining or bleaching and mechanical actions, especially those that induce curlation. The phenomenon has also been ascribed to longitudinal contraction of one fiber caused by transverse shrinkage of a crossing fiber by Page and coworkers [144,178]. Dumbleton [41] introduced such deformations into fibers artificially (see Section VII.B of Chapter 14). It has been observed that the load–elongation properties of sheets having a high proportion of such fibers are similar to those exhibited by the deformed fibers [41,142,143,178]. Microcompressions have been studied and measured thus far by light and electron microscopy in surface views. The morphological difference between a microcompression and a slip plane or dislocation (which see) is that the dislocation is localized within the microfibrillar structure (as in Fig. 59), whereas the microcompression takes the form of three closely spaced kinks of the fiber (see **Fiber Kink**, Section IV.B); both microcompressions and dislocations are phenomena associated with or induced by axial compression of the fibers. However, Nanko and Ohsawa [128] interpreted the fiber bond formation during the drying process by another mechanism instead of microcompressions on the basis of studies using the confocal scanning laser microscopy and TEM. In their work, they reported that no microcompression was observed. Further studies are required.

Moment of inertia Fiber moment of inertia and the computation of its magnitude are discussed in Chapter 14. The probability density or relative probability of finding fibers with a given moment of inertia in sheet cross sections was determined by using data obtained by digitizer, by Yang et al. [225]. Rothenberg and Fernandez [165] found this calculation useful in the interpretation of mechanical behavior of sheets, as shown, for example, in their data in Table 1. Its significance was also recognized by Duncker et al. [42], Kimura and Mark [102,103], and Schniewind et al. [176] in their studies on the stiffness properties of fibers.

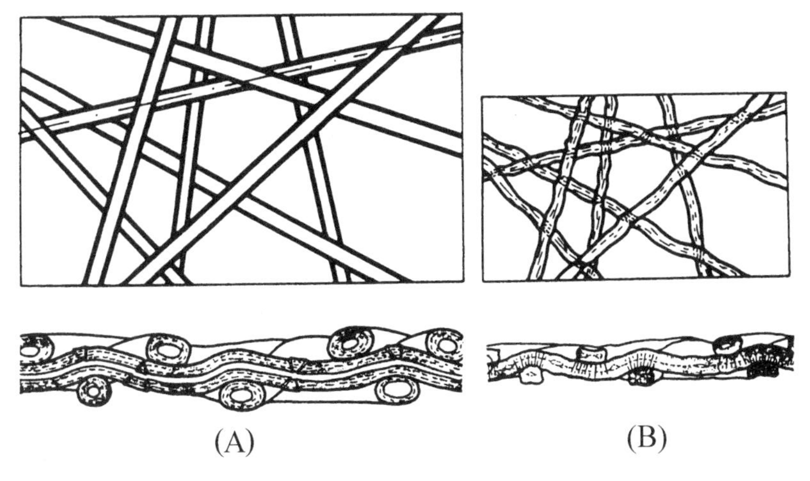

Fig. 61 Schematic of microcompression (A) Swollen fiber structure before drying; (B) fiber structure after drying. (From Ref. 59.)

Mühlsteph ratio This is the ratio of net to gross cross-sectional area of the fiber. It is principally used with respect to the dimensions of fibers as they exist (unmodified) in wood, as a means of evaluating the suitability of the wood for technical uses such as papermaking. It can be calculated by any of the techniques mentioned under **Area, Cross-sectional**.

Necking When certain dissimilar fibers such as wood pulp and rayon fibers bond together, a characteristic enlarged neck is created (Fig. 62). Page and Tydeman [145] assessed this phenomenon microscopically and evaluated the degree of necking N by the relation

$$N = \frac{2R_B}{R_1 + R_2} - 1 \tag{58}$$

where R_1, R_2, and R_B are as shown in Fig. 62.

Node (see Fiber Kink, Section IV.B.)

Perimeter The perimeter of a fiber viewed laterally is often difficult to measure because of the very high ratio of length to width in many fibers. If the fiber is viewed at a magnification small enough to contain the entire fiber, the width (which see) is small and difficult to measure. However, the boundary or surface perimeter and the convex perimeter (Figs. 42 and 44) can both be determined by planimetry, digitizing, and automatic image analysis under appropriate conditions—specifically, that boundary resolution is preserved as a low magnification image is projected and enlarged for measurements to be performed on it. The convex perimeter corresponds to the distance required to wrap a string around the fiber (see Fig. 42). As such, it can be used to derive the Feret diameter (which see) shown in Figs. 44 and 48. The boundary perimeter can be used to describe shape (for example, perimeter squared divided by gross area) of fiber surfaces in cross sections and also is useful in obtaining certain other derived size measurements as indicated in Fig. 48.

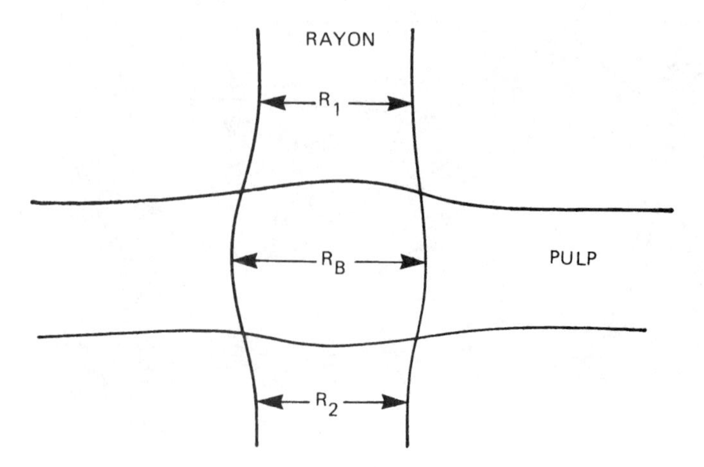

Fig. 62 Diagram to show measurements made to determine the degree of necking at a bonded crossing between a pulp and a rayon fiber. (From Ref. 145.)

Roundness The roundness factor of pulp fibers in cross section is defined by Isenberg [75] as the ratio of the gross area of the fibers to the area of their circumscribed circles. Thus, the roundness factor of the uncollapsed fiber on the left in Fig. 63 is greater than the collapsed fiber on the right. The factor has been used as a measure of relative collapse (which see) of the fibers of different pulps. Three other measures of fiber roundness are given by Isenberg: (1) the ratio of the gross fiber area to the area of a rectangle in which it is inscribed with the greatest width and thickness of a fiber parallel to the sides of the rectangle; (2) the ratio of the area of the cross section of the fiber to the area of a circle having a circumference equal to the perimeter (which see) of the fiber; (3) the ratio of the width to thickness of the fiber cross section. Definition (2) is identical to that for circularity (which see). Definition (3) closely approximates the definition of aspect ratio (which see), which is a highly useful measure of fiber collapse in sheets. The techniques of CLSM and image analysis permit any of these measurements to be made efficiently [78–80].

Runkel ratio This is the ratio of twice the fiber cell wall thickness to the lumen diameter. It is principally used with respect to the dimensions of fibers in wood for evaluating the suitability of a species for pulp wood. It is one of the eight basic wood characteristics reported by the FAO [45] in regard to the properties of plantation wood for papermaking. In general, values less than 1 are sought, although it has been pointed out [35] that there is no material improvement in using this ratio instead of the density of the wood as a predictor, the Runkel ratio and the basic density should have a roughly constant proportionality. Densitometry [55] and fluorescence [82] measurement techniques have been used in the past, but image analysis techniques will continue to displace slower microscopic methods [121].

Segment (see **Fiber segment.**)

Shive; Sliver The *Dictionary of Paper* [112] defines shive as "a bundle of incompletely separated fibers which may appear in the finished sheet as an imperfection" and a sliver as "a small splinter of wood, longer than a shive and of smaller diameter

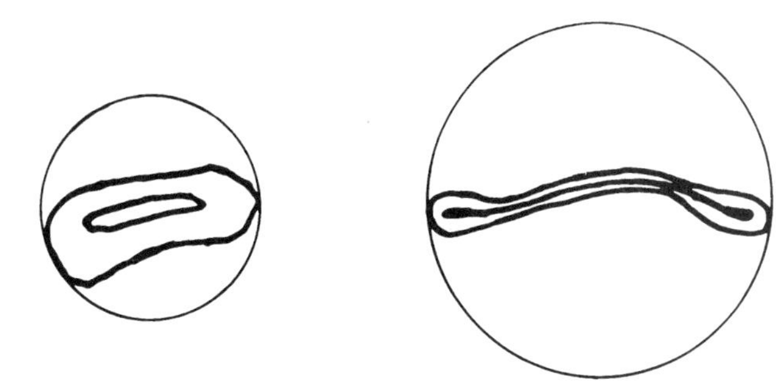

Fig. 63 Two fibers, one similar to one of the fibers marked *U* in Fig. 53A and the other similar to the fiber labeled *S*, in circumscribed circles.

in proportion to its length." Mechanical pulp usually contains shives and tends to have slivers much more often than does chemical pulp. Methods for their determination are given in the next section. They are readily observed in sheet cross sections, so data concerning them can be obtained and processed efficiently by using image analysis systems.

Slip planes (see **Dislocations.**)

Specific surface This is defined as the exposed area per unit mass of moisture-free fiber and thus provides a measure of the degree of fibrillation. Because its measurement is covered in Chapter 6 of Volume 2, it will not be discussed here. It is important to distinguish between this term, which refers to the external surface, and *specific internal surface*, which relates to the fine capillary system within the fiber wall.

Statistical diameter This is the statistical average of a specified parameter such as Feret or Martin diameter (which see) from microscopic or image analysis.

Thickness The thickness of the fiber cell wall in wood can be measured by a stage micrometer on a microscope [13], by densitometry [55], by fluorescence [82], or by CLSM and image analysis [111,221]. Isolated fiber wall thicknesses can be determined by pressing part of the fiber between transparent anvils as discussed in Section VII.A of Chapter 14, and making the measurement by variable permeance transducer [60] or interference microscopy [95,139]. In paper sheets, the measurement can be made by digitizing the fibers that lie along scan lines drawn on micrographs of sheet cross sections in a statistically acceptable sequence, as was done by Kimura and Mark [102,103].

Twist A freely dried pulp fiber undergoes extensive twisting as illustrated in Section VII.A of Chapter 14. However, bond formation in paper sheets prevents extensive twist. Therefore twist as measured in fibers in pulp is not appropriate as a measure of this parameter in paper. In Fig. 57 several fibers are seen to be twisted in a low density, thin sheet. These twists can be counted under the microscope or on the image, but the (rare) twists that occur in denser, thicker sheets are difficult to quantify. In a large number of serial sections, one can count twists of fibers, where they occur by noting the orientation of the major and minor axes (which see) of a given fiber in cross section, which rotate as the serial sections pass through the twist.

Unbonded fiber segment length (see **Free fiber length.**)

Width The width of a plant fiber generally changes substantially from its roughly polygonal shape in wood or other plant tissue to its generally more compacted shape in the paper sheet. The width and other dimensions change again upon recycling (see Table 2). Pulp fibers shrink laterally upon drying, as was demonstrated in an elegant microradiographic technique by Tydeman et al. [208]. If one is measuring original fiber diameter, examination of the wood or plant tissue in cross section is appropriate. Confocal laser scanning microscopy is being used more and more [121]. The techniques of stage micrometry [13], fluorescence [82], densitometry [55], and projection micrometry [75] also remain available together with image analysis. Individual

Table 2 Effect of Recycling on Fiber Dimensions: Scotch Pine Kraft Pulp Fiber

	Fiber diameter (μm)	Lumen diameter (μm)	Single wall thickness (μm)
Simple recycling			
Never dried	41.8	20.6	10.6
vs.			N.S.[a]
Dried	32.4	12.0	10.2
Never dried	41.8	20.6	10.6
vs.		N.S.	N.S.
Rewetted	34.0	12.6	10.7
Dried	32.4	12.0	10.2
vs.	N.S.	N.S.	N.S.
Rewetted	34.0	12.6	10.7
Deinking recycling			
Never dried	36.1	15.1	10.5
vs.			N.S.
Dried	29.4	9.5	10.0
Never dried	36.1	15.1	10.5
vs.			N.S.
Rewetted	32.6	12.4	10.0
Dried	29.4	9.5	10.0
vs.			N.S.
Rewetted	32.6	12.4	10.0

N.S. = no significant difference at 95% confidence interval. Each number represents the average of measurements on 40 fibers.
Source: From Ref. 120.

wet fiber widths have been determined by projection micrometry of undried stained fibers; the procedure is given by Isenberg [75]. Essentially the same procedure has been employed for cotton fibers [62]. Individual dry fibers can be measured as freely dried or compacted (if it is desired to ascertain the lumen-free cross-sectional area) by employing an image-splitting eyepiece [60]. Such an eyepiece was used to measure the fiber diameters in Table 2. The width of fibers in sheets can be determined by stage or projection micrometry, planimetry, semiautomatic image analysis (digitizing), and CLSM image analysis. The speed of data acquisition is generally in ascending order among those methods. Jordan and Page [84] made determinations of pulp fiber width by automatic image analysis, using two separate approaches designed to overcome the problem that fibers are generally too long to fit in the field of view if an adequate magnification is maintained for accurate width determination. The problem of width measurement of very fine fibers (asbestos fibers) has been dealt with by Taylor and Dixon [204] by taking brightness profiles at many points along the length of the fiber, always in a direction perpendicular to the local tangent to the fiber. Uncharacteristically large values (evidence of crossed fibers) are automatically deleted from the database. Width determination in the automatic image analyzer is further discussed in Section IV.A.

VI. DIMENSIONAL CHARACTERIZATION OF SHIVES, SLIVERS, AND FINES

This section deals with particles of fibrous nature that are present to some extent in most wood pulps and mixed pulps. Shives and slivers, which are defined in the glossary of the preceding section, consist of fiber bundles and their associated cellular elements (axial and ray parenchyma, vessels, etc.) that were present in the original wood or plant vascular tissue. They are accordingly always larger than fibers, whereas the fines fractions are smaller. At the present time, these fractions are principally described by size classes, numbers of particles present, and weight proportion.

A. Shives and Slivers

Several useful methods have been published by TAPPI that contain procedures for determining shives in pulp by screening or fractionation.

TAPPI Useful Method UM 240 [198] calls for a dispersion of about 1.13 kg (2.5 lb) of dry pulp in 300 L (66 gal) of water to be slurried onto a screen, which retains any shives and slivers present. The procedure takes over 30 min. The retained particles are oven dried and weighed to obtain the result as a percent of the oven-dry weight of the whole pulp.

TAPPI UM 241 [199] relates to a device that fractionates by varying the width of a slot or orifice through which stock at 0.3–1.0% consistency passes. A particle analyzer (digital counter) with a recorder monitors the stock flow. Although this device does not provide an exact measure of the shive and sliver content by either weight or number, it is widely used because it is relatively fast and gives the operator a control guideline for pulp quality.

TAPPI UM 242 [200] prescribes the dilution of about 100 g of pulp to 0.5% consistency. Samples that comprise about 25 g of oven-dried (OD) pulp are added to water that is passed under pressure through a porous screening plate, which retains the shives and slivers. The fibers and fines can be collected from the overflow. The procedure requires about 20 min. The shives are reported as a percent of OD weight of the whole pulp.

The shives content of a flash-dried pulp bale or other dry lap can be determined by counting the number of fiber bundles observed (by stereomicroscope) in two 60 g/m^2 handsheets prepared from a reslushed sample taken from the bale. The procedure is described in TAPPI UM 239 [197]. The number of shives per 100 g of moisture-free pulp is reported.

Several automatic shive analyzers have been developed using the image analysis technique. For example, the PQM-1000 laboratory analyzer produced by Sunds Defibrator AB measures the shape of shives within the range of 0.3–10.0 mm in length and 75–3000 μm in width. The pulp suspension passing through a glass cell (5×10 mm) is illuminated by a laser beam. Fibers and shives are projected onto a linear photosensor (an array of photodiodes) on the opposite side. The image data of shives are then treated statistically with a computer to provide a frequency distribution expressed as a matrix of length and width categories.

Because shives are readily observable in sheet surfaces (and to some extent beneath the surface) owing to their size and (frequently darker) color, there is certainly potential for rapid determination of their size and occurrence in sheet surfaces by automatic image analysis in the future if such procedures have not already been carried out. Dirt specks, halftone dots, and other dark particles can be observed in light-colored sheets easily this way.

In sheet cross sections, shives and other large particles are distinguishable without difficulty to a trained technician. The size, perimeter length, aspect ratio, moment of inertia, percent bonded area, and bonding state of such fiber bundles and aggregations can all be determined with the use of a graphic digitizer, as was done by Rothenberg and Fernandez [165] for various high yield pulps. The more classic methods of particle size and number determination by microscopy, microprojection, and hemacytometer particle counting have been described by Isenberg [75].

B. Fines

The Nature of Fines The dimensional characterization of the fines fraction in pulp or in paper sheets is a rather new endeavor. Until about 1960 there was little appreciation of the fact that the fines that arise from beating or refining action differ dramatically in shape, size, and contribution to sheet properties from the fines that result from pulping and defiberization. What are today known as primary fines were fairly well understood. These are mainly the axial and ray parenchyma cells, broken or cut ends of fibers, and miscellaneous storage, meristematic, protective, and so on, cells from the stems of trees and other plants. They are mostly separated from the fibers in pulping. Figure 64 shows the primary fines collected after passing a 200 mesh screen (opening width 76 μm) from a Southern pine (*Pinus* spp.) pulp that had not been refined (715 mL CSF). Fibers do not generally pass through a 200 mesh spacing.

Although most of the particles in Fig. 64 are cellular, some finer particles of a more threadlike or reticulate form are visible, especially near the upper right of the picture. These are more typical of the fines created by beating and other mechanical action, which Steenberg et al. [184] described and photographed, giving the fraction the name *crill*. Crill, now more frequently referred to as secondary fines, is essentially material that has been removed from the fiber walls by the rubbing and crushing action of beater bars and refiner plates and the rolling and rubbing of one fiber upon another. It is composed almost exclusively of slender fibrillar particles derived from the splitting and tearing off of microfibrillar strands, principally from the S1 layer. Steenberg et al. [184] observed such particles under the phase contrast microscope and noted that earlier characterizations of such material as flourlike or slimelike were erroneous. The early descriptions were probably based on the tendencies of secondary fines to retard drainage.

In fact, secondary fines or crill materials seldom, if ever, assume a compact shape such as exists in flour or meal, even with prolonged beating. It has been shown that very long beating times result in longer, not shorter, secondary fines as the S1 layer is stripped away and the beater begins to strip microfibrillar material from the S2 layer [97].

Figures 65–68 show secondary fines from pulp that was refined for 10 s. The primary fines and the fibers in the pulp had previously been removed. Figure 65 shows particles that pass 76 μm but are retained at 38 μm spacing. Figure 66 shows particles that pass 38 μm but are retained at 19 μm spacing. Figure 67 shows particles that pass

Fig. 64 Primary fines of unrefined, unbleached Southern pine pulp. (From Ref. 69.)

a 19 μm spacing screen and settle after 48 h. Figure 68 shows particles that pass a 19 μm screen but are still suspended after 48 h. Figures 69–72 show, respectively, secondary fines that were refined for 45 s and subsequently were fractionated in the same manner and order as the fines shown in Figs. 65–68.

What is evident in these eight micrographs is that the fibrillar shape is retained generally; the size is reduced as finer mesh openings are passed. Most of the cellular particles do not go through openings smaller than 38 μm, however.

Measurement of Fines In general, the way in which fines are reported is to express them as a percent by weight (mass) according to fraction. The following relation is used:

$$F_f = \left(1 - \frac{L_m}{S}\right) \times 100 \tag{59}$$

where

$F_f =$ fines fraction (%)

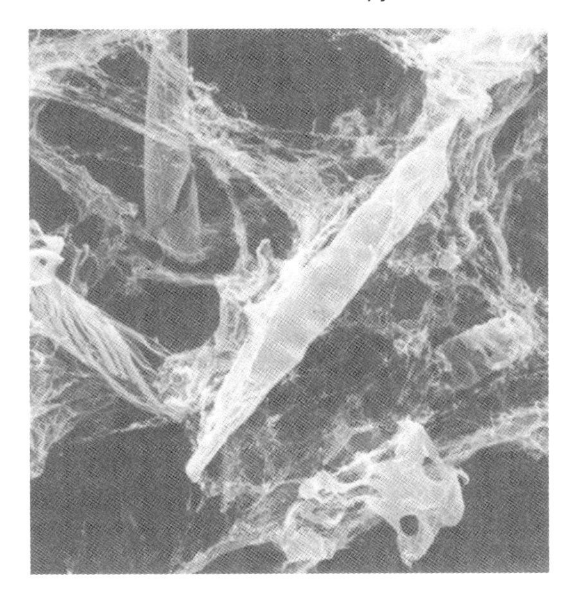

10 μm

Fig. 65 Secondary fines fraction 2° IA. Bar = 10 μm. (From Ref. 69.)

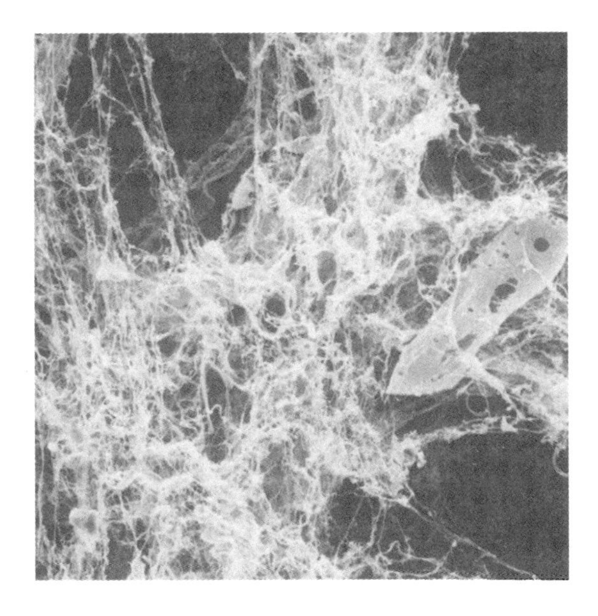

10 μm

Fig. 66 Secondary fines fraction 2° IB. Bar = 10 μm. (From Ref. 69.)

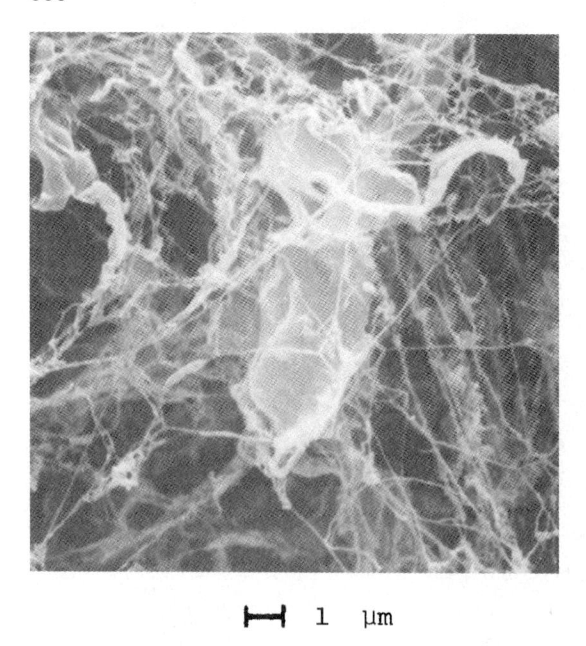

Fig. 67 Secondary fines fraction 2° IC. Bar = 1 μm. (From Ref. 69.)

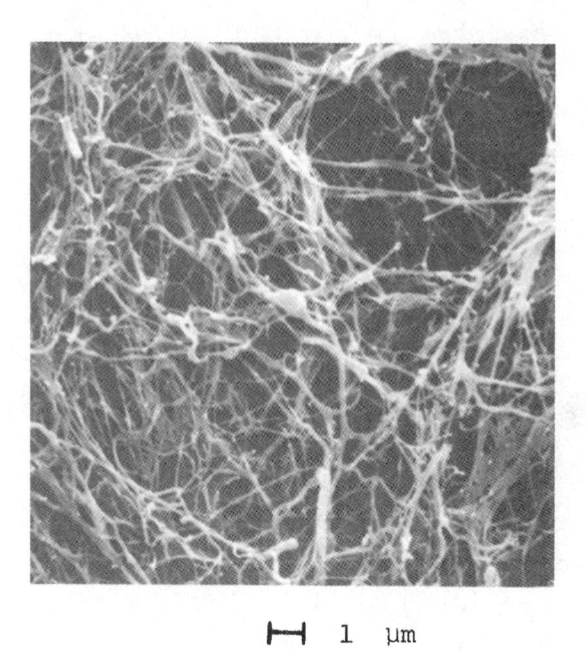

Fig. 68 Secondary fines fraction 2° ID. Bar = 1 μm. (From Ref. 69.)

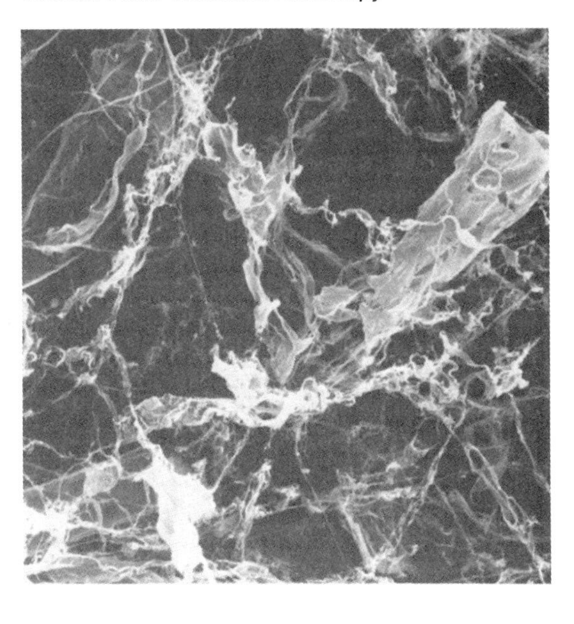

Fig. 69 Secondary fines fraction 2° IIA. Bar = 10 μm. (From Ref. 69.)

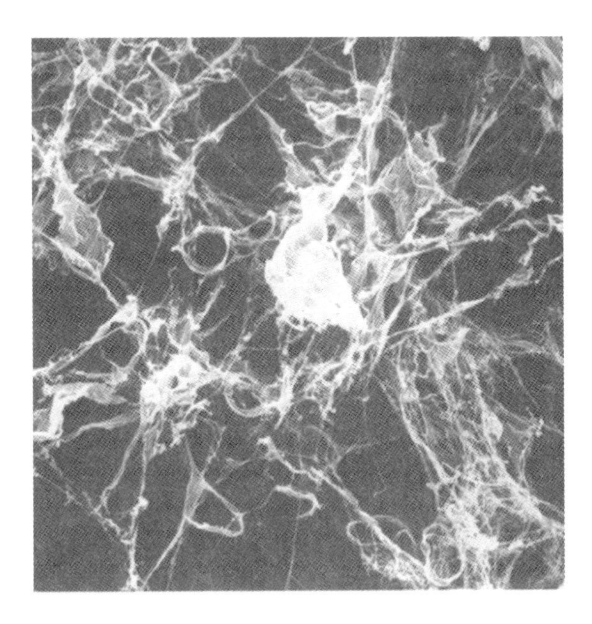

Fig. 70 Secondary fines fraction 2° IIB. Bar = 10 μm. (From Ref. 69.)

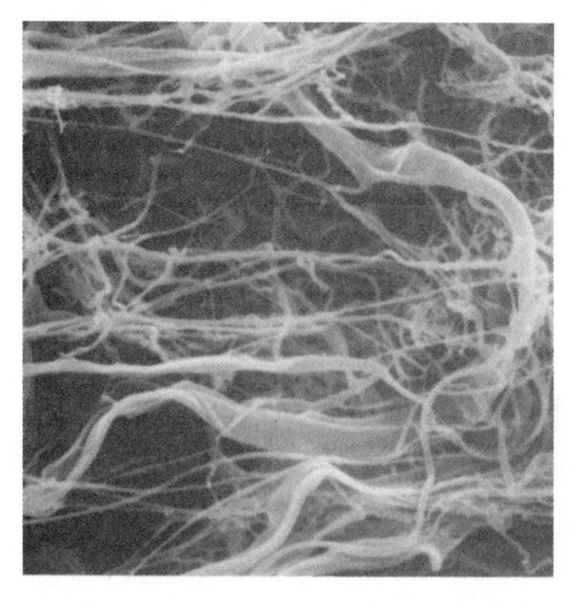

1 μm

Fig. 71 Secondary fines fraction 2° IIC. Bar = 1 μm. (From Ref. 69.)

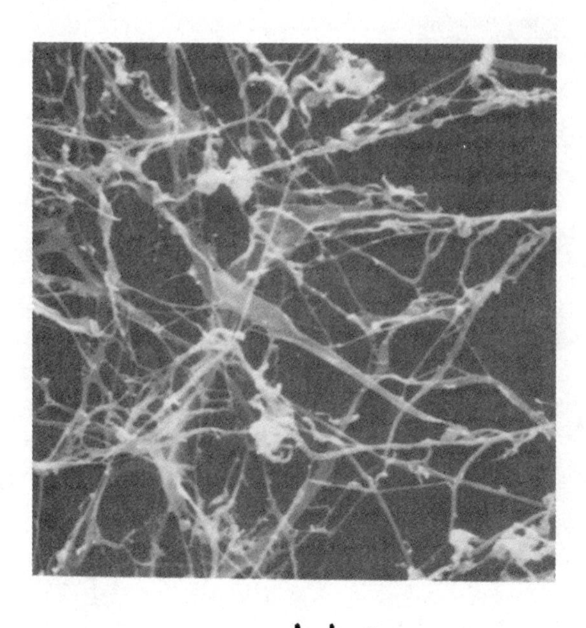

1 μm

Fig. 72 Secondary fines fraction 2° IID. Bar = 1 μm. (From Ref. 69.)

$L_m =$ mass of the long fiber in the pulp
$S =$ mass of all solids in the pulp

However, Kibblewhite and coworkers [98,101] developed a weighted *fines index* (FI), based on their examination of the effect (number of occurrences) of selected structural details of mechanical treatment on fiber surfaces that result from mechanical treatment, which may be useful for fundamental research purposes:

$$\mathrm{FI} = S_1 + 2S_{1-70} + 3S_{70-30} + 4S_2 \tag{60}$$

where

$S_1 =$ percentage of fibers in the long fiber fraction showing the primary wall largely removed to reveal the S1 layer with microfibrils mainly perpendicular to the fiber axes
$S_{1-70} =$ percentage of fibers showing the S1 layer partly removed to reveal microfibrils at angles of 90° to 70° to the fiber axes.
$S_{70-30} =$ percentage of fibers showing the S1 layer partly removed to reveal microfibrils at angles of 70° to 30° to the fiber axes.
$S_2 =$ percentage of fibers with some of the S1 layer removed to reveal the S2 layer

Access to scanning electron microscope facilities is needed to obtain the data for this calculation. The procedures used to fractionate fines for use in Eq. (59) have been developed. Kibblewhite et al. [101] used a Clark classifier [194].

Britt and Unbehend [20] and Unbehend [209] reported the development of a *dynamic retention/drainage jar* (DDJ) that is designed for wet screening of fibers, fines, and additives. Shown in Fig. 73, it consists of a barrel that screws into a base that has a recessed bottom. A perforated plate, a screen, and gaskets are positioned between the two. The screen holes are conical in both the retention and fractionation plates, which prevents binding of particles in the openings.

The procedure for determining the fines content in a sample of headbox or other stock is carried out using one 76 μm screen for operations such as quality control or several screen sizes for research purposes. Fibers and fines show good separation. However, the accuracy of the fines fractionation is critically dependent on knowing the exact consistency of the stock suspensions, because small errors in consistency can lead to large errors in the calculation of the fine fractions. Consistency should be determined by filtering a weighed quantity of stock through a weighed filter and drying in an oven. If there are fillers or other materials present, they must be accounted for precisely. Inorganic materials can be determined by ashing [192] and extractives content can be determined by following a standard such as TAPPI T 204 [191].

After fractionation, a size determination can be carried out with the use of a Coulter counter, described earlier in this chapter. Neither the wet screening in the DDJ nor the size determinations in the particle counter are without uncertainties due to the irregularity of the shape of fines particles. Some elongated particles will find their way through screen openings that are smaller than their long dimension. The Coulter counter yields results in terms of equivalent spherical diameters that hardly fit the extremely slender shape of secondary fines material. However, the DDJ particle counter procedure is decidedly an advance over previous procedures, which

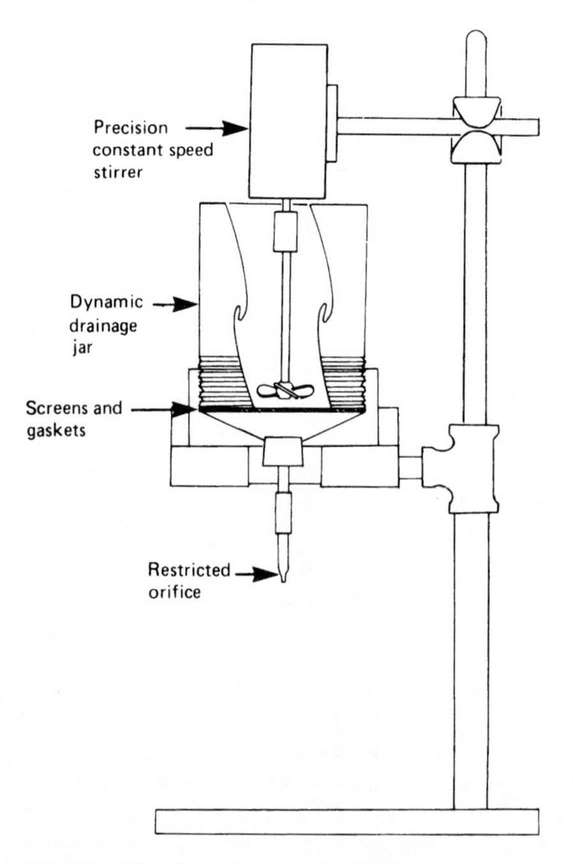

Fig. 73 The dynamic retention/drainage jar.

ranged from simple comparison of headbox and tray consistencies to tedious screening or sedimentation procedures.

The amount of fines in machine-made paper or handsheets can be determined by soaking the sheets in hot water for 30 min at approximately 0.5% solids. The suspension is carefully defibered at reduced speed for 5 min under mild agitation in a blender. Microscopic examination will tell the investigator if defibration and fines separation are complete. When this stage is reached, the fines determination is carried out by the same procedure as the one used for pulp [209]. The distribution of fines through the thickness can be determined by splitting the sheet into layers and making the determination on each layer separately.

REFERENCES

1. Aaltio, E. A., Prins, W., and Hermans, J. J. (1959). X-ray investigation into the orientation of cellulose fibers in paper with respect to the plane of the sheet. *Tappi* *42*(2):162A–163A.
2. Akai, T. J. (1993). *Applied Numerical Methods for Engineers*. Wiley, New York.

3. Algar, W. H. (1966). Effect of structure on the mechanical properties of paper. In: *Consolidation of the Paper Web*. F. Bolam, ed. British Paper and Board Makers Assoc., London, pp. 814–851.
4. Amidon, T. E. (1981). Effect of the wood properties of hardwoods on kraft paper properties. *Tappi 64*(3):123–126.
5. Anderson, J. G., and Edgar, R. F. (1987). Fiber-orientation anisotropy using a microwave moisture meter. *Tappi 70*(10):133–135.
6. Anon. (1979). Image analysis at the Institute. *Trend 29*:8–9.
7. Anon. (Ahlstrom Automation). (1985). Lippke fiber orientation sensor helps monitor sheet forming processes. *Fadum Ref. 5*(7):10.
8. Asunmaa, S., and Steenberg, B. (1958). Beaten pulps and fiber-to-fiber bond in paper. *Svensk Papperstidn. 61*:686–695.
9. Balodis, V., McKenzie, A. W., Harrington, K. J., and Higgins, H. G. (1966). Effects of hydrophilic colloids and other non-fibrous materials on fiber flocculation and network consolidation. In: *Consolidation of the Paper Web*. F. Bolam, ed. British Paper and Board Makers Assoc., London, pp. 639–691.
10. Barnea, D. I., and Silverman, H. F. (1972). A class of algorithms for fast digital image registration. *IEEE Trans. Comput. C-21*:179–186.
11. Bauer, W., and Stark, H. (1988). On the measurement of fiber orientation in paper sheet using laser light. *Wochenbl. Papierfabrik. 11/12*:461 (in German).
12. Baum, G. A. (1987). Polar diagrams of elastic stiffness, effect of machine variables. *Int. Paper Physics Conference*. Tech. Sect. CPPA, Montreal, pp. 161–166.
13. Berlyn, G. P., and Miksche, J. P. (1976). *Botanical Microtechnique and Cytochemistry*. Iowa State Univ. Press, Ames, IA.
14. Bernard, P., Boulay, R., Drouin, B., Gagnon, R., and Villeneuve, P. (1988). The role of absorbed water on the polarizing properties of paper at submillimetre wavelengths. *J. Pulp Paper Sci. 14*(5):J104–J108.
15. Borch, I., and Marchessault, R. H. (1969). Light scattering by cellulose. 1. Native cellulose films. *J. Polym. Sci. C 28*:153–167.
16. Boulay, R., Drouin, B., and Gagnon, R. (1986). Measurement of local fibre orientation variations. *J. Pulp Paper Sci. 12*(6):J177–J181.
17. Boulay, R., Drouin, B., and Gagnon, R. (1986). Paper fibre orientation measurement with a submillimetre laser. *J. Pulp Paper Sci. 12*(1):J26–J29.
18. Britt, K. W. (1965). Determination of fiber coarseness in wood samples. *Tappi 48*(1): 7–11.
19. Britt, K. W. (1980). Physical and chemical relationships in paper sheet formation. *Tappi 63*(5):105–108.
20. Britt, K. W., and Unbehend, J. E. (1976). New methods for monitoring retention. *Tappi 59*(2):67–70.
21. Bublitz, W. J., and Knutsen, D. P. (1980). Effects of deshive refining on high-yield kraft linerboard pulp. *Tappi 63*(5):109–113.
22. Butterfield, B. G., ed. (1998). *Microfibril Angle in Wood*. Proc. IAWA/IUFRO workshop on significance of microfibril angle to wood quality. B. G. Butterfield and Univ. Canterbury, Christchurch, NZ.
23. Carlsson, L. (1980). A study of the bending properties of paper and their relation to the layered structure. Ph.D. Thesis, Chalmers Univ. of Technology, Göteborg, Sweden.
24. Carroll, C. W. (1962). Joint probability function relating fibre segmental length and orientation. In: *Formation and Structure of Paper*. F. Bolam, ed. British Paper and Board Makers Assoc., London, pp. 243–245.
25. Charrier, J. M., and Marchessault, R. H. (1972). Light scattering by random and oriented anisotropic rods. *Fibre Sci. Technol. 5*:263–284.

26. Clark, J. d'A. (1942). The measurement and influence of fiber length. *Paper Trade J. 115*(26):36–42.
27. Clark, J. d'A. (1962). Effects of fiber coarseness and length. *Tappi 45*(8):628–634.
28. Clark, J. d'A. (1962). Weight average fiber length: A quick, visual method. *Pulp Paper Mag. Can 63*(2):T53–T60.
29. Corson, S. R., and Uprichard, J. M. (1972). Fiber length of Clark screen fractions. *Tappi 55*(11):1620.
30. Corte, H., and Kallmes, O. J. (1962). Statistical geometry of a fibrous network. In: *Formation and Structure of Paper*, Vol. 1. F. Bolam, ed. British Paper and Board Makers Assoc., London, pp. 13–52.
31. Cowan, W. F. (1975). Short Span Tensile Analysis. Pulmac Instruments Ltd., Montreal, Canada.
32. Cox, H. L. (1952). The elasticity and strength of paper and other fibrous materials. *Br. J. Appl. Phys. 3*:72–79.
33. Crosby, C. M., Eusufzai, A. R. K., Mark, R. E., Perkins, R. W., Chang, J. S., and Uplekar, N. V. (1981). A digitizing system for quantitative measurement of structural parameters in paper. *Tappi 64*(3):103–106.
34. Danielsen, R., and Steenberg, B. (1947). Quantitative determination of fibre orientation in paper. *Svensk Papperstidn. 50*:301–305.
35. Dinwoodie, J. M. (1965). The relationship between fiber morphology and paper properties: A review of literature. *Tappi 48*(8):440–447.
36. Dinwoodie, J. M. (1966). The influence of anatomical and chemical characteristics of softwood fibers on the properties of sulfate pulp. *Tappi 49*(2):57–67.
37. Dinwoodie, J. M. (1974). Failure in timber. 2. The angle of shear through the cell wall during longitudinal compression stressing. *Wood Sci. Technol. 8*:56–67.
38. Dodson, C. T. J., Serafino, L., and Yang, S. (1993). Interpretation of formation of paper. *Appita 46*(5):366–370.
39. Dodson, C. T. J., and Schaffnit, C. (1992). Flocculation and orientation effects on paper formation statistics. *Tappi 75*(1):167–171.
40. Donaldson, L. A., and Lausberg, M. J. F. (1998). Comparison of conventional transmitted light and confocal microscopy for measuring wood cell dimensions by image analysis. *IAWA J. 19*(3):321–336.
41. Dumbleton, D. P. (1972). Longitudinal compression of individual pulp fibers. *Tappi 55*(1):127–135.
42. Duncker, B., Hartler, N., and Samuelsson, L. G. (1966). Effect of drying on the mechanical properties of pulp fibers. In: *Consolidation of the Paper Web*. F. Bolam, ed. British Paper and Board Makers Assoc., London, pp. 529–537.
43. Evans, M., Hastings, N. A. J., and Peacock, J. B. (1993). *Statistical Distributions*. 2nd ed. Wiley, New York.
44. Fleishman, E. H., Baum, G. A., and Habeger, C. C. (1982). A study of elastic and dielectric anisotropy of paper. *Tappi 65*(10):115–118.
45. Food and Agriculture Organization, United Nations (1975). Pulping and Papermaking Properties of Fast Growing Plantation Wood Species. FO:MISC/75/31. FAO, Rome, Italy.
46. Forgacs, O. L. (1961). Structural weaknesses in softwood pulp tracheids. *Tappi 42*(2):112–119.
47. Forgacs, O. L., and Strelis, I. (1963). The measurement of the quantity and orientation of chemical pulp fibres in the surfaces of newsprint. *Pulp Paper Mag. Can. 64*(1):T3–T13.
48. Gagnon, R., Drouin, B., Bernard, P., and Bergeron, M. (1990). Far-infrared nondestructive testing of sheet material. *Tappi 73*(9):191–195.
49. Gagnon, R., Drouin, B., Dicaire, L.-G., and Bernard, P. (1990). High spatial resolution non-destructive testing of newspring. *J. Pulp Paper Sci. 16*(6):J179–J183.

50. Glynn, P., Jones, H. W. H., and Gallay, W. (1959). The fundamentals of curl in paper. *Pulp Paper Mag. Can. 60*(10):T316–T323.
51. Gonzales-Molina, J., Voillot, C., and Silvy, J. (1982). Effect of a mixture of softwood and hardwood fibers on the structure and physical properties of paper. *ATIP 36* (6–7):270,359–374 (in French).
52. Görres, J., Amiri, R., Wood, J. R., and Karnis, A. (1995). The shear bond strength of mechanical pulp fibers. *J. Pulp Paper Sci. 21*(5):J161–J164.
53. Graff, J. H., and Feavel, J. R. (1944). Projection arrangement for determination of fiber dimensions. *Paper Trade J. 118*(7):140–145.
54. Graminski, E. L., and Kirsch, R. A. (1977). Image analysis in paper manufacturing. *Proceedings IEEE Computer Society Conference on Pattern Recognition and Image Processing*, New York, pp. 137–143.
55. Green, H. V., and Worrall, J. (1964). Wood quality studies. 1. A scanning microphotometer for automatically measuring and recording certain wood characteristics. *Tappi 47*(7):419–427.
56. Habeger, C. C., and Baum, G. A. (1987). The use of microwave attenuation as a measure of fiber orientation anisotropy. *Tappi 70*(2):109–113.
57. Habeger, C. C., and Baum, G. A. (1983). Microwave dielectric constants of water-paper mixtures: Role of sheet structure and composition. *J. Appl. Polym. Sci. 28*(3):969–981.
58. Hagemeyer, R. W. (1960). The effect of pigment combination and solids concentration on particle packing and coated paper characteristics. 1. Relationship of particle shape to particle packing. *Tappi 43*(3):277–288.
59. Hansson, T., Fellers, C., and Htun, M. (1989). Drying strategies and a new restraint to improve cross directional properties of paper. In: *Fundamentals of Papermaking*. Vol. 2. C. F. Baker and V. W. Punton, eds. Mech. Eng. Pub., London, pp. 743–781.
60. Hardacker, K. W. (1969). Cross-sectional area measurement of individual wood pulp fibers by lateral compaction. *Tappi 52*(9):1742–1746.
61. Hardacker, K. W. (1970). Effects of loading rate, span, and beating on individual wood fiber tensile properties. In: *The Physics and Chemistry of Wood Pulp Fibers*. D. H. Page, ed. STAP 8. TAPPI, Atlanta, GA, pp. 201–216.
62. Harpham, J. A., and Hock, C. W. (1958). The fine paper properties of cotton linters. 1. The relationship of fiber composition of fine paper properties. *Tappi 41*(11):625–629.
63. Hasuike, M., Masuda, K., and Banbo, T. (1995). Characterization of three-dimensional sheet structure. *Mitshbishi Heavy Ind. Tech. Rev. 32*(3):127–131.
64. Hess, T. R., and Brodeur, P. H. (1996). Effects of wet straining and drying on fibre orientation and elastic stiffness orientation. *J. Pulp Paper Sci. 22*(5):J160–J164.
65. Hieta, K., Nanko, H., Mukoyoshi, S., and Ohsawa, J. (1990). Bonding ability of pulp fibers. *TAPPI Papermakers Conf. 1990*, pp. 123–131.
66. Higgins, H. G., de Yong, J., Balodis, V., Phillips, F. H., and Colley, J. (1973). The density and structure of hardwoods in relation to paper surface characteristics and other properties. *Tappi 56*(8):127–131.
67. Hoffman, J. (1992). *Numerical Methods for Engineers and Scientists*. McGraw-Hill, New York.
68. Hogg, R. V., and Craig, A. T. (1995). *Introduction to Mathematical Statistics*. 5th ed. Prentice-Hall, Englewood Cliffs, NJ.
69. Hsu, T. (1981). Classification and characterization of fines produced from Southern pine kraft pulp. M.S. Thesis, SUNY College of Environmental Science and Forestry, Syracuse, NY.
70. Htun, M. (1980). The influence of drying strategies on the mechanical properties of paper. Ph.D. Thesis, Royal Institute of Technology, Stockholm.
71. Htun, M., and de Ruvo, A. (1978). The implication of the fines fraction for the properties of bleached kraft sheet. *Svensk Papperstidn. 81*(16):507–510.

72. Huang, X.-C., and Bresee, R. R. (1993). Characterizing nonwoven web structure using image analysis techniques. Part II. Fiber orientation analysis in thin webs. *Nonwovens Res. 5*(2):14–21.
73. Huang, X.-C., and Bresee, R. R. (1993). Characterizing nonwoven web structure using image analysis technique. Part III. Web uniformity analysis. *Nonwoven Res. 5*(3):28–38.
74. Humphrey, K. (1995). *Image Analysis.* Pira Int., Leatherhead, UK.
75. Isenberg, I. H. (1967). *Pulp and Paper Microscopy.* 3rd ed. Inst. Paper Chemistry, Appleton, WI.
76. Jagels, R., Gardner, D. J., and Brann, T. B. (1982). Improved techniques for handling and staining wood fibers for digitizer-assisted measurement. *Wood Sci. 14*(4):165–167.
77. Jang, H. F., Robertson, A. G., and Seth, R. S. (1991). Optical sectioning of pulp fibers using confocal microscopy. *Tappi 74*(10):217–219.
78. Jang, H. F., Howard, R. C., and Seth, R. S. (1995). Fiber characterization using confocal microscopy: Effects of recycling. *Tappi 78*(12):131–137.
79. Jang, H. F., Amiri, R., Seth, R. S., and Karnis, A. (1996). Fiber characterization using confocal microscopy: Collapse behavior of mechanical pulp fibers. *Tappi 79*(4):203–210.
80. Jang, H. F., and Seth, R. S. (1998). Using confocal microscopy to characterize collapse behavior of fibers. *Tappi 81*(5):167–174.
81. Japan Industry Standards Committee JIS P-8207. Method of screening test of paper pulp.
82. Jayme, G., and Bauer, G. (1957). Differentiation of early- and late-wood fibers by secondary fluorescence. *Holzforschung 2*:16–18 (in German).
83. Jones, A. R. (1967). An experimental investigation of the in-plane elastic moduli of paper. Ph.D. Thesis, Institute of Paper Chemistry, Appleton, WI.
84. Jordan, B. D., and Page, D. H. (1980). Application of image analysis to pulp fibre characterization, 1. Presented at the Fiber Society Symposium, Baltimore, MD (September).
85. Jordan, B. D., and Page, D. H. (1983). Application of image analysis to pulp fibre characterization, In: *The Role of Fundamental Research in Paper Making.* J. Brander, ed. Mech. Eng. Pub., London, pp. 745–784.
86. Judt, M. (1958). Fiber alignment in paper. *Das Papier 12*(21/22):568–578 (in German).
87. Judt, M. (1959). The effect of the shake of paper machines on sheet formation and fiber orientation. *Das Papier 13*(3/4):46–54 (in German).
88. Kallmes, O. J. (1969). Technique for determining the fiber orientation distribution throughout the thickness of a sheet. *Tappi 52*(3):482–485.
89. Kallmes, O., and Bernier, G. (1962). The structure of paper. 3. The absolute, relative and maximum bonded areas of fiber networks. *Tappi 45*(11):867–872.
90. Kallmes, O., and Bernier, G. (1963). The structure of paper. 4. The free fiber length of a multiplanar sheet. *Tappi 46*(2):108–114.
91. Kallmes, O., and Bernier, G. (1964). The structure of paper. 8. Structure of idealized nonrandom networks. *Tappi 47*(11):694–703.
92. Kallmes, O., and Corte, H. (1960). The structure of paper. 1. The statistical geometry of an ideal two-dimensional fiber network. *Tappi 43*(9):737–752.
93. Kallmes, O., Corte, H., and Bernier, G. (1963). The structure of paper. 5. The bonding states of fibers in randomly formed paper. *Tappi 46*(8):493–502.
94. Kallmes, O., and Eckert, C. (1964). The structure of paper. 7. The application of the relative bonded area concept of paper evaluation. *Tappi 47*(9):540–548.
95. Kallmes, O. J., and Perez, M. (1966). Load/elongation properties of fibers. In: *Consolidation of the Paper Web.* F. Bolam, ed. British Paper and Board Makers Assoc., London, pp. 507–537.
96. Keith, C. T., and Côté, W. A., Jr. (1968). Microscopic characterization of slip lines and compression failures in wood cell walls. *Forest Products J. 18*(3):67–74.

97. Kibblewhite, R. P. (1972). Effect of beating on fibre morphology and fibre surface structure. *Appita 26*(3):196–202.
98. Kibblewhite, R. P. (1975). Interrelations between pulp refining treatments, fibre and fines quality, and pulp freeness. *Paperi Puu 57*(8):519–526.
99. Kibblewhite, R. P. (1977). Structural modifications to pulp fibers: Definitions and role in papermaking. *Tappi 60*(10):141–143.
100. Kibblewhite, R. P., and Brookes, D. (1975). Factors which influence the wet web strength of commercial pulps. *Appita 28*(4):227–231.
101. Kibblewhite, R. P., Brookes, D., and Allison, R. W. (1980). Effect of ozone on the fiber characteristics of thermomechanical pulps. *Tappi 63*(4):133–136.
102. Kimura, M., and Mark, R. E. (1986). Mechanical properties in relation to network structure for press-dried paper. *Sen-I Gakkaishi 42*(12): T539–T546.
103. Kimura, M., and Mark, R. E. (1987). Mechanical properties in relation to network geometry for press-dried paper. *Proc. 1987 Int. Paper Physics Conf.* Tech. Sect. CPPA, Montreal, pp. 167–172.
104. Kobar, L., Hajduczki, I., and Reinicz, E. (1975). The improvement of flat crush strength by the use of modified starch and lignin. *Zells. Papier 24*(9):269–270 (in German).
105. Kohl, A., and Hartig, W. (1985). Measurement of physical characteristics on running webs. *Das Papier 39*(10A):V172.
106. Koran, Z., Silvy, J., and Prud'homme, R. E. (1986). Network structure and fiber orientation in paper. *Tappi 69*(5):126–128.
107. Komori, T., Ujihara, Y., Matsunaga, Y., and Makishima, K. (1979). Crossings of curled fibers in two-dimensional assemblies. *Tappi 62*(3):93–95.
108. Koning, J. W., Jr., and Haskell, J. H. (1978). Papermaking Factors That Influence the Strength of Linerboard Weight Handsheets. U. S. Forest Products Laboratory, Madison, WI.
109. Ladell, J. L. (1959). A method of measuring the amount and distribution of cell wall material in transverse microscope sections of wood. *J. Inst. Wood Sci. 3*:43–46.
110. Lavigne, J. R. (1993). *Pulp and Paper Dictionary*, Miller Freeman, San Francisco.
111. Li, M., Johnston, R., Xu, L., Filonenko, Y., and Parker, I. (1999). Characterization of hydrocyclone-separated eucalypt fiber fractions. *J. Pulp Paper Sci. 25*(8):299–304.
112. Lim, Y. W., Sarko, A., and Marchessault, R. H. (1970). Light scattering by cellulose. 2. Oriented condenser paper. *Tappi 53*(12):2314–2319.
113. Luce, J. E. (1970). Transverse collapse of wood pulp fibers: Fiber models. In: *The Physics and Chemistry of Wood Pulp Fibers*. D. H. Page, ed. STAP 8. TAPPI, Augusta, GA, pp. 278–281.
114. Lynch, L. J., and Thomas, N. (1971). Optical diffraction profiles of single fibers. *Textile Res. J. 41*:568–572.
115. Marchessault, R. H. (1973). Light scattering by oriented native cellulose systems. In: *Structure and Properties of Polymer Films*. R. W. Lenz and A. R. Stein, eds. Plenum, New York, pp. 25–37.
116. Mardia, K. V., and Jupp, P. E. (2000). *Directional Statistics*. Wiley, Chichester, UK.
117. Mark, R. E. (1967). *Cell Wall Mechanics of Tracheids*. Yale Univ. Press, New Haven, CT.
118. Marton, J. (1974). Fines and wet end chemistry. *Tappi 57*(12):90–93.
119. Marton, R., Alexander, S. D., Brown, A. F., and Sherman, C. W. (1965). Morphological limitations to the quality of groundwood from hardwoods. *Tappi 48*(7):395–398.
120. Marton, R., Brown, A., Granzow, S., Koeppicus, R., and Tomlinson, S. (1974). Recycling and fiber structure. *Empire State Paper Res. Inst. Rep.* No. 61, pp. 39–63.
121. McMillin, C. W. (1981). Application of automatic image analysis to wood science. *Wood Sci. 14*(3):97–105.

122. Mercer, P. G. (1988). Online instrumentation for wet-end control. *Appita 41*(4):308–312.
123. Mignot, L. (1997). Development of a direct image analysis method to determine fiber orientation. Technological Research Diploma (Diplôme de Recherche Technologique), Ecole Française de Papeterie, Grenoble, France (in French.)
124. Morton, R. (Undated). An Introduction to Automatic Image Analysis. Bausch & Lomb Analytical Systems Div., Rochester, NY.
125. Morton, R. (Undated). Practical Considerations in Automatic Image Analysis. Bausch & Lomb Analytical Systems Div., Rochester, NY.
126. Moss, P. A., Retulainen, E., Paulapuro, H., and Aaltonen, P. (1993). Taking a new look at pulp and paper: Applications of confocal laser scanning microscopy (CSLM) to pulp and paper research. *Paperi Puu 75*(1–2):74–79.
127. Muggli, R., Marton, R., and Sarko, A. (1971). Light scattering by cellulose. 5. Anisotropic scattering by wood fibers. *J. Polym. Sci. C36*:121–139.
128. Nanko, H., and Ohsawa, J. (1989). Mechanisms of fiber bond formation. In: *Fundamentals of Papermaking*, Vol. 2. C. F. Baker and V. W. Punton, eds. Mech. Eng. Pub., London, pp. 783–830.
129. Nanko, H., and Ohsawa, J. (1989). How to see interfiber bonding in paper sheet. *J. Pulp Paper Sci. 15*(1):J17–J23.
130. Niskanen, K. J. (1987). Critical evaluation of some methods used to determine fiber orientation in paper. *International Paper Physics Conference*. Tech. Sect. CPPA, Montreal, pp. 107–111.
131. Niskanen, K. J. (1989). Distribution of fibre orientations in paper. In: *Fundamentals of Papermaking*, Vol. 1. C. F. Baker and V. W. Punton, eds. Mech. Eng. Pub., London, pp. 275–308.
132. Niskanen, K. J., and Sadowski, J. W. (1989). Evaluation of some fiber orientation measurements. *J. Pulp Paper Sci. 15*(6):J220–J224.
133. Niskanen, K. J., and Sadowski, J. W. (1990). Fiber orientation in paper by light diffraction. *J. Appl. Polym. Sci. 39*:483–486.
134. Nordman, L., Aaltonen, P., and Makkonen, T. (1966). Relationship between mechanical and optical properties of paper affected by web consolidation. In: *Consolidation of the Paper Web*. F. Bolam, ed. British Paper and Board Makers Assoc., London, pp. 909–927.
135. Öhrn, O. E. (1969). Fiber length measuring gauge with an "easy to use" measuring probe. *Svensk Papperstidn. 72*(20):667–668.
136. Olson, J. A., Robertson, A. G., Finnigan, T. D., and Turner, R. R. H. (1995). An analyzer for fibre shape and length. *J. Pulp Paper Sci. 21*(11):J367–J373.
137. Osaki, S. (1987). Microwaves quickly determine the fiber orientation of paper. *Tappi 70*(2):105–108.
138. Page, D. H. (1967). The collapse behavior of pulp fibers. *Tappi 50*(9):449–455.
139. Page, D. H., El-Hosseiny, F., Winkler, A., and Lancaster, A. P. S. (1977). Elastic modulus of single wood pulp fibers. *Tappi 60*(4):114–117.
140. Page, D. H., Sargeant, J. W., and Nelson, R. (1966). Structure of paper in cross-section. In: *Consolidation of the Paper Web*. F. Bolam, ed. British Paper and Board Makers Assoc., London, pp. 313–352.
141. Page, D. H., and Seth, R. S. (1980). The elastic modulus of paper. 2. The importance of fiber modulus, bonding and fiber length. *Tappi 63*(6):113–116.
142. Page, D. H., and Seth, R. S. (1980). The elastic modulus of paper. 3. The effects of dislocations, microcompressions, curl, crimps and kinks. *Tappi 63*(10):99–102.
143. Page, D. H., Seth, R. S., and DeGrâce, J. H. (1979). The elastic modulus of paper. 1. The controlling mechanisms. *Tappi 62*(9):99–102.
144. Page, D. H., and Tydeman, P. A. (1962). A new theory of the shrinkage, structure and properties of paper. In: *Formation and Structure of Paper*. F. Bolam, ed. British Paper

and Board Makers Assoc., London, pp. 397–421. (See also comments by J. G. Buchanan and O. V. Washburn, pp. 422–423.)

145. Page, D. H., and Tydeman, P. A. (1966). Physical processes occurring during the drying phase. In: *Consolidation of the Paper Web*. F. Bolam, ed. British Paper and Board Makers Assoc., London, pp. 371–396. (See also discussion, pp. 950–951.)
146. Page, D. H., Tydeman, P. A., and Hunt, M. (1962). A study of fibre-to-fibre bonding by direct observation. In: *Formation and Structure of Paper*. F. Bolam, ed. British Paper and Board Makers Assoc., London, pp. 171–193.
147. Page, D. H., Tydeman, P. A., and Hunt, M. (1962). The behavior of fiber-to-fiber bonds in sheets under dynamic conditions. In: *Formation and Structure of Paper*. F. Bolam, ed. British Paper and Board Makers Assoc., London, pp. 249–263.
148. Papermaking Additives Committee, TAPPI. (1975). Commercially available chemical agents for paper and board manufacture. Committee assignment report No. 60.
149. Parham, A. A., and Kaustinen, H. M. (1974). Papermaking materials. In: *An Atlas of Electron Micrographs*. Institute of Paper Chemistry, Appleton, WI.
150. Parker, J., and Mih, W. C. (1964). A new method for sectioning and analyzing paper in the transverse direction. *Tappi 47*(5):254–263.
151. Perez, M., and Kallmes, O. J. (1965). The role of fiber curl in paper properties. *Tappi 48*(10):601–606.
152. Perkins, R. W., Jr. (1978). Prediction of the elastic properties of paper from a knowledge of network geometric parameters. In: *General Constitutive Relations for Wood and Wood-Based Materials*. R. W. Perkins, B. A. Jayne, and J. A. Johnson, eds. Report of National Science Foundation Workshop, Syracuse Univ., pp. 1–16.
153. Perkins, R. W., Jr. (1980). Mechanical behavior of paper in relation to its structure. In: *The Cutting Edge*. Institute of Paper Chemistry, Appleton, WI, pp. 89–111.
154. Perkins, R. W., and Mark, R. E. (1983). Some new concepts of the relation between fibre orientation, fibre geometry and mechanical properties. In: *The Role of Fundamental Research in Paper Making*. J. Brander, ed. Mech. Eng. Pub., London, pp. 479–525.
155. Perkins, R. W., Mark, R. E., Silvy, J., Anderson, H., and Eusufzai, A. R. K. (1983). Effect of fiber orientation distribution on the mechanical properties of paper. *Int. Paper Physics Conference*. TAPPI Press, Atlanta, pp. 83–87.
156. Petkar, B. M., Oka, P. G., and Sundaram, V. (1980). The cross sectional shapes of a cotton fiber along its length. *Textile Res. J. 50*:541–543.
157. Press, W. H., Teukolsky, S. A., Vetterling, W. T., and Flannery, B. P. (1992). *Numerical Recipes: The Art of Scientific Computing*. 2nd ed. Cambridge Univ. Press, New York.
158. Prud'homme, R. E., Hien, N. V., Noah, J., and Marchessault, R. H. (1975). Determination of fiber orientation of cellulosic samples by X-ray diffraction. *J. Appl. Polym. Sci. 19*:2606–2620.
159. Prusas, Z. C. (1963). Laboratory study of the effects of fiber orientation on sheet anisotropy. *Tappi 46*(5):325–330.
160. Pye, I. T., Washburn, O. V., and Buchanan, I. G. (1966). Structural changes in paper on pressing and drying. In: *Consolidation of the Paper Web*. F. Bolam, ed. British Paper and Board Makers Assoc., London, pp. 353–370.
161. Qi, D. (1997). Microstructural model for a three-dimensional fiber network. *Tappi 80*(1):283–292.
162. Quackenbush, D. W. (1971). Faults in paper coatings and their relationship to base sheet structure. *Tappi 54*(1):47–52.
163. Radvan, B., and Dodson, C. T. J. (1966). Detection and cause of the layered structure of paper. In: *Consolidation of the Paper Web*. F. Bolam, ed. British Paper and Board Makers Assoc., London, pp. 189–214.

164. Ranger, A. E., and Hopkins, L. F. (1962). A new theory of the tensile behavior of paper. In: *Formation and Structure of Paper*, Vol. 1. F. Bolam, ed. British Paper and Board Makers Assoc., London, pp. 311–318.
165. Rothenberg, S., and Fernandez, J. M. (1979). Effect of chemical modification on the properties of lignin-containing fibers. 3. Geometrical parameters. *Empire State Paper Res. Inst. Rep.* No. 71, pp. 17–36.
166. Rothenberg, S., and Fernandez, J. M. (1980). Effect of chemical modification on the properties of lignin-containing fibers. 4. Geometrical parameters of TMP fibers treated chemically between refining stages. *Empire State Paper Res. Inst. Rep.* No. 72, pp. 65–85.
167. Rothenberg, S., and Fernandez, J. M. (1982). Effect of chemical modification on the properties of lignin-containing fibers. 5. Correlation of geometrical parameters and pulp properties of modified TMP. *Empire State Paper Res. Inst. Rep.* No. 76, pp. 77–96.
168. Ruck, H., and Krässig, H. (1958). The determination of fiber orientation in paper. *Pulp Paper Mag. Can. 59*(6):183–190.
169. Rudström, L., and Sjölin, U. (1970). A method for determining fibre orientation in paper using laser light. *Svensk Papperstidn. 73*(5):117–121.
170. Sadowski, I. (1976). Measurement of fibre orientation in paper by optical Fourier transform. Thesis, Helsinki Tech. Univ.
171. Saka, S., Thomas, A. I., and Gratzl, J. S. (1978). Lignin distribution: Determination by energy-dispersive analysis of X-rays. *Tappi 61*(1):73–76.
172. Sauret, G. (1963). Anisotropy in the paper sheet. 1. Influence of fiber orientation. *Tech. Rech. Papetières Bull. 2*:3–14 (in French).
173. Scandanavian Pulp, Paper and Board Testing Committee. Standard SCAN-M6:69. Fiber fractionation of mechanical pulp in the McNett apparatus.
174. Schafer, E. R., and Santaholma, M. (1933). Effect of different sized fibers on the physical properties of groundwood pulp. *Paper Trade J. (TAPPI Sect.) 97*: 224–229.
175. Schaffnit, C., Silvy, J., Dodson, C. T. J. (1992). Orientational density distributions of fibres in paper. *Nordic Pulp Paper Res. J. 3*:121–125.
176. Schniewind, A. P., Ifju, G., and Brink D. L. (1966). Effect of drying on the flexural rigidity of single fibers. In: *Consolidation of the Paper Web*. F. Bolam, ed. British Paper and Board Makers Assoc., London, pp. 538–543.
177. Schwalbe, H. C. (1966). Effects of sizing, adhesives and fillers on the formation and consolidation of paper webs. In: *Consolidation of the Paper Web*. F. Bolam, ed. British Paper and Board Makers Assoc., London, pp. 692–740.
178. Seth, R. S., and Page, D. H. (1983). The stress–strain curve of paper. In: *The Role of Fundamental Research in Paper Making*. J. Brander, ed. Mech. Eng. Pub., London, pp. 421–452.
179. Seth, R. S., Jang, H. F., Chan, B. K., and Wu, C. B. (1997). Transverse dimensions of wood pulp fibres and their implications for end use. In: *The Fundamentals of Paper Making Materials*. C. F. Baker, ed. Pira Int., London, pp. 473–503.
180. Setterholm, V., and Kuenzi, E. W. (1970). Fiber orientation and degree of restraint during drying. Effect on tensile anisotropy of paper handsheets. *Tappi 53*(10):1915–1920.
181. Schulgasser, K. (1985). Fiber orientation in machine-made paper. *J. Mater Sci. 20*:859–866.
182. Silvy, J. (1980). Structural study of fiber networks: The cellulosic fiber case. D.Sci. Thesis, Univ. Grenoble (in French).
183. Silvy, J., and Herve, P. (1989). Effects of flocculation, length and curvature of fibers on orientational distribution of fibers and the physical properties of papers. *ATIP 36* (6–7):270,343–357 (in French).

184. Steenberg, B., Sandgren, B., and Wahren, D. (1960). Studies on pulp crill. 1. Suspended fibrils in paper pulp fines. *Svensk Papperstidn. 63*(12):395–397.
185. Stein, A. S. (1973). Optical studies of the morphology of polymer films. In: *Structure and Properties of Polymer Films*. R. W. Lenz and R. S. Stein, eds. Plenum, New York, pp. 1–24.
186. Sugden, E. A. N. (1968). A semi-automated method for the determination of the average fiber length by projection. *Pulp Paper Mag. Can. 69*:T406–T411.
187. Sundararajan, P. R. (1981). X-ray studies on the orientation of inorganic materials in paper. *Tappi 64*(10):111–114.
188. Swanson, I. W. (1966). Effects of soluble non-fibrous materials on formation and consolidation of paper webs. In: *Consolidation of the Paper Web*. F. Bolam, ed. British Paper and Board Makers Assoc., London, pp. 741–776.
189. Takahashi, H., Suzuki, H., and Endoh, K. (1979). The effect of fiber shape on the mechanical strength of paper and board. *Tappi 62*(7):85–88.
190. TAPPI Provisional Method T 271 pm-91. Fiber length of pulp and paper by automated optical analyzer.
191. TAPPI T 204. Alcohol and dichloromethane solubles in wood and pulp.
192. TAPPI T 211. Ash in pulp.
193. TAPPI T 232. Fiber length of pulp by projection.
194. TAPPI T 233. Fiber length of pulp by classification.
195. TAPPI T 234. Coarseness of pulp fibers.
196. TAPPI T 261. Fines fraction of paper stock by wet screening.
197. TAPPI Useful Method UM 239. Fiber bundles in baled flash dried pulp.
198. TAPPI Useful Method UM 240. Shive content of mechanical pulp (laboratory flat screen).
199. TAPPI Useful Method UM 241. Shive content of mechanical pulp (von Alfthan shive analyzer).
200. TAPPI Useful Method UM 242. Shive content of mechanical pulp (Somerville Fractionator).
201. TAPPI Useful Method UM 808. Plybond peeling strength of linerboard and corrugated board.
202. Former TAPPI Suggested Method T 481 (withdrawn August 1972). Fiber orientation and squareness of paper (zero-span tensile strength).
203. Tasman, J. E. (1972). The fiber length of Bauer-McNett screen fractions. *Tappi 55*(1):136–138.
204. Taylor, C. J., and Dixon, R. N. (1983). Image analysis applied to fibre images. In: *The Role of Fundamental Research in Paper Making*. J. Brander, ed. Mech. Eng. Pub., London, pp. 777–784.
205. Thorpe, J. L., Mark, R. E., Eusufzai, A. R. K., and Perkins, R. W. (1976). Mechanical properties of fiber bonds. *Tappi 59*(5):96–100.
206. Thorpe, J. L. (1998). Determination of fiber orientation in z-directional layers of paper by resin Hough transformation. *Empire State Paper Res. Inst. Rep.* No. 108, pp. 80–89.
207. Toroi, M. (1959). The preparation of fibre oriented sheets on a laboratory scale. *Paperi Puu 41*(5):271–279.
208. Tydeman, P. A., Wembridge, D. R., and Page, D. H. (1966). Transverse shrinkage of individual fibres by microradiography. In: *Consolidation of the Paper Web*. F. Bolam, ed. British Paper and Board Makers Assoc., London, pp. 119–144.
209. Unbehend, J. E. (1977). The "dynamic retention/drainage jar." Increasing the credibility of retention measurements. *Tappi 60*(7):110–112.
210. Unger, E., and Unger, E. W. (1963). A contribution to the rapid determination of mean fiber length and fiber length distribution in paper stocks. *Zellstoff Papier 12*(1):4–10; *12*(2):40–45.

211. Unger, E. W., and Freund, F. (1975). New developments in fiber length analysis via projected images. *Zellstoff Papier 24*(5):143–146, 160 (in German).
212. Valley, R. B., and Morse, T. H. (1965). Measurement of fiber length using a modified Coulter particle counter. *Tappi 48*(6):372–376.
213. Vecchi, E. (1969). Quality control of poplar groundwood: Factors related to the structural composition of the pulp. *Tappi 52*(12):2390–2399.
214. Vepsäläinen, K. P., Seppä, H., and Varpula, T. (1998). Dielectric anisotropy of cellulose-water mixture based on microwave resonator method. *J. Pulp Paper Sci. 24*(6):J188–J196.
215. Visconti, S., Hien, N. V., Borch, J., and Marchessault, R. H. (1976). Light scattering by helical fiber structures: Experimental models. *J. Polym. Sci., Polym. Phys. 14*:631–641.
216. Walbaum, H. H., and Zak, H. (1976). Internal structure of paper and coatings in SEM cross sections. *Tappi 59*(3):102–105.
217. Wang, P. H., and McKimmy, M. D. (1977). The effect of pulping mixed species on paper properties. *Tappi 60*(7):140–143.
218. Ward, K., Jr., Voelker, M. H., and Maclaurin, D. J. (1965). Cotton linters as papermaking fibers: Comparative studies on rag, linters and cotton lint pulps. *Tappi 48*(11):657–650.
219. Watson, A. A., and Dadswell, H. E. (1961). Influence of fibre morphology on paper properties. 1. Fibre length. *Appita 14*(5):168–178.
220. Weibel, E. R. (1980). Anisotropic structures and stereology. In: *Stereological Methods.* Vol. 2. *Theoretical Foundations.* E. R. Weibel, ed. Academic Press, London, pp. 264–311.
221. Xu, L., Filonenko, Y., Li, M., and Parker, I. (1997). Measurement of cell wall thickness of fully collapsed fibers by confocal microscopy and image analysis. *Appita J. 50*(6):501–504.
222. Yan, J. F. (1975). A method for the interpretation of fiber length classification data. *Tappi 58*(8):191–192.
223. Yang, C. F. (1975). Plane modeling and analysis of fiber systems. Ph.D. Thesis, Univ. Washington, Seattle.
224. Yang, C. F., Crosby, C. M., Eusufzai, A. R. K., and Mark, R. E. (1987). Determination of paper sheet fiber orientation distribution by a laser optical diffraction method. *J. Appl. Polym. Sci. 34*:1145–1157.
225. Yang, C. F., Eusufzai, A. R. K., Sankar, R., Mark, R. E., and Perkins, R. W., Jr. (1978). Measurements of geometrical parameters of fiber networks. 1. Bonded surfaces, aspect ratios, fiber moments of inertia, bonding state probabilities. *Svensk Papperstidn. 81*(13):426–433.
226. Yang, C. F., Mark, R. E., Eusufzai, A. R. K., and Perkins, R. W., Jr. (1981). Measurements of geometrical parameters of fiber networks. 2. Mechanical properties in relation to network geometry for press-dried paper. *Svensk Papperstidn. 84*(9):R55–R60.
227. Yuhara, T., Hasuike, M., and Murakami, M. (1991). Fibre orientation measurement with the two-dimensional power spectrum of a high resolution soft X-ray image. *J. Pulp Paper Sci. 17*(4):J110–J114.

INDEX

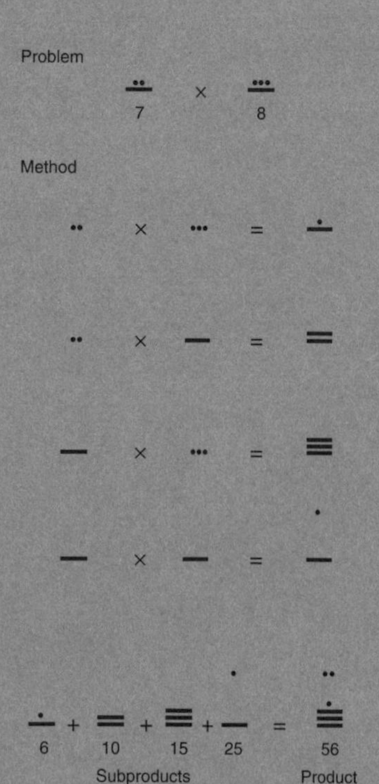

1.

• (n) × • (m) = • $(n + m - 1)$

2.

— (n) × • (m) = — $(n + m - 1)$

3.

— (n) × — (m) =

• $(n + m)$ + — $(n + m - 1)$

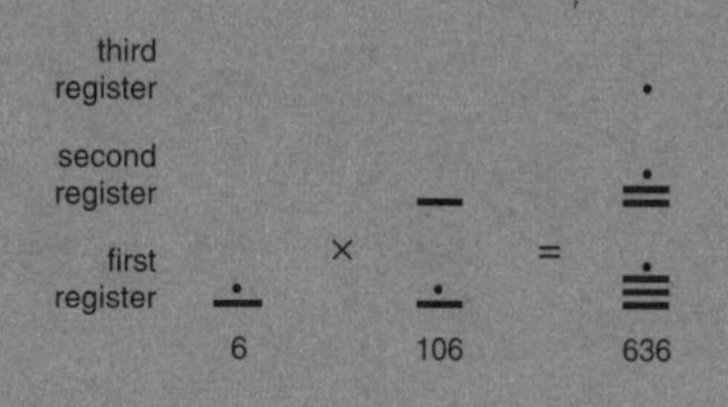

The procedures for multiplication and division described here are his, amplified by our codification of rules.

Th single-digit (0–9) Arabic multiplication table contains one hundred entries, but there are actually only thirty-eight distinct operations when regularities (0 times anything is 0; anything times 1 is itself) and commutations (2 × 3 = 3 × 2) are discounted. Since Maya notation contains only the dot and the bar, only three rules are needed to describe their interactions. These rules can be derived from the Arabic counterparts: (1) since 1 × 1 = 1, it follows that dot × dot = dot; (2) since 1 × 5 = 5, it follows that dot x bar = bar; and (3) since 5 × 5 = 25, then bar × bar = (bar + dot-in-the-next-higher-register)—i.e. 5 + 20, or 1.5 vigesimally. These three rules are equivalent to the thirty-eight operations of Arabic multiplication.

We must also have procedures for determining the "place," or order of magnitude of the product, which in Maya notation corresponds to the register into which the product is placed. To determine the order of magnitude, or highest place of the product, in Arabic notation, in multiplication problems that do not involve carrying, we can take the number of the highest place of the multiplicand and add 1 less than the highest place of the multiplier. Thus any number multiplied by a 1-digit multiplier will give a product with the same number of places as the multiplicand, since we add 1 – 1, or 0—e.g. 2 × 43 = 86; 43 and 86 each have 2 places. (We arbitrarily place the multiplier first and the multiplicand second here.) Multiplication by a 2-digit multiplier adds 1 to the number of places in the multiplicand—e.g. 16 × 324 = 5,184—and so on.

Maya multiplication is exactly analogous. A first-register multiplier gives a product in the same register as the multiplicand (1s times any order of magnitude gives the same order of magnitude). A second-register multiplier gives a product in the next register above that of the multiplicand (a multiplier in the 20s register gives a product 1 order of magnitude above the multiplicand). A third-register multiplier gives a product in the second register above that of the multiplicand, and so on. This procedure for determining registers is analogous to collecting exponents in base-10 multiplication. In all these examples, carrying may promote the product to a higher register whenever 4 bars collect in any given register.

Let us consider the simple problem 7 × 8. Each dot in the multiplicand (8) when multiplied by each dot in the multiplier (7) gives a dot, for a total of 6 (the first subproduct). The multiplicand bar when multiplied by the multiplier dots give 2 bars. The multiplicand dots when multiplied by the multiplier bar give 3 bars. Finally, the multiplicand bar when multiplied by the multiplier bar gives a bar in the same register and a dot in the next higher register. Collection of these four terms gives the product, 56 (2.16 vigesimally). The actual process is far less cumbersome than just outlined. The operation should be done by continuously summing the intermediate figures, as would be done on an abacus or with beans and pebbles on a flat surface. Each subproduct is added directly into the developing sum, and thus when the final operation is completed, the product results automatically.

The problem just given included only first-register numbers. For problems involving higher registers, the order of magnitude of the subproducts and products must be determined. This calculation can be made directly with three rules for the interaction of the symbols. In these rules, n and m refer to the register number; n and m are 1 for the units register, 2 for the 20s, etc. The rules are general for all registers. Thus a dot in any register n times a dot in any register m yields as the product a *dot* that goes into the register just below the one that corresponds to the sum of n and m $(n + m - 1)$. A bar times a dot yields a *bar* that goes into the $(n + m - 1)$ register. Finally, a bar times a bar gives a *dot* in the $(n + m)$ register and a *bar* in the $(n + m - 1)$ register.

Take, for instance, the problem 6 × 106. The dot in the multiplier (6) times the dot in the multiplicand (106) (both in the first register) gives a dot for the product in the first register (1 + 1 – 1 = 1). The multiplier dot times the multiplicand bar in the first register gives a bar in the first

Pages from a modern scientific article dealing with arithmetic operations with the Maya number system. Paper continues as an important medium for acquisition, storage and retrieval of human knowledge. Reprinted with permission from "Maya Arithmetic" by Joseph B. Lambert, Barbara Ownbey-McLaughlin, and Charles D. McLaughlin, *American Scientist 68*(3):249–255 (1980).